LPF	low-pass filter
LSSB	lower single sideband
MAP	maximum a posteriori (criterion)
MPSK	*M*-ary phase shift keying
MSK	minimum-shift keying
NBFM	narrowband frequency modulation
NCS	National Communications System of the federal government
NRZ	nonreturn-to-zero
OOK	on-off keying
OQPSK	offset quadrature phase shift keying
OSI	open system interconnection protocol model
PAM	pulse amplitude modulation
PCM	pulse code modulation
PD	phase detection
PDF	probability density function
PEP	peak envelope power
PLL	phase-locked loop
PM	phase modulation
PPM	pulse position modulation
PSD	power spectral density
PSK	phase shift keying
PTM	pulse time modulation
PWM	pulse width modulation
QAM	quadrature amplitude modulation
QPSK	quadrature phase shift
rms	root-mean-square
RT	remote telephone terminal
RZ	return-to-zero
SAW	surface acoustics wave
SDLC	synchronous data link control protocol
S/N	signal-to-noise ratio
SS	spread spectrum (system)
SSB	single sideband
TDM	time-division multiplexing
TDMA	time-division multiplex access
TELCO	telephone company
THD	total harmonic distortion
TTL	transistor transistor logic
TV	television
TVRO	TV receive only terminal
TWT	traveling-wave tube
UHF	ultra high frequency
USSB	upper single sideband
VCO	voltage-controlled oscillator
VF	voice frequency
VHF	very high frequency
VSB	vestigial sideband
WBFM	wideband frequency modulation
WER	word error rate

DIGITAL AND ANALOG COMMUNICATION SYSTEMS

FOURTH EDITION

DIGITAL AND ANALOG COMMUNICATION SYSTEMS

Leon W. Couch II

Professor of Electrical Engineering
University of Florida, Gainesville

Macmillan Publishing Company
New York

Maxwell Macmillan Canada
Toronto

Maxwell Macmillan International
New York Oxford Singapore Sydney

Editor(s): John Griffin
Production Supervisor: Margaret Comaskey
Production Manager: Su Levine
Text Designer: Eileen Burke
Cover Designer: Robert Vega

This book was set in Times Roman by York Graphic Services, printed and bound by Book Press. The cover was printed by Book Press.

Printed in the United States of America

Macmillan Publishing Company is part
of the Maxwell Communication Group of Companies.

Maxwell Macmillan Canada, Inc.
1200 Eglinton Avenue East
Suite 200
Don Mills, Ontario M3C 3N1

Macmillan Publishing Company
866 Third Avenue, New York, New York 10022

Library of Congress Cataloging in Publication Data
Couch, Leon W.
Digital and analog communication systems / Leon W. Couch II.—
4th ed.
p. cm.
Includes bibliographical references and index.
ISBN 0-02-325281-2
1. Telecommunication systems. 2. Digital communications.
I. Title.
TK5101.C69 1993
621.382—dc20 91-45917
CIP

Printing: 4 5 6 7 8 Year: 5 6 7 8 9 0 1

Preface

Continuing the tradition of the first, second, and third editions of this book, this new edition provides the latest up-to-date treatment of digital and analog communication systems. It includes a significant number of new homework problems, many of which require solutions via a personal computer. It is written as a textbook for junior or senior engineering students and is also appropriate for an introductory graduate course or as a modern technical reference for practicing electrical engineers. It covers *practical aspects* of communication systems developed from a sound *theoretical basis*.

The Theoretical Basis

- Representation of digital signals.
- Representation of analog signals.
- Magnitude and phase spectra.
- Fourier analysis.
- Power spectral density.
- Linear systems.
- Nonlinear systems.
- Simulation of communication systems.
- Intersymbol interference.
- Modulation theory.
- Random variables.
- Probability density.
- Random processes.
- Calculation of (S/N).
- Calculation of BER.
- Optimum systems.
- Block codes.
- Convolutional codes.

The Practical Applications

- PAM, PCM, DPCM, DM, PWM, and PPM baseband signaling.
- OOK, BPSK, QPSK, MPSK, MSK, and QAM bandpass digital signaling.
- AM, DSB-SC, SSB, VSB, PM, and FM bandpass analog signaling.
- Time-division multiplexing and the standards used.
- Digital line codes and spectra.

- Bit synchronizers.
- Frame synchronizers.
- Carrier synchronizers.
- Frequency-division multiplexing and the standards used.
- Telecommunication systems.
- Telephone systems.
- ISDN.
- Satellite communication systems.
- Effective input-noise temperature and noise figure.
- Link budget analysis.
- Fiber optic systems.
- Spread spectrum systems.
- Cellular radio telephone systems.
- Television systems.
- Technical standards for AM, FM, TV, and CATV.
- Computer communication systems.
- Protocols for computer communications.
- Technical standards for computer communications.
- Math tables.
- Illustrative examples.
- Over 540 homework problems with selected answers.
- Over 60 computer-solution homework problems with solution templates.
- Extensive references.
- Emphasis on design of communication systems.

The practical aspects are exhibited by describing the circuitry that is used in communication systems and summarizing the technical standards that have been adopted for digital, analog, and computer communication systems. The theoretical aspects are presented by using a sound mathematical basis that is made clear by the use of a definition, theorem, and proof format with worked examples. Many of the equations and homework problems are marked with a personal computer symbol, 💻, which indicates that there is a MathCAD† template solution for this equation or problem on a floppy disk (available to the instructor free of charge from the publisher).

†MathCAD is a trademark of MathSoft, Inc.

This book is an outgrowth of my teaching of graduate as well as undergraduate communication courses at the University of Florida and is tempered by my experiences as an amateur radio operator (K4GWQ). I believe that the reader does not fully understand the technical material unless he or she has the opportunity to work problems. Consequently, over 550 problems have been included. (Personal computer solutions are provided for over 60 of these problems on a floppy disk.) Some of them are easy, so that the beginning student will not become frustrated, and some are difficult enough to challenge the more advanced students. All the problems are designed to provoke thought about, and understanding of, communication systems.

The contents of this book may be divided into two main parts. Chapters 1 through 5 develop communication systems from a nonrandom signal viewpoint. This allows the reader to grasp some very important ideas without having to learn (or know) statistical concepts at the same time. In the second part of the book, Chapters 6 through 8 plus Appendix B, statistical concepts are developed and used. Statistical concepts are needed to analyze and design communication systems that are operating in the presence of noise. Some sections of Chapters 1 through 5 are marked with a star (★). This indicates that the material is optional or that mathematical developments of these sections are more difficult and generally use statistical concepts that are developed in later chapters. The starred sections are included because the results are significant, and it is hoped that these sections will motivate the beginning student to continue to study the exciting subject of communications.

The contents of the book are further subdivided as follows. Chapter 1 provides clear definitions of digital and analog signals as well as giving a comprehensive historical perspective. Coding theory and theoretical limits on the performance of communication systems are introduced. Chapter 2 develops the topics of spectra, orthogonal representations, and Fourier theory. Linear system concepts are reviewed. Chapter 3 covers baseband pulse and baseband digital signaling techniques, line codes and their spectra, and the prevention of intersymbol interference. Chapter 4 introduces bandpass communication circuits and develops the theory and practice of bandpass communication systems (amplitude modulation, frequency modulation, etc.). Chapter 5 provides examples of telephone, television, fiber optic, spread spectrum, digital, cellular mobile, and satellite communication systems. Summaries of technical standards are given. Chapter 6 develops the mathematical topic of random processes that is needed for analyzing the effects of noise. Chapter 7 describes the performance of digital and analog communication systems that are corrupted by noise. Chapter 8 is concerned with the design of optimum digital receivers that combat the effects of noise. A summary of useful mathematical techniques, identities, and tables is given in Appendix A. Appendix B covers the topic of probability and random variables. (This appendix is a prerequisite for Chapter 6 if the reader does not already have such knowledge.) Appendix C gives standards and terminology that are used in computer communication systems and Appendix D gives a brief introduction to MathCAD.

This book is written to be applicable to many different course structures. These are summarized in the accompanying table.

Course Length[a]	Chapters Covered	Course Title and Comments
Short Courses		
3 days	1, 3 (partially), 7 (partially)	Digital Communications and Coding
3 days	3, 5 (partially)	Telecommunications
3 days	4	Modulation, Transmitters, and Receivers
3 days	1, 5	Survey of Communication Systems
3 days	Appendix B	Probability and Random Variables
3 days	6	Random Processes
3 days	3 (partially), Appendix C	Computer and Data Communications
Undergraduate		
1 term 1 quarter	1, 2, 3, 4, 5 (partially) 1, 2, 3, 4	Introduction to Digital and Analog Communication Systems (student background in signals, networks, and statistics not required)
1 term 1 quarter	1, 2 (rapidly), 3, 4, 5 1, 2 (rapidly), 3, 4	Introduction to Digital and Analog Communication Systems (student background in statistics not required; knowledge of signals and networks required)
1 term	1, Appendix B, 6, 7	Digital and Analog Communication Systems in Noise (a course in probability, random variables, random processes, and applications to communication systems)
1 quarter	1, Appendix B, 6, Secs. 7-1 to 7-6	Digital Communication Systems in Noise
1 quarter	1, Appendix B, 6, Sec. 7-8	Analog Communication Systems in Noise
1 term	1, 2 (rapidly), 3, 4, 7	Digital and Analog Communication Systems (prior course in random processes required)
Two terms		
1st term	1, 2 (rapidly), 3, 4, 5	Communication I—Introduction to Communication Systems
2nd term	Appendix B, 6, 7	Communication II—Performance of Communication Systems in Noise
Three quarters		
1st quarter	1, 2 (rapidly), 3, 4	Communication I—Introduction to Communication Systems

Course Length[a]	Chapters Covered	Course Title and Comments
2nd quarter	5, Appendix B, 6	Communication II—Communication Systems and Noise
3rd quarter	7, 8	Communication III—Performance of Communication Systems and Optimum Digital Receivers
Graduate		
1 term	1, Appendix B, 6, 7, Secs. 8-1 to 8-4	Introduction to Communication Systems (some undergraduate knowledge of communications required)
1 term	1, 6, 7, 8	Introduction to Communication Systems with Optimum Digital Receivers (knowledge of probability required)

[a] One term is assumed to be equivalent to 3 class hours per week for a semester system or 4 class hours per week for a quarter system. One quarter is assumed to be 3 class hours per week.

I appreciate the help of the many persons who contributed to this book. In particular I appreciate the very helpful comments that have been provided by the Macmillan reviewers. For the first edition they were Ray W. Nettleton (Litton Amecon), James A. Cadzow (Arizona State University), Dean T. Davis (The Ohio State University), Jerry D. Gibson (Texas A&M University), and Gerald Lachs (Pennsylvania State University). For the second edition they were Jeff Burl (University of California, Irvine), Donald J. Healy (Georgia Institute of Technology), and Morris H. Mericle (Iowa State University). For the third and fourth editions they are Daniel Bukofzer (Naval Postgraduate School), Nasser Golshan (University of Wisconsin, Plattesville), Tim Healy (Santa Clara University), Theodore S. Rappaport (Virginia Polytechnic Institute and State University), and Dennis Silage (Temple University). Professor Thomas M. Scott of Iowa State University was especially helpful in providing marked pages with suggestions and comments as he was teaching the communications course there.

I also appreciate the help of my colleagues at the University of Florida, including Haniph Latchman, Scott Miller, and Peyton Z. Peebles. Special thanks are extended to the many University of Florida students who have been most helpful in making suggestions for this fourth edition. I also thank Lawrence K. Thompson and Charles S. Prewitt for assistance in preparing the solution manual. I am also grateful to the late Dr. T. S. George who taught me a great deal about communications while I was a graduate student under his direction. I am also very appreciative of the help of my wife, Dr. Margaret Couch, who typed the original and revised manuscripts and proofread the entire book.

Leon W. Couch II
Gainesville, Florida

Contents

CHAPTER 3

Baseband Pulse and Digital Signaling 133

CHAPTER 7
Performance of Communication Systems Corrupted by Noise 573

CHAPTER 8 Optimum Digital Receivers 641

APPENDIX A
Mathematical Techniques, Identities, and Tables 703

APPENDIX B
Probability and Random Variables 716

List of Symbols

There are not enough symbols in the English and Greek alphabets to allow the use of each letter only once. Consequently, some symbols may be used to denote more than one entity, but their use should be clear from the context. Furthermore, the symbols are chosen to be generally the same as those used in the associated mathematical discipline. For example, in the context of complex variables, x denotes the real part of a complex number (i.e., $c = x + jy$), whereas in the context of statistics x might denote a random variable.

Symbols

a_n	a constant
a_n	quadrature Fourier series coefficient
A_c	level of modulated signal of carrier frequency f_c
A_e	effective area of an antenna
b_n	quadrature Fourier series coefficient
B	baseband bandwidth
B_p	bandpass filter bandwidth
B_T	transmission (bandpass) bandwidth
c	a complex number where $c = x + jy$
c	a constant
c_n	complex Fourier series coefficient
C	channel capacity
C	capacitance
$\underline{\mathbf{C}}$	cost matrix
$\overline{C}$	average cost
°C	degrees Celsius
dB	decibel
D	dimensions/sec ($D = N/T_0$) or baud rate
D_f	frequency modulation gain constant
D_n	polar Fourier series coefficient
D_p	phase modulation gain constant

e	error
e	the natural number, 2.7183
E	modulation efficiency
E	energy
$\mathscr{E}(f)$	energy spectral density (ESD)
E_b/N_0	energy per bit/noise power spectral density ratio
f	frequency (Hz)
$f(x)$	probability density function (PDF)
f_c	carrier frequency
f_i	instantaneous frequency
f_0	a (frequency) constant; the fundamental frequency of a periodic waveform
f_s	sampling frequency
F	noise figure
$F(a)$	cumulative distribution function (CDF)
$g(t)$	complex envelope
$\tilde{g}(t)$	corrupted complex envelope
G	power gain
$G(f)$	power transfer function
h	Planck's constant, 6.2×10^{-34} joule-sec
$h(t)$	impulse response of a linear network
$h(x)$	mapping function of x into $h(x)$
H	entropy
$H(f)$	transfer function of a linear network
i	an integer
I_j	information in the jth message
j	the imaginary number $\sqrt{-1}$
j	an integer
k	Boltzmann's constant, 1.38×10^{-23} joule/K
k	an integer
$k(t)$	complex impulse response of a bandpass network
K	number of bits in a binary word that represents a digital message
K	degrees Kelvin (°C + 273)
l	an integer
ℓ	number of bits per dimension
L	inductance
L	number of levels permitted
m	an integer
m	mean value
$m(t)$	message (modulation) waveform
$\tilde{m}(t)$	corrupted (noisy received) message
M	an integer
M	number of messages permitted

n	an integer
n	number of bits in message
$n(t)$	noise waveform
N	an integer
N	number of dimensions used to represent a digital message
N	noise power
N_0	level of the power spectral density of white noise
$p(t)$	an absolutely time-limited pulse waveform
$p(t)$	instantaneous power
$p(m)$	probability density function of frequency modulation
P	average power
P_e	probability of bit error
$P(C)$	probability of correct decision
$P(E)$	probability of message error
$\mathcal{P}(f)$	power spectral density (PSD)
$Q(z)$	integral of Gaussian function
$Q(x_k)$	quantized value of the kth sample value, x_k
$r(t)$	received signal plus noise
R	data rate (bits/sec)
R	resistance
$R(t)$	real envelope
$R(\tau)$	autocorrelation function
$s(t)$	signal
$\tilde{s}(t)$	corrupted signal
S/N	signal power/noise power ratio
t	time
T	a time interval
T	absolute temperature (Kelvin)
T_b	bit period
T_e	effective input-noise temperature
T_0	duration of a transmitted symbol or message
T_0	period of a periodic waveform
T_0	standard room temperature (290 K)
T_s	sampling period
u_{11}	covariance
$v(t)$	a voltage waveform
$v(t)$	a bandpass waveform or a bandpass random process
$w(t)$	a waveform
$W(f)$	spectrum (Fourier transform) of $w(t)$
x	an input
x	a random variable

x	real part of a complex function or a complex constant
$x(t)$	a random process
y	an output
y	an output random variable
y	imaginary part of a complex function or a complex constant
$y(\bar{t})$	a random process
α	a constant
β	a constant
β_f	frequency modulation index
β_p	phase modulation index
δ	step size of delta modulation
δ_{ij}	Kronecker delta function
$\delta(t)$	impulse (Dirac delta function)
ΔF	peak frequency deviation (Hz)
$\Delta\theta$	peak-phase deviation
ϵ	a constant
ϵ	error
η	spectral efficiency [(bits/sec)/Hz]
$\theta(t)$	phase waveform
λ	dummy variable of integration
λ	wavelength
$\Lambda(r)$	likelihood ratio
π	3.14159
ρ	correlation coefficient
σ	standard deviation
τ	independent variable of autocorrelation function
τ	pulse width
$\varphi_j(t)$	orthogonal function
ϕ_n	polar Fourier series coefficient
ω_c	radian carrier frequency, $2\pi f_c$
$\equiv$	mathematical equivalence
$\triangleq$	mathematical definition of a symbol

Defined Functions

$J_n(\cdot)$	Bessel function of the first kind *n*th order
$\ln(\cdot)$	natural logarithm
$\log(\cdot)$	base 10 logarithm

$\log_2(\cdot)$	base 2 logarithm
$Q(z)$	integral of a Gaussian probability density function
$\mathrm{Sa}(z)$	$(\sin z)/z$
$u(\cdot)$	unit step function
$\Lambda(\cdot)$	triangle function
$\Pi(\cdot)$	rectangle function

Operator Notation

$\mathrm{Im}\{\cdot\}$	imaginary part of
$\mathrm{Re}\{\cdot\}$	real part of
$\overline{[\cdot]}$	ensemble average
$\langle[\cdot]\rangle$	time average
$[\cdot] * [\cdot]$	convolution
$[\cdot]^*$	conjugate
$\underline{/[\cdot]}$	angle operator or angle itself, see (2-108)
$\lvert[\cdot]\rvert$	absolute value
$\hat{[\cdot]}$	Hilbert transform
$\mathcal{F}[\cdot]$	Fourier transform
$\mathcal{L}[\cdot]$	Laplace transform
$[\cdot] \cdot [\cdot]$	dot product

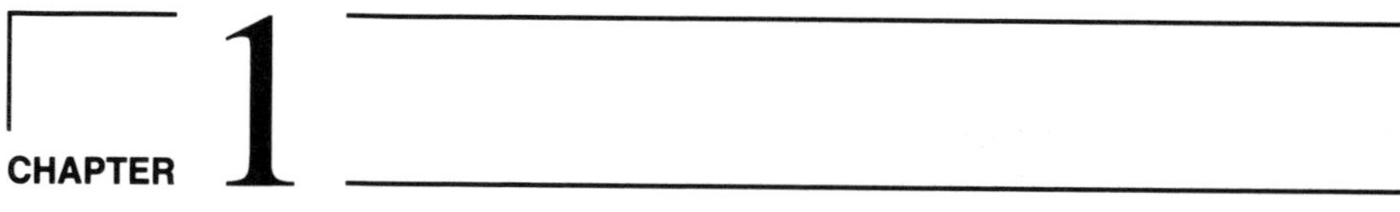

CHAPTER 1

Introduction

What is a communication system? To put the answer to this question in perspective, it is useful to ask a more fundamental question. What is electrical engineering (EE)? If one poses this question to 50 electrical engineers, one will probably get 50 different answers. However, it is clear that EE is concerned with solving problems of two types:

- Production and transmission of electrical energy
- Transmission of information

Communications systems are systems designed to transmit information.

Many areas within EE are needed to solve communication and electric energy problems. Some examples of these areas are computers, electrical and electronic circuits, electronic devices, digital signal processing, electromagnetics, and photonics.

It is important to realize that communication systems and electric energy systems have distinctly different sets of constraints. In electric energy systems, the waveforms are usually *known* and one is concerned with designing the system for *minimum energy loss.*

In communication systems, the waveform present at the sink (user) is *unknown* until after it is received—otherwise, no information would be transmitted and there would be no need for the communication system. More information is communicated to the sink when the user at the sink is "surprised" by the message that was transmitted. That is, the transmission of information implies the communication of messages that are not known ahead of time (a priori). On the other hand, in energy or power systems it is known a priori what the received waveform will be, and it is the energy imparted by the received waveform that is the desired quantity.

Noise limits our ability to communicate. If there were no noise, we could communicate messages electronically to the outer limits of the universe using an infinitely small amount of power. This has been obvious on an intuitive basis since the early days of radio. However, the theory that describes noise and the effect of noise on the transmission of information was not developed until the 1940s by such persons as D. O. North [1943], S. O. Rice [1944], C. E. Shannon [1948], and N. Wiener [1949].

1-1 Historical Perspective

A time chart showing the historical development of communications is given in Table 1-1. The reader is encouraged to spend some time studying this table in order to obtain an appreciation for the chronology of communications. Note that although the telephone was developed late in the nineteenth century, the first transatlantic telephone cable was not laid until 1954. Previous to this date, transatlantic calls were handled via shortwave radio. Similarly, although the British began television broadcasting in 1936, transatlantic television relay was not possible until 1962 when the *Telstar I* satellite was placed into orbit. Digital transmission systems—embodied by telegraph systems—were developed in the 1850s before analog systems—the telephone—in the twentieth century.

This book provides technical details for most of the communication inventions listed in the table. For example, pulse code modulation—used for modern digital transmission of telephone voice signals—was invented by Alex Reeves in 1937 and is described in detail in Chapter 3. Superheterodyne receivers and FM systems are discussed in Chapter 4. Recent developments such as fiber optics, the cellular mobile telephone, satellite, and spread spectrum systems are described in Chapter 5. The effects of noise on digital and analog systems are covered in Chapters 6 through 8. Computer communication system standards are tabulated in Appendix C.

More historical details about the engineers and their inventions that became the modern communication systems we enjoy today can be found in a book that was published in the centennial year for the Institute of Electrical and Electronic Engineers (IEEE) and its predecessors [Ryder and Fink, 1984].

1-2 Classification of Information Sources and Systems

Definition: A *digital information source* produces a finite set of possible messages.

A typewriter is a good example of a digital source. There is a finite number of characters (messages) that can be emitted by this source.

Definition: An *analog information source* produces messages that are defined on a continuum.

A microphone is a good example of an analog source. The output voltage describes the information in the sound, and it is distributed over a continuous range of values.

Definition: A *digital communication system* transfers information from a digital source to the sink.

TABLE 1-1 Important Dates in Communications

Year	Event
Before 3000 B.C.	Egyptians develop a picture language called *hieroglyphics*.
1500 B.C.	Semites devise an alphabet.
300 B.C.	Hindus invent numerals.
A.D. 800	Arabs adopt our present number system from India.
1440	Johannes Gutenberg invents movable metal type.
1622	European newsletters develop into printed sheets called *coranto*.
1752	Benjamin Franklin's kite shows that lightning is electricity.
1799	Alessandro Volta invents the first electric battery.
1820	Hans Christian Oersted shows that electric current produces a magnetic field.
1827	Georg Simon Ohm formulates his law ($I = E/R$).
1831	Michael Faraday discovers that electric fields are produced by changing magnetic fields.
1834	Carl F. Gauss and Ernst H. Weber build the electromagnetic telegraph.
1838	William F. Cooke and Sir Charles Wheatstone build the telegraph.
1839	Joseph Niepace and Louis Daguerre of France invent photography.
1844	Samuel F. B. Morse demonstrates the Baltimore, MD, and Washington, DC, telegraph line.
1850	Gustav Robert Kirchhoff first publishes his circuit laws.
1858	The first transatlantic cable is laid, and fails after 26 days.
1864	James C. Maxwell predicts electromagnetic radiation.
1866	The second transatlantic cable is installed.
1871	The Society of Telegraph Engineers is organized in London.
1876	Alexander Graham Bell develops and patents the telephone.
1877	Thomas A. Edison invents the phonograph.
1879	Thomas A. Edison demonstrates his practical electrical light.
1883	Thomas A. Edison discovers flow of electrons in a vacuum, called the "Edison effect," the foundation of the electron tube.
1884	The American Institute of Electrical Engineers (AIEE) is formed.
1885	Edward Branly invents the "coherer" radio-wave detector.
1887	Heinrich Hertz verifies Maxwell's theory.
1889	George Eastman develops practical photographic film.
1889	The Institute of Electrical Engineers (IEE) forms from the Society of Telegraph Engineers in London.
1894	Oliver Lodge demonstrates wireless communication over a distance of 150 yards.
1897	Guglielmo Marconi patents the wireless telegraph system.
1898	Valdemar Poulsen invents magnetic recording on steel wire.

TABLE 1-1 Important Dates in Communications (*continued*)

Year	Event
1900	Guglielmo Marconi transmits the first transatlantic wireless signal.
1904	John A. Fleming invents the vacuum-tube diode.
1905	Reginald Fessenden transmits speech and music by radio.
1906	Lee deForest invents the vacuum-tube triode amplifier.
1907	The Society of Wireless Telegraph Engineers is formed in the United States.
1908	A. A. Campbell-Swinton publishes basic ideas of television broadcasting.
1909	The Wireless Institute is established in the United States.
1912	The Institute of Radio Engineers (IRE) is formed in the United States from the Society of Wireless Telegraph Engineers and the Wireless Institute.
1915	Bell System completes a U.S. transcontinental telephone line.
1918	Edwin H. Armstrong invents the superheterodyne receiver circuit.
1920	KDKA, Pittsburgh, PA, begins the first scheduled radio broadcasts.
1920	J. R. Carson applies sampling to communications.
1923	Vladimir K. Zworykin devises the ''iconoscope'' television pickup tube.
1926	J. L. Baird, England, and C. F. Jenkins, United States, demonstrate television.
1927	The Federal Radio Commission is created in the United States.
1927	Harold Black develops the negative-feedback amplifier at Bell Laboratories.
1928	Philo T. Farnsworth demonstrates the first all-electronic television system.
1931	Teletypewriter service is initiated.
1933	Edwin H. Armstrong invents FM.
1934	The Federal Communication Commission (FCC) is created from the Federal Radio Commission in the United States.
1935	Robert A. Watson-Watt develops the first practical radar.
1935	Commercial color film with three emulsion layers (Kodachrome) is introduced.
1936	The British Broadcasting Corporation (BBC) begins the first television broadcasts.
1937	Alex Reeves conceives pulse code modulation (PCM).
1938	Chester Carlson develops ''xerography'' electrostatic copying.
1939	The Klystron microwave tube is developed by R. H. Varian, S. F. Varian, W. C. Hahn, and G. F. Metcalf.
1941	John V. Atanasoff invents the computer at Iowa State College.

TABLE 1-1 Important Dates in Communications (*continued*)

Year	Event
1941	The FCC authorizes television broadcasting in the United States.
1945	The ENIAC electronic digital computer is developed at the University of Pennsylvania by John W. Mauchly.
1947	Walter H. Brattain, John Bardeen, and William Shockley devise the transistor at Bell Laboratories.
1947	Steve O. Rice develops statistical representation for noise at Bell Laboratories.
1948	Claude E. Shannon publishes his work on information theory.
1950	Time-division multiplexing is applied to telephony.
1950s	Microwave telephone and communication links are developed.
1953	NTSC color television is introduced in the United States.
1953	The first transatlantic telephone cable (36 voice channels) is laid.
1954	J. P. Gordon, H. J. Zeiger, and C. H. Townes produce the first successful maser.
1955	J. R. Pierce proposes satellite communication systems.
1956	Videotape, by Ampex, is first used.
1957	First Earth satellite, *Sputnik I,* is launched by USSR.
1958	A. L. Schawlow and C. H. Townes publish the principles of the laser.
1958	Jack Kilby of Texas Instruments builds the first germanium integrated circuit (IC).
1958	Robert Noyce of Fairchild produces the first silicon IC.
1960	Theodore H. Marman produces the first successful laser.
1961	Stereo FM broadcasts begin in the United States.
1962	The first active satellite, *Telstar I,* relays television signals between the United States and Europe.
1963	Bell System introduces the touch-tone phone.
1963	The Institute of Electrical and Electronic Engineers (IEEE) is formed by merger of IRE and AIEE.
1963–66	Error-correction codes and adaptive equalization for high-speed error-free digital communications are developed.
1964	The electronic telephone switching system (No. 1 ESS) is placed into service.
1965	*Mariner IV* transmits pictures from Mars to Earth.
1965	The first commercial communications satellite, *Early Bird,* is placed into service.
1966	K. C. Kao and G. A. Hockham publish the principles of fiber optic communications.
1968	Cable television systems are developed.
1971	Intel Corporation develops the first single-chip microprocessor, the 4004.
1972	Motorola demonstrates the cellular telephone to the FCC.

TABLE 1-1 Important Dates in Communications (*continued*)

Year	Event
1973	Computerized axial tomography (CAT) scanners are introduced.
1976	Personal computers are developed.
1979	64 kb random access memory ushers in the era of very-large-scale integrated (VLSI) circuits.
1980	Bell System FT3 fiber optic communication is developed.
1980	Compact disk is developed by Philips and Sony.
1981	IBM PC is introduced.
1982	AT&T agrees to divest its 22 Bell System telephone companies.
1984	Macintosh computer is introduced by Apple.
1985	FAX machines become popular.
1989	''Pocket'' cellular telephone is introduced by Motorola.
1990–present	Era of digital signal processing with microprocessors, digital oscilloscopes, digitally tuned receivers, magaflop workstations, spread spectrum systems, integrated service digital networks (ISDNs), and high-definition television (HDTV).

DEFINITION: An *analog communication system* transfers information from an analog source to the sink.

Strictly speaking, a *digital waveform* is defined as a function of time that can have only a discrete set of values. If the digital waveform is a binary waveform, only two values are allowed. An *analog waveform* is a function of time that has a continuous range of values.

An electronic *digital* communication system usually has voltages and currents that have digital waveforms; however, it *may* have analog waveforms. For example, the information from a binary source may be transmitted to the sink by using a sine wave of 1000 Hz to represent a binary 1 and a sine wave of 500 Hz to represent a binary 0. Here the digital source information is transmitted to the sink by use of analog waveforms, but this is still called a digital communication system. Furthermore, this analog waveform is called a *digital signal* because it describes a digital source. Similarly, an *analog signal* describes an analog source. From this point of view, we see that a *digital* communication engineer needs to know how to analyze analog circuits as well as digital circuits.

Digital communication has a number of advantages:

- Relatively inexpensive digital circuits can be used.
- Privacy is preserved by using data encryption.
- Greater dynamic range (the difference between largest and smallest value) is possible.
- Data from voice, video, and data sources can be merged and transmitted over a common digital transmission system.

- In long-distance systems, noise does not accumulate from repeater to repeater.
- Errors in detected data are small, even when there is a large amount of noise on the received signal.
- Errors can often be corrected by the use of coding.

Digital communication also has disadvantages:

- Generally, more bandwidth is required than that for analog systems.
- Synchronization is required.

The advantages of digital communication systems usually outweigh their disadvantages. Consequently, digital systems are becoming more and more popular.

In this book we emphasize the *theory* of communication systems and the corresponding mathematical concepts since these ideas are basic to the understanding of future as well as present systems. This approach will give the graduate engineer a set of working tools that will be invaluable throughout his or her professional career and will not be hardware dependent. On the other hand, we will strive not to get bogged down in mathematical complexity. Optimum analog communication systems are often impossible to obtain, whereas it is possible to obtain and analyze some optimum digital communication systems without too much difficulty.

Practical applications are also of prime importance. These not only provide motivation and ease of understanding of theoretical concepts, but also give the young engineer a working knowledge of communication systems. We feel that this approach will provide an excellent basis for the development of future communication systems, as well as an understanding of present systems.

1-3 Deterministic and Random Waveforms

In communication systems we are concerned with two broad classes of waveforms: deterministic and random (or stochastic).

Definition: A *deterministic waveform* can be modeled as a completely specified function of time.

For example, if

$$w(t) = A\cos(\omega_0 t + \varphi_0) \tag{1-1}$$

describes a waveform where A, ω_0, and φ_0 are known constants, this waveform is said to be deterministic.

Definition: A *random waveform* (or stochastic waveform) cannot be completely specified as a function of time and must be modeled probabilistically.[†]

[†] A more complete definition of a random waveform, also called a *random process*, is given in Chapter 6.

Here we are faced immediately with a dilemma when analyzing communication systems. We know that the waveforms that represent the source cannot be deterministic. For example, in a digital communication system we might send information corresponding to any one of the letters of the English alphabet. Each letter might be represented by a deterministic waveform, but when we examine the waveform that is emitted from the source we find that it is a random waveform because we do not know exactly which characters will be transmitted. Consequently, we really need to design the communication system using a random signal waveform, and, of course, any noise that is introduced would also be described by a random waveform. This technique requires the use of probability and statistical concepts that make the design and analysis procedure more complicated. Fortunately, if we represent the signal waveform by a "typical" deterministic waveform, we can obtain most, but not all, of the results we are seeking. This is the approach taken in the first part of this book.

1-4 Organization of This Book

Chapters 1 through 5 use a deterministic approach in analyzing communication systems. This approach allows the reader to grasp some important concepts without the complications of statistical analysis. It also allows the reader who is not familiar with statistics to obtain a basic understanding of communication systems. However, the important topic of performance of communication systems in the presence of noise cannot be analyzed without the use of statistics. These topics are covered in Chapters 6 through 8 and in Appendix B.[†]

You will find that some portions of Chapters 1 through 5 are marked with a star (★). This indicates that the material is optional if time is limited, or that mathematical developments in these sections are more difficult and generally use statistical concepts. On the first reading, the beginning student should be concerned with the *results* of the starred sections and not labor over the mathematical details until he or she has gained more mathematical expertise as developed in Appendix B and Chapter 6. These starred sections are included because the results are significant and because it is hoped that these sections will motivate the beginning student to continue to study in this area in order to develop a sound theoretical basis that can be used to analyze and design practical communication systems after graduation.

This book can be used efficiently in a one-semester course, a two-semester course, or with the quarter system. (In a one-semester course, there is not enough time to teach communications from a statistical viewpoint and still provide a practical, broad overview of communication systems.) For a one-semester course, Chapters 1 through 5 can be taught to provide this broad perspective. In

[†] Appendix B covers the topic of probability and random variables and is a complete chapter in itself. This allows the reader who has not had a course on this topic to learn this material before Chapters 6 through 8 are studied.

the second semester of a two-semester course, Appendix B (if not covered in a prerequisite) and Chapters 6 through 7 or 8 can be taught to provide a thorough understanding of the effects of noise on communication systems. This book can be used as a textbook for short courses that cover various communication topics from a practical and/or theoretical viewpoint.

This book can also be used as a reference source for mathematics (Appendix A), statistics (Appendix B and Chapter 6), computer communication systems (Appendix C), and as a reference listing communication systems standards that have been adopted (Chapters 3, 4, and 5 and Appendix C).

Communications is an exciting area in which to work. There are the traditional areas such as AM radio, FM stereo, TV, and shortwave (around the world) radio. There are fast-developing areas such as computer communications, satellite systems, cellular telephone, spread spectrum, personal communication networks (PCNs), high-definition television (HDTV), and intelligent telecommunication networks. These topics and others are covered in this book—mainly in Chapter 5. The new reader is urged to browse through Chapter 5 looking at topics that are of special interest, such as telephone systems.

1-5 Use of a Personal Computer

There are over 540 homework problems in this book, of which over 60 are designed for computer solutions using a PC. These PC solution problems are marked with a PC symbol ( ). When one investigates possible ways of using the PC for solutions, one is faced with programming in Basic, Pascal, C, or FORTRAN, or one can use software packages that are designed to solve a particular class of problems. Obviously, it is better to use a good problem-specific package since it saves time and is less prone to mistakes in programming. For the PC, some of the popular EE packages are PSpice and Micro-CAP for circuit analysis, Touchstone for microwave circuits, and TESS for communications. For workstations, some communication packages are BDSS, SYSTID,† TOPSIM, CMSP, CHAMP, LINK, ICSSM, ICS, and MODEM. Most of the software that is designed for communication system work is not affordable by the average EE student. Consequently, the author has chosen to use MathCAD, which is available in a student version at a reasonable price. Although MathCAD is not designed specifically for communications, it provides user-friendly solutions to those problems that can be formulated in a mathematical equation. Consequently, it is suggested that the homework problems designated with be worked out using MathCAD. See Appendix D for an introduction to MathCAD and a sample problem with the PC solution. (An IBM PC-compatible floppy disk with MathCAD solution templates is made available to the instructor free of charge from the publisher.)

† See Sec. 4-2 for a digital computer simulation of a communications problem using SYSTID.

In addition, selected equations throughout this book are marked with the PC symbol (💻). This indicates that there is a MathCAD template for the designated equation on the floppy disk that is made available by your instructor. The PC can be used to plot results and see how they change with variations of the parameters. This template disk provides an excellent *design* tool and is especially helpful in solving nonlinear problems.

1-6 Block Diagram of a Communication System

Communication systems may be described by the block diagram shown in Fig. 1-1. Throughout this book we use the symbols indicated in the diagram so that the reader will not be confused about where the signals are located in the overall system. The purpose of a communication system is to transmit message information of the source, denoted by $m(t)$, to the sink. The corrupted message delivered to the sink is denoted by $\tilde{m}(t)$. The message information may be in analog or digital form, depending on the particular system, and it may represent audio, video, or some other type of information. In multiplexed systems there may be multiple input and output message sources and sinks. The spectra of $m(t)$ and $\tilde{m}(t)$ are concentrated about $f = 0$; consequently, they are said to be *baseband* signals.

The signal processing block at the transmitter conditions the source for more efficient transmission. For example, in a digital system the processor might consist of a minicomputer or microprocessor that provides redundancy reduction of the source. It may also provide channel encoding so that error detection and correction can be used at the signal processing unit at the receive site to reduce errors caused by noise in the channel. In an analog communication system the signal processor may be nothing more than an analog low-pass filter. In a hybrid system the signal processor might take samples of an analog input and digitize them to provide a pulse code modulation (PCM) "digital word" for transmission over the system. (We study PCM techniques in Chapter 3.) The signal at the output of the transmitter signal processor is also a baseband signal since it has frequencies concentrated near $f = 0$.

The transmitter carrier circuits convert the processed baseband signal into a frequency band that is appropriate for propagation over the transmission medium of the channel. For example, if the channel consists of a twisted-pair telephone

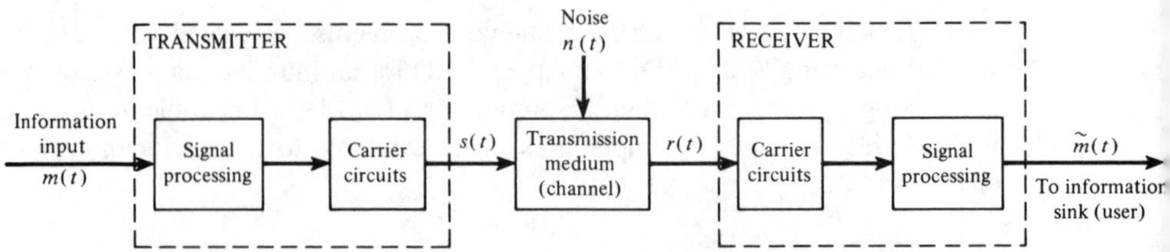

FIGURE 1-1 Communication system.

line, the spectrum of $s(t)$ will be in the audio range, say 300 to 3700 Hz, but if the channel consists of a fiber optic link, the spectrum of $s(t)$ will be at light frequencies. If the channel propagates baseband signals, no carrier circuits are necessary and $s(t)$ can be just the output of the processing circuit at the transmitter. Carrier circuits are needed when the channel can only propagate frequencies located in a band around f_c where $f_c \gg 0$. (The subscript denotes "carrier" frequency.) In this case $s(t)$ is said to be a *bandpass* signal since it is designed to have frequencies located in a band about f_c. For example, an amplitude modulated (AM) broadcasting station with an assigned frequency of 850 kHz has a carrier frequency of $f_c = 850$ kHz. The mapping of the baseband input information waveform $m(t)$ into the bandpass signal $s(t)$ is called *modulation*. [$m(t)$ is the audio signal in AM broadcasting.] In Chapter 4 it will be shown that any bandpass signal has the form

$$s(t) = R(t) \cos[\omega_c t + \theta(t)] \tag{1-2}$$

where $\omega_c = 2\pi f_c$. If $R(t) = 1$ and $\theta(t) = 0$, $s(t)$ would be a pure sinusoid of frequency $f = f_c$ with *zero* bandwidth. In the modulation process provided by the carrier circuits, the baseband input waveform $m(t)$ causes $R(t)$ and/or $\theta(t)$ to change as a function of $m(t)$. These fluctuations in $R(t)$ and $\theta(t)$ cause $s(t)$ to have a *finite* bandwidth that depends on the characteristics of $m(t)$ and on the mapping functions used to generate $R(t)$ and $\theta(t)$. In Chapter 5 practical examples of both digital and analog bandpass signaling are presented.

Channels may be classified into two categories: hardwire and softwire. Some examples of *hardwire* channels are twisted-pair telephone lines, coaxial cables, waveguides, and fiber optic cables. Some typical *softwire* channels are air, vacuum, and seawater. It should be realized that the general principles of digital and analog modulation apply to all types of channels, although channel characteristics may impose constraints that favor a particular type of signaling. In general, the channel medium attenuates the signal so that the noise level of the channel and/or the noise introduced by an imperfect receiver cause the delivered information $\tilde{m}$ to be deteriorated from that of the source. The channel noise may arise from natural electrical disturbances (e.g., lightning) or from man-made sources such as high-voltage transmission lines, ignition systems of cars, or even switching circuits of a nearby digital computer. The channel may contain active amplifying devices, such as repeaters in telephone toll systems or satellite transponders in space communication systems. These devices, of course, are necessary to help keep the signal above the noise level. In addition, the channel may provide *multiple paths* between its input and output that have different time delays and attenuation characteristics. Even worse, these characteristics may vary with time. This variation produces signal fading at the channel output. You have probably observed this type of fading when listening to distant shortwave stations.

The receiver takes the corrupted signal at the channel output and converts it to a baseband signal that can be handled by the receiver baseband processor. The baseband processor "cleans up" this signal and delivers an estimate of the source information $\tilde{m}(t)$ to the communication system output.

The goal of the engineer is to design communication systems so that the information is transmitted to the sink with as little deterioration as possible while satisfying design constraints such as allowable transmitted energy, allowable signal bandwidth, and cost. In digital systems the measure of deterioration is usually taken to be the probability of bit error of the delivered data $\tilde{m}$. In analog systems the performance measure is usually taken to be the signal-to-noise ratio at the receiver output.

1-7 Frequency Allocations

In communication systems that use the atmosphere for the transmission channel, interference and propagation conditions are strongly dependent on the transmission frequency. Theoretically, any type of modulation (e.g., amplitude modulation, frequency modulation, single sideband, phase-shift keying, frequency-shift keying, etc.) could be used at any transmission frequency. However, to provide some semblance of order and for political reasons, government regulations specify the modulation type, bandwidth, and type of information that can be transmitted over designated frequency bands. On an international basis, frequency assignments and technical standards are set by two committees: the International Telegraph and Telephone Consultative Committee (CCITT) and the International Radio Consultative Committee (CCIR). These committees function under the auspices of International Telecommunications Union (ITU), which in turn is one of the organizations of the Economic and Social Council of the United Nations. An excellent overview of international, regional, and national standards organizations is the subject of a special issue of the *IEEE Communications Magazine* [Ryan, 1985]. In the United States the Federal Communication Commission (FCC) regulates radio, telephone, television, and satellite transmission systems. The international frequency assignments are divided into subbands by the FCC to accommodate 70 categories of services and 9 million transmitters. Table 1-2 gives a general listing of frequency bands, their common designations, typical propagation conditions, and typical services assigned to these bands.

1-8 Propagation of Electromagnetic Waves

The propagation characteristics of electromagnetic waves used in softwire channels are highly dependent on the frequency. This situation is shown in Table 1-2, where users are assigned frequencies that have the appropriate propagation characteristics for the coverage needed. The propagation characteristics are the result of changes in the radio-wave velocity as a function of altitude and boundary conditions. The wave velocity is dependent on changes in air temperature, air density, and levels of air ionization.

Ionization (i.e., free electrons) of the rarified air at high altitudes has a dominant effect on wave propagation in the medium-frequency (MF) and high-

TABLE 1-2 Frequency Bands

Frequency Band[a]	Designation	Propagation Characteristics	Typical Uses
3–30 kHz	Very low frequency (VLF)	Ground wave; low attenuation day and night; high atmospheric noise level	Long-range navigation; submarine communication
30–300 kHz	Low frequency (LF)	Similar to VLF, slightly less reliable; absorption in daytime	Long-range navigation and marine communication radio beacons
300–3000 kHz	Medium frequency (MF)	Ground wave and night sky wave; attenuation low at night and high in day; atmospheric noise	Maritime radio, direction finding, and emergency frequencies; AM broadcasting
3–30 MHz	High frequency (HF)	Ionospheric reflection varies with time of day, season, and frequency; low atmospheric noise at 30 MHz	Amateur radio; international broadcasting, military communication, long-distance aircraft and ship communication, telephone, telegraph, facsimile
30–300 MHz	Very high frequency (VHF)	Nearly line-of-sight (LOS) propagation, with scattering because of temperature inversions, cosmic noise	VHF television, FM two-way radio, AM aircraft communication, aircraft navigational aids
0.3–3 GHz	Ultra high frequency (UHF)	LOS propagation, cosmic noise	UHF television, navigational aids, radar, microwave links
	Letter designation		
1.0–2.0	L		
2.0–4.0	S		
3–30 GHz	Super high frequency (SHF)	LOS propagation; rainfall attenuation above 10 GHz, atmospheric attenuation because of oxygen and water vapor; high water vapor absorption at 22.2 GHz	Satellite communication, radar microwave links
	Letter designation		
2.0–4.0	S		
4.0–8.0	C		
8.0–12.0	X		
12.0–18.0	Ku		
18.0–27.0	K		
27.0–40.0	Ka		
26.5–40.0	R		

TABLE 1-2 Frequency Bands (*continued*)

Frequency Band[a]	Designation	Propagation Characteristics	Typical Uses
30–300 GHz	Extremely high frequency (EHF)	Same; high water vapor absorption at 183 GHz and oxygen absorption at 60 and 119 GHz	Radar, satellite, experimental
	Letter designation		
27.0–40.0	Ka		
26.5–40.0	R		
33.0–50.0	Q		
40.0–75.0	V		
75.0–110.0	W		
110–300	mm (millimeter)		
10^3–10^7 GHz	Infrared, visible light, and ultraviolet	LOS propagation	Optical communications

[a]kHz = 10^3 Hz; MHz = 10^6 Hz; GHz = 10^9 Hz.

frequency (HF) bands. The ionization is caused by ultraviolet radiation from the sun, as well as cosmic rays. Consequently, the amount of ionization is a function of the time of day, season of the year, and the sun activity (sunspots). This results in several layers of varying ionization density located at various heights surrounding the Earth.

The dominant ionized regions are D, E, F_1, and F_2 layers. The D layer is located closest to the Earth's surface at an altitude of about 45 or 55 miles. For $f > 300$ kHz, the D layer acts as a radio-frequency (RF) sponge to absorb (or attenuate) these radio waves. The attenuation is inversely proportional to frequency and becomes small for frequencies above 4 MHz. For $f < 300$ kHz, the D layer provides refraction (bending) of RF waves. The D layer is most pronounced during the daylight hours, with maximum ionization when the sun is overhead, and almost disappears at night. The E layer has a height of 65 to 75 miles, has maximum ionization around noon (local time), and practically disappears after sunset. It provides reflection of HF frequencies during the daylight hours. The F layer ranges in altitude between 90 and 250 miles. It ionizes rapidly at sunrise, reaches its peak ionization in early afternoon, and decays slowly after sunset. The F region splits into two layers, F_1 and F_2, during the day and combines into one layer at night. The F region is the most predominant medium in providing reflection of HF waves. As shown in Fig. 1-2, the electromagnetic spectrum may be divided into three broad bands that have one of three dominant propagation characteristics: ground wave, sky wave, and line of sight (LOS).

Ground-wave propagation is illustrated in Fig. 1-2a. It is the dominant mode of propagation for frequencies below 2 MHz. Here the electromagnetic wave tends to follow the contour of the Earth. That is, diffraction of the wave causes it to propagate along the surface of the Earth. This is the propagation mode used in

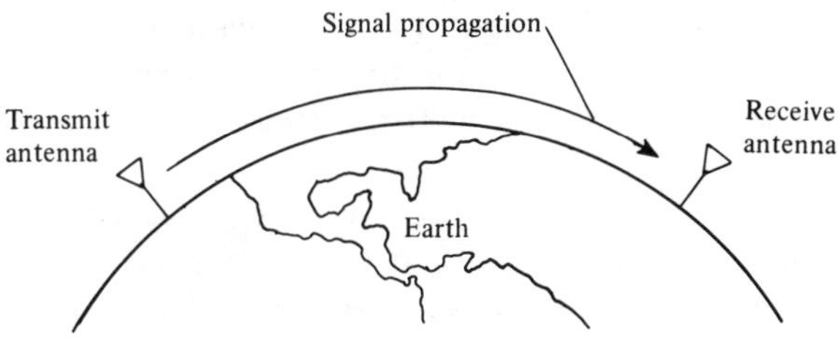

(a) Ground-Wave Propagation (Below 2 MHz)

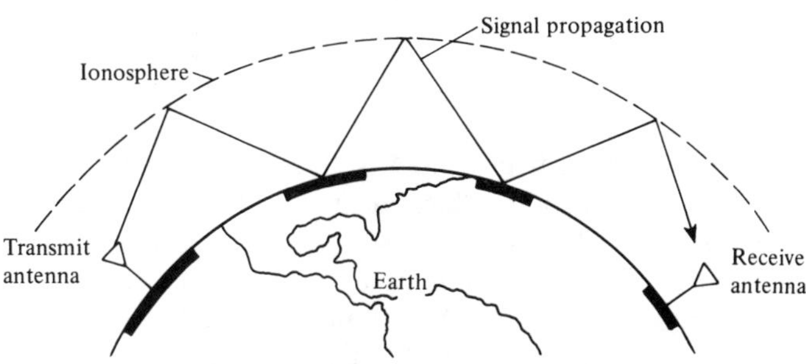

(b) Sky-Wave Propagation (2 to 30 MHz)

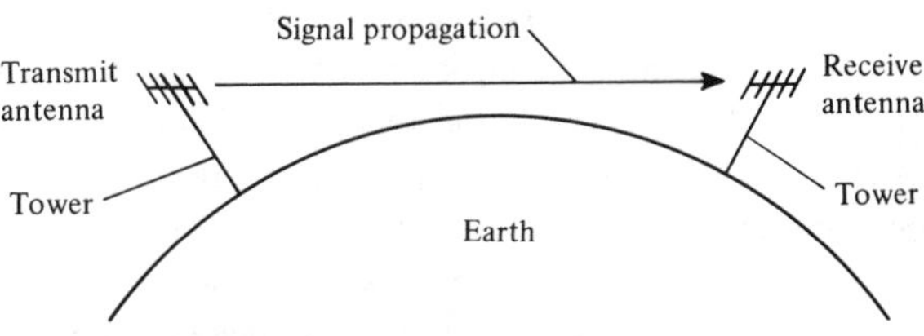

(c) Line-of-Sight (LOS) Propagation (Above 30 MHz)

FIGURE 1-2 Propagation of radio frequencies.

AM broadcasting, where the local coverage follows the Earth's contour and the signal propagates over the visual horizon. The question is often asked: What is the lowest radio frequency that can be used? The answer is that the value of the lowest useful frequency depends on how long you want to make the antenna. For efficient radiation, the antenna needs to be longer than $\frac{1}{10}$ of a wavelength. For example, for signaling with a carrier frequency of $f_c = 10$ kHz, the wavelength is

$$\lambda = \frac{c}{f_c} = \frac{(3 \times 10^8 \text{ m/sec})}{10^4} = 3 \times 10^4 \text{ m} \tag{1-3}$$

where c is the speed of light. (The formula $\lambda = c/f_c$ is distance = velocity × time, where the time needed to traverse one wavelength is $t = 1/f_c$.) Thus an antenna needs to be at least 3000 m in length for efficient electromagnetic radiation at 10 kHz.

Sky-wave propagation is illustrated in Fig. 1-2b. It is the dominant mode of propagation in the 2- to 30-MHz frequency range. Here long-distance coverage is obtained by reflecting the wave at the ionosphere and at the Earth's boundaries. Actually, in the ionosphere the waves are refracted (i.e., bent) gradually in an inverted U shape since the index of refraction varies with altitude as the ionization density changes. The refraction index of the ionosphere is given by [Griffiths, 1987; Jordan and Balmain, 1968]

$$n = \sqrt{1 - \frac{81N}{f^2}} \tag{1-4}$$

where n is the refractive index, N is the free electron density (number of electrons per cubic meter), and f is the frequency of the wave (in hertz). Typical N values range between 10^{10} and 10^{12} depending on the time of day, season, and the number of sunspots. In an ionized region $n < 1$, since $N > 0$, and outside the ionized region $n \approx 1$ since $N \approx 0$. In the ionized region, since $n < 1$, the waves will be bent according to Snell's law:

$$n \sin \varphi_r = \sin \varphi_i \tag{1-5}$$

where φ_i is the angle of incidence (between the wave direction and vertical) measured just below the ionosphere and φ_r is the angle of refraction for the wave (from vertical) measured in the ionosphere. Furthermore, the refraction index will vary with altitude within the ionosphere since N varies. For frequencies selected from the 2- to 30-MHz band the refraction index will vary with altitude over the appropriate range so that the wave will be bent back to Earth. Consequently, the ionosphere acts as a reflector. The transmitting station will have coverage areas as indicated in Fig. 1-2b by heavy black lines along the Earth's surface. The coverage near the transmit antenna is due to the ground-wave mode, and the other coverage areas are due to sky wave. Notice that there are areas of no coverage along the Earth's surface between the transmit and receive antennas. The angle of reflection and the loss of signal at an ionospheric reflection point depend on the frequency, the time of day, the season of the year, and the sunspot activity of the sun [Jordan, 1985, Chap. 33].

During the daytime (at the ionospheric reflection points), the electron density will be high so that $n < 1$. Consequently, sky waves from distant stations on the other side of the world will be heard on the shortwave bands. However, the D layer is also present during the day. This absorbs frequencies below 4 MHz. This is the case for AM broadcast stations, where distant stations cannot be heard during the day, but at night the layer disappears and distant AM stations can be heard via sky-wave propagation.

Sky-wave propagation is caused primarily by reflection from the F layer (90 to 250 miles in altitude). Because of this layer, international broadcast stations in

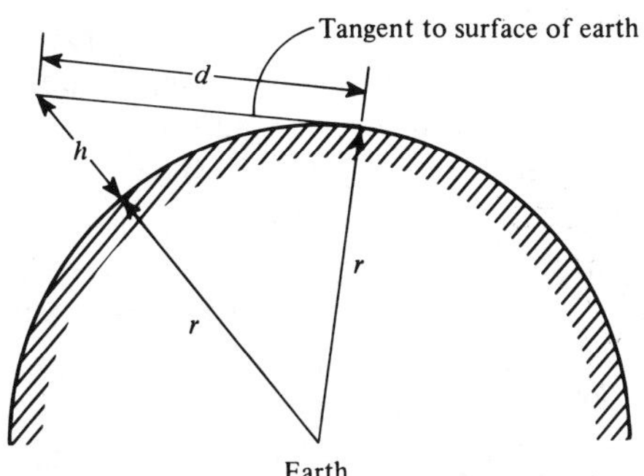

FIGURE 1-3 Calculation of distance to horizon.

the HF band can be heard from the other side of the world at almost any time during the day or night.

Line-of-sight (LOS) propagation is illustrated in Fig. 1-2c. This is the dominant mode for frequencies above 30 MHz. Here the electromagnetic wave propagates in a straight line. In this case $f^2 \gg 81N$, so that $n \approx 1$ and there is very little refraction by the ionosphere. In fact, the signal will propagate *through* the ionosphere. This property is used for satellite communications.

The LOS mode has the disadvantage that for communication between two terrestrial (Earth) stations, the signal path has to be above the horizon. Otherwise, the Earth will block the LOS path. Thus antennas need to be placed on tall towers so that the receiver antenna can "see" the transmitting antenna. A formula for the distance to the horizon, d, as a function of antenna height can be easily obtained by the use of Fig. 1-3. From this figure

$$d^2 + r^2 = (r + h)^2$$

or

$$d^2 = 2rh + h^2$$

where r is the radius of the Earth and h is the antenna height above the Earth's surface. In this application h^2 is negligible with respect to $2rh$. The radius of the Earth is 3960 statute miles. However, at LOS radio frequencies the effective Earth radius† is $\frac{4}{3}(3960)$ miles. Thus the distance to the radio horizon is

$$d = \sqrt{2h} \text{ miles} \tag{1-6}$$

where h is the antenna height measured in feet and d is in statute miles. For example, television stations have assigned frequencies above 30 MHz in the VHF or UHF range (see Table 1-2), and the fringe-area coverage of high-power stations is limited by the LOS radio horizon. For a television station with a 1000-ft tower, d is 44.7 miles. For a fringe-area viewer who has an antenna

† The refractive index of the atmosphere decreases slightly with height, which causes some bending of radio rays. This effect may be taken into account in LOS calculations by using an effective earth radius that is four thirds of the actual radius.

height of 30 ft, d is 7.75 miles. Thus for these transmitting and receiving heights, the television station would have fringe-area coverage out to a radius of 44.7 + 7.75 = 52.5 miles around the transmitting tower.

In addition to the LOS propagation mode, it is possible to have *ionospheric scatter propagation*. This mode occurs over the frequency range of 30 to 60 MHz, when the radio frequency signal is scattered because of irregularities in the refractive index of the lower ionosphere (about 50 miles above the earth's surface). Because of this scattering, communications can be carried out over path lengths of 1000 miles even though this is beyond the LOS distance. Similarly, *tropospheric scattering* (within 10 miles above the Earth's surface) can propagate radio frequency signals that are in the 40-MHz to 4-GHz range over paths of several hundred miles.

For more technical details about radio-wave propagation, the reader is referred to textbooks that include chapters on ground-wave and sky-wave propagation [Griffiths, 1987; Jordan and Balmain, 1968] and to a radio engineering handbook [Jordan, 1985]. A very readable description of this topic is also found in the ARRL handbook [ARRL, 1991], and a personal computer program, MINIMUF, is available that predicts sky-wave propagation conditions [Rose, 1982, 1984].

1-9 Information Measure

As we have seen, the purpose of communication systems is to transmit information from a source to a sink. However, what exactly is information, and how do we measure it? We know qualitatively that it is related to the surprise that is experienced when we receive the message. For example, the message "The ocean has been destroyed by a nuclear explosion" contains more information than the message "It is raining today."

DEFINITION: The *information* sent from a digital source when the jth message is transmitted is given by

$$I_j = \log_2\left(\frac{1}{P_j}\right) \text{ bits} \tag{1-7a}$$

where P_j is the probability of transmitting the jth message.†

From this definition we see that messages that are less likely to occur (smaller value for P_j) provide more information (larger value of I_j). We also observe that this information measure depends only on the likelihood of sending the message and does not depend on possible interpretation of the content as to whether or not it makes sense.

†The definition of probability is given in Appendix B.

The base of the logarithm determines the units used for the information measure. Thus for units of "bits," the base 2 logarithm is used. If the natural logarithm is used, the units are "nats"; and for base 10 logarithms, the unit is the "hartley," named after R. V. Hartley, who first suggested using the logarithm measure in 1928 [Hartley, 1948].

In this section the term *bit* denotes a unit of information as defined by (1-7a). In later sections, particularly in Chapter 3, *bit* is also used to denote a unit of binary data. Please do not confuse these two different meanings for the word *bit*. Some authors use *binit* to denote units of data and use *bit* exclusively to denote units of information. However, most engineers use the same word (*bit*) to denote both kinds of units, with the particular meaning understood from the context in which the word is used. This book follows that industry custom.

For ease of evaluating I_j on a calculator, (1-7a) can be written in terms of the base 10 logarithm or the natural logarithm.

$$I_j = -\frac{1}{\log_{10} 2} \log_{10} P_j = -\frac{1}{\ln 2} \ln P_j \tag{1-7b}$$

In general, the information content will vary from message to message since the P_j will not be equal. Consequently, we need an average information measure for the source, taking into account all the possible messages we can send.

DEFINITION: The *average information* measure of a digital source is

$$H = \sum_{j=1}^{m} P_j I_j = \sum_{j=1}^{m} P_j \log_2\left(\frac{1}{P_j}\right) \text{ bits} \tag{1-8}$$

where m is the number of possible different source messages and P_j is the probability of sending the jth message (m is finite since a digital source is assumed). The average information is called *entropy*.

EXAMPLE 1-1 EVALUATION OF INFORMATION AND ENTROPY

Find the information content of a message that consists of a digital word 12 digits long in which each digit may take on one of four possible levels. The probability of sending any of the four levels is assumed to be equal, and the level in any digit does not depend on the values taken on by previous digits.

In a string of 12 symbols (digits) where each symbol consists of one of four levels, there are $4 \cdot 4 \cdot \cdots \cdot 4 = 4^{12}$ different combinations (words) that can be obtained. Since each level is equally likely, all the different words are equally likely. Thus

$$P_j = \frac{1}{4^{12}} = \left(\frac{1}{4}\right)^{12}$$

or

$$I_j = \log_2\left(\frac{1}{(\frac{1}{4})^{12}}\right) = 12 \log_2(4) = 24 \text{ bits}$$

In this example we see that the information content in *any* of the possible messages equals that in any other possible message (24 bits). Thus the average information H is 24 bits.

Suppose that only two levels (binary) had been allowed for each digit and that all the words were equally likely. Then the information would be $I_j = 12$ bits for the binary words, and the average information would be $H = 12$ bits. Here, all the 12-bit words gave 12 bits of information since the words were equally likely. If they had not been equally likely, some of the 12-bit words would contain more than 12 bits of information and some would contain less, and the average information would have been less than 12 bits.

The rate of information is also important.

DEFINITION: The *source rate* is given by

$$R = \frac{H}{T} \text{ bits/sec} \tag{1-9}$$

where H is evaluated using (1-8) and T is the time required to send a message.

The definitions previously given apply to digital sources. The definitions for continuous sources are not presented here because the mathematics becomes more complicated and the physical meanings are more difficult to interpret. However, we do not need to be too concerned about the continuous case since, from a microscopic point of view, communication is inherently a discrete process, as pointed out by Shannon in 1948 and mentioned by Hartley in 1928. From a macroscopic point of view, analog sources can be approximated by digital sources with as much accuracy as we desire.

1-10 Ideal Communication Systems

A large number of criteria can be used to measure the effectiveness of a communication system to see if it is ideal or perfect. Some of these criteria are cost, channel bandwidth used, required transmitter power, signal-to-noise ratios at various points of the system, probability of bit error for digital systems, and time delay through the system.

In digital systems the optimum system might be defined as the system that minimizes the probability of bit error at the system output subject to constraints on transmitted energy and channel bandwidth. Thus methods for evaluation of bit error and signal bandwidth are of prime importance and are covered in subsequent chapters. This raises the question: Is it possible to invent a system with no bit error at the output even when we have noise introduced into the channel? This question was answered by Claude Shannon in 1948–1949 [Shannon, 1948, 1949]. The answer is yes, under certain assumptions. Shannon showed that (for the case of signal plus white Gaussian noise) a channel capacity C (bits/sec) could be calculated such that if the rate of information R (bits/sec) was less than C, the probability of error would approach zero. The equation for C is

$$C = B \log_2\left(1 + \frac{\mathrm{S}}{\mathrm{N}}\right) \tag{1-10}$$

where B is the channel bandwidth in hertz (Hz) and S/N is the signal-to-noise power ratio (watts/watts, not dB) at the input to the digital receiver. Shannon does not tell us how to build this system, but he proves that it is theoretically possible to have such a system. Thus Shannon gives us a theoretical performance bound that we can strive to achieve with practical communication systems. Systems that approach this bound usually incorporate error correction coding. The channel noise still causes errors at the input to the receiver decoder; however, enough redundancy has been added in the transmitted signal so that the decoder can detect and correct errors with its processing circuits.

In analog systems the optimum system might be defined as the one that achieves the largest signal-to-noise ratio at the receiver output subject to design constraints such as channel bandwidth and transmitted power. Here the evaluation of the output signal-to-noise ratio is of prime importance. We might ask the question: Is it possible to design a system with infinite signal-to-noise ratio at the output when noise is introduced by the channel? The answer, of course, is no. The performance of practical analog systems with respect to that of Shannon's ideal system is illustrated in Chapter 7 (Fig. 7-27).

The following section describes the improvement that can be obtained in digital systems when coding is used and how these coded systems compare with Shannon's ideal system.

1-11 Coding

If the data at the output of a digital communication system have errors that are too frequent for the desired use, the errors can often be reduced by the use of either of two main techniques.

- Automatic repeat request (ARQ).
- Forward error correction (FEC).

In an ARQ system, when a receiver circuit detects errors in a block of data, it requests that the data block be retransmitted. In an FEC system, the transmitted data are encoded so that the receiver can correct as well as detect errors. These procedures are also classified as *channel coding* because they are used to correct errors caused by channel noise. This is different from *source coding*, described in Chapter 3, where the purpose of the coding is to extract the essential informa-

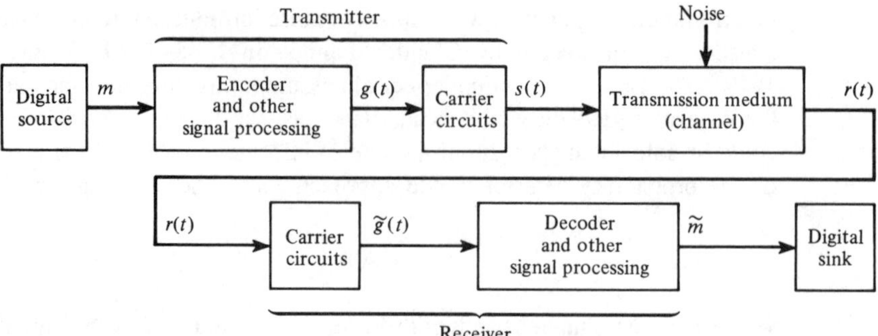

FIGURE 1-4 General digital communication system.

tion from the source and encode it into digital form so that it can be saved or transmitted using digital techniques.

The choice between using the ARQ or the FEC technique depends on the particular application. ARQ is generally used in computer communication systems because it is relatively inexpensive to implement and there is usually a duplex (two-way) channel so that the receiving end can transmit back an acknowledgment (ACK) for correctly received data or a request for retransmission (NAC) when the data are received in error. (See Appendix C, Section C-4, for examples of ARQ signaling.) FEC techniques are used to correct errors on simplex (one-way) channels where returning of an ACK/NAC indicator (required for the ARQ technique) is not feasible. FEC is preferred on systems with large transmission delays because if the ARQ technique were used, the effective data rate would be small; the transmitter would have long idle periods while waiting for the ACK/NAC indicator, which is retarded by the long transmission delay. Since ARQ systems are covered in Appendix C, we concentrate on FEC techniques in the remainder of this section.

Communication systems with FEC are illustrated in Fig. 1-4, where encoding and decoding blocks have been designated. From a *theoretical viewpoint*, Shannon's channel capacity theorem states that a finite value for S/N limits only the *rate* of transmission. That is, the probability of error, $P(E)$, can approach zero provided the information rate is less than the channel capacity. Of course, this theorem implies that coding is needed to make $P(E) \rightarrow 0$. The question is: *What $P(E)$ can be achieved when practical coding is used?*

The topic of coding is immense, and we will not even attempt to give an introduction to the numerous coding techniques that have been developed. However, we will summarize some of the basic concepts involved and cite results that indicate the practical improvement that can be achieved when coding is used. For further introduction to the topic of coding, an excellent textbook by Sweeney is recommended [Sweeney, 1991].

Coding provides improved performance because of the following phenomena:

- *Redundancy*. Redundant bits are added by the coder to accentuate the uniqueness of each (digital) message.

- *Noise averaging*. The code is designed so that the receiver can average out the noise over long time spans, T_0, where T_0 becomes very large.

The appropriate code is selected to provide the desired error performance within the allowed cost.

Codes may be classified into two broad categories:

- *Block codes*. A block code is a mapping of k input binary symbols into n output binary symbols. Consequently, the block coder is a *memoryless* device. Since $n > k$, the code can be selected to provide redundancy, such as *parity bits*, which are used by the decoder to provide some error detection and error correction. The codes are denoted by (n,k), where the code rate R is defined by $R = k/n$. Practical values of R range from $\frac{1}{4}$ to $\frac{7}{8}$, and k ranges from 3 to several hundred [Clark and Cain, 1981].
- *Tree codes*. A tree code is produced by a coder that has *memory*. *Convolutional codes* are a subset of tree codes. The convolutional coder accepts k binary symbols at its input and produces n binary symbols at its output, where the n output symbols are affected by $v + k$ input symbols. Memory is incorporated since $v > 0$. The code rate is defined by $R = k/n$. Typical values for k and n range from 1 to 8, and the values for v range from 2 to 60. The range of R is between $\frac{1}{4}$ and $\frac{7}{8}$ [Clark and Cain, 1981]. A small value for code rate R indicates a high degree of redundancy, which should provide more effective error control at the expense of reducing the information throughput.

★ Block Codes

Before discussing block codes, several definitions are needed. The *Hamming weight* of a code word is the number of binary 1 bits. For example, the code word 110101 has a Hamming weight of 4. The *Hamming distance* between two code words, denoted by d, is the number of positions by which they differ. For example, the code words 110101 and 111001 have a distance of $d = 2$. A received code word can be checked for errors. Some of the errors can be detected and corrected if $d \geq s + t + 1$, where s is the number of errors that can be detected and t is the number of errors that can be corrected ($s \geq t$). Thus a pattern of t or fewer errors can be both detected and corrected if $d \geq 2t + 1$.

A general code word can be expressed in the form

$$i_1 i_2 i_3 \cdots i_k p_1 p_2 p_3 \cdots p_r$$

where k is the number of information bits, r is the number of parity check bits, and n is the total word length in the (n,k) block code, where $n = k + r$. This arrangement of the information bits at the beginning of the code word followed by the parity bits is most common. Such a block code is said to be *systematic*. Other arrangements with the parity bits interleaved between the information bits are possible and are usually considered to be equivalent codes. For the parity bits to provide error detection and error correction, they must be functionally related to the information bits. This is accomplished with modulo 2 arithmetic:[†]

$$\begin{aligned} p_1 &= z_{11}i_1 \oplus z_{12}i_2 \oplus \cdots z_{1k}i_k \\ p_2 &= z_{21}i_1 \oplus z_{22}i_2 \oplus \cdots z_{2k}i_k \\ &\vdots \\ p_r &= z_{r1}i_1 \oplus z_{r2}i_2 \oplus \cdots z_{rk}i_k \end{aligned}$$

where

$$\mathbf{Z} = \begin{bmatrix} z_{11} & z_{12} & \cdots & z_{1k} \\ z_{21} & z_{22} & \cdots & z_{2k} \\ \vdots & & & \\ z_{r1} & z_{r2} & \cdots & z_{rk} \end{bmatrix}$$

The z_{ij} are binary numbers (0 or 1) that are specified to give the desired set of code words for the (n,k) block code. (The specific case of Hamming block codes is presented in the next section.)

The encoding process is conveniently represented by matrix notation using modulo 2 arithmetic. Let $\mathbf{c}$ be an n-dimensional row vector (binary-valued elements) that is the n-bit code word $i_1 i_2 i_3 \cdots i_k p_1 p_2 \cdots p_r$. That is, $\mathbf{c}$ is the result of encoding a k-bit information word where the information (message) word is the k-dimensional row vector $\mathbf{m}$. The encoder generates $\mathbf{c}$ from $\mathbf{m}$ by using[‡]

$$\mathbf{c} = \mathbf{mG} \tag{1-11a}$$

where $\mathbf{G}$ is the *generator* matrix. $\mathbf{G}$ is a $k \times n$ (k rows by n columns) matrix which is formed by combining a $k \times k$ identity matrix with the transpose of the $\mathbf{Z}$ matrix, $\mathbf{Z}^T$. That is,

$$\mathbf{G} = [\mathbf{I} \mid \mathbf{Z}^T] = \left[\begin{array}{ccccc|cccc} 1 & 0 & 0 & \cdots & 0 & z_{11} & z_{21} & \cdots & z_{r1} \\ 0 & 1 & 0 & \cdots & 0 & z_{12} & z_{22} & \cdots & z_{r2} \\ \vdots & & & & & & & & \\ 0 & 0 & 0 & \cdots & 1 & z_{1k} & z_{2k} & \cdots & z_{rk} \end{array}\right] \tag{1-11b}$$

For example, for a (7,4) code assume that

$$\mathbf{G} = \begin{bmatrix} 1 & 0 & 0 & 0 & 1 & 1 & 0 \\ 0 & 1 & 0 & 0 & 1 & 0 & 1 \\ 0 & 0 & 1 & 0 & 0 & 1 & 1 \\ 0 & 0 & 0 & 1 & 1 & 1 & 1 \end{bmatrix} \tag{1-11c}$$

[†]Modulo 2 multiplication is $0 \cdot 0 = 0$, $0 \cdot 1 = 0$, $1 \cdot 0 = 0$, and $1 \cdot 1 = 1$. Modulo 2 summation is $0 \oplus 0 = 0$, $0 \oplus 1 = 1$, $1 \oplus 0 = 1$, and $1 \oplus 1 = 0$. The summation is equivalent to an exclusive OR operation.

[‡]Some books use column matrices for code word representations. Either column or row representation is valid as long as all equations are consistent.

Then for the information word $\mathbf{m} = [1011]$, the corresponding (7,4) code word, $\mathbf{c}$, is

$$\mathbf{c} = [1011]\begin{bmatrix} 1 & 0 & 0 & 0 & 1 & 1 & 0 \\ 0 & 1 & 0 & 0 & 1 & 0 & 1 \\ 0 & 0 & 1 & 0 & 0 & 1 & 1 \\ 0 & 0 & 0 & 1 & 1 & 1 & 1 \end{bmatrix} = [1011010]$$

As expected, the first 4 bits of the code word are the original information word and the last 3 bits of the code word are the parity bits. That is, a systematic code word is obtained. (Nonsystematic code words can be obtained by exchanging columns of the identity matrix with the columns of the $\mathbf{Z}^T$ matrix.)

The forward error detection and correction (FEC) characteristics of block codes will now be examined. A received code word row vector will be denoted by $\tilde{\mathbf{c}}$. The tilde denotes that the received vector may have some errors (i.e., some of the binary 0's may be binary 1's, and vice versa). The errors can be caused by noise in the channel as well as other impairments, such as poor frequency response in the channel. The function of the receiver decoder is to try to recover the original information word, **m,** from $\tilde{\mathbf{c}}$. Consequently, the first task of the decoder is to determine if there are errors in $\tilde{\mathbf{c}}$ and, if so, to correct them to produce **c.** This is accomplished by computing the row vector, **s,** called the *syndrome.*† If $\mathbf{s} = \mathbf{0}$, no errors are detected, and if $\mathbf{s} \neq \mathbf{0}$, single errors can be corrected as demonstrated in Example 1-2, which follows. The syndrome is evaluated by

$$\mathbf{s} = \tilde{\mathbf{c}}\mathbf{H}^T \tag{1-12a}$$

where **H** is the *parity check* matrix. **H** is a $(n - k) \times n$-order matrix that is obtained by combining the matrix **Z** with a $(n - k) \times (n - k)$-order identity matrix.

$$\mathbf{H} = [\mathbf{Z} \mid \mathbf{I}] = \left[\begin{array}{cccc|ccccc} z_{11} & z_{12} & \cdots & z_{1k} & 1 & 0 & 0 & \cdots & 0 \\ z_{21} & z_{22} & \cdots & z_{2k} & 0 & 1 & 0 & \cdots & 0 \\ \vdots & & & & & & & & \\ z_{r1} & z_{r2} & \cdots & z_{rk} & 0 & 0 & 0 & \cdots & 1 \end{array}\right] \tag{1-12b}$$

In the next section, use of the syndrome for FEC is illustrated.

★Hamming Block Codes

To find good long codes is more easily said than done. The design of codes consists, to a large extent, of finding a suitable **Z** matrix which gives a parity check matrix that can be used not only to detect errors but also to correct errors. Hamming has given a procedure for designing block codes that have single-error-correction capability [Hamming, 1950].

†In medical terminology, a syndrome is a pattern of symptoms that are used to diagnose a disease. In this case, the "disease" consists of bit errors.

A Hamming code is a block code having a Hamming distance of 3. Since $d \geq 2t + 1$, $t = 1$ and a single error can be detected and corrected. For this code, the parity check matrix, **H**, and the **Z** matrix are easy to obtain. The **H** matrix is determined by choosing the **Z** matrix within **H** such that the columns of **H** contain all the possible r-bit binary words except for the all-zero word. This will allow generation of the (n,k) code word set (where $n = k + r$) by use of the generator matrix (1-11). However, only certain (n,k) codes are allowable. These allowable Hamming codes are

$$(n,k) = (2^m - 1,\ 2^m - 1 - m)$$

where m is an integer and $m \geq 3$. Thus some of the allowable codes are (7,4), (15,11), (31,26), (63,57), and (127,120). The code rate R approaches 1 as m becomes large, which indicates that the information throughput efficiency becomes very good for large m.

EXAMPLE 1-2 HAMMING CODE

Determine a parity check matrix for a (7,4) Hamming code and demonstrate that a single error can be detected and corrected.

As discussed above, it is possible for a (7,4) code to be designed such that it is a Hamming code. The procedure is to find a suitable $(n - k) \times n$-order parity check matrix, which is 3×7 order for the (7,4) code, where the columns of the **H** matrix are the binary-coded-decimal representations of the numbers 1 through $2^r - 1$. For this example these are the numbers 1 through 7. Using (1-12b), one possible **H** matrix is

$$\mathbf{H} = \begin{bmatrix} 1 & 1 & 0 & 1 & 1 & 0 & 0 \\ 1 & 0 & 1 & 1 & 0 & 1 & 0 \\ 0 & 1 & 1 & 1 & 0 & 0 & 1 \end{bmatrix}$$

because the columns of this matrix represent all the binary-coded-decimal numbers 1 through 7. (This is the same parity check matrix that was used in the preceding section, where the computation of the code word for an information word was illustrated by use of the corresponding code generation matrix.)

To demonstrate the FEC capability of this code, a syndrome will be evaluated. Assume that the message out of the information source is $\mathbf{m} = [1011]$. As shown in the preceding section, the transmitted (correct) code word is $\mathbf{c} = [1011010]$. But suppose that due to noise, the code word received is $\tilde{\mathbf{c}} = [1010010]$. Then, using (1-12a), the syndrome $\mathbf{s} = \tilde{\mathbf{c}}\mathbf{H}^T$ is

$$\mathbf{s} = [1010010] \begin{bmatrix} 1 & 1 & 0 \\ 1 & 0 & 1 \\ 0 & 1 & 1 \\ 1 & 1 & 1 \\ 1 & 0 & 0 \\ 0 & 1 & 0 \\ 0 & 0 & 1 \end{bmatrix} = [111]$$

Because this syndrome is nonzero, there is an error. Furthermore, the bit position of the error within the received code word is indicated by matching the syndrome to a column in the **H** matrix. Here **s** = [111] is identical to the fourth column (from the left) of the **H** matrix. This indicates that the fourth bit in the received code word is in error, so that $\tilde{\mathbf{c}} = [1010010]$ can be corrected to $\mathbf{c} = [1011010]$. Since the first 4 bits of this corrected code word are the decoded message bits, the receiver decoder would output the correct message $\mathbf{m} = [1011]$ and FEC has been achieved to correct the single error.

In addition to Hamming codes, there are many other types of block codes. One popular class consists of the cyclic codes. *Cyclic codes* are block codes such that another code word can be obtained by taking any one code word, shifting the bits to the right, and placing the dropped-off bits on the left. These types of codes have the advantage of being easily encoded from the message source by the use of inexpensive linear shift registers with feedback. This structure also allows these codes to be easily decoded. Examples of cyclic and related codes are Bose–Chaudhuri–Hocquenhem (BCH), Reed–Solomon, Hamming, maximal-length, Reed–Müller, and Golay codes. Some properties of block codes are given in Table 1-3 [Bhargava, 1983].

★ Convolutional Codes

A convolutional encoder is illustrated in Fig. 1-5. Here k bits (one input frame) are shifted in each time and, concurrently, n bits (one output frame) are shifted out, where $n > k$. Thus every k-bit input frame produces an n-bit output frame. Redundancy is provided in the output, since $n > k$. Also, there is memory in the coder, since the output frame depends on the previous K input frames where $K > 1$. The *code rate* is $R = k/n$, which is $\frac{3}{4}$ in this illustration. The *constraint*

TABLE 1-3 Properties of Block Codes

	Code[a]			
Property	**BCH**	**Reed–Solomon**	**Hamming**	**Maximal-Length**
Block length	$n = 2^m - 1$ $m = 3, 4, 5, \ldots$	$n = m(2^m - 1)$ bits	$n = 2^m - 1$	$n = 2^m - 1$
Number of parity bits		$r = m2t$ bits	$r = m$	
Minimum distance	$d \geq 2t + 1$	$d = m(2t + 1)$ bits	$d = 3$	$d = 2^m - 1$
Number of information bits	$k \geq n - mt$			$k = m$

[a] m is any positive integer unless otherwise indicated; n is the block length; k is the number of information bits.

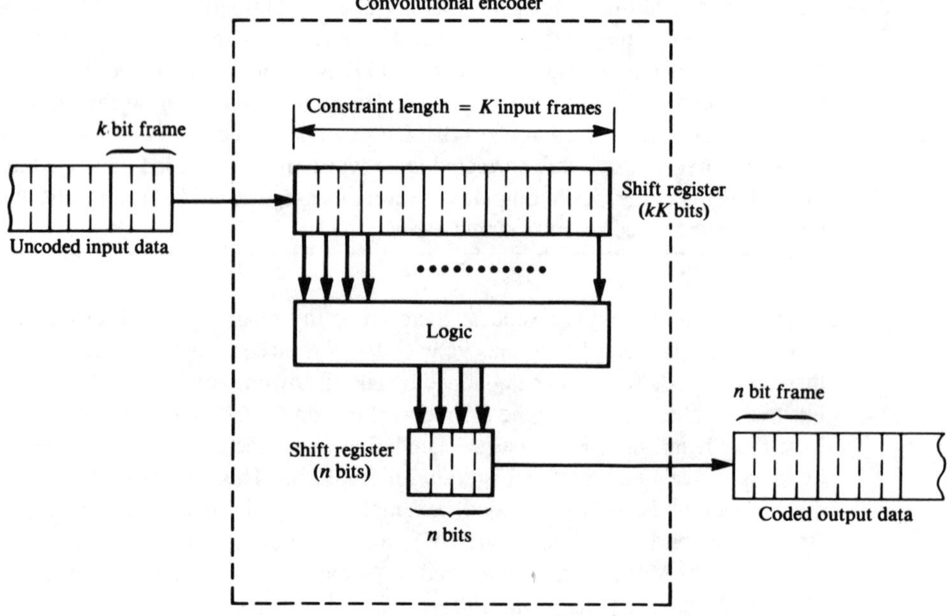

FIGURE 1-5 Convolutional encoding ($k = 3$, $n = 4$, $K = 5$, and $R = \frac{3}{4}$).

length, K, is the number of input frames that are held in the kK-bit shift register.[†] Depending on the particular convolutional code that is to be generated, data from the kK stages of the shift register are added (modulo 2) and used to set the bits in the n-stage output register.

For example, consider the convolutional coder shown in Fig. 1-6. Here $k = 1$, $n = 2$, $K = 3$, and a commutator with two inputs performs the function of a two-stage output shift register. The convolutional code is generated by inputting a bit of data and then giving the commutator a complete revolution. This process is repeated for successive input bits to produce the convolutionally encoded output. In this example, each $k = 1$ input bit produces $n = 2$ output bits, so the code rate is $R = k/n = \frac{1}{2}$. The code tree of Fig. 1-7 gives the encoded sequences for the convolutional encoder example of Fig. 1-6. To use the code tree, one moves up if the input is a binary 0 or down if the input is a binary 1. The corresponding encoded bits are shown in parentheses. For example, if the input sequence $x_{11} = 1010$ is fed into the input (most recent input bit on the right), the corresponding encoded output sequence is $y_{11} = 11010001$, as shown by path A in Fig. 1-7.

A convolutionally encoded signal is decoded by "matching" the encoded received data to the corresponding bit pattern in the code tree. In sequential decoding (a suboptimal technique), the path is found like that of a car driver who

[†] Several different definitions for constraint length are used in the literature [Blahut, 1983; Clark and Cain, 1981; Proakis, 1983].

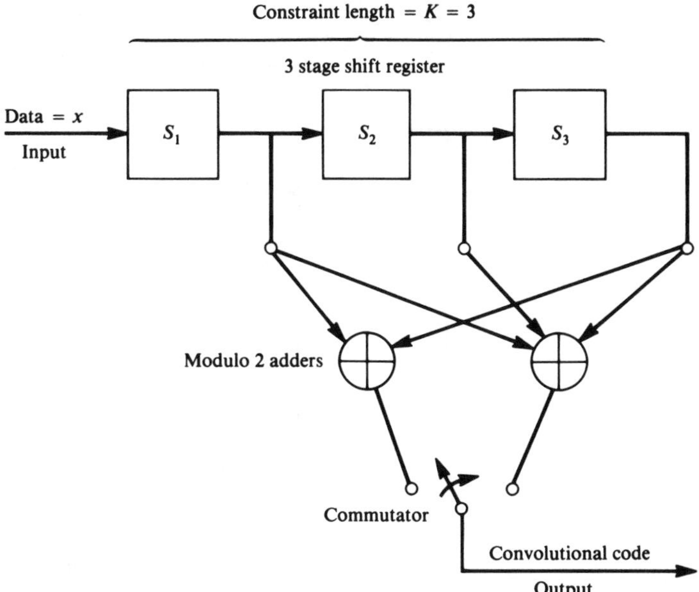

FIGURE 1-6 Convolutional encoder for a rate $\frac{1}{2}$, constraint length 3 code.

occasionally makes a wrong turn at a fork in a road but discovers the mistake, goes back, and tries another path. For example, if $y_{11} = 11010010$ was received, path A would be the closest match and the decoded data would be $x_{11} = 1010$. If noise was present in the channel, some of the received encoded bits might be in error, and then the paths would not match exactly. In this case, the match is found by choosing a path that will minimize the Hamming distance between the selected path sequence and the received encoded sequence.

An optimum decoding algorithm, called *Viterbi decoding*, uses a similar procedure. It examines the possible paths and selects the best ones based on some conditional probabilities [Forney, 1973]. The optimum Viterbi procedure uses so-called soft decisions. A *soft decision* algorithm first decides the result based on the test statistic[†] being above or below a decision threshold and then gives a "confidence" number that specifies how close the test statistic was to the threshold value. In *hard decisions* only the decision output is known, and it is not known if the decision was almost "too close to call" (because the test value was almost equal to the threshold value). The soft decision technique can translate into a 2-dB improvement (decrease) in the required receiver input E_b/N_0 [Clark and Cain, 1981]. E_b is the received signal energy over a one-bit time interval, and $N_0/2$ is the power spectral density (PSD) of the channel noise at the receiver input. Both E_b and N_0 will be defined in detail in later chapters—for example, see (5-91).

[†]The test statistic is a value that is computed at the receiver based on the receiver input during some specified time interval. See (7-4) and (8-48a) for examples of test statistics.

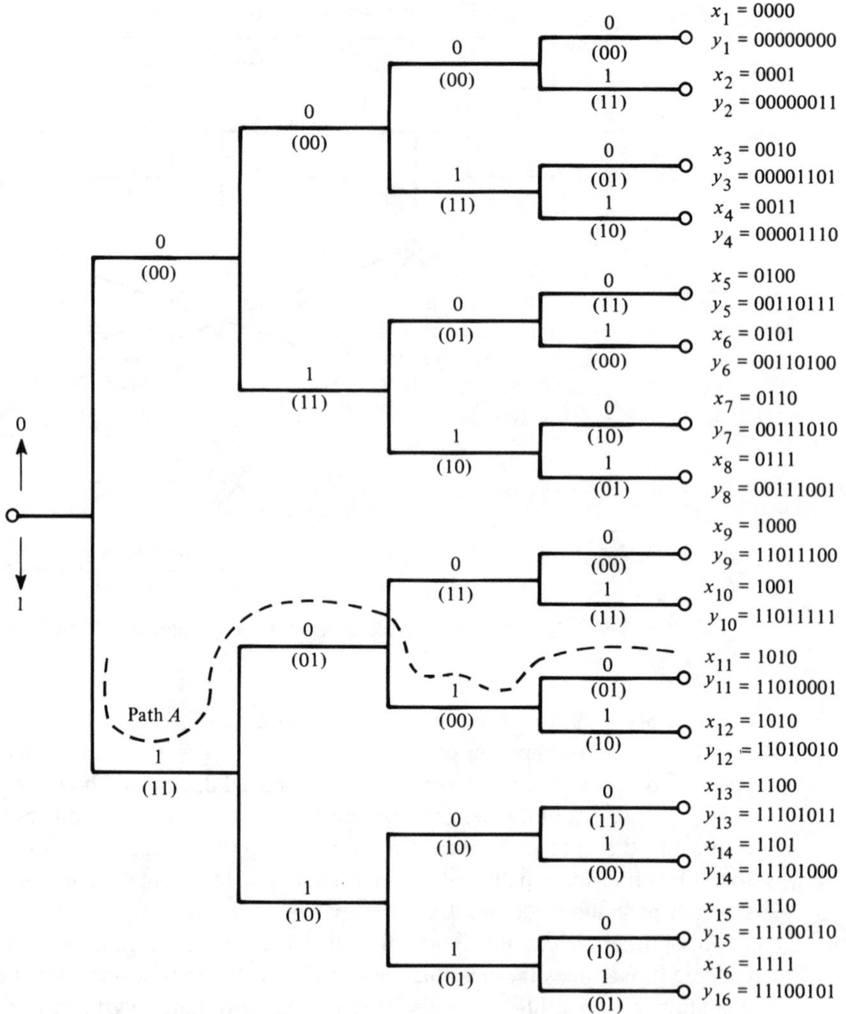

FIGURE 1-7 Code tree for convolutional encoder of Figure 1-6.

★ Code Interleaving

In the previous discussion it was assumed that if no coding was used, the channel noise would cause random bit errors at the receiver output that are more or less isolated, that is, not adjacent. When coding was added, the code redundancy allowed the receiver decoder to correct the errors so that the decoded output was almost error free. However, in some applications, large wide pulses of channel noise occur—such as in pulse jamming or in telephone channels with switching noise. If the usual coding techniques are used in these situations, bursts of errors will occur at the decoder output because the noise bursts are wider than the "redundancy time" of the code. This situation can be ameliorated by the use of code interleaving.

At the transmitting end the coded data are interleaved by shuffling (i.e., like shuffling a deck of cards) the coded bits over a time span of several block lengths (for block codes) or several constraint lengths (for convolutional codes). The required span length is several times the duration of the noise burst. At the receiver, before decoding, the data with error bursts are deinterleaved to produce coded data with isolated errors. The isolated errors are then corrected by passing the coded data through the decoder. This produces almost error-free output even when noise bursts occur at the receiver input. These are two classes of interleavers—block interleavers and convolutional interleavers [Sklar, 1988].

★Code Performance

The improvement in the performance of a digital communication system that can be achieved by the use of coding is illustrated in Fig. 1-8. It is assumed that a digital signal plus channel noise is present at the receiver input. The performance of a system that uses binary-phase-shift-keyed (BPSK) signaling is shown both for the case when coding is used and for the case when there is no coding. For the no code case, the optimum (matched filter) detector circuit is used in the receiver as derived in Chapter 7 and described by (7-38). For the coded case a (23,12)

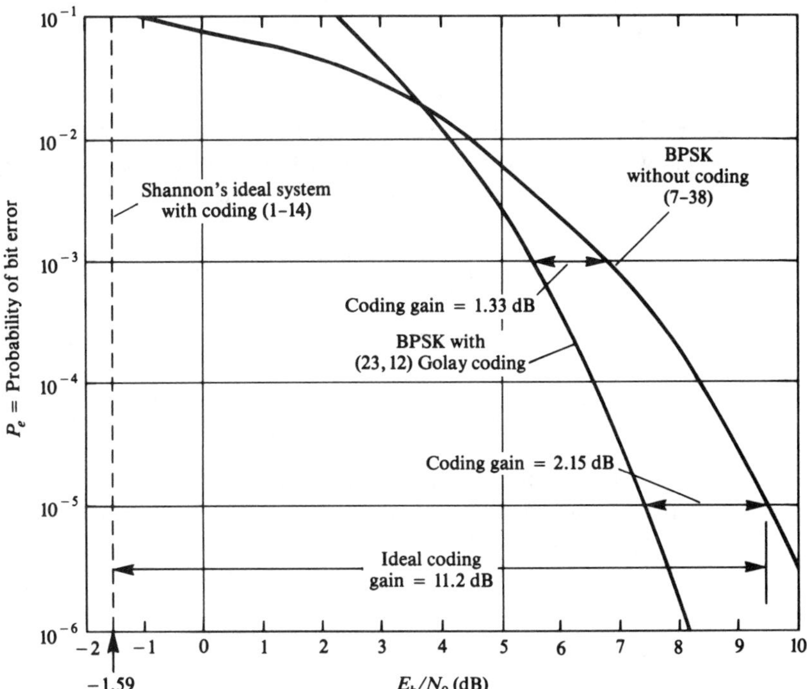

FIGURE 1-8 Performance of digital systems—with and without coding.

Golay code is used. P_e is the *probability of bit error*—also called the *bit error rate* (BER)—that is measured at the receiver output. E_b/N_0 is the energy-per-bit/noise-density ratio at the receiver input (as described in the preceding section). The *coding gain* is defined as the reduction in E_b/N_0 (in decibels) that is achieved when coding is used when compared to the E_b/N_0 required for the uncoded case at some specific level of P_e. For example, as can be seen in the figure, a coding gain of 1.33 dB is realized for a BER of 10^{-3}. The coding gain increases if the BER is smaller so that a coding gain of 2.15 dB is achieved when $P_e = 10^{-5}$. This improvement is significant in space communication applications where every decibel of improvement is valuable. It is also noted that there is a *coding threshold* in the sense that the coded system actually provides *poorer* performance than the uncoded system when E_b/N_0 is less than the threshold value. In this example the coding threshold is about 3.5 dB. A coding threshold is found in all coded systems.

For optimum coding, Shannon's channel capacity theorem, (1-10), gives the E_b/N_0 required. That is, if the source rate is below the channel capacity, the optimum code will allow the source information to be decoded at the receiver with $P_e \to 0$ (i.e., $10^{-\infty}$) even though there is some noise in the channel. We will now find the E_b/N_0 required so that $P_e \to 0$ with the optimum (unknown) code. The optimum encoded signal is not restricted in bandwidth, so, from (1-10),

$$C = \lim_{B\to\infty} \left\{ B \log_2\left(1 + \frac{S}{N}\right)\right\} = \lim_{B\to\infty} \left\{ B \log_2\left(1 + \frac{E_b/T_b}{N_0 B}\right)\right\}$$

$$= \lim_{x\to 0} \left\{ \frac{\log_2[1 + E_b/N_0 T_b)x]}{x} \right\}$$

where T_b is the time that it takes to send one bit and N is the noise power that occurs within the bandwidth of the signal. The power spectral density (PSD) is $\mathcal{P}_n(f) = N_0/2$, and, as shown in Chapter 2, the noise power is

$$N = \int_{-B}^{B} \mathcal{P}_n(f)\, df = \int_{-B}^{B} \left(\frac{N_0}{2}\right) df = N_0 B$$

where B is the signal bandwidth. L'Hospital's rule is used to evaluate this limit.

$$C = \lim_{x\to 0} \left\{ \frac{1}{1 + (E_b/N_0 T_b)x} \left(\frac{E_b}{N_0 T_b}\right) \log_2 e \right\} = \frac{E_b}{N_0 T_b \ln 2} \tag{1-13}$$

If we signal at a rate approaching the channel capacity, $P_e \to 0$ and we have the maximum information rate allowed for $P_e \to 0$ (i.e., the optimum system). Thus $1/T_b = C$, or, using (1-13),

$$\frac{1}{T_b} = \frac{E_b}{N_0 T_b \ln 2}$$

or

$$E_b/N_0 = \ln 2 = -1.59 \text{ dB} \tag{1-14}$$

Thus, if optimum coding is used, $P_e \rightarrow 0$ as long as $E_b/N_0 > -1.59$ dB. This is indicated in Fig. 1-8 by the dashed vertical line.

When the performance of the optimum encoded signal is compared to that of BPSK without coding (10^{-5} BER), it is seen that the optimum (unknown) coded signal has a coding gain of $9.61 - (-1.59) = 11.2$ dB. Using Fig. 1-8, compare this to the coding gain of 2.15 dB that is achieved when a (23,12) Golay-encoded BPSK signal is used. Table 1-4 shows the coding gains that can be obtained for some other codes [Bhargava, 1983].

All of the codes described above (excluding the ideal unknown code, predicted by Shannon) achieve their coding gains at the expense of *bandwidth expansion*. That is, when the redundant bits are added to provide the coding gain, the overall data rate and, consequently, the bandwidth are increased for the same information rate. Accordingly, we have to reduce the information rate when we

TABLE 1-4 Coding Gains with BPSK or QPSK

Coding Technique Used	Coding Gain (dB) at 10^{-5} BER	Coding Gain (dB) at 10^{-8} BER	Data Rate Capability
Ideal coding	11.2	13.6	
Concatenated[a] Reed–Solomon and convolution (Viterbi decoding)	6.5–7.5	8.5–9.5	Moderate
Convolutional with sequential decoding (soft decisions)	6.0–7.0	8.0–9.0	Moderate
Block codes (soft decisions)	5.0–6.0	6.5–7.5	Moderate
Concatenated[a] Reed–Solomon and short block	4.5–5.5	6.5–7.5	Very high
Convolutional with Viterbi decoding	4.0–5.5	5.0–6.5	High
Convolutional with sequential decoding (hard decisions)	4.0–5.0	6.0–7.0	High
Block codes (hard decisions)	3.0–4.0	4.5–5.5	High
Block codes with threshold decoding	2.0–4.0	3.5–5.5	High
Convolutional with threshold decoding	1.5–3.0	2.5–4.0	Very high

Source: Bhargava [1983].

[a]Two different encoders are used in series at the transmitter (see Fig. 1-4), and the corresponding decoders are used at the receiver.

attempt to implement these coding techniques on a bandlimited channel. Therefore, these techniques are not usually incorporated into modems that are designed for data communications over voice-frequency (VF) telephone channels that pass frequencies over a band in the 300- to 3000-Hz range. However, Gottfried Ungerboeck has invented a technique called *trellis-coded modulation* (TCM) that combines multilevel modulation and coding to achieve coding gain without bandwidth expansion [Ungerboeck, 1982, 1987; Biglieri, Divsalar, McLane, and Simon, 1991].† (TCM has been adopted for use in the CCITT V.32 modem that allows an information data rate of 9600 b/s (bits per second) to be transmitted over VF lines—see Table C-11.) As described earlier by (1-2), modulation is the process of mapping the baseband source information into amplitude and phase variations—$R(t)$ and $\theta(t)$, respectively—of a bandpass sinusoidal carrier signal, $s(t)$, which has a spectrum that is concentrated about the carrier frequency, f_c. The variations can be expressed mathematically as the complex envelope, $g(t)$, where

$$g(t) = R(t)e^{j\theta(t)} \tag{1-15}$$

(The complex envelope is indicated on the communication system block diagram in Fig. 1-4, and it will be described in detail in Chapters 4 and 5.) As seen in Fig. 1-9a, which describes conventional coded systems, the coding and modulation operations are carried out separately by first encoding the baseband data to produce a coded baseband waveform and then modulating this onto the carrier as amplitude and phase variations.

The combined modulation and coding operation of TCM is shown in Fig. 1-9b. Here, the serial data from the source, $m(t)$, are converted into parallel (m-bit) data, which are partitioned into k-bit and $(m - k)$-bit words where $k \le m$. The k-bit words (frames) are convolutionally encoded into $(n = k + 1)$-bit words so that the code rate is $R = k/(k + 1)$. The amplitude and phase are then set jointly on the basis of the coded n-bit word and the uncoded $(m - k)$-bit word. When a convolutional code of constraint length $K = 3$ is implemented, this TCM technique produces a coding gain of 3 dB relative to an uncoded signal that has the same bandwidth and information rate. Almost 6 dB of coding gain can be realized if coders of constraint length 9 are used. The larger constraint length codes are not too difficult to generate, but the corresponding decoder for a code of large constraint length is very complicated. However, very-high-speed integrated circuits (VHSIC) are making this feasible. As mentioned previously, the CCITT V.32 modem uses TCM. It has a coding gain of 4 dB and is described by Example 4 of Wei's paper [Wei, 1984; CCITT Study Group XVII, 1984].

For further study about coding, the reader is referred to several excellent books on the topic [Blahut, 1983; Clark and Cain, 1981; Gallagher, 1968; Lin and Costello, 1983; McEliece, 1977; Peterson and Weldon, 1972; Sweeney, 1991; Viterbi and Omura, 1979].

†Spectral conservation provided by multilevel signaling is discussed at the end of Secs. 3-4 and 5-2.

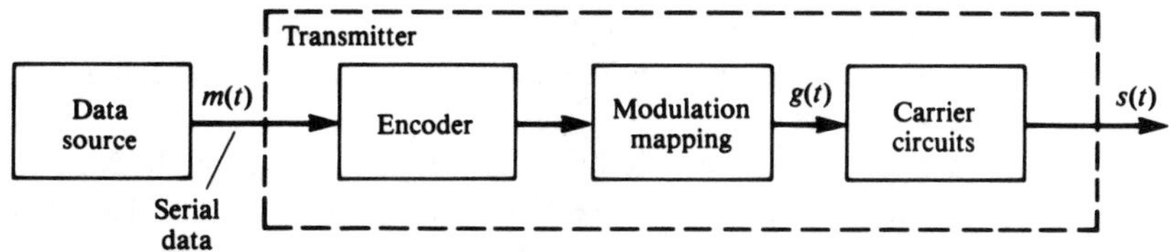

(a) Conventional Coding Technique

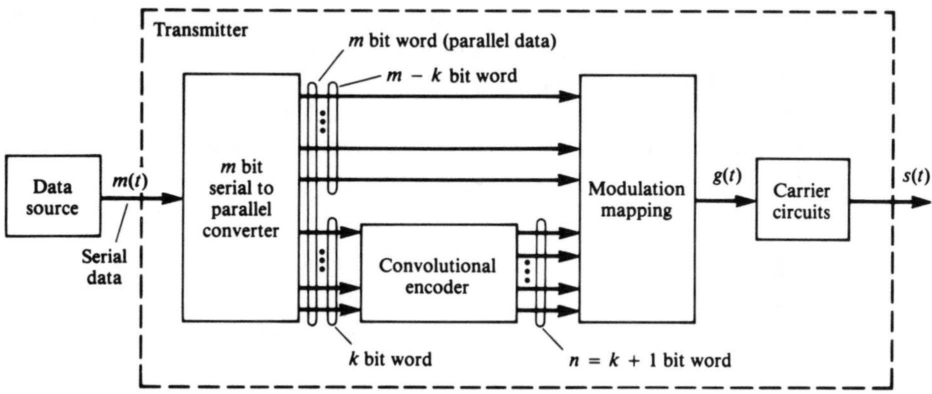

(b) Trellis-Coded Modulation Technique

FIGURE 1-9 Transmitters for conventional coding and for TCM.

1-12 Preview

From the previous discussions, we see the need for some basic tools to understand and design communication systems. Some prime tools needed are mathematical models to represent signals, noise, and linear systems. Chapter 2 provides these tools. It is divided into the broad categories of properties of signal and noise, orthogonal representations, bandlimited representations, and descriptions of linear systems. Measures of bandwidth are also defined. Baseband pulse techniques for both digital and analog signaling are discussed in Chapter 3. In Chapter 4 bandpass signaling techniques and modulated signal representations are explained. In addition, circuits that are used to generate and receive these signals are described. These tools are used in Chapter 5 to design and analyze communication systems. Examples of data, telephone, television, and satellite systems are given. A technique is developed for evaluating the data rate that a satellite communication system can support based on the specifications of the components used. Chapter 6 develops the theory of random processes. This is the tool used to evaluate the performance of both digital and analog communication systems that are operating in the presence of noise. Chapter 7 uses this theory to evaluate the probability of bit error for digital systems and the signal-to-noise ratio for analog systems. In Chapter 8 the performance of optimum digital systems is discussed in some detail. Optimum receiver structures are found, and the probability of error is evaluated. The additional complexity of fading channels is introduced and is solved.

Worked examples are scattered throughout this book to aid the reader. At the end of each chapter, ample exercise problems are given to help sharpen analytical skills and to provide insight in problem solving and communication system design. Answers for selected problems are provided at the end of the book.

Appendix A provides a review of general mathematical concepts. Appendix B provides a tutorial course on the topic of probability and random variables, including exercise problems. Appendix C summarizes the standards and terminology used in computer communication systems. Appendix D shows how to use MathCAD on a PC to solve communications problems. A bibliography of referenced books and papers is provided to give the reader a source of in-depth study material.

PROBLEMS

1-1 A high-power FM station of frequency 96.9 MHz has an antenna height of 1800 ft. If the signal is to be received 75 miles from the station, how high does a prospective listener need to mount his or her antenna in this fringe area?

1-2 Using geometry, prove that (1-6) is correct.

1-3 Referring to Sec. 1-8 concerning the propagation of electromagnetic waves, write a one-page discussion about radio-wave propagation for the 27-MHz citizen's-band (CB) radio service. Also, if you have experience as a CB operator, discuss how your prediction of CB signal propagation agrees with propagation conditions that you have observed.

1-4 A digital source emits -1.0- and 0.0-V levels with a probability of 0.2 each and $+3.0$- and $+4.0$-V levels with a probability of 0.3 each. Evaluate the average information of the source.

1-5 Prove that base 10 logarithms may be converted to base 2 logarithms by using the identity $\log_2(x) = [1/\log_{10}(2)] \log_{10}(x)$.

1-6 If all the messages emitted by a source are equally likely (i.e., $P_j = P$), show that (1-3) reduces to $H = \log_2(1/P)$.

1-7 For a binary source:

(a) Show that the entropy, H, is a maximum when the probability of sending a binary 1 is equal to the probability of sending a binary 0.

(b) Find the value of maximum entropy.

1-8 A single-digit, seven-segment liquid crystal display (LCD) emits a 0 with a probability of 0.25; a 1 and a 2 with a probability of 0.15 each; 3, 4, 5, 6, 7, and 8 with a probability of 0.07 each; and a 9 with a probability of 0.03. Find the average information for this source.

1-9 **(a)** A binary source sends a binary 1 with a probability of 0.3. Evaluate the average information for the source.

(b) For a binary source find the probability for sending a binary 1 and a binary 0 such that the average source information will be maximized.

1-10 A numerical keypad has the digits 0, 1, 2, 3, 4, 5, 6, 7, 8, and 9. Assume that the probability of sending any one digit is the same as that for sending any of the other digits. Calculate how often the buttons must be pressed in order to send out information at the rate of 2 bits/sec.

1-11 Refer to Example 1-1 and assume that words, each 12 digits in length, are sent over a system and that each digit can take on one of two possible values. Half of the possible words have a probability of being transmitted that is $(\frac{1}{2})^{13}$ for each word. The other half have probabilities equal to $3(\frac{1}{2})^{13}$. Find the entropy for this source.

1-12 Evaluate the channel capacity for a teleprinter channel that has a 300-Hz bandwidth and a signal-to-noise ratio of 30 dB.

1-13 Assume that a computer terminal has 110 characters (on its keyboard) and that each character is sent using binary words.

(a) What are the number of bits needed to represent each character?

(b) How fast can the characters be sent (characters/sec) over a telephone line channel having a bandwidth of 3.2 kHz and a signal-to-noise ratio of 20 dB?

(c) What is the information content of each character if each is equally likely to be sent?

1-14 **(a)** Suppose that a digitized television picture is to be transmitted from a source that uses a matrix of 480×500 picture elements (pixels) where each pixel can take on one of 32 intensity values. Assume that 30 pictures are sent per second. (This digital source is roughly equivalent to broadcast TV standards that have been adopted.) All picture elements are assumed to be independent, and all 32 intensity levels are assumed to be equally likely. Find the source rate R (bits/sec).

(b) Assume that the TV picture is to be transmitted over a channel with a 4.5-MHz bandwidth and a 35-dB signal-to-noise ratio. Find the capacity of the channel (bits/sec).

(c) Discuss how the parameters given in part (a) could be modified to allow transmission of color TV signals without increasing the required value for R.

1-15 Referring to (1-11), find a (7,4) Hamming code that is different from that shown in Example 1-2.

(a) Give the generator matrix for your (7,4) Hamming code.

(b) Find the code word that corresponds to the information word 1101.

(c) Demonstrate that a single error can be detected and corrected.

1-16 Design a (7,4) Hamming code such that each code word has the form

$$i_1 i_2 p_1 i_3 i_4 p_2 p_3$$

where i denotes information bits and p denotes parity check bits. In particular:

(a) Find the code generator matrix, **G.**
(b) Find the code parity check matrix, **H.**
(c) Show that a single error can be detected and corrected.

1-17 Using MathCAD, find all the code words for a (7,4) Hamming code.

1-18 Rework Example 1-2 for the case of a (15,11) Hamming code. That is:
(a) Find the parity check matrix.
(b) Find the encoded word if the input word is all binary 1's.
(c) Show that a single error can be detected and corrected.

1-19 Using the definitions for terms associated with convolutional coding, draw a block diagram for a convolutional coder that has rate $R = \frac{2}{3}$ and constraint length $K = 3$.

1-20 For the convolutional encoder shown in Fig. P1-20, compute the output coded data when the input data is $\mathbf{x} = [10111]$. (The first input bit is the leftmost element of the $\mathbf{x}$ row vector.)

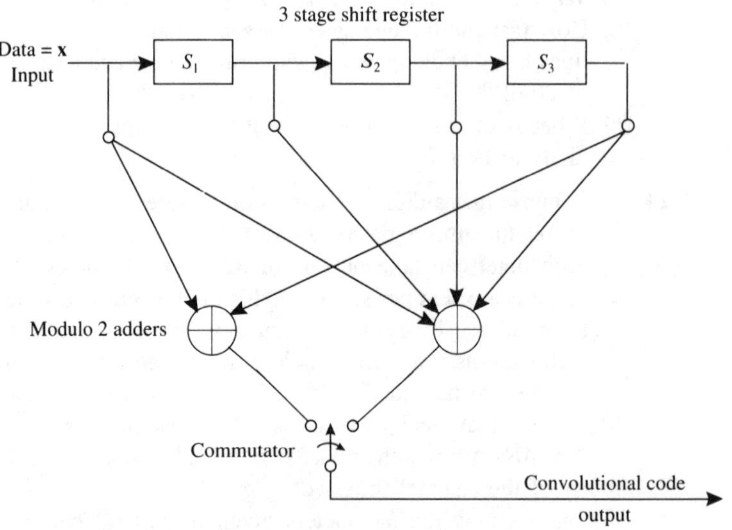

FIGURE P1-20

1-21 Design a code interleaver system. In particular, draw a block diagram for a:
(a) Code interleaver.
(b) Code deinterleaver.
Explain how your design works.

CHAPTER 2

Signals and Noise

2-1 Properties of Signals and Noise

In communication systems the received waveform is usually categorized into the desired part containing the information, and the extraneous or undesired part. The desired part is called the *signal* and the undesired part is called *noise*. In this text we will usually assume that the signal and the noise are added together to form the total received waveform. However, the multiplicative effect of noise that produces signal fading is discussed in Chapter 8.

This chapter develops mathematical tools that are used to describe signals and noise from a deterministic waveform point of view. (The random waveform approach is given in Chapter 6.) The waveforms will be represented by direct mathematical expressions or by the use of orthogonal series representations such as the Fourier series. Properties for describing these waveforms such as dc value, root mean square (rms) value, normalized power, magnitude spectrum, phase spectrum, power spectral density, and bandwidth will also be developed. In addition, techniques for evaluating the effect of linear filters on signals and noise will be studied.

The waveform of interest may be the voltage variation as a function of time, $v(t)$, or the current variation as a function of time, $i(t)$. Often the same mathematical techniques can be used when working with either type of waveform. Thus, for generality, waveforms will be denoted simply as $w(t)$ when the analysis applies to either case.

Physically Realizable Waveforms

Practical waveforms that are *physically realizable* (i.e., measurable in a laboratory) satisfy several conditions[†]:

1. The waveform has significant nonzero values over a composite time interval that is finite.
2. The spectrum of the waveform has significant values over a composite frequency interval that is finite.
3. The waveform is a continuous function of time.
4. The waveform has a finite peak value.
5. The waveform has only real values. That is, at any time, it cannot have a complex value $a + jb$ where b is nonzero.

[†] For an interesting discussion relative to the first and second conditions, see the paper by Slepian [1976].

The first condition is necessary because systems (and their waveforms) appear to exist for a finite amount of time. Physical signals also produce only a finite amount of energy. The second condition is necessary since any transmission medium—such as wires, coaxial cable, waveguides, or fiber optic cable—has a restricted bandwidth. The third condition is a consequence of the second, as will become clear from spectral analysis as developed in Sec. 2-2. The fourth condition is necessary because physical devices are destroyed if voltage or current of infinite value is present within the device. The fifth condition follows from the fact that only real waveforms can be observed in the real world, although *properties* of waveforms, such as spectra, may be complex. Later, in Chapter 4, it will be shown that complex waveforms can be very useful in representing real bandpass signals mathematically.

Mathematical models that violate some or all of the conditions listed above are often used, and for one main reason—to simplify the mathematical analysis. In fact, one often has to use a model that violates some of these conditions in order to calculate any type of answer. However, if one is careful with the mathematical model, the correct result can be obtained when the answer is properly interpreted. For example, consider the digital waveform shown in Fig. 2-1. The mathematical model waveform has discontinuities at the switching times. This situation violates the third condition—the physical waveform is continuous. The physical waveform is of finite duration (decays to zero before $t = \pm\infty$), but the duration of the mathematical waveform extends to infinity.

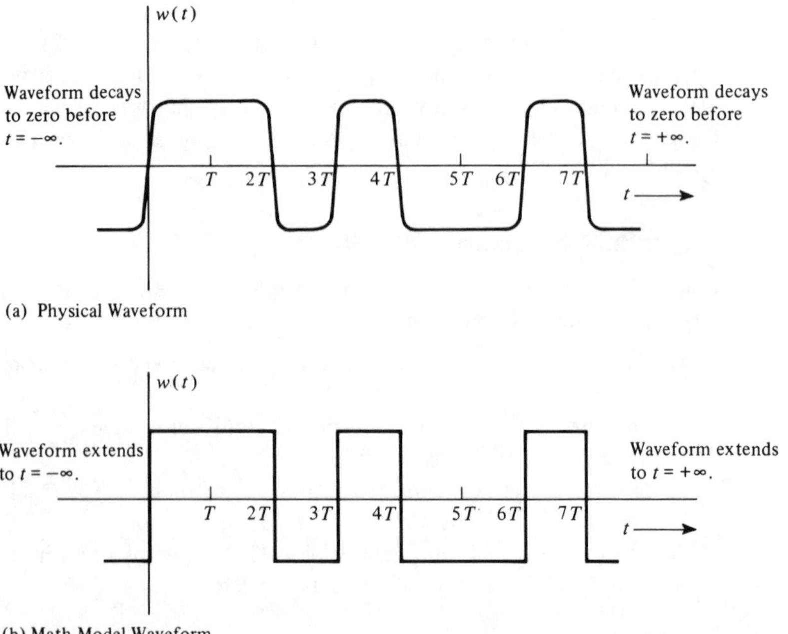

FIGURE 2-1 Physical and mathematical waveforms.

In other words, this mathematical model assumes that the physical waveform existed in its steady state condition for all time. Spectral analysis of the model will approximate the correct results except for the extremely high-frequency components. The average power that is calculated from the model will give the correct value for the average power of the physical signal that is measured over a time interval such that the physical signal exists in its steady state condition. The total energy of the mathematical model's signal will be infinity since it extends to infinite time, whereas that of the physical signal will be finite. Consequently, this model will not give the correct value for the total energy of the physical signal without using some additional information. However, the model can be used to evaluate the energy of the physical signal over some finite time interval in which the physical signal is in its steady state condition. This model is said to be a *power signal* because it has the property of finite power (and infinite energy), whereas the physical waveform is said to be an *energy signal* because it has finite energy. (Mathematical definitions for power and energy signals will be given in a subsequent section.) All physical signals are energy signals, although we generally use power signal mathematical models to simplify the analysis.

As described previously, waveforms may often be classified as signals or noise, digital or analog, deterministic or nondeterministic, physically realizable or nonphysically realizable, and belonging to the power or energy type. Additional classifications such as periodic and nonperiodic will be given in the next section.

Time Average Operator

Properties of signals and noise that are very useful to electrical engineers are dc (direct "current") value, average power, and rms (root mean square) value. Before these concepts are reviewed, the time average operator needs to be defined.

DEFINITION: The *time average operator*† is given by

$$\langle [\cdot] \rangle = \lim_{T\to\infty} \frac{1}{T} \int_{-T/2}^{T/2} [\cdot]\, dt \tag{2-1}$$

(This can also be called the *time mean* value of $[\cdot]$.)

It is seen that this operator is a *linear* operator since, from (2-1), the average of the sum of two quantities is the same as the sum of their averages‡:

$$\langle a_1 w_1(t) + a_2 w_2(t) \rangle = a_1 \langle w_1(t) \rangle + a_2 \langle w_2(t) \rangle \tag{2-2}$$

Equation (2-1) can be reduced to a simpler form if the operator is operating on a *periodic* waveform.

†In Appendix B the ensemble average operator is defined.

‡See (2-130) for the definition of linearity.

DEFINITION: A waveform $w(t)$ is *periodic* with period T_0 if

$$w(t) = w(t + T_0) \qquad \text{for all } t \tag{2-3}$$

where T_0 is the smallest positive number that satisfies this relationship.[†]

For example, as you know, a sinusoidal waveform of frequency $f_0 = 1/T_0$ hertz is periodic since it satisfies (2-3). From this definition it is clear that a periodic waveform will have significant values over an infinite time interval $(-\infty,\infty)$. Consequently, physical waveforms cannot be truly periodic, but they can have periodic values over a finite time interval. That is, (2-3) can be satisfied for t over some finite interval but not for all values of t.

THEOREM: *If the waveform involved is periodic, the time average operator can be reduced to*

$$\langle[\cdot]\rangle = \frac{1}{T_0}\int_{-T_0/2+a}^{T_0/2+a} [\cdot]\, dt \tag{2-4}$$

where T_0 is the period of the waveform and a is an arbitrary real constant, which may be taken to be zero.

It is seen that (2-4) readily follows from (2-1) since, referring to (2-1), integrals over successive time intervals T_0 seconds wide have identical areas because the waveshape is periodic with period T_0. As these integrals are summed, the total area and T are proportionally larger, resulting in a value for the time average that is the same as just integrating over one period and dividing by the width of that interval, T_0.

In summary, it is seen that (2-1) may be used to evaluate the time average of any type of waveform, whether or not it is periodic. Equation (2-4) is valid only for periodic waveforms.

Dc Value

DEFINITION: The *dc (direct "current")* value of a waveform $w(t)$ is given by its time average, $\langle w(t)\rangle$.

Thus

$$W_{dc} = \lim_{T\to\infty} \frac{1}{T}\int_{-T/2}^{T/2} w(t)\, dt \tag{2-5}$$

For any physical waveform, we are actually interested in evaluating the dc value only over a finite interval of interest, say from t_1 to t_2, so that the dc value is

$$\frac{1}{t_2 - t_1}\int_{t_1}^{t_2} w(t)\, dt$$

[†]Nonperiodic waveforms are called *aperiodic* waveforms by some authors.

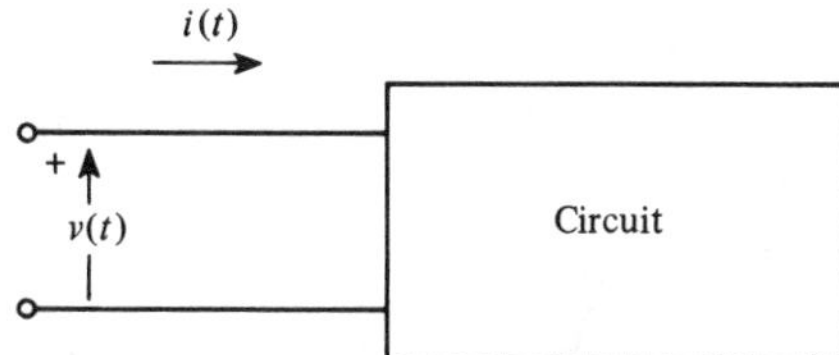

FIGURE 2-2 Polarity convention used for voltage and current.

However, if we use a mathematical model with a steady state waveform of infinite extent, we will obtain the correct result by using our definition (2-5), which involves the limit as $T \to \infty$. This will be demonstrated subsequently by Example 2-1. Moreover, as will be shown in Chapter 6, for the case of ergodic stochastic waveforms the time-averaging operator $\langle [\cdot] \rangle$ may be replaced by the ensemble average operator $\overline{[\cdot]}$.

Power

In communication systems, if the received (average) signal power is sufficiently large compared to the (average) noise power, information may be recovered. This concept was demonstrated by the Shannon channel capacity formula, (1-10), which indicates that information may be recovered even if the signal-to-noise power ratio is very small. Consequently, average power is an important concept that needs to be clearly understood. From physics it is known that power is defined as work per unit time, voltage is work per unit charge, and current is charge per unit time. Using these ideas we have the basis for the definition of power in terms of electrical quantities as given subsequently.

DEFINITION: Let $v(t)$ denote the voltage across a set of circuit terminals and let $i(t)$ denote the current into the terminal, as shown in Fig. 2-2. The *instantaneous power* (incremental work divided by incremental time) associated with the circuit is given by

$$p(t) = v(t)i(t) \tag{2-6}$$

where the instantaneous power flows into the circuit when $p(t)$ is positive and flows out of the circuit when $p(t)$ is negative. The *average power* is given by

$$P = \langle p(t) \rangle = \langle v(t)i(t) \rangle \tag{2-7}$$

As will be shown in Chapter 6, for the case of a stochastic $p(t)$ that is ergodic, the time average operator $\langle [\cdot] \rangle$ may be replaced by the ensemble average operator $\overline{[\cdot]}$.

EXAMPLE 2-1 EVALUATION OF POWER

Let the circuit of Fig. 2-2 contain a 120-V, 60-Hz fluorescent lamp wired in a high-power-factor configuration. Assume that the voltage and current are both

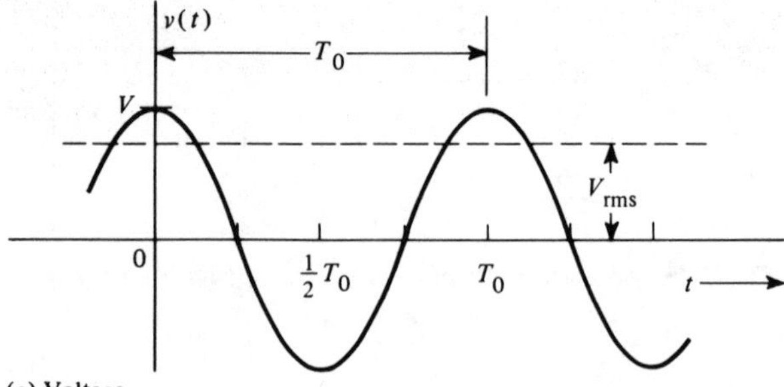

(a) Voltage

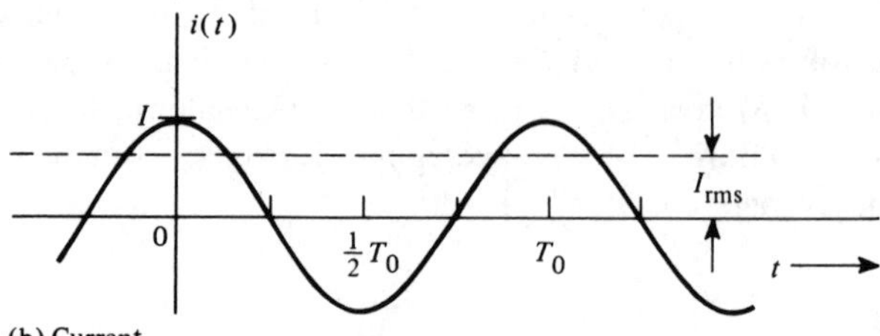

(b) Current

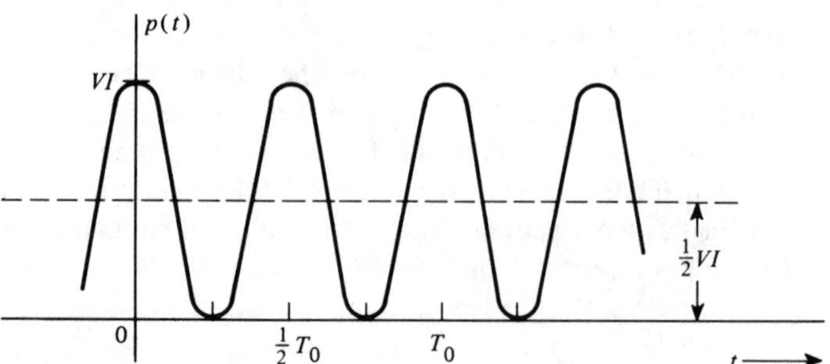

(c) Instantaneous Power

FIGURE 2-3 Steady state waveshapes for Example 2-1.

sinusoids and in phase (unity power factor), as shown in Fig. 2-3.[†] The dc value of this (periodic) voltage waveform is

$$V_{dc} = \langle v(t) \rangle = \langle V \cos \omega_0 t \rangle$$

$$= \frac{1}{T_0} \int_{-T_0/2}^{T_0/2} V \cos \omega_0 t \, dt = 0 \tag{2-8}$$

where $\omega_0 = 2\pi/T_0$ and $f_0 = 1/T_0 = 60$ Hz. Similarly, $I_{dc} = 0$. The instantaneous power is

$$p(t) = (V \cos \omega_0 t)(I \cos \omega_0 t) = \tfrac{1}{2}VI(1 + \cos 2\omega_0 t) \tag{2-9}$$

[†]Two-lamp fluorescent circuits can be realized with a high-power-factor ballast that gives an overall power factor greater than 90% [Fink and Beaty, 1978].

The average power is

$$P = \langle \tfrac{1}{2} VI(1 + \cos 2\omega_0 t) \rangle$$
$$= \frac{VI}{2T_o} \int_{-T_0/2}^{T_0/2} (1 + \cos 2\omega_0 t)\, dt$$
$$= \frac{VI}{2} \tag{2-10}$$

As can be seen from (2-9) and Fig. 2-3c, the power (i.e., light emitted) occurs in pulses at a rate of $2f_0 = 120$ pulses per second. (In fact, this lamp can be used as a stroboscope to "freeze" mechanically rotating objects.) The peak power is VI, and the average power is $\frac{1}{2}VI$, where V is the peak voltage and I is the peak current. Furthermore, for this case of sinusoidal voltage and sinusoidal current we see that the average power could be obtained by multiplying $V/\sqrt{2}$ with $I/\sqrt{2}$.

This subject brings up the concept of root mean square (rms) value.

Rms Value and Normalized Power

DEFINITION: The *root mean square (rms)* value of $w(t)$ is given by

$$W_{\text{rms}} = \sqrt{\langle w^2(t) \rangle} \tag{2-11}$$

THEOREM: *If a load is resistive (i.e., unity power factor), the average power is given by*

$$P = \frac{\langle v^2(t) \rangle}{R} = \langle i^2(t) \rangle R = \frac{V_{\text{rms}}^2}{R}$$
$$= I_{\text{rms}}^2 R = V_{\text{rms}} I_{\text{rms}} \tag{2-12}$$

where R is the value of the resistive load.

Equation (2-12) follows from (2-7) by the use of Ohm's law, which is $v(t) = i(t)R$, and (2-11).

Continuing Example 2-1, $V_{\text{rms}} = 120$ V. It is seen from (2-11), when sinusoidal waveshapes are used, that $V_{\text{rms}} = V/\sqrt{2}$ and $I_{\text{rms}} = I/\sqrt{2}$. Thus, using (2-12), we see that the average power is $\frac{1}{2}VI$, which is the same value that was obtained in the previous discussion.

The concept of *normalized* power is often used by communications engineers. In this concept R is assumed to be 1 Ω, although it may be another value in the actual circuit. Another way of expressing this concept is to say that the power is given on a per-ohm basis. In the signal-to-noise power ratio calculations R will automatically cancel out, so that normalized values of power may be used to obtain the correct ratio. If the actual value for the power is needed, say at the end of a long set of calculations, it can always be obtained by "denormalization" of the normalized value. From (2-12) it is also realized that the *square root of the normalized power is the rms value*.

DEFINITION: The *average normalized power* is given by

$$P = \langle w^2(t) \rangle = \lim_{T\to\infty} \frac{1}{T} \int_{-T/2}^{T/2} w^2(t)\, dt \tag{2-13}$$

where $w(t)$ represents a real voltage or current waveform.

Energy Waveforms and Power Waveforms†

DEFINITION: $w(t)$ is a *power waveform* if and only if the normalized average power, P, is finite and nonzero (i.e., $0 < P < \infty$).

DEFINITION: The *total normalized energy* is given by

$$E = \lim_{T\to\infty} \int_{-T/2}^{T/2} w^2(t)\, dt \tag{2-14}$$

DEFINITION: $w(t)$ is an *energy waveform* if and only if the total normalized energy is finite and nonzero (i.e., $0 < E < \infty$).

From these definitions it is seen that if a waveform is classified as either one of these types, it cannot be of the other type. That is, if $w(t)$ has finite energy, the power averaged over infinite time is zero, and if the power (averaged over infinite time) is finite, the energy is infinite. On the other hand, mathematical functions can be found that have both infinite energy and infinite power and, consequently, cannot be classified into either of these two categories. One example is $w(t) = e^{-t}$. Physically realizable waveforms are of the energy type, but we will often model them by infinite-duration waveforms of the power type. Laboratory instruments that measure average quantities—such as dc value, rms value, and average power—average over a finite time interval. That is, T of (2-1) remains finite instead of approaching some large number. Thus nonzero average quantities for finite energy (physical) signals can be obtained. For example, when the dc value is measured with a conventional volt-ohm-milliamp meter containing a meter movement, the time-averaging interval is established by the mass of the meter movement that provides damping. Thus the average quantities calculated from a power-type mathematical model (averaged over infinite time) will give the results that are measured in the laboratory (averaged over finite time).

†This concept is also called *energy signals* and *power signals* by some authors, but it applies to noise as well as signal waveforms.

Decibel

The *decibel* is a base 10 logarithmic measure of power ratios. For example, the ratio of the power level at the output of a circuit compared to that at the input is often specified by the decibel gain instead of the actual ratio.

Definition: The *decibel gain* of a circuit is given by[†]

$$\text{dB} = 10 \log \left(\frac{\text{average power out}}{\text{average power in}} \right) = 10 \log \left(\frac{P_{\text{out}}}{P_{\text{in}}} \right) \tag{2-15}$$

This definition gives a number that indicates the *relative* value of the *power out* with respect to the *power in* and does not indicate the actual magnitude of the power levels involved. If resistive loads are involved, (2-12) may be used to reduce (2-15) to

$$\text{dB} = 20 \log \left(\frac{V_{\text{rms out}}}{V_{\text{rms in}}} \right) + 10 \log \left(\frac{R_{\text{in}}}{R_{\text{load}}} \right) \tag{2-16}$$

or

$$\text{dB} = 20 \log \left(\frac{I_{\text{rms out}}}{I_{\text{rms in}}} \right) + 10 \log \left(\frac{R_{\text{load}}}{R_{\text{in}}} \right) \tag{2-17}$$

Note that the same value for decibels is obtained regardless of whether power, voltage, or current [(2-15), (2-16), or (2-17)] was used to obtain that value. That is, decibels are defined in terms of the logarithm of a power ratio but may be evaluated from voltage and/or current ratios.

If normalized powers are used,

$$\text{dB} = 20 \log \left(\frac{V_{\text{rms out}}}{V_{\text{rms in}}} \right) = 20 \log \left(\frac{I_{\text{rms out}}}{I_{\text{rms in}}} \right) \tag{2-18}$$

This equation does not give the true value for decibels unless $R_{\text{in}} = R_{\text{load}}$; however, it is common engineering practice to use (2-18) even if $R_{\text{in}} \neq R_{\text{load}}$ and even if the number that is obtained is not strictly correct. Engineers understand that if the correct value is needed, it may be calculated from this pseudovalue if R_{in} and R_{load} are known.

If the dB value is known, the power ratio or the voltage ratio can be easily obtained by inversion of the appropriate equations just given. For example, if the power ratio is desired, (2-15) can be inverted to obtain

$$\frac{P_{\text{out}}}{P_{\text{in}}} = 10^{\text{dB}/10} \tag{2-19}$$

[†]Logarithms to the base 10 will be denoted by $\log(\cdot)$, and logarithms to the base e will be denoted by $\ln(\cdot)$. Note that both dB and the ratio $(P_{\text{out}}/P_{\text{in}})$ are dimensionless quantities.

The decibel measure can also be used to express a measure of the signal power to noise power ratio, as measured at some point in a circuit.

DEFINITION: The *decibel signal-to-noise ratio* is†

$$(\mathrm{S/N})_{\mathrm{dB}} = 10 \log \left(\frac{P_{\text{signal}}}{P_{\text{noise}}} \right) = 10 \log \left(\frac{\langle s^2(t) \rangle}{\langle n^2(t) \rangle} \right) \tag{2-20}$$

Since the signal power is $\langle s^2(t)\rangle/R = V^2_{\text{rms signal}}/R$ and the noise power is $\langle n^2(t)\rangle/R = V^2_{\text{rms noise}}/R$, this definition is equivalent to

$$(\mathrm{S/N})_{\mathrm{dB}} = 20 \log \left(\frac{V_{\text{rms signal}}}{V_{\text{rms noise}}} \right) \tag{2-21}$$

The decibel measure may also be used to indicate absolute levels of power with respect to some reference level.

DEFINITION: The *decibel power level* with respect to 1 mW is given by

$$\begin{aligned} \mathrm{dBm} &= 10 \log \left(\frac{\text{actual power level (watts)}}{10^{-3}} \right) \\ &= 30 + 10 \log[\text{actual power level (watts)}] \end{aligned} \tag{2-22}$$

where the "m" in the dBm denotes a milliwatt reference.

Laboratory radio frequency (RF) signal generators are usually calibrated in terms of dBm.

Other decibel measures of absolute power levels are also used. When a 1-watt reference level is used, the decibel level is denoted dBW; when a 1-kilowatt reference level is used, the decibel level is denoted dBk. For example, a power of 5 W could be specified as +36.99 dBm, 6.99 dBW, or −23.0 dBk. The telephone industry uses a decibel measure with a "reference noise" level of 1 picowatt (10^{-12} W) [Jordan, 1985]. This decibel measure is denoted dBrn. A level of 0 dBrn corresponds to −90 dBm. The cable television (CATV) industry uses a 1-millivolt rms level across a 75-Ω load as a reference. This decibel measure is denoted dBmV, and it is defined as follows:

$$\mathrm{dBmV} = 20 \log \left(\frac{V_{\text{rms}}}{10^{-3}} \right) \tag{2-23}$$

0 dBmV corresponds to −48.75 dBm.

It should be emphasized that the general expression for evaluating power is given by (2-7). That is, (2-7) can be used to evaluate average power for any type of waveshape and load condition, whereas (2-12) is useful only for resistive

†This definition involves the ratio of the *average signal* power to the *average noise* power. An alternative definition that is also useful for some applications involves the ratio of the *peak* signal power to the average noise power—see Sec. 6-8.

loads. In other books, especially in the power area, equations are given that are valid only for sinusoidal waveshapes.

Phasors

Sinusoidal test signals occur so often in electrical engineering problems that a shorthand notation called *phasor notation* is often used.

DEFINITION: A complex number c is said to be a *phasor* if it is used to represent a *sinusoidal* waveform. That is,

$$w(t) = |c| \cos[\omega_0 t + \underline{/c}] = \text{Re}\{ce^{j\omega_0 t}\} \tag{2-24}$$

where the phasor $c = |c|e^{j\underline{/c}}$ and $\text{Re}\{\cdot\}$ denotes the real part of the complex quantity $\{\cdot\}$.

We will refer to $ce^{j\omega_0 t}$ as the *rotating* phasor, as distinguished from the phasor c. When c is given to represent waveform, it is *understood* that the actual waveform that appears in the circuit is a sinusoid as specified by (2-24). Since the phasor is a complex number, it can be written in either Cartesian form or polar form:

$$c \triangleq x + jy = |c|e^{j\phi} \tag{2-25}$$

where x and y are real numbers along the Cartesian coordinates and $|c|$ and $\underline{/c} = \phi = \tan^{-1}(y/x)$ are the length and angle (real numbers) in the polar coordinate system. Shorthand notation for the polar form on the right-hand side of (2-25) is $|c| \underline{/\phi}$.

For example, $25 \sin(2\pi 500t + 45°)$ could be denoted by the phasor $25\underline{/-45°}$ since

$$25 \sin(\omega_0 t + 45°) = 25 \cos(\omega_0 t - 45°) = \text{Re}\{(25e^{-j\pi/4})e^{j\omega_0 t}\}$$

where $\omega_0 = 2\pi f_0$ and $f_0 = 500$ Hz. Similarly, $10 \cos(\omega_0 t + 35°)$ could be denoted by $10\underline{/35°}$.

Some other authors may use a slightly different definition for the phasor. For example, $w(t)$ may be expressed in terms of the imaginary part of a complex quantity instead of the real part as defined in (2-24). In addition, the phasor may denote the rms value of $w(t)$ instead of the peak value [Kaufman and Seidman, 1979]. In this case, $10 \sin(\omega_0 t + 45°)$ should be denoted by the phasor $7.07\underline{/45°}$. However, throughout this book, the definition as given by (2-24) will be used, and it is obvious that phasors can represent only *sinusoidal* waveshapes.

2-2 Fourier Transform and Spectra

Definition

In electrical engineering problems the signal, the noise, or the combined signal plus noise usually consists of a voltage or current waveform that is a function of time. Let $w(t)$ denote the waveform of interest (either voltage or current). If desired, we could look at the waveform on an oscilloscope. The value of the voltage (or current) fluctuates as a function of time; consequently, certain frequencies or a frequency range is one of the main properties of interest to the electrical engineer. Theoretically, in order to evaluate the frequencies that are present, one needs to view the waveform over all time (i.e., $-\infty < t < \infty$) to be sure that the measurement is accurate and to guarantee that none of the frequency components is neglected. The relative level of one frequency as compared to another is given by the *voltage* (or *current*) *spectrum*. This is obtained by taking the *Fourier* transform of the voltage (or current) waveform.

DEFINITION: The *Fourier transform* (FT) of a waveform $w(t)$ is

$$W(f) = \mathscr{F}[w(t)] = \int_{-\infty}^{\infty} [w(t)]e^{-j2\pi ft}\, dt \tag{2-26}$$

where $\mathscr{F}[\cdot]$ denotes the Fourier transform of $[\cdot]$ and f is the frequency parameter with units of Hz (i.e., 1/sec).† This *defines* the term *frequency;* frequency is the parameter f in the Fourier transform.

This is also called a *two-sided spectrum* of $w(t)$ since both positive and negative frequency components are obtained from (2-26).

In general, since $e^{-j2\pi ft}$ is complex, $W(f)$ is a complex function of frequency. $W(f)$ may be decomposed into two real functions $X(f)$ and $Y(f)$ such that

$$W(f) = X(f) + jY(f) \tag{2-27}$$

This is identical to writing a complex number in terms of pairs of real numbers that can be plotted in a two-dimensional Cartesian coordinate system. For this reason, (2-27) is sometimes called the *quadrature* form or *Cartesian* form. Similarly, (2-26) can be written equivalently in terms of a polar coordinate system where the pair of real functions denotes the magnitude and phase

$$W(f) = |W(f)|e^{j\theta(f)} \tag{2-28}$$

where

$$|W(f)| = \sqrt{X^2(f) + Y^2(f)} \quad \text{and} \quad \theta(f) = \tan^{-1}\left(\frac{Y(f)}{X(f)}\right) \tag{2-29}$$

This is called the *magnitude-phase* form or *polar* form. To determine if certain frequency components are present, one would examine the magnitude spectrum $|W(f)|$, and sometimes engineers loosely call this just the *spectrum*.

It should be clear that the spectrum of a voltage (or current) waveform is obtained by a mathematical calculation and that it does not appear physically in

†Some authors define the FT in terms of the frequency parameter ω (radians per second), where $\omega = 2\pi f$. That definition is identical to (2-26) when ω is replaced by $2\pi f$. I prefer (2-26) since spectrum analyzers are usually calibrated in units of hertz, not radians per second.

an actual circuit. f is just a parameter (called *frequency*) that determines which point of the spectral function is to be evaluated. For example, the frequency $f = 10$ Hz is present in the waveform $w(t)$ if and only if $|W(10)| \neq 0$. From (2-26) it is realized that an exact spectral value can be obtained only if the waveform is observed over the infinite time interval $(-\infty,\infty)$. However, a special instrument called a *spectrum analyzer* may be used to obtain an approximation (i.e., finite time integral) for the magnitude spectrum $|W(f)|$.

If one happens to have the mathematical function for the spectrum $W(f)$, the time waveform may be calculated by using the inverse Fourier transform

$$w(t) = \int_{-\infty}^{\infty} W(f)e^{j2\pi ft}\, df \tag{2-30}$$

The functions $w(t)$ and $W(f)$ are said to constitute a *Fourier transform pair* where $w(t)$ is the time domain description and $W(f)$ is the frequency domain description. In this book, the time domain function is usually denoted by a lowercase letter and the frequency domain function by an uppercase letter. Shorthand notation for the pairing between the two domains will be denoted by a double arrow: $w(t) \leftrightarrow W(f)$. If the signal or noise description is known in one domain, the corresponding description in the other domain is known theoretically by using (2-26) or (2-30), as appropriate.

The waveform $w(t)$ is Fourier transformable (i.e., sufficient conditions) if it satisfies the *Dirichlet conditions*:

- Over any time interval of finite width, the function $w(t)$ is single valued with a finite number of maxima and minima and the number of discontinuities (if any) is finite.
- $w(t)$ is absolutely integrable. That is,

$$\int_{-\infty}^{\infty} |w(t)|\, dt < \infty \tag{2-31}$$

Although these conditions are sufficient, they are not necessary. In fact, some of the examples given subsequently do not satisfy the Dirichlet conditions and yet the Fourier transform can be found. The second Dirichlet condition is obtained by taking the absolute value of (2-26).

$$|W(f)| = \left|\int w(t)e^{-j2\pi ft}\, dt\right| \le \int_{-\infty}^{\infty} |w(t)e^{-j2\pi ft}\, dt| = \int_{-\infty}^{\infty} |w(t)|\, dt$$

Thus $|W(f)|$ is finite. Consequently, the Fourier transform exists if

$$\int_{-\infty}^{\infty} |w(t)|\, dt < \infty$$

Furthermore, a weaker sufficient condition for the existence of the Fourier transform is

$$E = \int_{-\infty}^{\infty} |w(t)|^2\, dt < \infty \tag{2-32}$$

where E is the normalized energy [Goldberg, 1961]. This is the finite energy condition that is satisfied by all physically realizable waveforms. Thus all physical waveforms encountered in engineering practice are Fourier transformable.

It should also be noted that mathematicians sometimes use other definitions, rather than (2-26), for the Fourier transform. However, in these cases, the corresponding inverse transforms, equivalent to (2-30), would also be different, so that when the transform and its inverse are used together, the original $w(t)$ would be recovered. This is a consequence of the Fourier integral theorem, which is

$$w(t) = \int_{-\infty}^{\infty} \int_{-\infty}^{\infty} w(\lambda) e^{j2\pi f(t-\lambda)}\, d\lambda\, df \tag{2-33}$$

Equation (2-33) may be decomposed into (2-26) and (2-30), as well as other definitions for Fourier transform pairs. The Fourier integral theorem is strictly true only for well-behaved functions (i.e., physical waveforms). For example, if $w(t)$ is an ideal square wave, then at a discontinuous point of $w(\lambda)$, denoted by λ_0, $w(t)$ will have a value that is the average of the two values that are obtained for $w(\lambda)$ on each side of the discontinuous point λ_0.

EXAMPLE 2-2 SPECTRUM OF AN EXPONENTIAL PULSE

Let $w(t)$ be a decaying exponential pulse that is switched on at $t = 0$.

$$w(t) = \begin{cases} e^{-t}, & t > 0 \\ 0, & t < 0 \end{cases}$$

The spectrum of this pulse is given by the Fourier transform

$$W(f) = \int_0^{\infty} e^{-t} e^{-j2\pi f t}\, dt$$

or

$$W(f) = \frac{1}{1 + j2\pi f}$$

In other words, this Fourier transform pair is

$$\left.\begin{cases} e^{-t}, & t > 0 \\ 0, & t < 0 \end{cases}\right\} \leftrightarrow \frac{1}{1 + j2\pi f} \tag{2-34}$$

The spectrum can also be expressed in terms of the quadrature functions by rationalizing the denominator of (2-34); thus

$$X(f) = \frac{1}{1 + (2\pi f)^2} \quad \text{and} \quad Y(f) = \frac{-2\pi f}{1 + (2\pi f)^2}$$

The magnitude-phase form is

$$|W(f)| = \sqrt{\frac{1}{1 + (2\pi f)^2}} \quad \text{and} \quad \theta(f) = -\tan^{-1}(2\pi f)$$

More examples will be given after some helpful theorems are developed.

Properties of Fourier Transforms

Many useful and interesting theorems follow from the definition of the spectrum as given by (2-26). One of particular interest is the consequence of working with real-time waveforms. In any physical circuit that can be built, the voltage (or current) waveforms are real functions (as opposed to complex functions) of time.

Theorem: *Spectral symmetry of real signals. If $w(t)$ is real, then*

$$W(-f) = W^*(f) \tag{2-35}$$

The asterisk superscript denotes the conjugate operation.

Proof. From (2-26)

$$W(-f) = \int_{-\infty}^{\infty} w(t)e^{j2\pi ft}\,dt \tag{2-36}$$

and taking the conjugate of (2-26) yields

$$W^*(f) = \int_{-\infty}^{\infty} w^*(t)e^{j2\pi ft}\,dt \tag{2-37}$$

But since $w(t)$ is real, $w^*(t) = w(t)$ and (2-35) follows since the right sides of (2-36) and (2-37) are equal. It can also be shown that if real $w(t)$ happens to be an even function of t, $W(f)$ is *real*. Similarly, if real $w(t)$ is an odd function of t, $W(f)$ is *imaginary*.

Another useful corollary of (2-35) is that for $w(t)$ real, the magnitude spectrum is even about the origin (i.e., $f = 0$):

$$|W(-f)| = |W(f)| \tag{2-38}$$

and the phase spectrum is odd about the origin:

$$\theta(-f) = -\theta(f) \tag{2-39}$$

This can easily be demonstrated by writing the spectrum in polar form:

$$W(f) = |W(f)|e^{j\theta(f)}$$

Then

$$W(-f) = |W(-f)|e^{j\theta(-f)}$$

and

$$W^*(f) = |W(f)|e^{-j\theta(f)}$$

Using (2-35), we see that (2-38) and (2-39) are true.

In summary, from the discussion above, some properties of the Fourier transform are:

- f, called frequency and having units of hertz, is just a parameter of the FT that specifies what frequency we are interested in looking for in the waveform $w(t)$.

- The FT looks for the frequency f in the $w(t)$ over *all* time, that is, over $-\infty < t < \infty$.
- $W(f)$ can be complex even though $w(t)$ is real.
- If $w(t)$ is real, then $W(-f) = W^*(f)$.

Parseval's Theorem and Energy Spectral Density

Parseval's Theorem

$$\int_{-\infty}^{\infty} w_1(t)w_2^*(t)\,dt = \int_{-\infty}^{\infty} W_1(f)W_2^*(f)\,df \tag{2-40}$$

If $w_1(t) = w_2(t) = w(t)$, this reduces to[†]

$$E = \int_{-\infty}^{\infty} |w(t)|^2\,dt = \int_{-\infty}^{\infty} |W(f)|^2\,df \tag{2-41}$$

Proof. Work with the left side of (2-40), using (2-30) to replace $w_1(t)$.

$$\int_{-\infty}^{\infty} w_1(t)w_2^*(t)\,dt = \int_{-\infty}^{\infty}\left[\int_{-\infty}^{\infty} W_1(f)e^{j2\pi ft}\,df\right]w_2^*(t)\,dt$$

$$= \int_{-\infty}^{\infty}\int_{-\infty}^{\infty} W_1(f)w_2^*(t)e^{j2\pi ft}\,df\,dt$$

Interchanging the order of integration on f and t[‡]

$$\int_{-\infty}^{\infty} w_1(t)w_2^*(t)\,dt = \int_{-\infty}^{\infty} W_1(f)\left[\int_{-\infty}^{\infty} w_2(t)e^{-j2\pi ft}\,dt\right]^* df$$

Using (2-26) produces the result of (2-40). Parseval's theorem gives an alternative method for evaluating the energy by using the frequency domain description instead of the time domain definition. This leads to the concept of the energy spectral density function.

DEFINITION: The *energy spectral density (ESD)* is defined for energy waveforms by

$$\mathscr{E}(f) = |W(f)|^2 \tag{2-42}$$

where $w(t) \leftrightarrow W(f)$. $\mathscr{E}(f)$ has units of joules per hertz.

[†]This is also known as *Rayleigh's energy theorem*.

[‡]Fubini's theorem states that the order of integration may be exchanged if all of the integrals are absolute convergent. That is, these integrals are finite valued when the integrands are replaced by their absolute values. We assume that this condition is satisfied.

By using Parseval's theorem, we see that the total normalized energy is given by the area under the ESD function:

$$E = \int_{-\infty}^{\infty} \mathscr{E}(f)\, df \tag{2-43}$$

For power waveforms, a similar function called the *power spectral density (PSD)* can be defined. This is developed in Sec. 2-3 and in Chapter 6. Later we will find that the PSD function is a very useful tool for solving communication problems.

There are many other Fourier transform theorems in addition to Parseval's theorem. Some of them are summarized in Table 2-1. These theorems can be proved by substituting the corresponding time function into the definition for the Fourier transform and reducing the result to that given in the rightmost column of the table. For example, the scale change theorem is proved by substituting $w(at)$ into (2-26). We get

$$\mathscr{F}[w(at)] = \int_{-\infty}^{\infty} w(at)e^{-j2\pi ft}\, dt$$

Letting $t_1 = at$ and assuming that $a > 0$, we get

$$\mathscr{F}[w(at)] = \int_{-\infty}^{\infty} \frac{1}{a} w(t_1)e^{-j2\pi(f/a)t_1}\, dt_1 = \frac{1}{a} W\left(\frac{f}{a}\right)$$

For $a < 0$ this becomes

$$\mathscr{F}[w(at)] = \int_{-\infty}^{\infty} \frac{-1}{a} w(t_1)e^{-j2\pi(f/a)t_1}\, dt_1 = \frac{1}{|a|} W\left(\frac{f}{a}\right)$$

Thus, for either $a > 0$ or $a < 0$, we get

$$w(at) \leftrightarrow \frac{1}{|a|} W\left(\frac{f}{a}\right)$$

The other theorems in Table 2-1 are proved in a similar straightforward manner, except for the integral theorem. The integral theorem is more difficult to derive since the transform result involves a Dirac delta function, $\delta(f)$. This theorem may be proved by the use of the convolution theorem, as illustrated by Prob. 2-36. The bandpass signal theorem will be discussed in more detail in Chapter 4. It is the basis for the digital and analog modulation techniques covered in Chapters 4 and 5. In Sec. 2-7 the relationship between the Fourier transform and the discrete Fourier transform (DFT) will be studied.

As we will see in the following examples, these theorems can greatly simplify the calculations required to solve Fourier transform problems. Consequently, the reader should study Table 2-1 and be prepared to use it when needed. After the Fourier transform is evaluated, one should check to see that the easy-to-verify properties of Fourier transforms are satisfied; otherwise, there is a mistake. For example, if $w(t)$ is real,

TABLE 2-1 Some Fourier Transform Theorems[a]

Operation	Function	Fourier Transform
Linearity	$a_1 w_1(t) + a_2 w_2(t)$	$a_1 W_1(f) + a_2 W_2(f)$
Time delay	$w(t - T_d)$	$W(f) e^{-j\omega T_d}$
Scale change	$w(at)$	$\frac{1}{\lvert a \rvert} W\left(\frac{f}{a}\right)$
Conjugation	$w^*(t)$	$W^*(-f)$
Duality	$W(t)$	$w(-f)$
Real signal frequency translation [$w(t)$ is real]	$w(t)\cos(\omega_c t + \theta)$	$\frac{1}{2}[e^{j\theta} W(f - f_c) + e^{-j\theta} W(f + f_c)]$
Complex signal frequency translation	$w(t) e^{j\omega_c t}$	$W(f - f_c)$
Bandpass signal	$\text{Re}\{g(t) e^{j\omega_c t}\}$	$\frac{1}{2}[G(f - f_c) + G^*(-f - f_c)]$
Differentiation	$\frac{d^n w(t)}{dt^n}$	$(j2\pi f)^n W(f)$
Integration	$\int_{-\infty}^{t} w(\lambda)\, d\lambda$	$(j2\pi f)^{-1} W(f) + \frac{1}{2} W(0)\, \delta(f)$
Convolution	$w_1(t) * w_2(t) = \int_{-\infty}^{\infty} w_1(\lambda) \cdot w_2(t - \lambda)\, d\lambda$	$W_1(f) W_2(f)$
Multiplication[b]	$w_1(t) w_2(t)$	$W_1(f) * W_2(f) = \int_{-\infty}^{\infty} W_1(\lambda)\, W_2(f - \lambda)\, d\lambda$
Multiplication by t^n	$t^n w(t)$	$(-j2\pi)^{-n} \frac{d^n W(f)}{df^n}$

[a] $\omega_c = 2\pi f_c$.

[b] $*$ denotes convolution as described in detail by (2-62).

- $W(-f) = W^*(f)$, or $\lvert W(f) \rvert$ is even and $\theta(f)$ is odd.
- $W(f)$ is real when $w(t)$ is even.
- $W(f)$ is imaginary when $w(t)$ is odd.

EXAMPLE 2-3 SPECTRUM OF A DAMPED SINUSOID

Let the damped sinusoid be given by

$$w(t) = \begin{cases} e^{-t/T} \sin \omega_0 t, & t > 0,\ T > 0 \\ 0, & t < 0 \end{cases}$$

The spectrum of this waveform is obtained by evaluating the Fourier transform. This is easily accomplished by using the result of the previous example plus some of the Fourier theorems. From (2-34), and using the scale change theorem, where $a = 1/T$,

$$\begin{Bmatrix} e^{-t/T}, & t > 0 \\ 0, & t < 0 \end{Bmatrix} \leftrightarrow \frac{T}{1 + j(2\pi fT)}$$

Using the real signal frequency translation theorem with $\theta = -\pi/2$, we get

$$W(f) = \frac{1}{2}\left\{e^{-j\pi/2}\frac{T}{1 + j2\pi T(f - f_0)} + e^{j\pi/2}\frac{T}{1 + j2\pi T(f + f_0)}\right\}$$
$$= \frac{T}{2j}\left\{\frac{1}{1 + j2\pi T(f - f_0)} - \frac{1}{1 + j2\pi T(f + f_0)}\right\} \qquad (2\text{-}44)$$

where $e^{\pm j\pi/2} = \cos(\pi/2) \pm j\sin(\pi/2) = \pm j$. This spectrum is complex (i.e., neither real nor imaginary) since $w(t)$ does not have even or odd symmetry about $t = 0$.

As expected, (2-44) shows that the peak of the magnitude spectrum for the damped sinusoid occurs at $f = \pm f_0$. Compare this to the peak of the magnitude spectrum for the exponential decay (Example 2-2) that occurs at $f = 0$. That is, the $\sin \omega_0 t$ factor caused the spectral peak to move from $f = 0$ to $f = \pm f_0$.

Dirac Delta Function and Unit Step Function

Another type of function that often occurs in communication problems is the Dirac delta function.

DEFINITION: The *Dirac delta function* $\delta(x)$ is defined by

$$\int_{-\infty}^{\infty} w(x)\,\delta(x)\,dx = w(0) \qquad (2\text{-}45)$$

where $w(x)$ is any function that is continuous at $x = 0$.

In this definition the variable x could be time or frequency depending on the application. An alternative definition for $\delta(x)$ is

$$\int_{-\infty}^{\infty} \delta(x)\,dx = 1 \qquad (2\text{-}46\text{a})$$

and

$$\delta(x) = \begin{cases} \infty, & x = 0 \\ 0 & x \neq 0 \end{cases} \qquad (2\text{-}46\text{b})$$

where both (2-46a) and (2-46b) need to be satisfied. It is obvious that the Dirac delta function is not a true function, so it is said to be a singular function. However, $\delta(x)$ can be defined as a function in a more general sense and treated as such in a branch of mathematics called the *theory of distributions*.

From (2-45), the *sifting property* of the δ function is

$$\int_{-\infty}^{\infty} w(x)\,\delta(x - x_0)\,dx = w(x_0) \tag{2-47}$$

That is, the δ function sifts out the value $w(x_0)$ from the integral.

In some problems it is also useful to use the equivalent integral for the δ function, which is

$$\delta(x) = \int_{-\infty}^{\infty} e^{\pm j2\pi xy}\,dy \tag{2-48}$$

where either the + or the − sign may be used as needed. This assumes that $\delta(x)$ is an even function: $\delta(-x) = \delta(x)$. Equation (2-48) may be verified by taking the Fourier transform of a delta function

$$\int_{-\infty}^{\infty} \delta(t)e^{-j2\pi ft}\,dt = e^0 = 1$$

and then taking the inverse Fourier transform of both sides of this equation; (2-48) follows. For additional properties of the delta function, see Appendix A.

Another function that is closely related to the Dirac delta function is the unit step function.

DEFINITION: The *unit step function* $u(t)$ is

$$u(t) = \begin{cases} 1, & t > 0 \\ 0, & t < 0 \end{cases} \tag{2-49}$$

Since $\delta(\lambda)$ is zero except at $\lambda = 0$, the Dirac delta function is related to the unit step function by

$$\int_{-\infty}^{t} \delta(\lambda)\,d\lambda = u(t) \tag{2-50}$$

and consequently,

$$\frac{du(t)}{dt} = \delta(t) \tag{2-51}$$

EXAMPLE 2-4 SPECTRUM OF A SINUSOID

Find the spectrum of a sinusoidal voltage waveform that has a frequency f_0 and a peak value of A volts.

$$v(t) = A \sin \omega_0 t \qquad \text{where } \omega_0 = 2\pi f_0$$

From (2-26), the spectrum is

$$V(f) = \int_{-\infty}^{\infty} A\left(\frac{e^{j\omega_0 t} - e^{-j\omega_0 t}}{2j}\right) e^{-j\omega t}\, dt$$
$$= \frac{A}{2j}\int_{-\infty}^{\infty} e^{-j2\pi(f-f_0)t}\, dt - \frac{A}{2j}\int_{-\infty}^{\infty} e^{-j2\pi(f+f_0)t}\, dt$$

By (2-48), these integrals are equivalent to Dirac delta functions. That is,

$$V(f) = j\frac{A}{2}\left[\delta(f+f_0) - \delta(f-f_0)\right]$$

Note that this spectrum is imaginary, as expected, since $v(t)$ is real and odd. In addition, a meaningful expression was obtained for the Fourier transform, although $v(t)$ was of the infinite energy type and not absolutely integrable. That is, this $v(t)$ does not satisfy the sufficient (but not necessary) Dirichlet conditions as given by (2-31) and (2-32).

The magnitude spectrum is

$$|V(f)| = \frac{A}{2}\,\delta(f-f_0) + \frac{A}{2}\,\delta(f+f_0)$$

where A is a positive number. Since only two frequencies ($f = \pm f_0$) are present, $\theta(f)$ is strictly defined only at these two frequencies. That is, $\theta(f_0) = \tan(-1/0) = -90°$ and $\theta(-f_0) = \tan(1/0) = +90°$. However, because $|V(f)| = 0$ for all frequencies except $f = \pm f_0$, $\theta(f)$ can be taken to be any convenient set of values for $f \neq \pm f_0$. Thus the phase spectrum is taken to be

$$\theta(f) = \begin{Bmatrix} -\pi/2, & f>0 \\ +\pi/2, & f<0 \end{Bmatrix} \text{radians} = \begin{Bmatrix} -90°, & f>0 \\ 90°, & f<0 \end{Bmatrix}$$

Plots of these spectra are shown in Fig. 2-4. It is seen that the magnitude spectrum is even and the phase spectrum is odd, as expected from (2-38) and (2-39).

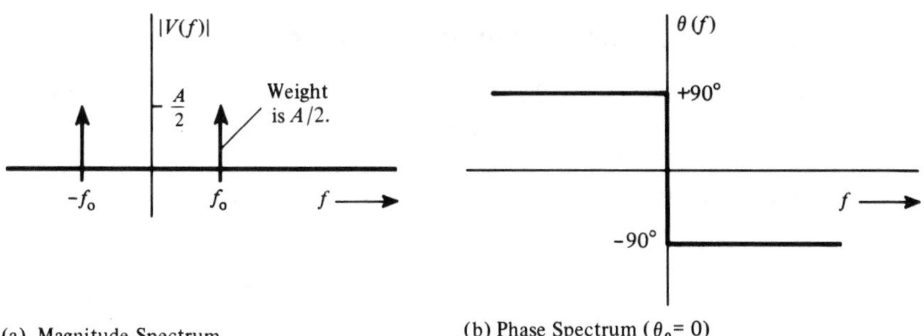

(a) Magnitude Spectrum

(b) Phase Spectrum ($\theta_0 = 0$)

FIGURE 2-4 Spectrum of a sine wave.

Now let us generalize the sinusoidal waveform to one with an arbitrary phase angle θ_0. Then

$$w(t) = A \sin(\omega_0 t + \theta_0) = A \sin[\omega_0(t + \theta_0/\omega_0)]$$

and, by using the time delay theorem, the spectrum becomes

$$W(f) = j\frac{A}{2}e^{j\theta_0(f/f_0)}[\delta(f + f_0) - \delta(f - f_0)]$$

The resulting magnitude spectrum is the same as that obtained before for the $\theta_0 = 0$ case. The new phase spectrum is the sum of the old phase spectrum plus the linear function $(\theta_0/f_0)f$. However, since the overall spectrum is zero except at $f = \pm f_0$, the value for the phase spectrum can be arbitrarily assigned at all frequencies except $f = \pm f_0$. At $f = f_0$ the phase is $(\theta_0 - \pi/2)$ radians, and at $f = -f_0$ the phase is $-(\theta_0 - \pi/2)$ radians.

From a *mathematical* viewpoint, Fig. 2-4 demonstrates that two frequencies are present in the sine wave, one at $f = +f_0$ and another at $f = -f_0$. This can also be seen from the time waveform; that is,

$$v(t) = A \sin \omega_0 t = \frac{A}{j2}e^{j\omega_0 t} - \frac{A}{j2}e^{-j\omega_0 t}$$

which implies that the sine wave consists of two rotating phasors, one rotating with a frequency $f = +f_0$ and another rotating with $f = -f_0$. From the *engineering* point of view it is said that one frequency is present, namely, $f = f_0$, since for any physical (i.e., real) waveform, (2-35) shows that for any positive frequency present, there is also a mathematical negative frequency present. The phasor associated with $v(t)$ is $c = 0 - jA = A\underline{/-90^\circ}$. Another interesting observation is that the magnitude spectrum consists of *lines* (i.e., Dirac delta functions). Later, it will be shown that the lines are a consequence of $v(t)$ being a periodic function. If the sinusoid is switched on and off, the spectrum is continuous as demonstrated by Example 2-9. A damped sinusoid also has a continuous spectrum, as demonstrated by Example 2-3.

Rectangular and Triangular Pulses

Some additional waveshapes frequently occur in communications problems, so some special symbols will be defined to shorten the notation.

Definition: Let $\Pi(\cdot)$ denote a single *rectangular pulse*

$$\Pi\left(\frac{t}{T}\right) \triangleq \begin{cases} 1, & |t| \le \dfrac{T}{2} \\ 0, & |t| > \dfrac{T}{2} \end{cases} \tag{2-52}$$

Definition: Let Sa(·) denote the function[†]

$$\text{Sa}(x) = \frac{\sin x}{x} \tag{2-53}$$

Definition: Let Λ(·) denote the triangular function

$$\Lambda\left(\frac{t}{T}\right) \triangleq \begin{cases} 1 - \dfrac{|t|}{T}, & |t| \leq T \\ 0, & |t| > T \end{cases} \tag{2-54}$$

These waveshapes are shown in Fig. 2-5. A tabulation of Sa(x) is given in Sec. A-9 (Appendix A).

EXAMPLE 2-5 SPECTRUM OF A RECTANGULAR PULSE

The spectrum is obtained by taking the Fourier transform of $w(t) = \Pi(t/T)$.

$$W(f) = \int_{-T/2}^{T/2} 1 e^{-j\omega t}\, dt = \frac{e^{-j\omega T/2} - e^{j\omega T/2}}{-j\omega}$$
$$= T \frac{\sin(\omega T/2)}{\omega T/2} = T\,\text{Sa}(\pi T f)$$

Thus,

$$\Pi\left(\frac{t}{T}\right) \leftrightarrow T\,\text{Sa}(\pi T f) \tag{2-55}$$

This Fourier transform pair is shown in Fig. 2-6a. Note the inverse relationship between the pulse width T and the spectral zero crossing $1/T$. Also, by use of the duality theorem, the spectrum of a $(\sin x)/x$ pulse is a rectangle. That is, realizing that $\Pi(x)$ is an even function, and applying the duality theorem to (2-55), we get

$$T\,\text{Sa}(\pi T t) \leftrightarrow \Pi\left(-\frac{f}{T}\right) = \Pi\left(\frac{f}{T}\right)$$

Replacing the parameter T by $2W$, we obtain the following Fourier transform pair.

$$2W\,\text{Sa}(2\pi W t) \leftrightarrow \Pi\left(\frac{f}{2W}\right) \tag{2-56}$$

where W is the absolute bandwidth in hertz. This Fourier transform pair is also shown in Fig. 2-6b.

[†]This is related to the sinc function by $\text{Sa}(x) = \text{sinc}(x/\pi)$ since $\text{sinc}(\lambda) \triangleq (\sin \pi\lambda)/\pi\lambda$. The notation Sa($x$) and sinc($x$) represent the same concept but can be confused because of scaling. In this book $(\sin x)/x$ will often be used because it avoids confusion and does not take much more text space.

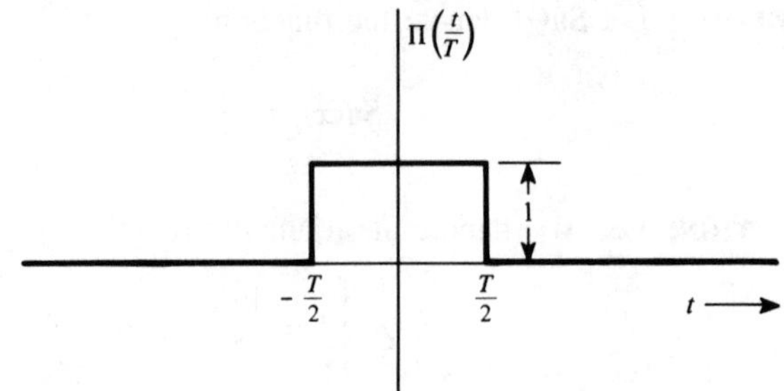

(a) Rectangular Pulse

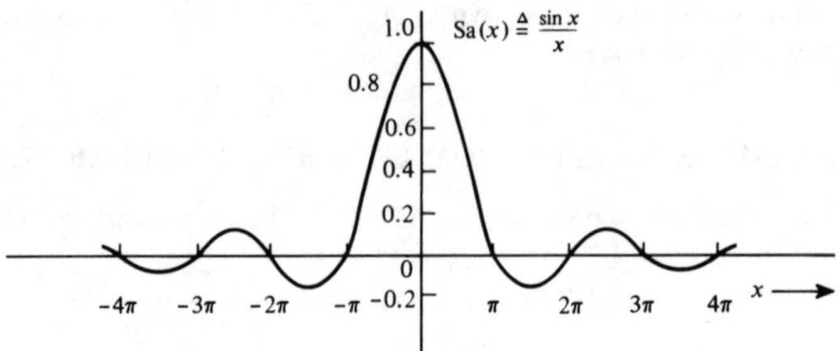

(b) Sa(x) Function

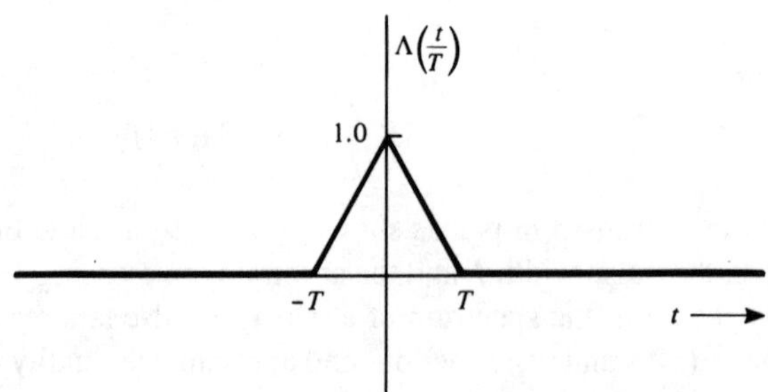

(c) Triangular Function

FIGURE 2-5 Waveshapes and corresponding symbolic notation.

The spectra shown in Fig. 2-6 are real since the time domain pulses are real and even. If the pulses are offset in time to destroy the even symmetry, the spectra will be complex. For example, let

$$v(t) = \begin{cases} 1, & 0 < t < T \\ 0, & t \text{ elsewhere} \end{cases} = \Pi\left(\frac{t - T/2}{T}\right)$$

Then, using the time delay theorem and (2-55), we get the spectrum

$$V(f) = Te^{-j\pi fT}\,\text{Sa}(\pi Tf) \tag{2-57}$$

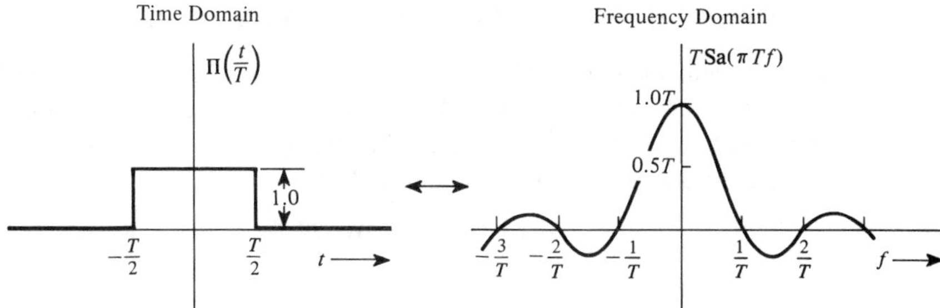

(a) Rectangular Pulse and Its Spectrum

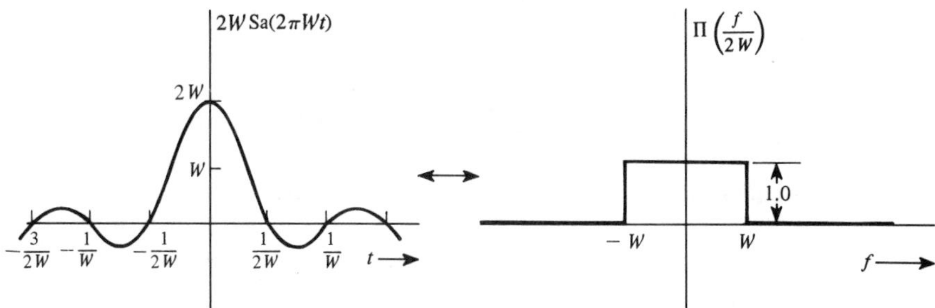

(b) Sa(x) Pulse and Its Spectrum

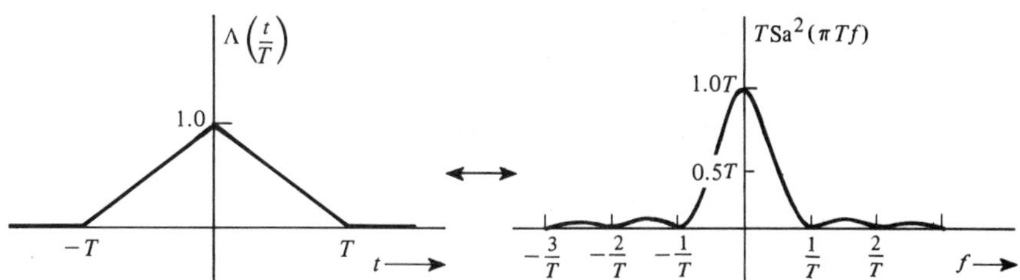

(c) Triangular Pulse and Its Spectrum

FIGURE 2-6 Spectra of rectangular, (sin x)/x, and triangular pulses.

In terms of quadrature notation, (2-57) becomes

$$V(f) = \underbrace{[T\,\mathrm{Sa}(\pi f T)\cos(\pi f T)]}_{X(f)} + j\underbrace{[-T\,\mathrm{Sa}(\pi f T)\sin(\pi f T)]}_{Y(f)} \tag{2-58}$$

The magnitude spectrum is

$$|V(f)| = T\left|\frac{\sin \pi f T}{\pi f T}\right| \tag{2-59}$$

and the phase spectrum is

$$\theta(f) = \tan^{-1}\left[\frac{-\sin(\omega T/2)}{\cos(\omega T/2)}\right] = -\pi f T \tag{2-60}$$

EXAMPLE 2-6 SPECTRUM OF A TRIANGULAR PULSE

The spectrum may be evaluated directly by taking the Fourier transform of $\Lambda(t/T)$ where the triangle is described by piecewise linear lines as shown in Fig. 2-6c. Another approach is to use the integral theorem. We will take the latter approach to demonstrate this technique. Let

$$w(t) = \Lambda(t/T)$$

Then

$$\frac{dw(t)}{dt} = \frac{1}{T}\,u(t+T) - \frac{2}{T}\,u(t) + \frac{1}{T}\,u(t-T)$$

and

$$\frac{d^2w(t)}{dt^2} = \frac{1}{T}\,\delta(t+T) - \frac{2}{T}\,\delta(t) + \frac{1}{T}\,\delta(t-T)$$

Consequently, we have the Fourier transform pair

$$\frac{d^2w(t)}{dt^2} \leftrightarrow \frac{1}{T}\,e^{j\omega T} - \frac{2}{T} + \frac{1}{T}\,e^{-j\omega T}$$

or

$$\frac{d^2w(t)}{dt^2} \leftrightarrow \frac{1}{T}(e^{j\omega T/2} - e^{-j\omega T/2})^2 = \frac{-4}{T}(\sin \pi f T)^2$$

Applying the integral theorem twice, we get

$$w(t) \leftrightarrow \frac{-4}{T}\,\frac{(\sin \pi f T)^2}{(j2\pi f)^2}$$

Thus

$$w(t) = \Lambda\left(\frac{t}{T}\right) \leftrightarrow T\,\mathrm{Sa}^2(\pi f T) \tag{2-61}$$

This is illustrated in Fig. 2-6c.

Convolution

The convolution operation, as first described in Table 2-1, is very useful. Later, we will show that the convolution operation can be used to describe the waveform at the output of a linear system.

DEFINITION: The *convolution* of a waveform $w_1(t)$ with a waveform $w_2(t)$ to produce a third waveform $w_3(t)$ is

$$w_3(t) = w_1(t) * w_2(t) \triangleq \int_{-\infty}^{\infty} w_1(\lambda) w_2(t-\lambda)\, d\lambda \tag{2-62}$$

where $w_1(t) * w_2(t)$ is shorthand notation that delineates this integral operation and $*$ is read "convolved with."

When this integral is examined, we realize that t is a parameter and λ is the variable of integration. Furthermore, the integrand is obtained by three operations:

1. Time reversal of w_2 to obtain $w_2(-\lambda)$.
2. Time shifting of w_2 to obtain $w_2(-(\lambda - t))$.
3. Multiplying this result by w_1 to form the integrand $w_1(\lambda)w_2(-(\lambda - t))$.

These three operations are illustrated in the following examples.

EXAMPLE 2-7 CONVOLUTION OF A RECTANGLE WITH AN EXPONENTIAL

Let

$$w_1(t) = \Pi\left(\frac{t - \frac{1}{2}T}{T}\right) \quad \text{and} \quad w_2(t) = e^{-t/T}\,u(t)$$

as shown in Fig. 2-7. Implementing step 3 above with the help of the figure, the convolution of $w_1(t)$ with $w_2(t)$ is

$$w_3(t) = w_1(t) * w_2(t) = \begin{cases} 0, & t < 0 \\ \int_0^t 1e^{(\lambda - t)/T}\,d\lambda, & 0 < t < T \\ \int_0^T 1e^{(\lambda - t)/T}\,d\lambda, & T < t \end{cases}$$

or

$$w_3(t) = \begin{cases} 0, & t < 0 \\ T(1 - e^{-t/T}), & 0 < t < T \\ T(e - 1)e^{-t/T}, & t > T \end{cases}$$

This result is plotted in Fig. 2-7.

EXAMPLE 2-8 SPECTRUM OF A TRIANGULAR PULSE BY CONVOLUTION

In Example 2-6 the spectrum of a triangular pulse was evaluated by using the integral theorem. The same result can be obtained by using the convolution theorem of Table 2-1. If the rectangular pulse of Fig. 2-6a is convolved with itself and then scaled (i.e., multiplied) by the constant $1/T$, the resulting time waveform is the triangular pulse of Fig. 2-6c. Applying the convolution theorem, we obtain the spectrum for the triangular pulse by multiplying the spectrum of the rectangular pulse (of Fig. 2-6a) with itself and scaling with a constant $1/T$. As expected, the result is the spectrum shown in Fig. 2-6c.

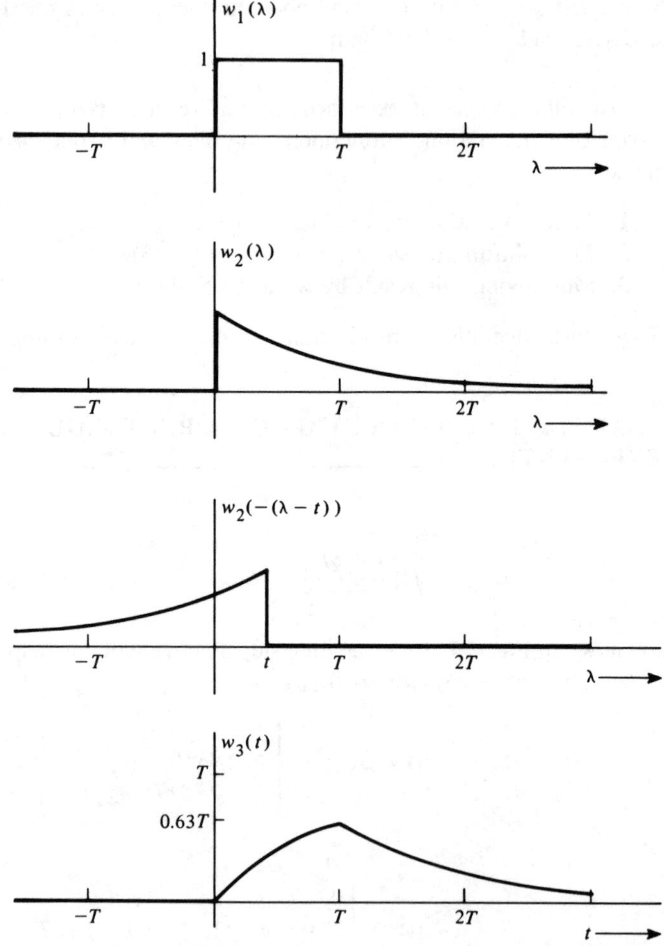

FIGURE 2-7 Convolution of a rectangle and an exponential.

EXAMPLE 2-9 SPECTRUM OF A SWITCHED SINUSOID

In Example 2-4 a continuous sinusoid was found to have a line spectrum with the lines located at $f = \pm f_0$. In this example we will see how the spectrum changes when the sinusoid is switched on and off. The switched sinusoid can be represented by

$$w(t) = \Pi\left(\frac{t}{T}\right) A \sin \omega_0 t = \Pi\left(\frac{t}{T}\right) A \cos\left(\omega_0 t - \frac{\pi}{2}\right)$$

Using (2-55) and the real signal translation theorem of Table 2-1, we see that the spectrum of this switched sinusoid is

$$W(f) = j\frac{A}{2}T[\text{Sa}(\pi T(f + f_0)) - \text{Sa}(\pi T(f - f_0))] \tag{2-63}$$

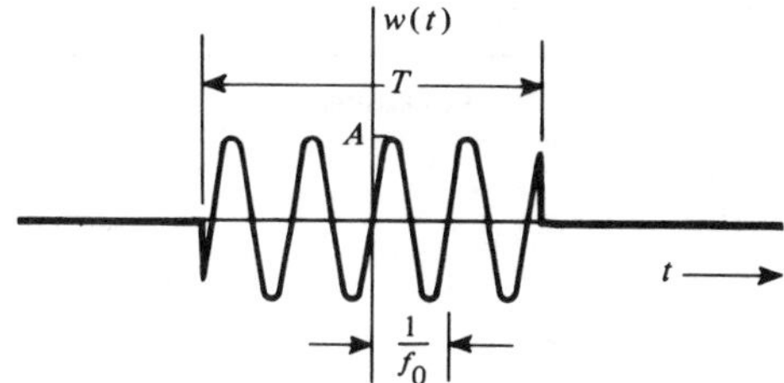

(a) Time Domain

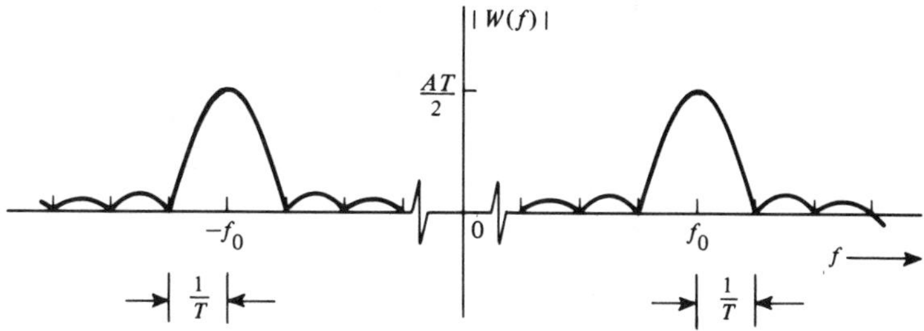

(b) Frequency Domain (Magnitude Spectrum)

FIGURE 2-8 Waveform and spectrum of a switched sinusoid.

This spectrum is continuous and imaginary. The magnitude spectrum is shown in Fig. 2-8. Compare this continuous spectrum of Fig. 2-8 with the discrete spectrum obtained for the continuous sine wave, as shown in Fig. 2-4a. In addition, note that if the duration of the switched sinusoid is allowed to become very large (i.e., $T \to \infty$), the continuous spectrum of Fig. 2-8 becomes the discrete spectrum of Fig. 2-4a by the use of (A-103) in Appendix A.

The spectrum for the switched sinusoid may also be evaluated using a different approach. The multiplication theorem of Table 2-1 may be used where $w_1(t) = \Pi(t/T)$ and $w_2(t) = A \sin \omega_0 t$. Here the spectrum of the switched sinusoid is obtained by working a convolution problem in the frequency domain:

$$W(f) = W_1(f) * W_2(f) = \int_{-\infty}^{\infty} W_1(\lambda) W_2(f - \lambda)\, d\lambda$$

This convolution integral is easy to evaluate since the spectrum of $w_2(t)$ consists of two delta functions. The details of this approach are left as a homework exercise for the reader.

A summary of Fourier transform pairs is given in Table 2-2. Numerical techniques using the discrete Fourier transform are discussed in Sec. 2-7.

TABLE 2-2 Some Fourier Transform Pairs

Function	Time Waveform $w(t)$	Spectrum $W(f)$
Rectangular	$\Pi\left(\frac{t}{T}\right)$	$T[\mathrm{Sa}(\pi f T)]$
Triangular	$\Lambda\left(\frac{t}{T}\right)$	$T[\mathrm{Sa}(\pi f T)]^2$
Unit step	$u(t) \triangleq \begin{cases} +1, & t>0 \\ 0, & t<0 \end{cases}$	$\frac{1}{2}\,\delta(f) + \dfrac{1}{j2\pi f}$
Signum	$\mathrm{sgn}(t) \triangleq \begin{cases} +1, & t>0 \\ -1, & t<0 \end{cases}$	$\dfrac{1}{j\pi f}$
Constant	1	$\delta(f)$
Impulse at $t = t_0$	$\delta(t - t_0)$	$e^{-j2\pi f t_0}$
Sinc	$\mathrm{Sa}(2\pi W t)$	$\dfrac{1}{2W}\,\Pi\left(\dfrac{f}{2W}\right)$
Phasor	$e^{j(\omega_0 t + \phi)}$	$e^{j\phi}\,\delta(f - f_0)$
Sinusoid	$\cos(\omega_c t + \phi)$	$\frac{1}{2}e^{j\phi}\,\delta(f - f_c) + \frac{1}{2}e^{-j\phi}\,\delta(f + f_c)$
Gaussian	$e^{-\pi(t/t_0)^2}$	$t_0 e^{-\pi(f t_0)^2}$
Exponential, one-sided	$\begin{cases} e^{-t/T}, & t>0 \\ 0, & t<0 \end{cases}$	$\dfrac{T}{1 + j2\pi f T}$
Exponential, two-sided	$e^{-\lvert t\rvert/T}$	$\dfrac{2T}{1 + (2\pi f T)^2}$
Impulse train	$\sum_{k=-\infty}^{k=\infty} \delta(t - kT)$	$f_0 \sum_{n=-\infty}^{n=\infty} \delta(f - n f_0)$, where $f_0 = 1/T$

2-3 Power Spectral Density and Autocorrelation Function

Power Spectral Density

The normalized power of a waveform will now be related to its frequency domain description by the use of a function called the *power spectral density (PSD)*. The PSD is very useful in describing how the power content of signals and noise is affected by filters and other devices in communication systems. In (2-42) the energy spectral density (ESD) was defined in terms of the magnitude squared version of the Fourier transform of the waveform. The PSD will be defined in a similar way. The PSD is more useful than the ESD since power-type models are generally used in solving communication problems.

First, define the truncated version of the waveform by

$$w_T(t) = \begin{cases} w(t), & -T/2 < t < T/2 \\ 0, & t \text{ elsewhere} \end{cases} = w(t)\Pi\left(\frac{t}{T}\right) \tag{2-64}$$

Using (2-13), we obtain the average normalized power

$$P = \lim_{T\to\infty} \frac{1}{T}\int_{-T/2}^{T/2} w^2(t)\, dt = \lim_{T\to\infty} \frac{1}{T}\int_{-\infty}^{\infty} w_T^2(t)\, dt$$

By the use of Parseval's theorem, (2-41), this becomes

$$P = \lim_{T\to\infty} \frac{1}{T}\int_{-\infty}^{\infty} |W_T(f)|^2\, df = \int_{-\infty}^{\infty} \left(\lim_{T\to\infty} \frac{|W_T(f)|^2}{T}\right) df \tag{2-65}$$

where $W_T(f) = \mathcal{F}[w_T(t)]$. The integrand of the right-hand integral has units of watts/hertz (or, equivalently, volts^2/hertz or amperes^2/hertz, as appropriate) and can be defined as the PSD.

Definition: The *power spectral density (PSD)* for a deterministic power waveform is†

$$\mathcal{P}_w(f) \triangleq \lim_{T\to\infty} \left(\frac{|W_T(f)|^2}{T}\right) \tag{2-66}$$

where $w_T(t) \leftrightarrow W_T(f)$ and $\mathcal{P}_w(f)$ has units of watts per hertz.

Note that the PSD is always a real nonnegative function of frequency. In addition, the PSD is not sensitive to the phase spectrum of $w(t)$ since that is lost because of the absolute value operation used in (2-66). From (2-65), the normalized average power is†

$$P = \langle w^2(t)\rangle = \int_{-\infty}^{\infty} \mathcal{P}_w(f)\, df \tag{2-67}$$

That is, the area under the PSD function is the normalized average power.

Autocorrelation Function

A related function called the *autocorrelation*, $R(\tau)$, can also be defined.‡

† Equations (2-66) and (2-67) give normalized (with respect to 1 Ω) PSD and average power, respectively. For unnormalized (i.e., actual) values, $\mathcal{P}_w(f)$ is replaced by the appropriate expression as follows. If $w(t)$ is a voltage waveform that appears across a resistive load of R ohms, the unnormalized PSD is $\mathcal{P}_w(f)/R$ watts/Hz, where $\mathcal{P}_w(f)$ has units of volts^2/Hz. Similarly, if $w(t)$ is a current waveform passing through a resistive load of R ohms, the unnormalized PSD is $\mathcal{P}_w(f)R$ watts/Hz, where $\mathcal{P}_w(f)$ has units of amperes^2/Hz.

‡ Here a time average is used in the definition of the autocorrelation function. In Chapter 6 an ensemble (statistical) average is used in the definition of $R(\tau)$. As shown in Chapter 6, these two definitions are equal if $w(t)$ is ergodic. Throughout this chapter, we assume that $w(t)$ is ergodic.

DEFINITION: The *autocorrelation* of a real (physical) waveform is[†]

$$R_w(\tau) \triangleq \langle w(t)w(t+\tau)\rangle = \lim_{T\to\infty} \frac{1}{T}\int_{-T/2}^{T/2} w(t)w(t+\tau)\,dt \qquad (2\text{-}68)$$

Furthermore, it can be shown that the PSD and the autocorrelation function are Fourier transform pairs.

$$R_w(\tau) \leftrightarrow \mathcal{P}_w(f) \qquad (2\text{-}69)$$

where $\mathcal{P}_w(f) = \mathcal{F}[R_w(\tau)]$. This is called the *Wiener–Khintchine theorem*. This theorem, along with properties for $R(\tau)$ and $\mathcal{P}(f)$, are developed in detail in Chapter 6 using the more general theory of random processes that is applicable to nondeterministic waveforms.

In summary, the PSD can be evaluated by either of two methods:

1. Direct evaluation using the definition, (2-66).
2. Indirect evaluation by first evaluating the autocorrelation function and then taking the Fourier transform, $\mathcal{P}_w(f) = \mathcal{F}[R_w(\tau)]$.

Furthermore, the total average normalized power for the waveform $w(t)$ can be evaluated using any of the *four* techniques:

$$P = \langle w^2(t)\rangle = W_{\text{rms}}^2 = \int_{-\infty}^{\infty} \mathcal{P}_w(f)\,df = R_w(0) \qquad (2\text{-}70)$$

EXAMPLE 2-10 PSD OF A SINUSOID

Let

$$w(t) = A\sin\omega_0 t$$

The PSD will be evaluated using the indirect method. The autocorrelation is

$$\begin{aligned} R_w(\tau) &= \langle w(t)w(t+\tau)\rangle \\ &= \lim_{T\to\infty}\frac{1}{T}\int_{-T/2}^{T/2} A^2 \sin\omega_0 t \sin\omega_0(t+\tau)\,dt \end{aligned}$$

Using a trigonometric identity, (A-12), we obtain

$$R_w(\tau) = \frac{A^2}{2}\cos\omega_0\tau \lim_{T\to\infty}\frac{1}{T}\int_{-T/2}^{T/2} dt - \frac{A^2}{2}\lim_{T\to\infty}\frac{1}{T}\int_{-T/2}^{T/2}\cos(2\omega_0 t + \omega_0\tau)\,dt$$

which reduces to

$$R_w(\tau) = \frac{A^2}{2}\cos\omega_0\tau \qquad (2\text{-}71a)$$

The PSD is then

$$\mathcal{P}_w(f) = \mathcal{F}\left[\frac{A^2}{2}\cos\omega_0\tau\right] = \frac{A^2}{4}[\delta(f-f_0) + \delta(f+f_0)] \qquad (2\text{-}71b)$$

[†]The autocorrelation of a complex waveform is $R_w(\tau) \triangleq \langle w^*(t)w(t+\tau)\rangle$.

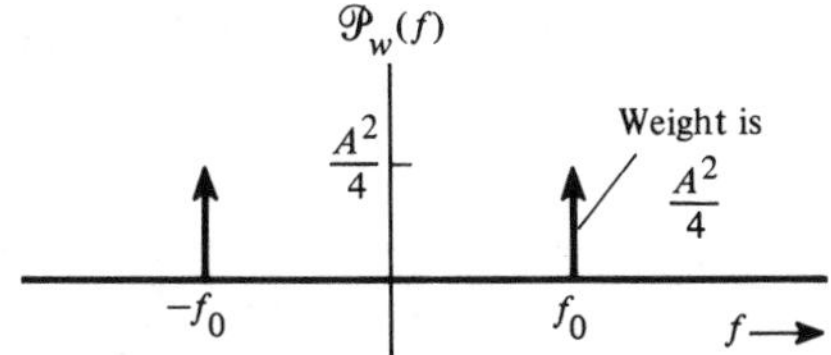

FIGURE 2-9 Power spectrum of a sinusoid.

as shown in Fig. 2-9. It may be compared to the "voltage" spectrum for a sinusoid that was obtained in Example 2-4 and shown in Fig. 2-4.

The average normalized power may be obtained by using (2-67):

$$P = \int_{-\infty}^{\infty} \frac{A^2}{4} [\delta(f - f_0) + \delta(f + f_0)]\, df = \frac{A^2}{2}$$

This value, $A^2/2$, checks with the known result for the normalized power of a sinusoid:

$$P = \langle w^2(t) \rangle = W_{\text{rms}}^2 = (A/\sqrt{2})^2 = A^2/2$$

It is also realized that $A \sin \omega_0 t$ and $A \cos \omega_0 t$ have exactly the same PSD (and autocorrelation function) since the phase has no effect on the PSD. This can be verified by evaluating the PSD for $A \cos \omega_0 t$, using the same procedure that was used above to evaluate the PSD for $A \sin \omega_0 t$.

Thus far, properties of signals and noise such as spectrum, average power, and rms value have been studied, but how do we represent the signal or noise waveform itself? The direct approach, of course, is to write a closed-form mathematical equation for the waveform itself. Other equivalent ways of modeling the waveform are often found to be very useful. One, which the reader has already studied in calculus, is to represent the waveform by the use of a Taylor series (i.e., power series) expansion about a point a:

$$w(t) = \sum_{n=0}^{\infty} \frac{w^{(n)}(a)}{n!} (t - a)^n \tag{2-72}$$

where

$$w^{(n)}(a) = \left. \frac{d^n w(t)}{dt^n} \right|_{t=a}$$

In this case, if the derivatives at $t = a$, $\{w^{(n)}(a)\}$, are known, this algorithm can be used to reconstruct the waveform. Another type of series representation that is especially useful in communication problems is the orthogonal series expansion, which is discussed in the next section.

2-4 Orthogonal Series Representation of Signals and Noise

An orthogonal series representation of signals and noise is a general representation that has many significant applications in communication problems such as Fourier series, sampling function series, and representation of digital signals. These specific cases are so important that they will be studied in some detail in separate sections that follow.

★ Orthogonal Functions

Before the orthogonal series is studied, a definition for orthogonal functions is needed.

DEFINITION: Functions $\varphi_i(t)$ in the set $\{\varphi_i(t)\}$ are said to be *orthogonal* with respect to each other over the interval $a < t < b$ if they satisfy the conditions

$$\int_a^b \varphi_i(t)\varphi_j^*(t)\,dt = \begin{Bmatrix} K_i, & i = j \\ 0, & i \neq j \end{Bmatrix} = K_i \delta_{ij} \tag{2-73}$$

where

$$\delta_{ij} \triangleq \begin{cases} 1, & i = j \\ 0, & i \neq j \end{cases} \tag{2-74}$$

δ_{ij} is called the *Kronecker delta function*. If the constants K_i are all equal to 1, the $\varphi_i(t)$ are said to be *orthonormal functions*.†

In other words, (2-73) is used to test pairs of functions to see if they are orthogonal. They are orthogonal over the interval (a,b) if the integral of their product is zero. The zero result implies that these functions are ''independent'' or in ''disagreement.'' If the result is not zero, they are not orthogonal, and consequently, the two functions have some ''dependence'' or ''alikeness'' to each other.

Equation (2-73) can also be used to evaluate normalized energy if $\varphi_j(t)$ corresponds to a (voltage or current) waveform component associated with a 1-Ω resistive load (i.e., the normalized condition). Using (2-14), we find that the normalized energy dissipated in the load in $(b - a)$ seconds as a result of $\varphi_j(t)$ is

$$E_j = \int_a^b |\varphi_j(t)|^2\,dt = K_j \tag{2-75}$$

Consequently, K_j is the normalized energy of $\varphi_j(t)$ in the interval (a,b).

† To normalize a set of functions, take each old $\varphi_i(t)$ and divide it by $\sqrt{K_i}$ to form the normalized $\varphi_i(t)$.

★ Orthogonal Series

Assume that $w(t)$ represents some practical waveform (signal, noise, or signal–noise combination) that we wish to represent over the interval $a < t < b$. Then we can obtain an equivalent orthogonal series representation by using the following theorem.

THEOREM: *$w(t)$ can be represented over the interval (a,b) by the series*

$$w(t) = \sum_j a_j \varphi_j(t) \tag{2-76}$$

where the orthogonal coefficients are given by

$$a_j = \frac{1}{K_j} \int_a^b w(t) \varphi_j^*(t)\, dt \tag{2-77}$$

and the range of j is over the integer values that correspond to the subscripts that were used to denote the orthogonal functions in the complete orthogonal set.

In order for (2-76) to be a valid representation for any physical signal (i.e., one with finite energy), the orthogonal set has to be complete. This implies that the set $\{\varphi_j(t)\}$ can be used to represent any function with an arbitrarily small error [Wylie, 1960]. In practice, it is usually difficult to prove that a given set of functions is complete. It can be shown that the complex exponential set and the harmonic sinusoidal sets that are used for the Fourier series in Sec. 2-5 are complete [Courant and Hilbert, 1953]. Many other useful sets are also complete such as Bessel functions, Legendre polynomials, and the $(\sin x)/x$-type sets [described by (2-161)].

Proof of Theorem. Assume that the set $\{\varphi_j(t)\}$ is sufficient to represent the waveform. Then in order for (2-76) to be correct, we only need to show that we can evaluate the a_j value by using (2-77). Using

$$w(t) = \sum_i a_i \varphi_i(t)$$

we operate on both sides of this equation with the integral operator

$$\int_a^b [\cdot]\, \varphi_j^*(t)\, dt \tag{2-78}$$

$$\begin{aligned} \int_a^b [w(t)]\, \varphi_j^*(t)\, dt &= \int_a^b \left[\sum_i a_i \varphi_i(t)\right] \varphi_j^*(t)\, dt \\ &= \sum_i a_i \int_a^b \varphi_i(t)\, \varphi_j^*(t)\, dt = \sum_i a_i K_i\, \delta_{ij} \\ &= a_j K_j \end{aligned} \tag{2-79}$$

Thus (2-77) follows.

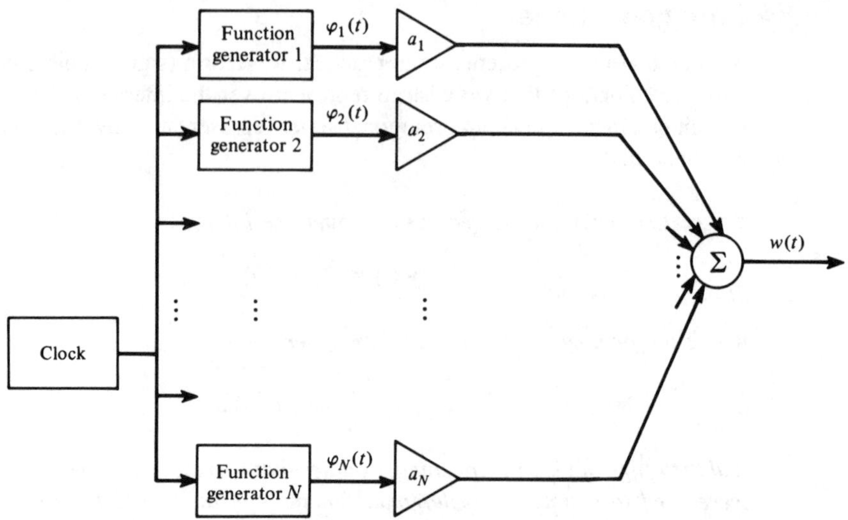

FIGURE 2-10 Waveform synthesis using orthogonal functions.

As mentioned previously, the orthogonal series is very useful in representing a signal, noise, or a signal–noise combination. The orthogonal functions $\varphi_j(t)$ are deterministic. Furthermore, if the waveform $w(t)$ is deterministic, the constants $\{a_j\}$ are also deterministic and may be evaluated using (2-77). In Chapter 6 we will see that if $w(t)$ is stochastic (e.g., in a noise problem), the $\{a_j\}$ are a set of random variables that give the desired random process $w(t)$.

It is also possible to use (2-76) to generate $w(t)$ from the $\varphi_j(t)$ functions and the coefficients a_j. In this case $w(t)$ is approximated by using a reasonable number of the $\varphi_j(t)$ functions. As shown in Fig. 2-10, for the case of real values for a_j and real functions for $\varphi_j(t)$, $w(t)$ can be synthesized by adding up weighted versions of $\varphi_j(t)$ where the weighting factors are given by $\{a_j\}$. The summing and gain weighting operation may be conveniently realized by using an operational amplifier with multiple inputs.

★ Vector Representations of Digital Signals†

Often in engineering problems we need to represent only a certain set (class) of signals, such as digital signals, that require the use of only a finite number of orthogonal functions, $\varphi_i(t)$. In this case, $j = 1, 2, \ldots, N$, where N is said to be the *dimension* of the waveform set. Furthermore, if the a_j and $\varphi_j(t)$ happen to be *real*, and the number of orthogonal functions is N, the mathematics can be simplified by representing the waveform $w(t)$ with an N-dimensional vector $\mathbf{w}$.‡

†This section may be skipped on the first reading. It is useful for the material covered in Chapter 8.

‡It is assumed that a real waveform is being represented by $\mathbf{w}$ and the $\varphi_j(t)$ are real; consequently, the a_j are real.

That is, the orthogonal function space representation of (2-76) corresponds to the orthogonal vector space represented by

$$\mathbf{w} = \sum_{j=1}^{N} w_j \boldsymbol{\varphi}_j \tag{2-80}$$

where the boldface type denotes a vector, $w_j \triangleq a_j$ of (2-76), and $\{\boldsymbol{\varphi}_j\}$ is an orthogonal vector set. $\mathbf{w}$ is an N-dimensional vector in Euclidean vector space, and $\{\boldsymbol{\varphi}_j\}$ is a set of N directional vectors that become a unit vector set if the K_j's are all unity. A shorthand notation for the vector $\mathbf{w}$ of (2-80) is given by a row vector:

$$\mathbf{w} = (w_1, w_2, w_3, \ldots, w_N) \tag{2-81}$$

The normalized energy of $w(t)$ over the interval $a < t < b$ is given by

$$E_w = \int_a^b w^2(t)\, dt = \int_a^b \left[\sum_i a_i \varphi_i(t)\right]^2 dt$$

$$= \sum_i \sum_j a_i a_j \int_a^b \varphi_i(t)\varphi_j(t)\, dt = \sum_i \sum_j a_i a_j K_i \delta_{ij} \tag{2-82}$$

or

$$E_w = \sum_{j=1}^{N} a_j^2 K_j \tag{2-83}$$

where $w(t)$, $\varphi_i(t)$, and $\varphi_j(t)$ are assumed to be real. This result is a special case of Parseval's theorem, and it relates the integral of the square of the time function $w(t)$ to the sum of the square of the orthogonal series coefficients. The general case of Parseval's theorem as related to orthogonal expansions is given by (8-56). It is obtained from (2-40) for the case of real (element) vectors.

If orthonormal functions and their corresponding unit directional vectors are used (i.e., $K_j = 1, j = 1, 2, \ldots, N$), the normalized energy over the interval (a, b) is given by the dot product of $\mathbf{w}$ with itself, which is equivalent to the square of the length of the vector.†

$$E_w = \mathbf{w} \cdot \mathbf{w} = \sum_{j=1}^{N} w_j^2 = |\mathbf{w}|^2 \tag{2-84}$$

This, of course, is an application of the Pythagorean theorem in N dimensions.

†The dot product of two real vectors $\mathbf{v} = (v_1, v_2, \ldots, v_N)$ and $\mathbf{w} = (w_1, w_2, \ldots, w_N)$ is $\mathbf{v} \cdot \mathbf{w} = \Sigma_{j=1}^{N} v_j w_j$. If $\mathbf{v}$ and $\mathbf{w}$ are column vectors, $\Sigma_{j=1}^{N} v_j w_j = \mathbf{v}^T\mathbf{w}$ where T denotes the vector transpose. Also, $\mathbf{v} \cdot \mathbf{w} = \mathbf{w} \cdot \mathbf{v}$, and if $\mathbf{v} \cdot \mathbf{w} = 0$, $\mathbf{v}$ and $\mathbf{w}$ are orthogonal.

EXAMPLE 2-11 REPRESENTATION OF A DIGITAL SIGNAL

Examine the representation for the waveform of a 3-bit (binary) digital word shown in Fig. 2-11a. This signal could be directly represented by

$$s(t) = \sum_{j=1}^{N=3} d_j p[t - (j - \tfrac{1}{2})T] = \sum_{j=1}^{N=3} d_j p_j(t) \tag{2-85}$$

where $p(t)$ is shown in Fig. 2-11b and $p_j(t) \triangleq p[t - (j - \frac{1}{2})T]$.

The $\{p_j(t)\}$ is a set of orthogonal functions that are not normalized.

$$\mathbf{d} = (d_1,d_2,d_3) = (1,0,1) \tag{2-86}$$

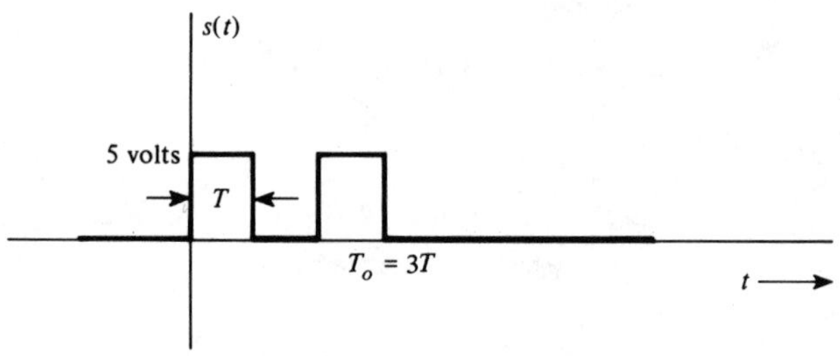

(a) A Three-Bit Signal Waveform

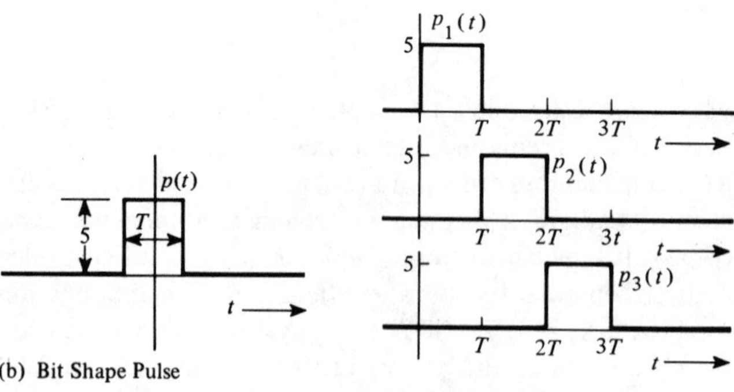

(b) Bit Shape Pulse

(c) Orthogonal Function Set

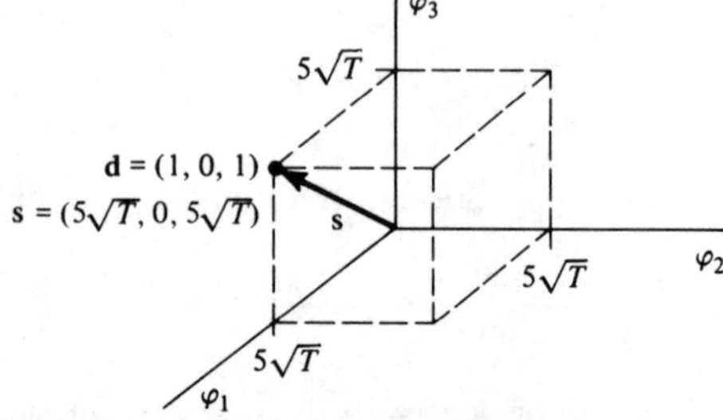

(d) Vector Representation of the Three-Bit Signal

FIGURE 2-11 Representation for a 3-bit binary digital signal.

is the binary word with 1 representing a binary 1 and 0 representing a binary 0. $p(t)$ is the pulse shape for each bit.

Using the orthogonal function approach, we can represent the waveform by

$$s(t) = \sum_{j=1}^{N=3} s_j \varphi_j(t) \tag{2-87}$$

Let $\{\varphi_j(t)\}$ be the corresponding set of ortho*normal* functions so that (2-84) will apply. Then

$$\varphi_j(t) = \frac{p_j(t)}{\sqrt{K_j}} = \frac{p_j(t)}{\sqrt{\int_0^{T_o} p_j^2(t)\, dt}} = \frac{p_j(t)}{\sqrt{25T}} \tag{2-88}$$

or

$$\varphi_j(t) = \begin{cases} \dfrac{1}{\sqrt{T}}, & (j-1)T < t < jT \\ 0, & t \text{ otherwise} \end{cases} \tag{2-89}$$

where $j = 1$, 2, or 3. Using (2-77) where $a = 0$ and $b = 3T$, we find that the orthonormal series coefficients for the digital signal shown in Fig. 2-11a are

$$\mathbf{s} = (s_1, s_2, s_3) = (5\sqrt{T}, 0, 5\sqrt{T}) \tag{2-90}$$

and the normalized energy, using (2-84), is†

$$E_s = 50T \text{ joules} \tag{2-91}$$

The vector representation for $s(t)$ using (2-90) is shown in Fig. 2-11d. From this figure, note that the square of the length of the vector ($|\mathbf{s}|^2 = 50T$) gives the normalized energy, in agreement with (2-91). In addition, all the other possible 3-bit words would be represented by points on the other vertices of the cube, and they might have different energy values depending on their distance from the origin (i.e., vector length).

It is also realized that a 3-bit digital signal could be represented by an analog waveform instead of a digital waveform. For example, let

$$\varphi_j(t) = \begin{cases} \sqrt{\dfrac{2}{T_0}} \cos j\omega_0 t, & 0 \le t \le T_0 \\ 0, & t \text{ elsewhere} \end{cases}$$

where $f_0 = 1/T_0$, $T_0 = 3T$, and $j = 1$, 2, and 3. The 3-bit digital signal is still given by (2-86), (2-90), and (2-87), but now $s(t)$ is an analog waveform since

†The energy can also be calculated directly from the $s(t)$ in Fig. 2-11a by evaluating

$$\int_0^{3T} s^2(t)\, dt.$$

$\varphi_j(t)$ is a function that takes on a continuous set of values between $-\sqrt{2/T_0}$ and $+\sqrt{2/T_0}$. The vector representation of this 3-bit signal is still identical to that shown in Fig. 2-11d and the energy is $50T$ joules, just as before for the case of the rectangular digital signaling waveform. As we will see later, the bandwidth of the analog waveform is less than that of the digital waveform, although either one could be used to represent the same digital signal.

2-5 Fourier Series

The Fourier series is a particular type of orthogonal series representation that is very useful in solving engineering problems, especially communication problems. The orthogonal functions that are used are either sinusoids, or, equivalently, complex exponential functions.†

Complex Fourier Series

The complex Fourier series uses the exponential functions.

$$\varphi_n(t) = e^{jn\omega_0 t} \tag{2-92}$$

where n ranges over all possible integer values, negative, positive, and zero; $\omega_0 = 2\pi/T_0$, where $T_0 = (b - a)$ is the length of the interval over which the series, (2-76), is valid; and $K_n = T_0$. Using (2-76), the Fourier series theorem follows.

THEOREM. *Any physical waveform (i.e., finite energy) may be represented over the interval $a < t < a + T_0$ by the complex exponential Fourier series*

$$w(t) = \sum_{n=-\infty}^{n=\infty} c_n e^{jn\omega_0 t} \tag{2-93}$$

where the complex (phasor) Fourier coefficients c_n are given by

$$c_n = \frac{1}{T_0}\int_a^{a+T_0} w(t)e^{-jn\omega_0 t}\,dt \tag{2-94}$$

and $\omega_0 = 2\pi f_0 = 2\pi/T_0$.

If the waveform $w(t)$ is periodic with period T_0, this Fourier series representation is valid *over all time* (i.e., over the interval $-\infty < t < +\infty$) since the $\varphi_n(t)$ are periodic functions that have a common fundamental period T_0. For this case of periodic waveforms, the choice of a value for the parameter a is arbitrary, and it is usually taken to be $a = 0$ or $a = -T_0/2$ for mathematical convenience. The frequency $f_0 = 1/T_0$ is said to be the *fundamental* frequency, and the frequency nf_0 is said to be the nth harmonic frequency, when $n > 1$. The Fourier coefficient

†Mathematicians generally call any orthogonal series a Fourier series.

c_0 is equivalent to the dc value of the waveform $w(t)$, since the integral is identical to that of (2-4) when $n = 0$.

c_n is, in general, a complex number. Furthermore, it is a phasor since it is the coefficient of a function of the type $e^{j\omega t}$. Consequently, (2-93) is said to be a complex or phasor Fourier series.

Some properties of the complex Fourier series are as follows.

1. If $w(t)$ is real,
$$c_n = c_{-n}^* \tag{2-95a}$$
2. If $w(t)$ is real and even [i.e., $w(t) = w(-t)$],
$$\text{Im}[c_n] = 0 \tag{2-95b}$$
3. If $w(t)$ is real and odd [i.e., $w(t) = -w(-t)$],
$$\text{Re}\,[c_n] = 0$$
4. Parseval's theorem is
$$\frac{1}{T_0}\int_a^{a+T_0} |w(t)|^2\,dt = \sum_{n=-\infty}^{n=\infty} |c_n|^2 \tag{2-95c}$$
See (2-125) for the proof.
5. The complex Fourier series coefficients of a real waveform are related to the quadrature Fourier series coefficients by
$$c_n = \begin{cases} \frac{1}{2}a_n - j\frac{1}{2}b_n, & n > 0 \\ a_0, & n = 0 \\ \frac{1}{2}a_{-n} + j\frac{1}{2}b_{-n}, & n < 0 \end{cases} \tag{2-95d}$$
See (2-97) and (2-98).
6. The complex Fourier series coefficients of a real waveform are related to the polar Fourier series coefficients by
$$c_n = \begin{cases} \frac{1}{2}D_n \underline{/\varphi_n}, & n > 0 \\ D_0, & n = 0 \\ \frac{1}{2}D_{-n}\ \underline{/-\varphi_{-n}}, & n < 0 \end{cases} \tag{2-95e}$$
See (2-106) and (2-107).

Note that these properties for the complex Fourier series coefficients are similar to those of the Fourier transform as given in Sec. 2-2.

Quadrature Fourier Series

The *quadrature* form of the Fourier series representing any physical waveform $w(t)$ over the interval $a < t < a + T_0$ is
$$w(t) = \sum_{n=0}^{n=\infty} a_n \cos n\omega_0 t + \sum_{n=1}^{n=\infty} b_n \sin n\omega_0 t \tag{2-96}$$

where the orthogonal functions are cos $n\omega_0 t$ and sin $n\omega_0 t$. Using (2-77), we find that these Fourier coefficients are given by

$$a_n = \begin{cases} \dfrac{1}{T_0}\displaystyle\int_a^{a+T_0} w(t)\, dt, & n = 0 \\ \dfrac{2}{T_0}\displaystyle\int_a^{a+T_0} w(t)\cos n\omega_0 t\, dt, & n \geq 1 \end{cases} \tag{2-97a}$$

and

$$b_n = \frac{2}{T_0}\int_a^{a+T_0} w(t)\sin n\omega_0 t\, dt, \qquad n > 0 \tag{2-97b}$$

Once again, since these sinusoidal orthogonal functions are periodic, this series is periodic with the fundamental period T_0, and if $w(t)$ is periodic with period T_0, the series will represent $w(t)$ over the whole real line (i.e., $-\infty < t < \infty$).

The complex Fourier series, (2-93), and the quadrature Fourier series, (2-96), are equivalent representations. This can be demonstrated by expressing the complex number c_n in terms of its conjugate parts, x_n and y_n. That is, using (2-94),

$$\begin{aligned} c_n &= x_n + jy_n \\ &= \left[\frac{1}{T_0}\int_a^{a+T_0} w(t)\cos n\omega_0 t\, dt\right] + j\left[\frac{-1}{T_0}\int_a^{a+T_0} w(t)\sin n\omega_0 t\, dt\right] \end{aligned} \tag{2-98}$$

for all integer values of n. Thus

$$x_n = \frac{1}{T_0}\int_a^{a+T_0} w(t)\cos n\omega_0 t\, dt \tag{2-99}$$

and

$$y_n = \frac{-1}{T_0}\int_a^{a+T_0} w(t)\sin n\omega_0 t\, dt \tag{2-100}$$

Using (2-97a) and (2-97b), we obtain the identities

$$a_n = \begin{Bmatrix} c_0, & n = 0 \\ 2x_n, & n \geq 1 \end{Bmatrix} = \begin{Bmatrix} c_0, & n = 0 \\ 2\ \mathrm{Re}\{c_n\}, & n \geq 1 \end{Bmatrix} \tag{2-101}$$

and

$$b_n = -2y_n = -2\ \mathrm{Im}\{c_n\}, \qquad n \geq 1 \tag{2-102}$$

where Re{·} denotes the real part of {·} and Im{·} denotes the imaginary part of {·}.

Polar Fourier Series

The quadrature Fourier series, (2-96), may be rearranged and written in a polar (amplitude-phase) form. The *polar* form is

$$w(t) = D_0 + \sum_{n=1}^{n=\infty} D_n \cos(n\omega_0 t + \phi_n) \tag{2-103}$$

where $w(t)$ is real and

$$a_n = \begin{cases} D_0, & n = 0 \\ D_n \cos \phi_n, & n \geq 1 \end{cases} \tag{2-104}$$

$$b_n = -D_n \sin \phi_n, \quad n \geq 1 \tag{2-105}$$

The latter two equations may be inverted, and we obtain

$$D_n = \begin{Bmatrix} a_0, & n = 0 \\ \sqrt{a_n^2 + b_n^2}, & n \geq 1 \end{Bmatrix} = \begin{cases} c_0, & n = 0 \\ 2|c_n|, & n \geq 1 \end{cases} \tag{2-106}$$

and

$$\phi_n = -\tan^{-1}\left(\frac{b_n}{a_n}\right) = \underline{/c_n}, \quad n \geq 1 \tag{2-107}$$

where the angle operator is defined by

$$\underline{/[\cdot]} = \tan^{-1}\left(\frac{\text{Im}[\cdot]}{\text{Re}[\cdot]}\right) \tag{2-108}$$

It should be clear from the context whether $\underline{/\quad}$ denotes the angle operator or the angle itself. For example, $\underline{/90^\circ}$ denotes an angle of 90°, but $\underline{/[1 + j2]}$ denotes the angle operator and is equal to 63.4° when evaluated.

The equivalence between the Fourier series coefficients is demonstrated geometrically in Fig. 2-12. It is seen that, in general, when a physical (real) waveform $w(t)$ is represented by a Fourier series, c_n is a complex number with a real

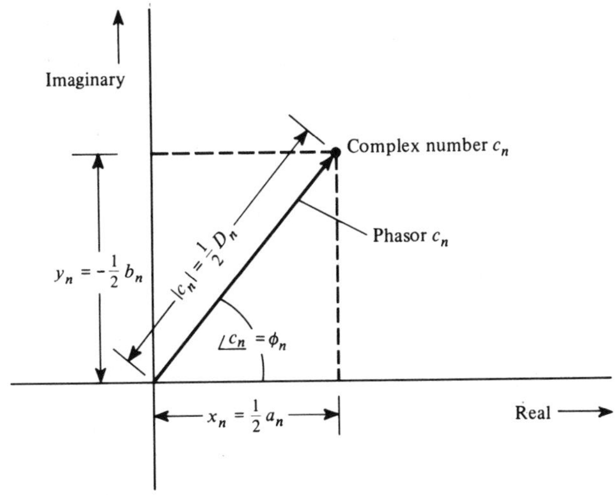

FIGURE 2-12 Fourier series coefficients, $n \geq 1$.

part x_n and an imaginary part y_n (which are both real numbers) and, consequently, a_n, b_n, D_n, and ϕ_n are real numbers. Furthermore, D_n is a nonnegative number for $n \geq 1$.

As we have seen, the complex, quadrature, and polar form of the Fourier series are all equivalent, but the question is: *Which is the best form to use?* The answer is that it depends on the particular problem being solved! If the problem is being solved analytically, the complex coefficients are *usually* easier to evaluate. On the other hand, if measurements of a waveform are being made in a laboratory, the polar form is usually more convenient since measuring instruments, such as voltmeters, scopes, vector voltmeters, and wave analyzers, usually give magnitude and phase readings. Using the laboratory results, engineers often draw *one-sided* spectral plots where lines are drawn corresponding to each D_n value at $f = nf_0$, where $n \geq 0$ (i.e., positive frequencies only). Of course, this one-sided spectral plot may be converted to the two-sided spectrum given by the c_n plot by using (2-106). It is understood that the actual spectrum is defined to be the Fourier transform of $w(t)$ and consists of two two-sided plots, one plot for $|c_n|$ and the other for $\angle c_n$. This is demonstrated by Fig. 2-4 where (2-109), from the following theorem, may be used.

Line Spectra for Periodic Waveforms

For *periodic* waveforms, the Fourier series representations are valid over all time (i.e., $-\infty < t < \infty$). Consequently, the (two-sided) spectrum, which depends on the waveshape from $t = -\infty$ to $t = \infty$, may be evaluated in terms of Fourier coefficients.

THEOREM: *If a waveform is periodic with period T_0, the spectrum of the waveform $w(t)$ is*

$$W(f) = \sum_{n=-\infty}^{n=\infty} c_n\, \delta(f - nf_0) \tag{2-109}$$

where $f_0 = 1/T_0$ and c_n are the phasor Fourier coefficients of the waveform as given by (2-94).

Proof

$$w(t) = \sum_{n=-\infty}^{n=\infty} c_n e^{jn\omega_0 t}$$

Taking the Fourier transform of both sides, we obtain

$$W(f) = \int_{-\infty}^{\infty} \left(\sum_{n=-\infty}^{n=\infty} c_n e^{jn\omega_0 t} \right) e^{-j\omega t}\, dt$$

$$= \sum_{n=-\infty}^{n=\infty} c_n \int_{-\infty}^{\infty} e^{-j2\pi(f-nf_0)t}\, dt = \sum_{n=-\infty}^{n=\infty} c_n\, \delta(f - nf_0)$$

where the integral representation for a delta function, (2-48), was used.

This theorem indicates that a periodic function *always* has a line (delta function) spectrum with the lines being at $f = nf_0$ and having weights given by the c_n values. An illustration of this property was given by Example 2-4, where $c_1 = -jA/2$ and $c_{-1} = jA/2$ and the other c_n's were zero. It is also obvious that there is no dc component since there is no line at $f = 0$ (i.e., $c_0 = 0$). On the other hand, if a function does not contain any periodic component, the spectrum will be continuous (no lines) except for a line at $f = 0$ when the function has a dc component.

It is also possible to evaluate the Fourier coefficients by sampling the Fourier transform of a pulse corresponding to $w(t)$ over a period. This is shown by the following theorem.

THEOREM: *If $w(t)$ is a periodic function with period T_0 and is represented by*

$$w(t) = \sum_{n=-\infty}^{n=\infty} h(t - nT_0) = \sum_{n=-\infty}^{n=\infty} c_n e^{jn\omega_0 t} \tag{2-110}$$

where

$$h(t) = \begin{cases} w(t), & |t| < \dfrac{T_0}{2} \\ 0, & t \text{ elsewhere} \end{cases} \tag{2-111}$$

then the Fourier coefficients are given by

$$c_n = f_0 H(nf_0) \tag{2-112}$$

where $H(f) = \mathcal{F}[h(t)]$ and $f_0 = 1/T_0$.

Proof

$$w(t) = \sum_{n=-\infty}^{\infty} h(t - nT_0) = \sum_{n=-\infty}^{\infty} h(t) * \delta(t - nT_0) \tag{2-113}$$

where $*$ denotes the convolution operation. Thus

$$w(t) = h(t) * \sum_{n=-\infty}^{\infty} \delta(t - nT_0) \tag{2-114}$$

But the impulse train may itself be represented by its Fourier series†

$$\sum_{n=-\infty}^{\infty} \delta(t - nT_0) = \sum_{n=-\infty}^{\infty} c_n e^{jn\omega_0 t} \tag{2-115}$$

where all the Fourier coefficients are just $c_n = f_0$. Substituting (2-115) into

†This is called the *Poisson sum formula*.

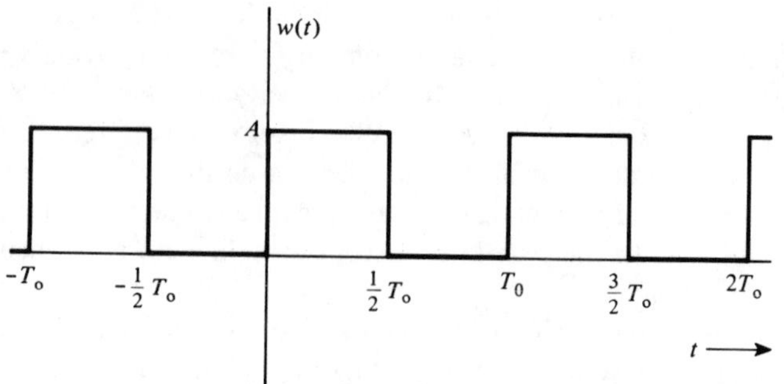

FIGURE 2-13 Periodic square wave used in Example 2-12.

(2-114), we obtain

$$w(t) = h(t) * \sum_{n=-\infty}^{\infty} f_0 e^{jn\omega_0 t} \tag{2-116}$$

Taking the Fourier transform of both sides of (2-116), we have

$$W(f) = H(f) \sum_{n=-\infty}^{\infty} f_0 \, \delta(f - nf_0)$$

$$= \sum_{n=-\infty}^{\infty} [f_0 H(nf_0)] \, \delta(f - nf_0) \tag{2-117}$$

Comparing (2-117) with (2-109), we see that (2-112) follows.

This theorem is useful for evaluating the Fourier coefficients c_n when the Fourier transform of the fundamental pulse shape $h(t)$ for the periodic waveform is known or can be obtained easily (such as from a Fourier transform table, e.g., Table 2-2).

EXAMPLE 2-12 FOURIER COEFFICIENTS FOR A RECTANGULAR WAVE

Find the Fourier coefficients for the rectangular wave shown in Fig. 2-13. The complex Fourier coefficients, using (2-94), are

$$c_n = \frac{1}{T_0} \int_0^{T_0/2} A e^{-jn\omega_0 t} \, dt = j \frac{A}{2\pi n} (e^{-jn\pi} - 1) \tag{2-118}$$

which reduce, using l'Hospital's rule for evaluating the indeterminant form for $n = 0$, to

$$c_n = \begin{cases} \dfrac{A}{2}, & n = 0 \\ -j\dfrac{A}{n\pi}, & n = \text{odd} \\ 0, & n \text{ otherwise} \end{cases} \tag{2-119}$$

This result may be verified by using (2-112) and the result of Example 2-5, where $T_0 = 2T$. That is,

$$c_n = \frac{A}{T_0} V(nf_0) = \frac{A}{2} e^{-jn\pi/2} \frac{\sin(n\pi/2)}{n\pi/2} \tag{2-120}$$

which is identical to (2-119). It is realized that the dc value of the waveform is $c_0 = A/2$, which checks with the result by using (2-4). The spectrum of the square wave is easily obtained by using (2-109).

The other types of Fourier coefficients may also be obtained. Using (2-101) and (2-102), we obtain the quadrature Fourier coefficients

$$a_n = \begin{cases} \dfrac{A}{2}, & n = 0 \\ 0, & n > 0 \end{cases} \tag{2-121a}$$

$$b_n = \begin{cases} \dfrac{2A}{\pi n}, & n = \text{odd} \\ 0, & n = \text{even} \end{cases} \tag{2-121b}$$

Here all the $a_n = 0$, except for $n = 0$, because, if the dc value were suppressed, the waveform would be an odd function about the origin. Using (2-106) and (2-107), we find that the polar Fourier coefficients are

$$D_n = \begin{cases} \dfrac{A}{2}, & n = 0 \\ \dfrac{2A}{n\pi}, & n = 1, 3, 5, \ldots \\ 0, & n \text{ otherwise} \end{cases} \tag{2-122}$$

and

$$\phi_n = -90^\circ \qquad \text{for } n \geq 1 \tag{2-123}$$

In communication problems the normalized average power is often needed, and, for the case of periodic waveforms, it can be evaluated using Fourier series coefficients.

THEOREM: *For a periodic waveform $w(t)$, the normalized power is given by*

$$P_w = \langle w^2(t) \rangle = \sum_{n=-\infty}^{n=\infty} |c_n|^2 \tag{2-124}$$

where $\{c_n\}$ are the complex Fourier coefficients for the waveform.

Proof. For periodic $w(t)$, the Fourier series representation is valid over all time and may be substituted into (2-12) to evaluate the normalized (i.e., $R = 1$) power.

$$P_w = \left\langle \left(\sum_n c_n e^{jn\omega_0 t} \right)^2 \right\rangle = \left\langle \sum_n \sum_m c_n c_m^* e^{jn\omega_0 t} e^{-jm\omega_0 t} \right\rangle$$

$$= \sum_n \sum_m c_n c_m^* \langle e^{j(n-m)\omega_0 t} \rangle = \sum_n \sum_m c_n c_m^* \, \delta_{nm} = \sum_n c_n c_n^*$$

or

$$P_w = \sum_n |c_n|^2 \tag{2-125}$$

Equation (2-124) is a statement for a special case of Parseval's theorem as applied to power signals. See (2-83) for the equivalent theorem as applied to energy signals. Furthermore, (2-124) checks with (2-83) since the $\{e^{jn\omega_0 t}\}$ orthogonal function set has constants $K_n = T_0$ and $P_w = E_w/T_0$, where E_w is the normalized energy dissipated in T_0 seconds.

If a large number of the Fourier coefficients are nonzero, then calculating the average power by summing the magnitude squared values of the Fourier coefficients is tedious. Then it is usually easier to evaluate the normalized average power directly by use of the time average, $\langle w^2(t) \rangle$.

Power Spectral Density for Periodic Waveforms

THEOREM: *For a periodic waveform, the power spectral density (PSD) is given by*

$$\mathcal{P}(f) = \sum_{n=-\infty}^{n=\infty} |c_n|^2 \, \delta(f - nf_0) \tag{2-126}$$

where $T_0 = 1/f_o$ is the period of the waveform and the $\{c_n\}$ are the corresponding Fourier coefficients for the waveform.

Proof. Let $w(t) = \sum_{-\infty}^{\infty} c_n e^{jn\omega_0 t}$. Then the autocorrelation function of $w(t)$ is

$$R(\tau) = \langle w^*(t) w(t + \tau) \rangle$$

$$= \left\langle \sum_{n=-\infty}^{\infty} c_n^* e^{-jn\omega_0 t} \sum_{m=-\infty}^{\infty} c_m e^{jm\omega_0 (t+\tau)} \right\rangle$$

or

$$R(\tau) = \sum_{n=-\infty}^{\infty} \sum_{m=-\infty}^{\infty} c_n^* c_m e^{jm\omega_0\tau} \langle e^{j\omega_0(m-n)t} \rangle \tag{2-127}$$

But $\langle e^{j\omega_0(n-m)t} \rangle = \delta_{nm}$, so (2-127) reduces to

$$R(\tau) = \sum_{n=-\infty}^{\infty} |c_n|^2 e^{jn\omega_0\tau} \tag{2-128}$$

The PSD is then

$$\mathcal{P}(f) = \mathcal{F}[R(\tau)] = \mathcal{F}\left[\sum_{-\infty}^{\infty} |c_n|^2 e^{jn\omega_0\tau}\right]$$

$$= \sum_{-\infty}^{\infty} |c_n|^2 \, \mathcal{F}[e^{jn\omega_0\tau}] = \sum_{-\infty}^{\infty} |c_n|^2 \, \delta(f - nf_0) \tag{2-129}$$

Equation (2-126) not only gives a way to evaluate the PSD for periodic waveforms but also can be used to evaluate the bandwidths of the waveforms. For example, the frequency interval could be found where 90% of the waveform power was concentrated.

2-6 Review of Linear Systems

Linear Time-Invariant Systems

A system is *linear* when *superposition* holds. That is,

$$y(t) = \mathcal{L}[a_1x_1(t) + a_2x_2(t)] = a_1\mathcal{L}[x_1(t)] + a_2\mathcal{L}[x_2(t)] \tag{2-130}$$

where $y(t)$ is the output and $x(t)$ is the input as shown in Fig. 2-14. $\mathcal{L}[\,\cdot\,]$ denotes the linear (differential equation) system operator acting on $[\cdot]$. The system is said to be time invariant if, for any delayed input $x(t - t_0)$, the output is just delayed by the same amount $y(t - t_0)$. That is, the *shape* of the response is the same no matter when the input is applied to the system.

A detailed discussion of the theory and practice of linear systems is beyond the scope of this book. That would require a book in itself [Irwin, 1984 or 1987]. However, some basic ideas that are especially relevant to communication problems will be reviewed here.

Impulse Response

The linear time-invariant system is described by a linear ordinary differential equation with constant coefficients and may be characterized by its impulse response, $h(t)$. The *impulse response* is the solution to the differential equation

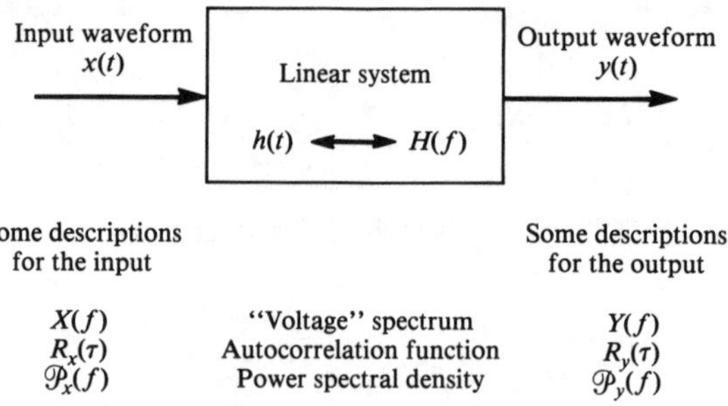

Some descriptions for the input		Some descriptions for the output
$X(f)$	"Voltage" spectrum	$Y(f)$
$R_x(\tau)$	Autocorrelation function	$R_y(\tau)$
$\mathcal{P}_x(f)$	Power spectral density	$\mathcal{P}_y(f)$

FIGURE 2-14 Linear system.

when the forcing function is a Dirac delta function. That is, $y(t) = h(t)$ when $x(t) = \delta(t)$. In physical networks the impulse response has to be *causal*. That is, $h(t) = 0$ for $t < 0$.†

This impulse response for the system may be used to obtain the system output when the input is *not* an impulse. In this case, a general waveform at the input may be approximated by‡

$$x(t) = \sum_{n=0}^{\infty} x(n\,\Delta t)[\delta(t - n\,\Delta t)]\,\Delta t \tag{2-131}$$

which indicates that samples of the input are taken at Δt-second intervals. Then using the time-invariant and superposition properties, the output is approximately

$$y(t) = \sum_{n=0}^{\infty} x(n\,\Delta t)[h(t - n\,\Delta t)]\,\Delta t \tag{2-132}$$

This expression becomes the exact result as Δt becomes zero. Letting $n\,\Delta t = \lambda$, we obtain

$$y(t) = \int_{-\infty}^{\infty} x(\lambda)h(t - \lambda)\,d\lambda \equiv x(t) * h(t) \tag{2-133}$$

An integral of this type is called the *convolution* operation, as first described by (2-62) in Sec. 2-2. That is, the output waveform for a time-invariant network can be obtained by convolving the input waveform with the impulse response for the system. Consequently, the impulse response can be used to characterize the response of the system in the time domain (see Fig. 2-14).

†The Paley–Wiener criterion gives the frequency domain equivalence for the causality condition of the time domain. It is that $H(f)$ must satisfy the condition

$$\int_{-\infty}^{\infty} \frac{|\ln |H(f)||}{1 + f^2}\,df < \infty$$

‡Δt corresponds to dx of (2-47).

Transfer Function

The spectrum of the output signal is obtained by taking the Fourier transform of both sides of (2-133). Using the convolution theorem of Table 2-1, we get

$$Y(f) = X(f)H(f) \tag{2-134}$$

or

$$H(f) = \frac{Y(f)}{X(f)} \tag{2-135}$$

where $H(f) = \mathcal{F}[h(t)]$ is said to be the *transfer function* or *frequency response* of the network. That is, the impulse response and frequency response are a Fourier transform pair:

$$h(t) \leftrightarrow H(f)$$

Of course, the transfer function $H(f)$ is, in general, a complex quantity and can be written in polar form

$$H(f) = |H(f)|e^{j\underline{/H(f)}} \tag{2-136}$$

where $|H(f)|$ is the *amplitude* (or *magnitude*) response and

$$\theta(f) = \underline{/H(f)} = \tan^{-1}\left[\frac{\text{Im}\{H(f)\}}{\text{Re}\{H(f)\}}\right] \tag{2-137}$$

is the phase response of the network. Furthermore, since $h(t)$ is a real function of time (for real networks), it follows from (2-38) and (2-39) that $|H(f)|$ is an even function of frequency and $\underline{/H(f)}$ is an odd function of frequency.

The transfer function of a linear time-invariant network can be measured by using a sinusoidal testing signal (that is swept over the frequency of interest) since the spectrum of a sinusoid is a line at the frequency under test. For example, if

$$x(t) = A \cos \omega_0 t$$

then the output of the network will be

$$y(t) = A|H(f_0)| \cos\left[\omega_0 t + \underline{/H(f_0)}\right] \tag{2-138}$$

where the amplitude and phase may be evaluated on an oscilloscope or by the use of a vector voltmeter.

If the input to the network is a periodic signal with a spectrum given by

$$X(f) = \sum_{n=-\infty}^{n=\infty} c_n\, \delta(f - nf_0) \tag{2-139}$$

where, using (2-109), $\{c_n\}$ are the complex Fourier coefficients of the input signal, the spectrum of the periodic output signal, using (2-134), is

$$Y(f) = \sum_{n=-\infty}^{n=\infty} c_n H(nf_0)\, \delta(f - nf_0) \tag{2-140}$$

We can also obtain the relationship between the power spectral density (PSD) at the input, $\mathcal{P}_x(f)$, and the output, $\mathcal{P}_y(f)$ of a linear time-invariant network.† From (2-66) we know that

$$\mathcal{P}_y(f) = \lim_{T\to\infty} \frac{1}{T} |Y_T(f)|^2 \tag{2-141}$$

Using (2-134) in a formal sense, we obtain

$$\mathcal{P}_y(f) = |H(f)|^2 \lim_{T\to\infty} \frac{1}{T} |X_T(f)|^2$$

or

$$\mathcal{P}_y(f) = |H(f)|^2 \mathcal{P}_x(f) \tag{2-142}$$

Consequently, the *power transfer function* of the network is

$$G_h(f) = \frac{\mathcal{P}_y(f)}{\mathcal{P}_x(f)} = |H(f)|^2 \tag{2-143}$$

A rigorous proof of this theorem is given in Chapter 6.

EXAMPLE 2-13 *RC* LOW-PASS FILTER

An *RC* low-pass filter is shown in Fig. 2-15, where $x(t)$ and $y(t)$ denote the input and output voltage waveforms, respectively. Using Kirchhoff's law for the sum of voltages around a loop, we get

$$x(t) = Ri(t) + y(t)$$

where $i(t) = C\,dy(t)/dt$ or

$$RC \frac{dy}{dt} + y(t) = x(t) \tag{2-144}$$

Using Table 2-1, we find that the Fourier transform of this differential equation is

$$RC(j2\pi f)Y(f) + Y(f) = X(f)$$

Thus the transfer function for this network is

$$H(f) = \frac{Y(f)}{X(f)} = \frac{1}{1 + j(2\pi RC)f} \tag{2-145}$$

Using the Fourier transform Table 2-2, we obtain the impulse response

$$h(t) = \begin{cases} \dfrac{1}{\tau_0} e^{-t/\tau_0}, & t \geq 0 \\ 0, & t < 0 \end{cases} \tag{2-146}$$

†The relationship between the input and output autocorrelation functions $R_x(\tau)$ and $R_y(\tau)$ can also be obtained as shown by (6-82).

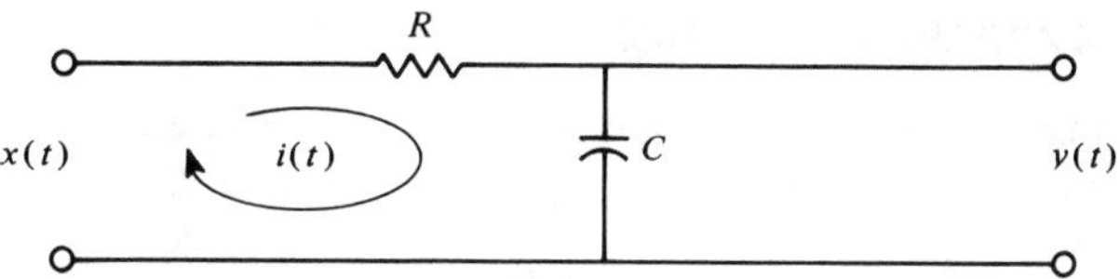

(a) RC Low-Pass Filter

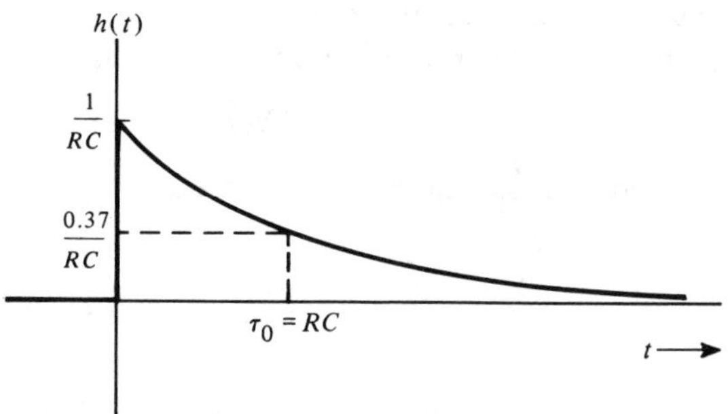

(b) Impulse Response

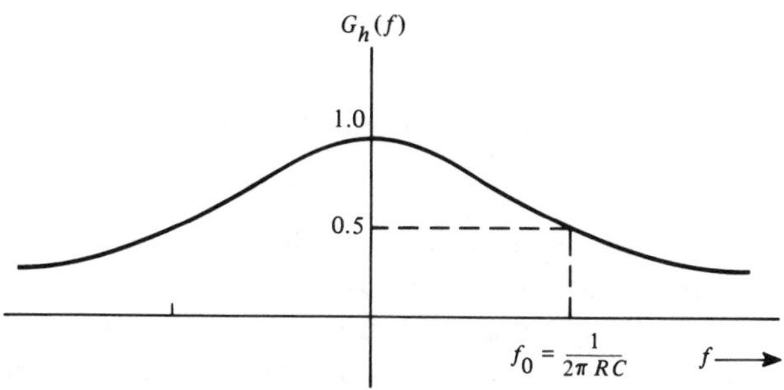

(c) Power Transfer Function

FIGURE 2-15 Characteristics of an *RC* low-pass filter.

where $\tau_0 = RC$ is called the *time constant*. When we combine (2-143) and (2-145), the power transfer function is

$$G_h(f) = |H(f)|^2 = \frac{1}{1 + (f/f_0)^2} \tag{2-147}$$

where $f_0 = 1/(2\pi RC)$.

The impulse response and the power transfer function are shown in Fig. 2-15. Note that the value of the power gain at $f = f_0$ (called the *corner frequency*) is $G_h(f_0) = \frac{1}{2}$. That is, the frequency component in the output waveform at $f = f_0$ is attenuated by 3 dB as compared to that at $f = 0$. Consequently, $f = f_0$ is said to be the 3-dB bandwidth of this filter. The topic of bandwidth is discussed in more detail in Sec. 2-8.

★ Distortionless Transmission

In communication systems a *distortionless channel* is often desired. This implies that the channel output is just proportional to a delayed version of the input

$$y(t) = Ax(t - T_d) \tag{2-148}$$

where A is the gain (which may be less than one) and T_d is the delay.

The corresponding requirement in the frequency domain specification is obtained by taking the Fourier transform of both sides of (2-148).

$$Y(f) = AX(f)e^{-j2\pi fT_d}$$

Thus, for distortionless transmission, we require that the transfer function of the channel be given by

$$H(f) = \frac{Y(f)}{X(f)} = Ae^{-j2\pi fT_d} \tag{2-149}$$

This implies that for no distortion at the output of a linear time-invariant system, two requirements must be satisfied:

1. The amplitude response is flat. That is,

$$|H(f)| = \text{constant} = A \tag{2-150a}$$

2. The phase response is a *linear* function of frequency. That is,

$$\theta(f) = \underline{/H(f)} = -2\pi fT_d \tag{2-150b}$$

When the first condition is satisfied, it is said that there is no *amplitude distortion*. When the second condition is satisfied, there is no *phase distortion*. For distortionless transmission both conditions must be satisfied.

The second requirement is often specified in an equivalent way using the time delay. Define the *time delay* of the system by

$$T_d(f) = -\frac{1}{2\pi f}\theta(f) = -\frac{1}{2\pi f}\underline{/H(f)]} \tag{2-151}$$

By (2-149), it is required that

$$T_d(f) = \text{constant} \tag{2-152}$$

for distortionless transmission. If $T_d(f)$ is not a constant as a function of frequency, there is phase distortion because the phase response, $\theta(f)$, is not a linear function of frequency.

EXAMPLE 2-14 DISTORTION CAUSED BY A FILTER

Let us examine the distortion effect caused by the *RC* low-pass filter studied in Example 2-13. From (2-145) the amplitude response is

$$|H(f)| = \frac{1}{\sqrt{1 + (f/f_0)^2}} \tag{2-153}$$

and the phase response is

$$\theta(f) = \underline{/H(f)} = -\tan^{-1}(f/f_0) \tag{2-154}$$

The corresponding time delay function is

$$T_d(f) = \frac{1}{2\pi f} \tan^{-1}(f/f_0) \tag{2-155}$$

These results are plotted in Fig. 2-16, as indicated by the solid lines. Obviously, this filter will produce some distortion since (2-150a) and (2-150b) are not satisfied. The dashed lines give the equivalent results for the distortionless filter. Several observations can be made. First, if the signals involved have spectral components at frequencies below $0.5f_0$, the filter will provide almost distortionless transmission since the error in the magnitude response (with respect to the distortionless case) is less than 0.5 dB and the error in the phase is less than 2.1° (8%). For $f < f_0$, the magnitude error is less than 3 dB and the phase error is less than 12.3° (27%). In engineering practice, this type of error is often considered to be tolerable. Waveforms with spectral components below $0.50f_0$ would be delayed by $1/(2\pi f_0)$ seconds. That is, if the cutoff frequency of the filter was $f_0 = 1$ kHz, the delay would be 0.2 msec. For wideband signals the higher-frequency components would be delayed less than the lower-frequency components.

The distortion effects previously discussed were based on the assumption that the system was linear and time invariant. A linear time-invariant system will produce amplitude distortion if the amplitude response is not flat and it will produce phase (i.e., time delay) distortion if the phase response is not a linear function of frequency. Moreover, if the system is nonlinear and/or time varying, other types of distortion will be produced. As a result, there are new frequency components at the output that are not present at the input. In some communication applications, these new frequencies are actually a desired result. The reader is referred to Sec. 4-3 for a study of these effects.

2-7 Bandlimited Signals and Noise

Often we deal with signal and noise waveforms that may be considered to be *bandlimited,* which implies that their spectra are nonzero only within a certain frequency band. In this case we can apply some very powerful theorems, namely, the sampling theorem to process the waveform. As we will show, these ideas are *especially applicable to digital communication* problems.

First, we examine some properties of bandlimited signals. Then the sampling and dimensionality theorems are developed.

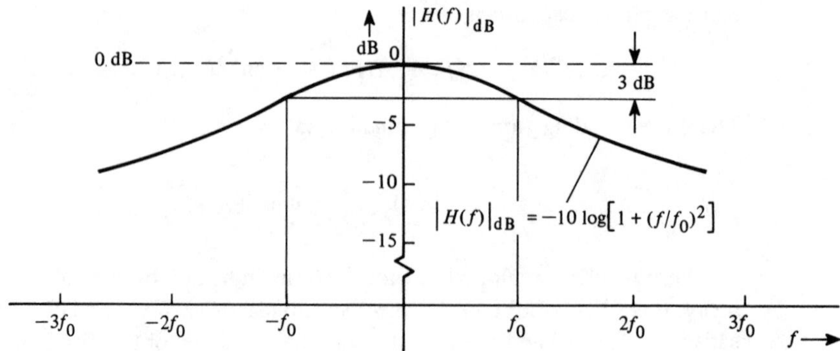

(a) Magnitude Response

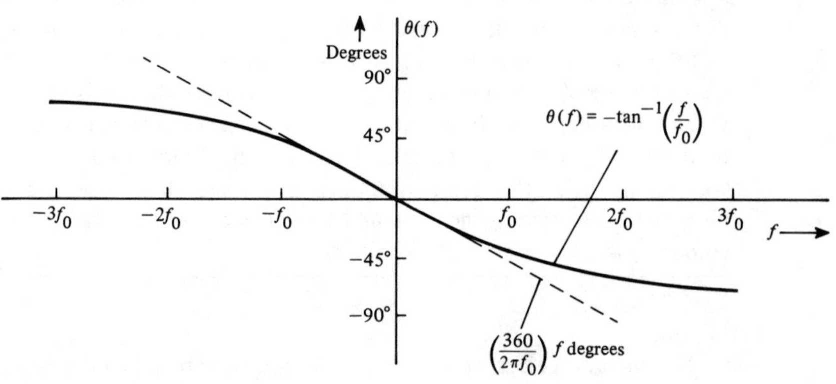

(b) Phase Response

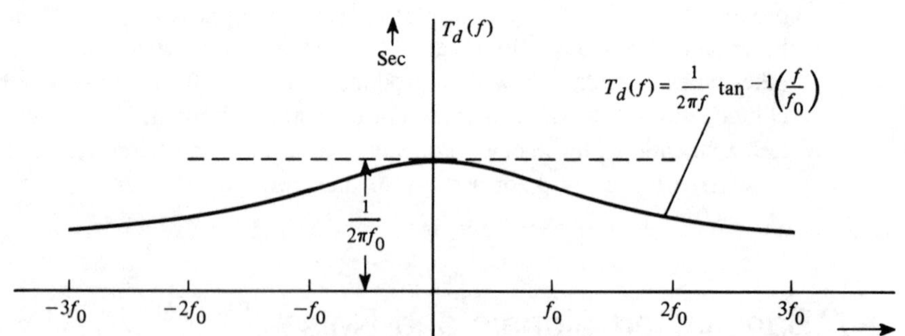

(c) Time Delay

FIGURE 2-16 Distortion caused by an *RC* low-pass filter.

Bandlimited Waveforms

Definition: A waveform $w(t)$ is said to be (absolutely) *bandlimited* to B hertz if

$$W(f) = \mathcal{F}[w(t)] = 0 \qquad \text{for } |f| \geq B \tag{2-156}$$

Definition: A waveform $w(t)$ is (absolutely) *time limited* if

$$w(t) = 0 \qquad \text{for } |t| > T \tag{2-157}$$

Theorem: *An absolutely bandlimited waveform cannot be absolutely time limited, and vice versa.*

The result of this theorem is illustrated by examining the spectrum for the rectangular pulse waveform of Example 2-5. This time-limited waveform is shown to have a spectrum that is not bandlimited. A relatively simple proof of this theorem can be obtained by contradiction [Wozencraft and Jacobs, 1965].

This theorem raises an engineering paradox. If we observe that a waveform is bandlimited, for example, by viewing its spectrum on a spectrum analyzer, the waveform cannot be time limited. However, we believe that this physical waveform *is* time limited because the device that generates the waveform was built at some finite past time and the device will decay at some future time (thus producing a time-limited waveform). This paradox is resolved by realizing that we are modeling a physical process with a mathematical model and perhaps the assumptions in the model are not satisfied, although we believe them to be satisfied. That is, there is some uncertainty as to what the actual time waveform and corresponding spectrum look like, especially at extreme values of time and frequency, owing to the inaccuracies of our measuring devices. This relationship between the physical process and our mathematical model is discussed in an interesting paper by David Slepian [1976]. In our problem, although the signal may not be absolutely bandlimited, it may be bandlimited for all practical purposes in the sense that the amplitude spectrum has a negligible level above a certain frequency. Of course, as engineers, we use mathematical models to help us predict the correct results. It should be realized that as measurement techniques improve, we may find that the model may need to be changed in order to predict what we perceive to be the correct answer. In short, common sense has to be used in formulating the mathematical model and interpreting the correct result.

Another interesting theorem states that if $w(t)$ is absolutely *bandlimited,* it is an *analytic* function. An analytic function is a function that possesses finite valued derivatives when they are evaluated for any finite value of t. This makes common sense. That is, a bandlimited function will not have frequency components above B hertz; consequently, the values of the function cannot fluctuate too rapidly as a function of time. This theorem may be proved by using a Taylor series expansion [Wozencraft and Jacobs, 1965].

Sampling Theorem

As indicated earlier, the sampling theorem is certainly one of the most useful theorems since it applies to digital communication systems, which are proliferating constantly. The sampling theorem is another application of an orthogonal series expansion.

SAMPLING THEOREM: *Any physical waveform may be represented over the interval* $-\infty < t < \infty$ *by*

$$w(t) = \sum_{n=-\infty}^{n=\infty} a_n \frac{\sin\{\pi f_s[t - (n/f_s)]\}}{\pi f_s[t - (n/f_s)]} \tag{2-158}$$

where

$$a_n = f_s \int_{-\infty}^{\infty} w(t) \frac{\sin\{\pi f_s[t - (n/f_s)]\}}{\pi f_s[t - (n/f_s)]} dt \tag{2-159}$$

and f_s *is a parameter that is assigned some convenient value greater than zero. Furthermore, if* $w(t)$ *is bandlimited to B hertz and* $f_s \geq 2B$, *then* (2-158) *becomes the sampling function representation, where*

$$a_n = w(n/f_s) \tag{2-160}$$

That is, for $f_s \geq 2B$, the orthogonal series coefficients are simply the values of the waveform that are obtained when the waveform is sampled every $1/f_s$ seconds.

The series given by (2-158) and (2-160) is sometimes called the *cardinal series*. It has been known by mathematicians since at least 1915 [Whittaker, 1915] and to engineers since the pioneering work of Shannon, who connected this series with information theory [Shannon, 1949]. An excellent tutorial paper on this topic has been published in the *Proceedings of the IEEE* [Jerri, 1977].

Proof. We need to show that

$$\varphi_n(t) = \frac{\sin\{\pi f_s[t - (n/f_s)]\}}{\pi f_s[t - (n/f_s)]} \tag{2-161}$$

form a set of the orthogonal functions. From (2-73) we need to demonstrate that (2-161) satisfies

$$\int_{-\infty}^{\infty} \varphi_n(t)\varphi_m^*(t)\, dt = K_n\, \delta_{nm} \tag{2-162}$$

Using Parseval's theorem, (2-40), we see that the left side becomes

$$\int_{-\infty}^{\infty} \varphi_n(t)\varphi_m^*(t)\, dt = \int_{-\infty}^{\infty} \Phi_n(f)\Phi_m^*(f)\, df \tag{2-163}$$

where

$$\Phi_n(f) = \mathcal{F}[\varphi_n(t)] = \frac{1}{f_s}\,\Pi\left(\frac{f}{f_s}\right) e^{-j2\pi(nf/f_s)} \tag{2-164}$$

We have

$$\int_{-\infty}^{\infty} \varphi_n(t)\varphi_m^*(t)\,dt = \frac{1}{(f_s)^2}\int_{-\infty}^{\infty}\left[\Pi\left(\frac{f}{f_s}\right)\right]^2 e^{-j2\pi(n-m)f/f_s}\,df$$

$$= \frac{1}{(f_s)^2}\int_{-f_s/2}^{f_s/2} e^{-j2\pi(n-m)(f/f_s)}\,df = \frac{1}{f_s}\,\delta_{nm} \tag{2-165}$$

Thus $\varphi_n(t)$ as given by (2-161) are orthogonal functions with $K_n = 1/f_s$. Using (2-77), we see that (2-159) follows. Furthermore, we will show that (2-160) follows for the case of $w(t)$ being absolutely bandlimited to B hertz with $f_s \geq 2B$. Using (2-77) and Parseval's theorem, (2-40), we get

$$a_n = f_s\int_{-\infty}^{\infty} w(t)\varphi_n^*(t)\,dt$$

$$= f_s\int_{-\infty}^{\infty} W(f)\,\Phi_n^*(f)\,df \tag{2-166}$$

Substituting (2-164),

$$a_n = \int_{-f_s/2}^{f_s/2} W(f)e^{j2\pi f(n/f_s)}\,df \tag{2-167}$$

But since $W(f)$ is zero for $|f| > B$, where $B \leq f_s/2$, the limits on the integral may be extended to $(-\infty,\infty)$ without changing the value of the integral. This integral with infinite limits is just the inverse Fourier transform of $W(f)$ evaluated at $t = n/f_s$. Thus $a_n = w(n/f_s)$, which is (2-160).

From (2-167) it is obvious that the minimum sampling rate allowed to reconstruct a bandlimited waveform without error is given by

$$(f_s)_{\min} = 2B \tag{2-168}$$

This is called the *Nyquist frequency*.

Since this sampling theorem is a special case of an orthogonal series expansion, orthogonal series properties hold such as describing the waveform by the use of an N-dimensional vector and the corresponding relationship for the energy of the waveform over a T_0-second interval in terms of the square of the elements for that vector.

For example, let us now examine the problem of reproducing a bandlimited waveform by using N sample values. Suppose that we are interested in reproducing the waveform over a T_0-second interval as shown in Fig. 2-17a. Then we can truncate the sampling function series of (2-158) so that we include only N of the

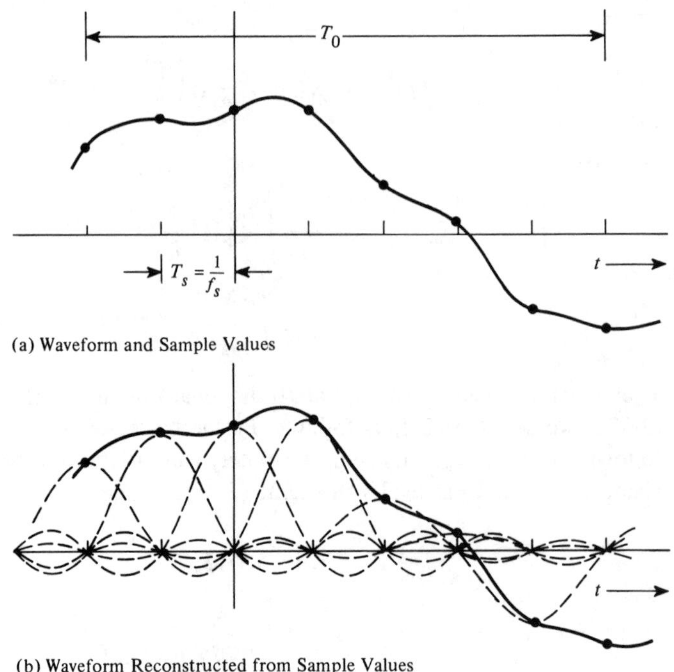

FIGURE 2-17 Sampling theorem.

$\varphi_n(t)$ functions that have their peaks within the T_0 interval of interest. That is, the waveform can be approximately reconstructed using N samples:

$$w(t) \approx \sum_{n=n_1}^{n=n_1+N} a_n \varphi_n(t)$$

where the $\{\varphi_n(t)\}$ are described by (2-161). Figure 2-17b shows the reconstructed waveform (solid line) which is obtained by the weighted sum of time-delayed $(\sin x)/x$ waveforms (dashed lines), where the weights are the sample values, $a_n = w(nf_s)$, denoted by the dots. The waveform is bandlimited to B hertz with the sampling frequency $f_s \geq 2B$. The sample values may be saved, for example, in the memory of a digital computer, so that the waveform may be reconstructed at a later time or the values may be transmitted over a communication system for waveform reconstruction at the receiving end. In either case, the waveform may be reconstructed from the sample values by the use of the equation above. That is, each sample value is multiplied by the appropriate $(\sin x)/x$ function, and these weighted $(\sin x)/x$ functions are summed to give the original waveform. This procedure is illustrated in Fig. 2-17b. The minimum number of sample values that are needed for waveform reconstruction is

$$N = \frac{T_0}{1/f_s} = f_s T_0 \geq 2BT_0 \tag{2-169}$$

and there are N orthogonal functions in the reconstruction algorithm. Using our geometric interpretation, we can say that N is the number of dimensions needed to reconstruct the T_0-second approximation of the waveform. Using (2-83), we find that the normalized energy in the T_0-second portion of the bandlimited waveform is approximately

$$E_w = \frac{1}{f_s} \sum_{n=n_1}^{n=n_1+N} \left[w\left(\frac{n}{f_s}\right) \right]^2 \tag{2-170}$$

where $N = f_s T_0 \geq 2BT_0$ and n_1 is adjusted so that the n/f_s sampling times fall within the T_0 interval.

Impulse Sampling

If one wants to reconstruct a waveform on a real-time basis, an alternative approach, called *impulse sampling*, is often used. The waveform is sampled with a unit-weight impulse train. This is accomplished by multiplying the waveform with a unit-weight impulse train. The resulting sampled waveform is

$$\begin{aligned} w_s(t) &= w(t) \sum_{n=-\infty}^{\infty} \delta(t - nT_s) \\ &= \sum_{n=-\infty}^{\infty} w(nT_s)\, \delta(t - nT_s) \end{aligned} \tag{2-171}$$

where $T_s = 1/f_s$. This situation is illustrated in Fig. 2-18.† In the figure the weight (area) of each impulse, $w(nT_s)$, is indicated by the height. The spectrum for the impulse sampled waveform, $w_s(t)$, can be evaluated by substituting the Fourier series of the (periodic) impulse train into (2-171).

$$w_s(t) = w(t) \sum_{n=-\infty}^{\infty} \frac{1}{T_s} e^{jn\omega_s t} \tag{2-172}$$

Taking the Fourier transform of both sides of this equation, we get

$$\begin{aligned} W_s(f) &= \frac{1}{T_s} W(f) * \mathcal{F}\left[\sum_{n=-\infty}^{\infty} e^{jn\omega_s t} \right] = \frac{1}{T_s} W(f) * \sum_{n=-\infty}^{\infty} \mathcal{F}[e^{jn\omega_s t}] \\ &= \frac{1}{T_s} W(f) * \sum_{n=-\infty}^{\infty} \delta(f - nf_s) \end{aligned}$$

or

$$W_s(f) = \frac{1}{T_s} \sum_{n=-\infty}^{\infty} W(f - nf_s) \tag{2-173}$$

†For illustrative purposes, assume that $W(f)$ is real.

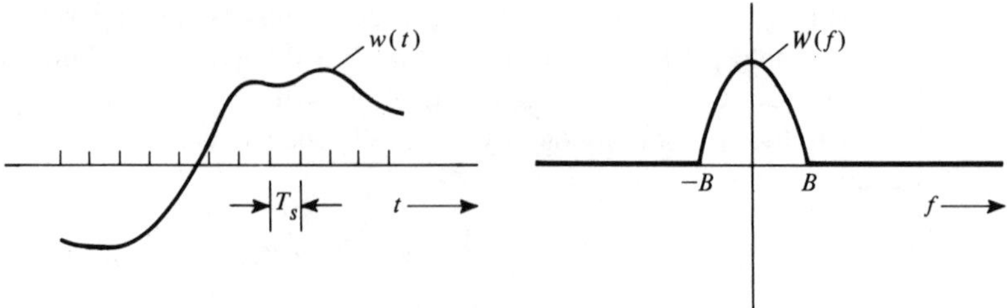

(a) Waveform and Its Spectrum

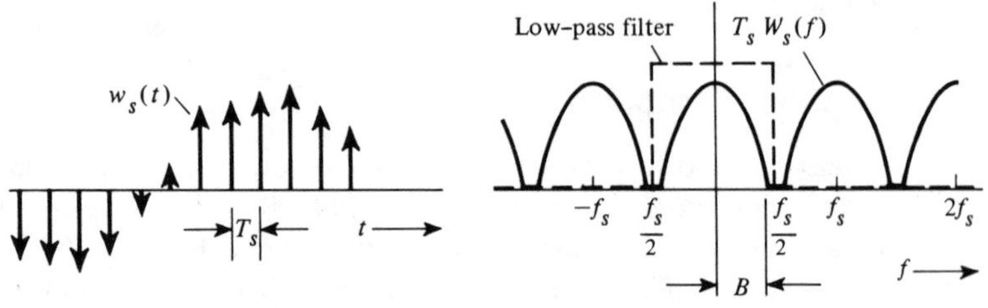

(b) Impulse Sampled Waveform and Its Spectrum ($f_s > 2B$)

FIGURE 2-18 Impulse sampling.

The spectrum of the impulse sampled waveform consists of the spectrum of the unsampled waveform that is replicated every f_s hertz.† This is shown in Fig. 2-18b. Note that this technique of impulse sampling may be used to translate the spectrum of a signal to another frequency band that is centered on some harmonic of the sampling frequency. A more general circuit that can translate the spectrum to any desired frequency band is called a mixer. Mixers are discussed in Sec. 4-3.

If $f_s \geq 2B$, as illustrated in Fig. 2-18, the replicated spectra do not overlap and the original spectrum can be regenerated by chopping $W_s(f)$ off above $f_s/2$. Thus $w(t)$ can be reproduced from $w_s(t)$ by simply passing $w_s(t)$ through an ideal low-pass filter that has a cutoff frequency of $f_c = f_s/2$ where $f_s \geq 2B$.

If $f_s < 2B$ (i.e., the waveform is undersampled), the spectrum of $w_s(t)$ will consist of overlapped, replicated spectra of $w(t)$, as illustrated in Fig. 2-19.‡ The spectral overlap or tail inversion is called *aliasing* or *spectral folding*.§ In this case, the low-pass filtered version of $w_s(t)$ will not be exactly $w(t)$. The recovered $w(t)$ will be distorted because of the aliasing. The alias distortion can be eliminated by *prefiltering* the original $w(t)$ before sampling so that the prefiltered

†In Chapter 3 this result is generalized to that of instantaneous sampling with a pulse train consisting of pulses with finite width and arbitrary shape (instead of impulses). The resulting spectrum is given by (3-10).

‡For illustrative purposes, assume that $W(f)$ is real.

§$f_s/2$ is the folding frequency where f_s is the sampling frequency. For no aliasing, $f_s > 2B$ is required, where $2B$ is the Nyquist frequency.

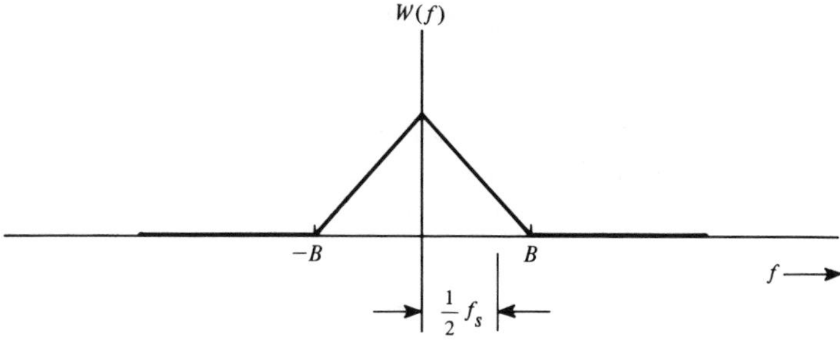

(a) Spectrum of Unsampled Waveform

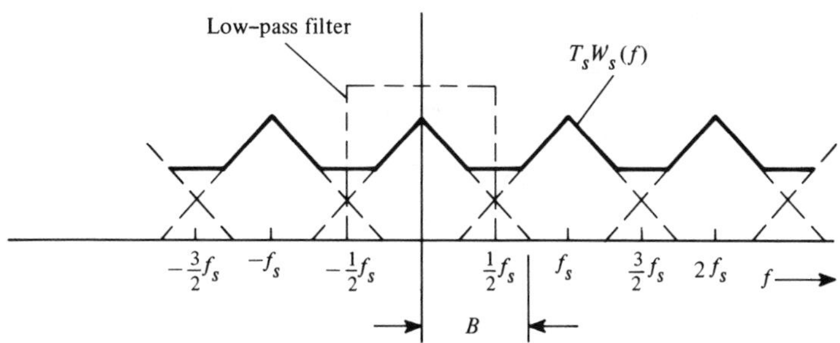

(b) Spectrum of Impulse Sampled Waveforn ($f_s < 2B$)

FIGURE 2-19 Undersampling and aliasing.

$w(t)$ has no spectral components above $|f| = f_s/2$. The prefiltering still produces distortion on the recovered waveform because the prefilter chops off the spectrum of the original $w(t)$ above $f = f_s/2$. However, from Fig. 2-19, it can be shown that if a prefilter is used, the recovered waveform obtained from the low-pass version of the sample signal will have one half of the error energy compared to the error energy that would be obtained without using the presampling filter.

A physical waveform, $w(t)$, has finite energy. From (2-42) and (2-43) it follows that the magnitude spectrum of the waveform, $|W(f)|$, has to be negligible for $|f| > B$ where B is an appropriately chosen positive number. Consequently, from a practical standpoint, the physical waveform is essentially bandlimited to B Hz where B is chosen to be large enough so that the error energy is below some specified amount.

Dimensionality Theorem

The sampling theorem may be restated in a more general way called the *dimensionality theorem.*

THEOREM: *A real waveform may be completely specified by*

$$N = 2BT_0 \tag{2-174}$$

independent pieces of information that will describe the waveform over a T_0 interval. N is said to be the number of dimensions required to specify the waveform, and B is the absolute bandwidth of the waveform.

The N sample values do not necessarily have to be periodic samples; they just have to be independent quantities. This can be demonstrated for the case of nonperiodic sampling by use of the complex Fourier series. We will represent the real $w(t)$ over a T_0-second interval by the Fourier series

$$w(t) = \sum_n c_n e^{jn\omega_0 t}$$

where $\omega_0 = 2\pi/T_0$, and it is known that $c_n^* = c_{-n}$ for any real signal. If $w(t)$ is bandlimited to B hertz, the range of n needed to represent $w(t)$ is $-B \le nf_0 \le B$, and since $f_0 = 1/T_0$, we have†

$$w(t) = \sum_{n=-BT_0}^{n=BT_0} c_n e^{jn\omega_0 t}$$

The c_0 term is the dc level for $w(t)$, and a dc level does not impart any information (see Chapter 1). Since $c_n = |c_n|e^{j\angle c_n}$ and $c_{-n} = c_n^*$, there are $N = 2BT_0$ independent (real) values required. Note that the c_n's in this example are not time samples of $w(t)$, and each c_n depends on the waveshape of $w(t)$ over the whole T_0 interval, as seen from (2-94).

The dimensionality theorem, (2-174), says simply that the information that can be conveyed by a bandlimited waveform or a bandlimited communication system is proportional to the product of the bandwidth of that system and the time allowed for transmission of the information. The *dimensionality theorem has profound implications in the design and performance of all types of communication systems*. For example, in radar systems it is well known that the time–bandwidth product of the received signal needs to be large for superior performance.

There are *two distinct ways in which the dimensionality theorem can be applied. First,* if any bandlimited waveform is given and we want to store some numbers in a table (or a computer memory bank) that could be used to reconstruct the waveform over a T_0-second interval, *at least* N numbers must be stored and, furthermore, the average sampling rate must be at least the Nyquist rate. That is,‡

† Actually, for $-B \le nf_0 \le B$ to hold with equality, we need an integer multiple of f_0 equal to B (which may be any real positive number). Thus, in general, T_0 needs to be large.

‡ If the spectrum of the waveform being sampled has a line at $f = \pm B$, there is some ambiguity as to whether or not the line is included within bandwidth B. The line is included by letting $f_s > 2B$ (i.e., dropping the equality sign).

$$f_s \geq 2B \tag{2-175}$$

Thus, in this first type of application, the dimensionality theorem is used to calculate the number of storage locations (amount of memory) required to represent a waveform.

The *second* type of application is the inverse type of problem. Here the dimensionality theorem is used to estimate the bandwidth of waveforms. For example, let M be the number of possible messages of a digital source that are transmitted by using M distinct waveforms. Suppose that the messages are represented by binary words, each word being n bits long. Thus $M = 2^n$. Let the information of a message be transmitted by a waveform having a duration of T_0 seconds. The transmitted waveform could be generated by using the orthogonal series

$$w(t) = \sum_{k=n_1}^{k=n_1+N} a_k \varphi_k(t)$$

where each bit of the binary word is encoded onto one dimension. That is, a_k would be, say $+1$ for a binary one and 0 for a binary zero so that $N = n = \log_2 M$. Note that in this example the number of dimensions used, N, is equal to the number of bits, n, that represents the message. Since $a_k = \pm 1$, this is called *binary encoding* or *binary signaling*. Furthermore, the transmitted waveform could be either analog or digital, depending on the type of orthogonal functions selected. For example, for generating a digital type of waveform, one could use the rectangular type of $\varphi_k(t)$ functions as described by Example 2-11 for TTL (transistor transistor logic) signaling. A *lower bound* for the bandwidth of this TTL digital waveform can be obtained by applying the dimensionality theorem (2-174)

$$B > \frac{N}{2T_0} \tag{2-176a}$$

On the other hand, an analog waveform would be generated if the $(\sin x)/x$ type of $\varphi_k(t)$ functions of (2-161) were used. Let the parameter f_s be $f_s = 2B = N/T_0$. Thus in this situation, the absolute bandwidth of the waveform is

$$B = \frac{N}{2T_0} \tag{2-176b}$$

This is identical to the bandwidth estimated by the dimensionality theorem as given by (2-174). In other words, if the source rate was $R = n/T_0 = 2400$ bits/sec, the bandwidth of the analog waveform [generated using $(\sin x)/x$ functions] would be $B = 1200$ Hz. The corresponding digital waveform (generated using rectangular functions) would have a much larger bandwidth, although either waveform could be used to convey the information of the M-message digital source. Thus in this second application, the bandwidth estimated by the dimensionality theorem is

$$B \geq \frac{N}{2T_0} \text{ hertz} \tag{2-176c}$$

where N is the number of dimensions used to generate a waveform of T_0-second duration.

What is even more interesting is that the dimensionality theorem shows that the required bandwidth can be reduced if N can be reduced. The required N is a function of the encoding scheme used. Specifically, N can be reduced by letting the number of allowable values for each of the a_k's to be more than two. This is called *multilevel encoding* or *multilevel signaling* and is illustrated in the following example.[†]

In this example of multilevel signaling, let the a_k for each dimension have eight possible levels instead of only two as the binary case. Then $L = 8 = 2^3$, and consequently, each dimension is carrying the equivalent of 3 bits of data since $M = 2^n = (L)^N = (2^3)^N$ or $n = 3N$. If the bit rate $R = n/T_0$ is 2400 bits/sec, the dimension rate $D = N/T_0$ is 800 dimensions per second. Thus, using the dimensionality theorem, the bandwidth for the $L = 8$ multilevel waveform is $B = 400$ Hz. Compare this to the bandwidth of $B = 1200$ Hz for the binary encoded waveform that was discussed earlier. All of these waveforms support the source data rate of $R = 2400$ bits/sec.

Definition: The *baud (symbol rate)* is

$$D = N/T_0 \text{ symbols/sec} \tag{2-177}$$

and is the number of dimensions or symbols sent per second.[‡]

Definition: The *bit rate* is

$$R = n/T_0 \text{ bits/sec} \tag{2-178}$$

and is the number of data bits sent per second.

The bit rate and the baud rate are related by

$$R = (\log_2 L)D \tag{2-179}$$

where L is the number of levels allowed in each dimension. For the eight-level signal discussed above, the symbol rate is 800 baud, which, using (2-179), corresponds to an equivalent bit rate of 2400 bits/sec. A more detailed discussion is given in Chapter 3 and is illustrated in Fig. 3-18.

From (2-176) it is clear that the baud rate (not the equivalent bit rate) determines the bandwidth of a multilevel signal. For a digital source with a given bit rate, the signaling bandwidth can be reduced if we design L to be larger so that the baud (rate) will be smaller. That is, the bandwidth of the eight-level waveform is one third the bandwidth of the binary waveform. Of course, it will cost more to build circuits of eight levels per dimension than of two levels per dimension (binary). In Chapter 7 we will find that multilevel signaling requires a larger

[†]Trellis-coded modulation, as discussed in Chapter 1, is an example of multilevel encoding.

[‡]In the technical literature, the term *baud rate* instead of baud is sometimes used even though baud rate is a misnomer, since the term *baud*, by definition, is the symbol rate (symbols/sec).

signal-to-noise ratio at the receiver input for error-free output, so it may not always be cost effective to choose the signaling scheme with the narrower bandwidth.

★ Discrete Fourier Transform

With the convenience of personal computers and the availability of digital signal processing integrated circuits, the spectrum of a waveform can be easily approximated by using the discrete Fourier transform (DFT). Here we show briefly how the DFT can be used to compute samples of the continuous Fourier transform (CFT), (2-26), and values for the complex Fourier series coefficients (2-94).

DEFINITION: The *discrete Fourier transform (DFT)* is defined by

$$X(n) = \sum_{k=0}^{k=N-1} x(k)e^{-j(2\pi/N)nk} \tag{2-180a}$$

where $n = 0, 1, 2, \ldots, N-1$, and the *inverse discrete Fourier transform (IDFT)* is defined by

$$x(k) = \frac{1}{N} \sum_{n=0}^{n=N-1} X(n)e^{j(2\pi/N)nk} \tag{2-180b}$$

where $k = 0, 1, 2, \ldots, N-1$.

Time and frequency do not appear explicitly since (2-180a) and (2-180b) are just definitions implemented on a digital computer to compute N values for the DFT and IDFT, respectively. Please be aware that other authors may use different (equally valid) definitions. For example, a $1/\sqrt{N}$ factor could be used on the right side of (2-180a) if the $1/N$ factor of (2-180b) is replaced by $1/\sqrt{N}$. This produces a different scale factor when the DFT is related to the CFT. Also, the signs of the exponents in (2-180a) and (2-180b) could be exchanged. This would reverse the spectral samples along the frequency axis. In fact, these changes in the definitions are used in the MathCAD fast Fourier transform (FFT) algorithm. Thus the MathCAD FFT algorithms icfft and cfft are related to the definitions of (2-180) by

$$\mathbf{X} = \sqrt{N}\,\text{icfft}\,(\mathbf{x}) \tag{2-181a}$$

and

$$\mathbf{x} = (1/\sqrt{N})\,\text{cfft}\,(\mathbf{X}) \tag{2-181b}$$

where $\mathbf{x}$ is an N-element array corresponding to samples of the waveform and $\mathbf{X}$ is the N-element DFT array. N is chosen to be a power of 2 (i.e., $N = 2^m$, where m is a positive integer). If other FFT software is used, the user should be aware of the specific definitions that are implemented so that the results can be interpreted properly.

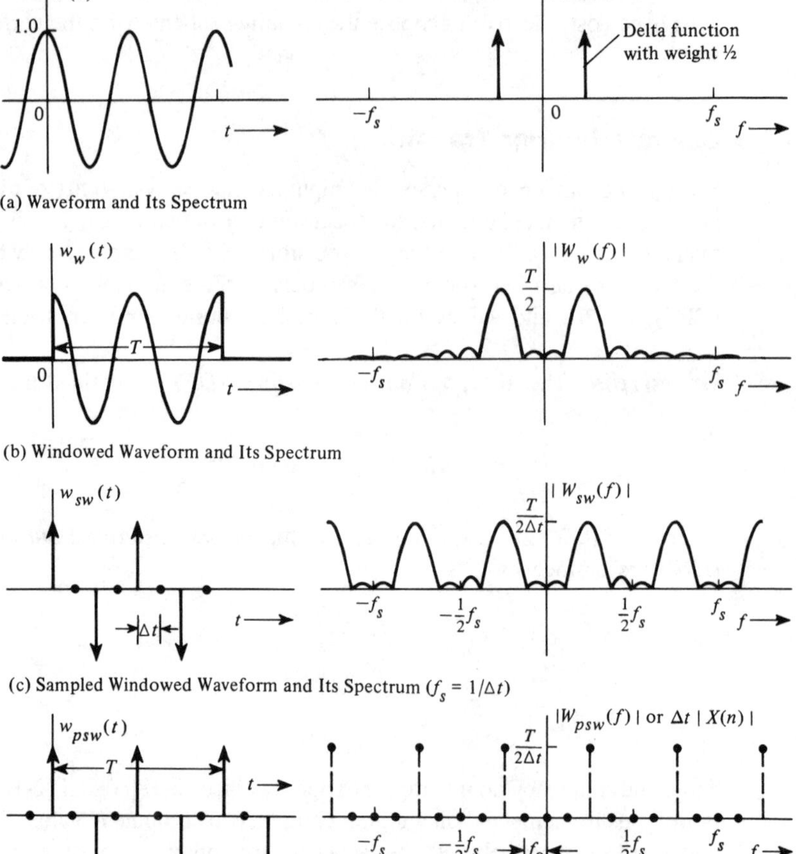

FIGURE 2-20 Comparison of CFT and DFT spectra.

The relationship between the DFT, as defined by (2-180), and the CFT will now be examined. It involves three concepts: windowing, sampling, and periodic sample generation. These are illustrated in Fig. 2-20, where the left side is time domain and the right side is the corresponding frequency domain. Suppose that the CFT of a waveform $w(t)$ is to be evaluated by use of the DFT. The time waveform is first windowed (truncated) over the interval $(0,T)$ so that only a finite number of samples, N, are needed. The windowed waveform, denoted by a w subscript, is

$$w_w(t) = \begin{Bmatrix} w(t), & 0 \le t \le T \\ 0, & t \text{ elsewhere} \end{Bmatrix} = w(t)\,\Pi\left(\frac{t-(T/2)}{T}\right) \tag{2-182}$$

The Fourier transform of the windowed waveform is

$$W_w(f) = \int_{-\infty}^{\infty} w_w(t)e^{-j2\pi ft}\,dt = \int_0^T w(t)e^{-j2\pi ft}\,dt \tag{2-183}$$

Now approximate the CFT by using a finite series to represent the integral where $t = k\,\Delta t$, $f = n/T$, $dt = \Delta t$, and $\Delta t = T/N$. Then

$$W_w(f)|_{f=n/T} \approx \sum_{k=0}^{N-1} w(k\,\Delta t)e^{-j(2\pi/N)nk}\,\Delta t \tag{2-184}$$

Comparing this with (2-180a), the relation between the CFT and DFT is obtained:

$$W_w(f)|_{f=n/T} \approx \Delta t\, X(n) \tag{2-185}$$

where $f = n/T$ and $\Delta t = T/N$. The sample values used in the DFT computation are $x(k) = w(k\,\Delta t)$, as shown in the left part of Fig. 2-20c. Also, since $e^{-j(2\pi/N)nk}$ of (2-184) is periodic in n—that is, the same values will be repeated for $n = N, N + 1, \ldots$ as were obtained for $n = 0, 1, \ldots$ —then $X(n)$ is periodic (although only the first N values are returned by DFT computer programs since the others are just repetitions). Another way of seeing that the DFT (and the IDFT) are periodic is to recognize that since samples are used, the discrete transform is an example of impulse sampling, and consequently, the spectrum must be periodic about the sampling frequency, $f_s = 1/\Delta t = N/T$ (as illustrated in Fig. 2-18 and again here in Fig. 2-20c and d). Furthermore, if the spectrum is desired for negative frequencies—the computer returns $X(n)$ for the positive n values of $0, 1, \ldots, N - 1$—(2-185) must be modified to give spectral values over the entire fundamental range of $-f_s/2 < f < f_s/2$. Thus for positive frequencies we use

$$W_w(f)|_{f=n/T} \approx \Delta t\, X(n), \qquad 0 \le n < \frac{N}{2} \tag{2-186a}$$

and for negative frequencies we use

$$W_w(f)|_{f=(n-N)/T} \approx \Delta t\, X(n), \qquad \frac{N}{2} < n < N \tag{2-186b}$$

In summary, four fundamental considerations need to be kept in mind when evaluating the CFT by use of the DFT:

1. Waveform windowing over $(0,T)$ is required so that a *finite* number of samples can be realized.
2. N sample values are used where the samples are taken at the times $t = k\,\Delta t$ $(k = 0, 1, \ldots, N - 1$ and $\Delta t = T/N)$.
3. The DFT gives the spectra of N frequencies $(f = n/T,\ n = 0, 1, \ldots, N - 1)$ over the frequency interval $(0, f_s)$, where $f_s = 1/\Delta t = N/T$.
4. The DFT and IDFT are periodic with periods f_s and T, respectively.

Figure 2-20 illustrates that if one is not careful, the DFT may give significant errors when used to approximate the CFT. The errors are due to a number of factors that may be categorized into three basic effects: *leakage, aliasing,* and the *picket-fence effect*.

The first effect is caused by windowing in the time domain. In the frequency domain this corresponds to convolving the spectrum of the unwindowed waveform with the spectrum (Fourier transform) of the window function. This spreads the spectrum of the frequency components of $w(t)$, as illustrated in Fig. 2-20b, and causes each frequency component to "leak" into adjacent frequencies. The leakage may produce errors when the DFT is compared with the CFT. This effect can be reduced by increasing the window width T or, equivalently, by increasing the number of points, N, that are used in the DFT. Window shapes other than the rectangle can also be used to reduce the sidelobes in the spectrum of the window function [Harris, 1978; Ziemer, Tranter, and Fannin, 1989]. Large periodic components in $w(t)$ cause more leakage, and if these components are known to be present, they might be eliminated before evaluating the DFT to reduce leakage.

From our previous study of sampling, it is known that the spectrum of a sampled waveform consists of replicating the spectrum of the unsampled waveform about harmonics of the sampling frequency. If $f_s < 2B$, where B is the highest significant frequency component in the unsampled waveform, aliasing errors will occur. The aliasing error can be decreased by using a higher sampling frequency or a presampling low-pass filter. Note that the highest frequency component that can be evaluated with an N-point DFT is $f = f_s/2 = N/(2T)$.

The third type of error, the picket-fence effect, occurs because the N-point DFT cannot resolve the spectral components any closer than the spacing $\Delta f = 1/T = f_s/N$. If the sampling rate is fixed, Δf can be decreased by increasing the number of points in the DFT. If the data length is limited to T_0 seconds, where $T_0 \le T$, T may be extended by adding additional zero-value sampling points. This is called *zero-padding* and will reduce Δf to produce better spectral resolution.

EXAMPLE 2-15 DFT FOR A RECTANGULAR PULSE

Table 2-3 demonstrates the calculation of the spectrum for a rectangular pulse using the MathCAD DFT algorithm. Note that (2-181a) and (2-185) are used to relate the MathCAD DFT results to the spectrum (i.e., the CFT). Also, the parameters M, tend, and T are selected so that the computed magnitude and phase spectral results compare favorably (i.e., little error) with the true spectrum for a rectangular pulse as given by (2-59) and (2-60) of Example 2-5. (T in Example 2-5 is equivalent to tend in Table 2-3). The MathCad template shown in Table 2-3 may be found on the computer disk for this book (as described in Chapter 1). The reader should try other values of M, tend, and T to see how leakage, aliasing, and picket-fence errors become large if the parameters are not carefully selected. Also, note that significant errors occur if tend = T. Why?

TABLE 2-3 Spectrum for a Rectangular Pulse Using the MathCAD DFT

Calculate DFT for a truncated step.

$M := 7 \qquad N := 2^M \qquad N = 128 \qquad k := 0 \,..N - 1$

Let xend be end of step.

$tend := 1 \qquad T := 10$

$dt := \frac{T}{N} \qquad t_k := k \cdot dt$

$w_k := 1.0 - \Phi(k \cdot dt - tend)$

$w_0 = 1 \qquad dt = 0.078$

WAVEFORM

1.5

w_k

-0.5

0 t_k T

$f4 := \frac{4}{tend}$

$n := 0 \,..N - 1$

$W := dt \cdot \left[\sqrt{N}\right] \cdot icfft(w)$

$f_n := \frac{n}{T} \qquad fs := \frac{1}{dt}$

$W_0 = 1.016 \qquad fs = 12.8$

$f_1 = 0.1 \qquad f4 = 4$

MAGNITUDE SPECTRUM out to 4th null

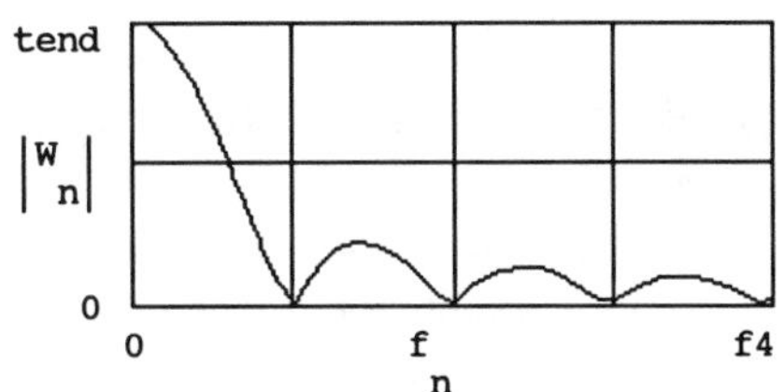

PHASE SPECTRUM

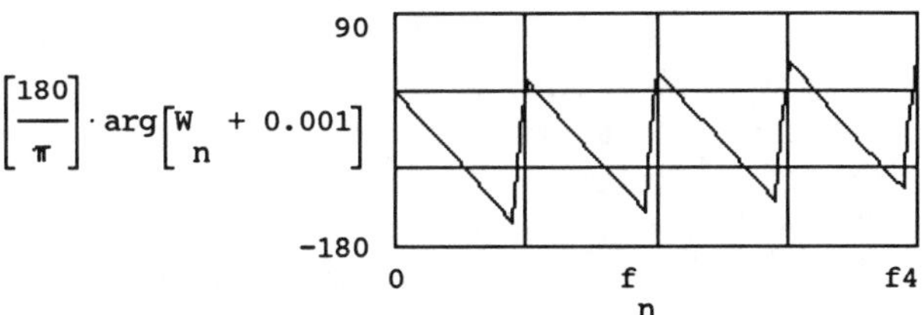

$\left[\frac{180}{\pi}\right] \cdot arg\left[W_n + 0.001\right]$

MAGNITUDE SPECTRUM over whole DFT frequency range

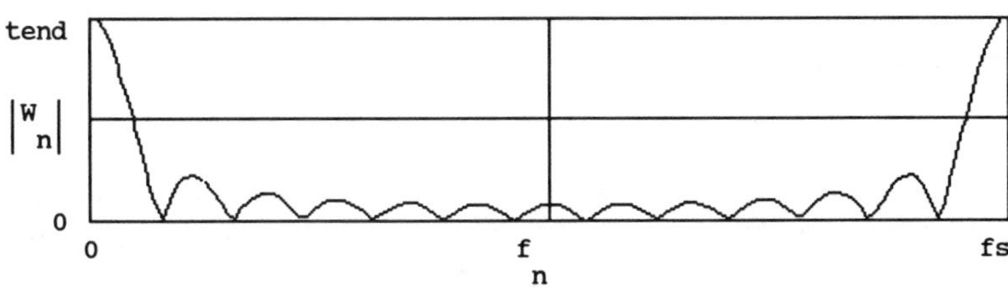

In a similar way the DFT may be also used to evaluate Fourier series coefficients. Using (2-94) yields

$$c_n = \frac{1}{T} \int_0^T w(t) e^{-j2\pi n f_0 t} \, dt \tag{2-187}$$

Approximate this integral by using a finite series where $t = k\,\Delta t$, $f_0 = 1/T$, $dt = \Delta t$, and $\Delta t = T/N$. Then

$$c_n \approx \frac{1}{T} \sum_{k=0}^{N-1} w(k\,\Delta t) e^{-j(2\pi/N)nk}\,\Delta t \tag{2-188}$$

Using (2-180a), the Fourier series coefficient is related to the DFT by

$$c_n \approx \frac{1}{N} X(n) \tag{2-189}$$

A discussion of the merits of different types of fast Fourier transform (FFT) algorithms for evaluating the DFT is beyond the scope of this book. However, there are many excellent books and papers on the subject [Bergland, 1969; Glisson, 1986; Harris, 1982; Ziemer, Tranter, and Fannin, 1989].

2-8 Bandwidth of Signals and Noise

Spectral width of signals and/or noise in communication systems is a very important concept, for two main reasons. First, more and more users are being assigned to increasingly crowded radio-frequency (RF) bands, so that the spectral width required for each one needs to be considered carefully. Second, the spectral width of signals/noise is important from the equipment design viewpoint since the circuits need to have enough bandwidth to accommodate the signal but reject the noise. The question is: What is *bandwidth?* As we will see, there are numerous definitions for bandwidth, just as there are numerous measures for length (centimeters, meters, feet, inches, etc.).

As long as we use the same definition when working with several signals and noise, we can compare their spectral widths by using the particular bandwidth definition that was selected. If we change definitions, "conversion factors" will be needed to compare the spectral widths that were obtained by using different definitions. Unfortunately, the conversion factors usually depend on the type of spectral *shape* involved [e.g., $(\sin x)/x$ type of spectrum or rectangular spectrum].

In engineering definitions, the bandwidth is taken to be the width of a *positive* frequency band. (We are describing the bandwidth of real signals or the bandwidth of a physical filter that has a real impulse response; consequently, the magnitude spectra of these waveforms are even about the origin $f = 0$.) In other words, the bandwidth would be $f_2 - f_1$, where $f_2 > f_1 \geq 0$ and f_2 and f_1 are determined by the particular definition that is used. For baseband waveforms or networks, f_1 is usually taken to be zero since the spectrum extends down to dc ($f = 0$). For bandpass signals, $f_1 > 0$ and the band $f_1 < f < f_2$ encompasses the carrier frequency f_c of the signal. It is recalled that as we increase the "signaling speed" of a signal (i.e., decrease T) the spectrum gets wider (see Fig. 2-6). Consequently, for engineering definitions of bandwidth, we require the bandwidth to vary as $1/T$.

We will give six engineering definitions and one legal definition of bandwidth that are often used. It is realized that there are also other definitions that might be useful, depending on the situation.

1. *Absolute bandwidth* is $f_2 - f_1$, where the spectrum is zero outside the interval $f_1 < f < f_2$ along the positive frequency axis.
2. *3-dB bandwidth* (or half-power bandwidth) is $f_2 - f_1$, where for frequencies inside the band $f_1 < f < f_2$, the magnitude spectra, say $|H(f)|$, fall no lower than $1/\sqrt{2}$ times the maximum value of $|H(f)|$, where the maximum value falls at a frequency inside the band.
3. *Equivalent noise bandwidth* is the width of a fictitious rectangular spectrum (height specified below) such that the power in that rectangular band is equal to the power associated with the actual spectrum over positive frequencies. For example, from (2-142), the PSD is proportional to the magnitude squared of the spectrum. Let f_0 be the frequency where the magnitude spectrum has a maximum; then the power in the equivalent rectangular band is proportional to

$$\text{equivalent power} = B_{eq}|H(f_0)|^2 \tag{2-190}$$

where B_{eq} is the equivalent bandwidth that is to be determined. The actual power for positive frequencies is proportional to

$$\text{actual power} = \int_0^\infty |H(f)|^2\, df \tag{2-191}$$

Setting (2-190) equal to (2-191), we have the formula that gives the *equivalent noise bandwidth,*

$$B_{eq} = \frac{1}{|H(f_0)|^2} \int_0^\infty |H(f)|^2\, df \tag{2-192}$$

4. *Null-to-null bandwidth* (or zero-crossing bandwidth) is $f_2 - f_1$, where f_2 is the first null in the envelope of the magnitude spectrum above f_0 and, for bandpass filters, f_1 is the first null in the envelope below f_0, where f_0 is the frequency where the magnitude spectrum is a maximum.† For baseband filters f_1 is usually zero.

5. *Bounded spectrum bandwidth* is $f_2 - f_1$ such that outside the band $f_1 < f < f_2$, the PSD, which is proportional to $|H(f)|^2$, must be down by at least a certain amount, say 50 dB, below the maximum value of the power spectral density.
6. *Power bandwidth* is $f_2 - f_1$, where $f_1 < f < f_2$ defines the frequency band in which 99% of the total power resides. This is similar to the FCC definition of *occupied bandwidth,* which states that the power above the upper band edge f_2 is $\frac{1}{2}$% and the power below the lower band edge is $\frac{1}{2}$%,

†In cases where there is no definite null in the magnitude spectrum, this definition is not applicable.

leaving 99% of the total power within the occupied band (FCC Rules and Regulations, Sec. 2.202).

7. *FCC bandwidth* is an authorized bandwidth parameter assigned by the FCC to specify the spectrum allowed in communication systems. When the FCC bandwidth parameter is substituted into the FCC formula, the minimum attenuation is given for the power level allowed in a 4-kHz band at the band edge with respect to the total average signal power. Quoting Sec. 21.106 of the FCC Rules and Regulations: "For operating frequencies below 15 GHz, in any 4 kHz band, the center frequency of which is removed from the assigned frequency by more than 50 percent up to and including 250 percent of the authorized bandwidth, as specified by the following equation, but in no event less than 50 dB:

$$A = 35 + 0.8(P - 50) + 10 \log_{10}(B) \tag{2-193}$$

(attenuation greater than 80 dB is not required)

where A = attenuation (in decibels) below the mean output power level,

P = percent removed from the carrier frequency,

B = authorized bandwidth in megahertz."

The FCC definition (as well as many other legal rules and regulations) is somewhat obscure, but it will be interpreted in Example 2-15. It actually defines a spectral mask. That is, the spectrum of the signal must be less than or equal to the values given by this spectral mask at all frequencies. This spectral mask is illustrated by the FCC envelope curve in Fig. 2-22. The FCC bandwidth parameter B for the spectral mask is assigned different values in different frequency bands: for example, B = 30 MHz in the 6-GHz band and 40 MHz in the 11-GHz band. It will be discussed later in the example associated with Fig. 2-22. The FCC bandwidth parameter is not consistent with the other definitions that are listed in the sense that it is not proportional to $1/T$, the "signaling speed" of the corresponding signal [Amoroso, 1980]. In this sense the FCC bandwidth parameter B is a *legal* definition instead of an engineering definition. The *rms bandwidth,* which is very useful in analytical problems, is defined in Chapter 6.

★ EXAMPLE 2-16 BANDWIDTHS FOR A BPSK SIGNAL

A binary-phase-shift-keyed (BPSK) signal will be used to illustrate how the bandwidth is evaluated for the different definitions just given.

The BPSK signal is described by

$$s(t) = m(t) \cos \omega_c t \tag{2-194}$$

where $\omega_c = 2\pi f_c$, f_c being the carrier frequency (hertz), and $m(t)$ is a serial binary (± 1 values) modulating waveform originating from a digital information source such as a digital computer and is illustrated in Fig. 2-21a. Let us evaluate the spectrum of $s(t)$ for the worst case (widest bandwidth).

The worst-case (widest-bandwidth) spectrum occurs when the digital modu-

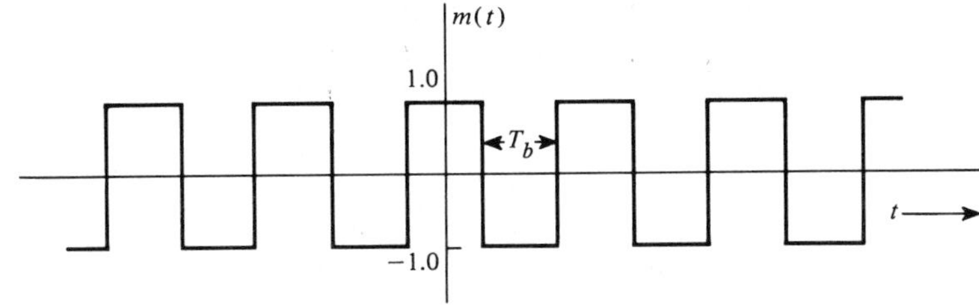

(a) Digital Modulating Waveform

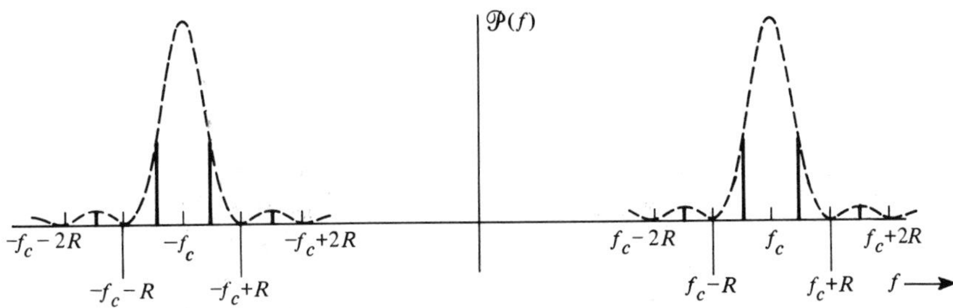

(b) Resulting BPSK spectrum

FIGURE 2-21 Spectrum of a BPSK signal.

lating waveform has transitions that occur most often. In this case $m(t)$ would be a square wave, as shown in Fig. 2-21a. Here a binary 1 is represented by a $+1$-V level and a binary 0 by a -1-V level, and the signaling rate is $R = 1/T_b$ bits/sec. The power spectrum of the square-wave modulation can be evaluated using Fourier series analysis. Using (2-126) and (2-120),

$$\mathcal{P}_m(f) = \sum_{n=-\infty}^{\infty} |c_n|^2\,\delta(f - nf_0) = \sum_{\substack{n=-\infty \\ n\neq 0}}^{\infty} \left[\frac{\sin(n\pi/2)}{n\pi/2}\right]^2 \delta\left(f - n\frac{R}{2}\right) \tag{2-195}$$

where $f_0 = 1/(2T_b) = R/2$. The PSD of $s(t)$ can be expressed in terms of the PSD for $m(t)$ by evaluating the autocorrelation of $s(t)$.

$$\begin{aligned} R_s(\tau) &= \langle s(t)s(t+\tau)\rangle \\ &= \langle m(t)m(t+\tau)\cos\omega_c t\cos\omega_c(t+\tau)\rangle \\ &= \tfrac{1}{2}\langle m(t)m(t+\tau)\rangle\cos\omega_c\tau + \tfrac{1}{2}\langle m(t)m(t+\tau)\cos(2\omega_c t + \omega_c\tau)\rangle \end{aligned}$$

or

$$R_s(\tau) = \tfrac{1}{2}R_m(\tau)\cos\omega_c\tau + \tfrac{1}{2}\lim_{T\to\infty}\frac{1}{T}\int_{-T/2}^{T/2} m(t)m(t+\tau)\cos(2\omega_c t + \omega_c\tau)\,dt \tag{2-196}$$

The integral is negligible since $m(t)m(t + \tau)$ is constant over small time intervals, but $\cos(2\omega_c t + \omega_c \tau)$ has many cycles of oscillation because $f_c \gg R$.† Any small area that is accumulated by the integral becomes negligible when divided by T, $T \rightarrow \infty$. Thus

$$R_s(\tau) = \tfrac{1}{2} R_m(\tau) \cos \omega_c \tau \tag{2-197}$$

The PSD is obtained by taking the Fourier transform of both sides of (2-197). Using the real-signal frequency transform theorem (Table 2-1), we get

$$\mathcal{P}_s(f) = \tfrac{1}{4}[\mathcal{P}_m(f - f_c) + \mathcal{P}_m(f + f_c)] \tag{2-198}$$

Substituting (2-195) into (2-198), we obtain the PSD for the BPSK signal:

$$\mathcal{P}_s(f) = \tfrac{1}{4} \sum_{\substack{n=-\infty \\ n \neq 0}}^{\infty} \left[\frac{\sin(n\pi/2)}{n\pi/2} \right]^2 \times \{\delta[f - f_c - n(R/2)] + \delta[f + f_c - n(R/2)]\} \tag{2-199}$$

This spectrum is shown in Fig. 2-21b.

The spectral shape that results from using this worst-case deterministic modulation is essentially the same as that obtained when random data are used; however, for the random case, the spectrum is continuous. The result for the random data case, as worked out in Chapter 3 where $\mathcal{P}_m(f)$ is given by (3-31), is

$$\mathcal{P}(f) = \tfrac{1}{4} T_b \left[\frac{\sin \pi T_b (f - f_c)}{\pi T_b (f - f_c)} \right]^2 + \tfrac{1}{4} T_b \left[\frac{\sin \pi T_b (f + f_c)}{\pi T_b (f + f_c)} \right]^2 \tag{2-200}$$

when the data rate is $R = 1/T_b$ bits/sec. This is shown by the dotted curve in Fig. 2-21b.

This demonstrates that we can often use (deterministic) square-wave test signals to help us analyze a digital communication system instead of using a more complicated random data model.

The bandwidth for the BPSK signal will now be evaluated for each of the bandwidth definitions given previously. To accomplish this, the shape of the PSD for the positive frequencies is needed. From (2-200) it is

$$\mathcal{P}(f) = \left[\frac{\sin \pi T_b (f - f_c)}{\pi T_b (f - f_c)} \right]^2 \tag{2-201}$$

Substituting this equation into the definitions, we obtain the resulting BPSK bandwidths as shown in Table 2-4, except for the FCC bandwidth.

The relationship between the spectrum and the FCC bandwidth parameter is a little more tricky to evaluate. We need to evaluate the decibel attenuation

$$A(f) = -10 \log_{10} \left[\frac{P_{4\text{kHz}}(f)}{P_{\text{total}}} \right] \tag{2-202}$$

†This is a consequence of the Riemann–Lebesque lemma from integral calculus [Olmsted, 1961].

TABLE 2-4 Bandwidths for BPSK Signaling Where the Bit Rate Is $R = 1/T_b$ Bits/Sec

Definition Used	Bandwidth	Bandwidth (kHz) for $R = 9600$ bits/sec
1. Absolute bandwidth	∞	∞
2. 3-dB bandwidth	$0.88R$	8.45
3. Equivalent noise bandwidth	$1.00R$	9.60
4. Null-to-null bandwidth	$2.00R$	19.20
5. Bounded spectrum bandwidth (50 dB)	$201.04R$	1930.0
6. Power bandwidth	$20.56R$	197.4

where $P_{4\text{kHz}}(f)$ is the power in a 4-kHz band centered at frequency f and P_{total} is the total signal power. The power in a 4-kHz band (assuming that the PSD is approximately constant across the 4-kHz bandwidth) is

$$P_{4\text{kHz}}(f) = 4000\, \mathcal{P}(f) \tag{2-203}$$

and, using the definition of equivalent bandwidth, we find that the total power is

$$P_{\text{total}} = B_{\text{eq}} \mathcal{P}(f_c) \tag{2-204}$$

where the spectrum has a maximum value at $f = f_c$. Using these two equations, (2-202) becomes

$$A(f) = -10 \log_{10}\left[\frac{4000\, \mathcal{P}(f)}{B_{\text{eq}} \mathcal{P}(f_c)}\right] \tag{2-205}$$

where $A(f)$ is the decibel attenuation of power measured in a 4-kHz band at frequency f compared to the total average power level of the signal. For the case of BPSK signaling, using (2-201), we find that the decibel attenuation is

$$A(f) = -10 \log_{10}\left\{\frac{4000}{R}\left[\frac{\sin \pi T_b(f - f_c)}{\pi T_b(f - f_c)}\right]^2\right\} \tag{2-206}$$

where $R = 1/T_b$ is the data rate. If we attempt to find the value of R such that $A(f)$ will fall below the specified FCC spectral envelope shown in Fig. 2-22 ($B = 30$ MHz), we will find that R is so small that there will be numerous zeros in the $(\sin x)/x$ function of (2-206) within the desired frequency range, $-50 \text{ MHz} < (f - f_c) < 50 \text{ MHz}$. This is difficult to plot, so we plot the envelope of $A(f)$ instead, by replacing $\sin \pi T(f - f_c)$ with its maximum value (unity). The resulting BPSK envelope curve for the decibel attenuation is shown in Fig. 2-22, where $R = 0.0171$ Mbits/sec.

It is obvious that the data rate allowed for a BPSK signal to meet the FCC $B = 30$ MHz specification is ridiculously low. This is because the FCC bandwidth parameter specifies an almost absolutely bandlimited spectrum. In order to

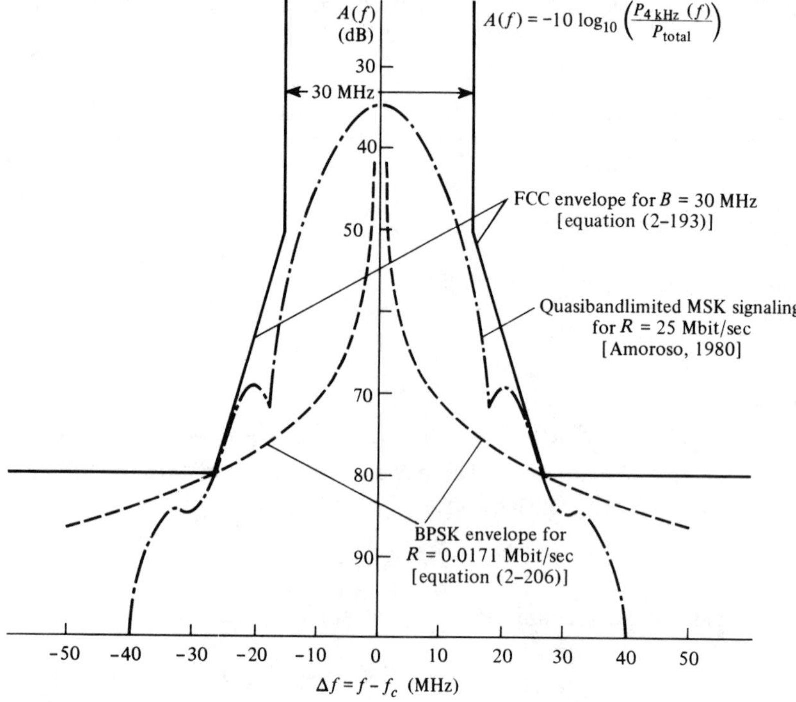

FIGURE 2-22 FCC-allowed envelope for $B = 30$ MHz. [From Frank Amoroso, "The bandwidth of digital data signals," *IEEE Communications Magazine,* Nov. 1980, p. 23, © 1980 IEEE.]

signal with a reasonable data rate, the pulse shape used in the transmitted signal must be modified from a rectangular shape (that gives the BPSK signal) to a rounded pulse such that the bandwidth of the transmitted signal will be almost absolutely bandlimited. Recalling our study of the sampling theorem, we realize that $(\sin x)/x$ pulse shapes are prime candidates since they have an absolutely bandlimited spectrum. However, the $(\sin x)/x$ pulses are not absolutely time-limited, so that this exact pulse shape cannot be used. Frank Amoroso and others have studied this problem, and a quasi-bandlimited version of the $(\sin x)/x$ pulse shape has been proposed [Amoroso, 1980]. The decibel attenuation curve for this type of signaling, shown in Fig. 2-22, is seen to fit very nicely under the allowed FCC spectral envelope curve for the case of $R = 25$ Mbits/sec. This allowable data rate of 25 Mbits/sec is a fantastic improvement over the $R = 0.0171$ Mbits/sec that was allowed for BPSK. It is also interesting to note that analog pulse shapes [$(\sin x)/x$ type] are required instead of a digital (rectangular) pulse shape. *This is another way of saying that it is vital for digital communication engineers to be able to analyze and design analog systems as well as digital systems.*

2-9 Summary

Signals and noise may be deterministic (waveform known), where the signal is the desired part and noise is the undesired part; or they may be stochastic (waveform unknown but statistics of waveform known). Several useful properties of signals and noise are their spectra, dc value, rms value, and associated power.

The properties of Fourier transforms and spectra were studied in detail. Examples illustrating spectral calculations were worked out. The autocorrelation function and the power spectral density function (PSD) were defined and examined.

Signals and noise may be represented by orthogonal series expansions. Illustrations were given for digital and analog signals. The Fourier series and sampling function series were found to be especially useful.

Linear systems were reviewed, and the condition for distortionless transmission was found.

The properties of bandlimited signals and noise were developed. This resulted in the sampling theorem and the dimensionality theorem. Discrete Fourier transforms (DFT) were studied.

The concept of bandwidth was discussed, and seven popular definitions were given. An example of a BPSK signal was used to illustrate how different measures of bandwidth resulted from the same signal. It was demonstrated that rounded pulse shapes are often required to meet FCC regulatory constraints.

In Chapter 4 we will study bandpass digital and analog communications in detail; but first, in Chapter 3, we will study how baseband analog waveforms are converted into baseband digital waveforms for transmission over digital communication systems.

PROBLEMS

2-1 For a sinusoidal waveform with a peak value of A and a frequency of f_0, use the time average operator to show that the rms value for this waveform is $A/\sqrt{2}$.

2-2 A function generator produces a periodic voltage waveform as shown in Fig. P2-2.

(a) Find the value for the dc voltage.

(b) Find the value for the rms voltage.

(c) If this voltage waveform is applied across a 1000-Ω load, what is the power dissipated in the load?

2-3 The voltage across a load is given by $v(t) = A_0 \cos \omega_0 t$, and the current through the load is a square wave,

$$i(t) = I_0 \sum_{n=-\infty}^{\infty} \left[\Pi\left(\frac{t - nT_0}{T_0/2}\right) - \Pi\left(\frac{t - nT_0 - (T_0/2)}{T_0/2}\right) \right]$$

where $\omega_0 = 2\pi/T_0$, $T_0 = 1$ sec, $A_0 = 10$ V, and $I_0 = 5$ mA.

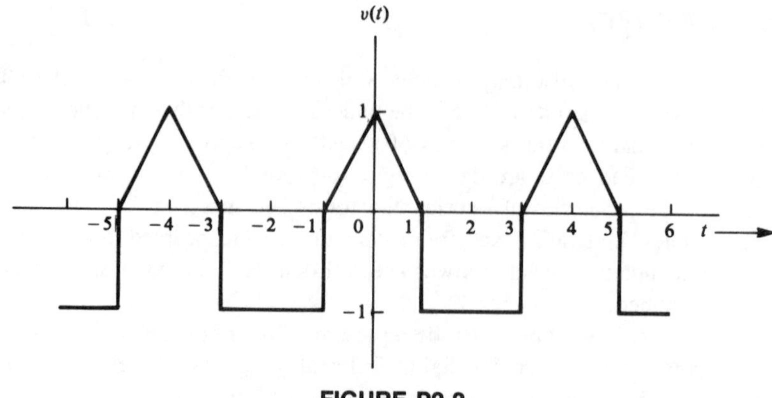

FIGURE P2-2

(a) Find the expression for the instantaneous power and sketch this result as a function of time.
(b) Find the value of the average power.

2-4 The voltage across a 50-Ω resistive load is the positive portion of a cosine wave. That is,

$$v(t) = \begin{cases} 10 \cos \omega_0 t, & |t - nT_0| < T_0/4 \\ 0, & t \text{ elsewhere} \end{cases}$$

where n is any integer.
(a) Sketch the voltage and current waveforms.
(b) Evaluate the dc values for the voltage and current.
(c) Find the rms values for the voltage and current.
(d) Find the total average power dissipated in the load.

2-5 For Prob. 2-4, find the energy dissipated in the load during a 1-hr interval if $T_0 = 1$ sec.

2-6 Determine whether each of the following signals is an energy signal or a power signal and evaluate the normalized energy or power, as appropriate.
(a) $w(t) = \Pi(t/T_0)$.
(b) $w(t) = \Pi(t/T_0) \cos \omega_0 t$.
(c) $w(t) = \cos^2 \omega_0 t$.

2-7 An average reading power meter is connected to the output circuit of a transmitter. The transmitter output is fed into a 75-Ω resistive load and the wattmeter reads 67 W.
(a) What is the power in dBm units?
(b) What is the power in dBk units?
(c) What is the value in dBmV units?

2-8 Assume that a waveform with a known rms value, V_{rms}, is applied across a 50-Ω load. Derive a formula that can be used to compute the dBm value from V_{rms}.

2-9 An amplifier is connected to a 50-Ω load and driven by a sinusoidal current source, as shown in Fig. P2-9. The output resistance of the amplifier is 10 Ω and the input resistance is 2 kΩ. Evaluate the true decibel gain of this circuit.

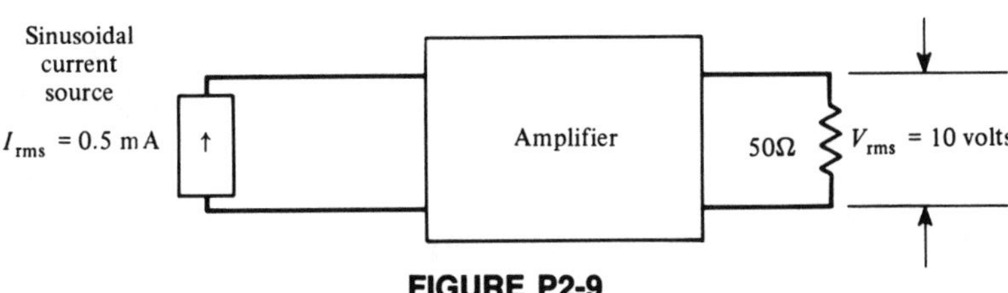

FIGURE P2-9

2-10 The voltage (rms) across the 300-Ω antenna input terminals of an FM receiver is 3.5 μV.
(a) Find the input power (watts).
(b) Evaluate the input power as measured in decibels below 1 mW (dBm).
(c) What would be the input voltage (in microvolts) for the same input power if the input resistance were 75 Ω instead of 300 Ω?

2-11 What is the value for the phasor that corresponds to the voltage waveform $v(t) = 12 \sin(\omega_0 t - 25°)$, where $\omega_0 = 2000\pi$?

2-12 A signal is $w(t) = 3 \sin(100\pi t - 30°) + 4 \cos(100\pi t)$. Find the corresponding phasor.

2-13 Evaluate the Fourier transform of

$$w(t) = \begin{cases} e^{-\alpha t}, & t \geq 1 \\ 0, & t < 1 \end{cases}$$

2-14 Find the spectrum for the waveform $w(t) = e^{-\pi(t/T)^2}$. What can we say about the width of $w(t)$ and $W(f)$ as T increases? [*Hint:* Use (A-75).]

2-15 Using the convolution property, find the spectrum for

$$w(t) = \sin 2\pi f_1 t \cos 2\pi f_2 t$$

2-16 Find the spectrum (Fourier transform) of the triangle waveform

$$s(t) = \begin{cases} At, & 0 < t < T_0 \\ 0, & t \text{ elsewhere} \end{cases}$$

in terms of A and T_0.

2-17 Find the spectrum for the waveform shown in Fig. P2-17.

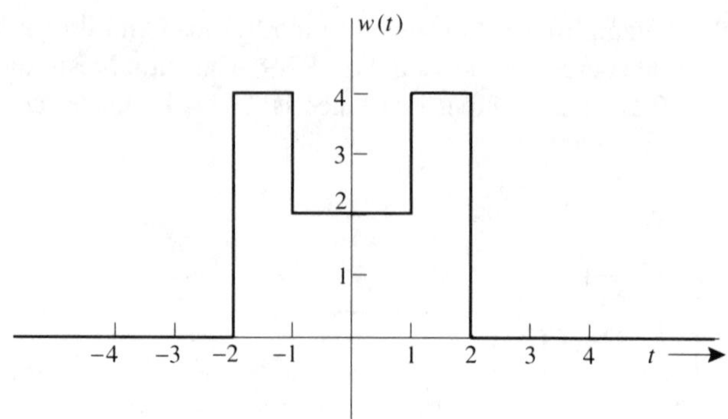

FIGURE P2-17

2-18 If $w(t)$ has the Fourier transform

$$W(f) = \frac{j2\pi f}{1 + j2\pi f}$$

find $X(f)$ for the following waveforms.
(a) $x(t) = w(2t + 2)$.
(b) $x(t) = e^{-jt}w(t - 1)$.
(c) $x(t) = 2\,\dfrac{dw(t)}{dt}$.
(d) $x(t) = w(1 - t)$.

2-19 Using (2-30), find $w(t)$ for $W(f) = A\Pi(f/2B)$ and verify your answer using the duality property.

2-20 Find the quadrature spectral functions $X(f)$ and $Y(f)$ for the damped sinusoidal waveform

$$w(t) = u(t)e^{-at} \sin \omega_0 t$$

where $u(t)$ is a unit step function, $a > 0$, and $W(f) = X(f) + jY(f)$.

2-21 Derive the spectrum of $w(t) = e^{-|t|/T}$.

2-22 Find the Fourier transforms for the following waveforms. Plot the waveforms and their magnitude spectra. [*Hint:* Use (2-186).]

(a) $\Pi\left(\dfrac{t-3}{4}\right)$.

(b) 2.

(c) $\Lambda\left(\dfrac{t-5}{5}\right)$.

2-23 Find the approximate Fourier transform for the following waveform using (2-186).

$$x(t) = \begin{cases} \sin(2\pi t/512) + \sin(70\pi t/512), & 5 < t < 75 \\ 0, & t \text{ elsewhere} \end{cases}$$

2-24 Evaluate the spectrum for the trapezoidal pulse shown in Fig. P2-24.

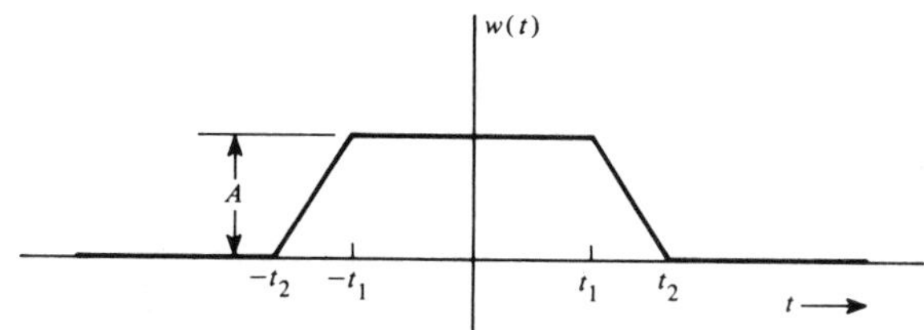

FIGURE P2-24

2-25 Show that

$$\mathscr{F}^{-1}\{\mathscr{F}[w(t)]\} = w(t)$$

[*Hint:* Use (2-33).]

2-26 Using the definition of the inverse Fourier transform, show that the value of $w(t)$ at $t = 0$ is equal to the area under $W(f)$. That is, show that

$$w(0) = \int_{-\infty}^{\infty} W(f)\, df$$

2-27 Prove that:
(a) If $w(t)$ is real and an even function of t, $W(f)$ is real.
(b) If $w(t)$ is real and an odd function of t, $W(f)$ is imaginary.

2-28 Suppose that the spectrum of a waveform as a function of frequency in hertz is

$$W(f) = \frac{1}{2}\delta(f - 4) + \frac{1}{2}\delta(f + 4) + \frac{j\pi f}{2 + j2\pi f}e^{j\pi f}$$

Find the corresponding spectrum as a function of radian frequency, $W(\omega)$.

2-29 The unit impulse can also be defined as follows:

$$\delta(t) = \lim_{a\to\infty}\left[Ka\left(\frac{\sin at}{at}\right)\right]$$

Find the value of K needed, and show that this definition is consistent with those given in the text. Give another example of an ordinary function such that in the limit of some parameter, the function becomes a Dirac delta function.

2-30 Use $v(t) = ae^{-at}$, $a > 0$, to approximate $\delta(t)$ as $a \to \infty$.
(a) Plot $v(t)$ for $a = 0.1$, 1, and 10.
(b) Plot $V(f)$ for $a = 0.1$, 1, and 10.

2-31 Show that

$$\operatorname{sgn}(t) \leftrightarrow \frac{1}{j\pi f}$$

[*Hint:* Use (2-30) and $\int_0^\infty (\sin x)/x \, dx = \pi/2$ from Appendix A.]

2-32 Show that

$$u(t) \leftrightarrow \tfrac{1}{2}\delta(f) + \frac{1}{j2\pi f}$$

[*Hint:* Use the linearity (superposition) theorem and the result of Prob. 2-31.]

2-33 Show that the sifting property of δ functions, (2-47), may be generalized to evaluate integrals that involve derivatives of the delta function. That is, show that

$$\int_{-\infty}^{\infty} w(x)\, \delta^{(n)}(x - x_0)\, dx = (-1)^n w(n)(x_0)$$

where the superscript (n) denotes the nth derivative. (*Hint:* Use integration by parts.)

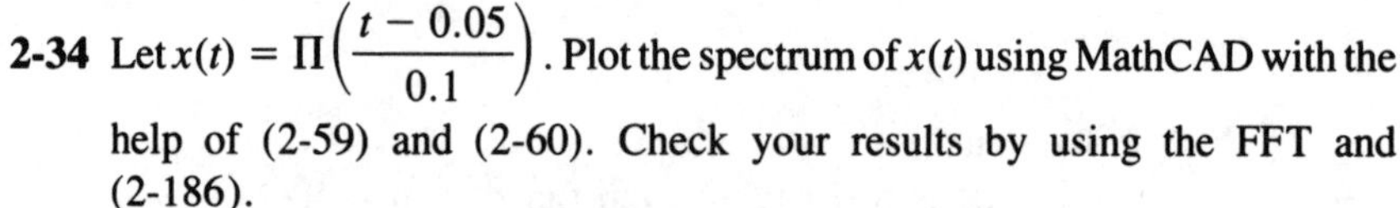

2-34 Let $x(t) = \Pi\left(\frac{t - 0.05}{0.1}\right)$. Plot the spectrum of $x(t)$ using MathCAD with the help of (2-59) and (2-60). Check your results by using the FFT and (2-186).

2-35 If $w(t) = w_1(t)w_2(t)$, show that

$$W(f) = \int_{-\infty}^{\infty} W_1(\lambda)W_2(f - \lambda)\, d\lambda$$

where $W(f) = \mathcal{F}[w(t)]$.

2-36 Show that:
(a) $\int_{-\infty}^{t} w(\lambda)\, d\lambda = w(t) * u(t)$.
(b) $\int_{-\infty}^{t} w(\lambda)\, d\lambda \leftrightarrow (j2\pi f)^{-1}W(f) + \tfrac{1}{2}W(0)\,\delta(f)$.
(c) $w(t) * \delta(t - a) = w(t - a)$.

2-37 Show that

$$\frac{dw(t)}{dt} \leftrightarrow (j2\pi f)W(f)$$

[*Hint:* Use (2-26) and integrate by parts. Assume that $w(t)$ is absolutely integrable.]

2-38 As discussed in Example 2-8, show that

$$\frac{1}{T}\Pi\left(\frac{t}{T}\right) * \Pi\left(\frac{t}{T}\right) = \Lambda\left(\frac{t}{T}\right)$$

2-39 Given the waveform $w(t) = A\Pi(t/T)\sin\omega_0 t$. Find the spectrum of $w(t)$ by using the multiplication theorem as discussed in Example 2-9.

2-40 Evaluate the following integrals.

(a) $\displaystyle\int_{-\infty}^{\infty} \frac{\sin 4\lambda}{4\lambda}\,\delta(t-\lambda)\,d\lambda.$

(b) $\displaystyle\int_{-\infty}^{\infty} (\lambda^3 - 1)\,\delta(2-\lambda)\,d\lambda.$

2-41 Prove that

$$M(f) * \delta(f - f_0) = M(f - f_0)$$

2-42 Evaluate $y(t) = w_1(t) * w_2(t)$, where

$$w_1(t) = \begin{cases} 1, & |t| < T_0 \\ 0, & t \text{ elsewhere} \end{cases}$$

$$w_2(t) = \begin{cases} [1 - 2|t|], & |t| < \frac{1}{2}T_0 \\ 0, & t \text{ elsewhere} \end{cases}$$

2-43 Given $w(t) = 5 + 12\cos\omega_0 t$ where $f_0 = 10$ Hz:

(a) Find $R_w(\tau)$.

(b) $\mathcal{P}_w(f)$.

2-44 Given the waveform

$$w(t) = A_1\cos(\omega_1 t + \theta_1) + A_2\cos(\omega_2 t + \theta_2)$$

where A_1, A_2, ω_1, ω_2, θ_1, and θ_2 are constants.

(a) Find the autocorrelation for $w(t)$ as a function of the constants.

(b) Find the PSD function for $w(t)$.

(c) Sketch the PSD for the case of $\omega_1 \neq \omega_2$.

(d) Sketch the PSD for the case of $\omega_1 = \omega_2$ and $\theta_1 = \theta_2 + 90°$.

(e) Sketch the PSD for the case of $\omega_1 = \omega_2$ and $\theta_1 = \theta_2$.

2-45 Given the periodic voltage waveform shown in Fig. P2-45:

(a) Find the dc value for this waveform.

(b) Find the rms value for this waveform.

(c) Find the complex exponential Fourier series.

(d) Find the voltage spectrum for this waveform.

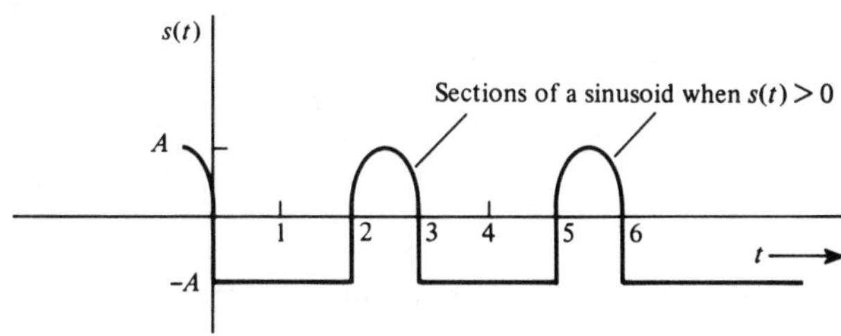

FIGURE P2-45

2-46 **(a)** Find a mathematical representation for a periodic triangular waveform that has a peak value of A volts, an average value of $\frac{1}{4}A$ volts, a period of T seconds, and a zero crossing at $t = 0$ with a positive slope. Assume that the slope of the sides of the triangle are equal with opposite signs.

(b) Find the spectrum for this triangular waveform.

(c) Find the rms value for this triangular waveform.

(d) Find the bandwidth required to pass 95% of the power. (*Hint:* Use the Fourier series.)

2-47 Determine if $s_1(t)$ and $s_2(t)$ are orthogonal over the interval $(-\frac{5}{2}T_2 < t < \frac{5}{2}T_2)$, where $s_1(t) = A_1 \cos(\omega_1 t + \varphi_1)$, $s_2(t) = A_2 \cos(\omega_2 t + \varphi_2)$, and $\omega_2 = 2\pi/T_2$ for the following cases.

(a) $\omega_1 = \omega_2$ and $\varphi_1 = \varphi_2$.

(b) $\omega_1 = \omega_2$ and $\varphi_1 = \varphi_2 + \pi/2$.

(c) $\omega_1 = \omega_2$ and $\varphi_1 = \varphi_2 + \pi$.

(d) $\omega_1 = 2\omega_2$ and $\varphi_1 = \varphi_2$.

(e) $\omega_1 = \frac{4}{5}\omega_2$ and $\varphi_1 = \varphi_2$.

(f) $\omega_1 = \pi\omega_2$ and $\varphi_1 = \varphi_2$.

2-48 Let $s(t) = A_1 \cos(\omega_1 t + \varphi_1) + A_2 \cos(\omega_2 t + \varphi_2)$. Determine the rms value of $s(t)$ in terms of A_1 and A_2 for the following cases.

(a) $\omega_1 = \omega_2$ and $\varphi_1 = \varphi_2$.

(b) $\omega_1 = \omega_2$ and $\varphi_1 = \varphi_2 + \pi/2$.

(c) $\omega_1 = \omega_2$ and $\varphi_1 = \varphi_2 + \pi$.

(d) $\omega_1 = 2\omega_2$ and φ_1 and φ_2.

(e) $\omega_1 = 2\omega_2$ and $\varphi_1 = \varphi_2 + \pi$.

2-49 Show that

$$\sum_{k=-\infty}^{\infty} \delta(t - kT_0) \leftrightarrow f_0 \sum_{n=-\infty}^{\infty} \delta(f - nf_0)$$

where $f_0 = 1/T_0$. [*Hint:* Expand $\Sigma_{k=-\infty}^{\infty} \delta(t - kT_0)$ into a Fourier series and then take the Fourier transform.]

2-50 Three functions are shown in Fig. P2-50.

(a) Show that these functions are orthogonal over the interval $(-4,4)$.

(b) Find the corresponding orthonormal set of functions.

(c) Expand the waveform

$$w(t) = \begin{cases} 1, & 0 \le t \le 4 \\ 0, & t \text{ elsewhere} \end{cases}$$

into an orthonormal series using this orthonormal set.

(d) Evaluate the mean square error for the orthogonal series obtained in part (c) by evaluating

$$\epsilon = \int_{-4}^{4} \left[w(t) - \sum_{j=1}^{3} a_j \varphi_j(t) \right]^2 dt$$

(e) Repeat parts (c) and (d) for the waveform

$$w(t) = \begin{cases} \cos(\frac{1}{4}\pi t), & -4 \le t \le 4 \\ 0, & t \text{ elsewhere} \end{cases}$$

Are the three orthonormal functions a complete orthonormal set?

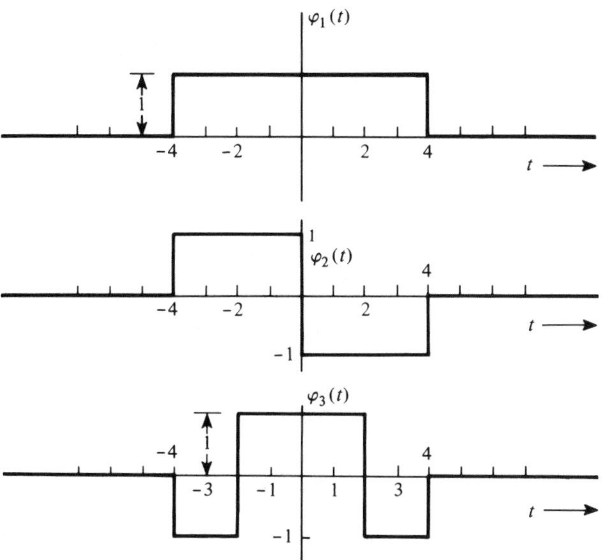

FIGURE P2-50

2-51 Refer to the signals shown in Fig. P2-51.

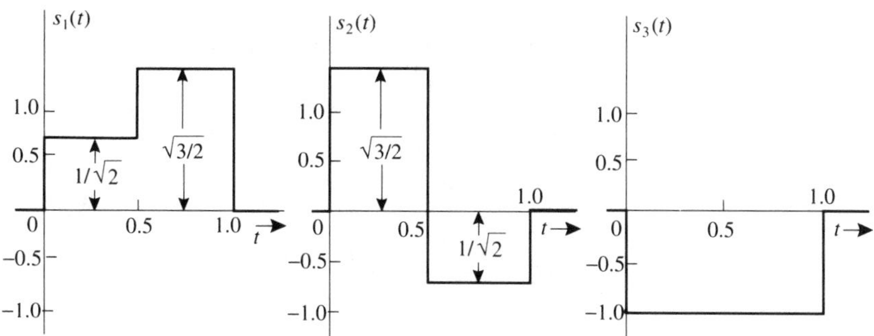

FIGURE P2-51

(a) What are the minimum number of orthogonal functions, N, required to represent this set of signals?
(b) Find N orthogonal basis functions.
(c) Find the corresponding signal vectors $\mathbf{s}_1$, $\mathbf{s}_2$, and $\mathbf{s}_3$.
[*Hint:* Refer to (8-21)–(8-26)].

2-52 Show that the complex Fourier series basis functions, as given by (2-92), are orthogonal.

2-53 Find expressions for the complex Fourier series coefficients that represent the waveform shown in Fig. P2-53.

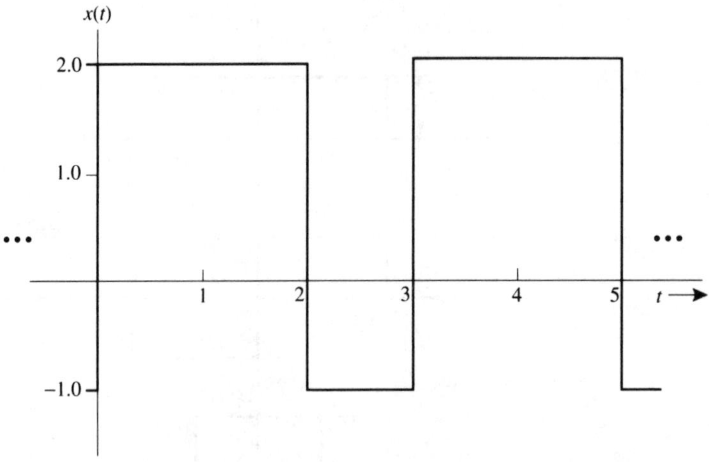

FIGURE P2-53

2-54 The periodic signal as shown in Fig. P2-53 is passed through a linear filter having the impulse response $h(t) = e^{-\alpha t}u(t)$, where $t > 0$ and $\alpha > 0$.
(a) Find expressions for the complex Fourier series coefficients associated with the output waveform $y(t) = x(t) * h(t)$.
(b) Find an expression for the normalized power of the output, $y(t)$.

2-55 Find the complex Fourier series for the periodic waveform given in Figure P2-2.

2-56 Find the complex Fourier series coefficients for the periodic rectangular waveform shown in Fig. P2-56 as a function of A, T, b, and τ_0. [*Hint:* The

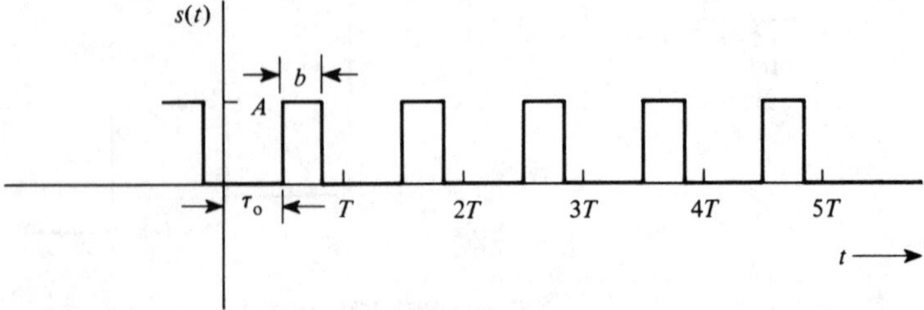

FIGURE P2-56

answer can be reduced to a $(\sin x)/x$ form multiplied by a phase-shift factor, $e^{j\theta_n(\tau_0)}$.]

2-57 For the waveform shown in Fig. P2-57, find:
(a) The complex Fourier series.
(b) The quadrature Fourier series.

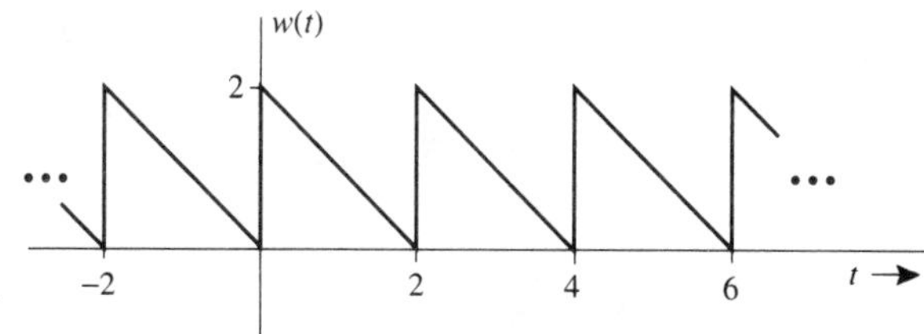

FIGURE P2-57

2-58 Given a periodic waveform $s(t) = \Sigma_{n=-\infty}^{\infty} p(t - nT_0)$, where

$$p(t) = \begin{cases} At, & 0 < t < T \\ 0, & t \text{ elsewhere} \end{cases}$$

and $T \le T_0$.
(a) Find the c_n Fourier series coefficients.
(b) Find the $\{x_n, y_n\}$ Fourier series coefficients.
(c) Find the $\{D_n, \phi_n\}$ Fourier series coefficients.

2-59 Prove that the polar form of the Fourier series, (2-103), can be obtained by rearranging the terms in the complex Fourier series, (2-93).

2-60 Prove that (2-95d) is correct.

2-61 Verify the relationships between the Fourier coefficients c_n, a_n, b_n, D_n, and ϕ_n as given by (2-101), (2-102), (2-104), (2-105), (2-106), and (2-107).

2-62 Let two complex numbers c_1 and c_2 be represented by $c_1 = x_1 + jy_1$ and $c_2 = x_2 + jy_2$, where x_1, x_2, y_1, and y_2 are real numbers. Show that $\text{Re}\{\cdot\}$ is a linear operator by demonstrating that

$$\text{Re}\{c_1 + c_2\} = \text{Re}\{c_1\} + \text{Re}\{c_2\}$$

2-63 Assume that $y(t) = s_1(t) + 2s_2(t)$, where $s_1(t)$ is given by Fig. P2-45 and $s_2(t)$ is given by Fig. P2-56. $T = 3$, $b = 1.5$, and $\tau_0 = 0$. Find the complex Fourier coefficients $\{c_n\}$ for $y(t)$.

2-64 Evaluate the PSD for the waveform shown in Fig. P2-2.

2-65 Assume that $v(t)$ is a triangular waveform, as shown in Fig. P2-65.
(a) Find the complex Fourier series for $v(t)$.
(b) Calculate the normalized average power.
(c) Calculate and plot the voltage spectrum.
(d) Calculate and plot the PSD.

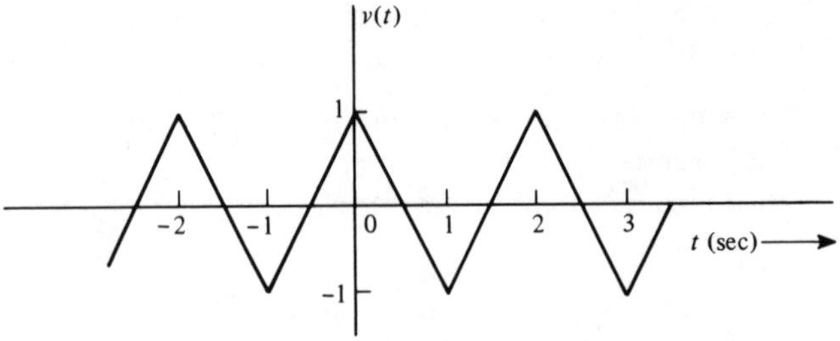

FIGURE P2-65

2-66 Let a complex number c be represented by $c = x + jy$ where x and y are real numbers. Show that:
(a) $\text{Re}\{c\} = \frac{1}{2}c + \frac{1}{2}c^*$.
(b) $\text{Im}\,\{c\} = \dfrac{1}{2j}\,c - \dfrac{1}{2j}\,c^*$.
where c^* is the complex conjugate of c. Note that when $c = e^{jz}$, (a) gives the definition of $\cos z$ and (b) gives the definition of $\sin z$.

2-67 The basic definitions for sine and cosine waves are

$$\sin z_1 \triangleq \frac{e^{jz_1} - e^{-jz_1}}{2j} \qquad \text{and} \qquad \cos z_2 \triangleq \frac{e^{jz_2} + e^{-jz_2}}{2}$$

Show that:
(a) $\cos z_1 \cos z_2 = \frac{1}{2}\cos(z_1 - z_2) + \frac{1}{2}\cos(z_1 + z_2)$.
(b) $\sin z_1 \sin z_2 = \frac{1}{2}\cos(z_1 - z_2) - \frac{1}{2}\cos(z_1 + z_2)$.
(c) $\sin z_1 \cos z_2 = \frac{1}{2}\sin(z_1 - z_2) + \frac{1}{2}\sin(z_1 + z_2)$.

2-68 Let two complex numbers c_1 and c_2 be represented by $c_1 = x_1 + jy_1$ and $c_2 = x_2 + jy_2$, where x_1, x_2, y_1, and y_2 are real numbers. Show that

$$\text{Re}\{c_1\}\,\text{Re}\{c_2\} = \tfrac{1}{2}\,\text{Re}\{c_1c_2^*\} + \tfrac{1}{2}\,\text{Re}\{c_1c_2\}$$

Note that this is a generalization of the $\cos z_1 \cos z_2$ identity of Prob. 2-67, where for the $\cos z_1 \cos z_2$ identity, $c_1 = e^{jz_1}$ and $c_2 = e^{jz_2}$.

2-69 Prove that the Fourier transform is a linear operator. That is, show that

$$\mathcal{F}[ax(t) + by(t)] = a\mathcal{F}[x(t)] + b\mathcal{F}[y(t)]$$

2-70 Plot the amplitude and phase response for the transfer function

$$H(f) = \frac{j10f}{5 + jf}$$

2-71 Given the filter shown in Fig. P2-71.
(a) Find the transfer function for this filter.
(b) Plot the magnitude and phase response.
(c) Find the power transfer function for this filter.
(d) Plot the power transfer function.

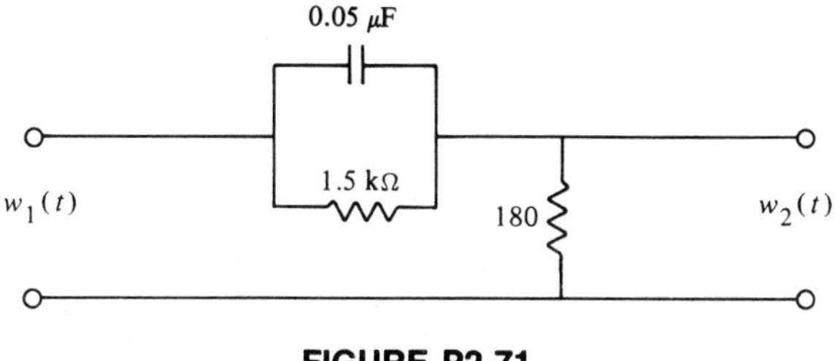

FIGURE P2-71

2-72 A signal with a PSD of

$$\mathcal{P}_x(f) = \frac{2}{(1/4\pi)^2 + f^2}$$

is applied to the network shown in Fig. P2-72.
(a) Find the PSD for $y(t)$.
(b) Find the average normalized power for $y(t)$.

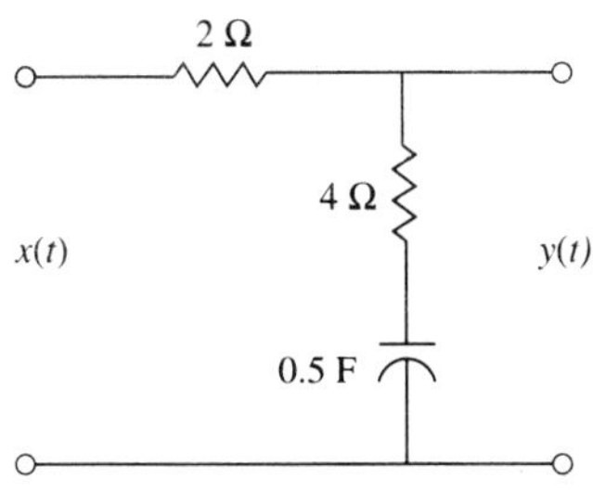

FIGURE P2-72

2-73 A signal $x(t)$ has a PSD

$$\mathcal{P}_x(f) = \frac{K}{[1 + (2\pi f/B)^2]^2}$$

where $K > 0$ and $B > 0$.
(a) Find the 3-dB bandwidth in terms of B.
(b) Find the equivalent noise bandwidth in terms of B.

2-74 The signal $x(t) = e^{-400\pi t}u(t)$ is applied to a brick-wall low-pass filter, $H(f)$, where

$$H(f) = \begin{cases} 1, & |f| \le B \\ 0, & |f| > B \end{cases}$$

Find the value of B such that the filter passes one-half the energy of $x(t)$.

2-75 Show that the average normalized power of a waveform can be found by evaluating the autocorrelation $R_w(\tau)$ at $\tau = 0$. That is, $P = R_w(0)$.

2-76 The signal $x(t) = 0.5 + 1.5\cos[(\frac{2}{3})\pi t] + 0.5\sin[(\frac{2}{3})\pi t)$ volts is passed through an RC low-pass filter (see Fig. 2-15a) where $R = 1\ \Omega$ and $C = 1$ F.

(a) What is the input PSD, $\mathcal{P}_x(f)$?
(b) What is the output PSD, $\mathcal{P}_y(f)$?
(c) What is the average normalized output power, P_y?

2-77 The input to the RC low-pass filter shown in Fig. 2-15 is

$$x(t) = 0.5 + 1.5\cos\omega_x t + 0.5\sin\omega_x t$$

Assume that the cutoff frequency is $f_0 = 1.5 f_x$.

(a) Find the input PSD, $\mathcal{P}_x(f)$.
(b) Find the output PSD, $\mathcal{P}_y(f)$.
(c) Find the normalized average power of the output, $y(t)$.

2-78 Using MathCAD, plot the frequency magnitude response and phase response for the low-pass filter shown in Fig. P2-78, where $R_1 = 7.5$ kΩ, $R_2 = 15$ kΩ, and $C = 0.1\ \mu$F.

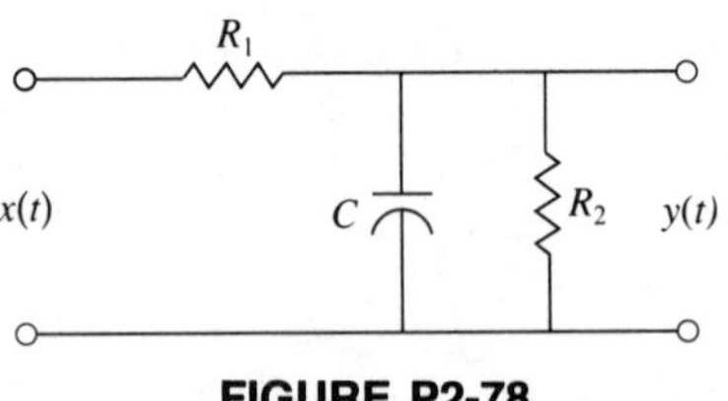

FIGURE P2-78

2-79 A comb filter is shown in Fig. P2-79. Let $T_d = 0.1$.

(a) Plot the magnitude of the transfer function for this filter.
(b) If the input is $x(t) = \Pi(t/T)$, where $T = 10$, plot the spectrum of the output $|Y(f)|$.

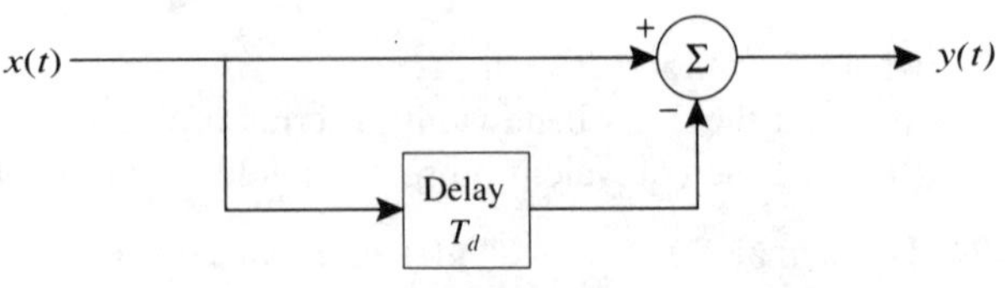

FIGURE P2-79

2-80 A signal $x(t) = \Pi(t - 0.5)$ passes through a filter that has the transfer function $H(f) = \Pi(f/B)$. Plot the output waveform when:

(a) $B = 0.6$ Hz.
(b) $B = 1$ Hz.
(c) $B = 50$ Hz.

2-81 Examine the distortion effect of an *RC* low-pass filter. Assume that a unity-amplitude square wave with a 50% duty cycle is present at the filter input and that the filter has a cutoff frequency of 1500 Hz. Using a computer or a programmable calculator, find the output waveshape if the square wave has a frequency of:
(a) 300 Hz.
(b) 500 Hz.
(c) 1000 Hz.
Sketch each of these results.

2-82 Given that the PSD of a signal is flat [i.e., $\mathcal{P}_s(f) = 1$], design an *RC* low-pass filter that will attenuate this signal by 20 dB at 15 kHz. That is, find the value for the *RC* of Fig. 2-15 such that the design specifications are satisfied.

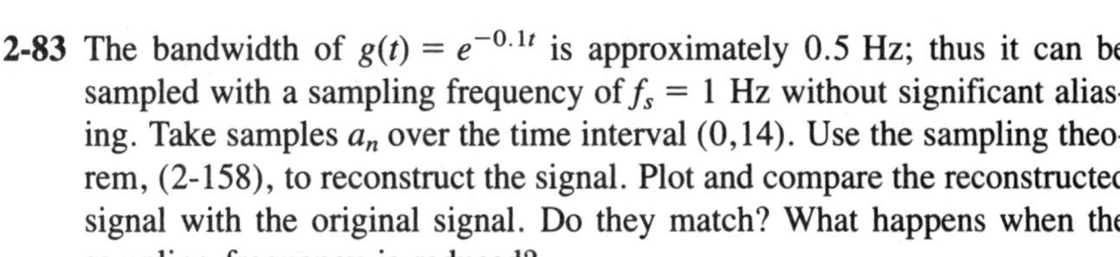

2-83 The bandwidth of $g(t) = e^{-0.1t}$ is approximately 0.5 Hz; thus it can be sampled with a sampling frequency of $f_s = 1$ Hz without significant aliasing. Take samples a_n over the time interval (0,14). Use the sampling theorem, (2-158), to reconstruct the signal. Plot and compare the reconstructed signal with the original signal. Do they match? What happens when the sampling frequency is reduced?

2-84 A waveform, $20 + 20\sin(500t + 30°)$, is to be sampled periodically and reproduced from these sample values.
(a) Find the maximum allowable time interval between sample values.
(b) How many sample values need to be stored in order to reproduce 1 sec of this waveform?

2-85 Determine the minimum (Nyquist) sampling frequency for the signal

$$s(t) = g_1^2(t) + g_2^2(t)$$

where $g_1(t)$ has a bandwidth of B_1 hertz and g_2 has a bandwidth of $3B_1$ hertz where $B_1 > 0$.

2-86 A baseband digital communication system sends one of 16 possible levels over the channel every 0.8 msec.
(a) What is the word length (number of bits) corresponding to each level?
(b) What is the data rate of the system?

2-87 A baseband digital communication system is to operate at a data rate of 9600 bits/sec.
(a) If 4-bit words are encoded into each level for transmission over the system, what is the minimum required bandwidth of the channel?
(b) Repeat part (a) for the case of 8-bit words.

2-88 Using a computer program, calculate the DFT of a rectangular pulse, $\Pi(t)$. Take five samples of the pulse and pad it with 59 zeros so that a 64-point FFT algorithm can be used. Sketch the resulting magnitude spectrum. Compare this result with the actual spectrum for the pulse. Try other combinations of number of pulse samples and zero-pads to see how the resulting FFT changes.

2-89 Given the low-pass filter shown in Fig. 2-15,
(a) Find the equivalent bandwidth in terms of R and C.
(b) Find the first zero-crossing (null) bandwidth of the filter.
(c) Find the absolute bandwidth of the filter.

2-90 Assume that the PSD of a signal is given by

$$\mathcal{P}_s(f) = \left[\frac{\sin(\pi f/B_n)}{\pi f/B_n}\right]^2$$

where B_n is the null bandwidth. Find the expression for the equivalent bandwidth in terms of the null bandwidth.

2-91 Table 2-4 gives the bandwidths of a BPSK signal using six different definitions. Using the six definitions and (2-201), show that the results given in the table are correct.

2-92 Given a triangular pulse signal

$$s(t) = \Lambda(t/T_0)$$

(a) Find the absolute bandwidth of this signal.
(b) Find the 3-dB bandwidth in terms of T_0.
(c) Find the equivalent bandwidth in terms of T_0.
(d) Find the zero-crossing bandwidth in terms of T_0.

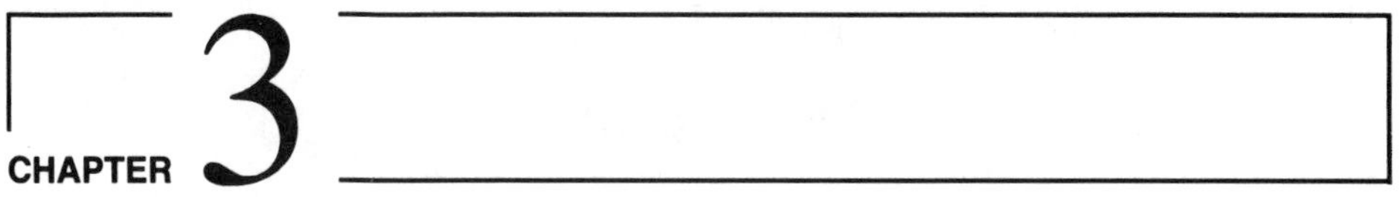

CHAPTER 3

Baseband Pulse and Digital Signaling

3-1 Introduction

In this chapter we are interested in studying how to encode analog waveforms (from analog sources) into baseband pulse signals and how to approximate analog signals with digital signals. As we will see, the digital approximation to the analog signal can be made very precise if we wish. In addition, we will learn how to process the digital baseband signals so that their bandwidth is minimized.

Digital signaling is popular and becoming even more popular because of the low cost of digital circuits and the flexibility of the digital approach. This flexibility arises, for example, because the digital data from digital sources may be merged with digitized data derived from analog sources to provide a general-purpose communication system.

The signals involved in the analog-to-digital conversion problem are *baseband* signals. We also realize that *bandpass* digital communication signals are produced by using baseband digital signals to modulate a carrier, as described in Chapter 1. For example, if a digital (baseband) signal phase modulates a carrier, the resulting digital bandpass signal is said to be a *phase-shift-keyed* (PSK) signal. Bandpass digital communication systems such as PSK are discussed in more detail in Chapter 5.

The three main goals of this chapter are:

1. To study how analog waveforms can be converted to digital waveforms that are acceptable for transmission over a digital communication system. The most popular technique is called *pulse code modulation* (PCM).
2. To examine how filtering of pulse signals affects our ability to recover the digital information at the receiver. This filtering can produce what is called *intersymbol interference* (ISI) in the recovered data signal.
3. To study how we can *multiplex* (combine) data from several digital bit streams into one high-speed digital stream for transmission over a digital system. One such technique, called *time-division multiplexing* (TDM), will be studied in this chapter. Another technique, called *frequency-division multiplexing* (FDM), is described in Chapter 4.

There is another very important problem in digital communication systems: the effect of noise in causing the digital receiver to produce some bit errors at the output. This problem will be studied in Chapter 7 since it involves the use of statistical concepts that are emphasized in the second part of this book.

3-2 Pulse Amplitude Modulation

Pulse amplitude modulation (PAM) is an engineering term that is used to describe the conversion of the analog signal to a pulse-type signal where the amplitude of the pulse denotes the analog information. As we will see in Sec. 3-3, this PAM signal can be converted into a PCM digital signal (baseband), which in turn is modulated onto a carrier in bandpass digital communication systems. Consequently, the analog-to-PAM conversion process is the first step in converting an analog waveform to a PCM (digital) signal. (In some applications the PAM signal is used directly, and conversion to PCM is not required.)

The sampling theorem, as studied in Chapter 2, provides a way to represent a bandlimited analog waveform by using sample values of that waveform and $(\sin x)/x$ orthogonal functions. The purpose of PAM signaling is to provide *another* waveform that looks like pulses, yet contains the information that was present in the analog waveform. Of course, since we are using pulses, we would expect the bandwidth of the PAM waveform to be wider than that of the analog waveform. However, the pulses are more practical to use in digital systems. It is also realized that $(\sin x)/x$ functions that are required for the sampling theorem representation are difficult to approximate and impossible to generate exactly since they approach zero only at $\pm\infty$ (noncausal). In summary, it is not required that the PAM signal "look" exactly like the original analog waveform; it is only required that an approximation to the original be *recovered* from the PAM signal. We will see, however, that the pulse rate, f_s, for PAM is the same as that required by the sampling theorem, namely, $f_s \geq 2B$, where B is the highest frequency in the analog waveform and $2B$ is called the *Nyquist rate*.

There are two classes of PAM signals: PAM that uses *natural sampling* (gating) and PAM that uses *instantaneous sampling* to produce a flat-top pulse. These signals are illustrated in Figs. 3-1 and 3-5, respectively. The flat-top type is more useful for conversion to PCM; however the naturally sampled type is easier to generate and is used in other applications.

★ Natural Sampling (Gating)

Definition: If $w(t)$ is an analog waveform bandlimited to B hertz, the *PAM* signal that uses *natural sampling* (gating) is

$$w_s(t) = w(t)S(t) \tag{3-1}$$

where

$$S(t) = \sum_{k=-\infty}^{\infty} \Pi\left(\frac{t - kT_s}{\tau}\right) \tag{3-2}$$

is a rectangular wave switching waveform and $f_s = 1/T_s \geq 2B$.

Theorem: *The spectrum for a naturally sampled PAM signal is*

$$W_s(f) = \mathcal{F}[w_s(t)] = d \sum_{n=-\infty}^{\infty} \frac{\sin \pi n d}{\pi n d} W(f - nf_s) \tag{3-3}$$

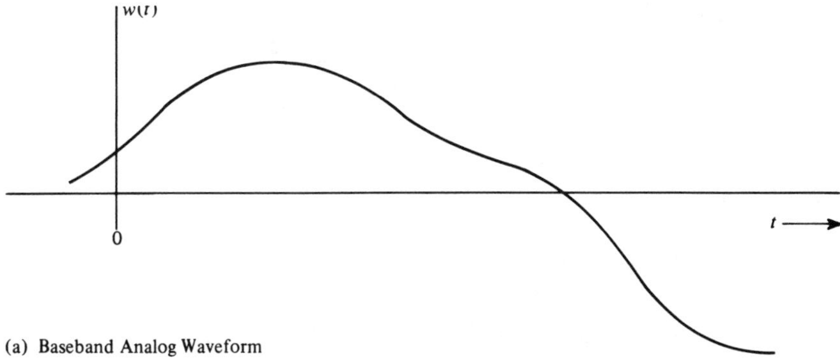

(a) Baseband Analog Waveform

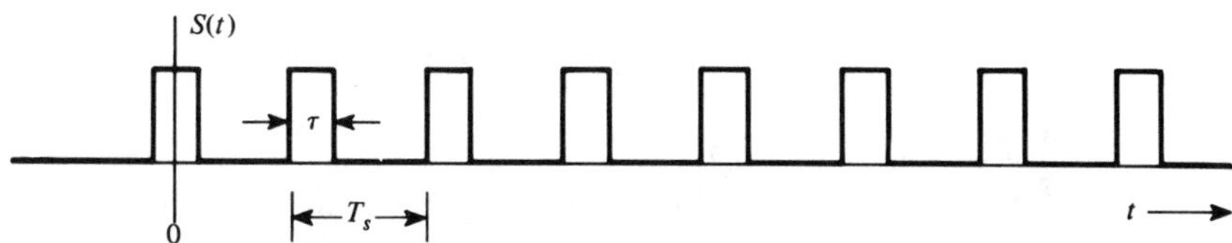

(b) Switching Waveform with Duty Cycle $d = \tau/T_s = 1/3$

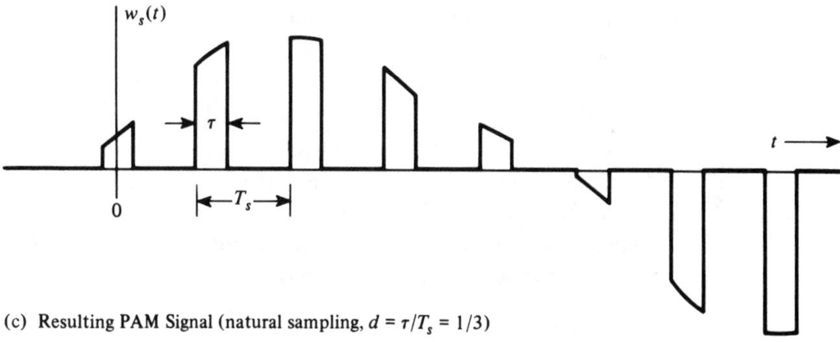

(c) Resulting PAM Signal (natural sampling, $d = \tau/T_s = 1/3$)

FIGURE 3-1 PAM signal with natural sampling.

where $f_s = 1/T_s$, $\omega_s = 2\pi f_s$, the duty cycle of $S(t)$ is $d = \tau/T_s$, and $W(f) = \mathcal{F}[w(t)]$ is the spectrum of the original unsampled waveform.

Proof. Taking the Fourier transform of (3-1), we get

$$W_s(f) = W(f) * S(f) \tag{3-4}$$

Since $S(t)$ is periodic, it may be represented by the Fourier series

$$S(t) = \sum_{n=-\infty}^{\infty} c_n e^{jn\omega_s t} \tag{3-5a}$$

where

$$c_n = d\frac{\sin n\pi d}{n\pi d} \tag{3-5b}$$

Thus

$$S(f) = \mathcal{F}[S(t)] = \sum_{n=-\infty}^{\infty} c_n\delta(f - nf_s) \tag{3-6}$$

and (3-4) becomes

$$W_s(f) = W(f) * \left(\sum_{n=-\infty}^{\infty} c_n\delta(f - nf_s)\right) = \sum_{n=-\infty}^{\infty} c_n W(f) * \delta(f - nf_s)$$

or

$$W_s(f) = \sum_{n=-\infty}^{\infty} c_n W(f - nf_s) \tag{3-7}$$

which becomes (3-3) upon substituting (3-5b).

The PAM waveform with natural sampling is relatively easy to generate since it only requires the use of an analog switch that is readily available in CMOS hardware (e.g., the 4016-quad bilateral switch). This is shown in Fig. 3-2, where the associated waveforms $w(t)$, $S(t)$, and $w_s(t)$ are as illustrated in Fig. 3-1.

The spectrum of the PAM signal with natural sampling is given by (3-3) in terms of the spectrum of the analog input waveform. This is illustrated in Fig. 3-3 for the case of an input waveform that has a rectangular spectrum where the duty cycle of the switching waveform is $d = \tau/T_s = \frac{1}{3}$ and the sampling rate is $f_s = 4B$. For this example, where $d = \frac{1}{3}$, the PAM spectrum is zero for $\pm 3f_s$, $\pm 6f_s$, and so on, since the spectrum in these harmonic bands is nulled out by the $(\sin x)/x$ function. From the figure it is obvious that the original analog waveform, $w(t)$, can be recovered from the PAM signal, $w_s(t)$, by passing the PAM signal through a low-pass filter where the cutoff frequency is $B < f_{\text{cutoff}} < f_s - B$. It is also obvious that the analog waveform cannot be recovered without

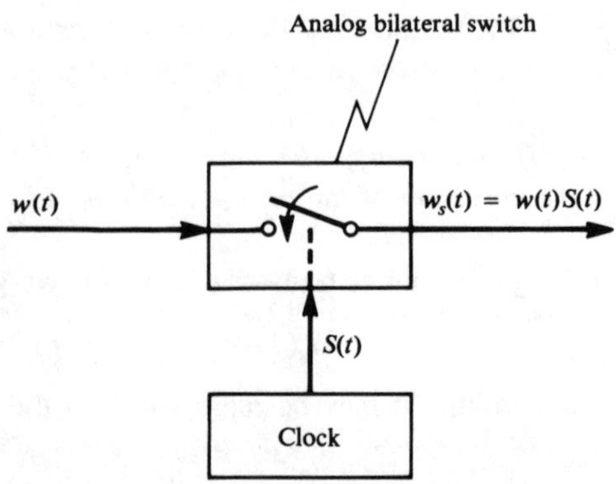

FIGURE 3-2 Generation of PAM with natural sampling (gating).

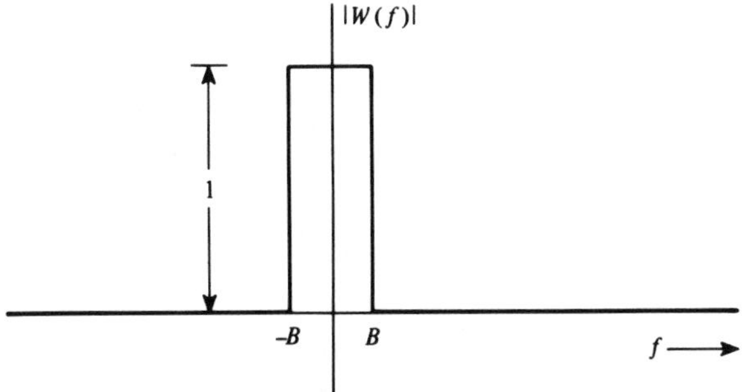

(a) Magnitude Spectrum of Input Analog Waveform

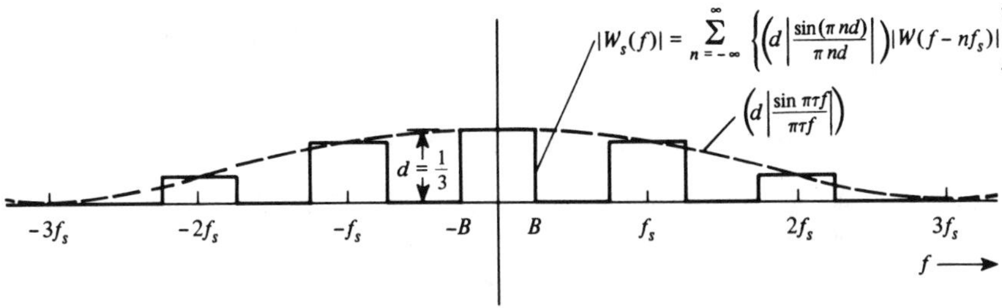

(b) Magnitude Spectrum of PAM (natural sampling) with $d = 1/3$ and $f_s = 4B$

FIGURE 3-3 Spectrum of a PAM waveform with natural sampling.

distortion unless $f_s \geq 2B$, since otherwise spectral components in harmonic bands (of f_s) would overlap. This is another illustration of the Nyquist sampling rate requirement. If the analog signal is *undersampled* ($f_s < 2B$), the effect of spectral overlapping is called *aliasing*. This results in a recovered analog signal that is distorted when compared to the original waveform. In practice, physical signals are usually considered to be time limited, so that (as we found in Chapter 2) they cannot be absolutely bandlimited. Consequently, there will be some aliasing in a PAM signal. Usually, if we prefilter the analog signal before it is introduced to the PAM circuit, we do not have to worry about this problem; however, the effect of aliasing noise has been studied [Spilker, 1977].

It can also be shown (see Prob. 3-4) that the analog waveform may be recovered from the PAM signal by using *product* detection, as shown in Fig. 3-4. Here the PAM signal is multiplied with a sinusoidal signal of frequency $\omega_o = n\omega_s$. This shifts the frequency band of the PAM signal that was centered about nf_s to baseband (i.e., $f = 0$) at the multiplier output. We will study the product detector in Chapter 4. For $n = 0$ this is identical to low-pass filtering, just discussed. Of course, you might ask: Why do you need to go to the trouble of using a product detector when a simple low-pass filter will work? The answer is: Because of noise on the PAM signal. Noise due to power supply hum or noise due to mechanical circuit vibration might fall in the band about $n = 0$, and other bands might be relatively noise free. In this case, a product detector might be used to get around the noise problem.

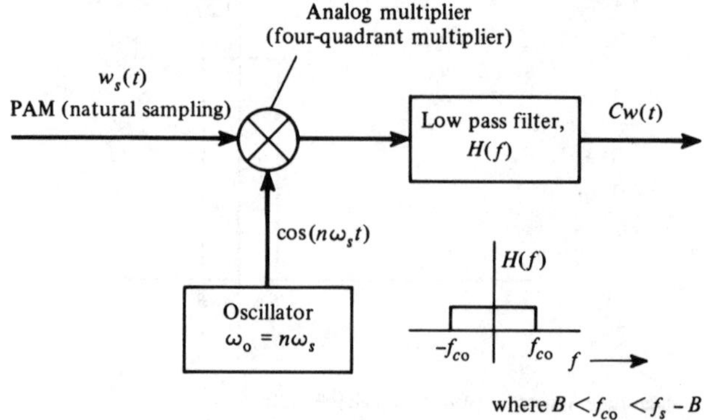

FIGURE 3-4 Demodulation of a PAM signal (naturally sampled).

Instantaneous Sampling (Flat-Top PAM)

Analog waveforms may also be converted to pulse signaling by the use of *flat-top* signaling with instantaneous sampling, as shown in Fig. 3-5. This is a generalization of the impulse train sampling technique that was studied in Sec. 2-7.

DEFINITION: If $w(t)$ is an analog waveform bandlimited to B hertz, the *instantaneous sampled* PAM signal is given by

$$w_s(t) = \sum_{k=-\infty}^{\infty} w(kT_s)h(t - kT_s) \tag{3-8}$$

where $h(t)$ denotes the sampling-pulse shape, and for flat-top sampling the pulse shape is

$$h(t) = \Pi\left(\frac{t}{\tau}\right) = \begin{cases} 1, & |t| < \tau/2 \\ 0, & |t| > \tau/2 \end{cases} \tag{3-9}$$

where $\tau \leq T_s = 1/f_s$ and $f_s \geq 2B$.

THEOREM: *The spectrum for a flat-top PAM signal is*

$$W_s(f) = \frac{1}{T_s} H(f) \sum_{k=-\infty}^{\infty} W(f - kf_s) \tag{3-10}$$

where

$$H(f) = \mathcal{F}[h(t)] = \tau\left(\frac{\sin \pi\tau f}{\pi\tau f}\right) \tag{3-11}$$

This type of PAM signal is said to consist of instantaneous samples since $w(t)$ is sampled at $t = kT_s$ and the sample values $w(kT_s)$ determine the amplitude of the flat-top rectangular pulses, as demonstrated in Fig. 3-5c. The flat-top PAM signal could be generated by using a sample-and-hold type of electronic circuit.

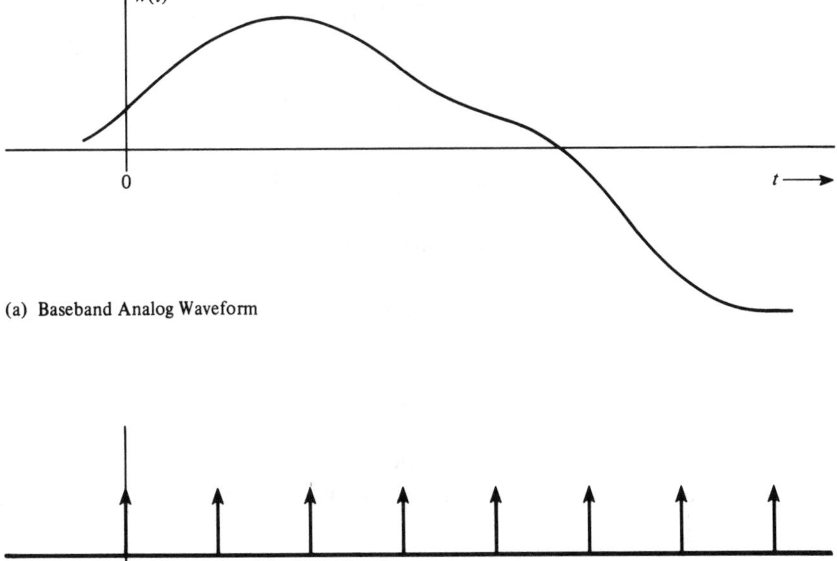

(a) Baseband Analog Waveform

(b) Impulse Train Sampling Waveform

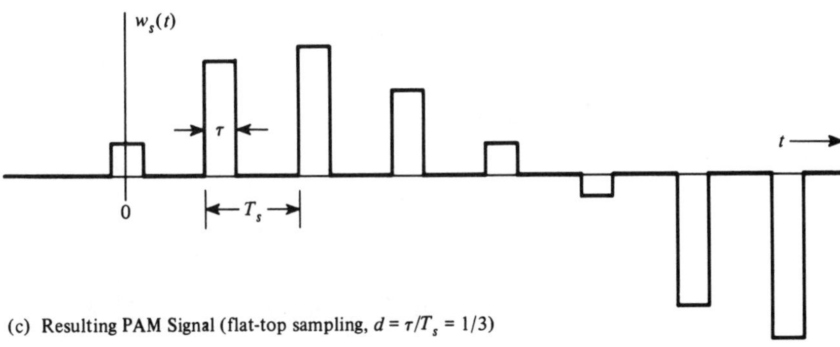

(c) Resulting PAM Signal (flat-top sampling, $d = \tau/T_s = 1/3$)

FIGURE 3-5 PAM signal with flat-top sampling.

Another pulse shape, rather than the rectangular shape, could be used in (3-8), but in this case the resulting PAM waveform would not be flat-topped. The only restriction on the pulse shape is that it be zero outside $T_s/2$ [i.e., $h(t) = 0$ for $|t| \geq T_s/2$] so that the pulses do not overlap. Note that if the $h(t)$ is of the $(\sin x)/x$ type with overlapping pulses, then (3-8) becomes identical to the sampling theorem of (2-158).

Proof. The spectrum for flat-top PAM can be obtained by taking the Fourier transform of (3-8). First, rewrite (3-8), using a more convenient form involving the convolution operation.

$$w_s(t) = \sum_k w(kT_s)h(t) * \delta(t - kT_s)$$

$$= h(t) * \sum_k w(kT_s)\, \delta(t - kT_s)$$

or

$$w_s(t) = h(t) * \left[w(t) \sum_k \delta(t - kT_s) \right]$$

The spectrum is

$$W_s(f) = H(f)\left[W(f) * \sum_k e^{-j2\pi f k T_s} \right] \tag{3-12}$$

But the sum of the exponential functions is equivalent to a Fourier series expansion (in the frequency domain), where the periodic function is an impulse train. That is,

$$\frac{1}{T_s} \sum_k \delta(f - kf_s) = \frac{1}{T_s} \sum_{n=-\infty}^{\infty} c_n e^{j(2\pi n T_s)f} \tag{3-13a}$$

where

$$c_n = \frac{1}{f_s} \int_{-f_s/2}^{f_s/2} \left[\sum_k \delta(f - kf_s) \right] e^{-j2\pi n T_s f}\, df = \frac{1}{f_s} \tag{3-13b}$$

Using (3-13a), we find that (3-12) becomes

$$W_s(f) = H(f)\left[W(f) * \frac{1}{T_s} \sum_k \delta(f - kf_s) \right]$$

$$= \frac{1}{T_s} H(f) \left[\sum_k W(f) * \delta(f - kf_s) \right]$$

which reduces to (3-10).

The spectrum of the flat-topped PAM signal is illustrated in Fig. 3-6 for the case of an analog input waveform that has a rectangular spectrum. The analog signal may be recovered from the flat-topped PAM signal by the use of a low-pass filter. However, it is noted that there is some high-frequency loss in the recovered analog waveform due to the filtering effect, $H(f)$, caused by the flat-top pulse shape. This loss, if significant, can be reduced by decreasing τ or by using some additional gain at the high frequencies in the low-pass filter transfer function. In this case, the low-pass filter would be called an *equalization* filter. The pulse width τ is also called the *aperture* since τ/T_s determines the gain of recovered analog signal, which is small if τ is small relative to T_s. It is also

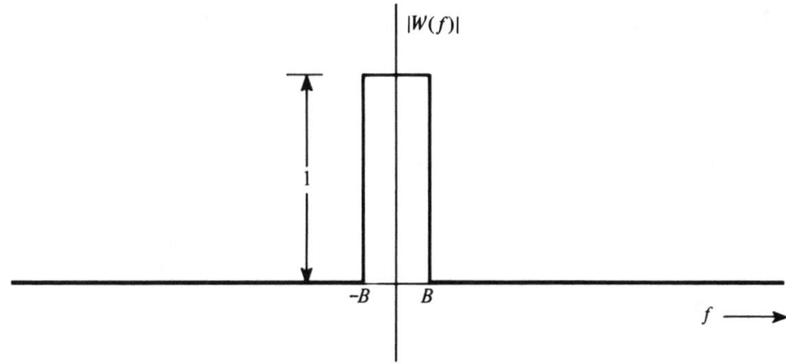

(a) Magnitude Spectrum of Input Analog Waveform

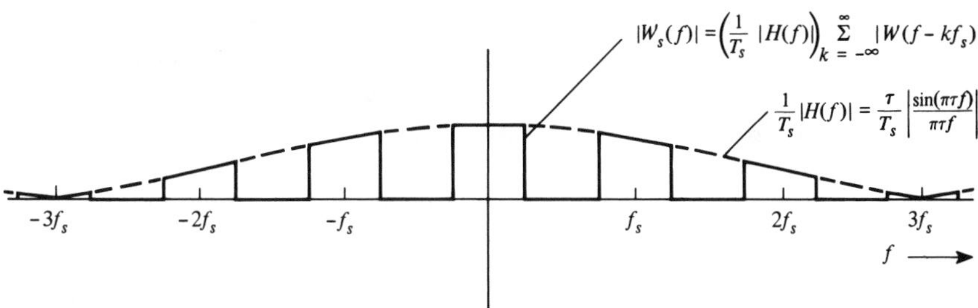

(b) Magnitude Spectrum of PAM (flat-top sampling), $\tau/T_s = 1/3$ and $f_s = 4B$

FIGURE 3-6 Spectrum of a PAM waveform with flat-top sampling.

possible to use product detection, similar to that shown in Fig. 3-4, except that now some prefilter might be needed before the multiplier (to make the spectrum flat in a band centered on $f = nf_s$), to compensate for the spectral loss due to the aperture effect.

In general, pulse modulation systems require *synchronization signals* at the receiver and at regenerative repeaters in order to process the pulse signal. As we have seen, PAM signals can be demodulated without a synchronization signal, although this is not true for most other pulse-type signals.

The transmission of either naturally or instantaneously sampled PAM over a channel requires a very wide frequency response because of the narrow pulse width. This imposes stringent requirements on the magnitude and phase response of the channel. The bandwidth required is much larger than that of the original analog signal, and the noise performance of the PAM system can never be better than that achieved by transmitting the analog signal directly. Consequently, PAM is not very good for long-distance transmission. It does provide a means for converting an analog signal to a PCM signal (as discussed later). PAM also provides a means for breaking a signal into time slots so that multiple PAM signals carrying information from different sources can be interleaved to transmit all of the information over a single channel. This is called time-division multiplexing and will be studied in Sec. 3-8.

3-3 Pulse Code Modulation

Definition: *Pulse code modulation* (PCM) is essentially analog-to-digital conversion of a special type where the information contained in the instantaneous samples of an analog signal is represented by digital words in a serial bit stream.

If we assume that each of the digital words has n binary digits, there are $M = 2^n$ unique code words that are possible, each code word corresponding to a certain amplitude level. However, each sample value from the analog signal can be any one of an infinite number of levels, so that the digital word that represents the amplitude closest to the actual sampled value is used. This is called *quantizing*. That is, instead of using the exact sample value of the analog waveform $w(kT_s)$, the sample is replaced by the closest allowed value, where there are M allowed values, and each allowed value corresponds to one of the code words. Other popular types of analog-to-digital conversion, such as delta modulation (DM) and differential pulse code modulation (DPCM), are discussed in later sections.

PCM is very popular because of the many advantages it offers. Some of these advantages are as follows.

- Relatively inexpensive digital circuitry may be used extensively in the system.
- PCM signals derived from all types of analog sources (audio, video, etc.) may be merged with data signals (e.g., from digital computers) and transmitted over a common high-speed digital communication system. This merging is called time-division multiplexing and is discussed in detail in a later section.
- In long-distance digital telephone systems requiring repeaters, a *clean* PCM waveform can be regenerated at the output of each repeater, where the input consists of a noisy PCM waveform. However, the noise at the input may cause bit errors in the regenerated PCM output signal.
- The noise performance of a digital system can be superior to that of an analog system. In addition, the probability of error for the system output can be reduced even further by the use of appropriate coding techniques such as adding parity check bits that may be used to detect and correct certain types of errors.

These advantages usually outweigh the main disadvantage of PCM: a much wider bandwidth than that of the corresponding analog signal.

Sampling, Quantizing, and Encoding

The PCM signal is generated by carrying out three basic operations: sampling, quantizing, and encoding (see Fig. 3-7). The sampling operation generates a flat-top PAM signal.

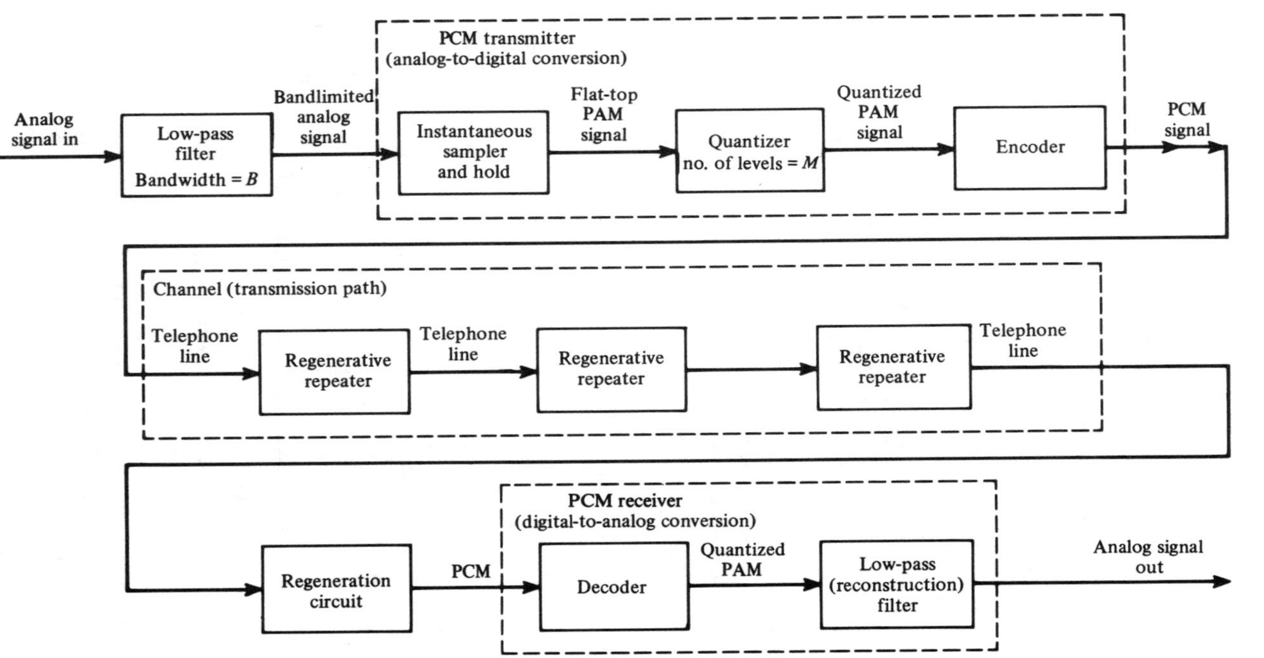

FIGURE 3-7 PCM transmission system.

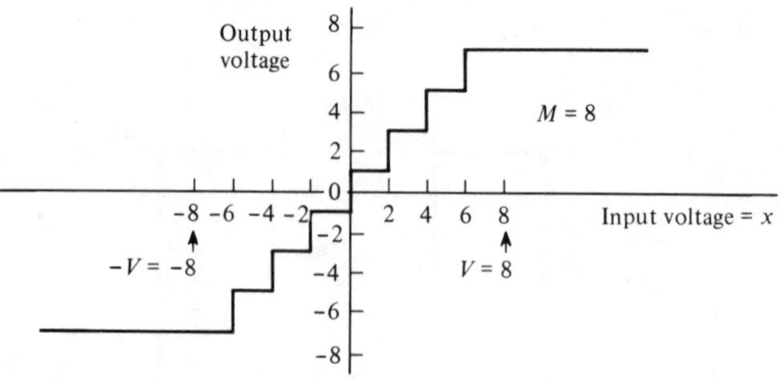

(a) Quantizer Output-Input Characteristics

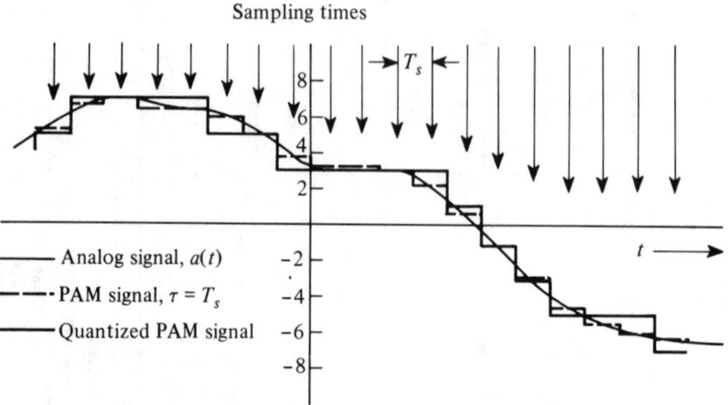

(b) Analog Signal, Flat-top PAM Signal, and Quantized PAM Signal

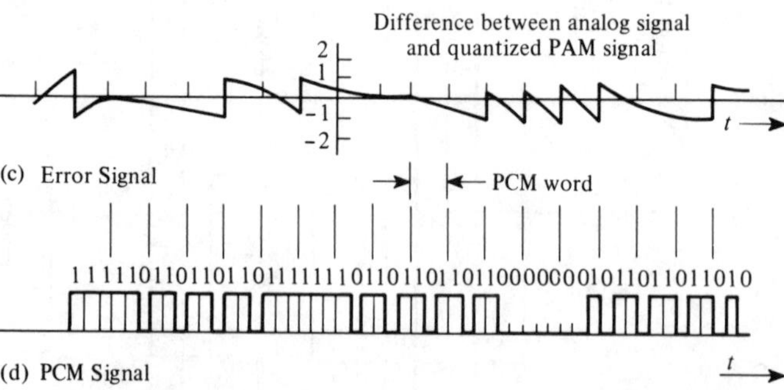

(c) Error Signal

(d) PCM Signal

FIGURE 3-8 Illustration of waveforms in a PCM system.

The quantizing operation is illustrated in Fig. 3-8 for the $M = 8$ level case. This quantizer is said to be *uniform* since all of the steps are of equal size. Since we are approximating the analog sample values by using a finite number of levels ($M = 8$ in this illustration), *error* is introduced into the recovered output analog signal because of the quantizing effect. The error waveform is illustrated in Fig. 3-8c. The quantizing error consists of the difference between the analog signal at the sampler input and the output of the quantizer. Note that the peak value of the error (± 1) is one half of the quantizer step size (2). If we sample at the Nyquist

TABLE 3-1 Three-Bit Gray Code for $M = 8$ Levels

Quantized Sample Voltage	Gray Code Word (PCM Output)	
+7	110	
+5	111	
+3	101	
+1	100	
		Mirror image except for sign bit
−1	000	
−3	001	
−5	011	
−7	010	

rate ($2B$) or faster, and there is negligible channel noise, there will still be noise, called *quantizing noise,* on the recovered analog waveform due to this error. The quantizing noise can also be thought of as a round-off error. In Sec. 7-7 statistics on this quantizing noise are evaluated, and a formula for the PCM system signal-to-noise ratio is developed. The quantizer output is a *quantized* PAM signal.

The PCM signal is obtained from the quantized PAM signal by encoding each quantized sample value into a digital word. It is up to the system designer to specify the exact code word that will represent a particular quantized level. If a Gray code of Table 3-1 is used, the resulting PCM signal is shown in Fig. 3-8d, where the PCM word for each quantized sample is strobed out of the encoder by the next clock pulse. The Gray code was chosen because it has only one bit change for each step change in the quantized level. Consequently, single errors in the received PCM code word will cause minimum errors in the recovered analog level, provided that the sign bit is not in error.

In concluding this section it should be pointed out that we have described PCM systems that represent the quantized analog sample values by *binary* code words. In general, of course, it is possible to represent the quantized analog samples by digital words using other than base 2. That is, for base q, the number of quantized levels allowed is $M = q^n$, where n is the number of q base digits in the code word. We will not pursue this topic since base q circuits, where $q > 2$, are not common in practice because they are usually too expensive in comparison with binary ($q = 2$) digital circuits.

Practical PCM Circuits

In terms of circuitry, three popular techniques are used to implement the analog-to-digital converter (ADC) encoding operation. These are the *counting or ramp, serial or successive approximation,* and *parallel or flash* encoders.

In the counting encoder, at the same time that the sample is taken, a ramp generator is energized and a binary counter is started. The output of the ramp generator is continuously compared to the sample value; when the value of the ramp becomes equal to the sample value, the binary value of the counter is read. This count is taken to be the PCM word. The binary counter and the ramp generator are then reset to zero and are ready to be reenergized at the next sampling time. This technique requires only a few components, but the speed of this type of ADC is usually limited by the speed of the counter. The Intersil ICL7126 CMOS ADC integrated circuit uses this technique.

The serial encoder compares the value of the sample with trial quantized values. Successive trials depend on whether the past comparator outputs are positive or negative. The trial values are chosen first in large steps and then in small steps so that the process will converge rapidly. The trial voltages are generated by a series of voltage dividers that are configured by (on-off) switches. These switches are controlled by digital logic. After the process converges, the value of the switch settings is read out as the PCM word. This technique requires more precision components (for the voltage dividers) than the ramp technique. The speed of the feedback ADC technique is determined by the speed of the switches. The National Semiconductor ADC0804 8-bit ADC uses this technique.

The parallel encoder uses a set of parallel comparators with reference levels that are the permitted quantized values. The sample value is fed into all of the parallel comparators simultaneously. The high or low level of the comparator outputs determines the binary PCM word with the aid of some digital logic. This is a fast ADC technique but requires more hardware than the other two methods. The RCA CA3318 8-bit ADC integrated circuit is an example of this technique.

All of the integrated circuits listed as examples above have parallel digital outputs that correspond to the digital word that represents the analog sample value. For generation of PCM, the parallel output (digital word) needs to be converted to serial form for transmission over a two-wire channel. This is accomplished by using a parallel-to-serial converter integrated circuit, which is also usually known as a *serial-input-output* (SIO) chip. The SIO chip includes a shift register that is set to contain the parallel data (from, usually, 8 or 16 input lines). Then the data are shifted out of the last stage of the shift register bit-by-bit onto a single output line to produce the serial format. Furthermore, the SIO chips are usually full duplex; that is, they have two sets of shift registers, one that functions for data flowing in each direction. One shift register converts parallel input data to serial output data for transmission over the channel, and, simultaneously, the other shift register converts received serial data from another input to parallel data that are available at another output. Three types of SIO chips are available: the *universal asynchronous receiver/transmitter* (UART), the *universal synchronous receiver/transmitter* (USRT), and the *universal synchronous/asynchronous receiver transmitter* (USART). The UART transmits and receives asynchronous serial data, the USRT transmits and receives synchronous serial data, and the USART combines both a UART and a USRT on one chip. (See Sec. 3-8 for a discussion of asynchronous and synchronous serial data lines.)

At the receiving end the PCM signal is decoded back into an analog signal by using a digital-to-analog converter (DAC) chip. If the DAC chip has a parallel data input, the received serial PCM data are first converted to a parallel form using a SIO chip as described above. The parallel data are then converted to an approximation of the analog sample value by the DAC chip. This conversion is usually accomplished by using the parallel digital word to set the configuration of electronic switches on a resistive current (or voltage) divider network so that the analog output is produced. This is called a *multiplying* DAC since the "analog" output voltage is directly proportional to the divider reference voltage multiplied by the value of the digital word. The Motorola MC1408 and the National Semiconductor DAC0808 8-bit DAC chips are examples of this technique. The DAC chip outputs samples of the quantized analog signal that approximates the analog sample values. Consequently, as the DAC chip is clocked, it generates a quantized PAM signal. This may be smoothed by a low-pass reconstruction filter to produce the analog output as illustrated in Fig. 3-7.

Bandwidth of PCM

Our study of the dimensionality theorem in Chapter 2 shows that the bandwidth of (serial) PCM signals depends on the bit rate and the pulse shape. From Fig. 3-8, the bit rate is

$$R = nf_s \tag{3-14}$$

where n is the number of bits in the PCM word ($M = 2^n$) and f_s is the sampling rate. For no aliasing we require $f_s \geq 2B$ where B is the bandwidth of the analog signal (that is to be converted to the PCM signal). Using (2-176), (2-178), and (3-14), we see that the dimensionality theorem shows that the bandwidth of the PCM waveform is bounded by

$$B_{\text{PCM}} \geq \tfrac{1}{2}R = \tfrac{1}{2}nf_s \tag{3-15a}$$

where equality is obtained if a $(\sin x)/x$ type of pulse shape is used to generate the PCM waveform. The exact spectrum for the PCM waveform will depend on the pulse shape that is used as well as on the type of line encoding. This will be studied in Sec. 3-4. For example, if one uses a rectangular pulse shape with polar NRZ line coding, Fig. 3-12(b) shows that the first null bandwidth is simply

$$B_{\text{PCM}} = R = nf_s \quad \text{(first null bandwidth)} \tag{3-15b}$$

Table 3-2 presents a tabulation of this result for the case of the minimum sampling rate, $f_s = 2B$. Note that the dimensionality theorem of (3-15a) demonstrates that the bandwidth of the PCM signal has a lower bound given by

$$B_{\text{PCM}} \geq nB \tag{3-15c}$$

where $f_s \geq 2B$ and B is the bandwidth of the corresponding analog signal. Thus, for reasonable values of n, *the bandwidth of the PCM signal will be significantly larger than the bandwidth of the corresponding analog signal* that it represents. For the example shown in Fig. 3-8 where $n = 3$, the PCM signal bandwidth will be at least three times wider than that of the corresponding analog signal. Furthermore, if the bandwidth of the PCM signal is reduced by improper filtering, or by passing the PCM signal through a system that has a poor frequency response,

TABLE 3-2 Performance of a PCM System with Uniform Quantizing and No Channel Noise

Number of Quantizer Levels Used, M	Length of the PCM Word, n (bits)	Bandwidth of PCM Signal (First Null Bandwidth)[a]	Recovered Analog Signal Power-to-Quantizing Noise Power Ratios (dB)	
			$(S/N)_{pk\ out}$	$(S/N)_{out}$
2	1	$2B$	10.8	6.0
4	2	$4B$	16.8	12.0
8	3	$6B$	22.8	18.1
16	4	$8B$	28.9	24.1
32	5	$10B$	34.9	30.1
64	6	$12B$	40.9	36.1
128	7	$14B$	46.9	42.1
256	8	$16B$	52.9	48.2
512	9	$18B$	59.0	54.2
1,024	10	$20B$	65.0	60.2
2,048	11	$22B$	71.0	66.2
4,096	12	$24B$	77.0	72.2
8,192	13	$26B$	83.0	78.3
16,384	14	$28B$	89.1	84.3
32,768	15	$30B$	95.1	90.3
65,536	16	$32B$	101.1	96.3

[a] B is the absolute bandwidth of the input analog signal.

the filtered pulses will be elongated (stretched in width) so that pulses corresponding to any one bit will smear into adjacent bit slots. If this condition becomes too serious, it will cause errors in the detected bits. This pulse smearing effect is called *intersymbol interference* (ISI). The filtering specifications for no ISI are discussed in Sec. 3-5.

Effects of Noise

The analog signal that is recovered at the PCM system output is corrupted by noise. Two main effects produce this noise or distortion:

1. Quantizing noise that is caused by the M-step quantizer at the PCM transmitter.
2. Bit errors in the recovered PCM signal. The bit errors are caused by *channel noise* as well as improper channel filtering, which causes ISI.

In addition, if the input analog signal that is sampled is not strictly bandlimited, there will be some aliasing noise on the recovered analog signal [Spilker, 1977].

In Chapter 7, under certain assumptions, it will be shown that the recovered analog *peak* signal power to total *average* noise power is given by[†]

$$\left(\frac{S}{N}\right)_{\text{pk out}} = \frac{3M^2}{1 + 4(M^2 - 1)P_e} \tag{3-16a}$$

and the *average* signal power to the average noise power is

$$\left(\frac{S}{N}\right)_{\text{out}} = \frac{M^2}{1 + 4(M^2 - 1)\, P_e} \tag{3-16b}$$

where M is the number of quantized levels used in the PCM system and P_e is the probability of bit error in the recovered binary PCM signal at the receiver DAC before it is converted back into an analog signal. In Chapter 7, P_e is evaluated for many different types of baseband and bandpass digital transmission systems. In Chapter 1 it was shown how channel coding could be used to correct some of the bit errors and, consequently, reduce P_e. Therefore, in many practical systems P_e is negligible. If we assume that there are no bit errors due to channel noise (i.e., $P_e = 0$) and no ISI, then, from (3-16a), the peak S/N due only to quantizing errors is

$$\left(\frac{S}{N}\right)_{\text{pk out}} = 3M^2 \tag{3-17a}$$

and from (3-16b) the average S/N due only to quantizing errors is

$$\left(\frac{S}{N}\right)_{\text{out}} = M^2 \tag{3-17b}$$

Numerical values for these S/N ratios are given in Table 3-2.

To realize these S/N ratios, one critical assumption is that the peak-to-peak level of the analog waveform at the input to the PCM encoder is set to the design level of the quantizer. For example, referring to Fig. 3-8a, this corresponds to the input traversing the range $-V$ to $+V$ volts where $V = 8$ volts is the design level of the quantizer. Equations (3-16) and (3-17) were derived for waveforms with equally likely values, such as a triangle waveshape, that have a peak-to-peak value of $2V$ and an rms value of $V/\sqrt{3}$, where V is the design peak level of the quantizer.

From a practical viewpoint, the quantizing noise at the output of the PCM decoder can be categorized into four types depending on the operating conditions. The four types are overload noise, random noise, granular noise, and hunting noise. As discussed earlier, the level of the analog waveform at the input of the PCM encoder needs to be set so that its peak level does not exceed the design peak of V volts. If the peak input does exceed V, then the recovered

[†]This derivation is postponed until Chapter 7 because a knowledge of statistics is needed to carry it out.

analog waveform at the output of the PCM system will have flat-tops near the peak values. This produces *overload noise*. The flat-tops are easily seen on an oscilloscope, and the recovered analog waveform sounds distorted since the flat-topping produces unwanted harmonic components. For example, this type of distortion can be heard on a PCM telephone system (e.g., the AT&T and SLC lines as described in Chapter 5) when there are high levels such as dial tones, busy signals, or off-hook signals.

The second type of noise, *random noise*, is produced by the random quantization errors in the PCM system under normal operating conditions when the input level is properly set. This type of condition is assumed in (3-17). Random noise has a "white" hissing sound. If the input level is not sufficiently large, the S/N will deteriorate from that given by (3-17) to one that will be described later by (3-28a) and (3-28b); however, the quantizing noise will still be more or less random.

If the input level is reduced further to a relatively small value with respect to the design level, the error values are not equally likely from sample to sample, and the noise has a harsh sound resembling gravel being poured into a barrel. This is called *granular noise*. This type of noise can be randomized (noise power decreased) by increasing the number of quantization levels, and, consequently, increasing the PCM bit rate. Alternatively, granular noise can be reduced by using a nonuniform quantizer, such as the μ-law or A-law quantizers that are described in the next section.

The fourth type of quantizing noise that may occur at the output of a PCM system is *hunting noise*. It can occur when the input analog waveform is nearly constant, including when there is no signal (i.e., zero level). For these conditions the sample values at the quantizer output (see Fig. 3-8) can oscillate between two adjacent quantization levels, causing an undesired sinusoidal type tone of frequency $\frac{1}{2}f_s$ at the output of the PCM system. Hunting noise can be reduced by filtering out the tone or by designing the quantizer so that there is no vertical step at the "constant" value of the inputs—such as at zero volts input for the no signal case. For the no signal case, the hunting noise is also called *idle channel noise*. Idle channel noise can be reduced by using a horizontal step at the origin of the quantizer output-input characteristic instead of a vertical step as shown in Fig. 3-8a.

Recalling that $M = 2^n$, we may express Eqs. (3-17a) and (3-17b) in decibels as

$$\left(\frac{S}{N}\right)_{dB} = 6.02n + \alpha \tag{3-18}$$

where n is the number of bits in the PCM word, $\alpha = 4.77$ for the peak S/N, and $\alpha = 0$ for the average S/N. This equation—called the *6 dB rule*—points out the significant performance characteristic for PCM: *an additional 6-dB improvement in S/N is obtained for each bit added to the PCM word*. This is illustrated in Table 3-2. Equation (3-18) is valid for a wide variety of assumptions (such as various types of input waveshapes and quantization characteristics), although the value of α will depend on these assumptions [Jayant and Noll, 1984]. Of course, it is assumed that there are no bit errors and that the input signal level is large enough to range over a significant number of quantizing levels.

EXAMPLE 3-1 DESIGN OF A PCM SYSTEM

Assume that an analog voice-frequency signal, which occupies a band from 300 to 3400 Hz, is to be transmitted over a binary PCM system. The minimum sampling frequency would be $2 \times 3.4 = 6.8$ kHz. In practice the signal would be oversampled, and in the United States a sampling frequency of 8 kHz is the standard used for voice-frequency signals in telephone communication systems. Assume that each sample value is represented by 7 information bits plus 1 parity bit; then the bit rate of the PCM signal is

$$R = (f_s \text{ samples/sec})(n \text{ bits/sample})$$
$$= (8\text{k samples/sec})(8 \text{ bits/sample}) = 64 \text{ kbits/sec} \qquad (3\text{-}19)$$

Since the signaling is binary (i.e., two possible levels), each dimension[†] (or symbol) is carrying one bit of data, so that the baud rate is equal to the bit rate, or $D = R = 64\text{k}$ symbols/sec. Referring to the dimensionality theorem (3-15a), we realize that the theoretically minimum absolute bandwidth of the PCM signal is

$$(B)_{\text{min}} = \tfrac{1}{2}D = 32 \text{ kHz} \qquad (3\text{-}20)$$

and this is realized if the PCM waveform consists of $(\sin x)/x$ pulse shapes. If rectangular pulse shaping is used, the absolute bandwidth is infinity and the first null bandwidth, as shown in Fig. 3-12b, is

$$(B)_{\text{null}} = R = \frac{1}{T_b} = 64 \text{ kHz} \qquad (3\text{-}21)$$

That is, we require a bandwidth of 64 kHz to transmit this digital voice PCM signal where the bandwidth of the original analog voice signal was, at most, 4 kHz. (In Sec. 3-5 we will see that the bandwidth of this binary PCM signal may be reduced somewhat by filtering without introducing ISI.) Using (3-17a), we observe that the peak signal-to-quantizing noise power ratio is

$$\left(\frac{S}{N}\right)_{\text{pk out}} = 3(2^7)^2 = 46.9 \text{ dB} \qquad (3\text{-}22)$$

where $M = 2^7$. Note that the inclusion of a parity bit does not affect the quantizing noise. However, coding (parity bits) may be used to decrease the decoded errors caused by channel noise or ISI. In this example, these effects were assumed to be negligible since P_e was assumed to be zero.

[†] See Fig. 2-11 for an illustration of a three-dimensional signal expansion. In this example we have an eight-dimensional signal expansion.

The performance of a PCM system for the most optimistic case (i.e., $P_e = 0$) can easily be obtained as a function of M, the number of quantizing steps used. The results are shown in Table 3-2. In obtaining the (S/N), no parity bits were used in the PCM word.

One might use this table to look at the design requirements in a proposed PCM system. For example, high-fidelity enthusiasts are turning to digital audio recording techniques. Here PCM signals are recorded instead of the analog audio signal to produce superb sound reproduction. For a dynamic range of 90 dB, it is seen that at least 15-bit PCM words would be required. Furthermore, if the analog signal had a bandwidth of 20 kHz, the first null bandwidth for rectangular bit-shape PCM would be 2×20 kHz $\times 15 = 600$ kHz. With an allowance for some oversampling and a wider bandwidth to minimize ISI, the bandwidth needed would be around 1.5 MHz. Consequently, video-type tape recorders are needed to record and reproduce high-quality digital audio signals. Although this type of recording technique might seem ridiculous at first, it is realized that expensive high-quality analog recording devices are hard pressed to reproduce a dynamic range of 70 dB. Thus digital audio is one way to achieve improved performance. This is being proven in the marketplace with the popularity of the digital compact disk (CD). The CD uses a 16-bit PCM word and a sampling rate of 44.1 kHz on each stereo channel [Miyaoka, 1984; Peek, 1985]. Reed–Solomon coding with interleaving is used to correct burst errors that occur as a result of scratches and fingerprints on the compact disk.

★ Nonuniform Quantizing: μ-Law and A-Law Companding

In practice, voice analog signals are more likely to have amplitude values near zero than at the extreme peak values allowed. For example, when digitizing voice signals, if the peak value allowed is 1 V, weak passages may have voltage levels on the order of 0.1 V (20 dB down). For signals such as these with nonuniform amplitude distribution, the granular quantizing noise will be a serious problem if the step size is not reduced for amplitude values near zero and increased for extremely large values. This is called *nonuniform quantizing* since a variable step size is used. An example of a nonuniform quantizing characteristic is shown in Fig. 3-9a.

The effect of nonuniform quantizing can be obtained by first passing the analog signal through a compression (nonlinear) amplifier and then into the PCM circuit that uses a uniform quantizer. In the United States a μ-*law* type of *compression* characteristic is used. It is defined [Smith, 1957] by

$$|w_2(t)| = \frac{\ln(1 + \mu|w_1(t)|)}{\ln(1 + \mu)} \tag{3-23}$$

where the allowed peak values of $w_1(t)$ are ± 1 (i.e., $|w_1(t)| \leq 1$), μ is a positive constant that is a parameter, and ln is the natural logarithm. This compression characteristic is shown in Fig. 3-9b, for several values of μ, and it is noted that

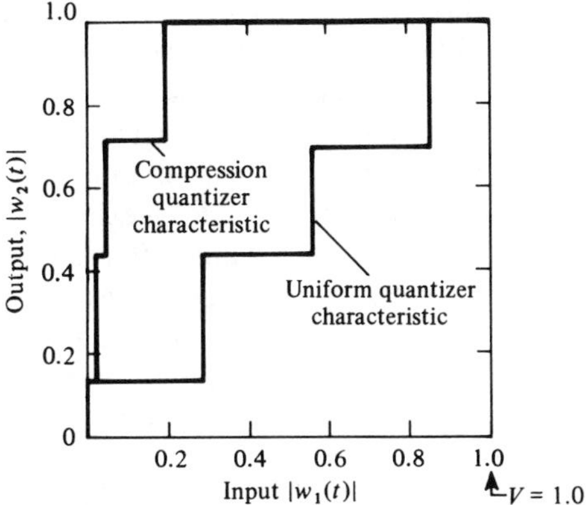

(a) $M = 8$ Quantizer Characteristic

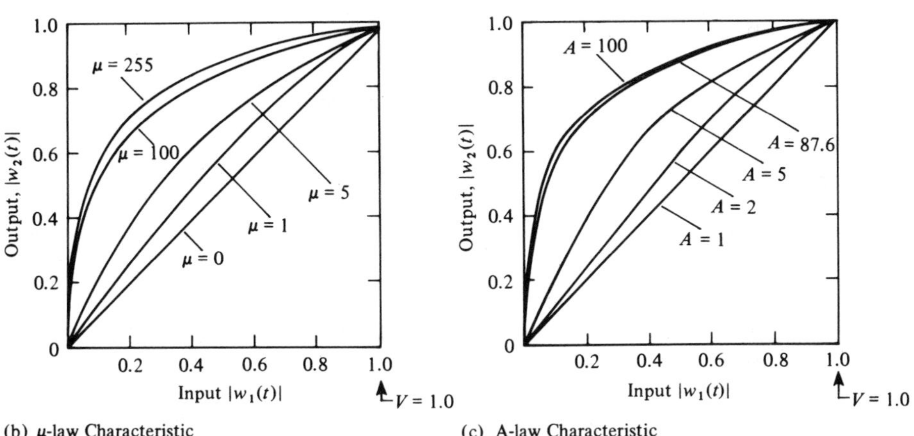

(b) μ-law Characteristic

(c) A-law Characteristic

FIGURE 3-9 Compression characteristics (first quadrant shown).

$\mu = 0$ corresponds to linear amplification (uniform quantization overall). In the United States, Canada, and Japan, the telephone companies use a $\mu = 255$ compression characteristic in their PCM systems [Dammann, McDaniel, and Maddox, 1972].

Another compression law, used mainly in Europe, is the *A-law* characteristic. It is defined [Cattermole, 1969] by

$$|w_2(t)| = \begin{cases} \dfrac{A|w_1(t)|}{1 + \ln A}, & 0 \le |w_1(t)| \le \dfrac{1}{A} \\[2ex] \dfrac{1 + \ln(A|w_1(t)|)}{1 + \ln A}, & \dfrac{1}{A} \le |w_1(t)| \le 1 \end{cases} \tag{3-24}$$

where $|w_1(t)| \leq 1$ and A is a positive constant. The A-law compression characteristic is shown in Fig. 3-9c. The typical value for A is 87.6.

When compression is used at the transmitter, *expansion* (i.e., decompression) must be used at the receiver output to restore signal levels to their correct relative values. The *expandor* characteristic is the inverse of the compression characteristic, and the combination of a compressor and an expandor is called a *compandor*.

We will now demonstrate that logarithmic companding gives a relatively constant output S/N over a broad range of input rms values, whereas a uniform quantizer (no companding) does not. Let us first study the results for the uniform quantizer. From Fig. 3-8c it is seen that the (quantizing) noise power remains the same, regardless of the rms value of the input waveform, as long as the input waveform has values that span several quantizing steps and does not exceed the peak design limit. Based on these observations, the expression for the average S/N, (3-17b), which is valid only for the maximum allowable input level, can be generalized to the case of an arbitrary input level. Consequently, the average S/N out of a PCM system with uniform quantization is

$$\left(\frac{S}{N}\right)_{\text{out}} = M^2\left(\frac{\sqrt{3}\,x_{\text{rms}}}{V}\right)^2 \tag{3-25}$$

where V is the peak design level of the quantizer and x_{rms} is the rms value of the input signal. It is assumed that x takes on equally likely values, and, consequently, $x_{\text{rms}} = x_p/\sqrt{3}$ where x_p is the peak value. This result is shown in Fig. 3-10 for the case of $M = 256$ (8-bit) quantization. The graph terminates at $20 \log(x_{\text{rms}}/V) = -4.77$ dB, which is the relative level when the peak value of the triangular input waveform is equal to V (the design peak value of the uniform quantizer).

For μ-law companding, the average output S/N is [Lathi, 1983]

$$\left(\frac{S}{N}\right)_{\text{out}} = \frac{M^2}{[\ln(1+\mu)]^2}\left[\frac{(\sqrt{3}\,x_{\text{rms}}/V)^2}{(x_{\text{rms}}/V)^2 + (2/\mu)(\langle|x|\rangle/V) + 1/\mu^2}\right] \tag{3-26}$$

For comparison purposes, assume that $M = 256$, $\mu = 255$, and we have a triangular waveform at the input. This gives $\langle|x|\rangle = \sqrt{3}\,x_{\text{rms}}/2$, and (3-26) becomes

$$\left(\frac{S}{N}\right)_{\text{out}} = 6394\left[\frac{(x_{\text{rms}}/V)^2}{(x_{\text{rms}}/V)^2 + 0.00679(x_{\text{rms}}/V) + 1.54 \times 10^{-5}}\right] \tag{3-27}$$

This result is also plotted in Fig. 3-10. Approximately the same result is obtained for other types of waveshapes. From this figure, the advantages of a nonuniform quantizer are obvious. The output S/N for a PCM system with no companding deteriorates rapidly as the input level decreases, but the output S/N for a PCM system with $\mu = 255$ companding remains around 35 dB over a 40-dB range of input level variation (8-bit encoding). Using $M = 2^n$, we observe that both (3-25) and (3-26) reduce to the form of (3-18), namely,

$$\left(\frac{S}{N}\right)_{\text{dB}} = 6.02n + \alpha \tag{3-28a}$$

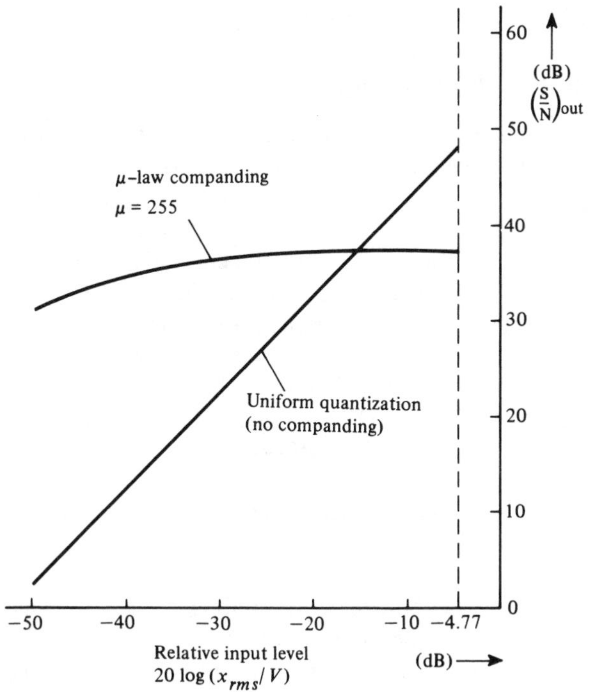

FIGURE 3-10 Output S/N of 8-bit PCM systems with and without companding.

where

$$\alpha = 4.77 - 20 \log(V/x_{\text{rms}}) \qquad \text{uniform quantizing} \qquad (3\text{-}28b)$$

and for sufficiently large input levels

$$\alpha \approx 4.77 - 20 \log[\ln(1 + \mu)] \qquad \mu\text{-law companding} \qquad (3\text{-}28c)$$

and [Jayant and Noll, 1984]

$$\alpha \approx 4.77 - 20 \log[1 + \ln A] \qquad A\text{-law companding} \qquad (3\text{-}28d)$$

Notice that α is a function of the input level for the uniform quantizing (no companding) case but is relatively constant for μ-law and A-law companding. The ratio V/x_{rms} is called the *loading factor*. The input level is often set for a loading factor of 4 = 12 dB to ensure that the overload quantizing noise will be negligible. In practice this gives $\alpha = -7.3$ for the case of uniform encoding as compared to $\alpha = 0$ that was obtained for the ideal conditions associated with (3-17b). All of these results give a 6-dB increase in the signal to quantizing noise ratio for each bit added to the PCM code word.

3-4 Digital Signaling Formats

Binary Line Coding

Binary 1's and 0's such as in PCM signaling may be represented in various serial-bit signaling formats called *line codes*. Some of the more popular line codes are shown in Fig. 3-11.† There are two major categories: *return-to-zero* (RZ) and *nonreturn-to-zero* (NRZ). With RZ coding, the waveform returns to a zero-volt level for a portion (usually one half) of the bit interval. The waveforms for the line code may be further classified according to the rule that is used to assign voltage levels to represent the binary data. Some examples follow.

Unipolar Signaling. In positive logic unipolar signaling the binary 1 is represented by a high level ($+A$ volts) and a binary 0 by a zero level. This type of signaling is also called *on-off keying*.

Polar Signaling. Binary 1's and 0's are represented by equal positive and negative levels.

Bipolar (Pseudoternary) Signaling. Binary 1's are represented by alternately positive or negative values. The binary 0 is represented by a zero level. The term *pseudoternary* refers to the use of three encoded signal levels to represent two-level (binary) data. This is also called *alternate mark inversion* (AMI) signaling.

Manchester Signaling. Each binary 1 is represented by a positive half-bit period pulse followed by a negative half-bit period pulse. Similarly, a binary 0 is represented by a negative half-bit period pulse followed by a positive half-bit period pulse. This type of signaling is also called *split-phase encoding*.

Later in this book we will often use shortened notations. *Unipolar NRZ will be denoted simply by unipolar, polar NRZ by polar, and bipolar RZ as bipolar*. It should also be pointed out that, unfortunately, the term *bipolar* has two different conflicting definitions as used in practice. The meaning is usually made clear by the context in which it is used: (1) In the space communication industry polar NRZ is sometimes called bipolar NRZ, or simply bipolar (this meaning will not be used in this book); and (2) in the telephone industry the term *bipolar* denotes pseudoternary signaling (used in this book) as in T1 bipolar RZ signaling described in Sec. 3-8.

The line codes shown in Fig. 3-11 are also known by alternate names [Deffeback and Frost, 1971; Sklar, 1988]. For example, polar NRZ is also called

†Strictly speaking, punched paper tape is a storage medium, not a line code. However, it is included for historical purposes to illustrate the origin of the terms *mark* and *space*. In punched paper tape the binary 1 corresponds to a hole (mark), and a binary 0 corresponds to no hole (space). In current loop signaling (see Appendix C), the presence of loop current denotes a mark and the absence of loop current denotes a space.

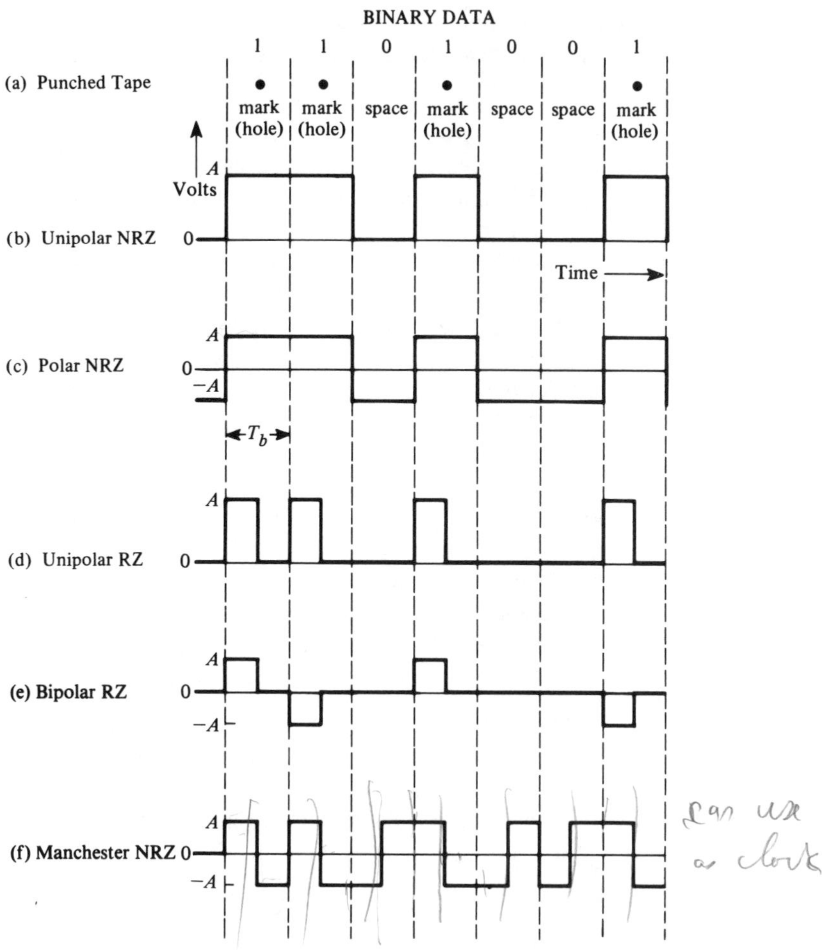

FIGURE 3-11 Binary signaling formats.

NRZ-L, where L denotes the normal logical level assignment. Bipolar RZ is also called RZ-AMI, where AMI denotes alternate mark space inversion. Bipolar NRZ is called NRZ-M, where M denotes inversion on mark. Negative logic bipolar NRZ is called NRZ-S, where S denotes inversion on space. Manchester NRZ is called Bi-φ-L for bi-phase with normal logic level.

Other line codes, too numerous to list here, can also be found in the literature. For example, the bipolar (pseudoternary) type may be extended into codes of several subclasses where the binary 0, which is usually represented by a zero-volt level, is replaced by some appropriate pulse sequence if there are n or more consecutive binary 0's where n is some designated positive integer [Benedetto, Biglieri, and Castellani, 1987]. This technique helps avoid long sequences of zero-volt levels so that bit synchronizers will be easier to build. It also allows the shape of the spectrum to be controlled (as discussed in the next section on PSD of codes) and permits the easy detection of single-bit errors.

Each of the line codes shown in Fig. 3-11 has advantages and disadvantages depending on its application. Some of the desirable properties of a line code are as follows:

- *Self-synchronization*. There is enough timing information built into the code so that bit synchronizers can be designed to extract the timing or clock signal. A long series of binary 1's and 0's should not cause a problem in time recovery.
- *Low probability of bit error*. Receivers can be designed that will recover the binary data with a low probability of bit error when the input data signal is corrupted by noise or ISI. The ISI problem is discussed in Sec. 3-5, and the effect of noise is covered in detail in Chapter 7.
- *A spectrum that is suitable for the channel*. For example, if the channel is ac coupled, the PSD of the line code signal should be negligible at frequencies near zero. In addition the signal bandwidth needs to be sufficiently small compared to the channel bandwidth so that ISI will not be a problem.
- *Transmission bandwidth*. It should be as small as possible.
- *Error detection capability*. It should be possible to implement this feature easily by the addition of channel encoders and decoders, or it should be incorporated into the line code.
- *Transparency*. The data protocol and line code are designed so that *every possible sequence of data* is faithfully and *transparently* received.

A protocol is not transparent if certain words are reserved for control sequences so that, for example, a certain word instructs the printer to give a carriage return. This feature causes a problem when a random data file (such as a machine language file) is transferred over the link since some of the words in the file might be control character sequences. These sequences would be intercepted by the receiving system, and the defined action would be carried out, instead of passing the word on to the intended destination. In addition, a code is not transparent if some sequence will result in a loss of clocking signal out of the bit synchronizer at the receiver. The bipolar format is not transparent since a string of zeros will result in a loss of the clocking signal.

The particular type of waveform used for binary signaling depends on the application. For example, unipolar or polar signaling is usually found in most digital circuits, but these waveshapes may have a nonzero dc level; furthermore, the dc level depends on the exact data being represented. On the other hand, bipolar and Manchester-type signaling will always have a zero dc level regardless of the data being represented. A more detailed discussion of the advantages and disadvantages of each signal format will be given after their spectra have been derived.

Power Spectra of Line Codes

The power spectral density (PSD) can be evaluated by using either deterministic or stochastic techniques. This was discussed first in Chapter 1 and later illustrated in Example 2-15. To evaluate the PSD using the deterministic technique, the waveform for a line code that results from a particular data sequence is used. The approximate PSD is then evaluated by using (2-66) or, if the line code is periodic, by (2-126). Work Prob. 3-21 to apply this deterministic approach. Alternatively, the PSD may be evaluated by using the stochastic approach that is developed in Chapter 6. The stochastic approach of Chapter 6 will be applied to obtain PSD of line codes that are shown in Fig. 3-11. The stochastic approach is used because it gives the PSD for the line code with a random data sequence (instead of that for a particular data sequence). Consequently, the PSD obtained by the stochastic approach is typical of those obtained for most deterministic calculations (where one calculation is obtained for each possible data sequence). The power spectral density (PSD) for the various line codes illustrated in Fig. 3-11 will now be evaluated.

Unipolar and Polar NRZ Signaling. The PSD for unipolar and polar signaling follows directly from the theory developed in Chapter 6. From (6-60) the PSD for a polar signal is

$$\mathcal{P}(f) = \frac{1}{T_b}|F(f)|^2 \tag{3-29}$$

where $F(f)$ is the Fourier transform of the basic pulse shape, $f(t)$. For rectangular NRZ pulse shapes, the Fourier transform pair is

$$f(t) = \Pi\left(\frac{t}{T_b}\right) \leftrightarrow F(f) = T_b \frac{\sin \pi f T_b}{\pi f T_b} \tag{3-30}$$

When we combine (3-29) and (3-30), the PSD for the polar signal with unit amplitude and rectangular pulse shape becomes

$$\mathcal{P}_{\text{polar}}(f) = T_b\left(\frac{\sin \pi f T_b}{\pi f T_b}\right)^2 \tag{3-31}$$

where the bit rate is $R = 1/T_b$ and the total signal power is unity. This PSD is plotted in Fig. 3-12b. The polar signal has the disadvantage of having a large PSD near dc. On the other hand, polar signals are relatively easy to generate, although positive and negative power supplies are required unless special-purpose integrated circuits are used that generate dual supply voltages from a single supply. The probability of bit error performance is superior to that of other signaling methods (see Fig. 7-14).

The unipolar signal consists of a polar signal plus a dc level. Thus the PSD for the unipolar signal is similar to that of the polar signal except for an added delta function at $f = 0$. The peak amplitude of the unipolar signal needs to be set to $\sqrt{2}$ for a total power of unity. This will allow it to be compared with the other signaling formats on an equal (unity) power basis. The PSD for the unipolar signal is then

$$\mathcal{P}_{\text{unipolar}}(f) = \tfrac{1}{2}T_b\left(\frac{\sin \pi f T_b}{\pi f T_b}\right)^2 + \tfrac{1}{2}\delta(f) \tag{3-32}$$

This is shown in Fig. 3-12a. The obvious disadvantage is the waste of power due to the dc level and the fact that the spectrum is not approaching zero near dc. Of course, dc coupled circuits are needed for this type of signaling. The advantages of unipolar signaling are that it is easy to generate (i.e., TTL and CMOS circuits) and it requires the use of only one power supply.

★ **Unipolar RZ.** The PSD for the unipolar RZ signal is more difficult to obtain and requires the use of the general formula for the PSD of digital signals. This is developed in Chapter 6. From (6-70a) the general result is

$$\mathcal{P}(f) = \frac{|F(f)|^2}{T_b}\left[R(0) + 2\sum_{k=1}^{\infty} R(k)\cos 2\pi k f T_b\right] \tag{3-33a}$$

where $F(f)$ is the spectrum of the pulse shape, T_b is the pulse width, and $R(k)$ is the autocorrelation of the data. This autocorrelation is given by

$$R(k) = \sum_{i=1}^{I} (a_n a_{n+k})_i P_i \tag{3-33b}$$

where a_n and a_{n+k} are the (voltage) levels of the data pulses at the nth and $n + k$th bit positions, respectively, and P_i is the probability of having the ith $a_n a_{n+k}$ product. For (unipolar) RZ, possible levels for the a's are $+A$ and 0 volts. For $k = 0$ the possible products of $a_n a_n$ are $A \times A = A^2$ and $0 \times 0 = 0$, and consequently, $I = 2$. For random data, the probability of having A^2 is $\frac{1}{2}$ and the probability of having 0 is $\frac{1}{2}$, so that

$$R(0) = \sum_{i=1}^{2} (a_n a_n)_i P_i = A^2 \cdot \tfrac{1}{2} + 0 \cdot \tfrac{1}{2} = \tfrac{1}{2}A^2 \tag{3-34}$$

For $k \neq 0$ there are $I = 4$ possibilities for the product values: $A \times A$, $A \times 0$, $0 \times A$, 0×0. They all occur with a probability of $\frac{1}{4}$. Thus, for $k \neq 0$,

$$R(k) = \sum_{i=1}^{4} (a_n a_{n+k}) P_i = A^2 \cdot \tfrac{1}{4} + 0 \cdot \tfrac{1}{4} + 0 \cdot \tfrac{1}{4} + 0 \cdot \tfrac{1}{4} = \tfrac{1}{4}A^2 \tag{3-35}$$

For RZ signaling the pulse duration is $T_b/2$. Using a rectangular pulse shape $f(t) = \Pi(2t/T_b)$, we obtain the corresponding pulse spectrum

$$F(f) = \frac{T_b}{2}\left[\frac{\sin(\pi f T_b/2)}{\pi f T_b/2}\right] \tag{3-36}$$

Substituting these results into (3-33a) and letting $A = 2$ for a unity power signal, we find that the PSD becomes

$$\begin{aligned}\mathcal{P}_{\text{RZ}}(f) &= \frac{T_b}{4}\left[\frac{\sin(\pi f T_b/2)}{\pi f T_b/2}\right]^2\left[2 + 2\sum_{k=1}^{\infty}\cos(2\pi k f T_b)\right] \\ &= \frac{T_b}{4}\left[\frac{\sin(\pi f T_b/2)}{\pi f T_b/2}\right]^2\left[1 + \sum_{k=-\infty}^{\infty} e^{jk\omega T_b}\right]\end{aligned} \tag{3-37}$$

$\mathcal{P}_{unipolar}(f)$

T_b

Weight $= \frac{1}{2}$

$0.5T_b$

(a) Unipolar NRZ

0 $0.5R$ R $1.5R$ $2R$ $f \longrightarrow$

$\mathcal{P}_{polar}(f)$

T_b

$0.5T_b$

(b) Polar NRZ

0 $0.5R$ R $1.5R$ $2R$ $f \longrightarrow$

$\mathcal{P}_{RZ}(f)$

$0.5T_b$

Weight $= \frac{1}{4}$

Weight $= 0.1$

(c) Unipolar RZ

0 $0.5R$ R $1.5R$ $2R$ $f \longrightarrow$

$\mathcal{P}_{bipolar}(f)$

$0.5T_b$

(d) Bipolar RZ

0 $0.5R$ R $1.5R$ $2R$ $f \longrightarrow$

$\mathcal{P}_{manchester}(f)$

$0.5T_b$

(e) Manchester NRZ

0 $0.5R$ R $1.5R$ $2R$ $f \longrightarrow$

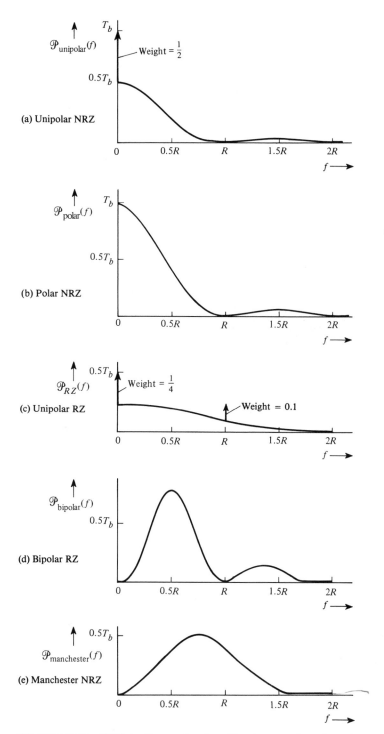

FIGURE 3-12 PSD for line codes (positive frequencies shown).

But[†]

$$\sum_{k=-\infty}^{\infty} e^{jk2\pi f T_b} = \frac{1}{T_b} \sum_{n=-\infty}^{\infty} \delta\left(f - \frac{n}{T_b}\right) \tag{3-38}$$

Then the PSD for the unipolar RZ with unity power reduces to

$$\mathcal{P}_{\text{RZ}}(f) = \frac{T_b}{4}\left[\frac{\sin(\pi f T_b/2)}{\pi f T_b/2}\right]^2 \left[1 + \frac{1}{T_b} \sum_{n=-\infty}^{\infty} \delta\left(f - \frac{n}{T_b}\right)\right] \tag{3-39}$$

This result is shown in Fig. 3-12c. As expected, the first null bandwidth is twice that for unipolar or polar signaling since the pulse width is half as wide. Note that there is a discrete (impulse) term at $f = R$. Consequently, this periodic component can be used for recovery of the clock signal. One disadvantage of this scheme is that it requires 3 dB more signal power than polar signaling for the same probability of bit error (see Chapters 7 and 8). Moreover, the spectrum is not negligible for frequencies near dc.

★ **Bipolar RZ Signaling.** The PSD for a bipolar signal can also be obtained from (3-33). The permitted values for a_n are $+A$, $-A$, and 0 where binary 1's are represented by alternating $+A$ and $-A$ values and a binary 0 is represented by $a_n = 0$. For $k = 0$ the products $a_n a_n$ are A^2 and 0 where each of these products occurs with a probability of $\frac{1}{2}$. Thus,

$$R(0) = \frac{A^2}{2} \tag{3-40}$$

For $k = 1$ (the adjacent bit case) and the data sequences (1,1), (1,0), (0,1), and (0,0), the possible $a_n a_{n+1}$ products are $-A^2$, 0, 0, 0. Each of these sequences occurs with a probability of $\frac{1}{4}$. Consequently,

$$R(1) = \sum_{i=1}^{4} (a_n a_{n+1})_i P_i = -\frac{A^2}{4} \tag{3-41}$$

For $k > 1$, the bits being considered are not adjacent, and the $a_n a_{n+k}$ products are $\pm A^2$, 0, 0, 0; these occur with a probability of $\frac{1}{4}$. Then

$$R(k > 1) = \sum_{i=1}^{5} (a_n a_{n+k})_i P_i = A^2 \cdot \tfrac{1}{8} - A^2 \cdot \tfrac{1}{8} = 0 \tag{3-42}$$

Using these autocorrelation values and substituting (3-36) into (3-33a) with $A = 2$ for a unity power signal, we obtain the PSD[‡]

$$\mathcal{P}_{\text{bipolar}}(f) = \frac{T_b}{4}\left[\frac{\sin(\pi f T_b/2)}{\pi f T_b/2}\right]^2 (2 - 2\cos 2\pi f T_b)$$

[†]Equation (3-38) is known as the Poisson sum formula as derived in (2-115).

[‡]Here it is assumed that the RZ pulse width is $T_b/2$. For an NRZ pulse width of T_b, replace $T_b/2$ by T_b in (3-36).

The PSD for the bipolar RZ signal reduces to

$$\mathcal{P}_{\text{bipolar}}(f) = T_b\left[\frac{\sin(\pi f T_b/2)}{\pi f T_b/2}\right]^2 \sin^2 \pi f T_b \qquad (3\text{-}$$

This PSD is plotted in Fig. 3-12d. Bipolar signaling has a spectral null at dc so that ac coupled circuits may be used in the transmission path.

The clock signal can be easily extracted from the bipolar waveform by converting the bipolar format to a unipolar RZ format by the use of full-wave rectification. The resulting (unipolar) RZ signal has a periodic component at the clock frequency (see Fig. 3-12c). Bipolar signals are not transparent. That is, a string of zeros will cause a loss in the clocking signal. This difficulty can be prevented by using ***high-density bipolar n*** (HDBn) signaling where a string of more than n consecutive zeros is replaced by a "filling" sequence that contains some pulses.† The calculation of the PSD for HBDn codes is difficult since the $R(k)$ have to be individually evaluated for large values of k [Davis and Barber, 1973]. There are many other popular line codes. They are too numerous to list here, but a list with appropriate references is available [Bylanski and Ingram, 1976].

Bipolar signals also have single-error detection capabilities built in since a single error will cause a violation of the bipolar line code rule. Any violations can easily be detected by receiver logic.

Some disadvantages of bipolar signals are that the receiver has to distinguish between three levels ($+A$, $-A$, and 0) instead of just two levels in the other signaling formats previously discussed. The bipolar signal also requires approximately 3 dB more signal power than a polar signal for the same probability of bit error. [It has $\frac{3}{2}$ the error of unipolar signaling as described by (7-28).]

★ **Manchester NRZ Signaling.** The Manchester signal uses the pulse shape

$$f(t) = \Pi\left(\frac{t + T_b/4}{T_b/2}\right) - \Pi\left(\frac{t - T_b/4}{T_b/2}\right) \qquad (3\text{-}44)$$

and the resulting pulse spectrum is

$$F(f) = \frac{T_b}{2}\left[\frac{\sin(\pi f T_b/2)}{\pi f T_b/2}\right] e^{j\omega T_b/4} - \frac{T_b}{2}\left[\frac{\sin(\pi f T_b/2)}{\pi f T_b/2}\right] e^{-j\omega T_b/4}$$

or

$$F(f) = jT_b\left[\frac{\sin(\pi f T_b/2)}{\pi f T_b/2}\right] \sin\left(\frac{\omega T_b}{4}\right) \qquad (3\text{-}45)$$

†For example, for HDB3, the filling sequences used to replace $n + 1 = 4$ zeros are the alternating sequences 000V and 100V, where the 1 bit is encoded according to the bipolar rule and the V is a 1 pulse of such polarity as to violate the bipolar rule. The alternating filling sequences are designed so that consecutive V pulses alternate in sign. Consequently, there will be a 0 dc value in the line code and the PSD will have a null at $f = 0$. To decode the HDB3 code, the bipolar decoder has to detect the bipolar violations and to count the number of zeros preceding each violation so that the substituted 1's can be deleted.

…g (3-45) into (3-29), we find the PSD for the Manchester signal:

$$\mathcal{P}_{\text{Manchester}}(f) = T_b\left[\frac{\sin(\pi f T_b/2)}{\pi f T_b/2}\right]^2 \sin^2\left(\frac{\pi f T_b}{2}\right) \tag{3-46}$$

This spectrum is plotted in Fig. 3-12e. The null bandwidth of the Manchester format is twice that of the bipolar bandwidth. However, the Manchester code has a zero dc level on a bit-by-bit basis. Moreover, a string of zeros will not cause a loss of the clocking signal.

In reviewing our study of PSD for digital signals, it should be emphasized that the spectrum is a function of the bit pattern (via the bit autocorrelation) as well as the pulse shape.

Differential Coding

When serial data are passed through many circuits along a communication channel, the waveform is often unintentionally inverted (i.e., data complemented). This result can occur in a twisted-pair transmission line channel just by switching the two leads at a connection point when a line code such as polar signaling is used. (Note that this would not affect the data on a bipolar signal.) To ameliorate this problem, *differential coding,* as illustrated in Fig. 3-13, is often employed. The encoded differential data are generated by

$$e_n = d_n \oplus e_{n-1} \tag{3-47}$$

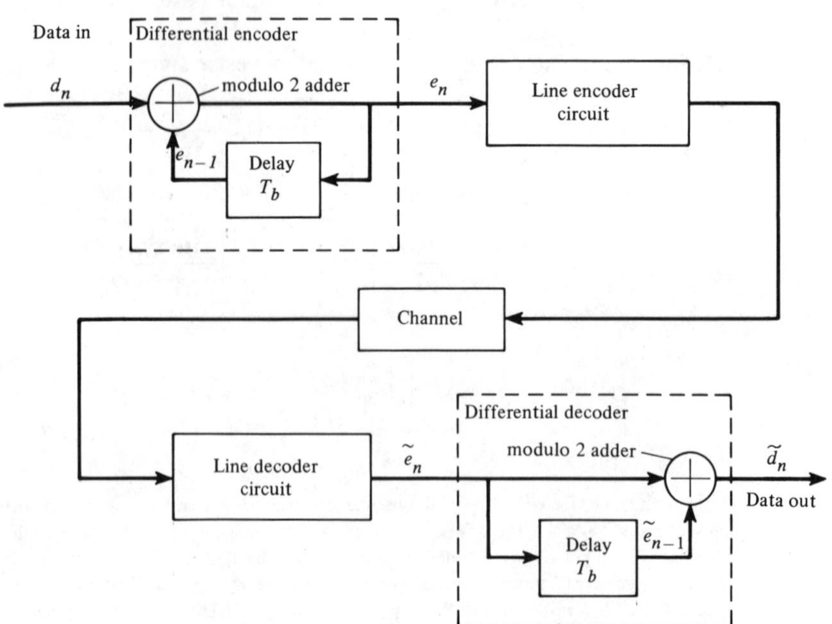

FIGURE 3-13 Differential coding system.

TABLE 3-3 Example of Differential Coding

Encoding									
Input sequence	d_n		1	1	0	1	0	0	1
Encoded sequence	e_n	1	0	1	1	0	0	0	1
Reference digit ⟶		↑							
Decoding (with correct channel polarity)									
Received sequence (correct polarity)	$\tilde{e}_n$	1	0	1	1	0	0	0	1
Decoded sequence	$\tilde{d}_n$		1	1	0	1	0	0	1
Decoding (with inverted channel polarity)									
Received sequence (inverted polarity)	$\tilde{e}_n$	0	1	0	0	1	1	1	0
Decoded sequence	$\tilde{d}_n$		1	1	0	1	0	0	1

where $\oplus$ is a modulo 2 adder or exclusive OR gate (XOR) operation. The received encoded data are decoded by

$$\tilde{d}_n = \tilde{e}_n \oplus \tilde{e}_{n-1} \tag{3-48}$$

where the tilde denotes receiving-end data.

Each digit in the encoded sequence is obtained by comparing the present input bit with the past encoded bit. A binary 1 is encoded if the present input bit and the past encoded bit are of opposite state, and a binary 0 is encoded if the states are the same. This is equivalent to the truth table of an XOR (exclusive OR) gate or a modulo 2 adder. An example of an encoded sequence is shown in Table 3-3, where the beginning reference digit is a binary 1. At the receiver the encoded signal is decoded by comparing the state of adjacent bits. If the present received encoded bit has the same state as the past encoded bit, a binary 0 is the decoded output. Similarly, a binary 1 is decoded for opposite states. As shown in the table, the polarity of the differentially encoded waveform may be inverted without affecting the decoded data. This is a great advantage when the waveform is passed through thousands of circuits in a communication system and the positive sense of the output is lost or changes occasionally as the network changes, such as switching between several data paths. As will be seen in Sec. 4-5, differential coding can be used to eliminate the problem of 180° phase ambiguity that occurs when a Costas loop or a squaring loop is turned on.

Eye Patterns

The effect of channel filtering and channel noise can be seen by observing the received line code on an oscilloscope. The left side of Fig. 3-14 shows received corrupted polar NRZ waveforms for the cases of (a) ideal channel filtering, (b) filtering that produces intersymbol interference (ISI), and (c) noise plus ISI. (ISI is described in Sec. 3-5.) On the right side of the figure, corresponding oscilloscope presentations of the corrupted signal are shown with multiple sweeps

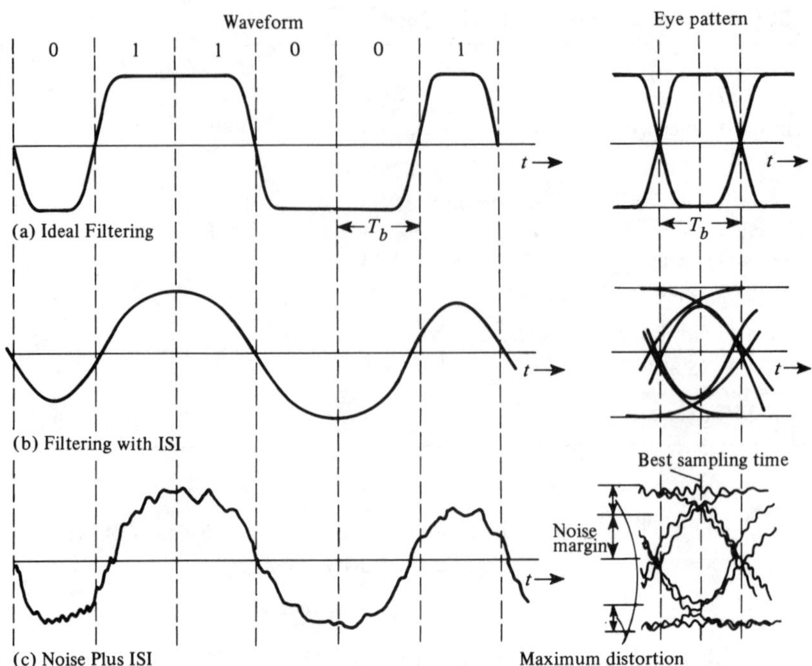

FIGURE 3-14 Distorted polar NRZ waveform and corresponding eye pattern.

where each sweep is triggered by a clock signal and the sweep width is slightly larger than T_b. These displays on the right are called *eye patterns* because they resemble the picture of a human eye. Under normal operating conditions (i.e., for no detected bit errors), the eye will be open. If there is a great deal of noise or ISI, the eye will close; this indicates that bit errors will be produced at the receiver output. The eye pattern provides an excellent way of assessing the quality of the received line code and the ability of the receiver to combat bit errors. As shown in the figure, the eye pattern provides the following information.

- The *timing error* allowed on the sampler at the receiver is given by the width inside the eye, called the *eye opening*. Of course, the preferred time for sampling is at the point where the vertical opening of the eye is the largest.
- The *sensitivity* to timing error is given by the slope of the open eye (evaluated at, or near, the zero-crossing point).
- The *noise margin* of the system is given by the height of the eye opening.
- The maximum distortion is given by the vertical width of the upper (or lower) portion of the eye at the sampling time.

Regenerative Repeaters

When a line code digital signal (such as PCM) is transmitted over a hardwire channel (such as a twisted-pair telephone line), it is attenuated, filtered, and corrupted by noise. Consequently, for long lines, the data cannot be recovered at the receiving end unless repeaters are placed in cascade along the line as illustrated in Fig. 3-7. These repeaters amplify and "clean up" the signal periodically. If the signal was analog instead of digital, only linear amplifiers with appropriate filters could be used since relative amplitude values would need to be preserved. In this case in-band distortion would accumulate from linear repeater to linear repeater. This is one of the disadvantages of analog signaling. However, with digital signaling, nonlinear processing can be used to regenerate a "noise-free" digital signal. This type of nonlinear processing is called a *regenerative repeater*. A simplified block diagram of a regenerative repeater for unipolar NRZ signaling is shown in Fig. 3-15. The amplifying filter increases the amplitude of the low-level input signal to a level that is compatible with the remaining circuitry and filters the signal in such a way as to minimize the effects of channel noise and ISI. (The filter that reduces ISI is called an *equalizing filter* and is discussed in Sec. 3-5.) The bit synchronizer generates a clocking signal at the bit rate that is synchronized so that the amplified distorted signal is sampled at a point where the eye opening is at maximum. (Bit synchronizers are discussed in detail in the next section.) The comparator produces a narrow pulse at its output only when the sample value is above some preselected threshold level, V_T. The pulse width is set by the width of the clocking pulse at the bit synchronizer output. V_T is usually selected to be one half of the expected peak-to-peak variation of the sample values. If the noise and ISI are small, the pulse at the comparator output will occur only when there is a binary 1 (i.e., high level) on the corrupted unipolar NRZ input. Consequently, the comparator—a threshold device—acts as a decision-making device. The monostable multivibrator outputs a pulse of width $T_b = 1/R$ for each narrow-width pulse that occurs at its input where R is the bit rate of the clocking signal that is produced by the bit synchronizer. Thus the unipolar NRZ line code is regenerated "noise free" except for bit errors that are caused when the input noise and ISI alter the sample values

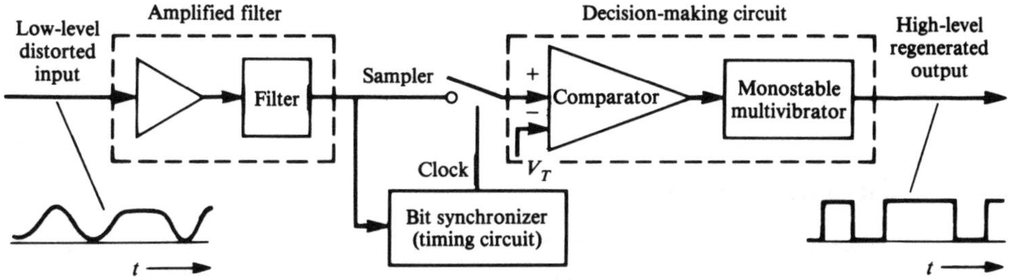

FIGURE 3-15 Regenerative repeater for unipolar NRZ signaling.

sufficiently so that the sample values occur on the wrong side of V_T. Chapter 7 shows how the probability of bit error is influenced by the S/N at the input to the repeater, the filter that is used, and the value of V_T that is selected.

In summary, the regenerative repeater consists of four main functional parts: *filtering with amplification, bit synchronization, sampling,* and *decision making*.

★ In long-distance digital communication systems, many repeaters may be used in cascade, as shown in Fig. 3-7. Of course, the spacing between the repeaters is governed by the path loss of the transmission medium and the amount of noise that is added. A repeater is required when the S/N at a point along the channel becomes lower than the value that is needed to maintain the overall probability-of-bit-error specification. The relationship between the repeater input S/N (or the equivalent energy-per-bit to noise PSD ratio) and the resulting probability of bit error is the topic of Chapter 7. The results are summarized in Fig. 7-14. For design purposes, assume that these results are known from that figure or that the repeater performance characteristics are supplied by the equipment vendor. Assume also that the repeaters are so spaced that each of them has the same probability of bit error, P_e, and that there are m repeaters in the system, including the end receiver. Then, for m repeaters in cascade, the overall probability of bit error, P_{me}, can be evaluated by the use of the binomial distribution from Appendix B. Using (B-33), we see that the probability of i out of the m repeaters producing a bit error is

$$P_i = \binom{m}{i} P_e^i (1 - P_e)^{m-i} \tag{3-49}$$

However, there is a bit error at the system output only when each of an *odd* number of the cascaded repeaters produces a bit error. Consequently, the overall probability of a bit error for m cascaded repeaters is

$$P_{me} = \sum_{\substack{i=1 \\ i=\text{odd}}}^{m} P_i = \sum_{\substack{i=1 \\ i=\text{odd}}}^{m} \binom{m}{i} P_e^i (1 - P_e)^{m-i}$$

$$= mP_e(1 - P_e)^{m-1} + \frac{m(m-1)(m-2)}{3!} P_e^3(1 - P_e)^{m-3} + \cdots \tag{3-50a}$$

Under useful operating conditions, $P_e \ll 1$, so only the first term of this series is significant. Thus the overall probability of bit error is approximated by

$$P_{me} \approx mP_e \tag{3-50b}$$

where P_e is the probability of bit error for a single repeater.

Bit Synchronization

As discussed earlier, synchronization signals are clock-type signals that are necessary within a receiver (or repeater) for detection (or regeneration) of the data from the corrupted input signal. These clock signals have a precise frequency and phase relationship with respect to the *received* input signal, and they are delayed when compared to the clock signals at the transmitter since there is propagation delay through the channel.

Digital communications usually need at least three types of synchronization signals: (1) *bit sync* to distinguish one bit interval from another, as discussed in this section; (2) *frame sync* to distinguish groups of data, as discussed in Sec. 3-8 in regard to time-division multiplexing; and (3) *carrier sync* for bandpass signaling with coherent detection, as discussed in Chapters 4 and 5. Systems are designed so that the sync is derived either directly from the corrupted signal or from a separate channel that is used only to transmit the sync information.

We will concentrate on systems with bit synchronizers that derive the sync directly from the corrupted signal since it is usually not economically feasible to send sync over a separate channel and since it is relatively easy to derive the bit sync for the separate channel case. The complexity of the bit synchronizer circuit depends on the sync properties of the line code. For example, the bit synchronizer for a unipolar RZ code with a sufficient number of alternating binary 1's and 0's is almost trivial since the PSD of that code has a delta function at the bit rate, $f = R$, as shown in Fig. 3-12c. Consequently, the bit sync clock signal can be obtained by passing the received unipolar RZ waveform through a narrowband bandpass filter that is tuned to $f_0 = R = 1/T_b$. This is illustrated by Fig. 3-16 if the square-law device is deleted. Alternatively, as described in Sec. 4-3, a phase-locked-loop (PLL) can be used to extract the sync signal from the unipolar RZ line code by locking the PLL to the discrete spectral line at $f = R$. For a polar NRZ line code, the bit synchronizer is slightly more complicated as shown in Fig. 3-16. Here the filtered polar NRZ waveform (Fig. 3-16b) is converted to a unipolar RZ waveform (Fig. 3-16c) by using a square-law (or, alternatively, a full-wave rectifier) circuit. As described previously, the clock signal is easily recovered by using a filter or a PLL since the unipolar RZ code has a delta function in its spectrum at $f = R$. All of the bit synchronizers discussed thus far use some technique for detecting the spectral line at $f = R$.

Another technique utilizes the symmetry property of the line code itself [Carlson, 1986]. Referring to Fig. 3-14 which illustrates the eye pattern for a polar NRZ code, we realize that a properly filtered line code has a pulse shape that is symmetrical about the optimum clocking (sampling) time, provided that the data are alternating between 1's and 0's. Referring to Fig. 3-17, let $w_1(t)$ denote the filtered polar NRZ line code and let $w_1(\tau_0 + nT_b)$ denote a sample value of the line code at the maximum (positive or negative) of the eye opening where n is an integer, $R = 1/T_b$ is the bit rate, τ is the relative clocking time (i.e., clock phase), and τ_0 is the optimum value corresponding to samples at the maximum of the eye opening. Since the pulse shape of the line code is approximately symmetrical about the optimum clocking time for alternating data,

$$|w_1(\tau_0 + nT_b - \Delta)| \approx |w_1(\tau_0 + nT_b + \Delta)|$$

where τ_0 is the optimum clocking phase and $0 < \Delta < \frac{1}{2}T_b$. $w_1(\tau + nT_b - \Delta)$ is called the *early sample* and $w_1(\tau + nT_b + \Delta)$ is called the *late sample*. These samples can be used to derive the optimum clocking signal as illustrated in the *early–late bit synchronizer* shown in Fig. 3-17. The control voltage, $w_3(t)$, for

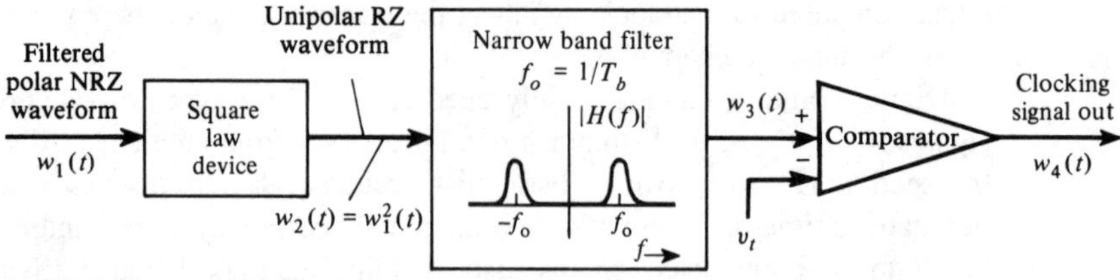

(a) Block Diagram of Bit Synchronizer

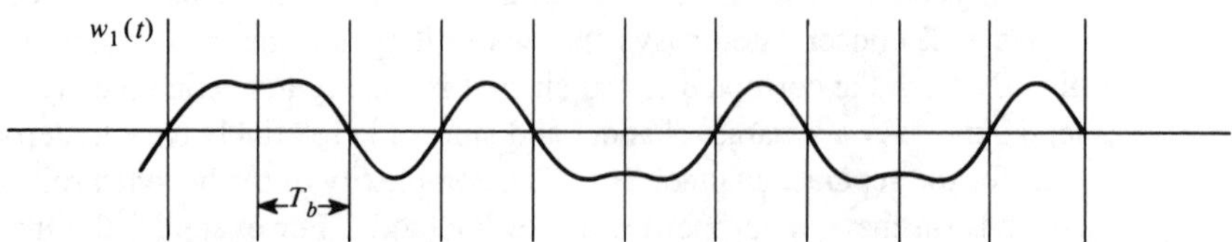

(b) Filtered Polar NRZ Input Waveform

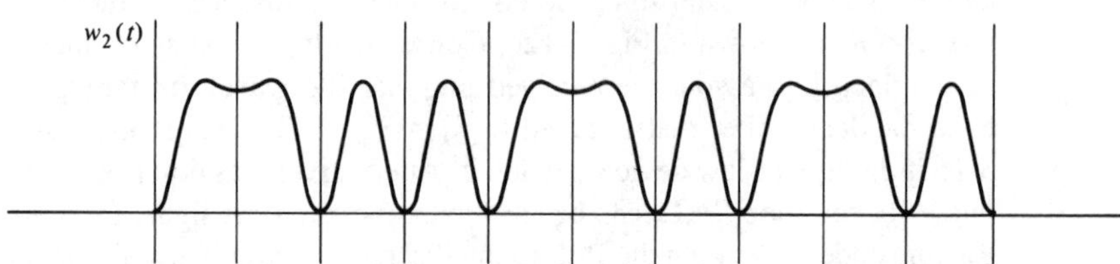

(c) Output of Square Law Device (Unipolar RZ)

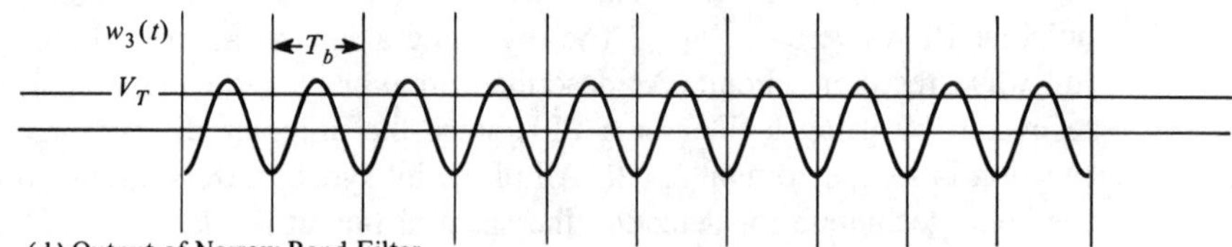

(d) Output of Narrow Band Filter

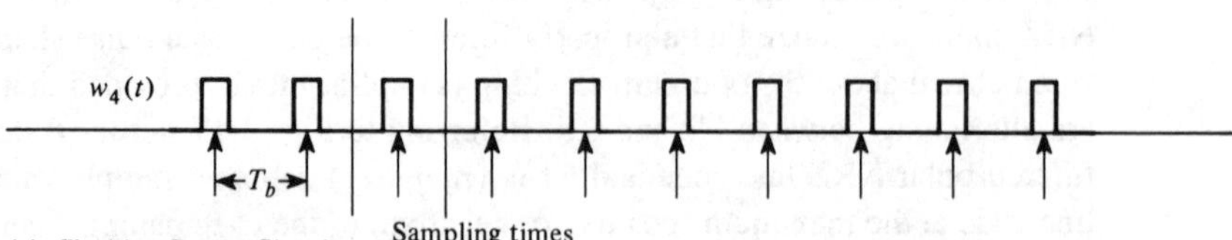

(e) Clocking Output Signal

FIGURE 3-16 Square-law bit synchronizer for polar NRZ signaling.

the voltage-controlled clock (VCC) is a smoothed (averaged) version of $w_2(t)$. That is,

$$w_3(t) = \langle w_2(t) \rangle \tag{3-51a}$$

where

$$w_2(t) = |w_1(\tau + nT_b - \Delta)| - |w_1(\tau + nT_b + \Delta)| \tag{3-51b}$$

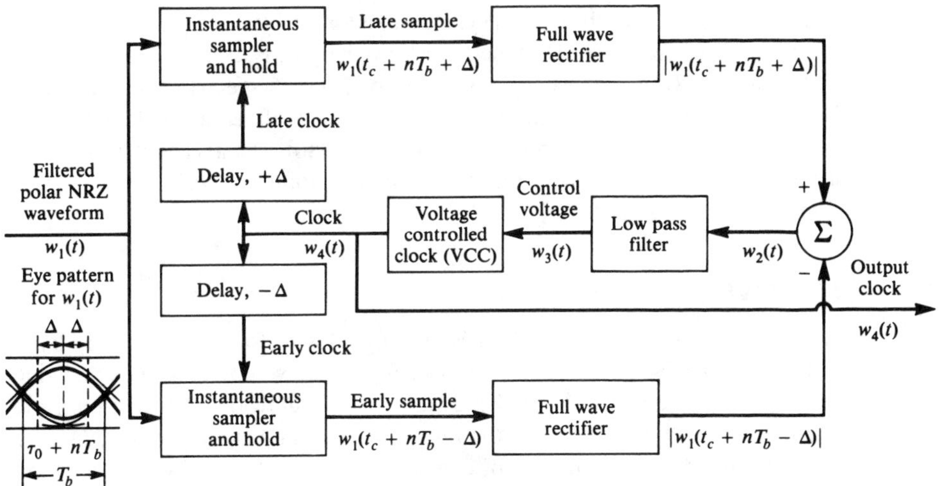

FIGURE 3-17 Early–late bit synchronizer for polar NRZ signaling.

(The averaging operation is needed so that the bit synchronizer will remain synchronized even if the data do not alternate for every bit interval.) If the VCC is producing clocking pulses with the optimum relative clocking time $\tau = \tau_0$ so that samples are taken at the maximum of the eye opening, equation (3-51) demonstrates that the control voltage $w_3(t)$ will be zero. If τ is late $w_3(t)$ will be a positive correction voltage, and if τ is early $w_3(t)$ will be negative. A positive (negative) control voltage will increase (decrease) the frequency of the VCC. Thus, the bit synchronizer will produce an output clock signal that is synchronized to the input data stream. That is, $w_4(t)$ will be a pulse train with narrow clock pulses occurring at the time $t = \tau + nT_b$, where n is any integer and τ approximates τ_0, the optimum clock phase that corresponds to sampling at the maximum of the eye opening. It is interesting to realize that the early–late bit synchronizer of Fig. 3-17 has the same canonical form as the Costas carrier synchronization loop of Fig. 4-34.

As explained in previous sections, unipolar, polar, and bipolar bit synchronizers will work only when there are a sufficient number of alternating 1's and 0's in the data. The loss of synchronization because of strings of all 1's or all 0's can be prevented by adopting one of two possible alternatives. One alternative, as discussed in Chapter 1, is to use bit interleaving (i.e., scrambling). In this case the source data with strings of 1's or 0's are scrambled to produce data with alternating 1's and 0's which are transmitted over the channel using a unipolar, polar, or bipolar line code. At the receiving end scrambled data are first recovered using the usual receiving techniques with bit synchronizers as just described; then the scrambled data are unscrambled. The other alternative is to use a completely different type of line code that does not require alternating data for bit synchronization. For example, Manchester NRZ encoding can be used, but it will require a channel with twice the bandwidth of that needed for a polar NRZ code.

For further study of the overall synchronization problem, the reader is referred to Chapter 7 of Feher [1981], which was written by L. E. Franks, and to Chapter 8 of Sklar [1988], which was written by M. A. King. A special issue of the *IEEE Transactions on Communications* gives tutorial as well as detailed analysis papers on synchronization [Gardner and Lindsey, 1980]. Details about bit synchronization can be found in a number of books [Holmes, 1982; Stiffler, 1971; Ziemer and Peterson, 1985].

Multilevel Signaling

As we have seen, the bandwidth of *binary* digital waveforms may be very large. In fact, the bandwidth of the binary waveform may well exceed the bandwidth allowed (e.g., the FCC-allowed bandwidth or the bandwidth available on a telephone line). If this is the case, the question is: What can we do to reduce the signal bandwidth? One answer, of course, is to convert to multilevel signaling. This type of signaling is illustrated in Fig. 3-18a, where an ℓ-bit digital-to-analog converter (DAC) is used to convert the binary signal with data rate R bits/sec to an $L = 2^{\ell}$ level multilevel digital signal.

For example, assume that an $\ell = 3$ bit DAC is used, so that $L = 2^3 = 8$ levels. Fig. 3-18b illustrates a typical input waveform, and Fig. 3-18c shows the corresponding eight-level multilevel output waveform where T_s is the time it takes to send one multilevel symbol. To obtain this waveform, the code shown in Table 3-4 was used. From the figure, we see that $D = 1/T_s = 1/(3T_b) = R/3$, or, in general, the baud rate is

$$D = \frac{R}{\ell} \tag{3-52}$$

The relationship between the output baud rate D and the associated bit rate R is identical to that discussed in conjunction with the dimensionality theorem in Chapter 2, where it was found that the bandwidth was constrained by $B \geq D/2$.

★ The PSD of the multilevel waveform shown in Fig. 3-18c can be obtained by the use of (3-33). Evaluating $R(k)$ for the case of equally likely levels a_n, as shown in Table 3-4, for the $k = 0$ case we have

$$R(0) = \sum_{i=1}^{8} (a_n)_i^2 P_i = 21$$

where $P_i = \frac{1}{8}$ for all of the eight possible values. For $k \neq 0$, $R(k) = 0$. Then, from (3-33a), the PSD for $w_2(t)$ is

$$\mathcal{P}_{w_2}(f) = \frac{|F(f)|^2}{T_s}(21 + 0)$$

where the pulse width (or symbol width) is now $T_s = 3T_b$. For the rectangular pulse shape of width $3T_b$ this becomes

$$\mathcal{P}_{w_2}(f) = 63T_b\left(\frac{\sin 3\pi f T_b}{3\pi f T_b}\right)^2 \tag{3-53}$$

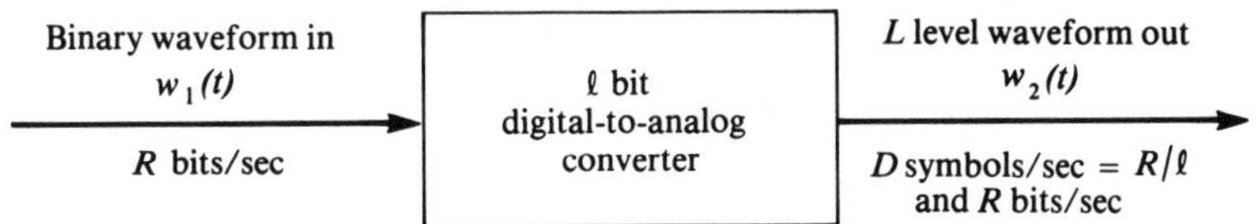

(a) ℓ Bit Digital-to-Analog Converter

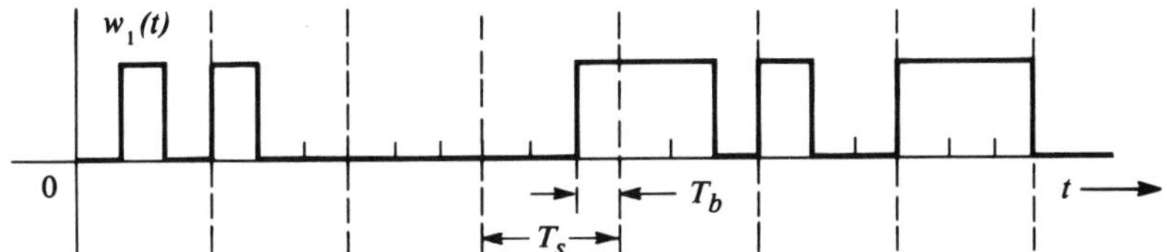

(b) Input Binary Waveform, $w_1(t)$

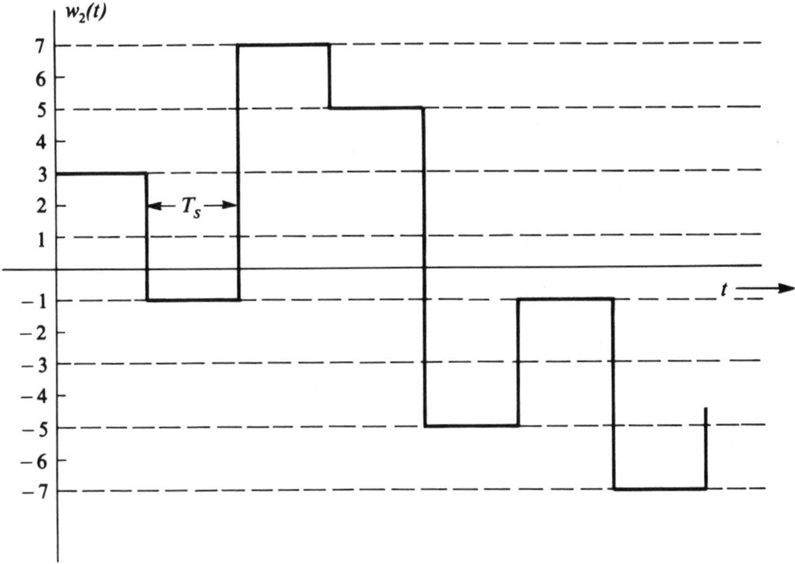

(c) $L = 8 = 2^3$ Level Waveform Out

FIGURE 3-18 Binary-to-multilevel digital signal conversion.

for this case where $l = 3$. Consequently, the first null bandwidth for this multilevel signal is $B_{\text{null}} = 1/(3T_b) = R/3$, or one third the bandwidth of the input binary signal.

For example, if the polar binary signal had a data rate of $R = 9600$ bits/sec, such as on an RS-232C-type line from a computer terminal (see Appendix C), the first null bandwidth of this binary signal would be 9.6 kHz, but the corresponding 8-level multilevel signal would have a bandwidth of only 3.2 kHz. Note that the multilevel signal would have a baud rate of $D = 3200$ symbols/sec and a bit rate of $R = 9600$ bits/sec, whereas the baud rate and bit rate of the binary signal are both equal to 9600.

TABLE 3-4 Three-Bit DAC Code

Digital Word	Output Level, $(a_n)_i$
000	+7
001	+5
010	+3
011	+1
100	−1
101	−3
110	−5
111	−7

In general, for multilevel signaling, the null bandwidth is

$$B_{\text{null}} = R/\ell \tag{3-54}$$

In summary, multilevel signaling, where $L > 2$, is used to reduce the bandwidth of a digital signal when compared to the bandwidth required for binary signaling. In practice, filtered multilevel signals are often used to modulate a carrier for transmission of digital information over a communication channel. This provides a relatively narrowband digital signal.

Spectral Efficiency

DEFINITION: The *spectral efficiency* of a digital signal is given by the number of bits per second of data that can be supported by each hertz of bandwidth:

$$\eta = \frac{R}{B} \qquad \text{(bits/sec)/Hz} \tag{3-55}$$

where R is the data rate and B is the bandwidth.

In applications where the bandwidth is limited by physical and regulatory constraints, the job of the communication engineer is to choose a signaling technique that gives the highest spectral efficiency while achieving given cost constraints and meeting specifications for low probability of bit error at the system output. Moreover, the maximum possible spectral efficiency is limited by the channel noise if the error is to be small. This maximum spectral efficiency is given by Shannon's channel capacity formula, (1-10),

$$\eta_{\max} = \frac{C}{B} = \log_2\left(1 + \frac{S}{N}\right) \tag{3-56}$$

Shannon's theory does not tell us how to achieve a system with this maximum theoretical spectral efficiency; however, practical systems that approach this spectral efficiency usually incorporate error correction coding and multilevel signaling.

TABLE 3-5 Spectral Efficiencies of Line Codes

Code Type	First-Null Bandwidth (Hz)	Spectral Efficiency $\eta = R/B$ [(bits/sec)/Hz]
Unipolar NRZ	R	1
Polar NRZ	R	1
Unipolar RZ	$2R$	$\frac{1}{2}$
Bipolar RZ	R	1
Manchester NRZ	$2R$	$\frac{1}{2}$
Multilevel NRZ	R/ℓ	ℓ

The spectral efficiency for multilevel NRZ signaling is obtained by substituting (3-54) into (3-55):

$$\eta = \ell \quad \text{(bits/sec)/Hz} \qquad \text{(multilevel NRZ signaling)} \tag{3-57}$$

where ℓ is the number of bits used in the DAC. Of course, ℓ cannot be increased without limit to an infinite efficiency because it is limited by the signal-to-noise ratio as given in (3-56).

The spectral efficiencies for all the line codes studied in the previous sections can be easily evaluated from their PSDs. The results are shown in Table 3-5. Unipolar NRZ, polar NRZ, and bipolar RZ are twice as efficient as unipolar or Manchester NRZ.

All of these binary line codes have $\eta \leq 1$.† Multilevel signaling can be used to achieve much greater spectral efficiency, but multilevel circuits are more costly. In practice, multilevel signaling is used in the T1G digital telephone lines as described in Sec. 3-8.

3-5 Intersymbol Interference

The absolute bandwidth of flat-top multilevel pulses is infinity. If these pulses are filtered improperly as they pass through a communication system, they will spread in time, and the pulse for each symbol may be smeared into adjacent time slots and cause *intersymbol interference* (ISI), as illustrated in Fig. 3-19. The question is: How can we restrict the bandwidth and still not introduce ISI? Of course, with restricted bandwidth, the pulses would have rounded tops (instead of flat tops). This problem was first studied by Nyquist in 1928 [Nyquist, 1928]. He discovered three different methods for pulse shaping that could be used to eliminate ISI. Each of these methods will be studied in the sections that follow.

Consider a digital signaling system as shown in Fig. 3-20. In general, multilevel signaling will be assumed where the flat-topped multilevel signal at the input is

$$w_{\text{in}}(t) = \sum_n a_n h(t - nT_s) \tag{3-58}$$

†Duobinary signaling, as discussed in the next section, has $\eta = 2$—see (3-89).

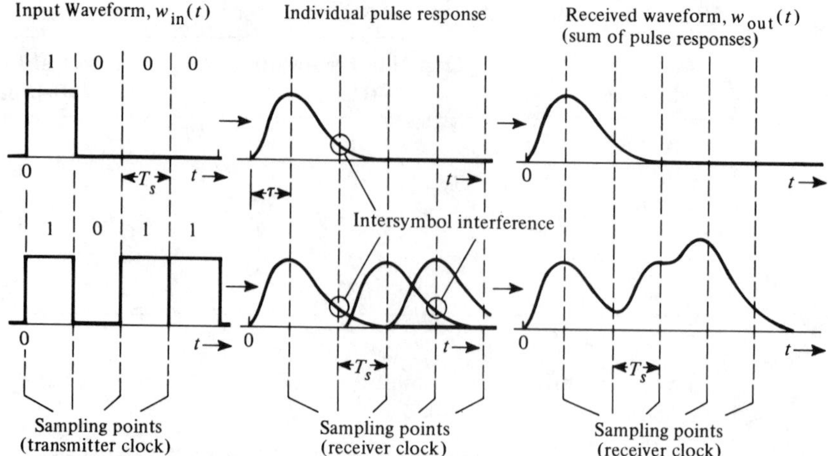

FIGURE 3-19 Examples of ISI on received pulses in a binary communication system.

where $h(t) = \Pi(t/T_s)$ and a_n may take on any of the allowed L multilevels ($L = 2$ for binary signaling). The symbol rate is $D = 1/T_s$ pulses/sec. Then (3-58) may be written as

$$w_{in}(t) = \sum_n a_n h(t) * \delta(t - nT_s)$$

$$= \left[\sum_n a_n \delta(t - nT_s)\right] * h(t) \tag{3-59}$$

The output of the linear system of Fig. 3-20 would be just the input impulse train convolved with the equivalent impulse response of the overall system

$$w_{out}(t) = \left[\sum_n a_n \delta(t - nT_s)\right] * h_e(t) \tag{3-60}$$

where the equivalent impulse response is

$$h_e(t) = h(t) * h_T(t) * h_C(t) * h_R(t) \tag{3-61}$$

Note that $h_e(t)$ is also the pulse shape that will appear at the output of the receiver filter when a single flat-top pulse is fed into the transmitting filter (Fig. 3-20).

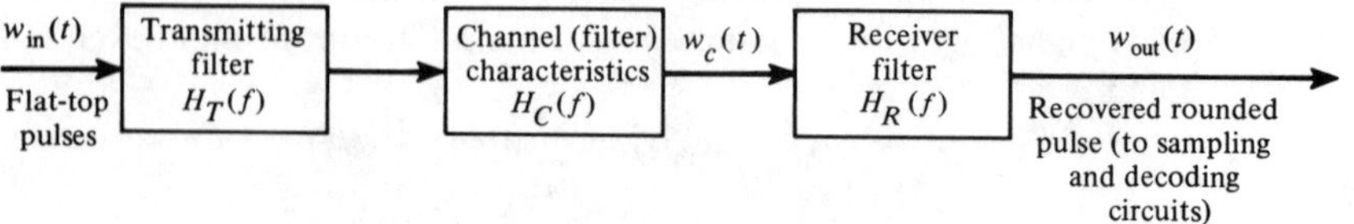

FIGURE 3-20 Baseband pulse transmission system.

The equivalent system transfer function is

$$H_e(f) = H(f)H_T(f)H_C(f)H_R(f) \tag{3-62}$$

where

$$H(f) = \mathcal{F}\left[\Pi\left(\frac{t}{T_s}\right)\right] = T_s\left(\frac{\sin \pi T_s f}{\pi T_s f}\right) \tag{3-63}$$

Equation (3-63) is used so that flat-top pulses will be present at the input to the transmitting filter. The receiving filter, $H_R(f)$, is given by

$$H_R(f) = \frac{H_e(f)}{H(f)H_T(f)H_C(f)} \tag{3-64}$$

where $H_e(f)$ is the overall filtering characteristic.

When $H_e(f)$ is chosen to minimize the ISI, then $H_R(f)$, obtained from (3-64), is called an *equalizing filter*. The equalizing filter characteristic depends on $H_C(f)$, the channel frequency response, as well as on the required $H_e(f)$. When the channel consists of dial-up telephone lines, the channel transfer function changes from call to call and the equalizing filter may need to be an *adaptive filter*. In this case, the equalizing filter adjusts itself by using feedback techniques to minimize the ISI. In some adaptive schemes, each communication session is preceded by a test bit pattern that is used to adapt the filter electronically for the maximum eye opening (i.e., minimum ISI). Such sequences are called *learning* or *training sequences* and *preambles*.

If we rewrite (3-60), the rounded pulse train at the output of the receiving filter is

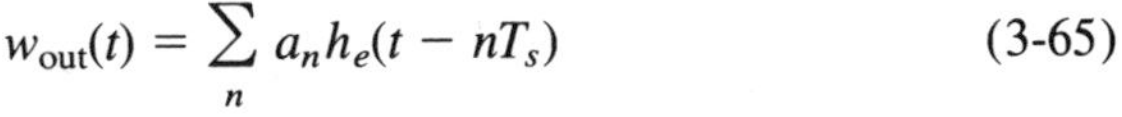

$$w_{\text{out}}(t) = \sum_n a_n h_e(t - nT_s) \tag{3-65}$$

The output pulse shape is affected by the input pulse shape (flat-topped in this case), the transmitter filter, the channel filter, and the receiving filter. Since, in practice, the channel filter is already specified, the problem is to determine the transmitting filter and the receiving filter that will minimize the ISI on the rounded pulse at the output of the receiving filter.

From an applications point of view, it is realized that when the required transmitter filter and/or receiving filter transfer functions are found, they can each be multiplied by $Ke^{-j\omega T_d}$, where K is a convenient gain factor and T_d is a convenient time delay. These parameters are chosen to make the filters easier to build. The $Ke^{-j\omega T_d}$ factor(s) would not affect the zero ISI result but, of course, would modify the level and delay of the output waveform.

Nyquist's First Method (Zero ISI)

Nyquist's first method for elimination of ISI is to use an equivalent transfer function, $H_e(f)$, such that the impulse response satisfies the following condition:

$$h_e(kT_s + \tau) = \begin{cases} C, & k = 0 \\ 0, & k \neq 0 \end{cases} \tag{3-66}$$

where k is an integer, T_s is the symbol (sample) clocking period, τ is the offset in the receiver sampling clock times when compared to the clocking times of the input symbols, and C is a nonzero constant. That is, for a single flat-top pulse of level a present at the input to the transmitting filter at $t = 0$, the received pulse would be $ah_e(t)$. It would have a value of aC at $t = \tau$ but would not cause interference at other sampling times because $h_e(kT_s + \tau) = 0$ for $k \neq 0$ (e.g., see Fig. 3-19).

Now suppose that we choose a $(\sin x)/x$ function for $h_e(t)$. In particular, let $\tau = 0$ and choose

$$h_e(t) = \frac{\sin \pi f_s t}{\pi f_s t} \tag{3-67}$$

where $f_s = 1/T_s$. This impulse response satisfies Nyquist's first criterion for zero ISI, (3-66). Consequently, if the transmit and received filters are designed so that the overall transfer function is

$$H_e(f) = \frac{1}{f_s} \Pi\left(\frac{f}{f_s}\right) \tag{3-68}$$

there will be no ISI. Furthermore, the absolute bandwidth of this transfer function is $B = f_s/2$. From our study of the sampling theorem and the dimensionality theorem in Chapter 2, we realize that this is the optimum filtering to produce a minimum bandwidth system. It will allow signaling at a baud rate of $D = 1/T_s = 2B$ pulses/sec, where B is the absolute bandwidth of the system. However, the $(\sin x)/x$ type of overall pulse shape has two practical difficulties.

1. The overall amplitude transfer characteristic $H_e(f)$ has to be flat over $-B < f < B$ and zero elsewhere. This is physically unrealizable (i.e., the impulse response is noncausal and of infinite duration). $H_e(f)$ is difficult to approximate because of the steep skirts in the filter transfer function at $f = \pm B$.
2. The synchronization of the clock in the decoding sampling circuit has to be almost perfect since the $(\sin x)/x$ pulse decays only as $1/x$ and is zero in adjacent time slots only when t is at the exactly correct sampling time. Thus, inaccurate sync will cause ISI.

Because of these difficulties, we are forced to consider other pulse shapes that have a slightly wider bandwidth. The idea is to find pulse shapes that go through zero at adjacent sampling points and yet have an envelope that decays much faster than $1/x$ so that clock jitter in the sampling times does not cause appreciable ISI. One solution for the equivalent transfer function, which has many desirable features, is the raised cosine-rolloff filter.

Raised Cosine-Rolloff Filtering

DEFINITION: The *raised cosine-rolloff filter* has the transfer function

$$H_e(f) = \begin{cases} 1, & |f| < f_1 \\ \frac{1}{2}\left\{1 + \cos\left[\frac{\pi(|f| - f_1)}{2f_\Delta}\right]\right\}, & f_1 < |f| < B \\ 0, & |f| > B \end{cases} \quad (3\text{-}69)$$

where B is the *absolute bandwidth* and the parameters f_1 and f_Δ are

$$f_\Delta = B - f_0 \quad (3\text{-}70)$$

$$f_1 \stackrel{\Delta}{=} f_0 - f_\Delta \quad (3\text{-}71)$$

f_0 is the *6-dB bandwidth* of the raised cosine-rolloff filter. The *rolloff factor* is defined to be

$$r = \frac{f_\Delta}{f_0} \quad (3\text{-}72)$$

This filter characteristic is illustrated in Fig. 3-21. The corresponding impulse response is

$$h_e(t) = \mathcal{F}^{-1}[H_e(f)] = 2f_0\left(\frac{\sin 2\pi f_0 t}{2\pi f_0 t}\right)\left[\frac{\cos 2\pi f_\Delta t}{1 - (4f_\Delta t)^2}\right] \quad (3\text{-}73)$$

Plots of the frequency response and the impulse response are shown in Fig. 3-22 for rolloff factors $r = 0$, $r = 0.5$, and $r = 1.0$. The $r = 0$ characteristic is the minimum bandwidth case where $f_0 = B$ and the impulse response is the $(\sin x)/x$ pulse shape. From this figure it is seen that *as the absolute bandwidth is increased* (e.g., $r = 0.5$ and $r = 1.0$), (1) *the filtering requirements are relaxed,* although $h_e(t)$ is still noncausal; and (2) *the clock timing requirements are relaxed* also since the envelope of the impulse response decays faster than $1/|t|$ (on the order of $1/|t|^3$ for large values of t).

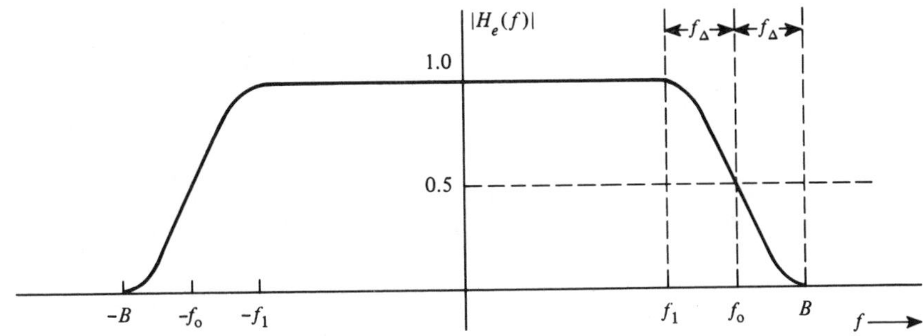

FIGURE 3-21 Raised cosine-rolloff filter characteristics.

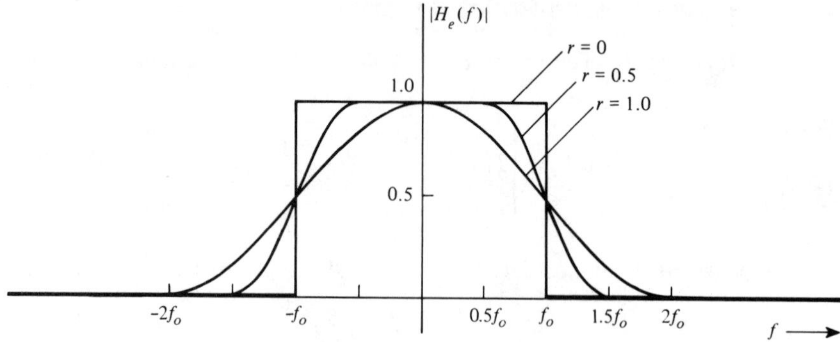

(a) Magnitude Frequency Response

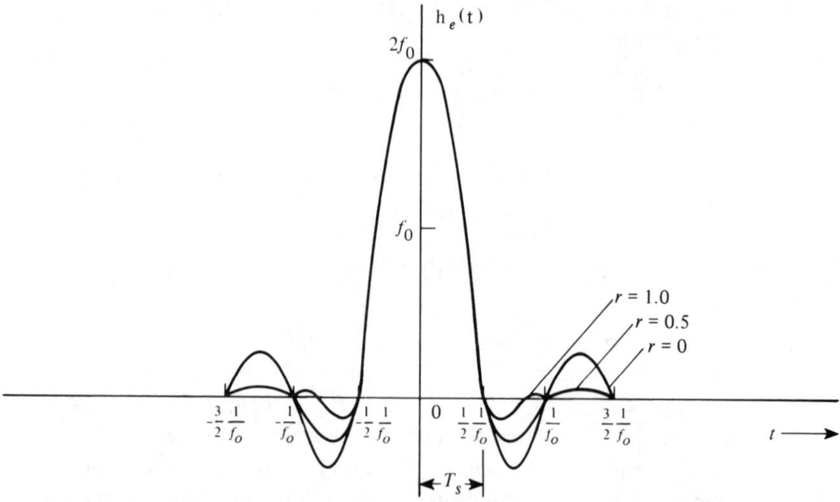

(b) Impulse Response

FIGURE 3-22 Frequency and time response for different rolloff factors.

The baud rate that the communication system can support without ISI can be related to the absolute bandwidth of the system and the rolloff factor of the raised cosine-rolloff filter characteristic. From Fig. 3-22b, the zeros in the system impulse response occur at $t = n/2f_0$, $n \neq 0$. Thus, the raised cosine-rolloff filter satisfies Nyquist's first criterion for no ISI as described by (3-66) if we use $\tau = 0$ and sample with a period of $T_s = 1/2f_0$. The corresponding baud rate is $D = 1/T_s = 2f_0$ symbols/sec. Using (3-70) and (3-72), we see that the baud rate that can be supported by the system is

$$D = \frac{2B}{1 + r} \tag{3-74}$$

where B is the absolute bandwidth of the system and r is the system rolloff factor.

EXAMPLE 3-1 (continued)

Assume that a binary PCM signal, with polar NRZ signaling, is passed through a communication system with a raised cosine-rolloff filtering characteristic and that the rolloff factor is 0.25. The bit rate of the PCM signal is 64 kbits/sec. Determine the absolute bandwidth of the filtered PCM signal.

Using (3-74), the absolute bandwidth is $B = 40$ kHz, which is less than the unfiltered PCM signal zero-crossing bandwidth of 64 kHz.

The raised cosine filter is only one of a more general class of filters that satisfy Nyquist's first criterion. This general class is described by the following theorem.

THEOREM: *The overall filter is said to be a Nyquist filter if the effective transfer function of a pulse system is*

$$H_e(f) = \begin{cases} \Pi\left(\dfrac{f}{2f_0}\right) + Y(f), & |f| < 2f_0 \\ 0, & f \text{ elsewhere} \end{cases} \tag{3-75}$$

where $Y(f)$ is a real function that is even symmetric about $f = 0$; that is,

$$Y(-f) = Y(f), \qquad |f| < 2f_0 \tag{3-76a}$$

and odd symmetric about $f = f_0$; that is,

$$Y(-f + f_0) = -Y(f + f_0), \qquad |f| < f_0 \tag{3-76b}$$

Then there will be no intersymbol interference at the system output if the symbol rate is

$$f_s = 2f_0 \tag{3-77}$$

This theorem is illustrated in Fig. 3-23. $Y(f)$ can be any real function that satisfies the symmetry conditions of (3-76). Thus an infinite number of filter characteristics can be used to produce zero ISI.

Proof. We need to show that the impulse response of this filter is 0 at $t = nT_s$, $n \neq 0$, where $T_s = 1/f_s = 1/(2f_0)$. Taking the inverse Fourier transform of (3-75), we have

$$h_e(t) = \int_{-2f_0}^{-f_0} Y(f)e^{j\omega t}\,df + \int_{-f_0}^{f_0} [1 + Y(f)]e^{j\omega t}\,df + \int_{f_0}^{2f_0} Y(f)e^{j\omega t}\,df$$

or

$$h_e(t) = \int_{-f_0}^{f_0} e^{j\omega t}\,df + \int_{-2f_0}^{2f_0} Y(f)e^{j\omega t}\,df$$

$$= 2f_0\left(\frac{\sin \omega_0 t}{\omega_0 t}\right) + \int_{-2f_0}^{0} Y(f)e^{j\omega t}\,df + \int_{0}^{2f_0} Y(f)e^{j\omega t}\,df$$

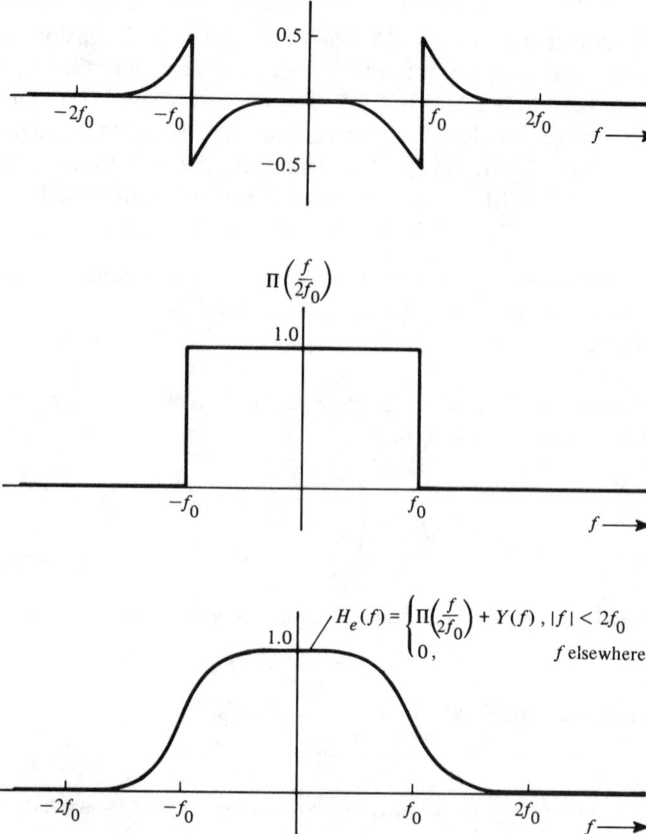

FIGURE 3-23 Nyquist filter characteristic.

Making a change in variable in the integrals, letting $f_1 = f + f_0$ in the first integral and $f_1 = f - f_0$ in the second integral, we obtain

$$h_e(t) = 2f_0\left(\frac{\sin \omega_0 t}{\omega_0 t}\right) + e^{-j\omega_0 t}\int_{-f_0}^{f_0} Y(f_1 - f_0)e^{j\omega_1 t}\,df_1$$

$$+\ e^{j\omega_0 t}\int_{-f_0}^{f_0} Y(f_1 + f_0)e^{j\omega_1 t}\,df_1$$

From (3-76a) and (3-76b) we know that $Y(f_1 - f_0) = -Y(f_1 + f_0)$; thus we have

$$h_e(t) = 2f_0\left(\frac{\sin \omega_0 t}{\omega_0 t}\right) + j2 \sin \omega_0 t \int_{-f_0}^{f_0} Y(f_1 + f_0)e^{j\omega_1 t}\,df_1$$

This impulse response is real since $H_e(-f) = H_e^*(f)$; furthermore, it satisfies Nyquist's first criterion. That is, $h_e(t)$ is zero at $t = n/(2f_0)$, $n \neq 0$, and $\tau = 0$.

Thus, if we sample at $t = n/(2f_0)$, there will be no ISI. However, the filter is noncausal.

In the preceding description of the Nyquist filter, we assumed that $H_e(f)$ was real. Of course, we could use a filter with a linear phase characteristic [i.e., $H_e(f)e^{-j\omega T_d}$], and there would be no ISI if we delayed the clocking by T_d sec, since the $e^{-j\omega T_d}$ factor is the transfer function of an ideal delay line. This would move the peak of the impulse response to the right (along the time axis), and then the filter would be approximately causal.

At the digital receiver, in addition to minimizing the ISI, we would like to minimize the effect of channel noise by proper receiver filtering. As will be shown in Chapter 6, the filter that minimizes the effect of channel noise is the *matched filter*. Unfortunately, if a matched filter is used for $H_R(f)$ at the receiver, the overall filter characteristic, $H_e(f)$, will usually not satisfy the Nyquist characteristic for minimum ISI. However, it can be shown that for the case of Gaussian noise into the receiver, the effects of both ISI and channel noise are minimized if the transmitter and receiver filters are designed so that [Shanmugan, 1979; Sunde, 1969; Ziemer and Peterson, 1985]

$$|H_T(f)| = \frac{\sqrt{|H_e(f)|}\,[\mathcal{P}_n(f)]^{1/4}}{\alpha\,|H(f)|\,\sqrt{|H_c(f)|}} \tag{3-78a}$$

and

$$|H_R(f)| = \frac{\alpha\sqrt{|H_e(f)|}}{\sqrt{H_c(f)}\,[\mathcal{P}_n(f)]^{1/4}} \tag{3-78b}$$

where $\mathcal{P}_n(f)$ is the PSD for the noise at the receiver input and α is an arbitrary positive constant (e.g., choose $\alpha = 1$ for convenience). $H_e(f)$ is selected from any appropriate frequency response characteristic that satisfies Nyquist's first criterion as discussed previously and $H(f)$ is given by (3-63). Any appropriate phase response can be used for $H_T(f)$ and $H_R(f)$ as long as the overall system phase response is linear. This results in a constant time delay versus frequency.

★Zero-Forcing Transversal Filter Equalizers

The equalizing (i.e., receiving) filter can be realized by using a transversal filter design, as shown in Fig. 3-24. As shown, the transversal filter consists of a tapped delay line where the tap outputs are multiplied by gain factors. The resulting products are summed to produce the filter output. Although other types of realizations are possible, the transversal structure is very popular for this application because (1) it is relatively easy to analyze and design and (2) it is easy to build using *charge coupled devices* (CCD) for the tapped analog delay line [Gersho, 1975]. As described in Fig. 3-20, the input, $w_c(t)$, is from the channel and the equalizer is designed to provide an output waveform, $w_{\text{out}}(t)$, that has zero ISI.

The design problem consists of finding the values for the tap gains such that there will be no ISI on the output waveform. This can be accomplished by

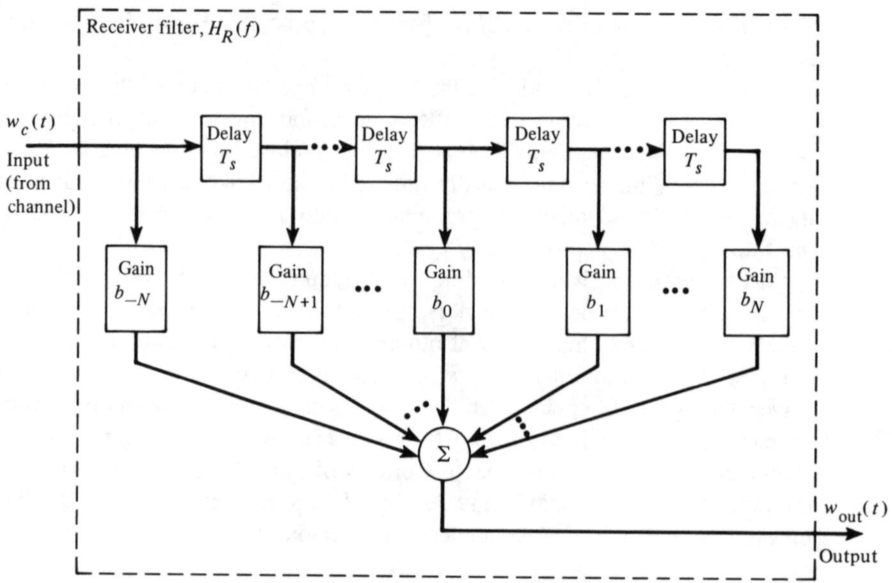

FIGURE 3-24 Transversal equalizing filter.

finding the b_n's that will satisfy Nyquist's first method for zero ISI. That is, the b_n's are found that give zeros for the impulse response, $h_e(t)$, at all sampling times except the present. To find the design equation, the system is analyzed using the impulse response technique described in the previous sections. As seen in Fig. 3-20, a single impulse into the pulse shaping filter $H(f)$ produces a single flat-topped pulse at the input to the transmit filter, $H_T(f)$. Then the corresponding output of the *channel* [i.e., the $H_c(f)$ filter] will be the response $w_c(t)$, and the overall system will have the equivalent impulse response of $h_e(t)$. For the transversal filter of Fig. 3-24, the response is related to its input by

$$h_e(t) = b_{-N}w_c(t) + b_{-N+1}w_c(t - T_s) + \cdots + b_N w_c(t - 2NT_s)$$

or

$$h_e(t) = \sum_{n=-N}^{N} b_n w_c(t - (n + N)T_s) \tag{3-79}$$

Evaluating the samples of $h_e(t)$ at the sampling times $t = (k + N)T_s = kT_s + \tau$ where the receiving clock offset time is $\tau = NT_s$, we get

$$h_e(kT_s + \tau) = \sum_{n=-N}^{N} b_n w_c((k + N)T_s - (n + N)T_s)$$

or

$$h_e(kT_s + \tau) = \sum_{n=-N}^{N} b_n w_c((k - n)T_s) \tag{3-80}$$

Using Nyquist's first method for zero ISI as described by (3-66), we realize that (3-80) becomes

$$\sum_{n=-N}^{N} b_n w_c((k-n)T_s) = \begin{cases} 1, & k = 0 \\ 0, & k = \pm 1, \pm 2, \ldots, \pm N \end{cases} \tag{3-81}$$

where, for convenience, $C = 1$ has been used. We can force only $2N$ points of the output impulse response to zero (i.e., N samples before and N samples after the peak of the pulse) because there are only $(2N + 1)$ taps on the transversal filter. However, the ISI beyond the $\pm N$ sample points is usually negligible since the time response has decayed to an insignificant amplitude over that range. Since k has the range $-N \le k \le N$, (3-81) represents a system of $(2N + 1)$ simultaneous equations with $(2N + 1)$ unknowns. These simultaneous equations may be solved for the $\{b_n\}$ unknowns by using matrix theory. Rewriting (3-81) in matrix form, we get

$$\begin{array}{l} k=-N \\ k=-N+1 \\ \vdots \\ k=-1 \\ k=0 \\ k=1 \\ \vdots \\ k=N \end{array} \begin{bmatrix} 0 \\ 0 \\ \vdots \\ 0 \\ 1 \\ 0 \\ \vdots \\ 0 \end{bmatrix} = \begin{bmatrix} w_c(0) & w_c(-T_s) & \cdots & & w_c(-2NT_s) \\ w_c(T_s) & w_c(0) & & & \\ \vdots & & & & \\ & & & & \\ & & & w_c(0) & w_c(-T_s) \\ w_c(2NT_s) & & & w_c(T_s) & w_c(0) \end{bmatrix} \begin{bmatrix} b_{-N} \\ b_{-N+1} \\ \vdots \\ b_{-1} \\ b_0 \\ b_1 \\ \vdots \\ b_N \end{bmatrix}$$

If we use matrix notation, this is

$$\boldsymbol{H}_e = \mathbf{W}_c \boldsymbol{B} \tag{3-82}$$

where the column vectors $\mathbf{H}_e$ and $\mathbf{B}$ are

$$\mathbf{H}_e = \begin{bmatrix} 0 \\ \vdots \\ 0 \\ 1 \\ 0 \\ \vdots \\ 0 \end{bmatrix}, \qquad \mathbf{B} = \begin{bmatrix} b_{-N} \\ \vdots \\ b_{-1} \\ b_0 \\ b_1 \\ \vdots \\ b_N \end{bmatrix} \tag{3-83a,b}$$

and the $(2N + 1) \times (2N + 1)$ matrix $\mathbf{W}_c$ is

$$\mathbf{W}_c = \begin{bmatrix} w_c(0) & w_c(-T_s) & \cdots & w_c(-2NT_s) \\ w_c(T_s) & w_c(0) & & \vdots \\ \vdots & & w_c(0) & w_c(-T_s) \\ w_c(2NT_s) & \cdots & w_c(T_s) & w_c(0) \end{bmatrix} \tag{3-83c}$$

Solving (3-82) for **B,** we find that the values for the tap coefficients of the transversal filter that gives zero ISI are

$$\mathbf{B} = \mathbf{W}_c^{-1}\,\mathbf{H}_e \tag{3-84}$$

Note that the b_n's that are specified by (3-84) depend on $H(f)$, $H_T(f)$, and $H_c(f)$ since $w_c(t)$ depends on all of these filters.

The transversal filter has the advantage that its transfer characteristic can be altered easily just by changing the tap coefficients. For example, in a "training mode," one or more flat-top test pulses could be fed into the transmit filter and, based on receiving the corresponding waveform $w_c(t)$ at the receiver input, the receiver could automatically compute the required tap gains for zero ISI as specified by (3-84). In the "run mode," the receiver could use these computed tap gains to detect the data with zero ISI. In practice, the channel transfer characteristic changes from time to time, so that this procedure could be repeated periodically to maintain the correct equalization. This type of automatic adjustment is called *preset equalization.* If we use another approach, it is possible to adjust the tap gains continuously by using a low-level reference signal or by computing the tap gains iteratively from the received data waveform, $w_c(t)$. This latter type of automatic equalization is called *adaptive equalization.* Both of these automatic equalization techniques can be used in practice [Lucky and Rudin, 1966; Lucky, Salz, and Weldon, 1968]. For medium-speed (2400 bits/sec) modems used for computer communications over switched (dial-up) telephone lines, it is usually adequate to use a fixed equalizer based on the expected average channel transfer characteristic. For higher speeds (larger than 2400 bits/sec), the variation in the characteristics from line to line is so large that some type of automatic equalization is usually required.

Equalizers can also be built that minimize the mean square error (MSE) on the equalizer output where the actual output due to a training signal is compared with the desired output. This type of equalizer has the advantage of minimizing the effect of channel noise as well as minimizing ISI. This approach is popular in applications of high-speed modems used on telephone lines because of the superior noise performance and because the tap gains for the adaptive equalizers can be made to converge rapidly to their desired values. A discussion of the theory of minimum MSE equalizers is beyond the scope of this book. For further study on this subject, the reader is referred to an excellent tutorial paper with its associated references [Qureshi, 1982].

★ Nyquist's Second and Third Methods for Control of ISI

Nyquist's *second method* of ISI control allows some ISI to be introduced in a *controlled* way so that it can be canceled out at the receiver and the data can be recovered without error if no noise is present. This technique also allows for the possibility of doubling the bit rate or, alternatively, halving the channel bandwidth. This phenomenon was observed by telegraphers in the 1900s and is known as "doubling the dotting speed" [Bennett and Davey, 1965]. Nyquist's

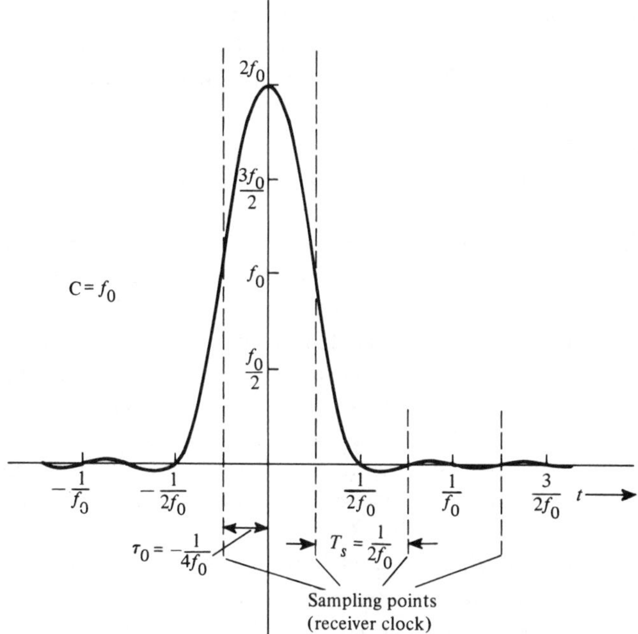

FIGURE 3-25 Nyquist's second criterion applied to the $r = 1$ raised cosine-rolloff filter characteristic.

second method requires that the received waveform be *correctly interpreted* since a controlled amount of ISI has been introduced. In fact, a vibrating relay that correctly interpreted such a waveform was described in the early technical literature [Gulstad, 1898, 1902], 30 years before Nyquist published his paper on the theory of elimination of ISI in 1928.

Nyquist's second method for control of ISI is to use an equivalent transfer function, $H_e(f)$, such that the impulse response satisfies the following condition:

$$h_e(kT_s + \tau_0) = \begin{cases} C, & k = 0 \\ C, & k = 1 \\ 0, & \text{other values of } k \end{cases} \tag{3-85}$$

where k is an integer, T_s is the clocking period, τ_0 is the offset in the receiver clocking times when compared to the transmitter clocking times, and C is a nonzero constant.

The raised cosine-rolloff characteristic $r = 1$ will now be examined to show that it satisfies Nyquist's second criterion and to study how the data are recovered.† The impulse response for the $r = 1$ raised cosine-rolloff filter, (3-73), is shown in Fig. 3-25. The offset of the receiver clock is $\tau_0 = -1/(4f_0)$ when

†Other raised cosine-rolloff impulse response characteristics (i.e., $r \neq 1$) do not satisfy Nyquist's second criterion.

compared to the timing of the transmitter impulse (clock). This characteristic satisfies Nyquist's second criterion of (3-85) since $h_e(0 - \tau_0) = f_0 = C$ for $k = 0$; $h_e(T_s + \tau_0) = f_0 = C$ for $k = 1$; and $h_e(kT_s + \tau_0) = 0$ for other values of k. The baud rate is

$$D = \frac{1}{T_s} = 2f_0 = B \tag{3-86}$$

Exactly this same baud rate was found when Nyquist's first criterion was applied—see (3-74). In other words, the $r = 1$ raised cosine-rolloff pulse does not provide speed doubling. (A pulse shape can be found that does provide speed doubling, as will be illustrated in the next section on duobinary signaling.) Referring to Fig. 3-25, note that the impulse response is *not* sampled at the peak of the pulse; instead, it is sampled symmetrically on each side of the peak. This is the distinguishing characteristic of Nyquist's second method, (3-85), when compared to the first method, (3-66). It also means that the receiver has to work harder to interpret the data on the received waveform because controlled ISI has been introduced (see Fig. 3-26).

Figure 3-26 shows several pulses that are received over a channel for the case of polar signaling. Each pulse is due to a bit that was transmitted. The received waveform would actually be a superposition (summation) of these pulses. The figure shows that there are several possibilities for the sample values of the superposition of the pulses: $+2V_0$ is obtained if the present bit is a binary 1 and the past bit is a binary 1; 0, if the present bit is a binary 1 and the past bit is a binary 0; 0, if the present bit is a binary 0 and the past bit is a binary 1; and $-2V_0$, if the present bit is a binary 0 and the past bit is a binary 0. Consequently, a receiver decoding circuit could be constructed to output the correct data bits based on the knowledge of previously decoded bits. However, if an error occurs (because of noise), a string of decoded errors would be produced. This effect of

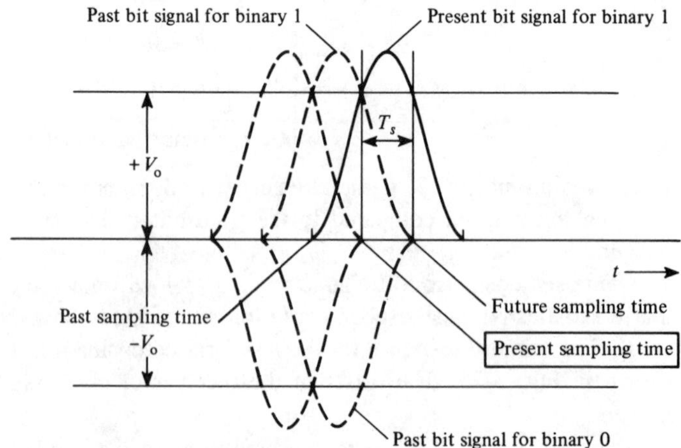

FIGURE 3-26 Received pulses for Nyquist's second method (controlled ISI).

error propagation can be eliminated by using differential encoding at the transmitter and decoding at the receiver.

Using Nyquist's second method, there are three decision levels ($-2V_0$, 0, and $+2V_0$) made at the receiver when sample values are distinguished in the presence of noise. In binary signaling using Nyquist's first method, there are only two decision levels. Consequently, the probability of bit error due to noise is slightly larger for the second method. In fact, for duobinary signaling (a three-decision-level type to be discussed in the next section), it can be shown that the transmitter power must be 2.1 dB greater than that for the direct binary case for the same probability of error [Ziemer and Peterson, 1985].

In *Nyquist's third method* of ISI control, the effect of ISI is eliminated by choosing $h_e(t)$ so that the area under the $h_e(t)$ pulse within the desired symbol interval, T_s, is not zero but the areas under $h_e(t)$ in adjacent symbol intervals are zero. For data detection, the receiver evaluates the area under the receiver waveform over each T_s interval. Pulses have been found that satisfy Nyquist's third criterion, but their performance in the presence of noise is inferior to the examples that have been discussed previously [Sunde, 1969].

★ Partial Response and Duobinary Signaling

The technique of introducing ISI in a controlled way so that receiver logic can cancel it out is called *partial response* or *correlative coding*. There is a whole class of partial response techniques [Kretzmer, 1966]. However, we will focus on the one that was invented by Lender, who pioneered partial response signaling [Lender, 1963]. This is called *duobinary signaling*, where "duo" implies the doubling of the transmission capacity—double the speed or half the bandwidth—when compared to the $r = 1$ raised cosine-rolloff filter. The equivalent transfer function for the duobinary system (channel) is

$$H_e(f) = \begin{cases} \dfrac{1}{B}\cos\dfrac{\pi f}{2B}, & |f| \le B \\ 0, & f \text{ elsewhere} \end{cases} \tag{3-87}$$

where B is the absolute bandwidth of the system. This is called a *cosine filter* characteristic (not to be confused with the raised cosine Nyquist filter). The corresponding impulse response is

$$h_e(t) = \frac{4}{\pi}\left[\frac{\cos 2\pi Bt}{1 - (4Bt)^2}\right] \tag{3-88}$$

Plots of these equations are shown in Fig. 3-27. It is seen that the impulse response satisfies Nyquist's second criterion, (3-85), where $\tau_0 = -1/(4B)$ and the sampling period is $T_s = 1/(2B)$. The baud rate and the bandwidth are related by

$$D = \frac{1}{T_s} = 2B \tag{3-89}$$

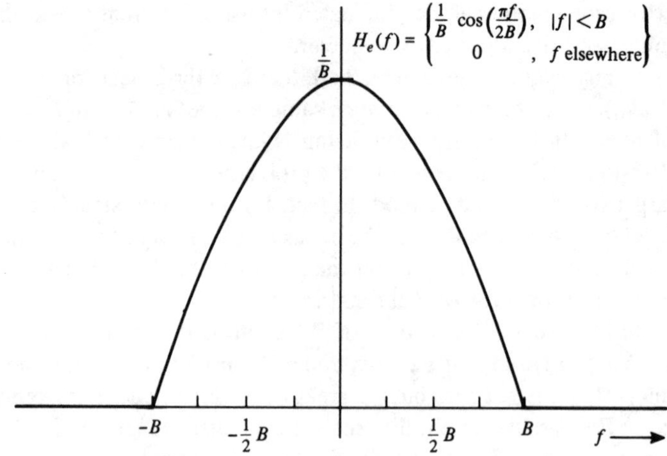

(a) Frequency Response

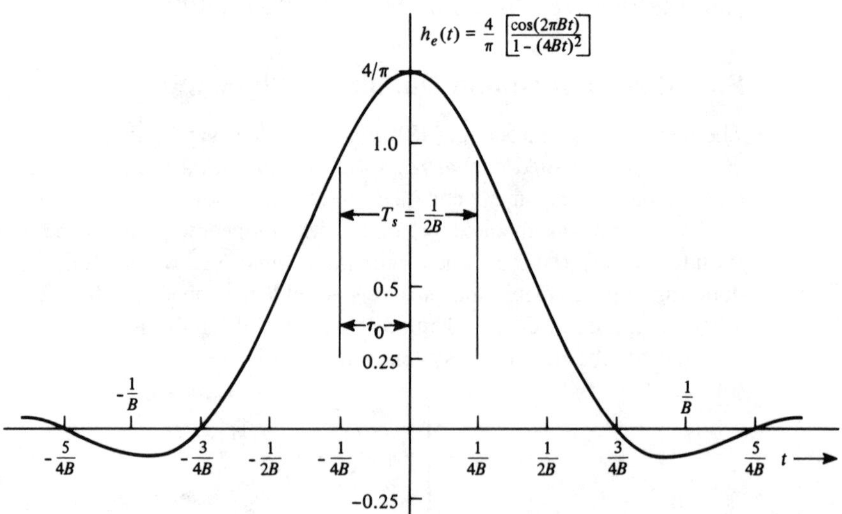

(b) Impulse Response

FIGURE 3-27 Cosine partial response.

This is exactly "double the speed" of the $r = 1$ raised cosine filter as specified by (3-74). By comparing (3-89) with the dimensionality theorem, (2-176), it is also realized that the duobinary technique provides the fastest speed that is theoretically possible for a given bandwidth. This maximum speed performance is also theoretically possible when an $r = 0$ raised cosine filter, $h_e(t) = (\sin 2\pi Bt)/(2\pi Bt)$, is used; however, the $r = 0$ system has two severe disadvantages, noted earlier in our study of Nyquist's first method.

A duobinary signaling system can also be realized by using a duobinary conversion filter, as shown in Fig. 3-28. Here a binary system is first designed for an

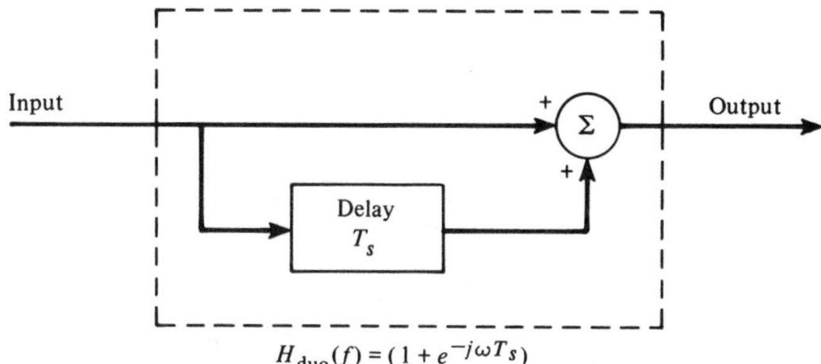

FIGURE 3-28 Duobinary conversion filter.

$r = 0$ raised cosine-rolloff (ideal rectangular) filtering characteristic. Then the duobinary conversion filter is inserted anywhere along the channel. The resulting equivalent transfer function is

$$H_e(f) = H_{duo}(f)\Pi\left(\frac{f}{2B}\right) \tag{3-90a}$$

where

$$H_{duo}(f) = 1 + e^{-j\omega T_s} \tag{3-90b}$$

Thus

$$H_e(f) = (1 + e^{-j\omega T_s})\Pi\left(\frac{f}{2B}\right) = e^{-j\omega T_s/2}(e^{j\omega T_s/2} + e^{-j\omega T_s/2})\Pi\left(\frac{f}{2B}\right)$$

With $T_s = 1/(2B)$, this reduces to

$$H_e(f) = (2Be^{-j\omega T_s/2})\begin{cases} \dfrac{1}{B}\cos\left(\dfrac{\pi f}{2B}\right), & |f| \le B \\ 0, & f \text{ elsewhere} \end{cases} \tag{3-91}$$

This is identical to the desired cosine filter characteristic of (3-87) except for a gain factor of $2B$ and a time delay of $T_s/2$.

The data on the received duobinary waveform can be recovered by the technique described previously for waveforms of the second Nyquist type. One disadvantage of the duobinary system is error propagation in the recovered data due to an error in some past bit. This effect can be eliminated by using differential encoding at the transmitter and differential decoding at the receiver. In a duobinary context, a differential encoder is called a *precoder*.

Another disadvantage of duobinary signaling is that the spectrum is not zero at $f = 0$. This problem can be corrected by using a *modified duobinary* technique, conceived by Lender [1966], that employs a modified duobinary conversion

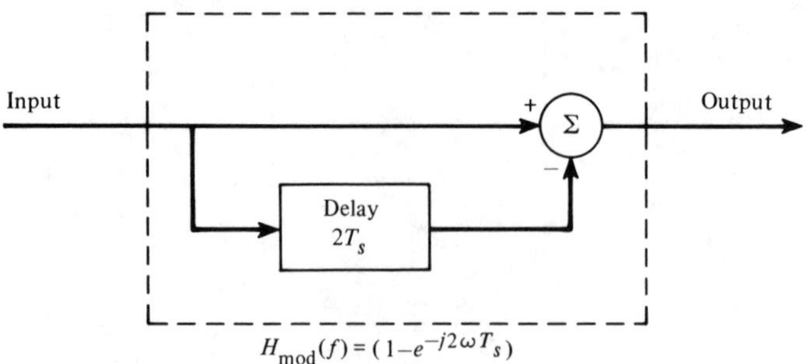

FIGURE 3-29 Modified duobinary conversion filter.

filter, as shown in Fig. 3-29. The resulting transfer function is

$$H_e(f) = H_{\text{mod}}(f)\Pi\left(\frac{f}{2B}\right) = (1 - e^{-j2\omega T_s})\Pi\left(\frac{f}{2B}\right)$$

$$= 2je^{-j\omega T_s}\left(\frac{e^{j\omega T_s} - e^{-j\omega T_s}}{2j}\right)\Pi\left(\frac{f}{2B}\right)$$

where

$$H_{\text{mod}}(f) = 1 - e^{-j2\omega T_s}$$

If we let $T_s = 1/(2B)$, the equivalent transfer function for modified duobinary signaling is

$$H_e(f) = (2je^{-j\omega T_s})\begin{cases} \sin\left(\dfrac{\pi f}{B}\right), & |f| \le B \\ 0, & f \text{ elsewhere} \end{cases} \tag{3-92}$$

The corresponding impulse response is

$$h_e(t) = \frac{4}{\pi}\,\frac{\sin[2\pi B(t - T_s)]}{(4B^2(t - T_s)^2 - 1)} \tag{3-93}$$

where $T_s = 1/(2B)$. These results are plotted in Fig. 3-30. This modification obviously eliminates the requirement for dc coupling in the channel.

From Figs. 3-29 and 3-30 it is realized that the present received sample value is correlated with (i.e., depends on) the present bit *and* the bit received two samples previously. If polar signaling is used, the received waveform samples have the possible values $+2V_0$, 0, and $-2V_0$. The value $+2V_0$ is obtained if the present received bit is a binary 1 and the second past received bit is a binary 0; 0 volt is obtained if the present bit is a binary 1 and the second past received bit is a binary 1, or if the present bit is a 0 and the second past received bit is a binary 0; and $-2V_0$ is obtained if the present bit is a binary 0 and the second past bit is a binary 1. Thus the data can be decoded when the present sample value and the second past received bit are known.

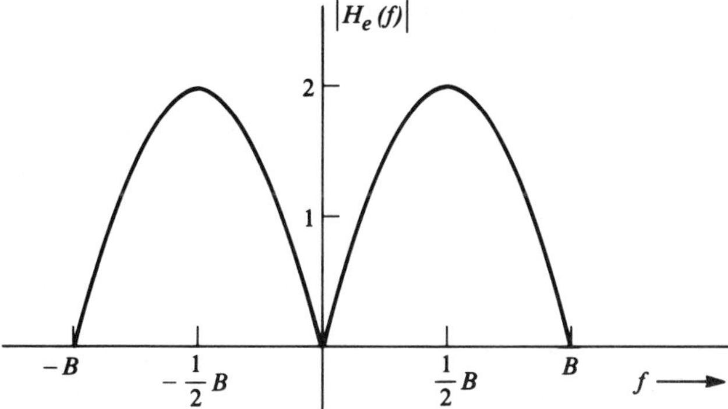

(a) Frequency Response

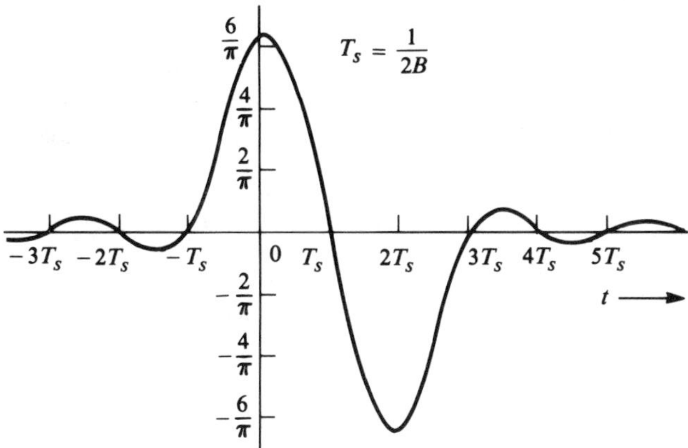

(b) Impulse Response

FIGURE 3-30 Modified duobinary partial response.

The modified binary signaling technique for controlled ISI can be decoded with zero error, although it does not satisfy (3-85). Equation (3-85) describes only one technique for signal transmission using controlled ISI. It could be generalized to cover the modified duobinary case, as well as other types of partial response signaling.

In summary, the modified duobinary technique has the advantages of not requiring a dc-coupled channel and does not require a sharp rolloff in the frequency response characteristic; yet the maximum (binary) signaling efficiency of $\eta = R/B = 2$ bits/Hz-sec is realized.

For additional study of partial response signaling, the reader is directed to a summary paper [Pasupathy, 1977] and to a paper that compares the performance of partial response signaling in several FM systems [Deshpande and Wittke, 1981].

3-6 Differential Pulse Code Modulation

When audio or video signals are sampled, it is usually found that adjacent samples are close to the same value. This means that there is a lot of redundancy in the signal samples and, consequently, that the bandwidth and the dynamic range of a PCM system are wasted when redundant sample values are retransmitted. One way to minimize redundant transmission and reduce the bandwidth is to transmit PCM signals corresponding to the difference in adjacent sample values. This, crudely speaking, is *differential pulse code modulation* (DPCM). At the receiver the present sample value is regenerated by using the past value plus the update differential value that is received over the differential system.

Moreover, the present value can be estimated from the past values by using a *prediction filter*. The filter may be realized by using a tapped delay line (bucket brigade device) to form a *transversal filter,* as shown in Fig. 3-31. When the tap gains $\{a_l\}$ are set so that the filter output will predict the present value from past values, the filter is said to be a *linear prediction filter* [Spilker, 1977]. The optimum tap gains are a function of the correlation properties of the audio or video signal [Jayant and Noll, 1984]. The output samples are

$$z(nT_s) = \sum_{l=1}^{K} a_l y(nT_s - lT_s) \tag{3-94a}$$

or, in simplified notation,

$$z_n = \sum_{l=1}^{K} a_l y_{n-l} \tag{3-94b}$$

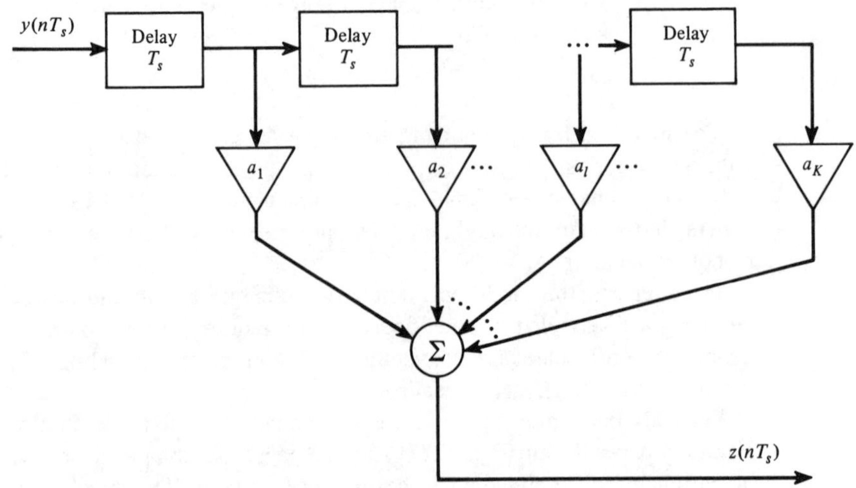

FIGURE 3-31 Transversal filter.

where y_{n-l} denotes the sample value at the filter input at time $t = (n - l)T_s$ and there are K delay devices in the transversal filter.

The linear prediction filter may be used in a differential configuration to produce DPCM. Two possible configurations will be examined.

The first DPCM configuration, shown in Fig. 3-32, uses the predictor to obtain a differential pulse amplitude modulated (DPAM) signal that is quantized and encoded to produce the DPCM signal. The recovered analog signal at the receiver output will be the same as that at the system input plus *accumulated* quantizing noise. We may eliminate the accumulation effect by using the transmitter configuration of Fig. 3-33.

In the second DPCM configuration shown in Fig. 3-33, the predictor operates on quantized values at the transmitter as well as at the receiver in order to minimize the quantization noise on the recovered analog signal. The analog output at the receiver is the same as the input analog signal at the transmitter except for quantizing noise; furthermore, the quantizing noise does not accumulate, as was the case in the first configuration.

It can be shown that DPCM, like PCM, follows the 6-dB rule [Jayant and Noll, 1984]

$$\left(\frac{S}{N}\right)_{dB} = 6.02n + \alpha \tag{3-95a}$$

where

$$-3 < \alpha < 15 \quad \text{for DPCM speech} \tag{3-95b}$$

and n is the number of quantizing bits ($M = 2^n$). Unlike companded PCM, the α for DPCM varies over a wide range depending on the properties of the input analog signal. Equation (3-95b) gives the range of α for voice-frequency (300 to 3400 Hz) telephone-quality speech. This DPCM performance may be compared to that for PCM. Equation (3-28) indicates that $\alpha = -10$ dB for μ-law companded PCM with $\mu = 255$. Thus there may be a S/N improvement as large as 25 dB when DPCM is used instead of $\mu = 255$ PCM. Alternatively, for the same S/N, DPCM could require 3 or 4 fewer bits per sample than companded PCM. This is why telephone DPCM systems often operate at a bit rate of $R = 32$ kbits/sec or $R = 24$ kbits/sec instead of the standard 64 kbits/sec as needed for companded PCM.

The CCITT has adopted a 32-kbits/sec DPCM standard that uses 4-bit quantization at an 8-ksample/sec rate for encoding 3.2-kHz bandwidth VF signals [Decina and Modena, 1988]. Moreover, a 64-kbits/sec DPCM CCITT standard (4-bit quantization and 16 ksamples/sec) has been adopted for encoding audio signals that have a 7-kHz bandwidth. A detailed analysis of DPCM systems is difficult and depends on the type of input signal present, the sample rate, the number of quantizing levels used, the number of stages in the prediction filter, and the predictor gain coefficients. This type of analysis is beyond the scope of this text, but for further study the reader is referred to published work on this topic [Flanagan et al., 1979; Jayant, 1974; Jayant and Noll, 1984; O'Neal, 1966b].

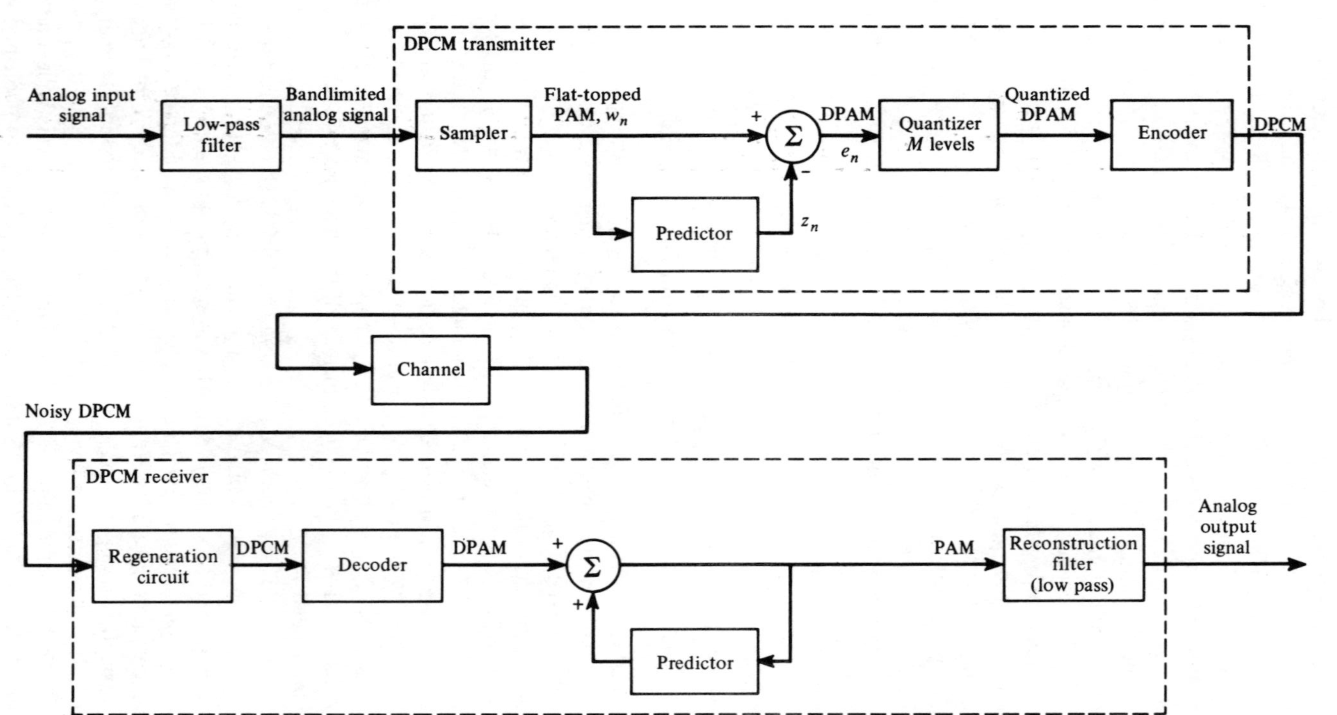

FIGURE 3-32 DPCM using prediction from samples of input signal.

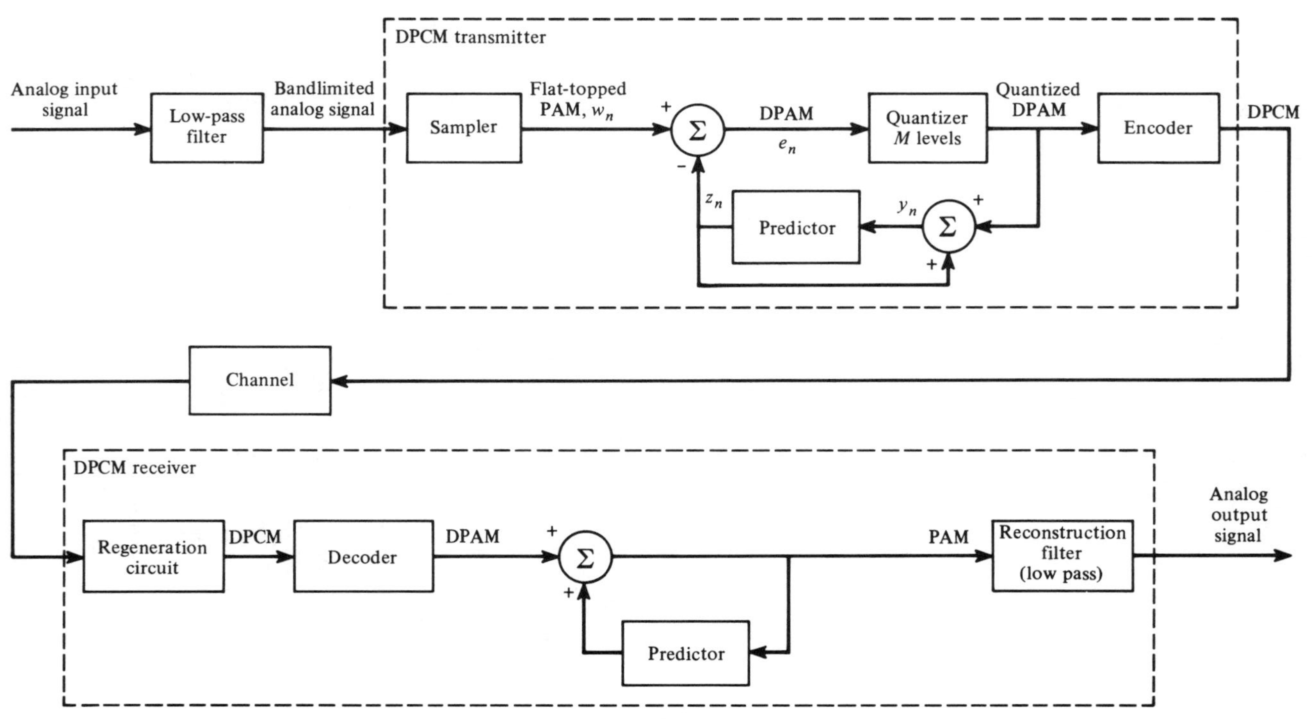

FIGURE 3-33 DPCM using prediction from quantized differential signal.

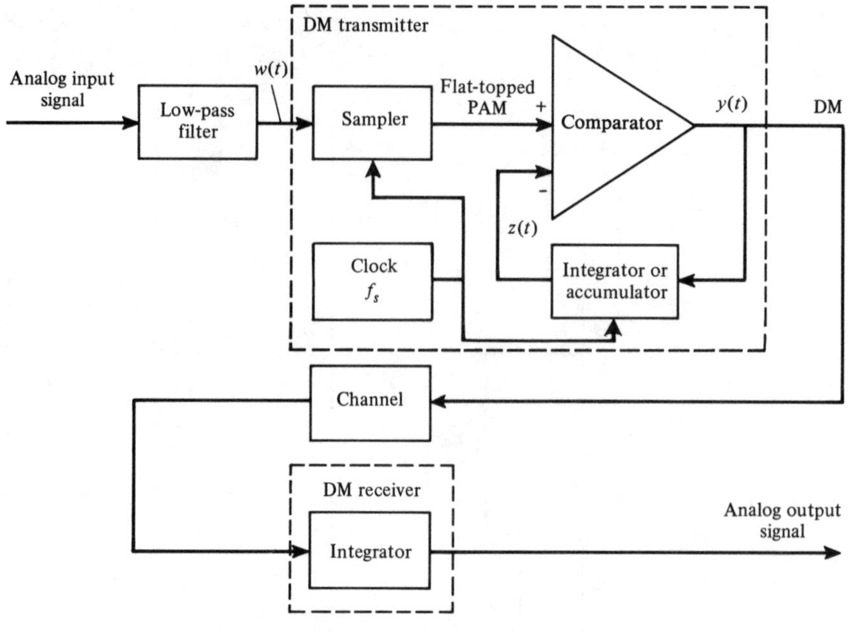

FIGURE 3-34 DM system.

3-7 Delta Modulation

From a block diagram point of view, *delta modulation* (DM) is a special case of DPCM where there are two quantizing levels. As shown in Fig. 3-33, for the case of $M = 2$, the quantized DPAM signal is *binary* and the encoder is not needed since the function of the encoder is to convert the multilevel DPAM signal to binary code words. For the case of $M = 2$, the DPAM signal is a DPCM signal where the code words are one bit long. The cost of a DM system is less than that of a DPCM system ($M > 2$) since the analog-to-digital converter (ADC) and digital-to-analog converter (DAC) are not needed. This is the main attraction of the DM scheme—it is relatively inexpensive. In fact, the cost may be further reduced by replacing the predictor by a low-cost integration circuit (such as an *RC* low-pass filter), as shown in Fig. 3-34.

In the DM circuit shown in Fig. 3-34, the operations of the subtractor and two-level quantizer are implemented by using a comparator so that the output is $\pm V_c$ (binary). In this case the DM signal is a polar signal. A set of waveforms associated with the delta modulator is shown in Fig. 3-35. In Fig. 3-35a an assumed analog input waveform is illustrated. If the instantaneous samples of the flat-topped PAM signal are taken at the beginning of each sampling period, the corresponding accumulator output signal is shown.† Here the integrator is as-

†The sampling frequency, $f_s = 1/T_s$, is selected to be within the range $2B_{\text{in}} < f_s < 2B_{\text{channel}}$, where B_{in} is the bandwidth of the input analog signal and B_{channel} is the bandwidth of the channel. The lower limit prevents aliasing of the analog signal and the upper limit prevents ISI in the DM signal at the receiver. See Example 3-2 for further restrictions on the selection of f_s.

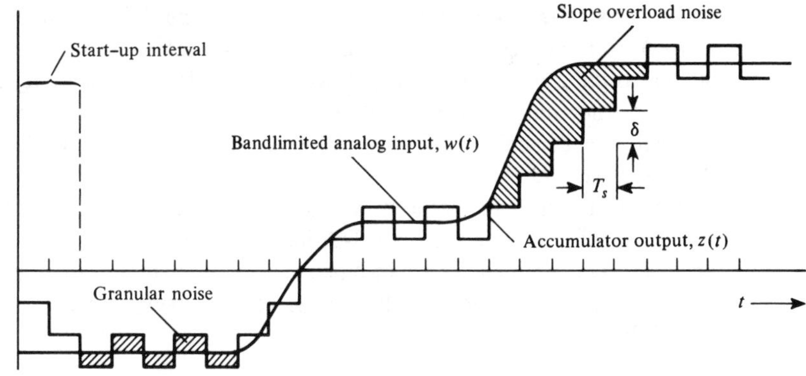

(a) Analog Input and Accumulator Output Waveforms

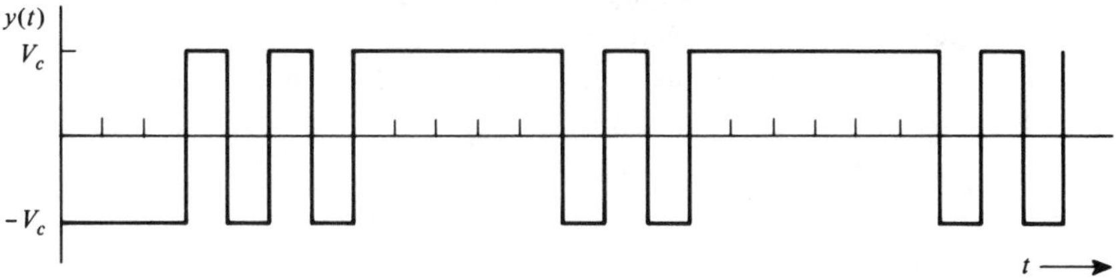

(b) Delta Modulation Waveform

FIGURE 3-35 DM system waveforms.

sumed to act as an accumulator (such as integrating impulses) so that the integrator output at time $t = nT_s$ is given by

$$z_n = \frac{1}{V_c} \sum_{i=0}^{n} \delta y_i \tag{3-96}$$

where $y_i = y(iT)$ and δ is the accumulator gain or step size. The corresponding DM output waveform is shown in Fig. 3-35b.

At the receiver the DM signal may be converted back to an analog signal approximation to the analog signal at the system input. This is accomplished by using an integrator for the receiver that produces a smoothed waveform corresponding to a smoothed version of the accumulator output waveform that is present in the transmitter (Fig. 3-35a).

Granular Noise and Slope Overload Noise

From Fig. 3-35a it is seen that the accumulator output signal does not always track the analog input signal. The quantizing noise error signal may be classified into two types of noise—*slope overload noise* and *granular noise*. Slope overload noise occurs when the step size δ is too small for the accumulator output to follow quick changes in the input waveform. Granular noise occurs for any step size but is smaller for a small step size. Thus we would like to have δ as small as possible to minimize the granular noise. The granular noise in a DM system is similar to the granular noise in a PCM system, whereas slope overload noise is a new phenomenon due to a differential signal being encoded (instead of the original signal itself). Both phenomena are also present in the DPCM system discussed earlier.

It is clear that there should be an optimum value for the step size, δ, since if δ is increased, the granular noise will increase but the slope overload noise will decrease. This optimum value of δ depends on the nature of the input signal and on the sampling frequency used. O'Neal has obtained results that are shown in Fig. 3-36 [O'Neal, 1966a].

EXAMPLE 3-2 DESIGN OF A DM SYSTEM

Find the step size δ required to prevent slope overload noise for the case when the input signal is a sine wave.

The maximum slope that can be generated by the accumulator output is

$$\frac{\delta}{T_s} = \delta f_s \tag{3-97}$$

For the case of the sine-wave input where $w(t) = A \sin \omega_a t$, the slope is

$$\frac{dw(t)}{dt} = A\omega_a \cos \omega_a t \tag{3-98}$$

and the maximum slope of the input signal is $A\omega_a$. Consequently, for no slope overload we require that $\delta f_s > A\omega_a$, or

$$\delta > \frac{2\pi f_a A}{f_s} \tag{3-99}$$

However, we do not want to make δ too much larger than this value or the granular noise will become too large.

★ The resulting detected signal-to-noise ratio can also be calculated. It has been determined experimentally that the spectrum of the granular noise is uniformly distributed over the frequency band $|f| \le f_s$. It can also be shown that the total granular quantizing noise is $\delta^2/3$ (see Sec. 7-7, where $\delta/2$ of PCM is replaced by δ for DM). Thus the PSD for the noise is $\mathcal{P}_n(f) = \delta^2/(6f_s)$. The granular noise power in the analog signal band, $|f| \le B$, is

$$N = \langle n^2 \rangle = \int_{-B}^{B} \mathcal{P}_n(f)\, df = \frac{\delta^2 B}{3f_s} \tag{3-100}$$

or, using (3-99), with equality

$$N = \frac{4\pi^2 A^2 f_a^2 B}{3f_s^3}$$

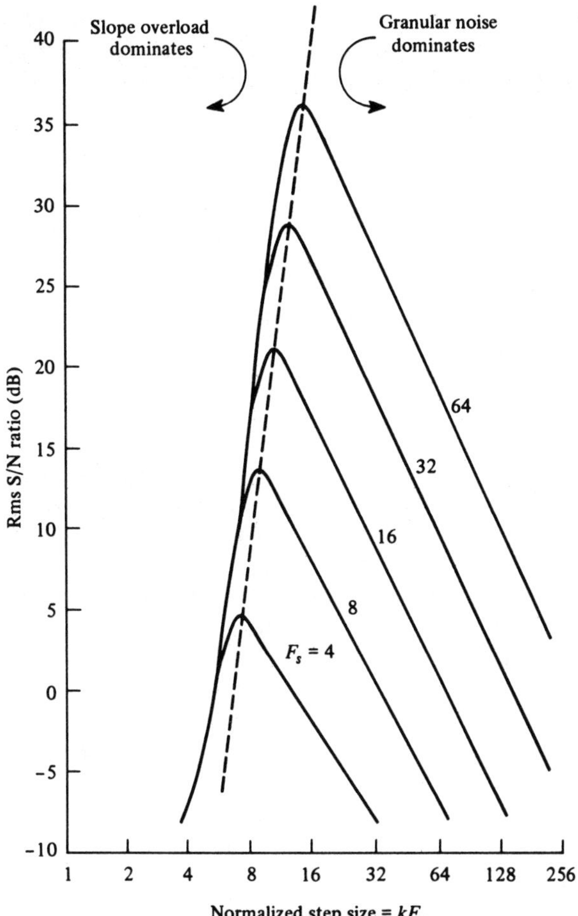

FIGURE 3-36 Signal-to-noise ratio out of a DM system as a function of step size for the case of an analog signal with a flat spectrum. $kF_s = (f_s\delta)\,(4\sigma B)$ where B is the absolute bandwidth of the signal, σ is the rms value of the signal, and $F_s = f_s/B$. [From J. B. O'Neal, Jr., "Delta modulation quantizing noise analytical and computer simulation results for Gaussian and television input signals," *Bell System Technical Journal*, Jan. 1966, p. 123. Copyright 1966, American Telephone and Telegraph Company. Reprinted by permission.]

The signal power is

$$S = \langle w^2(t) \rangle = \frac{A^2}{2} \tag{3-101}$$

The resulting average signal to quantizing noise ratio out of a DM system with a sine-wave test signal is

$$\left(\frac{S}{N}\right)_{\text{out}} = \frac{3}{8\pi^2}\frac{f_s^3}{f_a^2 B} \tag{3-102}$$

where f_s is the DM sampling frequency, f_a is the frequency of the sinusoidal input, and B is the bandwidth of the receiving system. It is emphasized that (3-102) was shown to be valid only for sinusoidal-type signals.

For voice-frequency (VF) audio signals, it has been shown that (3-99) is too restrictive if $f_a = 4$ kHz and that slope overload is negligible if [deJager, 1952]

$$\delta \geq \frac{2\pi 800 W_p}{f_s} \tag{3-103}$$

where W_p is the peak value of the input audio waveform, $w(t)$. (This is due to the fact that the midrange frequencies around 800 Hz dominate in the VF signal.) Combining (3-103) and (3-100), we obtain the S/N for the DM system with a VF-type signal

$$\left(\frac{S}{N}\right)_{\text{out}} = \frac{\langle w^2(t)\rangle}{N} = \frac{3f_s^3}{(1600\pi)^2 B}\left(\frac{\langle w^2(t)\rangle}{W_p^2}\right) \tag{3-104}$$

where B is the audio bandwidth and $(\langle w^2(t)\rangle/W^2)$ is the average-audio-power to peak-audio-power ratio.

This result can be used to design a VF DM system. For example, suppose that we desire an S/N of at least 30 dB. Assume that the VF bandwidth is 4 kHz and the average-to-peak audio power is $\frac{1}{2}$. Then (3-104) gives a required sampling frequency of 40.7 kbits/sec or $f_s = 10.2B$. It is also interesting to compare this DM system with a PCM system that has the same bandwidth (i.e., bit rate). The number of bits, n, required for each PCM word is determined by $R = (2B)n = 10.2B$ or $n \approx 5$. Then the average-signal to quantizing-noise ratio of the comparable PCM system is 30.1 dB (see Table 3-2). Thus, under these conditions, the PCM system with a comparable bandwidth has about the same S/N performance as the DM system. Furthermore, repeating the preceding procedure, it can be shown that if an S/N larger than 30 dB was desired, the PCM system would have a larger S/N than that of a DM system with comparable bandwidth; on the other hand, if an S/N less than 30 dB was sufficient, the DM system would outperform (i.e., have a larger S/N than) the PCM system of the same bandwidth. This is also shown by Fig. 3-38. Note that the S/N for DM increases as f_s^3 or 9 dB per octave increase of f_s.

It is also possible to improve the S/N performance of a DM system by using double integration instead of single integration, as was studied here. With a double integration system the S/N increases as f_s^5 or 15 dB per octave [Jayant and Noll, 1984].

Adaptive Delta Modulation and Continuously Variable Slope Delta Modulation

To minimize the slope overload noise while holding the granular noise at a reasonable value, *adaptive delta modulation* (ADM) is used. Here the step size is varied as a function of time as the input waveform changes. The step size is kept small to minimize the granular noise until the slope overload noise begins to

TABLE 3-6 Step-Size Algorithm

Data Sequence[a]				Number of Successive Binary 1's or 0's	Step-Size Algorithm, $f(d)$
×	×	0	1	1	δ
×	0	1	1	2	δ
0	1	1	1	3	2δ
1	1	1	1	4	4δ

[a] ×, don't care.

dominate. Then the step size is increased to reduce the slope overload noise. The step size may be adapted, for example, by examining the DM pulses at the transmitter output. When the DM pulses consist of a string of pulses with positive polarity, the step size is increased (see Fig. 3-35) until the DM pulses begin to alternate in polarity, then decreased, and so on. One possible algorithm for varying the step size is shown in Table 3-6, where the step size changes with discrete variation. Here the step size is normally set to a value δ when the ADM signal consists of data with alternating 1's and 0's or when two successive binary 1's occur. The step size is also δ for any number of binary 0's. However, if three successive binary 1's occur, the step size is increased to 2δ, and 4δ for four successive binary 1's. Figure 3-37 gives a block diagram for this ADM system.

Papers have been published with demonstration records that illustrate the quality of ADM and other digital techniques when voice signals are sent over digital systems and recovered [Flanagan et al., 1979; Jayant, 1974]. Another variation of ADM is *continuously variable slope delta modulation* (CVSD). Here the step size is made continuously variable instead of stepped in discrete increments.

The Motorola MC3417 is a CVSD integrated circuit which varies the slope (step size) as a function of the last three bits of the encoded data sequence; the MC3418 has a similar 4-bit algorithm [Motorola, 1985]. Both of these CVSD codec chips can be connected either as modulators (encoders) or as demodulators (decoders) and may be operated with bit rates ranging from 9.4 to 64 kbits/sec. Of course, a better S/N is obtained at the decoder output if a higher bit rate is used. For example, if MC3417 codecs are used for encoding and decoding in a 16-kbit/sec CVSD system, the S/N at the output will be 16 dB. This is called "communication quality." If MC3418s are used in a 37.7-kbit/sec system, the output S/N will be 30 dB. This performance is similar to that obtained for the DM and PCM systems discussed after equation (3-104).

The question might be asked: Which is better, PCM or DM? The answer, of course, depends on the criterion used for comparison and the type of message. If the objective is to have a relatively simple, low-cost system, delta modulation may be the best. However, the cost of ADCs is dropping, so this may not be a factor in the future. If high output S/N is the criterion, PCM is probably the best. This is illustrated by O'Neal's results when he compares the output S/N for PCM

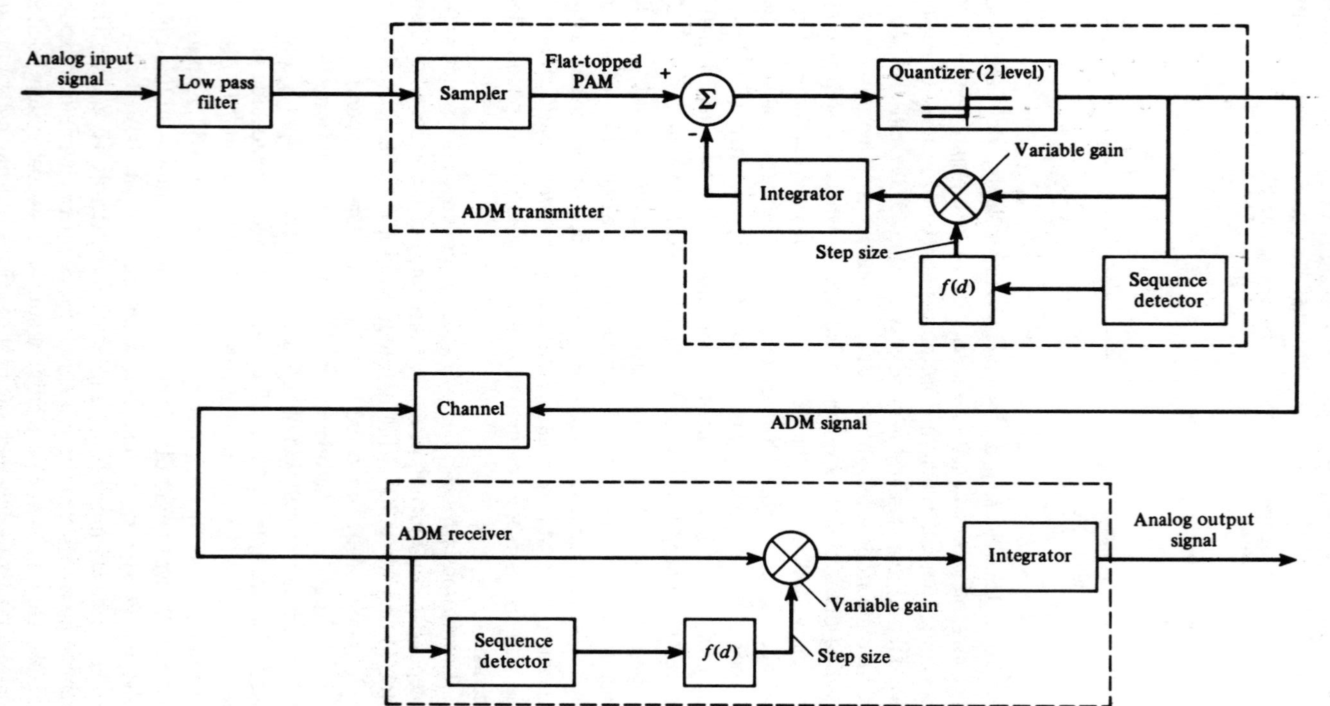

FIGURE 3-37 ADM system.

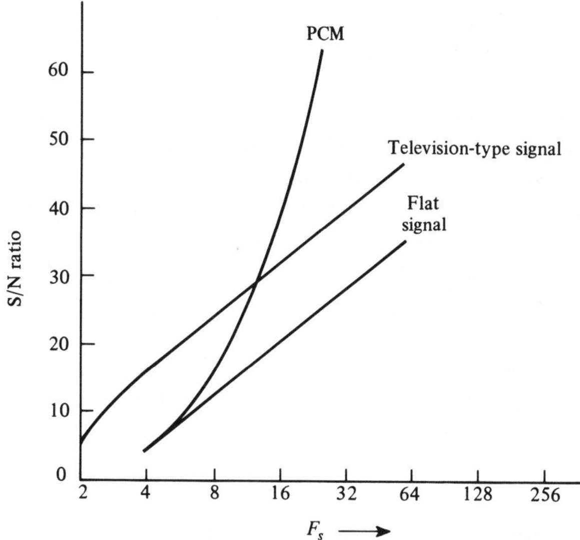

FIGURE 3-38 Comparison of $(S/N)_{out}$ for DM and PCM systems. The PCM curve is for the case of uniform quantizing steps. The television-type signal curve is for DM with an analog signal that has an *RC*-type low-pass filter rolloff (i.e., $\mathcal{P}(f) = 1/[1 + (f/f_{3dB})^2]$) but absolutely bandlimited to $B = 4f_{3dB}$ Hz. The flat signal curve is for a signal with a flat spectrum that is absolutely bandlimited to B. $F_s = f_s/B$ where f_s is the sampling frequency and B is the absolute bandwidth. [From J. B. O'Neal, Jr., "Delta modulation quantizing noise analytical and computer simulation results for Gaussian and television input signals," *Bell System Technical Journal,* Jan. 1966, p. 127. Copyright 1966, American Telephone and Telegraph Company. Reprinted by permission.]

and DM systems [O'Neal, 1966a] (see Fig. 3-38). If one is interfacing to existing equipment, compatibility, of course, would be a prime consideration. In this regard, PCM has the advantage because it was adopted first and is widely used.

Speech Coding

Digital speech coders can be classified into two categories: waveform coders and vocoders. *Waveform coders* use algorithms to encode and decode so that the system output is an approximation to the input waveform. *Vocoders* encode the speech by extracting a set of parameters. These parameters are digitized and transmitted to the receiver where they are used to set values for parameters in function generators and filters which, in turn, synthesize the output speech sound. Usually, the vocoder output waveform does not approximate the input waveform and has an artificial, unnatural sound. Although the words of the speaker may be clearly understandable, the speaker may not be identifiable. That is, the speaker may "sound like a robot." For this reason and because of space limitations, only waveform encoders (e.g., PCM, DPCM, DM, and CVSD) are covered in this book. It was demonstrated that VF-quality speech may be encoded at bit rates as low as 24 kbits/sec. More advanced techniques reduce the required bit rate to 16 or 8 kbits/sec, and some researchers believe that speech coding may be possible even at 2 kbits/sec. Some techniques that are available to achieve waveform coders at low bit rates are: linear prediction (already discussed as applied to DPCM), adaptive subband coding, and vector quantization. Adaptive subband coding allocates bits according to the input speech spectrum and the properties of hearing. With vector quantization, whole blocks of samples are encoded at a time instead of encoding on a sample-by-sample basis. An example is the vector sum excited linear prediction filter (VSELF) technique that is proposed for use in digital cellular telephones and described at the end of Sec. 5-8. Of course, these advanced techniques require more complex circuits and a larger delay while the output is being computed. For more details about this interesting topic of speech coders, the reader is referred to the literature [Aoyama, Daumer, and Modena, 1988; Flanagan et al., 1979; Jayant, 1986].

3-8 Time-Division Multiplexing†

Definition: *Time-division multiplexing* (TDM) is the time interleaving of samples from several sources so that the information from these sources can be transmitted serially over a single communication channel.

Figure 3-39 illustrates the TDM concept as applied to three analog sources that are multiplexed over a PCM system. For convenience, natural sampling is shown together with the corresponding gated TDM PAM waveform. In practice, an electronic switch is used for the commutation (sampler). In this example the pulse width of the TDM PAM signal is $T_s/3 = 1/(3f_s)$, and the pulse width of the TDM PCM signal is $T_s/(3n)$, where n is the number of bits used in the PCM word. Here $f_s = 1/T_s$ denotes the frequency of rotation for the commutator, and f_s satisfies the Nyquist rate for the analog source with the largest bandwidth. In some applications where the bandwidth of the sources is markedly different, the larger bandwidth sources may be connected to several switch positions on the sampler so that they will be sampled more often than the smaller bandwidth sources.

At the receiver the decommutator (sampler) has to be synchronized with the incoming waveform so that the PAM samples corresponding to source 1, for example, will appear on the channel 1 output. This is called frame synchronization. Low-pass filters are used to reconstruct the analog signals from the PAM samples. ISI resulting from poor channel filtering would cause PCM samples from one channel to appear on another channel, even though perfect bit and

†This is a method in which a common channel or system is shared by many users. Other methods for sharing a common communication system are discussed under the topic of multiple access techniques in Sec. 5-4.

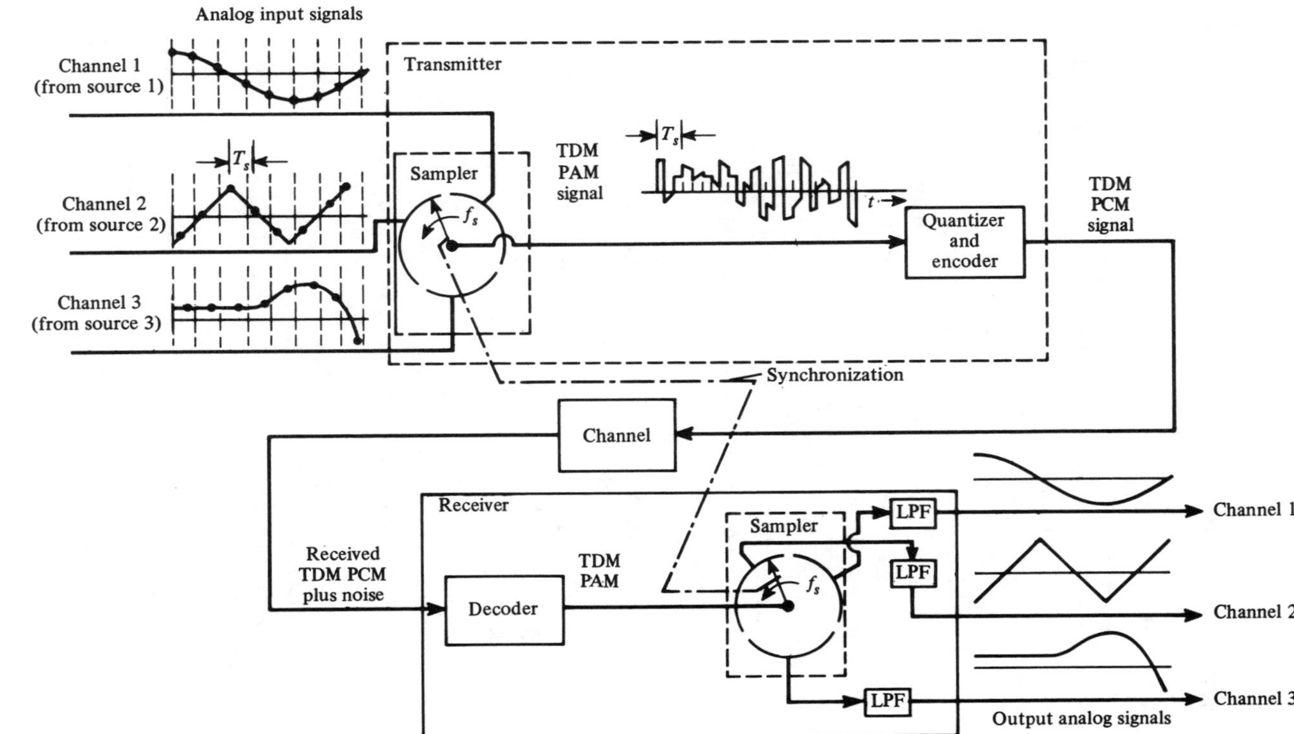

FIGURE 3-39 Three-channel TDM PCM system.

frame synchronization were maintained. This feedthrough of one channel's signal into another channel is called *crosstalk*.

★ Frame Synchronization

Frame synchronization is needed at the TDM receiver so that the received multiplexed data can be sorted and directed to the appropriate output channel. The frame sync can be provided to the receiver demultiplexer (demux) circuit either by sending a frame sync signal from the transmitter over a separate channel or by deriving the frame sync from the TDM signal itself. Because the implementation of the first approach is obvious, we will concentrate on implementation of the latter approach. In addition, the latter approach is usually more economical since a separate sync channel is not needed. As illustrated in Fig. 3-40, frame sync may be multiplexed along with the information words in an N channel TDM system by transmitting a unique K-bit sync word at the beginning of each frame. As illustrated in Fig. 3-41, the frame sync is recovered from the corrupted TDM signal by using a frame synchronizer circuit which crosscorrelates the regenerated TDM signal with the expected unique sync word $S = (s_1, s_2, \ldots, s_K)$ that is specified by the fixed tap gains $s_1, s_2, \ldots, s_K$. Here, assume that the bits of the sync word s_j's, are in a polar format (± 1). The present bit of the regenerated TDM signal is clocked into the first stage of the shift register and then shifted to the next stage on the next clock pulse so that the most immediate K bits are always stored in the shift register. The contents of the shift register are multiplied by the tap gains which correspond to the bits (± 1) of the sync word. These are *algebraically* summed and fed into the comparator input. When the unique K bit sync word appears in the data stream and is exactly aligned in the shift register so that s_1 is in stage 1, s_2 is in stage 2, and so on, then the voltage at the input to the comparator will be

$$v_c = s_1^2 + s_2^2 + \cdots + s_K^2 = K \tag{3-105}$$

K is also the maximum voltage that is possible for v_c for any possible data sequence that could be stored in the shift register. Furthermore, if the data in the shift register happen to differ in sign at any one-bit position compared to the bits in the sync word, then $v_c = K - 2$. Consequently, if the comparator threshold voltage, V_T, is set slightly smaller than K volts, say $V_T = K - 1$, then the comparator output will go high only during the T_b-second interval when the sync word is perfectly aligned in the shift register. Thus the frame synchronizer recovers the frame sync signal.

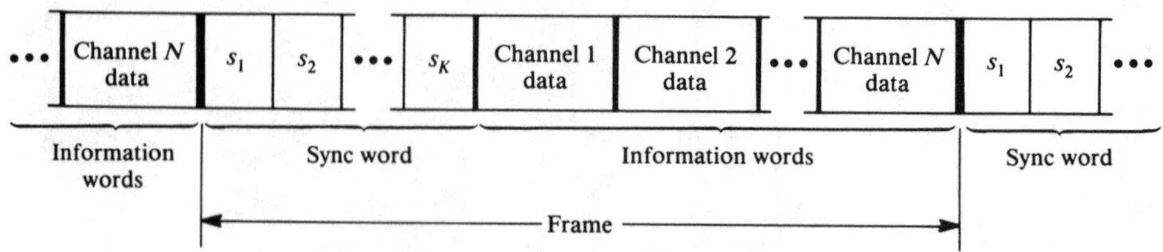

FIGURE 3-40 TDM frame sync format.

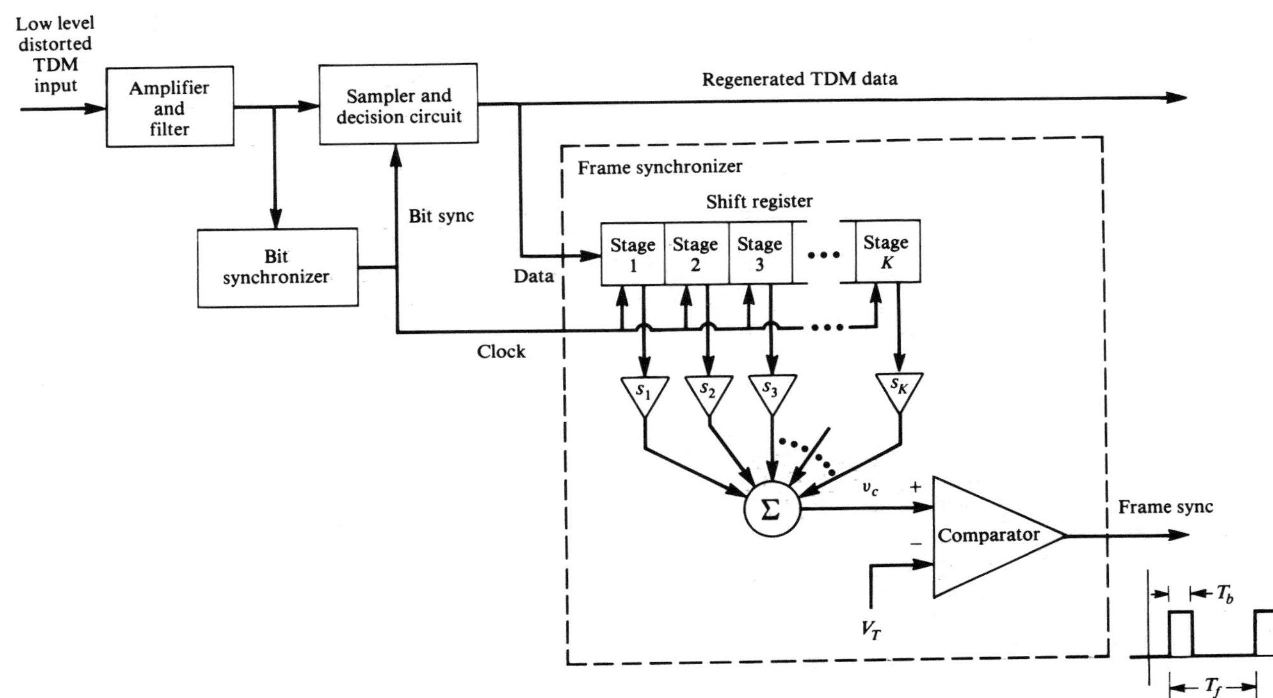

FIGURE 3-41 Frame synchronizer with TDM receiver front end.

False sync output pulses will occur if K successive information bits happen to match the bits in the sync word. For equally likely TDM data and $V_T = K - 1$, the probability of this false sync occurring is

$$P_f = \left(\frac{1}{2}\right)^K = 2^{-K} \tag{3-106}$$

In frame synchronizer design, this equation may be used to determine the number of bits, K, needed in the sync word so that the false lock probability will meet specifications. Alternatively, more sophisticated techniques such as aperture windows can be used to suppress false lock pulses [Ha, 1986]. The information words may also be encoded so that they are not allowed to have the bit strings that match the unique sync word.

Since the input to the comparator, v_c, is the crosscorrelation of the sync word with the passing K-bit word stored in the shift register, the sync word needs to be chosen so that its autocorrelation function, $R_s(k)$, has the desirable properties: $R_s(0) = 1$ and $R(k) \approx 0$ for $k \neq 0$. The PN codes (studied in Sec. 5-7) are almost ideal in this regard. For example, if $P_f = 4 \times 10^{-5}$ is the allowed probability of false sync, then from (3-106) a ($K = 15$)-bit sync word is required. Consequently, a 15-stage shift register is needed for the frame synchronizer in the receiver. The 15-bit PN sync word can be generated at the transmitter using a four-stage shift register. For further study of frame synchronizers the reader is referred to Ziemer and Peterson [1985] and Ha [1986].

Synchronous and Asynchronous Lines

For bit sync, data transmission systems are designed to operate with either synchronous or asynchronous serial data lines. In a *synchronous* system, each device is designed so that its internal clock is relatively stable for a long period of time, and it is synchronized to a system master clock. Each bit of data is clocked in synchronism with the master clock. The synchronizing signal may be provided by a separate clocking line or may be embedded in the data signal (e.g., by the use of Manchester line codes). In addition, synchronous transmission requires a higher level of synchronization to allow the receiver to determine the beginning and end of blocks of data. This is achieved by the use of frame sync as discussed previously and data link protocols as described in Appendix C.

In *asynchronous* systems, the timing is precise only for the bits within each character (or word). This is also called *start-stop* signaling because each character consists of a "start bit" that starts the receiver clock and concludes with one or two "stop bits" that terminate the clocking. Usually, two stop bits are used with terminals that signal at rates less than 300 bits/sec, and one stop bit is used if $R \geq 300$ bits/sec. Thus, with asynchronous lines, the receiver clock is started aperiodically and no synchronization with a master clock is required. This is ideal for keyboard terminals where the typist does not type at an even pace and the input rate is much slower than that of the data communication system. These asynchronous terminals often use a 7-bit ASCII code (see Appendix C), and the complete character consists of one start bit, 7 bits of the ASCII code, one parity bit, and one stop bit (for $R \geq 300$ bits/sec). This gives a total character length of 10 bits. In TDM of the asynchronous type, the different sources are multiplexed on a *character-interleaved* (i.e., character-by-character) basis instead of interleaving on a bit-by-bit basis. The synchronous transmission system is more efficient because start and stop bits are not required. However, the synchronous mode of transmission requires that the clocking signal be passed along with the data and that the receiver synchronize to the clocking signal.

''Intelligent'' TDMs may be used to *concentrate* data arriving from many different terminals or sources. They are capable of providing speed, code, and protocol conversion. At the input to a large mainframe computer, these devices are called *front-end processors*. The hardware in the intelligent TDM consists of microprocessors or minicomputers. Usually, they connect ''on the fly'' to the input data lines that have data present and momentarily disconnect the lines that do not have data present. For example, the user of a keyboard terminal is disconnected from the system while he or she is thinking (although to the user it does not appear to be disconnected) but is connected as each character or block of data is sent. Thus the output data rate of the multiplexer is much less than the sum of the data capacities of the input lines. This technique is called *statistical* multiplexing; it allows many more terminals to be connected on line to the system.

Multiplexers can also be classified into three general types. The first TDM type consists of those that connect to synchronous lines. The second TDM type consists of those that connect to quasi-synchronous lines. In this case, the individual clocks of the input data sources are not exactly synchronized in frequency. Consequently, there will be some variation in the bit rates between the data arriving from different sources. In addition, in some applications the clock rates of the input data streams are not related by a rational number. In these cases, the TDM output signal will have to be clocked at an increased rate above the nominal value to accommodate those inputs that are not synchronous. When a new input bit is not available at the multiplexer clocking time (due to nonsynchronization), *stuff bits,* which are dummy bits, are inserted in the TDM output data stream. This is illustrated by the bit-interleaved multiplexer shown in Fig. 3-42. The stuff bits may be binary 1's, 0's, or some alternating pattern, depending on the choice of the system designer. The third TDM type consists of those that operate with asynchronous sources and produce a high-speed asynchronous output (no stuff bits required) or high-speed synchronous output (stuff bits required).

EXAMPLE 3-3 DESIGN OF A TIME-DIVISION MULTIPLEXER

Design a time-division multiplexer that will accommodate 11 sources. Assume that the sources have the following specifications.

Source 1. Analog, 2-kHz bandwidth.
Source 2. Analog, 4-kHz bandwidth.
Source 3. Analog, 2-kHz bandwidth.
Sources 4–11. Digital, synchronous at 7200 bits/sec.

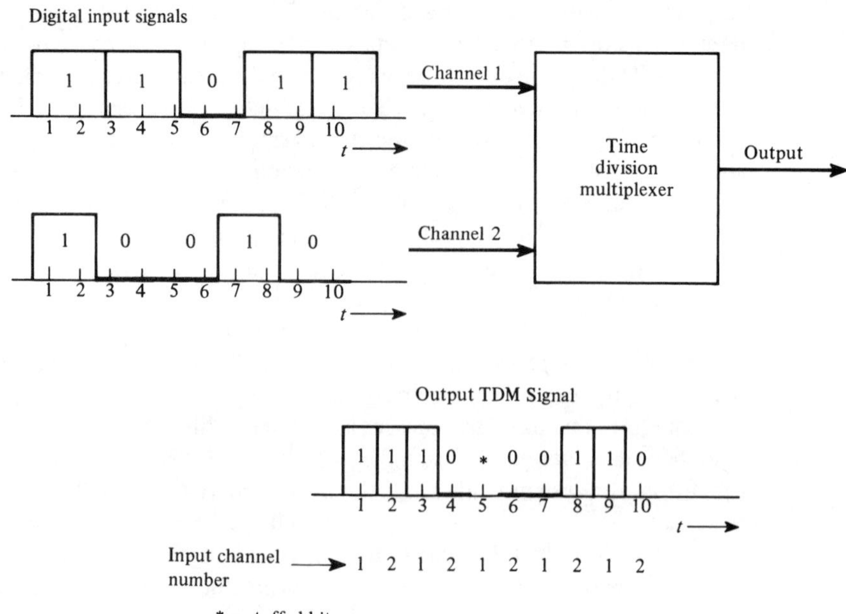

FIGURE 3-42 Two-channel bit-interleaved TDM with pulse stuffing.

Also assume that the analog sources will be converted into 4-bit PCM words and, for simplicity, that frame sync will be provided via a separate channel and synchronous TDM lines are used. To satisfy the Nyquist rate for the analog sources, sources 1, 2, and 3 need to be sampled at 4, 8, and 4 kHz, respectively. As shown in Fig. 3-43, this can be accomplished by rotating the first commutator at $f_1 = 4$ kHz and sampling source 2 twice on each revolution. This produces a 16-kilosamples/sec TDM PAM signal on the commutator output. Each of the analog sample values is converted into a 4-bit PCM word, so that the rate of the TDM PCM signal on the ADC output is 64 kbits/sec. The digital data on the ADC output may be merged with the data from the digital sources by using a second commutator rotating at $f_2 = 8$ kHz and wired so that the 64-kbits/sec PCM signal is present on 8 of 16 terminals. This provides an effective sampling rate of 64 kbits/sec. On the other eight terminals the digital sources are connected to provide a data transfer rate of 8 kbits/sec for each source. Since the digital sources are supplying a 7.2-kbit/sec data stream, pulse stuffing is used to raise the source rate to 8 kbits/sec.

This example illustrates the main advantage of TDM: it can easily accommodate both analog and digital sources. Unfortunately, when analog signals are converted to digital signals without redundancy reduction, they consume a great deal of digital system capacity.

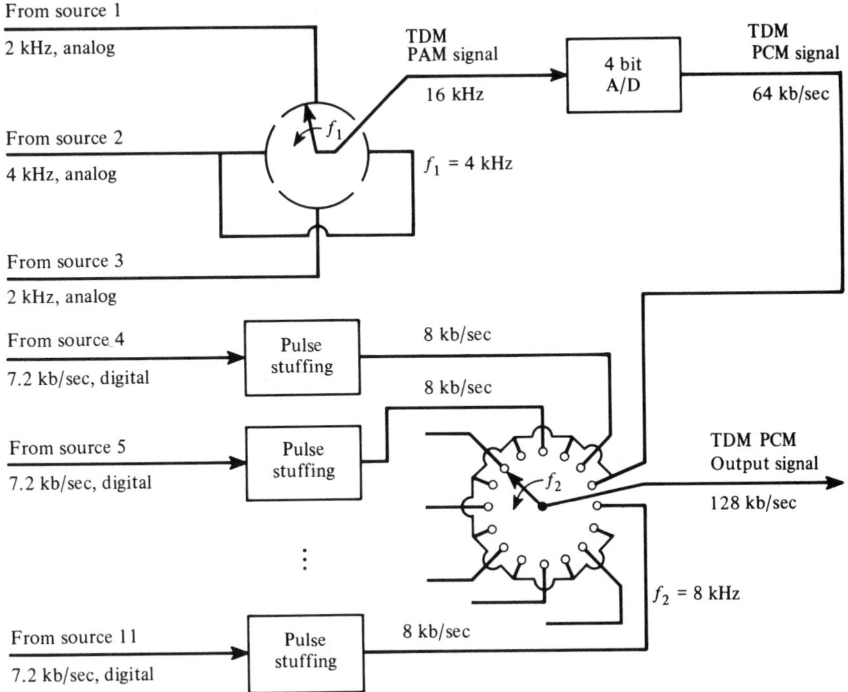

FIGURE 3-43 TDM with analog and digital inputs as described in Example 3-3.

TDM Hierarchy

In practice, TDMs may be grouped into two categories. The first category consists of multiplexers used in conjunction with digital computer systems to merge digital signals from several sources for TDM transmission over a high-speed line to a digital computer. The output rate of these multiplexers has been standardized to 1.2, 2.4, 3.6, 4.8, 7.2, 9.6, and 19.2 kbits/sec.

The second category of time-division multiplexers is used by the common carriers, such as the American Telephone and Telegraph Company (AT&T), to combine different sources into a high-speed digital TDM signal for transmission over toll networks. Unfortunately, the standards adopted by North America and Japan are different from those that have been adopted in other parts of the world. Of course, this leads to complications at the boundaries of two systems with different standards. The North America/Japan standards were first adopted by AT&T, and another set of standards has been adopted by the International Telegraph and Telephone Consultative Committee (CCITT) under the auspices of the International Telecommunications Union (ITU). The North American TDM

TABLE 3-7 TDM Standards for North America

Digital Signal Number	Bit Rate, R (Mbits/sec)	No. of 64 kbits/sec PCM VF Channels	Transmission Media Used
DS-0	0.064	1	Wire pairs
DS-1	1.544	24	Wire pairs
DS-1C	3.152	48	Wire pairs
DS-2	6.312	96	Wire pairs, fiber
DS-3	44.736	672	Coax., radio, fiber
DS-3C	90.254	1344	Radio, fiber
DS-4E	139.264	2016	Radio, fiber, coax.
DS-4	274.176	4032	Coax., fiber
DS-432	432.00	6048	Fiber
DS-5	560.160	8064	Coax., fiber

hierarchy is shown in Fig. 3-44 [James and Muench, 1972].† The telephone industry has standardized the bit rates to 1.544 Mbits/sec, 6.312 Mbits/sec, etc., and designates them as DS-1 for digital signal, type 1; DS-2 for digital signal, type 2; etc. as listed in Table 3-7. In Fig. 3-44 all input lines are assumed to be digital (binary) streams, and the number of voice-frequency (VF) analog signals that can be represented by these digital signals is shown in parentheses. The higher multiplexing level inputs are not always derived from lower level multiplexers. For example, one analog television signal can be converted directly to a DS-3 data stream (44.73 Mbits/sec). Similarly, the DS streams can carry a mixture of information from a variety of sources such as video, VF, and computers.

The transmission medium that is used for the multiplex levels depends on the DS level involved and on the economics of using a particular type of medium at a particular location (see Table 3-7). For example, higher DS levels may be transmitted over coaxial cables, fiber optic cable, microwave radio, or via satellite. A single DS-1 signal is usually transmitted over one pair of twisted wires. (One pair is used for each direction.) This type of DS-1 transmission over a twisted-pair medium is known (from its development in 1962 by AT&T) as the *T1 carrier system*. DS-1 signaling over a T1 system is very popular because of its relatively low cost and its excellent maintenance record. (T1 will be discussed in more detail in the following section.) Table 3-8 shows the specifications for the T-carrier digital baseband systems. Table 5-7 is a similar table for the capacity of common carrier bandpass systems. The corresponding CCITT TDM standard that is used throughout the world except for North America and Japan† is shown in Fig. 3-45 [Irmer, 1975].

†The Japanese TDM hierarchy is the same as that for North America for multiplex levels 1 and 2, but differs for levels 3, 4, and 5. For level 3 the Japanese standard is 32.064 Mbits/sec (480 VF), level 4 is 97.728 Mbits/sec (1440 VF), and level 5 is 397.200 Mbits/sec (5760 VF). Dissimilarities between standards used are briefly discussed and summarized by Jacobs [1986].

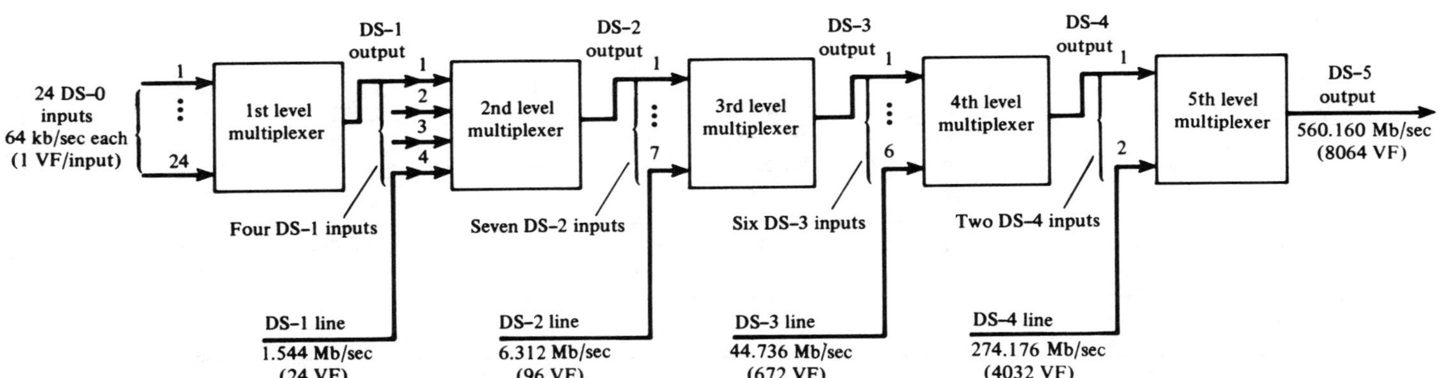

FIGURE 3-44 North American digital TDM hierarchy.

TABLE 3-8 Specifications for T-Carrier Baseband Digital Transmission Systems

		System Capacity							
System	Rate (Mbits/sec)	Digital Signal No.	Voice Channels	TV	Medium	Line Code	Repeater Spacing (miles)	Maximum System Length (miles)	System Error Rate
T1	1.544	DS-1	24	—	Wire pair	Bipolar RZ	1	50	10^{-6}
T1C	3.152	DS-1C	48	—	Wire pair	Bipolar RZ	1	—	10^{-6}
T1D	3.152	DS-1C	48	—	Wire pair	Duobinary NRZ	1	—	10^{-6}
T1G	6.443	DS-2	96	—	Wire pair	4-level NRZ	1	200	10^{-6}
T2	6.312	DS-2	96	—	Wire pair[a]	B6ZS[b] RZ	2.3	500	10^{-7}
T3	44.736	DS-3	672	1	Coax.	B3ZS[b] RZ	[c]	[c]	[c]
T4	274.176	DS-4	4032	6	Coax.	Polar NRZ	1	500	10^{-6}
T5	560.160	DS-5	8064	12	Coax.	Polar NRZ	1	500	4×10^{-7}

[a] Special two-wire cable is required for 12,000-ft repeater spacing. Since T2 cannot use standard exchange cables, it is not as popular as T1.

[b] BnZS denotes *binary n-zero substitution,* where a string of *n* zeros in the bipolar line code is replaced with a special three-level code word so that synchronization can be maintained [Fike and Friend, 1984].

[c] Used in central telephone office for building multiplex levels; not used for transmission from office to office.

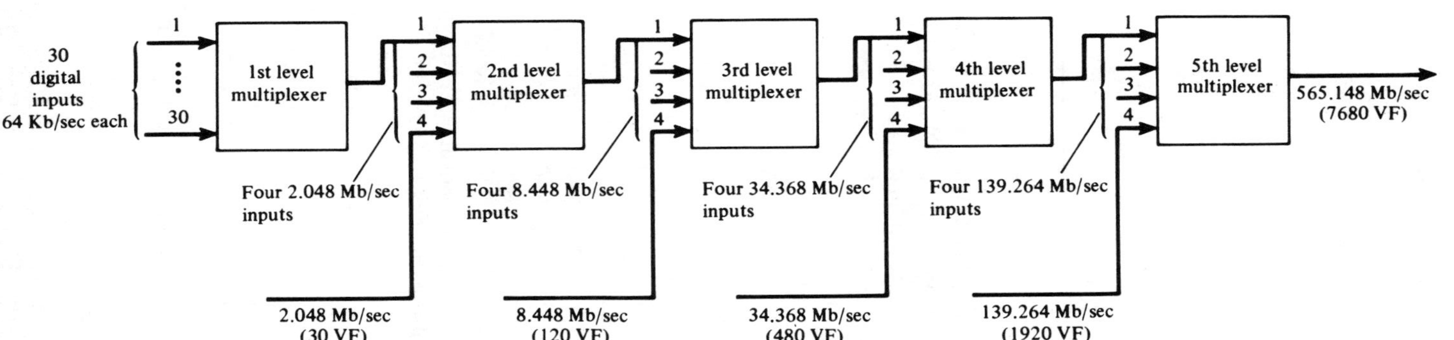

FIGURE 3-45 CCITT digital TDM hierarchy.

TABLE 3-9 SONET Signal Hierarchy

OC Level	Line Rate (MBits/sec)	Equivalent Number of: DS-3s	DS-1s	DS-0s
OC-1	51.84	1	28	672
OC-3	155.52	3	84	2,016
OC-9	466.56	9	252	6,048
OC-12	622.08	12	336	8,064
OC-18	933.12	18	504	12,096
OC-24	1,244.16	24	672	16,128
OC-36	1,866.24	36	1,008	24,192
OC-48	2,488.32	48	1,344	32,256

With the development of high-bit-rate fiber optic systems, it has become apparent that the original TDM standards are not adequate. A new TDM standard called SONET (Synchronous Optical Network) was proposed by Bellcore (Bell Communications Research) around 1985 and has evolved into an international standard that was adopted by the CCITT in 1989. This SONET standard is shown in Table 3-9. The OC-1 signal is an optical (light) signal that is turned on and off (modulated) by an electrical binary signal that has a line rate of 51.84 Mbits/sec. The electrical signal is called the STS-1 signal (Synchronous Transport Signal—level 1). Other OC-N signals have line rates of exactly N times the OC-1 rate and are formed by modulating a light signal with a STS-N electrical signal. The STS-N signal is obtained by byte-interleaving (scrambling) N STS-1 signals. (More details about fiber optic systems are given in Sec. 5-6.)

The telephone industry is moving toward an all-digital network that integrates voice and data over a single telephone line from each user to the telephone company equipment. This approach is called the *integrated service digital network* (ISDN). The standard implementation of *"basic rate"* ISDN service uses an overall data rate of 144 kbits/sec which is broken down into two B channels (64 kbits/sec each) and one D channel (16 kbits/sec). The two B channels carry PCM-encoded telephone conversations or data and the D channel is used for signaling to set up calls, disconnect calls, and route data for the B channels [Pandhi, 1987]. Basic rate ISDN is also called *narrowband ISDN* (N-ISDN). The *"primary rate"* ISDN service is in the development stage, but one implementation has an aggregate data rate of 1536 kbits/sec and consists of twenty-three 64-kbitsSsec B channels and one 64-kbits/sec D channel. Primary rate ISDN is also called *broadband ISDN* (B-ISDN). A more detailed discussion of ISDN will be delayed until Sec. 5-3 since material covered in the intervening chapters is useful for a better understanding of ISDN and its applications.

The T1 PCM System

For telephone voice service the first-level TDM multiplexer in Fig. 3-44 is replaced by a TDM PCM system, which will convert 24 VF analog telephone signals to a DS-1 (1.544 Mbits/sec) data stream. In AT&T terminology this is either a D-type channel bank or a digital carrier trunk (DCT) unit. ITT Telecommunications manufactures the T324 system, equivalent to the D3 channel bank, which also converts 24 VF signals into a DS-1, T1 type TDM PCM signal. Figures 3-46 and 3-47 display a functional block diagram of the transmit and receive portions of the T324 system. Here it is noticed that the VF and signaling for each of the 24 inputs are handled separately. In telephone terminology, *signaling* refers to address information (telephone number dialed); status information such as busy, dial tone, ringing, on hook and off hook; and metering information for billing purposes. *PAM Highway* denotes the 24 TDM PAM sampled signal that is encoded into a PCM signal. The *T1 line span* refers to a twisted-pair telephone line that is used to carry the TDM PCM (1.544 Mbits/sec) data stream. Two lines, one for transmit and one for receive, are used in the system. If the T1 line is connecting telephone equipment at different sites, repeaters are required about every mile.

The T1 system was developed by Bell Laboratories for short-haul digital communication of VF traffic up to 50 mi. The sampling rate used on each of the 24 VF analog signals is 8 kHz, which means that one frame length is $1/(8 \text{ kHz}) = 125$ μsec, as shown in Fig. 3-48. Presently, each analog sample is nominally encoded into an 8-bit PCM word so that there are $8 \times 24 = 192$ bits of data, plus one bit that is added for frame synchronization, yielding a total of 193 bits per frame. The T1 data rate is then (193 bits/frame)(8000 frames/sec) = 1.544 Mbits/sec, and the corresponding duration of each bit is 0.6477 μsec. The signaling is incorporated into the T1 format by replacing the eighth bit (the least significant bit) in each of the 24 channels of the T1 signal by a signaling bit in every sixth frame. Thus the signaling data rate for each of the 24 input channels is (1 bit/6 frames)(8000 frames/sec) = 1.333 kbits/sec. The framing bit used in the even-numbered frames follows the sequence 001110, and in the odd-numbered frames it follows the sequence 101010, so that the frames with the signaling information in the eighth bit position (for each channel) may be identified. The digital signaling on the T1 line is represented by a bipolar RZ waveform format (see Fig. 3-11) with peak levels of ± 3 V across a 100-Ω load. Consequently, there is no dc component on the T1 line regardless of the data pattern. In encoding the VF PAM samples, a $\mu = 255$ type compression characteristic is used, as described earlier in this chapter. Since the T1 data rate is 1.544 Mbits/sec and the line code is bipolar, the first zero-crossing bandwidth is 1.544 MHz and the spectrum peaks at 772 kHz, as shown in Fig. 3-12d. If the channel filtering transfer function was of the raised cosine-rolloff type with $r = 1$, the absolute channel bandwidth would be 1.544 MHz. In the past, when twisted-pair lines were used for analog transmission only, *loading* coils (inductors) were used to improve the frequency (amplitude) response. However, they cause the line to have a phase response that is not linear with frequency, resulting in ISI. Thus they must be removed when the line is used for T1 service.

T1G carrier, as described in Table 3-8, is the newest of the T-carrier type systems. Instead of binary levels, it uses $M = 4$ (quaternary) multilevel signaling, where $+3$ V represents the two binary bits 11, $+1$ V represents 01, -1 V

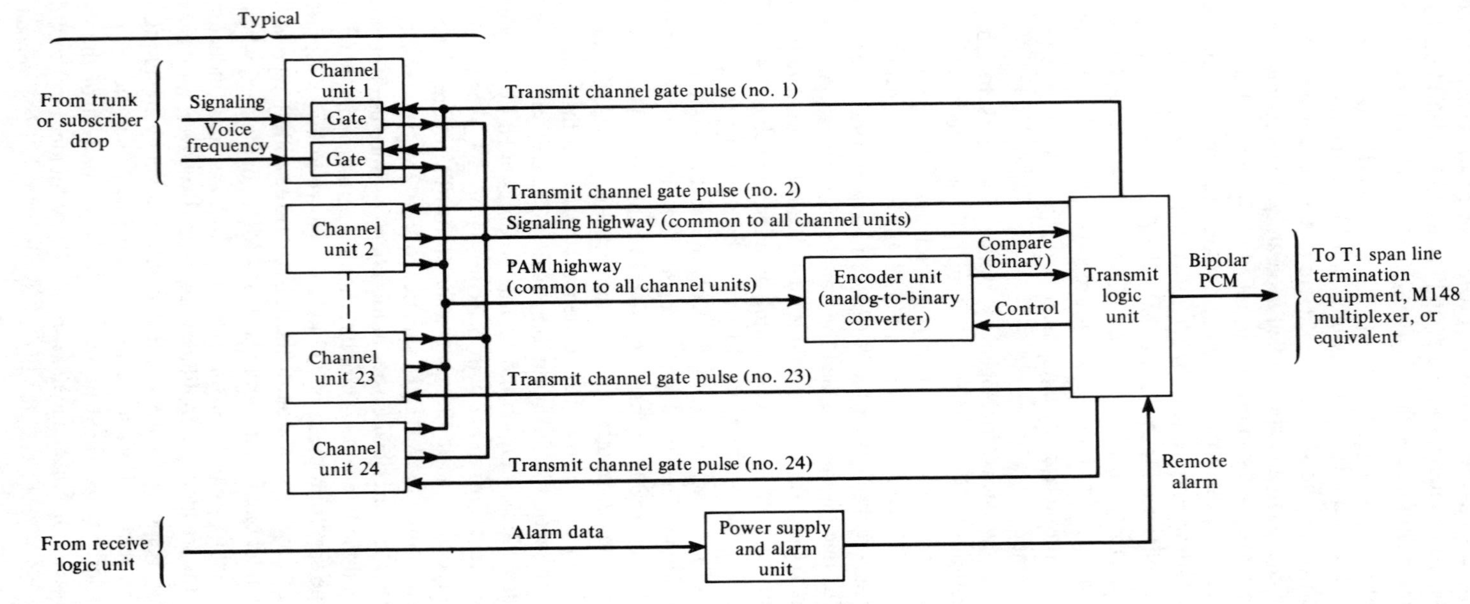

FIGURE 3-46 Transmit functional diagram of the ITT T324 PCM voice-frequency multiplexer. [From "Operation-Installation-Maintenance Manual for T324 PCM Carrier System," Pub. 650038-823-001, *ITT Telecommunications*, Transmissions Dept., Raleigh, N.C., 1977, p. 21, Sec. ITTR-1S-001. Reprinted by permission.]

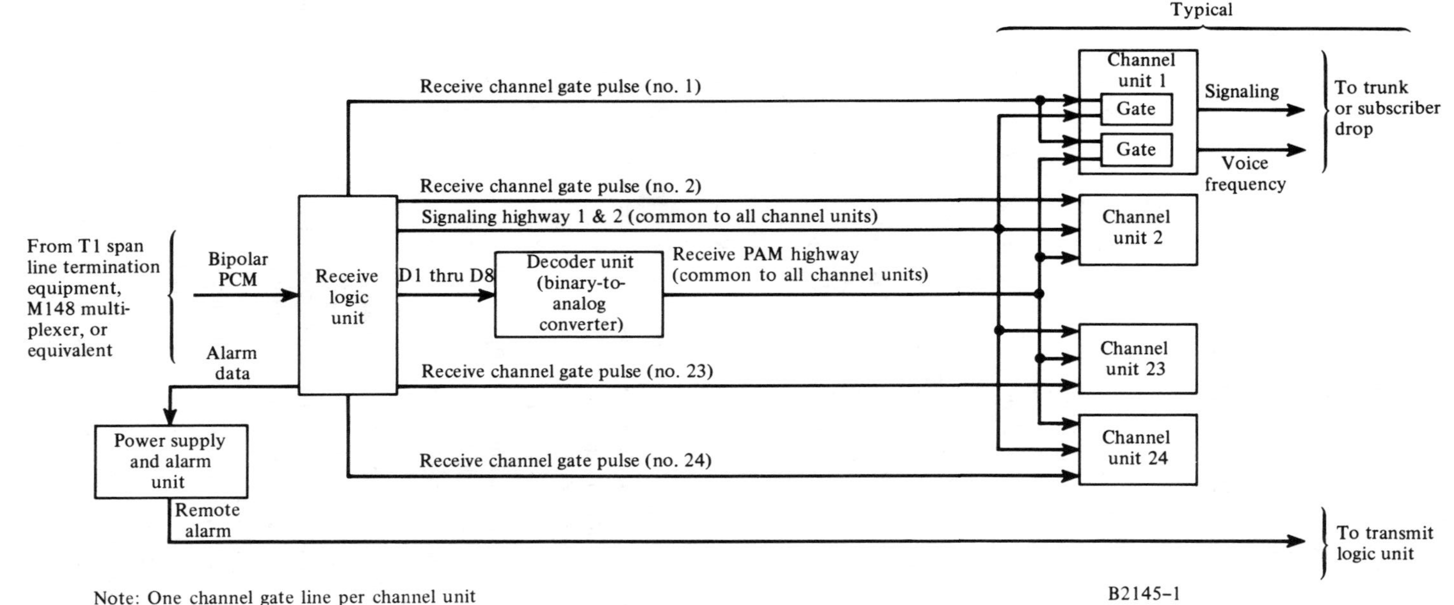

FIGURE 3-47 Receive functional diagram of the ITT T324 PCM voice-frequency multiplexer. [From "Operation-Installation-Maintenance Manual for T324 PCM Carrier System," Pub. 650038-823-001, *ITT Telecommunications,* Transmission Dept., Raleigh, N.C., 1977, p. 22, Sec. ITTR-15-001. Reprinted by permission.]

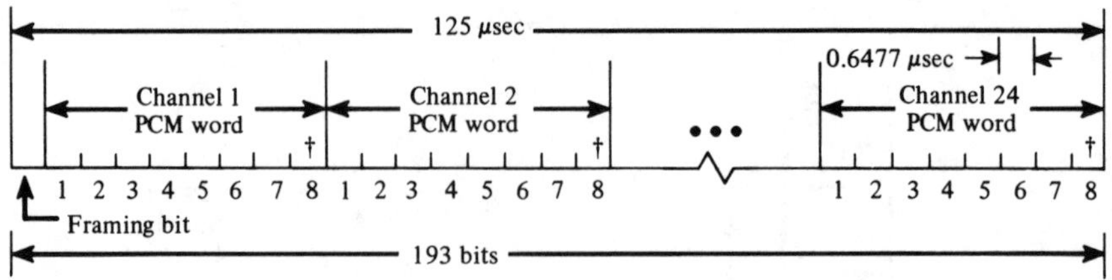

FIGURE 3-48 T1 TDM format for one frame.

represents 00, and −3 V represents 10 [Azaret et al., 1985]. Thus, a data rate of 6.443 MbitsSsec is achieved via 3.221-Mbaud signaling. This gives a 3.221-MHz zero-crossing bandwidth, which is close to the 3.152 zero-crossing bandwidth of the T1C system. This bandwidth can be supported by standard twisted-pair wires (one pair for each direction) with repeaters spaced at one-mile intervals. To achieve reliable signaling, the T1G equipment inserts additional parity and framing bits into the input stream DS-2 (6.312-Mbits/sec) signal so that a bit rate of 6.443 Mbits/sec is needed. For the first time the T1G system allows DS-2 (96 VF) capacity using standard twisted-pair cable that is common in existing telephone systems. In addition, old T1C-type systems can be upgraded to DS-2 capacity by replacing the old repeaters with T1G repeaters. The T1G system is also designed for economical operation, with the low-power repeaters having a mean time between failures of 1000 years and automated detection of faulty repeaters.

When considering the TDM approach for toll telephone transmission, one must consider the disadvantages as well as the advantages in comparison with frequency-division multiplexing (FDM). FDM is discussed in Chapter 4 (Fig. 4-47) and Chapter 5 (Fig. 5-19). It *was* the dominant multiplexing technique used in telephone toll service and for good reason—it *was* cheaper before the advent of low-cost ICs and fiber optic systems. The main difficulty with TDM is that it requires very wide bandwidth when compared to FDM, and it is difficult to obtain wide bandwidth with conventional wire transmission facilities. For example, the AT&T L5 coaxial cable transmission system can carry 10,800 VF channels in an FDM format using an overall bandwidth of 60 MHz. To transmit the same number of VF signals using TDM PCM, we need a bandwidth of about (10,800 VF channels)(64 kHz/VF channel) = 691 MHz, which is about 11 times the bandwidth required when using FDM. However, new fiber optic cable systems have phenomenal bandwidth and are relatively inexpensive on a per-channel basis. For example, the FTG1.7 fiber optic TDM system (see Table 5-7) has a capacity of 24,192 VF channels. Digital circuits are also much cheaper and more compact than analog circuits. Consequently, TDM systems are favored over FDM systems. Nevertheless, toll transmission facilities are very expensive, and it may be years before all FDM equipment can be economically replaced by TDM equipment.

For further information on digital transmission systems that are used by the telephone industry, the reader is referred to an excellent tutorial paper [Personick, 1980].

3-9 Pulse Time Modulation: Pulse Width Modulation and Pulse Position Modulation

Pulse time modulation (PTM) is a class of signaling techniques that encodes the sample values of an analog signal onto the *time axis* of a digital signal, and it is analogous to angle modulation techniques, which are introduced in Chapter 4. (As we have seen, PAM, PCM, and DM techniques encode the sample values into the amplitude characteristics of the digital signal.)

The two main types of pulse time modulation are *pulse width modulation* (PWM) and *pulse position modulation* (PPM) (see Fig. 3-49). In PWM, which is also called *pulse duration modulation* (PDM), sample values of the analog waveform are used to determine the *width* of the pulse signal. Either *instantaneous* sampling or *natural* sampling can be used. Figure 3-50 shows a technique for generating PWM signals with instantaneous sampling, and Fig. 3-51 displays

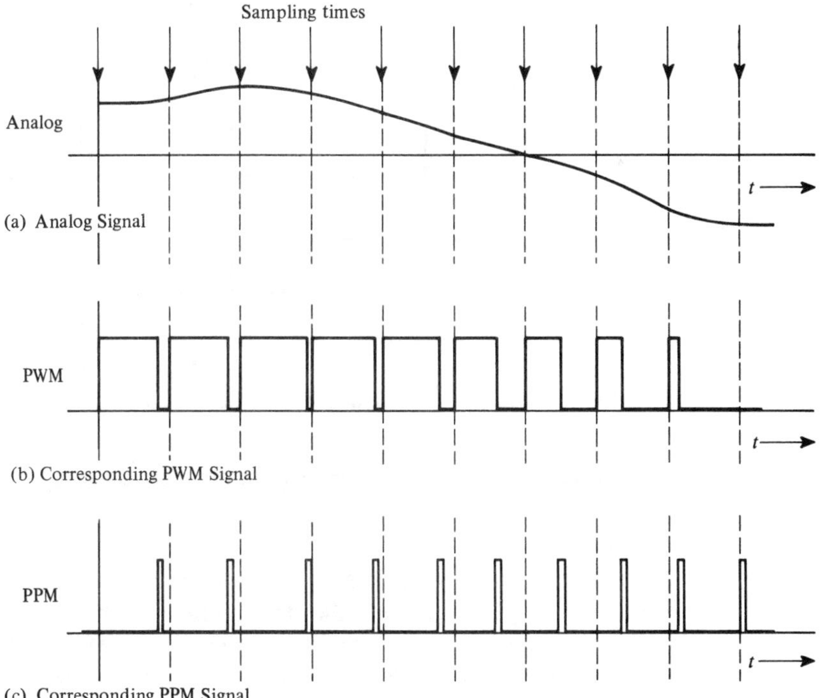

FIGURE 3-49 Pulse time modulation signaling.

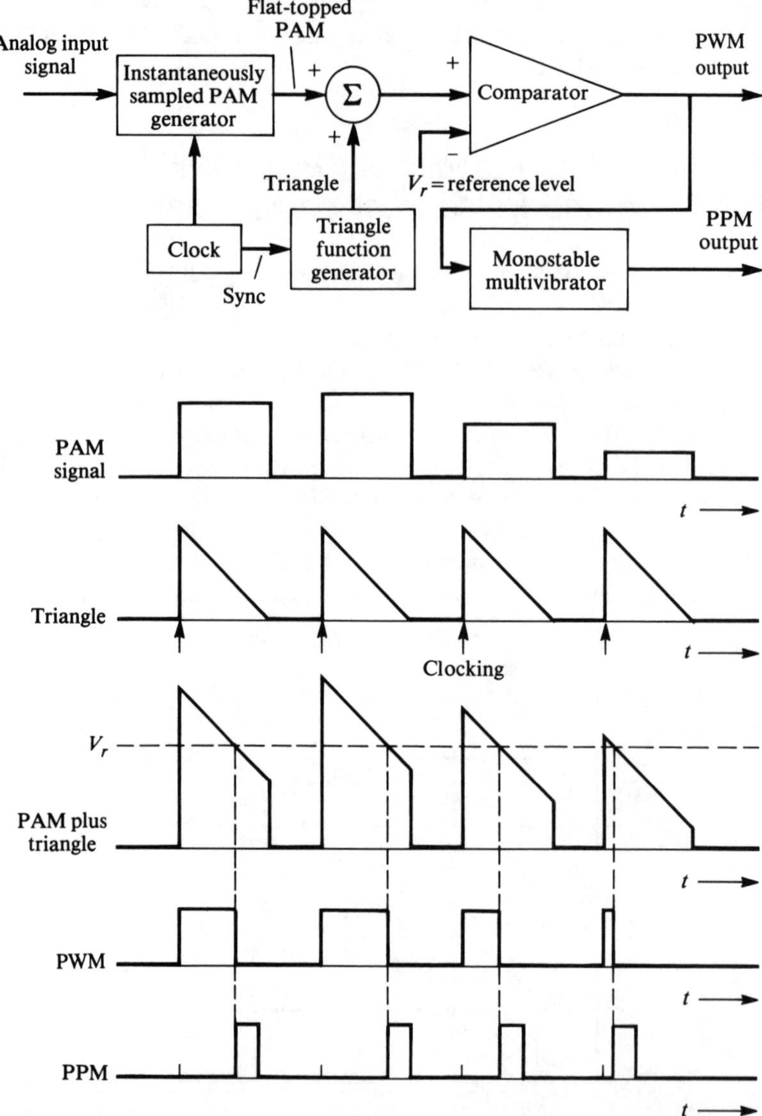

FIGURE 3-50 Technique for generating instantaneously sampled PTM signals.

PWM with natural sampling. In PPM the analog sample value determines the *position* of a narrow pulse relative to the clocking time. Techniques for generating PPM are also shown in the figures, and it is seen that PPM is easily obtained from PWM by using a monostable multivibrator circuit. In the literature on PTM signals, the comparator level V_r is often called the *slicing level.*

PWM or PPM signals may be converted back to the corresponding analog signal by a receiving system (Fig. 3-52). For PWM detection the PWM signal is

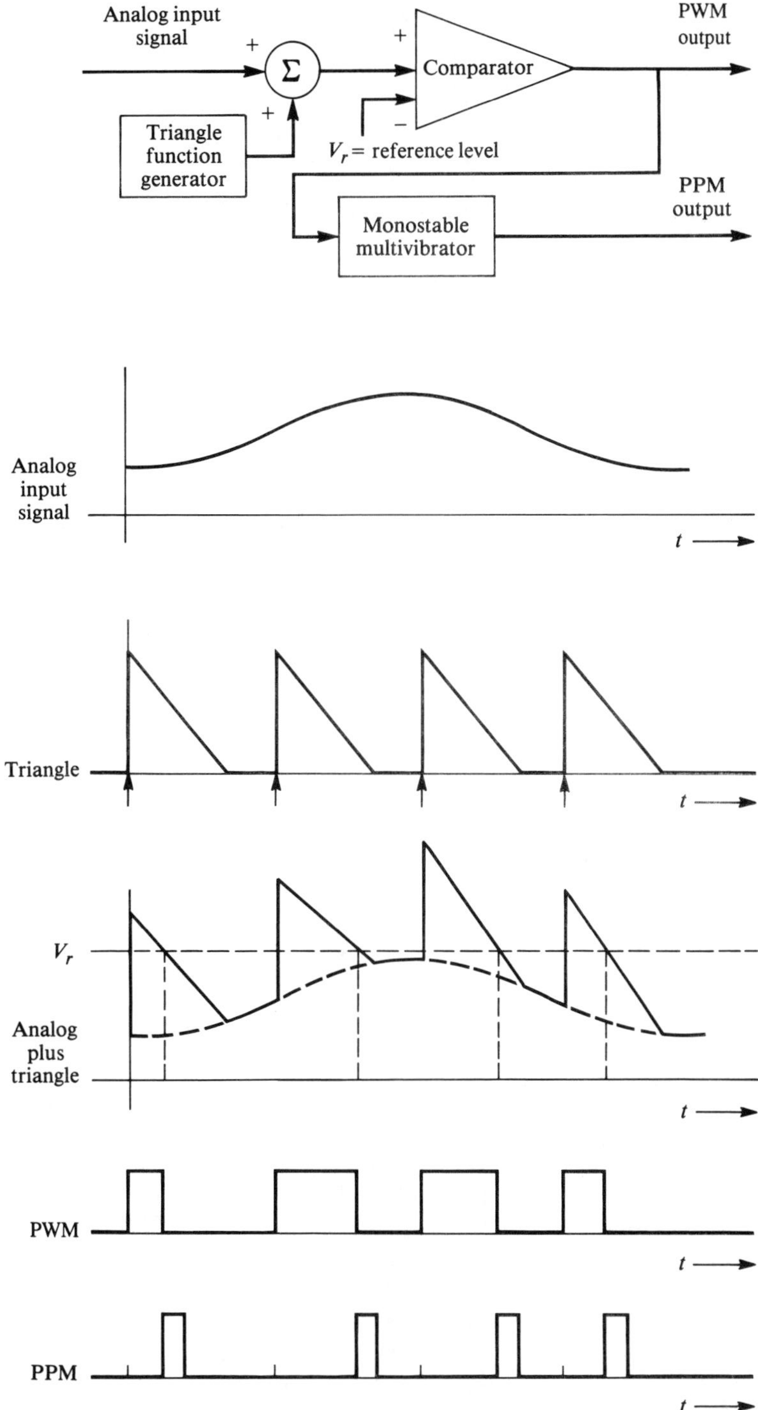

FIGURE 3-51 Technique for generating naturally sampled PTM signals.

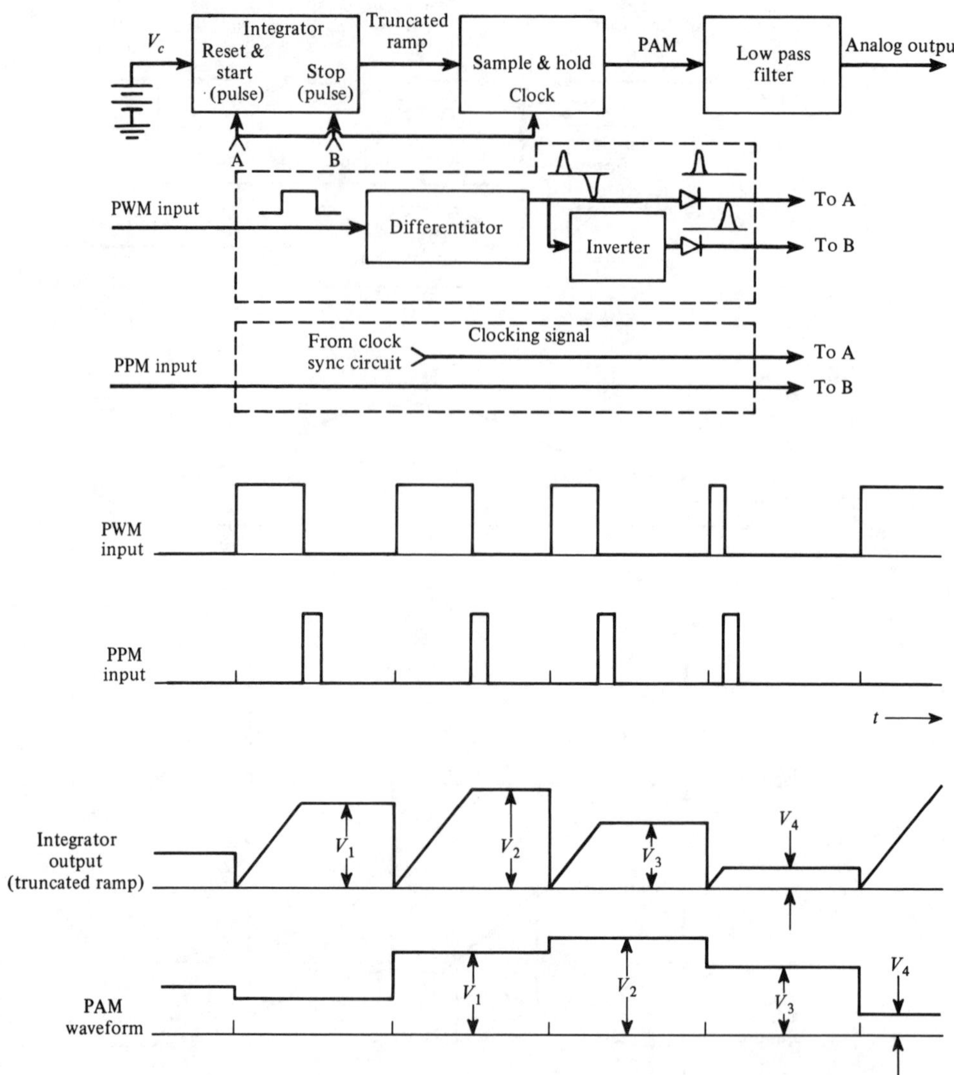

FIGURE 3-52 Detection of PWM and PPM signals.

used to start and stop the integration of an integrator; that is, the integrator is reset to zero, and integration is begun when the PWM pulse goes from a low level to a high level and the integrator integrates until the PWM pulse goes low. If the integrator input is connected to a constant voltage source, the output will be a truncated ramp. After the PWM signal goes low, the amplitude of the truncated ramp signal will be equal to the corresponding PAM sample value. At clocking time, the output of the integrator is then gated to a PAM output line (slightly before the integrator is reset to zero) using a sample-and-hold circuit. The PAM signal is then converted to the analog signal by low-pass filtering. In a similar way, PPM may be converted to PAM by using the clock pulse to reset the integrator to zero and start the integration. The PPM pulse is then used to stop the integration. The final value of the ramp is the PAM sample that is used to regenerate the analog signal.

Pulse time modulation signaling is not widely used to communicate across channels because a relatively wide bandwidth channel is needed, especially for PPM. However, PTM signals may be found internally in digital communications terminal equipment. The spectra of PTM signals are quite difficult to evaluate because of the nonlinear nature of the modulation [Rowe, 1965]. The main advantage of PTM signals is that they have a great immunity to additive noise when compared to PAM signaling, and they are easier to generate and detect than PCM, since PCM requires ADCs.

3-10 Summary

In this chapter we have studied baseband pulse and digital signaling techniques. It was found that, at least theoretically, the analog signal could be represented exactly by using pulse signaling techniques such as pulse amplitude modulation (PAM), pulse width modulation (PWM), or pulse position modulation (PPM). Furthermore, it was shown that analog signals could be approximated as accurately as desired by using digital signaling techniques such as pulse code modulation (PCM).

In our study of baseband digital signaling we concentrated on three major topics: (1) how the information in analog waveforms can be represented by digital signaling; (2) how filtering of the digital signal, due to the communication channel, affects our ability to recover the digital information at the receiver [i.e., the intersymbol interference (ISI) problem]; and (3) how we can merge the information from several sources into one digital signal by using time-division multiplexing (TDM). U.S. and worldwide standards for TDM telecommunications systems were given.

Digital signaling techniques were found to be very efficient because (1) binary digital circuitry could be used, (2) the signals could be easily regenerated, (3) uniform digital formats could be adopted that would handle information from both analog and digital sources, (4) many sources could be easily merged using TDM, and, as will be demonstrated in Chapter 7, (5) binary signaling is very noise resistant.

PCM is an analog-to-digital conversion scheme that involves three basic operations: (1) *sampling* a bandlimited analog signal, (2) *quantizing* the analog samples into M discrete values, and (3) *encoding* each sample value into an n-bit word where $M = 2^n$. There are two sources of noise in the signal that is recovered at the receiver output: (1) quantizing noise due to the approximation of the sample values using the M allowed values, and (2) noise due to receiver bit detection errors caused by channel noise or by ISI that arises because of improper channel frequency response. If the original analog signal is not strictly bandlimited, there will be a third noise component on the receiver output due to aliasing.

In studying the effect of improper channel filtering in producing ISI, the raised cosine-type rolloff filter was examined. Here it was found that the minimum bandwidth required to pass a PCM signal without ISI was equal to one half of the baud rate. However, this leads to technical difficulties since the overall filter characteristic is that for an ideal low-pass filter ($r = 0$) that is not causal. In addition, the sampling time at the receiver becomes very critical. A channel bandwidth equal to the baud rate ($r = 1$), which is also the zero-crossing bandwidth of the PCM signal, was found to be more realistic.

The channel bandwidth may be reduced if multilevel signal techniques are used (for a given data rate, R) or if differential pulse code modulation (DPCM) is used. In DPCM the difference between the actual and predicted sample values is transmitted over the system. Assuming that past values of the analog signal are related to the present value, we can obtain a bandwidth reduction over that needed when PCM is used. If equipment simplicity is of prime importance, delta modulation (DM) is the usual choice.

In this chapter we have been concerned with *baseband* signaling. In other words, the information is represented by a signal that has a spectrum concentrated about $f = 0$. In the next chapter we will be concerned with modulating baseband signals onto a carrier so that the spectrum will be concentrated about some desired frequency called the *carrier frequency*.

PROBLEMS

3-1 Demonstrate that the Fourier series coefficients for the switching waveform shown in Fig. 3-1b are given by (3-5b).

3-2 **(a)** Sketch the naturally sampled PAM waveform that results from sampling a 1-kHz sine wave at a 4-kHz rate.
(b) Repeat part (a) for the case of a flat-topped PAM waveform.

3-3 The spectrum of an analog audio signal is shown in Fig. P3-3. The waveform is to be sampled at a 10-kHz rate with a pulse width of $\tau = 50$ μsec.
(a) Find an expression for the spectrum of the naturally sampled PAM waveform. Sketch your result.
(b) Find an expression for the spectrum of the flat-topped PAM waveform. Sketch your result.

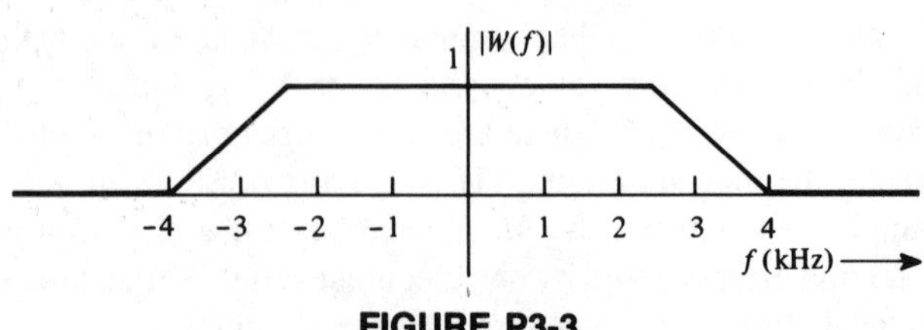

FIGURE P3-3

3-4 **(a)** Show that an analog output waveform (which is proportional to the original input analog waveform) may be recovered from a naturally sampled PAM waveform using the demodulation technique shown in Fig. 3-4.

(b) Find the constant of proportionality C that is obtained using this demodulation technique where $w(t)$ is the original waveform and $Cw(t)$ is the recovered waveform. Note that C is a function of n where the oscillator frequency is nf_s.

3-5 Figure 3-4 illustrates how a naturally sampled PAM signal can be demodulated to recover the analog waveform by use of a product detector. Show that the product detector can also be used to recover $w(t)$ from an instantaneously sampled PAM signal provided that the appropriate filter, $H(f)$, is used. Find the required $H(f)$ characteristic.

3-6 Starting with the PAM signal of Prob. 3-3:

(a) Design a PAM product demodulator with an oscillator operating at 30 kHz to recover the analog waveform from the naturally sampled PAM waveform.

(b) Design a PAM product demodulator with an oscillator operating at 30 kHz to demodulate the flat-topped PAM waveform.

(c) Discuss whether any synchronization signals are needed to operate these demodulators. Be sure to show the characteristics of any filters that are needed in your demodulator designs.

3-7 Assume that an analog signal with a spectrum shown in Fig. P3-3 is to be transmitted over a PAM system that has ac coupling. Thus a PAM pulse shape of the Manchester type as given by (3-44) is used. Find the spectrum for the Manchester encoded flat-topped PAM waveform. Sketch your result.

3-8 In a binary PCM system, if the quantizing noise is not to exceed $\pm P$ percent of the peak-to-peak analog level, show that the number of bits in each PCM word needs to be

$$n \geq [\log_2 10]\left[\log_{10}\left(\frac{50}{P}\right)\right] = 3.32 \log_{10}\left(\frac{50}{P}\right)$$

3-9 The information in an analog voltage waveform is to be transmitted over a PCM system with a $\pm 0.1\%$ accuracy (full scale). The analog waveform has an absolute bandwidth of 100 Hz and an amplitude range of -10 to $+10$ V.

(a) Determine the minimum sampling rate needed.

(b) Determine the number of bits needed in each PCM word.

(c) Determine the minimum bit rate required in the PCM signal.

(d) Determine the minimum absolute channel bandwidth required for transmission of this PCM signal.

3-10 A 40-Mbyte hard disk is used to store PCM data. Suppose that a VF (voice-frequency) signal is sampled at 8 ksamples/sec and the encoded PCM is to have an average S/N of at least 30 dB. How many minutes of VF conversation (i.e., PCM data) can be stored on the hard disk?

3-11 An analog signal with a bandwidth of 4.2 MHz is to be converted into binary PCM and transmitted over a channel. The peak-signal/quantizing-noise ratio at the receiver output must be at least 55 dB.

(a) If we assume $P_e = 0$ and no ISI, what will be the word length and the number of levels needed?

(b) What will be the equivalent bit rate?

(c) What will be the channel null bandwidth required if rectangular pulse shapes are used?

3-12 Modern compact disk (CD) players use 16-bit PCM, including one parity bit with 8 times oversampling of the analog signal. The analog signal bandwidth is 20 kHz.

(a) What is the null bandwidth of this PCM signal?

(b) Using (3-18), find the peak (S/N) in decibels.

3-13 Given an audio signal with spectral components in the frequency band 300 to 3000 Hz, assume that a sampling rate of 7 kHz will be used to generate a PCM signal. Design an appropriate PCM system, as follows.

(a) Draw a block diagram of the PCM system including the transmitter, channel, and receiver.

(b) Specify the number of uniform quantization levels needed and the zero-crossing channel bandwidth required, assuming that the peak signal-to-noise ratio at the receiver output needs to be at least 30 dB and that polar NRZ signaling is used.

(c) Discuss how nonuniform quantization can be used to improve the performance.

3-14 The S/N as given by (3-17a) and (3-17b) assume no ISI and no bit errors due to channel noise (i.e., $P_e = 0$). How large can P_e become before (3-17a) and (3-17b) are in error by 0.1% if $M = 4$, 8, or 16.

3-15 In a PCM system the bit error rate due to channel noise is 10^{-4}. Assume that the peak signal-to-noise ratio on the recovered analog signal needs to be at least 30 dB.

(a) Find the minimum number of quantizing levels that can be used to encode the analog signal into a PCM signal.

(b) If the original analog signal had an absolute bandwidth of 2.7 kHz, what is the zero-crossing bandwidth of the PCM signal for the polar NRZ signaling case?

3-16 Referring to Fig. 3-16 for a bit synchronizer using a square-law device, draw some typical waveforms that will appear in the bit synchronizer if a Manchester encoded PCM signal is present at the input. Discuss whether you would expect this bit synchronizer to work better for the Manchester encoded PCM signal or for a polar NRZ encoded PCM signal.

3-17 **(a)** Sketch the complete $\mu = 10$ compressor characteristic that will handle input voltages over the range -5 to $+5$ V.
(b) Plot the corresponding expandor characteristic.
(c) Draw a 16-level nonuniform quantizer characteristic that corresponds to the $\mu = 10$ compression characteristic.

3-18 For a 4-bit PCM system, calculate and sketch a plot of the output S/N (in decibels) as a function of the relative input level, $20 \log(x_{\text{rms}}/V)$ for
(a) A PCM system that uses $\mu = 10$ law companding.
(b) A PCM system that uses uniform quantization (no companding).
Assume that a triangular-type waveform is present at the input so that $\langle|x|\rangle = \sqrt{3}\, x_{\text{rms}}/2$.

Which of these systems is better to use in practice? Why?

3-19 The performance of a $\mu = 255$ law companded PCM system is to be examined when the input consists of a sine wave having a peak value of V volts. Assume that $M = 256$.
(a) Find an expression that describes the output S/N for this companded PCM system.
(b) Plot the $(S/N)_{\text{out}}$ (in decibels) as a function of the relative input level, $20 \log(x_{\text{rms}}/V)$. Compare this result with that shown in Fig. 3-10.

3-20 Assume a typical binary sequence and show that if the corresponding polar NRZ signal and unipolar NRZ signal have the same peak-to-peak amplitude, the polar signal has less power (an advantage) than the unipolar signal. If noise is added to these signals, how do the probabilities of bit errors compare for these two signaling techniques?

3-21 Consider a deterministic test pattern consisting of alternating binary 1's and binary 0's. Determine the magnitude spectra (not the PSD) for the following types of signaling formats as a function of T_b, where T_b is the time needed to send one bit of data.
(a) Unipolar NRZ signaling.
(b) Unipolar RZ signaling where the pulse width τ is $\tau = \frac{3}{4}T_b$.
How would each of these magnitude spectra change if the test pattern was changed to an alternating sequence of four binary 1s followed by four binary 0s?

3-22 Consider a random data pattern consisting of binary 1's and 0's where the probability of obtaining either a binary 1 or a binary 0 is $\frac{1}{2}$. Calculate the PSD function for the following types of signaling formats as a function of T_b, where T_b is the time needed to send one bit of data.
(a) Unipolar NRZ signaling.
(b) Unipolar RZ signaling where the pulse width τ is $\tau = \frac{3}{4}T_b$.
How do these PSDs for the random data cases compare to the magnitude spectra for the deterministic case of Prob. 3-21? What is the spectral efficiency for each of these cases?

3-23 Consider a deterministic data pattern consisting of alternating binary 1's and 0's. Determine the magnitude spectra (not the PSD) for the following types of signaling formats as a function of T_b, where T_b is the time needed to send one bit of data.

(a) Polar NRZ signaling.

(b) Manchester NRZ signaling.

How would each of these magnitude spectra change if the test pattern was changed to an alternating sequence of four binary 1's followed by two binary 0's?

3-24 Consider a random data pattern consisting of binary 1's and 0's, where the probability of obtaining either a binary 1 or a binary 0 is $\frac{1}{2}$. Calculate the PSD for the following types of signaling formats as a function of T_b, where T_b is the time needed to send 1 bit of data.

(a) Polar RZ signaling where the pulse width is $\tau = \frac{1}{2}T_b$.

(b) Manchester RZ signaling where the pulse width is $\tau = \frac{1}{4}T_b$. What is the first-null bandwidth of these signals? What is the spectral efficiency for each of these signaling cases?

3-25 Obtain the equations for the PSD of the bipolar NRZ and bipolar RZ (pulse width $\frac{1}{2}T_b$) line codes assuming peak values of $\pm A$ volts. Plot these PSD results as a function of the parameters A and $R = 1/T_b$.

3-26 In Fig. 3-12 the PSDs for several line codes are shown. These PSDs were derived assuming unity power for each signal so that the PSDs could be compared on an equal transmission power basis. Rederive the PSDs for these line codes assuming that the peak level is unity (i.e., $A = 1$). Plot these PSDs so that the spectra can be compared on an equal peak-signal-level basis.

3-27 Using (3-33), determine the conditions required so that there are delta functions in the PSD for line codes. Discuss how this affects the design of bit synchronizers for these line codes.

3-28 Consider a random data pattern consisting of binary 1's and 0's where the probability of obtaining either a binary 1 or a binary 0 is $\frac{1}{2}$. Assume that these data are encoded into a polar-type waveform where the pulse shape of each bit is given by

$$f(t) = \begin{cases} \cos\left(\dfrac{\pi t}{T_b}\right), & |t| < T_b/2 \\ 0, & |t| \text{ elsewhere} \end{cases}$$

T_b is the time needed to send one bit.

(a) Sketch a typical example of this waveform.

(b) Find the expression for the PSD of this waveform and sketch it.

(c) What is the spectral efficiency of this type of binary signal?

3-29 Create a practical block diagram for a differential encoding and decoding system. Explain how it works by showing the encoding and decoding for the sequence 001111010001. Assume that the reference digit is a binary 1. Show that error propagation cannot occur.

3-30 Design a regenerative repeater with its associated bit synchronizer for a polar RZ line code. Explain how your design works. (*Hint:* See Fig. 3-15 and the discussion on bit synchronizers.)

3-31 The approximate overall bit error for m cascaded repeaters is given by (3-50b), which is valid when $P_e \ll 1$. This approximation is the dominant (first) term in (3-50a). Assume that this approximation is considered to be valid when the first term of (3-50a) dominates the second term by three orders of magnitude (i.e., 1000). Under this assumption, how small does P_e need to be for (3-50b) to be valid? Give answers for $m = 5$, 10, and 30. Comment on the usefulness of (3-50b) when typical systems have overall error rates of 10^{-5} or less.

3-32 Design a bit synchronizer for a Manchester NRZ line code by completing the following steps:

(a) Give a simplified block diagram.

(b) Explain how it works.

(c) Specify filter or phase-locked loop (PLL) requirements.

(d) Explain the advantages and disadvantages of using this design for the Manchester NRZ line code as compared to using a polar NRZ line code and its associated bit synchronizer.

3-33 Figure 3-18c illustrates an eight-level multilevel signal. Assume that this line code is passed through a channel that filters this signal and adds some noise.

(a) Draw a picture of the eye pattern for the received waveform.

(b) Design a possible receiver with its associated symbol synchronizer for this line code.

(c) Explain how your receiver works.

3-34 The information in an analog waveform is to be transmitted over a multilevel PCM system where the number of multilevels is eight. Assume that an accuracy of $\pm 1\%$ is required (full scale) and that the absolute bandwidth is 2700 Hz. The analog voltage has a range of -1 to $+1$ V.

(a) Determine the minimum bit rate of the PCM signal.

(b) Determine the minimum baud rate of the PCM signal.

(c) Determine the minimum absolute channel bandwidth required for transmission of this PCM signal.

3-35 A binary waveform of 9600 bits/sec is converted into an octal (multilevel) waveform that is passed through a channel with a raised cosine-rolloff filter characteristic. The channel has a conditioned (equalized) phase response out to 2.4 kHz.

(a) What is the baud rate of the multilevel signal?

(b) What is the rolloff factor of the filter characteristic?

3-36 Assume that the spectral properties of an $L = 64$ level multilevel waveform with rectangular RZ-type pulse shapes are to be examined. The pulse shape is given by

$$f(t) = \Pi\left(\frac{2t}{T_s}\right)$$

T_s is the time needed to send one of the multilevel symbols.

(a) Determine the expression for the PSD for the case of equally likely levels where the peak signal levels for this multilevel waveform are ± 10 V.

(b) What is the null bandwidth?

(c) What is the spectral efficiency?

3-37 A binary communication system uses polar signaling. The overall impulse response is designed to be of the $(\sin x)/x$ type as given by (3-67) so that there will be no ISI. The bit rate is $R = f_s = 300$ bits/sec.

(a) What is the bandwidth of the polar signal?

(b) Plot the waveform of the polar signal at the system output when the input binary data is 01100101. Can you discern the data by looking at this polar waveform?

3-38 Equation (3-67) gives one possible noncausal impulse response for a communication system that will have no ISI. For a causal approximation, select

$$h_e(t) = \frac{\sin \pi f_s(t - 4 \times 10^{-3})}{\pi f_s(t - 4 \times 10^{-3})} \Pi\left(\frac{t - 4 \times 10^{-3}}{8 \times 10^{-3}}\right)$$

where $f_s = 1000$.

(a) Using a PC, calculate $H_e(f)$ by use of the Fourier transform integral and plot $|H_e(f)|$.

(b) What is the bandwidth of this causal approximation, and how does it compare to the bandwidth of the noncausal filter described by (3-67) and (3-68)?

3-39 Starting with (3-69), prove that the impulse response of the raised cosine-rolloff filter is given by (3-73).

3-40 Show that when a raised cosine-rolloff filter is used, the baud rate that the digital communication can support without ISI is $D = 2B/(1 + r)$.

3-41 Consider the raised cosine-rolloff filter given by (3-69) and (3-73).

(a) Plot $|H_e(f)|$ for the case of $r = 0.75$, indicating f_1, f_0, and B on your sketch, similar to Fig. 3-21.

(b) Plot $h_e(t)$ for the case of $r = 0.75$ in terms of $1/f_0$, similar to Fig. 3-22.

3-42 Find the PSD of the waveform out of an $r = 0.5$ raised cosine-rolloff channel when the input is a polar NRZ signal. Assume that equally likely binary signaling is used and the channel bandwidth is just large enough to prevent ISI.

3-43 Equation (3-66) gives the condition for no ISI (Nyquist's first method). Using (3-66) with $C = 1$ and $\tau = 0$, show that Nyquist's first method for no ISI is also satisfied if

$$\sum_{k=-\infty}^{\infty} H_e\left(f + \frac{k}{T_s}\right) = T_s \qquad \text{for } |f| \leq \frac{1}{2T_s}$$

3-44 Using the results of Prob. 3-43, demonstrate that the following filter characteristics do or do not satisfy Nyquist's criterion for no ISI ($f_s = 2f_0 = 2/T_0$).

(a) $H_e(f) = \frac{T_0}{2}\Pi\left(\frac{1}{2}fT_0\right)$.

(b) $H_e(f) = \frac{T_0}{2}\Pi\left(\frac{2}{3}fT_0\right)$.

3-45 Assume that a pulse transmission system has an overall raised cosine-rolloff filter characteristic as described by (3-69).

(a) Find the $Y(f)$ Nyquist function of (3-75) corresponding to the raised cosine-rolloff filter characteristic.

(b) Sketch $Y(f)$ for the case of $r = 0.75$.

(c) Sketch another $Y(f)$ that is not of the raised cosine-rolloff type and determine the absolute bandwidth of the resulting Nyquist filter characteristic.

3-46 An analog signal is to be converted into a polar PCM signal and transmitted over a channel that is absolutely bandlimited to 4 kHz. Assume that 16 quantization levels are used and that the overall equivalent system transfer function is of the raised cosine-rolloff type with $r = 0.5$.

(a) Find the maximum PCM bit rate that can be supported by this system without introducing ISI.

(b) Find the maximum bandwidth that can be permitted for the analog signal.

3-47 Rework Prob. 3-46 for the case of a multilevel PCM signal when the number of levels is four.

3-48 Find the tap coefficients required for a zero-forcing transversal equalizing filter for the case of a three-tap filter. For a single flat-topped pulse into the transmit filter the response of the channel at the sampling times is $w_c(-2T_s) = -0.08$, $w_c(-T_s) = 0.25$, $w_c(0) = 1.0$, $w_c(T_s) = 0.30$, and $w_c(2T_s) = -0.06$.

3-49 Find the three tap coefficients for a zero-forcing transversal filter equalizer with $w_c(-2T_s) = -0.01$, $w_c(-T_s) = 0.1$, $w_c(0) = 0.8$, $w_c(T_s) = -0.2$, and $w_c(2T_s) = 0.05$.

3-50 A communication channel is tested for ISI. The impulse response at the channel output is $w_c(t) = \Lambda(t/(2T_s))$. Design a three-tap zero forcing transversal filter to equalize the channel response.

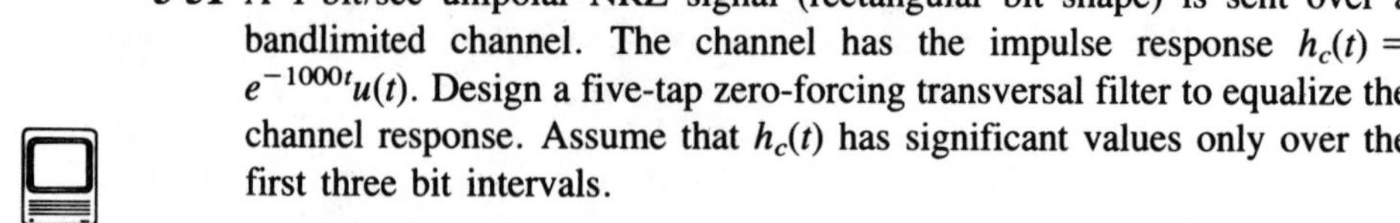

3-51 A 1-bit/sec unipolar NRZ signal (rectangular bit shape) is sent over a bandlimited channel. The channel has the impulse response $h_c(t) = e^{-1000t}u(t)$. Design a five-tap zero-forcing transversal filter to equalize the channel response. Assume that $h_c(t)$ has significant values only over the first three bit intervals.

3-52 A 1-bit/sec binary signal is to be transmitted over a channel. The equivalent impulse response is

$$h_e(t) = \begin{cases} e^{-t}, & t \geq 0 \\ e^{-t^2}, & t < 0 \end{cases}$$

(a) Plot the impulse response.

(b) Design a transversal filter to force four points (at the sampling times) to zero.

(c) Plot the impulse response that includes the zero-forcing equalizing filter.

3-53 The transfer function of the cosine-rolloff filter (not the raised cosine filter) is given by (3-87). Show that the corresponding impulse response is given by (3-88).

3-54 Find the PSD of the waveform out of a duobinary filtered channel (i.e., cosine-rolloff characteristic) when the input is a polar NRZ signal. Assume that equally likely binary signaling is used.

3-55 Assume that a unipolar binary signaling waveform is applied to the input of the digital communication system shown in Fig. P3-55. The channel filter, when driven by the polar differential signal, produces the response of a cosine partial response system. The data rate is $1/T_s$ bits/sec, and the abso-

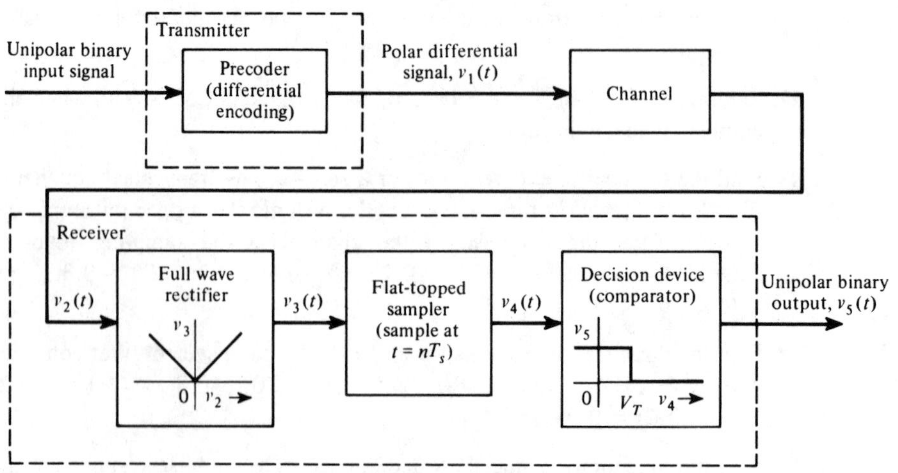

FIGURE P3-55

lute bandwidth of the channel is $1/(2T_s)$ Hz. At the receiver the full-wave rectifier and the decision device are nonlinear circuits having output-to-input characteristics as shown.

(a) Show whether or not the receiver gives the input sequence at the output (no channel noise).

(b) Assume that the input sequence is 1011010 and that the input and output binary levels are +1 V for a binary 1 and 0 V for a binary 0. Sketch the waveforms $v_1(t)$, $v_2(t)$, $v_3(t)$, $v_4(t)$, and $v_5(t)$.

(c) Suggest a value for the threshold voltage parameter V_T.

(d) Show that error propagation cannot occur in this system.

3-56 If a unipolar NRZ waveform is present at the input to a modified duobinary channel, determine

(a) The possible values (i.e., levels) for the received waveform at the sampling times (no-noise case).

(b) The decoding logic needed to recover the data from these received samples.

(*Hint:* Set up a table that gives the relationship between the possible sample values; the present bit, a_n; and the second previous bit, a_{n-2}.)

3-57 Assume that an analog signal is to be encoded into either PCM or DPCM. Discuss how the resulting spectra for the PCM and DPCM signals would be alike or different.

3-58 Assume that a PCM-type system is to be designed such that an audio signal can be delivered at the receiver output. This audio signal is to have a bandwidth of 3400 Hz and a S/N of at least 40 dB. Determine the bit rate requirements for a design that uses:

(a) $\mu = 255$ companded PCM signaling.

(b) DPCM signaling.

Discuss which of the systems above would be used in your design and why.

3-59 A DM binary sequence 10101110000101010110 is received. Referring to Fig. 3-34, plot the detected analog signal that appears at the output of the integrator.

3-60 Refer to Fig. 3-35, which shows typical DM waveforms. Draw an analog waveform that is different from the one shown in the figure. Draw the corresponding DM and integrator output waveforms. Denote the regions where slope overload noise dominates and where granular noise dominates.

3-61 A DM system is tested with a 10-kHz sinusoidal signal, 1 V peak to peak, at the input. It is sampled at 10 times the Nyquist rate.

(a) What is the step size required to prevent slope overload and to minimize granular noise?

(b) What is the PSD for the granular noise?

(c) If the receiver input is bandlimited to 200 kHz, what is the average-signal/quantizing-noise power ratio?

3-62 Assume that the input to a DM is $0.1t^8 - 5t + 2$. The step size of the DM is 1 V, and the sampler operates at 10 samples/sec. Over a time interval of 0 to 2 sec, sketch the input waveform, the delta modulator output, and the integrator output. Denote the granular noise and slope overload regions.

3-63 Rework Prob. 3-62 for the case of an adaptive delta modulator where the step size is selected according to the number of successive binary 1's or 0's on the DM output. Assume that the step size is 1.5 V when there are four or more binary digits of the same sign, 1 V for the case of three successive digits, and 0.5 V for the case of two or fewer successive digits.

3-64 A delta modulator is to be designed to transmit the information of an analog waveform that has a peak-to-peak level of 1 V and a bandwidth of 3.4 kHz. Assume that the waveform is to be transmitted over a channel where the frequency response is extremely poor above 1 MHz.

(a) Select the appropriate step size and sampling rate for a sine-wave test signal and discuss the performance of the system using the parameter values you have selected.

(b) If the DM system is to be used to transmit the information of a voice (analog) signal, select the appropriate step size when the sampling rate is 25 kHz. Discuss the performance of the system under these conditions.

3-65 A binary sequence 1110101000001010000000101 is transmitted by DM. Evaluate and sketch the resulting analog waveform that appears at the receiver output.

3-66 One analog waveform $w_1(t)$ is bandlimited to 3 kHz, and another, $w_2(t)$, is bandlimited to 9 kHz. These two signals are to be sent by TDM over a PAM-type system.

(a) Determine the minimum sampling frequency for each signal and design a TDM commutator and decommutator to accommodate these signals.

(b) Draw some typical waveforms for $w_1(t)$ and $w_2(t)$, and sketch the corresponding TDM PAM waveform.

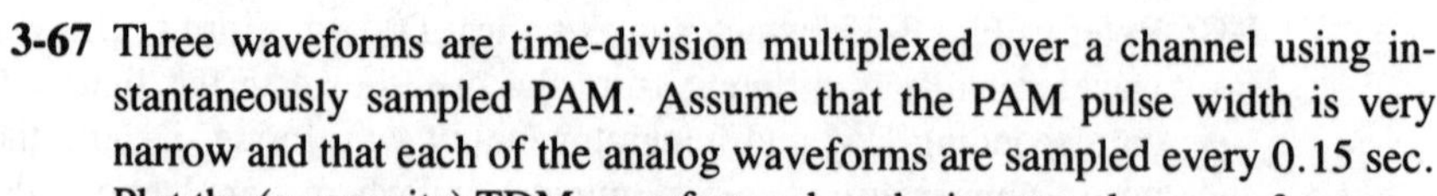

3-67 Three waveforms are time-division multiplexed over a channel using instantaneously sampled PAM. Assume that the PAM pulse width is very narrow and that each of the analog waveforms are sampled every 0.15 sec. Plot the (composite) TDM waveform when the input analog waveforms are

$$w_1(t) = 3\sin(2\pi t)$$

$$w_2(t) = \Pi\left(\frac{t-1}{2}\right)$$

$$w_3(t) = -\Lambda(t-1)$$

3-68 Twenty-three analog signals, each with a bandwidth of 3.4 kHz, are sampled at an 8-kHz rate and multiplexed together with a synchronization channel (8 kHz) into a TDM PAM signal. This TDM signal is passed through a channel with an overall raised cosine-rolloff filter characteristic of $r = 0.75$.

(a) Draw a block diagram for the system indicating the f_s of the commutator and the overall pulse rate of the TDM PAM signal.

(b) Evaluate the absolute bandwidth required for the channel.

3-69 Two flat-topped PAM signals are time-division multiplexed together to produce a composite TDM PAM signal that is transmitted over a channel. The first PAM signal is obtained from an analog signal that has a rectangular spectrum, $W_1(f) = \Pi(f/2B)$. The second PAM signal is obtained from an analog signal that has a triangular spectrum, $W_2(f) = \Lambda(f/B)$ where $B = 3$ kHz.

(a) Determine the minimum sampling frequency for each signal and design a TDM commutator and decommutator to accommodate these signals.

(b) Calculate and sketch the magnitude spectrum for the composite TDM PAM signal.

3-70 Rework Prob. 3-68 for a TDM pulse code modulation system where an 8-bit quantizer is used to generate the PCM words for each of the analog inputs and an 8-bit synchronization word is used in the synchronization channel.

3-71 Design a TDM PCM system that will accommodate four 300-bit/sec (synchronous) digital inputs and one analog input that has a bandwidth of 500 Hz. Assume that the analog samples will be encoded into 4-bit PCM words. Draw a block diagram for your design, analogous to Fig. 3-43, indicating the data rates at the various points on the diagram. Explain how your design works.

3-72 A character-interleaved TDM is used to combine the data streams of a number of 110-bit/sec asynchronous terminals for data transmission over a 2400-bit/sec digital line. Each terminal sends characters consisting of 7 data bits, 1 parity bit, 1 start bit, and 2 stop bits. Assume that one synchronization character is sent after every 19 data characters and, in addition, at least 3% of the line capacity is reserved for pulse stuffing to accommodate speed variations from the various terminals.

(a) Determine the number of bits per character.

(b) Determine the number of terminals that can be accommodated by the multiplexer.

(c) Sketch a possible framing pattern for the multiplexer.

3-73 Assume that two 600-bit/sec terminals, five 300-bit/sec terminals, and a number of 150-bit/sec terminals are to be time-multiplexed in a character-interleaved format over a 4800-bit/sec digital line. The terminals send 10 bits/character, and one synchronization character is inserted for every 99 data characters. All the terminals are asynchronous, and 3% of the line capacity is allocated for pulse stuffing to accommodate variations in the terminal clock rates.

(a) Determine the number of 150-bit/sec terminals that can be accommodated.
(b) Sketch a possible framing pattern for the multiplexer.

3-74 In designing a TDM system, a frame synchronizer identical to that shown in Fig. 3-41 is to be used. It is desired to choose a threshold so that frame sync can be detected even when one bit of the received sync word is in error. Thus, choose $V_T = K - 2$.
(a) Find the expression for the probability of false sync, P_f, assuming equally likely data into the synchronizer.
(b) Draw a block diagram for your design of the frame synchronizer including the number of shift register stages needed if $P_f \leq 10^{-3}$ is required. Choose a sync word that will satisfy this specification and show inverters or no inverters on your diagram, as needed, in place of the gains $s_1, s_2, \ldots, s_K$. Explain how your frame synchronizer works.
(c) Design a circuit (i.e., block diagram) that is used at the transmitter end to generate your unique frame sync word. Explain how your circuit works.

3-75 Find the number of the following devices that could be accommodated by a T1-type TDM line if 1% of the line capacity were reserved for synchronization purposes.
(a) 110-bit/sec teleprinter terminals.
(b) 300-bit/sec computer terminals.
(c) 1200-bit/sec computer terminals.
(d) 9600-bit/sec computer output ports.
(e) 64-kbit/sec PCM voice-frequency lines.
How would these numbers change if each of the sources were operational an average of 10% of the time?

3-76 A TDM system has four 300-bit/sec and one 1200-bit/sec computer terminals connected to its input, and 3% of the line capacity is used for pulse stuffing to accommodate speed variations among the terminals. No statistical multiplexing is used.
(a) Find the expression for the PSD of the polar NRZ signal at the TDM output. Assume that the binary 1's and 0's are equally likely and that the TDM signal levels are ± 12 V.
(b) Sketch the PSD of this TDM signal and compare it with the sketch of the PSD for the signals that are emitted from the computer terminals (polar signal with ± 12-V levels).

3-77 Still video data are sent over a basic rate ISDN system using 2B + D channels. The video data are sent over the two B channels, and the D channel is used for control and other data. Assume that each still picture consists of 480×500 pixels and that each pixel is encoded into one of eight allowed gray levels. How many pictures can be sent per second?

3-78 Assume that a sine wave is sampled at four times the Nyquist rate using instantaneous sampling.

(a) Sketch the corresponding PWM signal.

(b) Sketch the corresponding PPM signal.

3-79 Repeat Prob. 3-78, but assume that natural sampling is used.

3-80 Discuss why a PPM system requires a synchronizing signal, whereas PAM and PWM can be detected without the need for a synchronizing signal.

3-81 Assume that $w(t) = 10 \sin 2\pi t$ is instantaneously sampled every 0.1 sec, starting at $t = 0$.

(a) Evaluate the exact spectrum for the PWM signal, assuming that the maximum pulse width is obtained if the sample value is 10 V and the pulse width is zero if the sample value is -10 V.

(b) Evaluate the exact PPM spectrum, assuming that a pulse width of 0.01 sec is used.

3-82 Compare the bandwidth required to send a message by using PPM and PCM signaling. Assume that the digital source sends out 8 bits/character, so that the digital source can send 256 different messages (characters). Assume that the source rate is 10 characters/sec. Use the dimensionality theorem, $N/T_0 = 2B$, to determine the minimum bandwidth required, B.

(a) Determine the minimum bandwidth required for a PCM signal that will encode the source information.

(b) Determine the minimum bandwidth required for a PPM signal that will encode the source information.

CHAPTER 4

Bandpass Signaling Techniques and Components

4-1 Introduction

This chapter is concerned with *bandpass* signaling techniques. These are applicable to both digital and analog communication systems. As indicated in Chapter 1, the bandpass communication signal is obtained by modulating the baseband signal onto a carrier. The baseband signal might be an analog signal, such as that obtained directly from a microphone, or it might be digital, such as a PCM signal as discussed in Chapter 3. Here the classical bandpass signaling techniques of amplitude modulation, single-sideband, and angle modulation will be studied in detail. Classical modulation theory is directly applicable to the understanding of digital bandpass signaling techniques that are studied in Chapter 5.

For a better understanding of the implementation of communication systems, a description of communication component blocks such as filters, amplifiers, up-and-down converters, phase-locked loops, and detectors is covered in Sec. 4-3. (This section may be skipped if the material has been covered in electronics courses.)

Block diagrams for the various types of transmitters and receivers will be illustrated and analyzed. In addition to the theory, practical aspects of transmitter and receiver design will be emphasized. First, we will study the mathematical basis of modulation theory.

4-2 Representation of Bandpass Waveforms and Systems

What is a general representation for bandpass digital and analog signals? How do we represent a modulated signal? How do we represent bandpass noise? These are some of the questions that are answered in this section.

Definitions: Baseband, Bandpass, and Modulation

Definition: A *baseband* waveform has a spectral magnitude that is nonzero for frequencies in the vicinity of the origin (i.e., $f = 0$) and negligible elsewhere.

Definition: A *bandpass* waveform has a spectral magnitude that is nonzero for frequencies in some band concentrated about a frequency $f = \pm f_c$, where $f_c \gg 0$. The spectral magnitude is negligible elsewhere. f_c is called the *carrier frequency*.

For bandpass waveforms the value of f_c may be arbitrarily assigned for mathematical convenience in *some* problems. In others, namely, modulation problems, f_c is the frequency of an oscillatory signal in the transmitter circuit and is the assigned frequency of the transmitter, such as, for example, 850 kHz for an AM broadcasting station.

In communication problems, the information source signal is usually a baseband signal: for example, a transistor-transistor logic (TTL) waveform from a digital circuit, an audio signal from a microphone, or a video signal from a television camera. As described in Chapter 1, the communication engineer has the job of building a system that will transfer the information in this source signal $m(t)$ to the desired destination. As shown in Fig. 4-1, this usually requires the use of a bandpass signal, $s(t)$, which has a bandpass spectrum that is concentrated at $\pm f_c$ so that it will propagate across the communication channel (either a softwire or a hardwire channel).

Definition: *Modulation* is the process of imparting the source information onto a bandpass signal with a carrier frequency f_c by the introduction of amplitude and/or phase perturbations. This bandpass signal is called the *modulated* signal $s(t)$, and the baseband source signal is called the *modulating* signal $m(t)$.

Examples of exactly how modulation is accomplished are given later in this chapter. The definition indicates that modulation may be visualized as a mapping operation that maps the source information onto the bandpass signal $s(t)$ that will be transmitted over the channel.

As the modulated signal passes through the channel, noise corrupts it. The result is a bandpass signal-plus-noise waveform that is available at the receiver input, $r(t)$ (see Fig. 4-1). The receiver has the job of trying to recover the information that was sent from the source. $\tilde{m}$ denotes the corrupted version of m.

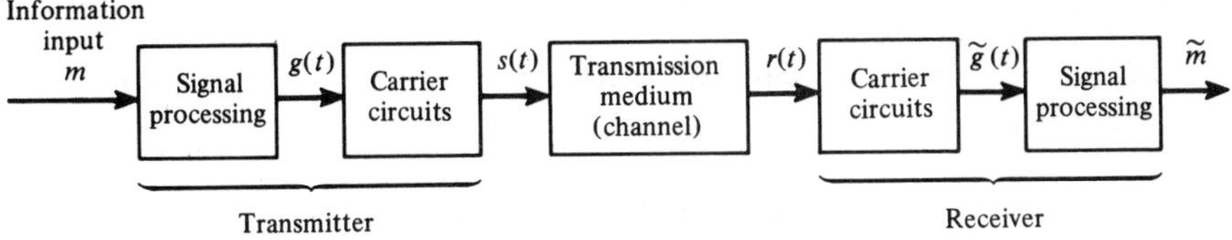

FIGURE 4-1 Communication system.

Complex Envelope Representation

All bandpass waveforms, whether they arise from a modulated signal, interfering signals, or noise, may be represented in a convenient form given by the following theorem. $v(t)$ will be used to denote the bandpass waveform canonically; specifically, it can represent the signal when $s(t) \equiv v(t)$, the noise when $n(t) \equiv v(t)$, the filtered signal plus noise at the channel output when $r(t) \equiv v(t)$, or any other type of bandpass waveform.

Theorem: *Any physical bandpass waveform can be represented by*

$$v(t) = \text{Re}\{g(t)e^{j\omega_c t}\} \tag{4-1a}$$

Re$\{\cdot\}$ *denotes the real part of* $\{\cdot\}$, $g(t)$ *is called the complex envelope of* $v(t)$, *and* f_c *is the associated carrier frequency (hertz) where* $\omega_c = 2\pi f_c$. *Furthermore, two other equivalent representations are*†

$$v(t) = R(t)\cos[\omega_c t + \theta(t)] \tag{4-1b}$$

and

$$v(t) = x(t)\cos\omega_c t - y(t)\sin\omega_c t \tag{4-1c}$$

where

$$g(t) = x(t) + jy(t) = |g(t)|e^{j\angle g(t)} \equiv R(t)e^{j\theta(t)} \tag{4-2}$$

$$x(t) = \text{Re}\{g(t)\} \equiv R(t)\cos\theta(t) \tag{4-3a}$$

$$y(t) = \text{Im}\{g(t)\} \equiv R(t)\sin\theta(t) \tag{4-3b}$$

$$R(t) \triangleq |g(t)| \equiv \sqrt{x^2(t) + y^2(t)} \tag{4-4a}$$

$$\theta(t) \triangleq \angle g(t) = \tan^{-1}\left(\frac{y(t)}{x(t)}\right) \tag{4-4b}$$

The waveforms $g(t)$, $x(t)$, $y(t)$, $R(t)$, and $\theta(t)$ are all baseband waveforms, and, except for $g(t)$, they are all real waveforms. $R(t)$ is a nonnegative real waveform. Thus, (4-1) is a low-pass-to-bandpass transformation.

Proof. Any physical waveform (it does *not* have to be periodic) may be represented over all time, $T_0 \to \infty$, by the complex Fourier series:

$$v(t) = \sum_{n=-\infty}^{n=\infty} c_n e^{jn\omega_0 t}, \qquad \omega_0 = 2\pi/T_0 \tag{4-5}$$

Furthermore, since the physical waveform is real $c_{-n} = c_n^*$ and using $\text{Re}\{\cdot\} = \frac{1}{2}\{\cdot\} + \frac{1}{2}\{\cdot\}^*$,

$$v(t) = \text{Re}\left\{c_0 + 2\sum_{n=1}^{\infty} c_n e^{jn\omega_0 t}\right\} \tag{4-6}$$

†The symbol $\equiv$ denotes an equivalence and the symbol $\triangleq$ denotes a definition.

When we introduce the arbitrary parameter f_c, this becomes†

$$v(t) = \text{Re}\left\{\left(c_0 e^{-j\omega_c t} + 2 \sum_{n=1}^{n=\infty} c_n e^{j(n\omega_0 - \omega_c)t}\right) e^{j\omega_c t}\right\} \tag{4-7}$$

so that (4-1a) follows where

$$g(t) \equiv c_0 e^{-j\omega_c t} + 2 \sum_{n=1}^{\infty} c_n e^{j(n\omega_0 - \omega_c)t} \tag{4-8}$$

Since $v(t)$ is a bandpass waveform with nonzero spectrum concentrated near $f = f_c$, the Fourier coefficients c_n are nonzero only for values of n in the range $\pm nf_0 \approx f_c$. Therefore, from (4-8) $g(t)$ has a spectrum that is concentrated near $f = 0$. That is, $g(t)$ is a *baseband* waveform. It is also obvious from (4-8) that $g(t)$ may be a complex function of time. Representing the complex envelope in terms of two real functions in Cartesian coordinates, we have

$$g(t) \equiv x(t) + jy(t)$$

where $x(t) = \text{Re}\{g(t)\}$ and $y(t) = \text{Im}\{g(t)\}$. $x(t)$ is said to be the *in-phase modulation* associated with $v(t)$, and $y(t)$ is said to be the *quadrature modulation* associated with $v(t)$. Alternatively, the polar form of $g(t)$, represented by $R(t)$ and $\theta(t)$, is given by (4-2), where the identities between Cartesian and polar coordinates are given by (4-3) and (4-4). $R(t)$ and $\theta(t)$ are real waveforms and, in addition, $R(t)$ is always nonnegative. $R(t)$ is said to be the *amplitude modulation* (AM) on $v(t)$, and $\theta(t)$ is said to be the *phase modulation* (PM) on $v(t)$. It is also realized that if $v(t)$ is a deterministic waveform, $x(t)$, $y(t)$, $R(t)$, and $\theta(t)$ are also deterministic. If $v(t)$ is stochastic, for example, representing bandpass noise, $x(t)$, $y(t)$, $R(t)$, and $\theta(t)$ are stochastic baseband processes. Thus, in general, bandpass noise includes both AM, $R(t)$, and PM, $\theta(t)$, noise components. This will be discussed further in Chapter 6. The usefulness of the complex envelope representation for bandpass waveforms cannot be overemphasized. In modern communication systems, the bandpass signal is often partitioned into two channels, one for $x(t)$ called the I (in-phase) channel and one for $y(t)$ called the Q (quadrature-phase) channel.

Bandpass Filtering

In Sec. 2-6 the general transfer function technique was described for the treatment of linear filter problems. Now a shortcut technique will be developed for modeling a bandpass filter by using an equivalent low-pass filter that has a complex valued impulse response (see Fig. 4-2a). $v_1(t)$ and $v_2(t)$ are the input and output bandpass waveforms with the corresponding complex envelopes $g_1(t)$

† Since the frequencies involved in the argument of Re{·} are all positive, it can be shown that the complex function $c_0 + 2\sum_{n=1}^{\infty} c_n e^{jn\omega_0 t}$ is analytic in the upper-half complex t plane. Many interesting properties result because this function is an analytic function of a complex variable.

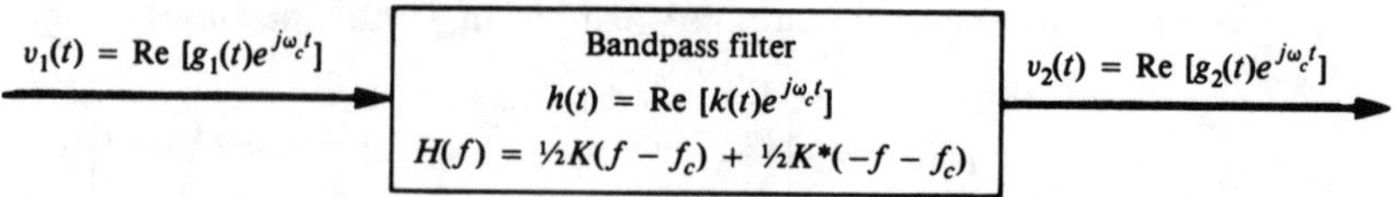

$v_1(t) = \text{Re}\,[g_1(t)e^{j\omega_c t}]$
Bandpass filter
$h(t) = \text{Re}\,[k(t)e^{j\omega_c t}]$
$H(f) = \frac{1}{2}K(f - f_c) + \frac{1}{2}K^*(-f - f_c)$
$v_2(t) = \text{Re}\,[g_2(t)e^{j\omega_c t}]$

(a) Bandpass Filter

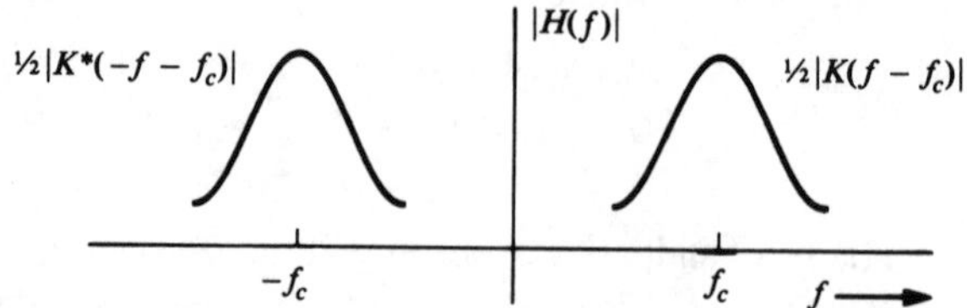

(b) Typical Bandpass Filter Frequency Response

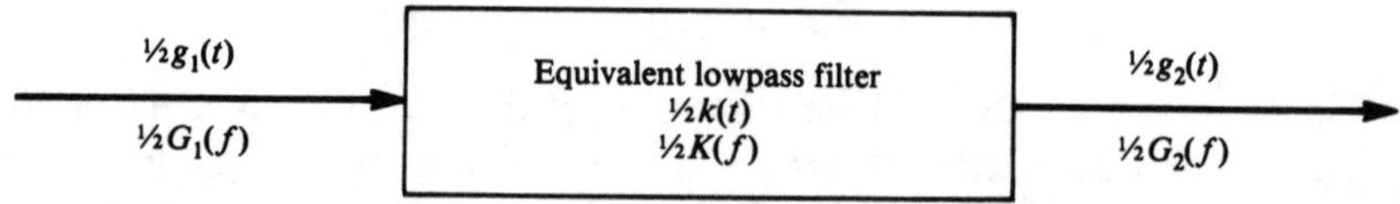

(c) Equivalent (Complex Impulse Response) Lowpass Filter

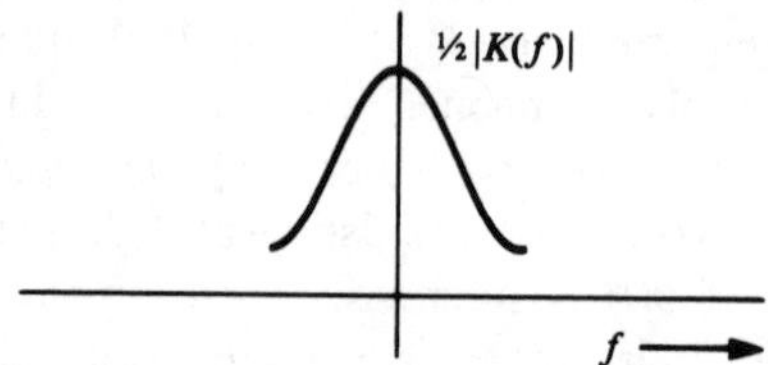

(d) Typical Equivalent Lowpass Filter Frequency Response

FIGURE 4-2 Bandpass filtering.

and $g_2(t)$. The impulse response of the bandpass filter, $h(t)$, can also be represented by its corresponding complex envelope $k(t)$. In addition, as shown in Fig. 4-2a, the frequency domain description, $H(f)$, can be expressed in terms of $K(f)$ with the help of (4-21) and (4-22). Figure 4-2b shows a typical bandpass frequency response characteristic $|H(f)|$.

Theorem: *The complex envelopes for the input, output, and impulse response of a bandpass filter are related by*

$$\tfrac{1}{2}g_2(t) = \tfrac{1}{2}g_1(t) * \tfrac{1}{2}k(t) \tag{4-9}$$

where $g_1(t)$ is the complex envelope of the input and $k(t)$ is the complex envelope of the impulse response. It also follows that

$$\tfrac{1}{2}G_2(f) = \tfrac{1}{2}G_1(f)\tfrac{1}{2}K(f) \tag{4-10}$$

These results may be proved by using the properties of linear systems (Sec. 2-6) and is given as Problem 4-2 at the end of this chapter.

This theorem indicates that any bandpass filter system may be described and analyzed by using an *equivalent low-pass filter* as shown in Fig. 4-2c. A typical equivalent low-pass frequency response characteristic is shown in Fig. 4-2d. Since equations for equivalent low-pass filters are usually much less complicated than those for bandpass filters, the equivalent low-pass filter system model is very useful. This is the basis for computer programs that simulate bandpass communication systems and for the bandpass sampling theorem. These topics will be examined in later sections.

It is realized that the linear bandpass filter can produce variations in the phase modulation at the output, $\theta_2(t)$, where $\theta_2(t) = \angle g_2(t)$, as a function of the amplitude modulation on the input complex envelope, $R_1(t)$, where $R_1(t) = |g_1(t)|$. This is called *AM-to-PM conversion*. Similarly, the filter can also cause variations on the AM at the output, $R_2(t)$, because of the PM on the input, $\theta_1(t)$. This is called *PM-to-AM conversion*.

Since $h(t)$ represents a linear filter, $g_2(t)$ will be a linear filtered version of $g_1(t)$; however, $\theta_2(t)$ and $R_2(t)$—the PM and AM components of $g_2(t)$—will be a *nonlinear* filtered version of $g_1(t)$ since $\theta_2(t)$ and $R_2(t)$ are nonlinear functions of $g_2(t)$. The analysis of the nonlinear distortion is very complicated. Although many analysis techniques have been published in the literature, none has been entirely satisfactory. Panter [1965] gives a three-chapter summary of some of these techniques, and a classical paper is also recommended [Bedrosian and Rice, 1968]. Furthermore, nonlinearities that occur in a practical system will also cause nonlinear distortion and AM-to-PM conversion effects. Nonlinear effects can be analyzed by several techniques, including power-series analysis; this is discussed in the section on amplifiers that follows later in this chapter. If a nonlinear effect in a *bandpass* system is to be analyzed, a Fourier series technique that uses the Chebyshev transform has been found to be useful [Spilker, 1977].

★ Distortionless Transmission

In Sec. 2-6 the general conditions were found for distortionless transmission. For linear *bandpass filters* (channels), a less restrictive set of conditions will now be shown to be satisfactory. For distortionless transmission of bandpass signals, the channel transfer function, $H(f) = |H(f)|e^{j\theta(f)}$, needs to satisfy the following requirements:

- The amplitude response is constant. That is,

$$|H(f)| = A \tag{4-11a}$$

  where A is a positive (real) constant.
- The derivative of the phase response is a constant. That is,

$$-\frac{1}{2\pi}\frac{d\theta(f)}{df} = T_g \tag{4-11b}$$

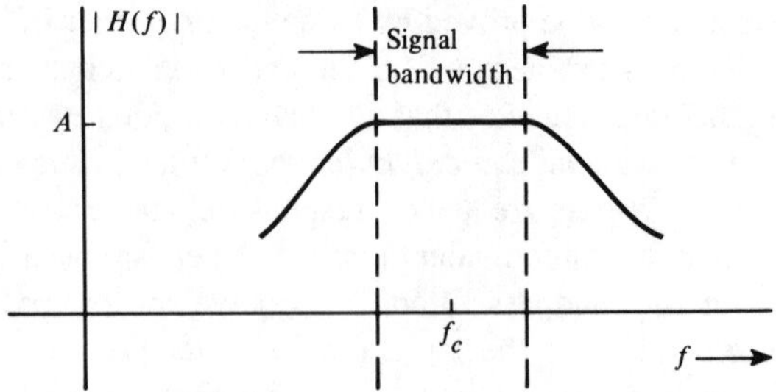

(a) Magnitude Response

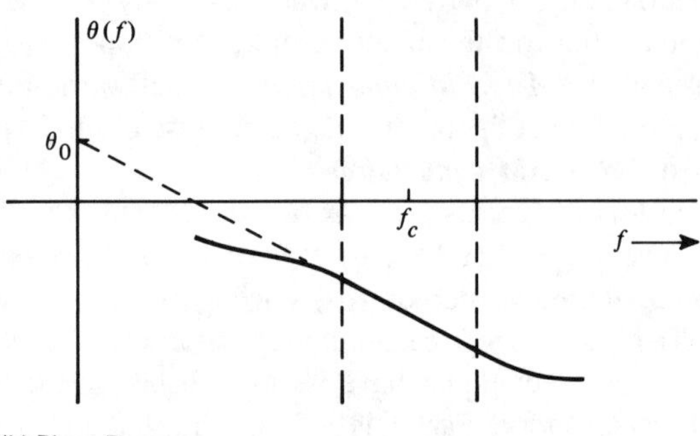

(b) Phase Response

FIGURE 4-3 Transfer characteristics of a distortionless bandpass channel.

where T_g is a constant called the complex *envelope delay* or, more concisely, the *group delay* and $\theta(f) = \underline{/H(f)}$.

This is illustrated in Fig. 4-3. Note that (4-11a) is identical to the general requirement of (2-150a), but (4-11b) is less restrictive than (2-150b). That is, if (2-150b) is satisfied, (4-11b) is satisfied where $T_d = T_g$; however, if (4-11b) is satisfied, (2-150b) is not necessarily satisfied because the integral of (4-11b) is

$$\theta(f) = -2\pi f T_g + \theta_0 \tag{4-12}$$

where θ_0 is a phase-shift constant, as shown in Fig. 4-3b. If θ_0 happens to be nonzero, (2-150b) is not satisfied.

Now it will be shown that (4-11a) and (4-11b) are sufficient requirements for distortionless transmission of bandpass signals. From (4-11a) and (4-12) the channel (or filter) transfer function is

$$H(f) = Ae^{j(-2\pi f T_g + \theta_0)} = (Ae^{j\theta_0})e^{-j2\pi f T_g} \tag{4-13}$$

over the bandpass of the signal. If the input to the bandpass channel is represented by

$$v_1(t) = x(t) \cos \omega_c t - y(t) \sin \omega_c t$$

then, using (4-13) and realizing that $e^{-j2\pi f T_g}$ causes a delay of T_g, the output of the channel is

$$v_2(t) = Ax(t - T_g)\cos[\omega_c(t - T_g) + \theta_0] - Ay(t - T_g)\sin[\omega_c(t - T_g) + \theta_0]$$

Using (4-12), we obtain

$$v_2(t) = Ax(t - T_g)\cos[\omega_c t + \theta(f_c)] - Ay(t - T_g)\sin[\omega_c t + \theta(f_c)]$$

where, by use of (2-150b) evaluated at $f = f_c$,

$$\theta(f_c) = -\omega_c T_g + \theta_0 = -2\pi f_c T_d$$

Thus the output bandpass signal can be described by

$$v_2(t) = Ax(t - T_g)\cos[\omega_c(t - T_d)] - Ay(t - T_g)\sin[\omega_c(t - T_d)] \quad (4\text{-}14)$$

where the modulation on the carrier (i.e., the x and y components) has been delayed by the *group time delay*, T_g, and the carrier has been delayed by the *carrier time delay*, T_d. Since $\theta(f_c) = -2\pi f_c T_d$, where $\theta(f_c)$ is the carrier phase shift, T_d is also called the *phase delay*. Equation (4-14) demonstrates that the bandpass filter delays the input complex envelope (i.e., the input information) by T_g seconds, whereas the carrier is delayed by T_d seconds. This is distortionless transmission, which is obtained when (4-11a) and (4-11b) are satisfied. Note that T_g will differ from T_d unless θ_0 happens to be zero.

In summary, the general requirements for distortionless transmission of either baseband or bandpass signals are given by (2-150a) and (2-150b). However, for the bandpass case (2-150b) is overly restrictive and may be replaced by (4-11b). (In this case $T_d \neq T_g$ unless $\theta_0 = 0$.) That is, for distortionless *bandpass* transmission, it is only necessary to have a transfer function with a *constant amplitude* and a *constant phase derivative* over the bandwidth of the signal.

Bandpass Sampling Theorem

Computer simulation is often used to analyze the performance of complicated communication systems, especially when coding schemes are used. This requires that the RF signal and noise for the system under test be sampled so that the data can be obtained for processing by digital computer simulation. If the sampling were carried out at the Nyquist rate or larger ($f_s \geq 2B$, where B is the highest frequency involved in the spectrum of the RF signal), the sampling rate would be ridiculous. For example, consider a satellite communication system with a carrier frequency of $f_c = 6$ GHz. The sampling rate required would be at least 12 GHz. Fortunately, for the signals of this type (bandpass signals), it can be shown that the sampling rate depends only on the *bandwidth* of the signal, not on the absolute frequencies involved. This is equivalent to saying that we can reproduce the signal from samples of the complex envelope.

BANDPASS SAMPLING THEOREM: *If a (real) bandpass waveform has a nonzero spectrum only over the frequency interval* $f_1 < |f| < f_2$ *where the transmission* *bandwidth B_T is taken to be the absolute bandwidth given by $B_T = f_2 - f_1$, then the waveform may be reproduced from sample values, if the sampling rate is*

$$f_s \geq 2B_T \tag{4-15}$$

This theorem may be demonstrated by using the quadrature bandpass representation

$$v(t) = x(t) \cos \omega_c t - y(t) \sin \omega_c t \tag{4-16}$$

Let f_c be the center of the bandpass so that $f_c = (f_2 + f_1)/2$. Then, from (4-8) it is seen that both $x(t)$ and $y(t)$ are baseband signals and absolutely bandlimited to $B = B_T/2$. From the baseband sampling theorem the sampling rate required to represent these baseband signals is $f_b \geq 2B = B_T$. Equation (4-16) becomes

$$v(t) = \sum_{n=-\infty}^{n=\infty} \left[x\left(\frac{n}{f_b}\right) \cos \omega_c t - y\left(\frac{n}{f_b}\right) \sin \omega_c t \right] \left[\frac{\sin\{\pi f_b[t - (n/f_b)]\}}{\pi f_b[t - (n/f_b)]} \right] \tag{4-17}$$

For the general case, where the $x(n/f_b)$ and $y(n/f_b)$ samples are independent, two real samples are obtained for each value of n so that the overall sampling rate for $v(t)$ is $f_s = 2f_b \geq 2B_T$. This is the bandpass sampling frequency requirement of (4-15). In most engineering applications, $f_c \gg B_T$. Thus the x and y samples can be obtained by sampling $v(t)$ at $t \approx (n/f_b)$, but adjusting t slightly, so that $\cos \omega_c t = 1$ and $\sin \omega_c t = 1$ at the *exact* sampling time for x and y, respectively. That is, for $t \approx n/f_s$, $v(n/f_b) = x(n/f_b)$ when $\cos \omega_c t = 1$ (i.e., $\sin \omega_c t = 0$) and $v(n/f_b) = y(n/f_b)$ when $\sin \omega_c t = 1$ (i.e., $\cos \omega_c t = 0$). If f_c is not large enough for the x and y samples to be obtained directly from $v(t)$, then $x(t)$ and $y(t)$ can first be obtained by the use of two quadrature product detectors as described by (4-71). The $x(t)$ and $y(t)$ baseband signals can then be individually sampled at a rate of f_b, and it is seen that the overall equivalent sampling rate is still $f_s = 2f_b \geq 2B_T$.

In the application of this theorem, it is assumed that the bandpass signal $v(t)$ is reconstructed by use of (4-17). This implies that *nonuniformly* spaced time samples of $v(t)$ are used since the samples are taken in pairs (for the x and y components) instead of being uniformly spaced T_s seconds apart. *Uniformly* spaced samples of $v(t)$ itself can be used with a minimum sampling frequency of $2B_T$, *provided* that either f_1 or f_2 is a harmonic of f_s [Taub and Schilling, 1986]. Otherwise, a minimum sampling frequency larger than $2B_T$ but not larger than $4B_T$ is required [Haykin, 1983; Peebles, 1976; Taub and Schilling, 1986]. This phenomenon also occurs with impulse sampling [eq. (2-173)] because f_s needs to be selected so that there is no spectral overlap in the $f_1 < f < f_2$ band when the bandpass spectrum is translated to harmonics of f_s.

BANDPASS DIMENSIONALITY THEOREM: *Assume that a bandpass waveform has a nonzero spectrum only over the frequency interval $f_1 < |f| < f_2$ where the transmission bandwidth B_T is taken to be the absolute bandwidth given by $B_T =$*

$f_2 - f_1$ and $B_T \ll f_1$. The waveform may be completely specified over a T_0-second interval by

$$N = 2B_T T_0 \tag{4-18}$$

independent pieces of information. N is said to be the number of dimensions required to specify the waveform.

The bandpass dimensionality theorem is similar to the dimensionality theorem discussed earlier, (2-174), and may be demonstrated by the use of the complex Fourier series to represent the bandpass waveform over a T_0-second interval. The number of nonzero Fourier coefficients required to produce a line spectrum over the B_T hertz frequency interval is B_T/f_0 since f_0 is the frequency spacing between adjacent lines. Since $c_n = |c_n|e^{j\angle c_n}$ and $c_{-n} = c_n^*$, the number of independent (real value) samples is $2B_T/f_0 = 2B_T T_0$.

As mentioned earlier, computer simulation is often used to analyze communication systems. The bandpass dimensionality theorem tells us that a bandpass signal B_T hertz wide can be represented over a T_0-second interval provided that at least $N = 2B_T T_0$ samples are used.

Representation of Modulated Signals

As defined earlier, modulation is the process of encoding the source information $m(t)$ (modulating signal) into a bandpass signal $s(t)$ (modulated signal). Consequently, the modulated signal is just a special application of the bandpass representation. The *modulated signal* is given by

$$s(t) = \text{Re}\{g(t)e^{j\omega_c t}\} \tag{4-19}$$

where $\omega_c = 2\pi f_c$. f_c is the carrier frequency. The complex envelope $g(t)$ is a function of the modulating signal $m(t)$. That is,

$$g(t) = g[m(t)] \tag{4-20}$$

Thus $g[\cdot]$ performs a mapping operation on $m(t)$. This was shown in Fig. 4-1.

Table 4-1 gives an overview on the *big picture* for the modulation problem. Examples of the mapping function $g[m]$ are given for amplitude modulation (AM), double-sideband suppressed carrier (DSB-SC), phase modulation (PM), frequency modulation (FM), single-sideband AM suppressed carrier (SSB-AM-SC), single-sideband PM (SSB-PM), single-sideband FM (SSB-FM), single-sideband envelope detectable (SSB-EV), single-sideband square-law detectable (SSB-SQ), and quadrature modulation (QM). Modulated signals are discussed in detail in Secs. 4-5, 4-6, and 4-7. Obviously, it is possible to use other $g[m]$ functions that are not listed in Table 4-1. The question is: Are they useful? $g[m]$ functions are desired that are easy to implement and that will give desirable spectral properties. Furthermore, in the receiver the inverse function $m[g]$ is required. The inverse should be single valued over the range used and should be easily implemented. The mapping should suppress as much noise as possible so that $m(t)$ can be recovered with little corruption.

TABLE 4-1 Complex Envelope Functions for Various Types of Modulation[a]

Type of Modulation	Mapping Functions $g[m]$	Corresponding Quadrature Modulation $x(t)$	$y(t)$
AM	$A_c[1 + m(t)]$	$A_c[1 + m(t)]$	0
DSB-SC	$A_c m(t)$	$A_c m(t)$	0
PM	$A_c e^{jD_p m(t)}$	$A_c \cos[D_p m(t)]$	$A_c \sin[D_p m(t)]$
FM	$A_c e^{jD_f \int_{-\infty}^{t} m(\sigma)\, d\sigma}$	$A_c \cos\left[D_f \int_{-\infty}^{t} m(\sigma)\, d\sigma\right]$	$A_c \sin\left[D_f \int_{-\infty}^{t} m(\sigma)\, d\sigma\right]$
SSB-AM-SC[b]	$A_c[m(t) \pm j\hat{m}(t)]$	$A_c m(t)$	$\pm A_c \hat{m}(t)$
SSB-PM[b]	$A_c e^{jD_p[m(t) \pm j\hat{m}(t)]}$	$A_c e^{\mp D_p \hat{m}(t)} \cos[D_p m(t)]$	$A_c e^{\mp D_p \hat{m}(t)} \sin[D_p m(t)]$
SSB-FM[b]	$A_c e^{jD_f \int_{-\infty}^{t} [m(\sigma) \pm j\hat{m}(\sigma)] d\sigma}$	$A_c e^{\mp D_f \int_{-\infty}^{t} \hat{m}(\sigma) d\sigma} \cos\left[D_f \int_{-\infty}^{t} m(\sigma)\, d\sigma\right]$	$A_c e^{\mp D_f \int_{-\infty}^{t} \hat{m}(\sigma) d\sigma} \sin\left[D_f \int_{-\infty}^{t} m(\sigma)\, d\sigma\right]$
SSB-EV[b]	$A_c e^{\{\ln[1 + m(t)] \pm j\hat{\ln}\|1 + m(t)\|\}}$	$A_c[1 + m(t)] \cos\{\hat{\ln}[1 + m(t)]\}$	$\pm A_c[1 + m(t)] \sin\{\hat{\ln}[1 + m(t)]\}$
SSB-SQ[b]	$A_c e^{(1/2)\{\ln[1 + m(t)] \pm j\hat{\ln}\|1 + m(t)\|\}}$	$A_c \sqrt{1 + m(t)} \cos\{\frac{1}{2} \hat{\ln}[1 + m(t)]\}$	$\pm A_c \sqrt{1 + m(t)} \sin\{\frac{1}{2} \hat{\ln}[1 + m(t)]\}$
QM	$A_c m_1(t) + j m_2(t)$	$A_c m_1(t)$	$A_c m_2(t)$

Type of Modulation	Corresponding Amplitude and Phase Modulation $R(t)$	$\theta(t)$	Linearity	Remarks
AM	$A_c\lvert 1 + m(t)\rvert$	$\begin{cases} 0, & m(t) > -1 \\ 180°, & m(t) < -1 \end{cases}$	L[c]	$m(t) > -1$ required for envelope detection
DSB-SC	$A_c\lvert m(t)\rvert$	$\begin{cases} 0, & m(t) > 0 \\ 180°, & m(t) < 0 \end{cases}$	L	Coherent detection required
PM	A_c	$D_p m(t)$	NL	D_p is the phase deviation constant (rad/volt)
FM	A_c	$D_f \int_{-\infty}^{t} m(\sigma)\, d\sigma$	NL	D_f is the frequency deviation constant (rad/volt-sec)
SSB-AM-SC[b]	$A_c\sqrt{[m(t)]^2 + [\hat{m}(t)]^2}$	$\tan^{-1}[\pm\hat{m}(t)/m(t)]$	L	Coherent detection required
SSB-PM[b]	$A_c e^{\mp D_p \hat{m}(t)}$	$D_p m(t)$	NL	
SSB-FM[b]	$A_c e^{\mp D_f \int_{-\infty}^{t} \hat{m}(\sigma) d\sigma}$	$D_f \int_{-\infty}^{t} m(\sigma)\, d\sigma$	NL	
SSB-EV[b]	$A_c[1 + m(t)]$	$\pm\widehat{\ln}[1 + m(t)]$	NL	$m(t) > -1$ is required so that the $\ln(\cdot)$ will have a real value
SSB-SQ[b]	$A_c\sqrt{1 + m(t)}$	$\pm\frac{1}{2}\widehat{\ln}[1 + m(t)]$	NL	$m(t) > -1$ is required so that the $\ln(\cdot)$ will have a real value
QM	$A_c\sqrt{m_1^2(t) + m_2^2(t)}$	$\tan^{-1}[m_2(t)/m_1(t)]$	L	Used in NTSC color television; requires coherent detection

[a] $A_c > 0$ is a constant that sets the power level of the signal as evaluated by use of (4-27); L, linear; NL. nonlinear; $\hat{[\cdot]}$ is the Hilbert transform (i.e., $-90°$ phase-shifted version) of $[\cdot]$ (see Sec. 4-6 and Sec. A-7, Appendix A).

[b] Use upper signs for upper sideband signals and lower signs for lower sideband signals.

[c] In the strict sense, AM signals are not linear because the carrier term does not satisfy the linearity (superposition) condition.

Spectrum of Bandpass Signals

The spectrum of a bandpass signal is directly related to the spectrum of its complex envelope.

THEOREM: *If a bandpass waveform is represented by*

$$v(t) = \text{Re}\{g(t)e^{j\omega_c t}\} \tag{4-21}$$

then the spectrum of the bandpass waveform is

$$V(f) = \tfrac{1}{2}[G(f - f_c) + G^*(-f - f_c)] \tag{4-22}$$

and the PSD of the waveform is

$$\mathcal{P}_v(f) = \tfrac{1}{4}[\mathcal{P}_g(f - f_c) + \mathcal{P}_g(-f - f_c)] \tag{4-23}$$

where $G(f) = \mathcal{F}[g(t)]$ and $\mathcal{P}_g(f)$ is the PSD of $g(t)$.

Proof

$$v(t) = \text{Re}\{g(t)e^{j\omega_c t}\} = \tfrac{1}{2}g(t)e^{j\omega_c t} + \tfrac{1}{2}g^*(t)e^{-j\omega_c t}$$

Thus

$$V(f) = \mathcal{F}[v(t)] = \tfrac{1}{2}\mathcal{F}[g(t)e^{j\omega_c t}] + \tfrac{1}{2}\mathcal{F}[g^*(t)e^{-j\omega_c t}] \tag{4-24}$$

If we use $\mathcal{F}[g^*(t)] = G^*(-f)$ from Table 2-1 and the frequency translation property of Fourier transforms from Table 2-1, this equation becomes

$$V(f) = \tfrac{1}{2}\{G(f - f_c) + G^*[-(f + f_c)]\} \tag{4-25}$$

which reduces to (4-22).

The PSD for $v(t)$ is obtained by first evaluating the autocorrelation for $v(t)$.

$$R_v(\tau) = \langle v(t)v(t + \tau)\rangle = \langle \text{Re}\{g(t)e^{j\omega_c t}\}\,\text{Re}\{g(t + \tau)e^{j\omega_c(t+\tau)}\}\rangle$$

Using the identity (see Prob. 2-68),

$$\text{Re}(c_2)\,\text{Re}(c_1) = \tfrac{1}{2}\,\text{Re}(c_2^* c_1) + \tfrac{1}{2}\,\text{Re}(c_2 c_1)$$

where $c_2 = g(t)e^{j\omega_c t}$ and $c_1 = g(t + \tau)e^{j\omega_c(t+\tau)}$, we get

$$R_v(\tau) = \langle \tfrac{1}{2}\,\text{Re}\{g^*(t)g(t + \tau)e^{-j\omega_c t}e^{j\omega_c(t+\tau)}\}\rangle + \langle \tfrac{1}{2}\,\text{Re}\{g(t)g(t + \tau)e^{j\omega_c t}e^{j\omega_c(t+\tau)}\}\rangle$$

Realizing that both $\langle\ \rangle$ and Re{ } are linear operators, we may exchange the order of the operators without affecting the result, and the autocorrelation becomes

$$R_v(\tau) = \tfrac{1}{2}\,\text{Re}\{\langle g^*(t)g(t + \tau)e^{j\omega_c \tau}\rangle\} + \tfrac{1}{2}\,\text{Re}\{\langle g(t)g(t + \tau)e^{j2\omega_c t}e^{j\omega_c \tau}\rangle\}$$

or

$$R_v(\tau) = \tfrac{1}{2}\,\text{Re}\{\langle g^*(t)g(t + \tau)\rangle e^{j\omega_c \tau}\} + \tfrac{1}{2}\,\text{Re}\{\langle g(t)g(t + \tau)e^{j2\omega_c t}\rangle e^{j\omega_c \tau}\}$$

But $\langle g^*(t)g(t + \tau)\rangle = R_g(\tau)$. The second term on the right is negligible because $e^{j2\omega_c t} = \cos 2\omega_c t + j \sin 2\omega_c t$ oscillates much faster than variations in $g(t)g(t + \tau)$. In other words, f_c is much larger than the frequencies in $g(t)$, so the integral is negligible. This is an application of the Riemann–Lebesque lemma from integral calculus [Olmsted, 1961]. Thus the autocorrelation reduces to

$$R_v(\tau) = \tfrac{1}{2}\,\mathrm{Re}\{R_g(\tau)e^{j\omega_c\tau}\} \tag{4-26}$$

The PSD is obtained by taking the Fourier transform of (4-26) (i.e., applying the Wiener–Khintchine theorem). Note that (4-26) has the same mathematical form as (4-21) when t is replaced by τ, so the Fourier transform has the same form as (4-22). Thus

$$\mathcal{P}_v(f) = \mathcal{F}[R_v(\tau)] = \tfrac{1}{4}[\mathcal{P}_g(f - f_c) + \mathcal{P}_g^*(-f - f_c)]$$

But $\mathcal{P}_g^*(f) = \mathcal{P}_g(f)$ since the PSD is a real function. Thus the PSD is given by (4-23).

Evaluation of Power

Theorem: *The total average normalized power of a bandpass waveform, $v(t)$, is*

$$P_v = \langle v^2(t)\rangle = \int_{-\infty}^{\infty} \mathcal{P}_v(f)\, df = R_v(0) = \tfrac{1}{2}\langle |g(t)|^2\rangle \tag{4-27}$$

where "normalized" implies that the load is equivalent to one ohm.

Proof. Substituting $v(t)$ into (2-67), we get

$$P_v = \langle v^2(t)\rangle = \int_{-\infty}^{\infty} \mathcal{P}_v(f)\, df$$

But $R_v(\tau) = \mathcal{F}^{-1}[\mathcal{P}_v(f)] = \int_{-\infty}^{\infty} \mathcal{P}_v(f)e^{j2\pi f\tau}\, df$, so

$$R_v(0) = \int_{-\infty}^{\infty} \mathcal{P}_v(f)\, df$$

Also, from (4-26),

$$R_v(0) = \tfrac{1}{2}\,\mathrm{Re}\{R_g(0)\} = \tfrac{1}{2}\,\mathrm{Re}\{\langle g^*(t)g(t+0)\rangle\}$$

or

$$R_v(0) = \tfrac{1}{2}\,\mathrm{Re}\{\langle |g(t)|^2\rangle\}$$

But $|g(t)|$ is always real, so

$$R_v(0) = \tfrac{1}{2}\,\langle |g(t)|^2\rangle$$

Another type of power rating, called the *peak envelope power* (PEP), is useful for transmitter specifications.

Definition: The *peak envelope power* (PEP) is the average power that would be obtained if $|g(t)|$ were to be held constant at its peak value.

This is equivalent to evaluating the average power in an unmodulated RF sinusoid that has a peak value of $A_p = \max[v(t)]$, as is readily seen from Fig. 4-32.

THEOREM: *The normalized PEP is given by*

$$P_{\text{PEP}} = \tfrac{1}{2}\,[\max|g(t)|]^2 \tag{4-28}$$

A proof of this theorem follows by applying the definition to (4-27). As described later in this chapter and in Chapter 5, the PEP is useful for specifying the power capability of AM, SSB, and television transmitters.

EXAMPLE 4-1 AMPLITUDE-MODULATED SIGNAL

Evaluate the magnitude spectrum for an amplitude-modulated (AM) signal. From Table 4-1, the complex envelope of an AM signal is

$$g(t) = A_c[1 + m(t)]$$

so that the spectrum of the complex envelope is

$$G(f) = A_c\delta(f) + A_cM(f) \tag{4-29}$$

Using (4-19), we obtain the AM signal waveform

$$s(t) = A_c[1 + m(t)]\cos\omega_c t$$

and, using (4-22), the AM spectrum

$$S(f) = \tfrac{1}{2}A_c[\delta(f - f_c) + M(f - f_c) + \delta(f + f_c) + M(f + f_c)] \tag{4-30a}$$

where, since $m(t)$ is real, $M^*(f) = M(-f)$, and $\delta(f) = \delta(-f)$ (the delta function was defined to be even) were used. Suppose that the magnitude spectrum of the modulation happens to be a triangular function, as shown in Fig. 4-4a. This spectrum might arise from an analog audio source where the bass frequencies are emphasized. The resulting AM spectrum, using (4-31), is shown in Fig. 4-4b. Note that since $G(f - f_c)$ and $G^*(-f - f_c)$ do not overlap, the magnitude spectrum is

$$|S(f)| = \begin{cases} \tfrac{1}{2}A_c\delta(f - f_c) + \tfrac{1}{2}A_c|M(f - f_c)|, & f > 0 \\ \tfrac{1}{2}A_c\delta(f + f_c) + \tfrac{1}{2}A_c|M(-f - f_c)|, & f < 0 \end{cases} \tag{4-30b}$$

The "1" in $g(t) = A_c[1 + m(t)]$ causes delta functions to occur in the spectrum at $f = \pm f_c$, where f_c is the assigned carrier frequency. Using (4-27), we obtain the total average signal power

$$P_s = \tfrac{1}{2}A_c^2\langle|1 + m(t)|^2\rangle = \tfrac{1}{2}A_c^2\langle 1 + 2m(t) + m^2(t)\rangle$$

$$= \tfrac{1}{2}A_c^2[1 + 2\langle m(t)\rangle + \langle m^2(t)\rangle]$$

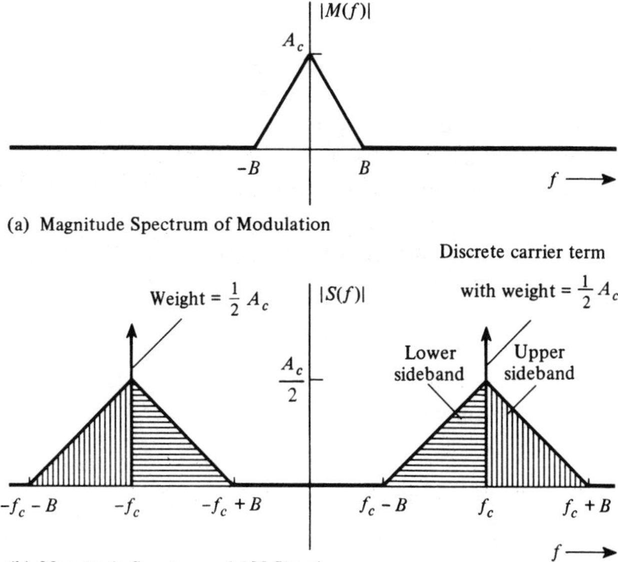

FIGURE 4-4 Spectrum of an AM signal.

If we assume that the dc value of the modulation is zero, as shown in Fig. 4-4a, the average signal power becomes

$$P_s = \tfrac{1}{2}A_c^2[1 + P_m] \tag{4-31}$$

where $P_m = \langle m^2(t) \rangle$ is the power in the modulation, $m(t)$, $\tfrac{1}{2}A_c^2$ is the carrier power, and $\tfrac{1}{2}A_c^2 P_m$ is the power in the sidebands of $s(t)$.

Received Signal Plus Noise

Using the representation of bandpass signals and including the effects of channel filtering, a model for the received signal plus noise will now be obtained. Referring to Fig. 4-1, the signal out of the transmitter is

$$s(t) = \text{Re}[g(t)e^{j\omega_c t}]$$

where $g(t)$ is the complex envelope for the particular type of modulation used (see Table 4-1).

A good mathematical model for the received signal plus noise, $r(t)$, may or may not be easy to obtain, depending on the complexity of the channel. If the channel is linear and time invariant, then

$$r(t) = s(t) * h(t) + n(t) \tag{4-32a}$$

where $h(t)$ is the impulse response of the channel and $n(t)$ is the noise at the receiver input. Furthermore, if the channel is distortionless, its transfer function is given by (4-13), and consequently, the signal plus noise at the receiver input is

$$r(t) = \text{Re}[Ag(t - T_g)e^{j\omega_c t + \theta(f_c))} + n(t)] \tag{4-32b}$$

where A is the gain of the channel (a positive number and usually less than 1), T_g is the channel group delay, and $\theta(f_c)$ is the carrier phase shift caused by the channel. In practice, the values for T_g and $\theta(f_c)$ are often not known, so that if values for T_g and $\theta(f_c)$ are needed by the receiver to detect the information that was transmitted, receiver circuits are devised to recover (i.e., estimate) the received carrier phase $\theta(f_c)$ and the group delay (e.g., a bit synchronizer for the case of digital signaling). To carry out a detailed analysis of the receiver performance when there are errors due to poor estimates of received carrier phase and group delay is beyond the scope of this book. (These topics are appropriate for advanced graduate courses.) Consequently, we will assume that the receiver circuits are designed to make these effects negligible, and therefore we can consider the signal plus noise at the receiver input to be

$$r(t) = \text{Re}[g(t)e^{j\omega_c t}] + n(t) \tag{4-33}$$

where the effects of channel filtering, if any, are taken into account by some modification of the complex envelope, $g(t)$, and the constant A_c that is implicit within $g(t)$ (see Table 4-1) is adjusted to reflect the effect of channel attenuation. Details of this approach are worked out in Sec. 5-5.

Given that (4-33) models the receiver input, the receiver output, $\tilde{m}(t)$, depends on the type of functional blocks used within the receiver, such as filters, limiters, and detectors. These blocks are described in Sec. 4-5 and their effects are summarized at the end of this section in Table 4-5.

★Digital Computer Simulation

Often communication systems are too complicated to analyze completely by using only analytical techniques, or a complete analytical approach would be too costly in terms of personnel and time. In these cases, *digital computer simulation* is used to provide a practical approach to system design and analysis. The complex envelope technique and sampling techniques are usually the basis of the computer algorithms. By using the complex envelope:

- Modulated signals can be represented (see Table 4-1).
- Noise can be represented (see Sec. 6-7).
- Filters can be simulated [see (4-9), (4-10), and Sec. 4-3].
- Detectors can be simulated (see Sec. 4-3).

Nonlinearities can be represented using time domain algorithms (see Sec. 4-3). The discrete Fourier transform (DFT) can be used for spectral analysis. Signal-to-noise ratios can be evaluated. Monte Carlo techniques can be used to estimate the probability of bit error for digital systems.

In Sec. 1-5 several computer programs that are designed for communication system simulation were listed and briefly described. Some of these programs are described in special issues of the *IEEE Journal on Selected Areas in Communica-*

tions, vol. SAC-2, January 1984, and SAC-6, January 1988. One of these simulation programs, SYSTID (system time domain), will now be studied in some detail [Fashano and Strodtbeck, 1984]. SYSTID has been in a continuous state of development and improvement by the Hughes Aircraft Company since 1967. Its development was supported under a National Aeronautics and Space Administration (NASA) contract during the early 1970s.

SYSTID uses the following techniques:

- Partitioning the communication system into basic functional blocks and processing the signal sequentially from the input to the output.
- Using complex envelope representation for modulated signals, noise, and filters.
- Using discrete time representation (sampling) of continuous time signals and systems.

The block diagram approach allows complicated communication systems to be built up from a set of interconnected functional blocks. The blocks are available from the SYSTID model library (see Table 4-2) or by calling user-defined models that are easy to program. The SYSTID package is written in FORTRAN code, and consists of a precompiler that takes SYSTID statements and generates the FORTRAN source code that simulates a particular communication system. For example, the SYSTID syntax for implementing a block in the simulation is

INPUT > MODEL NAME (PAR1, . . ., PARn, . . .) > OUTPUT

where MODEL NAME is the name of the model selected from the SYSTID library (or one defined by the user), PARn are pass parameters for the model, INPUT is the name assigned to the input node of the block that is being simulated, and OUTPUT is the output node name. Blocks are connected together by using a common node name for the output of a block that connects to the input of another block.

The communication system is simulated by executing the machine code that results from the compiled FORTRAN source code after it has been linked to the SYSTID library routines. The program output consists of data that are samples of the waveform (as a function of time) at each node of interest. A companion analysis program called RIP is used to look at the SYSTID output data and present a graphic output of the user-requested waveform properties. For example, the user may request plots of the waveform at any node, the PSD, the autocorrelation function, the eye pattern, and so on. An example of a SYSTID simulation for a simple communication system and the associated RIP graphic output is shown in Fig. 4-5.

The SYSTID package can be used to simulate very complicated communication systems. Some typical applications include bit error rate sensitivity, adjacent and cochannel interference, multicarrier frequency division multiple access (FDMA) frequency planning (see Chapter 5 for a description of FDMA), modulation and coding system performance, analog system performance (S/N and

TABLE 4-2 SYSTID Model Library

Signal Generators	Filters	Miscellaneous	Estimators
White Gaussian noise	Butterworth	Limiters	Power spectra
Arbitrary PDF sources	Chebyshev	TWT	Probability of bit error (MPSK, QASK)
Pulse generator	Bessel	Arbitrary nonlinearities	Noise power ratio
Transcendental functions	Butterworth–Thompson	TDA	Signal distortion
Square wave	Elliptic	Up/down converters	Signal to noise
Arbitrary tables	Quasi-elliptic	Differentiator	Noise bandwidth
PN sequence generator	Lead lag	Integration	Delay
Random bit stream generator	Integrate and dump	Time delays	Correlation
PN bit stream generator	Quadratic	Phase shifters	Statistics
MSK PN signal generator	Arbitrary poles and zeros	Latches and gates	Eye pattern analyzer
Raised cos PN signal generator	Transversal	Threshold detectors	Intelligible crosstalk
QPSK PN signal generator	Zonal (ideal)	Power meter	**Source Encoders**
FM noise loaded baseband signal generator	Measured response	Average meter	Analog-to-digital
Biphase L PN generator	Functional form	Rotary joint multipath	Digital-to-analog
ENB signal generator	Adaptive transversal equalizer	Recall file	Multilevel PCM
Impulse noise	**Mods and Demods**	Saver/loader	Orthogonal
Continuous wave	Amplitude		Fourier
	Frequency		Haar
	Phase		Hadamard
	Delta modulator		Ordered Hadamard
	FSK		Digital
	PSK		Interleaving and deinterleaving
	QASK, MPSK		Convolutional
	Digital echo modulation		
	Zero-crossing detector		

Source: Michael Fashano and Andrew L. Strodtbeck, "Communication System Simulation and Analysis with SYSTID," *IEEE Journal on Selected Areas in Communications,* vol. SAC-2, Janaury 1984, p. 11. © 1984 IEEE.

distortion), signal detection, adaptive signal processing, device modeling, and verification of channel specifications.

From the preceding discussion, it is clear that a study of communication system component blocks is needed before an overall study of communication systems (as presented in Chapter 5) is attempted. This study of the basic building block components is presented in the next section.

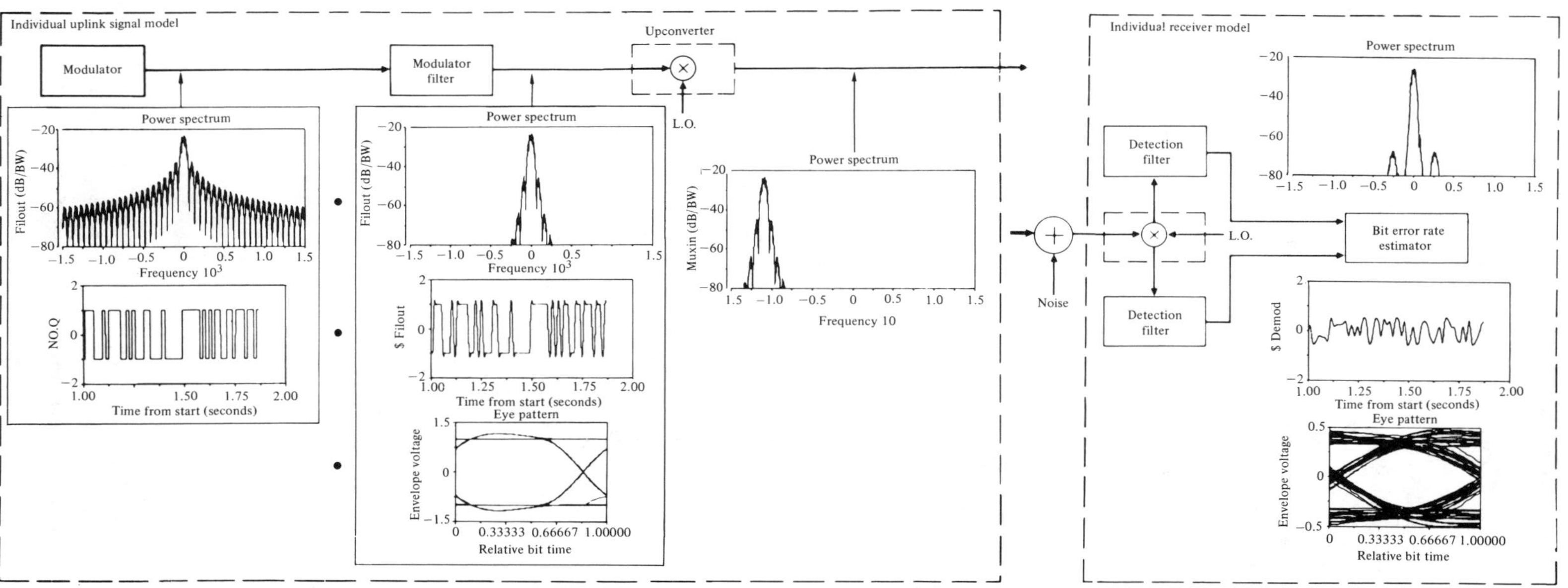

FIGURE 4-5 Transmitter and receiver simulation using SYSTID. [From Michael Fashano and Andrew L. Strodtbeck, "Communication System Simulation and Analysis with SYSTID," *IEEE Journal on Selected Areas in Communications,* vol. SAC-2, Jan. 1984, p. 17. © 1984 IEEE.]

4-3 Components of Communication Systems

Communication systems consist of component blocks such as amplifiers, filters, limiters, oscillators, mixers, frequency multipliers, phase-locked loops, detectors, and other analog and digital circuits. In most circuit and electronic courses, a great deal of time is spent on circuits that are linear and time invariant since they are relatively easy to analyze and form the basis for understanding other types of circuits. However, it is realized that all practical circuits are nonlinear, and we need to examine the distortion products that are produced by nonlinear effects. In addition, as we will see, nonlinear and time-varying circuits are often needed in communication systems to produce signals with new frequency components.

Filters

Filters are devices that take an input waveshape and modify the frequency spectrum to produce the output waveshape. Filters may be classified in several ways. One is by the type of construction used, such as *LC* elements or quartz crystal elements. Another is by the type of transfer function that is realized, such as the Butterworth or Chebyshev response (defined subsequently). These two topics of construction types and transfer function characteristics are discussed in this section.

Filters use energy storage elements to obtain frequency discrimination. In any physical filter the energy storage elements are imperfect. For example, a physical inductor has some series resistance as well as inductance, and a physical capacitor has some shunt (leakage) resistance as well as capacitance. A natural question, then, is: What is the quality, Q, of a circuit element or filter? Unfortunately, two different measures of filter quality are used in the technical literature. The first definition is concerned with the efficiency of *energy storage* in a circuit; it is [Ramo, Whinnery, and vanDuzer, 1967, 1984]

$$Q = \frac{2\pi(\text{maximum energy stored during one cycle})}{\text{energy dissipated per cycle}} \tag{4-34}$$

Of course, a larger value for Q corresponds to a more perfect storage element. That is, a perfect L or C element would have infinite Q. The second definition is concerned with the *frequency selectivity* of a circuit; it is

$$Q = \frac{f_0}{B} \tag{4-35}$$

where f_0 is the resonant frequency and B is the 3-dB bandwidth. Here the larger the value of Q, the better the frequency selectivity, since for a given f_0, the bandwidth would be smaller.

In general, the value of Q as evaluated using (4-34) is different from the value of Q obtained from (4-35). However, these two definitions give identical values for an *RLC* series resonant circuit driven by a voltage source or for an *RLC* parallel resonant circuit driven by a current source [Nilsson, 1990]. For bandpass filtering applications, frequency selectivity is the desired characteristic, so (4-35) is used. Also, (4-35) is easy to evaluate from laboratory measurements. If we are designing a passive filter (not necessarily a single-tuned circuit) of center frequency f_0 and 3-dB bandwidth B, the individual circuit elements will each need to have much larger Q's than f_0/B. Thus, for a practical filter design we first need to answer the question: What are the Q's needed for the filter elements, and what kind of elements will give these values of Q? This question is answered in Table 4-3.

Table 4-3 lists filters as classified by the type of energy storage elements used in their construction and gives typical values for the Q of the elements. Filters that use lumped† L and C elements become impractical to build above 300 MHz because the parasitic capacitance and inductance of the leads significantly affect the frequency response at high frequencies. Active filters, which use operational amplifiers with RC circuit elements, are practical only below 500 kHz since the op amps need to have a large open-loop gain over the operating band. For very-low-frequency filters, RC active filters are usually preferred to LC passive filters because the size of the LC components becomes large and the Q of the inductors becomes small in this frequency range.

Crystal filters are manufactured from quartz crystal elements. These act as a series resonant circuit in parallel with a shunt capacitance caused by the holder (mounting). Thus a parallel resonant as well as a series resonant mode of operation is possible. Above 100 MHz the quartz element becomes physically too small to manufacture, and below 1 kHz the size of the element becomes prohibitively large. Crystal filters have excellent performance because of the inherently high Q of the elements, but they are more expensive than LC and ceramic filters.

Mechanical filters use the vibrations of a resonant mechanical system to obtain the filtering action. The mechanical system usually consists of a series of disks spaced along a rod. Transducers are mounted on each end of the rod to convert the electrical signals to mechanical vibrations at the input, and vice versa at the output. Each disk is the mechanical equivalent of a high-Q electrical parallel resonant circuit. The mechanical filter usually has a high insertion loss resulting from the inefficiency of the input and output transducers.

Ceramic filters are constructed from piezoelectric ceramic disks with plated electrode connections on opposite sides of the disk. The behavior of the ceramic element is similar to that of the crystal filter element, as discussed earlier, except that the Q of the ceramic element is much lower. The advantage of the ceramic filter is that it often provides adequate performance at a cost that is low when compared to that for crystal filters.

Surface acoustic wave (SAW) filters utilize acoustic waves that are launched and travel on the surface of a piezoelectric substrate (slab). Metallic interleaved "fingers" have been deposited on the substrate. They help convert the electric

† A lumped element is a discrete R-, L-, or C-type element as compared to a continuously distributed RLC element such as that found in a transmission line.

TABLE 4-3 Filter Construction Techniques

Construction Type	Description of Elements or Filter	Center Frequency Range	Unloaded Q (Typical)	Filter Application[a]
LC (passive)	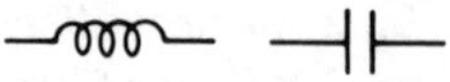	dc–300 MHz	100	Audio, video, IF, and RF
Active	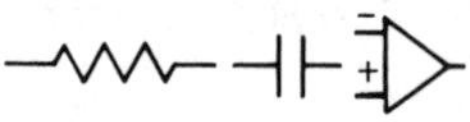	dc–500 kHz	200[b]	Audio
Crystal	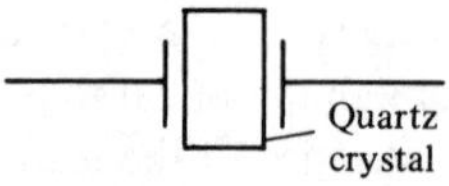 Quartz crystal	1 kHz–100 MHz	100,000	IF
Mechanical	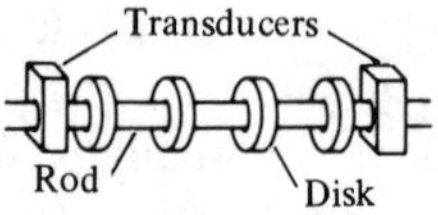 Transducers, Rod, Disk	50–500 kHz	1,000	IF
Ceramic	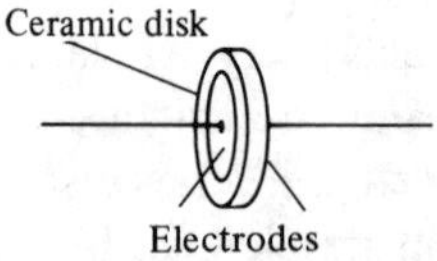 Ceramic disk, Electrodes	10 kHz–10.7 MHz	1,000	IF
Surface acoustic waves (SAW)	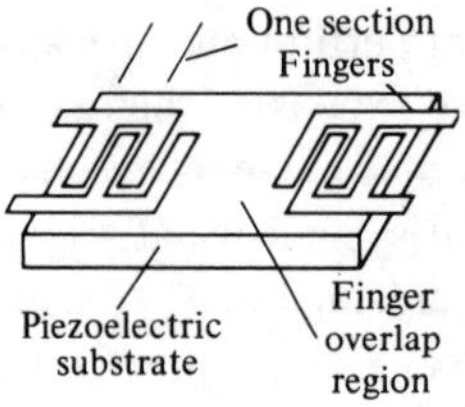 One section, Fingers, Piezoelectric substrate, Finger overlap region	10–800 MHz	c	IF and RF
Transmission line	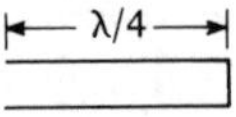 $\lambda/4$	UHF and microwave	1,000	RF
Cavity		Microwave	10,000	RF

[a]IF, intermediate frequency; RF, radio frequency (see Sec. 4-4).

[b]Bandpass Q's.

[c]Depends on design: $N = f_0/B$, where N is the number of sections, f_0 is the center frequency, and B is the bandwidth. Loaded Q's of 18,000 have been achieved.

signal to an acoustic wave (and vice versa) by providing an electric field pattern between the fingers that creates the acoustic wave as the result of the piezoelectric effect. The geometry of the fingers determines the frequency response of the filter as well as providing the input and output coupling. The insertion loss is somewhat larger than that for crystal or ceramic filters. However, the ease of shaping the transfer function and the wide bandwidth that can be obtained with controlled attenuation characteristics make the SAW filters very attractive. This new technology is being used to provide excellent IF amplifier characteristics in modern television sets.

SAW devices can also be tapped so that they are useful for transversal filter configurations operating in the RF range. At lower frequencies, charge transfer devices (CTDs) can be used to implement transversal filters [Gersho, 1975].

Transmission line filters utilize the resonant properties of open-circuited or short-circuited transmission lines. These are useful at UHF and microwave frequencies where wavelengths are small enough so that filters of reasonable size can be constructed. Similarly, the resonant effect of cavities is useful in building filters for microwave frequencies where the cavity size becomes small enough to be practical.

Filters are also characterized by the type of transfer function that is realized. The transfer function of a linear filter with lumped circuit elements may be written as the ratio of two polynomials

$$H(f) = \frac{b_0 + b_1(j\omega) + b_2(j\omega)^2 + \cdots + b_k(j\omega)^k}{a_0 + a_1(j\omega) + a_2(j\omega)^2 + \cdots + a_n(j\omega)^n}$$

where the constants a_i and b_i are functions of the element values and $\omega = 2\pi f$. The parameter n is said to be the *order* of the filter. By adjusting the constants to certain values, desirable transfer function characteristics can be obtained. Table 4-4 lists three different filter characteristics and the optimization criterion that defines each one. The Chebyshev filter is used when a sharp attenuation characteristic is required when using a minimum number of circuit elements. The Bessel filter is often used in data transmission when the pulse shape is to be preserved since it attempts to maintain a linear phase response in the passband. The Butterworth filter is often used as a compromise between the Chebyshev and Bessel characteristics.

The topic of filters is immense, and all aspects of filtering cannot possibly be covered here. For example, with the advent of inexpensive microprocessors, *digital filtering* and *digital signal processing* is becoming very important [Childers and Durling, 1975; Oppenheim and Schafer, 1975, 1989]. Here the analog waveform is sampled, and the sample values are processed by using digital techniques. Difference equations (as opposed to differential equations) describe this type of filtering operation. By changing the software, different filter characteristics can be realized. This technique of filter realization by software programming instead of hardware programming has many advantages. However, for real-time filtering, this technique is limited to realizing filters in the audio and video frequency range.

For additional reading on analog filters with emphasis on communication system applications, the reader is referred to the literature [Bowron and Stephenson, 1979].

TABLE 4-4 Some Filter Characteristics

Type	Optimization Criterion	Transfer Characteristic for the Low-Pass Filter[a]
Butterworth	Maximally flat: as many derivatives of $\lvert H(f)\rvert$ as possible go to zero as $f \to 0$	$\lvert H(f)\rvert = \dfrac{1}{\sqrt{1 + (f/f_b)^{2n}}}$
Chebyshev	For a given peak-to-peak ripple in the passband of the $\lvert H(f)\rvert$ characteristic, the $\lvert H(f)\rvert$ attenuates the fastest for any filter of nth order	$\lvert H(f)\rvert = \dfrac{1}{\sqrt{1 + \epsilon^2 C_n^2(f/f_b)}}$ ϵ = a design constant; $C_n(f)$ is the nth-order Chebyshev polynomial defined by the recursion relation $C_n(x) = 2xC_{n-1}(x) - C_{n-2}(x)$ where $C_0(x) = 1$ and $C_1(x) = x$
Bessel	Attempts to maintain linear phase in the passband	$H(f) = \dfrac{K_n}{B_n(f/f_b)}$ K_n is a constant chosen to make $H(0) = 1$ and the Bessel recursion relation is $B_n(x) = (2n - 1)\ B_{n-1}(x) - x^2 B_{n-2}(x)$ where $B_0(x) = 1$ and $B_1(x) = 1 + jx$

[a] f_b is the cutoff frequency of the filter.

Amplifiers

For analysis purposes, electronic circuits and, more specifically, amplifiers can be classified into two main categories:

- Nonlinear circuits.
- Linear circuits.

(Linearity was defined in Sec. 2-6.) In practice all circuits are nonlinear to some degree, even at low (voltage and current) signal levels, and become highly nonlinear for high signal levels. Linear circuit analysis is often used for the low signal level case since it greatly simplifies the mathematics and it gives accurate answers if the signal level is sufficiently small.

The main categories of nonlinear and linear analysis can be further classified into subcategories of circuits with

- Memory.
- No memory.

Circuits with memory contain inductive and capacitive effects which cause the present output value to be a function of previous input values as well as the present input value. If a circuit has no memory, its present output value is a function only of its present input value.

In introductory electrical engineering courses, it is first assumed that circuits are linear with no memory (resistive circuits) and, later, linear with memory (*RLC* circuits). It follows that linear amplifiers with memory may be described by a transfer function that is the ratio of the Fourier transform of the output signal divided by the Fourier transform of the input signal. As discussed in Chapter 2, the transfer function of a distortionless amplifier is given by $Ke^{-j\omega T_d}$, where K is the voltage gain of the amplifier and T_d is the delay between the output and input waveforms. If the transfer function of the linear amplifier is not of this form, the output signal will be a *linearly* distorted version of the input signal.

In addition to linear distortion, practical amplifiers also produce nonlinear distortion. In order to examine the effects of nonlinearity and yet keep a mathematical model that is tractable, we will assume that there is *no memory* in the following analysis. Thus, we will look at the present output as a function of the present input in the time domain. If the amplifier is linear, this relationship is

$$v_0(t) = Kv_i(t) \tag{4-36}$$

where K is the voltage gain of the amplifier. In practice, the output of the amplifier saturates at some value as the amplitude of the input signal is increased. This is illustrated by the nonlinear output-to-input characteristic shown in Fig. 4-6. The output-to-input characteristic may be modeled by a Taylor's expansion about $v_i = 0$ (i.e., a Maclaurin series):

$$v_0 = K_0 + K_1 v_i + K_2 v_i^2 + \cdots = \sum_{n=0}^{\infty} K_n v_i^n \tag{4-37}$$

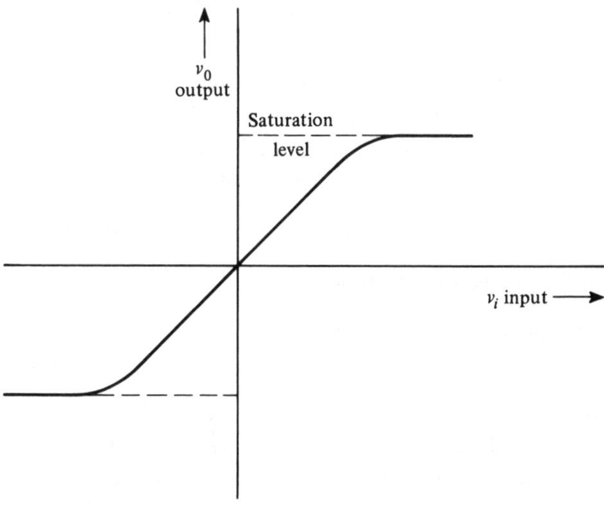

FIGURE 4-6 Nonlinear amplifier output-to-input characteristic.

where

$$K_n = \frac{1}{n!}\left(\frac{d^n v_0}{dv_i^n}\right)\bigg|_{v_i=0} \tag{4-38}$$

It is seen that there will be nonlinear distortion on the output signal if K_2, K_3, . . . , are not zero. K_0 is the output dc offset level, $K_1 v_i$ is the first-order (linear) term, $K_2 v_i^2$ is the second-order (square-law) term, and so on. Of course, K_1 will be larger than K_2, K_3, . . . , if the amplifier is anywhere near to being linear.

The *harmonic distortion* associated with the amplifier output is determined by applying a single sinusoidal test tone to the amplifier input. Let the input test tone be represented by

$$v_i(t) = A_0 \sin \omega_0 t \tag{4-39}$$

Then the second-order output term is

$$K_2(A_0 \sin \omega_0 t)^2 = \frac{K_2 A_0^2}{2}(1 - \cos 2\omega_0 t) \tag{4-40}$$

This indicates that the second-order distortion creates a dc level $K_2 A_0^2/2$ (in addition to any dc bias) and second harmonic distortion with amplitude $K_2 A_0^2/2$. In general, for a single tone input the output will be

$$v_{\text{out}}(t) = V_0 + V_1 \cos(\omega_0 t + \phi_1) + V_2 \cos(2\omega_0 t + \phi_2) + V_3 \cos(3\omega_0 t + \phi_3) + \cdots \tag{4-41}$$

where V_n is the peak value of the output at the frequency nf_0 hertz. Then the percentage of *total harmonic distortion* (THD) is defined by

$$\text{THD (\%)} = \frac{\sqrt{\sum_{n=2}^{\infty} V_n^2}}{V_1} \times 100 \tag{4-42}$$

The THD of an amplifier can be measured by using a distortion analyzer, or it can be evaluated by (4-42), with the V_n's obtained from a spectrum analyzer.

The *intermodulation distortion* (IMD) of the amplifier is obtained by using a two-tone test. If the input (tone) signals are

$$v_i(t) = A_1 \sin \omega_1 t + A_2 \sin \omega_2 t \tag{4-43}$$

then the second-order output term is

$$K_2(A_1 \sin \omega_1 t + A_2 \sin \omega_2 t)^2 = K_2(A_1^2 \sin^2 \omega_1 t + 2A_1A_2 \sin \omega_1 t \sin \omega_2 t + A_2^2 \sin^2 \omega_2 t)$$

The first and last terms on the right side of this equation produce harmonic distortion at frequencies $2f_1$ and $2f_2$. The cross-product term produces IMD. This term is present only when *both* input terms are present—thus the name "intermodulation distortion." Then the second-order IMD is

$$2K_2A_1A_2 \sin \omega_1 t \sin \omega_2 t = K_2A_1A_2 \{\cos[(\omega_1 - \omega_2)t] - \cos[(\omega_1 + \omega_2)t]\}$$

It is clear that IMD generates sum and difference frequencies.

The third-order term is

$$\begin{aligned} K_3 v_i^3 &= K_3(A_1 \sin \omega_1 t + A_2 \sin \omega_2 t)^3 \\ &= K_3(A_1^3 \sin^3 \omega_1 t + 3A_1^2 A_2 \sin^2 \omega_1 t \sin \omega_2 t \\ &\quad + 3A_1 A_2^2 \sin \omega_1 t \sin^2 \omega_2 t + A_2^3 \sin^3 \omega_2 t) \end{aligned} \tag{4-44}$$

The first and last terms on the right side of this equation will produce harmonic distortion and the second term, a cross product, becomes

$$\begin{aligned} 3K_3 A_1^2 A_2 \sin^2 \omega_1 t \sin \omega_2 t &= 3K_3 A_1^2 A_2 \sin \omega_2 t(1 - \cos^2 \omega_1 t) \\ &= \tfrac{3}{2} K_3 A_1^2 A_2 \{\sin \omega_2 t - \tfrac{1}{2}[\sin(2\omega_1 + \omega_2)t \\ &\quad - \sin(2\omega_1 - \omega_2)t]\} \end{aligned} \tag{4-45}$$

Similarly, the third term of (4-44) is

$$\begin{aligned} &3K_3 A_1 A_2^2 \sin \omega_1 t \sin^2 \omega_2 t \\ &\quad = \tfrac{3}{2} K_3 A_1 A_2^2 \{\sin \omega_1 t - \tfrac{1}{2}[\sin(2\omega_2 + \omega_1)t - \sin(2\omega_2 - \omega_1)t]\} \end{aligned} \tag{4-46}$$

The last two terms in (4-45) and (4-46) are intermodulation terms at nonharmonic frequencies. For the case of *bandpass* amplifiers where f_1 and f_2 are within the bandpass with f_1 close to f_2 (i.e., $f_1 \approx f_2 \gg 0$), the distortion products at $2f_1 + f_2$ and $2f_2 + f_1$ will usually fall outside the passband and, consequently, may not be a problem. However, the terms at $2f_1 - f_2$ and $2f_2 - f_1$ will fall inside the passband and will be close to the desired frequencies f_1 and f_2. These will be the main distortion products for bandpass amplifiers, such as those used for RF amplification in transmitters and receivers.

As (4-45) and (4-46) show, if either A_1 or A_2 is increased sufficiently, the IMD will become significant since the desired output varies linearly with A_1 or A_2 and the IMD output varies as $A_1^2 A_2$ or $A_1 A_2^2$. Of course, the exact input level required for the intermodulation products to be a problem depends on the relative values of K_3 and K_1. This may be specified by the amplifier third-order *intercept point,* which is evaluated by applying two equal test tones (i.e., $A_1 = A_2 = A$). The desired linearly amplified outputs will have amplitudes of $K_1 A$, and each of the third-order intermodulation products will have amplitudes of $3K_3 A^3/4$. The ratio of the desired output to the IMD output is then

$$R_{\text{IMD}} = \frac{4}{3}\left(\frac{K_1}{K_3 A^2}\right) \tag{4-47}$$

The input intercept point is defined as the input level that causes R_{IMD} to be unity (see Fig. 4-7). The solid curves are obtained by measurement using two sinusoidal signal generators to generate the tones and measuring the level of the desired output (at f_1 or f_2) and the IMD products (at $2f_1 - f_2$ or $2f_2 - f_1$) using a spectrum analyzer. The intercept point is a fictitious point that is obtained by extrapolation of the linear portion (decibel plot) of the desired output and IMD curves until they intersect. It is seen that the desired output (output at either f_1 or f_2)

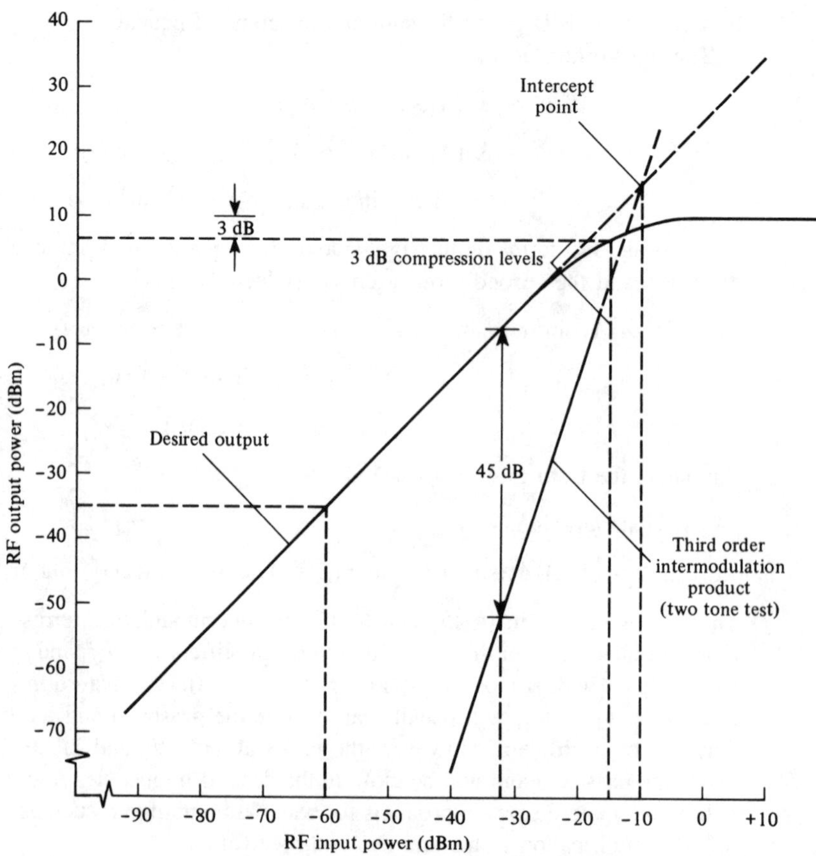

FIGURE 4-7 Amplifier output characteristics.

actually saturates when measurements are made since the higher order terms in the Taylor series have components at f_1 and f_2 that subtract from the linearly amplified output. For example, with K_3 being negative, the leading term in (4-46) occurs at f_1 and will subtract from the linearly amplified component at f_1, thus producing a saturated characteristic for the sinusoidal component at f_1. For an amplifier that happens to have the particular nonlinear characteristic shown in Fig. 4-7, the intercept point occurs for an RF input level of -10 dBm. Other properties of the amplifier are also illustrated in the figure. The gain of the amplifier is 25 dB in the linear region since a -60 dBm input produces a -35 dBm output level. The desired output is compressed by 3 dB for an input level of -15 dBm. Consequently, the amplifier might be considered to be linear only if the input level is less than -15 dBm. Furthermore, if the third-order IMD products are to be down by at least 45 dBm, the input level will have to be kept lower than -32 dBm.

Another term in the distortion products at the output of a nonlinear amplifier is called *cross modulation*. Cross-modulation terms are obtained when examining the third-order products resulting from a two-tone test. As shown in (4-45) and (4-46), the terms $\frac{3}{2}K_3A_1^2A_2 \sin \omega_2 t$ and $\frac{3}{2}K_3A_1A_2^2 \sin \omega_1 t$ are cross-modulation terms. Let us examine the term $\frac{3}{2}K_3A_1^2A_2 \sin \omega_2 t$. If we allow some amplitude variation on the input signal $A_1 \sin \omega_1 t$ so that it looks like an AM signal $A_1[1 + m_1(t)] \sin \omega_1 t$, where $m_1(t)$ is the modulating signal, a third-order distortion product becomes

$$\frac{3}{2}K_3A_1^2A_2[1 + m_1(t)]^2 \sin \omega_2 t \tag{4-48}$$

Thus the AM on the signal at carrier frequency f_1 will produce a signal at frequency f_2 with distorted modulation. That is, if two signals are passed through an amplifier having third-order distortion products in the output, and if either input signal has some AM, the amplified output of the *other* signal will be amplitude modulated to some extent by a distorted version of the modulation. This phenomenon is cross modulation.

Of course, passive as well as active circuits may have nonlinear characteristics and, consequently, will produce distortion products. For example, suppose that two AM broadcast stations have strong signals in the vicinity of a barn or house that has a metal roof with rusted joints. The roof may act as an antenna to receive and reradiate the RF energy, and the rusted joints may act as a diode (a nonlinear passive circuit). Signals at harmonics and intermodulation frequencies may be radiated and cause interference to other communication signals. In addition, cross-modulation products may be radiated. That is, distorted modulation of one station is heard on radios (located in the vicinity of the rusted roof) that are tuned to the other station's frequency.

When amplifiers are used to produce high-power signals, such as in transmitters, it is desirable to have amplifiers with high efficiency in order to reduce the costs of power supplies, cooling equipment, and energy consumed. The efficiency is the ratio of the output signal power to the dc input power. Amplifiers may be grouped into several categories, depending on the biasing levels and circuit configurations used. Some of these are Class A, B, C, D, E, F, G, H, and S [Krauss, Bostian, and Raab, 1980; Smith, 1986]. For Class A operation, the bias on the amplifier stage is adjusted so that current flows all during the complete cycle of an applied input test tone. For Class B operation, the amplifier is biased so that current flows for 180° of the applied signal cycle. Therefore, if a Class B amplifier is to be used for a baseband linear amplifier, such as an audio power amplifier in a hi-fi system, two devices are wired in push-pull configuration so that each one alternately conducts current over half of the input signal cycle. In *bandpass* Class B linear amplification, where the bandwidth is a small percentage of the operating frequency, only one active device is needed since tuned circuits may be used to supply the output current over the other half of the signal cycle. For Class C operation, the bias is set so that (collector or plate) current flows in pulses, each one having a pulse width that is usually much less than half of the input cycle. Unfortunately, with Class C operation, it is not possible to have linear amplification even if the amplifier is a bandpass RF amplifier with tuned circuits providing current over the nonconducting portion of the cycle. If one tries to amplify an AM signal with a Class C amplifier or other types of nonlinear amplifiers, the AM on the output will be distorted. However, RF signals with a constant real envelope, such as FM signals, may be amplified without distortion since a nonlinear amplifier preserves the zero-crossings of the input signal.

The efficiency of a Class C amplifier is determined essentially by the conduction angle of the active device since poor efficiency is caused by signal power being wasted in the device itself during the conduction time. The Class C amplifier is most efficient, having an efficiency factor of 100% for an ideal Class C amplifier. Class B amplifiers have an efficiency of $\pi/4 \times 100 = 78.5\%$ or less, and Class A amplifiers have an efficiency of 50% or less [Krauss, Bostian, and Raab, 1980]. Since Class C amplifiers are the most efficient, they are generally used to amplify constant envelope signals, such as FM signals used in FM broadcasting. Class D, E, F, G, H, and S amplifiers usually employ switching techniques in specialized circuits to obtain high efficiency [Krauss, Bostian, and Raab, 1980; Smith, 1986].

Many types of microwave amplifiers, such as traveling-wave tubes (TWT), operate on the velocity modulation principle. The input microwave signal is fed into a slow-wave structure. Here the velocity of propagation of the microwave signal is reduced so that it is slightly below the velocity of the dc electron beam. This enables a transfer of kinetic energy from the electron beam to the microwave signal, thereby amplifying the signal. In this type of amplifier, the electron current is *not* turned on and off to provide the amplifying mechanism; thus it is not classified in terms of Class B or C operation. The TWT is a linear amplifier when operated at the appropriate drive level. If the drive level is increased, the efficiency (RF output/dc input) is improved, but the amplifier becomes nonlinear. In this case constant envelope signals, such as PSK or PM, need to be used so that the intermodulation distortion will not cause a problem. This is often the mode of operation for satellite transponders (transmitters in communication satellites), where solar cells are costly and have limited power output. This subject is discussed in more detail in Chapter 5 in the section on satellite communications.

Limiters

A *limiter* is a nonlinear circuit with an output saturation characteristic. A soft saturating limiter characteristic is shown in Fig. 4-6. Figure 4-8 shows a hard (ideal) limiter characteristic, together with an illustration of the unfiltered output waveform obtained for an input waveform. The ideal limiter transfer function is essentially identical to the output-to-input characteristic of an ideal comparator with a zero reference level. The waveforms shown in Fig. 4-8 illustrate how amplitude variations on the input signal are eliminated on the output signal. A *bandpass limiter* is a nonlinear circuit with a saturating characteristic followed by a bandpass filter. For the case of an ideal bandpass limiter, the filter output waveform would be sinusoidal since the harmonics of the square wave would be filtered out. In general, any bandpass input (it might be a modulated signal plus

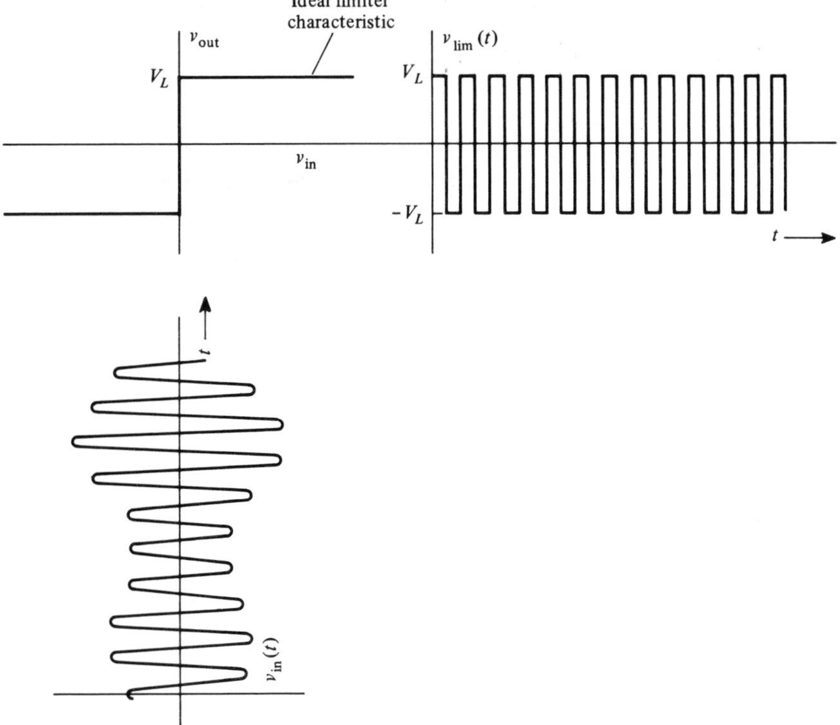

FIGURE 4-8 Ideal limiter characteristic with illustrative input and unfiltered output waveforms.

noise) can be represented, using (4-1b), by

$$v_{in}(t) = R(t)\cos[\omega_c t + \theta(t)] \tag{4-49}$$

where $R(t)$ is the equivalent real envelope and $\theta(t)$ is the equivalent phase function. The corresponding output of an ideal bandpass limiter becomes

$$v_{out}(t) = KV_L\cos[\omega_c t + \theta(t)] \tag{4-50}$$

where K is the level of the fundamental component of the square wave, $4/\pi$, multiplied by the gain of the output (bandpass) filter. This equation indicates that any AM that was present on the limiter input does not appear on the limiter output but that the phase function is preserved (i.e., the zero-crossings of the input are preserved on the limiter output). Limiters are often used in receiving systems designed for angle modulated signaling—such as PSK, FSK, and analog FM—to eliminate any variations in the real envelope of the receiver input signal that are caused by channel noise or signal fading. These amplitude variations could produce noise on the detected signal at the receiver output if the limiter was not used.

Mixers, Up Converters, and Down Converters

An *ideal mixer* is an electronic circuit that functions as a mathematical multiplier of two input signals. Usually one of these signals is a sinusoidal waveform produced by a local oscillator, as illustrated in Fig. 4-9.

Mixers are used to obtain frequency translation of the input signal. Assume that the input signal is a bandpass signal that has a nonzero spectrum in a band around or near $f = f_c$, then the signal is represented by

$$v_{in}(t) = \text{Re}\{g_{in}(t)e^{j\omega_c t}\} \tag{4-51}$$

where $g_{in}(t)$ is the complex envelope of the input signal. The signal out of the ideal mixer is then

$$\begin{aligned} v_1(t) &= [A_0 \,\text{Re}\{g_{in}(t)e^{j\omega_c t}\}] \cos \omega_0 t \\ &= \frac{A_0}{4}[g_{in}(t)e^{j\omega_c t} + g_{in}^*(t)e^{-j\omega_c t}](e^{j\omega_0 t} + e^{-j\omega_0 t}) \\ &= \frac{A_0}{4}[g_{in}(t)e^{j(\omega_c + \omega_0)t} + g_{in}^*(t)e^{-j(\omega_c + \omega_0)t} + g_{in}(t)e^{j(\omega_c - \omega_0)t} + g_{in}^*(t)e^{-j(\omega_c - \omega_0)t}] \end{aligned}$$

or

$$v_1(t) = \frac{A_0}{2}\text{Re}\{g_{in}(t)e^{j(\omega_c + \omega_0)t}\} + \frac{A_0}{2}\text{Re}\{g_{in}(t)e^{j(\omega_c - \omega_0)t}\} \tag{4-52}$$

This equation illustrates that the input bandpass signal with a spectrum near $f = f_c$ has been converted (i.e., frequency translated) into two output bandpass signals, one at the up-conversion frequency band, where $f_u = f_c + f_0$, and one at the down-conversion band, where $f_d = f_c - f_0$. A filter, as illustrated in Fig. 4-9, may be used to select either the up-conversion component or the down-conversion component. A *bandpass* filter is used to select the up-conversion component, but the down-conversion component is selected by either a *baseband* filter or a *bandpass* filter, depending on the location of $f_c - f_0$. For example, if $f_c - f_0 = 0$, a low-pass filter would be needed and the resulting output spectrum would be a baseband spectrum. If $f_c - f_0 > 0$, where $f_c - f_0$ was larger than

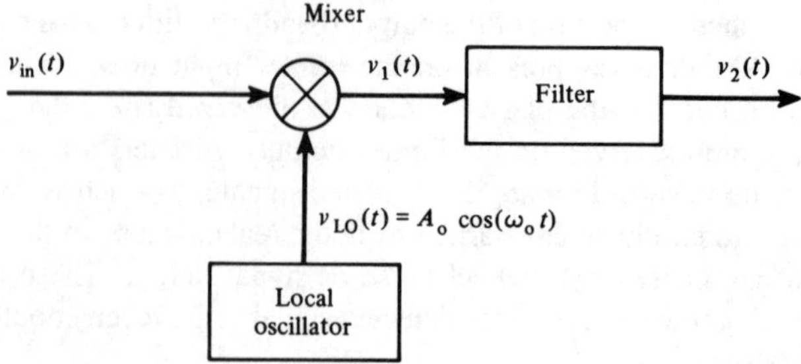

FIGURE 4-9 Mixer followed by a filter for either up or down conversion.

the bandwidth of $g_{in}(t)$, a bandpass filter would be used and the filter output would be

$$v_2(t) = \text{Re}\{g_2(t)e^{j(\omega_c - \omega_0)t}\} = \frac{A_0}{2}\text{Re}\{g_{in}(t)e^{j(\omega_c - \omega_0)t}\} \tag{4-53}$$

For this case of $f_c > f_0$, it is seen that the modulation on the mixer input signal $v_{in}(t)$ is preserved on the mixer up- or down-converted signals.

If $f_c < f_0$, we rewrite (4-52), obtaining

$$v_1(t) = \frac{A_0}{2}\text{Re}\{g_{in}(t)e^{j(\omega_c + \omega_0)t}\} + \frac{A_0}{2}\text{Re}\{g_{in}^*(t)e^{j(\omega_0 - \omega_c)t}\} \tag{4-54}$$

since the frequency in the exponent of the bandpass signal representation needs to be positive for easy physical interpretation of the location of spectral components. For this case of $f_0 > f_c$, the complex envelope of the down-converted signal has been conjugated when compared to the complex envelope of the input signal. This is equivalent to saying that the sidebands have been exchanged; that is, the upper sideband of the input signal spectrum becomes the lower sideband of the down-converted output signal, and so on. This is demonstrated mathematically by looking at the spectrum of $g^*(t)$. The spectrum is

$$\begin{aligned}\mathcal{F}[g_{in}^*(t)] &= \int_{-\infty}^{\infty} g_{in}^*(t)e^{-j\omega t}\,dt = \left[\int_{-\infty}^{\infty} g_{in}(t)e^{-j(-\omega)t}\,dt\right]^* \\ &= G_{in}^*(-f)\end{aligned} \tag{4-55}$$

The $-f$ indicates that the upper and lower sidebands have been exchanged, and the conjugate indicates that the phase spectrum has been inverted.

In summary, the complex envelope for the signal out of an *up converter* is

$$g_2(t) = \frac{A_0}{2}g_{in}(t) \tag{4-56a}$$

where $f_u = f_c + f_0 > 0$. Thus the same modulation is on the output signal as was on the input signal, but the amplitude has been changed by the $A_0/2$ scale factor.

For the case of *down conversion* there are two possibilities. For $f_d = f_c - f_0 > 0$ where $f_0 > f_c$,

$$g_2(t) = \frac{A_0}{2}g_{in}(t) \tag{4-56b}$$

This is called down conversion with *low-side injection* since the LO frequency is below that of the incoming signal (i.e., $f_0 < f_c$). Here the output modulation is the same as that of the input except for the $A_0/2$ scale factor. The other possibility is $f_d = f_0 - f_c > 0$, where $f_0 > f_c$, which produces the output complex envelope

$$g_2 = \frac{A_0}{2}g_{in}^*(t)$$

This is down conversion with *high-side* injection since $f_0 > f_c$. This is the case where the sidebands on the down-converted output signal are reversed from those on the input (e.g., a LSSB input signal becomes a USSB output signal).

Ideal mixers act as linear time-varying circuit elements since

$$v_1(t) = (A \cos \omega_0 t) v_{in}(t)$$

where $A \cos \omega_0 t$ is the time-varying gain of the linear circuit. It should also be recognized that mixers used in communication circuits are essentially mathematical multipliers. They should not be confused with the audio mixers that are used in radio and TV broadcasting studios. An *audio mixer* is a summing amplifier with multiple inputs so that several inputs from several sources—such as microphones, tape decks, and turntables—can be "mixed" (added) to produce one output signal. Unfortunately, the term *mixer* means entirely different things depending on the context used. As used in transmitters and receivers, it means a multiplying operation that produces a frequency translation of the input signal. In audio systems it means a summing operation to combine several inputs into one output signal.

In practice, the multiplying operation needed for mixers may be realized by using one of the following.

1. A continuously variable transconductance device such as a dual-gate FET.
2. A nonlinear device.
3. A linear device with time-varying discrete gain.

In the first method, when a dual-gate FET is used to obtain multiplication, $v_{in}(t)$ is usually connected to gate 1 and the local oscillator is connected to gate 2. The resulting output is

$$v_1(t) = K v_{in}(t) v_{LO}(t) \tag{4-57}$$

over the operative region, where $v_{LO}(t)$ is the local oscillator voltage. The multiplier is said to be of a *single-quadrant* type if the multiplier action of (4-57) is obtained only when both the input waveforms, v_{in} and $v_{LO}(t)$, have either nonnegative or nonpositive values [i.e., a plot of the values of $v_{in}(t)$ versus $v_{LO}(t)$ falls within a single quadrant]. The multiplier is of the *two-quadrant* type if multiplier action is obtained when either $v_{in}(t)$ or $v_{LO}(t)$ is nonnegative or nonpositive and the other is arbitrary. The multiplier is said to be of the *four-quadrant* type when multiplier action is obtained regardless of the signs of $v_{in}(t)$ and $v_{LO}(t)$.

In the second technique, a nonlinear device can be used to obtain multiplication by summing the two inputs as illustrated in Fig. 4-10. Looking at the square-law component at the output, we have

$$\begin{aligned} v_1(t) &= K_2(v_{in} + v_{LO})^2 + \text{other terms} \\ &= K_2(v_{in}^2 + 2v_{in}v_{LO} + v_{LO}^2) + \text{other terms} \end{aligned} \tag{4-58}$$

The cross-product term gives the desired multiplier action:

$$2K_2 v_{in} v_{LO} = 2K_2 A_0 v_{in}(t) \cos \omega_0 t \tag{4-59}$$

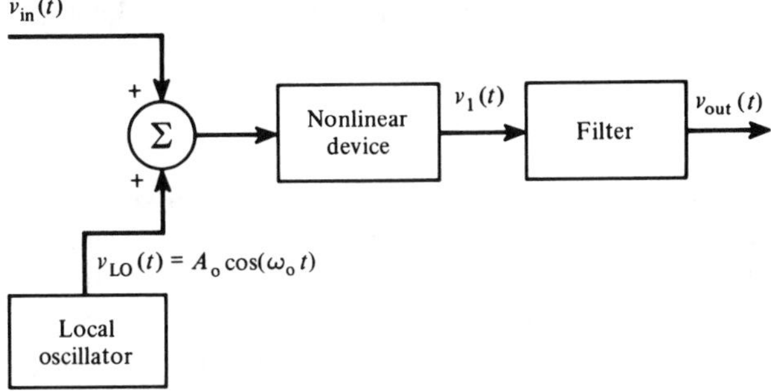

FIGURE 4-10 Nonlinear device used as a mixer.

If we assume that $v_{in}(t)$ is a bandpass signal, the filter can be used to pass either the up- or down-conversion terms. However, some distortion products may also fall within the output passband if ω_c and ω_0 are not chosen carefully.

In the third method, a linear device with time-varying discrete gain is used to obtain multiplier action. This is demonstrated in Fig. 4-11, where the time-varying device is an analog switch (such as a CMOS 4016 integrated circuit) that is activated by a square-wave oscillator signal, $v_0(t)$. The discrete gain of the switch is either unity or zero. The waveform at the output of the analog switch is

$$v_1(t) = v_{in}(t)S(t) \tag{4-60}$$

where $S(t)$ is a unipolar switching square wave that has unity peak amplitude. (This is analogous to PAM with natural sampling that was studied in Chapter 3.) Thus (4-60) becomes

$$v_1(t) = v_{in}(t)\left[\frac{1}{2} + \sum_{n=1}^{\infty} \frac{2 \sin(n\pi/2)}{n\pi} \cos n\omega_0 t\right] \tag{4-61}$$

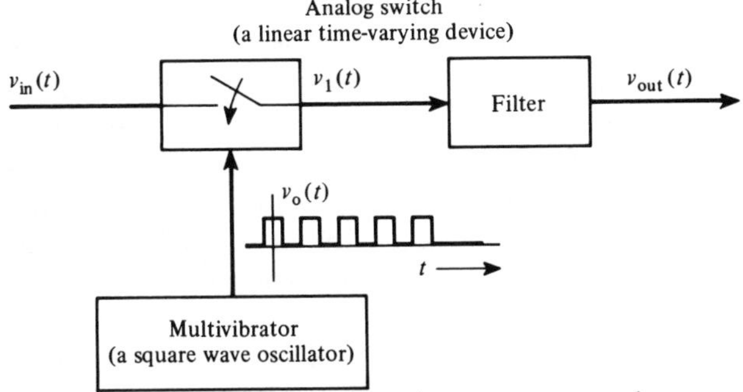

FIGURE 4-11 Linear time-varying device used as a mixer.

The multiplying action is obtained from the $n = 1$ term, which is

$$\frac{2}{\pi} v_{in}(t) \cos \omega_0 t \tag{4-62}$$

This would generate up- and down-conversion signals at $f_c + f_0$ and $f_c - f_0$ if $v_{in}(t)$ were a bandpass signal with a nonzero spectrum in the vicinity of $f = f_c$. However, (4-61) shows that other frequency bands are also present in the output signal, namely, at frequencies $f = |f_c \pm nf_0|$, $n = 3, 5, 7, \ldots$ and, in addition, there is the feed-through term $\frac{1}{2}v_{in}(t)$ appearing at the output. Of course, a filter may be used to pass either the up- or down-conversion component appearing in (4-61).

Mixers are often classified as being *unbalanced, single balanced,* or *double balanced*. That is, in general we obtain at the output of mixer circuits

$$v_1(t) = C_1 v_{in}(t) + C_2 v_0(t) + C_3 v_{in}(t) v_0(t) + \text{other terms} \tag{4-63}$$

When C_1 and C_2 are not zero, the mixer is said to be *unbalanced* since $v_{in}(t)$ and $v_0(t)$ feed through to the output. An example of an unbalanced mixer was illustrated in Fig. 4-10, where a nonlinear device was used to obtain mixing action. In the Taylor's expansion of the nonlinear device output-to-input characteristics, the linear term would provide feedthrough of both $v_{in}(t)$ and $v_0(t)$. A *single-balanced mixer* has feedthrough for only one of the inputs; that is, either C_1 or C_2 of (4-63) is zero. An example of a single-balanced mixer is shown in Fig. 4-11, which uses sampling to obtain mixer action. In this example, (4-61) demonstrates that $v_0(t)$ is balanced out (i.e., $C_2 = 0$) and $v_{in}(t)$ feeds through with a gain of $C_1 = \frac{1}{2}$. A *double-balanced mixer* has no feedthrough from either input; that is, both C_1 and C_2 of (4-63) are zero. An example of a double-balanced mixer is discussed in the next paragraph.

Figure 4-12a exhibits the circuit for a double-balanced mixer. This circuit is popular because it is relatively inexpensive and it has excellent performance. The third-order IMD is typically down at least 50 dB compared to the desired output components. It is usually designed for source and load impedances of 50 Ω and has broadband input and output ports. The RF [i.e., $v_{in}(t)$] port and the LO (local oscillator) port are often usable over a frequency range of 1000:1, say 1 to 1000 MHz; and the IF (intermediate frequency) output port, $v_1(t)$, is typically usable from dc to 600 MHz. The transformers are made by using small toroidal cores, and the diodes are matched hot carrier diodes. The input signal level at the RF port is relatively small, usually less than −5 dBm, and the local oscillator level at the LO port is relatively large, say +5 dBm. The LO signal is large, so that this signal, in effect, turns the diodes on and off so that the diodes will act as switches. The LO provides the switching control signal. This circuit thus acts as a time-varying linear circuit (with respect to the RF input port), and the analysis is very similar to that used for the analog-switch mixer of Fig. 4-11. During the portion of the cycle when $v_{LO}(t)$ has a positive voltage, the output voltage is proportional to $+v_{in}(t)$ as seen from the equivalent circuit shown in Fig. 4-12b.

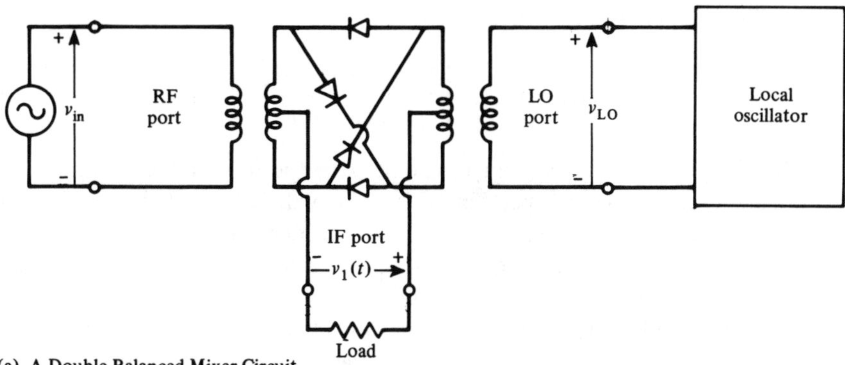

(a) A Double Balanced Mixer Circuit

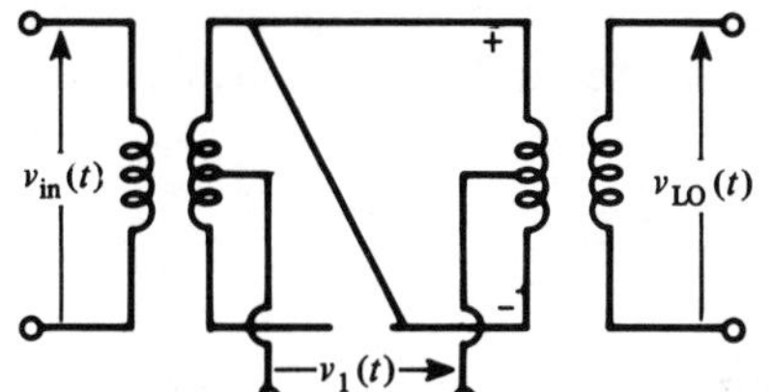

(b) Equivalent Circuit When $v_{LO}(t)$ Is Positive

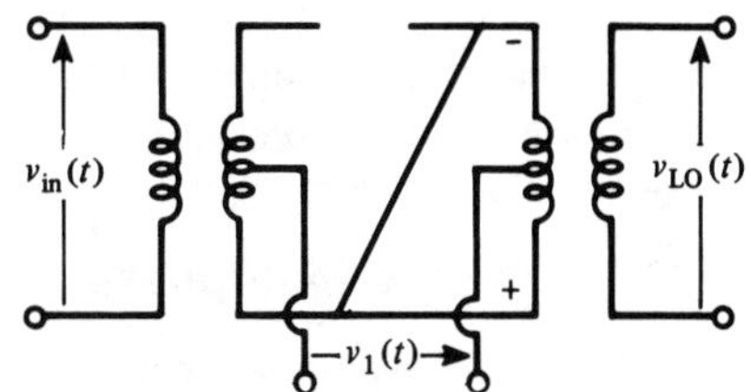

(c) Equivalent Circuit When $v_{LO}(t)$ Is Negative

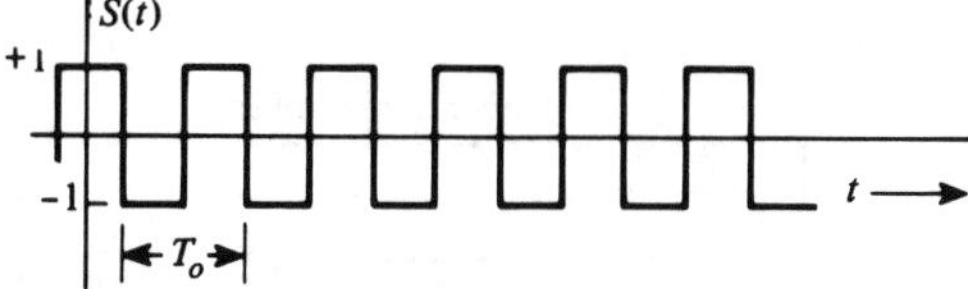

(d) Switching Waveform Due to the Local Oscillator Signal

FIGURE 4-12 Analysis of a double-balanced mixer circuit.

When $v_{LO}(t)$ is negative, the output voltage is proportional to $-v_{in}(t)$, as seen from the equivalent circuit shown in Fig. 4-12c. Thus the output of this double-balanced mixer is

$$v_1(t) = Kv_{in}(t)S(t) \tag{4-64}$$

where $S(t)$ is a bipolar switching waveform as shown in Fig. 4-12d. Since the switching waveform arises from the LO signal, its period is $T_0 = 1/f_0$. The switching waveform is described by

$$S(t) = 4 \sum_{n=1}^{\infty} \frac{\sin(n\pi/2)}{n\pi} \cos n\omega_0 t \tag{4-65}$$

so that the mixer output is

$$v_1(t) = [v_{in}(t)]\left[4K \sum_{n=1}^{\infty} \frac{\sin(n\pi/2)}{n\pi} \cos n\omega_0 t\right] \tag{4-66}$$

If the input is a bandpass signal with nonzero spectrum in the vicinity of f_c, this equation shows that the spectrum of the input will be translated to the frequencies $|f_c \pm nf_0|$, where $n = 1, 3, 5, \ldots$. In practice, the value K is such that the conversion gain (which is defined as the desired output level divided by the input level) at the frequency $|f_c \pm f_0|$ is about -6 dB. Of course, an output filter may be used to select the up-converted or down-converted frequency band.

In addition to up- or down-conversion applications, mixers (i.e., multipliers) may be used for modulators to translate a baseband signal to an RF frequency band, and mixers may be used as product detectors to translate RF signals to baseband. These applications will be discussed in later sections that deal with transmitters and receivers.

Frequency Multipliers

Frequency multipliers consist of a nonlinear circuit followed by a tuned circuit, as illustrated in Fig. 4-13. If a bandpass signal is fed into a frequency multiplier, the output will appear in a frequency band at the nth harmonic of the input frequency. Since the device is nonlinear, the bandwidth of the nth harmonic

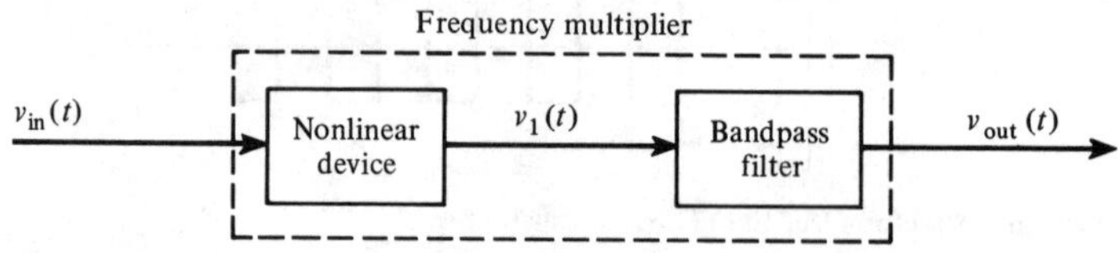

(a) Block Diagram of a Frequency Multiplier

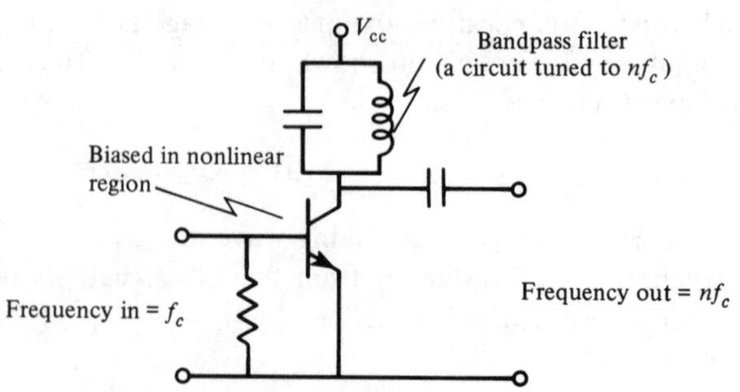

(b) Circuit Diagram for a Frequency Multiplier

FIGURE 4-13 Frequency multiplier.

output is larger than that of the input signal. In general, the bandpass input signal is represented by

$$v_{\text{in}}(t) = R(t)\cos[\omega_c t + \theta(t)] \tag{4-67}$$

The transfer function of the nonlinear device may be expanded in a Taylor's series, so that the *n*th order output term is

$$v_1(t) = K_n v_{\text{in}}^n(t) = K_n R^n(t)\cos^n[\omega_c t + \theta(t)]$$

or†

$$v_1(t) = CR^n(t)\cos[n\omega_c t + n\theta(t)] + \text{other terms} \tag{4-68}$$

This illustrates that the input *amplitude* variation $R(t)$ *appears distorted* on the output signal since the envelope on the output is $R^n(t)$. The waveshape of the *angle* variation, $\theta(t)$, is *not distorted* by the frequency multiplier, but the frequency multiplier does increase the magnitude of the angle variation by a factor of n. Thus frequency multiplier circuits are not used on signals where the AM is to be preserved; but as we will see, the frequency multiplier is very useful in PM and FM problems since it effectively "amplifies" the angle variation waveform, $\theta(t)$. The $n = 2$ multiplier is called a *doubler stage,* and the $n = 3$ frequency multiplier is said to be a *tripler stage*.

The frequency multiplier should not be confused with a mixer. The frequency multiplier acts as a nonlinear device. The mixer circuit (which uses a mathematical multiplier operation) acts as a linear circuit with time-varying gain (due to the LO signal). The bandwidth of the signal at the output of a frequency multiplier is larger than that of the input signal, and it appears in a frequency band located at the *n*th harmonic of the input. The bandwidth of a signal at the output of a mixer is the same as that of the input, but the input spectrum has been translated either up or down depending on the LO frequency and the bandpass of the output filter. Of course, it is realized that a frequency multiplier is essentially a nonlinear amplifier followed by a bandpass filter that is designed to pass the *n*th harmonic.

Envelope Detector

An *ideal envelope detector* is a circuit that produces a waveform at its output that is proportional to the real envelope of its input. From (4-1b) the bandpass input may be represented by $R(t)\cos[\omega_c t + \theta(t)]$, where $R(t) \geq 0$; then the output of the ideal envelope detector is

$$v_{\text{out}}(t) = KR(t) \tag{4-69}$$

where K is the proportionality constant.

A simple diode detector circuit that approximates an ideal envelope detector is shown in Fig. 4-14a. The diode current occurs in pulses proportional to the positive part of the input waveform. The current pulses charge the capacitor to

† *m*th-order output terms, where $m > n$, may also contribute to the *n*th harmonic output provided that K_m is sufficiently large with respect to K_n. This is illustrated by (A-18), where $m = 4$ and $n = 2$.

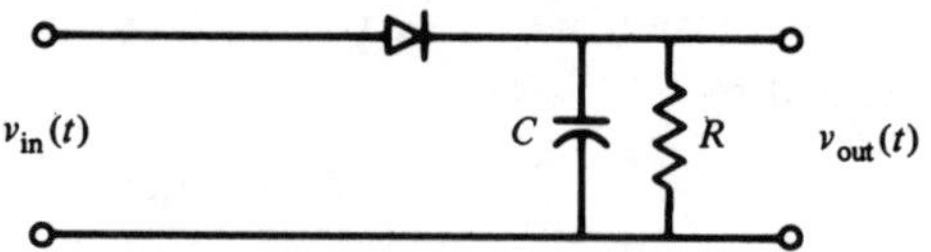

(a) A Diode Envelope Detector

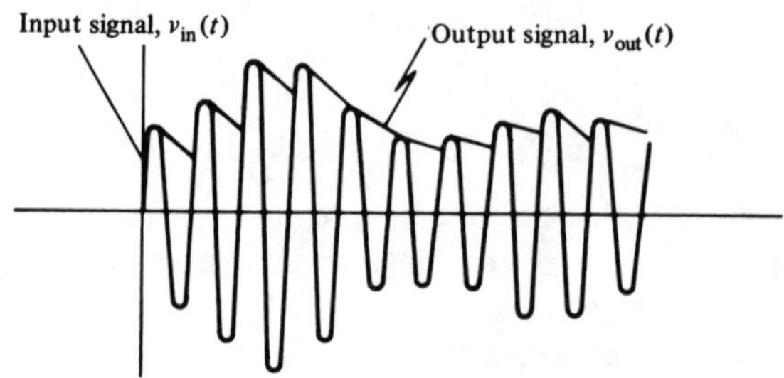

(b) Waveforms Associated with the Diode Envelope Detector

FIGURE 4-14 Envelope detector.

produce the output voltage waveform, as illustrated in Fig. 4-14b. The *RC* time constant is chosen so that the output signal will follow the real envelope, $R(t)$, of the input signal. Consequently, the cutoff frequency of the low-pass filter needs to be much smaller than the carrier frequency f_c and much larger than the bandwidth of the (detected) modulation waveform, B. That is,

$$B \ll \frac{1}{2\pi RC} \ll f_c \tag{4-70}$$

where *RC* is the time constant of the filter.

Product Detector

A *product detector* (Fig. 4-15) is a mixer circuit that down-converts the input (bandpass signal plus noise) to baseband. The output of the multiplier is

$$\begin{aligned} v_1(t) &= R(t)\cos[\omega_c t + \theta(t)]A_0\cos(\omega_c t + \theta_0) \\ &= \tfrac{1}{2}A_0 R(t)\cos[\theta(t) - \theta_0] + \tfrac{1}{2}A_0 R(t)\cos[2\omega_c t + \theta(t) + \theta_0] \end{aligned}$$

where the frequency of the oscillator is f_c and the phase is θ_0. The low-pass filter passes only the down-conversion term, so that the output is

$$v_{out}(t) = \tfrac{1}{2}A_0 R(t)\cos[\theta(t) - \theta_0] = \tfrac{1}{2}A_0 \text{Re}\{g(t)e^{-j\theta_0}\} \tag{4-71}$$

where the complex envelope of the input is denoted by

$$g(t) = R(t)e^{j\theta(t)} = x(t) + jy(t)$$

and $x(t)$ and $y(t)$ are the quadrature components [see (4-2)]. Since the frequency of the oscillator is the same as the carrier frequency of the incoming signal, the

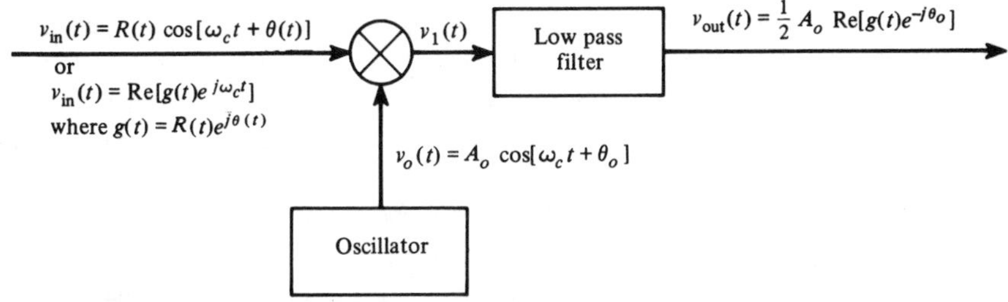

FIGURE 4-15 Product detector.

oscillator has been *frequency synchronized* with the input signal. Furthermore, if, in addition, $\theta_0 = 0$, the oscillator is said to be *phase synchronized* with the in-phase component and the output becomes

$$v_{out}(t) = \tfrac{1}{2}A_0 x(t) \tag{4-72a}$$

If $\theta_0 = 90°$,

$$v_{out} = \tfrac{1}{2}A_0 y(t) \tag{4-72b}$$

Equation (4-71) also indicates that a product detector is sensitive to AM and/or PM. For example, if the input contains no angle modulation so that $\theta(t) = 0$ and if the reference phase is set to zero (i.e., $\theta_0 = 0$), then

$$v_{out}(t) = \tfrac{1}{2}A_0 R(t) \tag{4-73a}$$

which implies that $x(t) \geq 0$ and the real envelope is obtained on the product detector output, just as for the case of the envelope detector discussed previously. However, if an angle-modulated signal $A_c \cos[\omega_c t + \theta(t)]$ is present at the input and $\theta_0 = 90°$, the product detector output is

$$v_{out}(t) = \tfrac{1}{2}A_0 \,\text{Re}\{A_c e^{j[\theta(t) - 90°]}\}$$

or

$$v_{out}(t) = \tfrac{1}{2}A_0 A_c \sin\theta(t) \tag{4-73b}$$

In this case, the product detector acts like a *phase detector* with a *sinusoidal characteristic* since the output voltage is proportional to the sine of the phase difference between the input signal and the oscillator signal. Phase detector circuits are also available that yield triangle and sawtooth characteristics [Krauss, Bostian, and Raab, 1980]. Referring to (4-73b) for the phase detector with a sinusoidal characteristic, and assuming that the phase difference is small [i.e., $|\theta(t)| \ll \pi/2$], we see that $\sin\theta(t) \approx \theta(t)$ and

$$v_{out}(t) \approx \tfrac{1}{2}A_0 A_c \theta(t) \tag{4-74}$$

which is a linear characteristic (for small angles). Thus the output of this phase detector is directly proportional to the phase differences when the difference angle is small, as illustrated in Fig. 4-21.

It should be realized that the product detector acts as a linear time-varying device with respect to the input $v_{in}(t)$, in contrast to the envelope detector, which is a nonlinear device. The property of being either linear or nonlinear significantly affects the results when two or more components, such as a signal plus noise, are applied to the input. This topic will be studied in Chapter 7.

Detectors may also be classified as being either coherent or noncoherent. A *coherent* detector has two inputs—one for a reference signal, such as the synchronized oscillator signal, and one for the modulated signal that is to be demodulated. The product detector is an example of a coherent detector. A *noncoherent* detector has only one input, namely, the modulated signal port. The envelope detector is an example of a noncoherent detector.

Frequency Modulation Detector

An *ideal frequency modulation (FM) detector* is a device that produces an output that is proportional to the instantaneous frequency of the input. That is, if the bandpass input is represented by $R(t) \cos[\omega_c t + \theta(t)]$, the output of the ideal FM detector is

$$v_{out}(t) = \frac{Kd[\omega_c t + \theta(t)]}{dt} = K\left[\omega_c + \frac{d\theta(t)}{dt}\right] \tag{4-75}$$

Usually, the FM detector is *balanced*. This means that the dc voltage $K\omega_c$ does not appear on the output if the detector is tuned to (or designed for) the carrier frequency f_c. In this case the output is

$$v_{out}(t) = K\frac{d\theta(t)}{dt} \tag{4-76}$$

There are many ways to build FM detectors, but almost all of them are based on one of three principles:

- FM-to-AM conversion.
- Phase-shift or quadrature detection.
- Zero-crossing detection.

A *slope detector* is one example of the FM-to-AM conversion principle. A block diagram is shown in Fig. 4-16. A bandpass limiter is needed to suppress any amplitude variations on the input signal since these would distort (produce noise in) the desired output signal.

The slope detector may be analyzed as follows. Assume that the input is a fading signal with frequency modulation. From Table 4-1, this FM signal may be

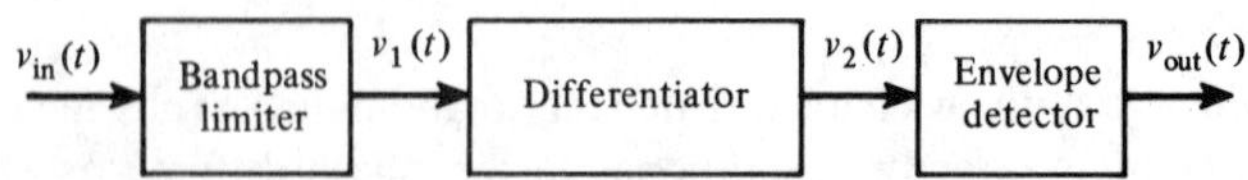

FIGURE 4-16 Frequency demodulation using slope detection.

represented by

$$v_{\text{in}}(t) = A(t) \cos[\omega_c t + \theta(t)] \tag{4-77}$$

where

$$\theta(t) = K_f \int_{-\infty}^{t} m(t_1)\, dt_1 \tag{4-78}$$

$A(t)$ represents the envelope that is fading and $m(t)$ is the modulation (e.g., audio) signal. It follows that the limiter output is

$$v_1(t) = V_L \cos\,[\omega_c t + \theta(t)] \tag{4-79}$$

and the output of the differentiator becomes

$$v_2(t) = -V_L\left[\omega_c + \frac{d\theta(t)}{dt}\right] \sin[\omega_c t + \theta(t)] \tag{4-80}$$

The output of the envelope detector is the magnitude of the complex envelope for $v_2(t)$:

$$v_{\text{out}}(t) = \left| -V_L\left[\omega_c + \frac{d\theta(t)}{dt}\right]\right|$$

and since $\omega_c \gg d\theta/dt$ in practice, this becomes

$$v_{\text{out}}(t) = V_L\left[\omega_c + \frac{d\theta(t)}{dt}\right]$$

Using (4-78), we obtain

$$v_{\text{out}}(t) = V_L\omega_c + V_L K_f m(t) \tag{4-81}$$

which indicates that the output consists of a dc voltage, $V_L\omega_c$, plus the ac voltage $V_L K_f m(t)$, which is proportional to the modulation on the FM signal. Of course, a capacitor could be placed in series with the output so that only the ac voltage would be passed to the load.

The differentiation operation can be obtained by any circuit that acts like a frequency-to-amplitude converter. For example, a single-tuned resonant circuit can be used as illustrated in Fig. 4-17, where the magnitude transfer function is $|H(f)| = K_1 f + K_2$ over the linear (useful) portion of the characteristic. A balanced FM detector, which is also called a *balanced discriminator,* is shown in Fig. 4-18. Two tuned circuits are used to balance out the dc when the input has a carrier frequency of f_c and to provide an extended linear frequency-to-voltage conversion characteristic.

Balanced discriminators can also be built that function because of the phase-shift properties of a double-tuned RF transformer circuit with primary and secondary windings [Stark, Tuteur, and Anderson, 1988]. In practice, discriminator circuits have been replaced by integrated circuits that operate on the quadrature principle.

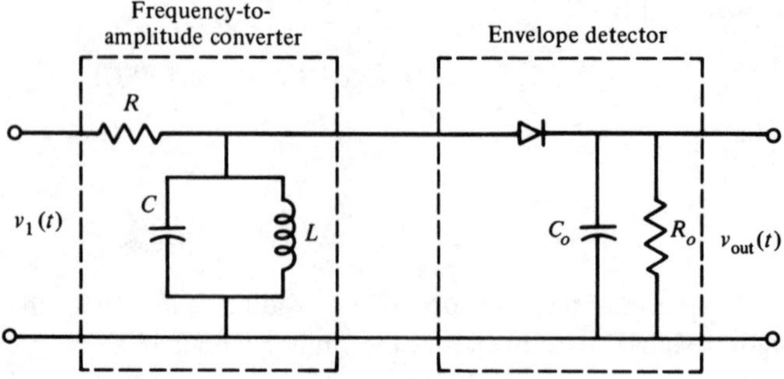

(a) Circuit Diagram of a Slope Detector

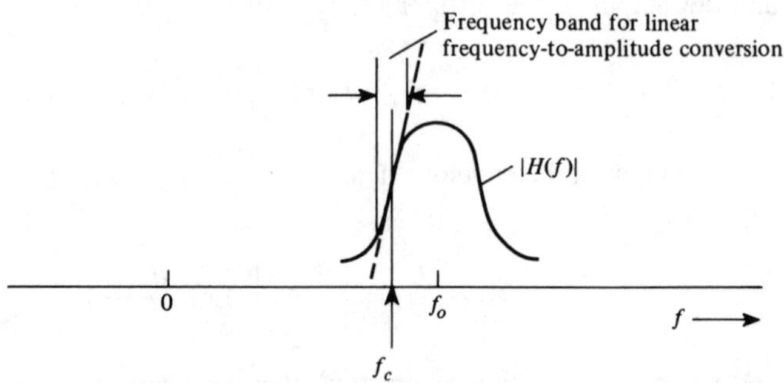

(b) Magnitude of Filter Transfer Function

FIGURE 4-17 Slope detection using a single-tuned circuit for frequency-to-amplitude conversion.

The *quadrature detector* is described as follows: A quadrature signal is first obtained from the FM signal; then, using a product detector, the quadrature signal is multiplied with the FM signal to produce the demodulated signal, $v_{out}(t)$. The quadrature signal can be produced by passing the FM signal through a capacitor (large) reactance that is connected in series with a parallel resonant circuit which is tuned to f_c. The quadrature signal voltage appears across the parallel resonant circuit. The series capacitance provides a 90° phase shift and the resonant circuit provides an additional phase shift that is proportional to the instantaneous frequency deviation (from f_c) of the FM signal. Using (4-79) and (4-78), the FM signal is

$$v_{in}(t) = V_L \cos[\omega_c t + \theta(t)] \tag{4-82a}$$

and the quadrature signal is

$$v_{quad}(t) = K_1 V_L \sin\left[\omega_c t + \theta(t) + K_2 \frac{d\theta(t)}{dt}\right] \tag{4-82b}$$

where K_1 and K_2 are constants that depend on component values used for the series capacitor and in the parallel resonant circuit. These two signals, (4-82a)

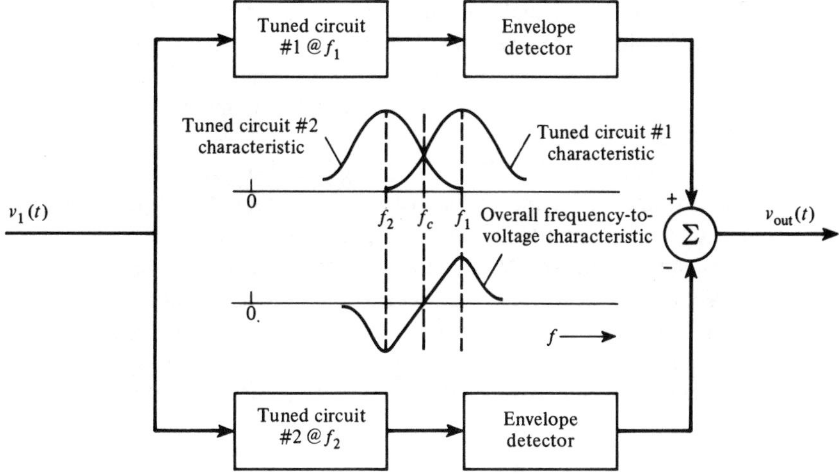

(a) Block Diagram

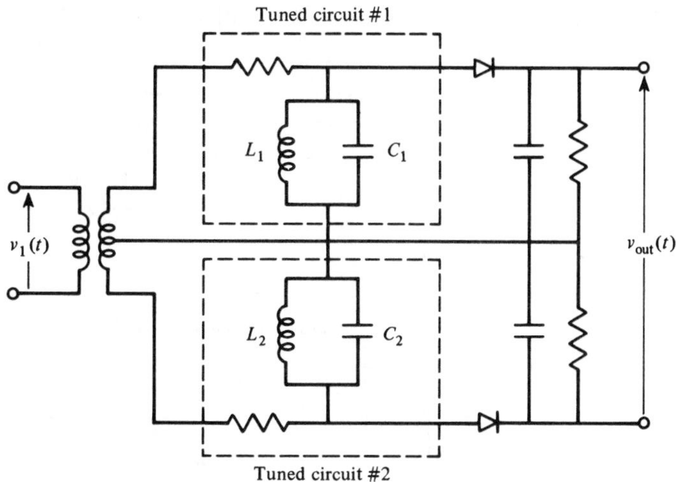

(b) Circuit Diagram

FIGURE 4-18 Balanced discriminator.

and (4-82b), are multiplied together by a product detector, for example, see Fig. 4-15, to produce the output signal

$$v_{out}(t) = \tfrac{1}{2}K_1 V_L^2 \sin\left[K_2 \frac{d\theta(t)}{dt}\right]$$

where the sum-frequency term is eliminated by the low-pass filter. For K_2 sufficiently small, $\sin x \approx x$, and by use of (4-78), the output becomes

$$v_{out} = \tfrac{1}{2}K_1 K_2 V_L^2 K_f m(t) \tag{4-83}$$

This demonstrates that the quadrature detector detects the modulation on the input FM signal. An example of the quadrature detector is shown on Fig. 4 of the MC3363 data sheets in the appendix of this chapter, Sec. 4-9. The quadrature detector principle is also used by phase-locked loops that are configured to detect FM—see (4-98e).

As indicated by (4-75), the output of an ideal FM detector is directly proportional to the instantaneous frequency of the input. This linear frequency-to-voltage characteristic may be obtained directly by counting the zero-crossings of the input waveform. An FM detector utilizing this technique is called a *zero-crossing detector*. A hybrid circuit (i.e., consisting of both digital and analog devices) that is a balanced FM zero-crossing detector is shown in Fig. 4-19. The limited (square-wave) FM signal, denoted by $v_1(t)$, is shown in Fig. 4-19b. For illustration purposes, it is assumed that $v_1(t)$ is observed over that portion of the modula-

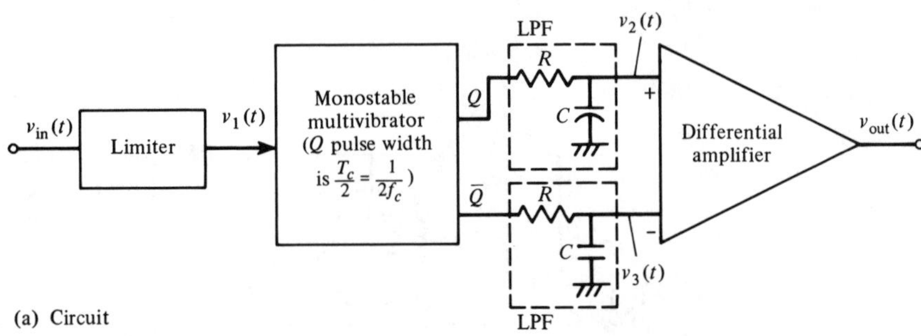

(a) Circuit

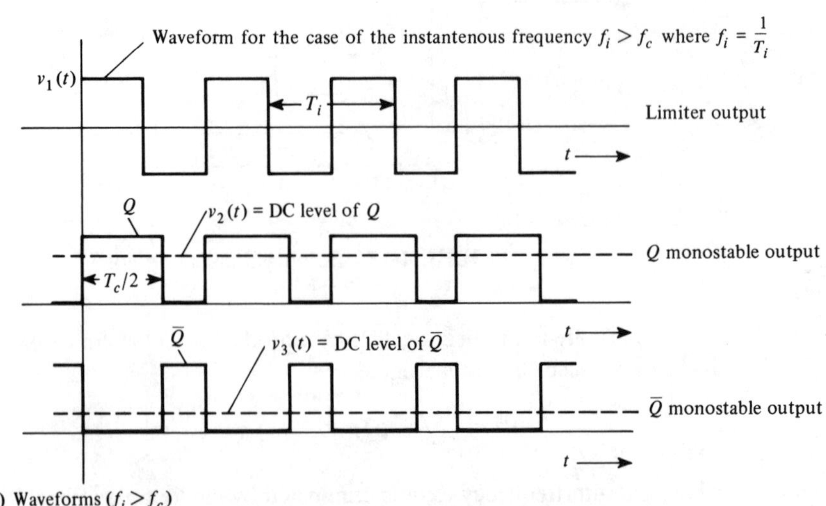

(b) Waveforms ($f_i > f_c$)

FIGURE 4-19 Balanced zero-crossing FM detector.

tion cycle when the instantaneous frequency

$$f_i(t) = f_c + \frac{1}{2\pi}\frac{d\theta(t)}{dt}$$

is larger than the carrier frequency, f_c. That is, $f_i > f_c$ in the illustration. Since the modulation voltage varies slowly with respect to the input FM signal oscillation, $v_1(t)$ appears (in the figure) to have a constant frequency, although it is actually varying in frequency according to $f_i(t)$. The monostable multivibrator (one shot) is triggered on the positive slope zero-crossings of $v_1(t)$. For balanced FM detection, the pulse width of the Q output is set to $T_c/2 = 1/2f_c$, where f_c is the carrier frequency of the FM signal at the input. Thus the differential amplifier output voltage is zero if $f_i = f_c$. For $f_i > f_c$ [as illustrated by the $v_1(t)$ waveform in the figure] the output voltage is positive and for $f_i < f_c$ the output voltage will be negative. Thus a linear frequency-to-voltage characteristic, $C[f_i(t) - f_c]$, is obtained where, for an FM signal at the input, $f_i(t) = f_c + (1/2\pi)K_f m(t)$.

Another circuit that can be used for FM demodulation as well as for other purposes is the phase-locked loop.

Phase-Locked Loops

A phase-locked loop (PLL) consists of three basic components: (1) a phase detector, (2) a low-pass filter, and (3) a voltage-controlled oscillator (VCO) (see Fig. 4-20). The VCO is an oscillator that produces a periodic waveform, the frequency of which may be varied about some free-running frequency, f_0, according to the value of the applied voltage $v_2(t)$. The free-running frequency, f_0, is the frequency of the VCO output when the applied voltage $v_2(t)$ is zero. The phase detector produces an output signal $v_1(t)$ that is a function of the phase difference between the incoming signal $v_{in}(t)$ and the oscillator signal $v_0(t)$. The filtered signal $v_2(t)$ is the control signal that is used to change the frequency of the VCO output. The PLL configuration may be designed so that it acts as a narrowband tracking filter when the low-pass filter (LPF) is a narrowband filter. In this operating mode the frequency of the VCO will become that of one of the line components of the input signal spectrum, so that, in effect, the VCO output

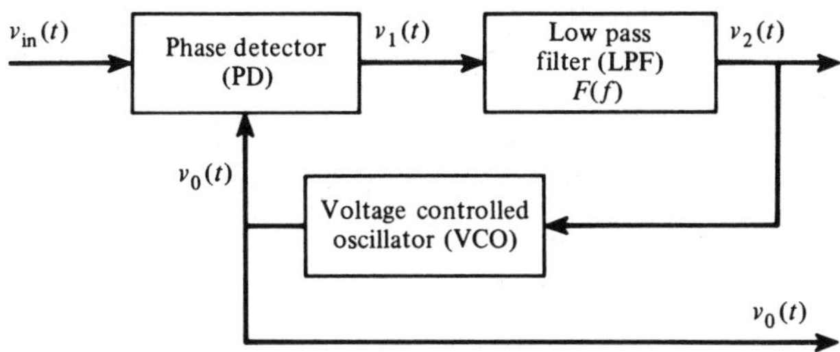

FIGURE 4-20 Basic PLL.

signal is a periodic signal with a frequency equal to the average frequency of this input signal component. Once the VCO has acquired this frequency component, the frequency of the VCO will track this input signal component if it changes slightly in frequency. In another mode of operation, the bandwidth of the LPF is wider so that the VCO can track the instantaneous frequency of the input signal. When the PLL tracks the input signal in either of these ways, the PLL is said to be "locked."

If the applied signal has an initial frequency of f_0, the PLL will acquire lock and the VCO will track the input signal frequency over some range, provided that the input frequency changes slowly. However, the loop will remain locked only over some finite range of frequency shift. This range is called the *hold-in* (or *lock*) *range*. The hold-in range depends on the overall dc gain of the loop, which includes the dc gain of the LPF. On the other hand, if the applied signal has an initial frequency not equal to f_0, the loop may not acquire lock even though the input frequency is within the hold-in range. The frequency range over which the applied input will cause the loop to lock is called the *pull-in* (or *capture*) *range*. This range is primarily determined by the loop filter characteristics, and it is never greater than the hold-in range. (This is illustrated in Fig. 4-24.) Another important PLL specification is the *maximum locked sweep rate*. It is defined as the maximum rate of change of the input frequency for which the loop will remain locked. If the input frequency changes faster than this rate, the loop will drop out of lock.

If the PLL is realized using analog circuit techniques, it is said to be an *analog phase-locked loop* (APLL). On the other hand, if digital circuits are used, it is said to be a *digital phase-locked loop* (DPLL). For example, the phase detection (PD) characteristic depends on the exact implementation used. Some PD characteristics are shown in Fig. 4-21. The sinusoidal characteristic is obtained if an (analog circuit) multiplier is used and the periodic signals are sinusoids. The multiplier may be implemented by using a double-balanced mixer. The triangle and sawtooth PD characteristics are obtained by using digital circuits. In addition to using digital VCO and PD circuits, the DPLL may incorporate a digital loop filter and signal processing techniques that use microprocessors. Gupta published a fine tutorial paper on analog phase-locked loops in the *IEEE Proceedings* [Gupta, 1975], and Lindsey and Chie followed with a survey paper on digital PLL techniques [Lindsey and Chie, 1981]. In addition, there are excellent books available [Blanchard, 1976; Gardner, 1979]. A detailed discussion of PLLs is beyond the scope of this book, but some of the significant results will be developed here.

The PLL may be studied by examining the APLL, as shown in Fig. 4-22. In this figure a multiplier (sinusoidal PD characteristic) is used. Assume that the input signal is

$$v_{\text{in}}(t) = A_i \sin[\omega_0 t + \theta_i(t)] \tag{4-84}$$

and that the VCO output signal is

$$v_0(t) = A_0 \cos[\omega_0 t + \theta_0(t)] \tag{4-85}$$

(a) Sinusoidal Characteristic

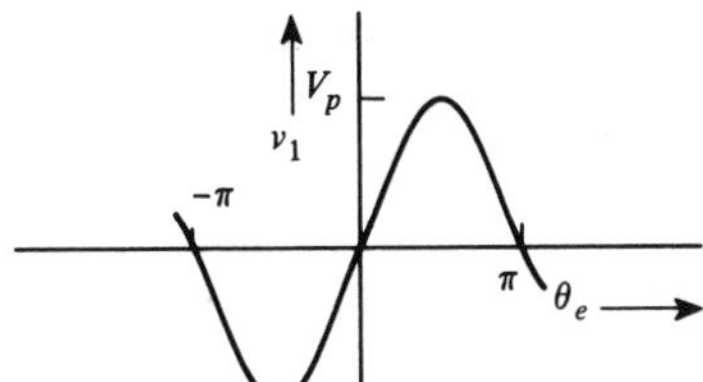

(b) Triangle Characteristic

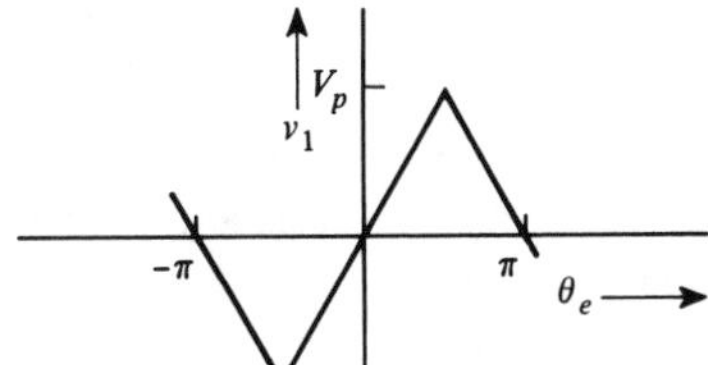

(c) Sawtooth Characteristic

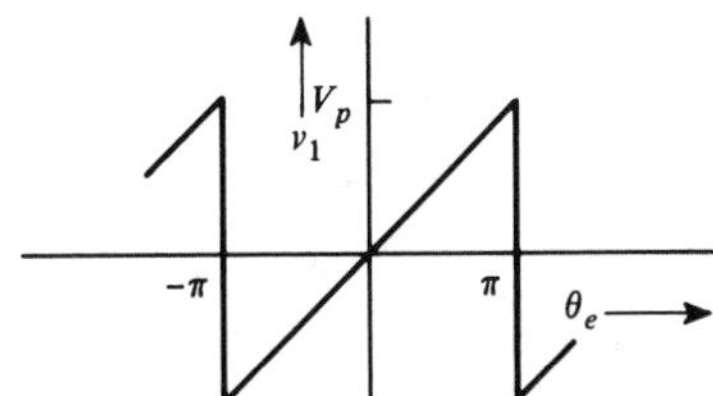

FIGURE 4-21 Some phase detector characteristics.

where

$$\theta_0(t) = K_v \int_{-\infty}^{t} v_2(\tau)\, d\tau \tag{4-86}$$

K_v is the VCO gain constant (rad/V-sec). The PD output is

$$\begin{aligned} v_1(t) &= K_m A_i A_0 \sin[\omega_0 t + \theta_i(t)] \cos[\omega_0 t + \theta_0(t)] \\ &= \frac{K_m A_i A_0}{2} \sin[\theta_i(t) - \theta_0(t)] + \frac{K_m A_i A_0}{2} \sin[2\omega_0 t + \theta_i(t) + \theta_0(t)] \end{aligned} \tag{4-87}$$

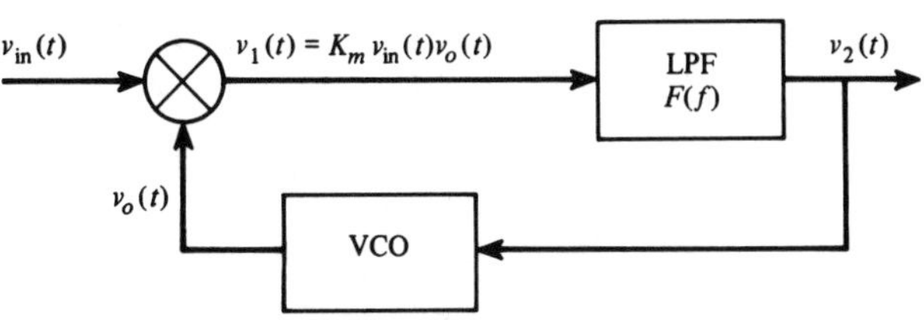

FIGURE 4-22 Analog PLL.

where K_m is the gain of the multiplier circuit. The sum frequency term does not pass through the LPF, so the LPF output is

$$v_2(t) = K_d[\sin \theta_e(t)] * f(t) \tag{4-88}$$

where

$$\theta_e(t) \triangleq \theta_i(t) - \theta_0(t) \tag{4-89}$$

$$K_d = \frac{K_m A_i A_0}{2} \tag{4-90}$$

and $f(t)$ is the impulse response of the LPF. $\theta_e(t)$ is called the *phase error; K_d* is the equivalent PD constant, which for the multiplier-type PD depends on the levels of the input signal, A_i, and the level of the VCO signal, A_0.

The overall equation describing the operation of the PLL may be obtained by taking the derivative of (4-86) and (4-89), and combining the result by use of (4-88). The resulting nonlinear equation that describes the PLL becomes

$$\frac{d\theta_e(t)}{dt} = \frac{d\theta_i(t)}{dt} - K_d K_v \int_0^t [\sin \theta_e(\lambda)] f(t-\lambda)\, d\lambda \tag{4-91}$$

where $\theta_e(t)$ is the unknown and $\theta_i(t)$ is the forcing function.

In general, this PLL equation is difficult to solve. However, it may be reduced to a linear equation if the gain K_d is large so that the error $\theta_e(t)$ is small. In this case, $\sin \theta_e(t) \approx \theta_e(t)$, and the resulting linear equation is

$$\frac{d\theta_e(t)}{dt} = \frac{d\theta_i(t)}{dt} - K_d K_v \theta_e(t) * f(t) \tag{4-92}$$

A block diagram that follows from this linear equation is shown in Fig. 4-23. It should be realized that in this linear PLL model, the *phase* of the input signal and the *phase* of the VCO output signal are used instead of actual signals themselves. The closed-loop transfer function $\Theta_0(f)/\Theta_i(f)$ is

$$H(f) = \frac{\Theta_0(f)}{\Theta_i(f)} = \frac{K_d K_v F(f)}{j2\pi f + K_d K_v F(f)} \tag{4-93}$$

where $\Theta_0(f) = \mathscr{F}[\theta_0(t)]$ and $\Theta_i(f) = \mathscr{F}[\theta_i(t)]$. Of course, the design and analysis techniques used to evaluate linear feedback control systems, such as Bode

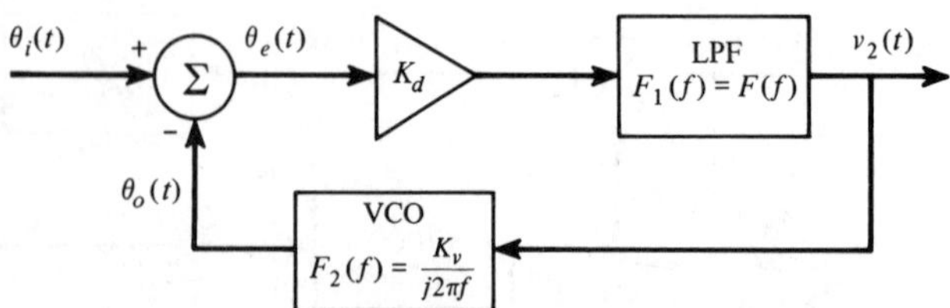

FIGURE 4-23 Linear model of the analog PLL.

plots, which will indicate phase gain and phase margins, are applicable. In fact, they are extremely useful in describing the performance of *locked* PLLs.

The equation for the *hold-in range* may be obtained by examining the nonlinear behavior of the PLL. From (4-86) and (4-88), the instantaneous frequency deviation of the VCO from ω_0 is

$$\frac{d\theta_0(t)}{dt} = K_v v_2(t) = K_v K_d[\sin \theta_e(t)] * f(t) \tag{4-94}$$

To obtain the hold-in range, the input frequency is changed very slowly from f_0. Here the dc gain of the filter is the controlling parameter, and (4-94) becomes

$$\Delta\omega = K_v K_d F(0) \sin \theta_e \tag{4-95}$$

The maximum and minimum values of $\Delta\omega$ give the hold-in range, and these are obtained when $\sin \theta_e = \pm 1$. Thus the maximum hold-in range (no noise case) is

$$\Delta f_h = \frac{1}{2\pi} K_v K_d F(0) \tag{4-96}$$

A typical lock-in characteristic is illustrated in Fig. 4-24. The solid curve shows the VCO control signal $v_2(t)$ as the sinusoidal testing signal is swept from a low frequency to a high frequency (with the free-running frequency of the VCO, f_0, being within the swept band). The dashed curve shows the result when sweeping from high to low. The hold-in range Δf_h is related to the dc gain of the PLL as described by (4-96).

The *pull-in range* Δf_p is determined primarily by the loop filter characteristics. For example, assume that the loop has not acquired lock and that the testing signal is swept slowly toward f_0. At the PD output there will be a beat (oscillatory) signal, and its frequency $|f_{\text{in}} - f_0|$ will vary from a large value to a small value as the test signal frequency sweeps toward f_0. As the testing signal frequency comes closer to f_0, the beat-frequency waveform will become nonsymmetrical so that it will have a nonzero dc value. This dc value tends to change the

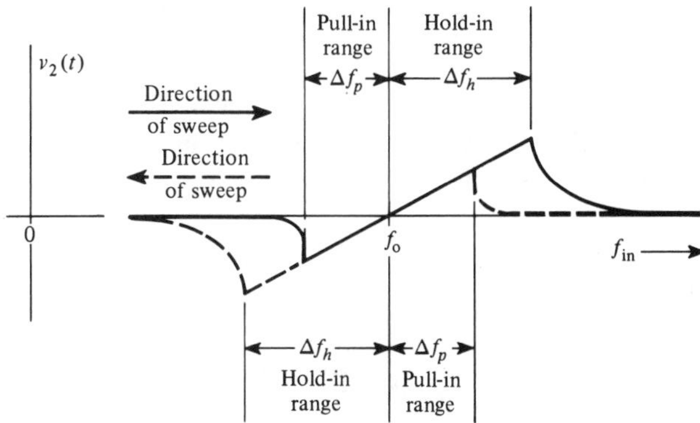

FIGURE 4-24 PLL VCO control voltage for a swept sinusoidal input signal.

frequency of the VCO to that of the input signal frequency so that the loop will tend to lock. The pull-in range, Δf_p, where the loop acquires lock will depend on exactly how the loop filter $F(f)$ processes the PD output to produce the VCO control signal. Furthermore, even if the input signal is within the pull-in range, it may take a fair amount of time for the loop to acquire lock since the LPF acts as an integrator and it takes some time for the control voltage (filter output) to build up to a value large enough for lock to occur. The analysis of the pull-in phenomenon is complicated. It is actually statistical in nature since it depends on the initial phase relationship of the input and VCO signals and on noise that is present in the circuit. Consequently, in the measurement of Δf_p, several repeated trials may be needed to obtain a typical value.

The locking phenomenon is not peculiar to PLL circuits but occurs in other types of circuits as well. For example, if an external signal is injected into the output port of an oscillator (i.e., a plain oscillator, not a VCO), the oscillator signal will tend to change frequency and will eventually lock onto the external signal frequency if the external signal is within the pull-in range of the oscillator. This phenomenon is called *injection locking* or *synchronization* of an oscillator. This injection-locking phenomenon may be modeled by a PLL model [Couch, 1971].

The PLL has numerous applications in communication systems. Some of these are (1) FM detection, (2) generation of highly stable FM signals, (3) coherent AM detection, (4) frequency multiplication, (5) frequency synthesis, and (6) use as a building block within complicated digital systems to provide bit synchronization and data detection.

Let us now find what conditions are required for the PLL to become an FM detector. Referring to Fig. 4-22, let the PLL input signal be an FM signal:

$$v_{\text{in}}(t) = A_{\text{i}}\sin\left[\omega_c t + D_f\int_{-\infty}^{t} m(\lambda)\, d\lambda\right] \tag{4-97a}$$

where

$$\theta_{\text{i}}(t) = D_f\int_{-\infty}^{t} m(\lambda)\, d\lambda \tag{4-97b}$$

or

$$\Theta_{\text{i}}(f) = \frac{D_f}{j2\pi f} M(f) \tag{4-97c}$$

and $m(t)$ is the baseband (e.g., audio) modulation that is to be detected. That is, we would like to find the conditions such that the PLL output, $v_2(t)$, is proportional to $m(t)$. Assume that f_c is within the capture (pull-in) range of the PLL; thus, for simplicity, let $f_0 = f_c$. Then the linearized PLL model, as shown in Fig. 4-23, can be used for analysis. Working in the frequency domain, we obtain the output

$$V_2(f) = \frac{\left(j\dfrac{2\pi f}{K_v}\right) F_1(f)}{F_1(f) + j\left(\dfrac{2\pi f}{K_v K_d}\right)} \Theta_{\text{i}}(f)$$

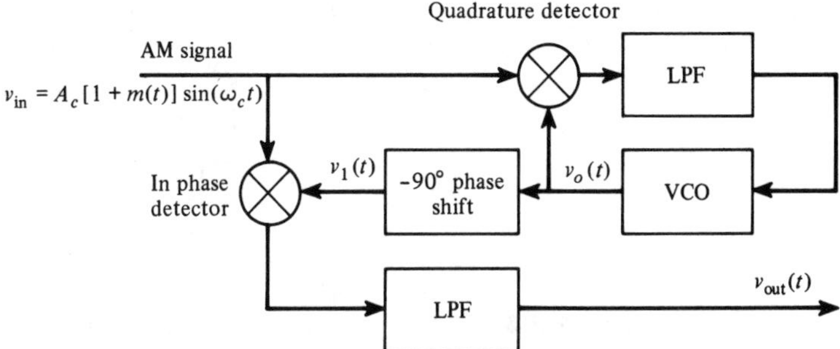

FIGURE 4-25 PLL used for coherent detection of AM.

which becomes, using (4-97c),

$$V_2(f) = \frac{\frac{D_f}{K_v} F_1(f)}{F_1(f) + j\left(\frac{2\pi f}{K_v K_d}\right)} M(f) \tag{4-98a}$$

We now need to find the conditions such that $V_2(f)$ is proportional to $M(f)$. Assume that the bandwidth of the modulation is B hertz, and let $F_1(f)$ be a low-pass filter. Thus

$$F(f) = F_1(f) = 1, \qquad |f| < B \tag{4-98b}$$

Also, let

$$\frac{K_v K_d}{2\pi} \gg B \tag{4-98c}$$

Then (4-98a) becomes

$$V_2(f) = \frac{D_f}{K_v} M(f) \tag{4-98d}$$

or

$$v_2(t) = Cm(t) \tag{4-98e}$$

where the constant of proportionality is $C = D_f/K_v$. Thus the PLL circuit of Fig. 4-22 will become a FM detector circuit where $v_2(t)$ is the detected FM output when the conditions of (4-98b) and (4-98c) are satisfied.

In another application, the PLL may be used to supply the coherent oscillator signal for product detection of an AM signal (see Fig. 4-25). Recall, from (4-84) and (4-85), that the VCO of a PLL locks 90° out of phase with respect to the incoming signal.† Then $v_0(t)$ needs to be shifted by $-90°$ so that it will be in

†This results from the characteristic of the phase detector circuit. This statement is correct for a PD that produces a zero dc output voltage when the two PD input signals are 90° out of phase (i.e., a multiplier-type PD). However, if the PD circuit produced a zero dc output when the two PD inputs were in phase, the VCO of the PLL would lock in phase with the incoming PLL signal.

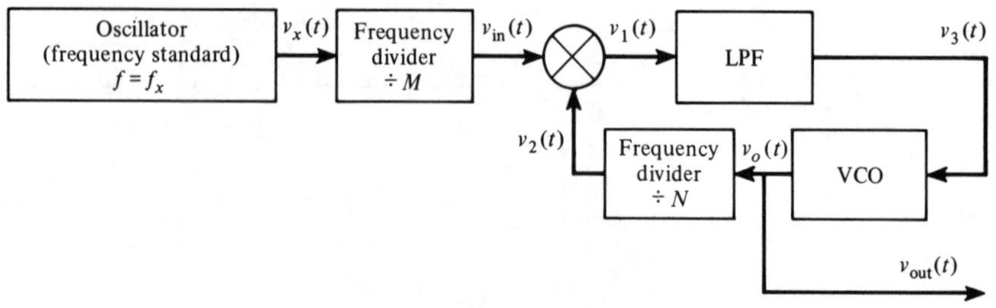

FIGURE 4-26 PLL used in a frequency synthesizer.

phase with the carrier of the input AM signal. This was found to be the requirement for coherent detection of AM as given by (4-72). In this application the bandwidth of the LPF needs to be just wide enough to provide the necessary pull-in range in order for the VCO to be able to lock onto the carrier frequency, f_c.

Figure 4-26 illustrates the use of a PLL in a frequency synthesizer. This frequency synthesizer generates a periodic signal of frequency

$$f_{\text{out}} = \left(\frac{N}{M}\right) f_x \tag{4-99}$$

where f_x is the frequency of the stable oscillator and N and M are the frequency-divider parameters. This result is verified by recalling that when the loop is locked, the dc control signal $v_3(t)$ shifts the frequency of the VCO so that $v_2(t)$ will have the same frequency as $v_{\text{in}}(t)$. Thus,

$$\frac{f_x}{M} = \frac{f_{\text{out}}}{N}$$

which is equivalent to (4-99). If programmable dividers are used, the synthesizer output frequency may be changed at will, according to (4-99), by selecting appropriate values of N and M. Of course, this could be carried out by using a microprocessor under software program control. It is noted that for the case of $M = 1$, this synthesizer configuration acts like a frequency multiplier. In addition, more complicated PLL synthesizer configurations can be built that incorporate mixers and additional oscillators.

In this section we have described the operating principles for communication system components such as filters, amplifiers, limiters, up/down converters, PLL, and detectors. The results are summarized in Table 4-5, which gives the output of each device as a function of the complex envelope for the input to that device. In succeeding sections of this chapter, we see how these devices are connected together to build communication systems. Space limitations do not permit further coverage of devices or detailed circuit analysis. The reader is referred to specialized books on these and related topics (such as oscillators and frequency synthesizers) [Manassewitsch, 1987; Sinnema and McPherson, 1991; Smith, 1986; Young, 1990].

TABLE 4-5 Effect of Communication System Components

Device Name	Device Parameter(s)[a]	Output Complex Envelope[b]	Output Waveform
Linear bandpass amplifier	Gain = A	$Ag(t)$	$\text{Re}\{Ag(t)e^{j\omega_c t}\}$
Linear bandpass filter	Impulse response = $\text{Re}\{k(t)e^{j\omega_c t}\}$	$\frac{1}{2}g(t) * k(t)$	$\frac{1}{2}\text{Re}\{g(t) * k(t)e^{j\omega_c t}\}$
Hard bandpass limiter	Limit level = V_L	$KV_L e^{j\theta(t)}$	$\text{Re}\{KV_L e^{j\theta(t)}e^{j\omega_c t}\}$
Down converter	Gain = A; $\omega_0 < \omega_c$	$Ag(t)$	$\text{Re}\{Ag(t)e^{j(\omega_c-\omega_0)t}\}$
Down converter[c]	Gain = A; $\omega_c < \omega_0$	$Ag^*(t)$	$\text{Re}\{Ag^*(t)e^{j(\omega_0-\omega_c)t}\}$
Up converter	Gain = A	$Ag(t)$	$\text{Re}\{Ag(t)e^{j(\omega_c+\omega_0)t}\}$
Frequency multiplier	Output level = V; multiply by n	$Ve^{jn\theta(t)}$	$\text{Re}\{Ve^{jn\theta(t)}e^{jn\omega_c t}\}$
Envelope detector	Gain = A	NA	$AR(t)$
Phase detector	Gain = A	NA	$A\theta(t)$
Frequency modulation detector	Gain = A	NA	$Ad\theta(t)/dt$
Product detector	Gain = A; reference = $K\cos(\omega_c t + \theta_0)$	NA	$A\,\text{Re}\{g(t)e^{-j\theta_0}\}$
Product detector	Gain = A; in-phase ref. ($\theta_0 = 0$)	NA	$Ax(t)$
Product detector	Gain = A; quad-phase ref. ($\theta_0 = 90°$)	NA	$Ay(t)$

[a] ω_0, LO frequency.

[b] NA, not applicable.

[c] For high-side injection (i.e., $\omega_c < \omega_0$), the upper and lower sidebands at the output are inverted compared to those at the input since $\mathcal{F}[g^*(t)] = G^*(-f)$.

4-4 Transmitters and Receivers

Generalized Transmitters

Transmitters generate the modulated signal at the carrier frequency f_c from the modulating signal $m(t)$. In Sec. 4-2 it was demonstrated that any type of modulated signal could be represented by

$$v(t) = \text{Re}\{g(t)e^{j\omega_c t}\} \tag{4-100}$$

or, equivalently,

$$v(t) = R(t)\cos[\omega_c t + \theta(t)] \tag{4-101}$$

and

$$v(t) = x(t)\cos\omega_c t - y(t)\sin\omega_c t \tag{4-102}$$

where the complex envelope

$$g(t) = R(t)e^{j\theta(t)} = x(t) + jy(t) \tag{4-103}$$

is a function of the modulating signal $m(t)$. The particular relationship that is chosen for $g(t)$ in terms of $m(t)$ defines the type of modulation that is used, such as AM, SSB, or FM (see Table 4-1). A generalized approach may be taken to obtain universal transmitter models that may be reduced to those used for a particular modulation type. We will also see that there are equivalent models that correspond to different circuit configurations, yet they may be used to produce the same type of modulated signal at their outputs. It is up to communications engineers to select an implementation method that will maximize performance, yet minimize cost based on the state of the art in circuit development.

There are two canonical forms for the generalized transmitter, as indicated by (4-101) and (4-102). Equation (4-101) describes an AM-PM type circuit as shown in Fig. 4-27. The baseband signal processing circuit generates $R(t)$ and $\theta(t)$ from $m(t)$. The R and θ are functions of the modulating signal $m(t)$ as given in Table 4-1 for the particular modulation type desired. The signal processing

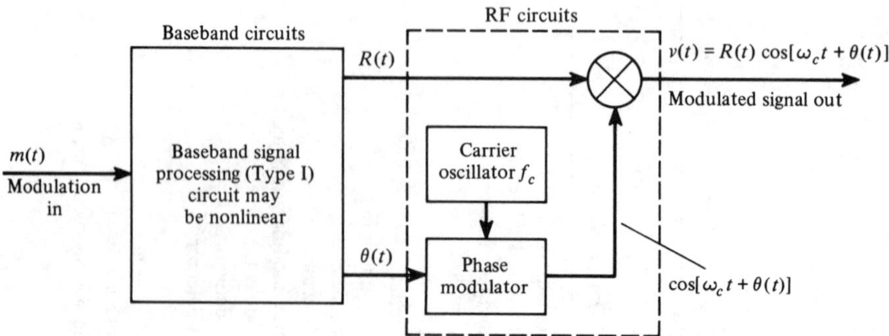

FIGURE 4-27 Generalized transmitter using the AM–PM generation technique.

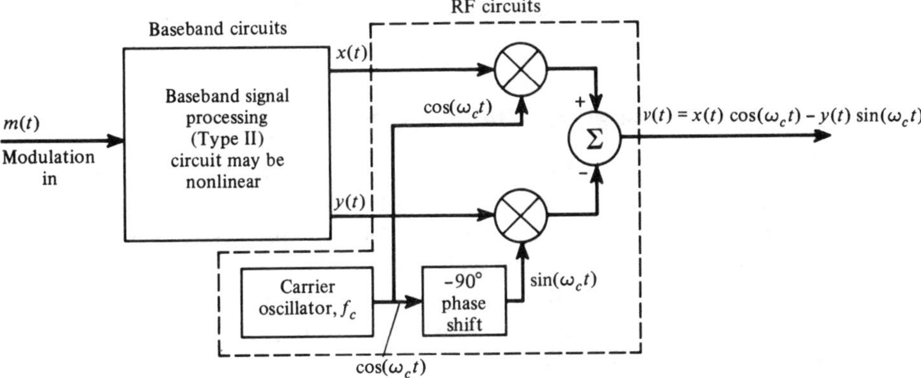

FIGURE 4-28 Generalized transmitter using the quadrature generation technique.

may be implemented either by using nonlinear analog circuits or a digital computer that incorporates the R and θ algorithms under software program control. In the implementation using a digital computer, one ADC will be needed at the input and two DACs will be needed at the output. The remainder of the AM-PM canonical form requires RF circuits, as indicated in the figure.

Figure 4-28 illustrates the second canonical form for the generalized transmitter. This uses in-phase and quadrature-phase (IQ) processing. Similarly, the formulas relating $x(t)$ and $y(t)$ to $m(t)$ are shown in Table 4-1, and the baseband signal processing may be implemented by using either analog hardware or digital hardware with software. The remainder of the canonical form utilizes RF circuits as indicated.

Once again, it is stressed that any type of signal modulation (AM, FM, SSB, QPSK, etc.) may be generated by using either of these two canonical forms. Both of these forms conveniently separate baseband processing from RF processing. Digital techniques are especially useful to realize the baseband processing portion. Furthermore, if digital computing circuits are used, any desired modulation type can be obtained by selecting the appropriate software program.

Most of the practical transmitters in use today are special variations on these canonical forms. Practical transmitters may perform the RF operations at some convenient lower RF frequency and then up-convert to the desired operating frequency. In the case of RF signals that contain no AM, frequency multipliers may be used to arrive at the operating frequency. Of course, power amplifiers are usually required to bring the output power level up to the specified value. If the RF signal contains no amplitude variations, Class C amplifiers (relatively high efficiency) may be used; otherwise, Class B amplifiers are used.

Students often ask: Can I build a transmitter or receiver for a laboratory project? The answer is yes, but it will be quite time consuming if only discrete components are used. However, integrated circuits make this type of project suitable and allow excellent performance to be achieved. Two inexpensive ICs that are manufactured to be used in 49-MHz cordless phones are recommended and described in Sec. 4-9 (data sheets for the MC 2831 FM Transmitter and the MC 3363 FM Receiver).

Generalized Receiver: The Superheterodyne Receiver

The receiver has the job of extracting the source information from the received modulated signal that has been corrupted by noise. Often it is desired that the receiver output be a replica of the modulating signal that was present at the transmitter input. There are two main classes of receivers: the *tuned radio frequency* (TRF) receiver and the *superheterodyne* receiver.

The TRF receiver consists of a number of cascaded high-gain RF bandpass stages that are tuned to the carrier frequency, f_c, followed by an appropriate detector circuit (envelope detector, product detector, FM detector, etc.). The TRF is not very popular because it is difficult to design tunable RF stages so that the desired station can be selected and yet have a narrow bandwidth so that adjacent channel stations are rejected. In addition, it is difficult to obtain high gain at radio frequencies and to have sufficiently small stray coupling between the output and input of the RF amplifying chain so that the chain will not become an oscillator at f_c. The "crystal set" that is built by the Cub Scouts is an example of a single-RF-stage TRF receiver that has no gain in the RF stage. TRF receivers are often used to measure time-dispersive (multipath) characteristics of radio channels [Rappaport, 1989].

Most receivers employ the *superheterodyne* receiving technique. This technique consists of either down-converting or up-converting the input signal to some convenient frequency band, called the *intermediate frequency* (IF) band, and then extracting the information (or modulation) by using the appropriate detector (see Fig. 4-29).† This basic receiver structure is used for the reception of all types of bandpass signals, such as television, FM, AM, satellite, and radar signals. The RF amplifier has a bandpass characteristic that passes the desired signal and provides amplification to override additional noise that is generated in the mixer stage. The RF filter characteristic also provides some rejection of adjacent channel signals and noise, but the main adjacent channel rejection is accomplished by the IF filter.

The IF filter is a bandpass filter that selects either the up-conversion or down-conversion component (whichever is chosen by the receiver designer). When up conversion is selected, it should be recalled that the complex envelope of the IF (bandpass) filter output is the same as the complex envelope for the RF input, except for RF filtering, $H_1(f)$, and IF filtering $H_2(f)$. However, if down conversion is used with $f_{LO} > f_c$, the complex envelope at the IF output will be the conjugate of that for the RF input [see (4-54)]. This means that the sidebands of the IF output will be inverted (i.e., the upper sideband on the RF input will become the lower sideband, etc., on the IF output). If $f_{LO} < f_c$, the sidebands are not inverted. In some special applications the LO frequency is selected to be the

† *Dual-conversion* superheterodyne receivers can also be built. Here a second mixer and a second IF stage follow the first IF stage shown in Fig. 4-29. (See the MC3363 data sheets in Sec. 4-9.)

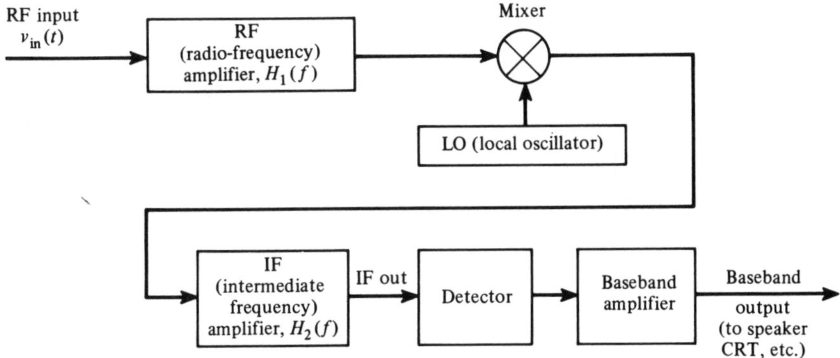

FIGURE 4-29 Superheterodyne receiver.

carrier frequency ($f_{LO} = f_c$), and then the superheterodyne receiver becomes a direct-conversion receiver.† In this case the IF filter is replaced by a low-pass filter so that the mixer-LPF combination becomes a product detector and the detector stage of Fig. 4-29 is deleted. Thus, the synchrodyne receiver is the same as a TRF receiver with product detection.

The center frequency selected for the IF amplifier is chosen based on three considerations.

1. The frequency should be such that a stable high-gain amplifier can be economically attained.
2. The frequency needs to be low enough so that with practical circuit elements in the IF filters, values of Q can be attained that will provide a steep attenuation characteristic outside the bandwidth of the IF signal. This decreases the noise and minimizes the interference from adjacent channels.
3. The frequency needs to be high enough so that the receiver image response can be made acceptably small.

The *image response* is the reception of an unwanted signal located at the image frequency due to insufficient attenuation of the image signal by the RF amplifier filter. The image response is best illustrated by an example.

EXAMPLE 4-2 AM BROADCAST SUPERHETERODYNE RECEIVER

Assume that an AM broadcast band radio is tuned to receive a station at 850 kHz and that the LO frequency is on the high side of the carrier frequency. If the IF frequency is 455 kHz, the LO frequency will be 850 + 455 = 1305 kHz (see Fig. 4-30). Furthermore, assume that other signals are present at the RF input of the radio and, particularly, that there is a signal at 1760 kHz; this signal will be down-converted by the mixer to 1760 − 1305 = 455 kHz. That is, the undesired (1760-kHz) signal will be translated to 455 kHz and will be added at the mixer

†A direct-conversion receiver is also called a *homodyne* or a *synchrodyne* receiver.

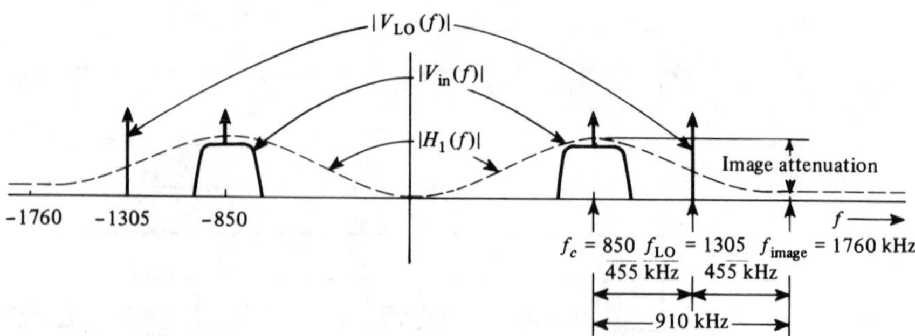

FIGURE 4-30 Spectra of signals and transfer function of an RF amplifier in a superheterodyne receiver.

output to the desired (850-kHz) signal, which was also down-converted to 455 kHz. This *undesired* signal that has been converted to the IF band is called the *image signal*. If the gain of the RF amplifier is down by, say, 25 dB at 1760 kHz as compared to 850 kHz, and if the undesired signal is 25 dB stronger at the receiver input than the desired signal, both signals will have the same level when translated to the IF. This undesired signal will definitely interfere with the desired signal in the detection process.

For down converters (i.e., $f_{IF} = |f_c - f_{LO}|$) the image frequency is

$$f_{image} = \begin{cases} f_c + 2f_{IF}, & \text{if } f_{LO} > f_c \quad \text{(high-side injection)} \\ f_c - 2f_{IF}, & \text{if } f_{LO} < f_c \quad \text{(low-side injection)} \end{cases} \tag{4-104a}$$

where f_c is the desired RF frequency, f_{IF} is the IF frequency, and f_{LO} is the local oscillator frequency. For up converters (i.e., $f_{IF} = f_c + f_{LO}$) the image frequency is

$$f_{image} = f_c + 2f_{LO} \tag{4-104b}$$

From Fig. 4-30 it is seen that the image response will usually be reduced if the IF frequency is increased, since f_{image} will occur farther away from the main peak (or lobe) of the RF filter characteristic, $|H_1(f)|$.

Recalling our earlier discussion on mixers, we also realize that other spurious responses (in addition to the image response) will occur in practical mixer circuits. These must also be taken into account in good receiver design.

Table 4-6 illustrates some typical IF frequencies that have become de facto standards. For the intended application, the IF frequency is low enough that the IF filter will provide good adjacent channel signal rejection when circuit elements with realizable Q are used; yet the IF frequency is large enough to provide adequate image-signal rejection by the RF amplifier filter.

The type of detector selected for use in the superheterodyne receiver depends on the intended application. For example, a product detector may be used in a PSK (digital) system, and an envelope detector is used in AM broadcast receivers. If the complex envelope $g(t)$ is desired for generalized signal detection or for

TABLE 4-6 Some Popular IF Frequencies in the United States

IF Frequency	Application
262.5 kHz	AM broadcast radios (in automobiles)
455 kHz	AM broadcast radios
10.7 MHz	FM broadcast radios
21.4 MHz	FM two-way radios
30 MHz	Radar receivers
43.75 MHz (video carrier)	TV sets
60 MHz	Radar receivers
70 MHz	Satellite receivers

optimum reception in digital systems, the $x(t)$ and $y(t)$ quadrature components, where $x(t) + jy(t) = g(t)$, may be obtained by using quadrature product detectors, as illustrated in Fig. 4-31. $x(t)$ and $y(t)$ could be fed into a signal processor to extract the modulation information. Disregarding the effects of noise, the signal processor could recover $m(t)$ from $x(t)$ and $y(t)$ (and, consequently, demodulate the IF signal) by using the inverse of the complex envelope generation functions given in Table 4-1.

The superheterodyne receiver has many advantages and some disadvantages. The main advantage is that extraordinarily high gain can be obtained without instability (self-oscillation). The stray coupling between the output of the receiver and the input does not cause oscillation because the gain is obtained in disjoint frequency bands—RF, IF, and baseband. The receiver is easily tunable to another frequency by changing the frequency of the LO signal (which may be supplied by a frequency synthesizer) and by tuning the bandpass of the RF amplifier to the desired frequency. Furthermore, high-Q elements—which are needed (to produce steep filter skirts) for adjacent channel rejection—are needed only in the fixed tuned IF amplifier. The main disadvantage of the superheterodyne receiver is the response to spurious signals that will occur if one is not careful with the design.

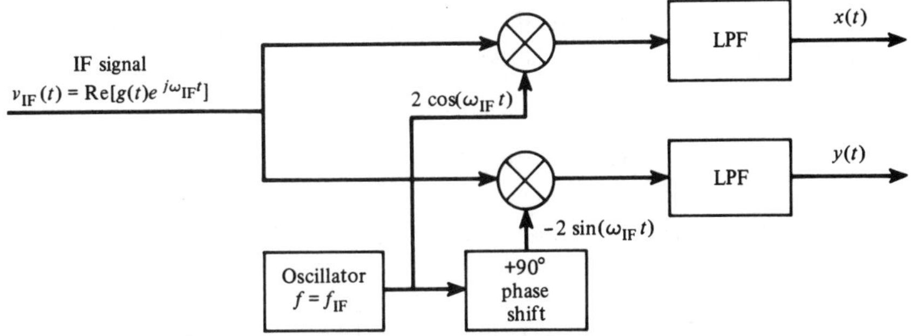

FIGURE 4-31 IQ (in-phase and quadrature-phase) detector.

As indicated previously, it is feasible to build a receiver for a laboratory project provided that ICs are used. Data sheets for the MC 3363 (FM receiver) are given in Sec. 4-9.

A discussion of receivers would not be complete without considering some of the causes of *interference*. Often the receiver owner thinks that a certain signal, such as an amateur radio signal, is causing the difficulty. This may or may not be the case. The origin of the interference may be at any of three locations.

1. At the interfering signal *source*, the transmitter may generate out-of-band signal components (such as harmonics) that fall in the band of the desired signal.
2. At the *receiver* itself, the front end may overload or produce spurious responses. Front-end overload occurs when the RF and/or mixer stage of the receiver is driven into the nonlinear range by the interfering signal and the nonlinearity causes cross modulation on the desired signal at the output of the receiver RF amplifier.
3. In the *channel*, a nonlinearity in the transmission medium may cause undesired signal components in the band of the desired signal.

For more discussion of receiver design and examples of practical receiver circuits, the reader is referred to the *ARRL Handbook* [ARRL, 1991].

4-5 Amplitude Modulation and Double-Sideband Suppressed Carrier

Amplitude Modulation

From Sec. 4-2 it is recalled that any type of modulated signal can be represented by

$$s(t) = \mathrm{Re}\{g(t)e^{j\omega_c t}\} \tag{4-105}$$

and that the complex envelope of an AM signal is given by

$$g(t) = A_c[1 + m(t)] \tag{4-106}$$

where the constant A_c has been included to specify the power level and $m(t)$ is the modulating signal (which may be analog or digital). These equations reduce to the representation for the AM signal:

$$s(t) = A_c[1 + m(t)] \cos \omega_c t \tag{4-107}$$

A waveform illustrating the AM signal, as seen on an oscilloscope, is shown in Fig. 4-32. For convenience it is assumed that the modulating signal $m(t)$ is a sinusoid. $A_c[1 + m(t)]$ corresponds to the in-phase component $x(t)$ of the complex envelope; it also corresponds to the real envelope $|g(t)|$ when $m(t) \geq -1$ (the usual case).

If $m(t)$ has a peak positive value of $+1$ and a peak negative value of -1, the AM signal is said to be 100% modulated.

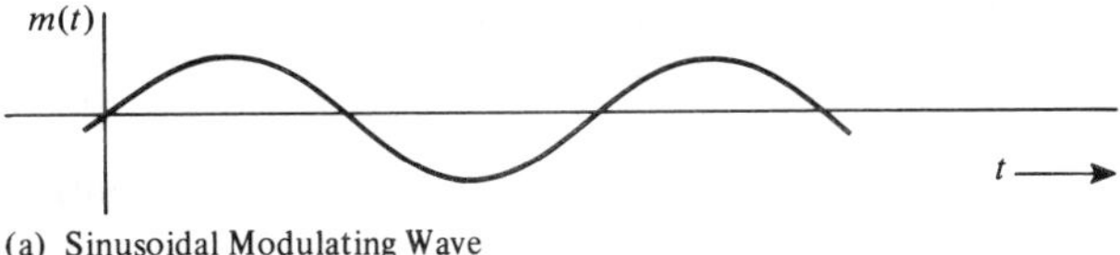

(a) Sinusoidal Modulating Wave

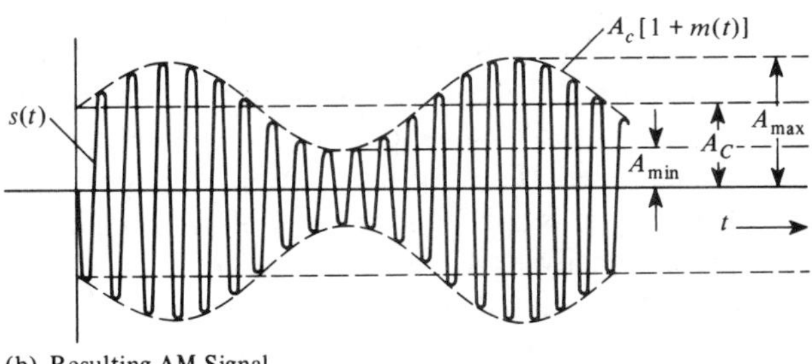

(b) Resulting AM Signal

FIGURE 4-32 AM signal waveform.

DEFINITION: The *percentage of positive modulation* on an AM signal is

$$\% \text{ positive modulation} = \frac{A_{\text{max}} - A_c}{A_c} \times 100 = \max[m(t)] \times 100 \quad (4\text{-}108)$$

and the *percentage of negative modulation* is

$$\% \text{ negative modulation} = \frac{A_c - A_{\text{min}}}{A_c} \times 100 = -\min[m(t)] \times 100 \quad (4\text{-}109)$$

The overall modulation percentage is

$$\% \text{ modulation} = \frac{A_{\text{max}} - A_{\text{min}}}{2A_c} \times 100 = \frac{\max[m(t)] - \min[m(t)]}{2} \times 100 \quad (4\text{-}110)$$

where A_{max} is the maximum value of $A_c[1 + m(t)]$, A_{min} is the minimum value, and A_c is the level of the AM envelope under the condition of no modulation [i.e., $m(t) = 0$].

Equation (4-110) may be obtained by averaging the positive and negative modulation as given by (4-108) and (4-109). A_{max}, A_{min}, and A_c are illustrated in Fig. 4-32b, where, in this example, $A_{\text{max}} = 1.5A_c$ and $A_{\text{min}} = 0.5A_c$, so that the percentages of positive and negative modulation are both 50% and the overall modulation is 50%.

The percentage of modulation can be over 100% (A_{min} will have a negative value) provided that a four-quadrant multiplier is used to generate the product of $A_c[1 + m(t)]$ and $\cos \omega_c t$ so that the true AM waveform, as given by (4-107), is obtained.[†] However, if the transmitter uses a two-quadrant multiplier that produces a zero output when $A_c[1 + m(t)]$ is negative, the output signal will be

$$s(t) = \begin{cases} A_c[1 + m(t)] \cos \omega_c t, & \text{if } m(t) \geq -1 \\ 0, & \text{if } m(t) < -1 \end{cases} \tag{4-111}$$

which is a distorted AM signal. The bandwidth of this signal is much wider than that of the undistorted AM signal, as is easily demonstrated by spectral analysis. This is the overmodulated condition that the FCC does not allow. An AM transmitter that uses plate modulation is an example of a circuit that acts as a two-quadrant multiplier. Here, for high-power AM signal generation, the unmodulated carrier signal is applied to the grid of the tube, and the dc plate voltage is varied proportionally to $A_c[1 + m(t)]$, where $A_c[1 + m(t)] \geq 0$. This produces the product $A_c[1 + m(t)] \cos \omega_c t$, provided that $m(t) \geq -1$, but produces no output when $m(t) < -1$.

If the percentage of negative modulation is less than 100%, an envelope detector may be used to recover the modulation without distortion since the envelope, $|g(t)| = |A_c[1 + m(t)]|$, is identical to $A_c[1 + m(t)]$. If the percentage of negative modulation is over 100%, undistorted modulation can still be recovered provided that the proper type of detector—a product detector—is used. This is seen from (4-71) with $\theta_0 = 0$. Furthermore, the product detector may be used for any percentage of modulation. In Chapter 7 it will be found that a product detector is superior to an envelope detector when the input signal-to-noise ratio is small.

Using (4-27), we realize that the *normalized average power* of the AM signal is

$$\begin{aligned} \langle s^2(t) \rangle &= \tfrac{1}{2}\langle |g(t)|^2 \rangle = \tfrac{1}{2} A_c^2 \langle [1 + m(t)]^2 \rangle \\ &= \tfrac{1}{2} A_c^2 \langle 1 + 2m(t) + m^2(t) \rangle \\ &= \tfrac{1}{2} A_c^2 + A_c^2 \langle m(t) \rangle + \tfrac{1}{2} A_c^2 \langle m^2(t) \rangle \end{aligned} \tag{4-112}$$

If the modulation contains no dc level, $\langle m(t) \rangle = 0$ and the *normalized* power of the AM signal is

$$\langle s^2(t) \rangle = \underbrace{\tfrac{1}{2} A_c^2}_{\substack{\text{discrete} \\ \text{carrier power}}} + \underbrace{\tfrac{1}{2} A_c^2 \langle m^2(t) \rangle}_{\text{sideband power}} \tag{4-113}$$

Note that this result is identical to that given by (4-33) if $A_c = 1$.

Definition: The *modulation efficiency* is the percentage of the total power of the modulated signal that conveys information.

[†]If the percentage of modulation becomes very large (approaching infinity), the AM signal becomes the double-sideband suppressed carrier signal that is described in the next section.

In AM signaling, only the sideband components convey information, so the modulation efficiency is

$$E = \frac{\langle m^2(t)\rangle}{1 + \langle m^2(t)\rangle} \times 100\% \tag{4-114}$$

The highest efficiency that could be obtained for a 100% AM signal would be 50%, which is attained only for square-wave modulation.

Using (4-28), we obtain the normalized peak envelope power (PEP) of the AM signal:

$$P_{\text{PEP}} = \frac{A_c^2}{2}\{1 + \max[m(t)]\}^2 \tag{4-115}$$

The voltage spectrum of the AM signal was obtained in (4-30a) of Example 4-1 and is

$$S(f) = \frac{A_c}{2}[\delta(f - f_c) + M(f - f_c) + \delta(f + f_c) + M(f + f_c)] \tag{4-116}$$

(The AM spectrum is given in Fig. 4-4.) The AM spectrum is relatively easy to obtain since it is just a translated version of the modulation spectrum plus delta functions that give the carrier line spectral component. The *bandwidth* is *twice* that of the modulation. As we will see shortly, the spectrum for an FM signal is much more complicated since the modulation mapping function $g(m)$ is nonlinear.

EXAMPLE 4-3 POWER OF AN AM SIGNAL

The FCC rates AM broadcast band transmitters by their average *carrier* power; this is common in other AM audio applications as well. Suppose that a 5000-W AM transmitter is connected to a 50-Ω load; then the constant A_c is given by $\frac{1}{2}A_c^2/50 = 5000$. Thus the peak voltage across the load will be $A_c = 707$ V during the times of no modulation. If the transmitter is then 100% modulated by a 1000-Hz test tone, the total (carrier plus sideband) average power will be, from (4-113),

$$1.5\left[\frac{1}{2}\left(\frac{A_c^2}{50}\right)\right] = (1.5) \times (5000) = 7500 \text{ W}$$

since $\langle m^2(t)\rangle = \frac{1}{2}$ for a sinusoidal modulation waveshape of unity (100%) amplitude. Note that 7500 W is the actual power, not the normalized power. The peak voltage (100% modulation) is (2)(707) = 1414 V across the 50-Ω load. From (4-115), the PEP is

$$4\left[\frac{1}{2}\left(\frac{A_c^2}{50}\right)\right] = (4)(5000) = 20{,}000 \text{ W}$$

The modulation efficiency would be 33% since $\langle m^2(t)\rangle = \frac{1}{2}$.

There are many ways of building AM transmitters. One might first consider generating the AM signal at a low power level (by using a multiplier) and then amplifying it. This, however, requires the use of linear amplifiers (such as Class A or B amplifiers, as discussed in Sec. 4-3) so that the AM will not be distorted. Because these linear amplifiers are not very efficient in converting the power-supply power to an RF signal, much of the energy is wasted in heat.[†] Consequently, high-power AM broadcast transmitters are built by amplifying the carrier oscillator signal to a high power level with efficient Class C or Class D amplifiers and then amplitude modulating the last high-power stage. This is called *high-level* modulation, and an example is shown in Fig. 4-33a. For high conversion efficiency, a pulse width modulation (PWM) technique is used to achieve the AM [DeAngelo, 1982], as shown in this figure. The audio input is converted to a PWM signal that is used to control a high-power switch (tube or transistor) circuit. The switch circuit output consists of a high-level PWM signal that is filtered by a low-pass filter to produce the "dc" component that is used as the power supply for the power amplifier (PA) stage. The PWM switching frequency is usually chosen in the range of 70 to 80 kHz so that the fundamental and harmonic components of the PWM signal can be easily suppressed by the low-pass filter, and yet the "dc" can vary at an audio rate as high as 12 or 15 kHz for good AM audio frequency response. This technique provides excellent frequency response and low distortion since no high-power audio transformers are needed.

AM Broadcast Technical Standards

Some of the FCC technical standards for AM broadcast stations are shown in Table 4-7. In the United States, assigned carrier frequencies are classified according to the station's intended coverage: clear-channel, regional, or local. Clear-channel stations, also called *Class I* or *Class II* stations, are licensed to operate with a carrier power as large as 50 kW. These stations are intended to cover large areas. Moreover, to accommodate as many stations as possible, nonclear-channel stations may be assigned to operate on clear-channel frequencies when they can be implemented without interfering with the dominant clear-channel station. Often, to prevent interference, these secondary stations have to operate with directional antenna patterns such that there is a null in the direction of the dominant station. This is especially true for nighttime operation when sky-wave propagation allows the clear-channel stations to cover most of the United States. Other secondary stations are allowed to operate only in the daytime since any nighttime operation would interfere with the clear-channel station. Other frequencies are set aside for regional stations, also called *Class III* stations, which are intended to cover medium-sized areas. These stations have allowed carrier power levels as large as 5 kW. A third class of frequencies is assigned to local (also called *Class IV*) stations, which are intended to cover one city or a county area and are licensed for a maximum power of 1 kW. Detailed listings of

[†]Do not confuse this conversion efficiency with modulation efficiency, which was defined by (4-114).

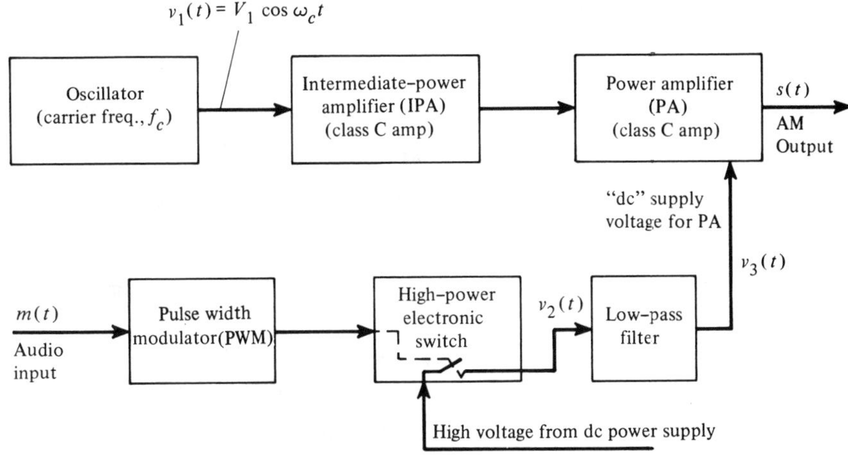

(a) Block Diagram

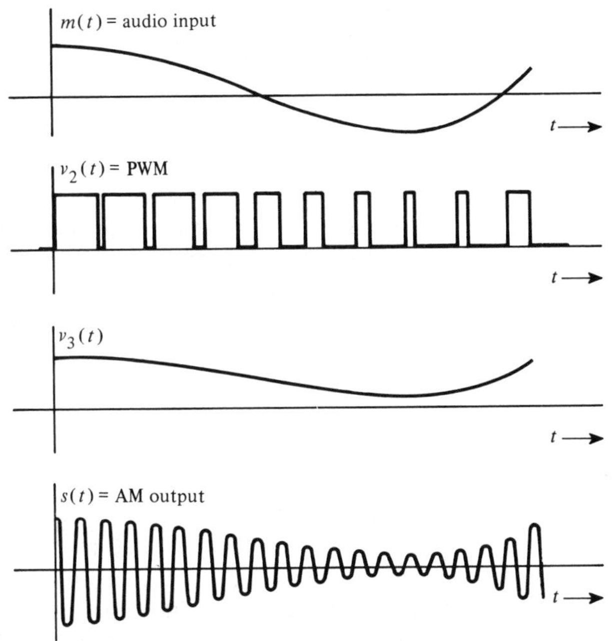

(b) Waveforms

FIGURE 4-33 Generation of high-power AM by the use of PWM.

the frequency assignments for these four classes of stations and their licensed powers are available. International broadcast AM stations, which operate in the shortwave bands (3 to 30 MHz), generally operate with higher power levels. Some of these feed 500 kW of carrier power into directional antennas. These antennas produce effective radiated power levels in the megawatt range (i.e., when the gain of the directional antenna is taken into account).

TABLE 4-7 AM Broadcast Station Technical Standards

Item	FCC Technical Standard
Assigned frequency, f_c	In 10-kHz increments from 540 to 1700 kHz
Channel bandwidth	10 kHz
Carrier frequency stability	±20 Hz of the assigned frequency
% modulation	Maintain 85–95%; max.: 100% neg., 125% pos.
Audio-frequency response[a]	±2 dB from 100 Hz to 5 kHz with 1 kHz being the 0-dB reference
Harmonic distortion[a]	Less than 5% for up to 85% modulation; less than 7.5% for modulation between 85% and 95%
Noise and hum	At least 45 dB below 100% modulation in the band 30 Hz to 20 kHz
Maximum power licensed	50 kW

[a] Under the new FCC deregulation policy, these requirements are deleted from the FCC rules, although broadcasters still use them as guidelines for minimum acceptable performance.

For many years, AM broadcasters have dreamed of offering stereo transmissions to their listeners. Technically, this advance has been possible for a long time, but for political reasons it was not implemented until recently. In 1977 the FCC conducted evaluations of five different AM stereo systems that were proposed by the Kahn Communication/Hazeltine Corporation, Belar Electronics, the Magnavox Corporation, the Harris Corporation, and Motorola [Mennie, 1978]. Each of these proposed systems had advantages and disadvantages, and the FCC decided to allow each AM broadcaster to implement any one of them. This led to confusion in the marketplace since each system required a different type of stereo decoding circuit in the AM receiver. Receiver manufacturers did not want to build all five decoding circuits into each receiver in order to accommodate all the possibilities AM broadcasters might offer. However, it appears that the Motorola system is the choice of most broadcasters, since about 440 out of 540 AM stereo stations use the Motorola technique. Nonetheless, it is claimed that the Kahn system is superior during multipath interference conditions that often occur at night when simultaneous sky waves and ground waves are received [Murphy, 1988].

The Kahn system implements AM stereo by transmitting the right-channel audio on an upper SSB-AM signal and the left-channel audio on a lower SSB-AM signal (see Table 4-1), with both SSB signals having the same carrier frequency, f_c. This is called *independent sideband* (ISB) transmission. A mono (conventional) AM receiver tuned to f_c receives both sidebands for mono (left plus right channel) reception. Two mono receivers can be used to receive stereo by tuning one slightly higher than f_c (right-channel audio output) and one slightly lower than f_c (left-channel audio output). In addition, stereo AM receivers can be built that give much better left- and right-channel separation than that obtained by using two mono receivers.

The Motorola C-QUAM[†] system uses quadrature modulation (QM) (see Table 4-1). The monaural audio, consisting of the sum of the left and right channels, is used to modulate a cosine carrier. That is, $m_1(t) = V_0 + m_L(t) + m_R(t)$, where V_0 is the dc offset used to produce the discrete carrier term in the AM spectrum, $m_L(t)$ is the left-channel audio, and $m_R(t)$ is the right-channel audio. The quadrature carrier is modulated by $m_2(t) = m_L(t) - m_R(t)$. The resulting real envelope $R(t)$ consists of a dc term and a distorted audio term. The distortion is reduced by using a limiter and a filter on the QM output and then amplitude modulating the resulting constant-amplitude signal with $m_L(t) + m_R(t)$ [Mennie, 1978]. The resulting C-QUAM signal is compatible with a mono (conventional) AM receiver for mono reception. Stereo AM receivers use an envelope detector to recover $m_L(t) + m_R(t)$ and a quadrature product detector to recover $m_L(t) - m_R(t)$. Sum and difference networks can then be used to obtain the left- and right-channel audio, $m_L(t)$ and $m_R(t)$, respectively.

Double-Sideband Suppressed Carrier

A *double-sideband suppressed carrier* (DSB-SC) signal is essentially an AM signal that has a suppressed discrete carrier. The DSB-SC signal is given by

$$s(t) = A_c m(t) \cos \omega_c t \tag{4-117a}$$

where $m(t)$ is assumed to have a zero dc level for the suppressed carrier case. The spectrum is identical to that for AM given by (4-116), except that the delta functions at $\pm f_c$ are missing. That is, the spectrum for DSB-SC is

$$S(f) = \frac{A_c}{2}[M(f - f_c) + M(f + f_c)] \tag{4-117b}$$

When compared with an AM signal, the percentage of modulation on a DSB-SC signal is infinite since there is no carrier line component. Furthermore, the modulation efficiency of a DSB-SC signal is 100% since no power is wasted in a discrete carrier. However, a product detector (which is more expensive than an envelope detector) is required for demodulation of the DSB-SC signal. If transmitting circuitry restricts the modulated signal to a certain peak value, say A_p, it can be demonstrated (see Prob. 4-49) that the sideband power of a DSB-SC signal is *four* times that of a comparable AM signal that has the same peak level. In this sense the DSB-SC signal has a fourfold power advantage over that of an AM signal.

It is also realized that if $m(t)$ is a polar binary data signal (instead of an audio signal), then (4-117a) is a BPSK signal, discussed in Example 2-15. As shown in Table 4-1, a QM signal can be generated by adding two DSB signals where there are two modulating signals, $m_1(t)$ and $m_2(t)$, modulating cosine and sine carriers, respectively.

[†]C-QUAM is the acronym used by Motorola for compatible quadrature amplitude modulation.

★ Costas Loop and Squaring Loop

The coherent reference for product detection cannot be obtained by the use of an ordinary phase-locked tracking loop since there are no spectral line components at $\pm f_c$. However, since the DSB-SC signal has a spectrum that is symmetrical with respect to the (suppressed) carrier frequency, either one of two types of carrier recovery loops shown in Fig. 4-34 may be used to demodulate the DSB-SC signal. Figure 4-34a shows the *Costas PLL* and Fig. 4-34b shows the *squaring loop*. It can be shown that the noise performance of these two loops are equivalent [Ziemer and Peterson, 1985], so the choice of which loop to implement depends on the relative cost of the loop components and the accuracy that can be realized when each component is built.

As shown in Fig. 4-34a, the Costas PLL is analyzed by assuming that the VCO is locked to the input suppressed carrier frequency, f_c, with a constant

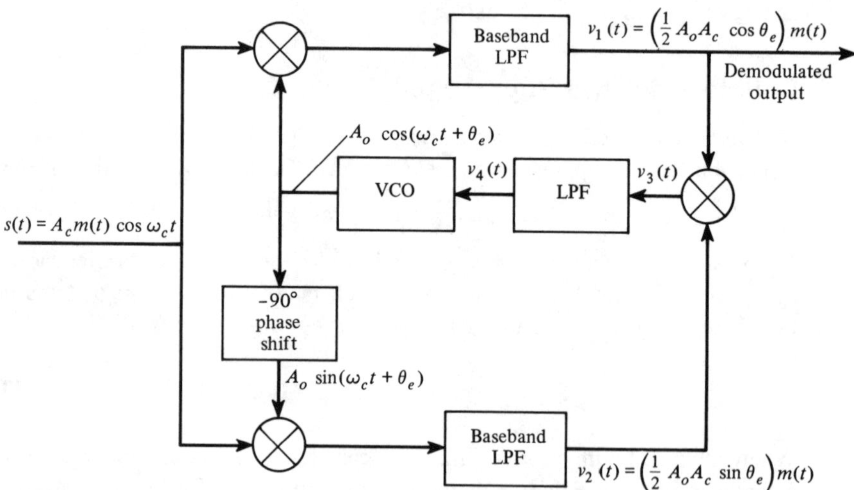

(a) Costas Phase-Locked Loop

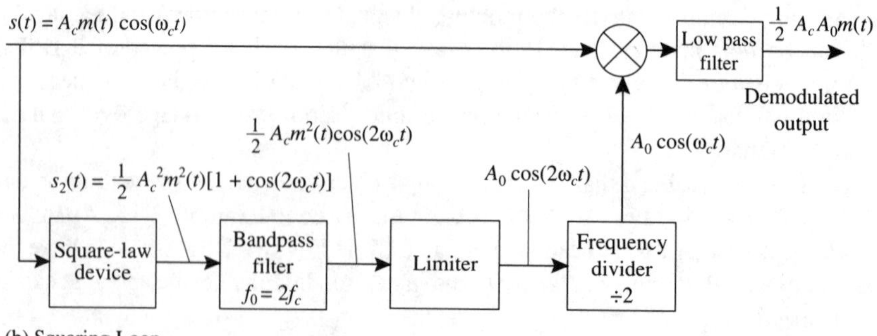

(b) Squaring Loop

FIGURE 4-34 Carrier recovery loops for DSB-SC signals.

phase error of θ_e. Then the voltages $v_1(t)$ and $v_2(t)$ are obtained at the output of the baseband low-pass filters as shown. Since θ_e is small, the amplitude of $v_1(t)$ is relatively large compared to that of $v_2(t)$ (i.e., $\cos \theta_e \gg \sin \theta_e$). Furthermore, $v_1(t)$ is proportional to $m(t)$, so it is the demodulated (product detector) output. The product voltage $v_3(t)$ is

$$v_3(t) = \tfrac{1}{2}(\tfrac{1}{2}A_0 A_c)^2 m^2(t) \sin 2\theta_e$$

The voltage $v_3(t)$ is filtered with a LPF that has a cutoff frequency near dc so that this filter acts as an integrator to produce the dc VCO control voltage

$$v_4(t) = K \sin 2\theta_e$$

where $K = \tfrac{1}{2}(\tfrac{1}{2}A_0 A_c)^2 \langle m^2(t) \rangle$ and $\langle m^2(t) \rangle$ is the dc level of $m^2(t)$. This dc control voltage is sufficient to keep the VCO locked to f_c with a small phase error, θ_e.

The squaring loop, shown in Fig. 4-34b, is analyzed by evaluating the expression for the signal out of each component block as shown on this figure. As shown by Fig. 4-34, either the Costas PLL or the squaring loop can be used to demodulate a DSB-SC signal, since for each case the output is $Cm(t)$, where C is a constant. Furthermore, either of these loops can be used to recover (i.e., demodulate) a binary-phase-shift keyed (BPSK) signal since the BPSK signal has the same mathematical form as a DSB-SC signal, where $m(t)$ is a polar NRZ signal as given in Fig. 3-11c and the BPSK waveform is shown in Fig. 5-1d.

Both the Costas PLL and the squaring loop have one major disadvantage—a 180° phase ambiguity. For example, suppose that the input is $-A_c m(t) \cos \omega_c t$ instead of $+A_c m(t) \cos \omega_c t$. Retracing the steps in the preceding analysis, we see that the output would be described by *exactly* the same equation that was obtained before. Then, whenever the loop is energized, it is just as likely to phase lock such that the demodulated output is proportional to $-m(t)$ as it is to $m(t)$. Thus we cannot be sure of the polarity of the output. This is no problem if $m(t)$ is a monaural audio signal since $-m(t)$ sounds the same to our ears as $m(t)$. However, if $m(t)$ is a polar data signal, binary 1's might come out as binary 0's after the circuit is energized, or vice versa. As we found in Chapter 3, there are two ways of abrogating this 180° phase ambiguity: (1) a known test signal can be sent over the system after the loop is turned on so that the sense of the polarity can be determined, or (2) differential coding and decoding may be used.

4-6 Asymmetric Sideband Signals

Single Sideband

Definition: An *upper single sideband* (USSB) signal has a zero-valued spectrum for $|f| < f_c$, where f_c is the carrier frequency.

A *lower single sideband* (LSSB) signal has a zero-valued spectrum for $|f| > f_c$, where f_c is the carrier frequency.

There are numerous ways in which the modulation $m(t)$ may be mapped into the complex envelope $g[m]$ such that an SSB signal will be obtained. Table 4-1 lists some of these methods.

- SSB-AM, which is detected by using a product detector.
- SSB-PM, which may be detected by using a phase detector.
- SSB-FM, which may be detected by using an FM detector.
- SSB-EV, which may be demodulated by using an envelope detector.
- SSB-SQ, which may be demodulated by using a square-law detector that squares the SSB-SQ signal and passes the result through a baseband (low-pass) filter.

SSB-AM is by far the most popular type. It is widely used by the military and by radio amateurs in high-frequency (HF) communication systems, and it is popular because the bandwidth is the same as that of the modulating signal (which is half the bandwidth of an AM or DSB-SC signal). For this reason, we will concentrate on this type of SSB signal. In the usual application, the term *SSB* refers to the SSB-AM type of signal unless otherwise denoted.

THEOREM: *An SSB signal (i.e., SSB-AM type) is obtained by using the complex envelope*

$$g(t) = A_c[m(t) \pm j\hat{m}(t)] \tag{4-118}$$

which results in the SSB signal waveform

$$s(t) = A_c[m(t) \cos \omega_c t \mp \hat{m}(t) \sin \omega_c t] \tag{4-119}$$

where the upper (−) sign is used for USSB and the lower (+) sign is used for LSSB. $\hat{m}(t)$ denotes the Hilbert transform of $m(t)$, which is given by[†]

$$\hat{m}(t) \triangleq m(t) * h(t) \tag{4-120}$$

where

$$h(t) = \frac{1}{\pi t} \tag{4-121}$$

and $H(f) = \mathcal{F}[h(t)]$ corresponds to a −90° phase shift network:

$$H(f) = \begin{cases} -j, & f > 0 \\ j, & f < 0 \end{cases} \tag{4-122}$$

Figure 4-35 illustrates this theorem. Assume that $m(t)$ has a magnitude spectrum that is of triangular shape, as shown in Fig. 4-35a. Then for the case of USSB (upper signs), the spectrum of $g(t)$ is zero for negative frequencies, as shown in Fig. 4-35b, and $s(t)$ has the USSB spectrum shown in Fig. 4-35c. This result is proved as follows.

Proof. We need to show that the spectrum of $s(t)$ is zero on the appropriate sideband (depending on the sign chosen). Taking the Fourier transform of

[†] A table of Hilbert transform pairs is given in Sec. A-7 (Appendix A).

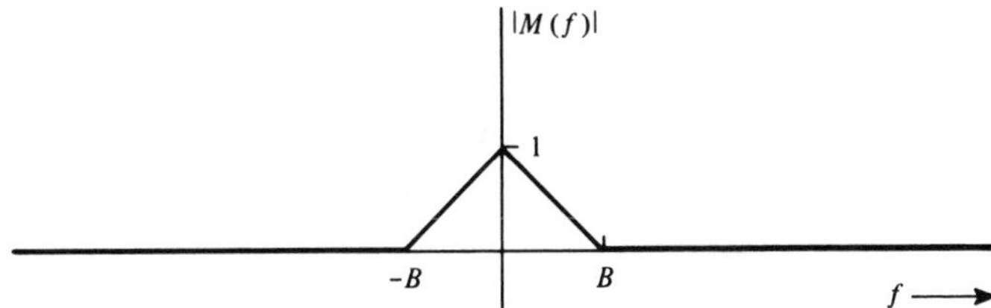

(a) Baseband Magnitude Spectrum

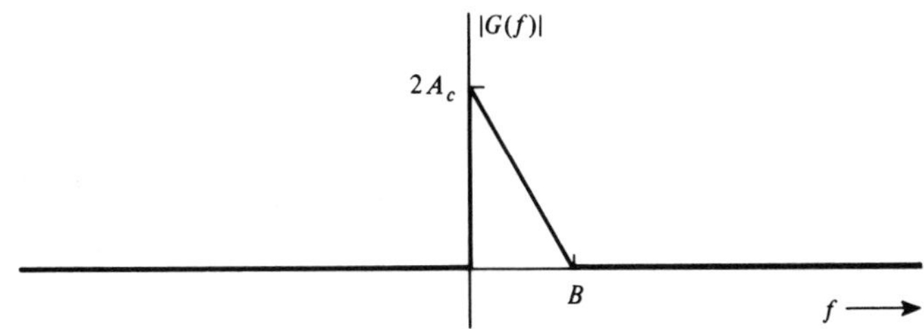

(b) Magnitude of Corresponding Spectrum of the Complex Envelope for USSB

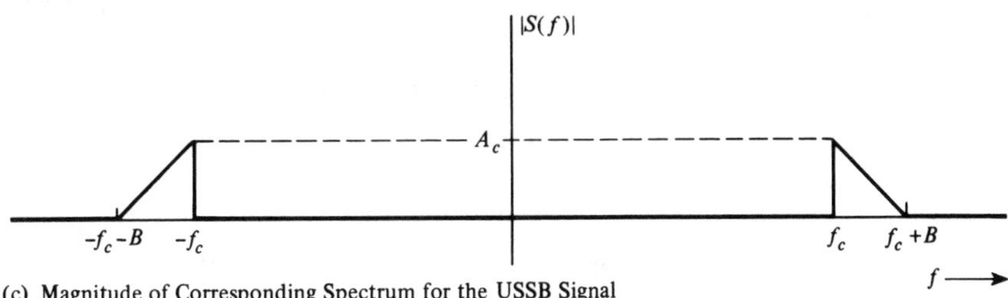

(c) Magnitude of Corresponding Spectrum for the USSB Signal

FIGURE 4-35 Spectrum for a USSB signal.

(4-119) and letting $\hat{m}(t) = n(t)$ for convenience, we have

$$S(f) = A_c \int_{-\infty}^{\infty} \left[m(t)\left(\frac{e^{j\omega_c t} + e^{-j\omega_c t}}{2}\right) \right.$$

$$\left. \mp\; n(t)\left(\frac{e^{j\omega_c t} - e^{-j\omega_c t}}{2j}\right)\right] e^{-j\omega t}\, dt$$

$$= \frac{A_c}{2} \int_{-\infty}^{\infty} [m(t) \pm jn(t)] e^{-j(\omega - \omega_c)t}\, dt$$

$$+ \frac{A_c}{2} \int_{-\infty}^{\infty} [m(t) \mp jn(t)] e^{-j(\omega + \omega_c)t}\, dt \qquad (4\text{-}123)$$

or

$$S(f) = \frac{A_c}{2}[M(f - f_c) \pm jN(f - f_c)] + \frac{A_c}{2}[M(f + f_c) \mp jN(f + f_c)]$$

Recalling that $n(t) = \hat{m}(t) = m(t) * h(t)$, we have

$$N(f) = M(f)H(f) = M(f)\begin{cases} -j, & f > 0 \\ j, & f < 0 \end{cases} \tag{4-124}$$

Using (4-124) in (4-123), we obtain

$$S(f) = \frac{A_c}{2}\left[M(f-f_c) \pm jM(f-f_c)\begin{Bmatrix} -j, & (f-f_c) > 0 \\ +j, & (f-f_c) < 0 \end{Bmatrix}\right]$$
$$+ \frac{A_c}{2}\left[M(f+f_c) \mp jM(f+f_c)\begin{Bmatrix} -j, & (f+f_c) > 0 \\ +j, & (f+f_c) < 0 \end{Bmatrix}\right] \tag{4-125}$$

For the USSB result choose the upper signs, and this equation reduces to

$$S(f) = A_c\begin{Bmatrix} M(f-f_c), & f > f_c \\ 0, & f < f_c \end{Bmatrix} + A_c\begin{Bmatrix} 0, & f > -f_c \\ M(f+f_c), & f < -f_c \end{Bmatrix} \tag{4-126}$$

which is indeed a USSB signal (see Fig. 4-35).

If the lower signs were chosen, an LSSB signal would have been obtained.

The *normalized average power* of the SSB signal is

$$\langle s^2(t)\rangle = \tfrac{1}{2}\langle |g(t)|^2\rangle = \tfrac{1}{2}A_c^2\langle m^2(t) + [\hat{m}(t)]^2\rangle \tag{4-127}$$

But $\langle \hat{m}(t)^2\rangle = \langle m^2(t)\rangle$, so that the SSB signal power is

$$\langle s^2(t)\rangle = A_c^2\langle m^2(t)\rangle \tag{4-128}$$

which is the power of the modulating signal $\langle m^2(t)\rangle$ multiplied by the power gain factor A_c^2.

The *normalized peak envelope power (PEP)* is

$$\tfrac{1}{2}\max\{|g(t)|^2\} = \tfrac{1}{2}A_c^2\max\{m^2(t) + [\hat{m}(t)]^2\} \tag{4-129}$$

Figure 4-36 illustrates two techniques for generating the SSB signal. The *phasing method* is identical to the IQ canonical form discussed earlier (Fig. 4-28) as applied to SSB signal generation. The *filtering method* is a special case where RF processing (by using the sideband filter) is used to form the equivalent $g(t)$, instead of using baseband processing to generate $g[m]$ directly. The filter method is the most popular method used since excellent sideband suppression can be obtained when a crystal filter is used for the sideband filter. Crystal filters are relatively inexpensive when produced in quantity at standard IF frequencies. In addition to these two techniques for generating SSB, there is a third technique called *Weaver's method*. This is described by Fig. P4-53 and Prob. 4-53. The practical SSB transmitter would incorporate an up converter to translate the SSB signal to the desired operating frequency and use a Class B amplifier to amplify the signal to the desired power level.

SSB signals have *both* AM and PM. Using (4-118), we have for the AM component (real envelope)

$$R(t) = |g(t)| = A_c\sqrt{m^2(t) + [\hat{m}(t)]^2} \tag{4-130}$$

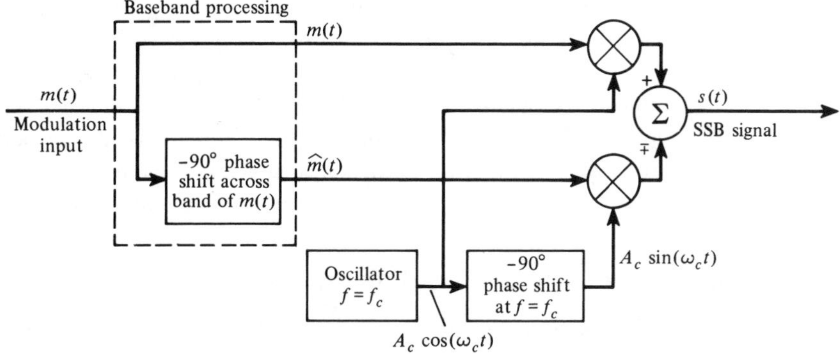

(a) Phasing Method

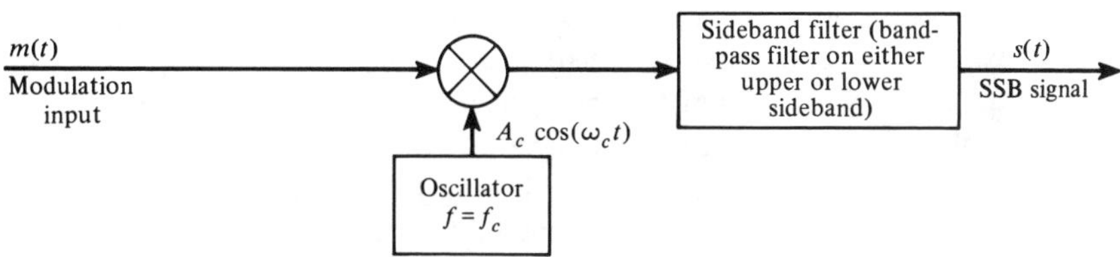

(b) Filter Method

FIGURE 4-36 Generation of SSB.

and for the PM component

$$\theta(t) = \underline{/g(t)} = \tan^{-1}\left[\frac{\pm\hat{m}(t)}{m(t)}\right] \tag{4-131}$$

SSB signals may be received by using a superheterodyne receiver that incorporates a product detector with $\theta_0 = 0$. Thus the receiver output is

$$v_{\text{out}}(t) = K\,\text{Re}\{g(t)e^{-j\theta_0}\} = KA_c m(t) \tag{4-132}$$

where K depends on the gain of the receiver and the loss in the channel. In detecting SSB signals with audio modulation, the reference phase θ_0 does not have to be zero because the same intelligence is heard regardless of the value of the phase used [although the $v_{\text{out}}(t)$ waveform will be drastically different depending on the value of θ_0]. Of course, for digital modulation, the phase has to be exactly correct if the digital waveshape is to be preserved. Furthermore, SSB is a poor modulation technique to use *if* the modulating data signal consists of a line code with a *rectangular* pulse shape. The rectangular shape (zero rise time) causes the value of the SSB-AM signal to be infinite adjacent to the switching times of the data because of the Hilbert transform operation. (This result will be demonstrated in a homework problem.) Thus an SSB signal with this type of modulation cannot be generated by any practical device since a device can produce only finite peak value signals. However, if rolled-off pulse shapes are used in the line code, such as (sin x)/x pulses, the SSB signal will have a reasonable peak value, and digital data transmission can then be accommodated via SSB.

SSB has many advantages such as the superior detected signal-to-noise ratio when compared to AM (see Chapter 7) and the fact that SSB has one half the bandwidth of AM or DSB-SC signals. Space limitations do not permit more discussion of SSB; for more information on this topic, the reader is referred to a book that is devoted wholly to SSB [Sabin and Schoenike, 1987].

Amplitude-Companded Single Sideband

One of the main disadvantages of SSB is that, for proper detection, the frequency of the oscillator signal at the product detector mixer, f_0, must be the same as that of the incoming SSB signal, f_c. That is, $f_0 = f_c$, as shown in Fig. 4-15. This is the assumption in (4-71) that leads to (4-132). However, the receiver operator usually does not know f_c; the operator knows only that he or she wants to tune in an SSB station. Consequently, if $f_0 \neq f_c$, the spectrum of the detected audio signal will be shifted by the frequency difference, $\Delta f = f_c - f_0$. This effect is observed when an amateur radio operator tunes in an SSB signal. That is, initially, the pitch of the audio signal out of the receiver is either too high or too low, and the operator keeps adjusting f_0 until the pitch of the output signal sounds natural (i.e., $f_0 \approx f_c$, so that $\Delta f \approx 0$). The assumption that $\Delta f \approx 0$, leading to (4-132), is now achieved, but it required the trial and error of a human operator. This trial-and-error approach is difficult, if not impossible to implement with an electronic circuit, so "automatic tuning" cannot be achieved. This problem of not being able to set $f_0 = f_c$ electronically can be avoided if a low-level sinusoidal signal of known frequency, called a pilot tone, is added to the modulating audio signal that is fed into the SSB transmitter. Then a phase-locked loop (PLL) (with appropriate frequency dividers) can be used at the receiver to recover the pilot tone and derive the oscillator signal where $f_0 = f_c$ for use by the product detector. This is the technique used in *amplitude-companded single sideband* (ACSSB). In 1985 the FCC approved the use of ACSSB for use in the land mobile service as an alternative to narrowband frequency modulation (as indicated in footnote c of Table 4-11). The ACSSB system uses a 3.1-kHz pilot tone that has a level 10 dB below the peak audio level [ARRL, 1991]. In addition, 6 dB of amplitude companding is used. (This is the same idea as using nonuniform quantizing, as described in Sec. 3-3.)

The ACSSB system gives a 12- to 15-dB improvement in postdetection signal-to-noise ratio when compared to a narrowband FM system for reception of fringe-area signals. In addition, this SSB technique allows four ACSSB stations to be assigned to the same bandwidth that was previously occupied by only one narrowband FM station.

Vestigial Sideband

In certain applications (such as television broadcasting) a DSB modulation technique takes too much bandwidth for the (television) channel and an SSB tech-

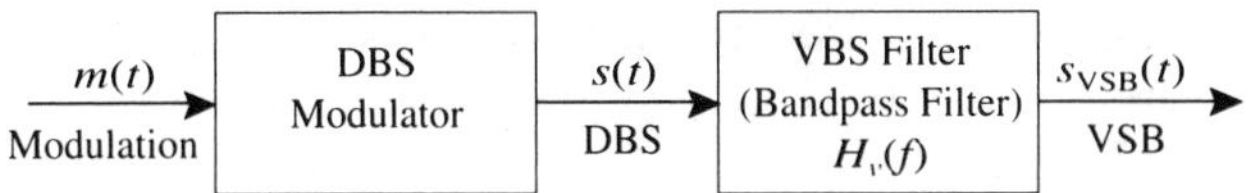

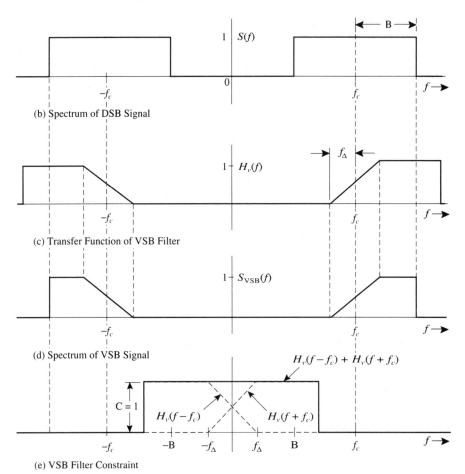

(a) Generation of VSB Signal

(b) Spectrum of DSB Signal

(c) Transfer Function of VSB Filter

(d) Spectrum of VSB Signal

(e) VSB Filter Constraint

FIGURE 4-37 VSB transmitter and spectra.

nique is too expensive to implement, although it takes only half the bandwidth. In this case a compromise between DSB and SSB, called *vestigial sideband* (VSB), is often chosen. VSB is obtained by partial suppression of one of the sidebands of a DSB signal. The DSB signal may be either an AM signal or a DSB-SC signal. This is illustrated by Fig. 4-37, where one sideband of the DSB signal is attenuated by using a bandpass filter, called a vestigial sideband filter, that has an asymmetrical frequency response about $\pm f_c$. The VSB signal is given by

$$s_{\text{VSB}}(t) = s(t) * h_v(t) \tag{4-133a}$$

where $s(t)$ is a DSB signal described by either (4-107) with carrier or (4-117a) with suppressed carrier, and $h_v(t)$ is the impulse response of the VSB filter. The spectrum of the VSB signal is

$$S_{\text{VSB}}(f) = S(f)H_v(f) \tag{4-133b}$$

as illustrated in Fig. 4-37d.

The modulation on the VSB signal can be recovered by a receiver that uses product detection or, if a large carrier is present, by use of envelope detection. For recovery of undistorted modulation, the transfer function for the VSB filter must satisfy the constraint

$$H_v(f - f_c) + H_v(f + f_c) = C, \qquad |f| \leq B \tag{4-134}$$

where C is a constant and B is the bandwidth of the modulation. Application of this constraint is shown in Fig. 4-37e, where it is seen that the condition specified by (4-134) is satisfied for the VSB filter characteristic shown in Fig. 4-37c.

The need for the constraint of (4-134) will now be proven. Assume that $s(t)$ is a DSB-SC signal. Then, by use of (4-117b) and (4-133b), the spectrum of the VSB signal is

$$S_{\text{VSB}}(f) = \frac{A_c}{2}[M(f - f_c)H_v(f) + M(f + f_c)H_v(f)]$$

Referring to Fig. 4-15, the product detector output is

$$v_{\text{out}}(t) = [A_0 s_{\text{VSB}}(t) \cos \omega_c t] * h(t)$$

where $h(t)$ is the impulse response of the low-pass filter having a bandwidth of B hertz. In the frequency domain this becomes

$$V_{\text{out}}(f) = A_0\{S_{\text{VSB}}(f) * [\tfrac{1}{2}\delta(f - f_c) + \tfrac{1}{2}\delta(f + f_c)]\}H(f)$$

where $H(f) = 1$ for $|f| < B$ and 0 for f elsewhere. Substituting for $S_{\text{VSB}}(f)$ and using the convolutional property $x(f) * \delta(f - a) = x(f - a)$, $V_{\text{out}}(f)$ is

$$V_{\text{out}}(f) = \frac{A_c A_0}{4}[M(f)H_v(f - f_c) + M(f)H_v(f + f_c)], \qquad |f| < B$$

or

$$V_{\text{out}}(f) = \frac{A_c A_0}{4}M(f)[H_v(f - f_c) + H_v(f + f_c)], \qquad |f| < B$$

If $H_v(f)$ satisfies the constraint of (4-134), this becomes

$$V_{\text{out}}(f) = KM(f)$$

or $v_{\text{out}}(t) = Km(t)$ where $K = A_c A_0/4$. This shows that the output of the product detector is undistorted when (4-134) is satisfied.

As discussed in Chapter 5, Sec. 5-9, broadcast television uses VSB to reduce the required channel bandwidth to 6 MHz. In this application, as shown in Fig. 5-52c, the frequency response of the visual TV transmitter is flat over the upper sideband out to 4.2 MHz above the visual carrier frequency and is flat over the lower sideband out to 0.75 MHz below the carrier frequency. The IF filter in the TV receiver has the VSB filter characteristic shown in Fig. 4-37c, where $f_\Delta =$ 0.75 MHz. This gives an overall frequency response characteristic that satisfies the constraint of (4-134) so that the video modulation on the visual VSB TV signal can be recovered at the receiver without distortion.

4-7 Phase Modulation and Frequency Modulation

Representation of PM and FM Signals

Phase modulation (PM) and *frequency modulation* (FM) are special cases of angle-modulated signaling. In angle modulated signaling the complex envelope is

$$g(t) = A_c e^{j\theta(t)} \tag{4-135}$$

Here the real envelope, $R(t) = |g(t)| = A_c$, is a constant, and the phase $\theta(t)$ is a linear function of the modulating signal $m(t)$. However, $g(t)$ is a *nonlinear* function of the modulation. Using (4-135), we find the resulting *angle-modulated signal* to be

$$s(t) = A_c \cos[\omega_c t + \theta(t)] \tag{4-136}$$

For PM the phase is directly proportional to the modulating signal:

$$\theta(t) = D_p m(t) \tag{4-137}$$

where the proportionality constant D_p is the phase sensitivity of the phase modulator, having units of radians per volt [assuming that $m(t)$ is a voltage waveform].

For FM the phase is proportional to the integral of $m(t)$:

$$\theta(t) = D_f \int_{-\infty}^{t} m(\sigma)\, d\sigma \tag{4-138}$$

where the frequency deviation constant D_f has units of radians/volt-second.

By comparing the last two equations, it is seen that if we have a PM signal modulated by $m_p(t)$, there is *also* FM on the signal corresponding to a *different* modulation waveshape that is given by

$$m_f(t) = \frac{D_p}{D_f}\left[\frac{dm_p(t)}{dt}\right] \tag{4-139}$$

where the subscripts f and p denote frequency and phase, respectively. Similarly, if we have an FM signal modulated by $m_f(t)$, the corresponding phase modulation on this signal is

$$m_p(t) = \frac{D_f}{D_p}\int_{-\infty}^{t} m_f(\sigma)\, d\sigma \tag{4-140}$$

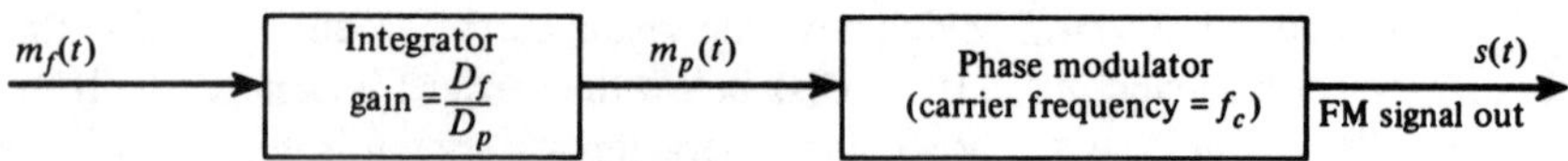

(a) Generation of FM Using a Phase Modulator

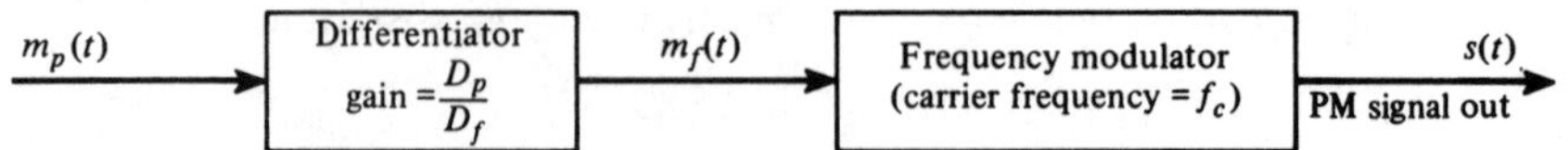

(b) Generation of PM Using a Frequency Modulator

FIGURE 4-38 Generation of FM from PM, and vice versa.

By using (4-140), a PM circuit may be used to synthesize an FM circuit by inserting an integrator in cascade with the phase modulator input (see Fig. 4-38a).

Direct PM circuits are realized by passing an unmodulated sinusoidal signal through a time-varying circuit that introduces a phase shift that varies with the applied modulating voltage (see Fig. 4-39a). D_p is the gain of the PM circuit (rad/V). Similarly, a direct FM circuit is obtained by varying the tuning of an oscillator tank (resonant) circuit according to the modulation voltage. This is shown in Fig. 4-39b, where D_f is the gain of the modulator circuit (which has units of radians per volt-second).

Definition: If a bandpass signal is represented by

$$s(t) = R(t) \cos \psi(t)$$

where $\psi(t) = \omega_c t + \theta(t)$, then the *instantaneous* frequency (hertz) of $s(t)$ is

$$f_i(t) = \frac{1}{2\pi} \omega_i(t) = \frac{1}{2\pi}\left[\frac{d\psi(t)}{dt}\right]$$

or

$$f_i(t) = f_c + \frac{1}{2\pi}\left[\frac{d\theta(t)}{dt}\right] \tag{4-141}$$

For the case of FM, using (4-138), the instantaneous frequency is

$$f_i(t) = f_c + \frac{1}{2\pi} D_f m(t) \tag{4-142}$$

Of course, this is the reason for calling this type of signaling *frequency modulation*—the instantaneous frequency varies about the assigned carrier frequency f_c directly proportional to the modulating signal $m(t)$. Figure 4-40b shows how the instantaneous frequency varies when a sinusoidal modulation (for illustrative purposes) is used. The resulting FM signal is shown in Fig. 4-40c.

The instantaneous frequency should not be confused with the term *frequency* as used in the spectrum of the FM signal. The spectrum is given by the Fourier transform of $s(t)$ and is evaluated by looking at $s(t)$ over the infinite time interval

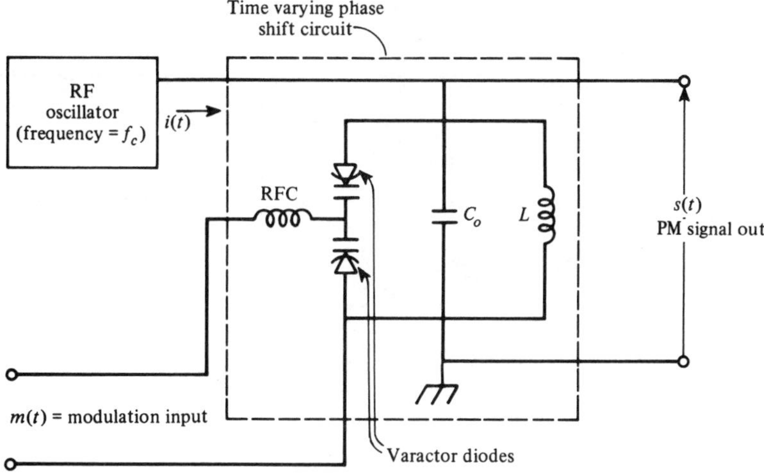

(a) A Phase Modulator Circuit

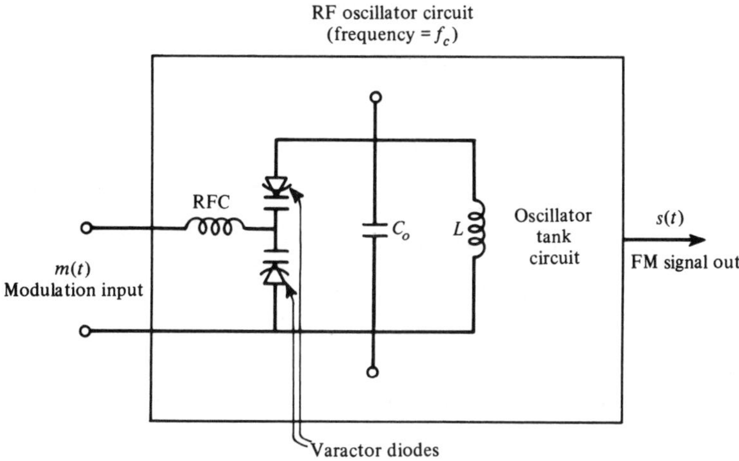

(b) A Frequency Modulator Circuit

FIGURE 4-39 Angle modulator circuits. RFC = radio-frequency choke.

$(-\infty < t < \infty)$. Thus the spectrum tells us what frequencies are present in the signal (on the average) *over all time*. The instantaneous frequency is the frequency that is present at a *particular* instant of time.

The *frequency deviation* from the carrier frequency is

$$f_d(t) = f_i(t) - f_c = \frac{1}{2\pi}\left[\frac{d\theta(t)}{dt}\right] \tag{4-143}$$

and the *peak frequency deviation* is

$$\Delta F = \max\left\{\frac{1}{2\pi}\left[\frac{d\theta(t)}{dt}\right]\right\} \tag{4-144}$$

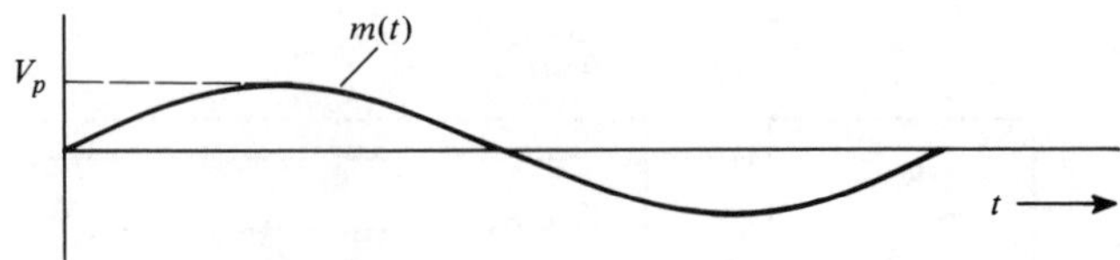

(a) Sinusoidal Modulating Signal

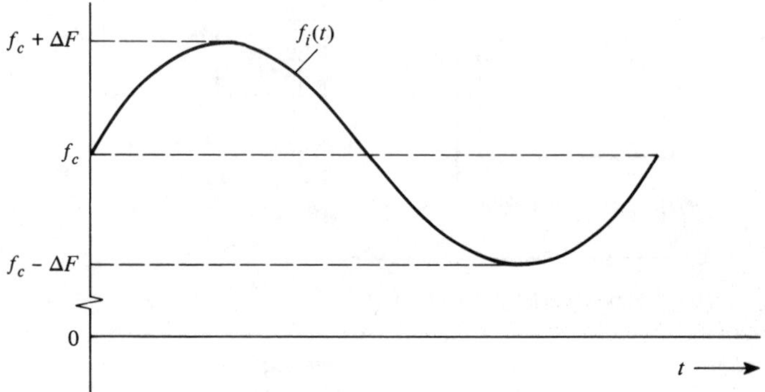

(b) Instantaneous Frequency of the Corresponding FM Signal

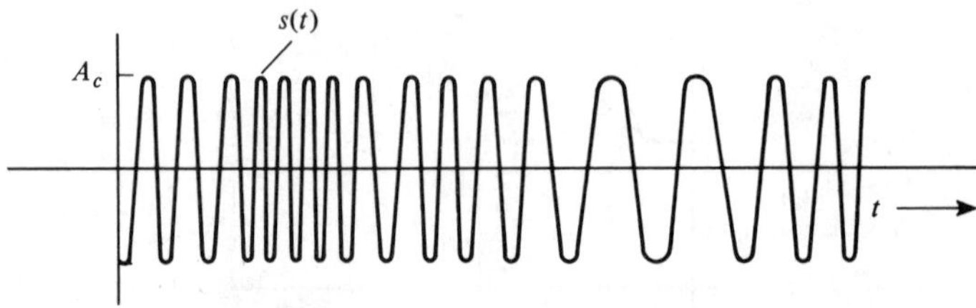

(c) Corresponding FM Signal

FIGURE 4-40 FM with a sinusoidal baseband modulating signal.

Note that ΔF is a nonnegative number. In some applications, such as (unipolar) digital modulation, the peak-to-peak deviation is used. This is defined by

$$\Delta F_{\text{pp}} = \max\left\{\frac{1}{2\pi}\left[\frac{d\theta(t)}{dt}\right]\right\} - \min\left\{\frac{1}{2\pi}\left[\frac{d\theta(t)}{dt}\right]\right\} \tag{4-145}$$

For FM signaling the peak frequency deviation is related to the peak modulating voltage by

$$\Delta F = \frac{1}{2\pi} D_f V_p \tag{4-146}$$

where $V_p = \max[m(t)]$, as illustrated in Fig. 4-40a.

From Fig. 4-40 it is obvious that an increase in the amplitude of the modulation signal V_p will increase ΔF. This will increase the bandwidth of the FM signal but will not affect the average power level of the FM signal, which is $A_c^2/2$. As V_p is increased, spectral components will appear farther and farther away from the carrier frequency, and the spectral components near the carrier frequency will decrease in magnitude since the total power in the signal remains constant (for specific details, see Example 4-4). This is distinctly different from AM signaling, where the level of the modulation affects the power in the AM signal but does not affect its bandwidth.

In a similar way, the *peak phase deviation* may be defined by

$$\Delta\theta = \max[\theta(t)] \tag{4-147}$$

and for PM this is related to the peak modulating voltage by

$$\Delta\theta = D_p V_p \tag{4-148}$$

where $V_p = \max[m(t)]$.

DEFINITION:[†] The *phase modulation index* is given by

$$\beta_p = \Delta\theta \tag{4-149}$$

where $\Delta\theta$ is the peak phase deviation.

The *frequency modulation index* is given by

$$\beta_f = \frac{\Delta F}{B} \tag{4-150}$$

where ΔF is the peak frequency deviation and B is the bandwidth of the modulating signal, which for the case of sinusoidal modulation is f_m, the frequency of the sinusoid.[‡]

For the case of PM or FM signaling with *sinusoidal* modulation such that the PM and FM signals have the same peak frequency deviation, β_p is identical to β_f.

Spectra of Angle-Modulated Signals

Using (4-22), we find that the spectrum of an angle modulated signal is given by

$$S(f) = \tfrac{1}{2}[G(f - f_c) + G^*(-f - f_c)] \tag{4-151}$$

where

$$G(f) = \mathcal{F}[g(t)] = \mathcal{F}[A_c e^{j\theta(t)}] \tag{4-152}$$

[†]For digital signals, an alternative definition of modulation index is sometimes used and is denoted by h in the literature. This digital modulation index is given by $h = 2\Delta\theta/\pi$, where $2\Delta\theta$ is the maximum peak-to-peak phase deviation change during the time that it takes to send one symbol, T_s (see Sec. 5-2).

[‡]Strictly speaking, the FM index is defined only for the case of single-tone (i.e., sinusoidal) modulation. However, it is often used for other waveshapes, where B is chosen to be the highest frequency or the dominant frequency in the modulating waveform.

When the spectra for AM, DSB-SC, and SSB were evaluated, we were able to obtain relatively simple formulas relating $S(f)$ to $M(f)$. For angle modulation signaling this is not the case because $g(t)$ is a nonlinear function of $m(t)$. Thus a general formula relating $G(f)$ to $M(f)$ cannot be obtained. This is unfortunate, but it is a fact of life. That is, to evaluate the spectrum for an angle-modulated signal, (4-152) must be evaluated on a case-by-case basis for the particular modulating waveshape of interest. Furthermore, since $g(t)$ is a nonlinear function of $m(t)$, superposition does not hold and the FM spectrum for the sum of two modulating waveshapes is not the same as summing the FM spectra that were obtained when the individual waveshapes were used.

One example of the spectra obtained for an angle-modulated signal is given in Chapter 2 (see Example 2-15). There a carrier was phase-modulated by a square wave where the peak-to-peak phase deviation was 180°. In this example the spectrum was easy to evaluate because this was the very special case where the PM signal reduces to a DSB-SC signal. In general, of course, the evaluation of (4-152) into a closed form is not easy, and one often has to use numerical techniques to approximate this Fourier transform integral. An example for the case of a sinusoidal modulating waveshape will now be worked out.

EXAMPLE 4-4 SPECTRUM OF A PM OR FM SIGNAL WITH SINUSOIDAL MODULATION

Assume that the modulation on the PM signal is

$$m_p(t) = A_m \sin \omega_m t \tag{4-153}$$

Then

$$\theta(t) = \beta \sin \omega_m t \tag{4-154}$$

where $\beta_p = D_p A_m = \beta$ is the phase modulation index.

The same phase function $\theta(t)$, as given by (4-154), could also be obtained if FM were used, where

$$m_f(t) = A_m \cos \omega_m t \tag{4-155}$$

and $\beta = \beta_f = D_f A_m/\omega_m$. The peak frequency deviation would be $\Delta F = D_f A_m/2\pi$.

The complex envelope is

$$g(t) = A_c e^{j\theta(t)} = A_c e^{j\beta \sin \omega_m t} \tag{4-156}$$

which is periodic with period $T_m = 1/f_m$. Consequently, $g(t)$ could be represented by a Fourier series that is valid over all time $(-\infty < t < \infty)$:

$$g(t) = \sum_{n=-\infty}^{n=\infty} c_n e^{jn\omega_m t} \tag{4-157}$$

where

$$c_n = \frac{A_c}{T_m} \int_{-T_m/2}^{T_m/2} (e^{j\beta \sin \omega_m t}) e^{-jn\omega_m t}\, dt \tag{4-158}$$

which reduces to

$$c_n = A_c\left[\frac{1}{2\pi}\int_{-\pi}^{\pi} e^{j(\beta \sin\theta - n\theta)}\, d\theta\right] = A_c J_n(\beta) \tag{4-159}$$

This integral—known as the *Bessel function of the first kind of the nth order*, $J_n(\beta)$—cannot be evaluated in closed form, but it has been evaluated numerically. Some tabulated four-place and two-place values for $J_n(\beta)$ are given in Tables 4-8 and 4-9, respectively. Extensive tables are available [Abramowitz and Stegun, 1964] or can be computed using MathCAD. The Bessel functions are standard function calls in mathematical personal computer programs such as MathCAD. Examination of the integral shows that (by making a change in variable)

$$J_{-n}(\beta) = (-1)^n J_n(\beta) \tag{4-160}$$

A plot of the Bessel functions for various orders, n, as a function of β is shown in Fig. 4-41.

Taking the Fourier transform of (4-157), we obtain

$$G(f) = \sum_{n=-\infty}^{n=\infty} c_n\, \delta(f - nf_m) \tag{4-161a}$$

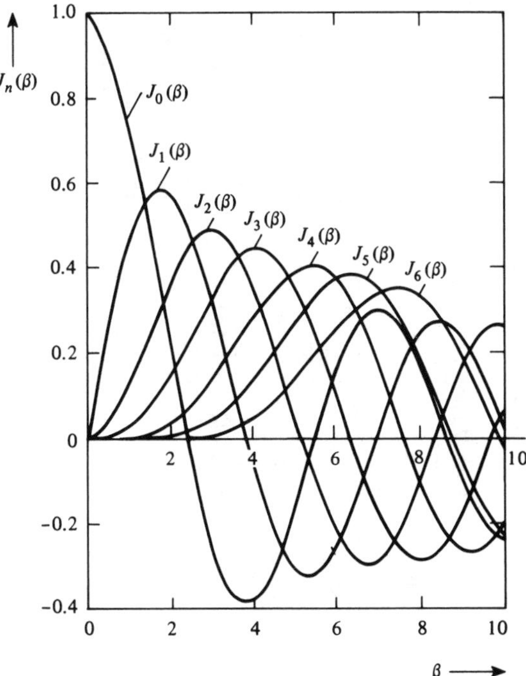

FIGURE 4-41 Bessel functions for $n = 0$ to $n = 6$.

TABLE 4-8 Four-Place Values of the Bessel Functions $J_{n(\beta)}$

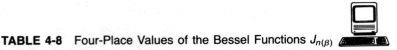

n \ β:	0.5	1	2	3	4	5	6	7	8	9	10
0	0.9385	0.7652	0.2239	−0.2601	−0.3971	−0.1776	0.1506	0.3001	0.1717	−0.09033	−0.2459
1	0.2423	0.4401	0.5767	0.3391	−0.06604	−0.3276	−0.2767	−0.004683	0.2346	0.2453	0.04347
2	0.03060	0.1149	0.3528	0.4861	0.3641	0.04657	−0.2429	−0.3014	−0.1130	0.1448	0.2546
3	0.002564	0.01956	0.1289	0.3091	0.4302	0.3648	0.1148	−0.1676	−0.2911	−0.1809	0.05838
4		0.002477	0.03400	0.1320	0.2811	0.3912	0.3576	0.1578	−0.1054	−0.2655	−0.2196
5			0.007040	0.04303	0.1321	0.2611	0.3621	0.3479	0.1858	−0.05504	−0.2341
6			0.001202	0.01139	0.04909	0.1310	0.2458	0.3392	0.3376	0.2043	−0.01446
7				0.002547	0.01518	0.05338	0.1296	0.2336	0.3206	0.3275	0.2167
8					0.004029	0.01841	0.05653	0.1280	0.2235	0.3051	0.3179
9						0.005520	0.02117	0.05892	0.1263	0.2149	0.2919
10						0.001468	0.006964	0.02354	0.06077	0.1247	0.2075
11							0.002048	0.008335	0.02560	0.06222	0.1231
12								0.002656	0.009624	0.02739	0.06337
13									0.003275	0.01083	0.02897
14									0.001019	0.003895	0.01196
15										0.001286	0.004508
16											0.001567

TABLE 4-9 Two-Place Values of the Bessel Functions $J_{n(\beta)}$

β \ *n*:	0	1	2	3	4	5	6	7	8	9	10	11	12
0.0	1.00	0											
0.2	0.99	0.10	0										
0.4	0.96	0.20	0.02										
0.6	0.91	0.29	0.04	0									
0.8	0.85	0.37	0.08	0.01									
1.0	0.77	0.44	0.11	0.02	0								
1.2	0.67	0.50	0.16	0.03	0.01								
1.4	0.57	0.54	0.21	0.05	0.01								
1.6	0.46	0.57	0.26	0.07	0.01								
1.8	0.34	0.58	0.31	0.10	0.02	0							
2.0	0.22	0.58	0.35	0.13	0.03	0.01							
2.2	0.11	0.56	0.40	0.16	0.05	0.01							
2.4	0	0.52	0.43	0.20	0.06	0.02	0						
2.6	−0.10	0.47	0.46	0.24	0.08	0.02	0.01						
2.8	−0.19	0.41	0.48	0.27	0.11	0.03	0.01						
3.0	−0.26	0.34	0.49	0.31	0.13	0.04	0.01						
3.2	−0.32	0.26	0.48	0.34	0.16	0.06	0.02	0					
3.4	−0.36	0.18	0.47	0.37	0.19	0.07	0.02	0.01					
3.6	−0.39	0.10	0.44	0.40	0.22	0.09	0.03	0.01					
3.8	−0.40	0.01	0.41	0.42	0.25	0.11	0.04	0.01					

TABLE 4-9 Two-Place Values of the Bessel Functions $J_{n(\beta)}$ (*continued*)

β \ n:	0	1	2	3	4	5	6	7	8	9	10	11	12
4.0	−0.40	−0.07	0.36	0.43	0.28	0.13	0.05	0.02	0				
4.2	−0.38	−0.14	0.31	0.43	0.31	0.16	0.06	0.02	0.01				
4.4	−0.34	−0.20	0.25	0.43	0.34	0.18	0.08	0.03	0.01				
4.6	−0.30	−0.26	0.18	0.42	0.36	0.21	0.09	0.03	0.01				
4.8	−0.24	−0.30	0.12	0.40	0.38	0.23	0.11	0.04	0.01	0			
5.0	−0.18	−0.33	0.05	0.36	0.39	0.26	0.13	0.05	0.02	0.01			
5.2	−0.11	−0.34	−0.02	0.33	0.40	0.29	0.15	0.07	0.02	0.01			
5.4	−0.04	−0.35	−0.09	0.28	0.40	0.31	0.18	0.08	0.03	0.01			
5.6	0.03	−0.33	−0.15	0.23	0.39	0.33	0.20	0.09	0.04	0.01	0		
5.8	0.09	−0.31	−0.20	0.17	0.38	0.35	0.22	0.11	0.05	0.02	0.01		
6.0	0.15	−0.28	−0.24	0.11	0.36	0.36	0.25	0.13	0.06	0.02	0.01		
6.2	0.20	−0.23	−0.28	0.05	0.33	0.37	0.27	0.15	0.07	0.03	0.01		
6.4	0.24	−0.18	−0.30	−0.01	0.29	0.37	0.29	0.17	0.08	0.03	0.01		
6.6	0.27	−0.12	−0.31	−0.06	0.25	0.37	0.31	0.19	0.10	0.04	0.01	0	
6.8	0.29	−0.07	−0.31	−0.12	0.21	0.36	0.33	0.21	0.11	0.05	0.02	0.01	
7.0	0.30	0	−0.30	−0.17	0.16	0.35	0.34	0.23	0.13	0.06	0.02	0.01	
7.2	0.30	0.05	−0.28	−0.21	0.11	0.33	0.35	0.25	0.15	0.07	0.03	0.01	
7.4	0.28	0.11	−0.25	−0.24	0.05	0.30	0.35	0.27	0.16	0.08	0.04	0.01	0
7.6	0.25	0.16	−0.21	−0.27	0	0.27	0.35	0.29	0.18	0.10	0.04	0.02	0.01
7.8	0.22	0.20	−0.16	−0.29	−0.06	0.23	0.35	0.31	0.20	0.11	0.05	0.02	0.01

8.0	0.17	0.23	−0.11	−0.29	−0.11	0.19	0.34	0.32	0.22	0.13	0.06	0.03	0.01
8.2	0.12	0.26	−0.06	−0.29	−0.15	0.14	0.32	0.33	0.24	0.14	0.07	0.03	0.01
8.4	0.07	0.27	0	−0.27	−0.19	0.09	0.30	0.34	0.26	0.16	0.08	0.04	0.02
8.6	0.01	0.27	0.05	−0.25	−0.22	0.04	0.27	0.34	0.28	0.18	0.10	0.04	0.02
8.8	−0.04	0.26	0.10	−0.22	−0.25	−0.01	0.24	0.34	0.29	0.20	0.11	0.05	0.02
9.0	−0.09	0.25	0.14	−0.18	−0.27	−0.06	0.20	0.33	0.31	0.21	0.12	0.06	0.03
9.2	−0.14	0.22	0.18	−0.14	−0.27	−0.10	0.16	0.31	0.31	0.23	0.14	0.07	0.03
9.4	−0.18	0.18	0.22	−0.09	−0.27	−0.14	0.12	0.30	0.32	0.25	0.16	0.08	0.04
9.6	−0.21	0.14	0.24	−0.04	−0.26	−0.18	0.08	0.27	0.32	0.27	0.17	0.10	0.05
9.8	−0.23	0.09	0.25	0.01	−0.25	−0.21	0.03	0.25	0.32	0.28	0.19	0.11	0.05
10.0	−0.25	0.04	0.25	0.06	−0.22	−0.23	−0.01	0.22	0.32	0.29	0.21	0.12	0.06
10.2	−0.25	−0.01	0.25	0.10	−0.19	−0.25	−0.06	0.18	0.31	0.30	0.22	0.14	0.07
10.4	−0.24	−0.06	0.23	0.14	−0.15	−0.26	−0.10	0.14	0.29	0.31	0.24	0.15	0.08
10.6	−0.23	−0.10	0.21	0.18	−0.11	−0.26	−0.14	0.10	0.28	0.31	0.26	0.17	0.10
10.8	−0.20	−0.14	0.18	0.21	−0.06	−0.25	−0.17	0.06	0.25	0.31	0.27	0.19	0.11
11.0	−0.17	−0.18	0.14	0.23	−0.02	−0.24	−0.20	0.02	0.22	0.31	0.28	0.20	0.12
11.2	−0.13	−0.20	0.10	0.24	0.30	−0.22	−0.22	−0.02	0.19	0.30	0.29	0.22	0.14
11.4	−0.09	−0.22	0.05	0.24	0.08	−0.19	−0.24	−0.06	0.16	0.29	0.30	0.23	0.15
11.6	−0.04	−0.23	0	0.23	0.12	−0.15	−0.25	−0.10	0.12	0.27	0.30	0.25	0.17
11.8	0	−0.23	−0.04	0.22	0.15	−0.12	−0.25	−0.14	0.08	0.25	0.30	0.26	0.18
12.0	0.05	−0.22	−0.08	0.20	0.18	−0.07	−0.24	−0.17	0.05	0.23	0.30	0.27	0.20
12.2	0.09	−0.21	−0.12	0.17	0.21	−0.03	−0.23	−0.20	0.01	0.20	0.29	0.28	0.21
12.4	0.13	−0.18	−0.16	0.13	0.22	0.01	−0.21	−0.22	−0.03	0.17	0.29	0.29	0.22
12.6	0.16	−0.15	−0.19	0.09	0.23	0.06	−0.18	−0.23	−0.07	0.14	0.27	0.29	0.24
12.8	0.19	−0.11	−0.21	0.05	0.23	0.10	−0.15	−0.24	−0.11	0.10	0.25	0.29	0.25

or

$$G(f) = A_c \sum_{n=-\infty}^{n=\infty} J_n(\beta)\, \delta(f - nf_m) \tag{4-161b}$$

Using this result in (4-151), we obtain the spectrum of the angle modulated signal. The magnitude spectrum for $f > 0$ is shown in Fig. 4-42 for the cases of $\beta = 0.2$, 1.0, 2.0, 5.0, and 8.0. It is noted that the discrete carrier term (at $f = f_c$) is proportional to $|J_0(\beta)|$; consequently, the level (magnitude) of the discrete carrier depends on the modulation index. It will be zero if $J_0(\beta) = 0$, which occurs if $\beta = 2.40$, 5.52, and so on, as shown in Table 4-10.

Figure 4-42 also shows that the bandwidth of the angle modulated signal depends on β and f_m. In fact, it can be shown that 98% of the total power is contained in the bandwidth

$$B_T = 2(\beta + 1)B \tag{4-162}$$

where β is either the phase modulating index or the frequency modulation index and B is the bandwidth of the modulating signal (which is f_m for sinusoidal modulation).† This formula gives a rule-of-thumb expression for evaluating the transmission bandwidth of PM and FM signals; it is called *Carson's rule*. B_T is shown in Fig. 4-42 for various values of β. Carson's rule, (4-162), is very important because it gives an easy way to compute the bandwidth of angle-modulated signals. Computation of the bandwidth using other definitions, such as 3-dB bandwidth, can be very difficult since the spectrum of the FM or PM signal must first be evaluated. This is a nontrivial task except for simple cases such as single-tone (sinusoidal) modulation or unless a digital computer is used to compute the approximate spectrum.

Since the exact spectrum of angle modulated signals is difficult to evaluate in general, formulas for the approximation of the spectra are extremely useful. Some relatively simple approximations may be obtained when the peak phase deviation is small and when the modulation index is large. These topics are discussed in the following sections on narrowband angle modulation and wide-band FM.

★ Narrowband Angle Modulation

When $\theta(t)$ is restricted to a small value, say $|\theta(t)| < 0.2$ rad, the complex envelope $g(t) = A_c e^{j\theta}$ may be approximated by a Taylor's series where only the first two terms are used. Thus, since $e^x \approx 1 + x$ for $|x| \ll 1$,

$$g(t) \approx A_c[1 + j\theta(t)] \tag{4-163}$$

†For the case of FM (not PM) with $2 < B < 10$, Carson's rule, (4-162), actually underestimates B_T somewhat. In this case a better approximation is $B_T = 2(\beta + 2)B$. Also, if the modulation signal contains discontinuities, such as square-wave modulation, both formulas may not be too accurate and the B_T should be evaluated by examining the spectrum of the angle-modulated signal. However, to avoid confusion in computing B_T, we will assume that (4-162) is approximately correct for all cases.

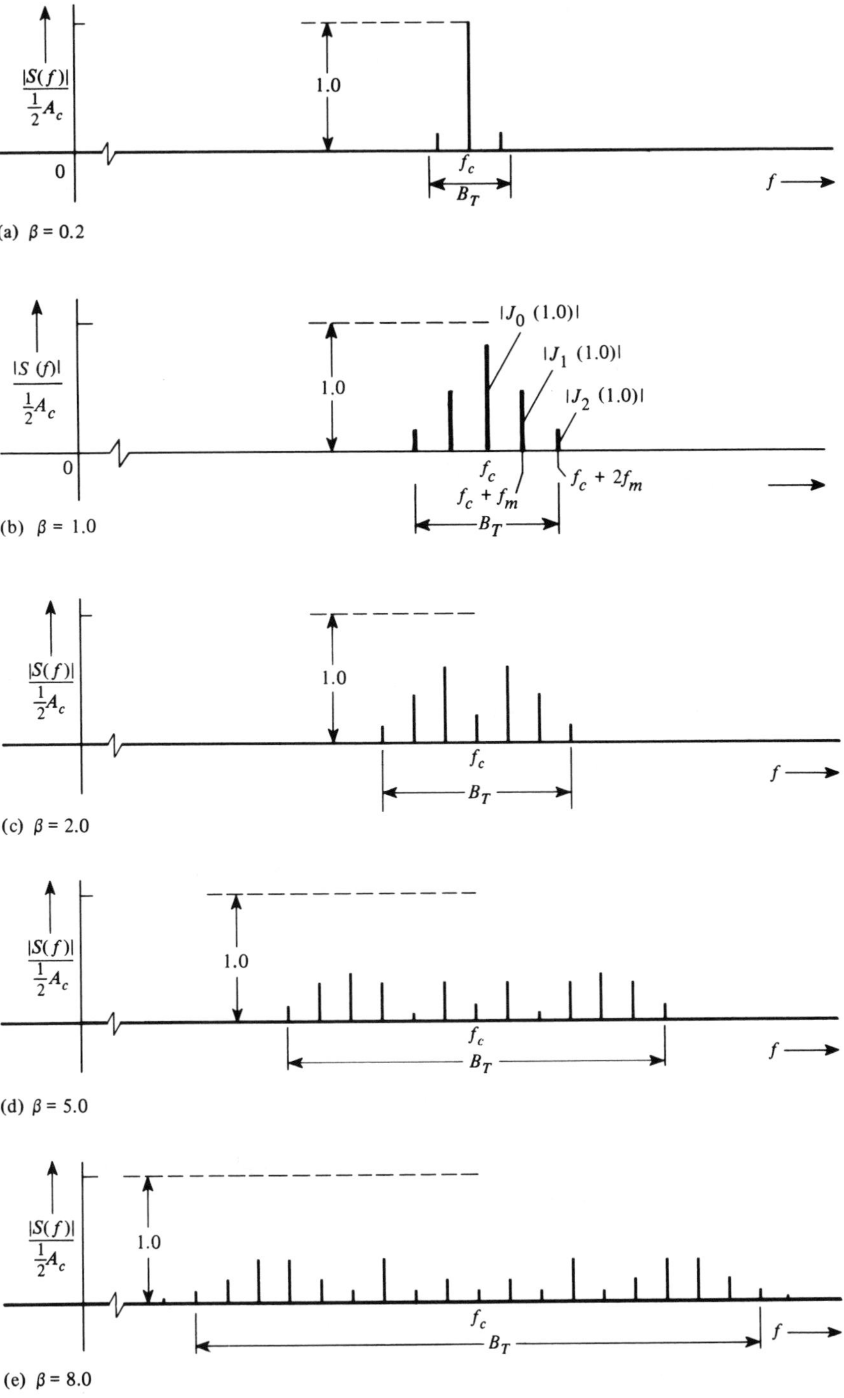

FIGURE 4-42 Magnitude spectra for FM or PM with sinusoidal modulation for various modulation indexes.

TABLE 4-10 Zeros of Bessel Functions: Values for β When $J_n(\beta) = 0$

	Order of Bessel Function, n						
	0	**1**	**2**	**3**	**4**	**5**	**6**
β for 1st zero	2.40	3.83	5.14	6.38	7.59	8.77	9.93
β for 2nd zero	5.52	7.02	8.42	9.76	11.06	12.34	13.59
β for 3rd zero	8.65	10.17	11.62	13.02	14.37	15.70	17.00
β for 4th zero	11.79	13.32	14.80	16.22	17.62	18.98	20.32
β for 5th zero	14.93	16.47	17.96	19.41	20.83	22.21	23.59
β for 6th zero	18.07	19.61	21.12	22.58	24.02	25.43	26.82
β for 7th zero	21.21	22.76	24.27	25.75	27.20	28.63	30.03
β for 8th zero	24.35	25.90	27.42	28.91	30.37	31.81	33.23

Using this approximation in (4-19), we obtain the expression for a *narrowband angle-modulated signal*

$$s(t) = \underbrace{A_c \cos \omega_c t}_{\substack{\text{discrete} \\ \text{carrier term}}} - \underbrace{A_c \theta(t) \sin \omega_c t}_{\text{sideband term}} \tag{4-164}$$

This result indicates that a narrowband angle modulated signal consists of two terms: a discrete carrier component (which does not change with modulating signal) and a sideband term. This is similar to AM-type signaling *except* that the sideband term is 90° out of phase with the discrete carrier term. The narrowband signal can be generated by using a balanced modulator (multiplier), as shown in Fig. 4-43a for the case of narrowband frequency modulation (NBFM). Furthermore, wideband frequency modulation (WBFM) may be generated from the

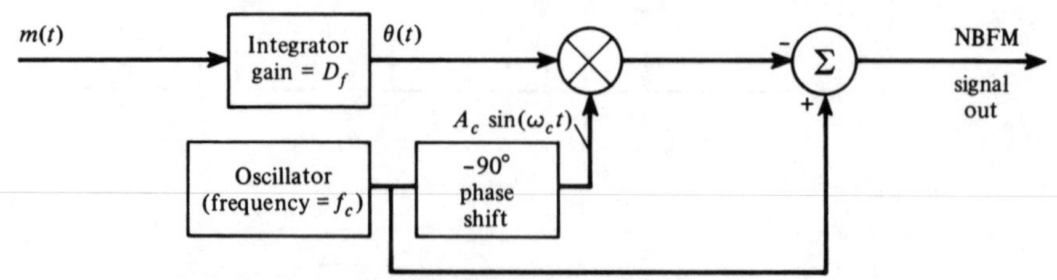

(a) Generation of NBFM Using a Balanced Modulator

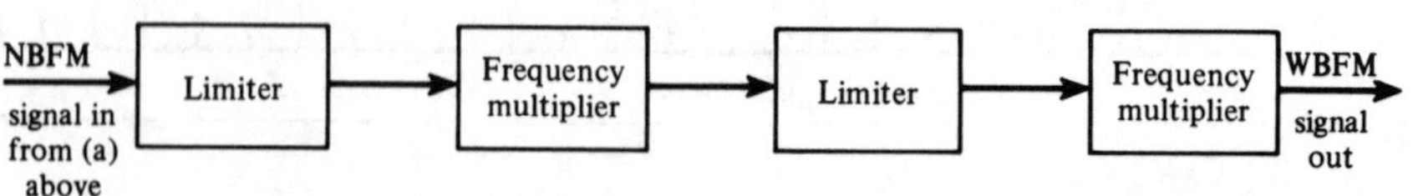

(b) Generation of WBFM From a NBFM Signal

FIGURE 4-43 Indirect method of generating WBFM (Armstrong method).

NBFM signal by using frequency multiplication, as shown in Fig. 4-43b. Limiter circuits are needed to suppress the incidental AM [which is $\sqrt{1+\theta^2(t)}$ as the result of the approximation (4-163)] that is present in the NBFM signal. This method of generating WBFM is called the *Armstrong method* or *indirect method*.

From (4-163) and (4-151), the spectrum of the narrowband angle-modulated signal is

$$S(f) = \frac{A_c}{2}\{[\delta(f-f_c) + \delta(f+f_c)] + j[\Theta(f-f_c) - \Theta(f+f_c)]\} \quad (4\text{-}165)$$

where

$$\Theta(f) = \mathscr{F}[\theta(t)] = \begin{cases} D_p M(f), & \text{for PM signaling} \\ \dfrac{D_f}{j2\pi f} M(f), & \text{for FM signaling} \end{cases} \quad (4\text{-}166)$$

★ Wideband Frequency Modulation

A *direct method* of generating wideband frequency modulation (WBFM) is to use a voltage-controlled oscillator (VCO), as illustrated in Fig. 4-39b. However, for VCOs that are designed for wide frequency deviation (ΔF large), the stability of the carrier frequency $f_c = f_0$ is not very good, so the VCO is incorporated into a PLL arrangement where the PLL is locked to a stable frequency source, such as a crystal oscillator (see Fig. 4-44). The frequency divider is needed to lower the modulation index of the WBFM signal to produce an NBFM signal ($\beta \approx 0.2$) so that a large discrete carrier term will always be present at frequency f_c/N to beat with the crystal oscillator signal and produce the dc control voltage [see Fig. 4-42a, (4-164), and (4-165)]. This dc control voltage holds the VCO on the assigned frequency with a tolerance determined by the crystal oscillator circuit.

The PSD of a WBFM signal may be *approximated* by using the probability density function (PDF) of the modulating signal. This is reasonable from an intuitive viewpoint, since the instantaneous frequency varies directly with the modulating signal voltage for the case of FM [$D_f/(2\pi)$ being the proportionality constant]. If the modulating signal spends more time at one voltage value than

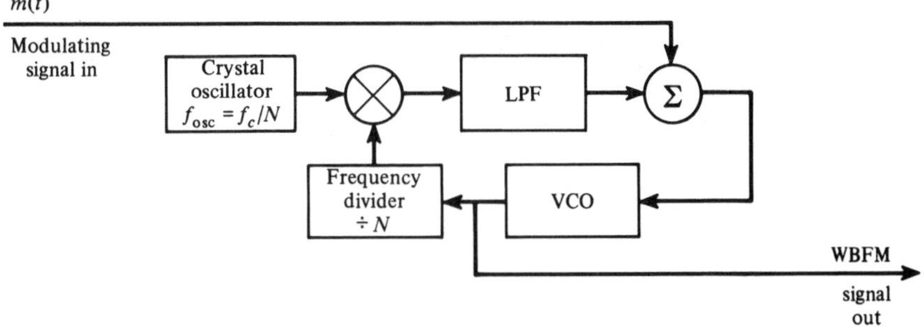

FIGURE 4-44 Direct method of generating WBFM.

another, the instantaneous frequency will dwell at the corresponding frequency, and the power spectrum will have a peak at this frequency. A discussion of the approximation involved—called the *quasi-static approximation*—is well documented [Rowe, 1965]. This result is stated in the following theorem.

THEOREM: *For WBFM signaling, where*

$$s(t) = A_c \cos\left[\omega_c t + D_f \int_{-\infty}^{t} m(\sigma)\, d\sigma\right]$$

$$\beta_f = \frac{D_f \max[m(t)]}{2\pi B} > 1$$

and B is the bandwidth of m(t), the normalized PSD of the WBFM signal is approximated by

$$\mathcal{P}(f) = \frac{\pi A_c^2}{2D_f}\left[f_m\left(\frac{2\pi}{D_f}(f - f_c)\right) + f_m\left(\frac{2\pi}{D_f}(-f - f_c)\right)\right] \qquad (4\text{-}167)$$

where $f_m(\cdot)$ is the PDF of the modulating signal.†

This theorem is proved in Prob. 6-44.

EXAMPLE 4-5 SPECTRUM FOR WBFM WITH TRIANGULAR MODULATION

The spectrum for a WBFM signal with a triangular modulating signal (Fig. 4-45a) will be evaluated. The associated PDF for triangular modulation is shown in Fig. 4-45b. The PDF is described by

$$f_m(m) = \begin{cases} \dfrac{1}{2V_p}, & |m| < V_p \\ 0, & m \text{ otherwise} \end{cases} \qquad (4\text{-}168)$$

where V_p is the peak voltage of the triangular waveform. Substituting this equation into (4-167), we obtain

$$\mathcal{P}(f) = \frac{\pi A_c^2}{2D_f}\left[\begin{Bmatrix} \dfrac{1}{2V_p}, & \left|\dfrac{2\pi}{D_f}(f - f_c)\right| < V_p \\ 0, & f \text{ otherwise} \end{Bmatrix} + \begin{Bmatrix} \dfrac{1}{2V_p}, & \left|\dfrac{2\pi}{D_f}(f + f_c)\right| < V_p \\ 0, & f \text{ otherwise} \end{Bmatrix}\right]$$

† See Appendix B for the definition of probability density function (PDF) and examples of PDFs for various waveforms. This topic may be deleted if the reader is not sufficiently familiar with PDFs. Do not confuse the PDF of the modulation, $f_m(\cdot)$, with the frequency variable, f.

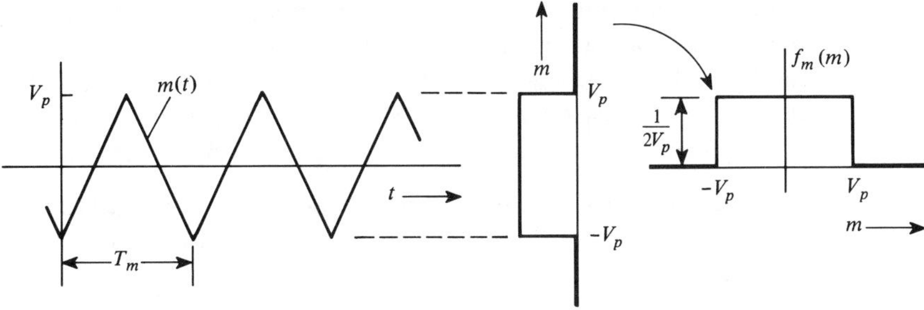

(a) Triangle Modulating Waveform

(b) PDF of Triangle Modulation

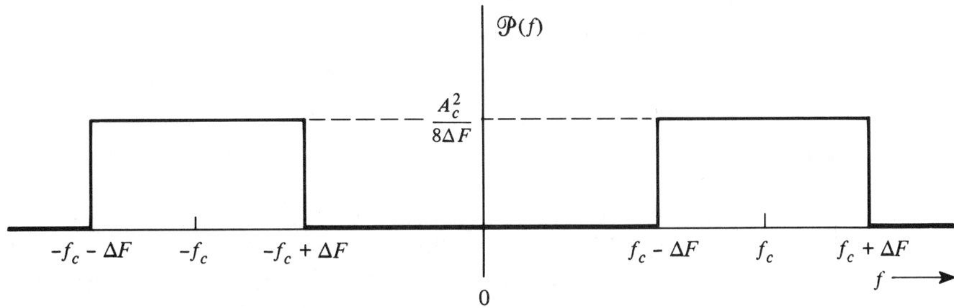

(c) PSD of WBFM Signal with Triangle Modulation, $\Delta F = D_f V_p / 2\pi$

FIGURE 4-45 Approximate spectrum of a WBFM signal with triangle modulation.

The PSD of the WBFM signal becomes

$$\mathcal{P}(f) = \begin{Bmatrix} \dfrac{A_c^2}{8\Delta F}, & (f_c - \Delta F) < f < (f_c + \Delta F) \\ 0, & f \text{ otherwise} \end{Bmatrix} + \begin{Bmatrix} \dfrac{A_c^2}{8\Delta F}, & (-f_c - \Delta F) < f < (-f_c + \Delta F) \\ 0, & f \text{ otherwise} \end{Bmatrix} \tag{4-169}$$

where the peak frequency deviation is

$$\Delta F = \frac{D_f V_p}{2\pi} \tag{4-170}$$

This PSD is plotted in Fig. 4-45c. It is recognized that this result is an *approximation* to the actual PSD. In this example the modulation is periodic with period T_m so that the actual spectrum is a line spectrum with the delta functions spaced $f_m = 1/T_m$ hertz apart from each other (like the spacing shown in Fig. 4-42 for the case of sinusoidal modulation). This approximation gives us the *approximate envelope* of the line spectrum. (If the modulation had been nonperiodic, the exact spectrum would be continuous and thus would not be a line spectrum.)

In the other example using sinusoidal (periodic) modulation, the exact PSD is known to contain line components with weights $[A_c J_n(\beta)]^2/2$ located at frequencies $f = f_c + nf_m$ (see Fig. 4-42). For the case of high index the envelope of these weights approximates the PDF of a sinusoid as given in Appendix B.

In summary, some important properties of angle modulated signals are

- It is a nonlinear function of the modulation, and, consequently, the bandwidth of the signal increases as the modulation index increases.
- The discrete carrier level changes, depending on the modulating signal, and is zero for certain types of modulating waveforms.
- The bandwidth of a narrowband angle modulated signal is twice the modulating signal bandwidth (same as that for AM signaling).
- The real envelope of an angle-modulated signal, $R(t) = A_c$, is constant and, consequently, does not depend on the modulating signal level.

Preemphasis and Deemphasis in Angle-Modulated Systems

In angle-modulated systems the signal-to-noise ratio at the output of the receiver can be improved if the level of the modulation (at the transmitter) is boosted at the top end of the (e.g., audio) spectrum—this is called *preemphasis*—and attenuated at high frequencies on the receiver output—this is called *deemphasis*. This gives an overall baseband frequency response that is flat while improving the signal-to-noise ratio at the receiver output (see Fig. 4-46). In the preemphasis characteristic the second corner frequency f_2 occurs much above the baseband spectrum of the modulating signal (say, 25 kHz for audio modulation). In FM broadcasting the time constant τ_1 is usually 75 μsec so that f_1 occurs at 2.12 kHz. The resulting overall system frequency response using preemphasis at the transmitter and deemphasis at the receiver is flat over the band of the modulating signal. It is interesting to note that in FM broadcasting with 75-μsec preemphasis, the signal transmitted is an FM signal for modulating frequencies up to 2.1 kHz but a *phase-modulated* signal for audio frequencies above 2.1 kHz since the preemphasis network acts as a differentiator for frequencies between f_1 and f_2. Hence, *preemphasized FM is actually a combination of FM and PM* and combines the advantages of both with respect to noise performance. In Chapter 7 it will be demonstrated that preemphasis–deemphasis improves the signal-to-noise ratio at the receiver output.

Frequency-Division Multiplexing and FM Stereo

Frequency-division multiplexing (FDM) is a technique for transmitting multiple messages simultaneously over a wideband channel by first modulating the message signals onto several subcarriers and forming a composite baseband signal that consists of the sum of these modulated subcarriers. This composite signal may then be modulated onto the main carrier, as shown in Fig. 4-47. Any type of modulation, such as AM, DSB, SSB, PM, FM, and so on, could be used. The modulation types used on the subcarriers, as well as the type used on the main

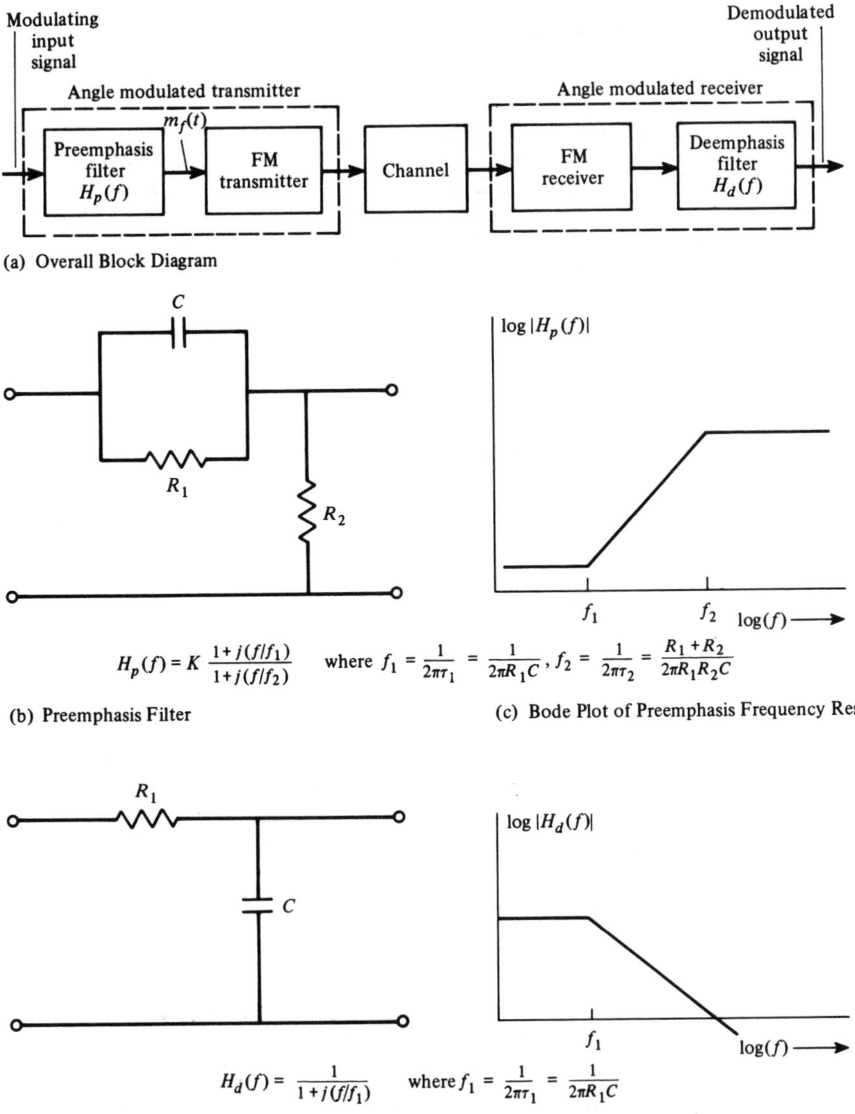

(a) Overall Block Diagram

$$H_p(f) = K\,\frac{1 + j(f/f_1)}{1 + j(f/f_2)} \quad \text{where } f_1 = \frac{1}{2\pi\tau_1} = \frac{1}{2\pi R_1 C},\; f_2 = \frac{1}{2\pi\tau_2} = \frac{R_1 + R_2}{2\pi R_1 R_2 C}$$

(b) Preemphasis Filter

(c) Bode Plot of Preemphasis Frequency Response

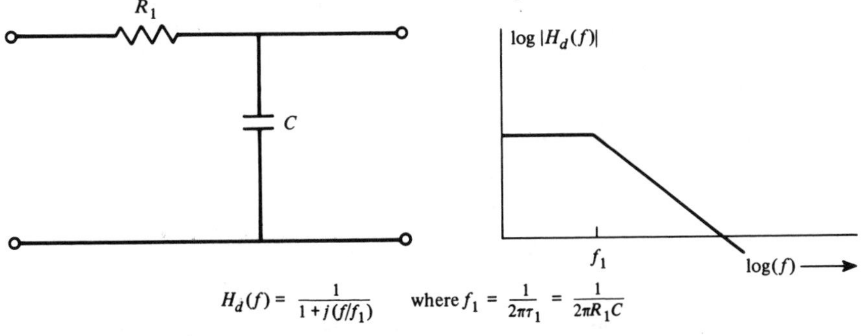

$$H_d(f) = \frac{1}{1 + j(f/f_1)} \quad \text{where } f_1 = \frac{1}{2\pi\tau_1} = \frac{1}{2\pi R_1 C}$$

(d) Deemphasis Filter

(e) Bode Plot of Deemphasis Characteristic

FIGURE 4-46 Angle-modulated system with preemphasis and deemphasis.

carrier, may be different. However, as shown in Fig. 4-47b, the composite signal spectrum must consist of modulated signals that do not have overlapping spectra; otherwise, crosstalk will occur between the message signals at the receiver output. The composite baseband signal then modulates a main transmitter to produce the FDM signal that is transmitted over the wideband channel.

The received FDM signal is first demodulated to reproduce the composite baseband signal that is passed through filters to separate the individual modulated

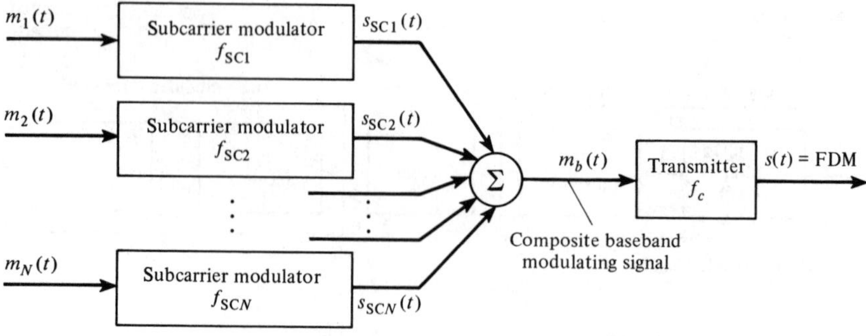

(a) Transmitter

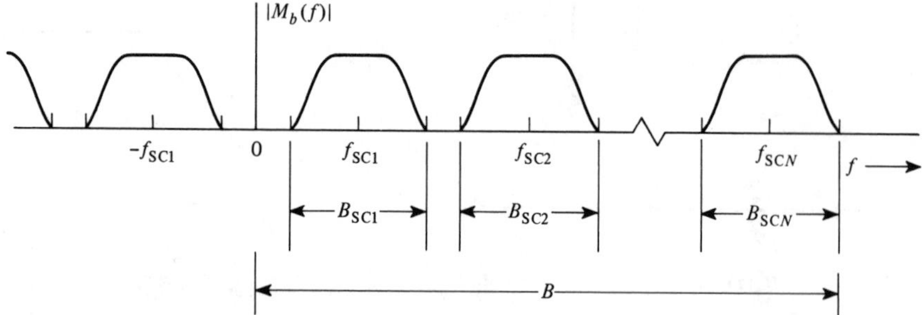

(b) Spectrum of Composite Baseband Signal

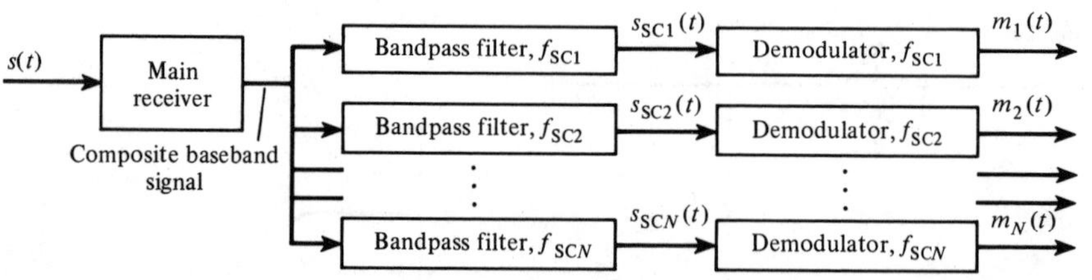

(c) Receiver

FIGURE 4-47 FDM system.

subcarriers. The subcarriers are then demodulated to reproduce the message signals $m_1(t)$, $m_2(t)$, and so on.†

The FM stereo broadcasting system that has been adopted in the United States is an example of an FDM system. Furthermore, it is compatible with the monaural FM system that has existed since the 1940s. That is, a listener with a conventional monaural FM receiver will hear the monaural audio (which consists of the left- plus the right-channel audio), while the listener with a stereo receiver will receive the left-channel audio on the left speaker and the right-channel audio on the right speaker (see Fig. 4-48). To obtain the compatibility feature, the left- and right-channel audios are combined (summed) to produce the monaural signal,

†Figure 5-19 shows a typical FDM hierarchy used by the telephone industry.

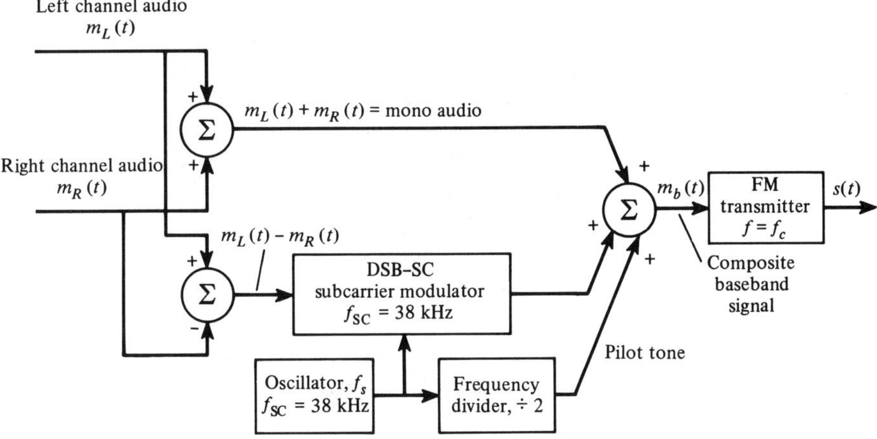

(a) FM Stereo Transmitter

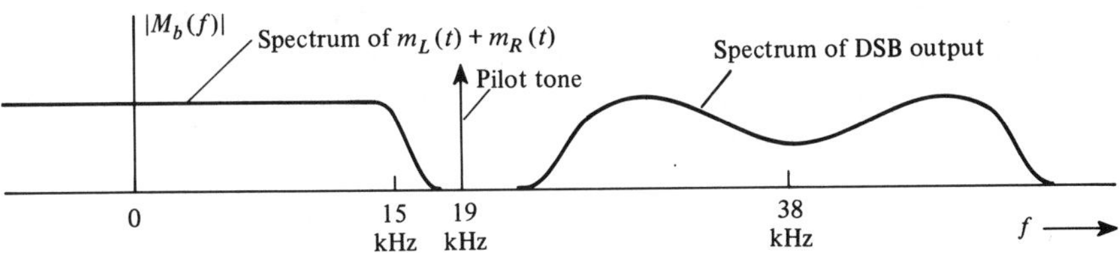

(b) Spectrum of Composite Baseband Signal

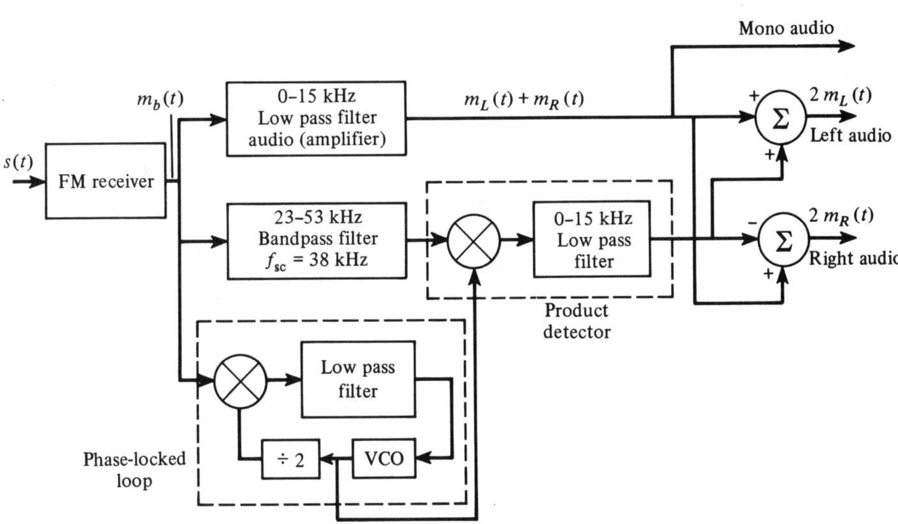

(c) FM Stereo Receiver

FIGURE 4-48 FM stereo system.

and the difference audio is used to modulate a 38-kHz DSB-SC signal. A 19-kHz pilot tone is added to the composite baseband signal $m_b(t)$ to provide a reference signal for coherent (product) subcarrier demodulation in the receiver. As seen from Fig. 4-48c, this system is compatible with existing FM monaural receivers. In Prob. 4-85 it will be found that a relatively simple switching (sampling) technique may be used to implement the demodulation of the subcarrier and the separation of the left and right signals in one operation.

The FM station may also be given *subsidiary communications authorization* (SCA) by the FCC. This allows the station to add a FM subcarrier to permit the transmission of private background music to business subscribers for use in their store or office. The SCA FM subcarrier frequency is usually 67 kHz, although this frequency is not specified by FCC rules. Moreover, up to four SCA subcarriers are permitted by the FCC, and each may carry either data or audio material for private subscribers.

FM Broadcast Technical Standards

Table 4-11 gives some of the FCC technical standards that have been adopted for FM systems. In the United States, FM stations are classified into one of three major categories that depend on their intended coverage area. Class A stations are local stations. They have a maximum effective radiated power (ERP) of 6 kW and a maximum antenna height of 300 ft above average terrain. The ERP is the average transmitter output power multiplied by the power gains of both the transmission line (a number less than 1) and the antenna (see Sec. 5-9 for some TV ERP calculations). Class B stations have a maximum ERP of 50 kW, with a maximum antenna height of 500 ft above average terrain. Class B stations are assigned to the northeastern part of the United States, southern California, Puerto Rico, and the Virgin Islands. Class C stations are assigned to the remainder of the United States. They have a maximum ERP of 100 kW and a maximum antenna height of 2000 ft above average terrain. As shown in Table 4-11, FM stations are further classified as commercial or noncommercial. Noncommercial stations operate in the 88.1 to 91.9 MHz segment of the FM band and provide educational programs with no commercials. Class D stations are limited to 10-W transmitter output and are for noncommercial service only. In the commercial segment of the FM band, 92.1 to 107.9 MHz, certain frequencies are reserved for Class A stations and the remainder are for Class B or Class C station assignment. A listing of these frequencies and the specific station assignments for each city is available [Broadcasting, 1991].

★ Dolby and DBX Noise Reduction Systems

As described previously, preemphasis and deemphasis are used in FM broadcasting for noise reduction. They are also used in audiotape recording for noise reduction. Other, more sophisticated noise reduction techniques have been developed that, in addition, use companding (see Sec. 3-3). This section will describe some of these techniques.

In audio cassette and reel-to-reel tape recording, noise occurs predominantly

TABLE 4-11 FCC FM Standards

Class of Service	Item	FCC Standard
FM broadcasting	Assigned frequency, f_c	In 200-kHz increments from 88.1 MHz (FM Channel 201) to 107.9 MHz (FM Channel 300)
	Channel bandwidth	200 kHz
	Noncommercial stations	88.1 MHz (Channel 201) to 91.9 MHz (Channel 220)
	Commercial stations	92.1 MHz (Channel 221) to 107.9 MHz (Channel 300)
	Carrier frequency stability	$\pm$2000 Hz of the assigned frequency
	100% modulation[a]	$\Delta F = 75$ kHz
	Audio frequency response[b]	50 Hz to 15 kHz following a 75-μsec preemphasis curve
	Modulation index	5 (for $\Delta F = 75$ kHz and $B = 15$ kHz)
	% harmonic distortion[b]	<3.5% (50–100 Hz) <2.5% (100–7500 Hz) <3.0% (7500–15,000 Hz)
	FM noise	At least 60 dB below 100% modulation at 400Hz
	AM noise	50 dB below the level corresponding to 100% AM in a band 50 Hz–15 kHz
	Maximum power licensed	100 kW in horizontal polarized plane plus 100 kW in vertical polarized plane
Two-way FM mobile radio	100% modulation	$\Delta F = 5$ kHz
	Modulation index	1 (for $\Delta F = 5$ kHz and $B = 5$ kHz)
	Carrier frequencies are within the frequency bands:	32–50 MHz (low VHF band) 144–148 MHz (2-m amateur band) 148–174 MHz (high VHF band)[c] 420–450 MHz ($\frac{3}{4}$-m amateur band) 450–470 MHz (UHF band) 470–512 MHz (UHF, T band) 806–928 MHz (900-MHz band)
TV aural (FM) signal	100% modulation	$\Delta F = 2$
	Modulation index	1.67 (fc

[a] For stereo transmission the 19-kHz pilot tone may contribute as much deviation. If SCA is used, each SCA subcarrier may also contribute up be 110% of 75 kHz provided that SCA subcarriers are present.

[b] Under the new FCC deregulation policy these requirements are deleted ers still use them as guidelines for minimum acceptable performance

[c] Amplitude-compressed SSB is also permitted in the 150- to 170-M

at the higher frequencies and is commonly called *hiss*. These noise reduction techniques use a combination of preemphasis filtering and dynamic range compression at the encoder before recording. The playback decoder circuit uses a combination of deemphasis and dynamic range expansion. If done properly, a high-quality audio signal can be recovered with low noise. However, if the encoder and decoders are not "matched," a "breathing" or "pumping" effect can result that can be more disturbing than the hiss on an unprocessed signal. Another disadvantage of these techniques is that if a tape with a processed signal is played back on a regular amplifier system without decoding, the reproduction is poor, with excessive high-frequency response.

Dolby Laboratories has developed a number of noise reduction techniques designated as Dolby-A, Dolby-B, and Dolby-C. *Dolby-A* is designed for commercial use and, consequently, is a relatively expensive system. The Dolby-A processor divides the audio spectrum into four frequency bands: low-pass, below 80 Hz; bandpass, 80 Hz to 3 kHz; high-pass, above 3 kHz; and high-pass, above 9 kHz. At the encoder, the gain in each of these bands is adaptively increased as the signal level in the associated band becomes lower. This provides an overall compression characteristic for the encoder. A maximum boost of 10 to 15 dB is used in each of the bands if the signal level is 45 dB below the maximum recording level (0 dB). For playback the decoder provides the appropriate expansion of the signal in each of the bands so that the overall frequency response is flat and the dynamic range of the original audio signal is preserved. If the recording and playback levels are properly set, the Dolby-A system provides a 10- to 15-dB increase in the signal-to-noise ratio.

The *Dolby-B* and *Dolby-C* systems are designed for consumer audio cassette and reel-to-reel tape recording. They are relatively inexpensive to implement and use a two-band instead of a four-band system, as in Dolby-A. As shown in Fig. 4-49, the Dolby-B encoder preemphasizes the higher frequencies and provides up to a 10-dB boost when the level in the high-frequency band is 45 dB below the reference level. The Dolby-C encoder provides as much as a 20-dB boost for the high-frequency components when their level is low. The Dolby-C encoder uses two cascaded stages of compression, one activated at high signal levels and another at low signal levels. Overall, the Dolby-B system reduces the tape hiss by about 10 dB for frequencies above 4 kHz, and the Dolby-C system reduces the hiss by about 20 dB for frequencies above 1 kHz. In FM broadcasting the Dolby-B system with a 6.4-kHz high-pass filter (25 μsec) is often used for additional signal-to-noise ratio improvement over the standard FM preemphasis–deemphasis system.

DBX, Inc. of Waltham, Massachusetts, has developed a noise reduction system that processes the whole audio signal in a single audio band. The *DBX system* uses voltage-controlled amplifiers with 130 dB of dynamic range at both the encoder and the decoder to provide the compression and expansion characteristics. The encoder also uses a 1.6-kHz preemphasis network that provides a 0-dB boost at the high end. The decoder has a matching deemphasis network. DBX system provides a 20- to 30-dB signal-to-noise ratio improvement and dle large signal overloads (above the 0-dB level) without distortion. This

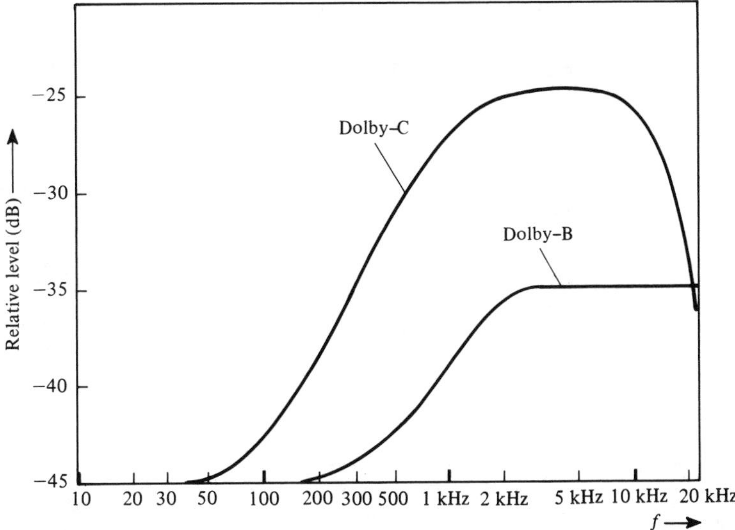

FIGURE 4-49 Low-level frequency response for Dolby-B and Dolby-C encoders.

system has the disadvantage, however, of single band processing. That is, a large-level component located anywhere within the frequency band will control the compression level of the whole signal.

4-8 Summary

The basic techniques used for bandpass signaling have been studied in this chapter. The complex envelope technique for representing bandpass signals and filters was found to be very useful. A description of communication circuits with performance analysis was presented for filters, amplifiers, limiters, mixers, frequency multipliers, phase-locked loops, and detector circuits. Nonlinear as well as linear circuit analysis techniques were used. The superheterodyne receiving circuit was found to be the fundamental technique used in communication receiver design, with the detector circuit determining whether the receiver is designed to detect digital or analog signals of a particular type. Generalized transmitters and receivers were developed, and practical aspects of their design, such as spurious signals, were evaluated.

AM, PM, and FM signaling techniques were developed in detail. Single-sideband and vestigial sideband systems were also studied. The spectra of these modulated RF signals were evaluated, and the transmission bandwidth was obtained. For example, in AM-type systems the transmission bandwidth is $B_T = 2B$, where B is the baseband bandwidth; and for FM-type systems, $B_T = 2(\Delta F + B)$, where ΔF is the peak frequency deviation of the carrier. Typical specifications for AM and FM broadcasting signals were also listed.

In the next chapter we will use these tools to analyze a number of digital and analog communication systems.

4-9 Appendix: Data Sheets for MC 2831A (FM Transmitter) and MC 3363 (FM Receiver)[†]

MOTOROLA
SEMICONDUCTOR
TECHNICAL DATA

MC2831A

LOW POWER FM TRANSMITTER SYSTEM

The MC2831A is a one-chip FM transmitter subsystem designed for cordless telephone and FM communication equipment. It includes a Microphone Amplifier, Pilot Tone Oscillator, Voltage Controlled Oscillator and Battery Monitor.

- Wide Range of Operating Supply Voltage (3.0 V–8.0 V)
- Low Drain Current (4.0 mA Typ Full Operation at V_{CC} = 4.0 V)
- Battery Checker (290 μA Typ at V_{CC} = 4.0 V)
- Low Number of External Parts Required

LOW POWER FM TRANSMITTER SYSTEM

SILICON MONOLITHIC INTEGRATED CIRCUIT

FIGURE 1 — FUNCTIONAL BLOCK DIAGRAM

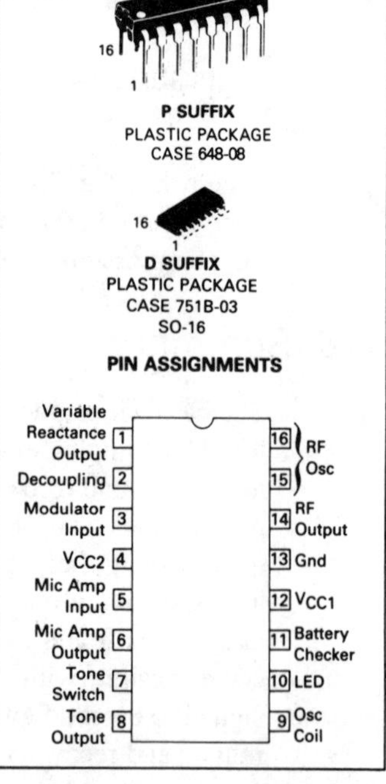

MAXIMUM RATINGS (T_A = 25°C, unless otherwise noted)

Rating	Pin	Symbol	Value	Unit
Power Supply Voltage	4, 12	V_{CC}	10	Vdc
Operating Supply Voltage Range	4, 12	V_{CC}	3.0 to 8.0	Vdc
Battery Checker Output Sink Current	10	I_{LED}	25	mA
Junction Temperature	—	T_J	+150	°C
Operating Ambient Temperature Range	—	T_A	−30 to +75	°C
Storage Temperature Range	—	T_{stg}	−65 to +150	°C

ELECTRICAL CHARACTERISTICS (V_{CC1} = 4.0 Vdc, V_{CC2} = 4.0 Vdc, T_A = 25°C, unless otherwise noted)

Characteristic	Symbol	Pin	Min	Typ	Max	Unit
Drain Current	I_{CC1}	12	150	290	420	μA
Drain Current	I_{CC2}	4	2.2	3.6	6.5	mA
BATTERY CHECKER						
Threshold Voltage (LED Off → On)	V_{TB}	11	1.0	1.2	1.4	Vdc
Output Saturation Voltage (Pin 11 = 0 V, Pin 10 Sink Current = 5.0 mA)	V_{OSAT}	10	—	0.15	0.5	Vdc
MIC AMPLIFIER						
Voltage Gain, Closed Loop (V_{in} = 1.0 mV_{rms}, f_{in} = 1.0 kHz)	—	5, 6	27	30	33	dB
Output dc Voltage	—	6	1.1	1.4	1.7	Vdc
Output Swing (V_{in} = 30 mV_{rms}, f_{in} = 1.0 kHz)	—	6	0.8	1.2	1.6	Vp-p
Total Harmonic Distortion (V_O = 31 mV_{rms}, f_{in} = 1.0 kHz)	THD	6	—	0.7	—	%
PILOT TONE OSCILLATOR (250 Ω LOADING)						
Output AF Voltage (f_o = 5.0 kHz)	—	8	—	50	—	mV_{rms}
Output dc Voltage	—	8	—	1.4	—	Vdc
Total Harmonic Distortion (f_o = 5.0 kHz, V_{AF} = 150 mV_{rms})	—	8	—	1.8	5.0	%
Tone Switch Threshold	—	7	1.1	1.4	1.7	Vdc
FM MODULATOR (120 Ω LOADING)						
Output RF Voltage (f_o = 16.6 MHz)	VRFO	14	—	40	—	mV_{rms}
Output dc Voltage	—	14	—	1.3	—	Vdc
Modulation Sensitivity (Note 1) (V_{in} = 1.0 V ± 0.2 V)	—	3, 14	6.0	10	18	Hz/mVdc
Maximum Deviation (Note 1) (V_{in} = 0 V to +2.0 V)	—	3, 14	±2.5	±5.0	±12.5	kHz
RF Frequency Range	—	14	—	—	60	MHz

Note 1. Modulation sensitivity and maximum deviation are measured at 49.815 MHz, which is the third harmonic of the crystal frequency.

FIGURE 2 — TEST CIRCUIT

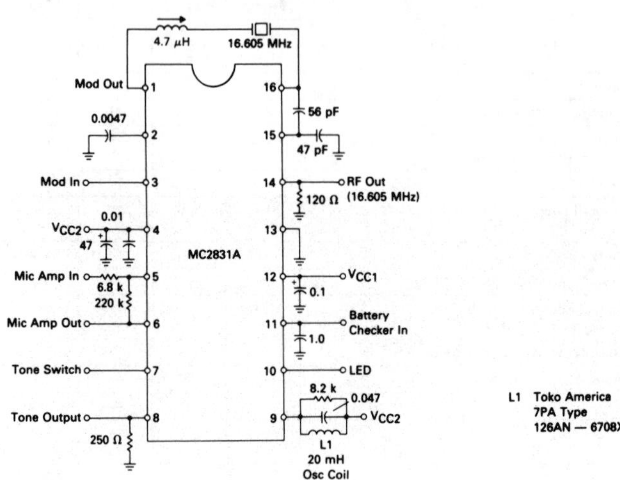

FIGURE 3 — SINGLE CHIP FM VHF TRANSMITTER AT 49.7 MHz

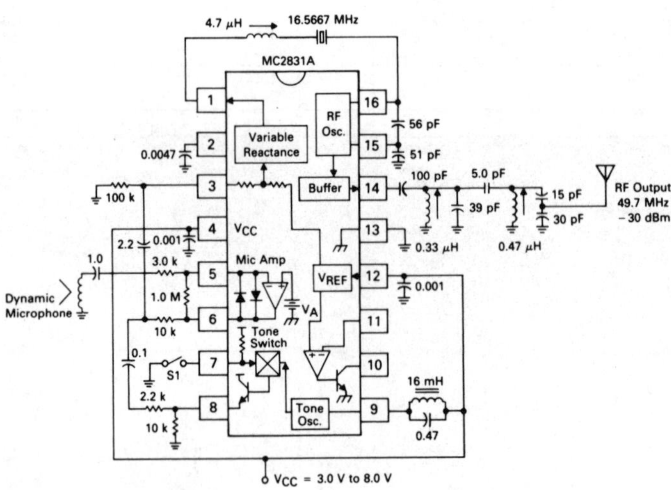

NOTES:

S1 is a normally closed push button type switch.

The crystal used is fundamental mode, calibrated for parallel resonance with a 32 pF load. The 49.7 MHz output is generated in the output buffer, which generates useful harmonics to 60 MHz.

The network on the output at Pin 14 provides output tuning and impedance matching to 50 Ω at 49.7 MHz. Harmonics are suppressed by more than 25 dB.

Battery checker circuit (Pins 10, 11) is not used in this application.

All capacitors in microfarads, inductors in Henries and resistors in Ohms, unless otherwise specified.

MC3363

Advance Information

LOW POWER DUAL CONVERSION FM RECEIVER

The MC3363 is a single chip narrowband VHF FM radio receiver. It is a dual conversion receiver with RF amplifier transistor, oscillators, mixers, quadrature detector, meter drive/carrier detect and mute circuitry. The MC3363 also has a buffered first local oscillator output for use with frequency synthesizers, and a data slicing comparator for FSK detection.

- Wide Input Bandwidth — 200 MHz Using Internal Local Oscillator
  — 450 MHz Using External Local Oscillator
- RF Amplifier Transistor
- Muting Operational Amplifier
- Complete Dual Conversion
- Low Voltage: V_{CC} = 2.0 V to 7.0 V
- Low Drain Current: I_{CC} = 3.6 mA (Typ) at V_{CC} = 3.0 V, Excluding RF Amplifier Transistor
- Excellent Sensitivity: Input 0.3 μV (Typ) for 12 dB SINAD Using Internal RF Amplifier Transistor
- Data Shaping Comparator
- Received Signal Strength Indicator (RSSI) with 60 dB Dynamic Range
- Low Number of External Parts Required
- Manufactured in Motorola's MOSAIC Process Technology
- See AN980 For Additional Design Information

LOW POWER DUAL CONVERSION FM RECEIVER

SILICON MONOLITHIC INTEGRATED CIRCUIT

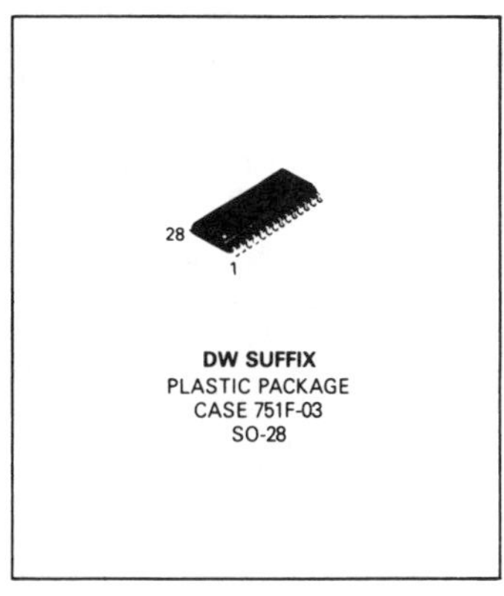

DW SUFFIX
PLASTIC PACKAGE
CASE 751F-03
SO-28

FIGURE 1 — PIN CONNECTIONS AND FUNCTIONAL BLOCK DIAGRAM

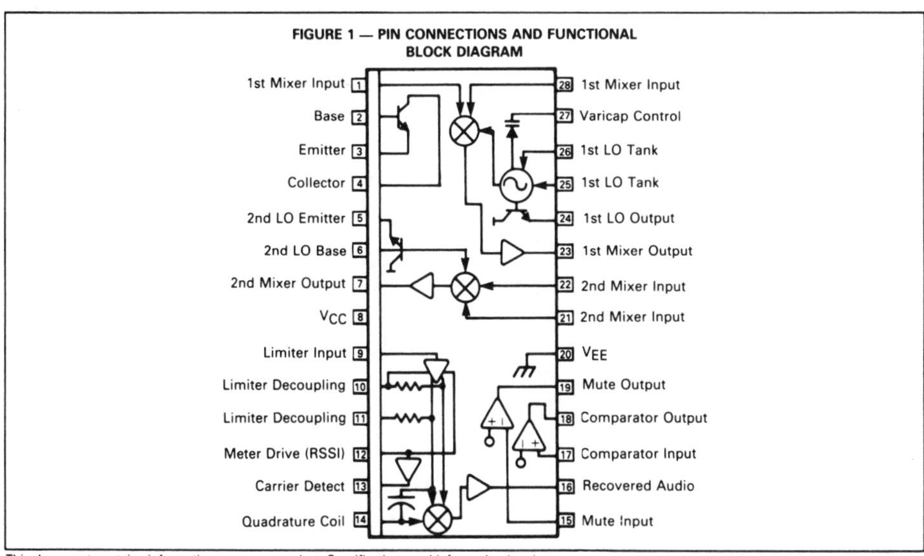

This document contains information on a new product. Specifications and information herein are subject to change without notice.

MAXIMUM RATINGS (T_A = 25°C unless otherwise noted)

Rating	Pin	Symbol	Value	Unit
Power Supply Voltage	8	$V_{CC(max)}$	8.0	Vdc
Operating Supply Voltage Range (Recommended)	8	V_{CC}	2.0 to 7.0	Vdc
Input Voltage (V_{CC} = 5.0 Vdc)	1, 28	V_{1-28}	1.0	Vrms
Mute Output Voltage	19	V_{19}	−0.7 to 8.0	Vpk
Junction Temperature	—	T_J	150	°C
Operating Ambient Temperature Range	—	T_A	−40 to +85	°C
Storage Temperature Range	—	T_{stg}	−65 to +150	°C

ELECTRICAL CHARACTERISTICS (V_{CC} = 5.0 Vdc, f_o = 49.7 MHz, Deviation = ±3.0 kHz, T_A = 25°C, Mod 1.0 kHz, Test Circuit of Figure 2 unless otherwise noted)

Characteristic	Pin	Min	Typ	Max	Unit
Drain Current (Carrier Detect Low)	8	—	4.5	8.0	mA
−3.0 dB Limiting Sensitivity (RF Amplifier Not Used)	—	—	0.7	2.0	μVrms
20 dB S/N Sensitivity (RF Amplifier Not Used)	—	—	1.0	—	μVrms
1st Mixer Input Resistance (Parallel — Rp)	1, 28	—	690	—	Ohm
1st Mixer Input Capacitance (Parallel — Cp)	1, 28	—	7.2	—	pF
1st Mixer Conversion Voltage Gain (A_{vc1}, Open Circuit)	—	—	18	—	dB
2nd Mixer Conversion Voltage Gain (A_{vc2}, Open Circuit)	—	—	21	—	dB
2nd Mixer Input Sensitivity (20 dB S/N) (10.7 MHz i/p)	21	—	10	—	μVrms
Limiter Input Sensitivity (20 dB S/N) (455 kHz i/p)	9	—	100	—	μVrms
RF Transistor DC Current Drain	4	1.0	1.5	2.5	mAdc
Recovered Audio (RF Signal Level = 1.0 mV)	16	120	200	—	mVrms
Noise Output Level (RF Signal = 0 mV)	16	—	70	—	mVrms
THD of Recovered Audio (RF Signal = 1.0 mV)	16	—	2%	—	%
Detector Output Impedance	16	—	400	—	Ohm
Data (Comparator) Output Voltage — High — Low	18 18	— 0.1	— 0.1	V_{CC} —	Vdc Vdc
Data (Comparator) Threshold Voltage Difference	17	70	110	150	mV
Meter Drive Slope	12	70	100	135	nA/dB
Carrier Detect Threshold (Below V_{CC})	12	0.53	0.64	0.77	Vdc
Mute Output Impedance — High — Low	19 19	— —	10 25	— —	Mohm Ohm

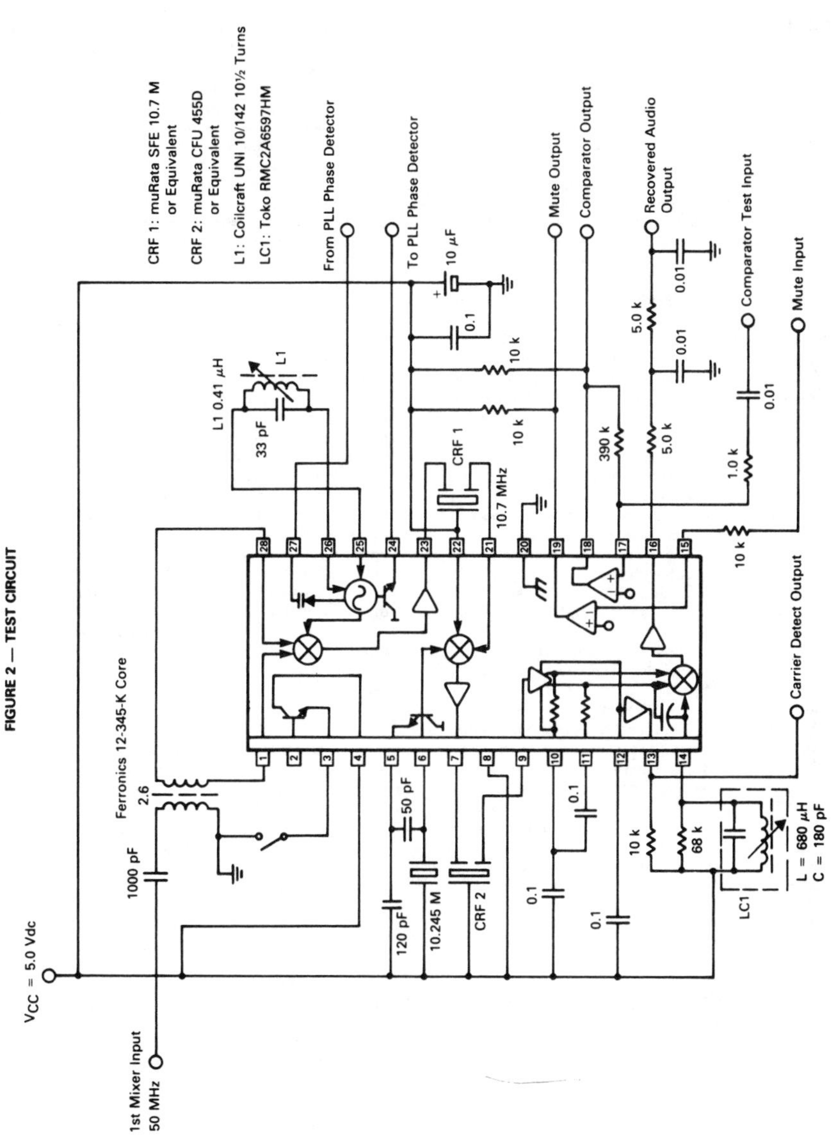

FIGURE 2 — TEST CIRCUIT

CIRCUIT DESCRIPTION

The MC3363 is a complete FM narrowband receiver from RF amplifier to audio preamp output. The low voltage dual conversion design yields low power drain, excellent sensitivity and good image rejection in narrowband voice and data link applications.

In the typical application, the input RF signal is amplified by the RF transistor and then the first mixer amplifies the signal and converts the RF input to 10.7 MHz. This IF signal is filtered externally and fed into the second mixer, which further amplifies the signal and converts it to a 455 kHz IF signal. After external bandpass filtering, the low IF is fed into the limiting amplifier and detection circuitry. The audio is recovered using a conventional quadrature detector. Twice-IF filtering is provided internally.

The input signal level is monitored by meter drive circuitry which detects the amount of limiting in the limiting amplifier. The voltage at the meter drive pin determines the state of the carrier detect output, which is active low.

APPLICATION

The first local oscillator is designed to serve as the VCO in a PLL frequency synthesized receiver. The MC3363 can operate together with the MC145166/7 to provide a two-chip ten channel frequency synthesized receiver in the 46/49 cordless telephone band. The MC3363 can also be used with the MC14515X series of CMOS PLL synthesizers and MC120XX series of ECL prescalers in VHF frequency synthesized applications to 200 MHz.

For single channel applications the first local oscillator can be crystal controlled. The circuit of Figure 4 has been used successfully up to 60 MHz. For higher frequencies an external oscillator signal can be injected into Pins 25 and/or 26 — a level of approximately 100 mVrms is recommended. The first mixer's transfer characteristic is essentially flat to 450 MHz when this approach is used (keeping a constant 10.7 MHz IF frequency). The second local oscillator is a Colpitts type which is typically run at 10.245 MHz under crystal control.

The mixers are doubly balanced to reduce spurious responses. The first and second mixers have conversion gains of 18 dB and 21 dB (typical), respectively. Mixer gain is stable with respect to supply voltage. For both conversions, the mixer impedances and pin layout are designed to allow the user to employ low cost, readily available ceramic filters.

Following the first mixer, a 10.7 MHz ceramic bandpass filter is recommended. The 10.7 MHz filtered signal is then fed into the second mixer input Pin 21, the other input Pin 22 being connected to V_{CC}.

The 455 kHz IF is filtered by a ceramic narrow bandpass filter then fed into the limiter input Pin 9. The limiter has 10 μV sensitivity for -3.0 dB limiting, flat to 1.0 MHz.

The output of the limiter is internally connected to the quadrature detector, including a quadrature capacitor. A parallel LC tank is needed externally from Pin 14 to V_{CC}. A 68 kOhm shunt resistance is included which determines the peak separation of the quadrature detector; a smaller value will lower the Q and expand the deviation range and linearity, but decrease recovered audio and sensitivity.

A data shaping circuit is available and can be coupled to the recovered audio output of Pin 16. The circuit is a comparator which is designed to detect zero crossings of FSK modulation. Data rates of 2000 to 35000 baud are detectable using the comparator. Best sensitivity is obtained when data rates are limited to 1200 baud maximum. Hysteresis is available by connecting a high-valued resistor from Pin 17 to Pin 18. Values below 120 kOhm are not recommended as the input signal cannot overcome the hysteresis.

The meter drive circuitry detects input signal level by monitoring the limiting of the limiting amplifier stages. Figure 5 shows the unloaded current at Pin 12 versus input power. The meter drive current can be used directly (RSSI) or can be used to trip the carrier detect circuit at a specified input power.

A muting op amp is provided and can be triggered by the carrier detect output (Pin 13). This provides a carrier level triggered squelch circuit which is activated when the RF input at the desired input frequency falls below a preset level. The level at which this occurs is determined by the resistor placed between the meter drive output (Pin 12) and V_{CC}. Values between 80–130 kOhms are recommended. This type of squelch is pictured in Figures 3 and 4.

Hysteresis is available by connecting a high-valued resistor Rh between Pins 12 and 13. The formula is:

$$\text{Hyst} = V_{CC}/\,(\text{Rh} \times 10^{-7})\ \text{dB}$$

The meter drive can also be used directly to drive a meter or to provide AGC. A current to voltage converter or other linear buffer will be needed for this application.

A second possible application of the op amp would be in a noise triggered squelch circuit, similar to that used with the MC3357/MC3359/MC3361 FM I.F.'s. In this case the op amp would serve as an active noise filter, the output of which would be rectified and compared to a reference on a squelch gate. The MC3363 does not have a dedicated squelch gate, but the NPN RF input stage or data shaping comparator might be used to provide this function if available. The op amp is a basic type with the inverting input and the output available. This application frees the meter drive to allow it to be used as a linear signal strength monitor.

The circuit of Figure 4 is a complete 50 MHz receiver from antenna input to audio preamp output. It uses few components and has good performance. The receiver operates on a single channel and has input sensitivity of <0.3 μV for 12 dB SINAD.

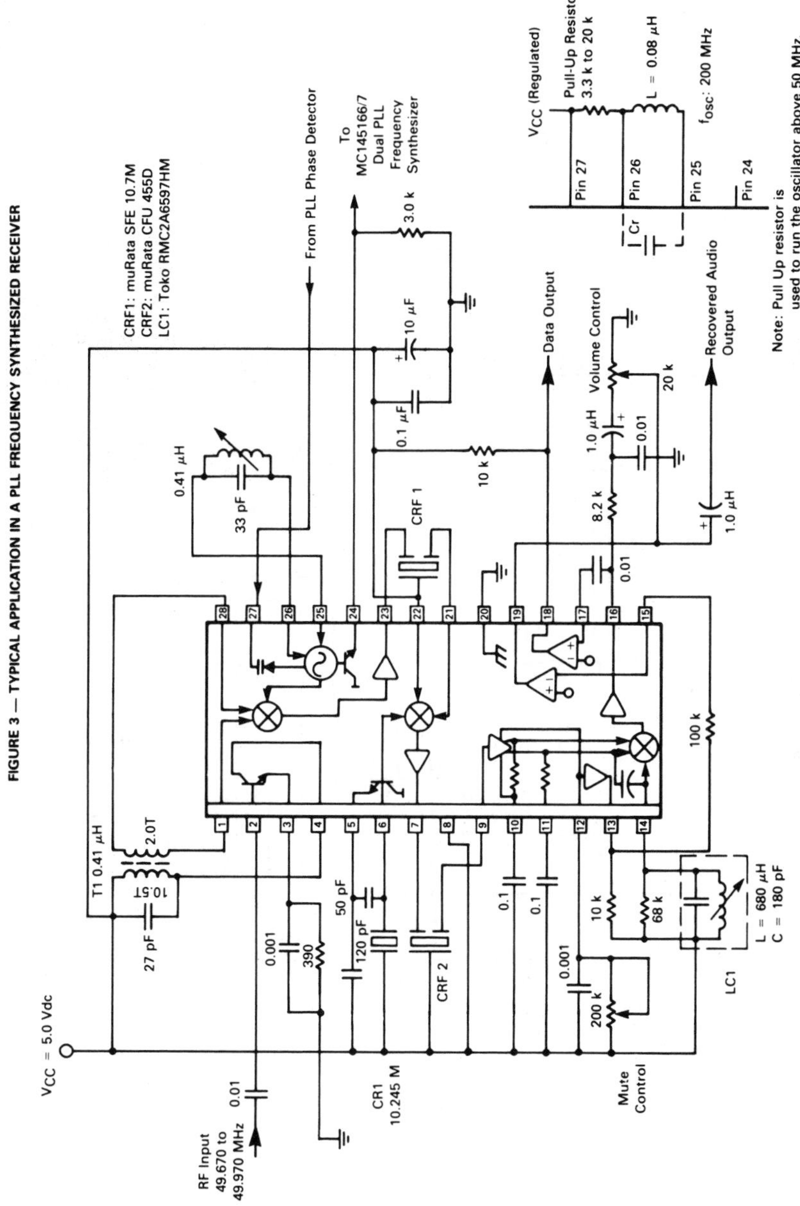

FIGURE 3 — TYPICAL APPLICATION IN A PLL FREQUENCY SYNTHESIZED RECEIVER

FIGURE 4 — SINGLE CHANNEL CRYSTAL CONTROLLED FM RECEIVER

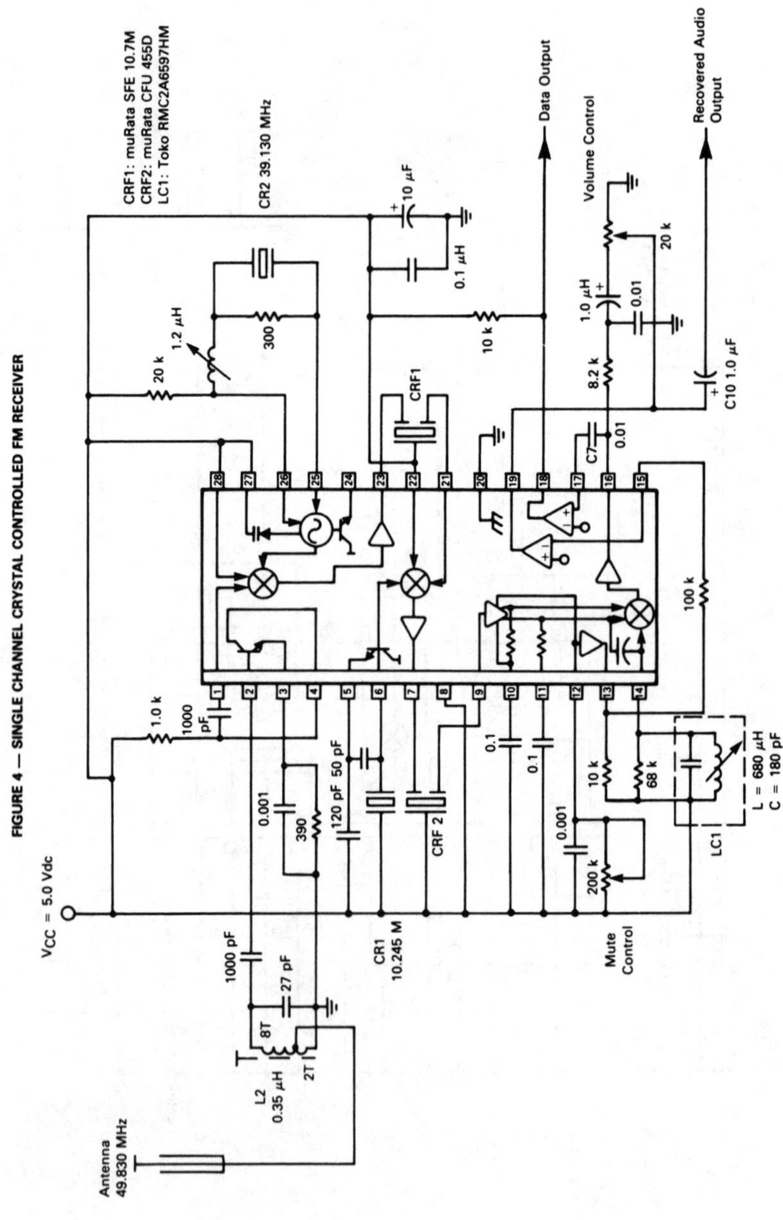

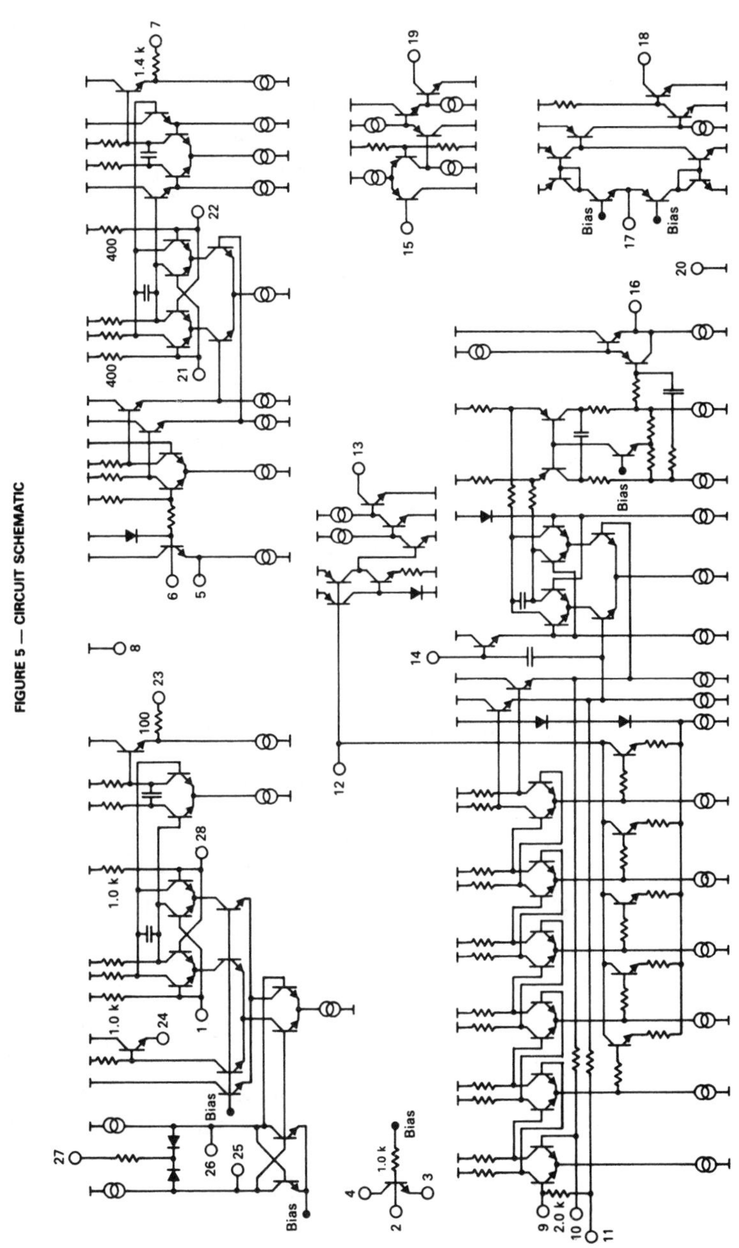

FIGURE 5 — CIRCUIT SCHEMATIC

PROBLEMS

4-1 If $v(t) = \text{Re}\{g(t)e^{j\omega_c t}\}$, show that (4-1b) and (4-1c) are correct where $g(t) = x(t) + jy(t) = R(t)e^{j\theta(t)}$.

4-2 Prove that (4-9) and (4-10) are correct.

4-3 A bandpass filter is shown in Fig. P4-3.
(a) Find the mathematical expression for the transfer function of this filter $H(f) = V_2(f)/V_1(f)$ as a function of R, L, and C. Sketch the magnitude transfer function $|H(f)|$.
(b) Find the expression for the equivalent low-pass filter transfer function and sketch the corresponding low-pass magnitude transfer function.

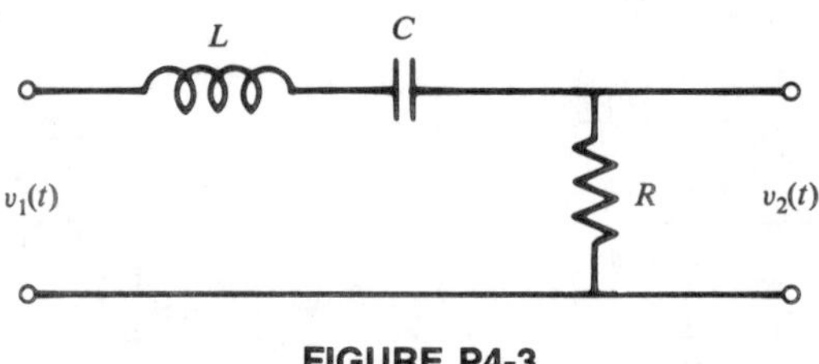

FIGURE P4-3

4-4 Let the transfer function of an ideal bandpass filter be given by

$$H(f) = \begin{cases} 1, & |f + f_c| < B_T/2 \\ 1, & |f - f_c| < B_T/2 \\ 0, & f \text{ elsewhere} \end{cases}$$

where B_T is the absolute bandwidth of the filter.
(a) Sketch the magnitude transfer function, $|H(f)|$.
(b) Find an expression for the waveform at the output, $v_2(t)$, if the input consists of the pulsed carrier:

$$v_1(t) = A\Pi(t/T)\cos(\omega_c t)$$

(c) Sketch the output waveform, $v_2(t)$, for the case when $B_T = 4/T$ and $f_c \gg B_T$.
Hint: Use the complex envelope technique and express the answer as a function of the sine integral where the sine integral is defined by

$$\text{Si}(u) = \int_0^u \frac{\sin \lambda}{\lambda} d\lambda$$

The sketch can be obtained by looking up values for the sine integral from published tables [Abramowitz and Stegun, 1964] or by numerically evaluating $\text{Si}(u)$.

4-5 Examine the distortion properties of an *RC* low-pass filter (shown in Fig. 2-15). Assume that the filter input consists of a bandpass signal that has a bandwidth of 1 kHz and a carrier frequency of 15 kHz. Let the time constant of the filter be $\tau_0 = RC = 10^{-5}$ sec.

(a) Find the phase delay for the output carrier.

(b) Determine the group delay at the carrier frequency.

(c) Evaluate the group delay for frequencies around and within the frequency band of the signal. Plot this delay as a function of frequency.

(d) Using the results of (a) through (c), explain why the filter does or does not distort the bandpass signal.

4-6 A bandpass filter as shown in Fig. P4-6 has the transfer function

$$H(s) = \frac{Ks}{s^2 + (\omega_0/Q)s + \omega_0^2}$$

where $Q = R\sqrt{C/L}$, the resonant frequency is $f_0 = 1/(2\pi\sqrt{LC})$, $\omega_0 = 2\pi f_0$, K is a constant, and values for R, L, and C are given in the figure. Assume that a bandpass signal with $f_c = 4$ kHz and a bandwidth of 200 Hz passes through the filter, where $f_0 = f_c$.

(a) What is the bandwidth of the filter?

(b) Plot the carrier delay as a function of f about f_0.

(c) Plot the group delay as a function of f about f_0.

(d) Explain why the filter does or does not distort the signal.

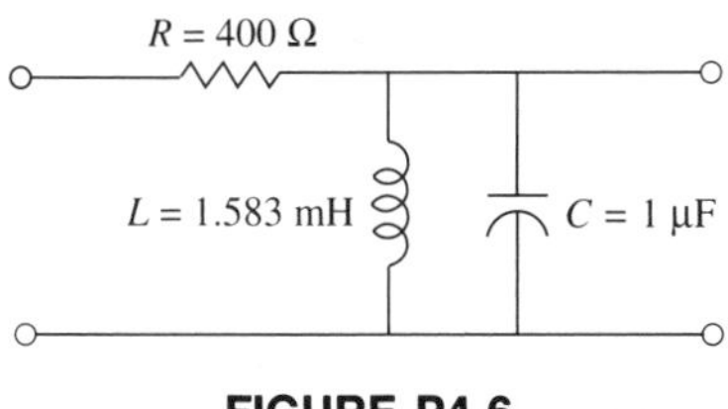

FIGURE P4-6

4-7 An FM signal is of the form

$$s(t) = \cos\left[\omega_c t + D_f \int_{-\infty}^{t} m(\sigma)\, d\sigma\right]$$

where $m(t)$ is the modulating signal and $\omega_c = 2\pi f_c$, f_c being the carrier frequency. Show that the functions $g(t)$, $x(t)$, $y(t)$, $R(t)$, and $\theta(t)$ as given for FM in Table 4-1 are correct.

4-8 SSB-PM has a complex envelope given by

$$g(t) = e^{jD_p[m(t) \pm j\hat{m}(t)]}$$

where $m(t)$ is the modulating signal and $\hat{m}(t)$ is the Hilbert transform of m. Show that the functions $x(t)$, $y(t)$, $R(t)$, and $\theta(t)$ as given for SSB-PM in Table 4-1 are correct.

4-9 Given a modulated signal

$$s(t) = 100 \sin(\omega_c + \omega_a)t + 500 \cos \omega_c t - 100 \sin(\omega_c - \omega_a)t$$

where the unmodulated carrier is 500 cos $\omega_c t$.

(a) Find the complex envelope for the modulated signal. What type of modulation is it? What is the modulating signal?

(b) Find the quadrature modulation components, $x(t)$ and $y(t)$, for this modulated signal.

(c) Find the magnitude and PM components, $R(t)$ and $\theta(t)$, for this modulated signal.

(d) Find the total average power where $s(t)$ is a voltage waveform that is applied across a 50-Ω load.

4-10 Find the spectrum for the modulated signal as given in Prob. 4-9 by two methods.

(a) By direct evaluation using the Fourier transform of $s(t)$.

(b) By the use of (4-22).

4-11 Given a pulse-modulated signal of the form

$$s(t) = e^{-at} \cos[(\omega_c + \Delta\omega)t]u(t)$$

where a, ω_c, and $\Delta\omega$ are positive constants and $\omega_c \gg \Delta\omega$ (ω_c is the carrier frequency),

(a) Find the complex envelope.

(b) Find the spectrum $S(f)$.

(c) Sketch the magnitude and phase spectra $|S(f)|$ and $\theta(f) = \underline{/S(f)}$.

4-12 In a digital computer simulation of a bandpass filter, the complex envelope of the impulse response is used where $h(t) = \text{Re}[k(t)e^{j\omega_c t}]$, as shown in Fig. 4-2. The complex impulse response can be expressed in terms of quadrature components: $k(t) = 2h_x(t) + j2h_y(t)$, where $h_x(t) = \frac{1}{2}\,\text{Re}[k(t)]$ and $h_y(t) = \frac{1}{2}\,\text{Im}\,[k(t)]$. The complex envelopes of the input and output are denoted, respectively, by $g_1(t) = x_1(t) + jy_1(t)$ and $g_2(t) = x_2(t) + jy_2(t)$. The bandpass filter simulation can be carried out by using four *real baseband* filters (i.e., filters having real impulse responses), as shown in Fig. P4-12. Note that although there are four filters, there are only two distinctly different impulse responses, $h_x(t)$ and $h_y(t)$.

(a) Using (4-9), show that Fig. P4-12 is correct.

(b) Show that $h_y(t) \equiv 0$ (i.e., no filter needed) if the bandpass filter has a transfer function with Hermitian symmetry about f_c, that is, if $H(-\Delta f + f_c) = H^*(\Delta f + f_c)$ where $|\Delta f| < B_T/2$ and B_T is the bounded spectral bandwidth of the bandpass filter. This Hermitian symmetry implies that the magnitude frequency response of the bandpass filter is even about f_c and the phase response is odd about f_c.

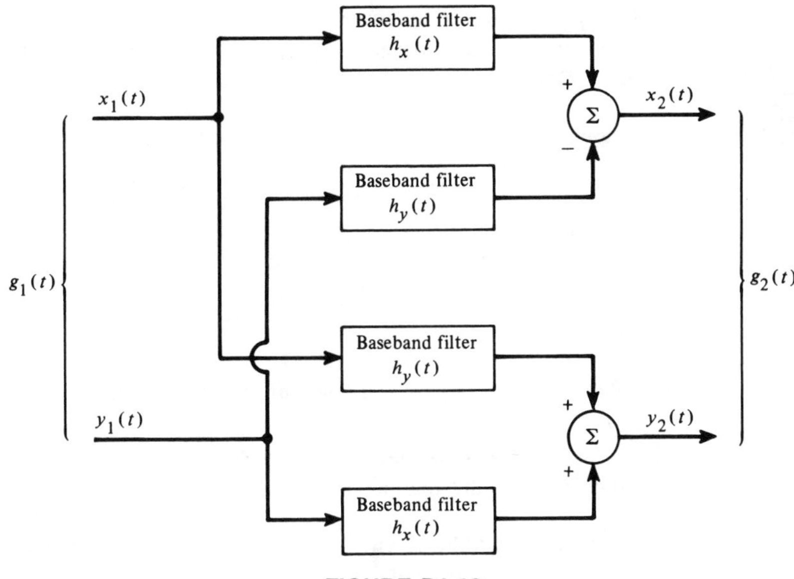

FIGURE P4-12

4-13 Evaluate and sketch the magnitude transfer function for **(a)** Butterworth, **(b)** Chebyshev, and **(c)** Bessel low-pass filters. Assume that $f_b = 10$ Hz and $\epsilon = 1$.

4-14 Plot the amplitude response, the phase response, and the phase delay as a function of frequency for the following low-pass filters where $B = 100$ Hz. Also, compare your results for these two filters.

(a) Butterworth filter, second order:

$$H(f) = \frac{1}{1 + \sqrt{2}(jf/B) + (jf/B)^2}$$

(b) Butterworth filter, fourth order:

$$H(f) = \frac{1}{[1 + 0.765(jf/B) + (jf/B)^2][1 + 1.848(jf/B) + (jf/b)^2]}$$

4-15 Assume that the output-to-input characteristic of a bandpass amplifier is described by (4-37) and that the linearity of the amplifier is being evaluated by using a two-tone test.

(a) Find the frequencies of the fifth-order intermodulation products that fall within the amplifier bandpass.

(b) Evaluate the levels for the fifth-order intermodulation products in terms of A_1, A_2, and the K's.

4-16 An amplifier is tested for total harmonic distortion (THD) by using a single-tone test. The output is observed on a spectrum analyzer. It is found that the peak values of the three measured harmonics decrease according to an exponential recursion relation, $V_{n+1} = V_n e^{-n}$, where $n = 1, 2, 3$. What is the THD?

4-17 The nonlinear output–input characteristic of an amplifier is

$$v_{out}(t) = 5v_{in}(t) + 1.5v_{in}^2(t) + 1.5v_{in}^3(t)$$

Assume that input signal consists of seven components

$$v_{in}(t) = \frac{1}{2} + \frac{4}{\pi^2}\sum_{k=1}^{6}\frac{1}{(2k-1)^2}\cos[(2k-1)\pi t]$$

(a) Plot the output signal and compare it with the linear output component $5v_{in}(t)$.

(b) Take the FFT of the output $v_{out}(t)$ and compare it with the spectrum for the linear output component.

4-18 For a bandpass limiter circuit, show that the bandpass output is given by (4-50), where $K = (4/\pi)A_0$. A_0 denotes the voltage gain of the bandpass filter, and it is assumed that the gain is constant over the frequency range of the bandpass signal.

4-19 Discuss whether or not the Taylor series nonlinear model is applicable for analysis of **(a)** soft limiters and **(b)** hard limiters.

4-20 Assume that an audio sine-wave testing signal is passed through an audio hard limiter circuit. Evaluate the total harmonic distortion (THD) on the signal at the limiter output.

4-21 Using the mathematical definition for linearity given in Chapter 2, show that the analog switch multiplier of Fig. 4-11 is a linear device.

4-22 An audio signal with a bandwidth of 10 kHz is transmitted over an AM transmitter with a carrier frequency of 1.0 MHz. The AM signal is received on a superheterodyne receiver with an envelope detector. What is the constraint on the *RC* time constant for the envelope detector?

4-23 Assume that an AM receiver with an envelope detector is tuned to an SSB-AM signal that has a modulation waveform given by $m(t)$. Find the mathematical expression for the audio signal that appears at the receiver output in terms of $m(t)$. Is the audio output distorted?

4-24 Evaluate the sensitivity of the zero-crossing FM detector shown in Fig. 4-19. Assume that the differential amplifier is described by $v_{out}(t) = A[v_2(t) - v_3(t)]$, where A is the voltage gain of the amplifier. In particular, show that $v_{out} = Kf_d$, where $f_d = f_i - f_c$ and find the value of the sensitivity constant K in terms of A, R, and C. Assume that the peak levels of the monostable outputs Q and $\overline{Q}$ are 4 V (TTL circuit levels).

4-25 **(a)** Using (4-92), show that the linearized block diagram model for a PLL is given by Fig. 4-23.

(b) Show that (4-93) describes the linear PLL model as given in Fig. 4-23.

4-26 Using the Laplace transform and the final value theorem, find an expression for the steady-state phase error, $\lim_{t\to\infty} \theta_e(t)$, for a PLL as described by (4-92). [*Hint:* The final value theorem is $\lim_{t\to\infty} f(t) = \lim_{s\to 0} sF(s)$.]

4-27 Assume that the loop filter of a PLL is a low-pass filter, as shown in Fig. P4-27.

(a) Evaluate the closed-loop transfer function $H(f)$ for a linearized PLL.

(b) Sketch the Bode plot $[|H(f)|]_{dB} \triangleq 20 \log |H(f)|$, for this PLL.

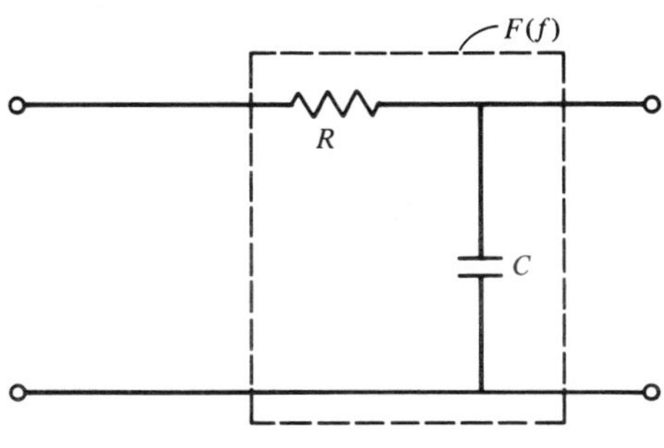

FIGURE P4-27

4-28 Assume that the phase noise characteristic of a PLL is being examined. The internal phase noise of the VCO is modeled by the input $\theta_n(t)$, as shown in Fig. P4-28.

(a) Find an expression for the closed-loop transfer function $\Theta_0(f)/\Theta_n(f)$, where $\theta_i(t) = 0$.

(b) If $F_1(f)$ is a low-pass filter given in Fig. P4-27, sketch the Bode plot $[|\Theta_0(f)/\Theta_n(f)|]_{dB}$ for the phase noise transfer function.

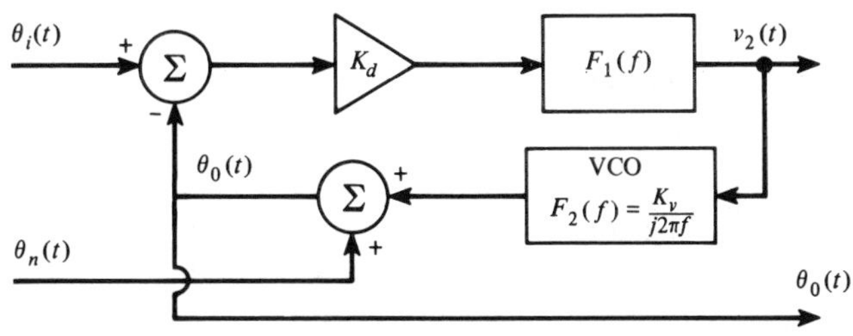

FIGURE P4-28

4-29 The input to a PLL is $v_{in}(t) = A \sin(\omega_0 t + \theta_i)$. The LPF has a transfer function $F(s) = (s + a)/s$.

(a) What is the steady-state phase error?

(b) What is the maximum hold-in range for the noiseless case?

4-30 **(a)** Refer to Fig. 4-26 for a PLL frequency synthesizer. Design a synthesizer that will cover a range of 144 to 148 MHz in 5-kHz steps starting at 144.000 MHz. Assume that the frequency standard operates at 5 MHz and that the M divider is fixed at some value and the N divider is programmable so that the synthesizer will cover the desired range. Sketch a diagram of your design, indicating the frequencies present at various points of the diagram.

(b) Modify your design so that the output signal can be frequency modulated with an audio-frequency input such that the peak deviation of the RF output is 5 kHz.

4-31 Assume that an SSB-AM transmitter is to be realized using the AM-PM generation technique, as shown in Fig. 4-27.

(a) Sketch a block diagram for the baseband signal processing circuit.

(b) Find expressions for $R(t)$, $\theta(t)$, and $v(t)$ when the modulation is $m(t) = A_1 \cos \omega_1 t + A_2 \cos \omega_2 t$.

4-32 Rework Prob. 4-31 for the case of generating an FM signal.

4-33 Assume that an SSB-AM transmitter is to be realized using the quadrature generation technique, as shown in Fig. 4-28.

(a) Sketch a block diagram for the baseband signal processing circuit.

(b) Find expressions for $x(t)$, $y(t)$, and $v(t)$ when the modulation is $m(t) = A_1 \cos \omega_1 t + A_2 \cos \omega_2 t$.

4-34 Rework Prob. 4-33 for the case of generating an FM signal.

4-35 An FM radio is tuned to receive an FM broadcasting station of frequency 96.9 MHz. The radio is of the superheterodyne type with the LO operating on the high side of the 96.9-MHz input and using a 10.7-MHz IF amplifier.

(a) Determine the LO frequency.

(b) If the FM signal has a bandwidth of 180 kHz, determine the filter characteristics for the RF and IF filters.

(c) Calculate the frequency of the image response.

4-36 A 4-GHz satellite downlink signal is received on an Earth station receiver. The IF frequency is 70 MHz. What is the image frequency for the case of:

(a) High-side injection?

(b) Low-side injection?

4-37 A superheterodyne receiver is tuned to a station at 20 MHz. The local oscillator frequency is 80 MHz and the IF is 100 MHz.

(a) What is the image frequency?

(b) If the LO has appreciable second-harmonic content, what two additional frequencies are received?

(c) If the RF amplifier contains a single resonant circuit with $Q = 50$ tuned to 20 MHz, what will be the image attentuation in dB?

4-38 An SSB-AM receiver is tuned to receive a 7.225-MHz lower SSB (LSSB) signal. The LSSB signal is modulated by an audio signal that has a 3-kHz bandwidth. Assume that the receiver uses a superheterodyne circuit with an SSB IF filter. The IF amplifier is centered on 3.395 MHz. The LO frequency is on the high (frequency) side of the input LSSB signal.

(a) Draw a block diagram of the single-conversion superheterodyne receiver, indicating frequencies present and typical spectra of the signals at various points within the receiver.

(b) Determine the required RF and IF filter specifications, assuming that the image frequency is to be attenuated by 40 dB.

4-39 **(a)** Draw a block diagram of a superheterodyne FM receiver that is designed to receive FM signals over a band from 144 to 148 MHz. Assume that the receiver is of the dual conversion type (i.e., a mixer and an IF amplifier followed by another mixer and a second IF amplifier) where the first IF frequency is 10.7 MHz and the second IF is 455 kHz. Indicate the frequencies of the signals at different points on the diagram and, in particular, show the frequencies involved when a signal at 146.82 MHz is being received.

(b) Replace the first oscillator by a frequency synthesizer such that the receiver can be tuned in 5-kHz steps from 144.000 to 148.000 MHz. Show the diagram of your synthesizer design and the frequencies involved.

4-40 An AM broadcast-band radio is tuned to receive a 1080-kHz AM signal and uses high-side LO injection. The IF is 455 kHz.

(a) Sketch the frequency response for the RF and IF filters.

(b) What is the image frequency?

4-41 Commercial AM broadcast stations operate in the 540- to 1700-kHz band with a transmission bandwidth limited to 10 kHz.

(a) What are the maximum number of stations that can be accommodated?

(b) If stations are not assigned to adjacent channels (in order to reduce interference on receivers that have poor IF characteristics), how many stations can be accommodated?

(c) For 455-kHz IF receivers, what band of image frequencies fall within the AM band for down-converter receivers with high-side injection?

4-42 An AM broadcast transmitter is tested by feeding the RF output into a 50-Ω (dummy) load. Tone modulation is applied. The carrier frequency is 850 kHz and the FCC licensed power output is 5000 W. The sinusoidal tone of 1000 Hz is set for 90% modulation.

(a) Evaluate the FCC power in dBk (dB above 1 kW) units.

(b) Write an equation for the voltage that appears across the 50-Ω load, giving numerical values for all constants.

(c) Sketch the spectrum of this voltage as it would appear on a calibrated spectrum analyzer.

(d) What is the average power that is being dissipated in the dummy load?

(e) What is the peak envelope power?

4-43 An AM transmitter is modulated with an audio testing signal given by $m(t) = 0.2 \sin \omega_1 t + 0.5 \cos \omega_2 t$, where $f_1 = 500$ Hz, $f_2 = 500\sqrt{2}$ Hz, and $A_c = 100$. Assume that the AM signal is fed into a 50-Ω load.
(a) Sketch the AM waveform.
(b) What is the modulation percentage?
(c) Evaluate and sketch the spectrum of the AM waveform.

4-44 For the AM signal given in Prob. 4-43:
(a) Evaluate the average power of the AM signal.
(b) Evaluate the PEP of the AM signal.

4-45 Assume that an AM transmitter is modulated with a video testing signal given by $m(t) = -0.2 + 0.6 \sin \omega_1 t$ where $f_1 = 3.57$ MHz. Let $A_c = 100$.
(a) Sketch the AM waveform.
(b) What is the percentage of positive and negative modulation?
(c) Evaluate and sketch the spectrum of the AM waveform about f_c.

4-46 A 50,000-W AM broadcast transmitter is being evaluated by means of a two-tone test. The transmitter is connected to a 50-Ω load and $m(t) = A_1 \cos \omega_1 t + A_1 \cos 2\omega_1 t$ where $f_1 = 500$ Hz. Assume that a perfect AM signal is generated.
(a) Evaluate the complex envelope for the AM signal in terms of A_1 and ω_1.
(b) Determine the value of A_1 for 90% modulation.
(c) Find the values for the peak current and average current into the 50-Ω load for the 90% modulation case.

4-47 An AM transmitter uses a two-quadrant multiplier so that the transmitted signal is described by (4-111). Assume that the transmitter is modulated by $m(t) = A_m \cos \omega_m t$, where A_m is adjusted so that 120% positive modulation is obtained. Evaluate the spectrum of this AM signal in terms of A_c, f_c, and f_m. Sketch your result.

4-48 A DSB-SC signal is modulated by $m(t) = \cos \omega_1 t + 2 \cos 2\omega_1 t$ where $\omega_1 = 2\pi f_1$, $f_1 = 500$ Hz, and $A_c = 1$.
(a) Write an expression for the DSB-SC signal and sketch a picture of this waveform.
(b) Evaluate and sketch the spectrum for this DSB-SC signal.
(c) Find the value of the average (normalized) power.
(d) Find the value of the PEP (normalized).

4-49 Assume that transmitting circuitry restricts the modulated output signal to a certain peak value, say A_P, because of power-supply voltages that are used and the peak voltage and current ratings of the components. If a DSB-SC signal with a peak value of A_P is generated by this circuit, show that the sideband power of this DSB-SC signal is four times the sideband power of a comparable AM signal having the same peak value, A_P, that could also be generated by this circuit.

4-50 A DSB-SC signal can be generated from two AM signals as shown in Fig. P4-50. Using mathematics to describe signals at each point on the figure, prove that the output is a DSB-SC signal.

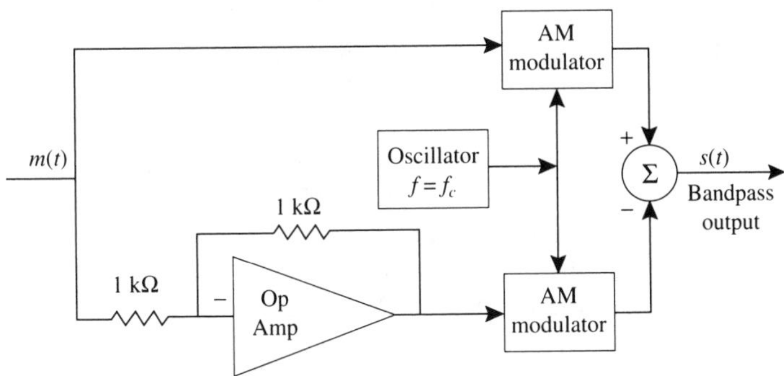

FIGURE P4-50

4-51 Show that the complex envelope $g(t) = m(t) - j\hat{m}(t)$ produces a lower SSB signal, provided that $m(t)$ is a real signal.

4-52 Show that the impulse response of a $-90°$ phase shift network (i.e., a Hilbert transformer) is $1/\pi t$. *Hint:*

$$H(f) = \lim_{\substack{\alpha \to 0 \\ \alpha > 0}} \begin{cases} -je^{-\alpha f}, & f > 0 \\ je^{\alpha f}, & f < 0 \end{cases}$$

4-53 SSB signals can be generated by the phasing method, Fig. 4-36a; the filter method, Fig. 4-36b; or by the use of Weaver's method as shown in Fig.

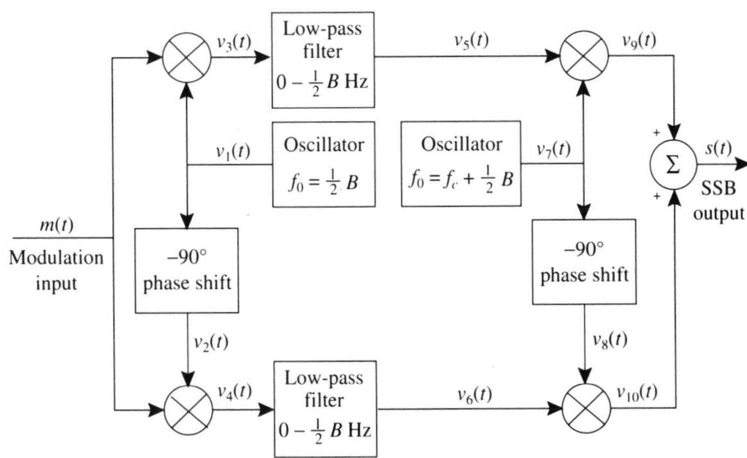

FIGURE P4-53 Weaver's method for generating SSB.

P4-53. For Weaver's method (Fig. P4-53) where B is the bandwidth of $m(t)$:

(a) Find a mathematical expression that describes the waveform out of each block on the block diagram.

(b) Show that $s(t)$ is an SSB signal.

4-54 An SSB-AM transmitter is modulated with a sinusoid $m(t) = 5 \cos \omega_1 t$, where $\omega_1 = 2\pi f_1$, $f_1 = 500$ Hz, and $A_c = 1$.

(a) Evaluate $\hat{m}(t)$.

(b) Find the expression for a lower SSB signal.

(c) Find the rms value of the SSB signal.

(d) Find the peak value of the SSB signal.

(e) Find the normalized average power of the SSB signal.

(f) Find the normalized PEP of the SSB signal.

4-55 An SSB-AM transmitter is modulated by a rectangular pulse such that $m(t) = \Pi(t/T)$ and $A_c = 1$.

(a) Prove that

$$\hat{m}(t) = \frac{1}{\pi} \ln \left| \frac{2t + T}{2t - T} \right|$$

as given in Table A-7.

(b) Find an expression for the SSB-AM signal, $s(t)$, and sketch $s(t)$.

(c) Find the peak value of $s(t)$.

4-56 For Prob. 4-55:

(a) Find the expression for the spectrum of a USSB-AM signal.

(b) Sketch the magnitude spectrum, $|S(f)|$.

4-57 A USSB transmitter is modulated with the pulse

$$m(t) = \frac{\sin \pi a t}{\pi a t}$$

(a) Prove that

$$\hat{m}(t) = \frac{\sin^2[(\pi a/2)t]}{(\pi a/2)t}$$

(b) Plot the corresponding USSB signal waveform for the case of $A_c = 1$, $a = 2$, and $f_c = 20$ Hz.

4-58 A USSB-AM signal is modulated by a rectangular pulse train:

$$m(t) = \sum_{n=-\infty}^{\infty} \Pi[(t - nT_0)/T]$$

where $T_0 = 2T$.

(a) Find the expression for the spectrum of the SSB-AM signal.

(b) Sketch the magnitude spectrum, $|S(f)|$.

4-59 A phasing-type SSB-AM detector is shown in Fig. P4-59. This circuit is attached to the IF output of a conventional superheterodyne receiver to provide SSB reception.

(a) Determine whether this detector is sensitive to LSSB or USSB signals. How would the detector be changed to receive SSB signals with alternate (opposite type of) sidebands?

(b) Assume that the signal at point A is a USSB signal with $f_c = 455$ kHz. Find the mathematical expressions for the signals at points B through I.

(c) Repeat part (b) for the case of an LSSB-AM signal at point A.

(d) Discuss the IF and LP filter requirements if the SSB signal at point A has a 3-kHz bandwidth.

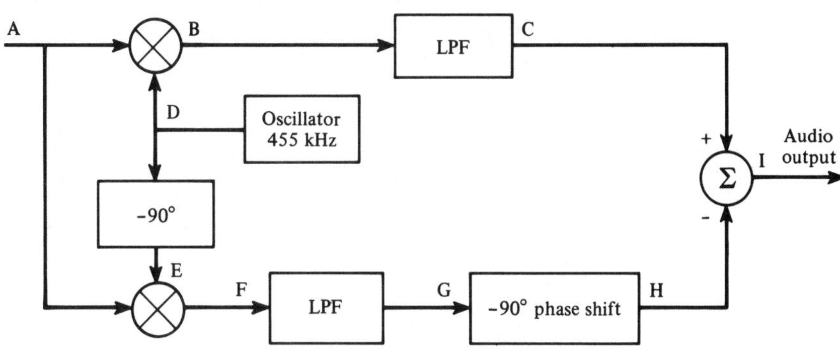

FIGURE P4-59

4-60 Can a Costas loop, as shown in Fig. 4-34, be used to demodulate an SSB-AM signal? Demonstrate that your answer is correct by using mathematics.

4-61 A modulated signal is described by the equation

$$s(t) = 10 \cos[(2\pi \times 10^8)\, t + 10 \cos(2\pi \times 10^3 t)]$$

Find each of the following.

(a) Percentage of AM.

(b) Normalized power of the modulated signal.

(c) Maximum phase deviation.

(d) Maximum frequency deviation.

4-62 A sinusoidal signal, $m(t) = \cos 2\pi f_m t$, is the input to an angle-modulated transmitter where the carrier frequency is $f_c = 1$ Hz and $f_m = f_c/4$.

(a) Plot $m(t)$ and the corresponding PM signal where $D_p = \pi$.

(b) Plot $m(t)$ and the corresponding FM signal where $D_f = \pi$.

4-63 A sinusoidal modulating waveform of amplitude 4 V and a frequency of 1 kHz is applied to an FM exciter that has a modulator gain of 50 Hz/V.

(a) What is the peak frequency deviation?

(b) What is the modulation index?

4-64 An FM signal has sinusoidal modulation with a frequency of $f_m = 15$ kHz and modulation index of $\beta = 2.0$.

(a) Find the transmission bandwidth using Carson's rule.

(b) What percentage of the total FM signal power lies within the Carson rule bandwidth?

4-65 An FM transmitter has a block diagram as shown in Fig. P4-65. The audio-frequency response is flat over the 20-Hz to 15-kHz audio band. The FM output signal is to have a carrier frequency of 103.7 MHz and a peak deviation of 75 kHz.

(a) Find the bandwidth and center frequency required for the bandpass filter.

(b) Calculate the frequency f_0 of the oscillator.

(c) What is the required peak deviation capability of the FM exciter?

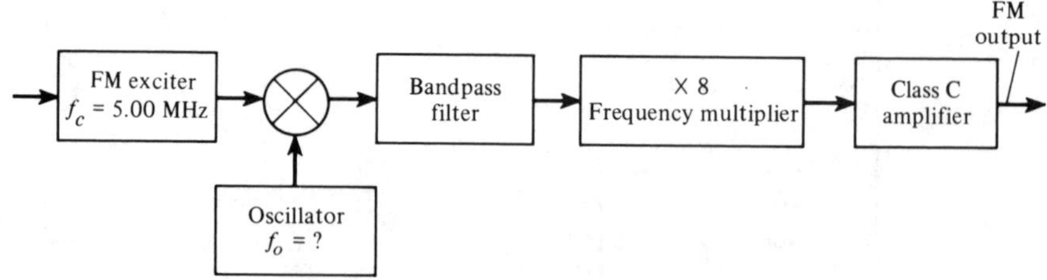

FIGURE P4-65

4-66 Analyze the performance of the FM circuit of Fig. 4-39b. Assume that the voltage appearing across the reversed-biased diodes, which provide the voltage variable capacitance, is $v(t) = 5 + 0.05m(t)$, where the modulating signal is a test tone, $m(t) = \cos \omega_1 t$, $\omega_1 = 2\pi f_1$, and $f_1 = 1$ kHz. The capacitance of each of the biased diodes is $C_d = 100/\sqrt{1 + 2v(t)}$ pF. Assume that $C_0 = 180$ pF and that L is chosen to resonate at 5 MHz.

(a) Find the value of L.

(b) Show that the resulting oscillator signal is an FM signal. For convenience, assume that the peak level of the oscillator signal is 10 V. Find the parameter D_f.

4-67 A modulated RF waveform is given by $500 \cos[\omega_c t + 20 \cos \omega_1 t]$, where $\omega_1 = 2\pi f_1$, $f_1 = 1$ kHz, and $f_c = 100$ MHz.

(a) If the phase deviation constant is 100 rad/V, find the mathematical expression for the corresponding phase modulation voltage $m(t)$. What is its peak value and its frequency?

(b) If the frequency deviation constant is 1×10^6 rad/V-sec, find the mathematical expression for the corresponding FM voltage, $m(t)$. What is its peak value and its frequency?

(c) If the RF waveform appears across a 50-Ω load, determine the average power and the PEP.

4-68 Given the FM signal $s(t) = 10 \cos [\omega_c t + 100 \int_{-\infty}^{t} m(\sigma)\, d\sigma]$, where $m(t)$ is a polar square-wave signal with a duty cycle of 50%, a period of 1 sec, and a peak value of 5 V.
(a) Sketch the instantaneous frequency waveform and the waveform of the corresponding FM signal (see Fig. 4-40).
(b) Plot the phase deviation $\theta(t)$ as a function of time.
(c) Evaluate the peak frequency deviation.

4-69 A carrier $s(t) = 100 \cos(2\pi \times 10^9 t)$ of an FM transmitter is modulated with a tone signal. For this transmitter a 1-V (rms) tone produces a deviation of 30 kHz. Determine the amplitude and frequency of all FM signal components (spectral lines) that are greater than 1% of the unmodulated carrier amplitude for the following modulating signals:
(a) $m(t) = 2.5 \cos(3\pi \times 10^4 t)$.
(b) $m(t) = 1 \cos(6\pi \times 10^4 t)$.

4-70 Referring to (4-160), show that

$$J_{-n}(\beta) = (-1)^n J_n(\beta)$$

4-71 Consider an FM exciter with the output $s(t) = 100 \cos[2\pi 1000t + \theta(t)]$. The modulation is $m(t) = 5 \cos(2\pi 8t)$ and the modulation gain of the exciter is 8 Hz/V. The FM output signal is passed through an ideal (brick-wall) bandpass filter which has a center frequency of 1000 Hz, a bandwidth of 56 Hz, and a gain of unity. Determine the normalized average power:
(a) At the bandpass filter input.
(b) At the bandpass filter output.

4-72 A l-kHz sinusoidal signal phase modulates a carrier at 146.52 MHz with a peak phase deviation of 45°. Evaluate the exact magnitude spectra of the PM signal if $A_c = 1$. Sketch your result. Using Carson's rule, evaluate the approximate bandwidth of the PM signal and see if it is a reasonable number when compared with your spectral plot.

4-73 A 1-kHz sinusoidal signal frequency modulates a carrier at 146.52 MHz with a peak deviation of 5 kHz. Evaluate the exact magnitude spectra of the FM signal if $A_c = 1$. Sketch your result. Using Carson's rule, evaluate the approximate bandwidth of the FM signal and see if it is a reasonable number when compared with your spectral plot.

4-74 The calibration of a frequency deviation monitor is to be verified by using a Bessel function test. An FM test signal with a calculated frequency deviation is generated by frequency modulating a sine wave onto a carrier. Assume that the sine wave has a frequency of 2 kHz and that the amplitude of the sine wave is slowly increased from zero until the discrete carrier term (at f_c) of the FM signal reduces to zero, as observed on a spectrum analyzer. What is the peak frequency deviation of the FM test signal when the discrete carrier term is zero? Suppose that the amplitude of the sine wave is increased further until this discrete carrier term appears, reaches a maxi-

mum, and then disappears again. What is the peak frequency deviation of the FM test signal now?

4-75 A frequency modulator has a modulator gain of 10 Hz/V and the modulating waveform is

$$m(t) = \begin{cases} 0, & t < 0 \\ 5, & 0 < t < 1 \\ 15, & 1 < t < 3 \\ 7, & 3 < t < 4 \\ 0, & 4 < t \end{cases}$$

(a) Plot the frequency deviation in hertz over the time interval $0 < t < 5$.
(b) Plot the phase deviation in radians over the time interval $0 < t < 5$.

4-76 A square-wave (digital) test signal of 50% duty cycle phase modulates a transmitter where $s(t) = 10 \cos[\omega_c t + \theta(t)]$. The carrier frequency is 60 MHz and the peak phase deviation is 45°. Assume that the test signal is of the unipolar NRZ type with a period of 1 msec and that it is symmetrical about $t = 0$. Find the exact spectrum for $s(t)$.

4-77 Two sinusoids, $m(t) = A_1 \cos \omega_1 t + A_2 \cos \omega_2 t$, phase modulate a transmitter. Derive a formula that gives the exact spectrum for the resulting PM signal in terms of the signal parameters A_c, ω_c, D_p, A_1, A_2, ω_1, and ω_2. [*Hint:* Use $e^{ja(t)} = (e^{ja_1(t)})(e^{ja_2(t)})$, where $a(t) = a_1(t) + a_2(t)$.]

4-78 Plot the magnitude spectrum centered on $f = f_c$ for an FM signal where the modulating signal is

$$m(t) = A_1 \cos 2\pi f_1 t + A_2 \cos 2\pi f_2 t$$

Assume that $f_1 = 10$ Hz, $f_2 = 17$ Hz, and that A_1 and A_2 are adjusted so that each tone contributes a peak deviation of 20 Hz.

4-79 For small values of β, $J_n(\beta)$ can be approximated by $J_n(\beta) = \beta^n/(2^n n!)$. Show that for the case of FM with sinusoidal modulation, $\beta = 0.2$ is sufficiently small to give NBFM.

4-80 A polar square wave with a 50% duty cycle frequency modulates an NBFM transmitter such that the peak phase deviation is 10°. Assume that the square wave has a peak value of 5 V, a period of 10 msec, and a zero-crossing at $t = 0$ with a positive-going slope.
(a) Determine the peak frequency deviation of this NBFM signal.
(b) Evaluate and sketch the spectrum of this signal, using the narrowband analysis technique. Assume that the carrier frequency is 30 MHz.

4-81 Design a wideband FM transmitter that uses the indirect method for generating a WBFM signal. Assume that the carrier frequency of the WBFM signal is 96.9 MHz and that the transmitter is capable of producing a high-quality FM signal with a peak deviation of 75 kHz when modulated by a 1-V (rms) sinusoid of frequency 20 Hz. Show a complete block diagram of your design, indicating the frequencies and peak deviations of the signals at various points.

4-82 An FM signal, $100 \cos[\omega_c t + D_f \int_{-\infty}^{t} m(\sigma)\, d\sigma]$, is modulated by the waveform shown in Fig. P4-82. Let $f_c = 420$ MHz.

(a) Determine the value of D_f so that the peak-to-peak frequency deviation is 25 kHz.

(b) Evaluate and sketch the approximate PSD.

(c) Determine the bandwidth of this FM signal such that spectral components are down at least 40 dB from the unmodulated carrier level for frequencies outside this bandwidth.

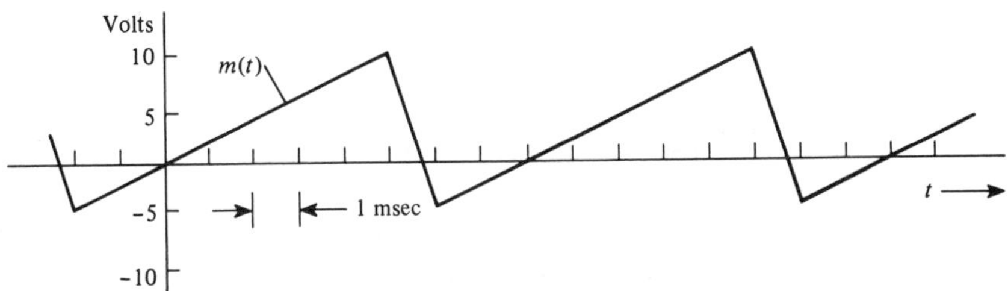

FIGURE P4-82

4-83 A multilevel digital test signal as shown in Fig. P4-83 modulates a WBFM transmitter. Evaluate and sketch the approximate power spectrum of this WBFM signal if $A_c = 5$, $f_c = 2$ GHz, and $D_f = 10^5$.

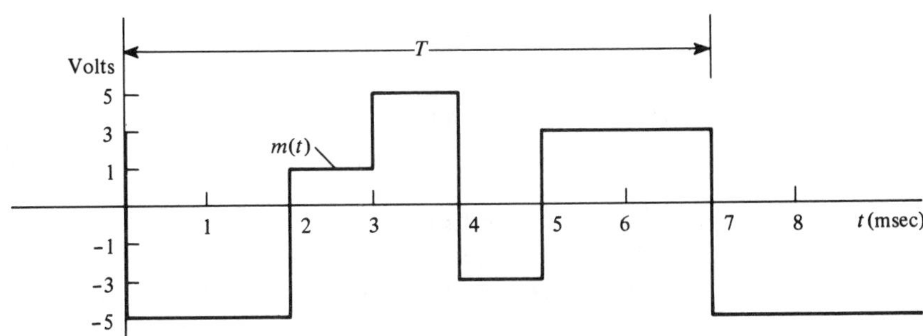

FIGURE P4-83

4-84 Refer to Fig. 4-46b, which displays the circuit diagram for a commonly used preemphasis filter network.

(a) Show that $f_1 = 1/(2\pi R_1 C)$.

(b) Show that $f_2 = (R_1 + R_2)/(2\pi R_1 R_2 C) \approx 1/(2\pi R_2 C)$.

(c) Evaluate K in terms of R_1, R_2, and C.

4-85 The composite baseband signal for FM stereo transmission is given by

$$m_b(t) = [m_L(t) + m_R(t)] + [m_L(t) - m_R(t)] \cos(\omega_{sc}t) + K \cos(\tfrac{1}{2}\omega_{sc}t)$$

A stereo FM receiver using a switching type of demultiplexer for recovery of $m_L(t)$ and $m_R(t)$ is shown in Fig. P4-85.

(a) Determine the switching waveforms at points C and D that activate the analog switches. Be sure to specify the correct phasing for each one. Sketch these waveforms.

(b) Draw a more detailed block diagram showing the blocks inside the PLL.

(c) Write equations for the waveforms at points A through F and explain how this circuit works by sketching typical waveforms at each of these points.

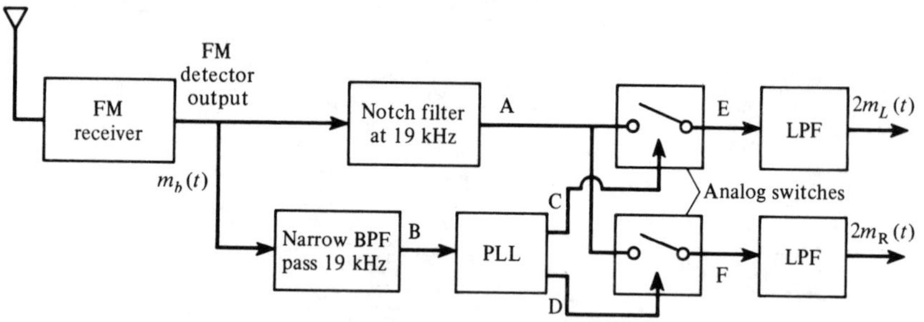

FIGURE P4-85

4-86 In a communication system two baseband signals (they may be analog or digital) are transmitted simultaneously by generating the RF signal

$$s(t) = m_1(t) \cos \omega_c t + m_2(t) \sin \omega_c t$$

The carrier frequency is 7.250 MHz. The bandwidth of $m_1(t)$ is 5 kHz and the bandwidth of $m_2(t)$ is 10 kHz.

(a) Evaluate the bandwidth of $s(t)$.

(b) Derive an equation for the spectrum of $s(t)$ in terms of $M_1(f)$ and $M_2(f)$.

(c) $m_1(t)$ and $m_2(t)$ can be recovered (i.e., detected) from $s(t)$ by using a superheterodyne receiver with two switching detectors. Draw a block diagram for the receiver and sketch the digital waveforms that are required to operate the samplers. Describe how they can be obtained. Show that the Nyquist sampling criterion is satisfied. (*Hint:* See Fig. P4-85.)

4-87 Suppose that a low-level audio signal that has a flat spectrum from 30 Hz to 20 kHz is tape recorded using Dolby-B preemphasis.

(a) Sketch the spectrum of the Dolby-B preemphasized signal.

(b) If the recording is mistakenly played back on a machine using Dolby-C deemphasis, sketch the spectrum of the signal played back.

(*Hint:* Use Fig. 4-49.)

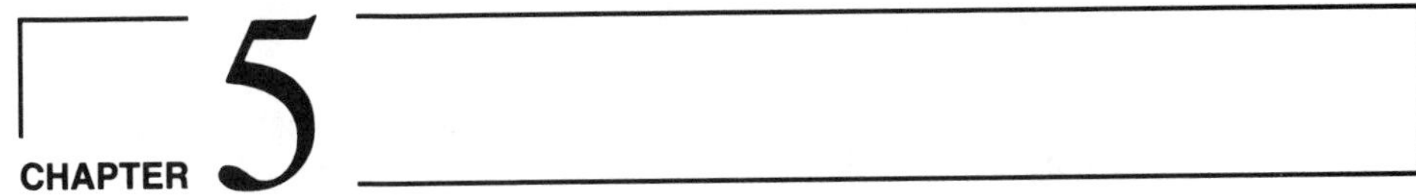

CHAPTER 5

Bandpass Communication Systems

5-1 Introduction

The goal of this chapter is to obtain a *working knowledge* of bandpass communication *systems*. To do so we will consider examples of data, telephone, cellular radio, television, satellite, spread spectrum, and fiber optic systems. The tools developed in previous chapters will be utilized here.

Theoretical system concepts are also emphasized. As indicated earlier, the goodness of a communication system is measured in terms of the required transmission power or energy, required system bandwidth, system cost, and the quality of information produced at the receiver output. In the first part of this chapter, we will study bandpass digital signals and their associated spectrum and bandwidth. For digital systems, the quality of the recovered information is usually specified by the probability of error on the output digital signal. For analog systems the quality of the recovered information is usually specified by the signal-to-noise ratio associated with the output analog waveform. In the latter part of this chapter, a technique will be developed for evaluating the signal-to-noise ratio at the output of an analog system and the probability of bit error for the output of a digital system. This technique does not require the use of statistics, *provided* that the noise characteristics of the receiver detector circuit are known. If the receiver is purchased from a vendor, these characteristics are included in the receiver specifications. They can also be obtained by measurement or by theoretical evaluation. The important topic of the theoretical evaluation of the detector characteristics is deferred until Chapters 6 through 8. In those chapters we will concentrate on the statistical concepts involved, yet have a practical understanding of their application that was gained in Chapters 1 through 5.

5-2 Digital Communication Systems

Since digital communication systems are becoming more significant each day, a whole chapter, Chapter 3, was devoted to *baseband* digital signaling. As stated in Chapter 1, the source information usually needs to be mapped into a *bandpass* signal in order to communicate over a channel. *Bandpass* digital communication systems consist of modulating a *baseband* digital signal (Chapter 3) onto a carrier using AM, PM, FM, or some type of generalized modulation technique (Chapter

4). In this section we extend our results of baseband digital signaling to bandpass digital signaling. From (4-23), the PSD for the bandpass signal, $s(t)$, is given by

$$\mathcal{P}_s(f) = \tfrac{1}{4}[\mathcal{P}_g(f - f_c) + \mathcal{P}_g(-f - f_c)] \tag{5-1}$$

where f_c is the carrier frequency and $\mathcal{P}_g(f)$ is the PSD of the complex envelope. [It is recalled that the complex envelope is a mapping of the baseband signal $m(t)$ onto the complex plane and that the particular type of mapping function used determines the modulation type (AM, FM, etc.).] In this section we will determine $\mathcal{P}_g(f)$ for various types of digital signaling so that the PSD of the bandpass digital signal can be obtained.

Bandpass digital communication systems may be divided into two main categories: binary digital systems and multilevel (i.e., more than two levels) digital systems.

Binary Signaling

The most common binary bandpass signaling techniques are illustrated in Fig. 5-1. They are

- *On–off keying* (OOK), also called *amplitude shift keying* (ASK), which consists of keying (switching) a carrier sinusoid on and off with a unipolar binary signal. It is identical to *unipolar* binary modulation on a DSB-SC signal, (4-117). Morse code radio transmission is an example of this technique. Consequently, OOK was one of the first modulation techniques to be used and precedes analog communication systems.
- *Binary-phase shift keying* (BPSK), which consists of shifting the phase of a sinusoidal carrier 0° or 180° with a unipolar binary signal. It is equivalent to PM signaling with a digital waveform and is also equivalent to modulating a DSB-SC signal with a polar digital waveform.
- *Frequency-shift keying* (FSK), which consists of shifting the frequency of a sinusoidal carrier from a mark frequency (corresponding, for example, to sending a binary 1) to a space frequency (corresponding to sending a binary 0) according to the baseband digital signal. It is identical to modulating a FM carrier with a binary digital signal.

As indicated in Sec. 3-5, usually the bandwidth of the digital signal needs to be minimized to achieve spectral conservation as one transmits information over a channel. This may be accomplished by using a premodulation raised cosine-rolloff filter to minimize the bandwidth of the digital signal and yet not introduce ISI. The shaping of the baseband digital signal produces an analog baseband waveform that modulates the transmitter. Figure 5-1f illustrates the resulting DSB-SC signal when a premodulation filter is used. Thus the BPSK signal of Fig. 5-1d becomes a DSB-SC signal (Fig. 5-1f) when premodulation filtering is used.

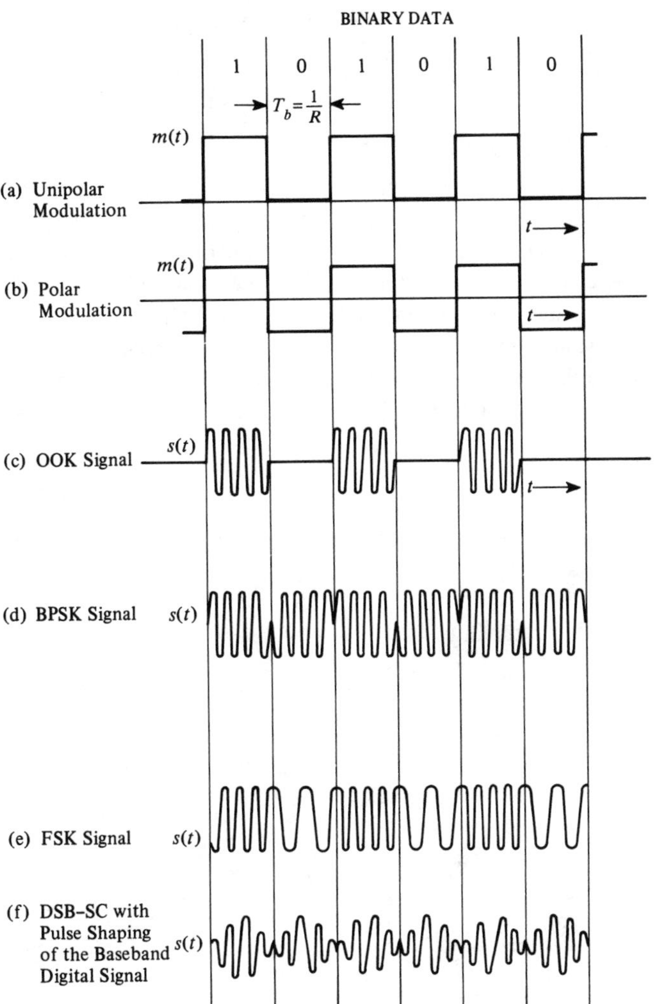

FIGURE 5-1 Bandpass digitally modulated signals.

On–Off Keying (OOK). The OOK signal is represented by

$$s(t) = A_c m(t) \cos \omega_c t \tag{5-2}$$

where $m(t)$ is a unipolar baseband data signal, as shown in Fig. 5-1a. Consequently, the complex envelope is simply

$$g(t) = A_c m(t) \qquad \text{for OOK} \tag{5-3}$$

and the PSD of this complex envelope is proportional to that for the unipolar signal. Using (3-32), we find that this PSD is

$$\mathcal{P}_g(f) = \frac{A_c^2}{2}\left[\delta(f) + T_b\left(\frac{\sin \pi f T_b}{\pi f T_b}\right)^2\right] \quad \text{for OOK} \tag{5-4}$$

where $m(t)$ has a peak value of $\sqrt{2}$ so that $s(t)$ has an average normalized power of $A_c^2/2$. The PSD for the corresponding OOK signal is then obtained by substituting (5-4) into (5-1). The result is shown for positive frequencies in Fig. 5-2a, where $R = 1/T_b$ is the bit rate. It is seen that the null-to-null bandwidth is $2R$. That is, the transmission bandwidth of the OOK signal is $B_T = 2B$ where B is the baseband bandwidth since OOK is AM-type signaling.

If raised cosine-rolloff filtering is used (to conserve bandwidth), the absolute bandwidth of the filtered binary signal is related to the bit rate R by (3-74), where $D = R$ for binary digital signaling. Thus

$$B = \tfrac{1}{2}(1 + r)R \tag{5-5}$$

where r is the rolloff factor of the filter. This gives an absolute transmission bandwidth of

$$B_T = (1 + r)R \tag{5-6}$$

for OOK signaling.

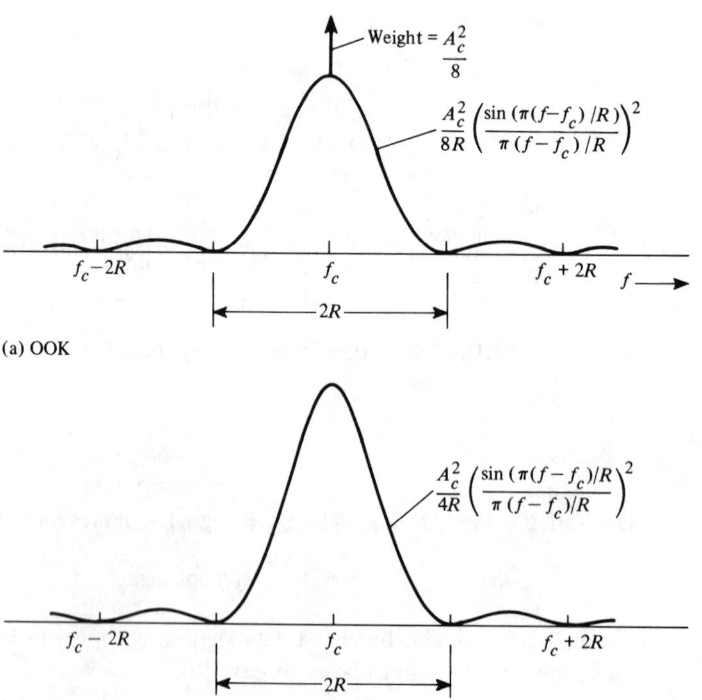

FIGURE 5-2 PSD of bandpass digital signals (positive frequencies shown).

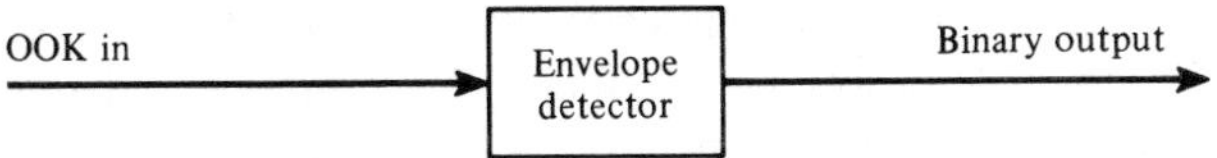

FIGURE 5-3 Noncoherent detection of OOK.

OOK may be detected using either an envelope detector (noncoherent detection) or a product detector (coherent detection) since it is a form of AM signaling. These detectors are shown in Figs. 5-3 and 5-4. For product detection the carrier reference must be provided. This may be accomplished by using a PLL to extract the carrier reference since there is a discrete carrier term in the OOK signal. In Chapters 7 and 8 the optimum detector that gives the lowest probability of bit error is found to be a coherent detector. It uses a specific type of filter, called a *matched filter*.

Binary-Phase Shift Keying (BPSK). The BPSK signal is represented by

$$s(t) = A_c \cos[\omega_c t + D_p m(t)] \tag{5-7}$$

where $m(t)$ is a polar baseband data signal. For convenience, let $m(t)$ have peak values of ± 1.

It will now be shown that BPSK is also a form of AM-type signaling. Expanding (5-7), we get

$$s(t) = A_c \cos(D_p m(t)) \cos \omega_c t - A_c \sin(D_p m(t)) \sin \omega_c t$$

Recalling that $m(t)$ has values of ± 1 and that $\cos(x)$ and $\sin(x)$ are even and odd functions of x, the representation of the BPSK signal reduces to

$$s(t) = \underbrace{(A_c \cos D_p) \cos \omega_c t}_{\text{pilot carrier term}} - \underbrace{(A_c \sin D_p) m(t) \sin \omega_c t}_{\text{data term}} \tag{5-8}$$

The level of the pilot carrier term is set by the value of the peak deviation, $\Delta\theta = D_p$.

For digital angle modulated signals, the *digital modulation index, h,* is defined by

$$h = \frac{2\Delta\theta}{\pi} \tag{5-9}$$

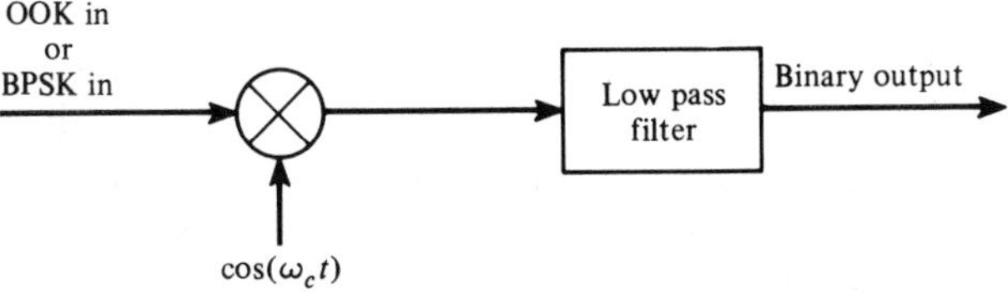

FIGURE 5-4 Coherent detection of OOK or BPSK.

where $2\Delta\theta$ is the maximum peak-to-peak phase deviation (radians) during the time required to send one symbol, T_s. For binary signaling, the symbol time is equal to the bit time, $T_s = T_b$.

The level of the pilot carrier term is set by the value of the peak deviation, which is $\Delta\theta = D_p$ for $m(t) = \pm 1$. If D_p is small, the pilot carrier term has a relatively large amplitude compared to the data term; consequently, there is very little power in the data term (which contains the source information). To maximize the signaling efficiency (low probability of error), the power in the data term needs to be maximized. This is accomplished by letting $\Delta\theta = D_p = 90° = \pi/2$ radians, which corresponds to a digital modulation index of $h = 1$. For this optimum case of $h = 1$, the BPSK signal becomes

$$s(t) = -A_c m(t) \sin \omega_c t \tag{5-10}$$

Throughout this text, we will assume that $\Delta\theta = 90°$, $h = 1$, is used for BPSK signaling (unless otherwise stated). Equation (5-10) shows that BPSK is equivalent to DSB-SC signaling with a polar baseband data waveform. The complex envelope for this BPSK signal is

$$g(t) = jA_c m(t) \qquad \text{for BPSK} \tag{5-11}$$

Using (3-31), we obtain the PSD for the complex envelope,

$$\mathcal{P}_g(f) = A_c^2 T_b \left(\frac{\sin \pi f T_b}{\pi f T_b} \right)^2 \qquad \text{for BPSK} \tag{5-12}$$

where $m(t)$ has values of ± 1 so that $s(t)$ has an average normalized power of $A_c^2/2$. The PSD for the corresponding BPSK signal is readily evaluated by translating the baseband spectrum to the carrier frequency, as specified by substituting (5-12) into (5-1). The resulting BPSK spectrum is shown in Fig. 5-2b. The null-to-null bandwidth for BPSK is also $2R$, the same as that found for OOK.

To detect BPSK, synchronous detection must be used, as illustrated in Fig. 5-4. Since there is no discrete carrier term in the BPSK signal, a PLL may be used to extract the carrier reference *only* if a low-level pilot carrier is transmitted together with this BPSK signal. Otherwise, a Costas loop or a squaring loop (Fig. 4-34) may be used to synthesize the carrier reference from this DSB-SC (i.e., BPSK) signal and to provide coherent detection. However, the 180° phase ambiguity must be resolved as discussed in Sec. 4.5. This can be accomplished by using differential coding at the transmitter input and differential decoding at the receiver output, as illustrated previously in Fig. 3-13.

In Chapters 6 through 8 it will be shown that optimum detection of BPSK consists of product detection as illustrated in Fig. 5-2, except that the low-pass filter is replaced by a resettable integrator that is turned on at the beginning of each bit period and integrates to the end of the bit. Then the integrator output is sampled and mapped into the digital value (e.g., a binary 1 if the sample value is a positive voltage and a binary 0 if the sample value is negative). The integrator is then reset to zero initial condition, and this procedure is repeated for detection of the next bit. This optimum detector, called an *integrate-and-dump detector,* requires knowledge of bit timing (called *bit synchronization*) as well as carrier sync.

Frequency-Shift Keying (FSK). The FSK signal can be characterized as one of two different types, depending on the method used to generate the FSK signal. One type is generated by switching the transmitter output line between two different oscillators, as shown in Fig. 5-5a. This type generates an output waveform that is discontinuous at the switching times. It is called *discontinuous-phase FSK* since $\theta(t)$ is discontinuous at the switching times. The discontinuous-phase FSK signal is represented by

$$s(t) = \begin{cases} A_c \cos(\omega_1 t + \theta_1), & \text{for } t \text{ in the time interval when a binary 1 is being sent} \\ A_c \cos(\omega_2 t + \theta_2), & \text{for } t \text{ in the time interval when a binary 0 is sent} \end{cases} \tag{5-13}$$

where f_1 is called the *mark* (binary 1) frequency and f_2 is called the *space* (binary 0) frequency. Since FSK transmitters are not usually built this way, we will turn to the second type, shown in Fig. 5-5b.

The *continuous-phase FSK* signal is generated by feeding the data signal into a frequency modulator, as shown in Fig. 5-5b. This FSK signal is represented by (referring to Sec. 4-7)

$$s(t) = A_c \cos\left[\omega_c t + D_f \int_{-\infty}^{t} m(\lambda)\, d\lambda\right]$$

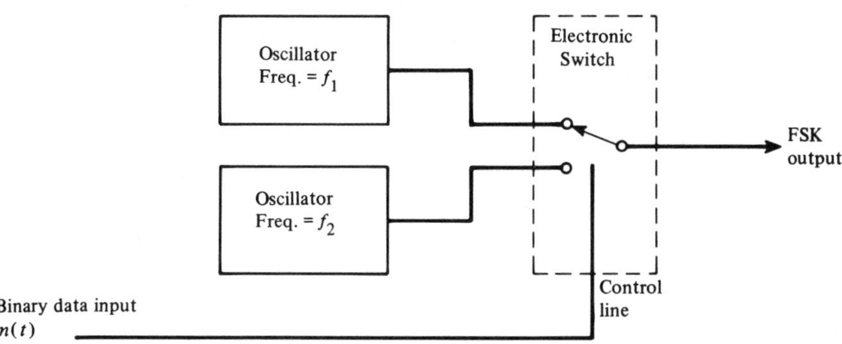

(a) Discontinuous-Phase FSK

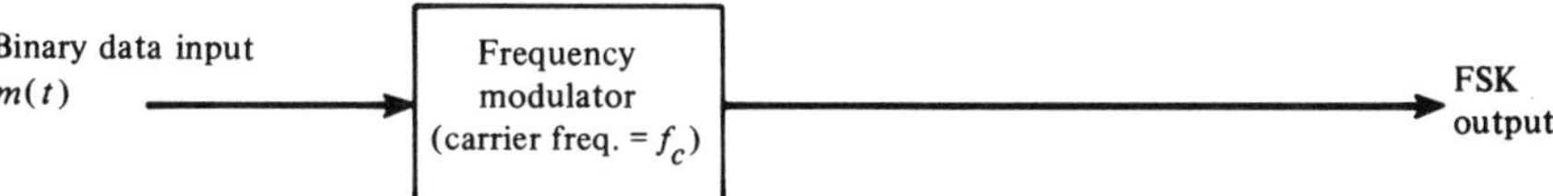

(b) Continous-Phase FSK

FIGURE 5-5 Generation of FSK.

or

$$s(t) = \text{Re}\{g(t)e^{j\omega_c t}\} \tag{5-14a}$$

where

$$g(t) = A_c e^{j\theta(t)} \tag{5-14b}$$

$$\theta(t) = D_f \int_{-\infty}^{t} m(\lambda)\, d\lambda \qquad \text{for FSK} \tag{5-14c}$$

and $m(t)$ is a baseband digital signal. Although $m(t)$ is discontinuous at the switching time, the phase function $\theta(t)$ is continuous because $\theta(t)$ is proportional to the integral of $m(t)$. If the serial data input waveform is binary, such as a polar baseband signal, the resulting FSK signal is called a *binary FSK* signal. Of course, a multilevel input signal would produce a multilevel FSK signal. We will assume that the input is a binary signal in this section and examine the properties of binary FSK signals.

In general, the spectra of FSK signals are difficult to evaluate since the complex envelope, $g(t)$, is a nonlinear function of $m(t)$. However, the techniques that were developed in Sec. 4-7 are applicable, as we show in the following example.

EXAMPLE 5-1 SPECTRUM OF THE BELL-TYPE 103 FSK MODEM

Keyboard-type computer terminals are often used for communication with a remote computer via analog dial-up telephone lines. *Dial-up* means that the computer terminal user calls the computer facility on a telephone and uses the telephone connection for data communication. To accomplish this, modems (i.e., a modulator and a demodulator) are connected to the phone line at each end, either by direct hardwire connection or by acoustical coupling with the telephone handset. This is illustrated in Fig. 5-6 with an FSK-type modem. (In Appendix C, specifications are also given for modems with other types of modulation.) The Bell Telephone System was the first to adopt this technique, with their Model 103 modem, which uses FSK with a signaling speed of 300 bits/sec. Each modem contains both an FSK transmitter and an FSK receiver so that the computer terminal can both "talk" and "listen." Two FSK frequency bands are used (one around 1 kHz and another around 2 kHz) so that it is possible to talk

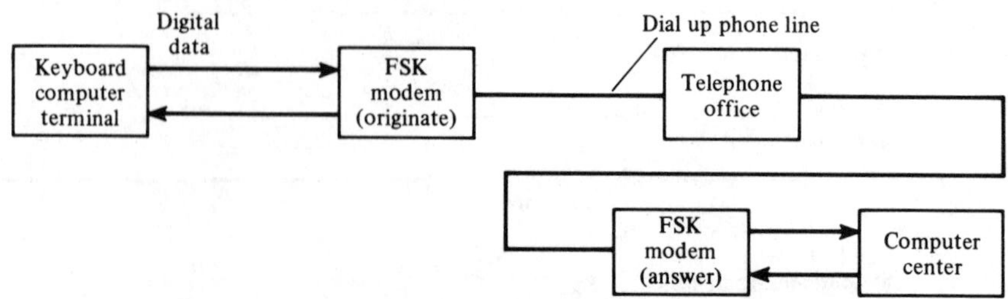

FIGURE 5-6 Computer communication using FSK signaling.

TABLE 5-1 Mark and Space Frequencies for the Bell-Type 103 Modem

	Originate Modem	Answer Modem
Transmit frequencies		
Mark (binary 1)	$f_1 = 1270$ Hz	$f_1 = 2225$ Hz
Space (binary 0)	$f_2 = 1070$ Hz	$f_2 = 2025$ Hz
Receive frequencies		
Mark (binary 1)	$f_1 = 2225$ Hz	$f_1 = 1270$ Hz
Space (binary 0)	$f_2 = 2025$ Hz	$f_2 = 1070$ Hz

and listen simultaneously. This is called *full-duplex capability*. (In *half-duplex* one cannot listen while talking, and vice versa; in *simplex* one can only talk or only listen.) The standard mark and space frequencies for the two bands are shown in Table 5-1. From this table it is seen that the peak-to-peak deviation is $2\Delta F = 200$ Hz.

The spectrum for a Bell-type 103 modem will now be evaluated for the case of the widest-bandwidth FSK signal. It is obtained when the input data signal consists of a (periodic) square wave corresponding to an alternating data pattern (i.e., 10101010). This is illustrated in Fig. 5-7a, where T_b is the time interval for

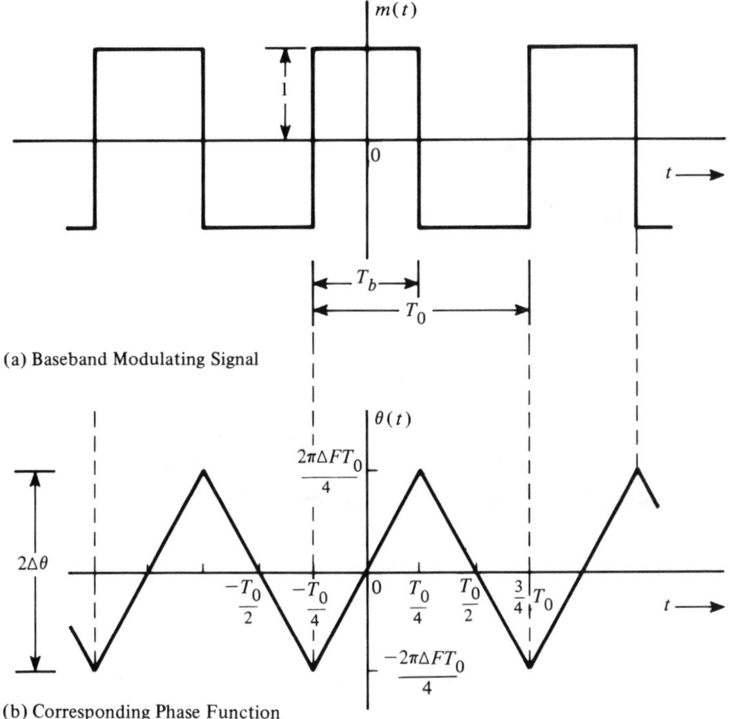

FIGURE 5-7 Input data signal and FSK signal phase function.

one bit and $T_0 = 2T_b$ is the period of the data modulation. The spectrum is obtained by using the Fourier series technique developed in Sec. 4-7 and Example 4-4. Since the modulation is periodic, we would expect the FSK spectrum to be a line spectrum (i.e., delta functions). Using (5-14c) and (4-144), we find that the peak frequency deviation is $\Delta F = D_f/(2\pi)$ for $m(t)$ having values of ± 1. This results in the triangular phase function shown in Fig. 5-7b. From this figure the digital modulation index is

$$h = \frac{2\Delta\theta}{\pi} = \Delta F T_0 = \frac{2\Delta F}{R} \tag{5-15}$$

where the bit rate is $R = 1/T_b = 2/T_0$. In this application, note that the digital modulation index as given by (5-15) is identical to the FM modulation index as defined by (4-150) since

$$h = \frac{\Delta F}{1/T_0} = \frac{\Delta F}{B} = \beta_f$$

provided that the bandwidth of $m(t)$ is defined as $B = 1/T_0$.

The Fourier series for the complex envelope is

$$g(t) = \sum_{-\infty}^{\infty} c_n e^{jn\omega_0 t} \tag{5-16}$$

where

$$\begin{aligned} c_n &= \frac{A_c}{T_0} \int_{-T_0/2}^{T_0/2} e^{j\theta(t)} e^{-jn\omega_0 t}\, dt \\ &= \frac{A_c}{T_0} \left[\int_{-T_0/4}^{T_0/4} e^{j\Delta\omega t - jn\omega_0 t}\, dt + \int_{T_0/4}^{3T_0/4} e^{-j\Delta\omega(t-(T_0/2))} e^{-jn\omega_0 t}\, dt \right] \end{aligned} \tag{5-17}$$

and $\Delta\omega = 2\pi\Delta F = 2\pi h/T_0$. This reduces to

$$c_n = \frac{A_c}{2} \left[\left(\frac{\sin[(\pi/2)(h-n)]}{(\pi/2)(h-n)} \right) + (-1)^n \left(\frac{\sin[(\pi/2)(h+n)]}{(\pi/2)(h+n)} \right) \right] \tag{5-18}$$

where the digital modulation index is $h = 2\Delta F/R$. $2\Delta F$ is the peak-to-peak frequency shift and R is the bit rate. Using (4-151) and (4-161a), we see that the spectrum of this FSK signal with alternating data is

$$S(f) = \tfrac{1}{2}[G(f - f_c) + G^*(-f - f_c)] \tag{5-19a}$$

where

$$G(f) = \sum_{-\infty}^{\infty} c_n\, \delta(f - nf_0) = \sum_{-\infty}^{\infty} c_n\, \delta\left(f - \frac{nR}{2}\right) \tag{5-19b}$$

and c_n is given by (5-18).

TABLE 5-2 BASIC Program for FSK Spectrum with 10101010 Data

```
10 PRINT:PRINT
20 PRINT "****** SPECTRUM FOR A FSK SIGNAL WITH 10101010 DATA
*******"
30 PRINT "L. W. COUCH
40 PRINT:PRINT
50 CLEAR
60 INPUT "THE BIT RATE (in bits/sec) IS";R
70 INPUT "THE FREQUENCY OF THE LOW FSK TONE IS";FL
80 INPUT "THE FREQUENCY SHIFT (between Mark and Space) IS";S
90 INPUT "DO YOU WANT A PRINT LISTING (Y or N)";PRT$
100 PI2=3.1415926#/2
110 DEF FNSA(X)=(SIN(PI2*X))/(PI2*X)
120 PRINT "N";TAB(6) "FREQ", "SPECTRUM";TAB(30) "DB SPECTRUM"
130 IF PRT$="Y" THEN LPRINT "BIT RATE=";R,"LOW FREQ=";FL,"FREQ
SHIFT=";S :LPRINT :LPRINT: LPRINT "N";TAB(6) "FREQ",
"SPECTRUM";TAB(30) "DB SPECTRUM"
140 FOR I=1 TO 21
150 N=I-11
160 F=FL+(S/2)+(N*R/2)
170 X=(S/R)-N
180 Y=(S/R)+N
190 IF ABS(X)<1E-08 THEN XX=1 ELSE XX=FNSA(X)
200 IF ABS(Y)<1E-08 THEN YY=1 ELSE YY=FNSA(Y)
210 SPEC=(ABS(XX+((-1)^N)*YY))/2
220 IF SPEC<.000001 THEN SPEC=.000001
230 DB=20*(LOG(SPEC))/LOG(10)
240 PRINT N;TAB(6) F,SPEC;TAB(30) DB
250 IF PRT$="Y" THEN LPRINT N;TAB(6) F,SPEC;TAB(30) DB
260 NEXT I
270 IF PRT$="Y" THEN LPRINT:LPRINT:LPRINT
280 END
290 IF PRT$="Y" THEN LPRINT N;TAB(6) F,SPEC TAB(30) DB
300 GOTO 260
```

FSK spectra can be evaluated easily for cases of different frequency shifts ΔF and bit rates R if a personal computer is used. A BASIC program that evaluates (5-18) and (5-19) is shown in Table 5-2. The resulting spectral values for the case of Bell 103 modem parameters are shown in Table 5-3. A summary of three computer runs using different sets of parameters is shown in Fig. 5-8. Figure 5-8a gives the FSK spectrum for the Bell 103. For this case, where the Bell 103 parameters are used, the digital modulation index is $h = 0.67$. Note that for this Bell 103 case there are no spectral lines at the mark and space frequencies, f_1 and f_2. Figures 5-8b and 5-8c give the FSK spectra for $h = 1.82$ and $h = 3.33$. Note that as the modulation index is increased, the spectrum concentrates about f_1 and f_2. This is the same result predicted by the PSD theorem for wideband FM (4-167) studied in Chapter 4, since the PDF of the binary modulation consists of two delta functions (one for $+1$ and one for -1 of Fig. 5-7a).

TABLE 5-3 Spectrum for Bell 103 FSK Modem: Originate Mode, $f_2 = 1070$ Hz, $f_1 = 1270$ Hz, $2\Delta F = 200$ Hz for $h = 0.67$

N	Frequency	Spectrum	DB Spectrum
-10	-330	3.691949E-03	-48.65489
-9	-180	2.634233E-03	-51.58692
-8	-30	5.783174E-03	-44.75667
-7	120	4.370366E-03	-47.18965
-6	270	1.033743E-02	-39.71175
-5	420	8.641859E-03	-41.26786
-4	570	2.362843E-02	-32.5313
-3	720	2.480333E-02	-32.1098
-2	870	.1033742	-19.71175
-1	1020	.3819719	-8.359373
0	1170	.8269934	-1.64996
1	1320	.3819719	-8.359373
2	1470	.1033742	-19.71175
3	1620	2.480333E-02	-32.1098
4	1770	2.362843E-02	-32.5313
5	1920	8.641859E-03	-41.26786
6	2070	1.033743E-02	-39.71175
7	2220	4.370366E-03	-47.18965
8	2370	5.783174E-03	-44.75667
9	2520	2.634233E-03	-51.58692
10	2670	3.691949E-03	-48.65489

The approximate transmission bandwidth B_T for the FSK signal is given by Carson's rule: $B_T = 2(\beta + 1)B$, where $\beta = \Delta F/B$. This is equivalent to

$$B_T = 2\Delta F + 2B \tag{5-20}$$

B is the bandwidth of the digital (e.g., square-wave) modulation waveform. For our example of an alternating binary 1 and 0 test pattern waveform, the bandwidth of this square-wave modulating waveform (assuming that the first null type of bandwidth is used) is $B = R$ and, using (5-20), the FSK transmission bandwidth becomes

$$B_T = 2(\Delta F + R) \tag{5-21}$$

This result is illustrated in Fig. 5-8. If a raised cosine-rolloff premodulation filter is used, the transmission bandwidth of the FSK signal becomes

$$B_T = 2\Delta F + (1 + r)R \tag{5-22}$$

For wideband FSK, where $\beta \gg 1$, ΔF dominates in these equations and we have $B_T = 2\Delta F$. For narrowband FSK the transmission bandwidth is $B_T = 2B$.

The exact PSD for continuous-phase FSK signals is difficult to evaluate for the case of random data modulation. However, it can be done by using some

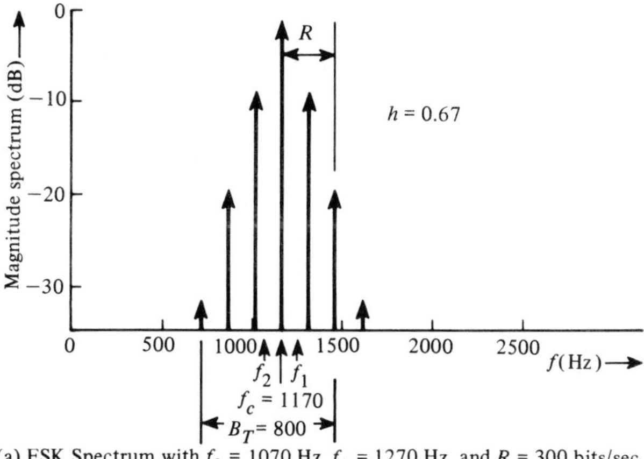

(a) FSK Spectrum with f_2 = 1070 Hz, f_1 = 1270 Hz, and R = 300 bits/sec (Bell 103 Parameters, Originate mode) for h = 0.67

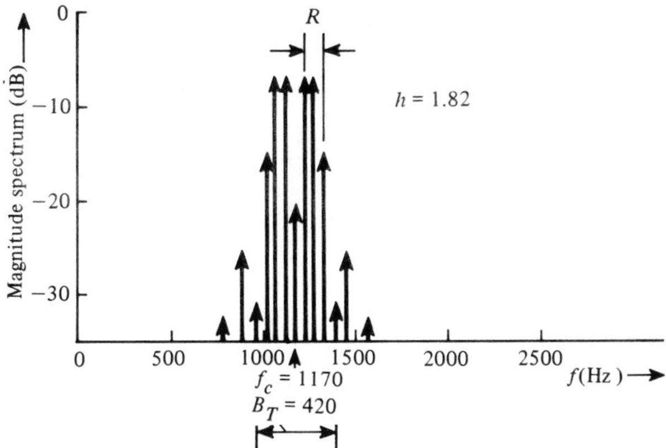

(b) FSK Spectrum with f_2 = 1070 Hz, f_1 = 1270 Hz, and R = 110 bits/sec for h = 1.82

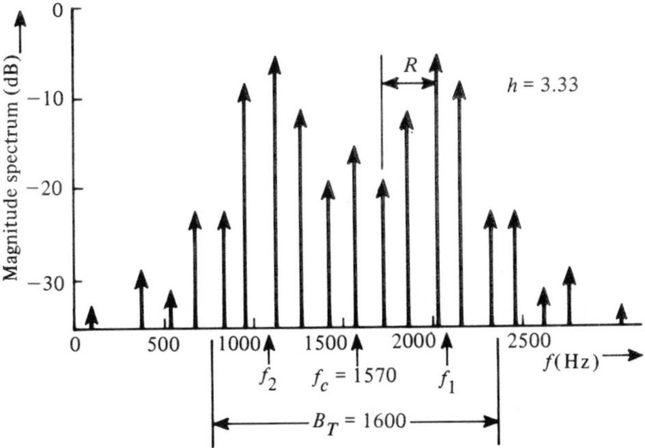

(c) FSK Spectrum with f_2 = 1070 Hz, f_1 = 2070 Hz, and R = 300 bits/sec for h = 3.33

FIGURE 5-8 FSK spectra for alternating data modulation (positive frequencies shown with one-sided magnitude values).

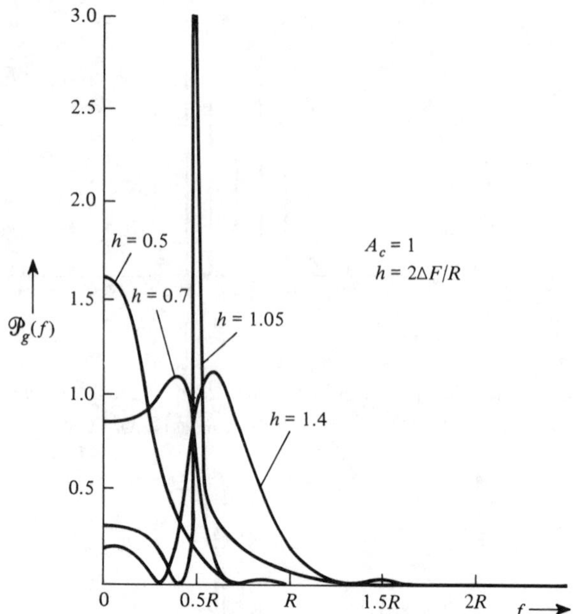

FIGURE 5-9 PSD for the complex envelope of FSK (positive frequencies shown).

elegant statistical techniques [Proakis, 1983, pp. 130–135]. The resulting PSD for the complex envelope of the FSK signal is[†]

$$\mathcal{P}_g(f) = \frac{A_c^2 T_b}{2} \times (A_1^2(f)[1 + B_{11}(f)] + A_2^2(f)[1 + B_{22}(f)] + 2B_{12}(f)A_1(f)A_2(f)) \quad \text{(5-23a)}$$

where

$$A_n(f) = \frac{\sin[\pi T_b(f - \Delta F(2n-3))]}{\pi T_b(f - \Delta F(2n-3))} \quad \text{(5-23b)}$$

and

$$B_{nm}(f) = \frac{\cos[2\pi f T_b - 2\pi \Delta F T_b(n+m-3)] - \cos(2\pi \Delta F T_b)\cos[2\pi \Delta F T_b(n+m-3)]}{1 + \cos^2(2\pi \Delta F T_b) - 2\cos(2\pi \Delta F T_b)\cos(2\pi f T_b)} \quad \text{(5-23c)}$$

where ΔF is the peak frequency deviation, $R = 1/T_b$ is the bit rate, and the digital modulation index is $h = 2\Delta F/R$. The PSD is plotted in Fig. 5-9 for several values of the digital modulation index. The curves in this figure were obtained by using the MathCAD computation shown in Appendix D.

[†]This result assumes that the carrier oscillator phase is not synchronized to any external source. That is, as discussed in Sec. 6-7, the startup phase θ_c is a random variable that is uniformly distributed over $(0, 2\pi)$. It is also assumed that $h \neq 0, 1, 2, \ldots$. When $h = 0, 1, 2, \ldots$, there are also discrete terms (delta functions) in the spectrum.

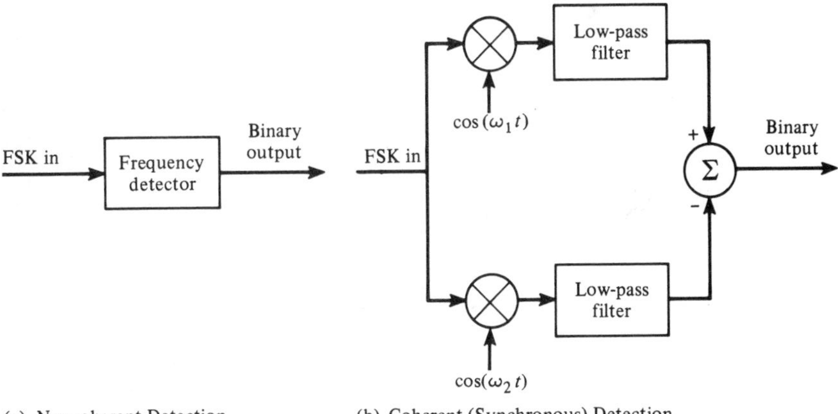

(a) Noncoherent Detection (b) Coherent (Synchronous) Detection

FIGURE 5-10 Detection of FSK.

FSK can be detected by using either a frequency (noncoherent) detector or two product detectors (coherent detection), as shown in Fig. 5-10. A detailed study of coherent and noncoherent detection of FSK is given in Secs. 7-3 and 7-4, where the probability of bit error is evaluated for OOK, BPSK, and FSK signaling.

Multilevel Signaling

With multilevel signaling, digital inputs with more than two modulation levels are allowed on the transmitter input. This is illustrated in Fig. 5-11, which shows how multilevel signals can be generated from a serial binary input stream using a digital-to-analog converter (DAC). For example, suppose that an $\ell = 2$-bit DAC is used. Then the number of levels (symbols) in the multilevel signal is $M = 2^\ell = 2^2 = 4$. The symbol rate (baud) of the multilevel signal is $D = R/\ell = \frac{1}{2}R$, where the bit rate is $R = 1/T_b$ bits/sec.

Quadrature Phase-Shift Keying (QPSK) and *M*-ary Phase-Shift Keying (MPSK). If the transmitter is a PM transmitter with an $M = 4$ level digital modulation signal, *M-ary phase-shift keying* (MPSK) is generated at the transmitter output. A plot of the permitted values of the complex envelope, $g(t) = A_c e^{j\theta(t)}$, would contain four points, one value of g (a complex number in general)

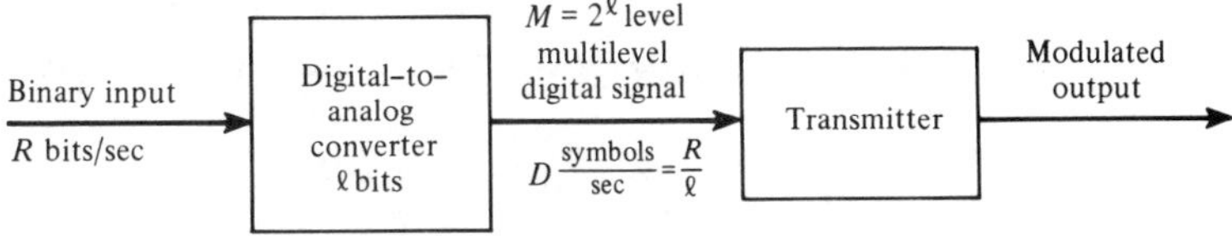

FIGURE 5-11 Multilevel digital transmission system.

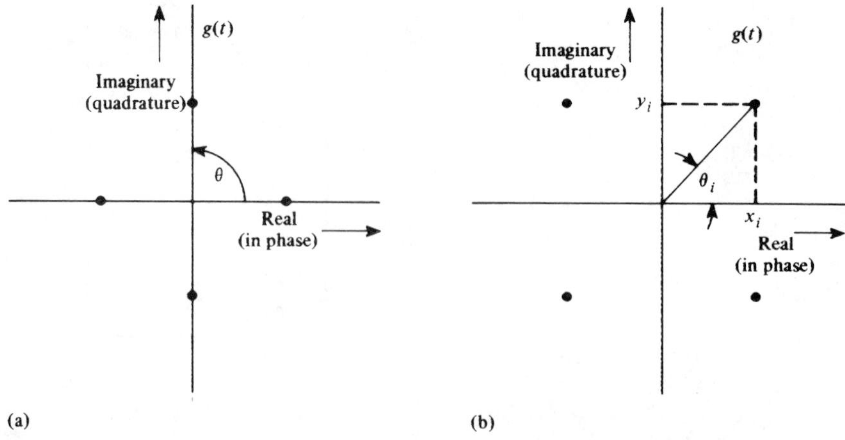

FIGURE 5-12 QPSK signal constellations (permitted values of the complex envelope).

for each of the four multilevel values, corresponding to the four phases that θ is permitted to have. A plot of two possible sets of $g(t)$ is shown in Fig. 5-12. For instance, suppose that the permitted multilevel values at the DAC are -3, -1, $+1$, and $+3$ V; then in Fig. 5-12a these multilevel values might correspond to PSK phases of 0, 90, 180, and 270°, respectively. In Fig. 5-12b these levels would correspond to carrier phases of 45, 135, 225, and 315°, respectively. These two signal constellations are essentially the same except for a shift in the carrier-phase reference.† This example of M-ary PSK where $M = 4$ is called *quadrature-phase-shift-keyed* (QPSK) signaling.

MPSK can also be generated by using two quadrature carriers modulated by the x and y components of the complex envelope (instead of using a phase modulator)

$$g(t) = A_c e^{j\theta(t)} = x(t) + jy(t) \tag{5-24}$$

where the permitted values of x and y are

$$x_i = A_c \cos \theta_i \tag{5-25}$$

$$y_i = A_c \sin \theta_i \tag{5-26}$$

for the permitted phase angles θ_i, $i = 1, 2, \ldots, M$, of the MPSK signal. This is illustrated by Fig. 5-13, where the signal processing circuit implements (5-25) and (5-26). Figure 5-12 gives the relationship between the permitted phase angles θ_i and the (x_i, y_i) components for two QPSK signal constellations. Note that this is identical to the quadrature method of generating modulated signals presented in Fig. 4-28.

†A constellation is an N-dimensional plot of the possible signal vectors corresponding to the possible digital signals (see Sec. 2-4).

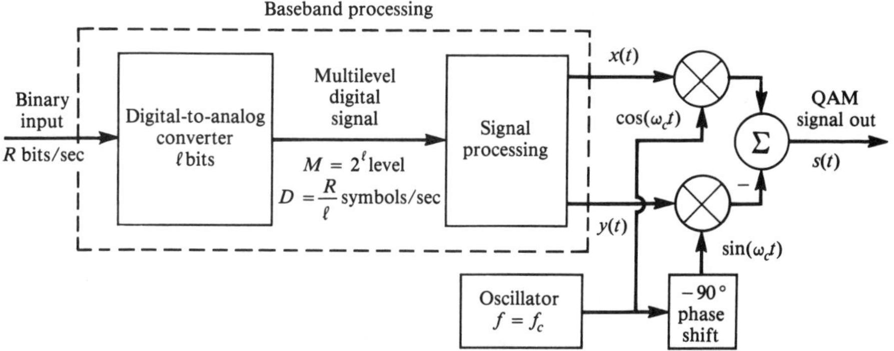

(a) Modulator for Generalized Signal Constellation

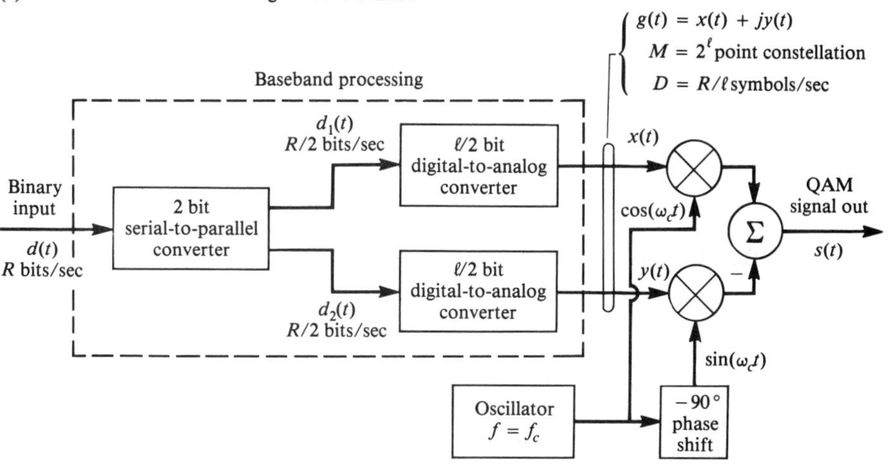

(b) Modulator for Rectangular Signal Constellation

FIGURE 5-13 Generation of QAM signals.

Quadrature Amplitude Modulation (QAM). Quadrature carrier signaling, as shown in Fig. 5-13, is called *quadrature amplitude modulation* (QAM). In general, QAM signal constellations are not restricted to having permitted signaling points only on a circle (of radius A_c, as was the case for MPSK). The general QAM signal is

$$s(t) = x(t) \cos \omega_c t - y(t) \sin \omega_c t \tag{5-27}$$

where

$$g(t) = x(t) + jy(t) = R(t)e^{j\theta(t)} \tag{5-28}$$

For example, a popular 16-symbol ($M = 16$ levels) QAM constellation is shown in Fig. 5-14, where the relationship between (R_i, θ_i) and (x_i, y_i) can readily be evaluated for each of the 16 signal values permitted. This type of signaling is used by 2400 bit/sec V.22 bis computer modems (see Table C-12). Here x_i and y_i

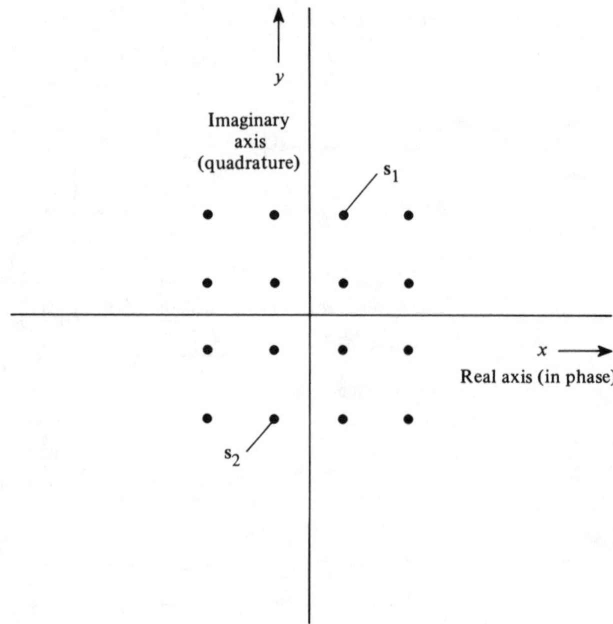

FIGURE 5-14 16-symbol QAM constellation (four levels per dimension).

are each permitted to have four levels per dimension. This 16-symbol QAM signal may be generated by using two ($\ell/2 = 2$)-bit digital-to-analog converters and quadrature balanced modulators as shown in Fig. 5-13b. The waveforms of x and y are represented by

$$x(t) = \sum_n x_n h_1\left(t - \frac{n}{D}\right) \tag{5-29}$$

and

$$y(t) = \sum_n y_n h_1\left(t - \frac{n}{D}\right) \tag{5-30}$$

where $D = R/\ell$ and (x_n, y_n) denotes one of the permitted (x_i, y_i) values during the symbol time that is centered on $t = nT = n/D$ seconds. (It takes T seconds to send each symbol.) $h_1(t)$ is the pulse shape that is used for each symbol. If the bandwidth of the QAM signal is not to be restricted, the pulse shape will be rectangular of T-second duration. In some applications the timing between the $x(t)$ and $y(t)$ components is offset by $T/2 = 1/(2D)$ seconds. That is, $x(t)$ would be described by (5-29), and the offset $y(t)$ would be described by

$$y(t) = \sum_n y_n h_1\left(t - \frac{n}{D} - \frac{1}{2D}\right) \tag{5-31}$$

One popular type of offset signaling is *offset QPSK (OQPSK),* which is offset QAM when $M = 4$. A special case of OQPSK when $h_1(t)$ has a sinusoidal type of pulse shape is called *minimum-shift keying* (MSK). This type will be studied immediately after the next section. Furthermore, a QPSK signal is said to be *unbalanced* if the $x(t)$ and $y(t)$ components have unequal powers and/or unequal data rates.

Power Spectral Density for MPSK and QAM. The PSD for MPSK and QAM signals is relatively easy to evaluate for the case of rectangular bit shape signaling. In this case, the PSD has the same spectral shape that was obtained for BPSK, *provided* that proper frequency scaling is used. For example, take the case of $M = 16$ QAM signaling illustrated in Fig. 5-14. The PSD can be evaluated by taking two *antipodal* signal vectors at a time and evaluating the PSD for the condition of signaling only with data modulation corresponding to those signal vectors. That is, referring to the figure, suppose that the vectors corresponding to the points $\mathbf{s}_1$ and $\mathbf{s}_2$ are used. This is equivalent to BPSK signaling. When we use (5-12), the *conditional* PSD of the complex envelope is of the form

$$\mathcal{P}_g(f|\mathbf{s}_1 \text{ or } \mathbf{s}_2) = A_1\left(\frac{\sin \pi f T}{\pi f T}\right)^2 \tag{5-32}$$

where the time that it takes to send one symbol is T and $T = 1/D = \ell/R = \ell T_b$. The resulting PSD, considering all possible antipodal signal pairs, is (using the probability theory of Appendix B)

$$\mathcal{P}_g(f) = \sum_{\substack{i=1\\ i=\text{odd}}}^{M} \mathcal{P}(f|\mathbf{s}_i \text{ or } \mathbf{s}_{i+1})P(\mathbf{s}_i \text{ or } \mathbf{s}_{i+1}) \tag{5-33}$$

where $i = 1, 3, 5, \ldots$ and $P(s_i \text{ or } s_{i+1})$ is the probability of sending the s_i or s_{i+1} antipodal vector pair. For equally likely random data (i.e., the usual situation), $P(s_i \text{ or } s_{i+1}) = 1/(M/2)$, so (5-33) becomes

$$\mathcal{P}_g(f) = \frac{2}{M}\sum_{\substack{i=1\\ i=\text{odd}}}^{M} A_i\left(\frac{\sin \pi f T}{\pi f T}\right)^2 = \frac{2}{M}\left(\frac{\sin \pi f T}{\pi f T}\right)^2 \sum_{\substack{i=1\\ i=\text{odd}}}^{M} A_i \tag{5-34}$$

But ΣA_i is just a constant. Thus the PSD for the complex envelope of MPSK or QAM signals with data modulation of rectangular bit shape is

$$\mathcal{P}_g(f) = K\left(\frac{\sin \pi f \ell T_b}{\pi f \ell T_b}\right)^2 \quad \text{for MPSK and QAM} \tag{5-35}$$

where $M = 2^\ell$ is the number of points in the signal constellation and the bit rate is $R = 1/T_b$. For a total transmitted power of P watts the value of K is $K = 2P\ell T_b$ since $\int_{-\infty}^{\infty} \mathcal{P}_s(f)\, df = P$. This PSD for the complex envelope is plotted in Fig. 5-15. The PSD of the MPSK or QAM signal is obtained by simply translating the PSD of Fig. 5-15 to the carrier frequency, as described by (5-1). Of course, for $\ell = 1$, the figure gives the PSD for BPSK (i.e., compare Fig. 5-15, $\ell = 1$, with Fig. 5-2b).

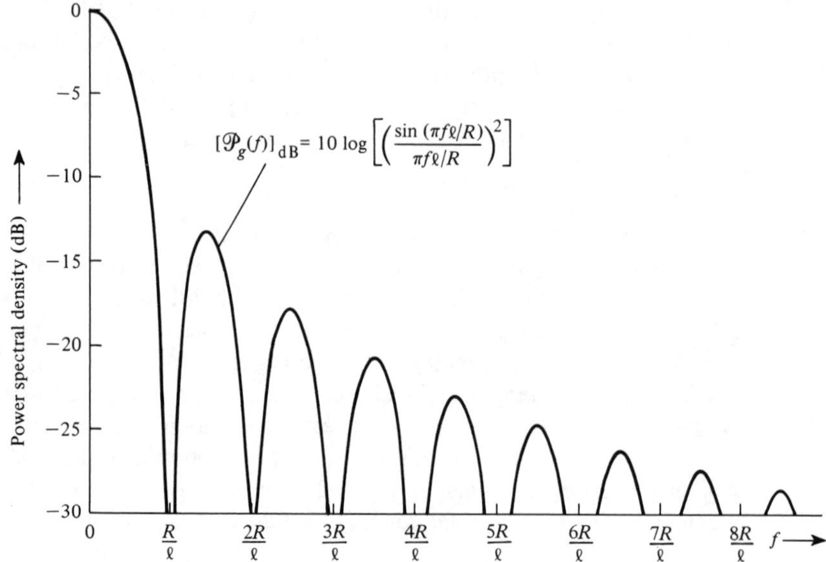

FIGURE 5-15 PSD for the complex envelope of MPSK and QAM where $M = 2^\ell$, R is bit rate, and $R/\ell = D$ is baud rate (positive frequencies shown).

From Fig. 5-15 we see that the null-to-null transmission bandwidth of MPSK or QAM is

$$B_T = 2R/\ell \tag{5-36}$$

when we use (3-55), the spectral efficiency of MPSK or QAM signaling with rectangular data pulses is

$$\eta = \frac{R}{B_T} = \frac{\ell}{2} \quad \frac{\text{bits/sec}}{\text{Hz}} \tag{5-37}$$

where $M = 2^\ell$ is the number of points in the signal constellation. For example, for the $M = 16$ QAM signal, the bandwidth efficiency is $\eta = 2$ bits/sec per hertz of bandwidth. For a 2400-bit/sec V.22bis modem ($M = 16$, QAM, see Table C-12), the null-to-null transmission bandwidth is $B_T = 1200$ Hz.

When QAM (or the QAM equivalent of MPSK) with a pulse shape of $h_1(t)$—see (5-29) and (5-30)—is transmitted over a bandlimited channel, the overall filter characteristic needs to be selected so that the intersymbol interference is negligible. $h_1(t) = h(t) * h_T(t)$ corresponds to the transmitter pulse shape, as described in Sec. 3-5. If the overall pulse shape satisfies the raised cosine-rolloff filter characteristic, then, by use of (3-74), the absolute bandwidth of the M-level modulating signal is

$$B = \tfrac{1}{2}(1 + r)D \tag{5-38}$$

where $D = R/\ell$ from Fig. 5-13 and r is the rolloff factor of the filter characteristic. Furthermore, from our study of AM (and DSB-SC to be specific) we know that the transmission bandwidth is related to the modulation bandwidth by $B_T = 2B$, so that the overall absolute transmission bandwidth of the QAM signal is

$$B_T = \left(\frac{1+r}{\ell}\right)R \tag{5-39}$$

Since $M = 2^\ell$, which implies that $\ell = \log_2 M = (\ln M)/(\ln 2)$, the spectral efficiency of QAM-type signaling with raised cosine filtering is

$$\eta = \frac{R}{B_T} = \frac{\ln M}{(1+r)\ln 2} \quad \frac{\text{bits/sec}}{\text{Hz}} \tag{5-40}$$

This result is an extremely important one since it tells us how fast we can signal for a prescribed bandwidth. Of course, this result holds for MPSK since this is a special case of QAM. This formula is used to generate Table 5-4, which illustrates the allowable bit rate per hertz of transmission bandwidth for QAM signaling. For example, suppose that we want to signal over a telephone line that has an available bandwidth of 2.4 kHz. If we used BPSK ($M = 2$) with a 50% rolloff factor, we could signal at a rate of $B_T \times \eta = 2.4 \times 0.677 = 1.60$ kbits/sec; but if we used QPSK ($M = 4$) with a 25% rolloff factor, we could signal at a rate of $2.4 \times 1.6 = 3.84$ kbits/sec.

In order to conserve bandwidth, the number of levels M in (5-40) cannot be increased too much since, for a given peak envelope power (PEP), the spacing between the signal points in the signal constellation will decrease and noise on the received signal will cause errors (since the noise moves the received signal vector to a new location that might correspond to a different signal level). This effect is discussed in more detail in Example 8-4 for QAM signaling. However, we know that R certainly has to be less than C, the channel capacity (Sec. 1-9) if the errors are to be kept small. Consequently, using (1-10), we require

$$\eta < \eta_{\max} \tag{5-41a}$$

where

$$\eta_{\max} = \log_2\left(1 + \frac{S}{N}\right) \tag{5-41b}$$

TABLE 5-4 Spectral Efficiency for QAM Signaling with Raised Cosine-Rolloff Pulse Shaping

Number of Levels, M (symbols)	Size of DAC, ℓ (bits)	$\eta = \frac{R}{B_T}\left(\frac{\text{bits/sec}}{\text{Hz}}\right)$					
		$r = 0.0$	$r = 0.1$	$r = 0.25$	$r = 0.5$	$r = 0.75$	$r = 1.0$
2	1	1.00	0.909	0.800	0.667	0.571	0.500
4	2	2.00	1.82	1.60	1.33	1.14	1.00
8	3	3.00	2.73	2.40	2.00	1.71	1.50
16	4	4.00	3.64	3.20	2.67	2.29	2.00
32	5	5.00	4.55	4.0	3.33	2.86	2.50

★ **Minimum-Shift Keying (MSK).** Minimum-shift keying is another bandwidth conservation technique that has been developed. It has the advantage of being a constant-amplitude signal and, consequently, can be amplified with Class C amplifiers without distortion. As we will see, MSK is equivalent to OQPSK with sinusoidal pulse shaping [for $h_i(t)$].

DEFINITION: *Minimum-shift keying* (MSK) is continuous-phase FSK with a minimum modulation index ($h = 0.5$) that will produce orthogonal signaling.

First, let us show that $h = 0.5$ is the minimum index allowed for orthogonal continuous-phase FSK. For a binary 1 to be transmitted over the bit interval $0 < t < T_b$, the FSK signal would be $s_1(t) = A_c \cos(\omega_1 t + \theta_1)$, and for a binary 0 to be transmitted, the FSK signal would be $s_2(t) = A_c \cos(\omega_2 t + \theta_2)$, where $\theta_1 = \theta_2$ for the continuous-phase condition at the switching time $t = 0$. For orthogonal signaling, from (2-73), we require the integral of the product of the two signals over the bit interval to be zero. Thus we require that

$$\int_0^{T_b} s_1(t)s_2(t)\, dt = \int_0^{T_b} A_c^2 \cos(\omega_1 t + \theta_1)\cos(\omega_2 t + \theta_2)\, dt = 0 \quad \text{(5-42a)}$$

This reduces to the requirement

$$\frac{A_c^2}{2}\left[\frac{\sin[(\omega_1 - \omega_2)T_b + (\theta_1 - \theta_2)] - \sin(\theta_1 - \theta_2)}{\omega_1 - \omega_2}\right] + \frac{A_c^2}{2}\left[\frac{\sin[(\omega_1 + \omega_2)T_b + (\theta_1 + \theta_2)] - \sin(\theta_1 + \theta_2)}{\omega_1 + \omega_2}\right] = 0 \quad \text{(5-42b)}$$

The second term is negligible because $\omega_1 + \omega_2$ is large,† so the requirement is that

$$\frac{\sin[2\pi h + (\theta_1 - \theta_2)] - \sin(\theta_1 - \theta_2)}{2\pi h} = 0 \quad \text{(5-43)}$$

where $(\omega_1 - \omega_2)T_b = 2\pi(2\Delta F)T_b$ and, from (5-15), $h = 2\Delta F T_b$. For the *continuous-phase* case $\theta_1 = \theta_2$, and (5-43) is satisfied for a minimum value of $h = 0.5$ or a peak frequency deviation of

$$\Delta F = \frac{1}{4T_b} = \frac{1}{4}R \qquad \text{for MSK} \quad \text{(5-44)}$$

Note that for *discontinuous-phase* FSK $\theta_1 \neq \theta_2$ and the minimum value for orthogonality for that case is $h = 1.0$.

Now we will demonstrate that the MSK signal (which is $h = 0.5$ continuous-phase FSK) is a form of OQPSK signaling with sinusoidal pulse shaping. First,

†If $\omega_1 + \omega_2$ is not sufficiently large to make the second term negligible, choose $f_c = \frac{1}{2}m/T_b = \frac{1}{2}mR$ where m is a positive integer. This will make the second term zero ($f_1 = f_c - \Delta F$ and $f_2 = f_c + \Delta F$).

consider the FSK signal over the signaling interval $(0,T_b)$. When we use (5-14), the complex envelope is

$$g(t) = A_c e^{j\theta(t)} = A_c e^{j2\pi\Delta F\int_0^t m(\lambda)\,d\lambda}$$

where $m(t) = \pm 1$, $0 < t < T_b$. Using (5-44), we find that the complex envelope becomes

$$g(t) = A_c e^{\pm j\pi t/(2T_b)} = x(t) + jy(t), \qquad 0 < t < T_b$$

where the $\pm$ signs denote the possible data during the $(0,\ T_b)$ interval. Thus

$$x(t) = A_c \cos\left(\pm 1 \frac{\pi t}{2T_b}\right), \qquad 0 < t < T_b \tag{5-45a}$$

$$y(t) = A_c \sin\left(\pm 1 \frac{\pi t}{2T_b}\right), \qquad 0 < t < T_b \tag{5-45b}$$

and the MSK signal is

$$s(t) = x(t)\cos\omega_c t - y(t)\sin\omega_c t \tag{5-45c}$$

A typical input data waveform, $m(t)$, and the resulting $x(t)$ and $y(t)$ quadrature modulation waveforms, are shown in Fig. 5-16. From (5-45a) and (5-45b), and realizing that $\cos[\pm\pi t/(2T_b)] = \cos[\pi t/(2T_b)]$ and $\sin[\pm\pi t/(2T_b)] = \pm\sin[\pi t/(2T_b)]$, we see that the ± 1 sign of $m(t)$ during the $(0,T_b)$ interval affects only $y(t)$ and not $x(t)$ over the signaling interval of $(0,2T_b)$. We also realize that the $\sin[\pi t/(2T_b)]$ pulse of $y(t)$ is $2T_b$ sec wide. Similarly, it can be seen that the ± 1 sign of $m(t)$ over the $(T_b,2T_b)$ interval affects only $x(t)$ and not $y(t)$ over the interval $(T_b,3T_b)$. In other words, the binary data of $m(t)$ *alternately* modulates the $x(t)$ and $y(t)$ components, and the pulse shape for the $x(t)$ and $y(t)$ symbols (which are $2T_b$ wide instead of T_b) is a sinusoid, as shown in the figure. Thus MSK is equivalent to OQPSK with sinusoidal pulse shaping.

The $x(t)$ and $y(t)$ waveforms as shown in Fig. 5-16 illustrate the so-called Type II MSK [Bhargava, Haccoun, Matyas, and Nuspl, 1981], where the basic pulse shape is always a positive half-cosinusoid. For Type I MSK, the pulse shape [for both $x(t)$ and $y(t)$] alternates between a positive and a negative half-cosinusoid. For both Type I and Type II MSK, it can be shown that there is *not* a one-to-one relationship between the input data $m(t)$ and the resulting mark and space frequencies, f_1 and f_2, in the MSK signal. This can be demonstrated by evaluating the instantaneous frequency, f_i, as a function of the data presented by $m(t)$ during the different bit intervals. The instantaneous frequency is $f_i = f_c + (1/2\pi)[d\theta(t)/dt] = f_c \pm \Delta F$, where $\theta(t) = \tan^{-1}[y(t)/x(t)]$. The $\pm$ sign is determined by the encoding technique (Type I or II) that is used to obtain the $x(t)$ and $y(t)$ waveforms in each bit interval, T_b, as well as the sign of the data on $m(t)$. To get a one-to-one frequency relationship between a Type I MSK signal and the corresponding $h = 0.5$ FSK signal, called a *fast frequency-shift keyed* (FFSK) signal, the data input to the Type I FSK modulator is first differentially encoded. Examples of the waveforms for MSK, Type I and II, and for FFSK can be found in the MathCAD solutions for Probs. 5-24, 5-25, and 5-26. Regardless of the

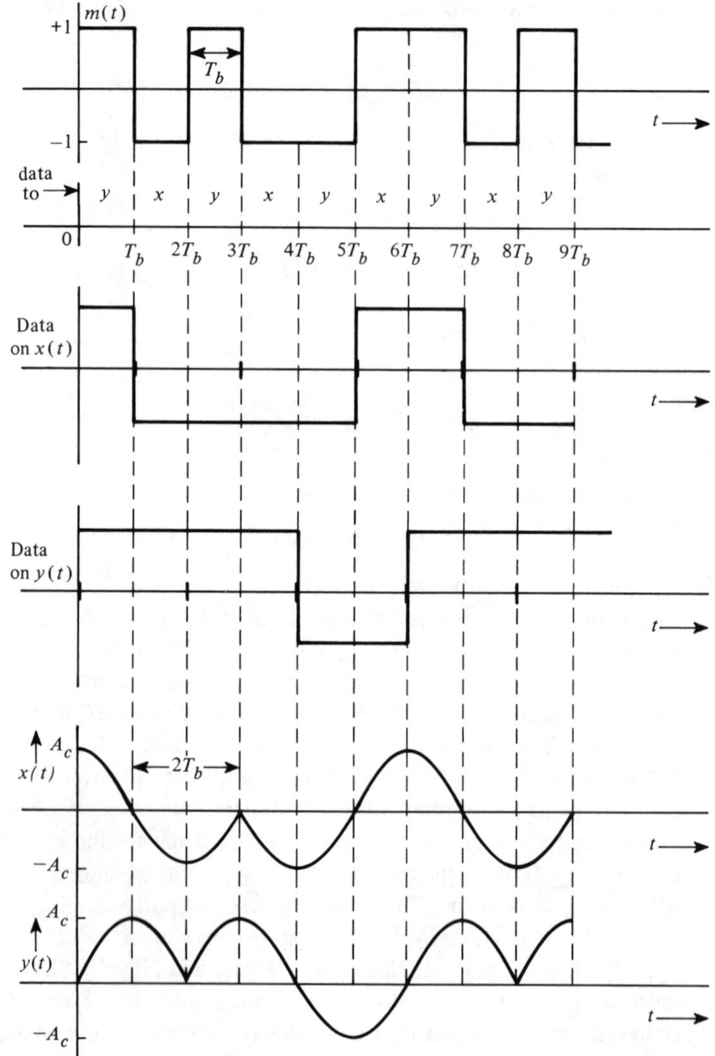

FIGURE 5-16 MSK quadrature component waveforms (Type II MSK).

differences noted, FFSK, MSK Type I, and MSK Type II are all constant-amplitude, continuous-phase FSK signals with a modulation index of $h = 0.5$.

The PSD for MSK (Type I and Type II) can be evaluated as follows. Since $x(t)$ and $y(t)$ have independent data and their dc value is zero, and since $g(t) = x(t) + jy(t)$, the PSD for the complex envelope is

$$\mathcal{P}_g(f) = \mathcal{P}_x(f) + \mathcal{P}_y(f) = 2\mathcal{P}_x(f)$$

where $\mathcal{P}_x(x) = \mathcal{P}_y(f)$ because $x(t)$ and $y(t)$ have the same type of pulse shape. When we use (3-29) or (6-60) with a pulse width of $2T_b$, this PSD becomes

$$\mathcal{P}_g(f) = \frac{2}{2T_b} |F(f)|^2 \tag{5-46}$$

where $F(f) = \mathcal{F}[f(t)]$ and $f(t)$ is the pulse shape. For the MSK half-cosinusoidal pulse shape, we have

$$f(t) = \begin{cases} A_c \cos\left(\dfrac{\pi t}{2T_b}\right), & |t| < T_b \\ 0, & t \text{ elsewhere} \end{cases} \tag{5-47a}$$

and the Fourier transform is

$$F(f) = \frac{4A_c T_b \cos 2\pi T_b f}{\pi[1 - (4T_b f)^2]} \tag{5-47b}$$

Thus, the PSD for the complex envelope of an MSK signal is

$$\mathcal{P}_g(f) = \frac{16A_c^2 T_b}{\pi^2}\left(\frac{\cos^2 2\pi T_b f}{[1 - (4T_b f)^2]^2}\right) \tag{5-48}$$

where the normalized power in the MSK signal is $A_c^2/2$. The PSD for MSK is readily obtained by translating this spectrum up to the carrier frequency, as described by (5-1). This complex envelope PSD for MSK is shown by the solid-line curve in Fig. 5-17. For comparison purposes, the corresponding PSD for OQPSK and (QPSK) complex envelopes is shown by the dashed-line curve. It is also known that there are other digital modulation techniques, such as *tamed frequency modulation* (TFM), that have even better spectral characteristics than MSK [DeJager and Dekker, 1978; Pettit, 1982; Taub and Schilling, 1986], and the optimum pulse shape for minimum spectral occupancy of FSK-type signals has been found [Campanella, LoFaso, and Mamola, 1984].

MSK signals can be generated by using any one of several methods, as illustrated in Fig. 5-18. Figure 5-18a shows the generation of FFSK (which is equivalent to Type I MSK with differential encoding of the input data). Here a simple FM-type modulator having a peak deviation of $\Delta F = 1/(4T_b) = (1/4)R$ is used. Figure 5-18b shows an MSK Type I modulator that is a realization of (5-45). This is called the *parallel* method of generating MSK since parallel in-phase (I) and quadrature-phase (Q) channels are used. Figure 5-18c shows the *serial* method of generating MSK. In this approach, BPSK is first generated at a carrier frequency of $f_2 = f_c - \Delta F$ and the bandpass filtered about $f_1 = f_c + \Delta F$ to produce an MSK signal with a carrier frequency of f_c. (See Prob. 5-27 to demonstrate that this technique is correct.)

This section on digital bandpass signaling techniques is summarized in Table 5-5, where the spectral efficiencies of various types of digital signals are shown, using two different bandwidth criteria—the null-to-null bandwidth and the 30-dB bandwidth. Of course, a larger value of η indicates better spectral efficiency. It is seen that MSK ranks better or worse than QPSK and 64 QAM depending on the bandwidth criteria used.

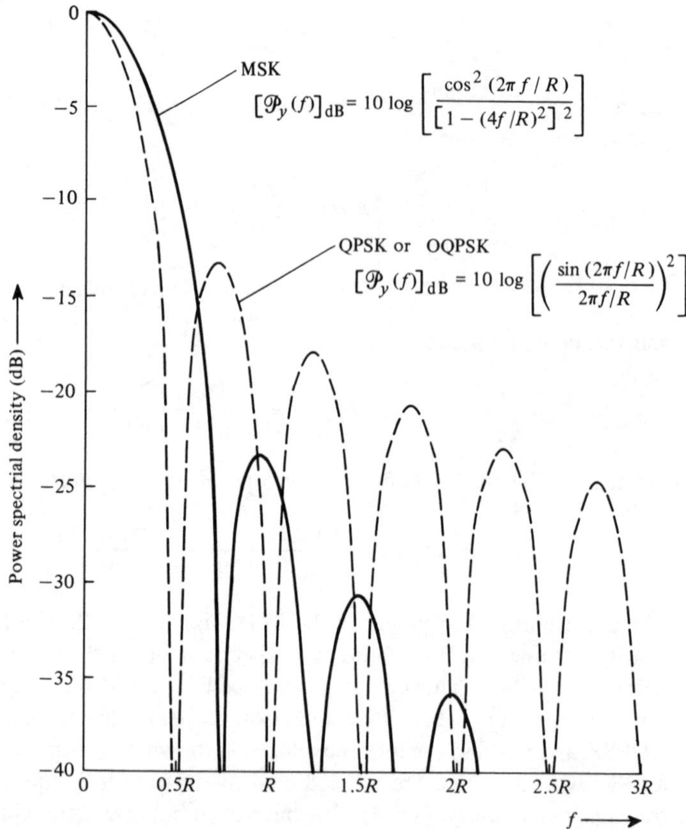

FIGURE 5-17 PSD for complex envelope of MSK, QPSK, and OQPSK where R is the bit rate (positive frequencies shown).

When designing a communication system, one is concerned with the cost and the error performance as well as the spectral occupancy of the signal. The topic of error performance is too immense to discuss in this section; it is covered in Chapters 7 and 8.

TABLE 5-5 Spectral Efficiency of Digital Signals

	Spectral Efficiency, $\eta = \frac{R}{B_T} \left(\frac{\text{bits/sec}}{\text{Hz}} \right)$	
Type of Signal	**Null-to-Null Bandwidth**	**30-dB Bandwidth**
OOK and BPSK	0.500	0.052
QPSK and OQPSK	1.00	0.104
MSK	0.667	0.438
16 QAM	2.00	0.208
64 QAM	3.00	0.313

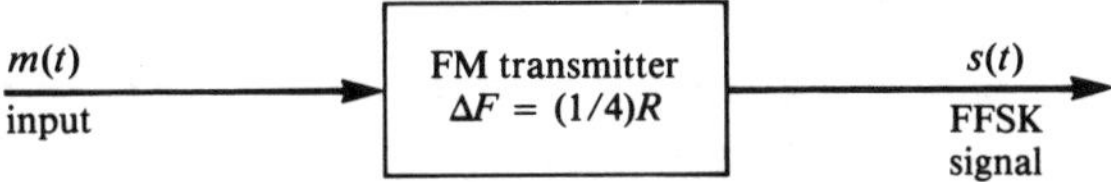

(a) Generation of Fast Frequency-Shift Keying (FFSK)

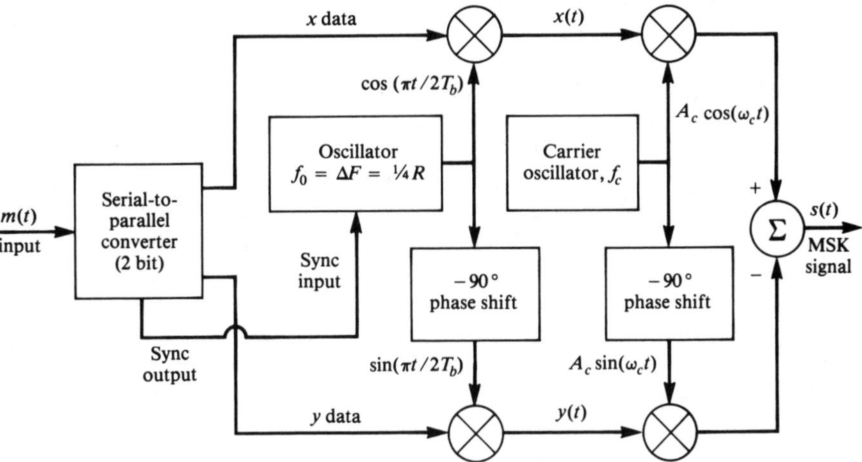

(b) Parallel Generation of Type I MSK (This will generate FFSK if a differential encoder is inserted at the input.)

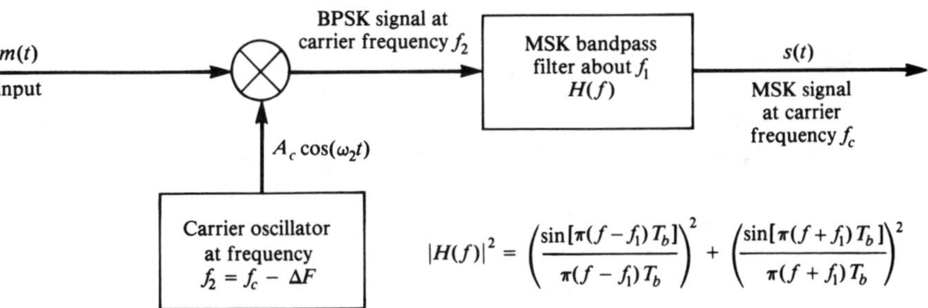

(c) Serial Generation of MSK

FIGURE 5-18 Generation of MSK signals.

5-3 Telecommunication Systems

Telecommunication systems are designed to transmit voice, data, or visual information over some distance. (The prefix *tele* is derived from a Greek word meaning distance.) Modern telecommunication systems have evolved from the telegraph and telephone systems of the 1800s. Telecommunication companies that provide services for a large number of individual users on a for-hire basis are known as *common carriers*. The term is applied to widely diverse businesses such as mail, airline, trucking, telephone, and data services. The common carriers are usually regulated by the government for the general welfare of the public, and, in some countries, certain common carrier services are provided by the government. In this section, we will study the common-carrier telecommunication systems with special emphasis given to telephone systems. The information from multiple users is transmitted over these systems primarily by using time-division multiplexing (TDM) and frequency-division multiplexing (FDM).

Time-Division Multiplexing

In Chapter 3, TDM techniques were first introduced. Examples of common carrier TDM techniques using digital signaling of the types DS-O, DS-1, DS-2, DS-3, DS-4, and DS-5 were studied. For a review of these techniques, the reader is referred to Sec. 3-3, specifically Figs. 3-43 and 3-44 and Tables 3-7 and 3-8. These figures and tables give the TDM standards that have been developed by the common carriers. For bandpass transmission, such as over microwave radio relay, fiber optic, or satellite systems, these TDM baseband signals are digitally modulated onto RF carriers using the OOK, QPSK, MPSK, or QAM techniques that were studied in Sec. 5-2. Fiber optic systems with OOK are described in Sec. 5-6.

TDM techniques are generally preferred to FDM techniques because of the ease of transmission of error-free data and voice information.

Frequency-Division Multiplexing

FDM techniques were introduced in Chapter 4, Fig. 4-47. This was followed by a discussion of the FDM technique used in FM stereo broadcasting, where the left and right audio signals are multiplexed onto one radio-frequency carrier. In toll telephone service, voice signals are transmitted over high-capacity channels using either TDM or FDM. The FDM technique was the first to be used, although TDM has become dominant, and FDM is being phased out.

The telephone FDM technique is illustrated by the American Telephone and Telegraph Company (AT&T) FDM hierarchy, which is given in Fig. 5-19. Here, 12 analog (0- to 4-kHz voice frequency) telephone signals are frequency-division-multiplexed using lower SSB modulators to form one *group* signal. Five group signals are frequency-division-multiplexed to form one *supergroup* signal. The supergroup signal contains the information of 60 voice-frequency (VF) telephone signals. The bandwidth of the group signal is 48 kHz, and any 48-kHz signal (analog or digital) with spectrum centered in the group passband may replace a group input in the FDM hierarchy. Similarly, a supergroup signal may be replaced by any 240-kHz bandwidth signal when needed. Ten supergroup signals are frequency-division-multiplexed to form one *mastergroup* signal, which contains 600 VF signals. In Fig. 5-19 the carrier frequencies are given for the AT&T Type U600 mastergroup [Bell Telephone Laboratories, 1970]. Six mastergroup signals may be frequency-division-multiplexed to form a *jumbogroup* containing the information of 3600-VF signals.

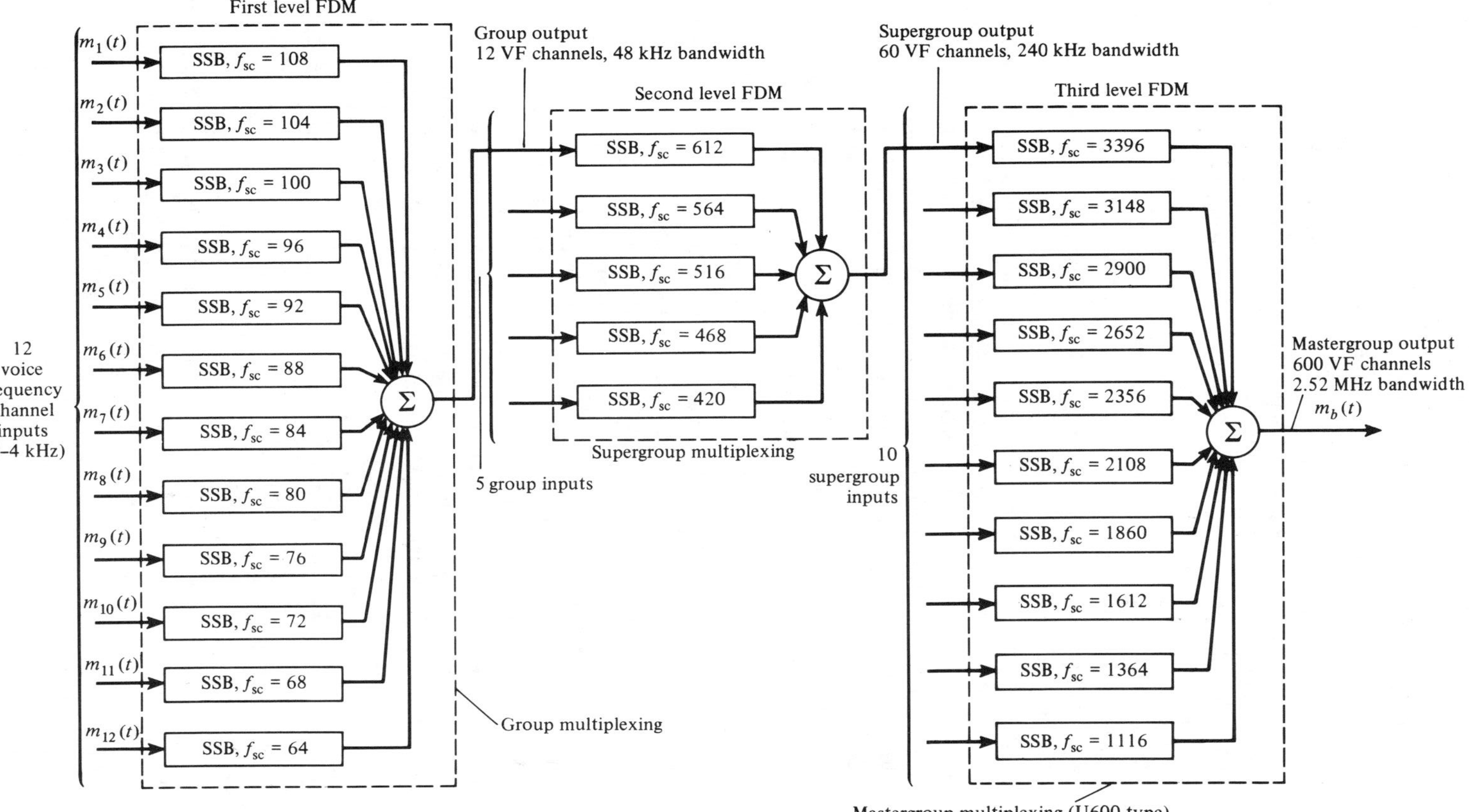

FIGURE 5-19 AT&T FDM hierarchy (f_{sc} given in kilohertz).

Telephone Systems

Telephone systems may be divided into two main categories:

1. The local subscriber-loop system, which connects each telephone to the equipment at the local telephone plant and the associated local switching equipment.
2. The toll system, which consists of long-distance circuits and the associated switching equipment.

When a local call is made, only the local-loop system is involved. The local switching office connects the two parties by making a hardwire connection between the two appropriate local loops. This is essentially a series connection, with the earphone and carbon microphone of each telephone handset wired in a series with a battery (located at the telephone plant). Figure 5-20 illustrates this analog local-loop system, which is said to supply *plain old telephone service* (POTS). The wire with positive voltage from the *central office* (CO) is called the *tip* lead and it is color coded *green*. The wire with negative voltage is called the *ring* lead and is color coded *red*. The terms *tip* and *ring* are derived from the era when a plug switchboard was used at the CO and these leads were connected to a plug with tip and ring contacts. This jack is similar to a stereo headphone jack that has tip, ring, and sleeve contacts. The Earth ground lead was connected to the sleeve contact.

The sequence of events that occur when placing a local phone call will now be described with the aid of Fig. 5-20 and Table 5-6. The calling party—the upper telephone set in Fig. 5-20—removes the telephone handset; this action closes the switchhook contact (off-hook) so that dc current flows through the caller's telephone line. The current, about 60 mA, is sensed at the CO and causes the CO to place a dial-tone signal (approximately 400 Hz) on the calling party's line. The calling party dials the number by using either pulse or touch-tone dialing. If pulse dialing is used, the dc line current is interrupted for the number of times equal to the digit dialed (at a rate of 10 pulses/sec). For example, there are five interruptions of the line current when the number 5 is dialed. Upon reception of the complete number sequence for the called party, the CO places the ringing generator (90 V, 20 Hz, on 2 sec, off 4 sec) on the line corresponding to the number dialed. This rings the phone. When the called party answers, dc current flows in that line to signal the CO to disconnect the ringing generator and to connect the two parties together via the CO circuit switch. Direct current is now flowing in the lines of both the called and the calling party, and there is a connection between the two parties via transformer coupling[†] as shown in Fig. 5-20. When either person speaks, the sound vibrations cause the resistance of the carbon microphone element to change in synchronization with the vibrations so that the dc line current is modulated (varied). This produces the ac audio signal that is coupled to the headphone of the other party's telephone. Note that both parties may speak and hear each other simultaneously. This is full-duplex operation.

[†] The transformers are called "repeat coils" in telephone parlance.

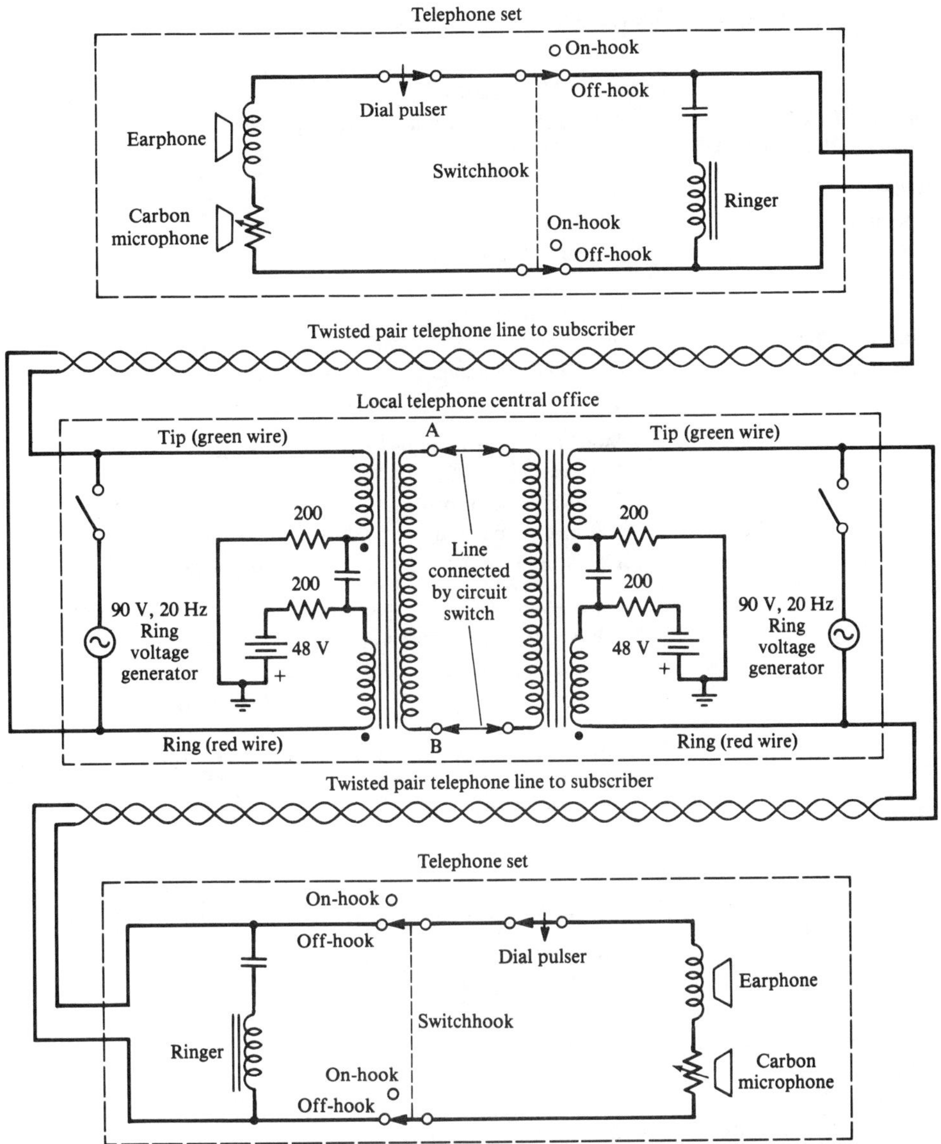

FIGURE 5-20 Local telephone system (simplified).

In some modern telephone sets, the carbon microphone element is replaced with an electret or dynamic microphone element and an associated IC amplifier that is powered by the battery voltage from the CO, but the principle of operation is the same as previously described.

The telephone system of Fig. 5-20 is satisfactory as long as the resistance of the twisted-pair loops is 1300 ohms or less. This limits the distance that telephones can be placed from this type of CO to about 15,000 ft for 26-gauge wire,

TABLE 5-6 Telephone Standards for the Subscriber Loop

Item	Standard
On-hook (idle status)	Line open circuit, minimum dc resistance 30 kΩ
Off-hook (busy status)	Line closed circuit, maximum dc resistance 200 Ω
Battery voltage	48 V
Operating current	20–80 mA
Subscriber-loop resistance	0–1300 Ω, 3600 Ω (max)
Loop loss	8 dB (typical), 17 dB (max)
Ringing voltage	90 V rms, 20 Hz (typical) (usually pulsed on 2 sec, off 4 sec)
Ringer equivalence number (REN)[a]	0.2 REN (minimum), 5.0 REN (maximum)
Pulse dialing	Momentary open-circuit loop
Pulsing rate	10 pulses/sec ±10%
Duty cycle	58–64% break (open)
Time between digits	600 msec minimum
Pulse code	1 pulse = 1, 2 pulses = 2, . . . , 10 pulses = 0
Touch-tone[b] dialing	Uses two tones, a low-frequency tone and a high-frequency tone, to specify each digit:

	High Tone (Hz)		
Low Tone	1209	1336	1477
697 Hz	1	2	3
770	4	5	6
852	7	8	9
941	*	0	#

Item	Standard
Level each tone	−6 to −4 dBm
Maximum difference in levels	4 dB
Maximum level (pair)	+2 dBm
Frequency tolerance	±1.5%
Pulse width	50 msec
Time between digits	45 msec minimum
Dial tone	350–440 Hz
Busy signal	480–620 Hz, with 60 interruptions per minute
Ringing signal tone	440–480 Hz, 2 sec on, 4 sec off

[a] Indicates the impedance loading caused by the telephone ringer. An REN of 1.0 equals about 8 kΩ, 0.2 REN is 40 kΩ, and 5.0 REN is 1.6 kΩ.

[b] Touch-tone was a registered trademark of AT&T. It is also known as *dual-tone multiple frequency* (DTMF) signaling.

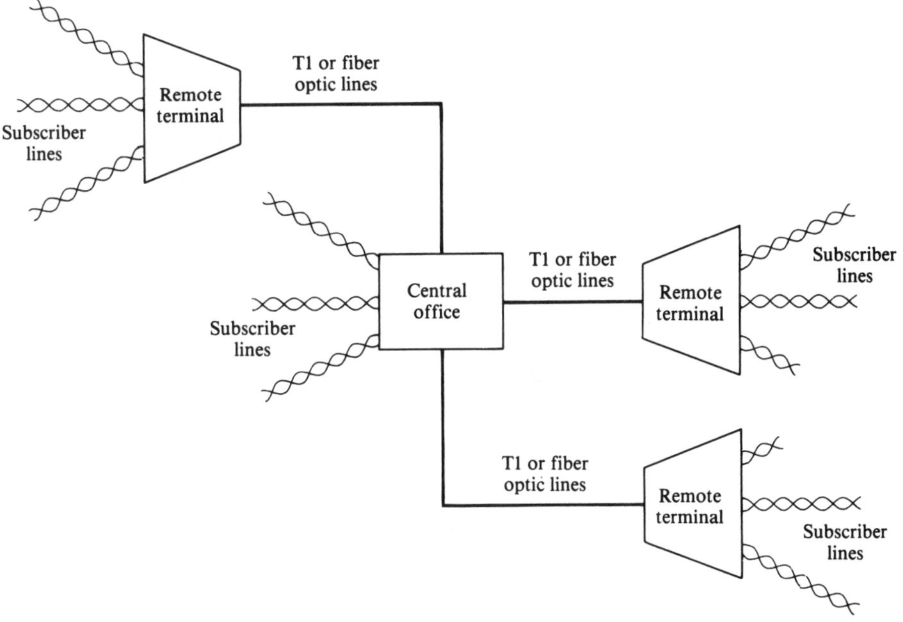

FIGURE 5-21 Telephone system with remote terminals.

24,000 ft for 24-gauge, and 38,000 ft or about 7 mi if 22-gauge twisted-pair wire is used. Sometimes 19-gauge wire is used in rural areas so that phones may be located up to 20 miles from the CO.

Supplying a dedicated wire pair from each subscriber all the way to the CO is expensive. In applications where a large number of subscribers are clustered some distance from the CO, costs can be substantially reduced by using *remote terminals* (RT) (see Fig. 5-21). Remote terminals also allow telephones to be placed at any distance from the CO, as we will describe subsequently.

The circuit for a typical RT is illustrated in Fig. 5-22. The RT supplies battery voltage and ringing current to the subscriber's telephone. The *2-wire circuit* that carries the duplex VF signals to and from the subscriber is converted into a *4-wire circuit* that carries two one-way (simplex) transmit and receive signals by the use of a *hybrid* circuit. The hybrid is a balanced transformer circuit (or its equivalent electronic circuit) that provides isolation for the transmit and receive signals, so that self-oscillation will not occur if the amplified transmitted signal is fed back via the receive line. As shown in the figure, the transmit VF signal is converted to a DS-0 PCM signal, which is time-division-multiplexed with the PCM signals from other subscribers attached to the RT. The TDM signal is sent over a DS-1 trunk to the CO. Similarly, the received DS-1 signal from the CO is demultiplexed and decoded to obtain the received VF audio for the subscriber. In a popular RT system manufactured by AT&T called SLC-96, 96 VF subscriber lines are digitized and multiplexed onto four T1 lines, and one additional T1 line is on standby in case one of the other T1 lines fails [Chapman, 1984]. From

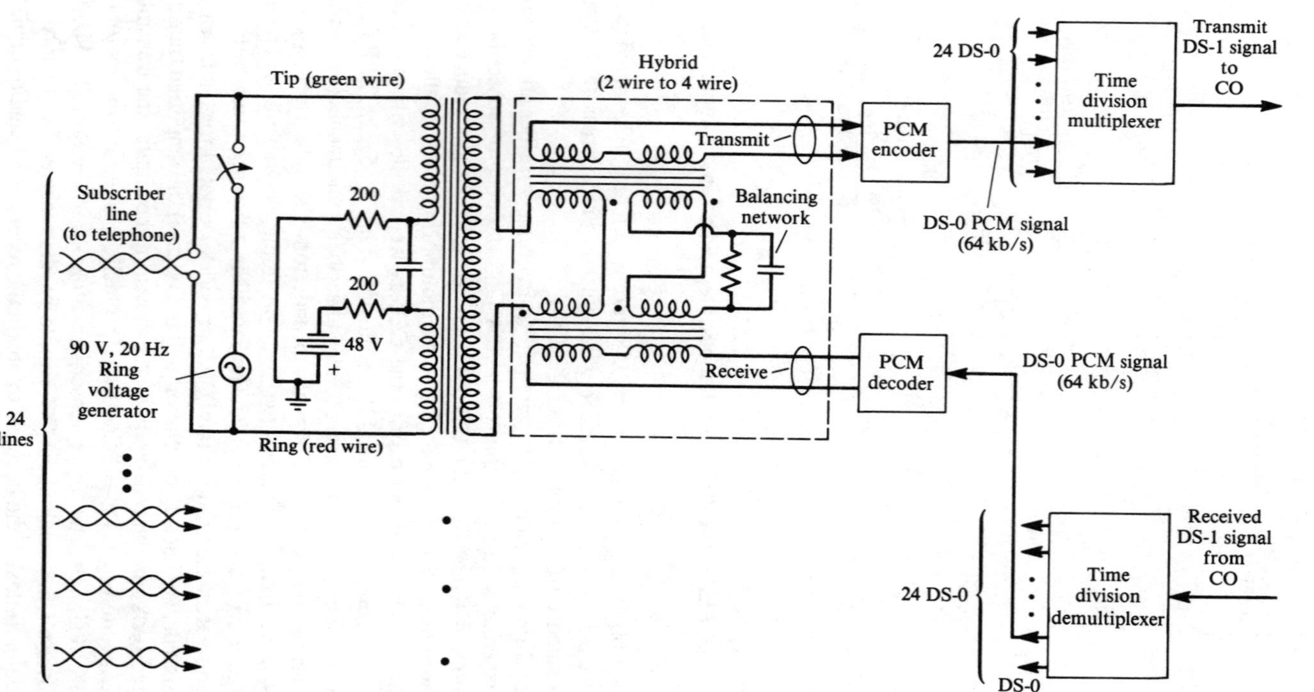

FIGURE 5-22 Remote terminal (RT).

Chapter 3 it is recalled that a T1 line requires two twisted pairs (one for the transmit data and one for the receive data) and that each T1 line (1.544 Mbits/sec) carries the equivalent of 24 VF signals.

Let us now compare the twisted-pair requirements for systems with and without RTs. If no RT is used, 96 pairs will be required to the CO for 96 subscribers, but if a SLC-96 RT is used, only 10 pairs (5 T1 lines) are needed to the CO. This provides a pair savings (also called pair gain) of 9.6 to 1. Furthermore, the RT may be located at any distance from the CO (there is no 1300-Ω limit) since the pairs are used for DS-1 signaling with repeaters spaced about every mile. Of course, fiber optic lines can also be used to connect the RT to the CO. For example, if two 560-Mbits/sec fiber optic lines are used (one for data sent in each direction), this has a capacity of 8064 VF channels for DS-5 signaling (see Table 3-7 and Fig. 3-44). Thus an RT with a 560-Mbits/sec fiber optic link to the CO could serve 8064 subscribers. Furthermore, in Sec. 5-8 we see that no optical repeaters will be required if the RT is located within 25 miles (typically) of the CO.

Telephone companies are replacing their analog switches at the CO with digital switches. Historically, at the CO a circuit switch was used to connect the calling and the called party (as shown in Fig. 5-20). These switches were controlled by hardwired relay logic. Now, modern telephone offices use *electronic switching* systems (ESS). With ESS, a digital computer controls the switching operation as directed by software called *stored program control*. Moreover, the latest ESS switches use *digital switching* instead of *analog switching*. Examples are AT&T's No. 4 ESS (electronic switching system) and No. 5 ESS, and Northern Telcom's DMS-100 switches. In a digitally switched CO, the customer's VF signal is converted to PCM and time-division-multiplexed with other PCM signals onto a high-speed digital line. (When RTs are used, the conversion of the VF signal to PCM is carried out at the RT.) The digital CO switches a call by placing the PCM data of the calling party into the TDM time slot that has been assigned for the destination (called) party. This is called *time-slot interchange* (TSI). The digital switch is less expensive on a per customer basis than the analog switch (for a large number of customers), and it allows switching of data and digitized video as well as PCM audio. However, many analog switch COs remain since it is not cost effective to scrap existing analog facilities, and, of course, analog VF signals must be converted to TDM PCM signaling before a digital switch can be used. On the other hand, there is the incentive to upgrade analog telephone systems to digital switching with remote terminals since this allows the ISDN service, as introduced in Chapter 3 and described in more detail in the next section. That is, the use of RTs permits the subscriber line to be short (about 1 mile). With short subscriber lines, ISDN service, and, if needed, DS-1 (1.544-Mbits/sec) service, can be provided directly over the existing twisted-pair lines that are already in place to the customer's premises.

For toll calls, the local CO switches the call to trunk lines that connect the local CO to a distant CO. The trunk lines usually carry either TDM or FDM signals, with TDM being the preferred technique. If the local CO uses analog switches, the two-wire circuit must be converted to a four-wire circuit (using a hybrid as described in Fig. 5-22) since unidirectional repeaters are used on the long-distance trunk lines. If the local CO uses a digital switch, it acts much like an RT with respect to the distant CO, and the two COs are connected by high-speed TDM digital trunk lines. Interestingly, long-distance lines are less than 5% of the total cost of the telephone network; most of the cost is in the switching equipment.

In summary, the analog telephone subscriber is connected to a CO or a RT via a 2-wire twisted-pair subscriber line. The CO or RT as a terminator of this subscriber loop provides the following functions for POTS:

1. Terminates the line with a 900-Ω ac load (balanced with respect to ground).
2. Supplies dc loop current from its battery through balanced 200-Ω resistors (for a total driven-point dc impedance of 400 Ω).
3. Monitors loop current to determine on-hook and off-hook conditions.
4. Receives dialing information from the subscriber via touch-tone or dial pulse.
5. Applies ringing voltage and call progress signaling (i.e., dial tone, busy signal, ringing tone, etc.).
6. Converts the 2-wire line to a 4-wire line (transmit and receive path) for use by digital COs and RTs or for use on toll calls. 2-wire to 4-wire conversion is usually not required for analog COs that switch local calls.
7. Provides analog-to-digital (i.e., PCM) and digital-to-analog conversion for digital systems.
8. Tests subscriber line and line termination (i.e., telephone loads connected to other end of the line).
9. Provide protective isolation of CO/RT equipment from unwanted line potentials such as lighting or cross connection to some ac power source.

Integrated Service Digital Network (ISDN)

As indicated previously, the telephone industry is evolving from an analog network to a digital network. Presently, the trend is to provide a digital CO and a digital network out to the RT; the "last mile" from the RT to the subscriber is usually analog. A new approach called the *integrated service digital network* (ISDN) converts the "last mile" analog subscriber line (ASL) to a *digital subscriber line* (DSL) so that digital data can be delivered directly to the subscriber's premises. The ISDN subscriber can demultiplex the data to provide any one or all of the following applications simultaneously: (1) decode data to produce VF signals for telephone handsets, (2) decode data for a video display, or (3) process data for telemetry and/or PC applications.

As indicated in Chapter 3, there are two categories of ISDN: (1) *narrowband* or "*basic rate*" ISDN, denoted by N-ISDN, and (2) *broadband* or "*primary rate*" ISDN, denoted by B-ISDN. B-ISDN is in the process of being developed; consequently, there are many proposed protocols and data rates. One proposal has an aggregate data rate of 1.536 Mbits/sec (approximately the same rate as for T1 lines) consisting of 23 B channels (64 kbits/sec each) and one D channel (64 kbits/sec). The B channels carry user (or bearer) data which may be PCM for encoded video or audio. The D channel is used for signaling to set up calls, disconnect calls, and/or route data for the 23 B channels.

Since B-ISDN standards are not yet fully developed, B-ISDN will be discussed only briefly. It is clear that twisted-pair copper lines must be used (presently) to provide B-ISDN for the "last mile" to the subscriber since it is not financially feasible to replace all copper lines already installed (about a $100 billion investment for U.S. copper line facilities) with fiber. Of course, fiber might be the economical solution for new installations. Furthermore, since 24-gauge copper lines have an attenuation of about 35 dB/mile at 1 MHz, the "last mile" copper line length from the RT to the subscriber will have to be limited to a mile or two, assuming that no repeaters are used. As discussed in Sec. 3-8, twisted-pair copper lines with repeaters spaced every mile can support a 1.5-Mbit/sec data rate (T1 line, which has a bipolar line code), or 6.4-Mbit/sec data rate (T1G line, which has a quarternary line code). Fiber optic (or coaxial) lines will be required for data rates on the order of 10 Mbits/sec or larger. For further reading on the developing B-ISDN technology, the reader is referred to the literature [Saltzberg et al., 1991].

The standard implementation of N-ISDN is shown in Fig. 5-23. The N-ISDN subscriber is connected to the RT of the telephone company by a 2-wire twisted

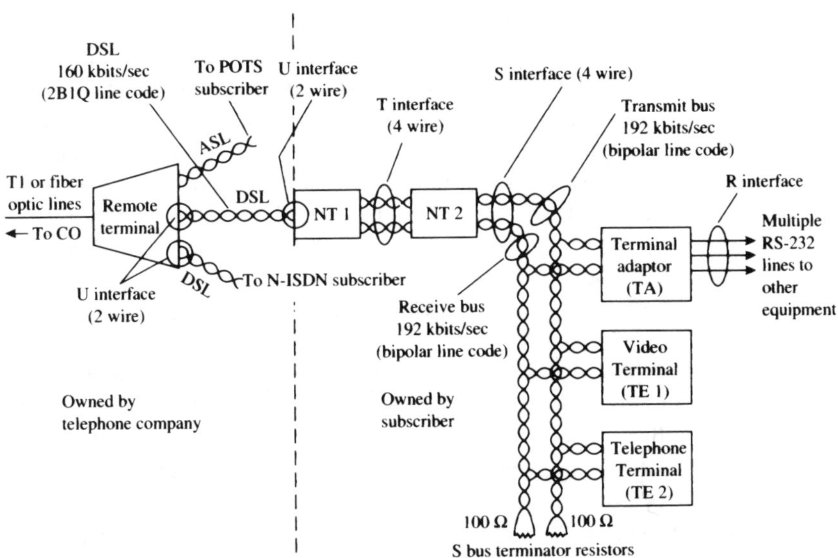

FIGURE 5-23 N-ISDN system with subscriber equipment attached.

pair (unloaded) telephone line. This allows existing copper pairs to be used for either POTS or N-ISDN simply by connecting the ends of the pairs to the appropriate terminating equipment. For N-ISDN service the line must be no longer than 18 kft (3.4 miles) for the 160-kbits/sec N-ISDN aggregate data rate. (If the subscriber is located within 18 kfeet of the CO, an RT is not necessary.) The data rate available to the N-ISDN subscriber is 144 kbits/sec, consisting of data partitioned into two B channels of 64 kbits/sec each and one D channel of 16 kbits/sec. In addition to the 2B + D data, the telephone company adds 12 kbits/sec for framing/timing plus 4 kbits/sec for overhead to support network operations. This gives an overall data rate of 160 kbits/sec on the DSL in both the transmit and receive directions (simultaneously) for full-duplex operation. The DSL is terminated on each end at the U interface as shown on Fig. 5-23. The NT1 (network termination) incorporates a hybrid circuit to convert the 2-wire U interface to a 4-wire T interface which has two 2-wire buses, one for the transmit and one for the receive data signals. The NT1 is a slave transceiver that derives its clocking signals from the DSL signal that is transmitted from the RT. The NT1 passes the 2B + D data between the U and T interfaces and also processes additional bits that appear on the transmit and receive lines at the T interface. The additional bits are needed for addressing, control, and supervision of terminal equipment. The data rate on each transmit and receive T line is 192 kbits/sec. The NT2 may or may not be needed. When used, it provides the appropriate higher-level protocols for a terminal cluster controller or for a local area network (LAN) access node. If not used, the NT1 4-wire interface becomes the S/T interface. The S and T interfaces are electrically identical and use a RJ45 8-pin modular connector. The S/T interface bus provided the connection medium for ISDN terminals (video, audio, and/or data) and/or a terminal adapter (TA). Up to eight terminals may be bridged onto the S/T bus. The TA allows non-ISDN terminals with RS-232 ports to be used on the ISDN network. For example, a personal computer (PC) can be connected to the N-ISDN network via its serial port and the R interface of the TA. Also, using an alternative approach, a PC adapter card with a built-in NT1 and a telephone codec can be used. That is, the DSL U interface modular connector is plugged directly into this adapter card on the PC. The PC then acts as an ISDN video terminal with a built-in telephone.

For details on the protocols used at the U, S, and T interfaces, the reader is referred to application notes for ISDN chips such as Motorola's MC 145472 U interface transceiver, MC 145474 S/T interface, and MC 145554 telephone codec.

N-ISDN service to the subscriber via 2-wire twisted-pair DSL up to 18 kft from a RT is made possible by the use of multilevel signaling. Referring to Fig. 5-15 with $R = 160$ bits/sec, a four-level (i.e., $\ell = 2$ bit) 80-kbaud line code has a null bandwidth of only 80 kHz instead of the 160-kHz null bandwidth for a binary ($\ell = 1$) line code. The 80-kHz bandwidth is supported by 26-gauge twisted-pair cable if it is less than about 18 kft in length. The particular four-level signal used is the 2B1Q line code (for two binary digits encoded into one quadrenary symbol) as shown in Fig. 5-24. Note that the 2B1Q line code is a

Previous 2B1Q level	Current binary word	Present 2B1Q level
+1 or +3	00	+1
	01	+3
	10	−1
	11	−3
−1 or −3	00	−1
	01	−3
	10	+1
	11	+3

(a) 2B1Q differential code table

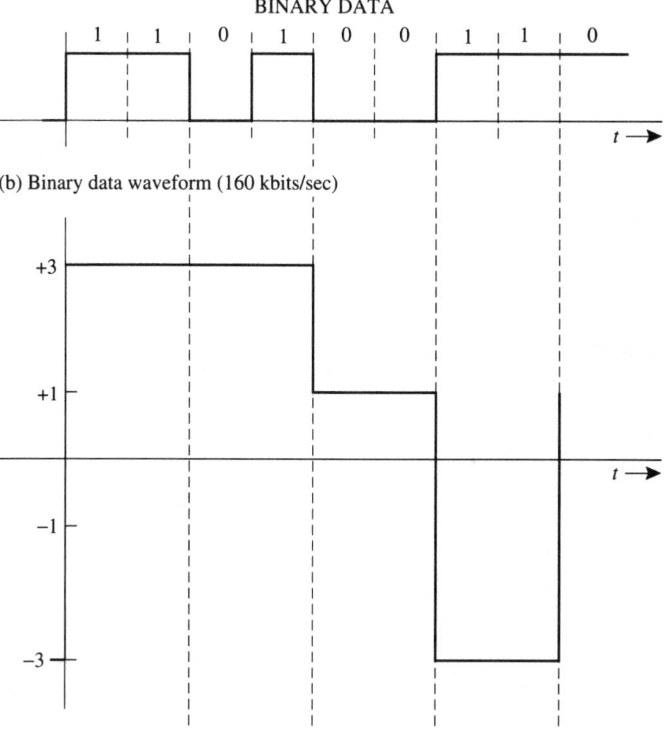

(b) Binary data waveform (160 kbits/sec)

(c) 2B1Q waveform (80 kbaud)

FIGURE 5-24 2B1Q Line code.

differential symbol code. Thus if (due to wiring error) the twisted pair is "turned over" so that the tip wire is connected to the ring terminal and the ring to the tip so that the 2B1Q signal polarity is inverted, the decoded binary data will still have the correct polarity (i.e., not complemented).

TABLE 5-7 Capacity of Some Common Carrier Links

Transmission Medium	Name	Developer	Year in Service	Number of Voice-Frequency Channels	Bit Rate (Mbits/sec)	Repeater Spacing (miles)	Operating Frequency (MHz)	Modulation[a]	
								D/A	Method
Open-wire pair	A	Bell	1918	4			0.005–0.025	A	SSB
	C	Bell	1924	3		150	0.005–0.030	A	SSB
		CCITT		12		50	0.036–0.140		
	J	Bell	1938	12		50	0.036–0.143	A	SSB
		CCITT		28			0.003–0.300	A	SSB
Twisted-pair cable	K	Bell	1938	12		17	0.012–0.060	A	SSB
		CCITT		12		19	0.012–0.060	A	SSB
	N1	Bell	1950	12			0.044–0.260	A	SSB
	N3	Bell	1964	24			0.172–0.268	A	SSB
	T1[b]	Bell	1962	24	1.544 (DS-1)	1		D	Bipolar
	T1G[b]	AT&T	1985	96	6.312 (DS-2)			D	4 level
	T2[b]	Bell		96	6.312 (DS-2)	2.3		D	B6ZS
Coaxial cable	L1	Bell	1941	600		8	0.006–2.79	A	SSB
	L3	Bell	1953	1,860		4	0.312–8.28	A	SSB
	L4	Bell	1967	3,600		2	0.564–17.55	A	SSB
	L5	Bell	1974	10,800		1	3.12–60.5	A	SSB
	T4[b]	Bell		4,032	274.176 (DS-4)	1		D	Polar
	T5[b]	Bell		8,064	560.16 (DS-5)	1		D	Polar
Fiber optic	FT3	Bell	1981	672	44.763 (DS-3)	4.4	0.82 μm	D	TDM/OOK
		Brit. Telcom	1984		140	6	1.3 μm	D	TDM/OOK
		Sask. Telcom	1985		45	6–18	0.84 and 1.3 μm	D	TDM/OOK
	F-400M	Nippon	1985		400	12	1.3 μm	D	TDM/OOK
	FT3C-90	AT&T	1985	1,344	90.254 (DS-3C)	16	1.3 μm	D	TDM/OOK
	FT4E-432	AT&T	1986	6,048	432 (DS-432)	16	1.3 μm	D	TDM/OOK
	LaserNet[d]	Microtel	1985	6,048	417.79 (9DS-3)	25	1.3 μm	D	TDM/OOK
	FTG1.7	AT&T	1987	24,192	1,668 (36DS-3)	29	1.3 μm	D	TDM/OOK

Transoceanic	TAT-1 (SB)	Bell	1956	48		20	0.024–0.168	A	SSB
	TAT-3 (SD)	Bell	1963	138		11	0.108–1.05	A	SSB
	TAT-5 (SF)	Bell	1970	845		6	0.564–5.88	A	SSB
	TAT-6 (SG)	Bell	1976	4,200		3	0.5–30	A	SSB
	TAT-8 (3 fibers)		1988	8,000	280		1.3 μm	D	TDM/OOK
	TAT-9 (3 fibers)		1991	16,000	565		1.55 μm	D	TDM/OOK
	TAT-10 (6 fibers)		1995	80,000[e]	565		1.55 μm	D	TDM/OOK
Troposphere scatter radio	DMEW		1961	240			UHF	A	FDM/FM
Microwave relay	TD-2	Bell	1948	600 (1954)	plus 1.544[c]	30	3700–4200	A	FDM/FM
	TH-1	Bell	1961	2,400 (1979)	plus 1.544[c]	30	5925–6245	A	FDM/FM
	TD-3	Bell	1967	1,800 (1979)	plus 1.544[c]	30	3700–4200	A	FDM/FM
	TN-1	Bell	1974	1,800	plus 1.544[c]	30	K band, 18 GHz	A	FDM/FM
	AR6A	Bell	1980	6,000		30	5925–6425	A	FDM/SSB
	18G274	NEC	1974	4,032	274.176 (DS-4)	7	18 GHz	D	4 PSK
	6G90	NEC	1979	1,344	90 (2DS-3)	30	6 GHz	D	16 QAM
	11G135	NEC	1980	2,016	135 (3 DS-3)	30	11 GHz	D	16 QAM
	6G135	NEC	1983	2,016	135 (3 DS-3)	30	6 GHz	D	64 QAM
	MDR-2306	Collins	1983	2,016	135 (3 DS-3)	30	6 GHz	D	64 QAM
	RD-6A	Northern Tel.	1984	2,016	135 (3 DS-3)	30	6 GHz	D	64 QAM
Communication satellite	Intelsat IV	COMSAT	1970	8,000		22,300	6 GHz up/4 GHz down	A/D	FDM/FM, QPSK/ SCPC
	Intelsat V	COMSAT	1980	25,000		22,300	6/4 and 14/11 GHz	A/D	FDM/FM, QPSK/ SCPC
	Intelsat VI	COMSAT	1986	80,000		22,300	6/4 and 14/11 GHz	A/D	FDM/FM, QPSK/ SCPC

[a] A-analog; D-digital plus modulation type. [b] See Table 3-8 for more details on the T-carrier system.

[c] Since 1974 data under voice were added to give a 1.544-Mbit/sec (DS-1) data channel in addition to the stated VF FDM capacity.

[d] See Example 5-7 in Sec. 5-6. [e] VF capacity with statistical multiplexing.

The primary advantage of ISDN is obvious—it provides TDM data channels directly to the subscriber. It also has disadvantages. For example, since the ISDN equipment is not powered by the DSL, it will not function when local ac power fails unless a backup power supply is provided by the subscriber. If only POTS is required, the conventional analog telephone is usually much cheaper and more reliable.

For study of additional N-ISDN topics, such as frame structure, timing recovery, synchronization, and echo cancellation, the reader is referred to a tutorial paper by [Huang and Valenti, 1991]. For additional study of telephone systems in general, the interested reader is urged to look at some of the numerous publications that are available [Briley, 1983; Chorafas, 1984; Fike and Friend, 1984; Jordan, 1985].

Capacities of Common Carrier Links

In FDM and TDM signaling, the wideband channels used to connect the toll offices consist of three predominant types: fiber optic cable, microwave radio relay systems, and (buried) coaxial cable systems. Table 5-7 lists some of the wideband systems that are used, or have been used in the past, and indicates the capacity of these systems in terms of the number of VF channels they can handle and their bit rate.

In 1980 the AR6A microwave relay system was introduced into service [Markle, 1978, 1983]. Until this time FDM/FM was the modulation method used on microwave links because the intermodulation distortion on klystron microwave power amplifiers was too large for SSB techniques to be used. The development of a low-distortion TWT (traveling-wave tube) amplifier has made FDM/SSB possible in the AR6A. The VF capacity of the AR6A is *three times* that of microwave systems using the FDM/FM technique, so that it will handle 10 mastergroups of 600 VF channels each. The performance of digital microwave systems is also shown.

Referring to Table 5-7 again, we realize that almost all of the open-wire pairs have been replaced by twisted-pair cable, but occasionally some can still be seen along railroad tracks. Fiber cable with TDM/OOK signaling is now rapidly overtaking twisted-pair and coaxial cable.

Tropospheric scatter radio is the transmission of UHF signals "over the horizon" by reflecting and refracting the signals in the troposphere. A ground range of 400 miles can be obtained by using high-power and large parabolic antennas. In most applications "tropo" systems are being replaced by satellite systems.

The capacity of satellite communication also depends on the type of modulation used. With FDM/FM the *Intelsat IV* and *V* series satellites can handle 300 VF channels on each 36-MHz bandwidth transponder (receiver–transmitter relay system). If quadrature phase-shift-keyed/single VF channel per carrier (QPSK/SCPC) is used, 800 VF channels can be accommodated. This subject is discussed in more detail in the following section, which describes satellite communication systems.

5-4 Satellite Communication Systems

The number of satellite communication systems has increased tremendously over the last few years. Satellite communications now relay the bulk of transoceanic telephone traffic, having greatly surpassed that sent by submarine cable. In addition, satellites have made transoceanic relaying of television signals possible. Satellite communications thus provide the relaying of data, telephone, and television signals, and it is now feasible to provide national direct-into-the-home television transmission via satellite.

Most communication satellites are placed in *geosynchronous* orbit. That is, they are located 22,300 miles above the equator so that their orbital period is the same as that of the Earth. Consequently, from Earth these satellites appear to be located at an approximately fixed point in the sky, as shown in Fig. 5-25. This enables the Earth station antennas to be simplified since they are pointed in a fixed direction and do not have to track a moving object. (For communication to the polar regions of the Earth, satellites in polar orbits are used, which require Earth stations with tracking antennas.)

The most desired, as well as the most popular, frequency band for satellite communication systems is 6 GHz on the uplink (Earth-to-satellite transmission) and 4 GHz on the downlink (satellite-to-Earth transmission). In this frequency range the equipment is relatively inexpensive, the cosmic noise is small, and the frequency is low enough so that rainfall does not appreciably attenuate the signals. Other losses, such as ionospheric scintillation and atmospheric absorption, are small at these frequencies [Spilker, 1977]. (Absorption occurs in specific frequency bands and is caused by the signal exciting atmospheric gases and water vapor.) However, existing terrestrial microwave radio relay links are already assigned to operate within the 6- and 4-GHz bands (see Table 5-7), so the FCC limits the power density on Earth from the satellite transmitters. One also has to be careful in locating the Earth station satellite receiving antennas so that they do

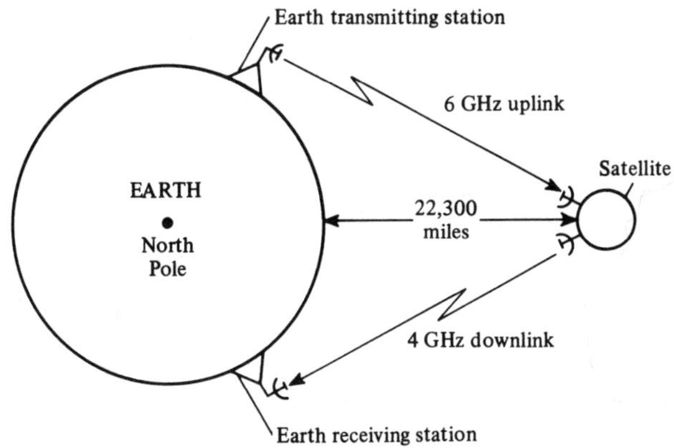

FIGURE 5-25 Communications satellite in geosynchronous orbit.

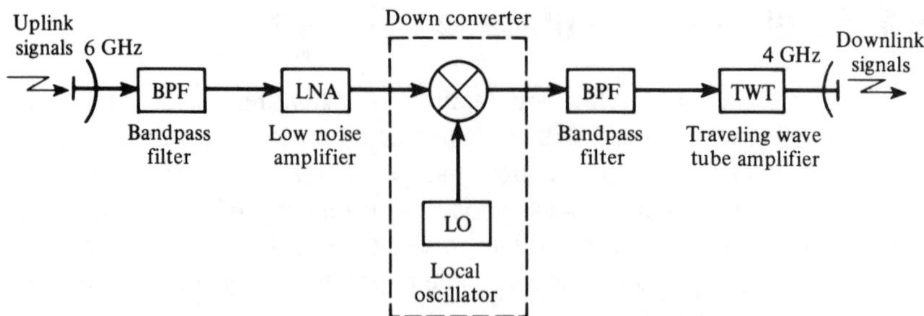

FIGURE 5-26 Simplified block diagram of a communications satellite transponder.

not receive interfering signals from the terrestrial microwave radio links that are using the same frequency assignment. In the 6/4-GHz band, synchronous satellites are assigned an orbital spacing of 2° (United States standard).

Newer satellites have to operate in higher-frequency bands because there are so few vacant spectral assignments in the 6/4-GHz band (C band). The Ku band satellites use 14 GHz on the uplink and 12 GHz on the downlink with an orbital spacing of 3 degrees. For direct-to-the-home broadcasting a 17-GHz uplink and 12-GHz downlink have been assigned [Pritchard and Kase, 1981], but do not appear to be commercially successful.

Each satellite has a number of *transponders* (receiver-to-transmitter) aboard to amplify the received signal from the uplink and to down-convert the signal for transmission on the downlink (see Fig. 5-26). This figure shows a "bent pipe transponder" that does not demodulate the received signal and perform signal processing but acts as a high-power-gain down converter. Most transponders are designed for a bandwidth of 36, 54, or 72 MHz, with 36 MHz being the standard used for C-band (6/4 GHz) television relay service. As technology permits, processing transponders will come into use since an improvement in error performance (for digital signaling) can be realized.

Each satellite is assigned a synchronous orbit position and a frequency band in which to operate. In the 6/4-GHz band, each satellite is permitted to use a 500-MHz-wide spectral assignment, and a typical satellite has 12 transponders aboard, with each transponder using 36 MHz of the 500-MHz bandwidth assignment. Some satellites reuse the same frequency band by having 12 transponders operating with vertically polarized radiated signals and 12 transponders with horizontally polarized signals.† The 6/4-GHz frequency assignment for satellites is shown in Fig. 5-27. These satellites use 24 transponders (12 using vertical polarization and 12 using horizontal polarization). The transponders are denoted by C1 for channel 1, C2 for channel 2, and so on. These satellites are used mainly to relay signals for cable antenna television (CATV) systems.

†A vertically polarized satellite signal has an E field that is oriented vertically (parallel with the axis of rotation of the Earth). With horizontal polarization the E field is horizontal.

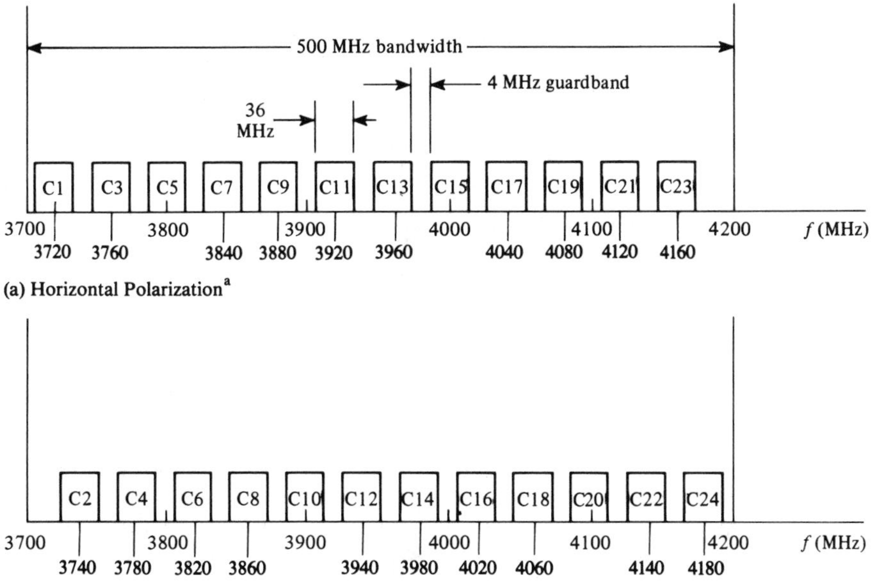

[a]These are the polarizations used for the Galaxy Satellites. Some of the other satellites use opposite polarization assignments.

FIGURE 5-27 6/4-GHz satellite transponder frequency plan for the downlink channels. (For the uplink frequency plan, add 2225 MHz to the numbers given above.)

Figure 5-28 shows an example of a 6/4-GHz satellite. This is the *Galaxy* series of satellites manufactured by the Hughes Aircraft Company, a member of their HS 376 family of satellites. The satellite is approximately 21 ft 8 in. high after deployment and has a diameter of 7 ft 1 in. The *Galaxy* mission's lifetime is 9 years and is limited by the amount of propellent stored onboard for firing thrusters that keep the satellite to within 0.1° of its assigned position (in both the north–south and east–west directions). The body of the satellite is spun at a rate of 50 rpm to provide gyroscopic stability. The solar cell/battery system supplies a dc power of at least 741 W to operate the electronics.

Television Signal Transmission

Broadcast-quality TV signals are relayed via satellite by using a single transponder (36-MHz bandwidth) for each TV signal. This is accomplished by modulating the composite 4.5-MHz bandwidth visual signal onto a 6-GHz FM carrier, as shown in Fig. 5-29. The composite visual signal consists of the black-and-white video signal, the color subcarrier signals, and the synchronizing pulse signal, as discussed in Sec. 5-9. The aural signal is also relayed over the same transponder by frequency modulating it onto a 6.8-MHz subcarrier that is frequency-division-multiplexed with the composite video signal. The resulting wideband signal frequency modulates the transmitter.

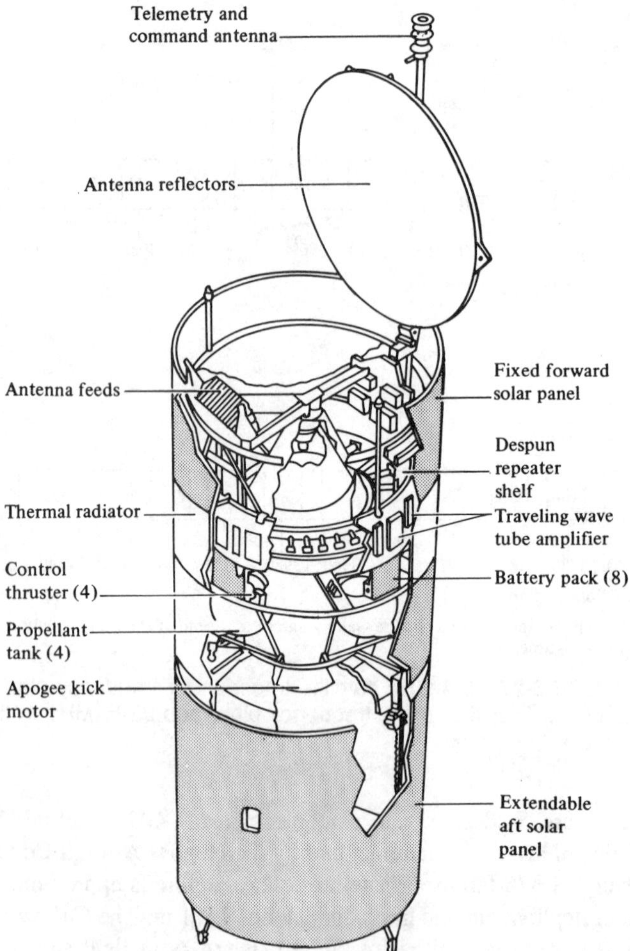

FIGURE 5-28 *Galaxy* satellite. [Courtesy of Hughes Communications, Inc., a wholly owned subsidiary of Hughes Aircraft Company.]

The bandwidth of the 6-GHz FM signal may be evaluated by using Carson's rule. The peak deviation of the composite video is 10.5 MHz and the peak deviation of the subcarrier is 2 MHz, giving an overall peak deviation of $\Delta F =$ 12.5 MHz. The baseband bandwidth is approximately 6.8 MHz. The transmission bandwidth is

$$B_T = 2(\Delta F + B) = 2(12.5 + 6.8) = 38.6 \quad \text{MHz} \tag{5-49}$$

which is accepted by the 36-MHz transponder.

This FDM/FM technique is also used by telephone companies for terrestrial relay of TV signals over microwave radio links that were characterized in Table 5-7 (except for the AR6A system, which cannot be used for TV relay). Each microwave link has the capacity for one TV channel (or for the number of VF

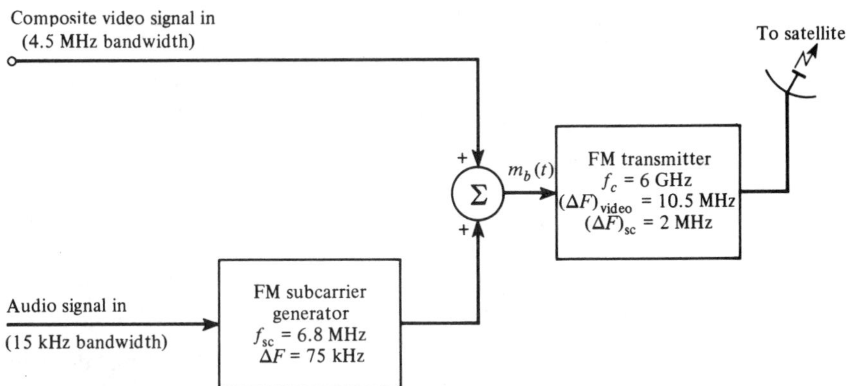

FIGURE 5-29 Transmission of broadcast-quality TV signals from a ground station.

channels, as indicated in the table). When video signals are relayed, wideband (0- to 15-kHz) aural signals can also be relayed by using FM subcarriers. Some typical subcarrier frequencies that are used are 5.58, 5.76, 6.2, 6.3, 6.48, 6.8, 7.38, and 7.56 MHz.

Data and Telephone Signal Multiple Access

Satellite relays provide a channel for data and VF (telephone) signaling similar to conventional terrestrial microwave radio links. That is, data may be time-division-multiplexed into DS-1, DS-2, and so on types of signals and (digitally) modulated onto a carrier for transmission via a satellite. On the other hand, the data may be frequency-division-multiplexed. In this case, modems are used to convert the data into signals that are compatible with VF group, supergroup, or mastergroup bandwidths, which are (analog) modulated onto the carrier for transmission to the satellite.

To relay telephone signals via satellite, the VF signals are frequency-division-multiplexed directly into groups, supergroups, and so on, or they may be time-division-multiplexed by converting the VF signal into a 64-kB/sec PCM signal that is merged with other digital signals to form DS-1, DS-2, and so on signals.

Satellite communication systems do differ from terrestrial microwave links in the techniques used for *multiple access* of a single transponder by multiple uplink and multiple downlink stations. Four main methods are used for multiple access.

1. *Frequency-division multiple access* (FDMA), which is similar to FDM.
2. *Time-division multiple access* (TDMA), which is similar to TDM.†

† *Satellite-switched time-division multiple access* (SS-TDMA) can also be used. With SS-TDMA satellites, different narrow-beam antennas are switched in at the appropriate time in the TDMA frame period to direct the transmit and receive beams to the desired direction. Thus, with multiple SS-TDMA beams the same transmit and receive frequencies may be reused simultaneously so that the capacity of the SS-TDMA satellite is much larger than that of a TDMA satellite. The *Intelsat VI* is a SS-TDMA satellite.

3. *Code-division multiple access* (CDMA) or *spread-spectrum multiple access* (SSMA).
4. *Space-division multiple access* (SDMA) where narrow-beam antenna patterns are switched from one direction to another.[†]

In addition, either of the following may be used:

1. A *fixed assigned multiple access* (FAMA) mode using either FDMA, TDMA, or CDMA techniques.
2. A *demand assigned multiple access* (DAMA) mode using either FDMA, TDMA, or CDMA.

In the FAMA mode the FDMA, TDMA, or CDMA format does not change, even though the traffic load of various Earth stations changes. For example, there is more telephone traffic between Earth stations during daylight hours (local time) than between these stations in the hours after midnight. In the FAMA mode a large number of satellite channels would be idle during the early morning hours since they are fixed assigned. In the DAMA mode, the FDMA or TDMA formats are changed as needed depending on the traffic demand of the Earth stations involved. Consequently, the DAMA mode uses the satellite capacity more efficiently, but it usually costs more to implement and maintain.

In CDMA the different users share the same frequency band simultaneously in time, as opposed to FDMA and TDMA, where users are assigned different frequency slots or time slots. With CDMA each user is assigned a particular digitally encoded waveform $\varphi_j(t)$ that is (nearly) orthogonal to the waveforms used by the others (see Secs. 2-4 and 5-7). Data may be modulated onto this waveform, transmitted over the communication system, and recovered. For example, if one bit of data, m_j, is modulated onto the waveform, the transmitted signal from the jth user might be $m_j\varphi_j(t)$ and the composite CDMA signal from all users would be $W(t) = \Sigma_j m_j \varphi_j(t)$. The data from the jth user could be recovered from the CDMA waveform by evaluating $\int_0^T W(t)\varphi_j(t)\, dt = m_j$, where T is the length of encoded waveform $\varphi_j(t)$. Gold codes are often used to obtain the encoded waveforms.

EXAMPLE 5-2 FIXED ASSIGNED MULTIPLE ACCESS MODE USING AN FDMA FORMAT

A good example of the fixed assigned multiple access technique is the FDM/FM format used with the *Intelsat IV* and *V* series satellites. Suppose that ground station A is sharing one 36-MHz transponder in the FDM mode with stations B, C, D, E, F, and G, which are all transmitting simultaneously through the

[†] *Satellite-switched time-division multiple access* (SS-TDMA) can also be used. With SS-TDMA satellites, different narrow-beam antennas are switched in at the appropriate time in the TDMA frame period to direct the transmit and receive beams to the desired direction. Thus, with multiple SS-TDMA beams the same transmit and receive frequencies may be reused simultaneously so that the capacity of the SS-TDMA satellite is much larger than that of a TDMA satellite. The *Intelsat VI* is a SS-TDMA satellite.

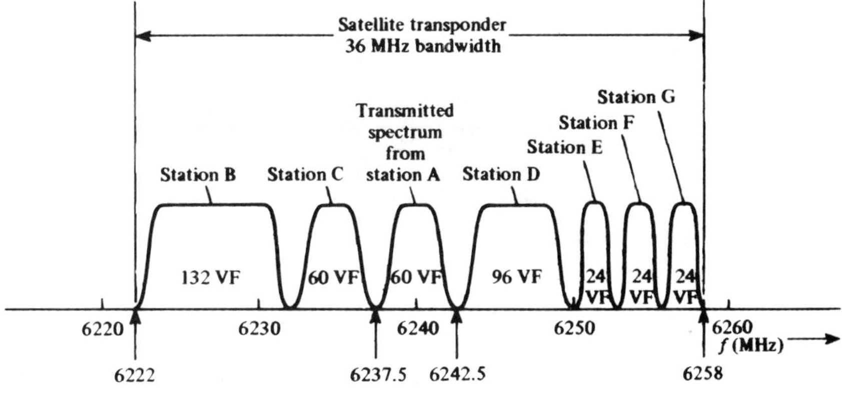

(a) Transponder Frequency Allocation

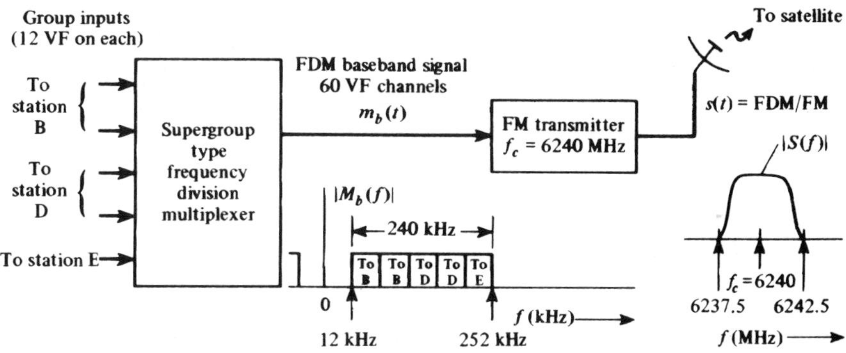

(b) Station A Ground Transmitting Equipment

FIGURE 5-30 Fixed assigned FDMA format for satellite communications.

transponder, as shown by their frequency assignments in Fig. 5-30a. Station A has an assigned band of 6237.5 to 6242.5 MHz in which it can transmit 60 VF channels using FDM/FM. This is the capacity of one supergroup. Suppose that station A has traffic as follows: 24 VF channels for station B, 24 VF channels for station D, and 12 VF channels for station E; then the equipment configuration would be as shown in Fig. 5-30b. In a similar way, the remaining spectrum of the 36-MHz transponder is divided among the other Earth stations according to their traffic needs on a fixed assignment basis.

EXAMPLE 5-3 SPADE SYSTEM

The *Intelsat IV* and *V* series satellites may also be operated in a DAMA mode using an FDMA format consisting of a single QPSK carrier for each telephone (VF) channel. This type of signaling is called *single channel per carrier* (SCPC), in which 800 QPSK signals may be accommodated in the 36-MHz bandwidth of

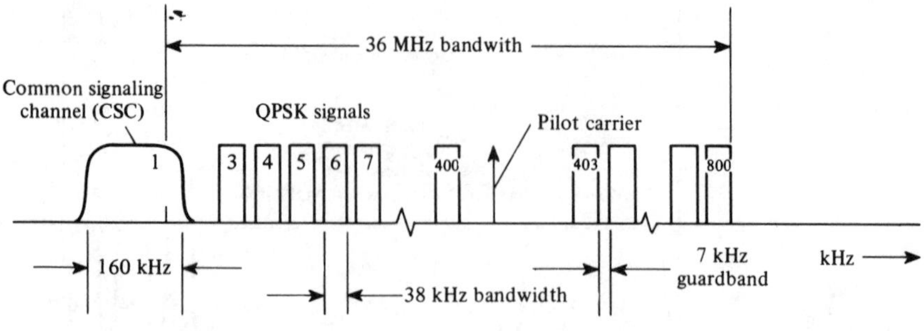

(a) Transponder Frequency Allocation

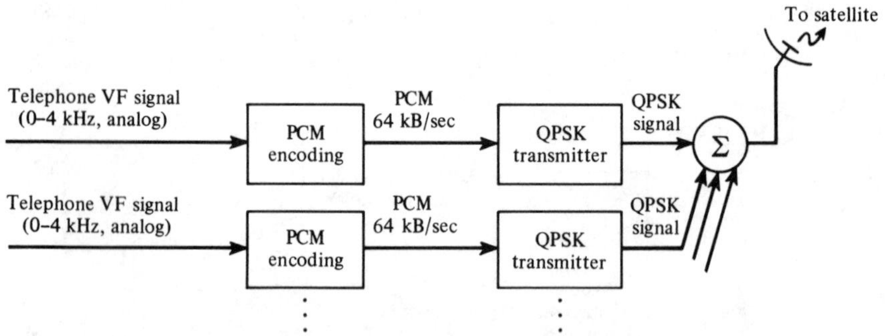

(b) Possible QPSK SCPC Transmitter Configuration

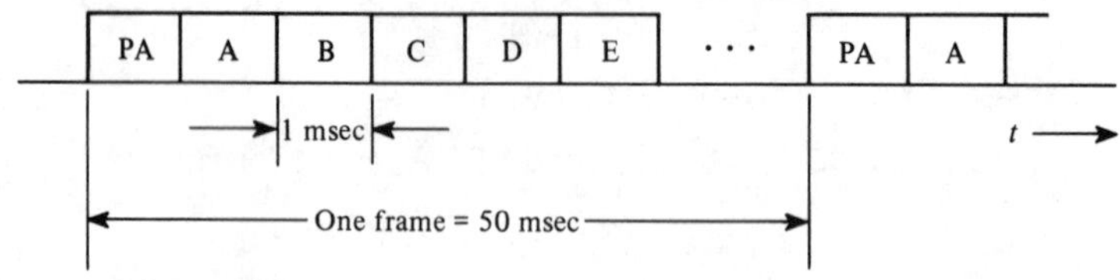

(c) TDMA CSC Signaling Format

FIGURE 5-31 SPADE satellite communication system for telephone VF message transmission.

the transponder, as shown in Fig. 5-31a. Thus 800 VF messages may be transmitted simultaneously through one satellite transponder, where each QPSK signal is modulated by a 64-kbit/sec PCM (digitized) voice signal (studied earlier in Example 3-1 and used as inputs to the North American digital hierarchy shown in Fig. 3-44). This SCPC-DAMA technique, illustrated in Fig. 5-31, is called the *SPADE system,* which is an acronym for *s*ingle channel per carrier, *p*ulse code modulation, multiple *a*ccess, *d*emand assignment *e*quipment [Edelson and Werth, 1972]!

The demand assignment of the carrier frequency for the QPSK signal to a particular Earth station uplink is carried out by TDM signaling in the *common signaling channel* (CSC) (see Fig. 5-31a). The CSC consists of a 128-kbit/sec PSK signal that is time shared among the Earth stations by using a TDMA format, as shown in Fig. 5-31c. PA denotes the synchronizing preamble that occurs at the beginning of each frame, and A, B, C, and so on denote the 1-msec time slots that are reserved for transmission by Earth stations A, B, C, and so on. In this way 49 different Earth stations may be accommodated in the 50-msec frame. For example, if Earth station B wishes to initiate a call to Earth station D, station B selects a QPSK carrier frequency randomly from among the idle channels that are available and transmits this frequency information along the address of station D (the call destination) in the station B TDMA time slot. If it is assumed that the frequency has not been selected by another station for another call, station D will acknowledge the request in its TDMA time slot. This acknowledgment would be heard by station B about 600 msec after its TDMA signaling request since the round-trip time delay to the satellite is 240 msec, plus equipment delays and the delay to the exact time slot assigned to station D with respect to that of station B. If another station, say station C, had selected the same frequency during the request period, a busy signal would be received at station B, and station B would randomly select another available frequency and try again. When the call is over, disconnect signals are transmitted in the TDMA time slot and that carrier frequency is returned for reuse. Since the CSC signaling rate is 128 kbits/sec and each time slot is 1 msec in duration, 128 bits are available for each accessing station to use for transmitting address information, frequency request information, and disconnect signaling.

In practice, since only 49 time slots are available for the TDMA mode, a number of the SCPC frequencies are assigned on a fixed basis.

In FDMA, such as the SPADE system, when the carriers are turned on and off on demand, there is amplitude modulation on the composite 36-MHz-wide signal that is being amplified by the transponder TWT. Consequently, the drive level of the TWT has to be "backed off" so that the amplifier is not saturated and will be sufficiently linear. Then the intermodulation products will be sufficiently low. On the other hand, if a single constant-amplitude signal (such as a single-wideband FM signal used in relaying television signals), had been used, IM products would not be a consideration and the amplifier could be driven harder to provide the full saturated power output level.

As indicated earlier, TDMA is similar to TDM, where the different Earth stations send up bursts of RF energy that contain packets of information. During the time slot designated for a particular Earth station, that station's signal uses the bandwidth of the whole transponder (see Fig. 5-32). Since the Earth stations use a constant envelope modulation technique, such as QPSK, and only one high-rate modulated signal is being passed through the transponder during any time interval, no inband IM products are generated (as in the FDMA technique described earlier). Thus the final TWT amplifier in the satellite may be driven into saturation for more power output. This advantage of TDMA over FDMA may be compared with the main disadvantage of the TDMA technique. The

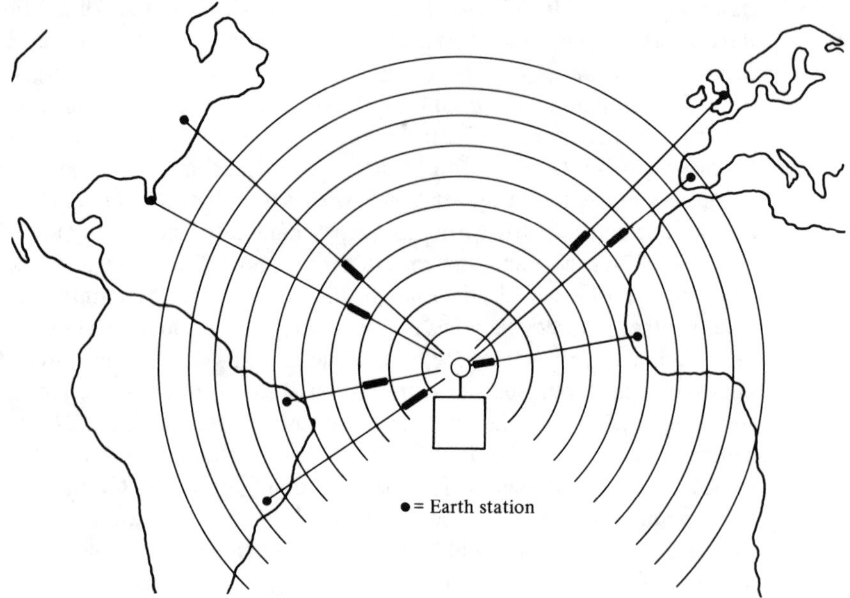

FIGURE 5-32 Interleaving of bursts in a TDMA satellite.

disadvantage of TDMA is that strict burst timing is required at the Earth station in order to prevent collision of the bursts at the satellite. In other words, the burst from a particular Earth station has to arrive at the satellite in the exact time slot designated for that station, so that its signal will not interfere with the bursts that are arriving from other Earth stations that are assigned adjacent time slots. Since the ground stations are located at different distances from the satellite and may use different equipment configurations, the time delay from each Earth station to the satellite will be different, and this must be taken into account when the transmission time for each Earth station is computed. In addition the satellite may be moving with respect to the Earth station, which means that the time delay is actually changing with time. Another disadvantage of TDMA is that the Earth stations are probably transmitting data that have arrived from synchronous terrestrial lines; consequently, the Earth station equipment must contain a large memory in which to buffer the data, which are read out at high speed when the packet of information is sent to the satellite.

A typical TDMA frame format for the data being relayed through a satellite is shown in Fig. 5-33. In this example, station B is sending data to stations A, E, G, and H. One frame consists of data arriving from each of the Earth stations. At any time, only one Earth station provides the time reference signal for the other Earth stations to use for computation of their transmitting time for their data bursts (frame synchronization). The burst lengths from the various stations could be different depending on the traffic volume. The second part of the figure shows an exploded view of a typical burst format that is transmitted from station B. This consists of two main parts, the preamble and data that are being sent to other

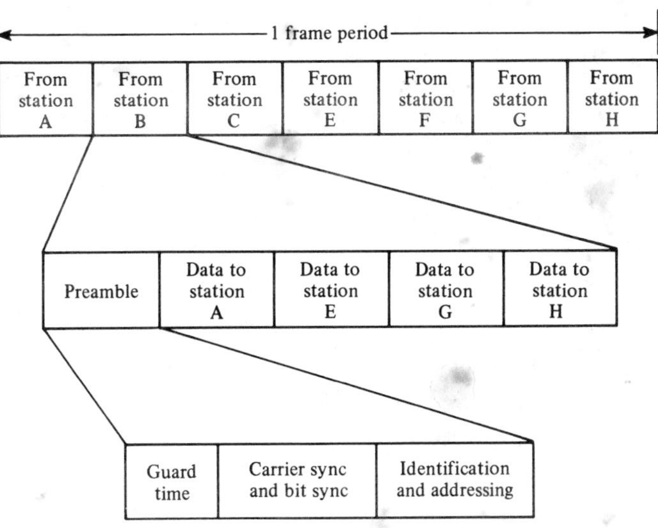

FIGURE 5-33 Typical TDMA frame format.

Earth stations from station B. The preamble includes a guard time before transmission is begun. Then a string of synchronization characters is transmitted that give the carrier sync recovery loops and bit timing recovery loops (in the ground station receivers) time to lock onto this burst from station B. The end of the preamble usually contains a unique word that identifies the burst as coming from station B and might indicate the addresses (stations) for which the data are intended.

Another multiple access method which is similar to TDMA is the *ALOHA* technique [Lam, 1979]. Here the multiple users send bursts of data, called *packets,* whenever they wish. If two or more of the bursts overlap in time, called a *collision,* the users involved retransmit their packets after a random time delay. It is hoped that a second collision will not occur. If it does, the retransmission process is repeated until each party is successful. This technique has the advantage of being relatively inexpensive to implement, but it will not work if there is heavy traffic loading on the satellite. Then the satellite becomes saturated with colliding packets, and the collisions can be avoided only by stopping all new transmissions and increasing the random delay required before retransmission. A more elaborate technique is called *slotted ALOHA*. Here the packets are transmitted at random, but only in certain time slots. This avoids collisions due to partial packet overlap.

Very small aperture terminals (VSAT) have become popular with the availability of Ku-band satellites and with the recent advances that have made low-cost terminals possible. "Very small aperture" implies that these systems use Earth-terminal antennas that are relatively small, which is about 1.8 meters in diameter. Solid-state power amplifiers (1 to 2 W), low-cost frequency converters, digital processing, and VLSI circuits have made VSATs feasible. The objective of VSAT systems is to provide low-cost data and voice transmission directly to users such as auto dealerships, banks (automatic teller machines), brokerage firms, pipeline companies (monitor and control), hotels and airlines (reservations), retail stores (data networks), and corporations (point-to-multi-point transmission). Typically, VSATs offer high-quality transmission (bit error rates of less than 10^{-7} for 99.5% of the time) at data rates from 100 bits/sec to 9.6 kbits/sec [Chakraborty, 1988]. Many users share a single satellite transponder via TDMA so that the user's cost can be substantially less than that for the same type of service provided by the long-distance public telephone network.

In the next section, a procedure will be developed for evaluating the performance of satellite systems in terms of the output probability of bit error for digital systems and the output signal-to-noise ratio for analog systems. More details can also be obtained from books devoted specifically to satellite systems [Ha, 1986, 1989; Pratt and Bostian, 1986; Spilker, 1977].

5-5 Receiver Signal-to-Noise Ratio

In this section, formulas will be developed for the signal-to-noise ratio at the detector input as a function of the transmitted *effective isotropic radiated power* (EIRP), the free-space loss, the receiver antenna gain, and the receiver system noise figure. These results will allow us to evaluate the quality of the receiver output (as measured by the probability of bit error for digital systems and output signal-to-noise ratio for analog systems), provided that the noise characteristics of the receiver detector circuit are known.

Signal Power Received

In communication systems the received signal power (as opposed to the voltage or current level) is the important quantity since it is the output signal power that ultimately drives the display device, which, for example, may be a loudspeaker or a cathode ray tube (CRT). Consequently, it is the *power* of the received signal that is of prime importance while trying to *minimize* the effect of *noise sources* that feed into the system and get amplified. The voltage gain, current gain, and impedance levels have to be such that the required power gain is achieved. In FET circuits it is accomplished by using relatively large voltage levels and small currents (high-impedance circuitry). With bipolar transistors power gain is achieved with relatively small voltages and large currents (low-impedance circuitry).

A block diagram of a communication system with a free-space transmission channel is shown in Fig. 5-34. The overall power gain (or power transfer function) of the channel is

$$\frac{P_{\text{Rx}}}{P_{\text{Tx}}} = G_{\text{AT}} G_{\text{FS}} G_{\text{AR}} \tag{5-50}$$

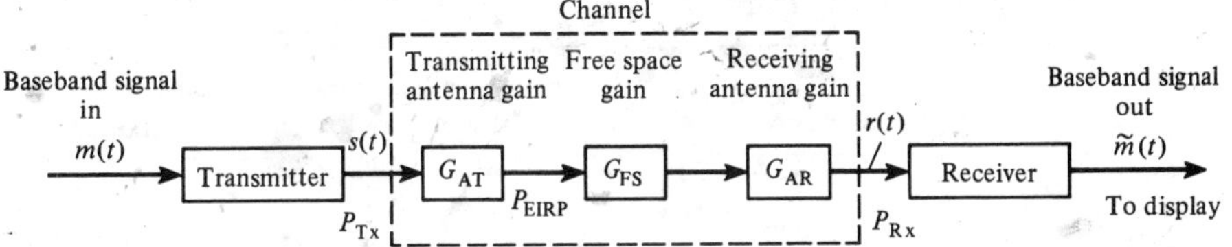

FIGURE 5-34 Block diagram of a communication system with a free-space transmission channel.

where P_{Tx} is the signal power into the transmitting antenna, G_{AT} is the transmitting antenna power gain, G_{FS} is the free-space power gain† (which is orders of magnitude less than one in typical communication systems), G_{AR} is the receiving antenna power gain, and P_{Rx} is the signal power into the receiver.

In order to use this relationship, these gains should be expressed in terms of useful antenna and free-space parameters [Kraus, 1986]. Here G_{AT} and G_{AR} are taken to be the power gains with respect to an isotropic antenna.‡ The EIRP is

$$P_{\text{EIRP}} = G_{\text{AT}} P_{\text{Tx}} \tag{5-51}$$

The antenna power gain is defined as

$$G_A = \frac{\begin{array}{c}\text{radiation power density of the actual antenna}\\ \text{in the direction of maximum radiation}\end{array}}{\begin{array}{c}\text{radiation power density of an isotropic antenna}\\ \text{with the same power input}\end{array}}$$

where the power density (W/m^2) is evaluated at the same distance, d, for both antennas. The gain for some practical antennas is given in Table 5-8.

The power density (W/m^2) of an isotropic antenna at a distance d from the antenna is

$$\text{power density at } d = \frac{\text{transmitted power}}{\text{area of a sphere with radius } d} = \frac{P_{\text{EIRP}}}{4\pi d^2} \tag{5-52a}$$

The FCC and others often specify the strength of an electromagnetic field by the field intensity, $\mathscr{E}$ (V/m), instead of power density (W/m^2). The two are related by

$$\text{power density} = \frac{\mathscr{E}^2}{377} \tag{5-52b}$$

where the power density and the field strength are evaluated at the same point in space and 377 Ω is the *free-space intrinsic impedance*. If the receiving antenna is

† A gain transfer function is the output quantity divided by the input quantity, whereas a loss transfer function is the input quantity divided by the output quantity.

‡ An isotropic antenna is a nonrealizable theoretical antenna that radiates equally well in all directions and is a useful reference for comparing practical antennas.

TABLE 5-8 Antenna Gains and Effective Areas

Type of Antenna	Power Gain, G_A (absolute units)	Effective Area, A_e (meters^2)
Isotropic	1	$\lambda^2/4\pi$
Infinitesimal dipole or loop	1.5	$1.5\lambda^2/4\pi$
Half-wave dipole	1.64	$1.64\lambda^2/4\pi$
Horn (optimized), mouth area, A	$10A/\lambda^2$	$0.81A$
Parabola or "dish" with area, A	$7.0A/\lambda^2$	$0.56A$
Turnstile (two crossed dipoles fed 90° out of phase)	1.15	$1.15\lambda^2/4\pi$

placed at d meters from the transmitting antenna, it will act like a "catcher's mitt" and intercept the power in an effective area of $(A_e)_{\text{Rx}}$ (meters^2), so that the received power will be

$$P_{\text{Rx}} = G_{\text{AT}}\left(\frac{P_{\text{Tx}}}{4\pi d^2}\right)(A_e)_{\text{Rx}} \tag{5-53}$$

where the gain of the transmitting antenna (with respect to an isotropic antenna), G_{AT}, has been included. Table 5-8 also gives the effective area for several types of antennas. The gain and the effective area of an antenna are related by

$$G_A = \frac{4\pi A_e}{\lambda^2} \tag{5-54}$$

where $\lambda = c/f$ is the wavelength, c being the speed of light (3×10^8 m/sec) and f the operating frequency (hertz). An antenna is a *reciprocal element*. That is, it has the same gain properties whether it is transmitting or receiving. Substituting (5-54) into (5-53), we obtain

$$\frac{P_{\text{Rx}}}{P_{\text{Tx}}} = G_{\text{AT}}\left(\frac{\lambda}{4\pi d}\right)^2 G_{\text{AR}} \tag{5-55}$$

where the free-space gain is

$$G_{\text{FS}} = \left(\frac{\lambda}{4\pi d}\right)^2 = \frac{1}{L_{\text{FS}}} \tag{5-56}$$

and L_{FS} is the free-space path loss (absolute units). Often the channel gain, expressed in decibel units, is obtained from (5-50):

$$(G_{\text{channel}})_{\text{dB}} = (G_{\text{AT}})_{\text{dB}} - (L_{\text{FS}})_{\text{dB}} + (G_{\text{AR}})_{\text{dB}} \tag{5-57}$$

where the free-space loss is

$$(L_{\text{FS}})_{\text{dB}} = 20 \log\left(\frac{4\pi d}{\lambda}\right) \text{ dB} \tag{5-58}$$

For example, the free-space loss at 4 GHz for the shortest path to a synchronous satellite from Earth (22,300 mi) is 195.6 dB.

Thermal Noise Sources

The noise power generated by a thermal noise source will be studied since the receiver noise is evaluated in terms of this phenomenon. A conductive element with two terminals may be characterized by its resistance, R ohms. This resistive or lossy element contains free electrons that have some random motion if the resistor has a temperature above absolute zero. This random motion causes a noise voltage to be generated at the terminals of the resistor. Although the noise is small, when the noise is amplified by a high-gain receiver it can become a problem. (If no noise were present, we could communicate to the edge of the universe with infinitely small power since the signal could always be amplified without having noise introduced.)

This physical lossy element, or physical resistor, can be modeled by an equivalent circuit that consists of a noiseless resistor in series with a noise voltage source (see Fig. 5-35). From quantum mechanics, it can be shown that the (normalized) power spectrum corresponding to the voltage source is [Van der Ziel, 1986]

$$\mathcal{P}_v(f) = 2R\left[\frac{h|f|}{2} + \frac{h|f|}{e^{h|f|/(kT)} - 1}\right] \tag{5-59}$$

where R = value of the physical resistor (ohms)
$h = 6.2 \times 10^{-34}$ J-sec is Planck's constant
$k = 1.38 \times 10^{-23}$ J/K is Boltzmann's constant, where K is kelvin
$T = (273 + C)$ is the absolute temperature of the resistor (kelvin)

At room temperature for frequencies below 1000 GHz, $[h|f|/(kT)] < 1/5$, so that $e^x = 1 + x$ is a good approximation. Then (5-59) reduces to

$$\mathcal{P}_v(f) = 2RkT \tag{5-60}$$

This equation will be used to develop other formulas in this text since the RF frequencies of interest are usually well below 1000 GHz and we are not dealing with temperatures near absolute zero.

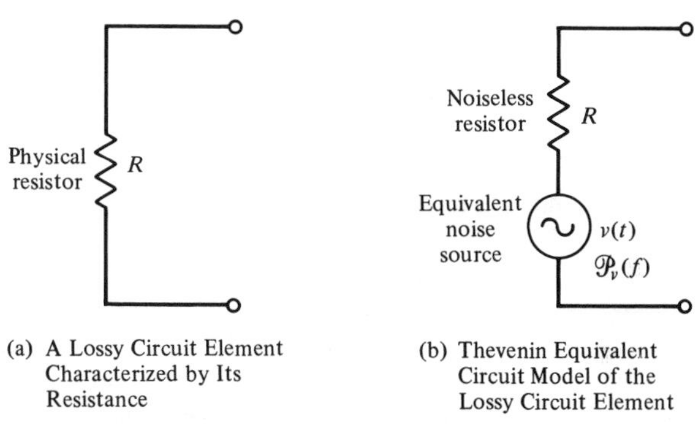

(a) A Lossy Circuit Element Characterized by Its Resistance

(b) Thevenin Equivalent Circuit Model of the Lossy Circuit Element

FIGURE 5-35 Thermal noise source.

If the open-circuit noise voltage that appears across a physical resistor is read by a true rms voltmeter that has a bandwidth of B hertz, then, using (2-67), the reading would be

$$V_{\text{rms}} = \sqrt{\langle v^2 \rangle} = \sqrt{2 \int_0^B \mathcal{P}_v(f)\, df} = \sqrt{4kTBR} \tag{5-61}$$

Characterization of Noise Sources

Noise sources may be characterized by the maximum amount of noise power or PSD that can be passed to a load.

Definition: The *available noise power* is the *maximum*† actual (i.e., not normalized) power that can be drawn from a source. The *available PSD* is the *maximum* actual (i.e., not normalized) PSD that can be obtained from a source.

For example, the available PSD for a thermal noise source is easily evaluated using Fig. 5-36 and (2-142).

$$\mathcal{P}_a(f) = \frac{\mathcal{P}_v(f)|H(f)|^2}{R} = \tfrac{1}{2}kT \quad \text{W/Hz} \tag{5-62}$$

where $H(f) = \frac{1}{2}$ for the resistor divider network. The available power from a thermal source in a bandwidth of B hertz is

$$P_a = \int_{-B}^{B} \mathcal{P}_a(f)\, df = \int_{-B}^{B} \frac{1}{2} kT\, df$$

or

$$P_a = kTB \tag{5-63}$$

This equation indicates that the available noise power from a thermal source does *not* depend on the value of R, even though the open-circuit rms voltage does.

The available noise power from a source (it does not have to be a thermal source) can be specified by a number called the noise temperature.

Definition: The *noise temperature* of a source is given by

$$T = \frac{P_a}{kB} \tag{5-64}$$

where P_a is the available power from the source in a bandwidth of B hertz.

In using this definition, it is noted that if the source happens to be of thermal origin, T will be the temperature of the device in kelvins, but if the source is not

†The maximum power or maximum PSD is obtained when $Z_L(f) = Z_s^*(f)$, where $Z_L(f)$ is the load impedance and $Z_s^*(f)$ is the conjugate of the source impedance.

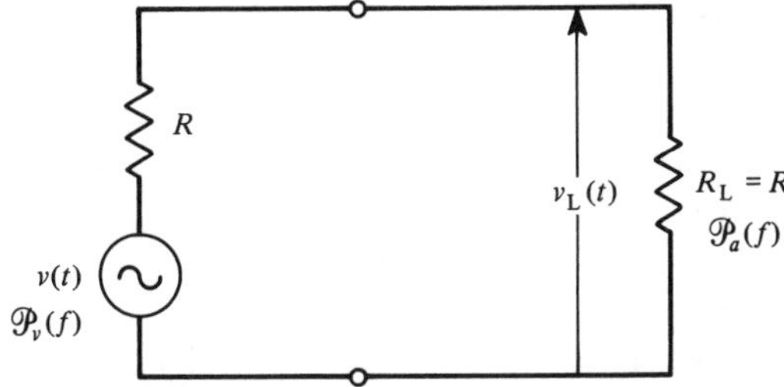

FIGURE 5-36 Thermal source with a matched load.

of thermal origin, the number obtained for T may have nothing to do with the physical temperature of the device.

Noise Characterization of Linear Devices

A linear device with internal noise generators may be modeled as shown in Fig. 5-37. Any device that can be built will have some internal noise sources. As shown in the figure, the device could be modeled as a noise-free device having a power gain $G_a(f)$ and an excess noise source at the output to account for the internal noise of the actual device. Some examples of linear devices that have to be characterized in receiving systems are lossy transmission lines, RF amplifiers, down converters, and IF amplifiers.

The power gain of the devices is the available power gain.

Definition: The *available power gain* of a linear device is

$$G_a(f) = \frac{\text{available PSD out of the device}}{\text{available PSD out of the source}} = \frac{\mathcal{P}_{ao}(f)}{\mathcal{P}_{as}(f)} \tag{5-65}$$

When $\mathcal{P}_{ao}(f)$ is measured to obtain $G_a(f)$, the source noise power is made large enough so that the amplified *source* noise that appears at the output dominates any other noise. In addition, note that $G_a(f)$ is defined in terms of actual

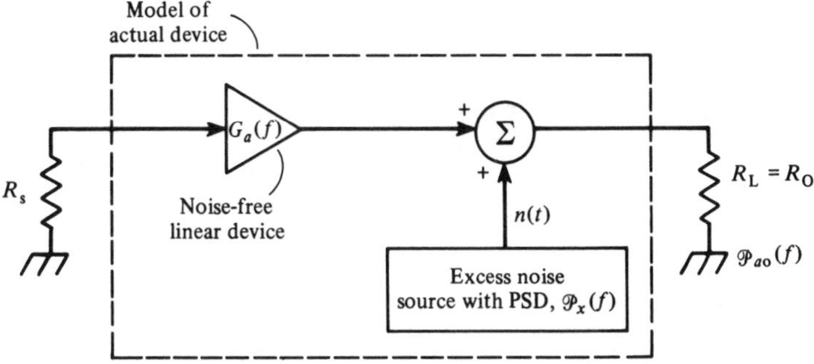

FIGURE 5-37 Noise model for an actual device.

(i.e., not normalized) PSD. In general, $G_a(f)$ will depend on the driving source impedance as well as on elements within the device itself, but it does *not* depend on the load impedance. If the source impedance and the output impedance of the device are equal, $G_a(f) = |H(f)|^2$, where $H(f)$ is the voltage or current transfer function of the linear device.

To characterize the goodness of a device, a figure of merit is needed that compares the actual (noisy) device with an ideal device (i.e., no internal noise sources). Two figures of merit, both of which tell us the same thing—namely, how bad the noise performance of the actual device is—are universally used. They are noise figure and effective input-noise temperature.

DEFINITION: The *spot noise figure* of a linear device is obtained by terminating the device with a thermal noise source of temperature T_0 on the input and a matched load on the output as indicated in Fig. 5-37. The spot noise figure is

$$F_s(f) = \frac{\begin{array}{c}\text{measured available PSD}\\ \text{out of the actual device}\end{array}}{\begin{array}{c}\text{available PSD out of an ideal device}\\ \text{with the same available gain}\end{array}}$$

or

$$F_s(f) = \frac{\mathcal{P}_{ao}(f)}{(kT_0/2)G_a(f)} = \frac{(kT_0/2)G_a(f) + \mathcal{P}_x(f)}{(kT_0/2)G_a(f)} > 1 \tag{5-66}$$

The value of R_s is the same as the source resistance that was used when $G_a(f)$ was evaluated. A standard temperature of $T_0 = 290$ K is used as adopted by the IEEE [Haus, 1963].

$F_s(f)$ is called the *spot noise figure* since it refers to the noise characterization at a particular "spot" or frequency in the spectrum. Note that $F_s(f)$ is always greater than unity for an actual device, but it is nearly unity if the device is almost an ideal device. $F_s(f)$ is a function of the source temperature, T_0. Consequently, when the noise figure is evaluated, a standard temperature of $T_0 = 290$ K is used. This corresponds to a room temperature of 62.3°F.

Often an average noise figure instead of a spot noise figure is desired. The average is measured over some bandwidth B.

DEFINITION: The *average noise figure* is

$$F = \frac{P_{ao}}{kT_0 \int_{f_0-B/2}^{f_0+B/2} G_a(f)\,df} \tag{5-67}$$

where

$$P_{ao} = 2\int_{f_0-B/2}^{f_0+B/2} \mathcal{P}_{ao}(f)\,df$$

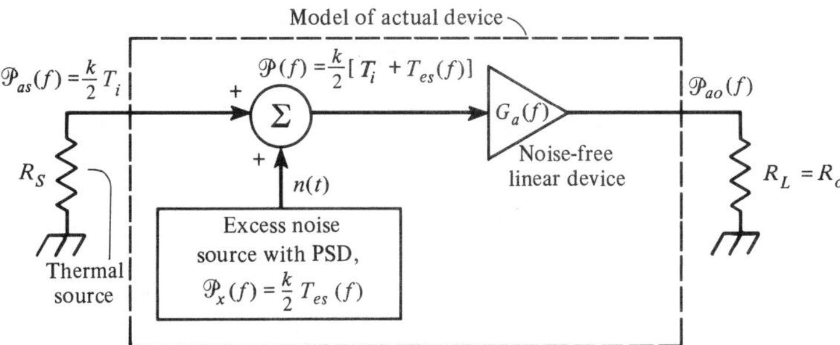

FIGURE 5-38 Another noise model for an actual device.

is the measured available output in a bandwidth B hertz wide centered on a frequency of f_0 and $T_0 = 290$ K.

If the available gain is constant over the band so that $G_a(f) = G_a$ over the frequency interval $(f_0 - B/2) \leq f \leq (f_0 + B/2)$, the noise figure becomes

$$F = \frac{P_{ao}}{kT_0 BG_a} \tag{5-68a}$$

The noise figure is often measured by using the Y-factor method. This technique is illustrated by Prob. 5-41. The Hewlett-Packard HP8970B Noise Figure Meter uses the Y-factor method for measuring the noise figure of devices.

The noise figure can also be specified in decibel units,†

$$F_{\text{dB}} = 10 \log(F) = 10 \log\left(\frac{P_{ao}}{kT_0 BG_a}\right) \tag{5-68b}$$

For example, suppose that the manufacturer of an RF preamplifier specifies that an RF preamp has a 2-dB noise figure. This means that the actual noise power at the output is 1.58 times the power that would occur because of amplification of noise from the input. The other figure of merit for evaluating the noise performance of a linear device is the effective input-noise temperature, which is illustrated in Fig. 5-38.

DEFINITION: The *spot effective input-noise temperature,* $T_{es}(f)$, of a linear device is the *additional* temperature required for an input source, which is driving an ideal (noise-free) device, to produce the same available PSD at the ideal device output as is obtained from the actual device when it is driven by the input source of temperature T_i kelvin. That is, $T_{es}(f)$ is defined by

$$\mathcal{P}_{ao}(f) = G_a(f)\,\frac{k}{2}\,[T_i + T_{es}(f)] \tag{5-69}$$

† Some authors call F the *noise factor* and F_{dB} the *noise figure*.

where $\mathcal{P}_{ao}(f)$ is the available PSD out of the actual device when driven by an input source of temperature T_i, and $T_{es}(f)$ is the spot effective input-noise temperature.

The *average effective input-noise temperature*, T_e, is defined by the equation

$$P_{ao} = k(T_i + T_e) \int_{f_0-B/2}^{f_0+B/2} G_a(f)\, df \tag{5-70}$$

where the measured available noise power out of the device in a band B hertz wide is

$$P_{ao} = 2 \int_{f_0-B/2}^{f_0+B/2} \mathcal{P}_{ao}(f)\, df \tag{5-71}$$

Since $G_a(f)$ depends on the source impedance as well as on the device parameters, $T_e(f)$ will depend on the source impedance used as well as on the characteristics of the device itself, but it is independent of the value of T_i used. In the definition of T_e, note that the IEEE standards do not specify that $T_i = T_0$ since the value of T_e obtained does not depend on the value of T_i that is used. However, $T_i = T_0 = 290$ may be used for convenience. The effective input-noise temperature can also be evaluated by the Y-factor method, as illustrated by Prob. 5-41.

When the gain is flat (constant) over the frequency band, $G_a(f) = G_a$, the effective input-noise temperature is simply

$$T_e = \frac{P_{ao} - kT_i G_a B}{kG_a B} \tag{5-72}$$

Note that $T_{es}(f)$ and T_e are greater than zero for an actual device, but if the device is nearly ideal (small internal noise sources), they will be very close to zero.

When T_e was evaluated by use of (5-70), an input source was used with some convenient value for T_i. However, when the device is used in a system, the available noise power from the source will be different if T_i is different. For example, suppose that the device is an RF preamplifier and the source is an antenna. The available power out of the amplifier when it is connected to the antenna is now[†]

$$P_{ao} = 2 \int_{f_0-B/2}^{f_0+B/2} \mathcal{P}_{as}(f) G_a(f)\, df + kT_e \int_{f_0-B/2}^{f_0+B/2} G_a(f)\, df \tag{5-73}$$

where $\mathcal{P}_{as}(f)$ is the available PSD out of the source (antenna). T_e is the average effective input temperature of the amplifier that was evaluated by using the input source T_i. If the gain of the amplifier is approximately constant over the band, this reduces to

$$P_{ao} = G_a P_{as} + kT_e B\, G_a \tag{5-74}$$

[†]The value of P_{ao} in (5-73) and (5-74) is different from that in (5-70), (5-71), and (5-72).

where the available power from the source (antenna) is

$$P_{as} = 2\int_{f_0-B/2}^{f_0+B/2} \mathcal{P}_{as}(f)\, df \tag{5-75}$$

Furthermore, the available power from the source might be characterized by its noise temperature, T_s, so that, using (5-64),

$$P_{as} = kT_sB \tag{5-76}$$

In satellite Earth station receiving applications, the antenna (source) noise temperature might be $T_s = 32$ K at 4 GHz for a parabolic antenna with a diameter of 5 m, where the noise from the antenna is due to cosmic radiation and to energy received from the ground as the result of the sidelobe beam pattern of the antenna. (The Earth acts as a blackbody noise source with $T = 280$ K.) Note that the $T_s = 32$ K of the antenna is "caused" by *radiation* resistance, which is not the same as a loss resistance (I^2R losses) associated with a thermal source and T_s has no relation to the physical temperature of the antenna.

In summary two figures of merit have been defined: noise figure and effective input-noise temperature. By combining (5-66) and (5-69) where $T_i = T_0$, a relationship between these two figures of merit for the spot measures is obtained:

$$T_{es}(f) = T_0[F_s(f) - 1] \tag{5-77a}$$

Here $T_i = T_0$ is required because $T_i = T_0$ is used in the definition for the noise figure that precedes (5-66).

Using (5-68a) and (5-72) where $T_i = T_0$, we obtain the same relationship for the average measures:

$$T_e = T_0(F - 1) \tag{5-77b}$$

EXAMPLE 5-4 T_e AND F FOR A TRANSMISSION LINE

The effective input-noise temperature, T_e, and the noise figure, F, for a lossy transmission line (a linear device) will now be evaluated.† This can be accomplished by terminating the transmission line with a source and a load resistance (all having the same physical temperature) that are both equal to the characteristic impedance of the line, as shown in Fig. 5-39. The gain of the transmission line is $G_a = 1/L$, where L is the transmission line loss (power in divided by power out) in absolute units (i.e., not dB units). Looking into the output port of the transmission line, one sees an equivalent circuit that is resistive (thermal source) with a value of R_0 ohms since the input of the line is terminated by a resistor of R_0 ohms (the characteristic impedance). Assume that the physical temperature of the transmission line is T_L, as measured on the Kelvin scale. Since the line acts as a thermal source, the available noise power at its output is

†These results also hold for the T_e and F of (impedance) matched attenuators.

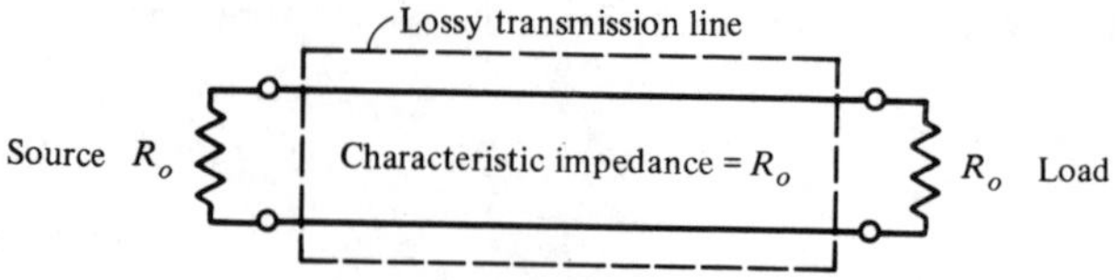

FIGURE 5-39 Noise figure measurement of a lossy transmission line.

$P_{ao} = kT_L B$. Using (5-72) where the source is at the same physical temperature, $T_i = T_L$, we get

$$T_e = \frac{kT_L B - kT_L G_a B}{kG_a B} = T_L\left(\frac{1}{G_a} - 1\right)$$

Thus the effective input-noise temperature for the transmission line is

$$T_e = T_L(L - 1) \tag{5-78a}$$

where T_L is the physical temperature (Kelvin) of the line and L is the line loss.

If the physical temperature of the line happens to be T_0, this becomes

$$T_e = T_0(L - 1) \tag{5-78b}$$

The noise figure for the transmission line is obtained by using (5-77b) to convert T_e to F. Thus, substituting (5-78a) into (5-77b), we get

$$T_L(L - 1) = T_0(F - 1)$$

Solving for F, we obtain the noise figure for the transmission line

$$F = 1 + \frac{T_L}{T_0}(L - 1) \tag{5-79a}$$

where T_L is the physical temperature (Kelvin) of the line, $T_0 = 290$, and L is the line loss. If the physical temperature of the line is 290 K (63°F), (5-79a) reduces to

$$F = \frac{1}{G_a} = L \tag{5-79b}$$

In decibel measure, this is $F_{\text{dB}} = L_{\text{dB}}$. In other words, if a transmission line has a 3-dB loss, it has a noise figure of 3 dB provided that it has a physical temperature of 63°F. If the temperature is 32°F (273 K), the noise figure, using (5-79a), will be 2.87 dB. Thus F_{dB} is approximately L_{dB}, if the transmission line is located in an environment (temperature range) that is inhabitable by humans.

Noise Characterization of Cascaded Linear Devices

In a communication system several linear devices, supplied by different vendors, are often cascaded together to form an overall system, as indicated in Fig. 5-40. In a receiving system these devices might be an RF preamplifier connected to a

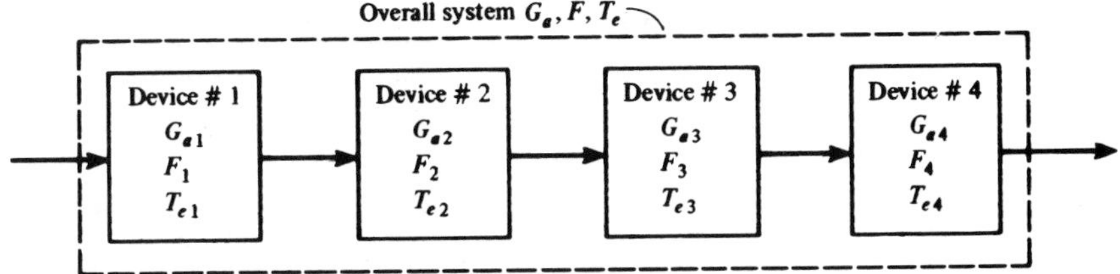

FIGURE 5-40 Cascade of four devices.

transmission line that feeds a down converter and an IF amplifier. (As discussed in Sec. 4-3, the down converter is a linear device and may be characterized by its conversion power gain and its noise figure.) For system performance calculations, we need to evaluate the overall power gain, G_a, and the overall noise characterization (which is given by the overall noise figure or the overall effective input-noise temperature) from the specifications for the individual devices provided by the vendors.

The overall available power gain is

$$G_a(f) = G_{a1}(f)G_{a2}(f)G_{a3}(f)G_{a4}(f)\cdots \tag{5-80}$$

since, for example, for a four-stage system,

$$G_a(f) = \frac{\mathcal{P}_{ao4}}{\mathcal{P}_{as}} = \left(\frac{\mathcal{P}_{ao1}}{\mathcal{P}_{as}}\right)\left(\frac{\mathcal{P}_{ao2}}{\mathcal{P}_{ao1}}\right)\left(\frac{\mathcal{P}_{ao3}}{\mathcal{P}_{ao2}}\right)\left(\frac{\mathcal{P}_{ao4}}{\mathcal{P}_{ao3}}\right)$$

THEOREM: *The overall noise figure for cascaded linear devices is*†

$$F = F_1 + \frac{F_2 - 1}{G_{a1}} + \frac{F_3 - 1}{G_{a1}G_{a2}} + \frac{F_4 - 1}{G_{a1}G_{a2}G_{a3}} + \cdots \tag{5-81}$$

as shown in Fig. 5-40 *(for a four-stage system).*

Proof. This result may be obtained by using an excess-noise model of Fig. 5-37 for each stage. We will prove the result for a two-stage system as modeled in Fig. 5-41. The overall noise figure is

$$F = \frac{P_{ao2}}{(P_{ao2})_{\text{ideal}}} = \frac{P_{x2} + P_{ao1}G_{a2}}{G_{a1}G_{a2}P_{as}}$$

which becomes

$$F = \frac{P_{x2} + G_{a2}(P_{x1} + G_{a1}P_{as})}{G_{a1}G_{a2}P_{as}} \tag{5-82}$$

where $P_{as} = kT_0B$ is the available power from the thermal source. P_{x1} and P_{x2} can be obtained from the noise figures of the individual devices by using Fig.

†This is known as Friis's noise formula.

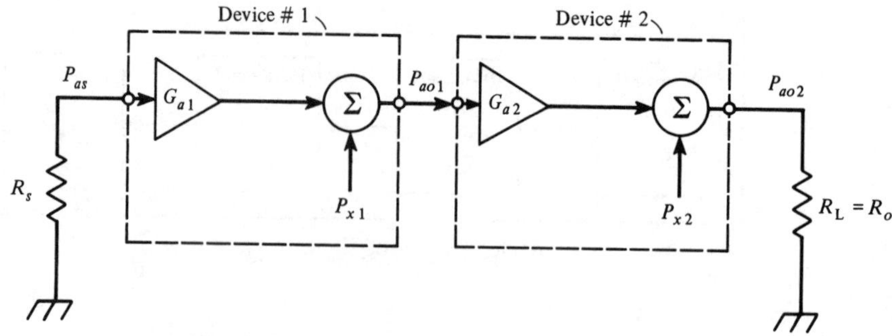

FIGURE 5-41 Noise model for two cascaded devices.

5-37, so that for the ith device

$$F_i = \frac{P_{aoi}}{G_{ai}P_{as}} = \frac{P_{xi} + G_{ai}P_{as}}{G_{ai}P_{as}}$$

or

$$P_{xi} = G_{ai}P_{as}(F_i - 1) \tag{5-83}$$

Substituting this equation into (5-82) for P_{x1} and P_{x2}, we obtain

$$F = F_1 + \frac{F_2 - 1}{G_{a1}}$$

which is identical to (5-81) for the case of two cascaded stages. In a similar way (5-81) can be shown to be true for any number of stages.

Looking at (5-81), we see that if the terms G_{a1}, $G_{a1}G_{a2}$, $G_{a1}G_{a2}G_{a3}$, and so on, are relatively large, F_1 will dominate the overall noise figure. Thus, in receiving system design, it is important that the first stage have a low noise figure and a large available gain so that the noise figure of the overall system will be as small as possible.

The overall effective input-noise temperature of several cascaded stages can also be evaluated.

THEOREM: *The* ***overall effective input-noise temperature*** *for cascaded linear devices is*

$$T_e = T_{e1} + \frac{T_{e2}}{G_{a1}} + \frac{T_{e3}}{G_{a1}G_{a2}} + \frac{T_{e4}}{G_{a1}G_{a2}G_{a3}} + \cdots \tag{5-84}$$

as shown in Fig. 5-40.

A proof of this result is left for the reader as an exercise.

For further study concerning the topics of effective input-noise temperature and noise figure, the reader is referred to an authoritative monograph [Mumford and Scheibe, 1968].

Link Budget Evaluation

The performance of a communication system depends on how large the signal-to-noise (S/N) ratio is at the detector input in the receiver. It is engineering custom to call the signal-to-noise ratio before detection the *carrier-to-noise* ratio (C/N). Thus, in this section, we will use C/N to denote the signal-to-noise ratio before detection (bandpass case) and S/N to denote the signal-to-noise ratio after detection (baseband case). Here we are interested in evaluating the detector input C/N as a function of the communication link parameters, such as transmitted EIRP, space loss, receiver antenna gain, and the effective input-noise temperature of the receiving system. The formula that relates these system link parameters to the C/N at the detector input is called the *link budget*.

The communication system may be described by the block diagram shown in Fig. 5-42. In this model, the receiving system from the receiving-antenna output to the detector input is modeled by one linear block that represents the cascaded stages in the receiving system, such as a transmission line, a low-noise amplifier (LNA), a down converter, and an IF amplifier. This linear block describes the overall available power gain and effective input-noise temperature of these cascaded devices, as described in the preceding section and modeled in Fig. 5-39.

As shown in Fig. 5-42, the C/N at the ideal amplifier input (with gain G_a) is identical to that at the detector input since the ideal amplifier adds no excess

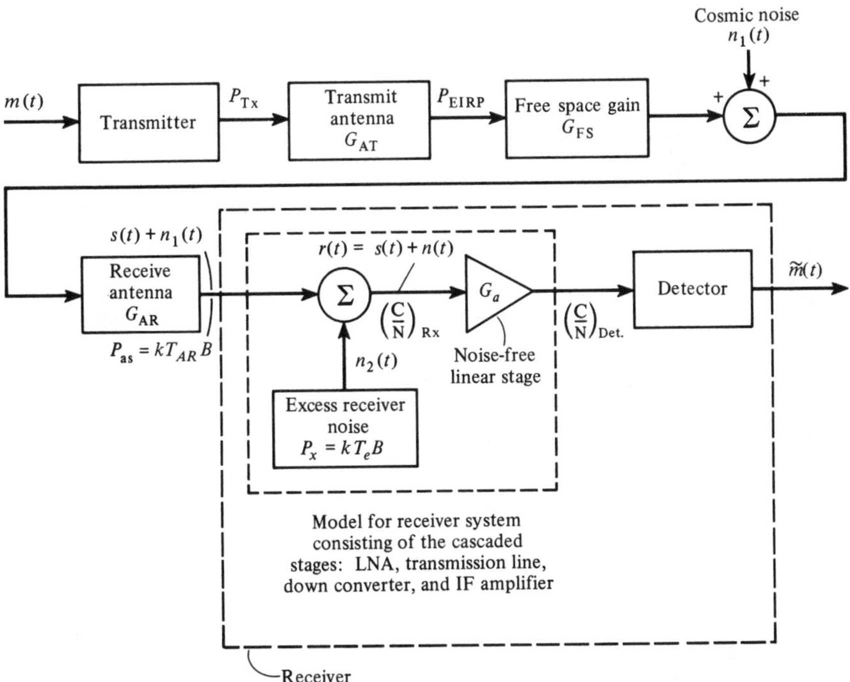

FIGURE 5-42 Communication system model for link budget evaluation.

noise and it amplifies the signal and noise equally well in a bandwidth of B hertz (the IF bandwidth). Thus, denoting this simply by C/N,

$$\left(\frac{\mathrm{C}}{\mathrm{N}}\right) \triangleq \left(\frac{\mathrm{C}}{\mathrm{N}}\right)_{Rx} = \left(\frac{\mathrm{C}}{\mathrm{N}}\right)_{\mathrm{det}} \tag{5-85}$$

where these carrier-to-noise power ratios are indicated on the figure.

Using (5-50) and (5-51), we obtain the received signal power

$$C_{\mathrm{Rx}} = (P_{\mathrm{EIRP}})G_{\mathrm{FS}}G_{\mathrm{AR}} \tag{5-86}$$

where P_{EIRP} is the EIRP from the transmitter, G_{FS} is the free-space gain, and G_{AR} is the receiving antenna power gain.

When we use (5-64), the available noise power at the input to the ideal amplifier in the model (Fig. 5-42) is

$$N = kT_{\mathrm{syst}}B \tag{5-87}$$

where B is the IF bandwidth. The receiving system noise temperature is

$$T_{\mathrm{syst}} = T_{\mathrm{AR}} + T_e \tag{5-88}$$

where T_{AR} is the noise temperature of the antenna (due to received cosmic noise and Earth blackbody radiation) and T_e is the effective input-noise temperature of the overall receiving system.

When (5-86) and (5-87) are combined, the carrier-to-noise ratio at the detector input is

$$\frac{\mathrm{C}}{\mathrm{N}} = \frac{P_{\mathrm{EIRP}}G_{\mathrm{FS}}G_{\mathrm{AR}}}{kT_{\mathrm{syst}}B} \tag{5-89}$$

For engineering applications, this formula is converted to decibel units. Using (5-56) in (5-89), the received carrier-to-noise ratio at the detector input in decibels is

$$\left(\frac{\mathrm{C}}{\mathrm{N}}\right)_{\mathrm{db}} = (P_{\mathrm{EIRP}})_{\mathrm{dBw}} - (L_{\mathrm{FS}})_{\mathrm{dB}} + \left(\frac{G_{\mathrm{AR}}}{T_{\mathrm{syst}}}\right)_{\mathrm{dB}} - k_{\mathrm{dB}} - B_{\mathrm{dB}} \tag{5-90}$$

where $(P_{\mathrm{EIRP}})_{\mathrm{dBw}} = 10\log(P_{\mathrm{EIRP}})$ is the EIRP of the transmitter in dB above 1 W

$(L_{\mathrm{FS}})_{\mathrm{dB}} = 20\log[(4\pi d)/\lambda]$ is the path loss

$k_{\mathrm{dB}} = 10\log(1.38 \times 10^{-23}) = -228.6$

$B_{\mathrm{dB}} = 10\log(B)$ (B is the IF bandwidth in hertz)

For analog communication systems, the S/N at the detector output can be related to the C/N at the detector input. The exact relationship depends on the type of detector as well as the modulation used. These relationships will be developed in Chapter 7, where statistical concepts are applied. Example 5-5 uses (5-90) to evaluate the performance of a CATV satellite receiving system.

E_b/N_0 Link Budget for Digital Systems

In digital communication systems the probability of bit error P_e for the digital signal at the detector output describes the quality of the recovered data. The P_e is a function of the ratio of the energy per bit to the noise PSD (E_b/N_0) as measured at the detector input. The exact relationship between P_e and E_b/N_0 depends on the type of digital signaling used, as will be developed in Chapter 7 using statistical concepts. In this section we evaluate the E_b/N_0 obtained at the detector input as a function of the communication link parameters.

The energy per bit is given by $E_b = CT_b$, where C is the signal power and T_b is the time required to send one bit. Using (5-64), we see that the noise PSD (one-sided) is $N_0 = kT_{\text{syst}}$. Thus

$$\frac{\text{C}}{\text{N}} = \frac{E_b/T_b}{N_0 B} = \frac{E_b R}{N_0 B} \tag{5-91}$$

where $R = 1/T_b$ is the data rate (bits/sec). Using (5-91) in (5-89), we have

$$\frac{E_b}{N_0} = \frac{P_{\text{EIRP}} G_{\text{FS}} G_{\text{AR}}}{kT_{\text{syst}} R} \tag{5-92}$$

In decibel units, the E_b/N_0 received at the detector input in a digital communications receiver is related to the link parameters by

$$\left(\frac{E_b}{N_0}\right)_{\text{dB}} = (P_{\text{EIRP}})_{\text{dBw}} - (L_{\text{FS}})_{\text{dB}} + \left(\frac{G_{\text{AR}}}{T_{\text{syst}}}\right)_{\text{dB}} - k_{\text{dB}} - R_{\text{dB}} \tag{5-93}$$

where $(R)_{\text{dB}} = 10 \log(R)$ and R is the data rate (bits/second).

For example, suppose that we have BPSK signaling and an optimum detector is used in the receiver; then $(E_b/N_0)_{\text{dB}} = 8.4$ dB is required for $P_e = 10^{-4}$ (see Fig. 7-14).† Using (5-93), the communication link parameters may be selected to give the required $(E_b/N_0)_{\text{dB}}$ of 8.4 dB. Note that as the bit rate is increased, the transmitted power has to be increased or the receiving system performance—denoted by $(G_{\text{AR}}/T_{\text{syst}})_{\text{dB}}$—has to be improved to maintain the required $(E_b/N_0)_{\text{dB}}$ of 8.4 dB.

EXAMPLE 5-5 LINK BUDGET EVALUATION FOR A TELEVISION RECEIVE-ONLY TERMINAL FOR SATELLITE SIGNALS

The link budget for a television receive-only (TVRO) terminal used for reception of TV signals from a satellite will be evaluated. It is assumed that this receiving terminal is located at Washington, D.C., and is receiving signals from a Hughes *Galaxy* satellite that is in a geostationary orbit at 134° W longitude above the equator. The specifications on the proposed receiving equipment are given in Table 5-9 together with the receiving site coordinates and satellite parameters.

†If, in addition, coding were used with a coding gain of 3 dB (see Sec. 1-11), an E_b/N_0 of 5.4 dB would be required for $P_e = 10^{-4}$.

TABLE 5-9 Typical Satellite Television Relay Parameters

Item	Parameter Value
Hughes *Galaxy I* satellite	
Orbit	Geostationary
Location (above equator)	134° W longitude
Uplink frequency band	6 GHz
Downlink frequency band	4 GHz
$(P_{\text{EIRP}})_{\text{dBw}}$	36 dBw
TVRO terminal	
Site location	Washington, D.C., 38.8° N latitude, 77° W longitude
Antenna	
Antenna type	5-m parabola
Noise temperature	32 K (at feed output) for 16.8° elevation
Feedline gain	0.98
Low-noise amplifier	
Noise temperature	120 K
Gain	50 dB
Receiver	
Manufacturer	Microdyne Corp., Model 1100-TVR (×24)
Noise temperature	2610 K
IF bandwidth	30 MHz
FM threshold	8 dB C/N

This TVRO terminal is typical of those used at the head end of CATV systems for reception of TV signals that are relayed via satellite. As discussed in Sec. 5-4, the composite NTSC baseband video signal is relayed via satellite by frequency-modulating this signal on a 6-GHz carrier that is radiated to the satellite. The satellite down-converts the FM signal to 4 GHz and retransmits this FM signal to the TVRO terminal.

The transmit antenna EIRP footprint for the *Galaxy I* satellite is shown in Fig. 5-43. From this figure it is seen that the EIRP directed toward Washington, D.C., is approximately 36 dBW, as obtained for inclusion in Table 5-9.

It can be shown that the *look angles* of the TVRO antenna can be evaluated using the following formulas [Davidoff, 1984, Chap. 9]. The elevation of the TVRO antenna is

$$E = \tan^{-1}\left[\frac{1}{\tan\beta} - \frac{R}{(R+h)\sin\beta}\right] \tag{5-94}$$

where

$$\beta = \cos^{-1}[\cos\phi\cos\lambda] \tag{5-95}$$

and λ is the difference between the longitude of the TVRO site and the longitude of the *substation point* (i.e., the point directly below the geostationary satellite along the equator). ϕ is the latitude of the TVRO site, $R = 3963$ statute miles is

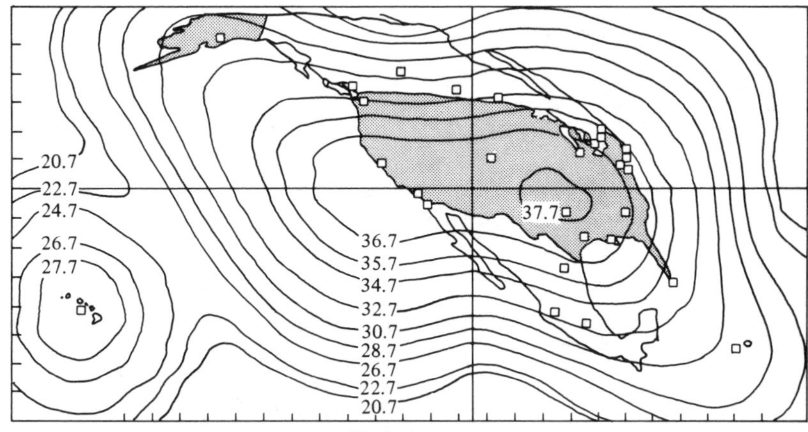

Vertical transmit

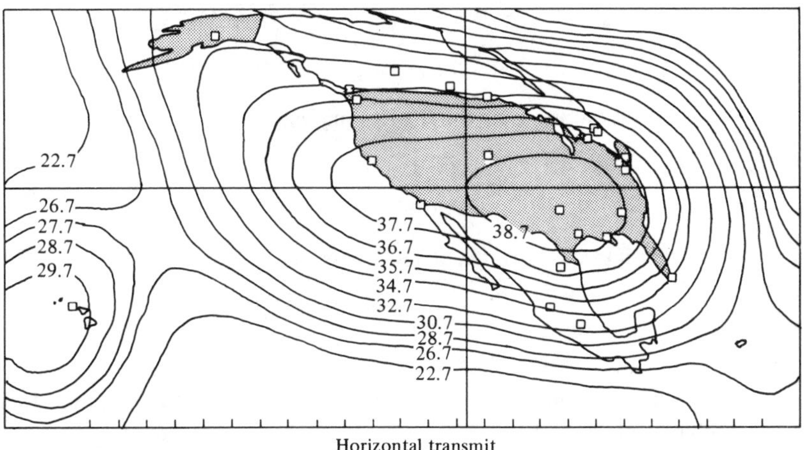

Horizontal transmit

Note: Actual values may vary ± 1 db.

FIGURE 5-43 *Galaxy I* (134° W longitude) antenna pattern EIRP footprint (contours given in dBW units). [Courtesy of Hughes Communications, Inc., a wholly owned subsidiary of Hughes Aircraft Company.]

the Earth radius, and $h = 22{,}242$ statute miles is the altitude of the synchronous satellite. The distance between the TVRO site and the satellite, called the *slant range*, is given by the cosine law:

$$d = \sqrt{(R + h)^2 + R^2 - 2R(R + h)\cos\beta} \tag{5-96}$$

The azimuth of the TVRO antenna is

$$A = \cos^{-1}\left(-\frac{\tan\phi}{\tan\beta}\right) \tag{5-97}$$

where the true azimuth as measured clockwise from the north is given by either A or $360° - A$, as appropriate, since a calculator gives a value between 0 and 180°

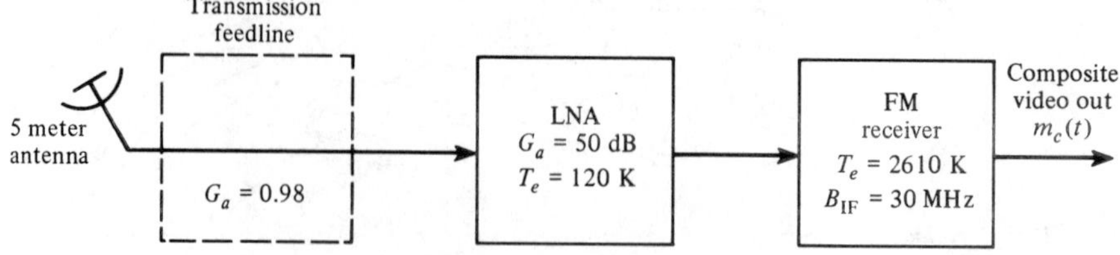

FIGURE 5-44 TVRO terminal.

for $\cos^{-1}(x)$. Using these formulas for the Washington, D.C., TVRO site,† we get $\lambda = 134 - 77 = 57°$ and

$$\beta = \cos^{-1}[\cos(38.8)\cos(57)] = 64.9° \tag{5-98a}$$

The slant range is

$$d = \sqrt{(26{,}205)^2 + (3963)^2 - 2(3963)(26{,}205)\cos(64.9)}$$
$$= 24{,}784 \quad \text{miles} \tag{5-98b}$$

The elevation angle of the TVRO antenna is

$$E = \tan^{-1}\left[\frac{1}{\tan(64.9)} - \frac{3963}{(26{,}205)\sin(64.9)}\right] = 16.8° \tag{5-98c}$$

The azimuth angle of the TVRO antenna is

$$A = \cos^{-1}\left[\frac{-\tan(38.8)}{\tan(64.9)}\right] = 247.9° \tag{5-98d}$$

Thus, the look angles of the TVRO antenna at Washington, D.C., to the *Galaxy I* satellite are $E = 16.8°$ and $A = 247.9°$.

A block diagram describing the receiving system is shown in Fig. 5-44. Using (5-90), the carrier-to-noise ratio at the detector input of the receiver is

$$\left(\frac{C}{N}\right)_{dB} = (P_{EIRP})_{dBw} - (L_{FS})_{dB} + \left(\frac{G_{AR}}{T_{syst}}\right)_{dB} - k_{dB} - B_{dB} \tag{5-99}$$

The corresponding path loss for $d = 24{,}784$ mi at a frequency of 4 GHz is

$$(L_{FS})_{dB} = 20\log\left(\frac{4\pi d}{\lambda}\right) = 196.5 \text{ dB}$$

For a 5-m parabola, using Table 5-8, the receiving antenna gain is

$$(G_{AR})_{dB} = 10\log\left[\frac{7\pi(\frac{5}{2})^2}{\lambda^2}\right] = 10\log(24{,}434) = 44 \text{ dB}$$

†Here for convenience and for ease in interpretation of answers, the calculator is set to evaluate all trigonometric functions in degrees (not radians).

Using (5-88) and (5-84), we obtain the system noise temperature

$$T_{syst} = T_{AR} + T_{feed} + \frac{T_{LNA}}{G_{feed}} + \frac{T_{receiver}}{G_{feed}G_{LNA}} \tag{5-100}$$

where the specified antenna noise temperature, including the feed, is $T_A = (T_{AR} + T_{feed}) = 32$ K.

The T_{AR} is *not* the ambient temperature of the antenna (which is about 290 K) because the effective resistance of the antenna does not consist of a thermal resistor but of a radiation resistor. That is, if RF power is fed into the antenna, it will not be dissipated in the antenna but instead will be radiated into space. Consequently, T_{AR} is called the *sky noise temperature* and is a measure of the amount of cosmic noise power that is received from outer space and noise caused by atmospheric attenuation, where this power is $P_{AR} = kT_{AR}B$. Graphs giving the measured sky noise temperature as a function of the RF frequency and the elevation angle of the antenna are available [Jordan, 1985, Chap. 27; Pratt and Bostian, 1986].

Substituting the receive system parameters into (5-100), we have

$$T_{syst} = 32 + \frac{120}{0.98} + \frac{2610}{(0.98)(100{,}000)} = 154.5 \text{ K}$$

Thus†

$$\left(\frac{G_{AR}}{T_{syst}}\right)_{dB} = 10 \log\left(\frac{24{,}434}{154.5}\right) = 22 \text{ dB/K}$$

Also,

$$(B)_{dB} = 10 \log(30 \times 10^6) = 74.8$$

Substituting these results into (5-99) yields

$$\left(\frac{C}{N}\right)_{dB} = 36 - 196.5 + 22 - (-228.6) - 74.8$$

Thus the carrier-to-noise ratio at the detector input (inside the FM receiver) is

$$\left(\frac{C}{N}\right)_{dB} = 15.3 \text{ dB} \tag{5-101}$$

A C/N of 15.3 dB does not seem to be a very good signal-to-noise ratio, does it? Actually, the question is: Does a C/N of 15.3 dB at the detector input give a high-quality video signal at the detector output? The theory that answers this question is quite involved. (We will analyze the performance of an FM detector in Chapter 7.) However, this question can be answered easily if we have some detector performance curves as shown in Fig. 5-45. Two curves are plotted in this figure. One is for the output of a receiver that uses an FM discriminator for

†Engineers specify dB/K as the units for $(G/T)_{dB}$, although it is a misnomer.

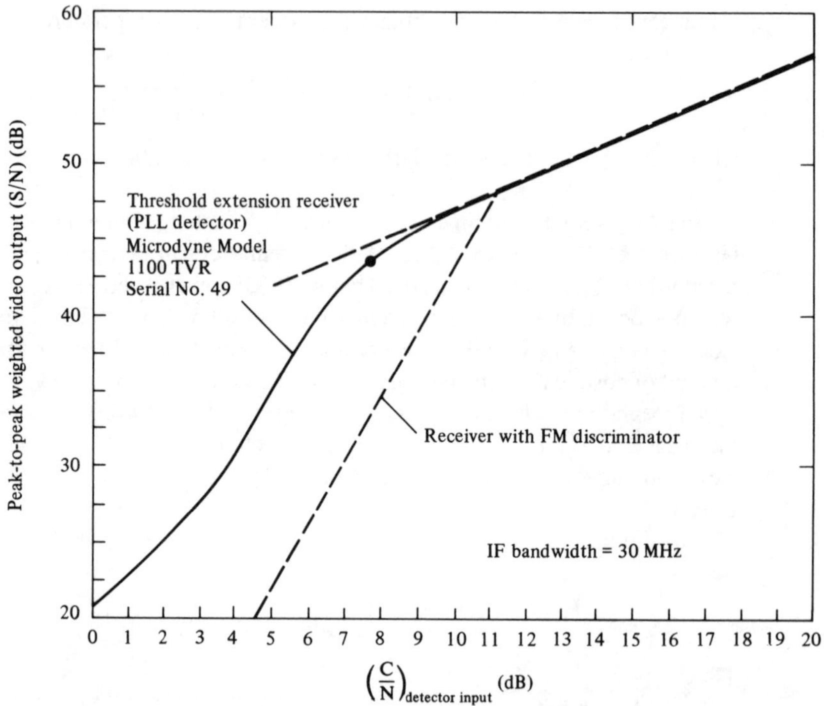

FIGURE 5-45 Output S/N as a function of FM detector input C/N. [Curve for the Microdyne receiver, Model 1100TVR, courtesy of the Microdyne Corporation, Ocala, FL, 1981.]

the FM detector, and the other is for a receiver (Microdyne 1100TVR) that uses a PLL FM detector to provide threshold extension.† For a C/N of 15.3 dB on the input, both receivers will give a 52-dB output S/N, which corresponds to a high-quality picture. However, if the receive signal from the satellite fades because of atmospheric conditions, and so on, by, say, 6 dB, the output S/N of the receiver with an FM discriminator will be only about 40 dB, whereas the receiver with threshold extension detection will have an output S/N of about 45 dB.

5-6 Fiber Optic Systems

Since 1980 fiber optic cable transmission systems have become commercially viable. This is indicated in Table 5-7, which shows that common carriers have installed extensive fiber optic systems in the United States, Japan, Britain, and

†The threshold point is the point where the knee occurs in the $(S/N)_{out}$ versus $(C/N)_{in}$ characteristic. In the characteristic for threshold extension receivers, this knee is extended to the left (improved performance) when compared to the knee for the corresponding receiver that uses a discriminator as a detector.

Canada. Fiber optic cable is the preferred underground transmission medium and is superseding twisted-pair cable and coaxial cable. Also, as shown in Table 5-7, transoceanic fiber optic cable systems are becoming popular. Digital fiber optic systems use simple OOK modulation of an optical source to produce a modulated light beam. The optical source has a wavelength in the range of 0.85 to 1.6 μm, which is about 77,000 GHz. (Analog AM fiber optic systems can also be built.)

The optical sources can be classified into two categories: (1) light-emitting diodes (LED) that produce noncoherent light and (2) solid-state lasers that produce coherent (single-carrier-frequency) light. The LED source is popular because it is rugged and inexpensive. It has a relatively low power output, about -15 dBm (including coupling loss), and a small modulation bandwidth of about 50 MHz. The laser diode has a relatively high power output, about $+5$ dBm (including coupling loss), and a large modulation bandwidth of about 1 GHz. The laser diode is preferred over the LED since it produces coherent light and has high output power, but it is high in cost and has a short lifetime. OOK modulation is obtained simply by switching the bias current on and off.

The light source is coupled into the fiber optic cable, and this cable is the channel transmission medium. Fiber cables can be classified into two categories: (1) multimodal and (2) single-mode. Multimodal fiber has a core diameter of 50 μm and a cladding diameter of 125 μm. The light is reflected off the core–cladding boundary as it propagates down the fiber to produce multiple paths of different lengths. This causes pulse dispersion of the OOK signal at the receiving end and thus severely limits the transmission bit rate that can be accommodated. The single-mode fiber has a core diameter of approximately 8 μm and propagates a single wave. Consequently, there is little dispersion of the received pulses of light. Because of its superior performance, single-mode fiber is preferred.

At the receiving end of the fiber, the receiver consists of a PIN diode, an avalanche photodiode (APD), or a GaAsMESFES transistor used as an optical detector. In 0.85-μm systems any of these devices may be used as a detector. In 1.3- and 1.55-μm systems the APD is usually used. The PIN diode is low in cost and reliable but is used only in 0.85-μm systems. These detectors act like a simple envelope detector. As shown in Chapter 7, better performance (lower probability of bit error) could be obtained by using a coherent detection system that has a product detector. This requires a coherent light source at the receiver that is mixed with the received OOK signal via the nonlinear action of the photodetector. Of course, it is theoretically possible to have coherent PSK and FSK systems, and researchers are working on these problems [Basch and Brown, 1985].

EXAMPLE 5-6 LINK BUDGET FOR A FIBER OPTIC SYSTEM

Tables 5-10 through 5-12 show the performance characteristics of a fiber optic common-carrier cable system that has been installed and operated by Microtel, Inc. in Florida [Siperko, 1985]. Called LaserNet, this system has been installed along the east coast between Jacksonville and Miami, with a connecting link

TABLE 5-10 Link Budget Analysis for LaserNet Operating at 1.3 μm

Description	Value
Average effective transmitter power	−3.5 dBm
Receiver sensitivity	−36.0 dBm
Available margin	32.5 dB
Losses in a fiber optic link	
Chromatic dispersion/reflection noise (40-km span)	1.5 dB
Fiber attenuation (40 km × 0.5 dB/km)	20.0 dB
15 splices at 0.1 dB/splice	1.5 dB
Connector losses	3.0 dB
Design margin	5.0 dB
Excess margin	1.5 dB
Total loss	32.5 dB

Source: Siperko [1985].

from Cocoa to Tampa via Orlando and a link from Tampa down to Naples on the west coast. From Table 5-10 it is seen that the average transmitted power is −3.5 dBm and the receiver sensitivity is −36.0 dBm. The link budget analysis shows that repeaters (demodulators and remodulators) are required every 40 km (or 25 miles). The bit rate is 417.792 Mbits/sec, as shown in Table 5-12, so that nine DS-3 signals can be accommodated on the TDM OOK light-wave signal. This is equivalent to a VF capacity of 6048 channels (see Table 3-7). In 1985, 480 miles of this system was constructed at a cost of \$3.50 per foot or \$6.98 per VF

TABLE 5-11 Single-Mode Optical Fiber Specifications

Core diameter	8.7 ± 1.3 μm
Reference surface diameter	125.0 ± 3 μm
Reference surface noncircularity	≤2.0%
Coating diameter	250.0 ± 15 μm
Spot size	5.0 ± 0.5 μm
Core/cladding concentricity (offset)	≤1.0 μm
Effective mode cutoff wavelength	1130–1270 nm
Refractive index difference	0.3 ± 0.5%
Zero dispersion wavelength	1310 nm nominal
Maximum dispersion at 1310 nm ± 20 nm	3.5 ps/nm-km
Dispersion at 1550 nm	17.0 ± 2.0 psec/nm-km
Mode field diameter	9.8 ± 1.0 μm
Attenuation at 1310 nm	0.5 dB/km maximum
Attenuation at 1550 nm[a]	0.4 dB/km maximum

Source: Charles M. Siperko, "LaserNet—a fiber optic intrastate network," *IEEE Communications Magazine*, May 1985, p. 41. © 1985 IEEE.

[a]Attenuation at 1550 nm typically 0.1 dB lower than attenuation at 1310 nm.

TABLE 5-12 Interface Specifications for an Optical Light Multiplexer

Line rate	417.792 Mbits/sec ± 15 ppm
Line code	Scrambled binary code, RZ
Wavelength	1310 nm ± 0.03 nm
Optical source	Indium gallium arsenide phosphide—ILD (InGaAsP)
Optical detector	Germanium, avalanche photodiode—APD
Optical fiber	8.7/125-μm single-mode step index
Transmitter power	Peak: +2 dBm ± 0.5 dB
	Average: −3.5 dBm ± 0.5 dB
Receiver sensitivity	Average: −36 dBm
Bit error rate (BER)	$\leq 10^{-11}$ for average −36 dBm receiver input

Source: Charles M. Siperko, "LaserNet—a fiber optic intrastate network," *IEEE Communications Magazine,* May 1985, p. 43. © 1985 IEEE.

channel mile. These figures are impressive. However, with newly developed optical termination equipment, the capacity can be increased to 1.3 Gbits/sec or 28 DS-3 channels, for a VF capacity of 18,816 channels when needed.

5-7 Spread Spectrum Systems

Thus far in our study of communication systems, we have been primarily concerned with the performance of communication systems in terms of bandwidth efficiency and energy efficiency (i.e., detected S/N or probability of bit error) with respect to natural noise. However, in some applications, we also need to consider antijam capability, interference rejection, multiple-access capability, and covert operation, or *low probability of intercept* (LPI) capability. These concerns are especially important in military applications. These performance objectives can be optimized by using spread spectrum techniques.

There are many types of *spread spectrum* (SS) systems. To be considered an SS system, a system must satisfy two criteria.

1. The bandwidth of the transmitted signal, $s(t)$, needs to be much greater than that of the message, $m(t)$.
2. The relatively wide bandwidth of $s(t)$ must be caused by an independent modulating waveform, $c(t)$, called the *spreading signal,* and this signal must be known by the receiver in order for the message signal, $m(t)$, to be detected.

Consequently, the complex envelope of the SS signal is a function of both $m(t)$ and $c(t)$. In most cases, a product function is used, so that

$$g(t) = g_m(t)g_c(t) \tag{5-102}$$

where $g_m(t)$ and $g_c(t)$ are the usual types of modulation complex envelope functions that generate AM, PM, FM, and so on, as given in Table 4-1. The SS signals are classified by the type of mapping functions that are used for $g_c(t)$.

Some of the most common types of SS signals are:

- *Direct Sequence* (DS). Here a DSB-SC type of spreading modulation is used [i.e., $g_c(t) = c(t)$] and $c(t)$ is a polar waveform.
- *Frequency Hopping* (FH). Here $g_c(t)$ is of the FM type where there are $M = 2^k$ hop frequencies determined by the k-bit words obtained from the spreading code waveform, $c(t)$.
- *Hybrid* techniques that include both DS and FH.

We will illustrate exactly how DS and FH systems work in the following sections.

Direct Sequence

Assume that the information waveform, $m(t)$, comes from a digital source and that $m(t)$ is a polar waveform having values of ± 1. Furthermore, let us examine the case of BPSK modulation where $g_m(t) = A_c m(t)$. Thus, for DS where $g_c(t) = c(t)$ is used in (5-102), the complex envelope for the SS signal becomes

$$g(t) = A_c m(t) c(t) \tag{5-103}$$

The resulting $s(t) = \text{Re}\{g(t)e^{j\omega_c t}\}$ is called a *binary-phase-shift keyed data, direct sequence spreading, spread spectrum signal* (BPSK-DS-SS) and $c(t)$ is a polar spreading signal. Furthermore, let this spreading waveform be generated by using a *pseudonoise* (PN) code generator, as shown in Fig. 5-46b, where the values of $c(t)$ are ± 1. The pulse width of $c(t)$ is denoted by T_c and is called a *chip* interval (as compared to a bit interval). The code generator uses a modulo 2 adder and r shift register stages that are clocked every T_c sec. It can be shown that $c(t)$ is periodic. Furthermore, feedback taps from the stage of the shift registers and modulo 2 adders are arranged so that the $c(t)$ waveform has a maximum period of N chips where $N = 2^r - 1$ [Ziemer and Peterson, 1985]. This type of PN code generator is said to generate a *maximum-length sequence* or *m-sequence* waveform.

Properties of Maximum-Length Sequences. Some properties of *m*-sequences are [Ziemer and Peterson, 1985]:

Property 1. In one period the number of 1s is always one more than the number of 0s.

Property 2. The modulo 2 sum of any *m*-sequence when summed chip by chip with a shifted version of the same sequence produces another shifted version of the same sequence.

Property 3. If a window of width r (where r is the number of stages in the shift register) is slid along the sequence for N shifts, then all possible r-bit words will appear exactly once, except for the all 0 r-bit word.

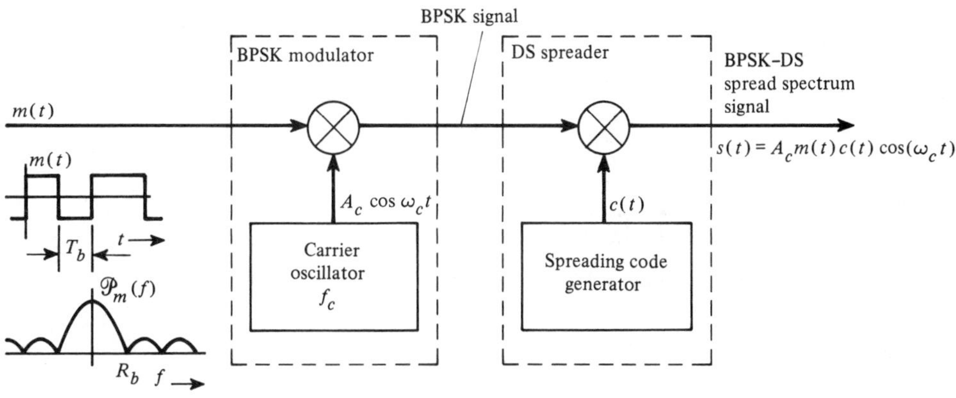

(a) BPSK–DS–SS Transmitter

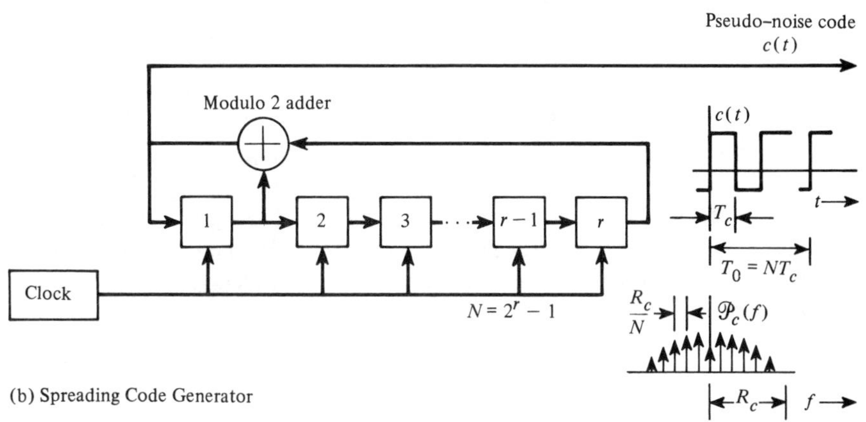

(b) Spreading Code Generator

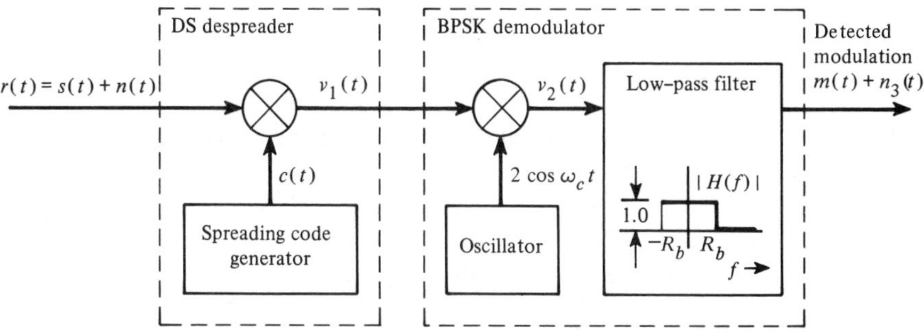

(c) BPSK–DS–SS Receiver

FIGURE 5-46 Direct sequence spread spectrum system (DS-SS).

Property 4. If the 0s and 1s are represented by -1 volt and $+1$ volt, the autocorrelation of the *sequence*, denoted by $R_c(k)$, is

$$R_c(k) = \begin{cases} 1, & k = \ell N \\ -\dfrac{1}{N}, & k \neq \ell N \end{cases} \tag{5-104a}$$

where $R_c(k) \triangleq (1/N)\sum_{n=0}^{N-1} c_n c_{n+k}$ and $c_n = \pm 1$.

The autocorrelation of the *waveform*, $c(t)$, denoted by $R_c(\tau)$, is

$$R_c(\tau) = \left(1 - \frac{\tau_\epsilon}{T_c}\right) R_c(k) + \frac{\tau_\epsilon}{T_c} R_c(k+1) \tag{5-104b}$$

where $R_c(\tau) = \langle c(t)c(t+\tau)\rangle$ and τ_ϵ is defined by

$$\tau = kT_c + \tau_\epsilon \qquad \text{and} \qquad 0 \leq \tau_\epsilon < T_c \tag{5-104c}$$

Equation (5-104b) reduces to

$$R_c(\tau) = \left[\sum_{\ell=-\infty}^{\ell=\infty} \left(1 + \frac{1}{N}\right) \Lambda\left(\frac{\tau - \ell N T_c}{T_c}\right)\right] - \frac{1}{N} \tag{5-104d}$$

This is plotted in Fig. 5-47a, where it is apparent that the autocorrelation function for the PN waveform is periodic with triangular pulses of width $2T_c$ repeated every NT_c seconds and that a correlation level of $-1/N$ occurs between these triangular pulses. Furthermore, since the autocorrelation function is periodic, the associated PSD is a line spectrum. That is, the autocorrelation is expressed as a Fourier series

$$R_c(\tau) = \sum_{n=-\infty}^{\infty} r_n e^{j2\pi n f_0 t} \tag{5-104e}$$

where $f_0 = 1/(NT_c)$ and $\{r_n\}$ is the set of Fourier series coefficients. Thus, using (2-109) yields

$$\mathcal{P}_c(f) = \mathcal{F}[R_c(\tau)] = \sum_{n=-\infty}^{\infty} r_n \delta(f - nf_0) \tag{5-104f}$$

where the Fourier series coefficients are evaluated and found to be

$$r_n = \begin{cases} \dfrac{1}{N^2}, & n = 0 \\ \left(\dfrac{N+1}{N^2}\right)\left(\dfrac{\sin(\pi n/N)}{\pi n/N}\right)^2, & n \neq 0 \end{cases} \tag{5-104g}$$

This PSD is plotted in Fig. 5-47b.

Now let us demonstrate that the bandwidth of the SS signal is relatively large compared to the data rate, R_b, and is determined primarily by the spreading

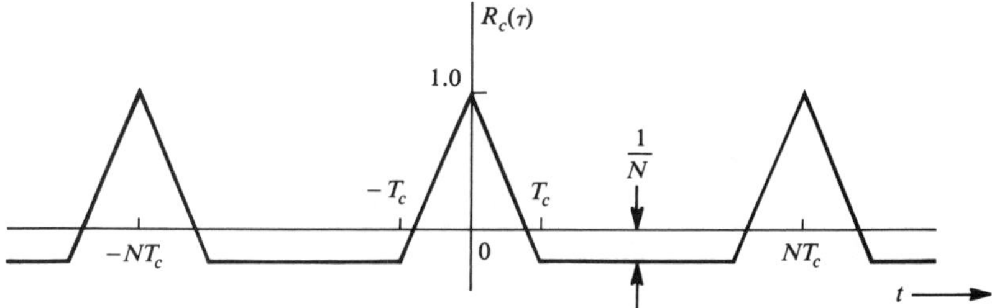

(a) Autocorrelation Function

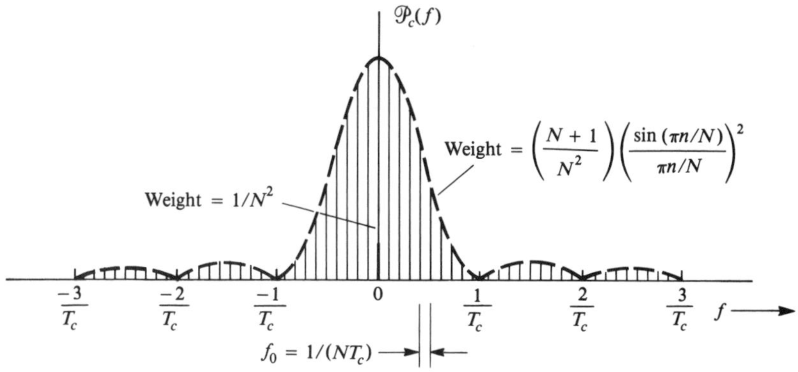

(b) Power Spectral Density (PSD)

FIGURE 5-47 Autocorrelation and PSD for an *m*-sequence PN waveform.

waveform, $c(t)$, and not by the data modulation, $m(t)$. Referring to Fig. 5-46, we see that the PSDs of both $m(t)$ and $c(t)$ are of the $[(\sin x)/x]^2$ type, where the bandwidth of $c(t)$ is much larger than that of $m(t)$ because it is assumed that the chip rate, $R_c = 1/T_c$ is much larger than the data rate, $R_b = 1/T_b$. That is, $R_c \gg R_b$. To simplify the mathematics, approximate these PSDs by rectangular spectra, as shown in Figs. 5-48a and 5-48b, where the heights of the PSD are selected so that the areas under the curves are unity because the powers of $m(t)$ and $c(t)$ are unity. (They both have only ± 1 values.) From (5-103), $g(t)$ is obtained by multiplying $m(t)$ and $c(t)$ in the time domain, and $m(t)$ and $c(t)$ are independent. Thus the PSD for the complex envelope of the BPSK–DS–SS signal is obtained by a convolution operation in the frequency domain.

$$\mathcal{P}_g(f) = A_c^2\,\mathcal{P}_m(f) * \mathcal{P}_c(f)$$

This result is shown in Fig. 5-48c for the approximate PSDs of $m(t)$ and $c(t)$. The bandwidth of the BPSK–DS–SS signal is determined essentially by the chip rate, R_c, since $R_c \gg R_b$. For example, let $R_b = 9.6$ kbits/sec and $R_c = 9.6$ Mchips/sec. Then the bandwidth of the SS signal is $B_T \approx 2R_c = 19.2$ MHz.

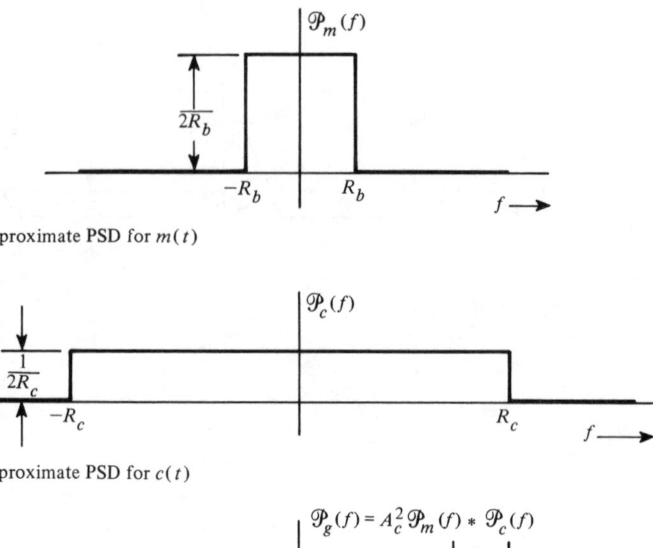

(a) Approximate PSD for $m(t)$

(b) Approximate PSD for $c(t)$

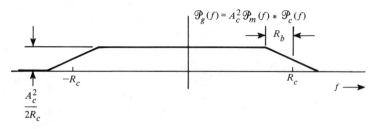

(c) Approximate PSD for Complex Envelope of the SS Signal

FIGURE 5-48 Approximate PSD of the BPSK–DS–SS signal.

From Fig. 5-48 we can also demonstrate that the spreading has made the signal less susceptible to detection by an eavesdropper. That is, this signal has LPI. Without spreading [i.e., if $c(t)$ was unity], the level of the in-band PSD would be proportional to $A_c^2/(2R_b)$, as seen in Fig. 5-48a, but with spreading the in-band spectral level drops to $A_c^2/(2R_c)$, as seen in Fig. 5-48c. This is a reduction of R_c/R_b. For example, for the values of R_b and R_c just cited, the reduction factor would be (9.6 Mchips/sec)/(9.6 kbits/sec) = 1000 or 30 dB. Often the eavesdropper detects the presence of a signal by using a spectrum analyzer, but when SS is used, this level will drop by 30 dB. This is often below the noise floor of the potential eavesdropper, and thus the SS signal will escape detection by the eavesdropper.

Figure 5-46c shows a receiver that recovers the modulation on the SS signal. The receiver has a despreading circuit that is driven by a PN code generator in synchronism with the transmitter spreading code. Assume that the input to the receiver consists of the SS signal plus a narrowband (sine wave) jammer signal. Then,

$$r(t) = s(t) + n(t) = A_c m(t)c(t) \cos \omega_c t + n_J(t) \tag{5-105a}$$

where the jamming signal is

$$n_J(t) = A_J \cos \omega_c t \tag{5-105b}$$

Here it is assumed that the jamming power is $A_J^2/2$ relative to the signal power of $A_c^2/2$ and that the jamming frequency is set to f_c for the worst-case jamming effect. Referring to Fig. 5-46c, we find that the output of the despreader is

$$v_1(t) = A_c m(t) \cos \omega_c t + A_J c(t) \cos \omega_c t \tag{5-106}$$

since $c^2(t) = (\pm 1)^2 = 1$. The BPSK–DS–SS signal has become simply a BPSK signal at the output of the despreader. That is, at the receiver input the SS signal has a bandwidth of $2R_c$, but at the despreader output the bandwidth of the resulting BPSK signal is $2R_b$, a 1000:1 reduction for the figures previously cited. The data on the BPSK despread signal are recovered by using a BPSK detector circuit as shown.

Now it will be shown that this SS receiver provides an antijam capability of 30 dB for the case of $R_b = 9.6$ kbits/sec and $R_c = 9.6$ Mchips/sec. From (5-106) it is seen that the narrowband jammer signal that was present at the receiver input has been *spread* by the despreader since it has been multiplied by $c(t)$. It is this spreading effect on the jamming signal that produces the antijam capability. Using (5-106) and referring to Fig. 5-46c, we obtain an input to the LPF of

$$v_2(t) = A_c m(t) + n_2(t) \tag{5-107a}$$

where

$$n_2(t) = A_J c(t) \tag{5-107b}$$

and the terms about $f = 2f_c$ have been neglected since they do not pass through the LPF. Referring to Fig. 5-46c, we note that the jammer power at the receiver output is

$$P_{n_3} = \int_{-R_b}^{R_b} \mathcal{P}_{n_2}(f)\, df = \int_{-R_b}^{R_b} A_J^2 \frac{1}{2R_c}\, df = \frac{A_J^2}{R_c/R_b}$$

and the jammer power at the input to the LPF is A_J^2. [$\mathcal{P}_{n_2}(f) = A_J^2/(2R_c)$ as seen from Fig. 5-48b and (5-107b).] For a conventional BPSK system (i.e., no spread spectrum), $c(t)$ would be unity and (5-107b) would become $n_2(t) = A_J$, so the jamming power out of the LPF would be A_J^2 instead of $A_J^2/(R_c/R_b)$ for the case of an SS system. [The output signal should be $A_c m(t)$ for both cases.] Thus, the SS receiver has reduced the effect of narrowband jamming by the factor of R_c/R_b. This factor, R_c/R_b, is called the *processing gain* of the SS receiver.[†] For our example of $R_c = 9.6$ Mchips/sec and $R_b = 9.6$ kbits/sec, the processing gain is 30 dB. This means that the narrowband jammer would have to have 30 dB more power to have the same jamming effect on this SS system, as compared to the

[†] The processing gain is defined as the ratio to the noise power out without SS divided by the noise power out with SS. This is equivalent to the ratio $(S/N)_{out}/(S/N)_{in}$ when $(S/N)_{in}$ is the signal-to-noise power into the receiver and $(S/N)_{out}$ is the signal-to-noise power out of the LPF.

conventional BPSK system (no SS). Thus this SS technique provides 30 dB of antijam capability for the R_c/R_b ratio cited in the example.

SS techniques can also be used to provide multiple access (as mentioned in Sec. 5-4). This is called *code division multiple access* (CDMA). Here each user is assigned a spreading code such that the signals are orthogonal. Thus multiple SS signals can be transmitted simultaneously in the same frequency band, and yet the data on a particular SS signal can be decoded by a receiver, *provided* that this receiver uses a PN code that is identical to that used on the particular SS signal that is to be decoded. Although CDMA provides a neat solution for the multiple access problem in an interference environment, it can be shown that TDMA and FDMA are more bandwidth efficient [Viterbi, 1985].

To accommodate more users in frequency bands that are now saturated with conventional narrowband users (such as the two-way radio bands), it is possible to assign in addition new SS stations. This is called *spread spectrum overlay*. The SS stations would operate with such a wide bandwidth that their PSD would appear to be negligible to narrowband receivers located sufficiently distant from the SS transmitters. On the other hand, to the SS receiver, the narrowband signals would have a minimal jamming effect because of the large coding gain of the SS receiver.

Frequency Hopping

As indicated previously, a frequency-hopped (FH) SS signal uses a $g_c(t)$ that is of the FM type where there are $M = 2^k$ hop frequencies controlled by the spreading code where k chip words are taken to determine each hop frequency. An FH-SS transmitter is shown in Fig. 5-49a. The source information is modulated onto a carrier using conventional FSK or BPSK techniques to produce an FSK or a BPSK signal. The frequency hopping is accomplished by using a mixer circuit where the LO signal is provided by the output of a frequency synthesizer that is hopped by the PN spreading code. The serial-to-parallel converter reads k serial chips of the spreading code and outputs a k chip parallel word to the programmable dividers in the frequency synthesizer. (See Fig. 4-26 and the related discussion of frequency synthesizers.) The k chip word specifies one of the possible $M = 2^k$ hop frequencies, $\omega_1, \omega_2, \ldots, \omega_M$.

The FH signal is decoded as shown in Fig. 5-49b. Here the receiver has the knowledge of the transmitter, $c(t)$, so that the frequency synthesizer in the receiver can be hopped in synchronism with that at the transmitter. This despreads the FH signal, and the source information is recovered from the dehopped signal by use of a conventional FSK or BPSK demodulator, as appropriate.

In 1985 the FCC opened up three shared frequency bands—902 to 928 MHz, 2400 to 2483.5 MHz, and 5725 to 5850 MHz—for commercial SS use with unlicensed 1-W transmitters. This has led to the production and use of SS equipment for telemetry systems, personal communication networks (PCNs), wireless local area networks for personal computers, and wireless fire and security systems [Schilling, Pickholtz, and Milstein, 1990]. Also, commercial SS applications have been proposed that have advantages over present systems. For exam-

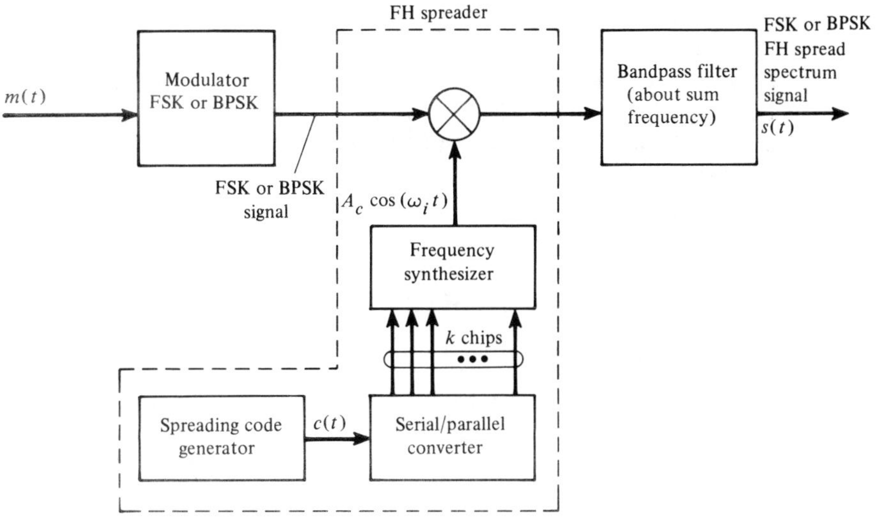

(a) Transmitter

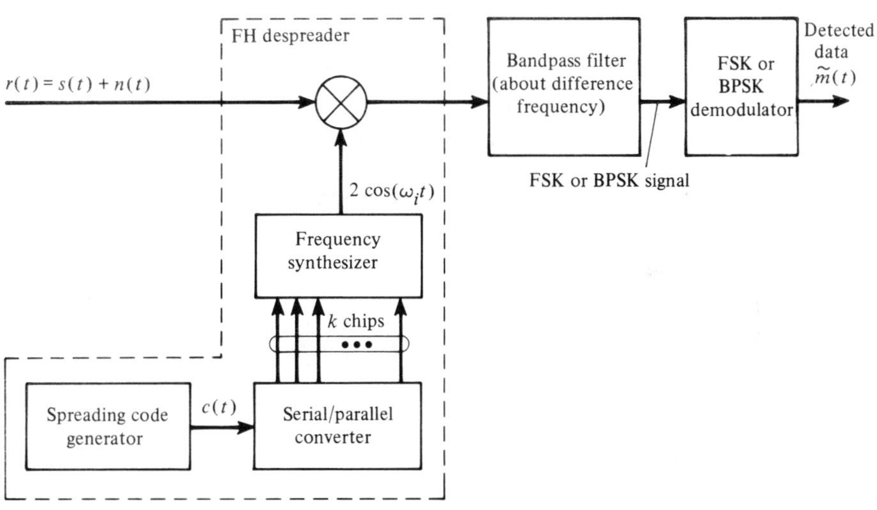

(b) Receiver

FIGURE 5-49 Frequency-hopped spread spectrum system (FH–SS).

ple, a digital SS cellular telephone system (i.e., CDMA) appears to be able to accommodate about 1000 users per cell as compared to the U.S. analog cellular system, which accommodates 55 users per cell [Schilling, Pickholtz, and Milstein, 1990]. In another application, digital stereo broadcasting using SS techniques appears to accommodate up to 75 stations in a region using the U.S. FM band, 88 to 108 MHz [Schilling et al., 1991]. This is a larger number of stations than found with present analog FM stereo assignments; moreover, digital stereo SS broadcasting will provide a much larger audio signal-to-noise ratio for the listener.

Limited space does not permit further discussion of SS systems. For further study on this interesting topic, the reader is referred to books that have been written on this subject [Cooper and McGillem, 1986; Dixon, 1984; Ziemer and Peterson, 1985].

5-8 Cellular Telephone Systems

Since the invention of radio systems, the goal of telephone engineers has been to provide personal telephone service to individuals by using radio systems to link phone lines with persons in their cars or on foot. In the past this type of personal telephone service was not possible because limited spectrum space did not permit the assignment of a "private-line" radio channel for each subscriber. In addition, the existing radio equipment was bulky and expensive. However, with the development of the integrated circuit technology, radio equipment can now be miniaturized, and relatively sophisticated operations can be implemented at low cost. This development allows the sharing of a number of radio channels on demand (FDMA) where a particular channel is assigned to a specific user only when a telephone call is in progress. An example is *trunked mobile radio systems* in which a number of radio channels are shared with different groups of users (i.e., statistical multiplexing on a FDMA basis). In these trunked systems, relatively high-power transmitters are used so that each radio channel covers a whole city or county. Consequently, when a channel is being used by anyone in that area, it is not available for use by others. Many more users can be accommodated simultaneously if the coverage of each transmitter is reduced to some small geographical area (a cell) so that the same radio frequency channel can be reused by another individual a few miles away in a distant cell.

This *cellular radio* concept is illustrated in Fig. 5-50. Each user communicates via radio from a cellular telephone set to the cell-site base station. This base station is connected via telephone lines to the *mobile telephone switching office* (MTSO). The MTSO connects the user to the called party. If the called party is land based, the connection is via the central office (CO) to the terrestrial telephone network. If the called party is mobile, the connection is made to the cell site that covers the area in which the called party is located, using an available radio channel in the cell associated with the called party. Theoretically, this cellular concept allows any number of mobile users to be accommodated for a given set of radio channels. That is, if more channels are needed, the existing cell sizes are decreased and additional small cells are inserted so that the existing channels can be reused more efficiently. The critical consideration is to design the cells for acceptable levels of cochannel interference [Lee, 1986]. As the mobile user travels from one cell to another, the MTSO automatically switches the user to an available channel in the new cell and the telephone conversation continues uninterrupted.

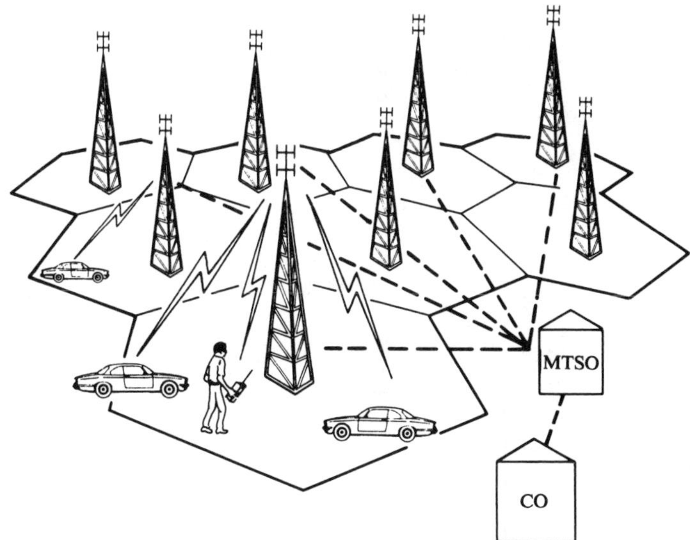

FIGURE 5-50 Cellular telephone system.

The cellular concept has the following advantages.

- Large subscriber capacity.
- Efficient use of the radio spectrum.
- Nationwide compatibility.
- Service to hand-held portables as well as vehicles.
- High-quality telephone and data service to the mobile user at relatively low cost.

Unfortunately, there are many cellular standards used throughout the world [Okumura and Shinji, 1986]. For example, as many as 15 different types of cellular phones are needed to remain in communication as one travels around Europe. Table 5-13 shows three major analog standards that are used around the world. Moreover, since cellular telephones are so popular, it is clear that these wideband FM analog systems do not have the capacity to accommodate the expected new subscribers. Consequently, narrowband FM analog systems with higher capacity, as shown in Table 5-14, are replacing wideband systems on a channel-by-channel basis as needed. For example, one wideband 30-kHz AMPS channel can be replaced by three narrowband 10-kHz NAMPS channels to triple the capacity. For larger capacity, new digital cellular systems, as shown in Table 5-15, will be used. The GSM digital system is being placed into service in Europe to provide one uniform standard that replaces the numerous analog systems that are being phased out. By 1996 it is estimated that the number of GSM subscribers will equal or surpass the number of European analog subscribers.

As shown in Table 5-13, the present cellular system used in the United States is the Advanced Mobile Phone System (AMPS), which was developed by AT&T

TABLE 5-13 Major Analog Cellular Telephone Systems

Item	System Name[a] and Where Used			
	AMPS North America			
	A Service (Nonwire Line)	B Service (Wire Line)	ETACS United Kingdom	JTACS Japan
Base cell station				
Transmit bands (MHz)	869–880, 890–891.5	880–890, 891.5–894	872–905	860–870
Mobile station				
Transmit bands (MHz)	824–835, 845–846.5	835–845, 846.5–849	917–950	915–925
Maximum power (watts)	3	3	3	2.8
Cell size, radius (km)	2–20	2–20	2–20	2–20
Number of duplex channels	416[b]	416[b]	1320	399
Channel bandwidth (kHz)	30	30	25	25
Modulation				
Voice	FM 12-kHz peak deviation	FM 12-kHz peak deviation	FM 9.5-kHz peak deviation	FM 9.5-kHz peak deviation
Control signal (Voice and paging channels)	FSK 8-kHz peak deviation 10 kbits/sec Manchester line code	FSK 8-kHz peak deviation 10 kbits/sec Manchester line code	FSK 6.4-kHz peak deviation 8 kbits/sec Manchester line code	FSK 6.4-kHz peak deviation 8 kbits/sec Manchester line code

Source: Most of the data in this table were assembled by Jim Worsham, Matsushita Communication Industrial Corporation of America/Panasonic, Alpharetta, Georgia.

[a] AMPS = Advanced Mobile Phone System; ETACS = Extended Total Access Communication System; JTACS = Japan Total Access Communication System.

[b] 21 of the 416 channels are used exclusively for paging.

TABLE 5-14 Narrowband Analog Cellular Telephone Systems

Item	System Name[a] and Where Used: NAMPS North America	NTACS Japan
Base cell station		
Transmit band (MHz)	869–894	843–870
Mobile station		
Transmit band (MHz)	824–849	898–925
Maximum power (watts)	3	2.8
Number of duplex channels	2496	2160
Channel bandwidth		
Voice (kHz)	10	12.5
Paging (kHz)	30	25
Modulation		
Voice	FM 5-kHz peak deviation	FM 5-kHz peak deviation
Signaling (voice channel)	FSK 0.7-kHz peak deviation 100 bits/sec, Manchester	FSK 0.7-kHz peak deviation 100 bits/sec, Manchester
Signaling (paging channel)	FSK 8-kHz peak deviation 10 kbits/sec, Manchester	FSK 6.4-kHz peak deviation 8 kbits/sec, Manchester

Source: Most of the data in this table were assembled by Jim Worsham, Matsushita Communication Industrial Corporation of America/Panasonic, Alpharetta, Georgia.

[a]NAMPS = Narrowband Advanced Mobile Phone System; NTACS = Narrowband Total Access Communication System.

and Motorola. To implement this cellular concept in the United States, the FCC had to find spectral space for assignment. It did so by using the 806- to 890-MHz band that was once assigned to TV channels 70 to 83 (see Table 5-17) but discontinued in 1974. Part of this band was assigned for cellular service as shown in Table 5-13. Furthermore, the FCC decided to license two competing call systems in each geographical area—one licensed to a conventional telephone company and another to a nontelephone company common carrier. The nontelephone system is called the *A service* or *nonwireline service,* and the telephone company system is called the *B service* or *wireline service*. Thus cell system subscribers have the option of renting service from the wireline company or the nonwireline company, or from both. In addition, as shown in the table, the standards provide for full duplex service. That is, one carrier frequency is used for mobile to cell base-station transmission, and another carrier frequency is used for base-station to mobile transmission. The telephone VF signals are frequency modulated onto the carriers, and control is provided by FSK signaling. Of the 416 channels that may be used by a cellular service licensee, 21 are used for paging and control with FSK signaling. FSK signaling is also used at the beginning and end of each call on a VF channel. When a cellular telephone is turned

TABLE 5-15 Major Digital Cellular Telephone Systems

Item	System Name[a] and Where Used		
	GSM Europe	**NADC (Proposed) North America**	**JDC (Proposed) Japan**
Base cell station			
Transmit bands (MHz)	935–960	869–894	810–826, 1477–1489, 1501–1513
Mobile station			
Transmit bands (MHz)	890–915	824–849	940–956, 1429–1441, 1453–1465
Maximum power (watts)	20	3	
Number of duplex channels	125	832	1600
Channel bandwidth (kHz)	200	30	25
Channel access method	TDMA	TDMA	TDMA
Users per channel	8	3	3
Modulation[b]			
Type	GMSK	$\pi/4$ DQPSK	$\pi/4$ DQPSK
Data rate	270.833 kbits/sec	48.6 kbits/sec	42 kbits/sec
Filter	0.3R Gaussian[c]	$r = 0.35$ raised cosine[d]	$r = 0.5$ raised cosine[d]
Speech coding[e]	RPE-LTP 13 kbits/sec	VSELF 8 kbits/sec	VSELF 8 kbits/sec

Source: Most of the data in this table were assembled by Jim Worsham, Matsushita Communication Industrial Corporation of America/Panasonic, Alpharetta, Georgia.

[a] GSM = Group Special Mobile; NADC = North American Digital Cellular; JDC = Japan Digital Cellular.

[b] GMSK = Gaussian MSK (i.e., $h = 0.5$ FSK, also called FFSK, with Gaussian premodulation filter; see Sec. 5-2); $\pi/4$ DQPSK = Differential QPSK with $\pi/4$ rotation of reference phase for each symbol transmitted.

[c] Baseband Gaussian filter with 3 dB bandwidth equal to 0.3 of the bit rate.

[d] Square root of raised cosine characteristic at transmitter and receiver.

[e] RPE-LTP = Regular Pulse Excitation—Long-Term Prediction; VSELF = Vector Sum Excited Linear Prediction Filter.

on, its receiver scans the paging channels looking for the strongest cell-site signal and then locks onto it. The cell site transmits FSK data continuously on a paging channel, and, in particular, it sends control information addressed to a particular cellular phone when a call comes in for it. That is, the FSK control signal tells the phone which voice-frequency (VF) channel to use for a call.

Each cellular telephone contains a PROM (programmable read-only memory) or EPROM (erasable programmable read-only memory)—called a *numeric assignment module* (NAM)—that is programmed to contain

- The telephone number—also called the *electronic service number* (ESN)—of the phone.
- The serial number of the phone as assigned by the manufacturer.
- Personal codes that can be used to prevent unauthorized use of the phone.

When the phone is "on-the-air" it automatically transmits its serial number to the MTSO. The serial number is used by the MTSO to lock out phone service to any phone that has been stolen. This feature, of course, discourages theft of the units. The MTSO uses the telephone number of the unit to provide billing information. When the phone is used in a remote city, it can be placed in the *roam* mode so that calls can be initiated or received and yet the service will be billed via the caller's "hometown" company.

To place a call the following sequence of events occurs.

1. The cellular subscriber initiates a call by keying in the telephone number of the called party and then presses the *send* key.
2. The MTSO verifies that the telephone number is valid and that the user is authorized to place a call.
3. The MTSO issues instructions to the user's cellular phone indicating which radio channel to use.
4. The MTSO sends out a signal to the called party to ring his or her phone. All of these operations occur within 10 sec of initiating the call.
5. When the called party answers, the MTSO connects the trunk lines for the two parties and initiates billing information.
6. When one party hangs up, the MTSO frees the radio channel and completes the billing information.

While a call is in progress, the cellular subscriber may be moving from one cell area to another, so the MTSO:

- Monitors the signal strength from the cellular telephone as received at the cell base station. If the signal drops below some designated level, the MTSO initiates a "hand-off" sequence.
- For "hand-off," the MTSO inquires about the signal strength as received at adjacent cell sites.
- When the signal level becomes sufficiently large at an adjacent cell site, the MTSO instructs the cellular radio to switch over to an appropriate channel for communication with that new cell site. This switching process takes less than 250 msec and usually is unnoticed by the subscriber.

As indicated by the tables, cellular systems and the standards for these systems are in a process of evolution throughout the world. The first commercial cellular service was implemented in the United States in 1983. There is a strong push for development of all digital cellular systems that have large user capacity, as illustrated by the proposed NADC and JDC systems listed in Table 5-15. For bandwidth conservation, these systems use low-bit-rate encoding of the VF signal. For example, the vector sum excited linear prediction filter (VSELF) encoder proposed for the NADC and JDC systems is a modified version of the codebook excited linear prediction (CELP) technique developed by B. Atal at the AT&T Bell Laboratories. This technique mimics the two main parts of the human speech system, the vocal cords and the vocal tract. The vocal tract is simulated by a time-varying linear prediction filter, and the vocal cord sounds used for the filter excitation are simulated with a database (i.e., a codebook) of possible excitations. The VSELF encoder (at the transmitter) partitions the talker's VF signal into 20-msec segments. The encoder sequences through the possible codebook excitation patterns and the possible values for the filter parameters to find the synthesized speech segment that gives the best match to the VF segment. The encoder parameters that specify this best match are then transmitted via digital data to the receiver, where the received data establish the parameters for the receiver speech synthesizer so that the VF signal may be reproduced for the listener.

The new digital cellular telephones need to be relatively inexpensive and compact in size. This, of course, requires the use of advanced integrated-circuit technology. For more details about cellular systems, the reader is referred to the technical literature, such as the *IEEE Transactions on Vehicular Technology* [Rhee and Lee, 1991] and a book [Lee, 1989].

5-9 Television

Television (TV) is a method of reproducing fixed or moving visual images by the use of electronic signals. There are numerous kinds of TV systems and different standards have been adopted around the world. Moreover, as discussed subsequently, TV standards are in a state of flux since there is a worldwide push to adopt a new *high-definition television* (HDTV) standard. First we will concentrate on black-and-white broadcast TV as it has developed in the United States since the 1930s and then color television will be examined.

Black-and-White Television

The black-and-white image of a scene is essentially the intensity of the light as a function of the x and y coordinates of the scene and of time. However, an electrical waveform is only a function of time, so that some means of encoding the x and y coordinates associated with the intensity of light from a single point in the scene must be used. In broadcast TV this is accomplished by using *raster*

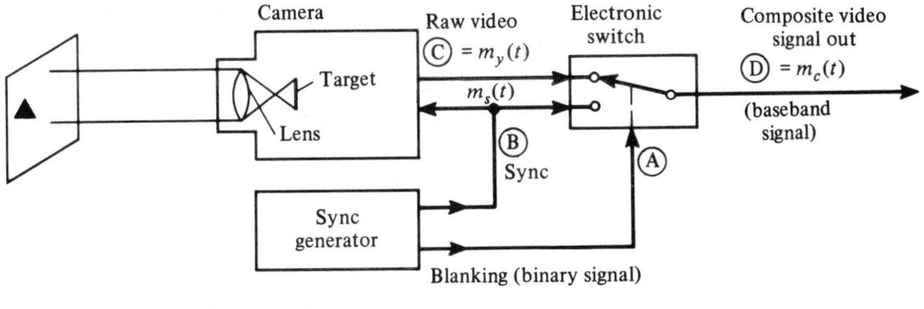

(a) Block Diagram of Camera System

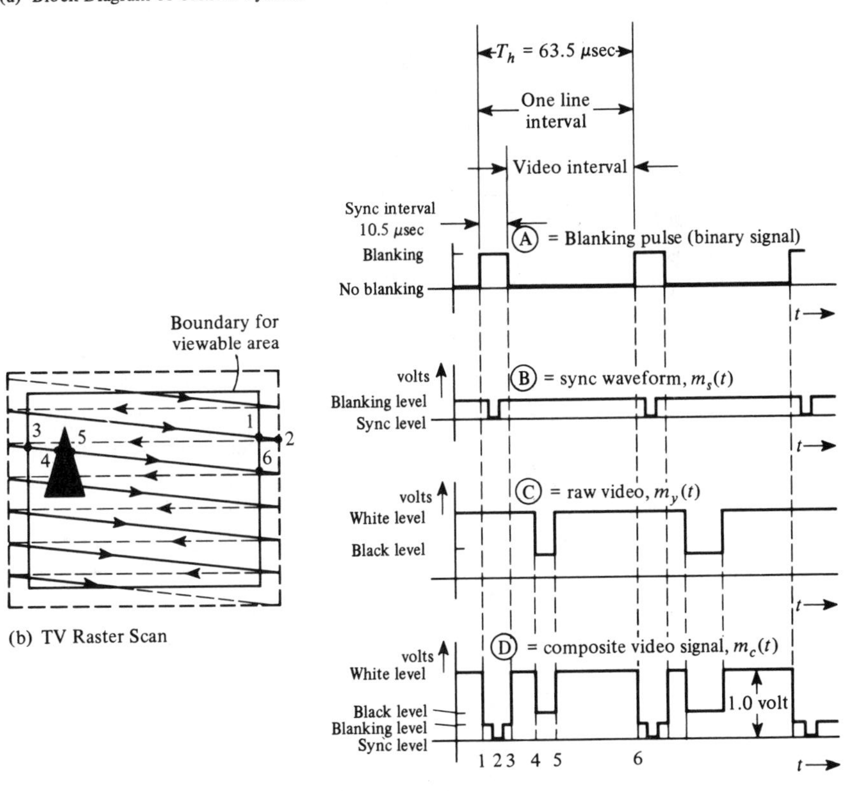

(b) TV Raster Scan

(c) TV Waveforms

FIGURE 5-51 Generation of a composite black-and-white video signal.

scanning, as shown in Fig. 5-51. It is assumed that the whole scene is scanned before any object in the scene has moved appreciably. Thus moving images can be transmitted by sending a series of still images, just as in motion pictures. As the scene is scanned, its intensity is signified by the amplitude of the video signal. A synchronizing pulse is inserted at the end of each scan line by the electronic switch, as shown in Fig. 5-51, to tell the TV receiver to start another scan line. For illustrative purposes, a black triangle is located in the upper left portion of the scene. In Figs. 5-51b and 5-51c, numbers are used to indicate the corresponding times on the video waveform, with the location of the point being scanned in the scene at that instant. The composite video signal is actually a hybrid signal that consists of a digital waveform during the synchronizing interval and an analog waveform during the video interval. Operating personnel monitor the quality of the video by looking at a *waveform monitor,* which displays the composite video waveform as shown in Fig. 5-51c.

The picture (as viewed) is generated by scanning left to right to create a line, and the lines move from the top to the bottom of the screen. During retrace, indicated by the dashed lines from right to left, the CRT is turned off by the video level that is "blacker than black" during the sync pulse interval.† The scene is actually covered by scanning alternate lines from the top to the bottom (one field) and returning to the top to scan the missed lines (the next field). This technique is called *interlacing* and reduces the flicker effect on moving pictures, giving two fields (passes down the screen) for every frame. In the United States there are approximately 30 frames/sec transmitted, which is equivalent to 60 fields/sec for the 2:1 interlacing that is used. When the scan reaches the bottom of a field, the sync pulse is widened out to cover the whole video interval plus sync interval. This occurs for 21 lines, and, consequently, a black bar is seen across the screen if the picture is "rolled up" by misadjusting the vertical hold control. In the receiver, the vertical sync (wide pulse) can be separated from the horizontal sync (narrow pulse) by using a differentiator circuit to recover the horizontal sync and an integrator to recover the vertical sync. The vertical interval is also used to send special signals, such as testing signals that can be used by TV engineers to test the frequency response and linearity of the equipment on-line [Solomon, 1979]. In addition, some TV programs have closed-caption information that can be decoded on special TV sets used by deaf viewers. These sets insert subtitles across the bottom of the screen. As shown in Table 5-16, this closed-captioning information is transmitted during line 21 of the vertical interval by inserting a 16-bit code word. Details of the exact sync structure are rigorously specified by the FCC (see Fig. 5-58).

The composite baseband signal consisting of the *luminance* (intensity) video signal $m_y(t)$ and the *synchronizing* signal $m_s(t)$ is described by

$$m_c(t) = \begin{cases} m_s(t) & \text{during the sync interval} \\ m_y(t) & \text{during the video interval} \end{cases} \tag{5-108}$$

The spectrum of $m_c(t)$ is very wide since $m_c(t)$ contains rectangular synchronizing pulses. In fact, the bandwidth would be infinite if the pulses had perfect rectangular shapes. In order to reduce the bandwidth, the pulses are rounded as specified by FCC standards (see Fig. 5-58). This allows $m_c(t)$ to be filtered to a bandwidth of $B = 4.2$ MHz (U.S. standard). For a still picture, the exact spectrum can be calculated using Fourier series analysis (as shown in Sec. 2-5). For a typical picture all fields are similar, thus $m_c(t)$ would be periodic with a period of

†In addition, most TV sets switch off the electron beam during retrace by using the sync pulse obtained from the horizontal output (flyback) transformer.

TABLE 5-16 U.S. Television Broadcasting Standards

Item	FCC Standard
Channel bandwidth	6 MHz
Visual carrier frequency	1.25 MHz ± 1000 Hz above the lower boundary of the channel
Aural carrier frequency	4.5 MHz ± 1000 Hz above the visual carrier frequency
Chrominance subcarrier frequency	3.579545 MHz ± 10 Hz
Aspect (width-to-height) ratio	Four units horizontally for every three units vertically
Modulation-type visual carrier	AM with negative polarity (i.e., a decrease in light level causes an increased real envelope level)
Aural carrier	FM with 100% modulation being $\Delta F = 25$ kHz with a frequency response of 50 to 15,000 Hz using 75-μsec preemphasis
Visual modulation levels	
Blanking level	75% ± 2.5% of peak real envelope level (sync tip level)
Reference black level	7.5% ± 2.5% (of the video range between blanking and reference white level) below blanking level; this is called the *setup* level by TV engineers[a]
Reference white level	12.5% ± 2.5% of sync tip level[a]
Scanning	
Number of lines	525 lines/frame, interlaced 2:1
Scanning sequence	Horizontally: left to right; vertically: top to bottom
Horizontal scanning frequency, f_h	15,734.264 ± 0.044 Hz (2/455 of chrominance frequency); 15,750 Hz may be used during monochrome transmissions
Vertical scanning frequency, f_v	59.94 Hz (2/525 of horizontal scanning frequency); 60 Hz may be used during monochrome transmissions; 21 equivalent horizontal lines occur during the vertical sync interval of each field
Vertical interval signaling	Lines 13, 14, 15, 16. Teletext Lines 17, 18. Vertical interval test signals (VITS) Line 19. Vertical interval reference (VIR) Line 20, Field 1. Station identification Line 21, Field 1. Captioning data Line 21, Field 2. Captioning framing code (½ line)

[a]See Fig. 5-52b.

approximately $T_0 = 1/60$ sec corresponding to the field rate of 60 fields/sec (see Table 5-16 for exact values). Thus, for a still picture, the spectrum would consist of lines spaced at 60-Hz intervals. However, from Fig. 5-51, we can observe that there are dominant intervals of width T_h corresponding to scanned lines of the frame. Furthermore, the adjacent lines usually have similar waveshapes. Consequently, $m_c(t)$ is quasiperiodic with period T_h, and the spectrum consists of clusters of spectral lines that are centered on harmonics of the scanning frequency, $nf_h = n/T_h$ where the spacing of the lines within these clumps is 60 Hz.

For moving pictures, the line structure of the spectrum "fuzzes out" to a continuous spectrum with spectral clusters centered about nf_h. Furthermore, between these clusters the spectrum is nearly empty. As we will see, these "vacant" intervals in the spectrum are used to transmit the color information for a color TV signal. This, of course, is a form of frequency-division multiplexing.

The resolution of TV pictures is often specified in terms of lines of resolution. The number of horizontal lines that can be distinguished from the top to the bottom of a TV screen for a horizontal line test pattern is called the *vertical line resolution*. The maximum number of distinguishable horizontal lines (vertical line resolution) is the total number of scanning lines in the raster less those not used for the image. That is, the vertical resolution is

$$n_v = (N_f - N_v) \quad \text{lines} \tag{5-109a}$$

where N_f is the total number of scanning lines per frame and N_v is the number of lines in the vertical interval (not image lines) for each frame. For U.S. standards (Table 5-16), the maximum vertical resolution is

$$n_v = 525 - 42 = 483 \quad \text{lines} \tag{5-109b}$$

The resolution in the horizontal direction is limited by the frequency response allowed for $m_c(t)$. For example, if a sine-wave test signal is used during the video interval, the highest sine-wave frequency that can be transmitted through the system will be $B = 4.2$ MHz (U.S. standard) where B is the system video bandwidth. For each peak of the sine wave a dot will appear along the horizontal direction as the CRT beam sweeps from left to right. Thus the horizontal resolution for 2:1 interlacing is

$$n_h = 2B(T_h - T_b) \quad \text{lines} \tag{5-109c}$$

where B is the video bandwidth, T_h is the horizontal interval, and T_b is the blanking interval (see Fig. 5-51). For U.S. standards, $B = 4.2$ MHz, $T_h = 63.5$ μs, and $T_b = 10.5\mu s$ so that the horizontal resolution is

$$n_h = 445 \quad \text{lines} \tag{5-109d}$$

Furthermore, because of poor video bandwidth and the poor interlacing characteristics of consumer TV sets, this horizontal resolution of 445 is usually not obtained in practice.[†] At best, the U.S. standard of 445 × 483 does not provide very good resolution, and this poor resolution is especially noticeable on large-screen TV sets. [For comparison the enhanced graphics adapter (EGA) computer monitor standard provides a resolution of 640 × 480.]

There is a great incentive to adopt a *high-definition television* (HDTV) standard in which the resolution is at least twice that of the present TV standard so that it approaches the quality of 35 mm film and which has an aspect ratio of horizontal picture width to vertical width of 16:9 instead of the present 4:3 ratio. There are many realities to consider such as (1) the desirability of adopting a

[†] Typically, a horizontal resolution of about 300 lines is obtained.

HDTV standard that is compatible with the present system so that standard TV sets may be used to view a HDTV signal, (2) the FCC requirement that HDTV must operate within the present bands allocated for broadcast TV, and (3) the cost, which must be reasonable. More than 20 HDTV standards have been proposed [Jurgen, 1988], but it is not clear which one, if any, will be universally accepted. Moreover, it is argued that any new HDTV standard should use digital instead of analog transmission [Jurgen, 1991]. Several digital standards are proposed [Zou, 1991].

For broadcast TV transmission (U.S. standard), the composite video signal of (5-108) is inverted and amplitude modulated onto an RF carrier so that the AM signal is

$$s_v(t) = A_c[1 - 0.875m_c(t)] \cos \omega_c t \tag{5-110}$$

This is shown in Fig. 5-52. The lower sideband of this signal is attenuated so that the spectrum will fit into the 6-MHz TV channel bandwidth. This attenuation can be achieved by using a vestigial sideband filter, which is a bandpass filter that attenuates most of the lower sideband. The resulting filtered signal is called a

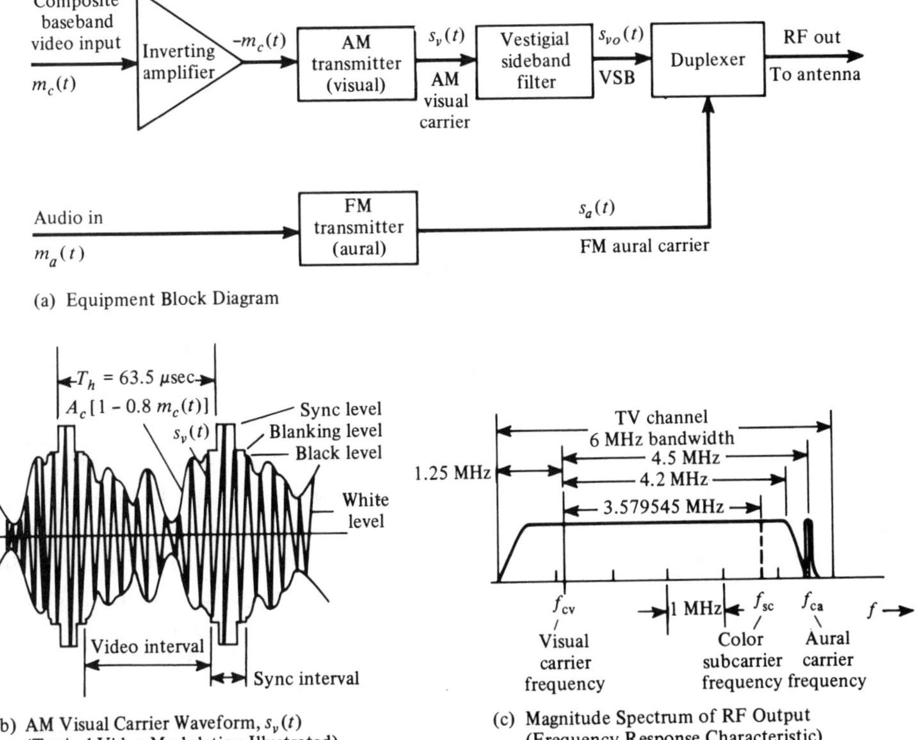

FIGURE 5-52 TV transmission system.

vestigial sideband (VSB) signal, $s_{v_0}(t)$ as described in Sec. 4-6. The discrete carrier term in the VSB signal changes when the picture is switched from one scene to another because the composite baseband signal has a nonzero dc level that depends on the particular scene being televised. The audio signal for the TV program is transmitted via a separate FM carrier located exactly 4.5 MHz above the visual carrier frequency.

The rated power of a TV visual signal is the effective isotropic radiated peak envelope power (EIRP, i.e., effective sync tip average power), and is usually called simply the *effective radiated power* (ERP). The EIRP is the power that would be required to be fed into an isotropic antenna to get the same field strength as that obtained from an antenna that is actually used as measured in the direction of its maximum radiation. The antenna pattern of an omnidirectional TV station is shaped like a doughnut with the tower passing through the hole. (The antenna structure is actually at the top of the tower for maximum line-of-sight coverage.) The ERP of the TV station (visual or aural) signal is

$$P_{\text{ERP}} = P_{\text{PEP}} G_{\text{A}} G_{\text{L}} \tag{5-111}$$

where P_{PEP} = peak envelope power out of the transmitter
G_{A} = power gain of the antenna (absolute units as opposed to decibel units) with respect to an isotropic antenna
G_{L} = total gain of the transmission line system from the transmitter output to the antenna (including the duplexer gain)

For a VHF TV station operating on channel 5, some typical values are $G_A = 5.7$, $G_L = 0.873$ (850 ft of transmission line plus duplexer), and a visual PEP of 20.1 kW. This gives an overall ERP of 100 kW, which is the maximum power licensed by the FCC for channels 2 to 6 (low-VHF band). For a UHF station operating on channel 20, some typical values are $P_{\text{PEP}} = 21.7$ kW, $G_A = 27$, $G_L = 0.854$ (850 ft of transmission line), which gives an ERP of 500 kW.

A simplified block diagram of a black-and-white TV receiver is shown in Fig. 5-53. The composite video, $m_c(t)$, as described earlier, plus a 4.5-MHz FM carrier containing the audio modulation, appears at the output of the envelope detector. The 4.5-MHz FM carrier is present because of the nonlinearity of the envelope detector and because the detector input (IF signal) contains, among other terms, the FM aural carrier plus a discrete (i.e., sinusoidal) term, which is the video carrier, located 4.5 MHz away from the FM carrier signal. The intermodulation product of these two signals produces the 4.5-MHz FM signal, which is called the *intercarrier signal,* and contains the aural modulation. Of course, if either of these two input signal components disappears, the 4.5-MHz output signal will also disappear. This will occur if the white level of the picture is allowed to go below, say, 10% of the peak envelope level (sync tip level) of the AM visual signal. This is why the FCC specifies that the white level cannot be lower than $(12.5 \pm 2.5)\%$ of the peak envelope level (see Table 5-16). When this occurs, a buzz will be heard in the sound signal since the 4.5-MHz FM carrier is disappearing at a 60-Hz (field) rate during the white portion of the TV scene. The sync separator circuit consists of a lossy integrator to recover the

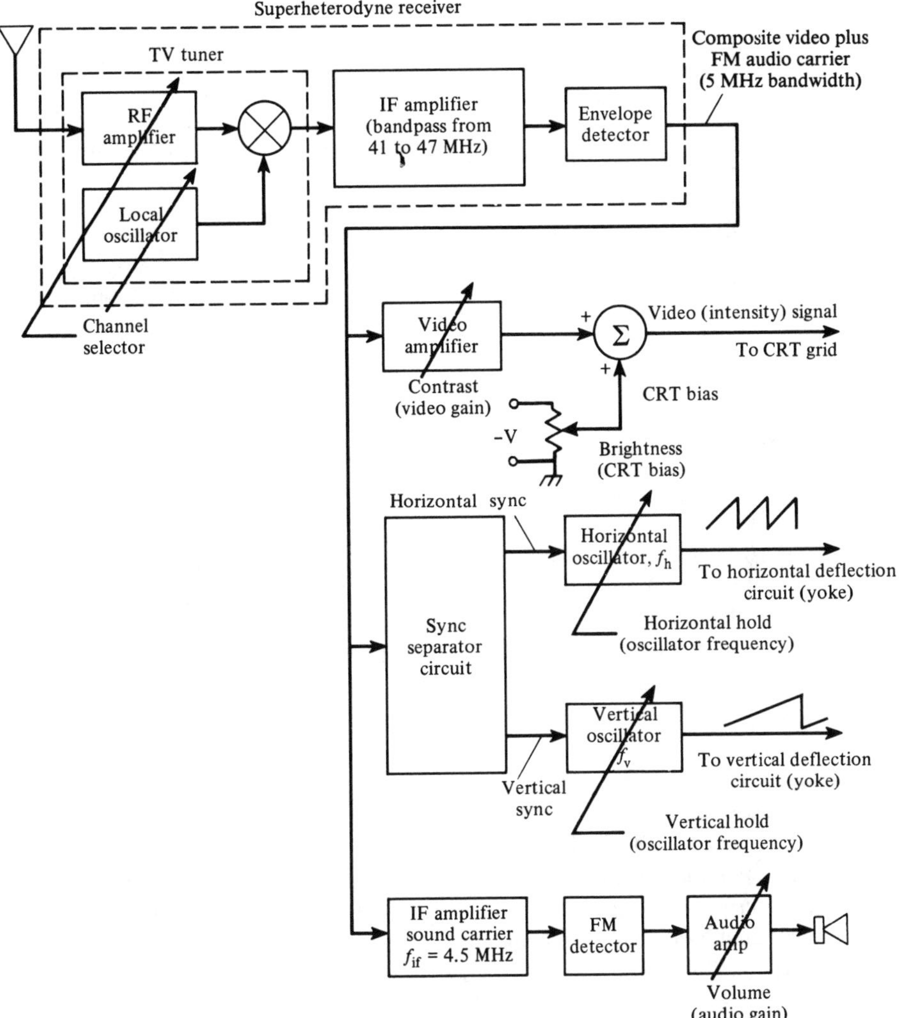

FIGURE 5-53 Black-and-white TV receiver.

(wide) vertical sync pulses and a lossy differentiator to recover the (narrow) horizontal pulses. The remainder of the block diagram is self-explanatory. The user-available adjustments for a black-and-white TV are also shown.

MTS Stereo Sound

Stereo audio, also known as *multichannel television sound* (MTS), is also available and is accomplished using the FDM technique shown in Fig. 5-54. This FDM technique is similar to that used in standard FM broadcasting, studied in Sec. 4-7 and illustrated by Fig. 4-48. For compatibility with monaural audio

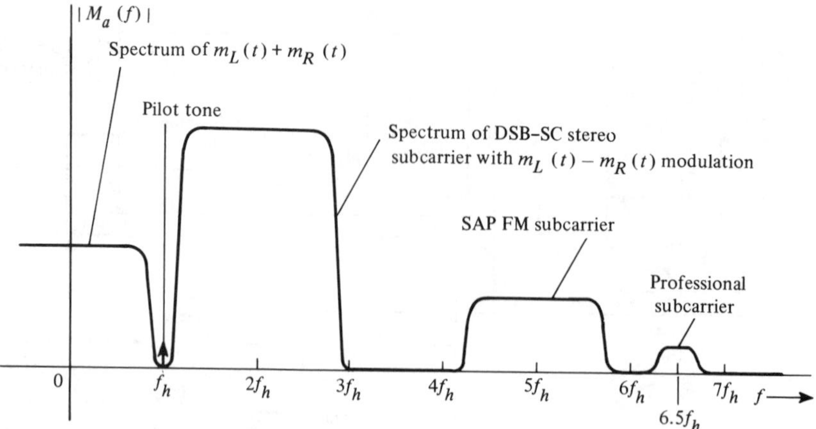

FIGURE 5-54 Spectrum of composite FDM baseband modulation on the FM aural carrier of U.S. TV stations.

television receivers, the left plus right audio, $m_L(t) + m_R(t)$, is frequency modulated directly onto the aural carrier of the television station. Stereo capability is obtained by using a DSB-SC subcarrier with $m_L(t) - m_R(t)$ modulation. The subcarrier frequency is $2f_h$, where $f_h = 15.734$ kHz is the horizontal sync frequency for the video (see Table 5-16). Capability is also provided for a *second audio program* (SAP), such as audio in a second language. This is accomplished by using an FM subcarrier of frequency $5f_h$. DBX audio noise reduction processing is used on the audio on both the DSB-SC stereo subcarrier and the SAP subcarrier that is frequency modulated by the second audio signal. (See Sec. 4-7 for a discussion of the DBX noise reduction technique.) Furthermore, a *professional channel* subcarrier at $6.5f_h$ is allowed for transmission of voice or data on a subscription service basis to paying customers. For further details on the FDM aural standards that have been adopted by television broadcasters and for other possible FDM formats, see a paper by Eilers [1985].

Color Television

Color television pictures are synthesized by combining red, green, and blue light in the proper proportion to produce the desired color. This is done by using a color CRT that has three types of phosphors—one type that emits red light, one that emits green, and a third type that emits blue light when they are struck by electrons. Therefore, three electronic video signals—$m_R(t)$, $m_G(t)$, and $m_B(t)$—are needed to drive the circuits that create the red, green, and blue light. Furthermore, a compatible color TV transmission system is needed such that black-and-white TV sets will receive the luminance video signal, $m_y(t)$, which corresponds to the gray scale of the scene. This compatibility between color and black-and-white transmission is accomplished by using a multiplexing technique analogous to that of FM stereo, except that three signals are involved in the problem instead of two.

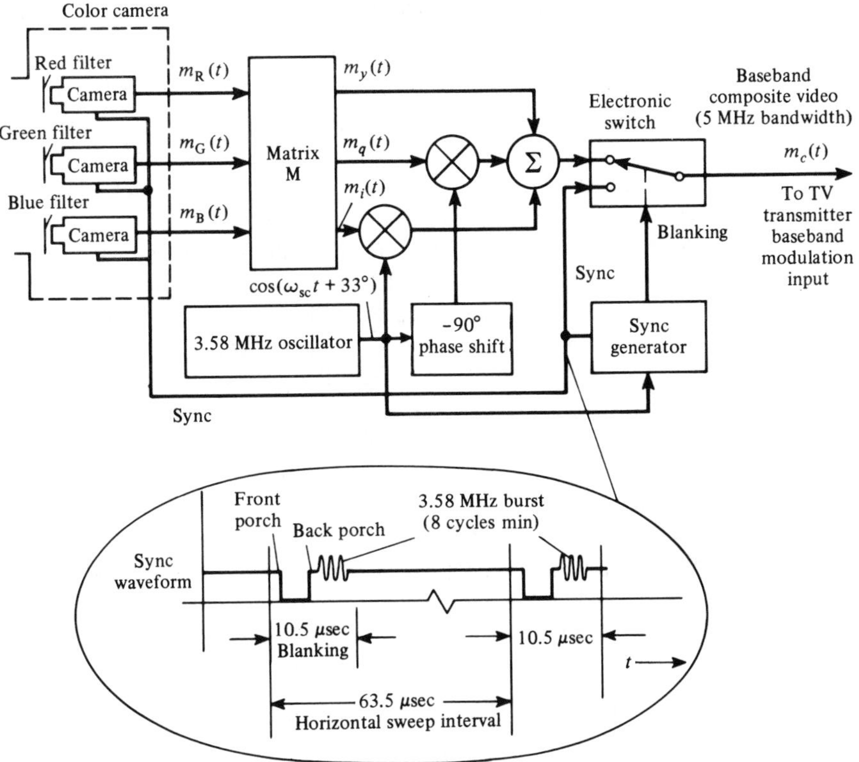

FIGURE 5-55 Generation of the composite NTSC baseband video signal. [Technically, $m_q(t)$ is low-pass filtered to an 0.5-MHz bandwidth and $m_i(t)$ is low-pass filtered to a 1.5-MHz bandwidth.]

The National Television System Committee (NTSC) compatible color TV system that was developed in the United States and adopted for U.S. transmissions in 1954 will be discussed. This technique is illustrated in Fig. 5-55. For compatible transmission, the waveforms that correspond to the intensity of the red, green, and blue components of the scene—$m_R(t)$, $m_G(t)$, and $m_B(t)$—are linearly combined to synthesize the equivalent black-and-white signal, $m_y(t)$, and two other independent video signals, called the *in-phase* and *quadrature* components, $m_i(t)$ and $m_q(t)$. (The names of these components arise from the quadrature modulation technique that is used to modulate them onto a subcarrier.) The exact equations used for these signals can be expressed in matrix form:

$$\begin{bmatrix} m_y(t) \\ m_i(t) \\ m_q(t) \end{bmatrix} = \underbrace{\begin{bmatrix} 0.3 & 0.59 & 0.11 \\ 0.6 & -0.28 & -0.32 \\ 0.21 & -0.52 & 0.31 \end{bmatrix}}_{M} \begin{bmatrix} m_R(t) \\ m_G(t) \\ m_B(t) \end{bmatrix} \tag{5-112}$$

where M is a 3×3 matrix that translates the red, green, and blue signals to the brightness, in-phase, and quadrature phase video (baseband) signals. For example, from (5-112), the equation for the *luminance* (black and white) signal is

$$m_y(t) = 0.3m_R(t) + 0.59m_G(t) + 0.11m_B(t) \tag{5-113}$$

Similarly, the equations for $m_i(t)$ and $m_q(t)$, which are called the *chrominance* signals, are easy to obtain. In addition, note that *if* $m_R(t) = m_G(t) = m_B(t) = 1$, which corresponds to maximum red, green, and blue, then $m_y(t) = 1$, which corresponds to the white level in the black-and-white picture.

The chrominance components are two other linearly independent components, and if they are transmitted to the color TV receiver together with the luminance signal, these three signals can be used to recover the red, green, and blue signals using an inverse matrix operation. The bandwidth of the luminance signal needs to be maintained at 4.2 MHz to preserve sharp (high-frequency) transition in the intensity of the light that occurs along edges of objects in the scene. However, the bandwidth of the chrominance signals does not need to be this large since the eye is not as sensitive to color transitions in a scene. In NTSC standards the bandwidth of the $m_i(t)$ signal is 1.5 MHz and the bandwidth of $m_q(t)$ is 0.5 MHz. When they are quadrature-modulated onto a subcarrier, the resulting composite baseband signal will then have a bandwidth of 4.2 MHz. (The upper sideband of the in-phase subcarrier signal is attenuated to keep a 4.2-MHz bandwidth.)

The composite NTSC baseband video signal is

$$m_c(t) = \begin{cases} m_s(t) & \text{during the sync interval} \\ m_y(t) + \text{Re}\{g_{sc}(t)e^{j\omega_{sc}t}\} & \text{during the video interval} \end{cases} \tag{5-114}$$

where

$$g_{sc}(t) = [m_i(t) - jm_q(t)]e^{j^{33°}} \tag{5-115}$$

Equation (5-115) indicates that the chrominance information is quadrature modulated onto a subcarrier (as described in Table 4-1). $m_s(t)$ is the sync signal, $m_y(t)$ is the luminance signal, and $g_{sc}(t)$ is the complex envelope of the subcarrier signal. The subcarrier frequency is $f_{sc} = 3.579545$ MHz $\pm$ 10 Hz, which, as we will see later, is chosen so that the subcarrier signal will not interfere with the luminance signal even though both signals fall within the 4.2-MHz passband.

In broadcast TV, this composite NTSC baseband color signal is amplitude-modulated onto the visual carrier as described by (5-110) and shown in Fig. 5-52. In microwave relay applications and satellite relay applications, this NTSC baseband signal is frequency-modulated onto a carrier, as described in Secs. 5-3 and 5-4. These color baseband signals can be transmitted over communication systems that are used for black-and-white TV, although the frequency response and linearity tolerances that are specified are tighter for color TV transmission.

As indicated before, the complex envelope for the subcarrier, as described by (5-115), contains the color information. The magnitude of the complex envelope, $|g_{sc}(t)|$, is the *saturation* or amount of color in the scene as it is scanned with time. The angle of the complex envelope, $\underline{/g_{sc}(t)}$, indicates hue or tint as a function of time. Thus $g_{sc}(t)$ is a vector that moves about in the complex plane as

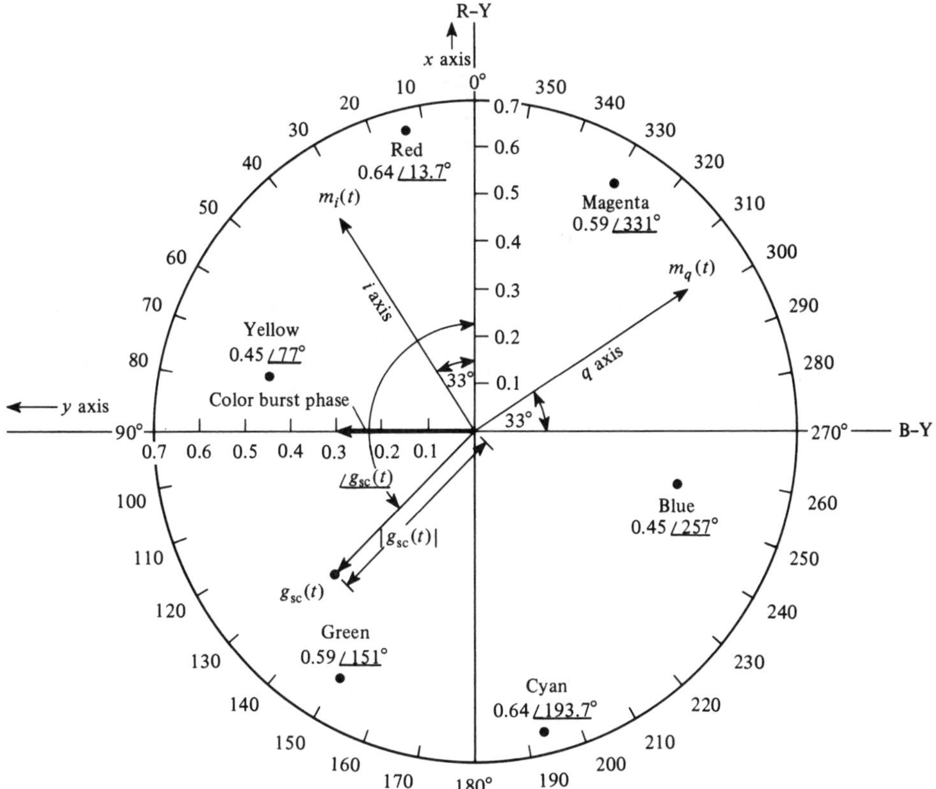

FIGURE 5-56 Vectorscope presentation of the complex envelope for the 3.58-MHz chrominance subcarrier.

a function of time as the lines of the picture are scanned. This vector may be viewed on an instrument called a *vectorscope,* which has a CRT that displays the vector $g_{sc}(t)$, as shown in Fig. 5-56. The vectorscope is a common piece of test equipment that is used to calibrate color TV equipment and to provide an on-line measure of the quality of the color signal being transmitted. The usual vectorscope presentation has the x axis located in the vertical direction, as indicated on the figure. Using (5-115), the positive axis directions for the $m_i(t)$ and $m_q(t)$ signals are also indicated on the vectorscope presentation, together with the vectors for the saturated red, green, and blue colors. For example, if saturated red is present, then $m_R = 1$ and $m_G = m_B = 0$. Using (5-112), we get $m_i = 0.6$ and $m_q = 0.21$. Thus, from (5-115), the saturated red vector is

$$(g_{sc})_{red} = (0.60 - j0.21)e^{j33^\circ} = 0.64 \underline{/13.7^\circ}$$

Similarly, the saturated green vector is $(g_{sc})_{green} = 0.59 \underline{/151^\circ}$, and the saturated blue vector is $(g_{sc})_{blue} = 0.45 \underline{/257^\circ}$. These vectors are shown in Fig. 5-56. It is interesting to note that the complementary colors have the opposite polarity on the vectorscope, where cyan is the complement of red, magenta is the complement of green, and yellow is the complement of blue.

For color subcarrier demodulation at the receiver, a phase reference is needed. This is provided by keying in an eight-cycle (or more) burst of the 3.58-MHz subcarrier sinusoid on the "back porch" of the horizontal sync pulse, as shown in Fig. 5-55. The phase of the reference sinusoid is +90° with respect to the x axis, as indicated on the vectorscope display (Fig. 5-56).

A color receiver is similar to a black-and-white TV receiver except that it has color demodulator circuits and a color CRT. This is shown in Fig. 5-57, where the color demodulator circuitry is indicated. The in-phase and quadrature signals, $m_i(t)$ and $m_q(t)$, are recovered (demodulated) by using product detectors where the phase reference signal is obtained from a PLL circuit that is locked onto the color burst (which was transmitted on the back porch of the sync pulse). M^{-1} is the inverse matrix of that given in (5-112) so that the three video waveforms corresponding to the red, green, and blue intensity signals are recovered. It is

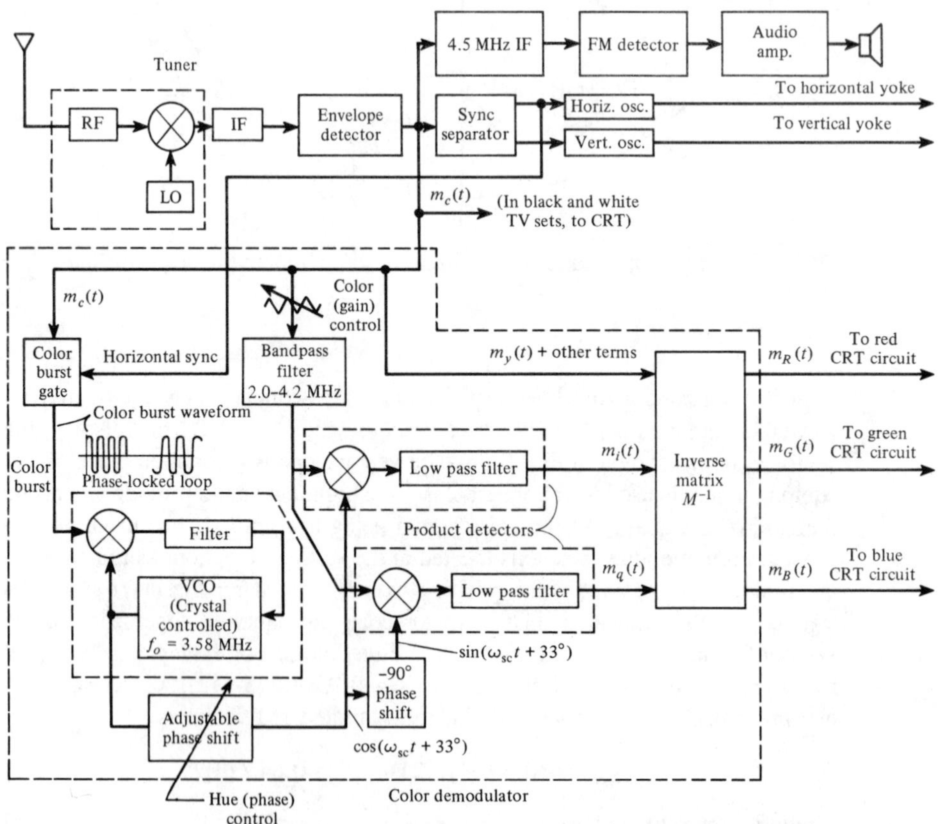

FIGURE 5-57 Color TV receiver with IQ demodulators.

realized that the hue control on the TV set adjusts the phase of the reference that sets the tint of the picture on the TV set. The color control sets the gain of the chrominance subcarrier signal. If the gain is reduced to zero, no subcarrier will be present at the input to the product detectors. Consequently, a black-and-white picture will be reproduced on the color TV CRT.

Of course, it is also possible to use different reference phases on the balanced detector inputs *provided* that the decoding matrix is changed to produce m_R, m_G, and m_B at its output. For example, one popular technique uses the R-Y and B-Y phases on the input to the balanced modulators [Grob, 1975]. Here the reference phase out of the hue control circuit is 0° instead of 33°, as shown in Fig. 5-57. This corresponds to detecting the subcarrier along the R-Y and B-Y axes, as indicated on the vectorscope presentation (Fig. 5-56). In this case $m_R(t) - m_Y(t)$ is obtained at the output of the upper product detector, and if a gain of two is used on the lower product detector, $m_B(t) - m_Y(t)$ is obtained at its output. These two outputs may be added in the proper proportions to obtain $m_G(t) - m_Y(t)$. These three signals, $m_R(t) - m_Y(t)$, $m_G(t) - m_Y(t)$, and $m_B(t) - m_Y(t)$, may be applied to the three control grids of a three-gun color CRT (one gun activating the red phosphor, one gun for the green, and one for the red). The $-m_Y(t)$ signal is applied to the common cathode of the three guns so that the effective signals on the guns are $m_R(t)$, $m_G(t)$, and $m_B(t)$, respectively, which are the desired control signals. Furthermore, for black-and-white operation, it is only necessary to remove the three signals from the control grid since the luminance drive signal is applied at the cathode.

As stated earlier, both the luminance and chrominance subcarrier signals are contained in the 4.2-MHz composite NTSC baseband signal. This gives rise to interfering signals in the receiver video circuits. For example, in Fig. 5-57 there is an interfering signal on the input to the inverse matrix circuit. Here $m_Y(t)$ plus interference is averaged out on a line-by-line scanning basis by the viewer's vision if the chrominance subcarrier frequency is an odd multiple of half of the horizontal scanning frequency. For example, from Table 5-16, $f_{sc} = 3{,}579{,}545$ Hz and $f_h = 15{,}734$ Hz. Consequently, there are 227.5 cycles of the 3.58-MHz chrominance interference across one scanned line of the picture. Since there is a half cycle left over, the interference on the *next* scanned line will be *180° out of phase* with that of the present line and the interference will be *canceled out* by the eye when one views the picture.

As we have seen, the NTSC color system is a very ingenious application of basic engineering principles, and this ingenuity is certainly not appreciated by the average color TV viewer.

Standards for TV and CATV Systems

A summary of some of the U.S. TV transmission standards as adopted by the FCC is shown in Table 5-16, and details of the synchronizing waveform are given in Fig. 5-58. A listing of the frequencies for the TV channels is given in Table 5-17. In addition, a comparison of some broadcast TV standards that are

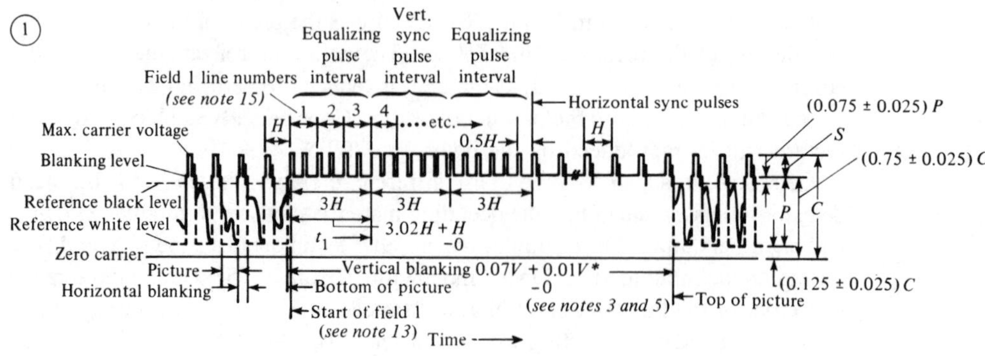

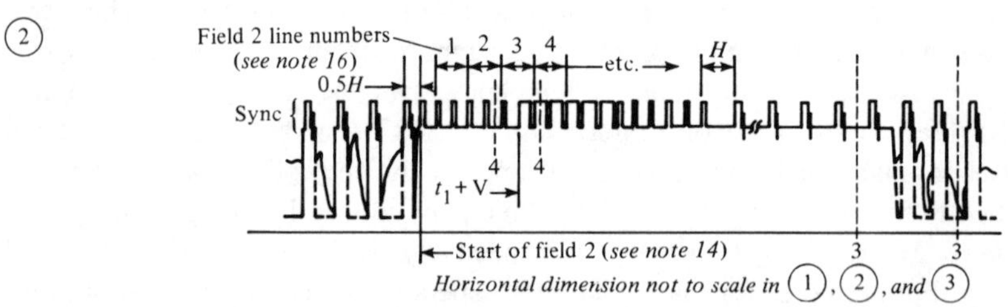

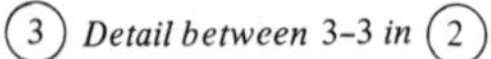

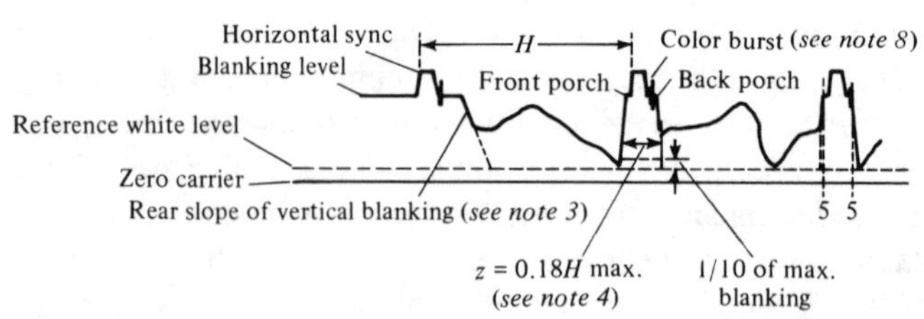

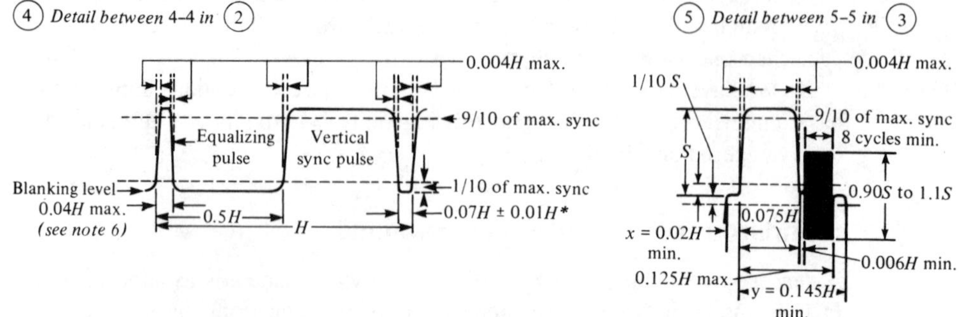

FIGURE 5-58 U.S. television synchronizing waveform standards for color transmission. (From Part 73 of FCC Rules and Regulations.)

used by various countries is shown in Table 5-18. The CCIR system B standard, which is a 625-lines/frame, 25-frames/sec, negative modulation system is the dominant European TV standard [Dhake, 1980]. As discussed previously, in addition to these standards for conventional TV as shown in this table, there are many proposed standards for HDTV [Jurgen, 1988]. There is also a push for adoption of an all-digital, instead of an analog, HDTV system [Jurgen, 1991; Zou, 1991]. When TV signals are relayed from one country to another via satellite, the standards of the originating country are used, and, if needed, the signal is converted to the standards of the receiving country at the receiving ground station using Digital Intercontinental Conversion Equipment (DICE).

U.S. CATV channel standards are presented in Table 5-19. CATV systems use coaxial cable with amplifiers to distribute TV and other signals from the "head end" to the subscriber. Since the cable signals are not radiated, the frequencies that are normally assigned to two-way radio and other on-the-air services can be used for additional TV channels on the cable. Consequently, the CATV channel assignments have been standardized as shown in this table. To receive these added channels, the subscriber has to have a TV set with CATV channel capability or use a down converter to convert the CATV channel to a

NOTES TO FIGURE 5-58

1 H = time from start of one line to start of next line = $1/f_h$ (see Table 5-16).
2 V = time from start of one field to start of next field = $1/f_v$ (see Table 5-16).
3 Leading and trailing edges of vertical blanking should be complete in less than 0.1 H.
4 Leading and trailing slopes of horizontal blanking must be steep enough to preserve minimum and maximum values of $(x + y)$ and (z) under all conditions of picture content.
*5 Dimensions marked with asterisk indicate that tolerances given are permitted only for long-time variations and not for successive cycles.
6 Equalizing pulse area shall be between 0.45 and 0.5 of area of a horizontal sync pulse.
7 Color burst follows each horizontal pulse, but is omitted following the equalizing pulses and during the broad vertical pulses.
8 Color bursts to be omitted during monochrome transmission.
9 The burst frequency shall be 3.579545 MHz. The tolerance on the frequency shall be ±10 Hz with a maximum rate of change of frequency not to exceed 1/10 Hz/sec.
10 The horizontal scanning frequency shall be 2/455 times the burst frequency.
11 The dimensions specified for the burst determine the times of starting and stopping the burst but not its phase. The color burst consists of amplitude modulation of a continuous sine wave.
12 Dimension P represents the peak excursion of the luminance signal from blanking level, but does not include the chrominance signal. Dimension S is the sync amplitude above blanking level. Dimension C is the peak carrier amplitude.
13 Start of field 1 is defined by a whole line between first equalizing pulse and preceding H sync pulses.
14 Start of field 2 is defined by a half line between first equalizing pulse and preceding H sync pulses.
15 Field 1 line numbers start with first equalizing pulse in field 1.
16 Field 2 line numbers start with second equalizing pulse in field 2.
17 Refer to Table 5-16 for further explanations and tolerances.

TABLE 5-17 U.S. Television Channel Frequency Assignments

	Channel Number	Band (MHz)	UHF Channel Number	Band	UHF Channel Number	Band
Radio astronomy, Aeronautical, Two-way radio (→ between 4 and 5)	2	54–60	29	560–566	57	728–734
	3	60–66	30	566–572	58	734–740
	4	66–72	31	572–578	59	740–746
	5	76–82	32	578–584	60	746–752
FM broadcasting, Aeronautical, Two-way radio (→ between 6 and 7)	6	82–88	33	584–590	61	752–758
	7	174–180	34	590–596	62	758–764
	8	180–186	35	596–602	63	764–770
	9	186–192	36	602–608	64	770–776
	10	192–198	37	608–614	65	776–782
	11	198–204	38	614–620	66	782–788
VHF	12	204–210	39	620–626	67	788–794
channels	13	210–216	40	626–632	68	794–800
			41	632–638	69	800–806
UHF channels	14	470–476	42	638–644	70[a]	806–812
	15	476–482	43	644–650	71[a]	812–818
	16	482–488	44	650–656	72[a]	818–824
	17	488–494	45	656–662	73[a]	824–830
	18	494–500	46	662–668	74[a]	830–836
	19	500–506	47	668–674	75[a]	836–842
	20	506–512	48	674–680	76[a]	842–848
	21	512–518	49	680–686	77[a]	848–854
	22	518–524	50	686–692	78[a]	854–860
	23	524–530	51	692–698	79[a]	860–866
	24	530–536	52	698–704	80[a]	866–872
	25	536–542	53	704–710	81[a]	872–878
	26	542–548	54	710–716	82[a]	878–884
	27	548–554	55	716–722	83[a]	884–890
	28	554–560	56	722–728		

[a]These frequencies are now allocated to two-way radio and cellular telephone service but were once allocated to UHF TV broadcasting.

standard channel such as channel 3 or channel 4. CATV systems can also be *bidirectional,* allowing some subscribers to transmit signals to the head end. In low-band split systems, frequencies below 50 MHz (T-band channels) are used for transmission to the head end. In mid-band, split system frequencies below 150 mHz are used for transmission to the head end.

Premium TV programs are often available to subscribers for an additional cost. The premium channels are usually scrambled by modifying the video sync or possibly by using digital encryption techniques. Subscribers to the premium services are provided with appropriate decoders that, in some systems, are addressable and remotely programmable by the CATV company.

More details about TV standards and engineering practice can be found in the *Television Engineering Handbook* [Benson and Whitaker, 1992].

TABLE 5-18 Comparison of Some Broadcast Television Standards

Item	Where Used: North America, South America, Japan	Spain, Italy, England, Germany, CCIR System B[a]	France		USSR
Lines/frame	525	625	625	819	625
Lines/second	15,750	15,625	15,625	20,475	15,625
Frames/second	30	25	25	25	25
Baseband video bandwidth (MHz)	4.2	5	6	10	6
Channel bandwidth (MHz)	6	7	8	14	8
Polarity of AM video modulation	Negative	Negative	Positive	Positive	Negative
Type of aural carrier	FM	FM	AM	AM	FM
Color system	NTSC[b]	PAL[c]	SECAM[d]	SECAM[d]	SECAM[d]
Color subcarrier frequency (MHz)	3.58	4.43	4.43	4.43	4.43

[a] The 625-line/frame, 25-frame/sec system is the CCIR (International Radio Consultative Committee) system B standard and is the dominant TV standard used on the European continent.

[b] NTSC, National Television System Committee (United States).

[c] PAL, Phase Alternation Line (Europe).

[d] SECAM, Sequential Couleur à Mémoire (French).

5-10 Summary

In this chapter, a wide range of practical communication systems were presented based on the theory developed in Chapters 1 through 4.

Basic bandpass digital signaling techniques such as OOK, BPSK, and FSK were developed from classical AM, PM, and FM theory. The spectra of these digital signals were evaluated in terms of the bit rate of the digital information source. Multilevel digital signaling techniques such as QPSK, MPSK, QAM, and MSK were also studied, and their spectra were evaluated.

Telephone and common carrier communication systems were examined. Specifications for voice frequency and data transmission were given for transmission over twisted-pair wire, coaxial cable, microwave radio relay, satellites, cellular radio, and fiber optic links. Examples of satellite frequency-division multiple access and time-division multiple access were given. Television signal relay via satellites was studied.

TABLE 5-19 CATV Channel Frequencies

	CATV Channel Number	Band (MHz)	Letter Channel Designator	CATV Channel Number	UHF Band (MHz)
Sub-VHF		5.75–11.75	T-7	37	300–306
		11.75–17.75	T-8	38	306–312
		17.75–23.75	T-9	39	312–318
		23.75–29.75	T-10	40	318–324
		29.75–35.75	T-11	41	324–330
		35.75–41.75	T-12	42	330–336
		41.75–47.55	T-13	43	336–342
Low-VHF	2	54–60		44	342–348
	3	60–66		45	348–354
	4	66–72		46	354–360
	5	76–82		47	360–366
	6	82–88		48	366–372
				49	372–378
	FM Broadcasting	88–108		50	378–384
High-VHF	7	174–180		51	384–390
	8	180–186		52	390–396
	9	186–192		53	396–402
	10	192–198		54	402–408
	11	198–204		55	408–414
	12	204–210		56	414–420
	13	210–216		57	420–426
				58	426–432
Midband	14	120–126	A	59	432–438
	15	126–132	B	60	438–444
	16	132–138	C	61	444–450
	17	138–144	D	.	
	18	144–150	E	.	
	19	150–156	F	.	
	20	156–162	G	89	612–618
	21	162–168	H	90	618–624
	22	168–174	I	91	624–630
Superband	23	216–222	J	92	630–636
	24	222–228	K	93	636–642
	25	228–234	L	94	642–648
	26	234–240	M	95	90–96
	27	240–246	N	96	96–102
	28	246–252	O	97	102–108
	29	252–258	P	98	108–114
	30	258–264	Q	99	114–120
	31	264–270	R		
	32	270–276	S		
	33	276–282	T		
	34	282–288	U		
	35	288–294	V		
	36	294–300	W		

A technique was developed for evaluating communication system performance. This required the calculation of the signal-to-noise ratio at the detector input of a receiver for cases of both digital and analog signaling. In developing this result the concepts of received signal power from a free-space channel, thermal noise sources, noise figure, and noise temperature were established. In Example 5-5 a link budget for the received signal-to-noise ratio at the output of a TV receive-only terminal for reception of satellite signals was presented. An example of a link budget for a fiber optic common carrier system was also given.

The antijam and low probability of intercept properties of spread spectrum systems were studied. This included the calculation of the spectrum for a spread spectrum signal and the examination of transmitter and receiver block diagrams for both direct sequence and frequency hopped spread spectrum systems.

Television transmission, both black-and-white and NTSC color, was discussed in detail. Standards for television and CATV systems were given.

The link budget analysis in this chapter pointed out the need for the use of statistics to evaluate the performance of digital and analog detector circuits. These techniques will be developed in succeeding chapters. In Chapter 6 the concepts of random processes are developed from probability theory (which is established in Appendix B), and in Chapter 7 the results are used to evaluate the performance of digital and analog receiving circuits for the case of noisy input signals.

PROBLEMS

5-1 A digital baseband signal consisting of rectangular binary pulses occurring at a rate of 2400 bits/sec is to be transmitted over a bandpass channel.
(a) Evaluate the magnitude spectrum for OOK signaling that is keyed by a baseband digital test pattern consisting of alternating 1's and 0's.
(b) Sketch the magnitude spectrum and indicate the value of the first null-to-null bandwidth. Assume a carrier frequency of 50 kHz.
(c) For a random data pattern, find the PSD and plot the result. Compare this result with that obtained in parts (a) and (b) for alternating data.

5-2 Repeat Prob. 5-1 for the case of BPSK signaling.

5-3 A carrier is angle modulated with a polar baseband data signal to produce a BPSK signal, $s(t) = 10\cos[\omega_c t + D_p m(t)]$, where $m(t) = \pm 1$ corresponds to the binary data {10010110}. $T_b = 0.0025$ sec and $\omega_c = 1000\pi$. Using MathCAD, plot the BPSK signal waveform and its corresponding FFT spectrum for the following digital modulation indices:
(a) $h = 0.2$.
(b) $h = 0.5$.
(c) $h = 1.0$.

5-4 Evaluate the magnitude spectrum for an FSK signal with alternating 1 and 0 data. Assume that the mark frequency is 50 kHz, the space frequency is

55 kHz, and the bit rate is 2400 bits/sec. Find the first null-to-null bandwidth.

5-5 Assume that 4800-bits/sec random data are sent over a bandpass channel by BPSK signaling. Find the transmission bandwidth B_T such that the spectral envelope is down at least 35 dB outside this band.

5-6 As indicated in Sec. 5-2, a BPSK signal can be demodulated by using a synchronous detector where the carrier reference is provided by a Costas loop for the case of $h = 1.0$. Alternatively, the carrier reference can be provided by a squaring loop that uses a $\times 2$ frequency multiplier. A block diagram for a squaring loop is shown in Fig. P5-6.

(a) Draw an overall block diagram for a BPSK receiver using the squaring loop.

(b) By using mathematics to represent the waveforms, show how the squaring loop recovers the carrier reference.

(c) Demonstrate that the squaring loop does or does not have a 180° phase ambiguity problem.

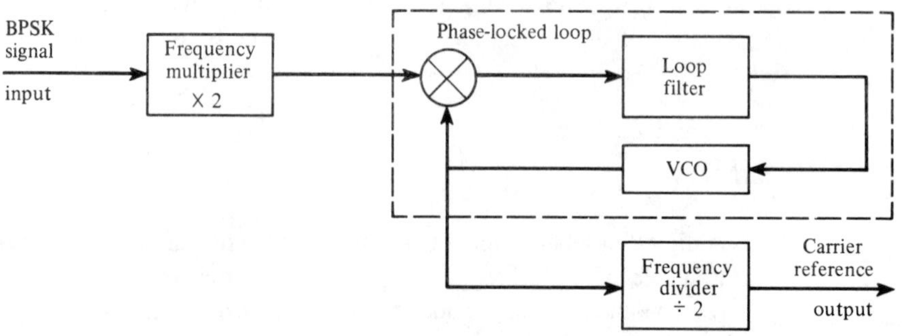

FIGURE P5-6

5-7 A binary data signal is differentially encoded and modulates a PM transmitter to produce a differentially encoded phase-shift-keyed signal (DPSK). The peak-to-peak phase deviation is 180° and f_c is harmonically related to the bit rate R.

(a) Draw a block diagram for the transmitter, including the differential encoder.

(b) Show typical waveforms at various points on the block diagram if the input data sequence is 01011000101.

(c) Assume that the receiver consists of a superheterodyne circuit. The detector that is used is shown in Fig. P5-7, where $T = 1/R$. If the DPSK IF signal $v_1(t)$ has a peak value of A_c volts, determine the appropriate value for the threshold voltage setting V_T.

(d) Sketch the waveforms that appear at various points of this detector circuit for the data sequence in part (b).

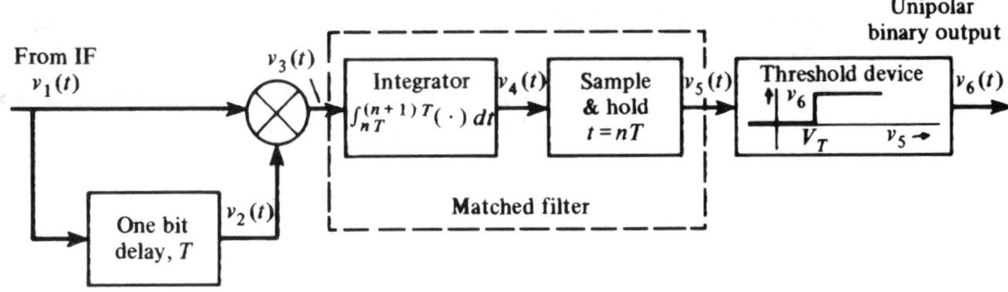

FIGURE P5-7

5-8 A binary baseband signal is passed through a raised cosine-rolloff filter with a 50% rolloff factor and is then modulated onto a carrier. The data rate is 2400 bits/sec. Evaluate:

(a) The absolute bandwidth of a resulting OOK signal.

(b) The approximate bandwidth of a resulting FSK signal when the mark frequency is 50 kHz and the space frequency is 55 kHz.

(*Note:* It is interesting to compare these bandwidths with those obtained in Probs. 5-1 and 5-4.)

5-9 Evaluate the exact magnitude spectrum of the FSK signal that is emitted by a Bell-type 103 modem operating in the answer mode at 300 bits/sec. Assume that the data are alternating 1's and 0's.

5-10 Repeat Prob. 5-9 if the data have the periodic pattern 111000111000.

5-11 Starting with (5-17), work out all of the mathematical steps to derive the result given in (5-18).

5-12 When FSK signals are detected using coherent detection as shown in Fig. 5-10b, it is assumed that $\cos \omega_1 t$ and $\cos \omega_2 t$ are orthogonal functions. This is approximately true if $f_1 - f_2 = 2\,\Delta F$ is sufficiently large. Find the *exact* condition required for the mark and space FSK signals to be orthogonal. (*Hint:* The answer relates f_1, f_2, and R.)

5-13 Show that the approximate transmission bandwidth for FSK is given by $B_T = 2R(1 + h/2)$, where h is the digital modulation index and R is the bit rate.

5-14 Assume that a QPSK modem is used to send data at a rate of 3.6 kbits/sec over a twisted pair telephone line. The line has a bandwidth of 2.4 kHz.

(a) If the line is equalized to have an equivalent raised cosine filter characteristic, what is the rolloff factor, r, required?

(b) Could a rolloff factor, r, be found so that a 4.8-kbit/sec data rate could be supported?

5-15 Show that two BPSK systems can operate simultaneously over the same channel by using quadrature (sine and cosine) carriers. Give a block diagram for the transmitters and receivers. What is the overall aggregate data rate, R, for this quadrature carrier multiplex system as a function of the channel null bandwidth, B_T? How does the aggregate data rate of this system compare to the data rate for a system that first time-division-multiplexes the two sources and transmits the TDM data via a QPSK carrier?

5-16 Assume that a telephone line channel is equalized to allow bandpass data transmission over a frequency range of 600 to 3000 Hz with raised cosine-rolloff filtering. The available channel bandwidth is 2400 Hz with a mid-channel frequency of 1800 Hz.

(a) Design a 2400-bit/sec QPSK transmission system with $f_c = 1800$ Hz. Show that the spectrum of this signal will fit into the channel when $r = 1$. Find the absolute and 6-dB bandwidths.

(b) Repeat part (a) for the case of 4800-bit/sec eight-phase PSK signaling.

5-17 Assume that a telephone line channel is equalized to allow bandpass data transmission over a frequency range of 400 to 3100 Hz so that the available channel bandwidth is 2700 Hz and the midchannel frequency is 1750 Hz. Design a 16-symbol QAM signaling scheme that will allow a data rate of 9600 bits/sec to be transferred over the channel. In your design, choose an appropriate rolloff factor r and indicate the absolute and 6-dB QAM signal bandwidth. Discuss why you selected the particular value of r that you used.

5-18 Assume that $R = 9600$ bits/sec. Calculate the *second* null-to-null bandwidth of BPSK, QPSK, MSK, 64PSK, and 64QAM. Discuss the advantages and disadvantages of using each of these signaling methods.

5-19 Referring to Fig. 5-13b, sketch the waveforms that appear at the output of each block assuming that the input is a TTL level signal with data 110100101 and $\ell = 4$. Explain how this QAM transmitter works.

5-20 Design a receiver (i.e., determine the block diagram) that will detect the data on a QAM waveform having an $M = 16$ point signal constellation as shown in Fig. 5-14. Explain how your receiver works. (*Hint:* Study Fig. 5-13b.)

5-21 Using MathCAD, plot the QPSK and OQPSK in-phase and quadrature modulation waveforms for the data stream

$$\{-1, -1, -1, +1, +1, +1, -1, -1, -1, +1, -1, -1, +1, -1, -1\}$$

Use a rectangular pulse shape. For convenience let $T_b = 1$.

5-22 Calculate and plot the PSD for a CCITT V.22bis 2400-bit/sec computer modem operating in the originate mode and sending random data. See Table C-12 for the modem specifications.

5-23 **(a)** Fig. 5-16 shows the $x(t)$ and $y(t)$ waveforms for Type II MSK. Redraw these waveforms for the case of Type I MSK.

(b) Show that (5-47b) is the Fourier transform of (5-47a).

5-24 Using MathCAD, plot the MSK Type I modulation waveforms $x(t)$, $y(t)$, and the MSK signal $s(t)$. Assume that the input data stream is

$$\{+1, -1, -1, +1, -1, -1, +1, -1, -1, +1, +1, -1, +1, +1, -1\}$$

Assume that $T_b = 1$ and that f_c has a value such that a good plot of $s(t)$ is obtained for a reasonable amount of computer time.

5-25 Repeat Prob. 5-24 but use the data stream

$$\{-1, -1, -1, +1, +1, +1, -1, -1, -1, +1, -1, -1, +1, -1, -1\}$$

This data stream is the differentially encoded version of the data stream in Prob. 5-24. The generation of FFSK is equivalent to Type I MSK with differential encoding of the data input. FFSK has a one-to-one relationship between the input data and the mark and space frequencies.

5-26 Using MathCAD, plot the MSK Type II modulation waveforms $x(t)$, $y(t)$, and the MSK signal $s(t)$. Assume that the input data stream is

$$\{-1, -1, -1, +1, +1, +1, -1, -1, -1, +1, -1, -1, +1, -1, -1\}$$

Assume that $T_b = 1$ and f_c has a value such that a good plot of $s(t)$ is obtained for a reasonable amount of computer time.

5-27 Show that MSK can be generated by the serial method of Fig. 5-18c. That is, show that the PSD for the signal at the output of the MSK bandpass filter is the MSK spectrum as described by (5-48) and (5-1).

5-28 Recompute the spectral efficiencies for all of the signals shown in Table 5-5 using a 40-dB bandwidth criterion.

5-29 The information in four analog signals is to be multiplexed and transmitted over a telephone channel that has a 400- to 3100-Hz bandpass. Each of the analog baseband signals is bandlimited to 500 Hz. Design a communication system (block diagram) that will allow the transmission of these four sources over the telephone channel using:
(a) FDM with SSB subcarriers.
(b) TDM using PAM with appropriate frequency translation and raised cosine-rolloff filtering.
Show the block diagrams of the complete system, including the transmission, channel, and reception portions. Include the bandwidths of the signals at the various points in the systems.

5-30 Give a more detailed block diagram of a group frequency-division multiplexer than that given on the left side of Fig. 5-19. Show a possible implementation of the SSB subcarrier generators and illustrate the spectra of each of the 12 SSB signals and the spectrum of the group FDM signal.

5-31 Assume that a twisted-pair telephone line acts as a transmission line for a VF signal. The line has a characteristic impedance of 500 Ω and it is terminated with a 500-Ω load. The line attenuation is 0.25 dB/1000 ft and the line is 150 miles long.

(a) If a 1-V (rms) VF signal is applied at the input and there are no amplifiers along the line, find the voltage (rms) that would appear across the load.

(b) Assume that a 1-V (rms) level is needed at the load and at the output of each amplifier when a 1-V (rms) signal is applied at the line input. Find the number of amplifiers needed and the amplifier spacing if each amplifier has a gain of 25 dB.

(c) In part (b), if each amplifier has 0.5% THD, estimate the worst-case THD for load signal when the input signal has no distortion.

5-32 A satellite with twelve 36-MHz bandwidth transponders operates in the 6/4-GHz bands with 500 MHz bandwidth and 4-MHz guardbands on the 4-GHz downlink, as shown in Fig. 5-27. Calculate the percentage bandwidth that is used for the guardbands.

5-33 An Earth station uses a 3-m-diameter parabolic antenna to receive a 4-GHz signal from a geosynchronous satellite. If the satellite transmitter delivers 10 W into a 3-m-diameter transmitting antenna and the satellite is located 36,000 km from the receiver, what is the power received?

5-34 Figure 5-30b shows an FDM/FM ground station for a satellite communication system. Find the peak frequency deviation needed to achieve the allocated spectral bandwidth for the 6240-MHz signal.

5-35 A microwave transmitter has an output of 0.1 W at 2 GHz. Assume that this transmitter is used in a microwave communication system where the transmitting and receiving antennas are parabolas, each 4 ft in diameter.

(a) Evaluate the gain of the antennas.

(b) Calculate the EIRP of the transmitted signal.

(c) If the receiving antenna is located 15 miles from the transmitting antenna over a free-space path, find the available signal power out of the receiving antenna in dBm units.

5-36 Using MathCAD, plot the PSD for a thermal noise source with a resistance of 10 kΩ over a frequency range of 10 to 100,000 GHz where $T = 300$ K.

5-37 Given the *RC* circuit of Fig. P5-37, where *R* is a physical resistor at temperature *T*, find the rms value of the noise voltage that appears at the output in terms of *k*, *T*, *R*, and *C*.

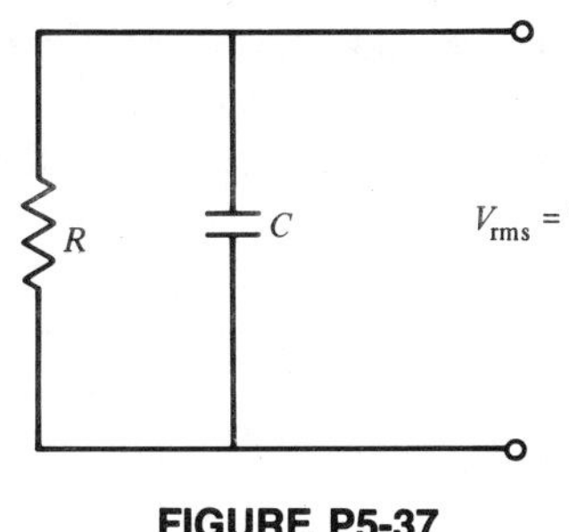

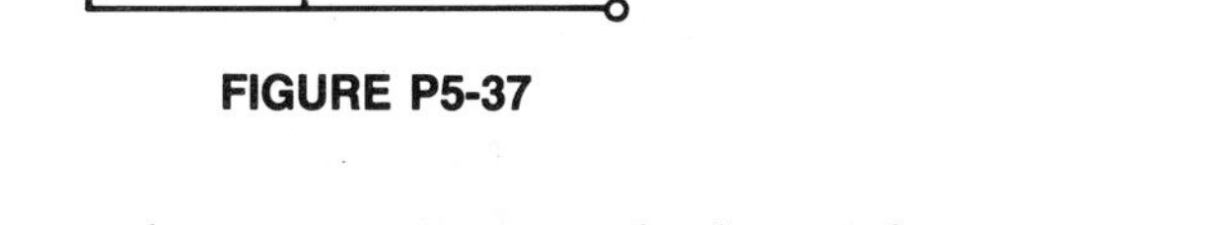

FIGURE P5-37

5-38 A receiver is connected to an antenna system that has a noise temperature of 100 K. Find the noise power that is available from the source over a 20-MHz band.

5-39 A bipolar transistor amplifier is modeled as shown in Fig. P5-39. Find a formula for the available power gain in terms of the appropriate parameters.

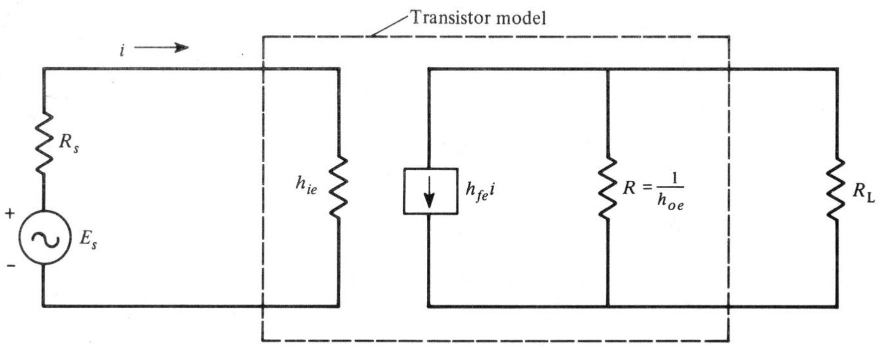

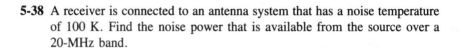

FIGURE P5-39

5-40 Using the definition for the available power gain, $G_a(f)$, as given by (5-65), show that $G_a(f)$ depends on the driving source impedance as well as the elements of the device and that $G_a(f)$ does not depend on the load impedance. [*Hint:* Calculate $G_a(f)$ for a simple resistive network.]

5-41 Show that the effective input-noise temperature and the noise figure can be evaluated from measurements that use the Y-factor method. With this method the device under test (DUT) is first connected to a noise source that has a relatively large output denoted by its source temperature, T_h, where the subscript h denotes "hot," and then the available noise power at the output of the DUT, P_{aoh} is measured with a power meter. Next, the DUT is connected to a source that has relatively low source temperature, T_c, where the subscript c denotes "cold," and noise power at the output of the DUT is measured, P_{aoc}. Show that

(a) The effective input noise temperature of the DUT is

$$T_e = \frac{T_h - YT_c}{Y - 1}$$

where $Y = P_{aoh}/P_{aoc}$ is obtained from the measurements.

(b) The noise figure of the DUT is

$$F = \frac{((T_h/T_0) - 1) - Y((T_c/T_0) - 1)}{Y - 1}$$

where $T_0 = 290$ K.

5-42 If a signal plus noise is fed into a linear device, show that the noise figure of that device is given by $F = (S/N)_{in}/(S/N)_{out}$. (*Hint:* Start with the basic definition of noise figure that is given in this chapter.)

5-43 An antenna is pointed in a direction such that it has a noise temperature of 30 K. It is connected to a preamplifier that has a noise figure of 1.6 dB and an available gain of 30 dB over an effective bandwidth of 10 MHz.

(a) Find the effective input noise temperature for the preamplifier.

(b) Find the available noise power out of the preamplifier.

5-44 A 10-MHz SSB-AM signal, which is modulated by an audio signal that is bandlimited to 5 kHz, is being detected by a receiver that has a noise figure of 10 dB. The signal power at the receiver input is 10^{-10} mW and the PSD of the input noise, $\mathcal{P}(f) = kT/2$, is 2×10^{-21}. Evaluate:

(a) The IF bandwidth needed.

(b) The signal-to-noise ratio at the receiver input.

(c) The signal-to-noise ratio at the receiver output, assuming that a product detector is used.

5-45 A phonograph cartridge that is used in a turntable to play phonograph records will produce 5 mV rms into a 47-kΩ matched load when a typical record is played. An audio preamplifier is needed to bring the audio signal up to a 1-V level with a 65-dB signal-to-noise ratio measured over the 20-kHz audio band. Determine the preamp gain and noise figure specifications that need to be met.

5-46 Prove that the overall effective input-noise temperature for cascaded linear devices is given by (5-84).

5-47 A TV set is connected to an antenna system as shown in Fig. P5-47. Evaluate:

(a) The overall noise figure.

(b) The overall noise figure if a 20-dB RF preamp with a 3-dB noise figure is inserted at point B.

(c) The overall noise figure if the preamp is inserted at point A.

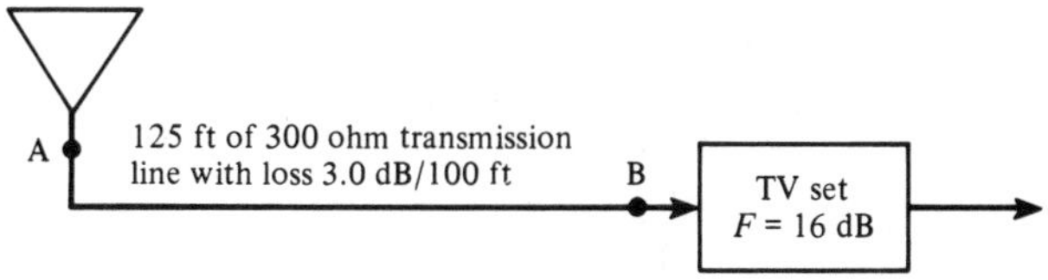

A — 125 ft of 300 ohm transmission line with loss 3.0 dB/100 ft — B → TV set $F = 16$ dB →

FIGURE P5-47

5-48 An Earth station receiving system consists of a 20-dB gain antenna with $T_{AR} = 80$ K, an RF amplifier with $G_a = 40$ dB and $T_e = 200$ K, and a down converter with $T_e = 15{,}000$ K. What is the overall effective input-noise temperature of this receiving system?

5-49 A low-noise amplifier (LNA), a down converter, and an IF amplifier are connected in cascade. The LNA has a gain of 40 dB and a T_e of 100°. The down converter has a noise figure of 8 dB and a conversion gain of 6 dB. The IF amplifier has a gain of 60 dB and a noise figure of 14 dB. Evaluate the overall effective input-noise temperature for this system.

5-50 A geosynchronous satellite transmits 13.5 dBw EIRP on a 4-GHz downlink to an Earth station. The receiving system has a gain of 60 dB, an effective input-noise temperature of 80 K, an antenna source noise temperature of 60 K, and an IF bandwidth of 36 MHz. If the satellite is located 24,500 miles from the receiver, what is the $(C/N)_{dB}$ at the input of the receiver detector circuit?

5-51 An antenna with $T_{AR} = 160$ K is connected to a receiver by means of a waveguide that has a physical temperature of 290 K and a loss of 2 dB. The receiver has a noise bandwidth of 1 MHz, an effective input-noise temperature of 800 K, and a gain of 120 dB from its antenna input to its IF output. Using MathCAD, find:

(a) The system noise temperature at the receiver input.
(b) The overall noise figure.
(c) The available noise power at the IF output.

5-52 The *Intelsat IV* satellite uses 36-MHz transponders with downlinks operating in the 4-GHz band. Each satellite transponder has a power output of 3.5 W and may be used with a 17° global coverage antenna that has a gain of 20 dB.

(a) For users located at the subsatellite point (i.e., directly below the satellite), show that $(C/N)_{dB} = (G_{AR}/T_{syst})_{dB} - 17.1$, where $(G_{AR}/T_{syst})_{dB}$ is the Earth receiving station figure of merit.
(b) Design a ground receiving station (block diagram) showing reasonable specifications for each block so that the IF output (C/N) will be 12 dB. Discuss the trade-offs that are involved.

5-53 The efficiency of a parabolic antenna is determined by the accuracy of the parabolic reflecting surface and other factors. The gain is $G_A = 4\pi\eta A/\lambda^2$, where η is the antenna efficiency. In Table 5-8 an efficiency of 56% was used to obtain the formula $G_A = 7A/\lambda^2$. In a ground receiving system for the *Intelsat IV* satellite, assume that a (G_{AR}/T_{syst}) of 40 dB is needed. Using a 30-m antenna, give the required antenna efficiency if the system noise temperature is 85 K. How does the required antenna efficiency change if a 25-m antenna is used?

5-54 Evaluate the performance of a TVRO system. Assume that the TVRO terminal is located in Miami, Florida (26.8° N latitude, 80.2° W longitude). A 10-ft-diameter parabolic receiving antenna is used and is pointed toward the *Galaxy I* satellite. Other TVRO parameters are given in Table 5-9, Fig. 5-43, and Fig. 5-45.

(a) Find the TVRO antenna look angles to the satellite and find the slant range.
(b) Find the overall receiver system temperature.
(c) Find the $(C/N)_{dB}$ into the receiver detector.
(d) Find the $(S/N)_{dB}$ out of the receiver.

5-55 Repeat Prob. 5-54 if the TVRO terminal is located at Anchorage, Alaska (61.2° N latitude, 149.8° W longitude), and a 5-m-diameter parabolic antenna is used.

5-56 A TVRO receive system consists of an 8-ft-diameter parabolic antenna that feeds a 50-dB, 85 K LNA. The sky noise temperature (with feed) is 32 K. The system is designed to receive signals from the *Galaxy I* satellite. The LNA has a post-mixer circuit that down-converts the satellite signal to 70 MHz. The 70-MHz signal is fed to the TVRO receiver via a 120-ft length of 75-Ω coaxial cable. The cable has a loss of 3 dB/100 ft. The receiver has a bandwidth of 36 MHz and a noise temperature of 3800 K. Assume that the TVRO site is located in Los Angeles, California (34° N latitude, 118.3° W longitude) and vertical polarization is of interest.

(a) Find the TVRO antenna look angles to the satellite and find the slant range.
(b) Find the overall system temperature.
(c) Find the $(C/N)_{dB}$ into the receiver detector.

5-57 An Earth station receiving system operates at 11.95 GHz and consists of a 20-m antenna with a gain of 65.53 dB and $T_{AR} = 60$ K, a waveguide with a loss of 1.28 dB and a physical temperature of 290 K, a LNA with $T_e =$ 150 K and 60 dB gain, and a downconverter with $T_e = 11{,}000$ K. Using MathCAD, find $(G/T)_{dB}$ for the receiving system evaluated at:

(a) The input to the LNA.
(b) The input to the waveguide.
(c) The input to the downconverter.

5-58 Rework Example 5-5 for the case of reception of TV signals from a direct broadcast satellite (DBS). Assume that the system parameters are similar to those given in Table 5-9, except that the satellite power is 316 kW EIRP radiated in the 12-GHz band. Furthermore, assume that a 1-m parabolic receiving antenna is used and the LNA has a 240-K noise temperature.

5-59 The most distant planet from our sun is Pluto, which is located at a (greatest) distance from the Earth of 7.5×10^9 km. Assume that an unmanned spacecraft with a 2-GHz, 10-W transponder is in the vicinity of Pluto. A receiving Earth station with a 64-m antenna is available that has a system noise temperature of 16 K at 2 GHz. Calculate the size of a spacecraft antenna that is required for a 300-bit/sec BPSK data link to Earth that has a 10^{-3} bit error rate (corresponding to E_b/N_0 of 9.88 dB). Allow 2 dB for incidental losses.

5-60 Using Lotus 1-2-3, SuperCalc, or some other spreadsheet program that will run on a personal computer (PC), design a spreadsheet that will solve Prob. 5-56. Run the spreadsheet on your PC and verify that it gives the correct answers. Print out your results. Also try other parameters, such as those appropriate for your location, and print out the results.

5-61 Assume that you want to analyze the overall performance of a satellite relay system that uses a "bent-pipe" transponder. Let $(C/N)_{\text{up}}$ denote the carrier-to-noise ratio for the transponder as evaluated in the IF band of the satellite transponder. Let $(C/N)_{\text{dn}}$ denote the IF carrier-to-noise ratio at the downlink receiving ground station when the satellite is sending a perfect (noise-free) signal. Show that the overall operating carrier-to-noise ratio at the IF of the receiving ground station, $(C/N)_{\text{ov}}$, is given by

$$\frac{1}{(C/N)_{\text{ov}}} = \frac{1}{(C/N)_{\text{up}}} + \frac{1}{(C/N)_{\text{dn}}}$$

5-62 Prove that (5-104b)—the autocorrelation for an m-sequence PN code—is correct. *Hint:* Use the definition for the autocorrelation function, $R_c(\tau) = \langle c(t)c(t+\tau)\rangle$ and (5-104a) where

$$c(t) = \sum_{-\infty}^{\infty} c_n p(t - nT_c)$$

and

$$p(t) = \begin{cases} 1, & 0 < t < T_c \\ 0, & t \text{ elsewhere} \end{cases}$$

5-63 Find an expression for the power spectral density (PSD) of an m-sequence PN code when the chip rate is 10 MHz and there are eight stages in the shift register. Sketch your result.

5-64 Referring to Fig. 4-47a, show that the complex Fourier series coefficients for the autocorrelation of an m-sequence PN waveform are given by (5-104g).

5-65 Refer to the BPSK–DS–SS system as described in Figs. 5-46 and 5-48. Find an expression for the processing gain of the receiver in terms of R_b

and R_c for the case of wideband jamming. Assume that the PSD of the jammer is given by

$$\mathcal{P}_{nJ}(f) = \begin{cases} N_o/2, & |f - f_c| < R_c \\ 0, & f \text{ otherwise} \end{cases}$$

and assume that $R_c \gg R_b$. [*Hint:* Use (6-129) and (6-133l).]

5-66 Assume that the modulator and demodulator for the FH–SS system of Fig. 5-49 are of the FSK type.

(a) Find a mathematical expression for the FSK–FH–SS signal, $s(t)$, at the transmitter output.

(b) Using your result for $s(t)$ from part (a) as the receiver input in Fig. 5-49b [i.e., $r(t) = s(t)$], show that the output of the receiver bandpass filter is an FSK signal.

5-67 Referring to Table 5-18, evaluate the maximum horizontal and vertical resolution (lines) for the CCIR System B TV standard. Assume that the horizontal blanking time is 10 μsec.

5-68 A VSB signal may be analyzed by using

$$g(t) = [m_1(t) + m_2(t)] + j[\hat{m}_2(t)]$$

as a model for the complex envelope. Explain how this model may be used to represent U.S. TV signals.

5-69 A low-power TV station is licensed to operate on TV Channel 35 with an effective radiated visual power of 1000 W. The tower height is 400 ft and the transmission line is 450 ft long. Assume that a $1\frac{5}{8}$-in.-diameter semi-rigid 50-Ω coaxial transmission line will be used that has a loss of 0.6 dB/100 ft (at the operating frequency). The antenna has a gain of 5.6 dB and the duplexer loss is negligible. Find the PEP required at the visual transmitter output.

5-70 A TV visual transmitter is tested by modulating it with a periodic test signal. The envelope of the transmitter output is viewed on an oscilloscope as shown in Fig. P5-70, where K is an unknown constant. The output of the transmitter is connected to a 50-Ω dummy load that has a calibrated average reading wattmeter. The wattmeter reads 6.9 kW. Find the PEP of the transmitter output.

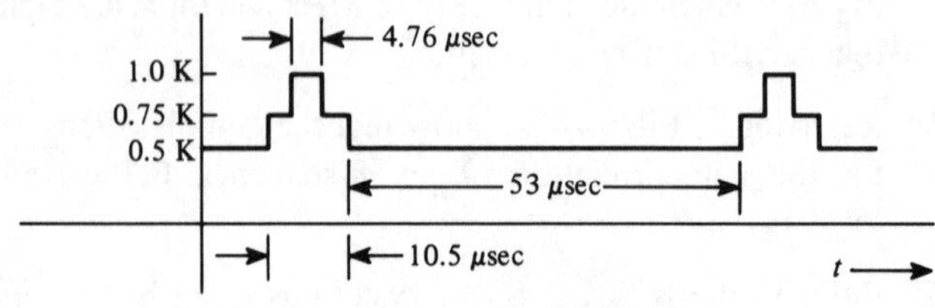

FIGURE P5-70

5-71 Specifications for the MTS stereo audio system for TV are given in Fig. 5-54. Using this figure:

(a) Design a block diagram for TV receiver circuitry that will detect the stereo audio signals and the second audio program (SAP) signal.

(b) Describe how the circuit works. That is, give mathematical expressions for the signal at each point on your block diagram and explain in words. Be sure to specify all filter transfer functions, VCO center frequencies, and so on.

5-72 A TV broadcaster wants to send digital data to subscribers concerning the stock market using the professional aural subcarrier. Propose two digital signaling techniques that will satisfy the spectral requirements of Fig. 5-54 where the available bandwidth is $0.5f_h$. Specifically, design and draw block diagrams of transmit and receive modems and give the specifications for transmitting data at a rate of **(a)** 1200 bits/sec and **(b)** 9600 bits/sec. Discuss why you think you have a good design.

5-73 In the R-Y, B-Y color TV subcarrier demodulation system, the G-Y signal is obtained from the R-Y and B-Y signals. That is,

$$[m_G(t) - m_Y(t)] = K_1[m_R(t) - m_Y(t)] + K_2[m_B(t) - m_Y(t)]$$

(a) Find the values for K_1 and K_2 that are required.

(b) Draw a possible block diagram for a R-Y, B-Y system and explain how it works.

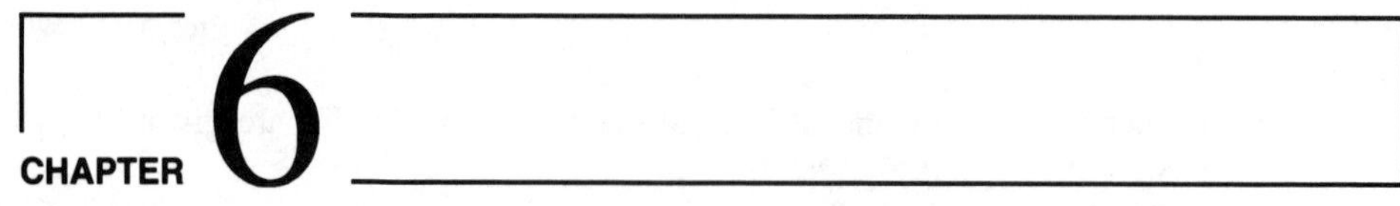

CHAPTER 6

Random Processes and Spectral Analysis

The mathematics used to describe random signals and noise will be developed in this chapter. From Chapter 1 it is recalled that random signals (in contrast to deterministic signals) are used to convey information. Noise is also described in terms of statistics. Thus the description of random signals and noise is fundamental to the understanding of communication systems.

This chapter is written with the assumption that the reader has a basic understanding of probability, random variables, and ensemble averages as developed in Appendix B. If the reader is not familiar with these ideas, Appendix B should be studied now just like a regular chapter. For the reader who has had courses in this area, Appendix B can be used as a quick review.

Random processes are extensions of the concepts associated with random variables when the time parameter is brought into the problem. As we will see, this enables frequency response to be incorporated into the statistical description.

6-1 Some Basic Definitions

Random Processes

DEFINITION: A *real random process* (or *stochastic process*) is an indexed set of real functions of some parameter (usually time) that has certain statistical properties.

Consider voltage waveforms that might be emitted from a noise source (see Fig. 6-1). One possible waveform is $v(t,E_1)$. Another is $v(t,E_2)$. In general, $v(t,E_i)$ denotes the waveform that is obtained when the event E_i of the sample space occurs. $v(t,E_i)$ is said to be a *sample* function of the sample space. The set of all possible sample functions $\{v(t,E_i)\}$ is called the *ensemble* and defines the *random process* $v(t)$ that describes the noise source. That is, the events $\{E_i\}$ are mapped into a set of time functions $\{v(t,E_i)\}$. This collection of functions is the random process $v(t)$. When we observe the voltage waveform generated by the noise source, we see one of the sample functions.

Sample functions can be obtained by observing simultaneously the outputs of a large number of identical noise sources. To obtain all of the sample functions in general, an infinite number of noise sources would be required.

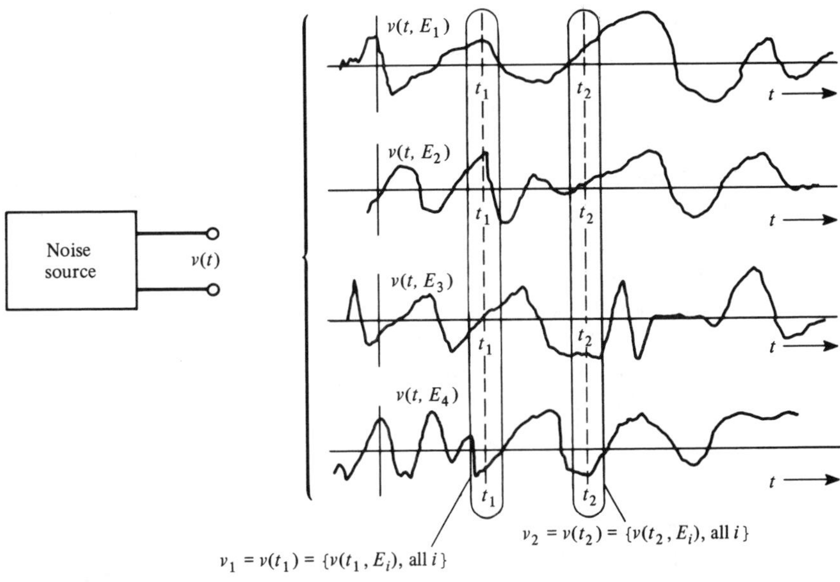

FIGURE 6-1 Random noise source and some sample functions of the random noise process.

The definition of a random process may be compared with that of a random variable. A random variable maps events into *constants,* whereas a random process maps events into *functions* of the parameter t.

THEOREM: *A **random process** may be described by an indexed set of random variables.*

Referring to Fig. 6-1, define a set of random variables $v_1 = v(t_1)$, $v_2 = v(t_2)$, . . . , where $v(t)$ is the random process. Here the random variable $v_j = v(t_j)$ takes on the values described by the set of constants $\{v(t_j,E_i), \text{ all } i\}$.

For example, suppose that the noise source has a Gaussian distribution. Then any of the random variables will be described by

$$f_{v_j}(v_j) = \frac{1}{\sqrt{2\pi}\sigma_j} e^{-(v_j - m_j)^2/(2\sigma_j^2)} \tag{6-1}$$

where $v_j \triangleq v(t_j)$. We realize that, in general, the probability density function (PDF) depends implicitly on time since m_j and σ_j correspond to the mean value and standard deviation measured at the time $t = t_j$. The $N = 2$ joint distribution for the Gaussian source for $t = t_1$ and $t = t_2$ is the bivariate Gaussian PDF $f_v(v_1,v_2)$ as given by (B-97), where $v_1 = v(t_1)$ and $v_2 = v(t_2)$.

To describe a general random process $x(t)$ completely, an N-dimensional PDF, $f_{\mathbf{x}}(\mathbf{x})$, is required where $\mathbf{x} = (x_i, x_2, \ldots, x_j, \ldots, x_N)$, $x_j \triangleq x(t_j)$, and

$N \to \infty$. Furthermore, the N-dimensional PDF is an implicit function of N time constants $t_1, t_2, \ldots, t_N$ since

$$f_{\mathbf{x}}(\mathbf{x}) = f_{\mathbf{x}}(x(t_1), x(t_2), \ldots, x(t_N)) \tag{6-2}$$

Random processes may be classified as being continuous or discrete. A *continuous random process* consists of a random process with associated continuously distributed random variables, $v_j = v(t_j)$. The Gaussian random process (described previously) is an example of a continuous random process. Noise in linear communication circuits is usually of the continuous type. (In many cases, noise in nonlinear circuits is also of the continuous type.) A *discrete random process* consists of random variables with discrete distributions. For example, the output of an ideal (hard) limiter is a binary (discrete with two levels) random process. Some sample functions of a binary random process are illustrated in Fig. 6-2.

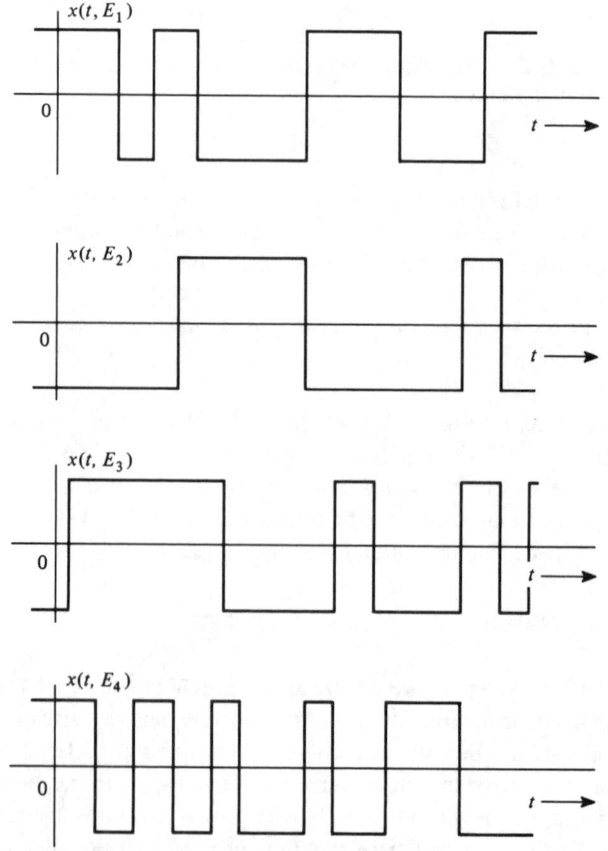

FIGURE 6-2 Sample functions of a discrete random process.

Stationarity and Ergodicity

DEFINITION: A random process $x(t)$ is said to be *stationary to the order N* if, for any $t_1, t_2, \ldots, t_N$,

$$f_{\mathbf{x}}(x(t_1), x(t_2), \ldots, x(t_N)) = f_{\mathbf{x}}(x(t_1 + t_0), x(t_2 + t_0), \ldots, x(t_N + t_0)) \quad (6\text{-}3)$$

where t_0 is any arbitrary real constant. Furthermore, the process is said to be *strictly stationary* if it is stationary to the order $N \to \infty$.

This definition implies that if a stationary process of order N is translated in time, then the Nth-order statistics do not change. Furthermore, the N-dimensional PDF depends on $N - 1$ time differences $t_2 - t_1, t_3 - t_1, \ldots, t_N - t_1$ since t_0 could be chosen to be $-t_1$.

EXAMPLE 6-1 FIRST-ORDER STATIONARITY

Examine a random process $x(t)$ to determine if it is first-order stationary. From (6-3), the requirement for first-order stationarity is that the first-order PDF not be a function of time. Let the random process be

$$x(t) = A \sin(\omega_0 t + \theta_0) \quad (6\text{-}4)$$

Case 1: A Stationary Result. First, assume that A and ω_0 are deterministic constants and θ_0 is a random variable. t is the time parameter. In addition, assume that θ_0 is uniformly distributed over $-\pi$ to π. Then $\psi \triangleq \theta_0 + \omega_0 t$ is a random variable uniformly distributed over the interval $\omega_0 t - \pi < \psi < \omega_0 t + \pi$. The first-order PDF for $x(t)$ can be obtained by the transformation technique developed in Sec. B-8 of Appendix B. This is essentially the same problem as the one worked out in Example B-5. From (B-71), the first-order PDF for $x(t)$ is

$$f(x) = \begin{cases} \dfrac{1}{\pi\sqrt{A^2 - x^2}}, & |x| \le A \\ 0, & x \text{ elsewhere} \end{cases} \quad (6\text{-}5a)$$

Since this PDF is not a function of t, $x(t)$is a first-order stationary process for the assumptions of Case 1 where θ_0 is a random variable. This result would be applicable to problems where θ_0 is the random start-up phase of an unsynchronized oscillator.

Case 2: A Nonstationary Result. Second, assume that A, ω_0, and θ_0 are deterministic constants. Then, at any time, the value of $x(t)$ is known with a probability of 1. Thus the first-order PDF of $x(t)$ is

$$f(x) = \delta(x - A \sin(\omega_0 t + \theta_0)) \quad (6\text{-}5b)$$

This PDF is a function of t; consequently, $x(t)$ is not first-order stationary for the assumption of Case 2, where θ_0 is a deterministic constant. This result will be applicable to problems where the oscillator is synchronized to some external source so that the oscillator start-up phase will have the known value θ_0.

DEFINITION: A random process is said to be *ergodic* if all time averages of any sample function are equal to the corresponding ensemble averages (expectations).

Two important averages in electrical engineering are dc and rms values. These values are defined in terms of time averages, but if the process is ergodic, they may be evaluated by the use of ensemble averages. The dc value of $x(t)$ is $x_{dc} \triangleq \langle x(t) \rangle$. When $x(t)$ is ergodic, the time average is equal to the ensemble average, so we obtain

$$x_{dc} \triangleq \langle x(t) \rangle \equiv \overline{[x(t)]} = m_x \tag{6-6a}$$

where the time average is

$$\langle [x(t)] \rangle = \lim_{T \to \infty} \frac{1}{T} \int_{-T/2}^{T/2} [x(t)] \, dt \tag{6-6b}$$

the ensemble average is

$$\overline{[x(t)]} = \int_{-\infty}^{\infty} [x] f_x(x) \, dx = m_x \tag{6-6c}$$

and m_x denotes the mean value. Similarly, the rms value is obtained:

$$X_{\text{rms}} \triangleq \sqrt{\langle x^2(t) \rangle} \equiv \sqrt{\overline{x^2}} = \sqrt{\sigma_x^2 + m_x^2} \tag{6-7}$$

where σ_x^2 is the variance of $x(t)$.

In summary, if a process is ergodic, all time and ensemble averages are interchangeable. The time average cannot be a function of time since the time parameter has been averaged out. Furthermore, the ergodic process must be stationary, since otherwise the ensemble averages (such as moments) would be a function of time. However, not all stationary processes are ergodic.

EXAMPLE 6-2 ERGODIC RANDOM PROCESS

Let the random process be given by

$$x(t) = A \cos(\omega_0 t + \theta) \tag{6-8}$$

where A and ω_0 are constants and θ is a random variable that is uniformly distributed over $(0, 2\pi)$.

First, evaluate some ensemble averages. The mean and the second moment are

$$\overline{x} = \int_{-\infty}^{\infty} [x(\theta)] f_\theta(\theta) \, d\theta = \int_0^{2\pi} [A \cos(\omega_0 t + \theta)] \frac{1}{2\pi} \, d\theta = 0 \tag{6-9}$$

and

$$\overline{x^2(t)} = \int_0^{2\pi} [A \cos(\omega_0 t + \theta)]^2 \frac{1}{2\pi}\, d\theta = \frac{A^2}{2} \tag{6-10}$$

In this example the time parameter t disappeared when the ensemble averages were evaluated. This would not be the case unless $x(t)$ was stationary.

Second, evaluate the corresponding time averages using a typical sample function of the random process. One sample function is $x(t,E_1) = A \cos \omega_0 t$, which occurs when $\theta = 0$. $\theta = 0$ corresponds to one of the events (outcomes). The time average for any of the sample functions can be evaluated by letting θ be the appropriate value between 0 and 2π. The time averages for the first and second moments are

$$\langle x(t) \rangle = \frac{1}{T_0} \int_0^{T_0} A \cos(\omega_0 t + \theta)\, dt = 0 \tag{6-11}$$

and

$$\langle x^2(t) \rangle = \frac{1}{T_0} \int_0^{T_0} [A \cos(\omega_0 t + \theta)]^2\, dt = \frac{A^2}{2} \tag{6-12}$$

where $T_0 = 1/f_0$ and the time-averaging operator for a periodic function, (2-4), has been used. In this example θ disappears when the time average is evaluated. This is a consequence of $x(t)$ being an ergodic process. However, in a nonergodic example, the time average would be a random variable.

Comparing (6-9) with (6-11) and (6-10) with (6-12), we see that the time average is equal to the ensemble average for the first and second moments. Consequently, we suspect that this process might be ergodic. However, we have not proven that the process is ergodic because we have not evaluated all the possible time and ensemble averages or all the moments. However, it appears as if the other time and ensemble averages would be equal, so we will *assume* that the process is ergodic. In general, it is difficult to prove that a process is ergodic, so we will assume that this is the case if the process appears to be stationary and some of the time averages are equal to the corresponding ensemble averages. An ergodic process has to be stationary since the time averages cannot be functions of time. However, if a process is known to be stationary, it may or may not be ergodic.

In Prob. 6-2 it will be shown that the random process as described by (6-8) would not be stationary (and, consequently, would not be ergodic) if θ were uniformly distributed over $(0,\pi/2)$ instead of $(0,2\pi)$.

Correlation Functions and Wide-Sense Stationarity

DEFINITION: The *autocorrelation function* of a real process $x(t)$ is

$$R_x(t_1,t_2) = \overline{x(t_1)x(t_2)} = \int_{-\infty}^{\infty}\int_{-\infty}^{\infty} x_1 x_2 f_x(x_1,x_2)\, dx_1\, dx_2 \tag{6-13}$$

where $x_1 = x(t_1)$ and $x_2 = x(t_2)$. If the process is stationary to the second order, the autocorrelation function is only a function of the time difference $\tau = t_2 - t_1$. That is,

$$R_x(\tau) = \overline{x(t)x(t + \tau)} \tag{6-14}$$

if $x(t)$ is second-order stationary.[†]

DEFINITION: A random process is said to be *wide-sense stationary* if

$$1.\ \overline{x(t)} = \text{constant} \tag{6-15a}$$

$$2.\ R_x(t_1,t_2) = R_x(\tau) \tag{6-15b}$$

where $\tau = t_2 - t_1$.

A process that is stationary to order 2 or greater is certainly wide-sense stationary. However, the converse is not necessarily true since only *certain* ensemble averages of second order or less, namely, those of (6-15), need to be satisfied for wide-sense stationarity.[‡] As indicated by (6-15), the mean and autocorrelation functions of a wide-sense stationary process do not change with a shift in the time origin. This implies that the associated circuit elements that generate the wide-sense stationary random process do not drift (age) with time.

The autocorrelation function gives us an idea of the frequency response that is associated with the random process. For example, if $R_x(\tau)$ remains relatively constant as τ is increased from zero to some positive number, it indicates that, on the average, sample values of x taken at $t = t_1$ and $t = t_1 + \tau$ are nearly the same. Consequently, $x(t)$ does not change rapidly with time (on the average) and we would expect $x(t)$ to have low frequencies present. On the other hand, if $R_x(\tau)$ decreases rapidly as τ is increased, we would expect $x(t)$ to change rapidly with time and, consequently, high frequencies would be present. We formulate this result rigorously in Sec. 6-2, where it will be shown that the PSD, which is a function of frequency, is the Fourier transform of the autocorrelation function.

Properties of the autocorrelation function of a real wide-sense stationary process are as follows:

$$1.\ R_x(0) = \overline{x^2(t)} = \text{second moment} \tag{6-16}$$

$$2.\ R_x(\tau) = R_x(-\tau) \tag{6-17}$$

$$3.\ R_x(0) \geq |R_x(\tau)| \tag{6-18}$$

The first two properties follow directly from the definition of $R_x(\tau)$ as given by (6-14). Furthermore, if $x(t)$ is ergodic, $R(0)$ is identical to the square of the rms

[†]The time average type of the autocorrelation function was defined in Chapter 2. The time average autocorrelation function, (2-68), is identical to the ensemble average autocorrelation function, (6-14), when the process is ergodic.

[‡]An exception occurs for the Gaussian random process. Here wide-sense stationarity does imply that the process is strict-sense stationary since the $N \to \infty$-dimensional Gaussian PDF is completely specified by the mean, variance, and covariance of $x(t_1)$, $x(t_2)$, . . . , $x(t_N)$.

value of $x(t)$. Property 3 is demonstrated as follows: $[x(t) \pm x(t+\tau)]^2$ is a nonnegative quantity, so that

$$\overline{[x(t) \pm x(t+\tau)]^2} \geq 0$$

or

$$\overline{x^2(t)} \pm \overline{2x(t)x(t+\tau)} + \overline{x^2(t+\tau)} \geq 0$$

This is equivalent to

$$R_x(0) \pm 2R_x(\tau) + R_x(0) \geq 0$$

which reduces to property 3.

The autocorrelation may be generalized to define an appropriate correlation function for two random processes.

Definition: The *cross-correlation function* for two real processes $x(t)$ and $y(t)$ is

$$R_{xy}(t_1,t_2) = \overline{x(t_1)y(t_2)} \tag{6-19}$$

Furthermore, if $x(t)$ and $y(t)$ are jointly stationary,† the cross-correlation function becomes

$$R_{xy}(t_1,t_2) = R_{xy}(\tau)$$

where $\tau = t_2 - t_1$.

Some *properties* of cross-correlation functions of jointly stationary real processes are

1. $R_{xy}(-\tau) = R_{yx}(\tau)$ (6-20)
2. $|R_{xy}(\tau)| \leq \sqrt{R_x(0)R_y(0)}$ (6-21)
3. $|R_{xy}(\tau)| \leq \frac{1}{2}[R_x(0) + R_y(0)]$ (6-22)

The first property follows directly from the definition, (6-19). Property 2 follows from the fact that

$$\overline{[x(t) + Ky(t+\tau)]^2} \geq 0 \tag{6-23}$$

†Similar to (6-3), $x(t)$ and $y(t)$ are said to be jointly stationary when

$$f_{xy}(x(t_1), x(t_2), \ldots, x(t_N), y(t_{N+1}), y(t_{N+2}), \ldots, y(t_{N+M}))$$
$$= f_{xy}(x(t_1 + t_0), x(t_2 + t_0), \ldots, x(t_N + t_0), y(t_{N+1} + t_0), y(t_{N+2} + t_0), \ldots, y(t_{N+M} + t_0))$$

$x(t)$ and $y(t)$ are strict-sense jointly stationary if this equality holds for $N \rightarrow \infty$ and $M \rightarrow \infty$ and are wide-sense jointly stationary for $N = 2$ and $M = 2$.

for any real constant K. Expanding (6-23), we obtain an equation that is a quadratic in K.

$$[R_y(0)]K^2 + [2R_{xy}(\tau)]K + [R_x(0)] \geq 0 \tag{6-24}$$

For K to be real, it can be shown that the discriminant of (6-24) has to be nonpositive.† That is,

$$[2R_{xy}(\tau)]^2 - 4[R_y(0)][R_x(0)] \leq 0 \tag{6-25}$$

This is equivalent to property 2 as described by (6-21). Property 3 follows directly from (6-24), where $K = \pm 1$. Furthermore, it is seen that

$$|R_{xy}(\tau)| \leq \sqrt{R_x(0)R_y(0)} \leq \tfrac{1}{2}[R_x(0) + R_y(0)] \tag{6-26}$$

since the geometric mean of two positive numbers, $R_x(0)$ and $R_y(0)$, does not exceed their arithmetic mean.

Note that the cross-correlation function of two random processes, $x(t)$ and $y(t)$, is a generalization of the concept of the joint mean of two random variables as defined by (B-91). Here x_1 is replaced by $x(t)$ and x_2 is replaced by $y(t + \tau)$. Thus two random processes $x(t)$ and $y(t)$ are said to be *uncorrelated* if

$$R_{xy}(\tau) = [\overline{x(t)}][\overline{y(t + \tau)}] = m_x m_y \tag{6-27}$$

for all values of τ. Similarly, two random processes $x(t)$ and $y(t)$ are said to be *orthogonal* if

$$R_{xy}(\tau) = 0 \tag{6-28}$$

for all values of τ.

As mentioned previously, it is realized that if $y(t) = x(t)$, the cross-correlation function becomes an autocorrelation function. In this sense the autocorrelation function is a special case of the cross-correlation function. Of course, when $y(t) \equiv x(t)$, all the properties of the cross-correlation function reduce to those of the autocorrelation function.

If the random processes $x(t)$ and $y(t)$ are jointly ergodic, the time average may be used to replace the ensemble average. For correlation functions this becomes

$$R_{xy}(\tau) \triangleq \overline{x(t)y(t + \tau)} \equiv \langle x(t)y(t + \tau)\rangle \tag{6-29}$$

where

$$\langle[\cdot]\rangle = \lim_{T\to\infty} \frac{1}{T}\int_{-T/2}^{T/2} [\cdot]\, dt \tag{6-30}$$

when $x(t)$ and $y(t)$ are jointly ergodic. In this case, cross-correlation functions and autocorrelation functions of voltage and/or current waveforms may be measured by using an electronic circuit that consists of a delay line, a multiplier, and an integrator. This measurement technique is illustrated in Fig. 6-3.

†The parameter K is a root of this quadratic only when the quadratic is equal to zero. When the quadratic is positive, the roots have to be complex.

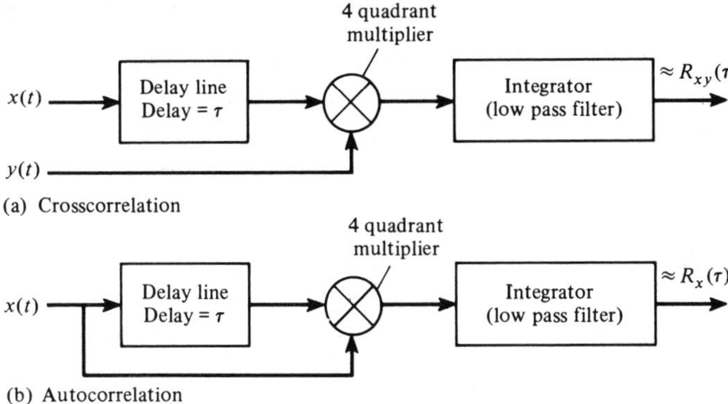

FIGURE 6-3 Measurement of correlation functions.

Complex Random Processes

In previous chapters, the complex envelope $g(t)$ was found to be extremely useful in describing bandpass waveforms. Bandpass *random* signals and noise can also be described in terms of the complex envelope, where $g(t)$ is a *complex* baseband *random process.*

DEFINITION: A *complex random process* is

$$g(t) \triangleq x(t) + jy(t) \tag{6-31}$$

where $x(t)$ and $y(t)$ are real random processes and $j = \sqrt{-1}$.

A complex process is strict-sense stationary if $x(t)$ and $y(t)$ are jointly strict-sense stationary; that is,

$$f_g(x(t_1), y(t_1'), x(t_2), y(t_2'), \ldots, x(t_N), y(t_N'))$$
$$= f_g(x(t_1 + t_0), y(t_1' + t_0), \ldots, x(t_N + t_0), y(t_N' + t_0)) \tag{6-32}$$

for any value of t_0 and any $N \to \infty$.

The definitions for the correlation functions can be generalized to cover *complex* random processes.

DEFINITION: The *autocorrelation function* for a *complex* random process is

$$R_g(t_1,t_2) = \overline{g^*(t_1)g(t_2)} \tag{6-33}$$

where the asterisk denotes the complex conjugate.

Furthermore, the complex random process is stationary in the wide sense if $\overline{g(t)}$ is a complex constant and $R_g(t_1,t_2) = R_g(\tau)$, where $\tau = t_2 - t_1$. The autocorrelation of a wide-sense stationary complex process has the Hermitian symmetry property

$$R_g(-\tau) = R_g^*(\tau) \tag{6-34}$$

Definition: The *cross-correlation function* for two *complex* random processes $g_1(t)$ and $g_2(t)$ is

$$R_{g_1g_2}(t_1,t_2) = \overline{g_1^*(t_1)g_2(t_2)} \tag{6-35}$$

When the complex random processes are jointly wide-sense stationary, the cross-correlation function becomes

$$R_{g_1g_2}(t_1,t_2) = R_{g_1g_2}(\tau)$$

where $\tau = t_2 - t_1$.

In Sec. 6-7 we will use these definitions in the statistical description of bandpass random signals and noise.

6-2 Power Spectral Density

Definition

A definition for the PSD $\mathcal{P}_x(f)$ was given in Chapter 2 for the case of deterministic waveforms (2-66). Here we develop a more general definition that is applicable to spectral analysis of random processes.

Suppose that $x(t,E_i)$ represents a sample function of a random process $x(t)$. A truncated version of this sample function can be defined:

$$x_T(t,E_i) = \begin{cases} x(t,E_i), & |t| < \frac{1}{2}T \\ 0, & t \text{ elsewhere} \end{cases} \tag{6-36}$$

where the subscript T denotes the truncated version. The corresponding Fourier transform is

$$X_T(f,E_i) = \int_{-\infty}^{\infty} x_T(t,E_i)e^{-j2\pi ft}\,dt$$

$$= \int_{-T/2}^{T/2} x(t,E_i)e^{-j2\pi ft}\,dt \tag{6-37}$$

This indicates that X_T is itself a random process since x_T is a random process. We will now simplify the notation and denote these functions simply by $X_T(f)$, $x_T(t)$, and $x(t)$ since it is clear that they are all random processes.

The normalized[†] energy in the time interval $(-T/2, T/2)$ is

$$E_T = \int_{-\infty}^{\infty} x_T^2(t)\,dt = \int_{-\infty}^{\infty} |X_T(f)|^2\,df \tag{6-38}$$

[†]If $x(t)$ is a voltage or current waveform, E_T is the energy on a per-ohm (i.e., $R = 1$) normalized basis.

Here Parseval's theorem was used to obtain the second integral. It is realized that E_T is a random variable since $x(t)$ is a random process. Furthermore, the mean normalized energy is obtained by taking the ensemble average of (6-38).

$$\overline{E_T} = \int_{-T/2}^{T/2} \overline{x^2(t)}\, dt = \int_{-\infty}^{\infty} \overline{x_T^2(t)}\, dt = \int_{-\infty}^{\infty} \overline{|X_T(f)|^2}\, df \tag{6-39}$$

The normalized average power is the energy expended per unit time, so that the normalized average power is

$$P = \lim_{T\to\infty} \frac{1}{T} \int_{-T/2}^{T/2} \overline{x^2(t)}\, dt = \lim_{T\to\infty} \frac{1}{T} \int_{-\infty}^{\infty} \overline{x_T^2(t)}\, dt$$

or

$$P = \int_{-\infty}^{\infty} \left[\lim_{T\to\infty} \frac{1}{T} \overline{|X_T(f)|^2} \right] df = \langle \overline{x^2(t)} \rangle \tag{6-40}$$

In the evaluation of the limit in (6-40), it is very important that the ensemble average be evaluated *before* the limit operation is carried out because we want to ensure that $X_T(f)$ is finite. [Since $x(t)$ is a power signal, $X(f) = \lim_{T\to\infty} X_T(f)$ may not exist.] Note that (6-40) indicates that for a random process, the average normalized power is given by the time average of the second moment. Of course, if $x(t)$ is wide-sense stationary, $\langle \overline{x^2(t)} \rangle = \overline{x^2(t)}$ since $\overline{x^2(t)}$ is a constant.

From the definition of the PSD in Chapter 2, we know that

$$P = \int_{-\infty}^{\infty} \mathcal{P}(f)\, df \tag{6-41}$$

Thus we see that the following definition of the PSD is consistent with that given in Chapter 2 (2-66).

Definition: The *power spectral density* (PSD) for a random process $x(t)$ is given by

$$\mathcal{P}_x(f) = \lim_{T\to\infty} \left(\frac{\overline{[|X_T(f)|^2]}}{T} \right) \tag{6-42}$$

where

$$X_T(f) = \int_{-T/2}^{T/2} x(t) e^{-j2\pi f t}\, dt \tag{6-43}$$

Wiener–Khintchine Theorem

Often the PSD is evaluated from the autocorrelation function for the random process by using the following theorem.

Wiener–Khintchine Theorem† *When $x(t)$ is a wide-sense stationary process, the PSD can be obtained from the Fourier transform of the autocorrelation function:*

$$\mathcal{P}_x(f) = \mathcal{F}[R_x(\tau)] = \int_{-\infty}^{\infty} R_x(\tau)e^{-j2\pi f\tau}\,d\tau \tag{6-44}$$

and, conversely,

$$R_x(\tau) = \mathcal{F}^{-1}[\mathcal{P}_x(f)] = \int_{-\infty}^{\infty} \mathcal{P}_x(f)e^{j2\pi f\tau}\,df \tag{6-45}$$

provided that $R(\tau)$ becomes sufficiently small for large values of τ so that

$$\int_{-\infty}^{\infty} |\tau R(\tau)|\,d\tau < \infty \tag{6-46}$$

Furthermore, this theorem is also valid for a nonstationary process, *provided* that we replace $R_x(\tau)$ by $\langle R_x(t,t+\tau)\rangle$.

Proof. From the definition of PSD,

$$\mathcal{P}_x(f) = \lim_{T\to\infty}\left(\frac{\overline{|X_T(f)|^2}}{T}\right)$$

where

$$\overline{|X_T(f)|^2} = \overline{\left|\int_{-T/2}^{T/2} x(t)e^{-j\omega t}\,dt\right|^2}$$

$$= \int_{-T/2}^{T/2}\int_{-T/2}^{T/2} \overline{x(t_1)x(t_2)}e^{-j\omega t_1}e^{j\omega t_2}\,dt_1dt_2$$

and $x(t)$ is assumed to be real. But $\overline{x(t_1)x(t_2)} = R_x(t_1,t_2)$. Furthermore, let $\tau = t_2 - t_1$ and make a change in variable from t_2 to $\tau + t_1$. We get

$$\overline{|X_T(f)|^2} = \int_{t_1=-T/2}^{t_1=T/2} \underbrace{\left[\int_{\tau=-T/2-t_1}^{\tau=T/2-t_1} R_x(t_1,t_1+\tau)e^{-j\omega\tau}\,d\tau\right] dt_1}_{①} \tag{6-47}$$

The area of this two-dimensional integration is shown on Fig. 6-4. On the figure, ① denotes the area covered by the inner integral times the differential width dt_1. To evaluate this two-dimensional integral easily, the order of integration will be

†Named after the American mathematician Norbert Wiener (1894–1964) and the German mathematician A. I. Khintchine (1894–1959). Other spellings of the German name are Khinchine and Khinchin.

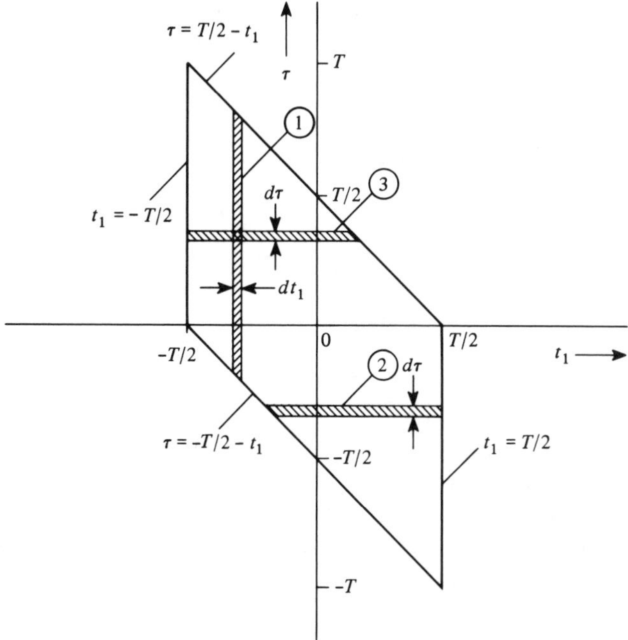

FIGURE 6-4 Region of integration for (6-47) and 6-48).

exchanged. As seen in the figure, this is accomplished by covering the total area by using ② when $\tau < 0$ and ③ when $\tau \geq 0$. Thus (6-47) becomes

$$\overline{|X_T(f)|^2} = \int_{-T}^{0} \underbrace{\left[\int_{t_1=-T/2-\tau}^{t_1=T/2} R_x(t_1,t_1+\tau) e^{-j\omega\tau}\, dt_1 \right] d\tau}_{②}$$

$$+ \int_{0}^{T} \underbrace{\left[\int_{t_1=-T/2}^{t_1=T/2-\tau} R_x(t_1,t_1+\tau) e^{-j\omega\tau}\, dt_1 \right] d\tau}_{③} \qquad (6\text{-}48)$$

Assume that $x(t)$ is stationary so that $R_x(t_1,t_1+\tau) = R_x(\tau)$, and factor $R_x(\tau)$ outside the inner integral. We get

$$\overline{|X_T(f)|^2} = \int_{-T}^{0} R_x(\tau) e^{-j\omega\tau} \left[t_1 \Big|_{-T/2-\tau}^{T/2} \right] d\tau + \int_{0}^{T} R(\tau) e^{-j\omega\tau} \left[t_1 \Big|_{-T/2}^{T/2-\tau} \right] d\tau$$

$$= \int_{-T}^{0} (T+\tau) R_x(\tau) e^{-j\omega\tau}\, d\tau + \int_{0}^{T} (T-\tau) R_x(\tau) e^{-j\omega\tau}\, d\tau$$

This equation can be written compactly as

$$\overline{|X_T(f)|^2} = \int_{-T}^{T} (T - |\tau|)R_x(\tau)e^{-j\omega\tau}\,d\tau \tag{6-49}$$

By substituting (6-49) into (6-42), we obtain

$$\mathcal{P}_x(f) = \lim_{T\to\infty} \int_{-T}^{T} \left(\frac{T - |\tau|}{T}\right) R_x(\tau)e^{-j\omega\tau}\,d\tau \tag{6-50a}$$

or

$$\mathcal{P}_x(f) = \int_{-\infty}^{\infty} R_x(\tau)e^{j\omega\tau}\,d\tau - \lim_{T\to\infty} \int_{-T}^{T} \frac{|\tau|}{T} R_x(\tau)\epsilon^{-j\omega\tau}\,d\tau \tag{6-50b}$$

Using the assumption of (6-46), we observe that the right-hand integral is zero and (6-50b) reduces to (6-44). Thus the theorem is proved. The converse relationship follows directly from the properties of Fourier transforms. Furthermore, if $x(t)$ is not stationary, (6-44) is still obtained if we replace $R_x(t_1,t_1 + \tau)$ in (6-48) by $\langle R_x(t_1,t_1 + \tau)\rangle = R_x(\tau)$.

Comparing the definition of the PSD with results of the Wiener–Khintchine theorem, we see that there are two distinctly different methods that may be used to evaluate the PSD of a random process.

1. The PSD is obtained directly from the definition as given by (6-42).
2. The PSD is obtained by evaluating the Fourier transform of $R_x(\tau)$, where $R_x(\tau)$ has to be obtained first.

Both methods will be demonstrated in Example 6-3.

Properties of the PSD

Some properties of the PSD are:

1. $\mathcal{P}_x(f)$ is always real. (6-51)
2. $\mathcal{P}_x(f) \geq 0$. (6-52)
3. When $x(t)$ is real, $\mathcal{P}_x(-f) = \mathcal{P}_x(f)$. (6-53)
4. $\int_{-\infty}^{\infty} \mathcal{P}_x(f)\,df = P =$ total normalized power. (6-54a)
   When $x(t)$ is wide-sense stationary,

$$\int_{-\infty}^{\infty} \mathcal{P}_x(f)\,df = P = \overline{x^2} = R_x(0) \tag{6-54b}$$

5. $\mathcal{P}_x(0) = \int_{-\infty}^{\infty} R_x(\tau)d\tau$. (6-55)

These properties follow directly from the definition and the use of the Wiener–Khintchine theorem.

EXAMPLE 6-3 EVALUATION OF THE PSD FOR A POLAR BASEBAND SIGNAL

Let $x(t)$ be a polar signal with random binary data. A sample function of this signal is illustrated in Fig. 6-5a. Assume that the data are independent from bit to bit and that the probability of obtaining a binary 1 during any bit interval is $\frac{1}{2}$. Find the PSD of $x(t)$.

The polar signal may be modeled by

$$x(t) = \sum_{n=-\infty}^{\infty} a_n f(t - nT_b) \tag{6-56}$$

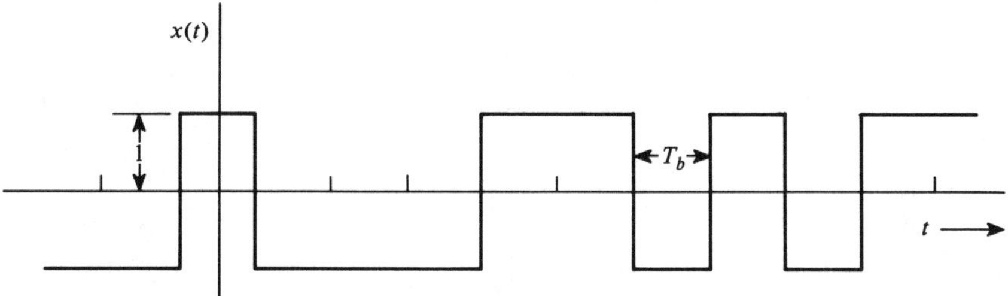

(a) Polar Signal

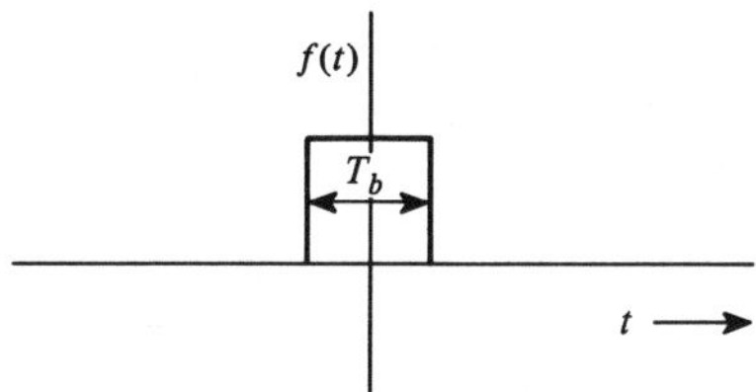

(b) Signaling Pulse Shape

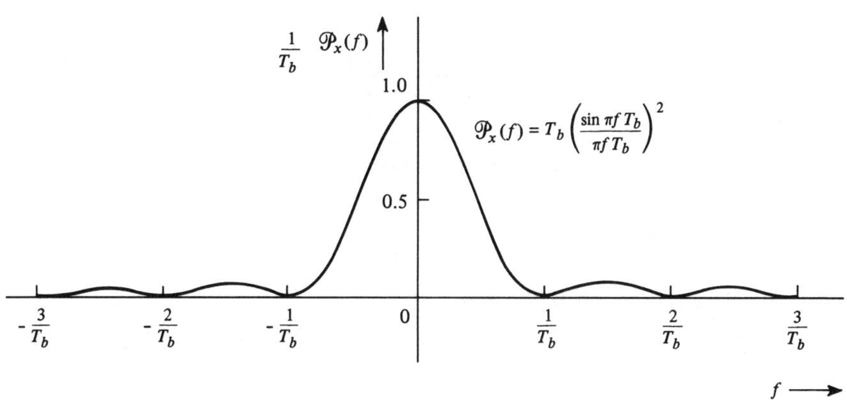

(c) Power Spectral Density of a Polar Signal

FIGURE 6-5 Random polar signal and its PSD.

where $f(t)$ is the signaling pulse shape, as shown in Fig. 6-5b, and T_b is the duration of one bit. $\{a_n\}$ is a set of random variables that represent the binary data. It is given that the random variables are independent. It is realized that each one is discretely distributed at $a_n = \pm 1$ and $P(a_n = 1) = P(a_n = -1) = \frac{1}{2}$, as described in the problem statement.

The PSD for $x(t)$ will be evaluated first by using method 1, which requires that $X_T(f)$ be obtained. We can obtain $x_T(t)$ by truncating (6-56).

$$x_T(t) = \sum_{n=-N}^{n=N} a_n f(t - nT_b)$$

where $T/2 = (N + \frac{1}{2})T_b$. Then

$$X_T(f) = \mathcal{F}[x_T(t)] = \sum_{n=-N}^{N} a_n \mathcal{F}[f(t - nT_b)] = \sum_{n=-N}^{N} a_n F(f) e^{-j\omega n T_b}$$

or

$$X_T(f) = F(f) \sum_{n=-N}^{N} a_n e^{-j\omega n T_b} \tag{6-57}$$

where $F(f) = \mathcal{F}[f(t)]$. When we substitute (6-57) into (6-42), the PSD is

$$\mathcal{P}_x(f) = \lim_{T\to\infty} \left(\frac{1}{T} |F(f)|^2 \overline{\left| \sum_{n=-N}^{N} a_n e^{-j\omega n T_b} \right|^2} \right)$$

$$= |F(f)|^2 \lim_{T\to\infty} \left(\frac{1}{T} \sum_{n=-N}^{N} \sum_{m=-N}^{N} \overline{a_n a_m} e^{j(m-n)\omega T_b} \right) \tag{6-58}$$

The average $\overline{a_n a_m}$ now needs to be evaluated for the case of polar signaling ($a_n = \pm 1$).

$$\overline{a_n a_m} = \begin{cases} \overline{a_n^2}, & n = m \\ \overline{a_n a_m}, & n \neq m \end{cases}$$

where $\overline{a_n a_m} = \overline{a_n}\,\overline{a_m}$ for $n \neq m$ since a_n and a_m are independent. Using the discrete distribution for a_n, we get

$$\overline{a_n} = (+1)\tfrac{1}{2} + (-1)\tfrac{1}{2} = 0$$

Similarly, $\overline{a_m} = 0$. Further,

$$\overline{a_n^2} = (+1)^2(\tfrac{1}{2}) + (-1)^2(\tfrac{1}{2}) = 1$$

Thus

$$\overline{a_n a_m} = \begin{cases} 1, & n = m \\ 0, & n \neq m \end{cases} \tag{6-59}$$

With this result, (6-58) becomes

$$\mathcal{P}_x(f) = |F(f)|^2 \lim_{T\to\infty} \left(\frac{1}{T} \sum_{n=-N}^{N} 1\right)$$

and, with $T = 2(N + \frac{1}{2})T_b$,

$$\mathcal{P}_x(f) = |F(f)|^2 \lim_{N\to\infty} \left[\frac{2N+1}{(2N+1)T_b}\right]$$

or

$$\mathcal{P}_x(f) = \frac{1}{T_b} |F(f)|^2 \qquad \text{(polar signaling)} \tag{6-60}$$

For the rectangular pulse shape shown in Fig. 6-5b,

$$F(f) = T_b\left(\frac{\sin \pi f T_b}{\pi f T_b}\right) \tag{6-61}$$

Thus the PSD for a polar signal with a rectangular pulse shape is

$$\mathcal{P}_x(f) = T_b\left(\frac{\sin \pi f T_b}{\pi f T_b}\right)^2 \tag{6-62}$$

This PSD is plotted in Fig. 6-5c.† The null bandwidth is $B = 1/T_b = R$, where R is the bit rate. Note that (6-62) satisfies the properties for a PSD function as listed above.

Method 2 will now be used to evaluate the PSD of the polar signal. This involves calculating the autocorrelation function and then evaluating the Fourier transform of $R_x(\tau)$ to obtain the PSD. When we use (6-56), the autocorrelation is

$$\begin{aligned} R_x(t, t+\tau) &= \overline{x(t)x(t+\tau)} \\ &= \overline{\sum_{n=-\infty}^{\infty} a_n f(t - nT_b) \sum_{m=-\infty}^{\infty} a_m f(t + \tau - mT_b)} \\ &= \sum_n \sum_m \overline{a_n a_m} f(t - nT_b) f(t + \tau - mT_b) \end{aligned}$$

By using (6-59), this equation reduces to

$$R_x(t,t+\tau) = \sum_{n=-\infty}^{\infty} f(t - nT_b) f(t + \tau - nT_b) \tag{6-63}$$

Obviously, $x(t)$ is *not* a wide-sense stationary process since the autocorrelation function depends on absolute time, t. To reduce (6-63) a particular type of pulse

†The PSD of this polar signal is purely continuous since the positive and negative pulses are assumed to be of equal amplitude.

shape needs to be designated. Assume, once again, the rectangular pulse shape

$$f(t) = \begin{cases} 1, & |t| \leq T_b/2 \\ 0, & t \text{ elsewhere} \end{cases}$$

The pulse product then becomes

$$f(t - nT_b)f(t + \tau - nT_b) = \begin{cases} 1, & \text{if } |t - nT_b| \leq T_b/2 \text{ and } |t + \tau - nT_b| \leq T_b/2 \\ 0, & \text{otherwise} \end{cases}$$

Working with the inequalities, we get a unity product only when

$$(n - \tfrac{1}{2})T_b \leq t \leq (n + \tfrac{1}{2})T_b$$

and

$$(n - \tfrac{1}{2})T_b - \tau \leq t \leq (n + \tfrac{1}{2})T_b - \tau$$

Assume that $\tau \geq 0$; then we will have a unity product when

$$(n - \tfrac{1}{2})T_b \leq t \leq (n + \tfrac{1}{2})T_b - \tau$$

provided that $\tau \leq T_b$. Thus, for $0 \leq \tau \leq T_b$,

$$R_x(t,t + \tau) = \sum_{n=-\infty}^{\infty} \begin{Bmatrix} 1, & (n - \tfrac{1}{2})T_b \leq t \leq (n + \tfrac{1}{2})T_b - \tau \\ 0, & \text{otherwise} \end{Bmatrix} \tag{6-64}$$

We know that the Wiener–Khintchine theorem is valid for nonstationary processes if we let $R_x(\tau) = \langle R_x(t,t + \tau)\rangle$. Using (6-64), we get

$$R_x(\tau) = \langle R_x(t,t + \tau)\rangle$$

$$= \lim_{T\to\infty} \frac{1}{T} \int_{-T/2}^{T/2} \sum_{n=-\infty}^{\infty} \begin{Bmatrix} 1, & (n - \tfrac{1}{2})T_b \leq t \leq (n + \tfrac{1}{2})T_b - \tau \\ 0, & \text{elsewhere} \end{Bmatrix} dt$$

or

$$R_x(\tau) = \lim_{T\to\infty} \frac{1}{T} \sum_{n=-N}^{N} \left(\int_{(n-1/2)T_b}^{(n+1/2)T_b-\tau} 1\, dt \right)$$

where $T/2 = (N - \tfrac{1}{2})T_b$. This reduces to

$$R_x(\tau) = \lim_{N\to\infty} \left[\frac{1}{(2N + 1)T_b} (2N + 1) \begin{Bmatrix} (T_b - \tau), & 0 \leq \tau \leq T_b \\ 0, & \tau > T_b \end{Bmatrix} \right]$$

or

$$R_x(\tau) = \begin{cases} \dfrac{T_b - \tau}{T_b}, & 0 \leq \tau \leq T_b \\ 0, & \tau > T_b \end{cases} \tag{6-65}$$

Similarly, for $\tau < 0$, results can be obtained. However, we know that $R_x(-\tau) = R_x(\tau)$, so (6-65) can be generalized for all values of τ. Thus

$$R_x(\tau) = \begin{cases} \dfrac{T_b - |\tau|}{T_b}, & |\tau| \le T_b \\ 0, & \text{otherwise} \end{cases} \tag{6-66}$$

which shows that $R_x(\tau)$ has a triangular shape. Evaluating the Fourier transform of (6-66), we obtain the PSD for the polar signal with a rectangular bit shape:

$$\mathcal{P}_x(f) = T_b\left(\frac{\sin \pi f T_b}{\pi f T_b}\right)^2 \tag{6-67}$$

Of course, this result, obtained by the use of method 2, is identical to (6-62), obtained by using method 1.

General Formula for the PSD of Baseband Digital Signals

A general formula for the PSD of baseband digital signals will now be obtained. The formulas for the PSD in Example 6-3 are valid only for polar signaling with $a_n = \pm 1$ and no correlation between the bits. A more general result can be obtained by starting with (6-58) and obtaining a result in terms of the autocorrelation of the data bits. Define the autocorrelation of the data bits by

$$R(k) = \overline{a_n a_{n+k}} \tag{6-68}$$

Make a change in the index in (6-58), letting $m = n + k$. Then, by using (6-68) and $T = (2N + 1)T_b$, (6-58) becomes

$$\mathcal{P}_x(f) = |F(f)|^2 \lim_{N\to\infty} \left[\frac{1}{(2N+1)T_b} \sum_{n=-N}^{n=N} \sum_{k=-N-n}^{k=N-n} R(k)e^{jk\omega T_b}\right]$$

Replacing the outer sum over the index n by $2N + 1$, we obtain the following expression. (This procedure is not strictly correct since the inner sum is also a function of n. The correct procedure would be to exchange the order of summation, similar to that used in (6-47) through (6-50) where the order of integration was exchanged. The result would be the same as that as obtained below when the limit is evaluated as $N \to \infty$.)

$$\begin{aligned}
\mathcal{P}_x(f) &= \frac{|F(f)|^2}{T_b} \lim_{N\to\infty} \left[\frac{(2N+1)}{(2N+1)} \sum_{k=-N-n}^{k=N-n} R(k)e^{jk\omega T_b}\right] \\
&= \frac{|F(f)|^2}{T_b} \sum_{k=-\infty}^{k=\infty} R(k)e^{jk\omega T_b} \\
&= \frac{|F(f)|^2}{T_b} \left[R(0) + \sum_{k=-\infty}^{-1} R(k)e^{jk\omega T_b} + \sum_{k=1}^{\infty} R(k)e^{jk\omega T_b}\right]
\end{aligned}$$

or

$$\mathcal{P}_x(f) = \frac{|F(f)|^2}{T_b}\left[R(0) + \sum_{k=1}^{\infty} R(-k)e^{-jk\omega T_b} + \sum_{k=1}^{\infty} R(k)e^{jk\omega T_b}\right] \quad (6\text{-}69)$$

But since $R(k)$ is an autocorrelation function, $R(-k) = R(k)$ and (6-69) becomes

$$\mathcal{P}_x(f) = \frac{|F(f)|^2}{T_b}\left[R(0) + \sum_{k=1}^{\infty} R(k)(e^{jk\omega T_b} + e^{-jk\omega T_b})\right]$$

Thus the general expression for the PSD of a baseband digital signal is

$$\mathcal{P}_x(f) = \frac{|F(f)|^2}{T_b}\left[R(0) + 2\sum_{k=1}^{\infty} R(k)\cos(2\pi k f T_b)\right] \quad (6\text{-}70a)$$

or, an equivalent expression is

$$\mathcal{P}_x(f) = \frac{|F(f)|^2}{T_b}\left[\sum_{k=-\infty}^{\infty} R(k)e^{-jk\omega T_b}\right] \quad (6\text{-}70b)$$

where the autocorrelation of the data is

$$R(k) = \overline{a_n a_{n+k}} = \sum_{i=1}^{I} (a_n a_{n+k})_i P_i \quad (6\text{-}70c)$$

P_i is the probability of getting the product $(a_n a_{n+k})_i$, and there are I possible values for the $a_n a_{n+k}$ product. $F(f)$ is the spectrum of the pulse shape of the digital symbol.

Note that the quantity in the brackets of (6-70b) is similar to the discrete Fourier transform (DFT) of the data autocorrelation function, $R(k)$, except that the frequency variable, ω, is continuous. Thus the PSD of the baseband digital signal is influenced by *both* the "spectrum" of the data and the spectrum of the pulse shape used for the line code. Furthermore, we will now demonstrate that the spectrum *may* also contain delta functions if the mean value of the data, $\overline{a}_n$, is nonzero. To do so, we first assume that the data symbols are uncorrelated; that is,

$$R(k) = \begin{Bmatrix} \overline{a_n^2}, & k = 0 \\ \overline{a}_n \overline{a}_{n+k}, & k \neq 0 \end{Bmatrix} = \begin{Bmatrix} \sigma_a^2 + m_a^2, & k = 0 \\ m_a^2, & k \neq 0 \end{Bmatrix}$$

where, as defined in Appendix B, the mean and the variance for the data are $m_a = \overline{a}_n$ and $\sigma_a^2 = \overline{(a_n - m_a)^2} = \overline{a_n^2} - m_a^2$. Substituting this into (6-70b), we get

$$\mathcal{P}_x(f) = \frac{|F(f)|^2}{T_b}\left[\sigma_a^2 + m_a^2 \sum_{k=-\infty}^{\infty} e^{-jk\omega T_b}\right]$$

From (2-115), recalling the Poisson sum formula,

$$\sum_{k=-\infty}^{\infty} e^{-jk\omega T_b} = R \sum_{n=-\infty}^{\infty} \delta(f - nR)$$

where $R = 1/T_b$, we see that the PSD becomes

$$\mathcal{P}_x(f) = \frac{|F(f)|^2}{T_b}\left[\sigma_a^2 + m_a^2 R \sum_{n=-\infty}^{\infty} \delta(f - nR)\right]$$

Thus, for the case of uncorrelated data, the PSD of the baseband digital signal is

$$\mathcal{P}_x(f) = \underbrace{\sigma_a^2 R|F(f)|^2}_{\text{Continuous spectrum}} + \underbrace{(m_a R)^2 \sum_{n=-\infty}^{\infty} |F(nR)|^2 \delta(f - nR)}_{\text{Discrete spectrum}} \tag{6-70d}$$

For the general case where there is correlation between the data, let the data autocorrelation function, $R(k)$, be expressed in terms of the *normalized data* autocorrelation function, $\rho(k)$. That is, let $\tilde{a}_n$ represent the corresponding data that have been normalized to have unity variance and zero mean. Thus

$$a_n = \sigma_a \tilde{a}_n + m_a$$

and, consequently,

$$R(k) = \sigma_a^2 \rho(k) + m_a^2$$

where

$$\rho(k) = \overline{[\tilde{a}_n \tilde{a}_{n+k}]}$$

Substituting this expression for $R(k)$ into (6-70b), we get

$$\mathcal{P}_x(f) = \frac{|F(f)|^2}{T_b}\left[\sigma_a^2 \sum_{k=-\infty}^{\infty} \rho(k) e^{-jk\omega T_b} + m_a^2 \sum_{k=-\infty}^{\infty} e^{-jk\omega T_b}\right]$$

Thus, for the general case of correlated data, the PSD of the baseband digital signal is

$$\mathcal{P}_x(f) = \underbrace{\sigma_a^2 R|F(f)|^2 \mathcal{W}_\rho(f)}_{\text{Continuous spectrum}} + \underbrace{(m_a R)^2 \sum_{n=-\infty}^{\infty} |F(nR)|^2 \delta(f - nR)}_{\text{Discrete spectrum}} \tag{6-70e}$$

where

$$\mathcal{W}_\rho(f) = \sum_{k=-\infty}^{\infty} \rho(k) e^{-j2\pi k f T_b} \tag{6-70f}$$

is a spectral weight function obtained from the Fourier transform of the normalized autocorrelation impulse train

$$\sum_{k=-\infty}^{\infty} \rho(k)\delta(\tau - kT_b)$$

This demonstrates that the PSD of the digital signal consists of a continuous spectrum that depends on the pulse-shape spectrum, $F(f)$, and the data correlation. Furthermore, if $m_a \neq 0$ and $F(nR) \neq 0$, the PSD will also contain spectral lines (delta functions) spaced at harmonics of the bit rate, R.

Examples of the application of these results are given in Sec. 3-4, where the PSD for unipolar RZ, bipolar, and Manchester line codes are evaluated. See Fig. 3-12 for a plot of the PSDs for these examples.

White Noise Processes

DEFINITION: A random process $x(t)$ is said to be a *white noise process* if the PSD is constant over all frequencies:

$$\mathcal{P}_x(f) = \frac{N_0}{2} \tag{6-71}$$

where N_0 is a positive constant.

The autocorrelation function for the white noise process is obtained by taking the inverse Fourier transform of (6-71). The result is

$$R_x(\tau) = \frac{N_0}{2}\,\delta(\tau) \tag{6-72a}$$

For example, the thermal noise process, as described in Sec. 5-5, can be considered to be a white noise process over the operating band where, from (6-71) and (5-62),

$$N_0 = kT \tag{6-72b}$$

Thermal noise also happens to have a Gaussian distribution. Of course, it is also possible to have white noise with other distributions.

Measurement of PSD

The PSD may be measured using analog or digital techniques. In either case, the measurement can only approximate the true PSD because the measurement is carried out over a finite time interval instead of the infinite interval as specified as (6-42).

Analog Techniques. The analog measurement technique consists of using either a bank of parallel narrowband bandpass filters with contiguous bandpass characteristics or by using a single bandpass filter that has a center frequency that is tunable. For the case of the filter bank, the waveform is fed simultaneously into the inputs of all the filters and the power at the output of each filter is evaluated. These output powers are divided (scaled) by the effective bandwidth of the corresponding filter so that an approximation for the PSD is obtained. That is, the PSD is evaluated for the frequency points corresponding to the center frequencies of the filters. Spectrum analyzers with this parallel type of analog processing are usually designed to cover the audio range of the spectrum where it is economically feasible to build a bank of bandpass filters.

Radio frequency (RF) spectrum analyzers are usually built using a single narrowband IF bandpass filter that is fed from the output of a mixer (up or down converter) circuit. The local oscillator (LO) of the mixer is swept slowly across an appropriate frequency band so that the RF spectrum analyzer is equivalent to a tunable narrowband bandpass filter where the center frequency of the tunable filter is swept across the desired spectral range. Once again, the PSD is obtained by evaluating the scaled power output of the narrowband filter as a function of the sweep frequency.

Numerical Computation of the PSD. The PSD is evaluated numerically in spectrum analyzers that use digital signal processing. One approximation for the PSD is

$$\mathcal{P}_T(f) = \frac{|X_T(f)|^2}{T} \tag{6-73a}$$

where the subscript T indicates that this approximation is obtained by viewing $x(t)$ over a T-second interval. T is called the *observation interval* or *observation length*. Of course, (6-73a) is an approximation to the true PSD as defined in (6-42) because T is finite and because only a sample of the ensemble is used. In more sophisticated spectrum analyzers, $\mathcal{P}_T(f)$ is evaluated for several $x(t)$ records and the average value of $\mathcal{P}_T(f)$ at each frequency is used to approximate the ensemble average required for a true $\mathcal{P}_x(f)$ as defined in (6-42). In evaluating $\mathcal{P}_T(f)$, the DFT is generally used to approximate $X_T(f)$. This, of course, brings the pitfalls of DFT analysis into play, as described in Sec. 2-7.

It should be stressed that (6-73a) is an approximation or *estimate* of the PSD. This estimate is called a *periodogram* because it was used historically to search for periodicities in data records that appear as delta functions in the PSD. [Delta functions are relatively easy to find in the PSD, and thus the periodicities in $x(t)$ are easily determined.] It is desirable that the estimate have an ensemble average that gives the true PSD. If this is the case, the estimator is said to be *unbiased*. We can easily check (6-73a) to see if it is unbiased.

$$\overline{\mathcal{P}_T(f)} = \overline{\left[\frac{|X_T(f)|^2}{T}\right]}$$

and, using (6-50a) for T finite, we obtain

$$\overline{\mathcal{P}_T(f)} = \mathcal{F}\left[R_x(\tau)\, \Lambda\left(\frac{\tau}{T}\right)\right]$$

and then, referring to Table 2-2,

$$\overline{\mathcal{P}_T(f)} = T\mathcal{P}_x(f) * \left(\frac{\sin \pi f T}{\pi f T}\right)^2 \tag{6-73b}$$

Since $\overline{\mathcal{P}_T(f)} \neq \mathcal{P}_x(f)$, the periodogram is a *biased* estimate. The bias is caused by the triangular window function, $\Lambda(\tau/T)$, that arises from the truncation of $x(t)$ to a T-second interval. Using (A-103) where $a = \pi T$, we see that $\lim_{T\to\infty} [\overline{\mathcal{P}_T(f)}] = \mathcal{P}_x(f)$ so that the periodogram becomes unbiased as $T \to \infty$. Consequently, the periodogram is said to be *asymptotically unbiased.*

In addition to being unbiased, it is desirable that an estimator be *consistent.* This means that the variance of the estimator should become small as $T \to \infty$. In this regard, it can be shown that (6-73a) gives an *inconsistent* estimate of the PSD when $x(t)$ is Gaussian [Bendat and Piersol, 1971].

Of course, window functions other than the triangle can be used to obtain different PSD estimators [Blackman and Tukey, 1958; Jenkins and Watts, 1968]. More modern techniques assume a mathematical model (i.e., form) for the autocorrelation function and estimate parameters for this model. The model is then checked to see if it is consistent with the data. Examples are the *moving average* (MA), *autoregressive* (AR), and the *autoregressive-moving average* (ARMA) models [Kay, 1986; Kay and Marple, 1981; Marple, 1986; Scharf, 1991; Shanmugan and Breipohl, 1988].

Spectrum analyzers that use microprocessor-based circuits often utilize numerical techniques to evaluate the PSD. These instruments can evaluate the PSD only for relatively low-frequency waveforms, say, over the audio or ultrasonic frequency range, since real-time digital signal-processing circuits cannot be built to process signals at RF rates.

6-3 Dc and Rms Values for Random Processes

In Chapter 2 the dc value, the rms value, and the average power were defined in terms of time average operations. For ergodic processes these time averages are equivalent to ensemble averages. Thus the dc value, the rms value, and the average power (which are all fundamental concepts in electrical engineering) can be related to the moments of an ergodic random process. A summary of these relationships follows. $x(t)$ is an ergodic random process that may correspond to either a voltage or a current waveform.

1. Dc value:

$$X_{\text{dc}} \triangleq \langle x(t)\rangle \equiv \overline{x} = m_x \tag{6-74}$$

2. Normalized dc power:

$$P_{\text{dc}} \triangleq [\langle x(t)\rangle]^2 \equiv (\overline{x})^2 \tag{6-75}$$

3. Rms value:

$$X_{\text{rms}} \triangleq \sqrt{\langle x^2(t)\rangle} \equiv \sqrt{\overline{(x^2)}} = \sqrt{R_x(0)} = \sqrt{\int_{-\infty}^{\infty} \mathcal{P}_x(f)\,df} \quad (6\text{-}76)$$

4. Rms value of the ac part:

$$\begin{aligned}(X_{\text{rms}})_{\text{ac}} &\triangleq \sqrt{\langle (x(t) - X_{\text{dc}})^2\rangle} \equiv \sqrt{\overline{(x - \bar{x})^2}} \\ &= \sqrt{\overline{x^2} - (\bar{x})^2} = \sqrt{R_x(0) - (\bar{x})^2} \\ &= \sqrt{\int_{-\infty}^{\infty} \mathcal{P}_x(f)\,df - (\bar{x})^2} = \sigma_x = \text{standard deviation}\end{aligned} \quad (6\text{-}77)$$

5. Normalized total average power:

$$P \triangleq \langle x^2(t)\rangle \equiv \overline{x^2} = R_x(0) = \int_{-\infty}^{\infty} \mathcal{P}_x(f)\,df \quad (6\text{-}78)$$

6. Normalized average power of the ac part:

$$\begin{aligned}P_{\text{ac}} &\triangleq \langle (x(t) - X_{\text{dc}})^2\rangle \equiv \overline{(x - \bar{x})^2} \\ &= \overline{x^2} - (\bar{x})^2 = R_x(0) - (\bar{x})^2 \\ &= \int_{-\infty}^{\infty} \mathcal{P}_x(f)\,df - (\bar{x})^2 = \sigma_x^2 = \text{variance}\end{aligned} \quad (6\text{-}79)$$

Furthermore, commonly available laboratory equipment may be used to evaluate the mean, second moment, and variance of an ergodic process. For example, if $x(t)$ is a voltage waveform, $\bar{x}$ can be measured by using a dc voltmeter and σ_x can be measured by using a "true rms" (ac coupled) voltmeter.† Using the measurements, we easily obtain the second moment from $\overline{x^2} = \sigma_x^2 + (\bar{x})^2$. At higher frequencies (e.g., radio, microwave, and optical), $\overline{x^2}$ and σ_x^2 can be measured by using a calibrated power meter. That is, $\sigma_x^2 = \overline{x^2} = RP$, where R is the load resistance of the power meter (usually 50 Ω) and P is the reading obtained from the power meter.

6-4 Linear Systems

Input–Output Relationships

As developed in Chapter 2, a linear time-invariant system may be described by its impulse response $h(t)$ or, equivalently, by its transfer function $H(f)$. This is

†Most "true rms" meters do not have a frequency response down to dc. Thus they do not measure the true rms value $\sqrt{\langle x^2(t)\rangle} = \sqrt{\overline{x^2}}$; instead, they measure σ.

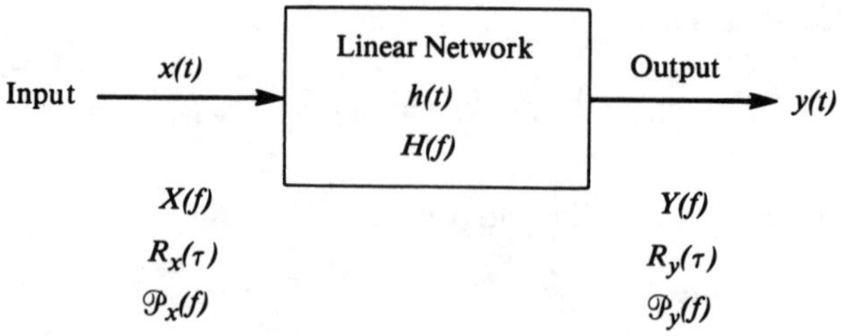

FIGURE 6-6 Linear system.

illustrated in Fig. 6-6, where $x(t)$ is the input and $y(t)$ is the output. The input–output relationships are

$$y(t) = h(t) * x(t) \tag{6-80}$$

The corresponding Fourier transform relationships are

$$Y(f) = H(f)X(f) \tag{6-81}$$

If $x(t)$ and $y(t)$ are random processes, these relationships are still valid (just as they were for the case of deterministic functions). In communication systems $x(t)$ might be a random signal plus (random) noise, or $x(t)$ might be noise alone when the signal is absent. For the case of random processes, autocorrelation functions and PSD functions may be used to describe the frequencies involved. Consequently, we need to address the question: What is the autocorrelation function and the PSD for the output process $y(t)$ when the autocorrelation and PSD for the input $x(t)$ are known?

THEOREM: *If a wide-sense stationary random process, $x(t)$, is applied to the input of a time-invariant linear network with impulse response $h(t)$, the output autocorrelation is*

$$R_y(\tau) = \int_{-\infty}^{\infty}\int_{-\infty}^{\infty} h(\lambda_1)h(\lambda_2)R_x(\tau - \lambda_2 + \lambda_1)\, d\lambda_1\, d\lambda_2 \tag{6-82a}$$

or

$$R_y(\tau) = h(-\tau) * h(\tau) * R_x(\tau) \tag{6-82b}$$

The output PSD is

$$\mathcal{P}_y(f) = |H(f)|^2\, \mathcal{P}_x(f) \tag{6-83}$$

where $H(f) = \mathcal{F}[h(t)]$.

It is also realized that (6-83) shows that the power transfer function of the network is

$$G_h(f) = \frac{\mathcal{P}_y(f)}{\mathcal{P}_x(f)} = |H(f)|^2 \tag{6-84}$$

as cited by (2-143).

Proof. From (6-80)

$$R_y(\tau) \triangleq \overline{y(t)y(t+\tau)}$$

$$= \overline{\left[\int_{-\infty}^{\infty} h(\lambda_1)x(t-\lambda_1)\,d\lambda_1\right]\left[\int_{-\infty}^{\infty} h(\lambda_2)x(t+\tau-\lambda_2)\,d\lambda_2\right]}$$

$$= \int_{-\infty}^{\infty}\int_{-\infty}^{\infty} h(\lambda_1)h(\lambda_2)\overline{x(t-\lambda_1)x(t+\tau-\lambda_2)}\,d\lambda_1 d\lambda_2 \tag{6-85}$$

But

$$\overline{x(t-\lambda_1)x(t+\tau-\lambda_2)} = R_x(t+\tau-\lambda_2-t+\lambda_1) = R_x(\tau-\lambda_2+\lambda_1)$$

so (6-85) is equivalent to (6-82a). Furthermore, (6-82a) may be written in terms of convolution operations as follows:

$$R_y(\tau) = \int_{-\infty}^{\infty} h(\lambda_1)\left\{\int_{-\infty}^{\infty} h(\lambda_2)R_x[(\tau+\lambda_1)-\lambda_2]\,d\lambda_2\right\}d\lambda_1$$

$$= \int_{-\infty}^{\infty} h(\lambda_1)\{h(\tau+\lambda_1) * R_x(\tau+\lambda_1)\}\,d\lambda_1$$

$$= \int_{-\infty}^{\infty} h(\lambda_1)\{h[-((-\tau)-\lambda_1)] * R_x[-((-\tau)-\lambda_1)]\}\,d\lambda_1$$

$$= h(-\tau) * h[-(-\tau)] * R_x[-(-\tau)]$$

which is equivalent to the convolution notation of (6-82b).

The PSD of the output is obtained by taking the Fourier transform of (6-82b).

$$\mathscr{F}[R_y(\tau)] = \mathscr{F}[h(-\tau)]\mathscr{F}[h(\tau) * R_x(\tau)]$$

or

$$\mathscr{P}_y(f) = H^*(f)H(f)\mathscr{P}_x(f)$$

where $h(t)$ is assumed to be real. This equation is equivalent to (6-83).

Of course, this theorem may be applied to cascaded linear systems. For example, two cascaded networks are shown in Fig. 6-7.

This theorem may also be generalized to obtain the cross-correlation or cross-spectrum of two linear systems, as illustrated in Fig. 6-8. $x_1(t)$ and $y_1(t)$ are the input and output processes of the first system, which has the impulse response

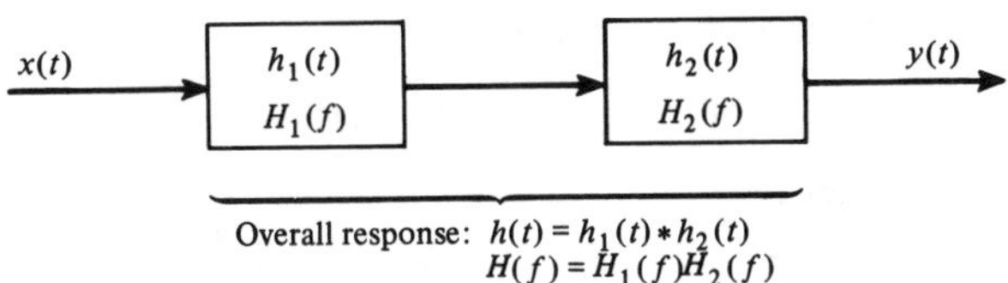

FIGURE 6-7 Two linear cascaded networks.

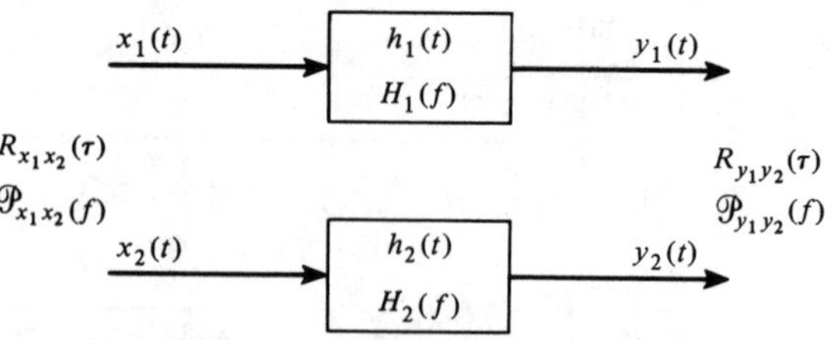

FIGURE 6-8 Two linear systems.

$h_1(t)$. Similarly, $x_2(t)$ and $y_2(t)$ are the input and output processes for the second system.

THEOREM: *Let $x_1(t)$ and $x_2(t)$ be wide-sense stationary inputs for two time-invariant linear systems, as shown in Fig. 6-8; then the output cross-correlation function is*

$$R_{y_1y_2}(\tau) = \int_{-\infty}^{\infty}\int_{-\infty}^{\infty} h_1(\lambda_1)h_2(\lambda_2)R_{x_1x_2}(\tau - \lambda_2 + \lambda_1)\, d\lambda_1 d\lambda_2 \qquad \text{(6-86a)}$$

or

$$R_{y_1y_2}(\tau) = h_1(-\tau) * h_2(\tau) * R_{x_1x_2}(\tau) \qquad \text{(6-86b)}$$

Furthermore, by definition the output cross power spectral density is the Fourier transform of the cross-correlation function; thus

$$\mathcal{P}_{y_1y_2}(f) = H_1^*(f)H_2(f)\mathcal{P}_{x_1x_2}(f) \qquad \text{(6-87)}$$

where $\mathcal{P}_{y_1y_2}(f) = \mathcal{F}[R_{y_1y_2}(\tau)]$, $\mathcal{P}_{x_1x_2}(f) = \mathcal{F}[R_{x_1x_2}(\tau)]$, $H_1(f) = \mathcal{F}[h_1(t)]$, and $H_2(f) = \mathcal{F}[h_2(t)]$.

The proof of this theorem is similar to that for the preceding theorem and will be left to the reader as an exercise.

EXAMPLE 6-4 OUTPUT AUTOCORRELATION AND PSD FOR AN *RC* LOW-PASS FILTER

An *RC* low-pass filter (LPF) is shown in Fig. 6-9. Assume that the input is an ergodic random process with a uniform PSD:

$$\mathcal{P}_x(f) = \tfrac{1}{2}N_0$$

Then the PSD of the output is

$$\mathcal{P}_y(f) = |H(f)|^2\mathcal{P}_x(f)$$

which becomes

$$\mathcal{P}_y(f) = \frac{\tfrac{1}{2}N_0}{1 + (f/B_{3\text{dB}})^2} \qquad \text{(6-88)}$$

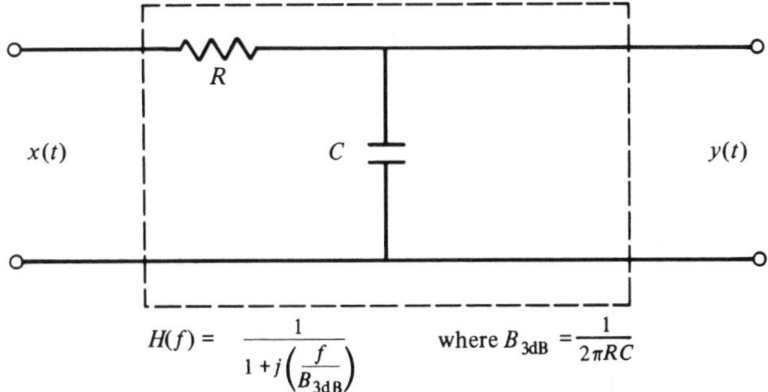

FIGURE 6-9 *RC* LPF.

where $B_{3db} = 1/(2\pi RC)$. Note that $\mathcal{P}_y(B_{3dB})/\mathcal{P}_y(0)$ is $\frac{1}{2}$, so $B_{3dB} = 1/(2\pi RC)$ is indeed the 3-dB bandwidth. Taking the inverse Fourier transform of $\mathcal{P}_y(f)$ as described by (6-88), we obtain the output autocorrelation function for the *RC* filter.

$$R_y(\tau) = \frac{N_0}{4RC}\, e^{-|\tau|/(RC)} \tag{6-89}$$

The normalized output power which is the second moment of the output is

$$P_y = \overline{y^2} = R_y(0) = \frac{N_0}{4RC} \tag{6-90}$$

Furthermore, the dc value of the output, which is the mean value, is zero since[†]

$$Y_{DC} = m_y = \sqrt{\lim_{\substack{\epsilon \to 0 \\ \epsilon > 0}} \int_{-\epsilon}^{\epsilon} \mathcal{P}_y(f)\, df} = 0 \tag{6-91}$$

The variance of the output is also given by (6-90) since $\sigma_y^2 = \overline{y^2} - m_y^2$, where $m_y = 0$.

EXAMPLE 6-5 SIGNAL-TO-NOISE RATIO AT THE OUTPUT OF AN *RC* LOW-PASS FILTER

Refer to Fig. 6-9 again and assume that $x(t)$ consists of a sine-wave (deterministic) signal plus white ergodic noise. Thus

$$x(t) = s_i(t) + n_i(t)$$

and

$$s_i(t) = A_0 \cos(\omega_0 t + \theta_0)$$

[†]When the integral is nonzero, the value obtained is equal or larger than the square of the mean value.

where A_0, ω_0, and θ_0 are known constants and $\mathcal{P}_{n_i}(f) = N_0/2$. The input signal power is

$$\langle s_i^2(t)\rangle = \frac{A_0^2}{2}$$

and the input noise power is

$$\langle n_i^2\rangle = \overline{n_i^2} = \int_{-\infty}^{\infty} \mathcal{P}_{n_i}(f)\, df = \int_{-\infty}^{\infty} \frac{N_0}{2}\, df = \infty$$

Thus the input signal-to-noise ratio is

$$\left(\frac{S}{N}\right)_{\text{in}} = \frac{\langle s_i^2(t)\rangle}{\langle n_i^2(t)\rangle} = 0 \tag{6-92}$$

The output consists of the sum of the filtered input signal plus the filtered input noise since the system is linear.

$$y(t) = s_0(t) + n_0(t)$$

The output signal is

$$\begin{aligned} s_0(t) &= s_i(t) * h(t) \\ &= A_0|H(f_0)| \cos[\omega_0 t + \theta_0 + \underline{/H(f_0)}\,] \end{aligned}$$

and the output signal power is

$$\langle s_0^2(t)\rangle = \frac{A_0^2}{2}\, |H(f_0)|^2 \tag{6-93}$$

From (6-90) of Example 6-4, the output noise power is

$$\overline{n_0^2} = \frac{N_0}{4RC} \tag{6-94}$$

The output signal-to-noise ratio is then

$$\left(\frac{S}{N}\right)_{\text{out}} = \frac{\langle s_0^2(t)\rangle}{\langle n_0^2(t)\rangle} = \frac{\langle s_0^2(t)\rangle}{\overline{n_0^2(t)}} = \frac{2A_0^2|H(f_0)|^2 RC}{N_0} \tag{6-95}$$

or

$$\left(\frac{S}{N}\right)_{\text{out}} = \frac{2A_0^2 RC}{N_0[1 + (2\pi f_0 RC)^2]}$$

6-5 Bandwidth Measures

A number of bandwidth measures were defined in Sec. 2-8, namely, absolute bandwidth, 3-dB bandwidth, equivalent bandwidth, null-to-null (zero-crossing) bandwidth, bounded bandwidth, power bandwidth, and the FCC bandwidth parameter. Of course, these definitions can be applied to evaluate the bandwidth of a wide-sense stationary process $x(t)$, where $\mathcal{P}_x(f)$ replaces $|H(f)|^2$ in the definitions. In this section we review the equivalent bandwidth and define a new bandwidth measure, rms bandwidth.

Equivalent Bandwidth

For a wide-sense stationary process $x(t)$, the equivalent bandwidth, as defined by (2-192), becomes

$$B_{\text{eq}} = \frac{1}{\mathcal{P}_x(f_0)} \int_0^{\infty} \mathcal{P}_x(f)\, df = \frac{R_x(0)}{2\mathcal{P}_x(f_0)} \tag{6-96}$$

f_0 is the frequency where $\mathcal{P}_x(f)$ is a maximum. This equation is valid for both bandpass and baseband processes (where $f_0 = 0$ is used for baseband processes).

Rms Bandwidth

The rms bandwidth is the square root of the second moment of the frequency with respect to a properly normalized PSD. In this case, f, although not random, may be *treated* as a random variable that has the "density function" $\mathcal{P}_x(f)/\int_{-\infty}^{\infty} \mathcal{P}_x(\lambda)\, d\lambda$. This is a nonnegative function in which the denominator provides the correct normalization so that the integrated value of this ratio is unity. Thus this function satisfies the properties of a PDF.

DEFINITION: If $x(t)$ is a *low-pass* wide-sense stationary process, the *rms bandwidth* is

$$B_{\text{rms}} = \sqrt{\overline{f^2}} \tag{6-97}$$

where

$$\overline{f^2} = \int_{-\infty}^{\infty} f^2 \left[\frac{\mathcal{P}_x(f)}{\int_{-\infty}^{\infty} \mathcal{P}_x(\lambda)\, d\lambda} \right] df = \frac{\int_{-\infty}^{\infty} f^2\, \mathcal{P}_x(f)\, df}{\int_{-\infty}^{\infty} \mathcal{P}_x(\lambda)\, d\lambda} \tag{6-98}$$

The rms measure of bandwidth is often used in theoretical comparisons of communication systems since the mathematics involved in the rms calculation is often easier to carry out than that for other bandwidth measures. However, the rms bandwidth is not easily measured with laboratory instruments.

THEOREM: *For a wide-sense stationary process $x(t)$, the mean-squared frequency $\overline{f^2}$ is*

$$\overline{f^2} = \left[-\frac{1}{(2\pi)^2 R(0)} \right] \left. \frac{d^2 R_x(\tau)}{d\tau^2} \right|_{\tau=0} \tag{6-99}$$

Proof. We know that

$$R_x(\tau) = \int_{-\infty}^{\infty} \mathcal{P}_x(f) e^{j2\pi f\tau}\, df$$

Taking the second derivative of this equation with respect to τ, we obtain

$$\frac{d^2R_x(\tau)}{d\tau^2} = \int_{-\infty}^{\infty} \mathcal{P}_x(f) e^{j2\pi f\tau}(j2\pi f)^2\, df$$

Evaluating both sides of this equation at $\tau = 0$ yields

$$\left.\frac{d^2R_x(\tau)}{d\tau^2}\right|_{\tau=0} = (j2\pi)^2 \int_{-\infty}^{\infty} f^2\mathcal{P}_x(f)\, df$$

Substituting this for the integral in (6-98), (6-98) becomes

$$\overline{f^2} = \frac{\displaystyle\int_{-\infty}^{\infty} f^2\mathcal{P}_x(f)\, df}{\displaystyle\int_{-\infty}^{\infty} \mathcal{P}_x(\lambda)\, d\lambda} = \frac{-\dfrac{1}{(2\pi)^2}\left[\dfrac{d^2R_x(\tau)}{d\tau^2}\right]_{\tau=0}}{R_x(0)}$$

which is identical to (6-99).

The rms bandwidth for a *bandpass* process can also be defined. Here we are interested in the square root of the second moment about the mean frequency of the *positive* frequency portion of the spectrum.

DEFINITION: If $x(t)$ is a *bandpass* wide-sense stationary process, the *rms bandwidth* is

$$B_{\text{rms}} = 2\sqrt{\overline{(f - f_o)^2}} \tag{6-100}$$

where

$$\overline{(f - f_0)^2} = \int_0^{\infty} (f - f_0)^2 \left(\frac{\mathcal{P}_x(f)}{\displaystyle\int_0^{\infty} \mathcal{P}_x(\lambda)\, d\lambda}\right) df \tag{6-101}$$

and

$$f_0 \triangleq \overline{f} = \int_0^{\infty} f \left(\frac{\mathcal{P}_x(f)}{\displaystyle\int_0^{\infty} \mathcal{P}_x(\lambda)\, d\lambda}\right) df \tag{6-102}$$

As a sketch of a typical bandpass PSD will verify, the quantity given by the radical in (6-100) is analogous to σ_f. Consequently, the factor of 2 is needed to give a reasonable definition for the bandpass bandwidth.

EXAMPLE 6-6 EQUIVALENT BANDWIDTH AND RMS BANDWIDTH FOR AN *RC* LPF

To evaluate the equivalent bandwidth and the rms bandwidth for a filter, white noise may be applied to the input. The bandwidth of the output PSD is the bandwidth of the filter since the input PSD is a constant.

For the *RC* LPF (Fig. 6-9) the output PSD, for white noise at the input, is given by (6-88). The corresponding output autocorrelation function is given by (6-89). When we substitute these equations into (6-96), the equivalent bandwidth for the *RC* LPF is

$$B_{eq} = \frac{R_y(0)}{2\mathcal{P}_y(0)} = \frac{N_0/4RC}{2(\frac{1}{2}N_0)} = \frac{1}{4RC} \quad \text{hertz} \tag{6-103}$$

Consequently, for an *RC* LPF,

$$B_{eq} = \frac{\pi}{2} B_{3dB} \tag{6-104}$$

The rms bandwidth is obtained by substituting (6-88) and (6-90) into (6-98).

$$B_{rms} = \sqrt{\frac{\int_{-\infty}^{\infty} f^2 \mathcal{P}_y(f)\, df}{R_y(0)}} = \sqrt{\frac{1}{2\pi^2 RC} \int_{-\infty}^{\infty} \frac{f^2}{(B_{3dB})^2 + f^2}\, df} \tag{6-105}$$

Examining the integral, we note that the integrand becomes *unity* as $f \to \pm\infty$, so that the value of the integral is infinity. Thus $B_{rms} = \infty$ for an *RC* LPF. For the rms bandwidth to be finite, the PSD needs to decay faster than $1/|f|^2$ as the frequency becomes large. Consequently, for the *RC* LPF, the rms definition is not very useful.

6-6 The Gaussian Random Process

DEFINITION: A random process $x(t)$ is said to be Gaussian if the random variables

$$x_1 = x(t_1),\ x_2 = x(t_2),\ \ldots,\ x_N = x(t_N) \tag{6-106}$$

have an N-dimensional Gaussian PDF for any N and any $t_1, t_2, \ldots, t_N$.

The N-dimensional Gaussian PDF can be written compactly by using matrix notation. Let $\mathbf{x}$ be the *column* vector denoting the N random variables,

$$\mathbf{x} = \begin{bmatrix} x_1 \\ x_2 \\ \vdots \\ x_N \end{bmatrix} = \begin{bmatrix} x(t_1) \\ x(t_2) \\ \vdots \\ x(t_N) \end{bmatrix} \tag{6-107}$$

The N-dimensional Gaussian PDF is

$$f_{\mathbf{x}}(\mathbf{x}) = \frac{1}{(2\pi)^{N/2}|\text{Det } \mathbf{C}|^{1/2}} e^{-(1/2)[(\mathbf{x}-\mathbf{m})^T\mathbf{C}^{-1}(\mathbf{x}-\mathbf{m})]} \tag{6-108}$$

where the mean vector is

$$\mathbf{m} = \overline{\mathbf{x}} = \begin{bmatrix} \overline{x}_1 \\ \overline{x}_2 \\ \vdots \\ \overline{x}_N \end{bmatrix} = \begin{bmatrix} m_1 \\ m_2 \\ \vdots \\ m_N \end{bmatrix} \tag{6-109}$$

$(\mathbf{x} - \mathbf{m})^T$ denotes the transpose of the column vector $(\mathbf{x} - \mathbf{m})$.

Det $\mathbf{C}$ is the determinant of the matrix $\mathbf{C}$, and $\mathbf{C}^{-1}$ is the inverse of the matrix $\mathbf{C}$. The covariance matrix $\mathbf{C}$ is defined by

$$\mathbf{C} = \begin{bmatrix} c_{11} & c_{12} & \cdots & c_{1N} \\ c_{21} & c_{22} & \cdots & c_{2N} \\ \vdots & \vdots & & \vdots \\ c_{N1} & c_{N2} & \cdots & c_{NN} \end{bmatrix} \tag{6-110}$$

where the elements of the matrix are

$$c_{ij} = \overline{(x_i - m_i)(x_j - m_j)} = \overline{[x(t_i) - m_i][x(t_j) - m_j]} \tag{6-111}$$

For a wide-sense stationary process $m_i = \overline{x(t_i)} = m_j = \overline{x(t_j)} = m$. The elements of the covariance matrix become

$$c_{ij} = R_x(t_j - t_i) - m^2 \tag{6-112}$$

If, in addition, the x_i happen to be uncorrelated, $\overline{x_i x_j} = \overline{x}_i \overline{x}_j$ for $i \neq j$, and the covariance matrix becomes

$$\mathbf{C} = \begin{bmatrix} \sigma^2 & 0 & \cdots & 0 \\ 0 & \sigma^2 & \cdots & 0 \\ \vdots & & \ddots & \vdots \\ 0 & 0 & \cdots & \sigma^2 \end{bmatrix} \tag{6-113}$$

where $\sigma^2 = \overline{x^2} - m^2 = R_x(0) - m^2$. That is, the covariance matrix becomes a diagonal matrix if the random variables are uncorrelated. Using (6-113) in (6-108), we can conclude that the Gaussian random variables are independent when they are uncorrelated.

Properties of Gaussian Processes

Some properties of Gaussian processes are now given.

1. $f_{\mathbf{x}}(\mathbf{x})$ depends only on $\mathbf{C}$ and on $\mathbf{m}$, which is another way of saying that the N-dimensional Gaussian PDF is completely specified by the first- and second-order moments (i.e., means, variances, and covariances).

2. Since the $\{x_i = x(t_i)\}$ are jointly Gaussian, the $x_i = x(t_i)$ are individually Gaussian.
3. When $\mathbf{C}$ is a diagonal matrix, the random variables are uncorrelated. Furthermore, the Gaussian random variables are independent when they are uncorrelated.
4. A linear transformation of a set of Gaussian random variables produces another set of Gaussian random variables.
5. A wide-sense stationary Gaussian process is also strict-sense stationary[†] [Papoulis, 1984, p. 222; Shanmugan and Breipohl, 1988, p. 141].

Property 4 is very useful in the analysis of linear systems. This property, as well as the following theorem, will be proved subsequently.

THEOREM: *If the input to a linear system is a Gaussian random process, the system output is also a Gaussian process.*

Proof. The output of a linear network having an impulse response $h(t)$ is

$$y(t) = h(t) * x(t) = \int_{-\infty}^{\infty} h(t - \lambda)x(\lambda)\, d\lambda$$

This can be approximated by

$$y(t) = \sum_{j=1}^{N} h(t - \lambda_j)x(\lambda_j)\, \Delta\lambda \tag{6-114}$$

which becomes exact as $N \to$ large, and $\Delta\lambda \to 0$.

The output random variables for the output random process are

$$y(t_1) = \sum_{j=1}^{N} [h(t_1 - \lambda_j)\, \Delta\lambda]x(\lambda_j)$$

$$y(t_2) = \sum_{j=1}^{N} [h(t_2 - \lambda_j)\, \Delta\lambda]x(\lambda_j)$$

$$\vdots$$

$$y(t_N) = \sum_{j=1}^{N} [h(t_N - \lambda_j)\, \Delta\lambda]x(\lambda_j)$$

In matrix notation this is

$$\begin{bmatrix} y_1 \\ y_2 \\ \vdots \\ y_N \end{bmatrix} = \begin{bmatrix} h_{11} & h_{12} & \cdots & h_{1N} \\ h_{21} & h_{22} & \cdots & h_{2N} \\ \vdots & \vdots & \ddots & \vdots \\ h_{N1} & h_{N2} & \cdots & h_{NN} \end{bmatrix} \begin{bmatrix} x_1 \\ x_2 \\ \vdots \\ x_N \end{bmatrix}$$

[†]This follows directly from (6-112) since the N-dimensional Gaussian PDF is a function only of τ and not the absolute times.

or

$$\mathbf{y} = \mathbf{Hx} \tag{6-115}$$

where the elements of the $N \times N$ matrix $\mathbf{H}$ are related to the impulse response of the linear network by

$$h_{ij} = [h(t_i - \lambda_j)]\,\Delta\lambda \tag{6-116}$$

We will now show that $\mathbf{y}$ is an N-dimensional Gaussian PDF when $\mathbf{x}$ is an N-dimensional Gaussian PDF. This may be accomplished by using the theory for a multivariate functional transformation as given in Appendix B. From (B-99) the PDF of $\mathbf{y}$ is

$$f_{\mathbf{y}}(\mathbf{y}) = \left.\frac{f_x(\mathbf{x})}{|J(\mathbf{y}/\mathbf{x})|}\right|_{\mathbf{x}=H^{-1}\mathbf{y}} \tag{6-117}$$

The Jacobian is

$$J\left(\frac{\mathbf{y}}{\mathbf{x}}\right) = \text{Det}\begin{bmatrix} \dfrac{dy_1}{dx_1} & \dfrac{dy_1}{dx_2} & \cdots & \dfrac{dy_1}{dx_N} \\ \dfrac{dy_2}{dx_1} & \dfrac{dy_2}{dx_2} & \cdots & \dfrac{dy_2}{dx_N} \\ \vdots & & & \vdots \\ \dfrac{dy_N}{dx_1} & \dfrac{dy_N}{dx_2} & \cdots & \dfrac{dy_N}{dx_N} \end{bmatrix} = \text{Det}[\mathbf{H}] \triangleq K \tag{6-118}$$

where K is a constant. In this problem $J(\mathbf{y}/\mathbf{x})$ is a *constant* (not a function of $\mathbf{x}$) since $\mathbf{y} = \mathbf{Hx}$ is a *linear* transformation. Thus

$$f_{\mathbf{y}}(\mathbf{y}) = \frac{1}{|K|} f_{\mathbf{x}}(\mathbf{H}^{-1}\mathbf{y})$$

or (6-119)

$$f_{\mathbf{y}}(\mathbf{y}) = \frac{1}{(2\pi)^{N/2}|K||\text{Det } \mathbf{C}_x|^{1/2}} e^{-(1/2)[(\mathbf{H}^{-1}\mathbf{y}-\mathbf{m}_x)^T\mathbf{C}_x^{-1}(\mathbf{H}^{-1}\mathbf{y}-\mathbf{m}_x)]}$$

where the subscript x has been appended to the quantities that are associated with $x(t)$. But we know that $\mathbf{m}_y = \mathbf{Hm}_x$ and from matrix theory we have the property $[\mathbf{AB}]^T = \mathbf{B}^T\mathbf{A}^T$, so that the exponent of (6-119) becomes

$$-\tfrac{1}{2}[(\mathbf{y} - \mathbf{m}_y)^T(\mathbf{H}^{-1})^T]\mathbf{C}_x^{-1}[\mathbf{H}^{-1}(\mathbf{y} - \mathbf{m}_y)] = -\tfrac{1}{2}[(\mathbf{y} - \mathbf{m}_y)^T\mathbf{C}_y^{-1}(\mathbf{y} - \mathbf{m}_y)] \tag{6-120}$$

where

$$\mathbf{C}_y^{-1} = (\mathbf{H}^{-1})^T\mathbf{C}_x^{-1}\mathbf{H}^{-1} \tag{6-121}$$

or

$$\mathbf{C}_y = \mathbf{H}\mathbf{C}_x\mathbf{H}^T \tag{6-122}$$

Thus the PDF for $\mathbf{y}$ is

$$f_{\mathbf{y}}(\mathbf{y}) = \frac{1}{(2\pi)^{N/2}|K||\text{Det } \mathbf{C}_x|^{1/2}} e^{-(1/2)(\mathbf{y}-\mathbf{m}_y)^T\mathbf{C}_y^{-1}(\mathbf{y}-\mathbf{m}_y)} \tag{6-123}$$

which is an N-dimensional Gaussian PDF. This completes the proof of this theorem.

If a linear system acts like an integrator or LPF, the output random variables (of the output random process) tend to be proportional to the sum of the input random variables. Consequently, by applying the central limit theorem (see Appendix B), the output of the integrator or LPF will tend toward a Gaussian random process when the input random variables are independent with non-Gaussian PDFs.

EXAMPLE 6-7 WHITE GAUSSIAN NOISE PROCESS

Assume that a Gaussian random process $n(t)$ is given that has a PSD of

$$\mathcal{P}_n(f) = \begin{cases} \frac{1}{2} N_0, & |f| \le B \\ 0, & f \text{ otherwise} \end{cases} \tag{6-124}$$

where B is a positive constant. This describes a *bandlimited white Gaussian* process as long as B is finite, but becomes a completely *white* (all frequencies present) *Gaussian process* as $B \to \infty$.

The autocorrelation function for the bandlimited white process is

$$R_n(\tau) = BN_0\left(\frac{\sin 2\pi B\tau}{2\pi B\tau}\right) \tag{6-125}$$

The total average power is $P = R_n(0) = BN_0$. The mean value of $n(t)$ is zero since there is no δ function in the PSD at $f = 0$. Furthermore, the autocorrelation function is zero for $\tau = k/(2B)$ when k is a nonzero integer. Therefore, the random variables $n_1 = n(t_1)$ and $n_2 = n(t_2)$ are uncorrelated when $t_2 - t_1 = \tau = k/(2B)$, $k \neq 0$. For other values of τ, the random variables are correlated. Since $n(t)$ is assumed to be Gaussian, n_1 and n_2 are jointly Gaussian random variables. Consequently, by property 3, the random variables are independent when $t_2 - t_1 = k/(2B)$. Of course, they are dependent for other values of t_2 and t_1. As $B \to \infty$, $R_n(\tau) \to \frac{1}{2}N_0\delta(\tau)$, and the random variables n_1 and n_2 become independent for all values of t_1 and t_2 provided that $t_1 \neq t_2$. Furthermore, as $B \to \infty$, the average power becomes infinite. Consequently, a white noise process is not physically realizable. However, it is a very useful mathematical idealization for system analysis, just as a deterministic impulse is useful for obtaining the impulse response of a linear system, although the impulse itself is not physically realizable.

6-7 Bandpass Processes†

Bandpass Representations

In Sec. 4-2 it was demonstrated that any bandpass waveform could be represented by

$$v(t) = \text{Re}\{g(t)e^{j\omega_c t}\} \tag{6-126a}$$

or, equivalently, by

$$v(t) = x(t)\cos\omega_c t - y(t)\sin\omega_c t \tag{6-126b}$$

and

$$v(t) = R(t)\cos[\omega_c t + \theta(t)] \tag{6-126c}$$

where $g(t)$ is the complex envelope, $R(t)$ is the real envelope, $\theta(t)$ is the phase, and $x(t)$ and $y(t)$ are the quadrature components. Thus the complex envelope is

$$g(t) = |g(t)|\, e^{j\angle g(t)} = R(t)e^{j\theta(t)} = x(t) + jy(t) \tag{6-127a}$$

This gives the relationships

$$R(t) = |g(t)| = \sqrt{x^2(t) + y^2(t)} \tag{6-127b}$$

$$\theta(t) = \angle g(t) = \tan^{-1}\left[\frac{y(t)}{x(t)}\right] \tag{6-127c}$$

$$x(t) = R(t)\cos\theta(t) \tag{6-127d}$$

$$y(t) = R(t)\sin\theta(t) \tag{6-127e}$$

Furthermore, the spectrum of $v(t)$ is related to that of $g(t)$ by (4-22), which is

$$V(f) = \tfrac{1}{2}[G(f - f_c) + G^*(-f - f_c)] \tag{6-128}$$

In Chapters 4 and 5 the bandpass representation was used to analyze communication systems from a *deterministic* viewpoint. Here we extend the representation to random processes. In communication systems the random processes may be random signals, noise, or signals corrupted by noise.

Of course, when $v(t)$ is a bandpass random process containing frequencies in the vicinity of $\pm f_c$, then $g(t)$, $x(t)$, $y(t)$, $R(t)$, and $\theta(t)$ will be baseband processes. In general, $g(t)$ is a complex process (as described in Sec. 6-1), and $x(t)$, $y(t)$, $R(t)$, and $\theta(t)$ are always real processes. This is readily seen from a Fourier series expansion of $v(t)$, as demonstrated by (4-5) through (4-8), where the Fourier series coefficients form a set of random variables since $v(t)$ is a random process. Furthermore, *if* $v(t)$ is a Gaussian random process, the Fourier series coefficients consist of a set of *Gaussian* random variables since they are obtained by linear operations on $v(t)$. Similarly, when $v(t)$ is a Gaussian process, $g(t)$, $x(t)$, and $y(t)$

†In some other texts this is called a *narrowband noise process*, which is a misnomer since it may be wideband or narrowband.

are Gaussian processes since they are linear functions of $v(t)$. However, $R(t)$ and $\theta(t)$ are not Gaussian because they are nonlinear functions of $v(t)$. The first-order PDF for these processes will be evaluated in Example 6-10.

We now need to address the topic of stationarity as applied to the bandpass representation.

THEOREM: *If $x(t)$ and $y(t)$ are jointly wide-sense stationary (WSS) processes, the real bandpass process*

$$v(t) = \text{Re}\{g(t)e^{j\omega_c t}\} = x(t)\cos\omega_c t - y(t)\sin\omega_c t \tag{6-129a}$$

will be WSS if and only if

$$1.\ \overline{x(t)} = \overline{y(t)} = 0 \tag{6-129b}$$

$$2.\ R_x(\tau) = R_y(\tau) \tag{6-129c}$$

$$3.\ R_{xy}(\tau) = -R_{yx}(\tau) \tag{6-129d}$$

Proof. The requirements for WSS are that $\overline{v(t)}$ be constant and $R_v(t,t+\tau)$ be only a function of τ. We see that $\overline{v(t)} = \overline{x(t)}\cos\omega_c t - \overline{y(t)}\sin\omega_c t$ is a constant for any value of t only if $\overline{x(t)} = \overline{y(t)} = 0$. Thus, condition 1 is required.

The conditions required to make $R_v(t, t+\tau)$ only a function of τ are found as follows:

$$\begin{aligned}
R_v&(t,t+\tau)\\
&= \overline{v(t)v(t+\tau)}\\
&= \overline{[x(t)\cos\omega_c t - y(t)\sin\omega_c t][x(t+\tau)\cos\omega_c(t+\tau) - y(t+\tau)\sin\omega_c(t+\tau)]}\\
&= \overline{x(t)x(t+\tau)}\cos\omega_c t\cos\omega_c(t+\tau) - \overline{x(t)y(t+\tau)}\cos\omega_c t\sin\omega_c(t+\tau)\\
&\quad - \overline{y(t)x(t+\tau)}\sin\omega_c t\cos\omega_c(t+\tau) + \overline{y(t)y(t+\tau)}\sin\omega_c t\sin\omega_c(t+\tau)
\end{aligned}$$

or

$$\begin{aligned}
R_v(t,t+\tau) &= R_x(\tau)\cos\omega_c t\cos\omega_c(t+\tau) - R_{xy}(\tau)\cos\omega_c t\sin\omega_c(t+\tau)\\
&\quad - R_{yx}(\tau)\sin\omega_c t\cos\omega_c(t+\tau) + R_y(\tau)\sin\omega_c t\sin\omega_c(t+\tau)
\end{aligned}$$

When we use trigonometric identities for products of sine and cosine, this equation reduces to

$$\begin{aligned}
R_v&(t,t+\tau)\\
&= \tfrac{1}{2}[R_x(\tau) + R_y(\tau)]\cos\omega_c\tau + \tfrac{1}{2}[R_x(\tau) - R_y(\tau)]\cos\omega_c(2t+\tau)\\
&\quad - \tfrac{1}{2}[R_{xy}(\tau) - R_{yx}(\tau)]\sin\omega_c\tau - \tfrac{1}{2}[R_{xy}(\tau) + R_{yx}(\tau)]\sin\omega_c(2t+\tau)
\end{aligned}$$

The autocorrelation for $v(t)$ can be made to be a function of τ only if the terms involving t are set equal to zero. That is, $[R_x(\tau) - R_y(\tau)] = 0$ and $[R_{xy}(\tau) + R_{yx}(\tau)] = 0$. Thus conditions 2 and 3 are required.

If conditions 1 through 3 of (6-129) are satisfied so that $v(t)$ is WSS, properties 1 through 5 of (6-133a) through (6-133e) are valid. Furthermore, the $x(t)$ and $y(t)$

components of $v(t) = x(t) \cos \omega_c t - y(t) \sin \omega_c t$ satisfy properties 6 through 14, as described by (6-133f) through (6-133n), when conditions 1 through 3 are satisfied. These properties are very useful in analyzing the random processes at various points in communication systems.

For a given bandpass waveform, the description of the complex envelope, $g(t)$, is not unique. This is easily seen in (6-126), where the choice of the value for the parameter f_c is left to our discretion. Consequently, for representation of a given bandpass waveform, $v(t)$, the frequency components that are present in the corresponding complex envelope, $g(t)$, depend on the value of f_c that is chosen in the model. Moreover, in representing random processes, one is often interested in having a representation for a WSS bandpass process with certain PSD characteristics. In this case, it can be shown that $\text{Re}\{(-j)g(t)e^{j\omega_c t}\}$ gives the same PSD as $\text{Re}\{g(t)e^{j\omega_c t}\}$ when $v(t)$ is a WSS process [Papoulis, 1984, pp. 314–322]. Consequently, $g(t)$ is not unique, and it can be chosen to satisfy some additional desired condition. Yet, properties 1 through 14 will still be satisfied when the conditions of (6-129) are satisfied.

In some applications, conditions 1 through 3 of (6-129) are not satisfied. This will be the case, for example, when the $x(t)$ and $y(t)$ quadrature components do not have the same power, as in an unbalanced quadrature modulation problem. Another example is the case in which $x(t)$ and/or $y(t)$ have a dc value. In these cases, the bandpass random process model described by (6-126) would be nonstationary. Consequently, one is faced with the question: Can another bandpass model be found that models a WSS $v(t)$, but yet does not require conditions 1 through 3 of (6-129)? The answer is, yes. Let the model of (6-126) be generalized to include a phase constant θ_c that is a random variable.Then we have the following theorem.

THEOREM: *If $x(t)$ and $y(t)$ are jointly WSS processes, the real bandpass process*

$$v(t) = \text{Re}\{g(t)e^{j(\omega_c t+\theta_c)}\} = x(t)\cos(\omega_c t + \theta_c) - y(t)\sin(\omega_c t + \theta_c) \quad (6\text{-}130)$$

will be WSS when θ_c is an independent random variable uniformly distributed over $(0, 2\pi)$.

This modification of the bandpass model should not worry us because we can argue that it is actually a better model for physically obtainable bandpass processes. That is, the constant θ_c is often called the *random start-up phase* since it depends on the "initial conditions" of the physical process. Any noise source or signal source starts up with a random-phase angle when it is turned on unless it is synchronized by injecting some external signal.

Proof. Using (6-130) to model our bandpass process, we now demonstrate that this $v(t)$ is wide-sense stationary when $g(t)$ is wide-sense stationary even though the conditions of (6-129) may be violated. To show that (6-130) is WSS, the first requirement is that $\overline{v(t)}$ be a constant.

$$\overline{v(t)} = \overline{\text{Re}\{g(t)e^{j(\omega_c t+\theta_c)}\}} = \text{Re}\{\overline{g(t)}e^{j\omega_c t}\,\overline{e^{j\theta_c}}\}$$

But $\overline{e^{j\theta_c}} = 0$, so we have $\overline{v(t)} = 0$, which is a constant. The second requirement is that $R_v(t,t+\tau)$ be only a function of τ.

$$R_v(t,\, t+\tau) = \overline{v(t)v(t+\tau)}$$

$$= \overline{\text{Re}\,\{g(t)e^{j(\omega_c t+\theta_c)}\}\,\text{Re}\{g(t+\tau)e^{j(\omega_c t+\omega_c\tau+\theta_c)}\}}$$

Using the identity $\text{Re}(c_1)\,\text{Re}(c_2) = \frac{1}{2}\text{Re}(c_1c_2) + \frac{1}{2}\text{Re}(c_1^*c_2)$ and recalling that θ_c is an independent random variable, we obtain

$$R_v(t,t+\tau) = \tfrac{1}{2}\,\text{Re}\{\overline{g(t)g(t+\tau)}\; e^{j(2\omega_c t+\omega_c\tau)}\overline{e^{j2\theta_c}}\}$$

$$+ \tfrac{1}{2}\,\text{Re}\{\overline{g^*(t)g(t+\tau)}e^{j\omega_c\tau}\}$$

But $\overline{e^{j2\theta_c}} = 0$ and $R_g(\tau) = \overline{g^*(t)g(t+\tau)}$, since $g(t)$ is assumed to be wide-sense stationary. Thus

$$R_v(t,t+\tau) = \tfrac{1}{2}\,\text{Re}\{R_g(\tau)e^{j\omega_c\tau}\} \tag{6-131}$$

The right-hand side of (6-131) is not a function of t, so $R_v(t,\, t+\tau) = R_{v(\tau)}$. Consequently, (6-130) gives a model for a wide-sense stationary bandpass process.

Furthermore, for this model as described by (6-130), properties 1 through 5 of (6-133a) through (6-133e) are valid, but all the properties 6 through 14, (6-133f) through (6-133h) are not valid for the $x(t)$ and $y(t)$ components of $v(t) = x(t)\cos(\omega_c t + \theta_c) - y(t)\sin(\omega_c t + \theta_c)$ unless all the conditions of (6-129) are satisfied. However, as will be proved later, the *detected* $x(t)$ and $y(t)$ components at the output of quadrature product detectors (see Fig. 6-11) satisfy properties 6 through 14, provided that the startup phase of the detectors, θ_0, is independent of $v(t)$. [Note that the $x(t)$ and $y(t)$ components associated with $v(t)$ at the input to the detector are not identical to the $x(t)$ and $y(t)$ quadrature output waveforms unless $\theta_c = \theta_0$; however, the PSDs may be identical.]

The complex envelope representation of (6-130) is very useful for evaluating the output of detector circuits. For example, if $v(t)$ is a signal plus noise process that is applied to a product detector, $x(t)$ is the output process if the reference is $2\cos(\omega_c t + \theta_c)$ and $y(t)$ is the output process if the reference is $-2\sin(\omega_c t + \theta_c)$ (see Chapter 4). Similarly, $R(t)$ is the output process for an envelope detector, and $\theta(t)$ is the output process for a phase detector.

Properties of WSS Bandpass Processes

Theorems giving the relationships between the autocorrelation functions and the PSD of $v(t)$, $g(t)$, $x(t)$, and $y(t)$ can be obtained. These and other theorems are listed subsequently as properties of bandpass random processes. These relationships assume that the bandpass process $v(t)$ is real and WSS.† The bandpass

† $v(t)$ also has zero mean value since it is a bandpass process.

nature of $v(t)$ is described mathematically with the aid of Fig. 6-10a. Here

$$\mathcal{P}_v(f) = 0 \qquad \text{for } f_2 < |f| < f_1 \tag{6-132}$$

where $0 < f_1 \le f_c \le f_2$. Furthermore, a positive constant B_0 is defined such that B_0 is the *largest* frequency interval between f_c and either band edge, as illustrated in Fig. 6-10a and $B_0 < f_c$.

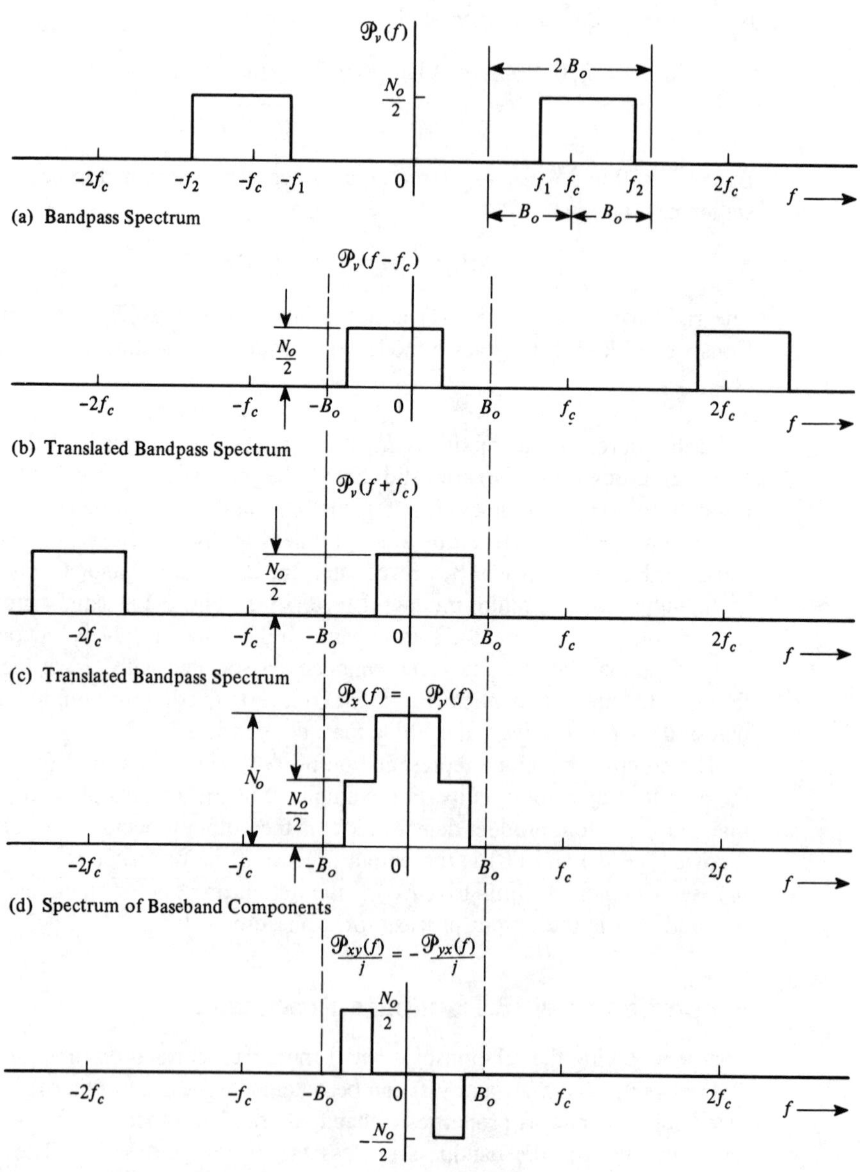

FIGURE 6-10 Spectra of random process for Example 6-8.

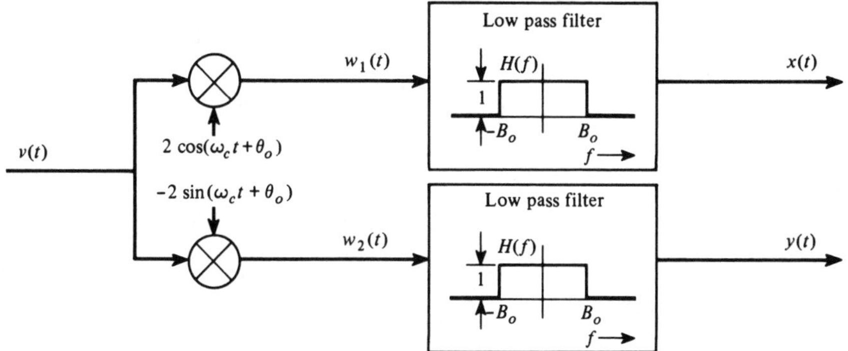

FIGURE 6-11 Recovery of $x(t)$ and $y(t)$ from $v(t)$.

The properties are:

1. $g(t)$ is a complex wide-sense-stationary baseband process. (6-133a)

2. $x(t)$ and $y(t)$ are real jointly wide-sense stationary baseband processes. (6-133b)

3. $R_v(\tau) = \frac{1}{2}\,\text{Re}\,\{R_g(\tau)e^{j\omega_c\tau}\}$. (6-133c)

4. $\mathcal{P}_v(f) = \frac{1}{4}[\mathcal{P}_g(f-f_c) + \mathcal{P}_g(-f-f_c)]$. (6-133d)

5. $\overline{v^2} = \frac{1}{2}\overline{|g(t)|^2} = R_v(0) = \frac{1}{2}R_g(0)$. (6-133e)

If the bandpass process $v(t)$ is WSS and if conditions 1 through 3 of (6-129) are satisified, $v(t)$ can be represented by (6-129a), where the $x(t)$ and $y(t)$ components satisfy properties 6 through 14 listed below. On the other hand, if $v(t)$ is WSS but does not satisfy all three conditions of (6-129), the representation of (6-129a) is not applicable because it will not be a WSS model, and consequently, properties 6 through 14 below are not applicable. However, for $v(t)$ WSS, the representation of (6-130) is always applicable regardless of whether none, some, or all of conditions 1 through 3 of (6-129) are satisfied; and, consequently, none, some, or all of properties 6 through 14 listed below are satisfied by the $x(t)$ and $y(t)$ components of (6-130). Furthermore, if the quadrature components of $v(t)$ are recovered by product detector circuits, as shown in Fig. 6-11, where θ_0 is a uniformly distributed random variable that is *independent* of $v(t)$, the *detected* $x(t)$ and $y(t)$ quadrature components satisfy properties 6 through 14 listed below.†

6. $\overline{x(t)} = \overline{y(t)} = 0$ (6-133f)

7. $\overline{v^2(t)} = \overline{x^2(t)} = \overline{y^2(t)} = \frac{1}{2}\overline{|g(t)|^2}$
$= R_v(0) = R_x(0) = R_y(0) = \frac{1}{2}R_g(0)$ (6-133g)

†Proofs of these properties will be given after Examples 6-8 and 6-9. Properties 6 through 14 also hold for the $x(t)$ and $y(t)$ components of (6-129a) provided that the conditions of (6-129b, c, and d) are satisfied.

8. $R_x(\tau) = R_y(\tau) = 2\int_0^{\infty} \mathcal{P}_v(f)\cos[2\pi(f-f_c)\tau]\,df$ (6-133h)

9. $R_{xy}(\tau) = 2\int_0^{\infty} \mathcal{P}_v(f)\sin[2\pi(f-f_c)\tau]\,df$ (6-133i)

10. $R_{xy}(\tau) = -R_{xy}(-\tau) = -R_{yx}(\tau)$ (6-133j)

11. $R_{xy}(0) = 0$ (6-133k)

12. $\mathcal{P}_x(f) = \mathcal{P}_y(f) = \begin{cases} [\mathcal{P}_v(f-f_c) + \mathcal{P}_v(f+f_c)], & |f| < B_0 \\ 0, & f \text{ elsewhere} \end{cases}$ (6-133l)

13. $\mathcal{P}_{xy}(f) = \begin{cases} j[\mathcal{P}_v(f-f_c) - \mathcal{P}_v(f+f_c)], & |f| < B_0 \\ 0, & f \text{ elsewhere} \end{cases}$ (6-133m)

14. $\mathcal{P}_{xy}(f) = -\mathcal{P}_{xy}(-f) = -\mathcal{P}_{yx}(f)$ (6-133n)

Additional properties can be obtained when $v(t)$ is a wide-sense stationary *single-sideband (SSB) random process*. If $v(t)$ is SSB about $f = \pm f_c$, from Chapter 4 we have

$$g(t) = x(t) \pm j\hat{x}(t) \tag{6-134}$$

where the upper sign is used for USSB and the lower sign is used for LSSB. $\hat{x}(t)$ is the Hilbert transform of $x(t)$. Using (6-134), we obtain some additional properties.

15. When $v(t)$ is a SSB process about $f = \pm f_c$,

$$R_g(\tau) = 2[R_x(\tau) \pm j\hat{R}_x(\tau)] \tag{6-135a}$$

where $\hat{R}_x(\tau) = [1/(\pi\tau)] * R_x(\tau)$

16. For USSB processes

$$\mathcal{P}_g(f) = \begin{cases} 4\mathcal{P}_x(f), & f > 0 \\ 0, & f < 0 \end{cases} \tag{6-135b}$$

17. For LSSB processes

$$\mathcal{P}_g(f) = \begin{cases} 0, & f > 0 \\ 4\mathcal{P}_x(f), & f < 0 \end{cases} \tag{6-135c}$$

From property 9 it is seen that if the PSD of $v(t)$ is even about $f = f_c, f > 0$, then $R_{xy}(\tau) \equiv 0$ for all τ. Consequently, $x(t)$ and $y(t)$ will be *orthogonal* processes when $\mathcal{P}_v(f)$ is even about $f = f_c, f > 0$. Furthermore, if in addition $v(t)$ is Gaussian, $x(t)$ and $y(t)$ will be independent Gaussian random processes.

EXAMPLE 6-8 SPECTRA FOR THE QUADRATURE COMPONENTS OF WHITE BANDPASS NOISE

Assume that $v(t)$ is an independent bandlimited white noise process. Let the PSD of $v(t)$ be $N_0/2$ over the frequency band $f_1 \le |f| \le f_2$, as illustrated in Fig. 6-10a.

Using property 12, we can evaluate the PSD for $x(t)$ and $y(t)$. This is obtained by summing the translated spectra $\mathcal{P}_v(f-f_c)$ and $\mathcal{P}_v(f+f_c)$ as illustrated in Fig. 6-10b and c to obtain $\mathcal{P}_x(f)$ shown in Fig. 6-10d. Note that the spectrum $\mathcal{P}_x(f)$ is zero for $|f| > B_0$. Similarly, the cross-spectrum, $\mathcal{P}_{xy}(f)$, can be obtained by using property 13. This is shown in Fig. 6-10e. It is interesting to note that over the frequency range where the cross-spectrum is nonzero, it is completely imaginary since $\mathcal{P}_v(f)$ is a real function. In addition, the cross-spectrum is always an odd function.

The total normalized power is

$$P = \int_{-\infty}^{\infty} \mathcal{P}_v(f)\, df = N_0(f_2 - f_1)$$

The same result is obtained if the power is computed from $\mathcal{P}_x(f) = \mathcal{P}_y(f)$ by using property 7.

$$P = R_x(0) = R_y(0) = \int_{-\infty}^{\infty} \mathcal{P}_x(f)\, df = N_0(f_2 - f_1)$$

EXAMPLE 6-9 PSD FOR A BPSK SIGNAL

The PSD for a BPSK signal that is modulated by random data will now be evaluated. In Chapters 2 and 5 it was demonstrated that the BPSK signal can be represented by

$$v(t) = x(t)\cos(\omega_c t + \theta_c) \tag{6-136}$$

where $x(t)$ represents the polar binary data (see Example 2-16) and θ_c is the random startup phase.

The PSD of $v(t)$ is found by using property 4, where $g(t) = x(t) + j0$. Thus

$$\mathcal{P}_v(f) = \tfrac{1}{4}[\mathcal{P}_x(f-f_c) + \mathcal{P}_x(-f-f_c)] \tag{6-137}$$

Now we need to find the PSD for the polar binary modulation $x(t)$. This was calculated in Example 6-3, where the PSD of a polar baseband signal with equally likely binary data was evaluated. Substituting (6-62) or, equivalently, (6-67) into (6-137), we obtain the PSD for the BPSK signal.

$$\mathcal{P}_v(f) = \frac{T_b}{4}\left\{\left[\frac{\sin \pi(f-f_c)T_b}{\pi(f-f_c)T_b}\right]^2 + \left[\frac{\sin \pi(f+f_c)T_b}{\pi(f+f_c)T_b}\right]^2\right\} \tag{6-138}$$

A sketch of this result is shown in Fig. 6-12. This result was cited earlier in (2-200), where bandwidths were also evaluated.

Proofs of Some Properties

Proofs for all 17 properties listed previously would be a long task. We therefore present detailed proofs for some of them. Proofs that involve similar mathematics will be left to the reader as exercise problems.

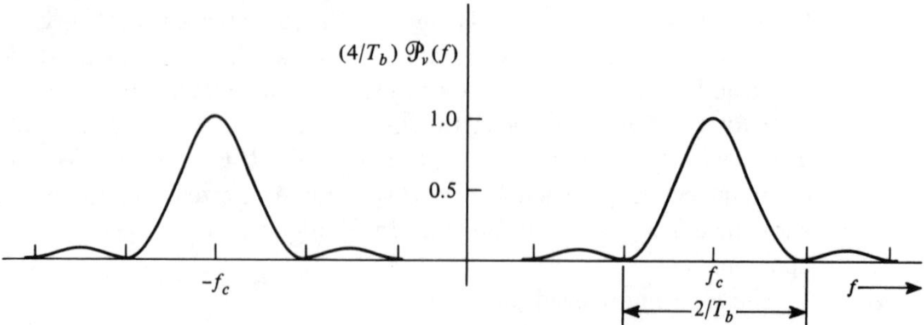

FIGURE 6-12 Power spectrum for a BPSK signal.

Properties 1 through 3 have already been given in the discussion preceding (6-131). Property 4 follows readily from property 3 by taking the Fourier transform of (6-133c). The mathematics is identical to that used in obtaining (4-25). Property 5 also follows directly from property 3. Property 6 will be shown subsequently. Property 7 follows from properties 3 and 8. As we will see later, properties 8 through 11 follow from properties 12 and 13.

Properties 6 and 12 are obtained with the aid of Fig. 6-11. That is, as shown by (4-71) in Sec. 4-3, $x(t)$ and $y(t)$ can be recovered by using product detectors. Thus

$$x(t) = [2v(t)\cos(\omega_c t + \theta_0)] * h(t) \tag{6-139}$$

and

$$y(t) = -[2v(t)\sin(\omega_c t + \theta_0)] * h(t) \tag{6-140}$$

where $h(t)$ is the impulse response of an ideal LPF that is bandlimited to B_0 hertz. θ_0 is an independent random variable that is uniformly distributed over $(0,2\pi)$ and corresponds to the random startup phase of a phase-incoherent receiver oscillator. Property 6 follows from (6-139) by taking the ensemble average since $\overline{\cos(\omega_c t + \theta_0)} = 0$ and $\overline{\sin(\omega_c t + \theta_0)} = 0$. Property 12, which is the PSD for $x(t)$, can be evaluated by first evaluating the autocorrelation for $w_1(t)$ of Fig. 6-11.

$$w_1(t) = 2v(t)\cos(\omega_c t + \theta_0)$$

$$R_{w_1}(\tau) = \overline{w_1(t)w_1(t+\tau)}$$

$$= \overline{4v(t)v(t+\tau)\cos(\omega_c t + \theta_0)\cos(\omega_c(t+\tau) + \theta_0)}$$

But since θ_0 is an independent random variable, using a trigonometric identity, we have

$$R_{w_1}(\tau) = 4\overline{v(t)v(t+\tau)}[\tfrac{1}{2}\cos\omega_c\tau + \tfrac{1}{2}\overline{\cos(2\omega_c t + \omega_c\tau + 2\theta_0)}]$$

But $\overline{\cos(2\omega_c t + \omega_c\tau + 2\theta_0)} = 0$, so

$$R_{w_1}(\tau) = 2R_v(\tau)\cos\omega_c\tau \tag{6-141}$$

The PSD of $w_1(t)$ is obtained by taking the Fourier transform of (6-141).

$$\mathcal{P}_{w_1}(f) = 2\mathcal{P}_v(f) * [\tfrac{1}{2}\delta(f-f_c) + \tfrac{1}{2}\delta(f+f_c)]$$

or

$$\mathcal{P}_{w_1}(f) = \mathcal{P}_v(f-f_c) + \mathcal{P}_v(f+f_c)$$

Finally, the PSD of $x(t)$ is

$$\mathcal{P}_x(f) = |H(f)|^2\mathcal{P}_{w_1}(f)$$

or

$$\mathcal{P}_x(f) = \begin{cases} [\mathcal{P}_v(f-f_c) + \mathcal{P}_v(f+f_c)], & |f| < B_0 \\ 0, & \text{otherwise} \end{cases}$$

which is property 12.

Property 8 follows directly from property 12 by taking the inverse Fourier transform.

$$\begin{aligned} R_x(\tau) &= \mathcal{F}^{-1}[\mathcal{P}_x(f)] \\ &= \int_{-B_0}^{B_0} \mathcal{P}_v(f-f_c)e^{j2\pi f\tau}\,df + \int_{-B_0}^{B_0} \mathcal{P}_v(f+f_c)e^{j2\pi f\tau}\,df \end{aligned}$$

Making changes in the variables, let $f_1 = -f + f_c$ in the first integral and let $f_1 = f + f_c$ in the second integral.

$$R_x(\tau) = \int_{f_c-B_0}^{f_c+B_0} \mathcal{P}_v(-f_1)e^{j2\pi(f_c-f_1)\tau}\,df_1 + \int_{f_c-B_0}^{f_c+B_0} \mathcal{P}_v(f_1)e^{j2\pi(f_1-f_c)\tau}\,df_1$$

But $\mathcal{P}_v(-f) = \mathcal{P}_v(f)$ since $v(t)$ is a real process. Furthermore, since $v(t)$ is bandlimited, the limits on the integrals may be changed to integrate over the interval $(0,\infty)$. Thus

$$R_x(\tau) = 2\int_0^\infty \mathcal{P}_v(f_1)\left[\frac{e^{j2\pi(f_1-f_c)\tau} + e^{-j2\pi(f_1-f_c)\tau}}{2}\right] df_1$$

which is identical to property 8.

In a similar way, properties 13 and 9 can be shown to be valid. Properties 10 and 14 follow directly from property 9.

For SSB processes we know that $y(t) = \pm\hat{x}(t)$. Property 15 is then obtained as follows.

$$\begin{aligned} R_{gg}(\tau) &= \overline{g^*(t)g(t+\tau)} \\ &= \overline{[x(t) \mp j\hat{x}(t)][x(t+\tau) \pm j\hat{x}(t+\tau)]} \\ &= [\overline{x(t)x(t+\tau)} + \overline{\hat{x}(t)\hat{x}(t+\tau)}] \\ &\quad \pm j[-\overline{\hat{x}(t)x(t+\tau)} + \overline{x(t)\hat{x}(t+\tau)}] \end{aligned} \tag{6-142}$$

Using the definition for crosscorrelation functions and using property 10, we have

$$R_{x\hat{x}}(\tau) = \overline{x(t)\hat{x}(t+\tau)} = -R_{\hat{x}x}(\tau) = -\overline{\hat{x}(t)x(t+\tau)} \tag{6-143}$$

Furthermore, knowing that $\hat{x}(t)$ is the convolution of $x(t)$ with $1/(\pi t)$, we can demonstrate (see Prob. 6-35) that

$$R_{\hat{x}}(\tau) = R_x(\tau) \tag{6-144}$$

and

$$R_{x\hat{x}}(\tau) = \hat{R}_{xx}(\tau) \tag{6-145}$$

Thus (6-142) reduces to property 15. Taking the Fourier transform of (6-135a), we obtain properties 16 and 17.

As demonstrated by property 6, the detected mean values $\overline{x(t)}$ and $\overline{y(t)}$ are zero when $v(t)$ is independent of θ_0. However, from (6-139) and (6-140) we realize that a less restrictive condition is required. That is, $\overline{x(t)}$ or $\overline{y(t)}$ will be zero if $v(t)$ is orthogonal to $\cos(\omega_c t + \theta_0)$ or $\sin(\omega_c t + \theta_0)$, respectively; otherwise, they will be nonzero. For example, suppose that $v(t)$ is

$$v(t) = 5\cos(\omega_c t + \theta_c) \tag{6-146}$$

where θ_c is a random variable uniformly distributed over $(0,2\pi)$. If the reference $2\cos(\omega_c t + \theta_0)$ of Fig. 6-11 is phase coherent with $\cos(\omega_c t + \theta_c)$ (i.e., $\theta_0 \equiv \theta_c$), the output of the upper LPF will have a mean value of 5. On the other hand, if the random variables θ_0 and θ_c are independent, then $v(t)$ will be orthogonal to $\cos(\omega_c t + \theta_0)$ and $\bar{x}$ of the output of the upper LPF of Fig. 6-11 will be zero. The output dc value (i.e., time average) will be $5\cos(\theta_0 - \theta_c)$ in either case.

No properties have been given pertaining to the autocorrelation or the PSD of $R(t)$ and $\theta(t)$ as related to the autocorrelation and PSD of $v(t)$. In general, this is a difficult problem because $R(t)$ and $\theta(t)$ are nonlinear functions of $v(t)$. This topic is discussed in more detail after Example 6-10.

As indicated previously, $x(t)$, $y(t)$, and $g(t)$ are Gaussian processes when $v(t)$ is Gaussian. However, $R(t)$ and $\theta(t)$ are not Gaussian processes when $v(t)$ is Gaussian, as we will demonstrate.

EXAMPLE 6-10 PDF FOR THE ENVELOPE AND PHASE FUNCTIONS OF A GAUSSIAN BANDPASS PROCESS

Assume that $v(t)$ is a wide-sense stationary *Gaussian* process with finite PSD that is symmetrical about $f = \pm f_c$. We want to find the one-dimensional PDF for the envelope process, $R(t)$. Of course, this is identical to the process that appears at the output of an envelope detector when the input is a Gaussian process, such as Gaussian noise. Similarly, the PDF for the phase, $\theta(t)$ (the output of a phase detector), will also be obtained.

This problem is solved by evaluating the two-dimensional random variable transformation of $x = x(t)$ and $y = y(t)$ into $R = R(t)$ and $\theta = \theta(t)$. This is illustrated in Fig. 6-13. Since $v(t)$ is Gaussian, we know that x and y are jointly

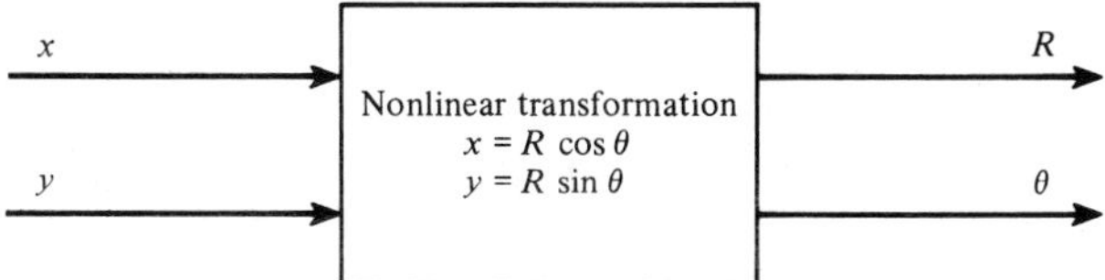

FIGURE 6-13 Nonlinear (polar) transformation of two Gaussian random variables.

Gaussian. For $v(t)$ having a finite PSD that is symmetrical about $f = \pm f_c$, the mean values for x and y are both zero and the variances of both are

$$\sigma^2 = \sigma_x^2 = \sigma_y^2 = R_v(0) \tag{6-147}$$

Furthermore, x and y are independent since they are uncorrelated Gaussian random variables [since the PSD of $v(t)$ is symmetrical about $f = \pm f_c$]. Therefore, the joint PDF of x and y is

$$f_{xy}(x,y) = \frac{1}{2\pi\sigma^2}\, e^{-(x^2+y^2)/(2\sigma^2)} \tag{6-148}$$

The joint PDF for R and θ is obtained by the two-dimensional transformation of x and y into R and θ.

$$f_{R\theta}(R,\theta) = \frac{f_{xy}(x,y)}{|J[(R,\theta)/(x,y)]|}\Bigg|_{\substack{x=R\cos\theta\\ y=R\sin\theta}}$$

$$= f_{xy}(x,y)\left|J\left(\frac{(x,y)}{(R,\theta)}\right)\right|\Bigg|_{\substack{x=R\cos\theta\\ y=R\sin\theta}} \tag{6-149}$$

We will work with $J[(x,y)/(R,\theta)]$ instead of $J[(R,\theta)/(x,y)]$ because in this problem the partial derivatives in the former are easier to evaluate than those in the latter.

$$J\left(\frac{(x,y)}{(R,\theta)}\right) = \text{Det}\begin{bmatrix} \dfrac{\partial x}{\partial R} & \dfrac{\partial x}{\partial \theta} \\[2ex] \dfrac{\partial y}{\partial R} & \dfrac{\partial y}{\partial \theta} \end{bmatrix}$$

where x and y are related to R and θ as shown in Fig. 6-13. Of course, $R \geq 0$, and θ falls in the interval $(0,2\pi)$.

$$J\left(\frac{(x,y)}{(R,\theta)}\right) = \text{Det}\begin{bmatrix} \cos\theta & -R\sin\theta \\ \sin\theta & R\cos\theta \end{bmatrix}$$

$$= R[\cos^2\theta + \sin^2\theta] = R \tag{6-150}$$

Substituting (6-148) and (6-150) into (6-149), the joint PDF of R and θ is

$$f_{R\theta}(R,\theta) = \begin{cases} \dfrac{R}{2\pi\sigma^2}\, e^{-R^2/2\sigma^2}, & R \geq 0 \text{ and } 0 \leq \theta \leq 2\pi \\[2ex] 0, & R \text{ and } \theta \text{ otherwise} \end{cases} \tag{6-151}$$

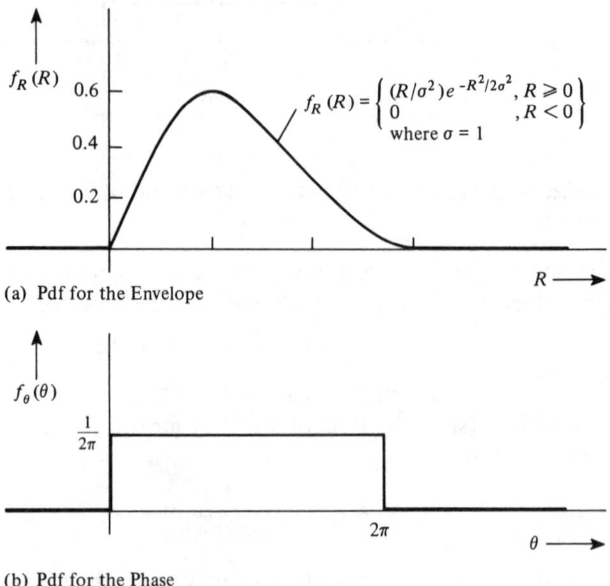

FIGURE 6-14 PDF for the envelope and phase of a Gaussian process.

The PDF for the envelope is obtained by calculating the marginal PDF:

$$f_R(R) = \int_{-\infty}^{\infty} f_R(R,\theta)\, d\theta = \int_0^{2\pi} \frac{R}{2\pi\sigma^2}\, e^{-R/(2\sigma^2)}\, d\theta, \qquad R \geq 0$$

or

$$f_R(R) = \begin{cases} \dfrac{R}{\sigma^2}\, e^{-R^2/(2\sigma^2)}, & R \geq 0 \\ 0, & R \text{ otherwise} \end{cases} \tag{6-152}$$

This is called a *Rayleigh* PDF. Similarly, the PDF of θ is obtained by integrating out R in the joint PDF:

$$f_\theta(\theta) = \begin{cases} \dfrac{1}{2\pi}, & 0 \leq \theta \leq 2\pi \\ 0, & \text{otherwise} \end{cases} \tag{6-153}$$

Of course, this is called a *uniform* PDF. Sketches of these PDFs are shown in Fig. 6-14.

The random variables $R = R(t_1)$ and $\theta = \theta(t_1)$ are independent since it is seen that $f_R(R,\theta) = f_R(R)f(\theta)$. However, $R(t)$ and $\theta(t)$ are *not* independent random processes since the random variables $R = R(t_1)$ and $\theta = \theta(t_1 + \tau)$ are not independent for all values of τ. To verify this statement, a four-dimensional transformation problem consisting of transforming the random variables $(x(t_1),x(t_2),y(t_1),y(t_2))$ into $(R(t_1),R(t_2),\theta(t_1),\theta(t_2))$ needs to be worked out where $t_2 - t_1 = \tau$ (Davenport and Root, 1958).

The evaluation of the autocorrelation function for the envelope $R(t)$ generally requires that the two-dimensional density function of $R(t)$ be known since $R_R(\tau) = \overline{R(t)R(t+\tau)}$. However, to obtain this joint density function for $R(t)$, the four-dimensional density function of $(R(t_1),R(t_2),\theta(t_1),\theta(t_2))$ must first be obtained by a four-dimensional transformation discussed in the preceding paragraph. A similar problem is worked to evaluate the autocorrelation function for the phase $\theta(t)$. These difficulties in evaluating the autocorrelation function for $R(t)$ and $\theta(t)$ arise because they are nonlinear functions of $v(t)$. The PSDs for $R(t)$ and $\theta(t)$ are obtained by taking the Fourier transform of the autocorrelation function.

6-8 Matched Filters

General Results

In preceding sections of this chapter, we have developed techniques for describing random processes and *analyzing* the effect of linear systems on these processes. Here we develop a technique for *designing* a linear filter to minimize the effect of the noise while maximizing the signal.

A general representation for a matched filter is illustrated in Fig. 6-15. The input signal is denoted by $s(t)$ and the output signal by $s_0(t)$. Similar notation is used for the noise. This filter is used in applications where the signal may or may not be present, but when the signal is present, *its waveshape is known*. (This has application to digital signaling and radar problems as will become clear in Examples 6-11 and 6-12.) The signal is assumed to be (absolutely) time limited to the interval $(0,T)$ and is zero otherwise. The PSD, $\mathcal{P}_n(f)$, of the additive input noise $n(t)$ is also known. We wish to determine the filter characteristic such that the instantaneous output signal power is maximized at a sampling time t_0, when compared with the average output noise power. That is, we want to find $h(t)$ or, equivalently, $H(f)$, so that

$$\left(\frac{S}{N}\right)_{\text{out}} = \frac{s_0^2(t_0)}{\overline{n_0^2(t)}} \tag{6-154}$$

is a maximum. This is the matched filter design criterion.

The matched filter does *not* preserve the input signal waveshape. This is not the objective. The objective is to distort the input signal waveshape and filter the

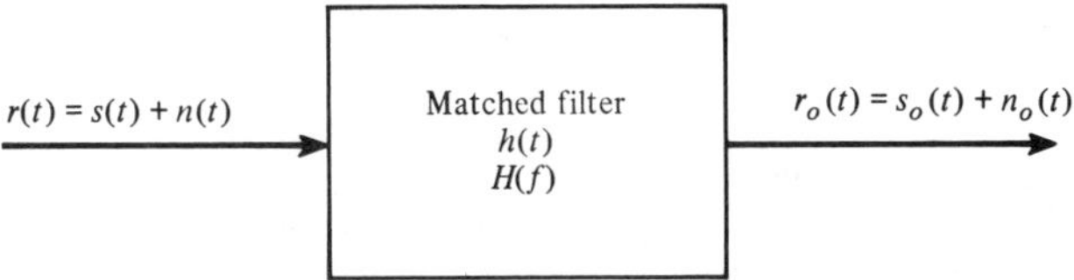

FIGURE 6-15 Matched filter.

noise so that at the sampling time t_0, the output signal level will be as large as possible with respect to the rms (output) noise level. In Chapters 7 and 8 we demonstrate that, under certain conditions, this minimizes the probability of error when receiving digital signals.

THEOREM: *The **matched filter** is the linear filter that maximizes* $(S/N)_{out} = s_0^2(t_0)/\overline{n_0^2(t)}$ *of Fig. 6-15 and has a transfer function given by*[†]

$$H(f) = K\frac{S^*(f)}{\mathcal{P}_n(f)}\, e^{-j\omega t_0} \tag{6-155}$$

where $S(f) = \mathcal{F}[s(t)]$ *is the Fourier transform of the known input signal* $s(t)$ *of duration T sec.* $\mathcal{P}_n(f)$ *is the PSD of the input noise,* t_0 *is the sampling time when* $(S/N)_{out}$ *is evaluated, and K is an arbitrary real nonzero constant.*

Proof. The output signal at time t_0 is

$$s_0(t_0) = \int_{-\infty}^{\infty} H(f)S(f)e^{j\omega t_0}\, df$$

The average power of the output noise is

$$\overline{n_0^2(t)} = R_{n_0}(0) = \int_{-\infty}^{\infty} |H(f)|^2 \mathcal{P}_n(f)\, df$$

Substituting these equations into (6-154), we get

$$\left(\frac{S}{N}\right)_{out} = \frac{\left|\int_{-\infty}^{\infty} H(f)S(f)e^{j\omega t_0}\, df\right|^2}{\int_{-\infty}^{\infty} |H(f)|^2 \mathcal{P}_n(f)\, df} \tag{6-156}$$

We wish to find the particular $H(f)$ that maximizes $(S/N)_{out}$. This can be obtained with the aid of the Schwarz inequality,[‡] which is

$$\left|\int_{-\infty}^{\infty} A(f)B(f)\, df\right|^2 \le \int_{-\infty}^{\infty} |A(f)|^2\, df \int_{-\infty}^{\infty} |B(f)|^2\, df \tag{6-157}$$

where $A(f)$ and $B(f)$ may be complex functions of the real variable f. Furthermore, equality is obtained only when

$$A(f) = KB^*(f) \tag{6-158}$$

[†] It appears that this formulation for the matched filter was first discovered independently by B. M. Dwork and T. S. George in 1950; the result for the white noise case was shown first by D. O. North in 1943 [Root, 1987].

[‡] A proof for the Schwarz inequality is given in the appendix at the end of this chapter.

where K is any arbitrary real constant. The Schwarz inequality may be used to replace the numerator on the right-hand side of (6-156) by letting

$$A(f) = H(f)\sqrt{\mathcal{P}_n(f)}$$

and

$$B(f) = \frac{S(f)e^{j\omega t_0}}{\sqrt{\mathcal{P}_n(f)}}$$

Then (6-156) becomes

$$\left(\frac{\text{S}}{\text{N}}\right)_{\text{out}} \leq \frac{\int_{-\infty}^{\infty} |H(f)|^2 \mathcal{P}_n(f)\,df \int_{-\infty}^{\infty} \frac{|S(f)|^2}{\mathcal{P}_n(f)}\,df}{\int_{-\infty}^{\infty} |H(f)|^2 \mathcal{P}_n(f)\,df}$$

where it is realized that $\mathcal{P}_n(f)$ is a nonnegative real function. Thus

$$\left(\frac{\text{S}}{\text{N}}\right)_{\text{out}} \leq \int_{-\infty}^{\infty} \frac{|S(f)|^2}{\mathcal{P}_n(f)}\,df \tag{6-159}$$

The maximum $(\text{S/N})_{\text{out}}$ is obtained when $H(f)$ is chosen such that equality is attained. This occurs when $A(f) = KB^*(f)$ or

$$H(f)\sqrt{\mathcal{P}_n(f)} = \frac{KS^*(f)e^{-j\omega t_0}}{\sqrt{\mathcal{P}_n(f)}}$$

which reduces to (6-155) of the theorem.

From a practical viewpoint it is realized that the constant K is arbitrary since both the input signal and the input noise would be multiplied by K and K cancels out when $(\text{S/N})_{\text{out}}$ is evaluated. However, both the output signal and noise levels depend on the value of the constant.

In this proof, no constraint was applied to assure that $h(t)$ would be causal. Thus the filter as specified by (6-155) may not be realizable (i.e., causal). However, the transfer function given by (6-155) can often be approximated by a realizable (causal) filter. If the causal constraint is included (in solving for the matched filter), the problem becomes more difficult. A linear integral equation is obtained that must be solved to obtain the unknown function $h(t)$ [Thomas, 1969].

Results for White Noise

For the case of white noise, the description of the matched filter is simplified as follows. For white noise $\mathcal{P}_n(f) = N_0/2$. Thus (6-155) becomes

$$H(f) = \frac{2K}{N_0} S^*(f)e^{-j\omega t_0}$$

From this equation we obtain the following theorem.

THEOREM: *When the input noise is white, the impulse response of the matched filter becomes*

$$h(t) = Cs(t_0 - t) \tag{6-160}$$

where C is an arbitrary real positive constant, t_0 is the time of the peak signal output, and s(t) is the known input-signal waveshape.

Proof

$$h(t) = \mathcal{F}^{-1}[H(f)] = \frac{2K}{N_0}\int_{-\infty}^{\infty} S^*(f)e^{-j\omega t_0}e^{j\omega t}\,df$$

$$= \frac{2K}{N_0}\left[\int_{-\infty}^{\infty} S(f)e^{j2\pi f(t_0-t)}\,df\right]^*$$

$$= \frac{2K}{N_0}[s(t_0 - t)]^*$$

But $s(t)$ is a real signal and let $C = 2K/N_0$, so that the impulse response is equivalent to (6-160).

Equation (6-160) shows that the impulse response of the matched filter (white noise case) is simply the known signal waveshape that is "played backward" and translated by an amount t_0 (as illustrated by Example 6-11). Thus the filter is said to be "matched" to the signal.

An important property is the actual value of $(S/N)_{out}$ that is obtained from the matched filter. From (6-159), using Parseval's theorem as given by (2-41), we obtain

$$\left(\frac{S}{N}\right)_{out} = \int_{-\infty}^{\infty} \frac{|S(f)|^2}{N_0/2}\,df = \frac{2}{N_0}\int_{-\infty}^{\infty} s^2(t)\,dt$$

But $\int_{-\infty}^{\infty} s^2(t)\,dt = E_s$ is the energy in the (finite-duration) input signal. Thus

$$\left(\frac{S}{N}\right)_{out} = \frac{2E_s}{N_0} \tag{6-161}$$

This is a very interesting result. It states that the $(S/N)_{out}$ depends on the signal *energy* and PSD level of the noise and not on the particular signal waveshape that is used. Of course, the signal energy can be increased, to improve $(S/N)_{out}$, by increasing the signal amplitude, the signal duration, or both.

Equation (6-161) can also be written in terms of a time–bandwidth product and the ratio of the input average signal power (over T seconds) to average noise power. Assume that the input noise power is measured in a band that is W hertz wide. We also know that the signal has a duration of T seconds. Then, from (6-161),

$$\left(\frac{S}{N}\right)_{out} = 2TW\frac{(E_s/T)}{(N_0W)} = 2(TW)\left(\frac{S}{N}\right)_{in} \tag{6-162}$$

where $(S/N)_{in} = (E_s/T)/(N_0W)$. From (6-162) we see that it is desirable to have an input signal and noise with a large time–bandwidth product, but (6-161) clearly shows that it is the input-signal energy with respect to N_0 that actually determines the $(S/N)_{out}$ that is attained [Turin, 1976].

EXAMPLE 6-11 INTEGRATE-AND-DUMP (MATCHED) FILTER

Suppose that the known signal is the rectangular pulse, as shown in Fig. 6-16a.

$$s(t) = \begin{cases} 1, & t_1 \leq t \leq t_2 \\ 0, & t \text{ otherwise} \end{cases} \tag{6-163}$$

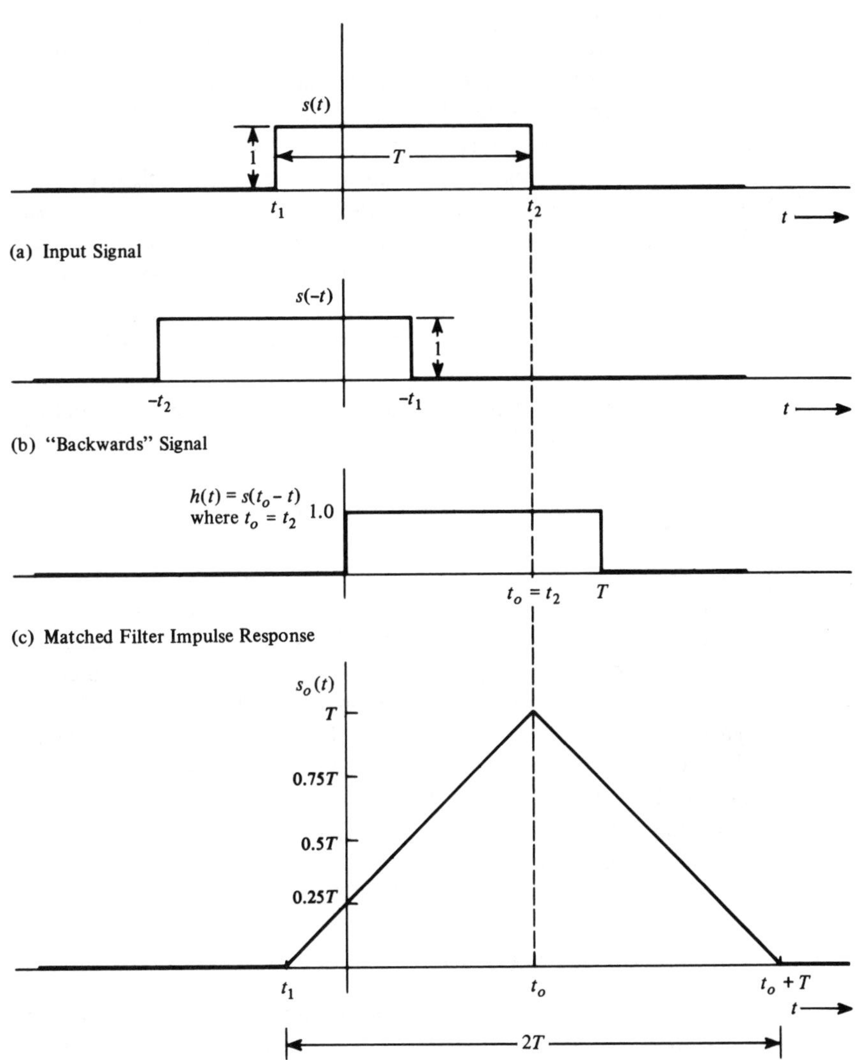

FIGURE 6-16 Waveforms associated with the matched filter of Example 6-11.

where the signal duration is $T = t_2 - t_1$. Then, for the case of white noise, the impulse response required for the matched filter is

$$h(t) = s(t_0 - t) = s(-(t - t_0)) \tag{6-164}$$

C was chosen to be unity for convenience. $s(-t)$ is shown in Fig. 6-16b. From this figure, it is obvious that for the impulse response to be causal we require

$$t_0 \geq t_2 \tag{6-165}$$

We will use $t_0 = t_2$ since this is the smallest allowed value that satisfies the causal condition, and we would like to minimize the time that we have to wait before the maximum signal level occurs at the filter output (i.e., $t = t_0$). A sketch of $h(t)$ for $t_0 = t_2$ is shown in Fig. 6-16c. The resulting output signal is shown in Fig. 6-16d. Note that the peak output signal level does indeed occur at $t = t_0$ and that the input signal waveshape has been distorted by the filter in order to peak up the output signal at $t = t_0$.

In applications to digital signaling with a rectangular bit shape, this matched filter is equivalent to an *integrate-and-dump filter,* as we now illustrate. Assume that we signal with one rectangular pulse and are interested in sampling the filter output when the signal level is maximum. Then the filter output at $t = t_0$ is

$$r_0(t_0) = r(t_0) * h(t_0) = \int_{-\infty}^{\infty} r(\lambda)h(t_0 - \lambda)\, d\lambda$$

When we substitute for the matched-filter impulse response shown in Fig. 6-16c, this equation becomes

$$r_0(t_0) = \int_{t_0-T}^{t_0} r(\lambda)\, d\lambda \tag{6-166}$$

Thus we need to integrate the digital input signal plus noise over one symbol period T (which is the bit period for binary signaling) and "dump" the integrator output at the end of the symbol period. This is illustrated in Fig. 6-17 for binary signaling. Note that for proper operation of this optimum filter, an external clocking signal called *bit sync* is required (see Chapter 3 for a discussion of bit synchronizers). In addition, the output signal is not binary since the output sample values are still corrupted by noise (although the noise has been minimized by the matched filter). Of course, the output could be converted into a binary signal by feeding it into a comparator. This is exactly what is done in digital receivers, as described in Chapters 7 and 8.

Correlation Processing

THEOREM: *For the case of white noise, the matched filter may be realized by correlating the input with $s(t)$.*

$$r_0(t_0) = \int_{t_0-T}^{t_0} r(t)s(t)\, dt \tag{6-167}$$

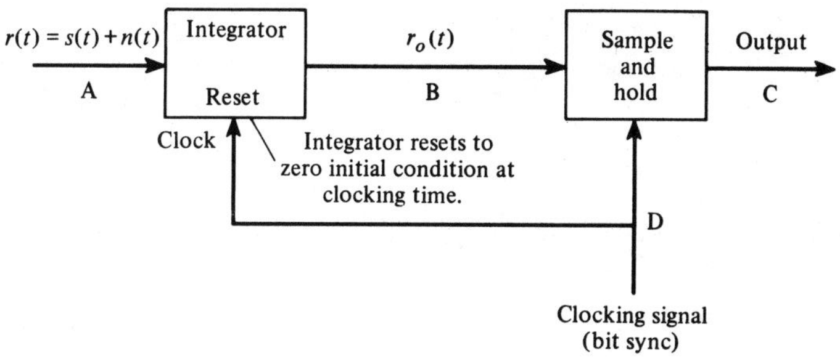

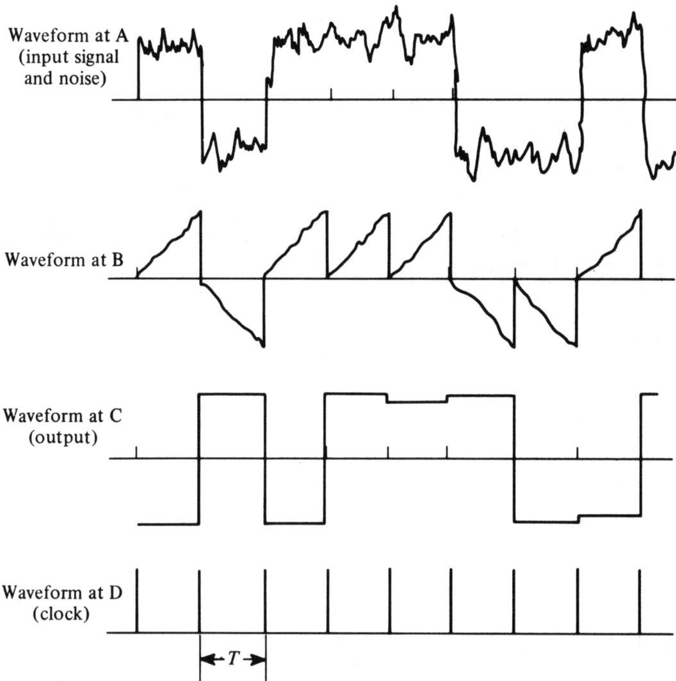

FIGURE 6-17 Integrate-and-dump realization of a matched filter.

where $s(t)$ is the known signal waveshape and $r(t)$ is the processor input, as illustrated in Fig. 6-18.

Proof. The output of the matched filter at time t_0 is

$$r_0(t_0) = r(t_0) * h(t_0) = \int_{-\infty}^{t_0} r(\lambda)h(t_0 - \lambda)\, d\lambda$$

But from (6-160),

$$h(t) = \begin{cases} s(t_0 - t), & 0 \le t \le T \\ 0, & \text{elsewhere} \end{cases}$$

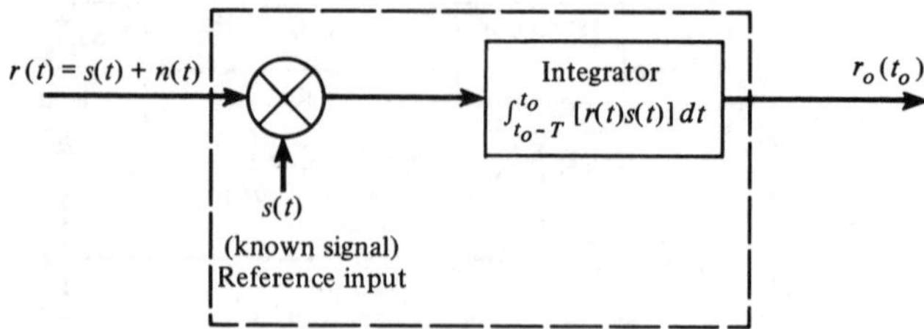

FIGURE 6-18 Matched-filter realization by correlation processing.

so

$$r_0(t_0) = \int_{t_0-T}^{t_0} r(\lambda)s[t_0 - (t_0 - \lambda)]\,d\lambda$$

which is identical to (6-167).

The correlation processor is often used as a matched filter for bandpass signals, as illustrated by Example 6-12.

EXAMPLE 6-12 MATCHED FILTER FOR DETECTION OF A BPSK SIGNAL

Referring to Fig. 6-18, let the filter input be a BPSK signal plus white noise. For example, this might be the IF output in a BPSK receiver. The BPSK signal can be written as

$$s(t) = \begin{cases} +A\cos\omega_c t, & nT < t \le (n+1)T \quad \text{for a binary 1} \\ -A\cos\omega_c t, & nT < t \le (n+1)T \quad \text{for a binary 0} \end{cases}$$

where f_c is the IF center frequency, T is the duration of one bit of data, and n is an integer. The reference input to the correlation processor should be either $+A\cos\omega_c t$ or $-A\cos\omega_c t$, depending on whether we are attempting to detect a binary 1 or a binary 0. Since these waveshapes are identical except for the ± 1 constants, we could just use $\cos\omega_c t$ for the reference and recognize that when a binary 1 BPSK signal is present at the input (no noise), a positive voltage, $\frac{1}{2}AT$, would be produced at the output. Similarly, a binary 0 BPSK signal would produce a negative voltage, $-\frac{1}{2}AT$, at the output. Thus, for BPSK signaling plus white noise, we obtain the matched filter shown in Fig. 6-19. Notice that this looks like the familiar product detector, except that the LPF has been replaced by a gated integrator that is controlled by the bit-sync clock. With this type of postdetection processing, the product detector becomes a matched filter. However, to implement this optimum detector, both bit sync and carrier sync are needed. It is also realized that this technique, shown in Fig. 6-19, could be classified as a more general form of an integrate-and-dump filter (first shown in Fig. 6-17).

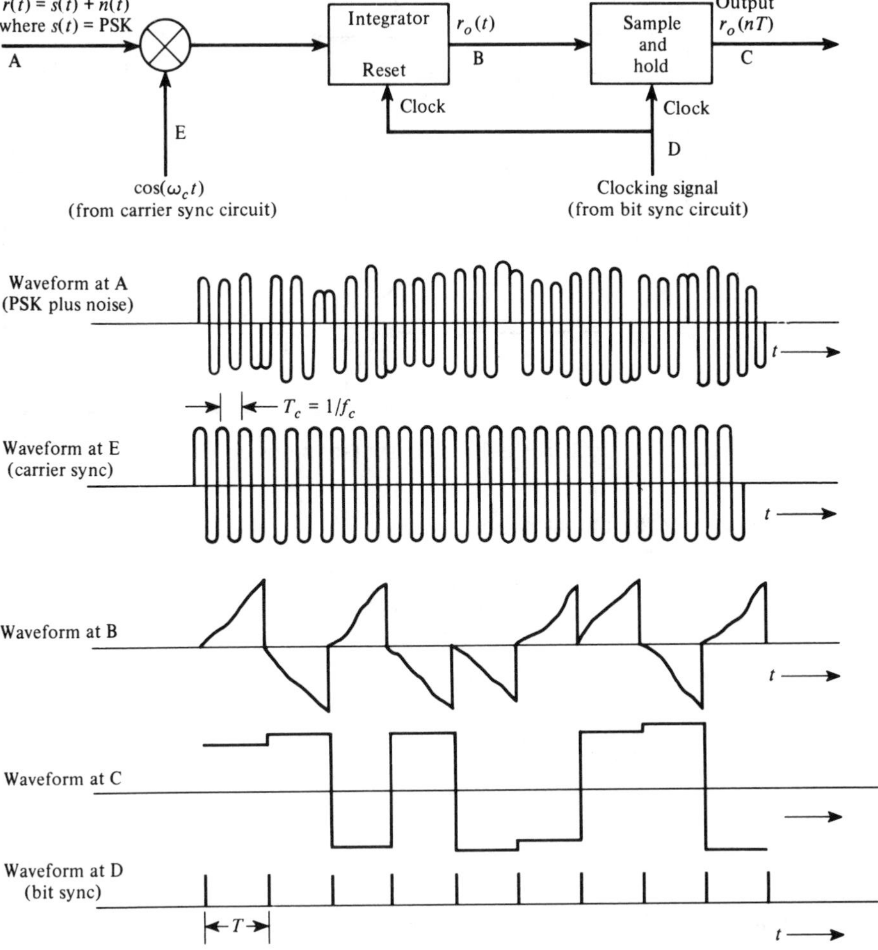

FIGURE 6-19 Correlation (matched-filter) detection for BPSK.

Transversal Matched Filter

A transversal filter can also be designed to satisfy the matched-filter criterion. Referring to Fig. 6-20, we wish to find the set of transversal filter coefficients $\{a_i;\ i = 1, 2, \ldots, N\}$ such that $s_0^2(t_0)/\overline{n_0^2(t)}$ is maximized.

The output signal at time $t = t_0$ is

$$s_0(t_0) = a_1 s(t_0) + a_2 s(t_0 - T) + a_3 s(t_0 - 2T) + \cdots + a_N s(t - (N-1)T)$$

or

$$s_0(t_0) = \sum_{k=1}^{N} a_k s(t_0 - (k-1)T) \qquad (6\text{-}168)$$

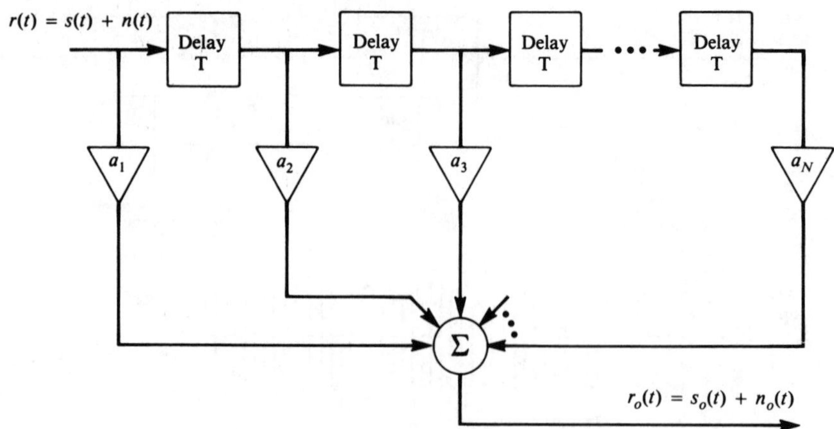

FIGURE 6-20 Transversal matched filter.

Similarly, for the output noise,

$$n_0(t) = \sum_{k=1}^{N} a_k n(t - (k-1)T) \tag{6-169}$$

The average noise power is

$$\overline{n_0^2(t)} = \sum_{k=1}^{N} \sum_{l=1}^{N} a_k a_l \overline{n(t - (k-1)T)n(t - (l-1)T)}$$

or

$$\overline{n_0^2(t)} = \sum_{k=1}^{N} \sum_{l=1}^{N} a_k a_l R_n(kT - lT) \tag{6-170}$$

where $R_n(\tau)$ is the autocorrelation of the input noise. Thus the output peak signal to average noise power ratio is

$$\frac{s_0^2(t_0)}{\overline{n_0^2(t)}} = \frac{\left[\sum_{k=1}^{N} a_k s(t_0 - (k-1)T)\right]^2}{\sum_{k=1}^{N} \sum_{l=1}^{N} a_k a_l R_n(kT - lT)} \tag{6-171}$$

We can find the a's that maximize this ratio by using Lagrange's method of maximizing the numerator while constraining the denominator to be a constant [Olmsted, 1961, p. 518]. Consequently, we need to maximize the function

$$M(a_1, a_2, \ldots, a_N) = \left[\sum_{k=1}^{N} a_k s(t_0 - (k-1)T)\right]^2 - \lambda \sum_{k=1}^{N} \sum_{l=1}^{N} a_k a_l R_n(kT - lT) \tag{6-172}$$

where λ is the Lagrange multiplier. The maximum occurs when $\partial M/\partial a_i = 0$ for all the $i = 1, 2, \ldots, N$. Thus

$$\frac{\partial M}{\partial a_i} = 0 = 2\left[\sum_{k=1}^{N} a_k s(t_0 - (k-1)T)\right] s(t_0 - (i-1)T)$$

$$- 2\lambda \sum_{k=1}^{N} a_k R_n(kT - iT) \tag{6-173}$$

for $i = 1, 2, \ldots, N$. But $\Sigma_{k=1}^{N} a_k s(t_0 - (k-1)T) = s_0(t_0)$, which is a constant. Furthermore, let $\lambda = s_0(t_0)$. Then we obtain the required condition,

$$s(t_0 - (i-1)T) = \sum_{k=1}^{N} a_k R_n(kT - iT) \tag{6-174}$$

for $i = 1, 2, \ldots, N$. This is a set of N simultaneous linear equations that must be solved to obtain the a's. We can obtain these coefficients conveniently by writing (6-174) in matrix form. Define the elements

$$s_i \triangleq s[t_0 - (i-1)T], \qquad i = 1, 2, \ldots, N \tag{6-175}$$

$$r_{ik} = R_n(kT - iT), \qquad i = 1, \ldots, N \tag{6-176}$$

In matrix notation (6-174) becomes

$$\mathbf{s} = \mathbf{Ra} \tag{6-177}$$

where the known signal vector is

$$\mathbf{s} = \begin{bmatrix} s_1 \\ s_2 \\ \vdots \\ s_N \end{bmatrix} \tag{6-178}$$

the known autocorrelation matrix for the input noise is

$$\mathbf{R} = \begin{bmatrix} r_{11} & r_{12} & \cdots & r_{1N} \\ r_{21} & r_{22} & \cdots & r_{2N} \\ \vdots & \vdots & & \vdots \\ r_{N1} & r_{N2} & \cdots & r_{NN} \end{bmatrix} \tag{6-179}$$

and the unknown transversal filter coefficient vector is

$$\mathbf{a} = \begin{bmatrix} a_1 \\ a_2 \\ \vdots \\ a_N \end{bmatrix} \tag{6-180}$$

The coefficients for the transversal matched filter are then given by

$$\mathbf{a} = \mathbf{R}^{-1}\mathbf{s} \tag{6-181}$$

where $\mathbf{R}^{-1}$ is the inverse of the autocorrelation matrix for the noise and $\mathbf{s}$ is the (known) signal vector.

6-9 Summary

A random process is the extension of the concept of a random variable to random waveforms. A random process $x(t)$ is described by an N-dimensional PDF where the random variables are $x_1 = x(t_1)$, $x_2 = x(t_2)$, . . . , $x_N = x(t_N)$. If this PDF is invariant for a shift in the time origin, as $N \to \infty$, the process is said to be *strict-sense stationary*.

The *autocorrelation function* of a random process $x(t)$ is

$$R_x(t_1,t_2) = \overline{x(t_1)x(t_2)} = \int_{-\infty}^{\infty}\int_{-\infty}^{\infty} x_1x_2f_{\mathbf{x}}(x_1,x_2)\,dx_1\,dx_2$$

In general a two-dimensional PDF of $x(t)$ is required to evaluate $R_x(t_1,t_2)$. If the process is stationary,

$$R_x(t_1,t_2) = \overline{x(t_1)x(t_1 + \tau)} = R_x(\tau)$$

where $\tau = t_2 - t_1$.

If $\overline{x(t)}$ is a constant and if $R_x(t_1,t_2) = R_x(\tau)$, the process is said to be *wide-sense stationary*. If a process is strict-sense stationary, it is wide-sense stationary, but the converse is not generally true.

A process is *ergodic* when the time averages are equal to the corresponding ensemble averages. If a process is ergodic, it is also stationary, but the converse is not generally true. For ergodic processes, the dc value (a time average) is also $X_{dc} = \overline{x(t)}$ and the rms value (a time average) is also $X_{rms} = \sqrt{\overline{x^2(t)}}$.

The *power spectral density* (PSD), $\mathcal{P}_x(f)$, is the Fourier transform of the autocorrelation function $R_x(\tau)$ (Wiener–Khintchine theorem).

$$\mathcal{P}_x(f) = \mathcal{F}[R_x(\tau)]$$

The PSD is a nonnegative real function and is even about $f = 0$ for real processes. The PSD can also be evaluated by the ensemble average of a function of the Fourier transforms of the truncated sample functions.

The autocorrelation function of a wide-sense stationary real random process is a real function, and it is even about $\tau = 0$. Furthermore, $R_x(0)$ gives the total normalized average power, and this is the maximum value of $R_x(\tau)$.

For *white noise* the PSD is a constant, and the autocorrelation function is a Dirac delta function located at $\tau = 0$. White noise is not physically realizable because it has infinite power, but it is a very useful approximation for many problems.

The *cross-correlation* function of two jointly stationary real processes $x(t)$ and $y(t)$ is

$$R_{xy}(\tau) = \overline{x(t)y(t+\tau)}$$

and the cross-PSD is

$$\mathcal{P}_{xy}(f) = \mathcal{F}[R_{xy}(\tau)]$$

The two processes are said to be *uncorrelated* if

$$R_{xy}(\tau) = [\overline{x(t)}][\overline{y(t)}]$$

for all τ. They are said to be orthogonal if

$$R_{xy}(\tau) = 0$$

for all τ.

Let $y(t)$ be the output process of a *linear system* and $x(t)$ the input process, where $y(t) = x(t) * h(t)$. Then

$$R_y(\tau) = h(-\tau) * h(\tau) * R_x(\tau)$$

and

$$\mathcal{P}_y(f) = |H(f)|^2 \mathcal{P}_x(f)$$

where $H(f) = \mathcal{F}[h(t)]$.

The *equivalent bandwidth* of a linear system is defined by

$$B = \frac{1}{|H(f_0)|^2} \int_0^\infty |H(f)|^2\, df$$

where $H(f)$ is the transfer function of the system and f_0 is usually taken to be the frequency, where $|H(f)|$ is a maximum. Similarly, the equivalent bandwidth of a random process $x(t)$ is

$$B = \frac{1}{\mathcal{P}_x(f_0)} \int_0^\infty \mathcal{P}_x(f)\, df = \frac{R_x(0)}{2\mathcal{P}_x(f_0)}$$

If the input to a linear system is a Gaussian process, the output is another *Gaussian* process.

A real stationary *bandpass random process* can be represented by

$$v(t) = \text{Re}\{g(t)e^{j(\omega_c t + \theta_c)}\}$$

where the complex envelope $g(t)$ is related to quadrature processes $x(t)$ and $y(t)$. Numerous properties of these random processes can be obtained as listed in Sec. 6-7. For example, $x(t)$ and $y(t)$ are independent Gaussian processes when the PSD of $v(t)$ is symmetrical about $f = f_c$, $f > 0$, and $v(t)$ is Gaussian. Properties for SSB bandpass processes are also obtained.

The *matched filter* is a linear filter that maximizes the instantaneous output-signal power to the average output-noise power for a given input-signal waveshape. For the case of white noise, the impulse response of the matched filter is

$$h(t) = Cs(t_0 - t)$$

where $s(t)$ is the known signal waveshape, C is a real constant, and t_0 is the time that the output signal power is a maximum. The matched filter can be realized in many forms, such as the integrate-and-dump, the correlator, and the transversal filter configurations.

6-10 Appendix: Proof of Schwarz's Inequality

Schwarz's inequality is

$$\left|\int_{-\infty}^{\infty} f(t)g(t)\,dt\right|^2 \le \int_{-\infty}^{\infty} |f(t)|^2\,dt \int_{-\infty}^{\infty} |g(t)|^2\,dt \qquad (6\text{-}182)$$

and becomes an *equality* if and only if

$$f(t) = Kg^*(t) \qquad (6\text{-}183)$$

where K is an arbitrary real constant. $f(t)$ and $g(t)$ may be complex valued. It is assumed that both $f(t)$ and $g(t)$ have finite energy. That is,

$$\int_{-\infty}^{\infty} |f(t)|^2\,dt < \infty \quad \text{and} \quad \int_{-\infty}^{\infty} |g(t)|^2\,dt < \infty \qquad (6\text{-}184)$$

Proof. Schwarz's inequality is equivalent to the inequality.

$$\left|\int_{-\infty}^{\infty} f(t)g(t)\,dt\right| \le \sqrt{\int_{-\infty}^{\infty} |f(t)|^2\,dt}\sqrt{\int_{-\infty}^{\infty} |g(t)|^2\,dt} \qquad (6\text{-}185)$$

Furthermore, it is realized that

$$\left|\int_{-\infty}^{\infty} f(t)g(t)\,dt\right| \le \int_{-\infty}^{\infty} |f(t)g(t)|\,dt = \int_{-\infty}^{\infty} |f(t)|\,|g(t)|\,dt \qquad (6\text{-}186)$$

and equality holds if (6-183) is satisfied. Thus, if we can prove that

$$\int_{-\infty}^{\infty} |f(t)|\,|g(t)|\,dt \le \sqrt{\int_{-\infty}^{\infty} |f(t)|^2\,dt}\sqrt{\int_{-\infty}^{\infty} |g(t)|^2\,dt} \qquad (6\text{-}187)$$

then we have proved Schwarz's inequality. To simplify the notation, replace $|f(t)|$ and $|g(t)|$ by the real-valued functions $a(t)$ and $b(t)$ where

$$a(t) = |f(t)| \qquad (6\text{-}188a)$$

$$b(t) = |g(t)| \qquad (6\text{-}188b)$$

Then, we need to show that

$$\int_{-\infty}^{\infty} a(t)b(t) < \sqrt{\int_{-\infty}^{\infty} a^2(t)\,dt}\sqrt{\int_{-\infty}^{\infty} b^2(t)\,dt} \qquad (6\text{-}189)$$

This can easily be shown by using an orthonormal functional series to represent $a(t)$ and $b(t)$. Let

$$a(t) = a_1\varphi_1(t) + a_2\varphi_2(t) \tag{6-190a}$$

$$b(t) = b_1\varphi_1(t) + b_2\varphi_2(t) \tag{6-190b}$$

where, as described in Chapter 2, $\mathbf{a} = (a_1,a_2)$ and $\mathbf{b} = (b_1,b_2)$ represent $a(t)$ and $b(t)$. This is illustrated in Fig. 6-21.

Then, using the figure,

$$\cos\theta = \cos(\theta_a - \theta_b) = \cos\theta_a \cos\theta_b + \sin\theta_a \sin\theta_b = \frac{a_1}{|\mathbf{a}|}\frac{b_1}{|\mathbf{b}|} + \frac{a_2}{|\mathbf{a}|}\frac{b_2}{|\mathbf{b}|}$$

or

$$\cos\theta = \frac{\mathbf{a}\cdot\mathbf{b}}{|\mathbf{a}|\,|\mathbf{b}|} \tag{6-191}$$

Furthermore, using (8-56), we see that the dot product is equivalent to the inner product

$$\mathbf{a}\cdot\mathbf{b} = \int_{-\infty}^{\infty} a(t)b(t) \tag{6-192a}$$

and the length (or norm) of the vector, using the Pythagorean theorem, (2-84), is equivalent to

$$|\mathbf{a}| = \sqrt{\mathbf{a}\cdot\mathbf{a}} = \sqrt{\int_{-\infty}^{\infty} a^2(t)\,dt} \tag{6-192b}$$

and

$$|\mathbf{b}| = \sqrt{\mathbf{b}\cdot\mathbf{b}} = \sqrt{\int_{-\infty}^{\infty} b^2(t)\,dt} \tag{6-192c}$$

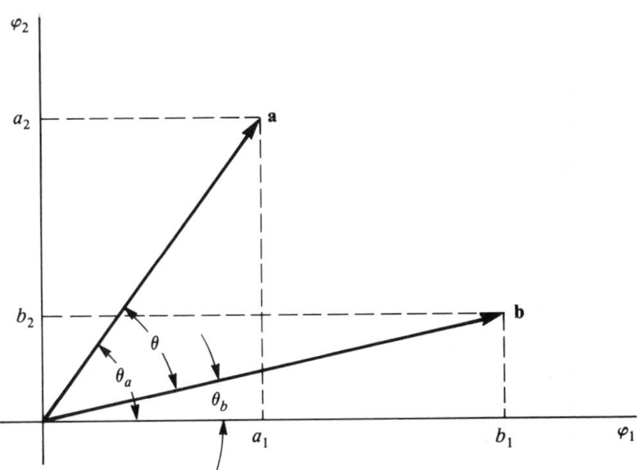

FIGURE 6-21 Vector representation for $a(t)$ and $b(t)$.

Since $|\cos\,\theta| \le 1$ we have

$$\left|\int_{-\infty}^{\infty} a(t)b(t)\,dt\right| \le \sqrt{\int_{-\infty}^{\infty} a^2(t)\,dt}\;\sqrt{\int_{-\infty}^{\infty} b^2(t)\,dt} \tag{6-193}$$

where equality is obtained when we substitute

$$\mathbf{a} = K\mathbf{b} \tag{6-194}$$

Using (6-188), when $f(t) = Kg^*(t)$, we find that (6-194) is also satisfied so that equality is obtained. Thus the proof of (6-193) proves Schwarz's inequality.

PROBLEMS

6-1 Let a random process $x(t)$ be defined by

$$x(t) = At + B$$

(a) If B is a constant and A is uniformly distributed between -1 and $+1$, sketch a few sample functions.

(b) If A is a constant and B is uniformly distributed between 0 and 2, sketch a few sample functions.

6-2 Let a random process be given by

$$x(t) = A\cos(\omega_0 t + \theta)$$

where

$$f(\theta) = \begin{cases} \dfrac{2}{\pi}, & 0 \le \theta \le \dfrac{\pi}{2} \\ 0, & \text{elsewhere} \end{cases}$$

(a) Evaluate $\overline{x(t)}$.

(b) From the result of part (a), what can be said about the stationarity of the process?

6-3 Using the random process described in Prob. 6-2:

(a) Evaluate $\langle x^2(t)\rangle$.

(b) Evaluate $\overline{x^2(t)}$.

(c) Using the results of parts (a) and (b), determine if the process is ergodic for these averages.

6-4 A conventional average reading ac voltmeter (volt-ohm multimeter) has a schematic diagram as shown in Fig. P6-4. The needle of the meter movement deflects proportionally to the average current flowing through the meter. The meter scale is marked to give the rms value of sine-wave voltages. Suppose that this meter is used to determine the rms value of a noise voltage. The noise voltage is known to be an ergodic Gaussian process having a zero mean value. What is the value of the constant that is multiplied by the meter reading to give the true rms value of the Gaussian noise?

(*Hint:* The diode is a short circuit when the input voltage is positive and an open circuit when the input voltage is negative.)

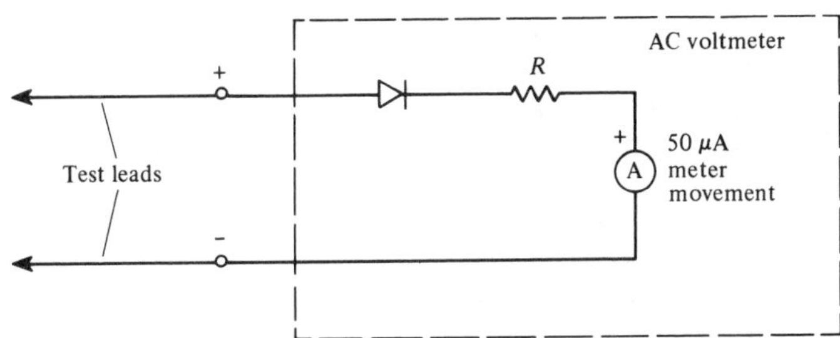

FIGURE P6-4

6-5 Let $x(t) = A_0 \sin(\omega_0 t + \theta)$ be a random process where θ is a random variable that is uniformly distributed between 0 and 2π, and A_0 and ω_0 are constants.
(a) Find $R_x(\tau)$.
(b) Show that $x(t)$ is wide-sense stationary.
(c) Verify that $R_x(\tau)$ satisfies the appropriate properties.

6-6 Let $r(t) = A_0 \cos \omega_0 t + n(t)$, where A_0 and ω_0 are constants. Assume that $n(t)$ is a wide-sense stationary random noise process with a zero mean value and an autocorrelation of $R_n(\tau)$.
(a) Find $\overline{r(t)}$ and determine if $r(t)$ is wide-sense stationary.
(b) Find $R_r(t_1,t_2)$.
(c) Evaluate $\langle R_r(t,t + \tau)\rangle$ where $t_1 = t$ and $t_2 = t + \tau$.

6-7 Let an additive signal plus noise process be described by $r(t) = s(t) + n(t)$.
(a) Show that $R_r(\tau) = R_s(\tau) + R_n(\tau) + R_{sn}(\tau) + R_{ns}(\tau)$.
(b) Simplify the result for part (a) for the case when $s(t)$ and $n(t)$ are independent and the noise has a zero mean value.

6-8 Consider the sum of two ergodic noise voltages

$$n(t) = n_1(t) + n_2(t)$$

The power of $n_1(t)$ is 5 W, the power of $n_2(t)$ is 10 W, the dc value of $n_1(t)$ is -2 V, and the dc value n_2 is $+1$ V. Find the power of $n(t)$ if
(a) $n_1(t)$ and $n_2(t)$ are orthogonal.
(b) $n_1(t)$ and $n_2(t)$ are uncorrelated.
(c) The cross-correlation of $n_1(t)$ and $n_2(t)$ is 2 for $\tau = 0$.

6-9 Assume that $x(t)$ is ergodic and let $x(t) = m_x + y(t)$, where $m_x = \overline{x(t)}$ is the dc value of $x(t)$ and $y(t)$ is the ac component of $x(t)$. Show that
(a) $R_x(\tau) = m_x^2 + R_y(\tau)$.
(b) $\lim_{\tau\to\infty} R_x(\tau) = m_x^2$.
(c) If $R_x(\tau)$ can be measured, how can the dc value of $x(t)$ be determined from $R_x(\tau)$?

6-10 Determine if the following functions satisfy the properties for autocorrelation functions.
(a) $\sin \omega_0 \tau$.
(b) $(\sin \omega_0 \tau)/(\omega_0 \tau)$.
(c) $\cos \omega_0 \tau + \delta(\tau)$.
(d) $e^{-a|\tau|}$, where $a < 0$.
(*Note:* $\mathcal{F}[R(\tau)]$ must also be a nonnegative function.)

6-11 A random process $x(t)$ has an autocorrelation function given by $R_x(\tau) = 5 + 8e^{-3|\tau|}$. Find:
(a) The rms value for $x(t)$.
(b) The PSD for $x(t)$.

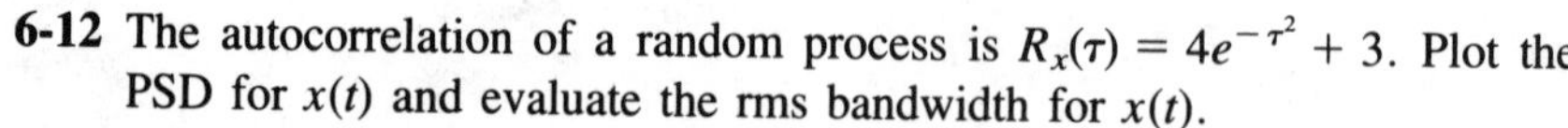

6-12 The autocorrelation of a random process is $R_x(\tau) = 4e^{-\tau^2} + 3$. Plot the PSD for $x(t)$ and evaluate the rms bandwidth for $x(t)$.

6-13 Show that two random processes $x(t)$ and $y(t)$ are uncorrelated (i.e., $R_{xy}(\tau) = m_x m_y$) if the processes are independent.

6-14 If $x(t)$ contains periodic components, show that:
(a) $R_x(\tau)$ contains periodic components.
(b) $\mathcal{P}_x(f)$ contains delta functions.

6-15 Find the PSD for the random process described in Prob. 6-2.

6-16 Determine if the following functions can be valid PSD functions for a real process.
(a) $2e^{-2\pi|f-45|}$.
(b) $4e^{-2\pi[f^2-16]}$.
(c) $25 + \delta(f - 16)$.
(d) $10 + \delta(f)$.

6-17 The PSD of an ergodic random process $x(t)$ is

$$\mathcal{P}_x(f) = \begin{cases} \dfrac{1}{B}(B - |f|), & |f| \leq B \\ 0, & f \text{ otherwise} \end{cases}$$

where $B > 0$. Find:
(a) The rms value of $x(t)$.
(b) $R_x(\tau)$.

6-18 Referring to the techniques described in Example 6-3, evaluate the PSD for a PCM signal that uses Manchester NRZ encoding (see Fig. 3-11). Assume that the data have values of $a_n = \pm 1$, which are equally likely, and that the data are independent from bit to bit.

6-19 The *magnitude* frequency response of a linear time-invariant network is to be determined from a laboratory setup as shown in Fig. P6-19. Discuss how $|H(f)|$ is evaluated from the measurements.

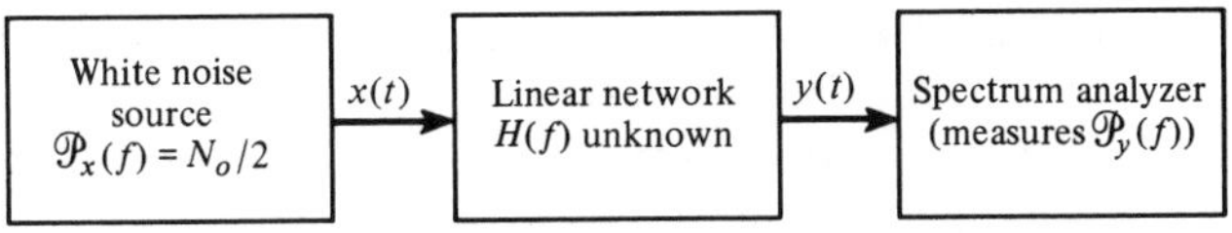

White noise source
$\mathcal{P}_x(f) = N_o/2$
x(t)
Linear network
H(f) unknown
y(t)
Spectrum analyzer
(measures $\mathcal{P}_y(f)$)

FIGURE P6-19

6-20 A linear time-invariant network with an unknown $H(f)$ is shown in Fig. P6-20.

(a) Find a formula for evaluating $h(t)$ in terms of $R_{xy}(\tau)$ and N_0.

(b) Find a formula for evaluating $H(f)$ in terms of $\mathcal{P}_{xy}(f)$ and N_0.

White noise source
$\mathcal{P}_x(f) = N_o/2$
x(t)
Linear network
H(f) unknown
y(t)
x(t)

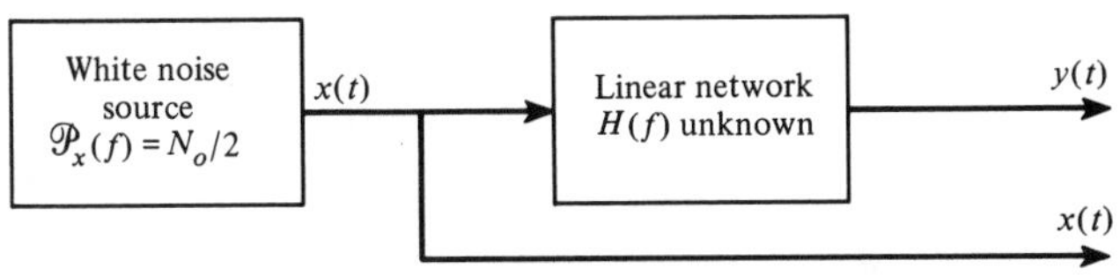

FIGURE P6-20

6-21 The output of a linear system is related to the input by $y(t) = h(t) * x(t)$, where $x(t)$ and $y(t)$ are jointly wide-sense stationary. Show that:

(a) $R_{xy}(\tau) = h(\tau) * R_x(\tau)$.

(b) $\mathcal{P}_{xy}(f) = H(f)\mathcal{P}_x(f)$.

(c) $R_{yx}(\tau) = h(-\tau) * R_x(\tau)$.

(d) $\mathcal{P}_{yx}(f) = H^*(f)\mathcal{P}_x(f)$.

[*Hint:* Use (6-86) and (6-87).]

6-22 Ergodic white noise with a PSD of $\mathcal{P}_n(f) = N_0/2$ is applied to the input of an ideal integrator with a gain of K (a real number) such that $H(f) = K/(j2\pi f)$.

(a) Find the PSD for the output.

(b) Find the rms value of the output noise.

6-23 A linear system has a power transfer function $|H(f)|^2$ as shown in Fig. P6-23. The input $x(t)$ is a Gaussian random process with a PSD given by

$$\mathcal{P}_x(f) = \begin{cases} \frac{1}{2} N_0, & |f| \le 2B \\ 0, & f \text{ elsewhere} \end{cases}$$

|H(f)|²
1.0
-B
B
f

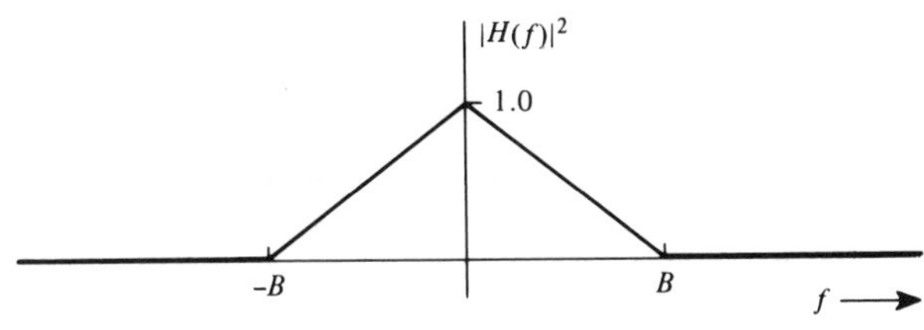

FIGURE P6-23

(a) Find the autocorrelation function for the output $y(t)$.
(b) Find the PDF for $y(t)$.
(c) When are the two random variables $y_1 = y(t_1)$ and $y_2 = y(t_2)$ independent?

6-24 A linear filter evaluates the T-second moving average of an input waveform where the filter output is

$$y(t) = \frac{1}{T} \int_{t-(T/2)}^{t+(T/2)} x(u)\, du$$

and $x(t)$ is the input.
(a) Show that the impulse response is $h(t) = (1/T)\ \Pi(t/T)$.
(b) Show that

$$R_y(\tau) = \frac{1}{T} \int_{-T}^{T} \left(1 - \frac{|u|}{T}\right) R_x(\tau - u)\, du$$

(c) If $R_x(\tau) = e^{-|\tau|}$ and $T = 1$ sec, plot $R_y(\tau)$ and compare it with $R_x(\tau)$.

6-25 As shown in Example 6-5, the output signal-to-noise ratio of an RC LPF is given by (6-95) when the input is a sinusoidal signal plus white noise. Find the value of the RC product such that the output signal-to-noise ratio will be a maximum.

6-26 Assume that a sine wave of peak amplitude A_0 and frequency f_0 plus white noise with $\mathcal{P}_n(f) = N_0/2$ is applied to a linear filter. The transfer function of the filter is

$$H(f) = \begin{cases} \dfrac{1}{B}(B - |f|), & |f| < B \\ 0, & f \text{ elsewhere} \end{cases}$$

where B is the absolute bandwidth of the filter. Find the signal-to-noise power ratio for the filter output.

6-27 For the random process $x(t)$ with the PSD shown in Fig. P6-27, determine
(a) The equivalent bandwidth.
(b) The rms bandwidth.

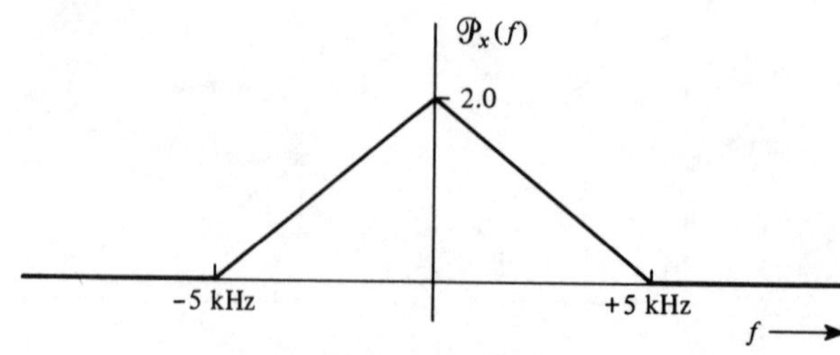

FIGURE P6-27

6-28 If $x(t)$ is a real bandpass random process that is wide-sense stationary, show that the definition for rms bandwidth, (6-100), is equivalent to

$$B_{\text{rms}} = 2\sqrt{\overline{f^2} - (f_0)^2}$$

where $\overline{f^2}$ is given by (6-98) or (6-99) and f_0 is given by (6-102).

6-29 In the definition for the rms bandwidth of a *bandpass* random process, f_0 is used. Show that

$$f_0 = \frac{1}{2\pi R_x(0)}\left(\frac{d\hat{R}_x(\tau)}{d\tau}\right)\Bigg|_{\tau=0}$$

where $\hat{R}_x(\tau)$ is the Hilbert transform of $R_x(\tau)$.

6-30 Two identical *RC* LPFs are coupled in cascade by an isolation amplifier that has a voltage gain of 10.

(a) Find the overall transfer function of the network as a function of R and C.

(b) Find the 3-dB bandwidth in terms of R and C.

6-31 Let $x(t)$ be a Gaussian process where two random variables are $x_1 = x(t_1)$ and $x_2 = x(t_2)$. The random variables have variances of σ_1^2 and σ_2^2 and means of m_1 and m_2. The correlation coefficient is

$$\rho = \overline{(x_1 - m_1)(x_2 - m_2)}/(\sigma_1\sigma_2)$$

Using the matrix notation for the $N = 2$ dimensional PDF, show that this reduces to the bivariate Gaussian PDF as given by (B-97).

6-32 A bandlimited white Gaussian random process has an autocorrelation function that is specified by (6-125). Show that as $B \to \infty$, the autocorrelation function becomes $R_n(\tau) = \frac{1}{2}N_0\delta(\tau)$.

6-33 Let two random processes $x(t)$ and $y(t)$ be jointly Gaussian with zero-mean values. That is, $(x_1, x_2, \ldots, x_N, y_1, y_2, \ldots, y_M)$ is described by an $(N + M)$-dimensional Gaussian PDF. The cross correlation is

$$R_{xy}(\tau) = \overline{x(t_1)y(t_2)} = 10\sin(2\pi\tau)$$

(a) When are the random variables $x_1 = x(t_1)$ and $y_2 = y(t_2)$ independent?

(b) Show that $x(t)$ and $y(t)$ are or are not independent random processes.

6-34 Starting with (6-121), show that

$$\mathbf{C}_y = \mathbf{H}\mathbf{C}_x\mathbf{H}^T$$

(*Hint:* Use the identity matrix property, $\mathbf{A}\mathbf{A}^{-1} = \mathbf{A}^{-1}\mathbf{A} = \mathbf{I}$, where $\mathbf{I}$ is the identity matrix.)

6-35 Given the random process

$$x(t) = A_0 \cos(\omega_c t + \theta)$$

where A_0 and ω_0 are constants and θ is a random variable that is uniformly distributed over the interval $(0, \pi/2)$.

(a) Determine if $x(t)$ is wide-sense stationary.
(b) Find the PSD for $x(t)$.
(c) If θ is uniformly distributed over $(0,2\pi)$, is $x(t)$ wide-sense stationary?

6-36 A bandpass WSS random process $v(t)$ is represented by (6-129a) where the conditions of (6-129) are satisfied. The PSD of $v(t)$ is shown in Fig. P6-36 where $f_c = 1$ MHz. Using MathCAD:
(a) Plot $\mathcal{P}_x(f)$.
(b) Plot $\mathcal{P}_{xy}(f)$.

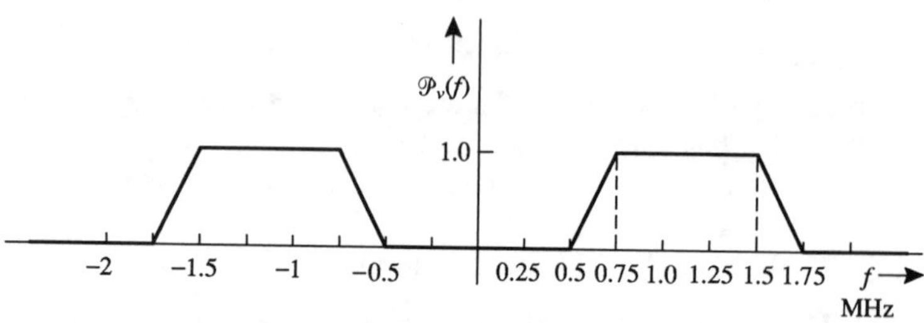

FIGURE P6-36

6-37 The PSD of a bandpass WSS process $v(t)$ is shown in Fig. P6-37. $v(t)$ is the input to a product detector and the oscillator signal (i.e., the second input to the multiplier) is $5\cos(\omega_c t + \theta_0)$, where $f_c = 1$ MHz and θ_0 is an independent random variable with a uniform distribution over $(0,2\pi)$. Using MathCAD, plot the PSD for the output of the product detector.

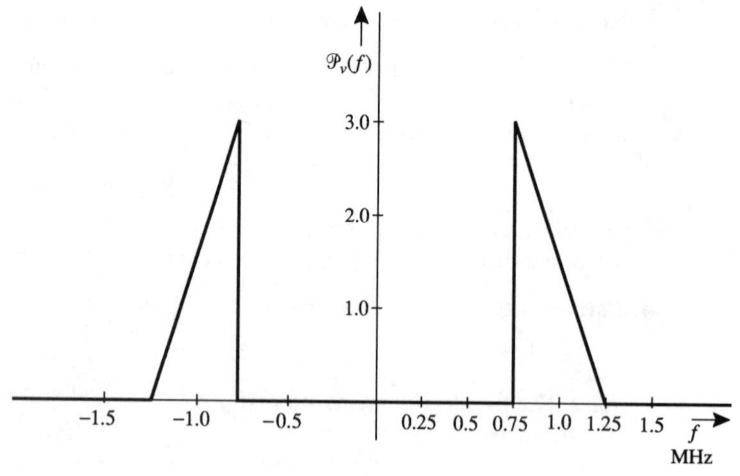

FIGURE P6-37

6-38 A WSS bandpass process $v(t)$ is applied to a product detector as shown in Fig. 6-11 where $\theta_c = 0$.

(a) Derive an expression for the autocorrelation of $w_1(t)$ in terms of $R_v(\tau)$. Is $w_1(t)$ WSS?
(b) Using $R_{w_1}(\tau)$ obtained in part (a), to find an expression for $\mathcal{P}_{w_1}(f)$. (*Hint:* Use the Wiener–Khintchine theorem.)

6-39 A USSB signal is

$$v(t) = 10\,\text{Re}\{[x(t) + j\hat{x}(t)]e^{j(\omega_c t + \theta_c)}\}$$

where θ_c is a random variable that is uniformly distributed over $(0,2\pi)$. The PSD for $x(t)$ is given in Fig. P6-27. Find:
(a) The PSD for $v(t)$.
(b) The total power of $v(t)$.

6-40 Show that:
(a) $R_{\hat{x}}(\tau) = R_x(\tau)$.
(b) $R_{x\hat{x}}(\tau) = \hat{R}_x(\tau)$.

6-41 A bandpass random signal can be represented by

$$s(t) = x(t)\cos(\omega_c t + \theta_c) - y(t)\sin(\omega_c t + \theta_c)$$

where the PSD of $s(t)$ is shown in Fig. P6-41. θ_c is an independent random variable that is uniformly distributed over $(0,2\pi)$. Assume that $f_3 - f_2 = f_2 - f_1$. Find the PSD for $x(t)$ and $y(t)$ when
(a) $f_c = f_1$. This is USSB signaling where $y(t) = \hat{x}(t)$.
(b) $f_c = f_2$. This represents independent USSB and LSSB signaling with two different modulations [neither one is $x(t)$ or $y(t)$].
(c) $f_1 < f_c < f_2$. This is vestigial sideband signaling.
(d) For which, if any, of these cases are $x(t)$ and $y(t)$ orthogonal?

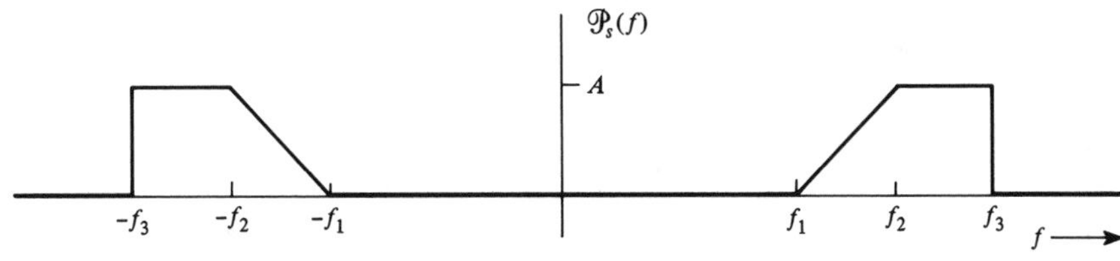

FIGURE P6-41

6-42 Referring to Prob. 6-41(b), how are the two modulations $m_1(t)$ and $m_2(t)$ for the independent sidebands related to $x(t)$ and $y(t)$? Give the PSD for $m_1(t)$ and $m_2(t)$, where $m_1(t)$ is the modulation on the USSB portion of the signal and $m_2(t)$ is the modulation on the LSSB portion of $s(t)$.

6-43 For the bandpass random process, show that:
(a) Equation (6-133m) is valid (property 13).
(b) Equation (6-133i) is valid (property 9).

6-44 Referring to Example 6-9, find the PSD for a BPSK signal with Manchester-encoded data (see Fig. 3-11). Assume that the data have values of $a_n = \pm 1$ that are equally likely and that the data are independent from bit to bit.

6-45 The input to an envelope detector is an ergodic bandpass Gaussian noise process. The rms value of the input is 2 V and the mean value is 0 V. The envelope detector has a voltage gain of 10. Find:
(a) The dc value of the output voltage.
(b) The rms value of the output voltage.

6-46 A narrowband signal plus noise process is

$$r(t) = A\cos(\omega_c t + \theta_c) + x(t)\cos(\omega_c t + \theta_c) - y(t)\sin(\omega_c t + \theta_c)$$

where $A\cos(\omega_c t + \theta_c)$ is a sinusoidal carrier and the remaining terms are the bandpass noise. Let the noise have an rms value of σ and a mean of 0. This signal plus noise process appears at the input to an envelope detector. Show that the PDF for the output of the envelope detector is

$$f(R) = \begin{cases} \dfrac{R}{\sigma^2} e^{-[(R^2+A^2)/(2\sigma^2)]} I_0\left(\dfrac{RA}{\sigma^2}\right), & R \geq 0 \\ 0, & R < 0 \end{cases}$$

where

$$I_0(z) \triangleq \frac{1}{2\pi}\int_0^{2\pi} e^{z\cos\theta}\, d\theta$$

is the modified Bessel function of the first kind of zero order. $f(R)$ is known as a *Rician* PDF in honor of the late S. O. Rice, who was an outstanding engineer at Bell Telephone Laboratories.

6-47 Assume that a known signal is

$$s(t) = \begin{cases} \dfrac{A}{T} t\cos\omega_c t, & 0 \leq t \leq T \\ 0, & t \text{ otherwise} \end{cases}$$

This signal plus white noise is present at the input to a matched filter.
(a) Design a matched filter for this signal. Sketch the waveforms for this problem analogous to those shown in Fig. 6-16.
(b) Sketch the waveforms for the correlation processor shown in Fig. 6-18.

6-48 A baseband digital communication system uses polar signaling with a bit rate of $R = 2000$ bits/sec. The transmitted pulses are rectangular and the frequency response of the channel filter is

$$H_c(f) = \frac{B}{B + jf}$$

where $B = 6000$ Hz. The filtered pulses are the input to a receiver that uses integrate-and-dump processing as illustrated in Fig. 6-17. Examine the integrator output for ISI. In particular:

(a) Plot the integrator output when a single binary "1" is sent.

(b) Plot the integrator output for an all-pass channel and compare this result with that obtained in part (a).

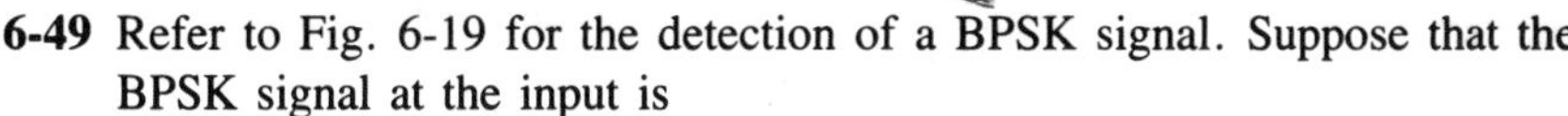

6-49 Refer to Fig. 6-19 for the detection of a BPSK signal. Suppose that the BPSK signal at the input is

$$r(t) = s(t) = \sum_{n=0}^{7} d_n p(t - nT)$$

where

$$p(t) = \begin{cases} e^{-t}\cos(\omega_c t), & 0 < t < T \\ 0, & t \text{ elsewhere} \end{cases}$$

and the binary data d_n is the 8 bit string $\{+1,-1,-1,+1,+1,-1,+1,-1\}$. Use MathCAD to:

(a) Plot the input waveform $r(t)$.

(b) Plot the integrator output waveform $r_0(t)$.

6-50 A matched filter is described in Fig. P6-50.

(a) Find the impulse response of the matched filter.

(b) Find the pulse shape to which this filter is matched (white noise case).

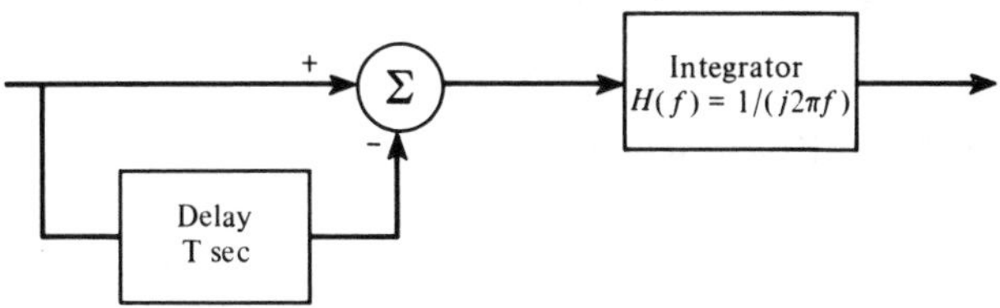

FIGURE P6-50

6-51 A FSK signal, $s(t)$, is applied to a matched-filter receiver circuit that is shown in Fig. P6-51. The FSK signal is

$$s(t) = \begin{cases} A\cos(\omega_1 t), & \text{when a binary 1 is sent} \\ A\cos(\omega_2 t), & \text{when a binary 0 is sent} \end{cases}$$

where $f_1 = f_c + \Delta F$ and $f_2 = f_c - \Delta F$. Let ΔF be chosen to satisfy the MSK condition, which is $\Delta F = 1/(4T)$. T is the time that it takes to send one bit and the integrator is reset every T seconds. Assume that $A = 1$, $f_c = 1000$ Hz, and $\Delta F = 50$ Hz.

(a) If a binary 1 is sent, plot $v_1(t)$, $v_2(t)$, and $r_0(t)$ over a T-second interval.

(b) Referring to Fig. 6-16, find an expression that describes the output of a filter that is matched to the FSK signal when a binary 1 is sent. Plot it.

(c) Discuss how the plots obtained for parts (a) and (b) agree or disagree.

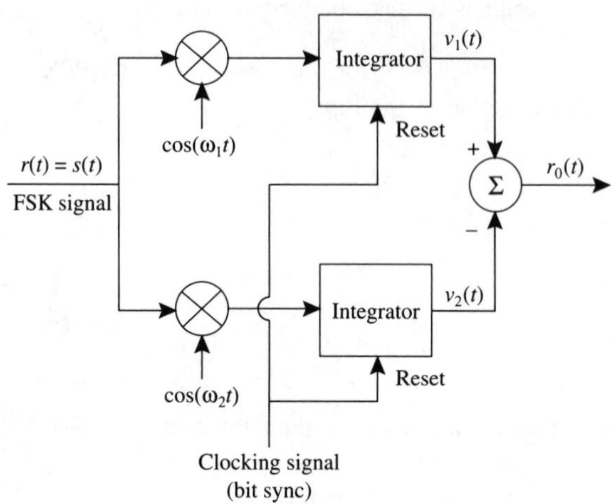

FIGURE P6-51

6-52 Let

$$s(t) = \text{Re}\{g(t)e^{j(\omega_c t + \theta_c)}\}$$

be a wideband FM signal where

$$g(t) = A_c e^{jD_f \int_{-\infty}^{t} m(\lambda)d\lambda}$$

and $m(t)$ is a random modulation process.

(a) Show that

$$R_g(\tau) = A_c^2[\overline{e^{jD_f \tau\, m(t)}}]$$

when the integral $\int_t^{t+\tau} m(\lambda)\, d\lambda$ is approximated by $\tau m(t)$.

(b) Using the results of part (a), prove that the PSD of the wideband FM signal is given by (4-167). That is, show that

$$\mathcal{P}_s(f) = \frac{\pi A_c^2}{2D_f}\left[f_m\left(\frac{2\pi}{D_f}(f - f_c)\right) + f_m\left(\frac{2\pi}{D_f}(-f - f_c)\right)\right]$$

CHAPTER 7

Performance of Communication Systems Corrupted by Noise

As indicated in Chapter 1, the two primary considerations in the design of a communication system are:

1. The *performance* of the system when it is corrupted by noise. The performance measure for a digital system is the probability of error of the output signal. For analog systems, the performance measure is the output signal-to-noise ratio.
2. The channel *bandwidth* that is required for transmission of the communication signal. This bandwidth was evaluated for various types of digital and analog signals in the previous chapters.

There are numerous ways in which the information can be demodulated (recovered) from the received signal that has been corrupted by noise. Some receivers provide optimum performance, but most do not. Often a suboptimum receiver will be used in order to lower the cost. In addition, some suboptimum receivers perform almost as well as optimum ones for all practical purposes. Here we will *analyze* the performance of some suboptimum as well as some optimum receivers. In Chapter 8 *design* techniques are developed that yield optimum receivers for digital signaling.

7-1 Error Probabilities for Binary Signaling

General Results

Figure 7-1 is a general block diagram for a binary communication system. The receiver input $r(t)$ consists of the transmitted signal $s(t)$ plus channel noise $n(t)$. For *baseband* signaling the processing circuits in the receiver consist of low-pass filtering with appropriate amplification. For *bandpass* signaling, such as OOK, BPSK, and FSK, discussed in Sec. 5-2, the processing circuits normally consist of a superheterodyne receiver containing a mixer, an IF amplifier, and a detector. These circuits produce a baseband analog output $r_0(t)$. (For example, when BPSK signaling is used, the detector might consist of a product detector and an integrator as described in Sec. 6-8 and illustrated in Fig. 6-19.)

The analog baseband waveform $r_0(t)$ is sampled at the clocking time $t = t_0 + nT$ to produce the samples $r_0(t_0 + nT)$, which are fed into a threshold de-

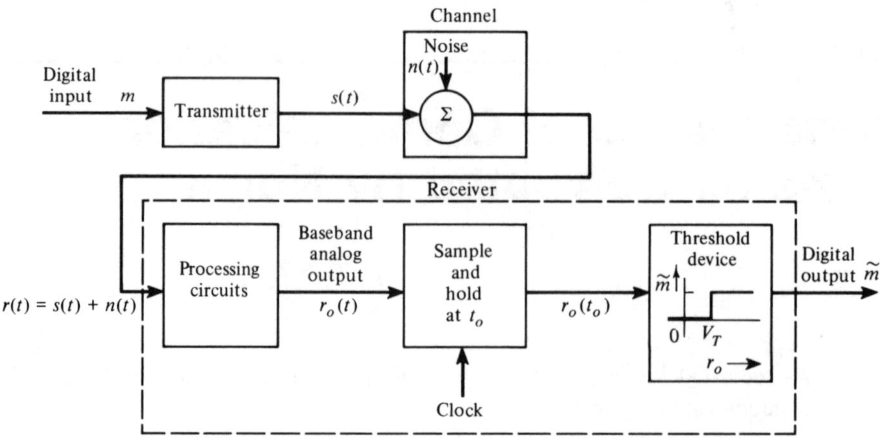

FIGURE 7-1 General binary communication system.

vice (a comparator). The threshold device produces the binary serial-data waveform $\tilde{m}(t)$.

In this subsection we develop a *general technique* for evaluating the probability of bit error, also called the *bit error rate* (BER), for binary signaling. In later sections, this technique will be used to obtain specific expressions for the BER of various binary signaling schemes such as OOK, BPSK, and FSK.

To develop a general formula for the BER of a detected binary signal, let T be the length of time that it takes to transmit one bit of data. The transmitted signal over a bit interval of $(0,T)$ is represented by

$$s(t) = \begin{cases} s_1(t), & 0 < t \le T \quad \text{for a binary 1} \\ s_2(t), & 0 < t \le T \quad \text{for a binary 0} \end{cases} \tag{7-1}$$

where $s_1(t)$ is the waveform that is used if a binary 1 is transmitted and $s_2(t)$ is the waveform that is used if a binary 0 is transmitted. If $s_1(t) = -s_2(t)$, $s(t)$ is called an *antipodal* signal. The binary signal plus noise at the receiver input produces a baseband analog waveform at the output of the processing circuits that is denoted by

$$r_0(t) = \begin{cases} r_{01}(t), & 0 < t \le T \quad \text{for a binary 1 sent} \\ r_{02}(t), & 0 < t \le T \quad \text{for a binary 0 sent} \end{cases} \tag{7-2}$$

where $r_{01}(t)$ is the output signal that is corrupted by noise for a binary 1 transmission and $r_{02}(t)$ is the output for the binary 0 transmission. (Note that if the receiver uses nonlinear processing circuits, such as an envelope detector, the superposition of the signal plus noise outputs are not valid operations.) This analog voltage waveform $r_0(t)$ is sampled at some time t_0 during the bit interval.

That is, $0 < t_0 \le T$. For matched-filter processing circuits, t_0 is usually T. The resulting sample is

$$r_0(t_0) = \begin{cases} r_{01}(t_0) & \text{for a binary 1 sent} \\ r_{02}(t_0) & \text{for a binary 0 sent} \end{cases} \tag{7-3}$$

It is realized that $r_0(t_0)$ is a *random variable* that has a *continuous* distribution because the channel noise has corrupted the signal. To shorten the notation, we will denote $r_0(t_0)$ simply by r_0. That is,

$$r_0 = r_0(t_0) = \begin{cases} r_{01} & \text{for a binary 1 sent} \\ r_{02} & \text{for a binary 0 sent} \end{cases} \tag{7-4}$$

$r_0 = r_0(t_0)$ is called the *test statistic.*

For the moment, let us assume that we can evaluate the PDFs for the two random variables $r_0 = r_{01}$ and $r_0 = r_{02}$. These PDFs are actually *conditional PDFs* since they depend, respectively, on a binary 1 or a binary 0 being transmitted. That is, when $r_0 = r_{01}$, the PDF is $f(r_0|s_1 \text{ sent})$ and when $r_0 = r_{02}$ the PDF is $f(r_0|s_2 \text{ sent})$. These conditional PDFs are shown in Fig. 7-2. For illustrative purposes Gaussian shapes are shown. The actual shapes of the PDFs depend on the characteristics of the channel noise, the specific types of filter and detector circuits used, and the types of binary signals transmitted. (In later sections, we obtain specific PDFs using the theory developed in Chapter 6.)

In our development of a general formula for the BER, assume that the *polarity of the processing circuits* of the receiver is such that if the *signal only* (no noise) were present at the receiver input, $r_0 > V_T$ when a *binary 1 is sent* and $r_0 < V_T$ when a *binary 0 is sent;* V_T is the threshold (voltage) setting of the comparator (threshold device).

When signal plus noise is present at the receiver input, errors can occur in two ways. An error occurs when $r_0 < V_T$ if a binary 1 is sent:

$$P(\text{error}|s_1 \text{ sent}) = \int_{-\infty}^{V_T} f(r_0|s_1)\, dr_0 \tag{7-5}$$

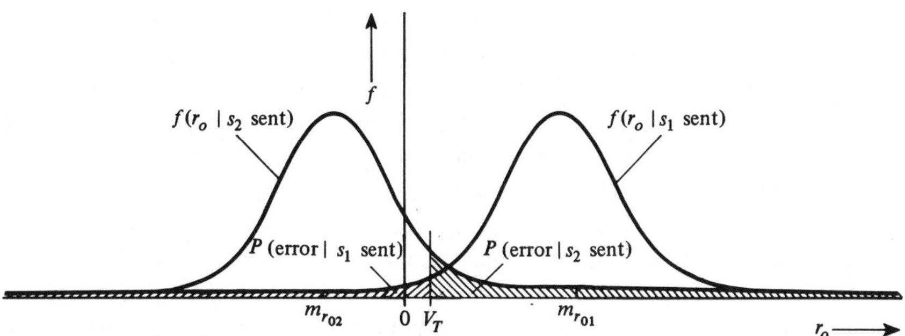

FIGURE 7-2 Error probability for binary signaling.

This is illustrated by a shaded area to the left of V_T in Fig. 7-2. Similarly, an error occurs when $r_0 > V_T$ if a binary 0 is sent:

$$P(\text{error}|s_2 \text{ sent}) = \int_{V_T}^{\infty} f(r_0|s_2)\, dr_0 \tag{7-6}$$

The bit error rate is then

$$P_e = P(\text{error}|s_1 \text{ sent})P(s_1\text{sent}) + P(\text{error}|s_2 \text{ sent})P(s_2\text{sent}) \tag{7-7}$$

This follows from probability theory (see Appendix B), where the probability of an event that consists of joint events is

$$P(E) = \sum_{i=1}^{2} P(E, s_i) = \sum_{i=1}^{2} P(E|s_i)P(s_i)$$

When we combine (7-5), (7-6), and (7-7), the general expression for the BER of any binary communication system is

$$P_e = P(s_1 \text{ sent}) \int_{-\infty}^{V_T} f(r_0|s_1)\, dr_0 + P(s_2 \text{ sent}) \int_{V_T}^{\infty} f(r_0|s_2)\, dr_0 \tag{7-8}$$

$P(s_1$ sent) and $P(s_2$ sent) are known as the *source statistics* or *a priori statistics*. In most applications, the source statistics are considered to be equally likely. That is,

$$P(\text{binary 1 sent}) = P(s_1 \text{ sent}) = \tfrac{1}{2} \tag{7-9a}$$

$$P(\text{binary 0 sent}) = P(s_2 \text{ sent}) = \tfrac{1}{2} \tag{7-9b}$$

In the results that we obtain throughout the remainder of this chapter we will assume that the source statistics are equally likely. As indicated earlier, the conditional PDFs depend on the signaling waveshapes involved, the channel noise, and the receiver processing circuits used. These are obtained subsequently for the case of Gaussian channel noise and linear processing circuits.

Results for Gaussian Noise

Assume that the channel noise is a zero-mean wide-sense stationary Gaussian process and that the receiver processing circuits, except for the threshold device, are linear. Then we know (see Chapter 6) that for Gaussian noise (only) at the input, the output of the linear processor, $r_0(t) = n_0(t)$, will also be a Gaussian process. For baseband signaling, the processing circuits would consist of linear filters with some gain. For bandpass signaling, as we demonstrated in Chapter 4, a superheterodyne circuit (consisting of a mixer, IF stage, and product detector) is a linear circuit. However, if automatic gain control (AGC) or limiters are used, the receiver will be nonlinear and the results of this section will not be applicable. In addition, if a nonlinear detector such as an envelope detector is used, the output noise will not be Gaussian. For the case of a linear processing receiver circuit with a binary signal plus noise at the input, the sampled output is

$$r_0 = s_0 + n_0 \tag{7-10}$$

Here the shortened notation $r_0(t_0) = r_0$ is used. $n_0(t_0) = n_0$ is a zero-mean Gaussian random variable, and $s_0(t_0) = s_0$ is a constant that depends on the signal being sent. That is,

$$s_0 = \begin{cases} s_{01} & \text{for a binary 1 sent} \\ s_{02} & \text{for a binary 0 sent} \end{cases} \tag{7-11}$$

where s_{01} and s_{02} are known constants for a given type of receiver with known input signaling waveshapes $s_1(t)$ and $s_2(t)$. Since the output noise sample n_0 is a zero-mean Gaussian random variable, the total output sample r_0 is a Gaussian random variable with a mean value of either s_{01} or s_{02}, depending on whether a binary 1 or a binary 0 was sent. This is illustrated in Fig. 7-2, where the mean value of r_0 is $m_{r_{01}} = s_{01}$ when a binary 1 is sent and the mean value of r_0 is $m_{r_{02}} = s_{02}$ when a binary 0 is sent. Thus the two conditional PDFs are

$$f(r_0|s_1) = \frac{1}{\sqrt{2\pi}\,\sigma_0} e^{-(r_0 - s_{01})^2/(2\sigma_0^2)} \tag{7-12}$$

and

$$f(r_0|s_2) = \frac{1}{\sqrt{2\pi}\,\sigma_0} e^{-(r_0 - s_{02})^2/(2\sigma_0^2)} \tag{7-13}$$

$\sigma_0^2 = \overline{n_0^2} = \overline{n_0^2(t_0)} = \overline{n_0^2(t)}$ is the average power of the *output* noise from the receiver processing circuit where the output noise process is wide-sense stationary.

Using equally likely source statistics and substituting (7-12) and (7-13) into (7-8), we find that the bit error rate becomes

$$P_e = \frac{1}{2}\int_{-\infty}^{V_T} \frac{1}{\sqrt{2\pi}\,\sigma_0} e^{-(r_0 - s_{01})^2/(2\sigma_0^2)}\, dr_0 + \frac{1}{2}\int_{V_T}^{\infty} \frac{1}{\sqrt{2\pi}\,\sigma_0} e^{-(r_0 - s_{02})^2/(2\sigma_0^2)}\, dr_0 \tag{7-14}$$

This can be reduced to the $Q(z)$ functions defined in Sec. B-7 (Appendix B) and tabulated in Sec. A-10 (Appendix A). Let $\lambda = -(r_0 - s_{01})/\sigma_0$ in the first integral and $\lambda = (r_0 - s_{02})/\sigma_0$ in the second integral; then

$$P_e = \frac{1}{2}\int_{-(V_T - s_{01})/\sigma_0}^{\infty} \frac{1}{\sqrt{2\pi}} e^{-\lambda^2/2}\, d\lambda + \frac{1}{2}\int_{(V_T - s_{02})/\sigma_0}^{\infty} \frac{1}{\sqrt{2\pi}} e^{-\lambda^2/2}\, d\lambda$$

or

$$P_e = \frac{1}{2} Q\left(\frac{-V_T + s_{01}}{\sigma_0}\right) + \frac{1}{2} Q\left(\frac{V_T - s_{02}}{\sigma_0}\right) \tag{7-15}$$

By using the appropriate value for the comparator threshold, V_T, this probability of error can be minimized. To find the V_T that minimizes P_e, we need to solve $dP_e/dV_T = 0$. Using Leibniz's rule, (A-36), for differentiating the integrals of (7-14), we obtain

$$\frac{dP_e}{dV_T} = \frac{1}{2}\frac{1}{\sqrt{2\pi}\sigma_0} e^{-(V_T - s_{01})^2/(2\sigma_0^2)} - \frac{1}{2}\frac{1}{\sqrt{2\pi}\sigma_0} e^{-(V_T - s_{02})^2/(2\sigma_0^2)} = 0$$

or

$$e^{-(V_T - s_{01})^2/(2\sigma_0^2)} = e^{-(V_T - s_{02})^2/(2\sigma_0^2)}$$

which implies the condition

$$(V_T - s_{01})^2 = (V_T - s_{02})^2$$

Consequently, for minimum P_e, the threshold setting of the comparator needs to be

$$V_T = \frac{s_{01} + s_{02}}{2} \tag{7-16}$$

Substituting (7-16) into (7-15), we obtain the expression for the minimum P_e. Thus, for binary signaling in Gaussian noise and with the optimum threshold setting as specified by (7-16), the BER is

$$P_e = Q\left(\frac{s_{01} - s_{02}}{2\sigma_0}\right) = Q\left(\sqrt{\frac{(s_{01} - s_{02})^2}{4\sigma_0^2}}\right) \tag{7-17}$$

where it is assumed that $s_{01} > V_T > s_{02}$.† So far, we have optimized only the threshold level, not the filters in the processing circuits.

Results for White Gaussian Noise and Matched-Filter Reception

If the receiving filter (in the processing circuits of Fig. 7-1) is optimized, the BER as given by (7-17) can be reduced. To *minimize* P_e we need to *maximize* the argument of Q, as is readily seen from Fig. B-7. Thus we need to find the linear filter that maximizes

$$\frac{[s_{01}(t_0) - s_{02}(t_0)]^2}{\sigma_0^2} = \frac{[s_d(t_0)]^2}{\sigma_0^2}$$

$s_d(t_0) \triangleq s_{01}(t_0) - s_{02}(t_0)$ is the *difference signal* sample value that is obtained by subtracting the sample s_{02} from s_{01}. The corresponding instantaneous power of the difference output signal at $t = t_0$ is $s_d^2(t_0)$. As derived in Sec. 6-8, the linear filter that maximizes the instantaneous output signal power at the sampling time $t = t_0$ when compared to the average output noise power $\sigma_0^2 = \overline{n_0^2(t)}$ is the *matched filter*. For the case of *white noise* at the receiver input, the matched filter needs to be matched to the difference signal $s_d(t) = s_1(t) - s_2(t)$. Thus the impulse response of the matched filter for binary signaling is

$$h(t) = C[s_1(t_0 - t) - s_2(t_0 - t)] \tag{7-18}$$

†If $s_{01} < V_T < s_{02}$, the result is

$$P_e = Q\left(\frac{s_{02} - s_{01}}{2\sigma_0}\right) = Q\left(\sqrt{\frac{(s_{02} - s_{01})^2}{4\sigma_0^2}}\right)$$

where the characteristic of the threshold device in Fig. 7-1 is altered so that a binary 0 is chosen when $r_0 > V_T$ and a binary 1 is chosen when $r_0 < V_T$.

where $s_1(t)$ is the signal (only) that appears at the receiver input when a binary 1 is sent, $s_2(t)$ is the signal that is received when a binary 0 is sent, and C is a real constant. Furthermore, by using (6-161) the output peak signal to average noise ratio that is obtained from the matched filter is

$$\frac{[s_d(t_0)]^2}{\sigma_0^2} = \frac{2E_d}{N_0}$$

$N_0/2$ is the PSD of the noise at the *receiver input,* and E_d is the difference signal energy at the *receiver input,* where

$$E_d = \int_0^T [s_1(t) - s_2(t)]^2 \, dt \tag{7-19}$$

Thus, for binary signaling corrupted by *white* Gaussian noise, *matched-filter* reception, and by using the optimum threshold setting, the bit error rate is

$$P_e = Q\left(\sqrt{\frac{E_d}{2N_0}}\right) \tag{7-20}$$

We will use this result to evaluate the P_e for various types of binary signaling schemes where matched-filter reception is used.

Results for Colored Gaussian Noise and Matched-Filter Reception

The technique that we just used to obtain the BER for binary signaling in white noise can be modified to evaluate the BER for the colored noise case. The modification used is illustrated in Fig. 7-3. Here a *prewhitening filter* is inserted ahead of the receiver processing circuits. The transfer function of the prewhitening filter is

$$H_p(f) = \frac{1}{\sqrt{\mathcal{P}_n(f)}} \tag{7-21}$$

so that the noise that appears at the filter output, $\tilde{n}(t)$, is white. We have now converted the colored noise problem into a white noise problem, so that the design techniques presented in the preceding section are applicable. The matched

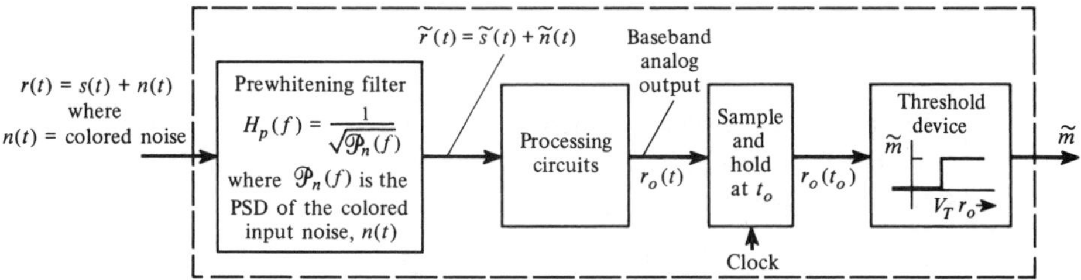

FIGURE 7-3 Matched-filter receiver for colored noise.

filter in the processing circuits is now matched to the *filtered waveshapes,*

$$\tilde{s}(t) = \tilde{s}_1(t) = s_1(t) * h_p(t) \qquad \text{(binary 1)} \tag{7-22a}$$

$$\tilde{s}(t) = \tilde{s}_2(t) = s_2(t) * h_p(t) \qquad \text{(binary 0)} \tag{7-22b}$$

where $h_p(t) = \mathcal{F}^{-1}[H_p(f)]$. Since the prewhitening will produce signals $\tilde{s}_1(t)$ and $\tilde{s}_2(t)$, which are spread beyond the T-sec signaling interval, two types of degradation will result.

1. The signal energy of the filtered signal that occurs beyond the T-sec interval will not be used by the matched filter in maximizing the output signal.
2. The portions of signals from previous signaling intervals that occur in the present signaling interval will produce ISI (see Chapter 3).

Both of these effects can be reduced if the duration of the original signal is made less than the T-sec signaling interval so that almost all of the spread signal will occur within that interval.

7-2 Performance of Baseband Binary Systems

Unipolar Signaling

As illustrated in Fig. 7-4b, the two baseband signaling waveforms are

$$s_1(t) = +A, \qquad 0 < t \le T \qquad \text{(binary 1)} \tag{7-23a}$$

$$s_2(t) = 0, \qquad 0 < t \le T \qquad \text{(binary 0)} \tag{7-23b}$$

where $A > 0$. This unipolar signal plus white Gaussian noise is present at the receiver input.

First, evaluate the performance of a receiver that uses an LPF $H(f)$ with unity gain. Assume that the equivalent bandwidth of this LPF is $B > 2/T$, so that the unipolar signaling waveshape is preserved (approximately) at the filter output and yet the noise will be reduced by the filter. Thus $s_{01}(t_0) \approx A$ and $s_{02}(t_0) \approx 0$. The noise power at the output of the filter is $\sigma_0^2 = (N_0/2)(2B)$, where B is the equivalent bandwidth of the filter. The optimum threshold setting is then $V_T = \frac{1}{2}A$. When we use (7-17), the BER is

$$P_e = Q\left(\sqrt{\frac{A^2}{4N_0B}}\right) \qquad \text{(low-pass filter)} \tag{7-24a}$$

for a receiver that uses an LPF with an equivalent bandwidth of B.

The performance of a matched-filter receiver is obtained by using (7-20), where the sampling time is $t_0 = T$. The energy in the difference signal is $E_d = A^2T$, so that the BER is

$$P_e = Q\left(\sqrt{\frac{A^2T}{2N_0}}\right) = Q\left(\sqrt{\frac{E_b}{N_0}}\right) \qquad \text{(matched filter)} \tag{7-24b}$$

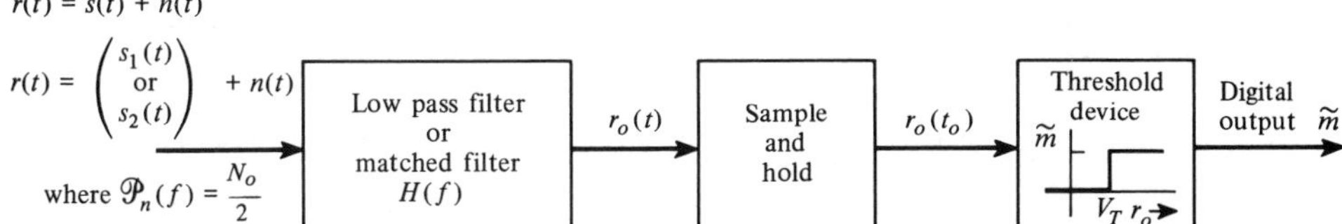

(a) Receiver

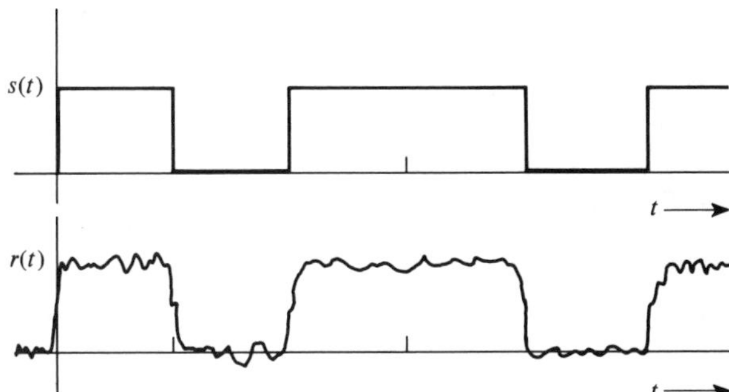

(b) Unipolar Signaling

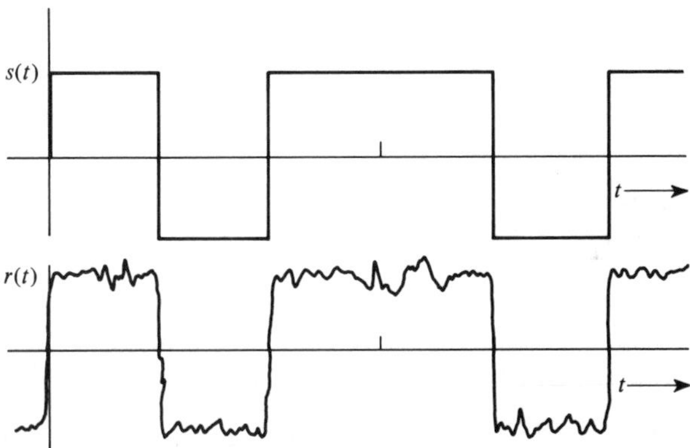

(c) Polar Signaling

FIGURE 7-4 Receiver for baseband binary signaling.

where the average energy per bit is $E_b = A^2T/2$. For the rectangular pulse shape, the matched filter is an integrator. Consequently, the optimum threshold value is

$$V_T = \frac{s_{01} + s_{02}}{2} = \frac{1}{2}\left(\int_0^T A\,dt + 0\right) = \frac{AT}{2}$$

Often it is desirable to express the BER in terms of E_b/N_0 since it indicates the average energy required to transmit one bit of data over a white (thermal) noise channel. By expressing the BER in terms of E_b/N_0, the performance of different signaling techniques can be easily compared.

A plot of (7-24b) is shown in Fig. 7-5.

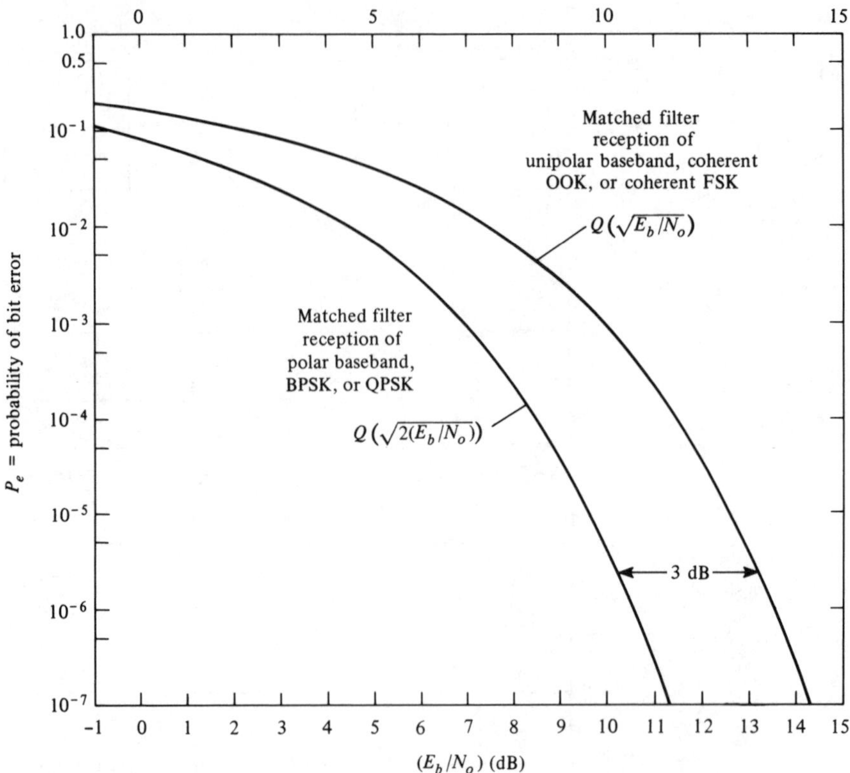

FIGURE 7-5 P_e for matched-filter reception of several binary signaling schemes.

Polar Signaling

As shown in Fig. 7-4c, the baseband polar signaling waveform is

$$s_1(t) = +A, \qquad 0 < t \le T \qquad \text{(binary 1)} \tag{7-25a}$$

$$s_2(t) = -A, \qquad 0 < t \le T \qquad \text{(binary 0)} \tag{7-25b}$$

The polar signal is an antipodal signal since $s_1(t) = -s_2(t)$.

The performance of an LPF receiver system is obtained by using (7-17). Assuming that the equivalent bandwidth of the LPF is $B \ge 2/T$, we realize that the output signal samples are $s_{01}(t_0) \approx A$ and $s_{02}(t_0) \approx -A$ at the sampling time $t = t_0$. In addition, $\sigma_0^2 = N_0 B$. The optimum threshold setting is now $V_T = 0$. Thus the BER for polar signaling is

$$P_e = Q\left(\sqrt{\frac{A^2}{N_0 B}}\right) \qquad \text{(low-pass filter)} \tag{7-26a}$$

where B is the equivalent bandwidth of the LPF.

The performance of the matched-filter receiver is obtained, once again, by using (7-20), where $t_0 = T$. (The integrate-and-dump matched filter for polar signaling was given in Fig. 6-17.) Since the energy of the difference signal is $E_d = (2A)^2 T$, the BER is

$$P_e = Q\left(\sqrt{\frac{2A^2T}{N_0}}\right) = Q\left(\sqrt{2\left(\frac{E_b}{N_0}\right)}\right) \qquad \text{(matched filter)} \qquad (7\text{-}26b)$$

where the average energy per bit is $E_b = A^2T$. The optimum threshold setting is $V_T = 0$.

A plot of the BER for unipolar and polar baseband signaling is shown in Fig. 7-5. It is apparent that polar signaling has a 3-dB advantage over unipolar signaling since unipolar signaling requires a E_b/N_0 that is 3 dB larger than that for polar signaling for the same P_e.

Bipolar Signaling

For bipolar NRZ signaling binary 1's are represented by alternating positive and negative values, and the binary 0's are represented by a zero level. Thus

$$s_1(t) = \pm A, \qquad 0 < t \le T \qquad \text{(binary 1)} \qquad (7\text{-}27a)$$

$$s_2(t) = 0, \qquad 0 < t \le T \qquad \text{(binary 0)} \qquad (7\text{-}27b)$$

where $A > 0$. This is similar to unipolar signaling except two thresholds, $+V_T$ and $-V_T$, are needed as shown in Fig. 7-6. Figure 7-6b illustrates the error probabilities for the case of additive Gaussian noise. The BER is

$$P_e = P(\text{error}| + A \text{ sent})P(+A \text{ sent}) + P(\text{error}| - A \text{ sent})P(-A \text{ sent}) + P(\text{error}|s_2 \text{ sent})P(s_2 \text{ sent})$$

and, by using Fig. 7-6b,

$$P_e \approx \left[Q\left(\frac{V_T}{\sigma_0}\right)\right]\frac{1}{4} + \left[Q\left(\frac{V_T}{\sigma_0}\right)\right]\frac{1}{4} + \left[2Q\left(\frac{V_T}{\sigma_0}\right)\right]\frac{1}{2}$$

which reduces to

$$P_e = \frac{3}{2}Q\left(\frac{V_T}{\sigma_0}\right)$$

or, by using the optimum threshold value, $V_T = \frac{1}{2}A$,

$$P_e = \frac{3}{2}Q\left(\frac{A}{2\sigma_0}\right)$$

For the case of a receiver with a low-pass filter that has a bipolar signal plus white noise at its input, $\sigma_0^2 = N_0 B$. Thus the BER is

$$P_e = \frac{3}{2}Q\left(\sqrt{\frac{A^2}{4N_0B}}\right) \qquad \text{(low-pass filter)} \qquad (7\text{-}28a)$$

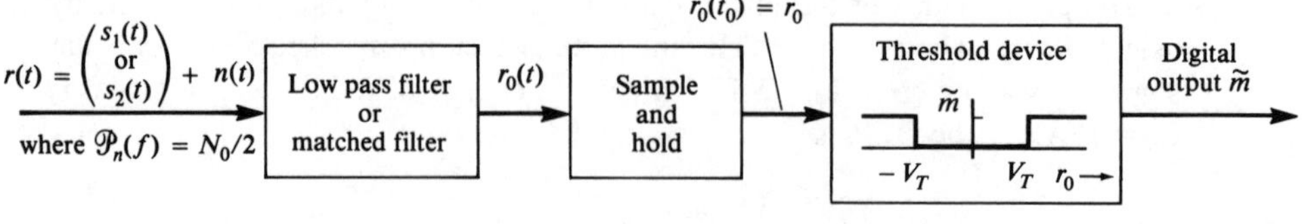

(a) Receiver

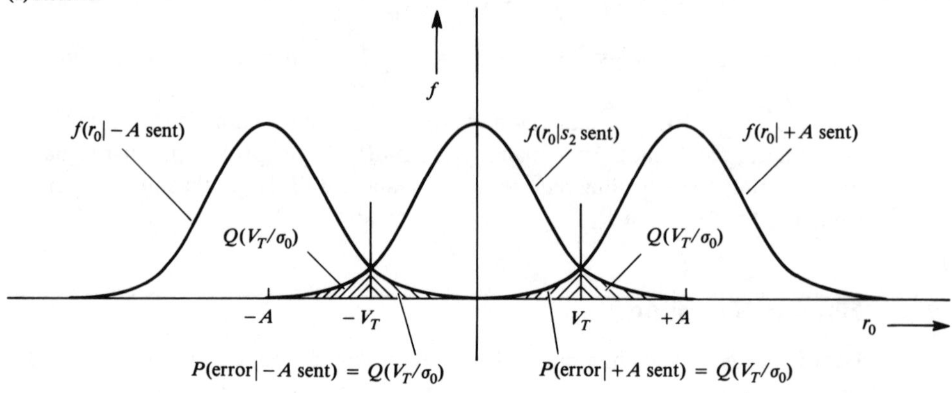

(b) Conditional PDFs

FIGURE 7-6 Receiver for bipolar signaling.

where the PSD of the input noise is $N_0/2$ and the equivalent bandwidth of the filter is B Hz. If a matched filter is used, its output (S/N) is, using (6-161),

$$\left(\frac{S}{N}\right)_{\text{out}} = \frac{A^2}{\sigma_0^2} = \frac{2E_d}{N_0}$$

For bipolar NRZ signaling the energy in the different signal is $E_d = A^2T = 2E_b$ where E_b is the average energy per bit. Thus for a matched-filter receiver, the BER is

$$P_e = \frac{3}{2}Q\left(\sqrt{\frac{E_b}{N_0}}\right) \qquad \text{(matched filter)} \qquad (7\text{-}28b)$$

For bipolar RZ signaling, $E_d = A^2T/4 = 2E_b$ so that the resulting BER formula is identical to (7-28b). These results show that the BER for bipolar signaling is just $\frac{3}{2}$ that for unipolar signaling as described by (7-24b).

7-3 Coherent Detection of Bandpass Binary Signals

On-Off Keying

From Fig. 5-1c, an on-off keying (OOK) signal is represented by

$$s_1(t) = A\cos(\omega_c t + \theta_c), \qquad 0 < t \leq T \qquad \text{(binary 1)} \qquad (7\text{-}29a)$$

or

$$s_2(t) = 0, \qquad 0 < t \leq T \qquad \text{(binary 0)} \qquad (7\text{-}29b)$$

For coherent detection, a product detector is used as illustrated in Fig. 7-7. Actually, in RF applications a mixer would convert the incoming OOK signal to an IF so that stable high-gain amplification could be conveniently achieved and then a product detector would translate the signal to baseband. Figure 7-7, equivalently, represents these operations by converting the incoming signal and noise directly to baseband.

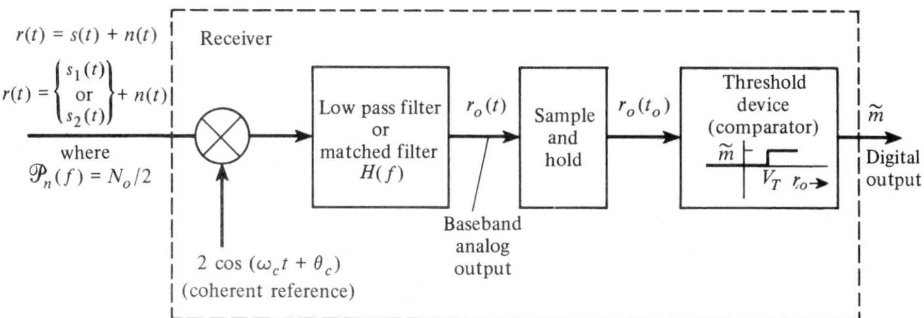

(a) Receiver

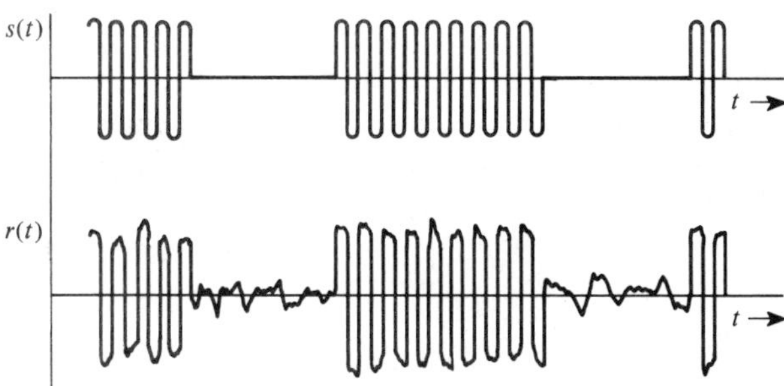

(b) OOK Signaling

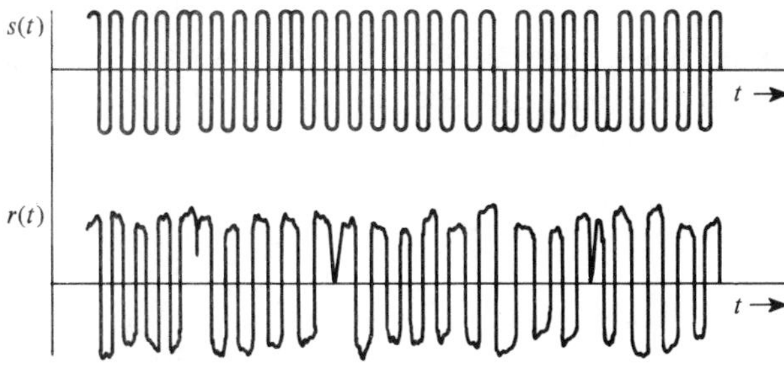

(c) BPSK Signaling

FIGURE 7-7 Coherent detection of OOK or BPSK signals.

Assume that the OOK signal plus white (over the equivalent bandpass) Gaussian noise is present at the receiver input. As developed in Chapter 6, this bandpass noise may be represented by

$$n(t) = x(t)\cos(\omega_c t + \theta_n) - y(t)\sin(\omega_c t + \theta_n)$$

where the PSD of $n(t)$ is $\mathcal{P}_n(f) = N_0/2$, and θ_n is a uniformly distributed random variable that is independent of θ_c.

The receiving filter $H(f)$ of Fig. 7-7 may be some convenient LPF or it may be a matched filter. Of course, the receiver would be optimized (for the lowest P_e) if a matched filter were used.

First, evaluate the performance of a receiver that uses an LPF where the filter has a dc gain of unity. Assume that the equivalent bandwidth of the filter is $B \geq 2/T$ so that the envelope of the OOK signal is (approximately) preserved at the filter output. The baseband analog output will be

$$r_0(t) = \begin{Bmatrix} A, & 0 < t \leq T, & \text{binary 1} \\ 0, & 0 < t \leq T, & \text{binary 0} \end{Bmatrix} + x(t) \qquad (7\text{-}30)$$

where $x(t)$ is the baseband noise. The noise power is $\overline{x^2(t)} = \sigma_0^2 = \overline{n^2(t)} = 2(N_0/2)(2B) = 2N_0B$. Since $s_{01} = A$ and $s_{02} = 0$, the optimum threshold setting is $V_T = A/2$. When we use (7-17), the BER is

$$P_e = Q\left(\sqrt{\frac{A^2}{8N_0B}}\right) \quad \text{(narrowband filter)} \qquad (7\text{-}31)$$

B is the equivalent bandwidth of the LPF. The equivalent bandpass bandwidth of this receiver is $B_p = 2B$.

The performance of a matched-filter receiver is obtained by using (7-20). The energy in the difference signal at the receiver input is†

$$E_d = \int_0^T [A\cos(\omega_c t + \theta_c) - 0]^2\, dt = \frac{A^2T}{2} \qquad (7\text{-}32)$$

Consequently, the BER is

$$P_e = Q\left(\sqrt{\frac{A^2T}{4N_0}}\right) = Q\left(\sqrt{\frac{E_b}{N_0}}\right) \quad \text{(matched filter)} \qquad (7\text{-}33)$$

where the average energy per bit is $E_b = A^2T/4$. For this case, where $s_1(t)$ has a rectangular (real) envelope, the matched filter is an integrator. Consequently, the optimum threshold value is

$$V_T = \frac{s_{01} + s_{02}}{2} = \frac{1}{2}s_{01} = \frac{1}{2}\left[\int_0^T 2A\cos^2(\omega_c t + \theta_c)\, dt\right]$$

† Strictly speaking, f_c needs to be an integer multiple of half the bit rate, $R = 1/T$, to obtain exactly $A^2T/2$ for E_d. However, since $f_c \gg R$, for all practical purposes $E_d = A^2T/2$ regardless of whether or not $f_c = nR/2$.

which reduces to $V_T = AT/2$ when $f_c \gg R$. Note that the performance of OOK is exactly the same as that for baseband unipolar signaling, as illustrated in Fig. 7-5.

Binary-Phase-Shift Keying

Referring to Fig. 7-7 once again, we see that the BPSK signal is

$$s_1(t) = A\cos(\omega_c t + \theta_c), \qquad 0 < t \le T \qquad \text{(binary 1)} \tag{7-34a}$$

and

$$s_2(t) = -A\cos(\omega_c t + \theta_c), \qquad 0 < t \le T \qquad \text{(binary 0)} \tag{7-34b}$$

BPSK signaling is also called *phase reversal keying* (PRK). The BPSK signal is an antipodal signal since $s_1(t) = -s_2(t)$.

Once again, first evaluate the performance of a receiver that uses an LPF having a gain of unity and an equivalent bandwidth of $B \ge 2/T$. The baseband analog output is

$$r_0(t) = \begin{Bmatrix} A, & 0 < t \le T & \text{binary } 1 \\ -A, & 0 < t \le T & \text{binary } 0 \end{Bmatrix} + x(t) \tag{7-35}$$

where $\overline{x^2(t)} = \sigma_0^2 = \overline{n^2(t)} = 2N_0B$. Since $s_{01} = A$ and $s_{02} = -A$, the optimum threshold is $V_T = 0$. When we use (7-17), the BER is

$$P_e = Q\left(\sqrt{\frac{A^2}{2N_0B}}\right) \qquad \text{(narrowband filter)} \tag{7-36}$$

When BPSK is compared with OOK on a *peak envelope power* (PEP) basis for a given value of N_0, 6 dB less (peak) signal power is required for BPSK signaling to give the same P_e as that for OOK. However, if the two are compared on an average (signal) power basis, the performance of BPSK has a 3-dB advantage over OOK since the average power of OOK is 3 dB below its PEP (equally likely signaling), but the average power of BPSK is equal to its PEP.

The performance of the matched-filter receiver is obtained by using (7-20). Recall that the matched filter for BPSK signaling was illustrated in Fig. 6-19, where a correlation processor using an integrate-and-dump filter was shown. The energy in the difference signal at the receiver input is

$$E_d = \int_0^T [2A\cos(\omega_c t + \theta_c)]^2\, dt = 2A^2T \tag{7-37}$$

Thus the BER is

$$P_e = Q\left(\sqrt{\frac{A^2T}{N_0}}\right) = Q\left(\sqrt{2\left(\frac{E_b}{N_0}\right)}\right) \qquad \text{(matched filter)} \tag{7-38}$$

where the average energy per bit is $E_b = A^2T/2$ and $V_T = 0$. Note that the performance of BPSK is exactly the same as that for baseband polar signaling; however, it is 3 dB superior to OOK signaling (see Fig. 7-5).

Frequency-Shift Keying

The FSK signal can be coherently detected by using two product detectors. This is illustrated in Fig. 7-8 where identical LPFs at the output of the product detectors have been replaced by only one of the filters since the order of linear operations may be exchanged without affecting the results. The mark (binary 1) and space (binary 0) signals are

$$s_1(t) = A\cos(\omega_1 t + \theta_c), \qquad 0 < t \le T \qquad \text{(binary 1)} \tag{7-39a}$$

$$s_2(t) = A\cos(\omega_2 t + \theta_c), \qquad 0 < t \le T \qquad \text{(binary 0)} \tag{7-39b}$$

where the frequency shift is $2\,\Delta F = f_1 - f_2$, assuming that $f_1 > f_2$. This FSK signal plus white Gaussian noise is present at the receiver input. The PSD for $s_1(t)$ and $s_2(t)$ is shown in Fig. 7-8b.

First, evaluate the performance of a receiver that uses an LPF $H(f)$ with a dc gain of 1. Once again, assume the equivalent bandwidth of the filter is $2/T \le B < \Delta F$. The LPF, when combined with the frequency translation produced by the product detectors, acts as dual bandpass filters—one centered at $f = f_1$ and the other at $f = f_2$, where each has an equivalent bandwidth of $B_p = 2B$. Thus the input noise that affects the output consists of two narrowband components $n_1(t)$ and $n_2(t)$, where the spectrum of $n_1(t)$ is centered at f_1 and the spectrum of $n_2(t)$ is centered at f_2, as shown in Fig. 7-8. Furthermore, $n(t) = n_1(t) + n_2(t)$, where

$$n_1(t) = x_1(t)\cos(\omega_1 t + \theta_c) - y_1(t)\sin(\omega_1 t + \theta_c) \tag{7-40a}$$

and

$$n_2(t) = x_2(t)\cos(\omega_2 t + \theta_c) - y_2(t)\sin(\omega_2 t + \theta_c) \tag{7-40b}$$

Assume that the frequency shift is $2\Delta F > 2B$, so that the mark and space signals may be separated by the filtering action. Then the input signal and noise that passes through the upper channel in Fig. 7-8a is

$$r_1(t) = \begin{Bmatrix} s_1(t), & \text{binary 1} \\ 0, & \text{binary 0} \end{Bmatrix} + n_1(t) \tag{7-41}$$

and the signal and noise that passes through the lower channel is

$$r_2(t) = \begin{Bmatrix} 0, & \text{binary 1} \\ s_2(t), & \text{binary 0} \end{Bmatrix} + n_2(t) \tag{7-42}$$

where $r(t) = r_1(t) + r_2(t)$. The noise power of $n_1(t)$ and $n_2(t)$ is $\overline{n_1^2(t)} = \overline{n_2^2(t)} = (N_0/2)(4B) = 2N_0B$. Thus the baseband analog output is

$$r_0(t) = \begin{Bmatrix} +A, & 0 < t \le T, & \text{binary 1} \\ -A, & 0 < t \le T, & \text{binary 0} \end{Bmatrix} + n_0(t) \tag{7-43}$$

where $s_{01} = +A$, $s_{02} = -A$, and $n_0(t) = x_1(t) - x_2(t)$. The optimum threshold setting is $V_T = 0$. Furthermore, the bandpass noise processes $n_1(t)$ and $n_2(t)$ are

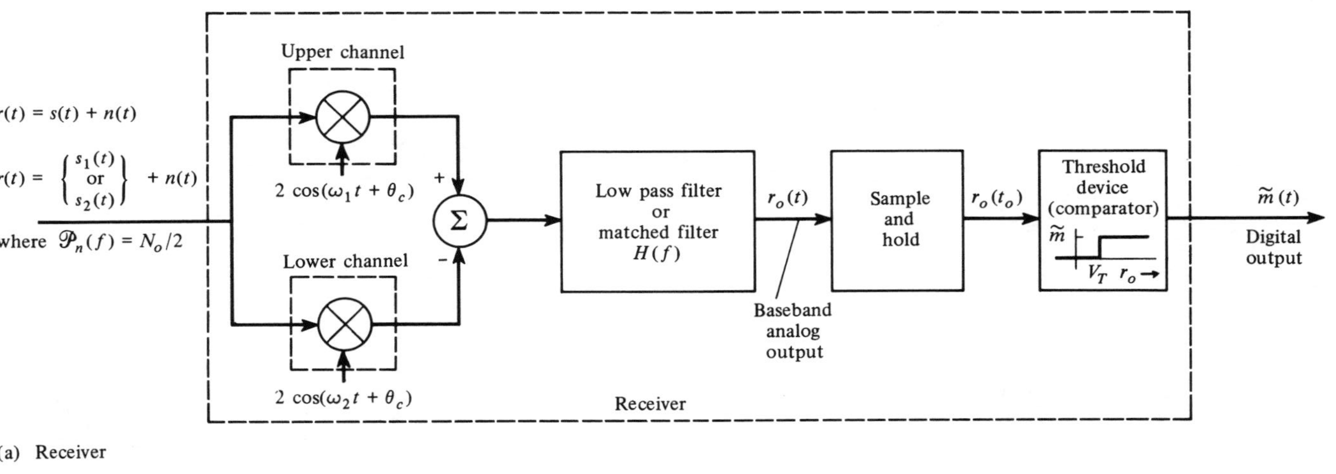

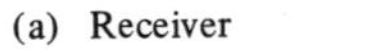

(a) Receiver

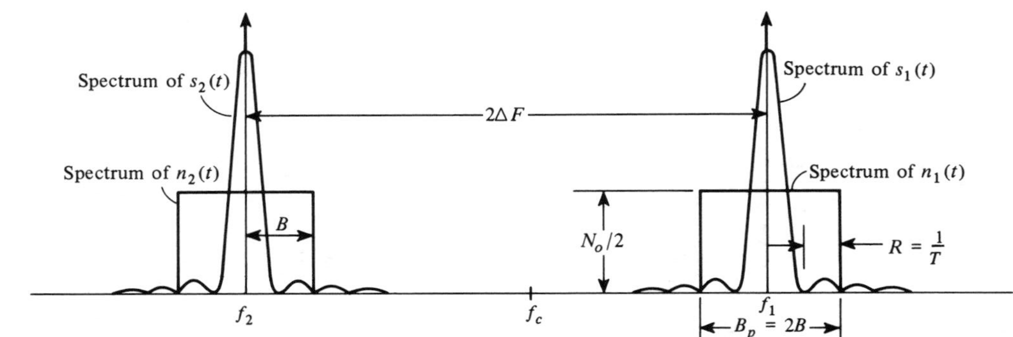

(b) Power Spectra for Positive Frequencies

FIGURE 7-8 Coherent detection of an FSK signal.

independent since they have spectra in nonoverlapping frequency bands (see Fig. 7-8) and they are white. (See Prob. 7-29 for the verification of this statement.) Consequently, the resulting baseband noise processes $x_1(t)$ and $x_2(t)$ are independent, and the output noise power is

$$\overline{n_0^2(t)} = \sigma_0^2 = \overline{x_1^2(t)} + \overline{x_2^2(t)} = \overline{n_1^2(t)} + \overline{n_2^2(t)} = 4N_0B \tag{7-44}$$

Substituting for s_{01}, s_{02}, and σ_0 into (7-17), we have

$$P_e = Q\left(\sqrt{\frac{A^2}{4N_0B}}\right) \quad \text{(bandpass filters)} \tag{7-45}$$

Comparing the performance of FSK with that of BPSK and OOK on a PEP basis, we see that FSK requires 3 dB more power than BPSK for the same P_e but 3 dB less power than OOK. Comparing the performance on an average power basis, we observe that FSK is 3 dB worse than BPSK but is equivalent to OOK (since the average power of OOK is 3 dB below its PEP).

The performance of FSK signaling with matched-filter reception is obtained from (7-20). The energy in the difference signal is

$$E_d = \int_0^T [A\cos(\omega_1 t + \theta_c) - A\cos(\omega_2 t + \theta_c)]^2\, dt$$

$$= \int_0^T [A^2\cos^2(\omega_1 t + \theta_c) - 2A^2\cos(\omega_1 t + \theta_c)$$

$$\times \cos(\omega_2 t + \theta_c) + A^2\cos^2(\omega_2 t + \theta_c)]\, dt$$

or†

$$E_d = \tfrac{1}{2}A^2T - A^2\int_0^T [\cos(\omega_1 - \omega_2)t]\, dt + \tfrac{1}{2}A^2T \tag{7-46}$$

Consider the case when $2\Delta F = f_1 - f_2 = n/(2T) = nR/2$. Under this condition the integral (i.e., the cross-product term) goes to zero. This condition is required for $s_1(t)$ and $s_2(t)$ to be orthogonal. Consequently, $s_1(t)$ will not contribute an output to the lower channel (see Fig. 7-8), and vice versa. Furthermore, if $(f_1 - f_2) \gg R$, $s_1(t)$ and $s_2(t)$ will be approximately orthogonal since the value of this integral will be negligible compared to A^2T. Assuming that one or both of these conditions is satisfied, then $E_d = A^2T$, and the BER for FSK signaling is

$$P_e = Q\left(\sqrt{\frac{A^2T}{2N_0}}\right) = Q\left(\sqrt{\frac{E_b}{N_0}}\right) \quad \text{(matched filter)} \tag{7-47}$$

where the average energy per bit is $E_b = A^2T/2$. The performance of FSK signaling is equivalent to that of OOK signaling (matched-filter reception) and is 3 dB inferior to BPSK signaling (see Fig. 7-5).

†For integrals of the type $\int_0^T A^2\cos^2(\omega_1 t + \theta_c)\, dt = \frac{1}{2}A^2\,[\int_0^T dt + \int_0^T \cos(2\omega_1 t + 2\theta_c)\, dt]$, the second integral on the right is negligible compared to the first integral on the right because of the oscillation in the second integral (Riemann–Lebesque lemma [Olmsted, 1961]).

As we demonstrate in the following section, coherent detection is superior to noncoherent detection. On the other hand, for coherent detection, the coherent reference must be available. This reference is often obtained from the noisy input signal so that the reference itself is also noisy. This, of course, increases P_e over those values given by the preceding formulas. The circuitry that extracts the carrier reference is usually complex and expensive. Often, one is willing to accept the poorer performance of a noncoherent system to simplify the circuitry and reduce the cost.

7-4 Noncoherent Detection of Bandpass Binary Signals

The *derivation* of the equations for the BER of noncoherent receivers is considerably more difficult than the derivation of the BER for coherent receivers. On the other hand, the *circuitry* for noncoherent receivers is *relatively simple* when compared to that used in coherent receivers. For example, *OOK with noncoherent reception is the most popular signaling technique used in fiber optic communication systems.*

In this section the BER will be computed for two noncoherent receivers—one for the reception of OOK signals and the other for the reception of FSK signals. As indicated in Chapter 5, BPSK cannot be detected noncoherently. However, as we will see, differential phase-shift-keyed (DPSK) signals may be demodulated by using a partially (quasi) coherent technique.

On-Off Keying

A noncoherent receiver for detection of OOK signals is shown in Fig. 7-9. Assume that an OOK signal plus white Gaussian noise is present at the receiver input. Then the noise at the filter output $n(t)$ will be bandlimited Gaussian noise, and the total filter output, consisting of signal plus noise, is

$$r(t) = \begin{cases} r_1(t), & 0 < t \leq T, \quad \text{binary 1 sent} \\ r_2(t), & 0 < t \leq T, \quad \text{binary 0 sent} \end{cases} \tag{7-48}$$

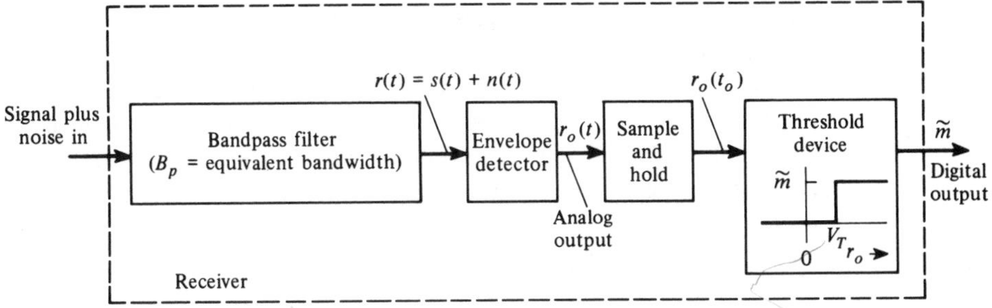

FIGURE 7-9 Noncoherent detection of OOK.

Let the bandwidth of the filter be B_p, where B_p is at least as large as the transmission bandwidth of the OOK signal so that the signal waveshape is preserved at the filter output. Then for the case of a binary 1, $s_1(t) = A\cos(\omega_c t + \theta_c)$, so

$$r_1(t) = A\cos(\omega_c t + \theta_c) + n(t), \qquad 0 < t \le T$$

or

$$r_1(t) = [A + x(t)]\cos(\omega_c t + \theta_c) - y(t)\sin(\omega_c t + \theta_c), \qquad 0 < t \le T \quad (7\text{-}49)$$

For a binary 0, $s_2(t) = 0$ and

$$r_2(t) = x(t)\cos(\omega_c t + \theta_c) - y(t)\cos(\omega_c t + \theta_c), \qquad 0 < t \le T \quad (7\text{-}50)$$

The BER is obtained by using (7-8), which, for the case of equally likely signaling, is

$$P_e = \tfrac{1}{2}\int_{-\infty}^{V_T} f(r_0|s_1)\,dr_0 + \tfrac{1}{2}\int_{V_T}^{\infty} f(r_0|s_2)\,dr_0 \quad (7\text{-}51)$$

We need to evaluate the conditional PDFs for the output of the envelope detector, $f(r_0|s_1)$ and $f(r_0|s_2)$. $f(r_0|s_1)$ is the PDF for $r_0 = r_0(t) = r_{01}$ that occurs when $r_1(t)$ is present at the input of the envelope detector, and $f(r_0|s_2)$ is the PDF for $r_0 = r_0(t_0) = r_{02}$ that occurs when $r_2(t)$ is present at the input of the envelope detector.

We will evaluate $f(r_0|s_2)$ first. When $s_2(t)$ is sent, the input to the envelope detector, $r_2(t)$, consists of bandlimited bandpass Gaussian noise as seen from (7-50). In Example 6-10 we demonstrated that for this case, the PDF of the envelope is Rayleigh distributed. Of course, the output of the envelope detector is the envelope, so $r_0 = \text{R} = r_{02}$. Thus the PDF for the case of noise alone is

$$f(r_0|s_2) = \begin{cases} \dfrac{r_0}{\sigma^2} e^{-r_0^2/(2\sigma^2)}, & r \ge 0 \\ 0, & r_0 \text{ otherwise} \end{cases} \quad (7\text{-}52)$$

The parameter σ^2 is the variance of the noise at the input of the envelope detector. Thus $\sigma^2 = (N_0/2)(2B_p) = N_0 B_p$, where B_p is the effective bandwidth of the bandpass filter and $N_0/2$ is the PSD of the white noise at the receiver input.

For the case of $s_1(t)$ being transmitted, the input to the envelope detector is given by (7-49). Since $n(t)$ is a Gaussian process (that has no delta functions in its spectrum at $f = \pm f_c$) the in-phase baseband component, $A + x(t)$, is also a Gaussian process with a mean value of A instead of a zero mean, as in (7-50). The PDF for the envelope $r_0 = R = r_{01}$ is evaluated using the same technique used in Example 6-10 and the result as cited in Prob. 6-46. Thus, for this case of a sinusoid plus noise at the envelope detector input,

$$f(r_0|s_1) = \begin{cases} \dfrac{r_0}{\sigma^2} e^{-(r_0^2+A^2)/(2\sigma^2)} I_0\!\left(\dfrac{r_0 A}{\sigma^2}\right), & r_0 \ge 0 \\ 0, & r_0 \text{ otherwise} \end{cases} \quad (7\text{-}53)$$

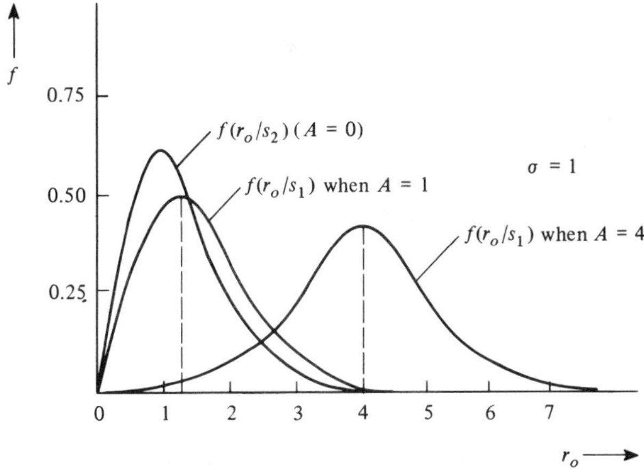

FIGURE 7-10 Conditional PDFs for noncoherent OOK reception.

which is a Rician PDF, where

$$I_0(z) \triangleq \frac{1}{2\pi}\int_0^{2\pi} e^{z\cos\theta}\,d\theta \tag{7-54}$$

is the modified Bessel function of the first kind of zero order.

The two conditional PDFs, $f(r_0|s_2)$ and $f(r_0|s_1)$, are shown in Fig. 7-10. Actually, $f(r_0|s_2)$ is a special case of $f(r_0|s_1)$ when $A = 0$ since for this condition we have noise only at the detector input and (7-53) becomes (7-52). Two other plots of (7-53) are also given, one for $A = 1$ and another for $A = 4$. It is seen that for $A/\sigma \gg 1$, the mode of the distribution [i.e., where $f(r_0|s_1)$ is a maximum] occurs at the value $r_0 = A$. In addition, note that for $A/\sigma \gg 1, f(r_0|s_1)$ takes on a Gaussian shape (as demonstrated subsequently).

The bit error rate for the noncoherent OOK receiver is obtained by substituting (7-52) and (7-53) into (7-51).

$$P_e = \frac{1}{2}\int_0^{V_T} \frac{r_0}{\sigma^2}\, e^{-(r_0^2+A^2)/(2\sigma^2)} I_0\left(\frac{r_0 A}{\sigma^2}\right) dr_0 + \frac{1}{2}\int_{V_T}^{\infty} \frac{r_0}{\sigma^2}\, e^{-r_0^2/(2\sigma^2)}\, dr_0 \tag{7-55}$$

The optimum threshold level is the value of V_T for which P_e is a minimum. For $A/\sigma \gg 1$, the optimum threshold is close to $V_T = A/2$, so we will use this value to simplify the mathematics.† The integral involving the Bessel function cannot be evaluated in closed form. However, $I_0(z)$ can be approximated by $I_0(z) = e^z/\sqrt{2\pi z}$, which is valid for $z \gg 1$. Then, for $A/\sigma \gg 1$, the left integral in (7-55) becomes

$$\frac{1}{2}\int_0^{V_T} \frac{r_0}{\sigma^2}\, e^{-(r_0^2+A^2)/(2\sigma^2)} I_0\left(\frac{r_0 A}{\sigma^2}\right) dr_0 \approx \frac{1}{2}\int_0^{A/2} \sqrt{\frac{r_0}{2\pi\sigma^2 A}}\; e^{(r_0-A)^2/(2\sigma^2)}\, dr_0$$

† Most practical systems operate with $A/\sigma \gg 1$.

Since $A/\sigma \gg 1$, the integrand is negligible except for values of r_0 in the vicinity of A, so the lower limit may be extended to $-\infty$ and $\sqrt{r_0/(2\pi\sigma^2 A)}$ can be replaced by $\sqrt{1/(2\pi\sigma^2)}$. Thus

$$\frac{1}{2}\int_0^{V_T} \frac{r_0}{\sigma^2} e^{-(r_0^2+A^2)/(2\sigma^2)} I_0\left(\frac{r_0 A}{\sigma^2}\right) dr_0 \approx \frac{1}{2}\int_{-\infty}^{V_T} \frac{1}{\sqrt{2\pi}\sigma} e^{-(r_0-A)^2/(2\sigma^2)} dr_0 \quad (7\text{-}56)$$

When we substitute (7-56) into (7-55), the BER becomes

$$P_e = \frac{1}{2}\int_{-\infty}^{A/2} \frac{1}{\sqrt{2\pi}\sigma} e^{-(r_0-A)^2/(2\sigma^2)} dr_0 + \frac{1}{2}\int_{A/2}^{\infty} \frac{r_0}{\sigma^2} e^{-r_0^2/(2\sigma^2)} dr_0$$

or

$$P_e = \tfrac{1}{2}Q\left(\frac{A}{2\sigma}\right) + \tfrac{1}{2}e^{-A^2/(8\sigma^2)} \quad (7\text{-}57)$$

Using $Q(z) = e^{-z^2/2}/\sqrt{2\pi z^2}$ for $z \gg 1$, we have

$$P_e = \frac{1}{\sqrt{2\pi}(A/\sigma)} e^{-A^2/(8\sigma^2)} + \tfrac{1}{2}e^{-A^2/(8\sigma^2)}$$

Since $A/\sigma \gg 1$, the second term on the right dominates over the first term. *Finally,* we obtain the approximation for the BER for noncoherent detection of OOK. It is

$$P_e = \tfrac{1}{2}e^{-A^2/(8\sigma^2)}, \qquad \frac{A}{\sigma} \gg 1$$

or

$$P_e = \tfrac{1}{2}e^{-[1/(2TB_p)](E_b/N_0)}, \qquad \frac{E_b}{N_0} \gg \frac{TB_p}{4} \quad (7\text{-}58)$$

where the average energy per bit is $E_b = A^2T/4$ and $\sigma^2 = N_0 B_p$. $R = 1/T$ is the bit rate of the OOK signal and B_p is the equivalent bandwidth of the bandpass filter that precedes the envelope detector.

Equation (7-58) indicates that the BER depends on the bandwidth of the bandpass filter and that P_e becomes smaller as B_p is decreased. Of course, this result is valid only when the ISI is negligible. Referring to (3-74), we realize that the minimum bandwidth allowed (i.e., for no ISI) is obtained when the rolloff factor is $r = 0$. This implies that the minimum bandpass bandwidth that is allowed is $B_p = 2B = R = 1/T$. A plot of the BER is given in Fig. 7-14 for this minimum bandwidth case of $B_p = 1/T$.

Frequency-Shift Keying

A noncoherent receiver for detection of frequency-shift-keyed (FSK) signals is shown in Fig. 7-11. The input consists of an FSK signal as described by (7-39) plus white Gaussian noise with a PSD of $N_0/2$. A sketch of the spectrum of the

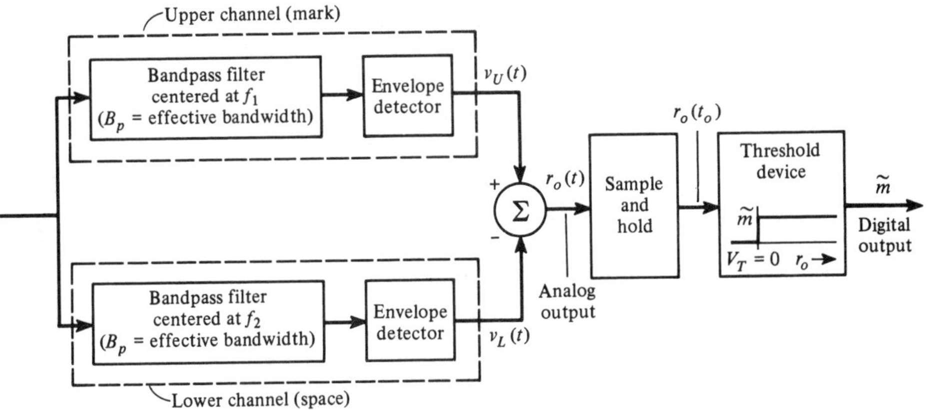

FIGURE 7-11 Noncoherent detection of FSK.

FSK signal and noise was given in Fig. 7-8b, where the bandpass filter bandwidth is B_p. It is assumed that the frequency shift, $2\Delta F = f_1 - f_2$, is sufficiently large so that the spectra of $s_1(t)$ and $s_2(t)$ have negligible overlap.

The BER for this receiver is obtained by evaluating (7-8). For signal alone at the receiver input, the output of the summing junction is $r_0(t) = +A$ when a mark (binary 1) is transmitted and $r_0(t) = -A$ when a space (binary 0) is transmitted. Because of this symmetry and because the noise out of the upper and lower receiver channels is similar, the optimum threshold is $V_T = 0$. Similarly, it is realized that the PDF of $r_0(t)$ conditioned on s_1 and the PDF of $r_0(t)$ conditioned on s_2 are similar. That is,

$$f(r_0|s_1) = f(-r_0|s_2) \tag{7-59}$$

Substituting (7-59) into (7-8), we find the BER to be

$$P_e = \tfrac{1}{2}\int_{-\infty}^{0} f(r_0|s_1)\,dr_0 + \tfrac{1}{2}\int_{0}^{\infty} f(r_0|s_2)\,dr_0$$

or

$$P_e = \int_{0}^{\infty} f(r_0|s_2)\,dr_0 \tag{7-60}$$

As shown in Fig. 7-11, $r_0(t)$ is positive when the upper channel output $v_U(t)$ exceeds the lower channel output $v_L(t)$. Thus

$$P_e = P(v_U > v_L|s_2) \tag{7-61}$$

For the case of a space signal plus noise at the receiver input, we know that the output of the upper bandpass filter is only Gaussian noise (no signal). Thus the output of the upper envelope detector v_U is noise having a Rayleigh distribution

$$f(v_U|s_2) = \begin{cases} \dfrac{v_U}{\sigma^2}\, e^{-v_U^2/(2\sigma^2)}, & v_U \ge 0 \\ 0, & v_U < 0 \end{cases} \tag{7-62}$$

where $\sigma^2 = N_0 B_p$. On the other hand, v_L has a Rician distribution since a sinusoid (the space signal) plus noise appears at the input to the lower envelope detector.

$$f(v_L|s_2) = \begin{cases} \dfrac{v_L}{\sigma^2}\, e^{-(v_L^2+A^2)/(2\sigma^2)} I_0\left(\dfrac{v_L A}{\sigma^2}\right), & v_L \geq 0 \\ 0, & v_L < 0 \end{cases} \tag{7-63}$$

where $\sigma^2 = N_0 B_p$ also. Using (7-62) and (7-63) in (7-61), we obtain

$$P_e = \int_0^\infty \frac{v_L}{\sigma^2}\, e^{-(v_L^2+A^2)/(2\sigma^2)} I_0\left(\frac{v_L A}{\sigma^2}\right)\left[\int_{v_L}^\infty \frac{v_U}{\sigma^2}\, e^{-v_U^2/(2\sigma^2)}\, dv_U\right] dv_L$$

When we evaluate the inner integral, the BER becomes

$$P_e = e^{-A^2/(2\sigma^2)} \int_0^\infty \frac{v_L}{\sigma^2}\, e^{-v_L^2/\sigma^2} I_0\left(\frac{v_L A}{\sigma^2}\right) dv_L \tag{7-64}$$

This integral can be evaluated by using (A-81a) from the integral table in Appendix A. Thus for noncoherent detection of FSK, the BER is

$$P_e = \tfrac{1}{2} e^{-A^2/(4\sigma^2)}$$

or

$$P_e = \tfrac{1}{2} e^{-[1/(2TB_p)](E_b/N_0)} \tag{7-65}$$

where the average energy per bit is $E_b = A^2T/2$ and $\sigma^2 = N_0 B_p$. $N_0/2$ is the PSD of the input noise, and B_p is the effective bandwidth of each of the bandpass filters (see Fig. 7-11). Comparing (7-65) with (7-58), we see that OOK and FSK are equivalent on an E_b/N_0 basis. A plot of (7-65) is given in Fig. 7-14 for the case of the minimum filter bandwidth allowed, $B_p = R = 1/T$, for no ISI. When comparing the error performance of noncoherently detected FSK with that of coherently detected FSK, it is seen that noncoherent FSK requires, at most, only 1 dB more E_b/N_0 than that for coherent FSK if P_e is 10^{-4} or less. The noncoherent FSK receiver is considerably easier to build since the coherent reference signals do not have to be generated. Thus, in practice, almost all of the FSK receivers use noncoherent detection.

Differential Phase-Shift Keying

Phase-shift-keyed signals cannot be detected incoherently. However, a partially coherent technique can be used whereby the phase reference for the present signaling interval is provided by a delayed version of the signal that occurred during the previous signaling interval. This is illustrated by the receivers shown in Fig. 7-12, where differential decoding is provided by the (one-bit) delay and the multiplier. If a BPSK signal (no noise) were applied to the receiver input, the output of the sample-and-hold circuit, $r_0(t_0)$, would be positive (binary 1) if the

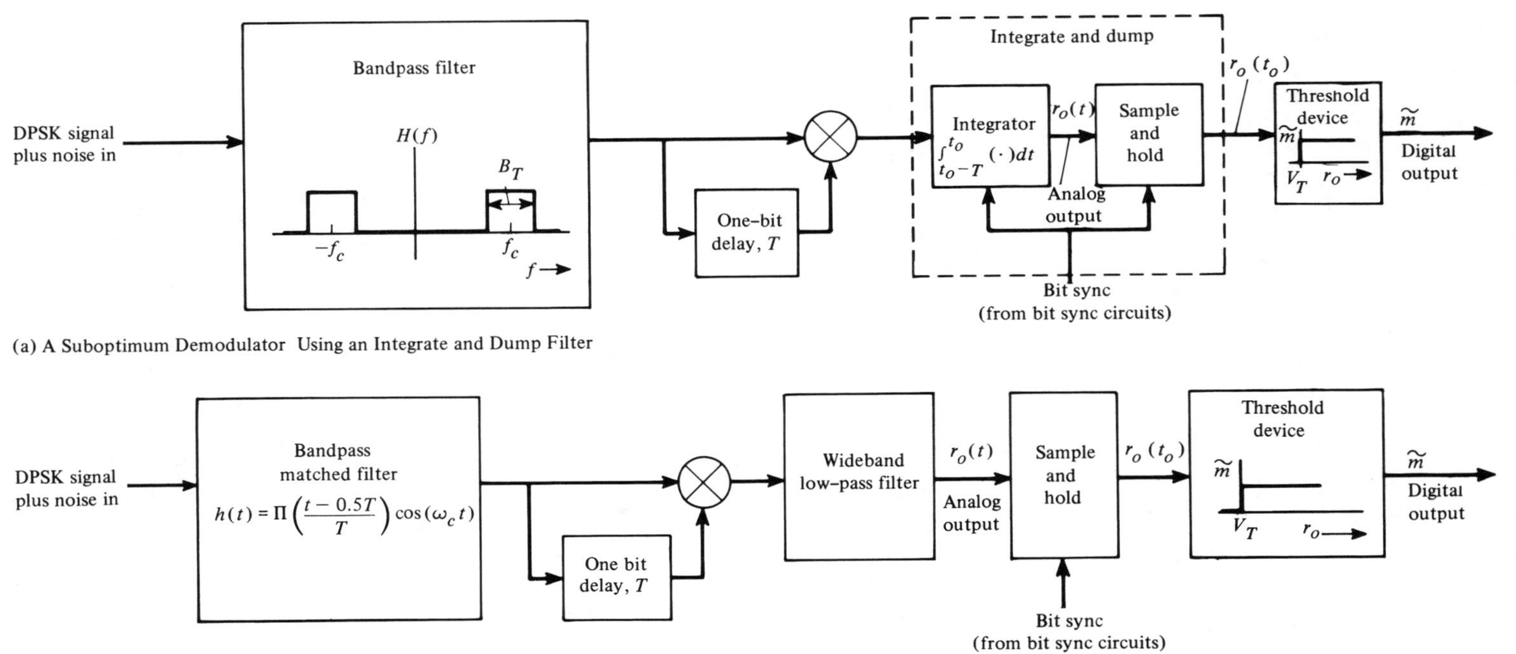

(a) A Suboptimum Demodulator Using an Integrate and Dump Filter

(b) An Optimum Demodulator Using a Bandpass Matched Filter

FIGURE 7-12 Demodulation of DPSK.

present data bit and the previous data bit were of the same sense; $r_0(t_0)$ would be negative (binary 0) if the two data bits were different. Consequently, if the data on the BPSK signal are differentially encoded (e.g., see the illustration in Table 3-3), the *decoded* data sequence will be recovered at the output of this receiver. This signaling technique consisting of transmitting a differentially encoded BPSK signal is known as *differential phase-shift keying* (DPSK).

The BER for these DPSK receivers can be derived under the following assumptions.

- The additive input noise is white and Gaussian.
- The phase perturbation of the composite signal plus noise varies slowly so that the phase reference is essentially a constant from the past signaling interval to the present signaling interval.
- The transmitter carrier oscillator is sufficiently stable so that the phase during the present signaling interval is the same as that from the past signaling interval.

The BER for the suboptimum demodulator of Fig. 7-12a has been obtained by J. H. Park for the case of a large input signal-to-noise ratio and for $B_T > 2/T$ but yet not too large. The result is [Park, 1978]

$$P_e = Q\left(\sqrt{\frac{(E_b/N_0)}{1 + [(B_T T/2)/(E_b/N_0)]}}\right) \tag{7-66a}$$

For typical values of B_T and E_b/N_0 in the range of $B_T = 3/T$ and $E_b/N_0 = 10$, this BER can be approximated by

$$P_e = Q\left(\sqrt{E_b/N_0}\right) \tag{7-66b}$$

Thus the performance of the suboptimum receiver of Fig. 7-12a is similar to that obtained for OOK and FSK as plotted in Fig. 7-14.

Figure 7-12b shows one form of an optimum DPSK receiver that is derived in Chapter 8 (see Fig. 8-25). From Chapter 8, the BER for optimum demodulation of DPSK is found to be

$$P_e = \tfrac{1}{2}e^{-(E_b/N_0)} \tag{7-67}$$

An alternative method of deriving this result for the optimum DPSK receiver of Fig. 7-12b is given by P. Z. Peebles [1987]. Other alternative forms of optimum DPSK receivers are possible [Lindsey and Simon, 1973; Simon, 1978].

A plot of this error characteristic for the case of optimum demodulation of DPSK, (7-67), is shown in Fig. 7-14. In comparing the error performance of BPSK and DPSK with optimum demodulation, it is seen that for the same P_e, DPSK signaling requires, at most, 1 dB more E_b/N_0 than BPSK provided that $P_e = 10^{-4}$ or less. In practice, DPSK is often used instead of BPSK since the DPSK receiver does not require a carrier synchronization circuit.

7-5 Quadrature Phase-Shift Keying and Minimum-Shift Keying

As described in Sec. 5-2, quadrature phase-shift keying (QPSK) is a multilevel signaling technique that uses $L = 4$ levels per symbol. Thus 2 bits are transmitted during each signaling interval (T seconds). The QPSK signal may be represented by

$$s(t) = (\pm A)\cos(\omega_c t + \theta_c) - (\pm A)\sin(\omega_c t + \theta_c), \qquad 0 < t \le T \quad (7\text{-}68)$$

where the $(\pm A)$ factor on the cosine carrier is one bit of data and the $(\pm A)$ factor on the sine carrier is another bit of data. The relevant input noise is represented by

$$n(t) = x(t)\cos(\omega_c t + \theta_n) - y(t)\sin(\omega_c t + \theta_n)$$

Obviously, the QPSK signal is equivalent to two BPSK signals—one using a cosine carrier and the other using a sine carrier. The QPSK signal is detected by using the coherent receiver shown in Fig. 7-13. (This is an application of the IQ detector that was first given in Fig. 4-31.) Since both the upper and lower channels of the receiver are BPSK receivers, the BER is the same as that for a BPSK system. Thus from (7-38) the BER for the QPSK receiver is

$$P_e = Q\left(\sqrt{2\left(\frac{E_b}{N_0}\right)}\right) \quad (7\text{-}69)$$

This is also shown in Fig. 7-14. The BERs for the BPSK and QPSK signaling are identical, but for the same bit rate R, the bandwidths of the two signals are *not* the same. The bandwidth of the QPSK signal is exactly one-half the bandwidth of the BPSK signal for a given bit rate. This result is given by (5-39), where the QPSK signal transmits one symbol for every 2 bits of data, whereas the BPSK signal transmits one symbol for each bit of data.

In Chapter 5 it was shown that minimum-shift keying (MSK) is essentially equivalent to QPSK except that the data on the $x(t)$ and $y(t)$ quadrature modulation components are offset and their equivalent data pulse shape is a positive part of a cosine function instead of a rectangular pulse. (This gives a PSD for MSK that rolls off faster than that for PSK.) The optimum receiver for detection of MSK is similar to that for QPSK (Fig. 7-13), except that a matched filter with a cosine pulse shape is used instead of the rectangular pulse shape that is synthesized by the integrate-and-dump circuit. Consequently, since the MSK and QPSK signal representations and the optimum receiver structures are identical except for the pulse shape, the probability of bit error for MSK and QPSK is identical, as described by (7-69). A plot of this BER is shown in Fig. 7-14.

If the data are properly encoded, the data on an MSK signal can also be detected by using FM-type detectors since the MSK signal is also an FSK signal with a minimum amount of frequency shift that makes $s_1(t)$ and $s_2(t)$ orthogonal signals. Thus, for suboptimum detection of MSK, the BER is given by the BER

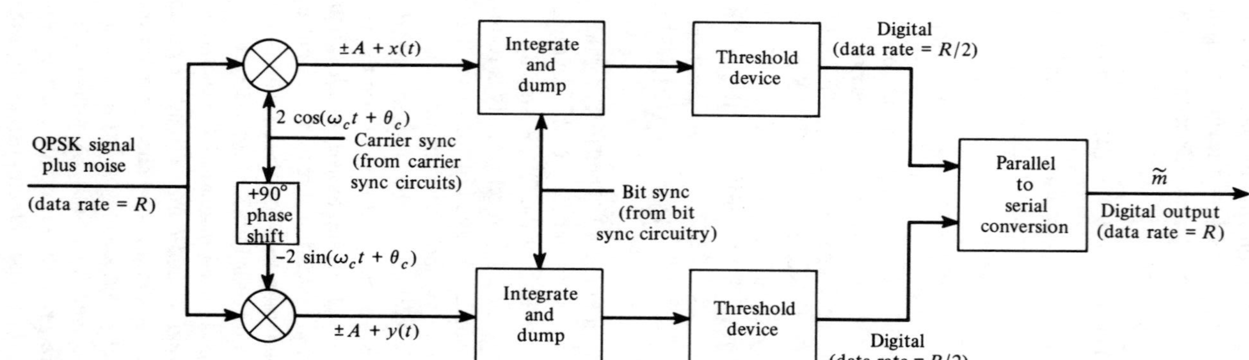

FIGURE 7-13 Matched-filter detection of QPSK.

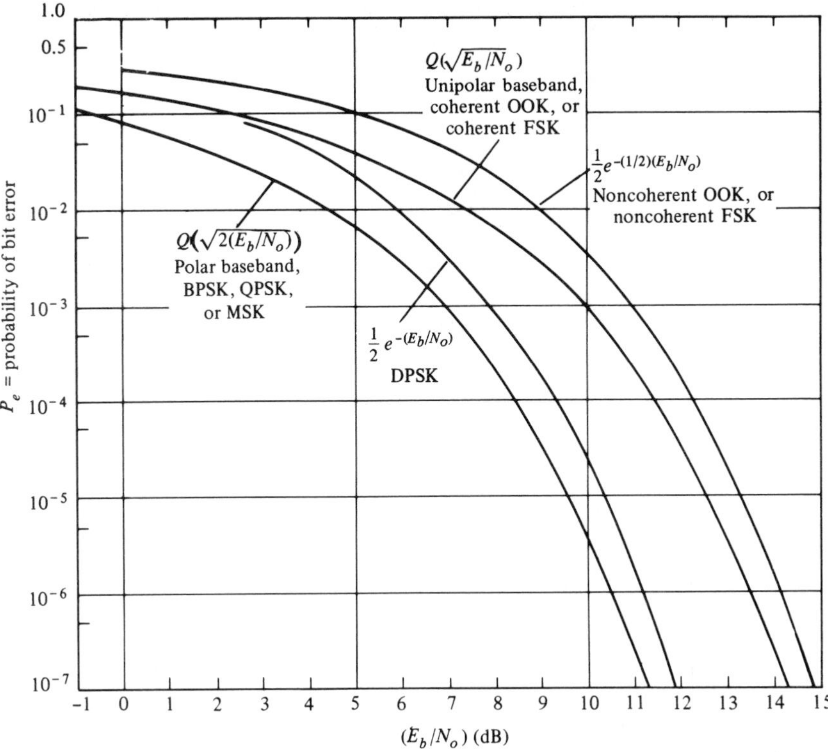

FIGURE 7-14 Comparison of the probability of bit error for several digital signaling schemes.

for FSK, as described by (7-47) for coherent FM detection and (7-65) for noncoherent FM detection.

7-6 Comparison of Digital Signaling Systems

Bit Error Rate and Bandwidth

A summary of the bit error rates that we obtained in the previous sections is given in Table 7-1. The minimum bandwidths for the signals are also tabulated using the equations developed in Chapter 5.

Zero-crossing bandwidths are not given in the table because they depend on the exact pulse shape used as well as the data statistics. The zero-crossing bandwidth can be evaluated from the PSD that describes the situation of interest. In Chapters 3 and 5 methods were given for evaluation of the PSD, and the PSD and corresponding zero-crossing bandwidth were worked out for many cases of practical interest.

TABLE 7-1 Comparison of Digital Signaling Methods

Type of Digital Signaling	Minimum Transmission Bandwidth Required[a] (Where R Is the Bit Rate)		Error Performance			
			Coherent detection		*Noncoherent detection*	
Baseband signaling						
Unipolar	$\frac{1}{2}R$	(5-38)			$Q\left[\sqrt{\left(\frac{E_b}{N_0}\right)}\right]$	(7-24b)
Polar	$\frac{1}{2}R$	(5-38)			$Q\left[\sqrt{2\left(\frac{E_b}{N_0}\right)}\right]$	(7-26b)
Bipolar	$\frac{1}{2}R$	(5-38)			$\frac{3}{2}Q\left[\sqrt{\left(\frac{E_b}{N_0}\right)}\right]$	(7-28b)
Bandpass signaling						
OOK	R	(5-39)	$Q\left[\sqrt{\left(\frac{E_b}{N_0}\right)}\right]$	(7-33)	$\frac{1}{2}e^{-(1/2)(E_b/N_0)}$, $\left(\frac{E_b}{N_0}\right) > \frac{1}{4}$	(7-58)
BPSK	R	(5-39)	$Q\left[\sqrt{2\left(\frac{E_b}{N_0}\right)}\right]$	(7-38)	Requires coherent detection	
FSK	$2\Delta F + R$ where $2\Delta F = f_2 - f_1$ is the frequency shift	(5-22)	$Q\left[\sqrt{\left(\frac{E_b}{N_0}\right)}\right]$	(7-47)	$\frac{1}{2}e^{-(1/2)(E_b/N_0)}$	(7-65)
DPSK	R	(5-39)	Not used in practice		$\frac{1}{2}e^{-(E_b/N_0)}$	(7-67)
QPSK	$\frac{1}{2}R$	(5-39)	$Q\left[\sqrt{2\left(\frac{E_b}{N_0}\right)}\right]$	(7-69)	Requires coherent detection	
MSK	$1.5R$ (null bandwidth)	(5-48)	$Q\left[\sqrt{2\left(\frac{E_b}{N_0}\right)}\right]$	(7-69)	$\frac{1}{2}e^{-(1/2)(E_b/N_0)}$	(7-65)

[a] Typical bandwidth specifications by the International Telecommunications Union (ITU) are larger that these minima [Jordan, 1985].

In Fig. 7-14 the BER curves are plotted using the equations presented in Table 7-1. Except for the curves describing the noncoherent detection cases, all of these results assume that the optimum filter—the matched filter—is used in the receiver. In practice, simpler filters work almost as well as the matched filter. For example, in a SYSTID computer simulation of a BPSK system with a three-pole Butterworth receiving filter having a bandwidth equal to the bit rate, the E_b/N_0 needs to be increased no more than 0.4 dB to obtain the same BER as that obtained when a matched filter is used (for error rates above 10^{-12}).

Comparing the various bandpass signaling techniques, we see that QPSK and MSK give the best overall performance in terms of the minimum bandwidth required for a given signaling rate and one of the smallest P_e for a given E_b/N_0. However, QPSK is relatively expensive to implement since it requires coherent detection. Figure 7-14 shows that for the same P_e, DPSK (using a noncoherent receiver) requires only about 1 dB more E_b/N_0 than that for QPSK or BPSK, for error rates of 10^{-4} or less. Since DPSK is much easier to receive than BPSK, it is often used in practice in preference to BPSK. Similarly, the BER performance of a noncoherent FSK receiver is very close to that of a coherent FSK receiver. Since noncoherent FSK receivers are simpler than coherent FSK receivers, they are often used in practice. In some applications there are multiple paths of different lengths between the transmitter and receiver (e.g., caused by reflection). This causes fading and noncoherence in the received signal (phase). In this case, it is very difficult to implement coherent detection, regardless of the complexity of the receiver, and one resorts to the use of noncoherent detection techniques.

Channel coding can be used to reduce the P_e below the values given in Fig. 7-14. This concept was developed in Chapter 1. By using Shannon's channel capacity formula, it was found that theoretically the BER would approach zero as long as E_b/N_0 was above -1.59 dB when optimum (unknown) coding was used on an infinite bandwidth channel. With practical coding it was found that coding gains as large as 9 dB could be achieved on BPSK and QPSK systems. That is, for a given BER the E_b/N_0 requirements in Fig. 7-14 could be reduced by as much as 9 dB when coding was used.

Synchronization

As we have seen, three levels of synchronization are needed in digital communication systems.

1. Bit synchronization.
2. Frame or word synchronization.
3. Carrier synchronization.

Bit and frame synchronizations were discussed in Chapter 3.

Carrier synchronization is required in receivers that use coherent detection. If the spectrum of the digital signal has a discrete line at the carrier frequency, such as in equally likely OOK signaling, a PLL can be used to recover the carrier reference from the received signal. This was described in Chapter 4 and shown in Fig. 4-25. In BPSK signaling there is no discrete carrier term, but the spectrum is symmetrical about the carrier frequency. Consequently, a Costas loop or a squaring loop may be used for carrier sync recovery. These loops were illustrated in Figs. 4-34 and P5-6. As indicated in Sec. 4-5, these loops may lock up with a 180° phase error, which must be resolved to ensure that the recovered data will not be complemented. For QPSK signals, the carrier reference may be obtained by a more generalized Costas loop or, equivalently, by a fourth-power loop [Spilker, 1977]. These loops have a four-phase ambiguity of 0, ±90, or 180°, which must also be resolved to obtain correctly demodulated data. These facts illustrate once again why noncoherent reception techniques (which can be used for OOK, FSK, and DPSK) are so popular.

Bit synchronizers are needed at the receiver to provide the bit sync signal for clocking the sample-and-hold circuit and, if used, the matched filter circuit. The bit synchronizer was illustrated in Fig. 3-16.

All the BER formulas that we have obtained assume that noise-free bit sync and carrier sync (for coherent detection) are available at the receiver. Of course, if these sync signals are obtained from noisy signals that are present at the receiver input, the reference signals are also noisy. Consequently, the P_e will be larger than that given for the ideal case when noise-free sync is assumed.

The third type of synchronization that is required by most digital systems is frame sync or word sync. In some systems this sync is used simply to demark the serial data into digital words or *bytes*. In other systems, block coding or convolutional coding is used at the transmitter so that some of the bit errors at the output of the threshold device of the receiver can be detected and corrected by using decoding circuits. In these systems word sync is needed to clock the receiver decoding circuits. In addition, frame sync is required in time-division multiplex (TDM) systems. This was described in Fig. 3-41. Higher levels of synchronization, such as network synchronization, may be required when data are received from several sources. For example, multiple-access satellite communication systems require network synchronization, as illustrated in Figs. 5-32 and 5-33.

For further reading on the interesting topic of synchronization, the reader is referred to a special issue of the *IEEE Transactions on Communications,* which presents tutorial as well as in-depth papers on the different aspects of synchronization problems [Gardner and Lindsey, 1980].

7-7 Output Signal-to-Noise Ratio for PCM Systems

In the previous sections we have analyzed how the P_e for various digital systems depends on the energy per bit E_b of the signal at the receiver input and on the level of the input noise spectrum $N_0/2$. Now suppose that we look at applications of these signaling techniques where an analog signal is encoded into a PCM signal that comprises the data that are transmitted over the digital system having a BER of P_e. This is illustrated in Fig. 7-15. The digital transmitter and receiver may be any one of those associated with the digital signaling systems studied in

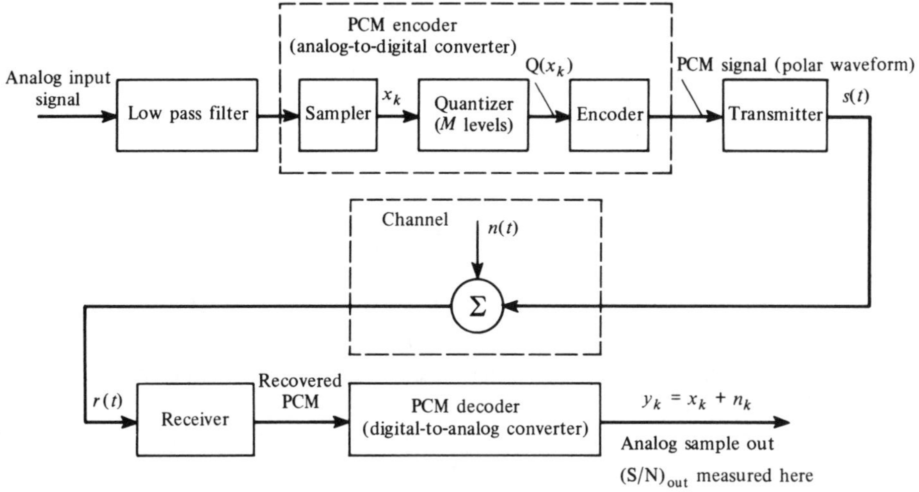

FIGURE 7-15 PCM communications system.

the previous sections. For example, $s(t)$ might be an FSK signal and the receiver would be an FSK receiver. The recovered PCM signal has some bit errors (caused by channel noise). Consequently, the decoded analog waveform at the output of the PCM decoder will have noise because of these bit errors as well as quantizing noise. The question is: What is the peak signal to average noise ratio $(S/N)_{out}$ for the analog output? If the input analog signal has a uniform PDF over $-V$ to $+V$ and there are M steps in the uniform quantizer, the answer is

$$\left(\frac{S}{N}\right)_{pk\ out} = \frac{3M^2}{1 + 4(M^2 - 1)P_e} \tag{7-70}$$

Applications of this result were first studied in Sec. 3-3.

Equation (7-70) will now be derived. As shown in Fig. 7-15, the analog sample x_k is obtained at the sampling time $t = kT_s$. This sample is quantized to a value $Q(t_k)$, which is one of the M possible levels, as indicated in Fig. 7-15. The quantized sample $Q(x_k)$ is encoded into an n-bit PCM word $(a_{k1}, a_{k2}, \ldots, a_{kn})$, where $M = 2^n$. If polar signaling is used, the a_k's take on values of $+1$ or -1. For simplicity, we assume that the PCM code words are related to the quantized values by

$$Q(x_k) = V \sum_{j=1}^{n} a_{kj} \left(\tfrac{1}{2}\right)^j \tag{7-71}$$

For example, if the PCM word for the kth sample happens to be $(+1, +1, \ldots, +1)$, then the value of the quantized sample will be

$$Q(x_k) = V\left(\frac{1}{2} + \frac{1}{4} + \cdots + \frac{1}{2^n}\right) = \frac{V}{2}\left(1 + \frac{1}{2} + \frac{1}{4} + \cdots + \frac{1}{2^{n-1}}\right)$$

Using (A-90), we find that the sum of this finite series is

$$Q(x_k) = \frac{V}{2}\left[\frac{(\frac{1}{2})^n - 1}{\frac{1}{2} - 1}\right] = V - \frac{V}{2^n}$$

or, since $V = (\delta/2)2^n$,

$$Q(x_k) = V - \frac{\delta}{2}$$

where δ is the step size of the quantizer (see Fig. 7-16). Thus the PCM word $(+1, +1, \ldots, +1)$ represents the maximum value of the quantizer, as illustrated in Fig. 7-16. Similarly, the level of the quantizer corresponding to the other code words can be obtained.

Referring to Fig. 7-15 again, we note that the analog sample output of the PCM system for the kth sampling time is

$$y_k = x_k + n_k$$

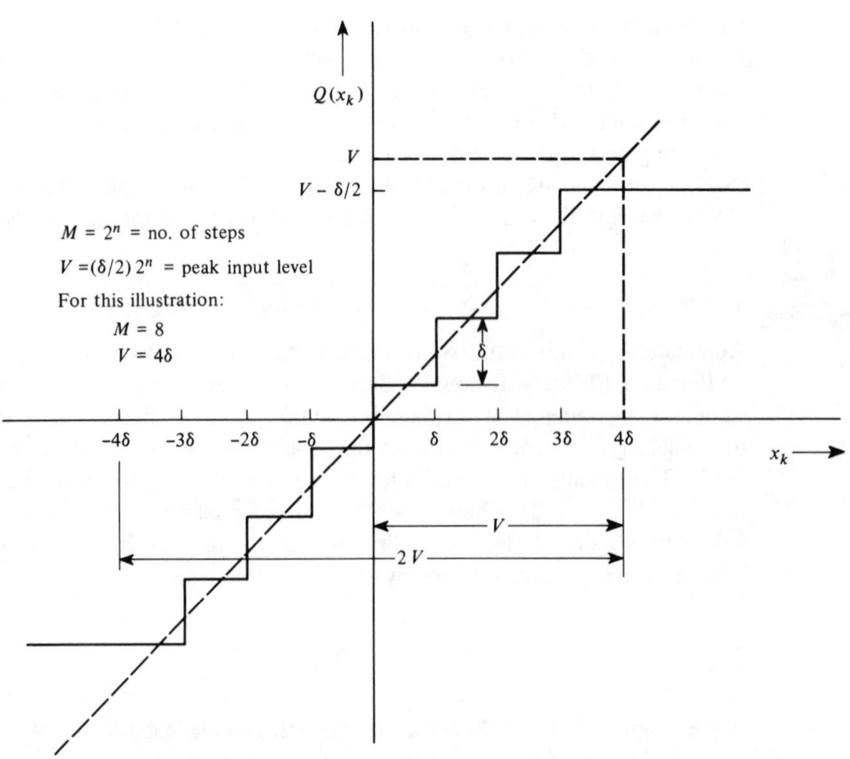

FIGURE 7-16 Uniform quantizer characteristic for $M = 8$ ($n = 3$ bits in each PCM word).

where x_k is the signal (same as the input sample) and n_k is noise. The output peak signal power to average noise power is then

$$\left(\frac{S}{N}\right)_{\text{pk out}} = \frac{[(x_k)_{\max}]^2}{\overline{n_k^2}} = \frac{V^2}{\overline{n_k^2}} \tag{7-72}$$

where $(x_k)_{\max} = V$, as is readily seen from Fig. 7-16. As indicated in Chapter 3, it is assumed that the noise n_k consists of two uncorrelated effects.

1. Quantizing noise that is due to the quantizing error:

$$e_q = Q(x_k) - x_k \tag{7-73}$$

2. Noise due to bit errors that are caused by the channel noise:

$$e_b = y_k - Q(x_k) \tag{7-74}$$

Thus

$$\overline{n_k^2} = \overline{e_q^2} + \overline{e_b^2} \tag{7-75}$$

First, evaluate the quantizing noise power. For a uniformly distributed signal, the quantizing noise is uniformly distributed. Furthermore, as indicated by Fig. 3-8c, the interval of the uniform distribution is $(-\delta/2, \delta/2)$, where δ is the step size ($\delta = 2$ in Fig. 3-8c). Thus, with $M = 2^n = 2V/\delta$ (from Fig. 7-16),

$$\overline{e_q^2} = \int_{-\infty}^{\infty} e_q^2 f(e_q)\, de_q = \int_{-\delta/2}^{\delta/2} e_q^2 \frac{1}{\delta}\, de_q = \frac{\delta^2}{12} = \frac{V^2}{3M^2} \tag{7-76}$$

The noise power due to bit errors is evaluated by the use of (7-74),

$$\overline{e_b^2} = \overline{[y_k - Q(x_k)]^2} \tag{7-77}$$

where $Q(x_k)$ is given by (7-71). The recovered analog sample y_k is reconstructed from the received PCM code word using the same algorithm as that of (7-71). Assuming that the received PCM word for the kth sample is $(b_{k1}, b_{k2}, \ldots, b_{kn})$, then we see that

$$y_k = V \sum_{j=1}^{n} b_{kj} (\tfrac{1}{2})^j \tag{7-78}$$

The b's will be different from the a's whenever there is a bit error in the recovered digital (PCM) waveform. By using (7-78) and (7-71), (7-77) becomes

$$\begin{aligned}\overline{e_b^2} &= V^2 \overline{\left[\sum_{j=1}^{n} (b_{kj} - a_{kj})(\tfrac{1}{2})^j\right]^2} \\ &= V^2 \sum_{j=1}^{n} \sum_{\ell=1}^{n} \left(\overline{b_{kj}b_{k\ell}} - \overline{a_{kj}b_{k\ell}} - \overline{b_{kj}a_{k\ell}} + \overline{a_{kj}a_{k\ell}}\right) 2^{-j-\ell}\end{aligned} \tag{7-79}$$

where b_{kj} and $b_{k\ell}$ are two bits in the received PCM word that occur at different bit positions when $j \neq \ell$. Similarly, a_{kj} and $b_{k\ell}$ are the transmitted (Tx) and received

(Rx) bits in two different bit positions, where $j \neq \ell$ (for the PCM word corresponding to the kth sample). The encoding process produces bits that are independent unless $j = \ell$. Furthermore, the bits have a zero-mean value. Thus $\overline{b_{kj}b_{k\ell}} = \overline{b_{kj}}\,\overline{b_{k\ell}} = 0$ for $j \neq \ell$. Similar results are obtained for the other averages on the right-hand side of (7-79). Equation (7-79) becomes

$$\overline{e_b^2} = V^2 \sum_{j=1}^{n} \left(\overline{b_{kj}^2} - 2\overline{a_{kj}b_{kj}} + \overline{a_{kj}^2}\right) 2^{-2j} \tag{7-80}$$

Evaluating the averages in (7-80), we obtain†

$$\begin{aligned}
\overline{b_{kj}^2} &= (+1)^2 P(+1\text{Rx}) + (-1)^2 P(-1\text{Rx}) = 1 \\
\overline{a_{kj}^2} &= (+1)^2 P(+1\text{Tx}) + (-1)^2 P(-1\text{Tx}) = 1 \\
\overline{a_{kj}b_{kj}} &= (+1)(+1)P(+1\text{Tx}, +1\text{Rx}) + (-1)(-1)P(-1\text{Tx}, -1\text{Rx}) \\
&\quad + (-1)(+1)P(-1\text{Tx}, +1\text{Rx}) + (+1)(-1)P(+1\text{Tx}, -1\text{Rx}) \\
&= [P(+1\text{Tx}, +1\text{Rx}) + P(-1\text{Tx}, -1\text{Rx})] \\
&\quad - [P(-1\text{Tx}, +1\text{Rx}) + P(+1\text{Tx}, -1\text{Rx})] \\
&= [1 - P_e] - [P_e] = 1 - 2P_e
\end{aligned}$$

Thus (7-80) reduces to

$$\overline{e_b^2} = 4V^2 P_e \sum_{j=1}^{n} \left(\tfrac{1}{4}\right)^j$$

which, by (A-90), becomes

$$\overline{e_b^2} = V^2 P_e \frac{\left(\frac{1}{4}\right)^n - 1}{\frac{1}{4} - 1} = \frac{4}{3} V^2 P_e \frac{(2^n)^2 - 1}{(2^n)^2}$$

or

$$\overline{e_b^2} = \tfrac{4}{3} V^2 P_e \frac{M^2 - 1}{M^2} \tag{7-81}$$

Substituting (7-76) and (7-81) into (7-72) with the help of (7-75), we have

$$\left(\frac{S}{N}\right)_{\text{pk out}} = \frac{V^2}{(V^2/3M^2) + (4V^2/3M^2)P_e(M^2 - 1)}$$

which reduces to (7-70).

Equation (7-70) is used to obtain the curves shown in Fig. 7-17. For a PCM system with M quantizing steps, $(S/N)_{\text{out}}$ is given as a function of the BER of the digital receiver. For $P_e < 1/(4M^2)$ the analog signal at the output is corrupted primarily because of quantizing noise. In fact, for $P_e = 0$, $(S/N)_{\text{out}} = 3M^2$,

†The notation $+1$Tx denotes a binary 1 transmitted, -1Tx denotes a binary 0 transmitted, $+1$Rx denotes a binary 1 received, and -1Rx denotes a binary 0 received.

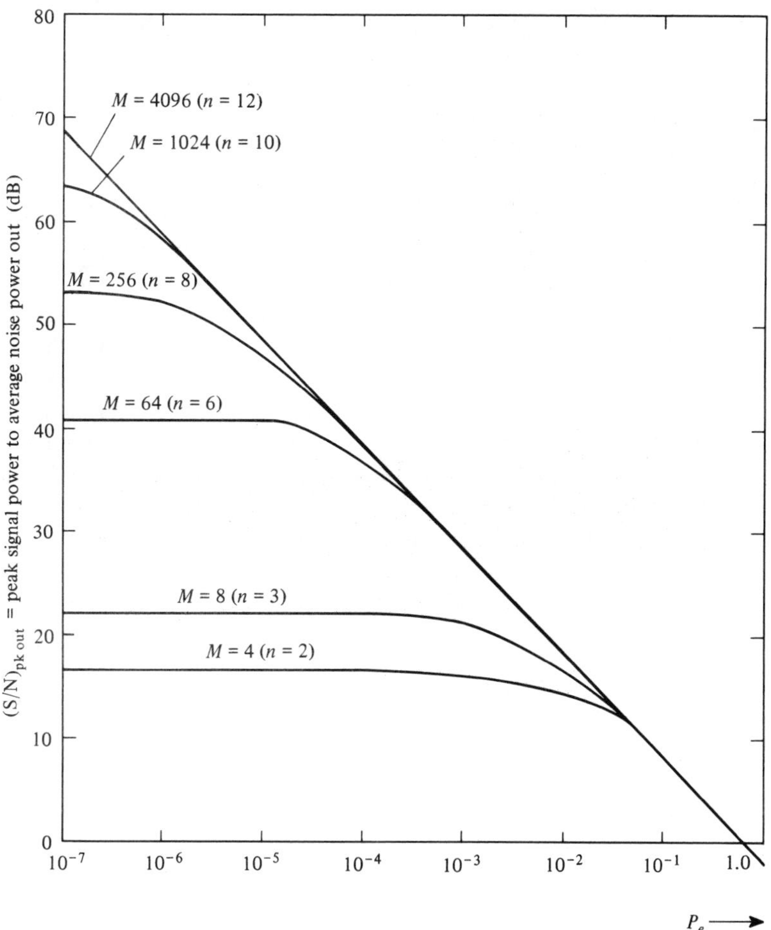

FIGURE 7-17 $(S/N)_{out}$ of a PCM system as a function of P_e and the number of quantizer steps, M.

where all the noise is quantizing noise. On the other hand, for $P_e > 1/(4M^2)$, the output is corrupted primarily because of channel noise.

It is also stressed that (7-70) is the *peak* signal to average noise ratio. The *average* signal to average noise ratio is obtained easily from the results given above. The average signal to average noise ratio is

$$\left(\frac{S}{N}\right)_{\text{out}} = \frac{\overline{(x_k)^2}}{\overline{n_k^2}} = \frac{V^2}{3\overline{n_k^2}} = \frac{1}{3}\left(\frac{S}{N}\right)_{\text{pk out}}$$

where $\overline{x_k^2} = V^2/3$ since x_k is uniformly distributed from $-V$ to $+V$. Thus, using (7-72) and (7-70),

$$\left(\frac{S}{N}\right)_{\text{out}} = \frac{M^2}{1 + 4(M^2 - 1)P_e} \tag{7-82}$$

when $(S/N)_{out}$ is the average signal to average noise power ratio at the output of the system.

7-8 Output Signal-to-Noise Ratios for Analog Systems

In Chapter 4 generalized modulation and demodulation techniques for analog and digital signals were studied. Specific techniques were shown for AM, DSB-SC, SSB, PM, and FM signaling, and the bandwidths of these signals were evaluated. Here we evaluate the output signal-to-noise ratios for these systems as a function of the input signal, noise, and system parameters. Once again, we will find that the mathematical analysis of noncoherent systems is more difficult than that for coherent systems and that approximations are often used to obtain simplified results. However, noncoherent systems are often found to be more prevalent in practice since the receiver cost is usually lower. This is certainly the case for overall system cost in applications involving one transmitter and thousands or millions of receivers such as FM, AM, and TV broadcasting.

For systems with additive noise channels the input to the receiver is

$$r(t) = s(t) + n(t)$$

For bandpass communication systems having a transmission bandwidth of B_T, this is

$$\begin{aligned} r(t) &= \text{Re}\{g_s(t)e^{j(\omega_c t + \theta_c)}\} + \text{Re}\{g_n(t)e^{j(\omega_c t + \theta_c)}\} \\ &= \text{Re}\{[g_s(t) + g_n(t)]e^{j(\omega_c t + \theta_c)}\} \end{aligned}$$

or

$$r(t) = \text{Re}\{g_T(t)e^{j(\omega_c t + \theta_c)}\} \tag{7-83a}$$

where

$$\begin{aligned} g_T(t) &\triangleq g_s(t) + g_n(t) \\ &= [x_s(t) + x_n(t)] + j[y_s(t) + y_n(t)] \\ &= x_T(t) + jy_T(t) \\ &= R_T(t)e^{j\theta_T}(t) \end{aligned} \tag{7-83b}$$

$g_T(t)$ denotes the total (i.e., composite) complex envelope at the receiver input; it consists of the complex envelope of the signal plus the complex envelope of the noise. The properties of the total complex envelope, as well as those of Gaussian noise, were given in Chapter 6. The complex envelopes for a number of different types of modulated signals $g_s(t)$ were given in Table 4-1.

Baseband Systems

To compare the output signal-to-noise ratio $(S/N)_{out}$ for various bandpass systems, we need a common measurement criterion for the receiver input. For

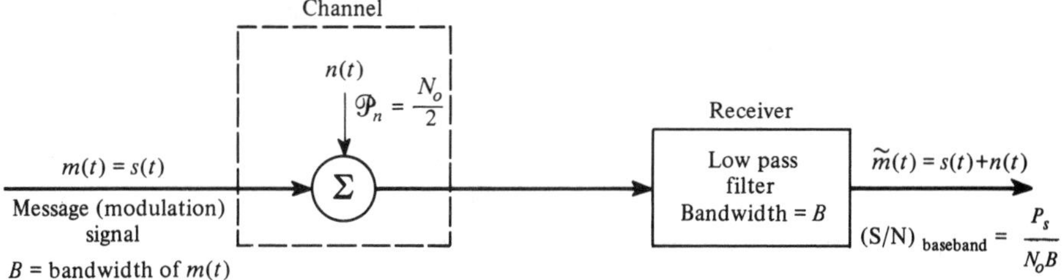

FIGURE 7-18 Baseband system.

example, E_b/N_0 was used for digital systems. For analog systems the criterion is the received signal power P_s divided by the amount of power in the white noise that is contained in a bandwidth equal to the message (modulation) bandwidth. This is equivalent to the $(S/N)_{out}$ of a baseband transmission system as illustrated in Fig. 7-18. That is,

$$\left(\frac{S}{N}\right)_{\text{baseband}} = \frac{P_s}{N_0 B} \tag{7-84}$$

Of course, a baseband system involves no modulation or demodulation.

We can compare the performance of different modulated systems by evaluating $(S/N)_{out}$ for each system for a given value of $(P_s/N_0B) = (S/N)_{\text{baseband}}$.

AM Systems with Product Detection

Figure 7-19 illustrates the receiver for an AM system with coherent detection. From (4-107), the complex envelope of the AM signal is

$$g_s(t) = A_c[1 + m(t)]$$

The complex envelope of the composite received signal plus noise is

$$g_T(t) = [A_c + A_c m(t) + x_n(t)] + jy_n(t) \tag{7-85}$$

For product detection, using (4-71), we find that the output is

$$\tilde{m}(t) = \text{Re}\{g_T(t)\} = A_c + A_c m(t) + x_n(t) \tag{7-86}$$

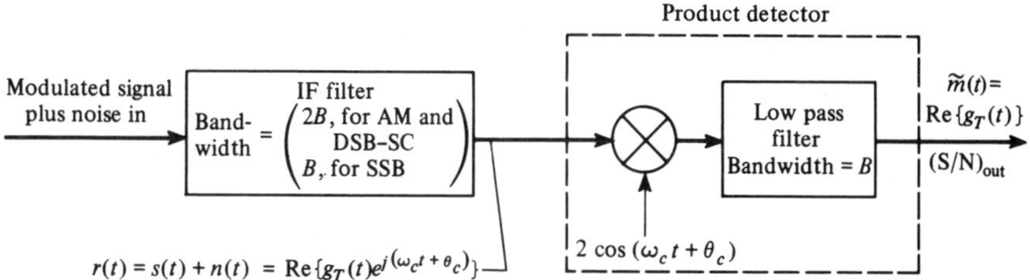

FIGURE 7-19 Coherent receiver.

Here A_c is the dc voltage at the detector output that occurs because of the discrete AM carrier,† $A_c m(t)$ is the detected modulation, and $x_n(t)$ is the detected noise. The output S/N is

$$\left(\frac{S}{N}\right)_{out} = \frac{A_c^2 \overline{m^2(t)}}{\overline{x_n^2(t)}} = \frac{A_c^2 \overline{m^2(t)}}{2N_0 B} \tag{7-87}$$

where, from (6-133g), $\overline{x_n^2} = \overline{n^2} = 2(N_0/2)(2B)$. The input signal power is

$$P_s = \frac{A_c^2}{2} \overline{[1 + m(t)]^2} = \frac{A_c^2}{2} [1 + \overline{m^2}]$$

where it is assumed that $\overline{m(t)} = 0$ (i.e., no dc on the modulating waveform). Thus the input S/N is

$$\left(\frac{S}{N}\right)_{in} = \frac{(A_c^2/2)(1 + \overline{m^2})}{2N_0 B} \tag{7-88}$$

Combining (7-87) and (7-88) yields

$$\frac{(S/N)_{out}}{(S/N)_{in}} = \frac{2\overline{m^2}}{1 + \overline{m^2}} \tag{7-89}$$

For 100% sine-wave modulation $\overline{m^2} = \frac{1}{2}$ and $(S/N)_{out}/(S/N)_{in} = \frac{2}{3}$.

For comparison purposes, (7-87) can be evaluated in terms of $(S/N)_{baseband}$ by substituting for P_s:

$$\frac{(S/N)_{out}}{(S/N)_{baseband}} = \frac{\overline{m^2}}{1 + \overline{m^2}} \tag{7-90}$$

For 100% sine-wave modulation, $(S/N)_{out}/(S/N)_{baseband} = \frac{1}{3}$. This illustrates that this AM system is 4.8 dB worse than a baseband system that uses the same amount of signal power. This is shown in Fig. 7-27.

AM Systems with Envelope Detection

The envelope detector produces $KR_T(t)$ at its output, where K is a proportionality constant. Then, for additive signal plus noise at the input, the output is

$$KR_T(t) = K|g_s(t) + g_n(t)|$$

Substituting (7-85) for $g_s(t) + g_n(t)$ gives

$$KR_T(t) = K|[A_c + A_c m(t) + x_n(t)] + j[y_n(t)]| \tag{7-91}$$

The power is

$$\overline{[KR_T(t)]^2} = K^2 A_c^2 \overline{\left\{\left[1 + m(t) + \frac{x_n(t)}{A_c}\right]^2 + \left[\frac{y_n(t)}{A_c}\right]^2\right\}} \tag{7-92}$$

†This dc voltage is often used to provide automatic gain control (AGC) of the preceding RF and IF stages.

For the case of large $(S/N)_{in}$, $(y_n/A_c)^2 \ll 1$, so we have

$$\overline{[KR_T(t)]^2} = (KA_c)^2 + K^2A_c^2\,\overline{m^2} + K^2\,\overline{x_n^2} \tag{7-93}$$

where $(KA_c)^2$ is the power of the AGC term, $K^2A_c^2\,\overline{m^2}$ is the power of the detected modulation (signal), and $K^2\,\overline{x_n^2}$ is the power of the detected noise. Thus, for large $(S/N)_{in}$,

$$\left(\frac{S}{N}\right)_{out} = \frac{A_c^2\,\overline{m^2}}{\overline{x_n^2}} = \frac{A_c^2\,\overline{m^2}}{2N_0B} \tag{7-94}$$

Comparing this with (7-87), we see that *for large $(S/N)_{in}$, the performance of the envelope detector is identical to that of the product detector.*

For small $(S/N)_{in}$ the performance of the envelope detector is much inferior to that of the product detector. Referring to (7-91), we note that the detector output is

$$KR_T(t) = K|g_T(t)| = K|A_c[1 + m(t)] + R_n(t)e^{j\theta_n(t)}|$$

For $(S/N)_{in} < 1$, as seen from Fig. 7-20, the magnitude of $g_T(t)$ can be approximated by

$$KR_T(t) \approx K\{A_c[1 + m(t)]\cos\theta_n(t) + R_n(t)\} \tag{7-95}$$

Thus, for the case of a Gaussian noise channel, the output of the envelope detector consists of Rayleigh distributed noise $R_n(t)$, plus a signal term that is *multiplied* by a random noise factor $\cos\theta_n(t)$. This multiplicative effect corrupts the signal to a much larger extent than the additive Rayleigh noise. This produces a *threshold* effect. That is, $(S/N)_{out}$ becomes *very small* when $(S/N)_{in} < 1$. In fact, it can be shown that $(S/N)_{out}$ is proportional to the *square* of $(S/N)_{in}$ for the case of $(S/N)_{in} < 1$ [Schwartz, Bennett, and Stein, 1966]. Although the envelope detector is greatly inferior to the product detector for small $(S/N)_{in}$, this deficiency is seldom noticed in practice for AM broadcasting applications. This is because the AM listener is usually interested in listening to stations only if they have reasonably good $(S/N)_{out}$, say 25 dB or more. Under these conditions the

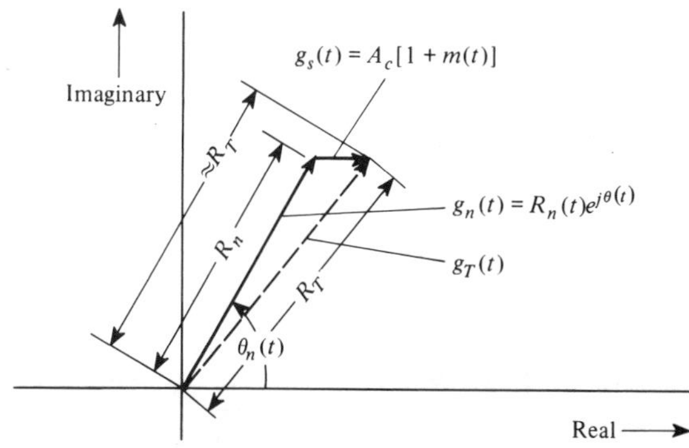

FIGURE 7-20 Vector diagram for AM, $(S/N)_{in} \ll 1$.

envelope detector performance is equivalent to that of the product detector. Moreover, the envelope detector is inexpensive and does not require a coherent reference. For these reasons the envelope detector is used almost exclusively in AM broadcast receivers. For other applications, such as listening to weak AM stations or for AM data transmission systems, product detection may be required to eradicate the multiplicative noise that would occur with envelope detection of weak systems.

DSB-SC Systems

As indicated in Chapter 4, the DSB-SC signal is essentially an AM signal in which the discrete carrier term has been suppressed (i.e., equivalent to infinite percent AM). The modulating waveform $m(t)$ is recovered from the DSB-SC signal by using coherent detection, as shown in Fig. 7-19. For DSB-SC,

$$g_s(t) = A_c m(t) \tag{7-96}$$

Following the development leading to (7-89), the S/N for DSB-SC is

$$\frac{(\text{S/N})_{\text{out}}}{(\text{S/N})_{\text{in}}} = 2 \tag{7-97}$$

Using (7-90), we obtain

$$\frac{(\text{S/N})_{\text{out}}}{(\text{S/N})_{\text{baseband}}} = 1 \tag{7-98}$$

Thus the noise performance of a DSB-SC system is the same as that of baseband signaling systems, although the bandwidth requirement is twice as large (i.e., $B_T = 2B$).

SSB Systems

The receiver for an SSB signal is also shown in Fig. 7-19, where the IF bandwidth is now $B_T = B$. The complex envelope for SSB is

$$g_s(t) = A_c[m(t) \pm j\hat{m}(t)] \tag{7-99}$$

where the upper sign is used for USSB and the lower sign is used for LSSB. The complex envelope for the (total) received SSB signal plus noise is

$$g_T(t) = [A_c m(t) + x_n(t)] + j[\pm A_c \hat{m}(t) + y_n(t)] \tag{7-100}$$

The output of the product detector is

$$\tilde{m}(t) = \text{Re}\{g_T(t)\} = A_c m(t) + x_n(t) \tag{7-101}$$

The corresponding S/N is

$$(\text{S/N})_{\text{out}} = \frac{A_c^2\, \overline{m^2(t)}}{\overline{x_n^2(t)}} = \frac{A_c^2\, \overline{m^2(t)}}{N_0 B} \tag{7-102}$$

where $\overline{x_n^2} = \overline{n^2} = 2(N_0/2)(B)$. Using (6-133g), we see that the input signal power is

$$P_s = \tfrac{1}{2}\overline{|g_s(t)|^2} = \frac{A_c^2}{2}\left[\overline{m^2} + \overline{(\hat{m})^2}\right] = A_c^2\,\overline{m^2} \tag{7-103}$$

and the input noise power is $P_n = \overline{n^2(t)} = N_0 B$. Thus

$$\frac{(S/N)_{out}}{(S/N)_{in}} = 1 \tag{7-104}$$

Similarly,

$$\frac{(S/N)_{out}}{(S/N)_{baseband}} = 1 \tag{7-105}$$

SSB is exactly equivalent to baseband signaling, in terms of both the noise performance and the bandwidth requirements (i.e., $B_T = B$).

PM Systems

As shown in Fig. 7-21, the modulation on a PM signal is recovered by a receiver that uses a (coherent) phase detector. (In Chapter 4 it was found that a phase detector could be realized by using a limiter that followed a product detector when β_p is small.) The PM signal has a complex envelope of

$$g_s(t) = A_c e^{j\theta_s(t)} \tag{7-106a}$$

where

$$\theta_s(t) = D_p m(t) \tag{7-106b}$$

The complex envelope of the composite signal plus noise at the detector input is

$$g_T(t) = |g_T(t)|e^{j\theta_T(t)} = [g_s(t) + g_n(t)] \tag{7-107}$$

$$= A_c e^{j\theta_s(t)} + R_n(t)e^{j\theta_n(t)}$$

When the input is Gaussian noise (only), $R_n(t)$ is Rayleigh distributed and $\theta_n(t)$ is uniformly distributed as demonstrated in Example 6-10.

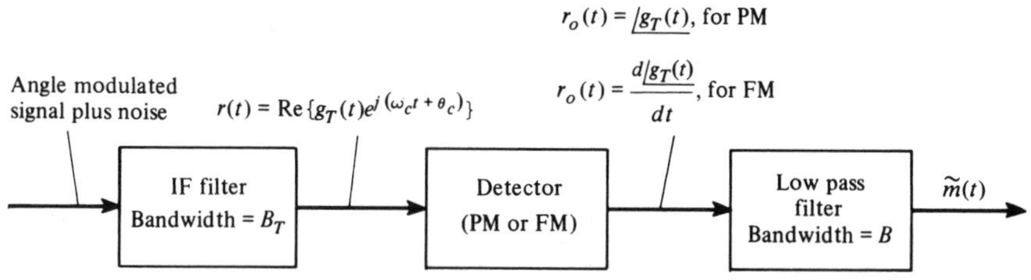

FIGURE 7-21 Receiver for angle-modulated signals.

The phase detector output is proportional to $\theta_T(t)$:

$$r_0(t) = K\underline{/g_T(t)} = K\theta_T(t)$$

where K is the gain constant of the detector. For large $(S/N)_{in}$, the phase angle of $g_T(t)$ can be approximated with the help of a vector diagram for $g_T = g_s + g_n$. This is shown in Fig. 7-22. Then, for $A_c \gg R_n(t)$, the composite phase angle is approximated by

$$r_0(t) = K\theta_T(t) \approx K\left\{\theta_s(t) + \frac{R_n(t)}{A_c}\sin[\theta_n(t) - \theta_s(t)]\right\} \tag{7-108}$$

For the case of no phase modulation (the carrier signal is still present), the equation reduces to

$$r_0(t) \approx \frac{K}{A_c}\, y_n(t), \qquad \theta_s(t) = 0 \tag{7-109}$$

where, using (6-127e), $y_n(t) = R_n(t)\sin\theta_n(t)$. This shows that the presence of an unmodulated carrier (at the input to the PM receiver) suppresses the noise on the output. This is called the *quieting effect,* and it occurs when the input signal power is above the *threshold* [i.e., when $(S/N)_{in} \gg 1$]. Furthermore, when phase modulation is present and $(S/N)_{in} \gg 1$, we can replace $R_n(t)\sin[\theta_n(t) - \theta_s(t)]$ by $R_n(t)\sin\theta_n(t)$. This can be done because $\theta_s(t)$ can be considered to be deterministic, and, consequently, $\theta_s(t)$ is a constant for a given value of t. Then $\theta_n(t) - \theta_s(t)$ will be uniformly distributed over some 2π interval since, from (6-153), $\theta_n(t)$ is uniformly distributed over $(0, 2\pi)$. That is, $\cos[\theta_n(t) - \theta_s(t)]$ will have the same PDF as $\cos[\theta_n(t)]$ so that the replacement can be made. Thus, for large $(S/N)_{in}$, the relevant part of the PM detector output is approximated by

$$r_0(t) \approx s_0(t) + n_0(t) \tag{7-110}$$

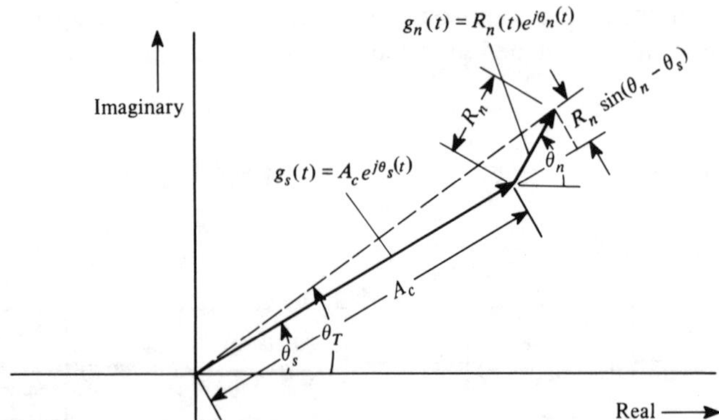

FIGURE 7-22 Vector diagram for angle modulation, $(S/N)_{in} \gg 1$.

where

$$s_0(t) = K\theta_s(t) = KD_p m(t) \tag{7-111a}$$

$$n_0(t) = \frac{K}{A_c}\, y_n(t) \tag{7-111b}$$

The PSD of this output noise, $n_0(t)$, is obtained by the use of (6-133l).

$$\mathcal{P}_{n_0}(f) = \begin{cases} \dfrac{K^2}{A_c^2} N_0, & |f| \le B_T/2 \\ 0, & f \text{ otherwise} \end{cases} \tag{7-112}$$

where the PSD of the bandpass input noise is $N_0/2$ over the bandpass of the IF filter and zero outside the IF bandpass. Equation (7-112) is plotted in Fig. 7-23a,

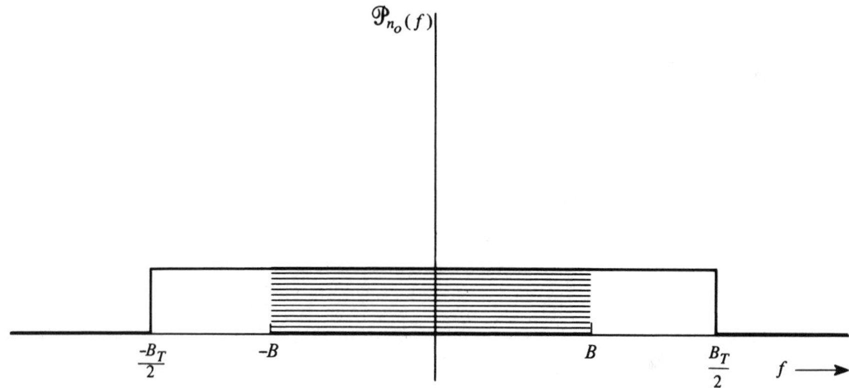

(a) PM Detector

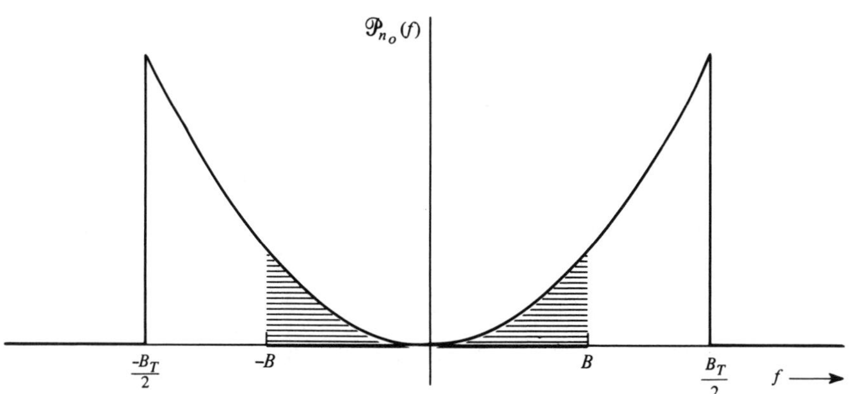

(b) FM Detector

FIGURE 7-23 PSD for noise out of detectors for receivers of angle-modulated signals.

and the portion of the spectrum that is passed by the LPF is denoted by the hatched lines.

The receiver output consists of the low-pass filtered version of $r_0(t)$. However, the spectrum of $s_0(t)$ is contained within the bandpass of the filter, so

$$\tilde{m}(t) = s_0(t) + \tilde{n}_0(t) \tag{7-113}$$

where $\tilde{n}_0(t)$ is bandlimited white noise of the hatched portion of Fig. 7-23a. The noise power of the receiver is

$$\overline{[\tilde{n}_0(t)]^2} = \int_{-B}^{B} \mathcal{P}_{n_0}(f)\, df = \frac{2K^2 N_0 B}{A_c^2} \tag{7-114}$$

The output S/N is now easily evaluated with the help of (7-111a) and (7-114):

$$\left(\frac{\text{S}}{\text{N}}\right)_{\text{out}} = \frac{\overline{s_0^2}}{\overline{\tilde{n}_0^2}} = \frac{A_c^2 D_p^2\, \overline{m^2}}{2N_0 B}$$

Using (4-148) and (4-149), we can write the sensitivity constant of the PM transmitter as

$$D_p = \frac{\beta_p}{V_p}$$

where β_p is the PM index and V_p is the peak value of $m(t)$. Thus the output S/N becomes

$$\left(\frac{\text{S}}{\text{N}}\right)_{\text{out}} = \frac{A_c^2 \beta_p^2 \overline{(m/V_p)^2}}{2N_0 B} \tag{7-115}$$

The input S/N is

$$\left(\frac{\text{S}}{\text{N}}\right)_{\text{in}} = \frac{A_c^2/2}{2(N_0/2)B_T} = \frac{A_c^2}{2N_0 B_T} \tag{7-116}$$

where B_T is the transmission bandwidth of the PM signal (and also the IF filter bandwidth).

The transmission bandwidth of the PM signal is given by Carson's rule, (4-162),

$$B_T = 2(\beta_p + 1)B \tag{7-117}$$

where β_p is the PM index. The PM index is identical to the peak angle deviation of the PM signal, as indicated in (4-149). Thus

$$\left(\frac{\text{S}}{\text{N}}\right)_{\text{in}} = \frac{A_c^2}{4N_0(\beta_p + 1)B} \tag{7-118}$$

Combining (7-115) and (7-118), we obtain the ratio of output to input S/N.

$$\frac{(\text{S/N})_{\text{out}}}{(\text{S/N})_{\text{in}}} = 2\beta_p^2(\beta_p + 1)\overline{\left(\frac{m}{V_p}\right)^2} \tag{7-119}$$

The output S/N can also be expressed in terms of the equivalent baseband system by substituting (7-84) into (7-115), where $P_s = A_c^2/2$.

$$\frac{(\mathrm{S/N})_{\mathrm{out}}}{(\mathrm{S/N})_{\mathrm{baseband}}} = \beta_p^2 \overline{\left(\frac{m}{V_p}\right)^2} \tag{7-120}$$

This equation shows that the improvement of a PM system over a baseband signaling system depends on the amount of phase deviation that is used. It seems to indicate that we can make the improvement as large as we wish simply by increasing β_p. This depends on the types of circuits used. If the peak phase deviation exceeds π radians, special "phase unwrapping" techniques have to be used in some circuits to obtain the true value (as compared to the relative value) of the phase at the output. Thus the maximum value of $\beta_p m(t)/V_p = D_p m(t)$ might be taken to be π. For sinusoidal modulation this would provide an improvement of $D_p^2 \overline{m^2} = \pi^2/2$ or 6.9 dB over baseband signaling.

It is emphasized that the results obtained previously for $(\mathrm{S/N})_{\mathrm{out}}$ are valid only when the input signal is above the threshold [i.e., $(\mathrm{S/N})_{\mathrm{in}} > 1$].

FM Systems

The procedure that we will use to evaluate the output S/N for FM systems is essentially the same as that used for PM systems, except that the output of the FM detector is proportional to $d\theta_T(t)/dt$, whereas the output of the PM detector is proportional to $\theta_T(t)$. The detector in the angle modulated receiver of Fig. 7-21 is now an FM detector. The complex envelope of the FM signal (only) is

$$g_s(t) = A_c e^{j\theta_s(t)} \tag{7-121a}$$

where

$$\theta_s(t) = D_f \int_{-\infty}^{t} m(\lambda)\, d\lambda \tag{7-121b}$$

It is assumed that an FM signal plus white noise is present at the receiver input.

The output of the FM detector is proportional to the derivative of the composite phase at the detector input:

$$r_0(t) = \left(\frac{K}{2\pi}\right)\frac{d\underline{/g_T(t)}}{dt} = \left(\frac{K}{2\pi}\right)\frac{d\theta_T(t)}{dt} \tag{7-122}$$

where K is the FM detector gain. Using the same procedure as that leading to (7-108), (7-110), and (7-111), we can approximate the detector output by

$$r_0(t) \approx s_0(t) + n_0(t) \tag{7-123}$$

where, for FM,

$$s_0(t) = \left(\frac{K}{2\pi}\right)\frac{d\theta_s(t)}{dt} = \left(\frac{KD_f}{2\pi}\right)m(t) \tag{7-124a}$$

and

$$n_0(t) = \left(\frac{K}{2\pi A_c}\right)\frac{dy_n(t)}{dt} \tag{7-124b}$$

This result is valid only when the input signal is above the threshold [i.e., $(S/N)_{in} \gg 1$]. The derivative of the noise, (7-124b), makes the PSD of the FM output noise different from that for the PM case. For FM we have

$$\mathcal{P}_{n_0}(f) = \left(\frac{K}{2\pi A_c}\right)^2 |j2\pi f|^2 \mathcal{P}_{y_n}(f)$$

or

$$\mathcal{P}_{n_0}(f) = \begin{cases} \left(\dfrac{K}{A_c}\right)^2 N_0 f^2, & |f| < B_T/2 \\ 0, & f \text{ otherwise} \end{cases} \tag{7-125}$$

This shows that the PSD for the noise out of the FM detector has a parabolic shape, as illustrated in Fig. 7-23b.

The receiver output consists of the low-pass filtered version of $r_0(t)$. The noise power for the filtered noise is

$$\overline{[\tilde{n}_0(t)]^2} = \int_{-B}^{B} \mathcal{P}_{n_0}(f)\, df = \frac{2}{3}\left(\frac{K}{A_c}\right)^2 N_0 B^3 \tag{7-126}$$

The output S/N is now easily evaluated using (7-124a) and (7-126):

$$\left(\frac{S}{N}\right)_{out} = \frac{\overline{s_0^2}}{\overline{[\tilde{n}_0]^2}} = \frac{3A_c^2[D_f/(2\pi B)]^2\, \overline{m^2}}{2N_0 B}$$

From (4-146) and (4-150), it is realized that

$$\frac{D_f}{2\pi B} = \frac{\beta_f}{V_p}$$

where V_p is the peak value of $m(t)$. Then the output S/N becomes

$$\left(\frac{S}{N}\right)_{out} = \frac{3A_c^2\beta_f^2\,\overline{(m/V_p)^2}}{2N_0 B} \tag{7-127}$$

The input S/N is

$$\left(\frac{S}{N}\right)_{in} = \frac{A_c^2}{4N_0(\beta_f + 1)B} \tag{7-128}$$

Combining (7-127) and (7-128), we obtain the ratio of output to input S/N,

$$\frac{(S/N)_{out}}{(S/N)_{in}} = 6\beta_f^2(\beta_f + 1)\overline{\left(\frac{m}{V_p}\right)^2} \tag{7-129}$$

where β_f is the FM index and V_p is the peak value of the modulating signal $m(t)$.

The output S/N can be expressed in terms of the equivalent baseband S/N by substituting (7-84) into (7-127), where $P_s = A_c^2/2$,

$$\frac{(\mathrm{S/N})_{\mathrm{out}}}{(\mathrm{S/N})_{\mathrm{baseband}}} = 3\beta_f^2 \overline{\left(\frac{m}{V_p}\right)^2} \tag{7-130}$$

For the case of sinusoidal modulation $\overline{(m/V_p)^2} = \frac{1}{2}$, and (7-130) becomes

$$\frac{(\mathrm{S/N})_{\mathrm{out}}}{(\mathrm{S/N})_{\mathrm{baseband}}} = \frac{3}{2}\,\beta_f^2 \qquad \text{(sinusoidal modulation)} \tag{7-131}$$

At first glance, these results seem to indicate that the performance of FM systems can be increased without limit simply by increasing the FM index β_f. However, as β_f is increased, the transmission bandwidth increases, and, consequently, $(\mathrm{S/N})_{\mathrm{in}}$ decreases. These equations for $(\mathrm{S/N})_{\mathrm{out}}$ are *valid only* when $(\mathrm{S/N})_{\mathrm{in}} \gg 1$ (i.e., input signal power above the threshold), so $(\mathrm{S/N})_{\mathrm{out}}$ *does not* increase to an excessively large value simply by increasing the FM index β_f. This threshold effect was first demonstrated by the measured output S/N characteristic shown in Fig. 5-45. Plots of (7-131) are given by the dashed lines in Fig. 7-24.

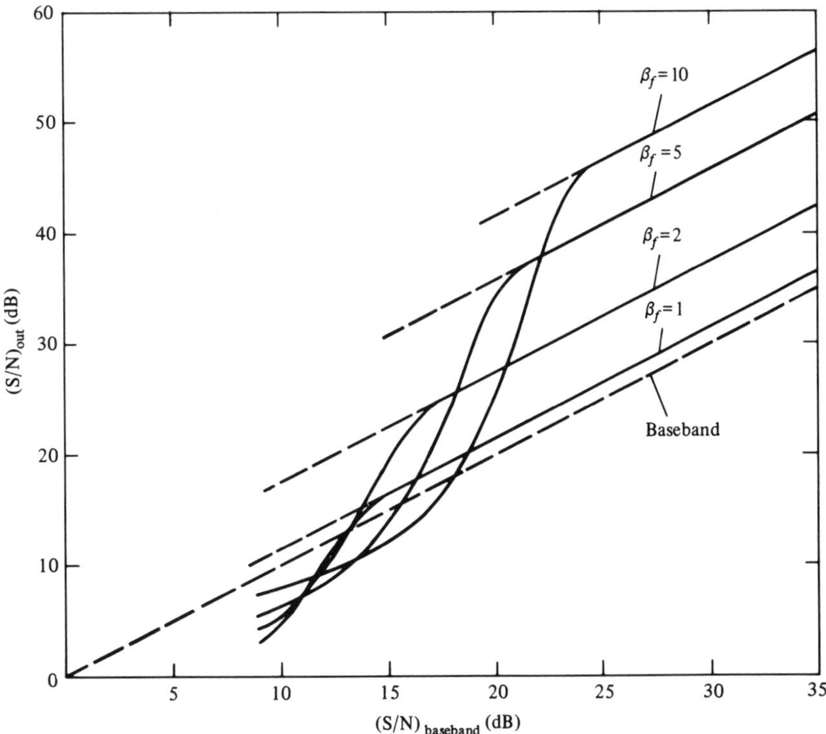

FIGURE 7-24 Noise performance of an FM discriminator for a sinusoidal modulated FM signal plus Gaussian noise (no deemphasis).

The threshold effect was first analyzed in 1948 [Rice, 1948; Stumpers, 1948]. An excellent tutorial treatment has been given by Taub and Schilling [1986]. They have shown that (7-131) can be generalized to describe $(S/N)_{out}$ near the threshold. For the case of sinusoidal modulation, the output S/N for a FM discriminator is shown to be

$$\left(\frac{S}{N}\right)_{out} = \frac{\frac{3}{2}\beta_f^2\,(S/N)_{baseband}}{1 + \left(\frac{12}{\pi}\,\beta_f\right)\left(\frac{S}{N}\right)_{baseband} e^{\left\{-\left[\frac{1}{2(\beta_f+1)}\left(\frac{S}{N}\right)_{baseband}\right]\right\}}} \tag{7-132}$$

(No deemphasis is used in obtaining the result.) This output S/N characteristic showing the threshold effect of an FM discriminator is plotted by the solid lines in Fig. 7-24. This figure illustrates that the FM noise performance can be substantially better than baseband performance. For example, for $\beta_f = 5$ and $(S/N)_{baseband} = 25$ dB, the FM performance is 15.7 dB better than the baseband performance. The performance can be improved even further by the use of deemphasis, as we will demonstrate in a later section.

FM Systems with Threshold Extension

Any one of several techniques may be used to lower the threshold below that provided by a receiver that uses only an FM discriminator. For example, Fig. 5-45 demonstrated (by measurement) that a PLL FM detector could be used to extend the threshold below that provided by an FM discriminator. However, when the input S/N is large, all the FM receiving techniques provide the same performance—namely, that as predicted by (7-129) or (7-130).

An *FM receiver with feedback* (FMFB) is shown in Fig. 7-25. This is another threshold extension technique. The FMFB receiver provides threshold extension by lowering the modulation index for the FM signal that is applied to the discriminator input. That is, the modulation index of $\tilde{e}(t)$ is smaller than that for $v_{in}(t)$, as we will show. Thus the threshold will be lower than that illustrated in Fig. 7-24. The calculation of the exact amount of threshold extension that is realized by an FMFB receiver is somewhat involved [Taub and Schilling, 1986]. However, we can easily show that the FMFB technique does indeed reduce the modulation

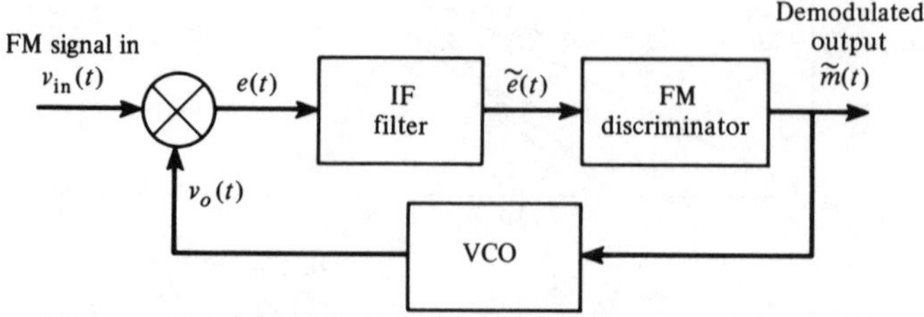

FIGURE 7-25 FMFB receiver.

index of the FM signal at the discriminator input. Referring to Fig. 7-25, we find that the FM signal at the receiver input is

$$v_{\text{in}}(t) = A_c \cos[\omega_c t + \theta_i(t)]$$

where

$$\theta_i(t) = D_f \int_{-\infty}^{t} m(\lambda)\, d\lambda$$

The output of the VCO is

$$v_0(t) = A_0 \cos[\omega_0 t + \theta_0(t)]$$

where

$$\theta_0(t) = D_v \int_{-\infty}^{t} \tilde{m}(\lambda)\, d\lambda$$

With these representations for $v_{\text{in}}(t)$ and $v_0(t)$, the output of the multiplier (mixer) is

$$\begin{aligned} e(t) &= A_c A_0 \cos[\omega_c t + \theta_i(t)] \cos[\omega_0 t + \theta_0(t)] \\ &= \tfrac{1}{2} A_c A_0 \cos[(\omega_c - \omega_0)t + \theta_i(t) - \theta_0(t)] \\ &\quad + \tfrac{1}{2} A_c A_0 \cos[(\omega_c + \omega_0)t + \theta_i(t) + \theta_0(t)] \end{aligned}$$

If the IF filter is tuned to pass the band of frequencies centered about $f_{\text{if}} \triangleq f_c - f_0$, the IF output is

$$\tilde{e}(t) = \frac{A_c A_0}{2} \cos[\omega_{\text{if}} t + \theta_i(t) - \theta_0(t)] \tag{7-133a}$$

or

$$\tilde{e}(t) = \frac{A_c A_0}{2} \cos\left\{\omega_{\text{if}} t + \int_{-\infty}^{t} [D_f m(\lambda) - D_v \tilde{m}(\lambda)]\, d\lambda\right\} \tag{7-133b}$$

The FM discriminator output is proportional to the derivative of the phase deviation.

$$\tilde{m}(t) = \frac{K}{2\pi} \frac{d\left\{\int_{-\infty}^{t} [D_f m(\lambda) - D_v\, \tilde{m}(\lambda)]\, d\lambda\right\}}{dt}$$

Evaluating the derivative and solving the resulting equation for $\tilde{m}(t)$, we obtain

$$\tilde{m}(t) = \left(\frac{KD_f}{2\pi + KD_v}\right) m(t)$$

Substituting this expression for $\tilde{m}(t)$ into (7-133), we get

$$\tilde{e}(t) = \frac{A_c A_0}{2} \cos\left[\omega_{\text{if}} t + \left(\frac{1}{1 + (K/2\pi)D_v}\right) D_f \int_{-\infty}^{t} m(\lambda)\, d\lambda\right] \tag{7-134}$$

This demonstrates that the modulation index of $\tilde{e}(t)$ is exactly $1/[1 + (K/2\pi)D_v]$ of the modulation index of $v_{\text{in}}(t)$. The threshold extension provided by the FMFB receiver is on the order of 5 dB, whereas that of a PLL receiver is on the order of 3 dB (when both are compared to the threshold of an FM discriminator). Although this is not a fantastic improvement, it can be quite significant for systems that operate near the threshold, such as satellite communication systems. For example, a system that uses a threshold extension receiver instead of a conventional receiver with an FM discriminator may be much less expensive than the system that requires a doubled-size antenna to provide the 3-dB signal gain. This was demonstrated in Example 5-5. Other threshold extension techniques have been described and analyzed in the literature [Klapper and Frankle, 1972].

FM Systems with Deemphasis

In Chapter 4 (see Fig. 4-46) it was illustrated that the noise performance of an FM system could be improved by preemphasizing the higher frequencies of the modulation signal at the transmitter input and deemphasizing the output of the receiver. This improvement occurs because the PSD of the noise at the output of the FM detector has a parabolic shape, as shown in Fig. 7-23b.

Referring to Fig. 7-21, we incorporate deemphasis into the receiver by including a deemphasis response in the LPF characteristic. Assume that the transfer function of the LPF is

$$H(f) = \frac{1}{1 + j(f/f_1)} \tag{7-135}$$

over the low-pass bandwidth of B hertz. For standard FM broadcasting, a 75-μsec deemphasis filter is used so that $f_1 = 1/[(2\pi)(75 \times 10^{-6})] = 2.1$ kHz. From (7-125) and (7-126), the noise power out of the receiver with deemphasis is

$$\overline{[\tilde{n}_0(t)]^2} = \int_{-B}^{B} |H(f)|^2 \mathcal{P}_{n_0}(f)\, df = \left(\frac{K}{A_c}\right)^2 N_0 \int_{-B}^{B} \left[\frac{1}{1 + (f/f_1)^2}\right] f^2\, df$$

or

$$\overline{[\tilde{n}_0(t)]^2} = 2\left(\frac{K}{A_c}\right)^2 N_0 f_1^3 \left[\frac{B}{f_1} - \tan^{-1}\left(\frac{B}{f_1}\right)\right] \tag{7-136}$$

In a typical application $B/f_1 \gg 1$, so $\tan^{-1}(B/f_1) \approx \pi/2$, which is negligible when compared to B/f_1. Thus (7-136) becomes

$$\overline{[\tilde{n}_0(t)]^2} = 2\left(\frac{K}{A_c}\right)^2 N_0 f_1^2 B \tag{7-137}$$

for $B/f_1 \gg 1$.

The output *signal* power for the preemphasis–deemphasis system is the same as that when preemphasis–deemphasis is not used because the overall frequency response of the system to $m(t)$ is flat (constant) over the bandwidth of B hertz.[†] Thus from (7-137) and (7-124a), the output S/N is

$$\left(\frac{S}{N}\right)_{out} = \frac{\overline{s_0^2}}{\overline{n_0^2}} = \frac{A_c^2[D_f/(2\pi f_1)]^2\overline{m^2}}{2N_0B}$$

In addition, $D_f/(2\pi B) = \beta_f/V_p$, so the output S/N reduces to

$$\left(\frac{S}{N}\right)_{out} = \frac{A_c^2\beta_f^2(B/f_1)^2\,\overline{(m/V_p)^2}}{2N_0B} \tag{7-138}$$

Using (7-128), we find that the output to input S/N is

$$\frac{(S/N)_{out}}{(S/N)_{in}} = 2\beta_f^2(\beta_f+1)\left(\frac{B}{f_1}\right)^2\overline{\left(\frac{m}{V_p}\right)^2} \tag{7-139}$$

where β_f is the FM index, B is the bandwidth of the baseband (modulation) circuits, f_1 is the 3-dB bandwidth of the deemphasis filter, V_p is the peak value of the modulating signal $m(t)$, and $\overline{(m/V_p)^2}$ is the square of the rms value of $m(t)/V_p$.

The output S/N is expressed in terms of the equivalent baseband S/N by substituting (7-84) into (7-138).

$$\frac{(S/N)_{out}}{(S/N)_{baseband}} = \beta_f^2\left(\frac{B}{f_1}\right)^2\overline{\left(\frac{m}{V_p}\right)^2} \tag{7-140}$$

When a sinusoidal test tone is transmitted over this FM system, $\overline{(m/V_p)^2} = \frac{1}{2}$ and (7-140) becomes

$$\frac{(S/N)_{out}}{(S/N)_{baseband}} = \tfrac{1}{2}\beta_f^2\left(\frac{B}{f_1}\right)^2 \quad \text{(sinusoidal modulation)} \tag{7-141}$$

Of course, all of these results are valid only when the FM signal at the receiver input is above the threshold.

It is interesting to compare the noise performance of commercial FM systems. As shown in Table 4-10, for standard FM broadcasting $\beta_f = 5$, $B = 15$ kHz, and $f_1 = 2.1$ kHz. Using these parameters in (7-141), we obtain the noise performance of an FM broadcasting system, as shown in Fig. 7-26 by a solid curve. The corresponding performance of the same system, but without preemphasis–deemphasis, is shown by a dashed curve [from (7-131)]. Similarly, results are shown for the performance of the TV FM aural transmission system where $\beta_f = 1.67$, $B = 15$ kHz, and $f_1 = 2.1$ kHz.

[†]For a fair comparison of FM systems with and without preemphasis, the peak deviation ΔF needs to be the same for both cases. With typical audio program signals, preemphasis does not increase ΔF appreciably because the low frequencies dominate in the spectrum of $m(t)$. Thus this analysis is valid. However, if $m(t)$ is assumed to have a flat spectrum over the audio passband, the gain of the preemphasis filter needs to be reduced so that the peak deviation will be the same with and without preemphasis [Peebles, 1976, pp. 266–271]. In the latter case, there is less improvement in performance when preemphasis is used.

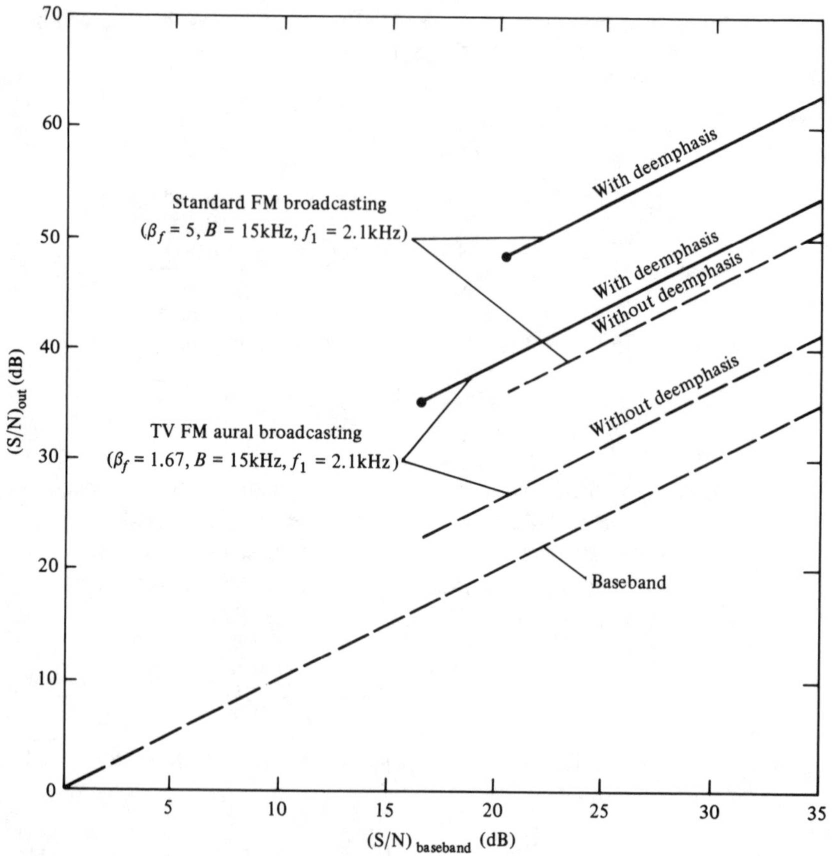

FIGURE 7-26 Noise performance of standard FM systems for sinusoidal modulation.

Figure 7-26 also illustrates that the FM noise performance with deemphasis can be substantially better than that of FM without deemphasis. For example, for standard FM broadcasting ($\beta_f = 5$, $B = 15$ kHz, and $f_1 = 2.1$ kHz) with $(S/N)_{\text{baseband}} = 25$ dB, the FM performance is superior by 13.3 dB.

7-9 Comparison of Analog Signaling Systems

Table 7-2 compares the analog systems that were analyzed in the previous sections. It is seen that the nonlinear modulation systems provide significant improvement in the noise performance, *provided* that the input signal is above the threshold. Of course, the improvement in the noise performance is obtained at the expense of having to use a wider transmission bandwidth. If the input S/N is very low, the linear systems outperform the nonlinear systems. SSB is best in

TABLE 7-2 Comparison of Analog Signaling Techniques[a]

Type	Linearity	Transmission Bandwidth Required[b]	$\frac{(S/N)_{out}}{(S/N)_{baseband}}$		Comments
Baseband	L	B	1	(7-84)	No modulation
AM	L[c]	$2B$	$\frac{\overline{m^2}}{1+\overline{m^2}}$	(7-90)	Valid for all $(S/N)_{in}$ with coherent detection; valid above threshold for envelope detection and $\lvert m(t)\rvert \le 1$
DSB-SC	L	$2B$	1	(7-98)	Coherent detection required
SSB	L	B	1	(7-105)	Coherent detection required; performance identical to baseband system
PM	NL	$2(\beta_p + 1)B$	$\beta_p^2 \overline{\left(\frac{m}{V_p}\right)^2}$	(7-120)	Coherent detection required; valid for $(S/N)_{in}$ above threshold
FM	NL	$2(\beta_f + 1)B$	$3\beta_f^2 \overline{\left(\frac{m}{V_p}\right)^2}$	(7-130)	Valid for $(S/N)_{in}$ above threshold
FM with deemphasis	NL	$2(\beta_f + 1)B$	$\beta_f^2 \left(\frac{B}{f_1}\right)^2 \overline{\left(\frac{m}{V_p}\right)^2}$	(7-140)	Valid for $(S/N)_{in}$ above threshold
PCM	NL	[d]	$M^2/(S/N)_{baseband}$	(7-82)	Valid for $(S/N)_{in}$ above threshold (i.e., $P_e \to 0$)

[a] B, absolute bandwidth of the modulating signal; f_1, 3-dB bandwidth of the deemphasis signal; L, linear; $m = m(t)$ is the modulating signal; $M = 2^n$, number of quantizing steps where n is the number of bits in the PCM word; NL, nonlinear; V_p, peak value of $m(t)$; β_p, PM index; β_f, FM index.

[b] Typical bandwidth specifications by the International Telecommunications Union (ITU) are larger than these minima [Jordan, 1985].

[c] In the strict sense, AM signaling is not linear because of the carrier term (see Table 4-1).

[d] This approximation of an analog signal depends on the type of digital system used (e.g., OOK, FSK).

terms of small bandwidth, and it has one of the best noise characteristics at low input S/N.

The selection of a particular system depends on the transmission bandwidth that is allowed and the available receiver input S/N. A comparison of the noise performance of these systems is given in Fig. 7-27, assuming that $V_p = 1$, $\overline{m^2} = \frac{1}{2}$ for the AM, PM, and FM systems. For the nonlinear systems a bandwidth spreading ratio of $B_T/B = 12$ is chosen for system comparisons. This corresponds to a $\beta_f = 5$ for the FM systems cited in the figure (which corresponds to commercial FM broadcasting).

Note that, except for the "wasted" carrier power in AM, all of the linear modulation methods have the same S/N performance as that for the baseband system. (SSB has the same performance as DSB because the coherence of the two sidebands in DSB compensates for the half noise power in SSB due to bandwidth reduction.) These comparisons are made on the basis of signals with equal *average* powers. If comparisons are made on equal *peak* powers (i.e., equal peak values for the signals), then SSB has a $(S/N)_{out}$ that is 3 dB better than

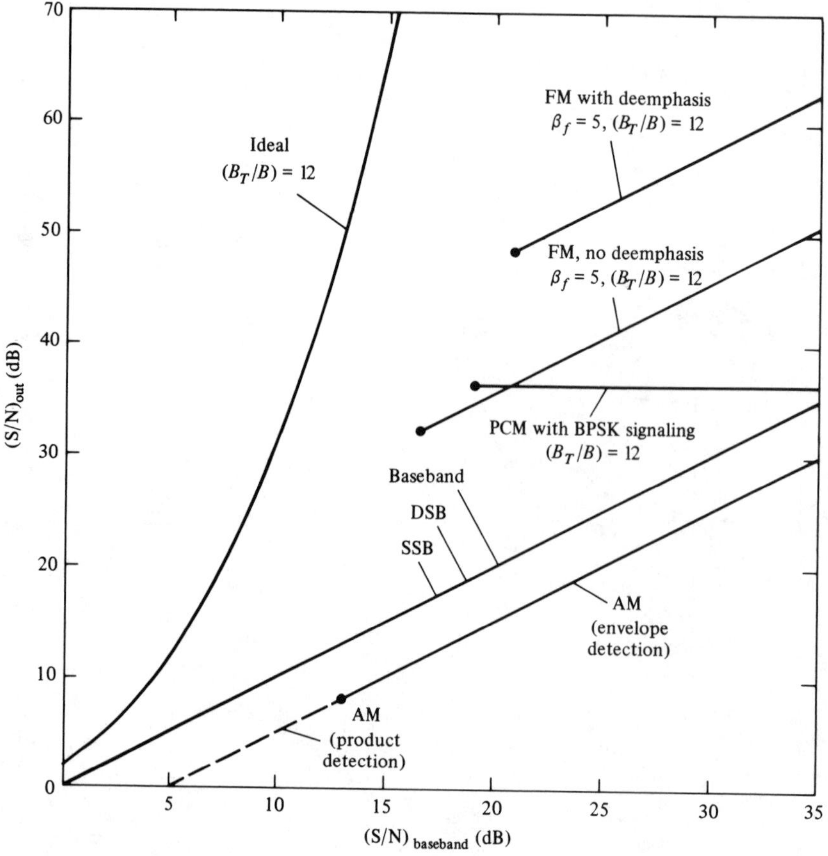

FIGURE 7-27 Comparison of the noise performance of analog systems.

DSB and 9 dB better than AM with 100% modulation (as demonstrated by Prob. 7-34). Of course, when operating above threshold, all of the nonlinear modulation systems have better S/N performance than linear modulation systems because the nonlinear systems have larger transmission bandwidths.

Ideal System Performance

It is interesting to ask the question: What is the best noise performance that is theoretically possible? Here a wide transmission bandwidth would be used to gain improved noise performance. The answer is given by Shannon's channel capacity theorem. The *ideal system* is defined as one that does not lose channel capacity in the detection process. Thus

$$C_{\text{in}} = C_{\text{out}} \tag{7-142}$$

where C_{in} is the bandpass channel capacity and C_{out} is the channel capacity after detection. Using (1-10) in (7-142), we have

$$B_T \log_2[1 + (\text{S/N})_{\text{in}}] = B \log_2[1 + (\text{S/N})_{\text{out}}] \tag{7-143}$$

where B_T is the transmission bandwidth of the bandpass signal at the receiver input and B is the bandwidth of the baseband signal at the receiver output. Solving for $(\text{S/N})_{\text{out}}$, we have

$$\left(\frac{\text{S}}{\text{N}}\right)_{\text{out}} = \left[1 + \left(\frac{\text{S}}{\text{N}}\right)_{\text{in}}\right]^{B_T/B} - 1 \tag{7-144}$$

But

$$\left(\frac{\text{S}}{\text{N}}\right)_{\text{in}} = \frac{P_s}{N_0 B_T} = \left(\frac{P_s}{N_0 B}\right)\left(\frac{B}{B_T}\right) = \left(\frac{B}{B_T}\right)\left(\frac{\text{S}}{\text{N}}\right)_{\text{baseband}} \tag{7-145}$$

where (7-84) was used. Thus (7-144) becomes

$$\left(\frac{\text{S}}{\text{N}}\right)_{\text{out}} = \left[1 + \left(\frac{B}{B_T}\right)\left(\frac{\text{S}}{\text{N}}\right)_{\text{baseband}}\right]^{B_T/B} - 1 \tag{7-146}$$

Equation (7-146), which describes the ideal system performance, is plotted in Fig. 7-27 for the case of $B_T/B = 12$. As expected, none of the practical signaling systems equals the performance of the ideal system. However, some of the nonlinear systems (near threshold) approach the performance of the ideal system.

7-10 Summary

The performance of digital systems has already been summarized in Sec. 7-6, and the performance of analog systems has been summarized in Sec. 7-9. The reader is invited to review these sections for a summary of this chapter. However, to condense these sections, we will simply state that there is no "best system" that provides a universal solution to all problems. The solution depends on the noise performance required, the transmission bandwidth available, and the state of the art in electronics that might favor the use of one type of communication system over another. In addition, the performance has been evaluated for the case of an additive white Gaussian noise channel. For other types of noise distributions or for multiplicative noise, the results would be different.

In the next chapter, we look at the problem of synthesizing optimum digital receivers for digital signaling over an additive white Gaussian noise channel.

PROBLEMS

7-1 In a binary communication system the receiver test statistic, $r_0(t_0) = r_0$, consists of a polar signal plus noise. The polar signal has values $s_{01} = +A$ and $s_{02} = -A$. Assume that the noise has a *Laplacian* distribution, which is

$$f(n_0) = \frac{1}{\sqrt{2}\,\sigma_0} e^{-\sqrt{2}|n_0|/\sigma_0}$$

where σ_0 is the rms value of the noise.

(a) Find the probability of error P_e as a function of A/σ_0 for the case of equally likely signaling and V_T having the optimum value.

(b) Plot P_e as a function of A/σ_0 decibels. Compare this result with that obtained for Gaussian noise as given by (7-26a).

7-2 Using (7-8), show that the optimum threshold level for the case of antipodal signaling with additive white Gaussian noise is

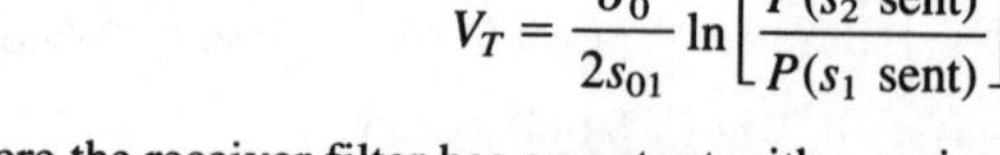

$$V_T = \frac{\sigma_0^2}{2s_{01}} \ln\left[\frac{P(s_2 \text{ sent})}{P(s_1 \text{ sent})}\right]$$

Here the receiver filter has an output with a variance of σ_0^2. s_{01} is the value of the sampled binary 1 signal at the filter output. $P(s_1$ sent) and $P(s_2$ sent) are the probabilities of transmitting a binary 1 and a binary 0, respectively.

7-3 A baseband digital communication system uses polar signaling with matcher filter in the receiver. The probability of sending a binary 1 is p and the probability of sending a binary zero is $1 - p$.

(a) For $E_b/N_0 = 10$ dB, plot P_e as a function of p using a log scale.

(b) Referring to (1-8), plot the entropy, H, as a function of p. Compare the shapes of these two curves.

7-4 A *whole* binary communication system can be modeled as an *information channel*, as shown in Fig. P7-4. Find equations for the four transition probabilities $P(\tilde{m}|m)$, where both $\tilde{m}$ and m can be binary 1's or binary 0's. Assume that the test statistic is a linear function of the receiver input and that additive white Gaussian noise appears at the receiver input. [*Hint:* Look at (7-15).]

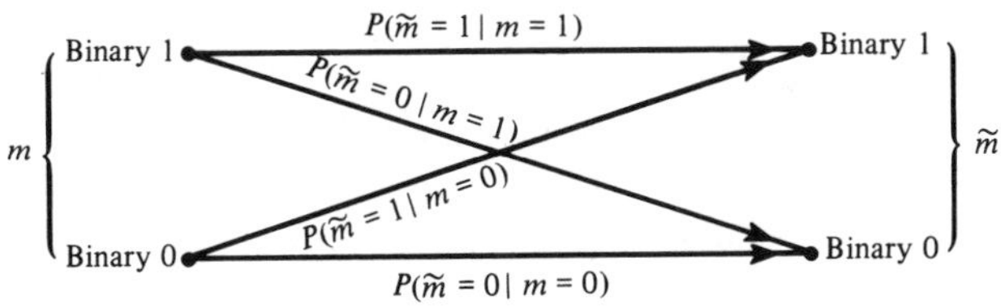

Binary 1 — $P(\tilde{m} = 1 \mid m = 1)$ — Binary 1

$P(\tilde{m} = 0 \mid m = 1)$

m $\tilde{m}$

$P(\tilde{m} = 1 \mid m = 0)$

Binary 0 — $P(\tilde{m} = 0 \mid m = 0)$ — Binary 0

FIGURE P7-4

7-5 A baseband digital communication system uses unipolar signaling (rectangular pulse shape) with matched filter detection. The data rate is $R = 9600$ bits/sec.

(a) Find an expression for bit error rate (BER), P_e, as a function of $(S/N)_{in}$. $(S/N)_{in}$ is the signal-to-noise power ratio at the receiver input where the noise is measured in a bandwidth corresponding to the equivalent bandwidth of the matched filter. [*Hint:* First find an expression of E_b/N_0 in terms of $(S/N)_{in}$.]

(b) Plot P_e vs. $(S/N)_{in}$ in dB units on a log scale over a range of $(S/N)_{in}$ from 0 to 15 dB.

7-6 Work Prob. 7-5 for the case of polar signaling.

7-7 Examine how the performance of a baseband digital communication system is affected by the receiver filter. Equation (7-26a) describes the BER when a low-pass filter is used and the bandwidth of the filter is sufficiently large so that the signal level at the filter output is $s_{01} = +A$ or $s_{02} = -A$. Instead, suppose that a RC low-pass filter with a restricted bandwidth is used where $T = 1/f_0 = 2\pi RC$. T is the duration (pulse width) of one bit and f_0 is the 3-dB bandwidth of the RC low-pass filter as described by (2-147). Assume that the initial conditions of the filter are reset to zero at the beginning of each bit interval.

(a) Derive an expression for P_e as a function of E_b/N_0.

(b) On a log scale, plot the BER obtained in part (a) for E_b/N_0 over a range of 0 to 15 dB.

(c) Compare this result with that for a matched-filter receiver (as shown in Fig. 7-5).

7-8 Consider a baseband digital communication system that uses polar signaling (rectangular pulse shape) where the receiver is shown in Fig. 7-4a. Assume that the receiver uses a second-order Butterworth filter with a 3-dB bandwidth of f_0. The filter impulse response and transfer function are

$$h(t) = \left[\sqrt{2}\omega_0 e^{-(\omega_0/\sqrt{2})t} \sin\left(\frac{\omega_0}{\sqrt{2}} t\right)\right] u(t)$$

$$H(f) = \frac{1}{(jf/f_0)^2 + \sqrt{2}(jf/f_0) + 1}$$

where $\omega_0 = 2\pi f_0$. Let $f_0 = 1/T$, where T is the bit interval (i.e., pulse width) and assume that the initial conditions of the filter are reset to zero at the beginning of each bit interval.

(a) Derive an expression for P_e as a function of E_b/N_0.

(b) On a log scale, plot the BER obtained in part (a) for E_b/N_0 over a range of 0 to 15 dB.

(c) Compare this result with that for a matched-filter receiver (as shown in Fig. 7-5).

7-9 Consider a baseband unipolar communication system with equally likely signaling. Assume that the receiver uses a simple RC LPF with a time constant of $RC = \tau$ where $\tau = T$ and $1/T$ is the bit rate. (By "simple" it is meant that the initial conditions of the LPF are *not* reset to zero at the beginning of each bit interval.)

(a) For signal alone at the receiver input, evaluate the approximate worst-case signal to ISI ratio (in decibels) out of the LPF at the sampling time $t = t_0 = nT$, where n is an integer.

(b) Evaluate the signal to ISI ratio (in decibels) as a function of the parameter K, where $t = t_0 = (n + K)T$ and $0 < K \leq 1$.

(c) What is the optimum sampling time to use to maximize the signal-to-ISI power ratio out of the LPF?

7-10 Consider a baseband polar communication system with equally likely signaling and no channel noise. Assume that the receiver uses a simple RC LPF with a time constant of $\tau = RC$. (By "simple" it is meant that the initial conditions of the LPF are *not* reset to zero at the beginning of each bit interval.) Evaluate the worst-case approximate signal to ISI ratio (in decibels) out of the LPF at the sampling time $t = t_0 = nT$, where n is an integer. This approximate result will be valid for $T/\tau > \frac{1}{2}$. Plot this result as a function of T/τ for $\frac{1}{2} \leq T/\tau \leq 5$.

7-11 Consider a baseband unipolar system described in Prob. 7-9. Assume that white Gaussian noise is present at the receiver input.

(a) Derive an expression for P_e as a function of E_b/N_0 for the case of sampling at the times $t = t_0 = nT$.

(b) Compare the BER obtained in part (a) with the BER characteristic that is obtained when a matched-filter receiver is used. Plot both of these BER characteristics as a function of $(E_b/N_0)_{\text{dB}}$ over the range 0 to 15 dB.

7-12 Work Prob. 7-11 for the case of *polar* baseband signaling.

7-13 Work Prob. 7-9 for the case of *polar* baseband signaling.

7-14 For unipolar baseband signaling as described by (7-23),

(a) Find the matched-filter frequency response and show how the filtering operation can be implemented by using an integrate-and-dump filter.

(b) Show that the equivalent bandwidth of the matched filter is $B_{\text{eq}} = 1/(2T) = R/2$.

7-15 Equally likely polar signaling is used in a baseband communication system. Gaussian noise having a PSD of $N_0/2$ W/Hz plus a polar signal with a peak level of A volts is present at the receiver input. The receiver uses a matched-filter circuit having a voltage gain of 1000.

(a) Find the expression for P_e as a function of A, N_0, T, and V_T, where $R = 1/T$ is the bit rate and V_T is the threshold level.

(b) Plot P_e as a function of V_T for the case of $A = 8 \times 10^{-3}$ V, $N_0/2 = 4 \times 10^{-9}$ W/Hz, and $R = 1200$ bits/sec.

7-16 Consider a baseband polar communication system with matched-filter detection. Assume that the channel noise is white and Gaussian with a PSD of $N_0/2$. The probability of sending a binary 1 is $P(1)$ and the probability of sending a binary 0 is $P(0)$. Find the expression for P_e as a function of the threshold level V_T when the signal level out of the matched filter is A, and the variance of the noise out of the matched filter is $\sigma^2 = N_0/(2T)$.

7-17 Design a receiver for detecting the data on a bipolar RZ signal that has a peak value of $A = 5$ volts. In your design assume that an RC low-pass filter will be used and the data rate is 2400 b/sec.

(a) Draw a block diagram of your design and explain how it works.

(b) Give the values for the design parameters R, C, and V_T.

(c) Calculate the PSD level for the noise N_0 that is allowed if P_e is to be less than 10^{-6}.

7-18 A BER of 10^{-5} or less is desired for an OOK communication system where the bit rate is $R = 2400$ bits/sec. The input to the receiver consists of the OOK signal plus white Gaussian noise.

(a) Find the minimum transmission bandwidth required.

(b) Find the minimum E_b/N_0 required at the receiver input for coherent matched-filter detection.

(c) Rework part (b) for the case of noncoherent detection.

7-19 Rework Prob. 7-18 for the case when FSK signaling is used where $2\Delta F = f_2 - f_1 = 1.5R$.

7-20 In this chapter the BER for a BPSK receiver was derived under the assumption that the coherent receiver reference (see Fig. 7-7) was exactly in phase with the received BPSK signal. Suppose that there is a phase error of θ_e between the reference signal and the incoming BPSK signal. Obtain new equations that give the P_e in terms of θ_e as well as the other parameters. In particular:

(a) Obtain a new equation that replaces (7-36).

(b) Obtain a new equation that replaces (7-38).

(c) Plot results from part (b) where the log plot of P_e is given as a function of θ_e over a range $-\pi < \theta_e < \pi$ for the case when $E_b/N_0 = 10$ dB.

7-21 Digital data are transmitted over a communication system that uses nine repeaters plus a receiver. BPSK signaling is used. The P_e for each of the

regenerative repeaters (see Sec. 3-4) is 5×10^{-8}, assuming additive Gaussian noise.
(a) Find the overall P_e for the system.
(b) If each repeater is replaced by an ideal amplifier (no noise or distortion), what is the P_e of the overall system?

7-22 Digital data are to be transmitted over a toll telephone system using BPSK. Regenerative repeaters are spaced 50 miles apart along the system. The total length of the system is 600 miles. The telephone lines between the repeater sites are equalized over a 300- to 2700-Hz band and provide an E_b/N_0 (Gaussian noise) of 15 dB to the repeater input.
(a) Find the largest bit rate R that can be accommodated with no ISI.
(b) Find the overall P_e for the system. (Be sure to include the receiver at the end of the system.)

7-23 A BPSK signal is given by

$$s(t) = A \sin[\omega_c t + \theta_c + (\pm 1)\beta_p], \qquad 0 < t \leq T$$

The binary data are represented by (± 1), where $(+1)$ is used to transmit a binary 1 and (-1) is used to transmit a binary 0. β_p is the phase modulation index as defined by (4-149).
(a) For $\beta_p = \pi/2$ show that this BPSK signal becomes the BPSK signal as described by (7-34).
(b) For $0 < \beta_p < \pi/2$ show that a discrete carrier term is present in addition to the BPSK signal as described by (7-34).

7-24 Referring to the BPSK signal as described in Prob. 7-23, find P_e as a function of the modulation index β_p, where $0 < \beta_p \leq \pi/2$.
(a) Find P_e as a function of A, β_p, N_0, and B for the receiver that uses a narrowband filter.
(b) Find P_e as a function of E_b, N_0, and β_p for the receiver that uses a matched filter. (E_b is the average BPSK signal energy that is received during one bit.)

7-25 Referring to the BPSK signal as described in Prob. 7-23, let $0 < \beta_p < \pi/2$.
(a) Show a block diagram for detection of the BPSK signal where a PLL is used to recover the coherent reference signal from the BPSK signal.
(b) Explain why Manchester-encoded data are often used when the receiver uses a PLL for carrier recovery, as in part (a). (*Hint:* Look at the spectrum of the Manchester-encoded PSK signal.)

7-26 In obtaining the P_e for FSK signaling with coherent reception, the energy in the difference signal E_d was needed, as shown in (7-46). For *orthogonal* FSK signaling the cross-product integral was found to be zero. Suppose that f_1, f_2, and T are chosen so that E_d is maximized.
(a) Find the relationship as a function of f_1, f_2, and T for maximum E_d.
(b) Find the P_e as a function of E_b and N_0 for signaling with this FSK signal.

(c) Sketch the P_e for this type of FSK signal and compare it with a sketch of the P_e for the orthogonal FSK signal that is given by (7-47).

7-27 An FSK signal with $R = 110$ bits/sec is transmitted over an RF channel that has white Gaussian noise. The receiver uses a noncoherent detector and has a noise figure of 6 dB. The impedance of the antenna input of the receiver is 50 Ω. The signal level at the receiver input is 0.05 μV, and the noise level is $N_0 = kT_0$, where $T_0 = 290$ K and k is Boltzmann's constant. Find the P_e for the digital signal at the output of the receiver.

7-28 Work Prob. 7-27 for the case of DPSK signaling.

7-29 An analysis of the noise associated with the two channels of an FSK receiver precedes (7-44). It is stated that $n_1(t)$ and $n_2(t)$ are independent since they arise from a common white Gaussian noise process and since $n_1(t)$ and $n_2(t)$ have nonoverlapping spectra. Prove that this statement is correct. [*Hint:* $n_1(t)$ and $n_2(t)$ can be modeled as the outputs of two linear filters that have nonoverlapping transfer functions and the same white noise process, $n(t)$, at their inputs.]

7-30 In most applications, communication systems are designed to have a BER of 10^{-5} or less. Find the minimum E_b/N_0 decibels required to achieve an error rate of 10^{-5} for the following types of signaling.
(a) Polar baseband.
(b) OOK.
(c) BPSK.
(d) FSK.
(e) DPSK.

7-31 Digital data are to be transmitted over a telephone line channel. Suppose that the telephone line is equalized over a 300- to 2700-Hz band and that the signal-to-Gaussian-noise (power) ratio at the output (receive end) is 25 dB.
(a) Of all the digital signaling techniques studied in this chapter, choose the one that will provide the largest bit rate for a P_e of 10^{-5}. What is the bit rate R for this system?
(b) Compare this result with the bit rate R that is possible when an ideal digital signaling scheme is used, as given by the Shannon channel capacity stated by (1-10).

7-32 An analog baseband signal has a uniform PDF and a bandwidth of 3500 Hz. This signal is sampled at an 8-kHz rate, uniformly quantized, and encoded into a PCM signal having 8-bit words. This PCM signal is transmitted over a DPSK communication system that contains additive white Gaussian channel noise. The signal-to-noise ratio at the receiver input is 8 dB.
(a) Find the P_e of the recovered PCM signal.
(b) Find the peak signal/average noise ratio (decibels) out of the PCM system.

7-33 An SS signal is often used to combat narrowband interference and for communication security. The SS signal with direct sequence spreading is (see Sec. 5-7)

$$s(t) = A_c c(t) m(t) \cos(\omega_c t + \theta_c)$$

where θ_c is the start-up carrier phase, $m(t)$ is the polar binary data baseband modulation, and $c(t)$ is a polar baseband spreading waveform that usually consists of a pseudonoise (PN) code. The PN code is a binary sequence that is N bits long. The "bits" are called *chips* since they do not contain data and since many chips are transmitted during the time that it takes to transmit 1 bit of the data [in $m(t)$]. The same N-bit code word is repeated over and over, but N is a large number, so the chip sequence in $c(t)$ looks like digital noise. The PN sequence may be generated by using a clocked r-stage shift register having feedback so that $N = 2^r - 1$. The autocorrelation of a long sequence is approximately

$$R_c(\tau) = \Lambda\left(\frac{\tau}{T_c}\right)$$

where T_c is the duration of one chip (the time it takes to send one chip of the PN code). $T_c \ll T_b$, where T_b is the duration of a data bit.

(a) Find the PSD for the SS signal $s(t)$. [*Hint:* Assume that $m(t)$, $c(t)$, and θ_c are independent. In addition, note that the PSD of $m(t)$ can be approximated by a delta function since the spectral width of $m(t)$ is very small when compared to that for the spreading waveform $c(t)$.]

(b) Draw a block diagram for an optimum coherent receiver. Note that $c(t)m(t)$ is first coherently detected and then the data, $m(t)$, are recovered by using a correlation processor.

(c) Find the expression for P_e.

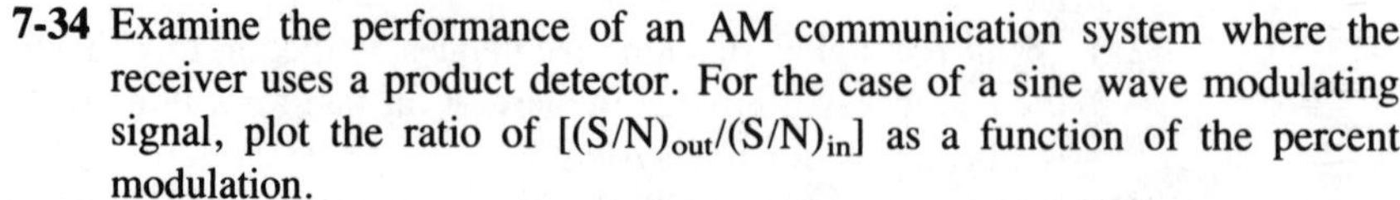

7-34 Examine the performance of an AM communication system where the receiver uses a product detector. For the case of a sine wave modulating signal, plot the ratio of $[(S/N)_{out}/(S/N)_{in}]$ as a function of the percent modulation.

7-35 An AM transmitter is modulated 40% by a sine-wave audio test tone. This AM signal is transmitted over an additive white Gaussian noise channel. Evaluate the noise performance of this system and determine by how many decibels this system is inferior to a DSB-SC system.

7-36 A phasing-type receiver for SSB signals is shown in Fig. P7-36. (This was the topic of Prob. 4-59.)

(a) Show that this receiver is or is not a linear system.

(b) Derive the equation for S/N out of this receiver when the input is an SSB signal plus white noise with a PSD of $N_0/2$.

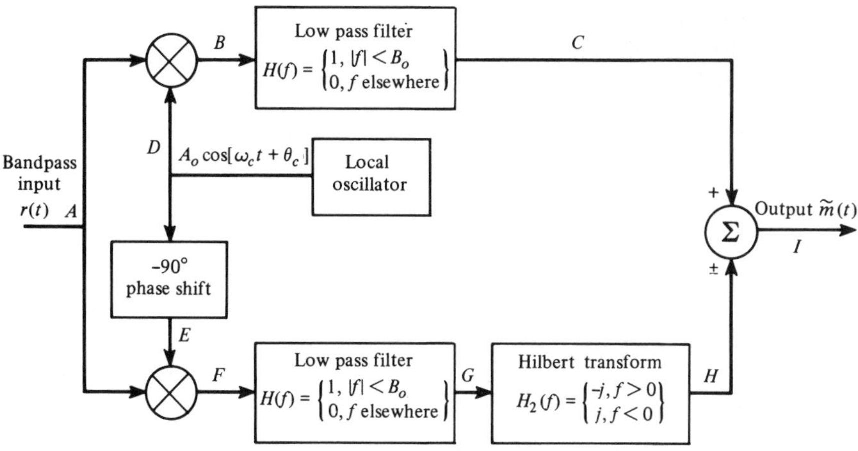

FIGURE P7-36

7-37 Referring to Fig. P7-36, suppose that the receiver consists only of the upper portion of the figure, so that point C is the output. Let the input be an SSB signal plus white noise with a PSD of $N_0/2$. Find $(S/N)_{\text{out}}$.

7-38 Consider the receiver shown in Fig. P7-38. Let the input be a DSB-SC signal plus white noise with a PSD of $N_0/2$. The mean value of the modulation is zero.

(a) For A_0 large, show that this receiver acts like a product detector.

(b) Find the equation for $(S/N)_{\text{out}}$ as a function of A_c, $\overline{m^2}$, N_0, A_0, and B_T when A_0 is large.

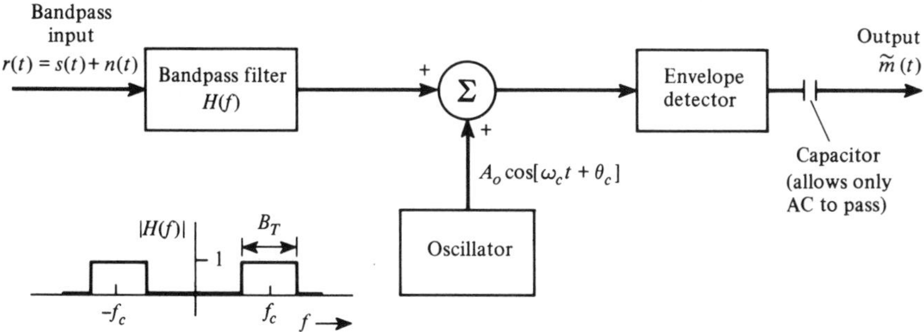

FIGURE P7-38

7-39 Compare the performance of AM, DSB-SC, and SSB systems when the modulating signal $m(t)$ is a Gaussian random process. Assume that the Gaussian modulation has a zero mean value and a peak value of $V_p = 1$ where $V_p \approx 4\sigma_m$. Compare the noise performance of these three systems by plotting $(S/N)_{\text{out}}/(S/N)_{\text{baseband}}$ for:

(a) The AM system.
(b) The DSB-SC system.
(c) The SSB system.

7-40 If linear modulation systems are to be compared on an equal *peak* power basis (i.e., all have equal peak signal values), show that:
(a) SSB has a $(S/N)_{out}$ that is 3 dB better than DSB.
(b) SSB has a $(S/N)_{out}$ that is 9 dB better than AM.
(*Hint:* See Prob. 4-49.)

7-41 Using (7-132), plot the output S/N threshold characteristic of a discriminator for the parameters of a TV FM aural system. ($\Delta F = 25$ kHz and $B =$ 15 kHz.) Compare the $(S/N)_{out}$ versus $(S/N)_{baseband}$ for this system with those shown in Fig. 7-24.

7-42 An FM receiver has an IF bandwidth of 25 kHz and a baseband bandwidth of 5 kHz. The noise figure of the receiver is 12 dB, and it uses a 75-μsec deemphasis network. An FM signal plus white noise is present at the receiver input, where the PSD of the noise is $N_0/2 = kT/2$. $T = 290$ K. Find the minimum input signal level (in dBm) that will give an S/N of 35 dB at the output when sine-wave test modulation is used.

7-43 Referring to Table 4-11, a two-way FM mobile radio system uses the parameters $\beta_f = 1$ and $B = 5$ kHz.
(a) Find $(S/N)_{out}$ for the case of no deemphasis.
(b) Find $(S/N)_{out}$ if deemphasis is used with $f_1 = 2.1$ kHz. It is realized that B is not much larger than f_1 in this application.
(c) Sketch the $(S/N)_{out}$ versus $(S/N)_{baseband}$ for this system and compare them with the result for FM broadcasting as shown in Fig. 7-26.

7-44 Compare the performance of two FM systems that use different deemphasis characteristics. Assume that $\beta_f = 5$, $\overline{(m/V_p)^2} = \frac{1}{2}$, $B = 15$ kHz, and an additive white Gaussian noise channel is used. Find and sketch $(S/N)_{out}/(S/N)_{baseband}$ for:
(a) 25-μsec deemphasis.
(b) 75-μsec deemphasis.

7-45 A baseband signal $m(t)$ that has a Gaussian (amplitude) distribution frequency modulates a transmitter. Assume that the modulation has a zero-mean value and a peak value of $V_p \approx 4\sigma_m$. The FM signal is transmitted over an additive white Gaussian noise channel. Let $\beta_f = 3$ and $B = 15$ kHz. Find and sketch the $(S/N)_{out}/(S/N)_{baseband}$ characteristic when:
(a) No deemphasis is used.
(b) 75-μsec deemphasis is used.

7-46 In FM broadcasting, a preemphasis filter is used at the audio input of the transmitter and a deemphasis filter is used at the receiver output to improve the output S/N. For 75-μsec emphasis, the 3-dB bandwidth of the receiver deemphasis LPF is $f_1 = 2.1$ kHz. The audio bandwidth is $B = 15$ kHz.

(a) Derive a formula for the improvement factor, I, as a function of B/f_1, where

$$I = \frac{(\text{S/N})_{\text{out}} \text{ for system with preemphasis–deemphasis}}{(\text{S/N})_{\text{out}} \text{ for system without preemphasis–deemphasis}}$$

(b) For $B = 15$ kHz, plot the decibel improvement that is realized as a function of the design parameter f_1 where 50 Hz $< f_1 <$ 15 kHz.

7-47 In FM broadcasting, preemphasis is used, and yet $\Delta F = 75$ kHz is defined as 100% modulation. Examine the incompatibility of these two standards. For example, assume that the amplitude of a 1-kHz audio test tone is adjusted to produce 100% modulation (i.e., $\Delta F = 75$ kHz).

(a) If the frequency is changed to 15 kHz, find the ΔF that would be obtained ($f_1 = 2.1$ kHz). What is the percent modulation?

(b) Explain why this phenomenon does not cause too much difficulty when typical audio program material (modulation) is broadcast.

7-48 Stereo FM transmission was studied in Sec. 4-7. At the transmitter, the left-channel audio, $m_L(t)$, and the right-channel audio, $m_R(t)$, are each preemphasized by an $f_1 = 2.1$-kHz network. These preemphasized audio signals are then converted into the composite baseband modulating signal $m_b(t)$, as shown in Fig. 4-48. At the receiver, the FM detector outputs the composite baseband signal that has been corrupted by noise. (Assume that the noise comes from a white Gaussian noise channel.) This corrupted composite baseband signal is demultiplexed into corrupted left- and right-channel audio signals, $\tilde{m}_L(t)$ and $\tilde{m}_R(t)$, each having been deemphasized by a 2.1-kHz filter. The noise on these outputs arises from the noise at the output of the FM detector that occurs in the 0- to 15-kHz and 23- to 53-kHz bands. The subcarrier frequency is 38 kHz. Assuming that the input S/N of the FM receiver is large, show that the stereo FM system is 22.2 dB more noisy than the corresponding monaural FM system.

7-49 An FDM signal, $m_b(t)$, consists of five 4-kHz-wide channels denoted by C1, C2, . . . , C5, as shown in Fig. P7-49. The FDM signal was obtained by modulating five audio signals (each with 4-kHz bandwidth) onto USSB (upper single-sideband) subcarriers. This FDM signal, $m_b(t)$, modulates a DSB-SC transmitter. The DSB-SC signal is transmitted over an additive white Gaussian noise channel. At the receiver the average power of the DSB-SC signal is P_s and the noise has a PSD of $N_0/2$.

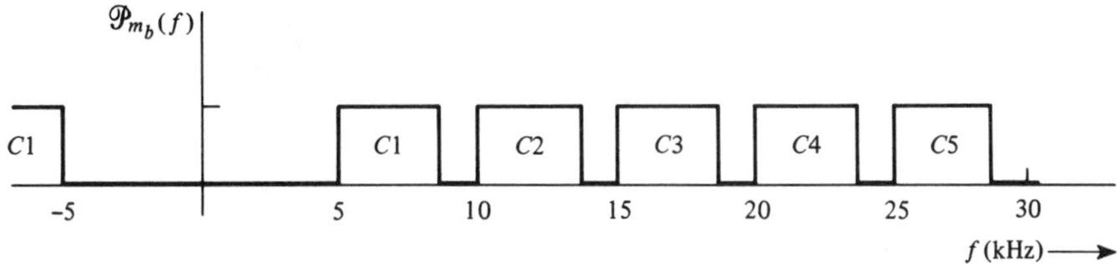

FIGURE P7-49

(a) Draw a block diagram for a receiving system with five outputs, one for each audio channel.

(b) Calculate the output S/N for each of the five audio channels.

7-50 Rework Prob. 7-49 when the FDM signal is frequency modulated onto the main carrier. Assume that the rms carrier deviation is denoted by ΔF_{rms} and that the five audio channels are independent. No deemphasis is used.

7-51 Refer to Fig. 7-27 and:

(a) Verify that the PCM curve is correct.

(b) Find the equation for the PCM curve for the case of QPSK signaling over the channel.

(c) Compare the performance of the PCM/QPSK system with that for PCM/BPSK and for the ideal case by sketching the $(S/N)_{\text{out}}$ characteristics.

CHAPTER 8

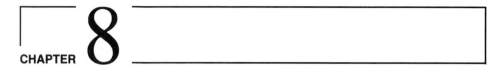

Optimum Digital Receivers

8-1 Introduction

Chapter 7 was concerned with the *analysis* of suboptimum as well as some optimum (i.e., matched-filter) receivers. In this chapter we concentrate on the *design* of optimum receivers for digital signals that have been corrupted by noise. We will find that the optimum receiver designer needs to answer the following questions:

- What is the criterion that will be used for the optimization?
- What is the resulting optimum receiver structure?
- What is the performance of the optimum receiver?

In digital problems it is assumed that one of M possible messages is transmitted over the system and the receiver is to decide which message was transmitted. Each message might be represented by a binary word that is K bits long, so that the total number of available messages from the source is

$$M = 2^K$$

We will find that the optimum receiver makes a decision on a message-by-message (i.e., word-by-word or symbol-by-symbol) basis as compared to one that makes decisions on a bit-by-bit basis, as studied in Chapter 7. (Of course, if the source has only two possible messages, message-by-message and bit-by-bit decisions are equivalent since $K = 1$ for $M = 2$.)

To describe the error performance, we will evaluate the probability of message error, $P(E)$, as compared to the bit error, P_e, that was used previously. The bit error, P_e, is also called the *bit error rate* (BER). The message error, $P(E)$, is also called the *message error rate* (MER), *symbol error rate* (SER), or the *word error rate* (WER). All three terms are used to describe the same thing, the probability of message error.

Word Error Rate (WER) for Bit-by-Bit Detection

As indicated above, this chapter is concerned with the detection of signals on a message-by-message basis. Before we get to the details of this topic, however, we will study how the WER, $P(E)$, differs for systems that detect on a bit-by-bit basis when compared to those that detect on a message-by-message basis. For bit-by-bit detection of a K-bit message

$$P(E) = 1 - P(C) = 1 - (1 - P_e)^K$$

where the probability of detecting a bit correctly is $P_c = 1 - P_e$. Using the approximation $(1 - P_e)^K \approx 1 - KP_e$ when $P_e \ll 1$, we get a WER of

$$P(E) = KP_b \qquad \text{(bit-by-bit detection)} \tag{8-1}$$

where P_e is the BER as evaluated on a bit-by-bit basis in Chapter 7. In this chapter techniques will be developed for evaluating the WER for systems that detect on a message-by-message basis.

Some examples will now be worked out to compare the WER for systems that use bit-by-bit detection (Chapter 7) with those that use message-by-message detection (Chapter 8). For the first example, suppose that we compare the WER for a K-bit PCM system that uses bit-by-bit detection with the WER of one that uses message-by-message detection. Assume that polar NRZ signaling is used. Substituting (7-26b) into (8-1), we get a WER of

$$P(E) = KQ\left(\sqrt{\frac{2E_b}{N_0}}\right) = KQ\left(\sqrt{\frac{2E_s}{KN_0}}\right)$$

for bit-by-bit detection where the energy of the K-bit signal is $E_s = KE_b$. For message-by-message detection, from (8-75), we get the same result. Thus, the WER performance for a K-bit PCM system with polar NRZ signaling is the same for either bit-by-bit or message-by-message detection.

For nonlinear modulation systems the results will differ. For example, compare a binary FSK system (bit-by-bit detection of K-bit words) with a M-ary FSK system (message-by-message detection) where $M = 2^K$. For the BFSK system, by substituting (7-47) into (8-1), the WER is

$$P(E) = KQ\left(\sqrt{\frac{E_b}{N_0}}\right) = KQ\left(\sqrt{\frac{E_s}{KN_0}}\right)$$

for bit-by-bit detection of K-bit words. For the MFSK system with orthogonal signaling,[†] from (8-55), the WER is

$$P(E) = (M - 1)Q\left(\sqrt{\frac{E_s}{N_0}}\right) = (2^K - 1)Q\left(\sqrt{\frac{E_s}{N_0}}\right)$$

for message-by-message detection ($M = 2^K$). When these two results are compared for E_s/N_0 sufficiently large, $P(E)$ for message-to-message detection (i.e., MFSK) is lower than that for bit-to-bit detection (i.e., BPSK, K-bit words).

Conversion of WER to BER

In M-ary systems, the signal is usually detected on a message-by-message basis, and the WER, $P(E)$, is evaluated using techniques developed in this chapter. The message is often expressed as a K-bit word where $M = 2^K$. For example, if the message is the letter A, it could be represented by the ASCII 7-bit code word

[†] See (7-46) for the frequency separation required for orthogonal signaling.

1000001 (see Appendix C). Consequently, one is often interested in obtaining the equivalent BER, P_e, from the WER, $P(E)$. Unfortunately, there is no general formula that can be used in all cases to convert WER to BER, except for the trivial case of one-bit messages (i.e., $M = 2$) where $P_e = P(E)$. However, upper and lower bounds on the conversion can be obtained by working on the results for limiting cases.

For the upper conversion bound, assume that all possible error messages are equally likely. That is, when a message is sent, there is one correct choice at the receiver and a message error will be made by selecting any of the other $M - 1$ possible messages. If we assume that all of the $M - 1$ error messages are equally likely, the probability of a particular message being in error is $P(E)/(M - 1)$. If error messages consist of K bits, the probability of k bits being in error is

$$P(k) = \binom{K}{k} \frac{P(E)}{M-1}$$

The average number of error bits is

$$\bar{k} = \sum_{k=1}^{K} kP(k) = \sum_{k=1}^{K} k \binom{K}{k} \frac{P(E)}{M-1} = 2^{K-1}K \frac{P(E)}{M-1}$$

Thus the average probability of bit error is $\bar{k}$ error bits out of K bits, or

$$P_e = \frac{\bar{k}}{K} = \frac{2^{K-1}}{M-1} P(E) = \frac{(M/2)}{M-1} P(E)$$

where $M = 2^K$. Here the worst-case assumption was used to obtain the upper bound value for P_e.

For the lower bound on the BER, there are still $M - 1$ possible ways to make a message error, but assume that the receiver makes an error by choosing a message corresponding to the signal vector (see Chapter 2 or, for example, Fig. 8-13) that is nearest to (but not identical to) the transmitted (correct) signal vector and the receiver does not choose any of the other $M - 1$ possible errors. This is a reasonable assumption in useful systems where the noise (and, consequently, the error) is reasonably small. Also assume that all messages are equally likely to be sent. Thus, when this "nearest neighbor" signal vector is selected (an error), and if a K-bit Gray code (see Chapter 3, Table 3-1) is used to represent the message, there will be only a one-bit error in the K-bit error message. These assumptions give a BER of

$$P_e = \frac{P(E)}{K}$$

which is a lower bound on the BER when compared to results that would be obtained for other assumptions.

Using these results for the conversion of WER to BER, we find that the BER is bounded by

$$\frac{P(E)}{K} \leq P_e \leq \frac{(M/2)}{M-1} P(E) \tag{8-2}$$

where $P(E)$ is the WER and $M = 2^K$. For example, if a $(K = 7)$-bit code is used, this would give bounds of

$$0.143P(E) \leq P_e \leq 0.504P(E)$$

Under the usual operating conditions of low error rates (say $P_e < 10^{-3}$), the "nearest neighbor" conditions would apply and the result would be near the lower bound (almost equal to the lower bound if the 7-bit code was a Gray code).

In the next section criteria for obtaining optimum receivers will be discussed. This discussion will be followed by development of the vector signal concept, first introduced in Chapter 2, that will lead to optimum receiver structures and to techniques for evaluation of $P(E)$ for these receivers.

8-2 Optimization Criteria

The statistical technique called *hypothesis testing* is used to obtain the optimum receiver. That is, using the noisy signal that is present at the input, the receiver has to *decide* which one of M hypotheses is correct:

Hypothesis 1: m_1 sent
Hypothesis 2: m_2 sent
⋮
Hypothesis M: m_M sent

Here $\{m_i, i = 1, 2, \ldots, M\}$ denotes the M possible messages that can be sent by the source.

Any one of a number of optimization criteria can be used to decide which message was sent. There are numerous books written on the subject of detection theory [Kazakos and Papantoni-Kazakos, 1990; Melsa and Cohn, 1978; Scharf, 1991; Whalen, 1971; Wozencraft and Jacobs, 1965]. The most important detection criteria are summarized on the following pages. Some criteria require more a priori knowledge about the communication system than others. The criteria that use more a priori knowledge yield optimum receivers with better performance, *provided* that the source and channel actually have the specifications that are the design a priori conditions.

Bayes Criterion

The *Bayes criterion* minimizes the *average cost*. The average cost is given by

$$\overline{C} = \sum_{i=1}^{M} \overline{C}(D|m_i)P(m_i) \tag{8-3a}$$

where $P(m_i)$ is the probability of sending the message m_i and $\overline{C}(D|m_i)$ is the conditional average cost.

$$\overline{C}(D|m_i) = \sum_{j=1}^{M} P(\tilde{m}_j|m_i)c_{ji} \tag{8-3b}$$

where $P(\tilde{m}_j|m_i)$ is the probability of deciding that m_j was transmitted when m_i was actually sent. As we will see, this conditional probability is a function of the channel noise and receiver parameters. $\{c_{ji}\}$ are the elements in the cost matrix

$$C = \begin{bmatrix} c_{11} & c_{12} & \cdots & c_{1M} \\ c_{21} & c_{22} & \cdots & c_{2M} \\ \vdots & \vdots & & \vdots \\ c_{M1} & c_{M2} & \cdots & c_{MM} \end{bmatrix} \tag{8-4}$$

where c_{ji} is the cost of deciding that m_j was transmitted when m_i was actually sent. The values for c_{ji} are assigned by the receiver designer based on the weight that he or she gives to making this particular correct or incorrect decision. For example, c_{22} would be the cost of deciding that m_2 was transmitted when m_2 was actually transmitted. Since this is a correct decision, a zero value or a negative value (i.e., a reward) would be assigned. c_{12} is the cost of deciding that m_1 was transmitted when m_2 was actually transmitted. It would be assigned a positive value since this is an error. If the value assigned for c_{12} were larger than that assigned for c_{21}, the resulting Bayes receiver would strive to make fewer errors in deciding that m_1 was transmitted when m_2 was transmitted than in deciding that m_2 was transmitted when m_1 was transmitted. In this way the Bayes criterion allows the receiver designer to rank the different types of errors and correct decisions that are made.

In summary, the Bayes receiver minimizes the average cost. To implement the Bayes criterion, the source statistics $\{P(m_i)\}$ must be known and the designer must be willing to specify a cost matrix C. In addition, the statistical characteristics of the noisy received signal must be known so that $\{P(\tilde{m}_j|m_i)\}$ can be evaluated. This is illustrated in the following example.

EXAMPLE 8-1 BAYES CRITERION FOR A BINARY SOURCE

Assume that the source is capable of sending only two possible messages, m_1 and m_2. Thus $M = 2$. Using (8-3), we have

$$\begin{aligned} \overline{C} = {} & P(m_1)[c_{11}P(\tilde{m}_1|m_1) + c_{21}P(\tilde{m}_2|m_1)] \\ & + P(m_2)[c_{12}P(\tilde{m}_1|m_2) + c_{22}P(\tilde{m}_2|m_2)] \end{aligned} \tag{8-5}$$

Let the noisy signal at the receiver input produce a one-dimensional random variable r. [This could be $r_0(t_0)$ of Fig. 7-1.] Furthermore, let the conditional PDFs of r be similar to those illustrated in Fig. 8-1. Then we can relate $P(\tilde{m}_j|m_i)$ to the appropriate areas under the conditional probabilities. Assume that the

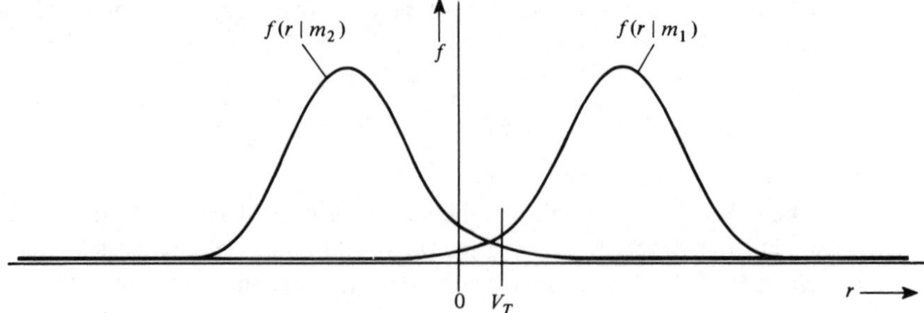

FIGURE 8-1 Conditional probabilities for a binary message problem.

receiver outputs $\tilde{m} = m_1$ if $r > V_T$ and $\tilde{m} = m_2$ if $r < V_T$. In this case (8-5) becomes

$$\overline{C} = P(m_1)\left[c_{11} \int_{V_T}^{\infty} f(r|m_1)\,dr + c_{21} \int_{-\infty}^{V_T} f(r|m_1)\,dr\right] + P(m_2)\left[c_{12} \int_{V_T}^{\infty} f(r|m_2)\,dr + c_{22} \int_{-\infty}^{V_T} f(r|m_2)\,dr\right] \tag{8-6}$$

We need to design the receiver such that the minimum cost is obtained. This is achieved by finding the value of V_T that minimizes $\overline{C}$. Thus, evaluating $\partial\overline{C}/\partial V_T = 0$, using Leibniz's rule for differentiating integrals (A-36),

$$\begin{aligned}\frac{\partial\overline{C}}{\partial V_T} &= P(m_1)[-c_{11}f(V_T|m_1) + c_{21}f(V_T|m_1)] \\ &\quad + P(m_2)[-c_{12}f(V_T|m_2) + c_{22}f(V_T|m_2)] \\ &= 0\end{aligned}$$

or

$$(c_{21} - c_{11})P(m_1)f(V_T|m_1) - (c_{12} - c_{22})P(m_2)f(V_T|m_2) = 0$$

Thus V_T has to satisfy the following condition:

$$\frac{f(V_T|m_1)}{f(V_T|m_2)} = \frac{(c_{12} - c_{22})P(m_2)}{(c_{21} - c_{11})P(m_1)} \tag{8-7}$$

In other words, the Bayes receiver biases the threshold value according to the values of the elements in the cost matrix as well as according to the source probabilities.

Another way of describing the decision algorithm for the Bayes receiver is to use the *likelihood ratio* $\Lambda(r)$. The likelihood ratio is given by

$$\Lambda(r) = \frac{f(r|m_1)}{f(r|m_2)} \tag{8-8}$$

Then the Bayes receiver chooses $\tilde{m} = m_1$ if

$$\Lambda(r) > \frac{(c_{12} - c_{22})P(m_2)}{(c_{21} - c_{11})P(m_1)} \tag{8-9}$$

and chooses $\tilde{m} = m_2$ otherwise. It is realized that $\Lambda(r)$ may be used as the test statistic (instead of r). In this case, for a given value of r, $\Lambda(r)$ is evaluated by the receiver processor circuits and $\tilde{m} = m_1$ is the output decision if (8-9) is satisfied.

Minimax Criterion

The minimax criterion strives to minimize the average cost when the source probabilities $\{P(m_i)\}$ are unknown. The receiver is designed such that the average cost $\overline{C}$ is minimized while using the worst-case set for $\{P(m_i)\}$. Thus the $\{P(m_i)\}$ probabilities that maximize $\overline{C}$ are assumed when the Bayes problem (obtaining the receiver with minimum cost) is solved.

Neyman–Pearson Criterion

The *Neyman–Pearson* criterion assumes that both the source probabilities and the cost matrix are unknown. The receiver is designed for acceptable values of $P(\tilde{m}_j|m_i)$.

For example, in a radar problem there are two hypotheses: m_1 could denote noise alone present at the receiver, and m_2 could denote signal plus noise present. In this radar example $P(\tilde{m} = m_2|m_1)$ is called the probability of *false alarm*, $P(\tilde{m} = m_2|m_2)$ is called the probability of *detection*, and $P(\tilde{m} = m_1|m_2)$ is called the probability of *miss*. Using a Neyman–Pearson criterion, we can design the radar to give acceptable values for the probability of detection and the probability of false alarm.

Maximum a Posteriori Criterion

The maximum a posteriori (MAP) criterion maximizes the a posteriori probability $P(m_i|r)$. Given a received value r, the receiver chooses $\tilde{m} = m_k$ when $P(m_i|r)$, $i = 1, 2, \ldots, M$, is maximized for the value $i = k$.

This criterion can also be formulated in terms of a likelihood ratio. For example, to test whether m_1 or m_2 was sent, choose $\tilde{m} = m_1$ if

$$P(m_1|r) > P(m_2|r) \tag{8-10}$$

But

$$P(m_1|r) = \frac{f(m_1,r)}{f(r)} = \frac{P(m_1)f(r|m_1)}{f(r)} \tag{8-11}$$

and a similar result can be obtained for $P(m_2|r)$. Equation (8-10) becomes

$$\frac{P(m_1)f(r|m_1)}{f(r)} > \frac{P(m_2)f(r|m_2)}{f(r)}$$

where $f(r)$ is a nonnegative quantity.

Therefore, the MAP criterion is to choose $\tilde{m} = m_1$ when

$$P(m_1)f(r|m_1) > P(m_2)f(r|m_2) \tag{8-12}$$

or when

$$\Lambda(r) = \frac{f(r|m_1)}{f(r|m_2)} > \frac{P(m_2)}{P(m_1)} \tag{8-13}$$

Comparing (8-13) with (8-9), we see that the MAP receiver is equivalent to a Bayes receiver when the cost matrix is

$$C = \begin{bmatrix} c_{11} & c_{12} \\ c_{21} & c_{22} \end{bmatrix} = \begin{bmatrix} 0 & 1 \\ 1 & 0 \end{bmatrix} \tag{8-14}$$

Using this result in (8-6), we realize that the *MAP receiver minimizes the probability of error* since $\overline{C}$ becomes $P(E)$ when the cost matrix of (8-14) is used.

A *maximum likelihood* receiver is an MAP receiver where $P(m_i)$, $i = 1, 2, \ldots, M$, are equally likely [i.e., $P(m_i) = 1/M$].

The MAP criterion will be used in later sections to obtain optimum digital receivers. Moreover, in Chapter 7, we obtained some MAP receivers. That is, we used matched filters to minimize $P(E)$ for the case of binary messages ($M = 2$). In this chapter we find optimum receivers for $M \geq 2$.

In this section we have used one-dimensional PDFs to simplify the mathematics (and thereby provide a lucid presentation). However, to develop the MAP concept further, it will be necessary to describe signals and noise in terms of an N-dimensional vector-space representation.

8-3 *N*-Dimensional Vector Space

Signal Representation

For each possible source message there is a unique waveform that is transmitted over the channel. That is, the M possible messages $\{m_1, m_2, \ldots, m_i, \ldots, m_M\}$ have the corresponding transmitted waveforms $\{s_1(t), s_2(t), \ldots, s_i(t), \ldots, s_M(t)\}$. The waveform $s_i(t)$ has a duration (at most) of T_0 seconds, which is the time required to send the message. By using the theory that was developed in Chapter 2, the transmitted waveform $s_i(t)$ can be expanded over an orthogonal set of basis functions. In this chapter we *assume* that these basis functions are *normalized* for unity energy (over T_0 seconds). That is,

$$s_i(t) = \sum_{j=1}^{N} s_{ij}\, \varphi_j(t) \tag{8-15}$$

where $\{\varphi_j(t)\}$ is a set of N *orthonormal* functions that are time limited to a T_0-second duration. Thus $s_i(t)$ is represented by an N-dimensional vector

$$\mathbf{s}_i = (s_{i1}, s_{i2}, \ldots, s_{iN}) \tag{8-16}$$

s_{ij} are the elements of the signal vector as described by (8-15). When a message m_i emanates from the source, the waveform $s_i(t)$ is transmitted over the channel. This is equivalent to sending the N-dimensional vector $\mathbf{s}_i$. It is also realized that *if* the elements s_{ij} happen to have only binary values (e.g., $s_{ij} = \pm 1$), then (8-16) would be an equivalent binary word that represents the message $s_i(t)$.

From the theory developed in Chapter 2, the energy in the signal $s_i(t)$ is

$$E_{s_i} = \int_0^{T_0} s_i^2(t)\, dt = \sum_{j=1}^{N} s_{ij}^2 = |\mathbf{s}_i|^2 \tag{8-17}$$

This is obtained from (2-83), where $K_j = 1$ ($j = 1, 2, \ldots, N$) since $\{\varphi_j(t)\}$ is an orthonormal set. Furthermore, the dimensionality theorem (2-176) may be used to obtain a lower bound on the bandwidth of $s_i(t)$.

The system designer selects the waveform set $\{s_i(t)\}$ to represent the messages $\{m_i\}$. Furthermore, for a given set of signal vectors $\{\mathbf{s}_i\}$, different waveform sets can be obtained by using different sets of basis functions $\{\varphi_j(t)\}$. Two distinctly different types of orthonormal sets are shown in Figs. 8-2 and 8-3. They are of the time-division and frequency-division types.

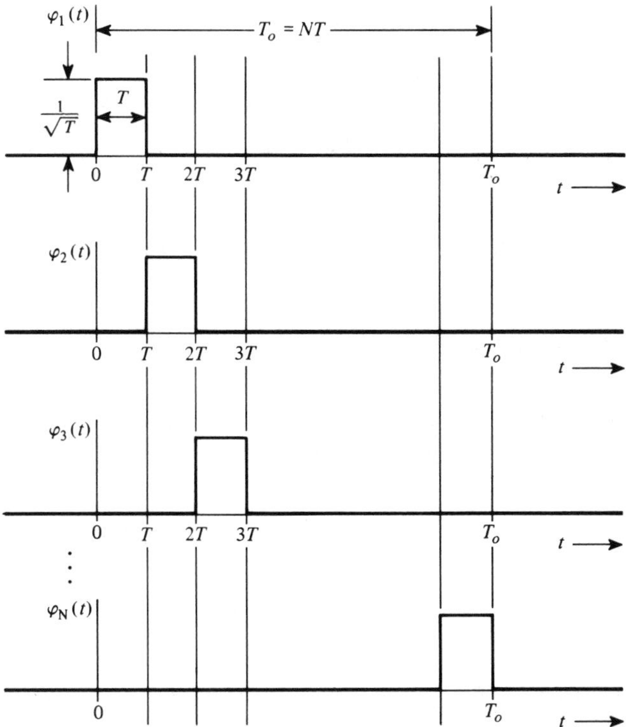

FIGURE 8-2 Type I—rectangular pulse orthonormal function set.

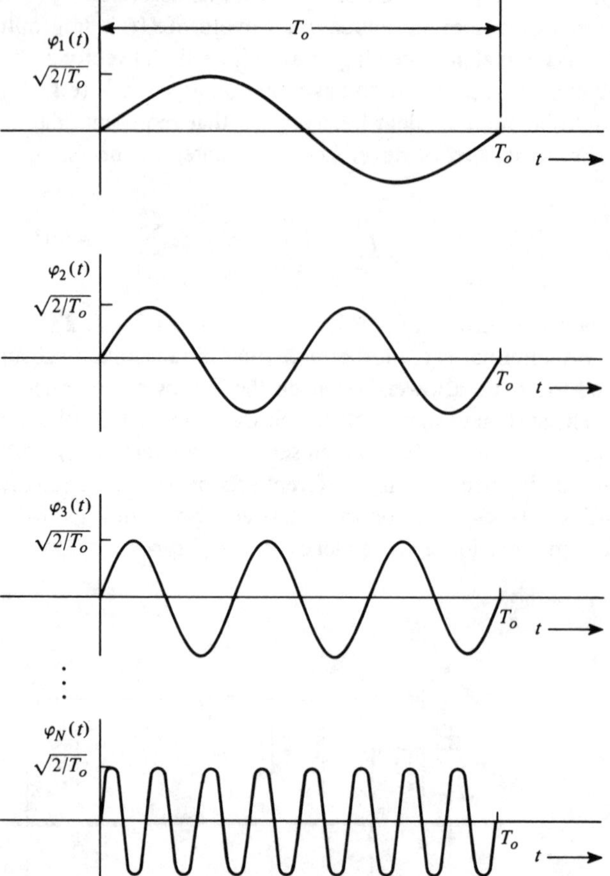

FIGURE 8-3 Type II—sinusoidal pulse orthonormal function set.

Type I: Time-Division Basis Functions. Let $\varphi_j(t)$, $j = 1, 2, \ldots, N$, be nonoverlapping pulses such that the whole set is nonzero over the interval $(0, T_0)$, and each $\varphi_j(t)$ is nonzero over an interval T_0/N.

The nonoverlapping rectangular pulse set of Fig. 8-2 is one example. It is described by

$$\varphi_j(t) = \begin{cases} \dfrac{1}{\sqrt{T}}, & (j-1)T < t \leq jT \\ 0, & t \text{ otherwise} \end{cases} \tag{8-18}$$

where $T_0 = NT$ and $j = 1, 2, \ldots, N$. In this case $\varphi_i(t)$ does not overlap $\varphi_j(t)$ (in the time domain) when $i \neq j$.

Type II: Frequency-Division Basis Functions. $\varphi_j(t), j = 1, 2, \ldots, N,$ are absolutely time limited to $(0, T_0)$ and there is little overlap of $\mathcal{F}[\varphi_i(t)]$ and $\mathcal{F}[\varphi_j(t)]$, $i \neq j$, in the frequency domain.

We know that if the $\varphi_j(t), j = 1, 2, \ldots, N$, are absolutely time limited to $(0, T_0)$, then the $\varphi_j(t)$ cannot be absolutely frequency limited (i.e., absolutely no overlap in the frequency domain). (This theorem was studied in Sec. 2-7.) However, orthonormal sets $\varphi_j(t)$ can be obtained where there is very little overlap in the frequency domain. One example is the sinusoidal pulse orthonormal set illustrated in Fig. 8-3. These orthonormal functions are described by

$$\varphi_j(t) = \begin{cases} \sqrt{\dfrac{2}{T_0}} \sin\left[\left(\dfrac{2\pi j}{T_0}\right)t\right], & 0 < t \leq T_0 \\ 0, & t \text{ elsewhere} \end{cases} \tag{8-19}$$

where $j = 1, 2, \ldots, N$. When $|i - j|$ becomes large, there is little overlap in the spectra of these functions.

Another interesting set of orthonormal functions are those of the $(\sin x)/x$ type illustrated in Fig. 8-4. They are defined as

$$\varphi_j(t) = \sqrt{2B}\, \frac{\sin\{2\pi B[t - j/(2B)]\}}{2\pi B[t - j/(2B)]} \tag{8-20}$$

where $B = 1/(2T)$. Here the $\varphi_j(t), j = 1, 2, \ldots, N$, overlap in the frequency domain. In fact, all the spectra have a constant spectral magnitude over the band $-B$ to $+B$ hertz and are zero outside this range. However, in the time domain the $\varphi_j(t)$ pulse is concentrated in an interval around jT seconds. In this sense this orthonormal set tends to be of the Type I class. Although these functions are not absolutely time limited (since they are absolutely bandlimited), they are approximately of the time-limited class.

Of course, orthonormal sets in general do not satisfy either the Type I or Type II classifications.

Gram–Schmidt Orthogonalization Procedure

If we are designing a receiver for a given waveform set, we need to answer the questions: What is the minimum number of dimensions required in the vector-space representation of the signals, and what is an associated set of basis functions, $\varphi_j(t), j = 1, 2, \ldots, N$, that can be used?

Theorem: *If M signals are given, $s_i(t)$, $i = 1, 2, \ldots, M$, then the number of dimensions, N, required to generate these waveforms is $N \leq M$. Furthermore, the minimum number of basis functions is obtained by the Gram–Schmidt orthogonalization procedure.*

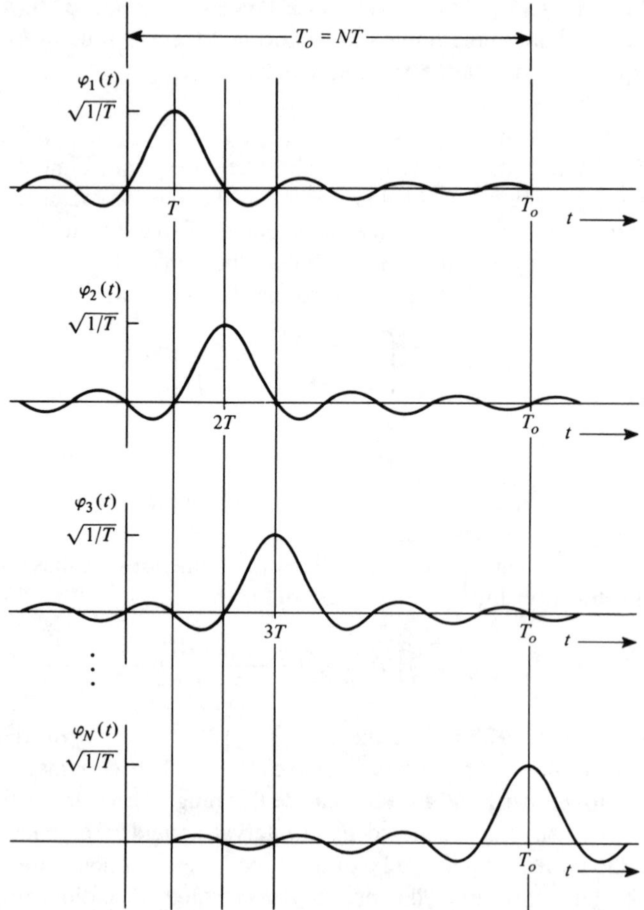

FIGURE 8-4 (sin *x*)/*x* orthonormal function set.

The set of M signals spans $N = M$ dimensional space only if the $s_i(t)$ are linearly independent [i.e., none of the $s_i(t)$ can be expressed as a linear combination of the remaining $s_i(t)$]. If the $s_i(t)$ are not linearly independent, then $N < M$.

The number of dimensions, N, required to represent the signal set [consisting of the M waveforms $s_1(t)$, $s_2(t)$, . . . , $s_M(t)$] and a corresponding set of basis functions $\{\varphi_j(t)\}$ can be obtained by use of the *Gram–Schmidt* procedure. Roughly speaking, this procedure consists of taking any one of the signal waveforms and normalizing it for unity energy. This normalized waveform is taken to be the basis function, $\varphi_1(t)$, which defines the first dimension. The representation for another signal is obtained by finding the component that is projected along the first dimension, and any remaining term is used to define the basis function for the second dimension. This process is repeated until all of the M signals have been decomposed into N linearly independent components, as described by (8-15). A detailed description of the Gram–Schmidt procedure is given by the following steps.[†]

1. Index the given set of functions by designating one waveform as $s_1(t)$, another as $s_2(t)$, and so on, to obtain $s_1(t), s_2(t), \ldots, s_M(t)$. (By the proper choice of index in some problems, the mathematical procedure is simplified.)
2. Let $\psi_1(t) \triangleq s_1(t)$. Find the first orthonormal function. It is

$$\varphi_1(t) = \frac{\psi_1(t)}{\sqrt{E_{\psi_1}}} \tag{8-21a}$$

where

$$E_{\psi_1} = \int_0^{T_0} \psi_1^2(t)\, dt \tag{8-21b}$$

Then the vector corresponding to the waveform $s_1(t)$ is

$$\mathbf{s}_1 = (s_{11}, 0, 0, \ldots, 0) \tag{8-22a}$$

where

$$s_{11} = \sqrt{E_{\psi_1}} \tag{8-22b}$$

3. Find $\varphi_2(t)$. Let

$$\psi_2(t) = s_2(t) - s_{21}\varphi_1(t) \tag{8-23a}$$

where, by (2-77),

$$s_{ij} = \int_0^{T_0} s_i(t)\varphi_j(t)\, dt \tag{8-23b}$$

If this $\psi_2(t) \neq 0$, then

$$\varphi_2(t) = \frac{\psi_2(t)}{\sqrt{E_{\psi_2}}} \tag{8-24a}$$

where

$$E_{\psi_j} = \int_0^{T_0} \psi_j^2(t)\, dt \tag{8-24b}$$

Thus,

$$\mathbf{s}_2 = (s_{21}, s_{22}, 0, 0, \ldots, 0) \tag{8-24c}$$

where

$$s_{22} = \sqrt{E_{\psi_2}} \tag{8-24d}$$

[†] Here it is assumed that both sets $\{s_i(t)\}$ and $\{\varphi_j(t)\}$ are described by *real* waveforms.

If the $\psi_2(t) = 0$ when $s_2(t)$ is used, then

$$\mathbf{s}_2 = (s_{21}, 0, 0, \ldots, 0) \tag{8-24e}$$

and another $\psi_2(t)$ is found by repeating this procedure where

$$\psi_2(t) = s_3(t) - s_{31}\varphi_1(t) \tag{8-25}$$

Then $\mathbf{s}_3$ is evaluated similar to the procedure leading to (8-24c). $\theta_2(t)$ will be found when $\psi_2(t) \neq 0$.

4. Find $\varphi_3(t)$. Choose the next waveform, say $s_k(t)$, and evaluate

$$\psi_3(t) = s_k(t) - s_{k1}\ \varphi_1(t) - s_{k2}\ \varphi_2(t) \tag{8-26a}$$

where

$$s_{kj} = \int_0^{T_0} s_k(t)\ \varphi_j(t)\ dt \tag{8-26b}$$

Then, if $\psi_3(t) \neq 0$.

$$\varphi_3(t) = \frac{\psi_3(t)}{\sqrt{E_{\psi_3}}} \tag{8-26c}$$

5. The procedure is continued until all the $s_i(t)$, $i = 1, 2, \ldots, M$, have been used. When the procedure is completed, all the $\varphi_j(t)$, $j = 1, 2, \ldots, N$, will have been obtained. Similarly, the vectors $\mathbf{s}_i$, $i = 1, 2, \ldots, M$, will have been evaluated. Since some of the $\psi(t)$'s may be zero, as in (8-24e), we have $N \leq M$.

The Gram–Schmidt procedure is not difficult to remember once you master the logic involved.

EXAMPLE 8-2 VECTORS FOR A RECTANGULAR WAVEFORM SET

Let the $M = 4$ waveforms signal set be given as shown on the left side of Fig. 8-5.

The orthonormal basis set and the signal vectors will now be evaluated using the Gram–Schmidt procedure.

The waveforms are already indexed, so we will proceed to step 2. The first nonnormalized orthogonal function is $\psi_1(t) = s_1(t)$, and it has the energy

$$E_{\psi_1} = \int_0^3 s_1^2(t)\ dt = 3 \tag{8-27a}$$

Then the first orthonormal function is

$$\varphi_1(t) = \frac{s_1(t)}{\sqrt{E_{\psi_1}}} = \frac{s_1(t)}{\sqrt{3}} \tag{8-27b}$$

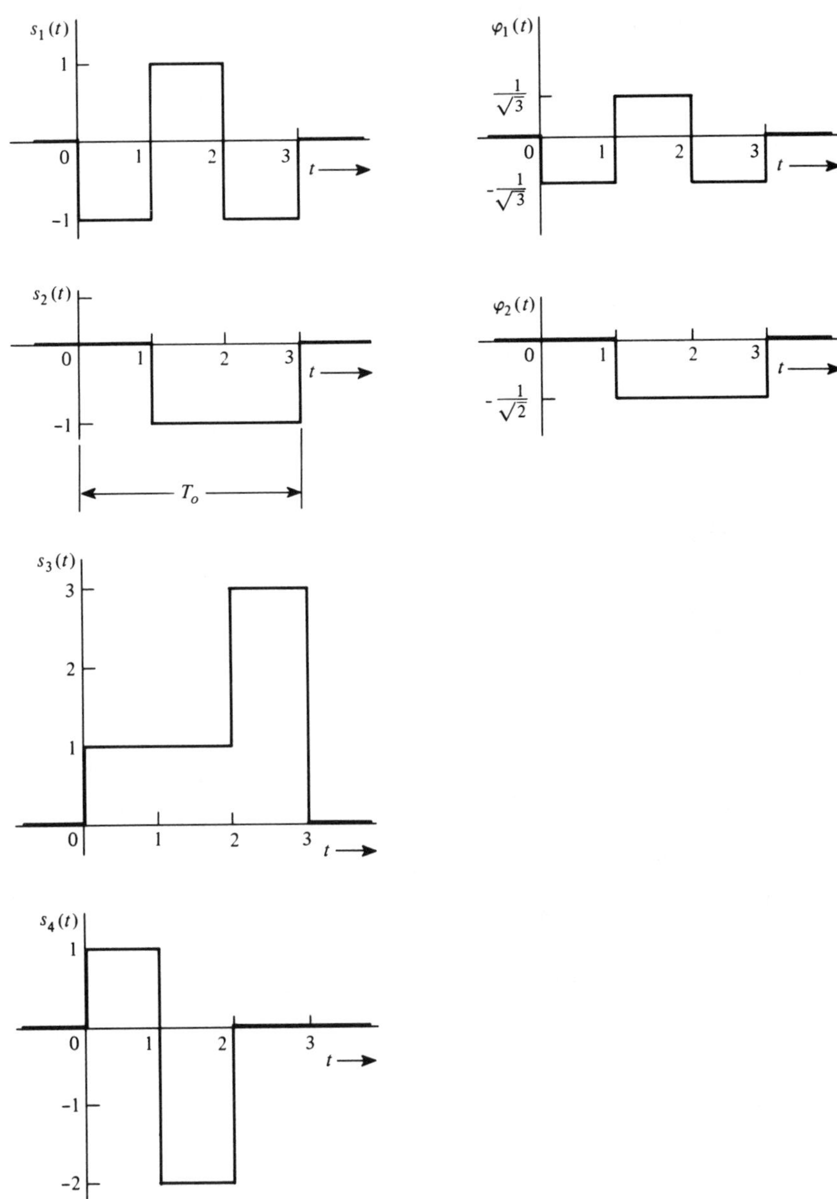

FIGURE 8-5 Example of the Gram–Schmidt procedure.

This basic function is shown in the top right corner of Fig. 8-5. The vector for $s_1(t)$ is

$$\mathbf{s}_1 = (1.73, 0, 0, 0) \tag{8-27c}$$

where $s_{11} = \sqrt{3} \approx 1.73$.

Step 3 is followed to evaluate $\varphi_2(t)$.

$$s_{21} = \int_0^3 s_2(t)\,\varphi_1(t)\,dt$$

$$= (0)\left(\frac{-1}{\sqrt{3}}\right) + (-1)\left(\frac{1}{\sqrt{3}}\right) + (-1)\left(\frac{-1}{\sqrt{3}}\right) = 0 \qquad \text{(8-28a)}$$

Then

$$\psi_2(t) = s_2(t) - s_{21}\,\varphi_1(t) = s_2(t) \qquad \text{(8-28b)}$$

and

$$\varphi_2(t) = \frac{\psi_2(t)}{\sqrt{E_{\psi_2}}} = \frac{s_2(t)}{\sqrt{2}} \qquad \text{(8-28c)}$$

The second orthonormal basis function is also shown in Fig. 8-5. The vector for $s_2(t)$ is

$$\mathbf{s}_2 = (0,\ 1.41,\ 0,\ 0) \qquad \text{(8-28d)}$$

Next, $\varphi_3(t)$ is evaluated by step 4.

$$s_{31} = \int_0^3 s_3(t)\,\varphi_1(t)\,dt$$

$$= (1)\left(\frac{-1}{\sqrt{3}}\right) + (1)\left(\frac{1}{\sqrt{3}}\right) + (3)\left(\frac{-1}{\sqrt{3}}\right) = -\sqrt{3} \qquad \text{(8-29a)}$$

$$s_{32} = \int_0^3 s_3(t)\varphi_2(t)\,dt$$

$$= (1)(0) + (1)\left(\frac{-1}{\sqrt{2}}\right) + (3)\left(\frac{-1}{\sqrt{2}}\right) = -\frac{4}{\sqrt{2}} \qquad \text{(8-29b)}$$

Then

$$\psi_3(t) = s_3(t) - s_{31}\varphi_1(t) - s_{32}\varphi_2(t)$$

$$= s_3(t) - (-\sqrt{3})\varphi_1(t) - \left(\frac{-4}{\sqrt{2}}\right)\varphi_2(t) = 0 \qquad \text{(8-29c)}$$

Since $\psi_3(t) = 0$, we do not need $\varphi_3(t)$ to describe $s_3(t)$, and the signal vector is

$$\mathbf{s}_3 = (-1.73,\ -2.83,\ 0) \qquad \text{(8-29d)}$$

Since $\psi_3(t) = 0$ when $s_3(t)$ is used, we will continue to look for a third orthonormal function by using $s_4(t)$.

$$\psi_3(t) = s_4(t) - s_{41}\varphi_1(t) - s_{42}\varphi_2(t) \qquad \text{(8-30a)}$$

where

$$s_{41} = \int_0^3 s_4(t)\varphi_1(t)\,dt$$

$$= (+1)\left(\frac{-1}{\sqrt{3}}\right) + (-2)\left(\frac{1}{\sqrt{3}}\right) + (0)\left(\frac{-1}{\sqrt{3}}\right) = -\sqrt{3} \qquad \text{(8-30b)}$$

and

$$s_{42} = \int_0^3 s_4(t)\varphi_2(t)\,dt$$

$$= (+1)(0) + (-2)\left(\frac{-1}{\sqrt{2}}\right) + (0)\left(\frac{-1}{\sqrt{2}}\right) = \sqrt{2} \qquad \text{(8-30c)}$$

Then

$$\psi_3(t) = s_4(t) - (-\sqrt{3})\varphi_1(t) - (\sqrt{2})\varphi_2(t) = 0 \qquad \text{(8-30d)}$$

Since $\psi_3(t) = 0$ when $s_4(t)$ is used, we do not need $\varphi_3(t)$ to describe $s_4(t)$ either. The signal vector for $s_4(t)$ is then

$$\mathbf{s}_4 = (-1.73, 1.41) \qquad \text{(8-30e)}$$

This completes the Gram–Schmidt procedure since we have used all of our $s_i(t)$ waveforms. It is interesting to note that the length of the vectors that we obtained can be verified as being correct by using (8-17), which relates the vector length to the energy of the corresponding waveform. Since we found that only two basis functions are needed, the minimum number of dimensions required is $N = 2$ in this problem, where the number of waveforms is $M = 4$. The vector-space representation for the signal set is shown in Fig. 8-6.

Vector Channels and Relevant Noise

A communication system with an additive white noise channel is shown in Fig. 8-7a. We will show that this system can be described by an N-dimensional vector model as illustrated in Fig. 8-7b.

The noisy signal at the input of the receiver is

$$r(t) = s_i(t) + n_w(t) \qquad \text{(8-31)}$$

where $s_i(t)$ is the waveform that is transmitted for the message m_i and $n_w(t)$ is the white channel noise. We know that the signal $s_i(t)$ may be represented by an N-dimensional orthogonal series. To represent the white noise waveform in a series, an infinite (complete) set of orthogonal functions is required. Then (8-31)

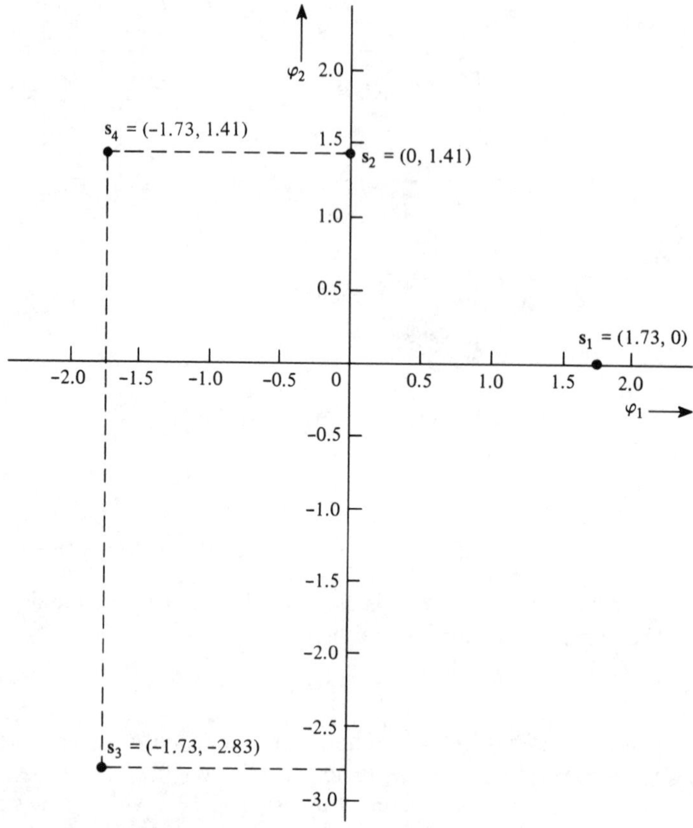

FIGURE 8-6 Vector signals for the waveform signals of Fig. 8-5.

becomes

$$r(t) = \sum_{j=1}^{N} s_{ij}\varphi_j(t) + \sum_{j=1}^{\infty} n_j\varphi_j(t)$$

or

$$r(t) = \sum_{j=1}^{N} (s_{ij} + n_j)\varphi_j(t) + \sum_{j=N+1}^{\infty} n_j\varphi_j(t) \tag{8-32}$$

[Note that $\{n_j\}$ is a set of random variables since $n_w(t)$ is a random process.]

The optimum receiver is sensitive only to the dimensions in which signal energy is present (Wozencraft and Jacobs, 1965). Thus the relevant part of the received waveform is

$$r(t) = \sum_{j=1}^{N} (s_{ij} + n_j)\varphi_j(t) = \sum_{j=1}^{N} r_j\varphi_j(t) \tag{8-33a}$$

where

$$r_j = s_{ij} + n_j, \qquad j = 1, 2, \ldots, N \tag{8-33b}$$

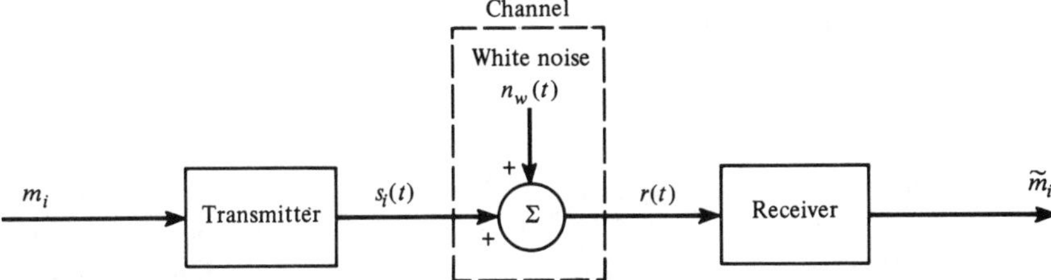

(a) Waveform Communication System

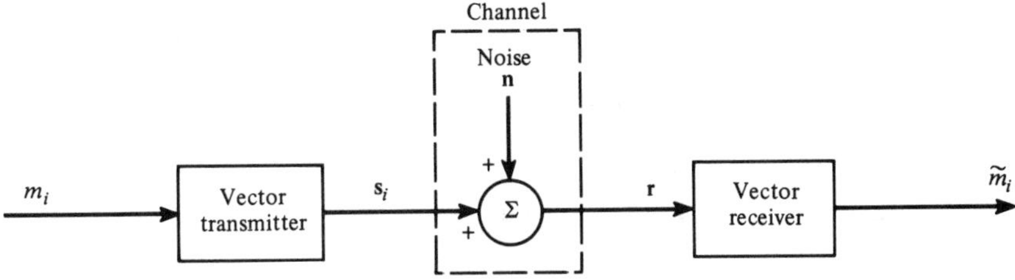

(b) Vector Model of System

FIGURE 8-7 Digital communication system.

Using (8-33b), we see that the received vector corresponding to the waveform $r(t)$ is

$$\mathbf{r} = \mathbf{s}_i + \mathbf{n} \tag{8-34a}$$

where

$$\mathbf{s}_i = (s_{i1}, s_{i2}, \ldots, s_{iN}) \tag{8-34b}$$

and

$$\mathbf{n} = (n_1, n_2, \ldots, n_N) \tag{8-34c}$$

The transmitted signal $s_i(t)$ has the vector representation $\mathbf{s}_i$ and the relevant channel noise has the vector representation $\mathbf{n}$. This is the vector model illustrated in Fig. 8-7b.

PDF for the Noise Vector

The N-dimensional joint PDF for the noise vector $\mathbf{n}$ will now be evaluated for the case of white Gaussian noise in the waveform channel. We know that the representation for the white channel noise is

$$n_w(t) = \sum_{j=1}^{\infty} n_j \varphi_j(t) \tag{8-35a}$$

where the waveform for the relevant noise is

$$n(t) = \sum_{j=1}^{N} n_j \varphi_j(t) \tag{8-35b}$$

and, using (2-77), we get

$$n_j = \int_0^{T_0} n_w(t)\varphi_j(t)\, dt \tag{8-35c}$$

The random variables, n_j, $j = 1, 2, \ldots, N$, are jointly Gaussian because they are linear functions of $n_w(t)$, which is assumed to be a white Gaussian process.

Furthermore, it is easy to demonstrate that the random variables, n_j, $j = 1, 2, \ldots, N$, are uncorrelated. Using (8-35c),

$$\overline{n_j n_k} = \overline{\left[\int_0^{T_0} n_w(t_1)\varphi_j(t_1)\, dt_1\right]\left[\int_0^{T_0} n_w(t_2)\varphi_k(t_2)\, dt_2\right]}$$

$$= \int_0^{T_0}\int_0^{T_0} \overline{n_w(t_1)n_w(t_2)}\, \varphi_j(t_1)\varphi_k(t_2)\, dt_1 dt_2 \tag{8-36}$$

Since the channel noise is white,

$$\overline{n_w(t_1)n_w(t_2)} = R_{n_w}(t_2 - t_1) = \frac{N_0}{2}\,\delta(t_2 - t_1) \tag{8-37a}$$

where

$$\mathcal{P}_{n_w}(f) = \frac{N_0}{2} \tag{8-37b}$$

Then (8-36) becomes

$$\overline{n_j n_k} = \frac{N_0}{2}\int_0^{T_0} \varphi_j(t_1)\varphi_k(t_1)\, dt_1 \tag{8-38}$$

Using the definition for orthonormal functions, as given by (2-73) where $K_j = 1$, and realizing that the orthonormal functions are real, we observe that (8-38) becomes

$$\overline{n_j n_k} = \frac{N_0}{2}\,\delta_{jk} \tag{8-39}$$

Also,

$$\overline{n_j} = 0, \qquad j = 1, 2, \ldots, N \tag{8-40}$$

since

$$\overline{n_j} = \int_0^{T_0} \overline{n_w(t)}\varphi_j(t)\, dt \qquad \text{and} \qquad \overline{n_w(t)} = 0$$

Since the random variables n_j are uncorrelated and jointly Gaussian with zero-mean values, they are independent (see Appendix B). That is,

$$f_{\mathbf{n}}(\mathbf{n}) = f(n_1, n_2, \ldots, n_N) = f(n_1)f(n_2)\cdots f(n_N)$$

$$= \left(\frac{1}{\sqrt{2\pi}\sigma_1}\, e^{-n_1^2/(2\sigma_1^2)}\right)\left(\frac{1}{\sqrt{2\pi}\sigma_2}\, e^{-n_2^2/(2\sigma_2^2)}\right)\cdots\left(\frac{1}{\sqrt{2\pi}\sigma_N}\, e^{-n_N^2/(2\sigma_N^2)}\right) \tag{8-41}$$

Using (8-39) and (8-40), we see that

$$\sigma_j^2 = [n_j^2 - (n_j)^2] = \frac{N_0}{2} \stackrel{\Delta}{=} \sigma^2 \tag{8-42}$$

Then the N-dimensional PDF for the noise vector is

$$f_{\mathbf{n}}(\mathbf{n}) = \frac{1}{(2\pi\sigma^2)^{N/2}} e^{-\sum_{j=1}^{N} n_j^2/(2\sigma^2)}$$

or

$$f_{\mathbf{n}}(\mathbf{n}) = \frac{1}{(2\pi\sigma^2)^{N/2}} e^{-|\mathbf{n}|^2/(2\sigma^2)} \tag{8-43a}$$

where

$$|\mathbf{n}|^2 = \sum_{j=1}^{N} n_j^2 \tag{8-43b}$$

and

$$\sigma^2 = \frac{N_0}{2} \tag{8-43c}$$

$\mathcal{P}_{n_w}(f) = N_0/2$ is the PSD for the white channel (waveform) noise.

8-4 MAP Receiver

Decision Rule for the Additive White Gaussian Noise Channel

In this section we find the MAP decision rule for the case of an *additive white Gaussian noise* (AWGN) channel. This decision rule will be used to obtain optimum receiver structures in the succeeding section.

The MAP receiver chooses $\tilde{m} = m_k$ when

$$P(m_i|\mathbf{r}), \qquad i = 1, 2, \ldots, M \tag{8-44a}$$

is maximized for the value $i = k$. $\mathbf{r}$ is the N-dimensional vector representing the received waveform that is present at the receiver input. Equation (8-44a) can be written as

$$\frac{f(m_i,\mathbf{r})}{f(\mathbf{r})} = \frac{P(m_i)f(\mathbf{r}|m_i)}{f(\mathbf{r})}, \qquad i = 1, 2, \ldots, M \tag{8-44b}$$

But $f(\mathbf{r})$ is a constant, so it may be neglected when finding the value of i for which (8-44b) is maximized. Thus the general *MAP decision rule* is to choose $\tilde{m} = m_k$ when

$$P(m_i)f(\mathbf{r}|m_i), \qquad i = 1, 2, \ldots, M \tag{8-45}$$

is maximized for $i = k$. For an additive noise channel $\mathbf{r} = \mathbf{s}_i + \mathbf{n}$, so (8-45) becomes

$$P(m_i)f_{\mathbf{n}}(\mathbf{r} - \mathbf{s}_i), \qquad i = 1, 2, \ldots, M \tag{8-46}$$

where $f_{\mathbf{n}}(\mathbf{n})$ is the N-dimensional PDF for the channel noise vector. Using (8-43a), the decision rule for an AWGN channel is to choose $\tilde{m} = m_k$ when

$$P(m_i)\frac{1}{(2\pi\sigma^2)^{N/2}}\, e^{-|\mathbf{r}-\mathbf{s}_i|^2/(2\sigma^2)}, \qquad i = 1, 2, \ldots, M \tag{8-47}$$

is maximized for $i = k$. This result can be reduced to a more convenient form as stated by the following theorem.

THEOREM: *If the received waveform consists of a signal with AWGN, $r(t) = s(t) + n_w(t)$, the* ***MAP decision rule*** *chooses $\tilde{m} = m_k$ when the test statistic*

$$D_i = \mathbf{r} \cdot \mathbf{s}_i + c_i, \qquad i = 1, 2, \ldots, M \tag{8-48a}$$

is maximized for $i = k$. The constants $\{c_i\}$ are

$$c_i = \left\{\frac{N_0}{2}\ln[P(m_i)] - \tfrac{1}{2}E_{s_i}\right\}, \qquad i = 1, 2, \ldots, M \tag{8-48b}$$

E_{s_i} is the energy in the waveform $s_i(t)$ over its duration of T_0 sec, $N_0/2$ is the PSD of the white noise $n_w(t)$, $\{P(m_i)\}$ is the set of source (message) probabilities, and $\mathbf{r} \cdot \mathbf{s}_i$ is the dot product of the received vector $\mathbf{r}$ with the known signal vector $\mathbf{s}_i$.

The *dot product* is given by

$$\mathbf{r} \cdot \mathbf{s}_i = \sum_{j=1}^{N} r_j s_{ij} \tag{8-49}$$

In applying (8-48), $\mathbf{r}$ is the input to the receiver algorithm. $\{\mathbf{s}_i\}$ and $\{c_i\}$ are known and stored in the receiver. The $\{c_i\}$ can be precomputed from the known quantities N_0, $\{P(m_i)\}$, and E_{s_i} by using (8-48b).

An interesting fact is realized when the decision rule of (8-48a) is examined. Suppose we want to compare the error performance of a number of different digital communication systems. Furthermore, suppose that the different systems have the *same vector signal set* and the same source statistics $\{P(m_i)\}$. That is, it is possible to have entirely different waveform signal sets $\{s_i(t)\}$ (one set for each system) by using different sets of basis functions $\{\varphi_j(t)\}$. In these cases, the $\{s_i(t)\}$ could be different and yet the $\{\mathbf{s}_i\}$ would be identical for the different systems. For systems of this type, the error performance with MAP receivers would be identical because the D_i, $i = 1, 2, \ldots, M$, that are obtained in (8-48) are identical for the different systems. For example, in Chapter 7 we found that a baseband system with polar signaling has the same performance as a BPSK system (same E_b/N_0). We now realize that this occurs because the *vector signal sets* of the two systems are *identical*.

Proof of Theorem (8-48). Maximizing (8-47) is the same as minimizing the reciprocal:

$$\frac{e^{+|\mathbf{r}-\mathbf{s}_i|^2/(2\sigma^2)}}{P(m_i)} \tag{8-50}$$

The positive constant $(2\pi\sigma^2)^{N/2}$ has been neglected since it does not affect the minimization operation. Since $\ln x$ is a monotonically *increasing* function for $x > 0$, the minimization of (8-50) is the same as minimizing

$$\ln\left[\frac{e^{+|\mathbf{r}-\mathbf{s}_i|^2/(2\sigma^2)}}{P(m_i)}\right] = \frac{|\mathbf{r}-\mathbf{s}_i|^2}{2\sigma^2} - \ln P(m_i) \tag{8-51}$$

From (8-43c), $2\sigma^2 = N_0$ is a positive number. Then minimizing (8-51) is equivalent to minimizing

$$|\mathbf{r}-\mathbf{s}_i|^2 - N_0 \ln P(m_i) \tag{8-52}$$

or

$$|\mathbf{r}|^2 - 2\mathbf{r}\cdot\mathbf{s}_i + |\mathbf{s}_i|^2 - N_0 \ln P(m_i)$$

This is equivalent to minimizing

$$-\mathbf{r}\cdot\mathbf{s}_i + \tfrac{1}{2}|\mathbf{s}_i|^2 - \frac{N_0}{2}\ln P(m_i) \tag{8-53}$$

Here $|\mathbf{r}|^2$ has been neglected since it does not have any effect on finding the value of i that produces the minimum. Minimizing (8-53) is equivalent to *maximizing*

$$\mathbf{r}\cdot\mathbf{s}_i - \tfrac{1}{2}|\mathbf{s}_i|^2 + \frac{N_0}{2}\ln P(m_i)$$

This is identical to (8-48) since, from (8-17), $E_{s_i} = |\mathbf{s}_i|^2$.

Receiver Structures

We will now find optimum digital receiver structures that implement the MAP decision rule as given by (8-48). Four canonical receiver structures are possible, and they provide identical performance. They are receivers that incorporate

1. N correlators where the correlators have reference waveforms: $\varphi_j(t)$, $j = 1, 2, \ldots, N$.
2. N matched filters that are matched to the waveforms: $\varphi_j(t)$, $j = 1, 2, \ldots, N$.
3. M correlators where the correlators have reference waveforms: $s_i(t)$, $i = 1, 2, \ldots, M$.
4. M matched filters that are matched to the waveforms: $s_i(t)$, $i = 1, 2, \ldots, M$.

These receiver structures are shown in Figs. 8-8 to 8-11, respectively.

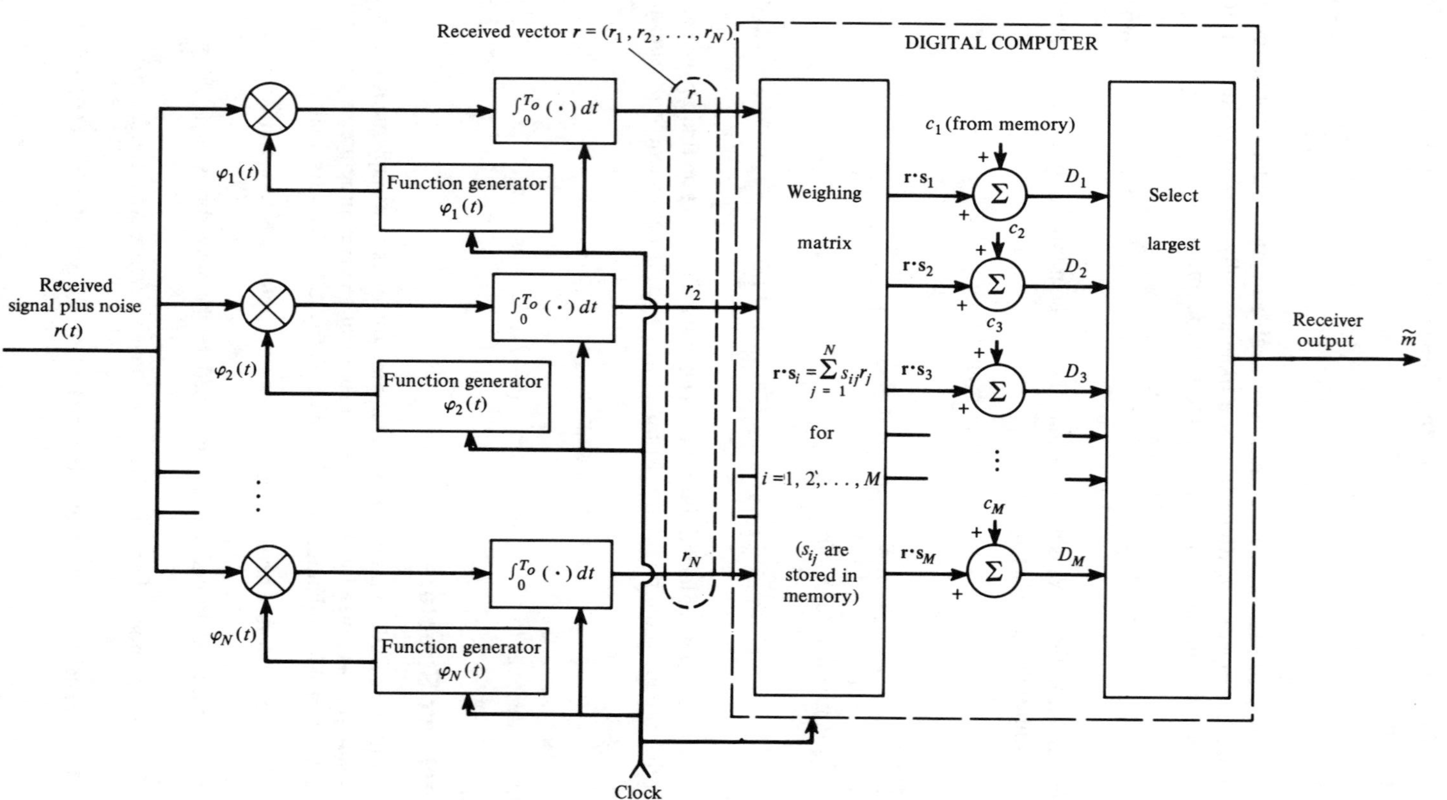

FIGURE 8-8 MAP receiver using correlation with orthonormal functions.

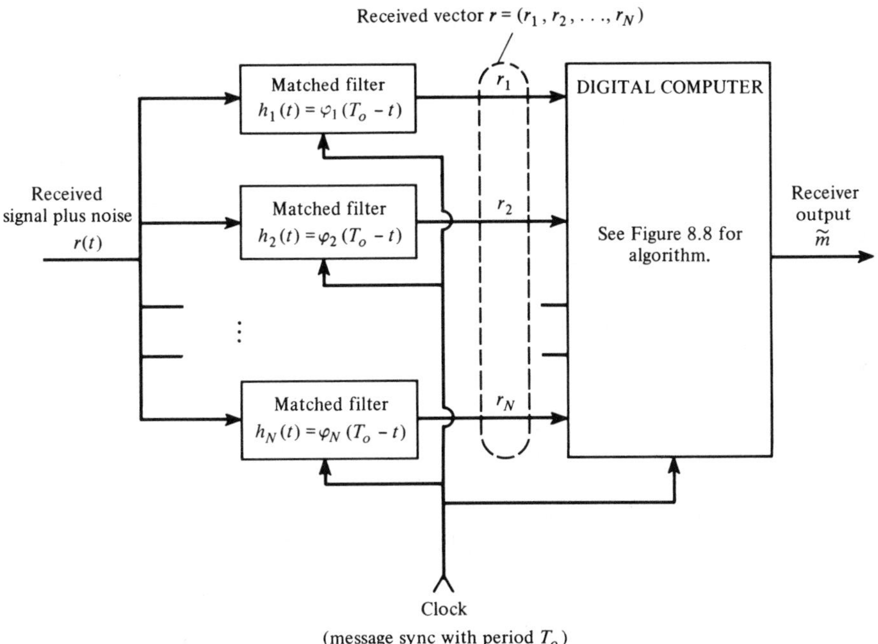

FIGURE 8-9 MAP receiver using filters matched to the orthonormal functions.

We will first demonstrate that the receiver illustrated in Fig. 8-8 is an optimum receiver that satisfies the MAP criterion. The elements of the received vector, $\mathbf{r} = (r_1, r_2, \ldots, r_N)$, are evaluated by using (2-77):

$$r_j = \int_0^{T_0} r(t)\varphi_j(t)\, dt \tag{8-54}$$

where $\varphi_j(t)$ are real orthonormal ($K_j = 1$) functions. Thus the elements r_j are obtained by multiplying the received waveform $r(t)$ with $\varphi_j(t)$ and integrating the product over the interval of T_0 seconds. Now that $\mathbf{r}$ has been obtained from the front end of the receiver, the remainder of the receiver (i.e., the detector) consists of implementing the MAP decision algorithm, as given by (8-48). This can be implemented by using a digital computer or microprocessor, as illustrated in Fig. 8-8.

If we are describing a receiver for a baseband communication system, the product operation is obtained by a four-quadrant multiplier circuit. On the other hand, if we are describing a receiver for an RF (i.e., bandpass) communication system, the multiplier is an RF mixer. Furthermore for the bandpass case, the $\varphi_j(t)$ are bandpass waveforms and, consequently, carrier synchronization must be applied to the function generators in addition to the word synchronization clock. For the case of binary message signaling, $M = 2$, this receiver, which is shown in Fig. 8-8, is exactly equivalent to the optimum (matched-filter) receiv-

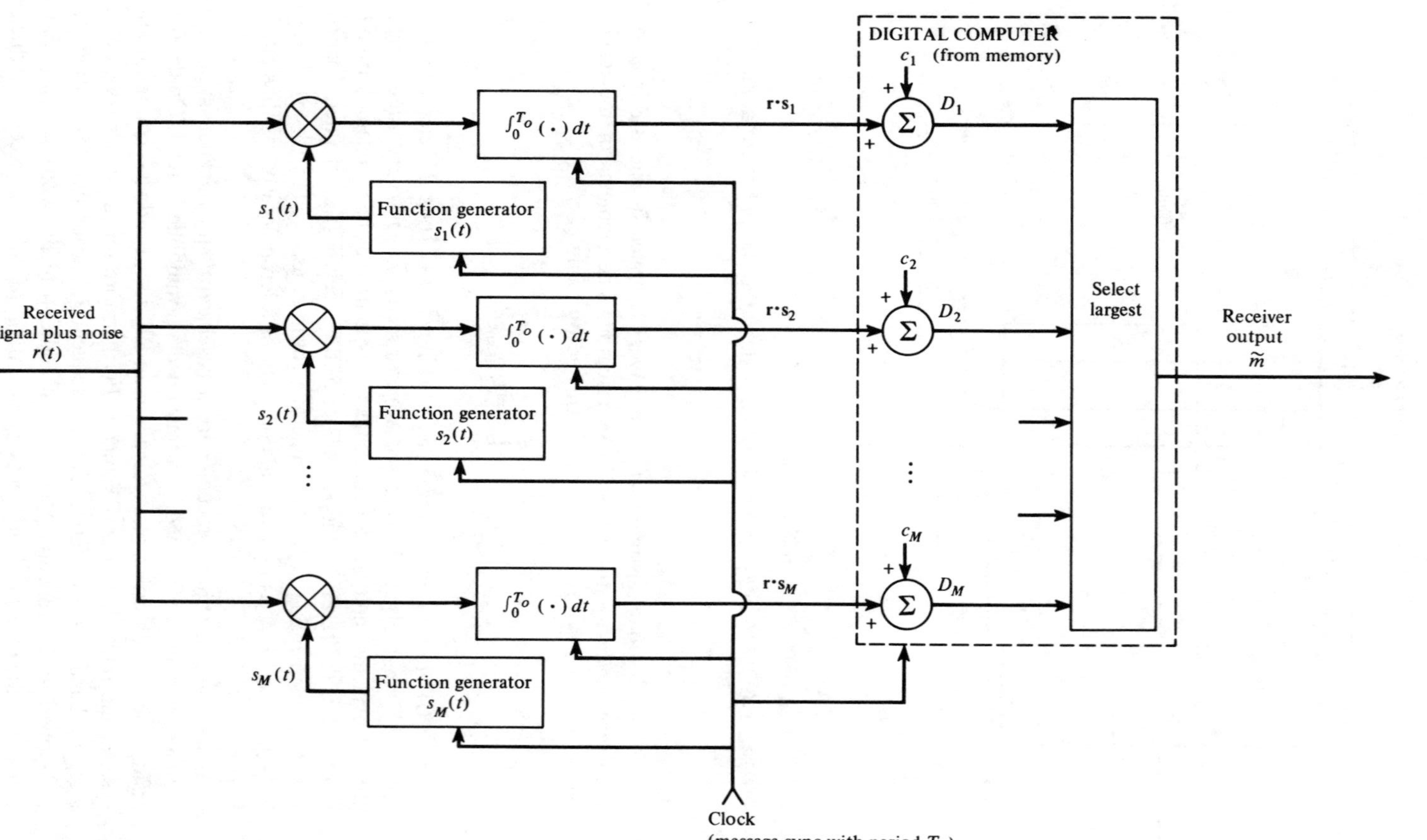

FIGURE 8-10 MAP receiver using correlation with signal waveforms.

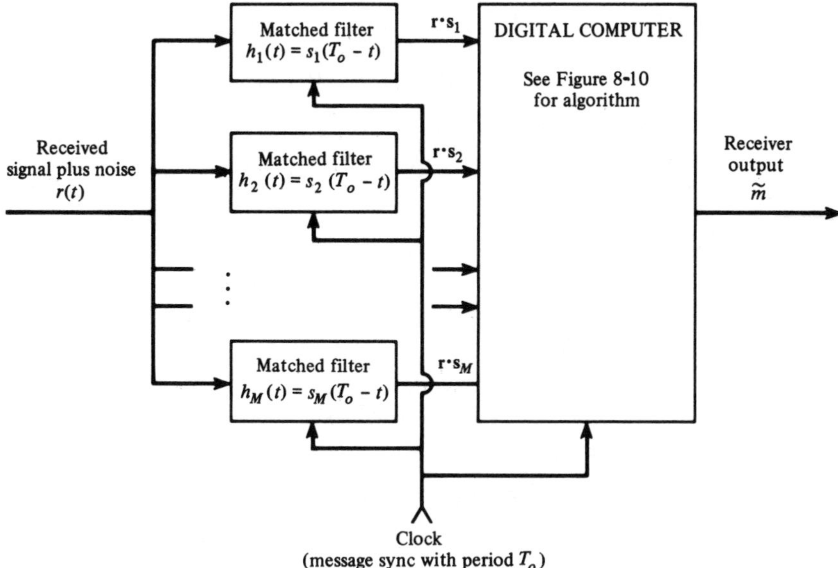

FIGURE 8-11 MAP receiver using filters matched to the signal waveforms.

ers that were studied in Chapter 7. For example, see Fig. 7-1, where the matched filter of Fig. 6-18 is used for the processing and sample-and-hold circuits.

Another equivalent MAP receiver uses N matched filters. This is shown in Fig. 8-9. The impulse response of the jth matched filter is

$$h_j(t) = \varphi_j(T_0 - t)$$

so that the output of the jth matched filter is

$$\int_0^{T_0} r(\lambda)h_j(T_0 - \lambda)\,d\lambda = \int_0^{T_0} r(\lambda)\varphi_j[T_0 - (T_0 - \lambda)]\,d\lambda$$

$$= \int_0^{T_0} r(\lambda)\varphi_j(\lambda)\,d\lambda = r_j \tag{8-55}$$

Thus the output of the jth integrator is r_j, which is identical to (8-54). The received vector has been obtained. Consequently, the remainder of the receiver is the same as that of Fig. 8-8.

To obtain the MAP receivers of Figs. 8-10 and 8-11, we need the following theorem.

PARSEVAL'S THEOREM: *Assume that* **v** *and* **w** *are vectors corresponding to the waveforms* $v(t)$ *and* $w(t)$, *where*

$$v(t) = \sum_{j=1}^{N} v_j\varphi_j(t) \qquad \text{and} \qquad w(t) = \sum_{j=1}^{N} w_j\varphi_j(t)$$

then

$$\int_{-\infty}^{\infty} v(t)w(t)\,dt = \int_{-\infty}^{\infty} V(f)W^*(f)\,df = \mathbf{v}\cdot\mathbf{w} \tag{8-56}$$

where $V(f) = \mathcal{F}[v(t)]$ *and* $W(f) = \mathcal{F}[w(t)]$.

Proof

$$\int_{-\infty}^{\infty} v(t)w(t)\,dt = \int_{-\infty}^{\infty} \sum_{j=1}^{N} v_j\varphi_j(t) \sum_{k=1}^{N} w_k\,\varphi_k(t)\,dt$$

$$= \sum_{j=1}^{N}\sum_{k=1}^{N} v_j w_k \int_{-\infty}^{\infty} \varphi_j(t)\varphi_k(t)\,dt$$

But $\varphi_j(t)$ and $\varphi_k(t)$ are real orthonormal functions, so, using (2-73), we get

$$\int_{-\infty}^{\infty} v(t)w(t)\,dt = \sum_{j=1}^{N}\sum_{k=1}^{N} v_j w_k \delta_{jk} = \sum_{j=1}^{N} v_j w_j = \mathbf{v}\cdot\mathbf{w}$$

Furthermore, it can be shown [see Parseval's theorem, (2-40)] that

$$\int_{-\infty}^{\infty} v(t)w(t)\,dt = \int_{-\infty}^{\infty} V(f)W^*(f)\,df$$

by substituting $\int_{-\infty}^{\infty} V(f_1)e^{j2\pi f_1 t}\,df_1$ for $v(t)$ and $\int_{-\infty}^{\infty} W(f_2)e^{j2\pi f_2 t}\,df_2$ for $w(t)$.

To obtain Figs. 8-10 and 8-11, use (8-56) in (8-48a). Then the MAP decision rule is to choose $\tilde{m} = m_k$ when the test statistic

$$D_i = \left[\left[\int_0^{T_0} r(t)s_i(t)\,dt\right]\right] + c_i, \qquad i = 1, 2, \ldots, M \tag{8-57}$$

is maximized for $i = k$. c_i is given by (8-48b). This is implemented by the correlation receiver shown in Fig. 8-10. In Fig. 8-11 the correlators are replaced by equivalent matched filters. The output of the ith matched filter is

$$\int_0^{T_0} r(\lambda)h_i(T_0 - \lambda)\,d\lambda = \int_0^{T_0} r(\lambda)s_i[T_0 - (T_0 - \lambda)]\,d\lambda$$

$$= \int_0^{T_0} r(\lambda)s_i(\lambda)\,d\lambda = \mathbf{r}\cdot\mathbf{s}_i \tag{8-58}$$

In comparing the cost of the four equivalent receivers (Figs. 8-8 to 8-11), the receivers of Fig. 8-8 and 8-9 are usually less costly since $N \ll M$ in most problems. This is illustrated in the examples shown in the following section.

8-5 Probability of Error for MAP Receivers

The probability of error $P(E)$ is evaluated on an example-by-example basis. As seen from the decision rule for the AWGN channel, (8-48), the decision depends on the vector signal set used, $\{\mathbf{s}_i\}$; the source probabilities, $\{P(m_i)\}$; and the level of the AWGN, $N_0/2$. Consequently, the $P(E)$ also depends on all of these quantities. The following theorem is very helpful in evaluating $P(E)$.

THEOREM: *For an AWGN channel and a given set of source probabilities $\{P(m_i)\}$, the $P(E)$ of an optimum (MAP) receiver is unaffected if the vector signal set $\{\mathbf{s}_i\}$ is rotated and/or translated.*

Proof. This result follows directly from the MAP decision rule, (8-48). If the signal set $\{\mathbf{s}_i\}$ is rotated and/or translated the *relative* position of $\mathbf{s}_i$ with respect to $\mathbf{s}_j$, $j = 1, 2, \ldots, M$, is unaffected. Thus the same $\tilde{m} = m_k$ is obtained from the decision rule regardless of the amount of rotation and/or translation. Consequently, the probability of error remains the same. (However, the signal energy changes since the length from the signal vector tip to the origin is $\sqrt{E_{s_i}}$.)

EXAMPLE 8-3 BINARY SIGNALING

An arbitrary signal set for binary signaling (i.e., $M = 2$) is illustrated in Fig. 8-12a. The distance between the signal vectors is denoted by the parameter d.

Using the foregoing theorem, we can rotate and translate the signal set without affecting $P(E)$. To simplify the mathematics, rotate and translate the signal set so that the two-dimensional problem of Fig. 8-12a beomes a one-dimensional problem of 8-12b. The resulting conditional PDFs along the φ_1 axis [i.e., $r(t) = r_1\varphi_1(t)$] are shown in Fig. 8-12c. These are obtained by using (8-43):

$$f(\mathbf{r}|\mathbf{s}_1) = f_{\mathbf{n}}(\mathbf{r} - \mathbf{s}_1) = \frac{1}{\sqrt{\pi N_0}} e^{-[r_1 + (d/2)]^2/N_0} \qquad \text{(8-59a)}$$

and

$$f(\mathbf{r}|\mathbf{s}_2) = f_{\mathbf{n}}(\mathbf{r} - \mathbf{s}_2) = \frac{1}{\sqrt{\pi N_0}} e^{-[r_1 - (d/2)]^2/N_0} \qquad \text{(8-59b)}$$

V_T is the threshold level (i.e., decision boundary) given by the MAP decision rule. Thus, using (8-52), which is equivalent to (8-48), we can give the value of V_T by

$$|\mathbf{V}_T - \mathbf{s}_1|^2 - N_0 \ln P(m_1) = |\mathbf{V}_T - \mathbf{s}_2|^2 - N_0 \ln P(m_2) \qquad \text{(8-60)}$$

In a one-dimensional vector space this becomes

$$\left(V_T + \frac{d}{2}\right)^2 - N_0 \ln P(m_1) = \left(V_T - \frac{d}{2}\right)^2 - N_0 \ln P(m_2) \qquad \text{(8-61)}$$

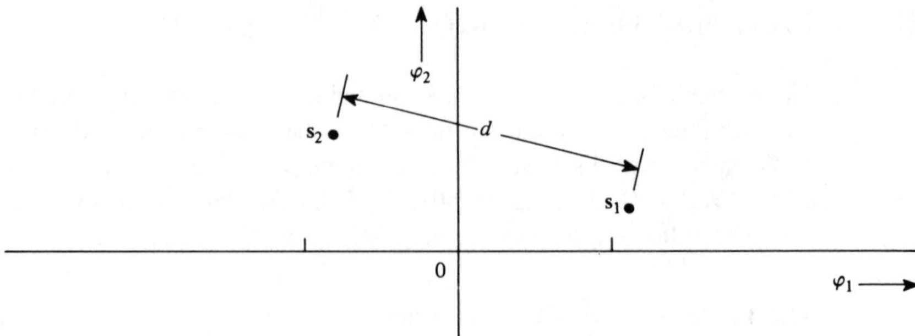

(a) A Binary Signal Set

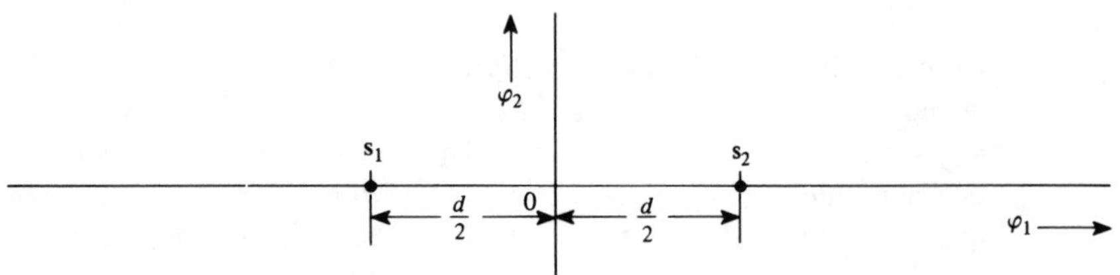

(b) Translated and Rotated Signal Set

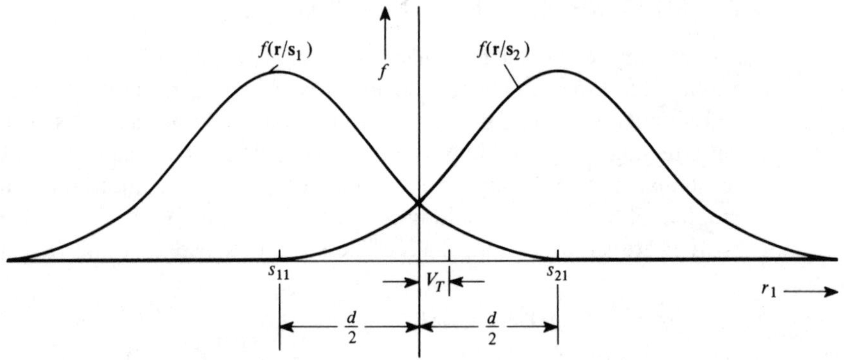

(c) Conditional PDF for Received Vector

FIGURE 8-12 Binary signaling problem for Example 8-3.

Solving for V_T, we get

$$V_T = \frac{N_0}{2d} \ln \left[\frac{P(m_1)}{P(m_2)} \right] \tag{8-62}$$

The resulting probability of error is

$$P(E) = P(m_1)P(E|m_1) + P(m_2)P(E|m_2)$$

or

$$P(E) = P(m_1) \int_{V_T}^{\infty} \frac{1}{\sqrt{\pi N_0}} e^{-[r_1+(d/2)]^2/N_0}\, dr_1 + P(m_2) \int_{-\infty}^{V_T} \frac{1}{\sqrt{\pi N_0}} e^{-[r_1-(d/2)]^2/N_0}\, dr_1$$

or

$$P(E) = P(m_1)Q\left[\frac{V_T + (d/2)}{\sqrt{N_0/2}}\right] + P(m_2)Q\left[\frac{(d/2) - V_T}{\sqrt{N_0/2}}\right]$$

where, from (B-56),

$$Q(z) \triangleq \frac{1}{\sqrt{2\pi}}\int_z^{\infty} e^{-\lambda^2/2}\,d\lambda$$

A plot of $Q(z)$ is shown in Fig. B-7. [A tabulation of $Q(z)$ is given in Sec. A-10 (Appendix A), and an approximation for $Q(z)$ is given by (B-64).] Thus $P(E)$ reduces to

$$P(E) = P(m_1)Q\left(\frac{d + 2V_T}{\sqrt{2N_0}}\right) + P(m_2)Q\left(\frac{d - 2V_T}{\sqrt{2N_0}}\right) \tag{8-63}$$

For the case of equally likely signaling $P(m_1) = P(m_2) = \frac{1}{2}$ and (8-63) becomes

$$P(E) = Q\left(\frac{d}{\sqrt{2N_0}}\right) \tag{8-64}$$

EXAMPLE 8-4 QUADRATURE AMPLITUDE MODULATION SIGNALING

Quadrature amplitude modulation (QAM) digital signaling was introduced in Sec. 5-2. It is a popular type of digital signaling technique used in computer modem systems.

Referring to Fig. 8-13a, assume that we have an $M = 6$ signal set as illustrated. Furthermore, assume that all the symbols (i.e., signal vectors) are equally likely to be transmitted. In this case the boundaries for the decision regions, D_1, $D_2, \ldots, D_6$, are given by the perpendicular bisectors (the dashed lines) of the invisible lines connecting adjacent signal vectors. This result was found in Example 8-3 for the case of equally likely signaling. In this problem, however, $P(E)$ is a little more difficult to evaluate since two dimensions are involved. It is easier first to evaluate the probability of a correct decision $P(C)$ and then obtain $P(E)$.

$$P(E) = 1 - P(C) \tag{8-65a}$$

where

$$P(C) = \sum_{i=1}^{6} P(C|m_i)P(m_i) = \sum_{i=1}^{6} \tfrac{1}{6}P(C|m_i) \tag{8-65b}$$

$P(C|m_i)$ is the probability of a correct decision given that $\mathbf{s}_i$ (corresponding to m_i) is transmitted.

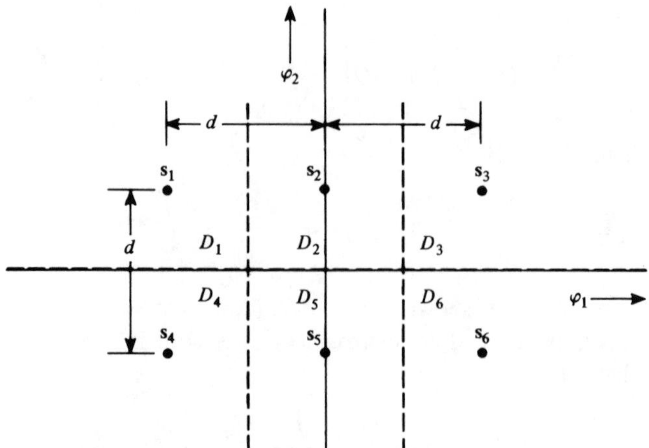

(a) $M = 6$ Rectangular Signal Set

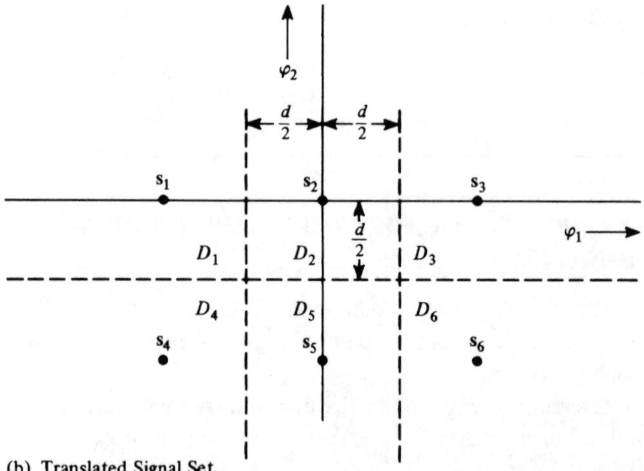

(b) Translated Signal Set

FIGURE 8-13 QAM vector signal set for Example 8-4.

Each of the $P(C|m_i)$, $i = 1, 2, \ldots, 6$, can be individually evaluated by translating the vector $\mathbf{s}_i$ to the origin. For example, to evaluate $P(C|m_2)$, translate $\mathbf{s}_2$ to the origin, as shown in Fig. 8-13b. Then

$$P(C|m_2) = \int_{-d/2}^{d/2} f(\lambda_1)\, d\lambda_1 \int_{-d/2}^{\infty} f(\lambda_2)\, d\lambda_2 \tag{8-66a}$$

where

$$f(\lambda) = \frac{1}{\sqrt{\pi N_0}}\, e^{-\lambda^2/N_0} \tag{8-66b}$$

Thus (8-66a) reduces to

$$P(C|m_2) = (1 - 2p)(1 - p) \tag{8-67}$$

where

$$p \triangleq Q\left(\frac{d}{\sqrt{2N_0}}\right) \tag{8-68}$$

Similarly,

$$P(C|m_5) = (1 - 2p)(1 - p) \tag{8-69}$$

In addition,

$$\begin{aligned} P(C|m_1) &= P(C|m_3) = P(C|m_4) = P(C|m_6) \\ &= \int_{-d/2}^{\infty} f(\lambda_1)\,d\lambda_1 \int_{-d/2}^{\infty} f(\lambda_2)\,d\lambda_2 = (1 - p)^2 \end{aligned} \tag{8-70}$$

Substituting (8-67), (8-69), and (8-70) into (8-66) and (8-65), we obtain the probability of symbol error. Thus for this $M = 6$ QAM signal set,

$$P(E) = 1 - \tfrac{4}{6}(1 - p)^2 - \tfrac{2}{6}(1 - 2p)(1 - p) \tag{8-71}$$

where p is given by (8-68).

EXAMPLE 8-5 PULSE CODE MODULATION SIGNALING

Assume that we have a binary PCM signaling system that uses polar NRZ signaling in which each PCM word has a length of K bits. An example of a ($K = 3$)-bit PCM signal was shown in Fig. 3-8d. There are $M = 2^K$ possible messages. The orthonormal functions are of the rectangular pulse type shown in Fig. 8-2. Then, for PCM signaling we have the relationship

$$M = 2^K = 2^N \tag{8-72}$$

or $K = N$. The signal vectors terminate on the *vertices of a hypercube,* as illustrated in Fig. 8-14 for the case of $K = N = 3$.

If we assume that each of the signals (i.e., PCM words) is equally likely to occur, the probability of a correct decision is

$$P(C) = \sum_{i=1}^{M} \frac{1}{M} P(C|m_i) \tag{8-73}$$

where

$$P(C|m_i) = \underbrace{\int_{-d/2}^{\infty} f(\lambda_1) d\lambda_1 \int_{-d/2}^{\infty} f(\lambda_2)\,d\lambda_2 \cdots \int_{-d/2}^{\infty} f(\lambda_N)\,d\lambda_N}_{N \text{ integrals}}$$

or

$$P(C|m_i) = (1 - p)^N \tag{8-74}$$

Both $f(\lambda)$ and p have been defined previously [see (8-66b) and (8-68)]. Note that all the $P(C|m_i)$, $i = 1, 2, \ldots, M$, are identical because of the symmetry in-

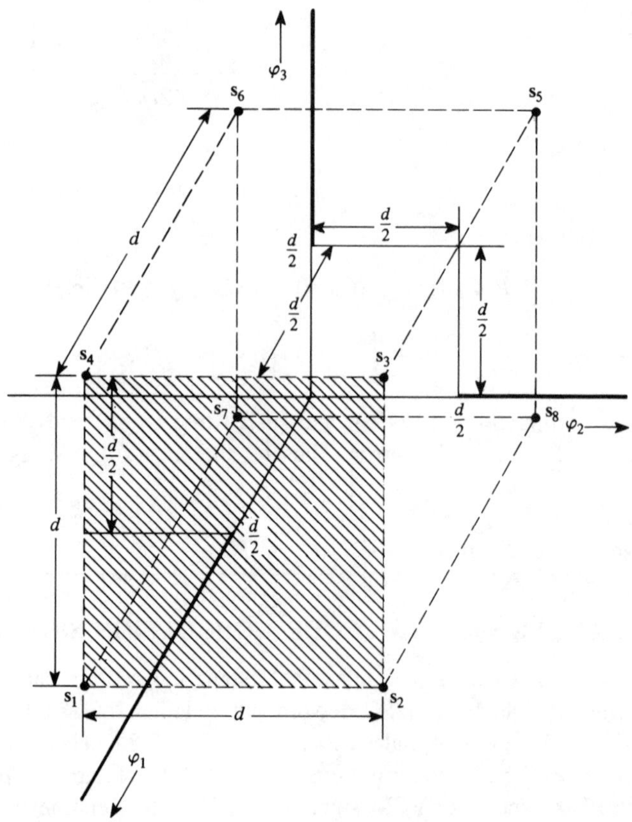

FIGURE 8-14 Vector signal set for PCM signals of Example 8-5 ($N = 3$).

volved. Using (8-74) in (8-73), we find that the probability of word error for this PCM system is

$$P(E) = 1 - (1 - p)^N \tag{8-75}$$

where p is given by (8-68). Furthermore,

$$E_{s_i} = |\mathbf{s}_i|^2 = \sum_{j=1}^{N} s_{ij}^2 = N\left(\frac{d}{2}\right)^2 = E_s \tag{8-76}$$

or

$$d = 2\sqrt{\frac{E_s}{N}} \tag{8-77}$$

so

$$p = Q\left(\sqrt{\frac{2E_s}{NN_0}}\right) \tag{8-78}$$

in the PCM (i.e., hypercube) problem.

Union Bound

The mathematical manipulations involved in the preceding examples were relatively easy because of the symmetry in those problems. In general, the mathematics becomes tedious when $P(E)$ is evaluated. Often we are happy to obtain a *bound* on the $P(E)$. Moreover, for practical purposes, it is often satisfactory to know that the $P(E)$ will be below some upper bound. One upper bound that is very useful is the union bound. It is articulated in the following theorem.

THEOREM: *For equally likely signaling over an AWGN channel, the MAP decision rule gives the following bound on the conditional probability of error:*

$$P(E|m_i) \le \sum_{\substack{k=1 \\ k\neq i}}^{M} Q\left(\frac{d_{ik}}{\sqrt{2N_0}}\right) \tag{8-79}$$

where $d_{ik} \triangleq |\mathbf{s}_i - \mathbf{s}_k|$ *is the distance between the signal vectors* $\mathbf{s}_i$ *and* $\mathbf{s}_k$. *This bound is called the* ***union bound.***

Proof. Let E_{ik} be the error event that $\mathbf{r}$ is closer to $\mathbf{s}_k$ than to $\mathbf{s}_i$ when $\mathbf{s}_i$ is transmitted. Then

$$P(E|m_i) = P(E_{i1} \cup E_{i2} \cup \cdots \cup E_{ih} \cup E_{ij} \cup \cdots \cup E_{iM}) \tag{8-80}$$

Note that E_{ii} does *not* appear on the right side of this equation.

Furthermore, we know that for the union of two events [see (B-4)],

$$P(A \cup B) = P(A) + P(B) - P(A \cap B) \le P(A) + P(B)$$

If we extend this inequality for the case of the union of $M - 1$ events, (8-80) becomes

$$P(E|m_i) \le \sum_{\substack{k=1 \\ k\neq i}}^{M} P(E_{ik}) \tag{8-81}$$

But $P(E_{ik})$ is just the probability of error for a binary signaling problem, such as Example 8-3, so using (8-64) yields

$$P(E_{ik}) = Q\left(\frac{d_{ik}}{\sqrt{2N_0}}\right) \tag{8-82}$$

Substituting (8-82) into (8-81), we obtain (8-79).

EXAMPLE 8-6 QUANTIZED PULSE POSITION MODULATION SIGNALING

Pulse position modulation (PPM) signaling was studied in Sec. 3-9. Now assume that only M possible pulse positions are allowed over the T_0 signaling interval so that digital (i.e., quantized) signaling is obtained. The signal set $s_i(t)$, $i = 1, 2, \ldots, M$, is described by

$$s_i(t) = s_{ii}\varphi_i(t), \qquad i = 1, 2, \ldots, M \tag{8-83}$$

where the orthonormal functions are of the rectangular pulse type, as shown in Fig. 8-2. The signal energy is $E_s = |\mathbf{s}_i|^2$ for all $i = 1, 2, \ldots, M$. Thus

$$M = 2^K = N \tag{8-84}$$

for quantized PPM signaling. The corresponding *vector set* $\{\mathbf{s}_i\}$ *is orthogonal,* as illustrated in Fig. 8-15 for the case of $M = N = 3$.

The probability of error can be described by the union bound. The distance between any of the signal vectors is

$$d_{ik} = \sqrt{2E_s} \qquad \text{for all } i \text{ and } k \text{ where } i \neq k$$

From (8-79) we have a bound on the conditional probability of error:

$$P(E|m_i) \leq (M-1)Q\left(\sqrt{\frac{E_s}{N_0}}\right), \qquad i = 1, 2, \ldots, M$$

Furthermore,

$$P(E) = \sum_{i=1}^{M} \frac{1}{M} P(E|m_i)$$

Thus the probability of error for PPM signaling (with M quantized positions) is bounded by

$$P(E) \leq (M-1)Q\left(\sqrt{\frac{E_s}{N_0}}\right) \tag{8-85}$$

A comparison of this union bound result with the exact result for $P(E)$ is shown in Fig. 8-16. The exact result is obtained from a published set of tabulated

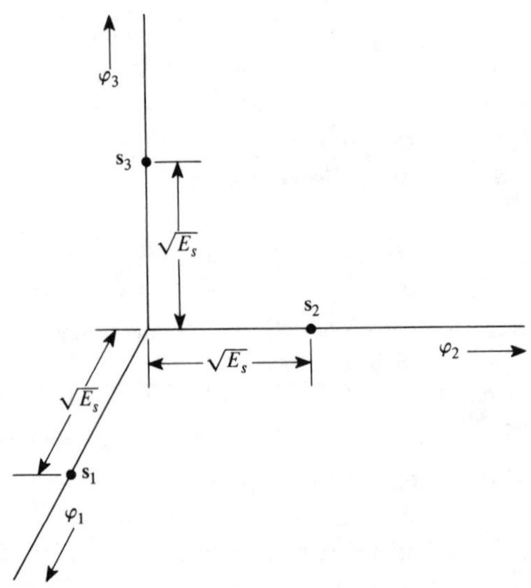

FIGURE 8-15 $M = 3$ vector signal set for quantized PPM signaling of Example 8-6.

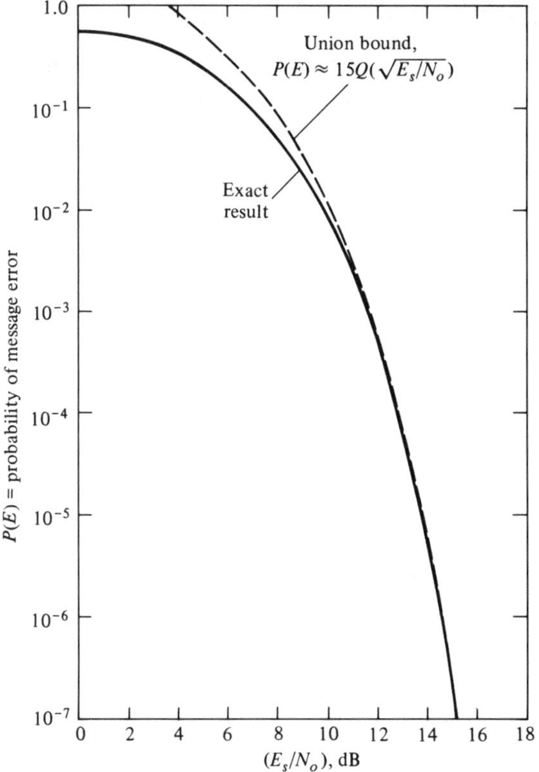

FIGURE 8-16 *P*(*E*) for $M = 16$ quantized PPM signal of Example 8-6.

values [Golomb, 1964]. For this case of $M = 16$ ($K = 4$), it is seen that the union bound is a very good approximation to the exact result when $P(E) < 10^{-3}$. In general, the union bound becomes a good approximation to the exact result when $P(E)$ is small.

8-6 Random-Phase and Fading Channels

In the previous sections of this chapter the waveforms, $s_i(t)$, $i = 1, 2, \ldots, M$, could be either bandpass or baseband signals, depending on the choice of the $\{\varphi_j(t)\}$ used. In this section we study *bandpass signals* that have an unknown phase. To distinguish the bandpass waveform set $\{s_i(t)\}$ from the modulation (baseband) waveform set, we will use $\{m_i(t)\}$ to denote the baseband waveforms. We will first study the random-phase problem and then generalize the result to model fading signals (random amplitude and phase).

Double-Sideband Transmission with Random Phase

Refer to Fig. 8-17 and let $\{m_i(t)\}$ be the baseband modulation waveform set and $\{s_i(t)\}$ be the corresponding double-sideband (bandpass) signal set: θ is a random variable that models the unknown carrier phase at the receiver. Let θ be uniformly distributed:

$$f(\theta) = \begin{cases} \dfrac{1}{2\pi}, & |\theta| \leq \pi \\ 0, & \theta \text{ elsewhere} \end{cases}$$

Note that θ is assumed to be a random variable (as opposed to a random process), so we are assuming that the phase is a constant over the signaling interval of T_0 sec. As related to practice, θ models the slow (with respect to f_c) random change in the phase because of, for example, the change in the propagation time between the transmitter and receiver or because of oscillator drift at the transmitter and/or receiver.

Using the theory of Sec. 6-7 for bandpass processes, we find that the quadrature detector outputs of Fig. 8-17 are

$$x(t) = m(t) \cos \theta + x_n(t) \tag{8-86a}$$

$$y(t) = -m(t) \sin \theta + y_n(t) \tag{8-86b}$$

The PSD of the baseband noise processes are

$$\mathcal{P}_{x_n}(f) = \mathcal{P}_{x_y}(f) = N_0 \tag{8-87}$$

for the case of white channel noise with a PSD of $\mathcal{P}_{n_N}(f) = N_0/2$.

The baseband signal set may be represented by the orthonormal expansion

$$m_i(t) = \sum_{j=1}^{N} m_{ij}\varphi_j(t), \qquad i = 1, 2, \ldots, M \tag{8-88}$$

Then the vector representation for (8-86) is

$$\mathbf{x} = \mathbf{m} \cos \theta + \mathbf{x}_n \tag{8-89a}$$

$$\mathbf{y} = -\mathbf{m} \sin \theta + \mathbf{y}_n \tag{8-89b}$$

Furthermore, assume that $n_w(t)$ is a Gaussian random process. Then, using the results of Secs. 6-7 and 8-3, we see that the PDF of $\mathbf{x}_n$ and $\mathbf{y}_n$ is

$$f_{\mathbf{n}}(\mathbf{x}_n, \mathbf{y}_n) = f(\mathbf{x}_n) f(\mathbf{y}_n)$$
$$= \left[\frac{1}{(2\pi N_0)^{N/2}} e^{-(1/2N_0)|\mathbf{x}_n|^2}\right]\left[\frac{1}{(2\pi N_0)^{N/2}} e^{-(1/2N_0)|\mathbf{y}_n|^2}\right]$$

or

$$f_{\mathbf{n}}(\mathbf{x}_n, \mathbf{y}_n) = \frac{1}{(2\pi N_0)^N} e^{-(1/2N_0)(|\mathbf{x}_n|^2 + |\mathbf{y}_n|^2)} \tag{8-90}$$

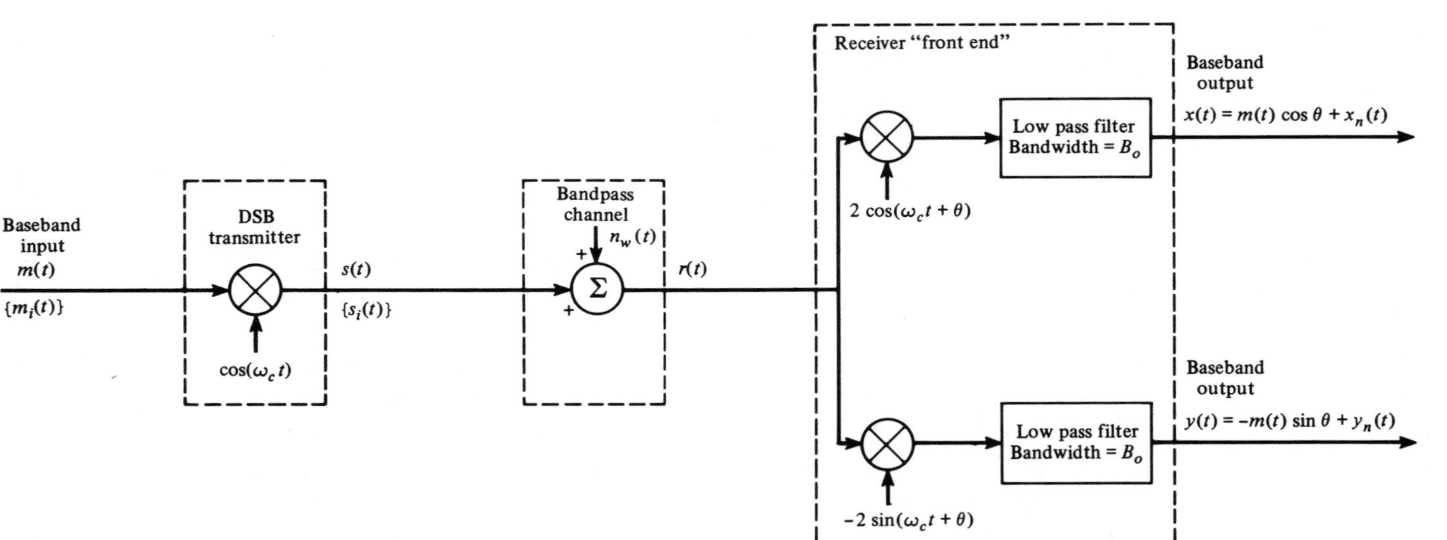

FIGURE 8-17 DSB transmission system with random phase.

The MAP receiver is found by applying the MAP decision rule of (8-46). Assume that the messages are equally likely; then $P(m_i) = 1/M$, a constant, may be neglected in the maximization procedure. The MAP decision rule is to choose $\tilde{m} = m_k$ when

$$\overline{f_{\mathbf{n}}(\mathbf{x} - \mathbf{m}_i \cos\theta, \mathbf{y} + \mathbf{m}_i \sin\theta)} = \overline{\frac{1}{(2\pi N_0)^N} e^{-(1/2N_0)(|\mathbf{x}-\mathbf{m}_i\cos\theta|^2 + |\mathbf{y}+\mathbf{m}_i\sin\theta|^2)}},$$

$$i = 1, 2, \ldots, M \quad \text{(8-91a)}$$

is maximized for $i = k$. The expectation operator, averaging with respect to θ, is needed since the PDF is actually conditioned on the value of θ and we want the MAP criteria to be satisfied regardless of the value of θ. Equation (8-91a) reduces to maximizing

$$\overline{\exp\left[\frac{1}{N_0}(\mathbf{x}\cdot\mathbf{m}_i\cos\theta - \mathbf{y}\cdot\mathbf{m}_i\sin\theta)\right]} \times \exp\left[-\left(\frac{E_{m_i}}{2N_0}\right)\right] \quad \text{(8-91b)}$$

where $|\mathbf{x}|^2$, $|\mathbf{y}|^2$, and $1/(2\pi N_0)^N$ have been neglected since they do not affect the maximizing operation with respect to i. $E_{m_i} = |\mathbf{m}_i|^2$ is the energy in the signal $m_i(t)$ over its duration T_0 sec. To evaluate the average in (8-91) conveniently, define the Cartesian-to-polar coordinate transformation:

$$D_i \triangleq \sqrt{(\mathbf{x}\cdot\mathbf{m}_i)^2 + (\mathbf{y}\cdot\mathbf{m}_i)^2} \quad \text{(8-92a)}$$

and

$$\phi_i \triangleq \tan^{-1}\left(\frac{\mathbf{y}\cdot\mathbf{m}_i}{\mathbf{x}\cdot\mathbf{m}_i}\right) \quad \text{(8-92b)}$$

where, from Fig. 8-18,

$$\mathbf{x}\cdot\mathbf{m}_i = D_i\cos\phi_i \quad \text{(8-92c)}$$

$$\mathbf{y}\cdot\mathbf{m}_i = D_i\sin\phi_i \quad \text{(8-92d)}$$

By using (8-92), (8-91) becomes

$$\overline{\exp\left[\frac{1}{N_0}(D_i\cos\phi_i\cos\theta - D_i\sin\phi_i\sin\theta)\right]} \times \exp\left[-\left(\frac{E_{m_i}}{2N_0}\right)\right]$$

$$= \overline{\exp\left[\frac{D_i}{N_0}\cos(\phi_i + \theta)\right]} \times \exp\left[-\left(\frac{E_{m_i}}{2N_0}\right)\right]$$

$$= \left\{\int_{-\pi}^{\pi} \exp\left[\frac{D_i}{N_0}\cos(\phi_i + \theta)\right]\frac{1}{2\pi}\,d\theta\right\} \times \exp\left[-\left(\frac{E_{m_i}}{2N_0}\right)\right] \quad \text{(8-93)}$$

But the modified Bessel function of the first kind, zero order, is

$$I_0(z) = \frac{1}{2\pi}\int_{-\pi}^{\pi} e^{z\cos(\theta+\alpha)}\,d\theta \quad \text{(8-94)}$$

where α may be any constant since the cosine is periodic over the 2π interval of integration. Thus, for the case of a random-phase variation over an AWGN

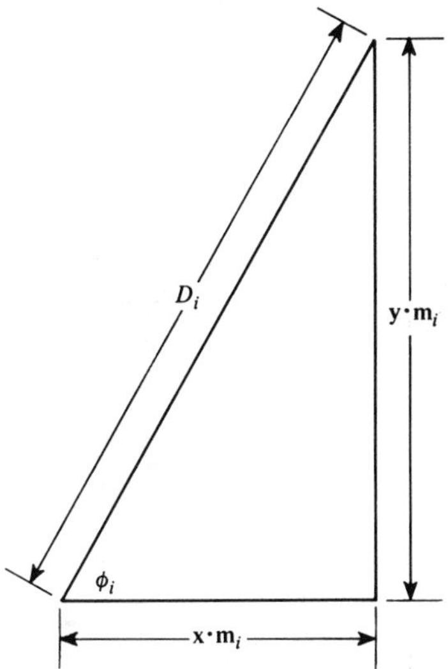

FIGURE 8-18 Cartesian-to-polar transformation.

channel with equally likely signaling, the MAP decision rule is to choose $\tilde{m} = m_k$ when

$$I_0\left(\frac{D_i}{N_0}\right) \times \exp\left(\frac{E_{m_i}}{2N_0}\right), \qquad i = 1, 2, \ldots, M \tag{8-95}$$

is maximized for $i = k$.

If we also assume that all the $m_i(t)$, $i = 1, 2, \ldots, M$, have the same energy $E_{m_i} = E_m$, then (8-95) can be further simplified. $I_0(z)$ is a monotonically increasing function for $z \geq 0$, so the MAP receiver needs to choose $\tilde{m} = m_k$ when

$$D_i = \sqrt{(\mathbf{x}\cdot\mathbf{m}_i)^2 + (\mathbf{y}\cdot\mathbf{m}_i)^2}, \qquad i = 1, 2, \ldots, M \tag{8-96}$$

is maximized for $i = k$.

The decision rule of (8-96) can be realized in a number of ways. One canonical form is the correlation receiver shown in Fig. 8-19. By using Parseval's theorem, (8-56), the integrator outputs are obtained.

$$\int_0^{T_0} x(t)m_i(t)\,dt = \mathbf{x}\cdot\mathbf{m}_i \tag{8-97a}$$

and

$$\int_0^{T_0} y(t)m_i(t)\,dt = \mathbf{y}\cdot\mathbf{m}_i \tag{8-97b}$$

These are the inputs to the decision rule algorithm of (8-96). Thus the optimum receiver structure of Fig. 8-19 is obtained.

Figure 8-20 illustrates another receiver structure that implements the MAP decision rule of (8-96). This receiver incorporates M matched filters and M

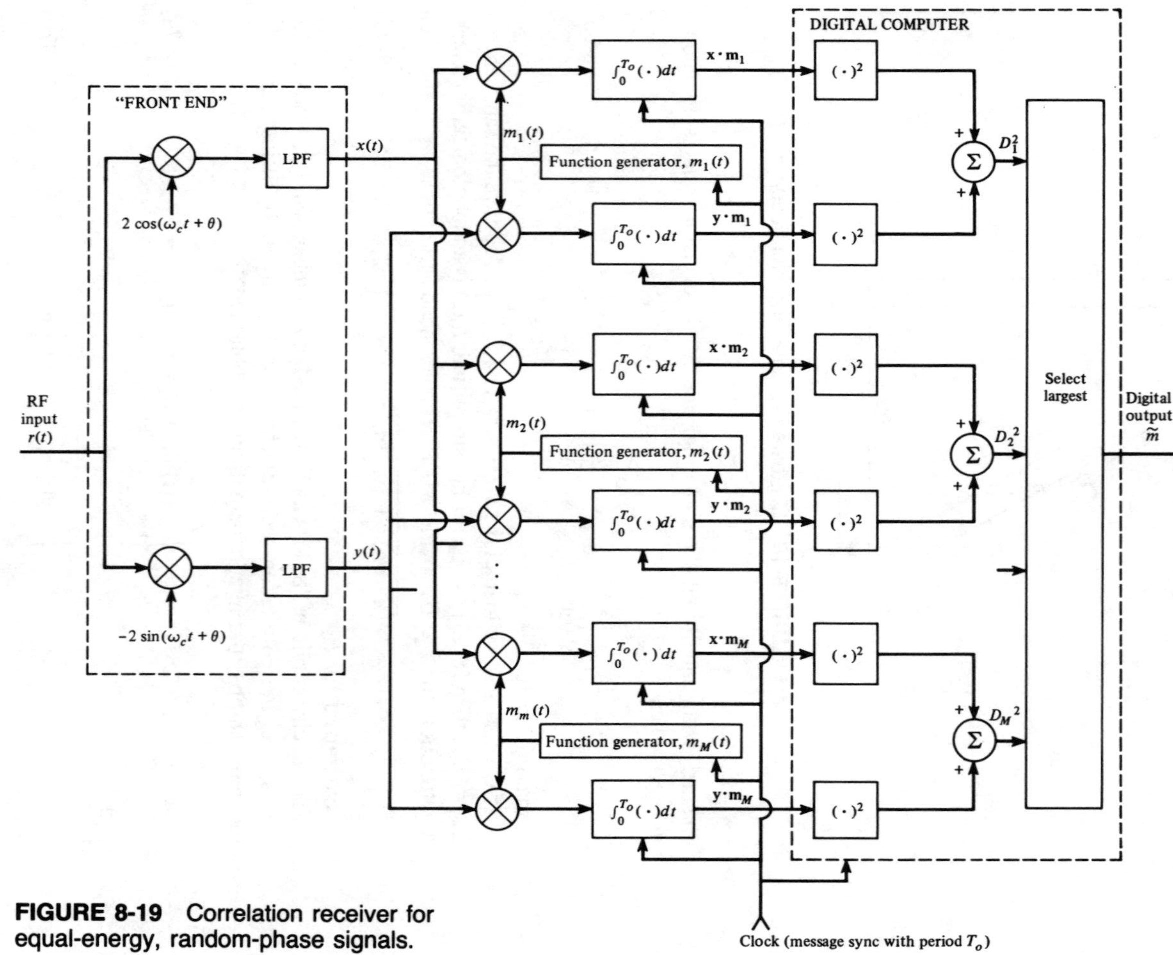

FIGURE 8-19 Correlation receiver for equal-energy, random-phase signals.

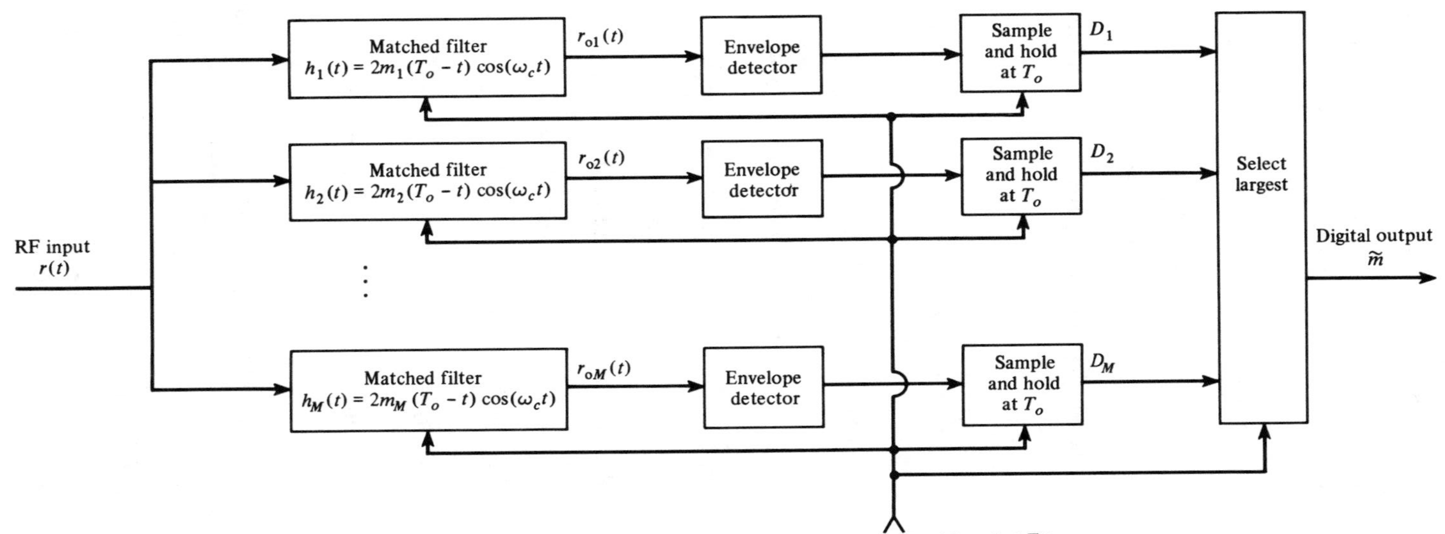

FIGURE 8-20 Matched-filter, envelope detector receiver for equal-energy, random-phase signals.

envelope detectors. The message clock signal is needed to reset the initial conditions of the matched filters to zero at the beginning of each message period and to activate the sample-and-hold circuit at the end of each message period.

We will now demonstrate that this receiver structure, as described in Fig. 8-20, implements the decision rule of (8-96). The output of the ith matched filter may be represented by

$$r_{0i}(t) = \text{Re}\{g_{0i}(t)e^{j\omega_c t}\}, \qquad i = 1, 2, \ldots, M \tag{8-98a}$$

where, from (4-9), the output complex envelope is

$$g_{0i}(t) = \tfrac{1}{2}\, g(t) * k_i(t) \tag{8-98b}$$

$k_i(t)$ is the complex envelope of the impulse response of the ith matched filter. The input complex envelope is

$$g(t) = x(t) + jy(t) \tag{8-98c}$$

where $x(t)$ and $y(t)$ are given by (8-86). At the sampling time $t = T_0$, the output complex envelope is

$$\begin{aligned} g_{0i}(T_0) &= \tfrac{1}{2}[x(T_0) + jy(T_0)] * k_i(T_0) \\ &= \tfrac{1}{2}\int_0^{T_0} x(\lambda)k_i(T_0 - \lambda)\, d\lambda + j\,\tfrac{1}{2}\int_0^{T_0} y(\lambda)k_i(T_0 - \lambda)\, d\lambda \end{aligned} \tag{8-99}$$

Now assume that the complex envelope of the ith matched filter is

$$k_i(t) = 2m_i(T_0 - t) \tag{8-100a}$$

which implies that the impulse response of the ith filter is assumed to be

$$h_i(t) = 2m_i(T_0 - t)\cos \omega_c t \tag{8-100b}$$

By using (8-100a), (8-99) becomes

$$g_{0i}(T_0) = \int_0^{T_0} x(\lambda)m_i(\lambda)\, d\lambda + j\int_0^{T_0} y(\lambda)m_i(\lambda)\, d\lambda \tag{8-101a}$$

By Parseval's theorem, (8-56), (8-101a) is equivalent to

$$g_{0i}(T_0) = \mathbf{x}\cdot\mathbf{m}_i + j\mathbf{y}\cdot\mathbf{m}_i \tag{8-101b}$$

From the theory developed in Chapter 4, we know that the output of an envelope detector is the magnitude of the complex envelope that is present at its input. Thus at the sampling time T_0, the output of the ith envelope detector is

$$|g_{0i}(T_0)| = |\mathbf{x}\cdot\mathbf{m}_i + j\mathbf{y}\cdot\mathbf{m}_i| = \sqrt{(\mathbf{x}\cdot\mathbf{m}_i)^2 + (\mathbf{y}\cdot\mathbf{m}_i)^2} \tag{8-102}$$

This is identical to the decision rule of (8-96), so Fig. 8-20 is a realization of a MAP receiver for equal-energy, random-phase signaling.

Probability of Error for Orthogonal Binary Random-Phase Signaling

The probability of error is evaluated on a case-by-case basis since it depends on the vector modulation set $\{\mathbf{m}_i\}$. In this section the probability of error will be evaluated for a binary ($M = 2$) orthogonal signal set. A random-phase communication system will be used. Furthermore, the signals $\mathbf{m}_1$ and $\mathbf{m}_2$ are assumed to have the same energy E_m and $P(m_i) = \frac{1}{2}$, $i = 1, 2$.

The waveforms and their corresponding vectors are represented by

$$m_1(t) = \sqrt{E_m}\,\varphi_1(t), \qquad \mathbf{m}_1 = \sqrt{E_m}\,\boldsymbol{\varphi}_1 \tag{8-103a}$$

and

$$m_2(t) = \sqrt{E_m}\,\varphi_2(t), \qquad \mathbf{m}_2 = \sqrt{E_m}\,\boldsymbol{\varphi}_2 \tag{8-103b}$$

where $\boldsymbol{\varphi}_1$ and $\boldsymbol{\varphi}_2$ are unit vectors along the first and second dimensional axes. The MAP receiver chooses $\tilde{m} = m_1$ when

$$D_1^2 = (\mathbf{x} \cdot \mathbf{m}_1)^2 + (\mathbf{y} \cdot \mathbf{m}_1)^2 > (\mathbf{x} \cdot \mathbf{m}_2)^2 + (\mathbf{y} \cdot \mathbf{m}_2)^2 = D_2^2 \tag{8-104}$$

Substituting (8-103) into (8-104), we choose $\tilde{m} = m_1$ when

$$(\mathbf{x} \cdot \boldsymbol{\varphi}_1)^2 + (\mathbf{y} \cdot \boldsymbol{\varphi}_1)^2 > (\mathbf{x} \cdot \boldsymbol{\varphi}_2)^2 + (\mathbf{y} \cdot \boldsymbol{\varphi}_2)^2 \tag{8-105a}$$

where

$$\mathbf{x} = x_1\boldsymbol{\varphi}_1 + x_2\boldsymbol{\varphi}_2 \tag{8-105b}$$

and

$$\mathbf{y} = y_1\boldsymbol{\varphi}_1 + y_2\boldsymbol{\varphi}_2 \tag{8-105c}$$

First, evaluate the probability of error for the MAP receiver conditioned on m_1 being transmitted, $P(E|m_1)$. As we will see, the evaluation of $P(E|m_1)$ is tedious, but the procedure is straightforward. For the case of $\mathbf{m} = \mathbf{m}_1$, then, $\mathbf{x}$ and $\mathbf{y}$ become

$$\mathbf{x} = \mathbf{m}_1 \cos\theta + \mathbf{x}_n = (\sqrt{E_m}\cos\theta + x_{n_1})\boldsymbol{\varphi}_1 + x_{n_2}\boldsymbol{\varphi}_2 \tag{8-106a}$$

and

$$\mathbf{y} = -\mathbf{m}_1 \sin\theta + \mathbf{y}_n = (-\sqrt{E_m}\sin\theta + y_{n_1})\boldsymbol{\varphi}_1 + y_{n_2}\boldsymbol{\varphi}_2 \tag{8-106b}$$

or

$$x_1 = (\sqrt{E_m}\cos\theta + x_{n_1}), \qquad x_2 = x_{n_2} \tag{8-106c}$$

and

$$y_1 = (-\sqrt{E_m}\sin\theta + y_{n_1}), \qquad y_2 = y_{n_2} \tag{8-106d}$$

When (8-106) is substituted into (8-105a), the receiver decision rule is to choose $\tilde{m} = m_1$ when

$$x_1^2 + y_1^2 > x_2^2 + y_2^2 \tag{8-107}$$

But we realize that $x_2 = x_{n_2}$ and $y_2 = y_{n_2}$ are independent zero-mean Gaussian random variables with $\sigma^2 = N_0$ for the case of an AWGN channel. An error is made when the right side of (8-107) is larger than the left side (m_1 transmitted). Thus

$$P(E|m_1,\theta,R_1) = P(R_1^2 \triangleq x_1^2 + y_1^2 < R_2^2 \triangleq x_2^2 + y_2^2)$$
$$= \iint_{R_1^2 \le x_2^2 + y_2^2} \frac{1}{2\pi N_0} e^{-[1/(2N_0)](x_2^2+y_2^2)}\, dx_2\, dy_2$$

Converted to polar coordinates, this becomes

$$P(E|m_1,\theta,R_1) = \int_{R_2=R_1}^{R_2=\infty} \int_{\theta_2=0}^{\theta_2=2\pi} \frac{1}{2\pi N_0} e^{-[1/(2N_0)]R_2^2} R_2\, dR_2\, d\theta_2$$

or

$$P(E|m_1,\theta,R_1) = e^{-R_1^2/(2N_0)} \tag{8-108}$$

Referring to (8-106c), we see that x_1 is a Gaussian random variable with variance $\sigma_{x_1}^2 = N_0$ and mean $m_{x_1} = \sqrt{E_m} \cos \theta$. Similarly, from (8-106d), y_1 is an independent Gaussian random variable with variance $\sigma_{y_1}^2 = N_0$ and mean $m_{y_1} = -\sqrt{E_m} \sin \theta$. Thus, since $R_1^2 = x_1^2 + y_1^2$,

$$P(E|m_1,\theta) = \int_{-\infty}^{\infty} \int_{-\infty}^{\infty} P(E|m_1,\theta,R_1) f(x_1) f(y_1)\, dx_1\, dy_1$$
$$= \int_{-\infty}^{\infty} e^{-x_1^2/(2N_0)} f(x_1)\, dx_1 \int_{-\infty}^{\infty} e^{-y_1^2/(2N_0)} f(y_1)\, dy_1 \tag{8-109}$$

The integrals on the right side of (8-109) can be conveniently evaluated by using the following identity. If $f(z)$ is a Gaussian PDF with mean m_z and variance σ_z^2,

$$\int_{-\infty}^{\infty} e^{az^2} f(z)\, dz = \frac{e^{am_z^2/(1-2a\sigma_z^2)}}{\sqrt{1 - 2a\sigma_z^2}} \qquad \text{for } a < \frac{1}{2\sigma_z^2} \tag{8-110}$$

Substituting (8-110) into (8-109), we have

$$P(E|m_1,\theta) = \left(\frac{1}{\sqrt{2}} e^{[E_m/(4N_0)]\cos^2\theta}\right) \left(\frac{1}{\sqrt{2}} e^{-[E_m/(4N_0)]\sin^2\theta}\right)$$

or

$$P(E|m_1,\theta) = \tfrac{1}{2} e^{-E_m/(4N_0)} \tag{8-111}$$

It is interesting that the right side of (8-111) does not happen to be a function of m_1 or θ. Thus

$$P(E) = \tfrac{1}{2} e^{-E_m/(4N_0)} \tag{8-112}$$

However, $P(E)$ is obtained from $P(E|m_i,\theta)$, $i = 1, 2$, by using

$$P(E) = \sum_i \int_{-\pi}^{\pi} P(E|m_i,\theta) P(m_i) f(\theta)\, d\theta \tag{8-113}$$

Moreover, since $s(t) = m(t) \cos \omega_c t$,

$$E_s = \tfrac{1}{2} E_m \tag{8-114}$$

Consequently, the probability of error for binary equally likely orthogonal signaling over an AWGN random-phase channel is

$$P(E) = \tfrac{1}{2} e^{-E_s/(2N_0)} \tag{8-115}$$

when an optimum (i.e., MAP) receiver is used.

EXAMPLE 8-7 FSK ORTHOGONAL SIGNALING

In Sec. 7-3 it was demonstrated that FSK is an orthogonal binary signaling technique when the frequency shift is much larger than the bit rate, R [e.g., see (7-46)]. Thus (8-115) describes the $P(E)$ for noncoherent (random-phase) reception of FSK. Furthermore, for this example of $M = 2$, $P(E)$ is identical to the probability of bit error P_e and E_s is identical to the energy per bit, E_b. Thus (8-115) becomes

$$P_e = \tfrac{1}{2} e^{-(1/2)(E_b/N_0)} \tag{8-116}$$

But this is the same result we obtained in Table 7-1 when the "minimum bandwidth" filter was used. We *now realize* that the optimum filter that is needed for noncoherent detection of FSK is the *matched filter*. That is, the filters shown in Fig. 8-20 are matched to the FSK signals of (7-39). For optimum noncoherent reception of FSK, the receiver described in Fig. 7-11 is identical to that given in Fig. 8-20 (for the case of $M = 2$). A comparison of the error performance of the optimum noncoherent (random-phase) FSK system with that for the optimum coherent FSK system was given in Fig. 7-14.

Fading Channels

We have just found the optimum receiver for the case of a random-phase channel. We will now extend this result to model a fading channel. The fading channel produces a random amplitude variation as well as a random-phase variation on the received signal. This is illustrated by the scattering channel model as shown in Fig. 8-21. The signal reflected from the kth scatterer is

$$s_k(t) = A_k m(t - \tau_k) \cos[\omega_c (t - \tau_k)] \tag{8-117}$$

If we assume that the relative delays τ_k are small compared to the signal duration T_0, then

$$s_k(t) = A_k m(t) \cos(\omega_c t - \theta_k)$$

where $\theta_k = \omega_c \tau_k$. Then the composite received signal (no noise) is

$$s(t) = \sum_k s_k(t)$$

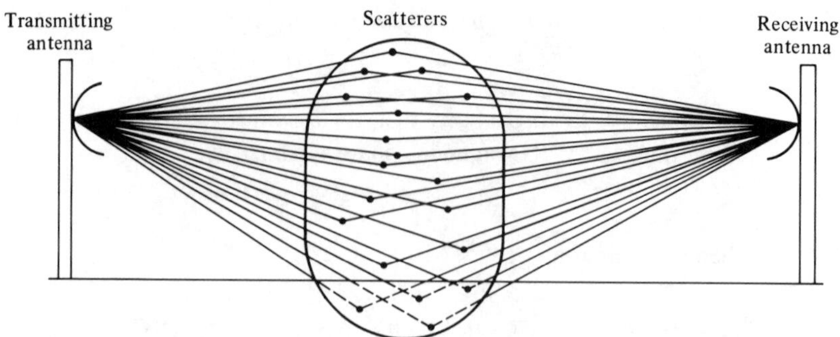

FIGURE 8-21 Scattering model for a fading channel.

or

$$s(t) = Am(t)\cos(\omega_c t - \theta) \tag{8-118}$$

where A and θ are random variables that depend on the statistics of A_k and θ_k.

The composite received signal plus noise is

$$r(t) = Am(t)\cos(\omega_c t - \theta) + n_w(t) \tag{8-119}$$

We will assume that $n_w(t)$ is AWGN as used previously. We will also assume that A and θ are independent where A has a Rayleigh PDF and θ has a uniform PDF. Under these assumptions, the channel is said to be a *Rayleigh fading channel.* The θ is identical to the random-phase variable that was used in the random-phase model studied in the previous sections.

The Rayleigh fading channel is valid for the case when each path contributes the same amount of energy to the composite received signal. For the case where one path dominates (e.g., when the direct path contributes much more energy), the PDF of A would be Rician (see Prob. 6-46).

The optimum MAP receiver is obtained by observing that if A were known, the decision rule would be identical to that for the random-phase problem where the modulation signal set, $\{m_i(t)\}$, is replaced by $\{Am_i(t)\}$. Thus, using (8-95), the MAP decision rule is to choose $\tilde{m} = m_k$ when

$$I_0\left(\frac{AD_i}{N_0}\right) \times \exp\left[-\frac{A^2E_{m_i}}{2N_0}\right] \tag{8-120}$$

is maximized for $i = k$. Furthermore, when we assume that the energy in all the $m_i(t)$ signals is the same [i.e., $E_m = E_{m_i}$, $i = 1, 2, \ldots, N$], (8-120) becomes

$$I_0\left(\frac{AD_i}{N_0}\right) \tag{8-121}$$

Since $I_0(z)$ is a monotonically increasing function for $z \geq 0$, the MAP decision rule is to choose $\tilde{m} = m_k$ when AD_i/N_0 is maximized. Furthermore, since $A > 0$ (because A is a Rayleigh distributed random variable) and $N_0 > 0$, this is equivalent to maximizing D_i, regardless of the value of the unknown A. Thus *for the*

Rayleigh fading channel with AWGN, the MAP receiver is identical to that obtained for the random-phase case for equal-energy signaling (see Figs. 8-19 and 8-20).

EXAMPLE 8-8 PROBABILITY OF ERROR FOR A FADING CHANNEL WITH FSK SIGNALING

The probability of error will now be evaluated for orthogonal binary signaling over a Rayleigh fading channel. Of course, orthogonal FSK signaling is an example of a binary orthogonal signaling scheme.

The probability of error is

$$P(E) = \int_0^{\infty} P(E|A)f(A)\, dA \tag{8-122}$$

where, from (8-112),

$$P(E|A) = \tfrac{1}{2}e^{-A^2E_m/(4N_0)} \tag{8-123}$$

For Rayleigh fading, the PDF of A is

$$f(A) = \frac{A}{\sigma^2}\, e^{-A^2/(2\sigma^2)}, \qquad A > 0 \tag{8-124}$$

where σ^2 is a parameter that will be evaluated later. Substituting (8-123) and (8-124) into (8-122), we find that the probability of error becomes

$$P(E) = \frac{1}{2}\int_0^{\infty} \frac{A}{\sigma^2} e^{-[E_m/(4N_0)+1/(2\sigma^2)]A^2}\, dA$$

which reduces to

$$P(E) = \frac{1}{2}\left[\frac{1}{1 + \frac{1}{2}(\sigma^2E_m)(1/N_0)}\right] \tag{8-125}$$

The parameter σ^2 will now be related to the received signal energy. From (8-118), the received signal energy is

$$E_s = \frac{\overline{A^2}E_m}{2} = \frac{E_m}{2}\int_0^{\infty} A^2f(A)\, dA$$

Using (8-124), we evaluate the integral and get

$$E_s = \sigma^2E_m \tag{8-126}$$

By substituting (8-126) into (8-125), the probability of error for binary orthogonal signaling over a Rayleigh fading channel with AWGN is

$$P(E) = \frac{1}{2}\left[\frac{1}{1 + \frac{1}{2}(E_s/N_0)}\right] \tag{8-127}$$

when the optimum (MAP) receiver is used. For binary signaling $P(E)$ is identical to the BER.

In Fig. 8-22 the error performances of several optimum receiving systems are compared for the case of binary orthogonal signals. The curve for the Rayleigh fading channel is a plot of (8-127). The BER, P_e, is also shown for a nonfading channel for both cases of noncoherent (random-phase) detection and coherent (known-phase) detection. Equations (8-116) and (7-47) were used to obtain these curves for the noncoherent and coherent cases. Obviously, Rayleigh fading is very detrimental to the performance of a communication system, even one with an optimum receiver.

Diversity reception is used to combat the detrimental effect of fading. Diversity reception embodies any or all of the following.

- *Space diversity* uses multiple receiver sites to combine the received signals. When the signal at one site fades, there is a high probability that the S/N at some of the other sites will be large so that the composite $P(E)$ can be kept small.
- *Time diversity* uses the retransmission of the message a number of times. The effect of a low received S/N at one time is counterbalanced by high received S/N at other times, thus providing a low composite $P(E)$.
- *Frequency diversity* uses multiple frequency channels. When the signal fades in one channel, it is counterbalanced by a high received S/N in other channels, thus providing a low composite $P(E)$.

Optimum diversity (MAP) receivers can also be obtained [Wozencraft and Jacobs, 1965].

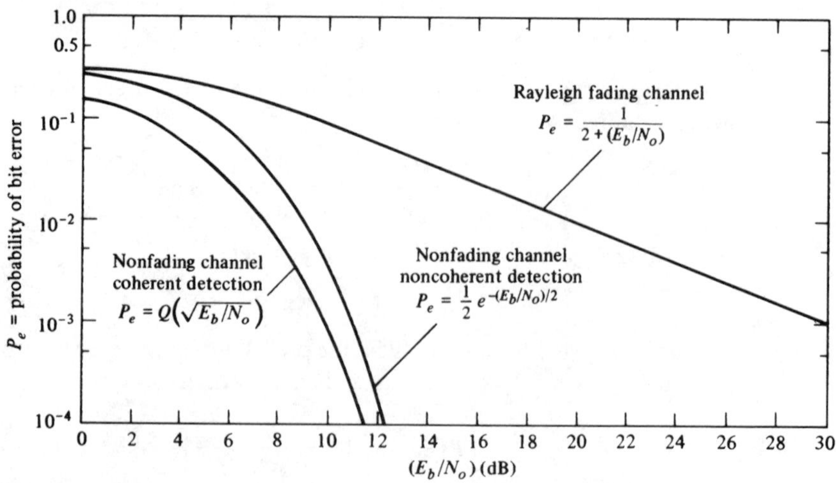

FIGURE 8-22 Comparison of optimum receiver performance for FSK signaling over fading and nonfading channels.

8-7 Bit Error Rate for DPSK Signaling

In Sec. 7-4 the differential phase-shift-keying (DPSK) technique was discussed, and there it was indicated that the equation for the BER would be derived in this chapter. This derivation has been postponed until now because the theory that has just been developed in Sec. 8-6 will be used to obtain the BER for DPSK signaling.

We will now show that DPSK signaling is another example of binary orthogonal signaling with an unknown carrier phase. As indicated in Chapter 7, DPSK uses the past data bit as a reference for detection of the present data bit—that is, differential encoding is used. The transmitted signal is

$$s_i(t) = m_i(t) \cos \omega_c t, \qquad i = 1, 2 \tag{8-128a}$$

where the differentially encoded message signals are

$$m_1(t) = p(t) + m_0(t) \qquad \text{(binary 1)} \tag{8-128b}$$

$$m_2(t) = p(t) - m_0(t) \qquad \text{(binary 0)} \tag{8-128c}$$

This is illustrated by the rectangular signaling waveforms shown in Fig. 8-23. $p(t)$ is the past data bit and $\pm m_0(t)$ is the present data bit. Using (8-128b), the waveform representing a binary 1, $m_1(t)$, is shown in Fig. 8-23c, and the signal representing a binary 0, $m_2(t)$, is shown in Fig. 8-23d. T_0 is the time it takes to transmit the whole message $m_i(t)$ (which includes the receiver reference). T_b is the time it takes to send one bit, $\pm m_0(t)$. We see that $p(t)$ and $m_0(t)$ are orthogonal functions since they are absolutely time limited to an interval of T_0 sec and the intervals for $p(t)$ and $m_0(t)$ do not overlap. Of course, in practice $p(t)$ and $m_0(t)$ may not have rectangular pulse shapes; in fact, they may be "rounded" to reduce the bandwidth of the transmitted signal, $s(t)$. Assume that the energy in one bit of the transmitted signal $s(t)$ is E_b; then the corresponding orthonormal functions for the series expansion of $m_i(t)$ are

$$\varphi_1(t) = \frac{1}{\sqrt{2E_b}} p(t) \tag{8-129a}$$

and

$$\varphi_2(t) = \frac{1}{\sqrt{2E_b}} m_0(t) \tag{8-129b}$$

From (8-128) and (8-129), the corresponding vectors are

$$\mathbf{p} = \left(\sqrt{2E_b}, 0\right) \tag{8-130a}$$

$$\mathbf{m}_0 = \left(0, \sqrt{2E_b}\right) \tag{8-130b}$$

$$\mathbf{m}_1 = \left(\sqrt{2E_b}, \sqrt{2E_b}\right) \tag{8-130c}$$

$$\mathbf{m}_2 = \left(\sqrt{2E_b}, -\sqrt{2E_b}\right) \tag{8-130d}$$

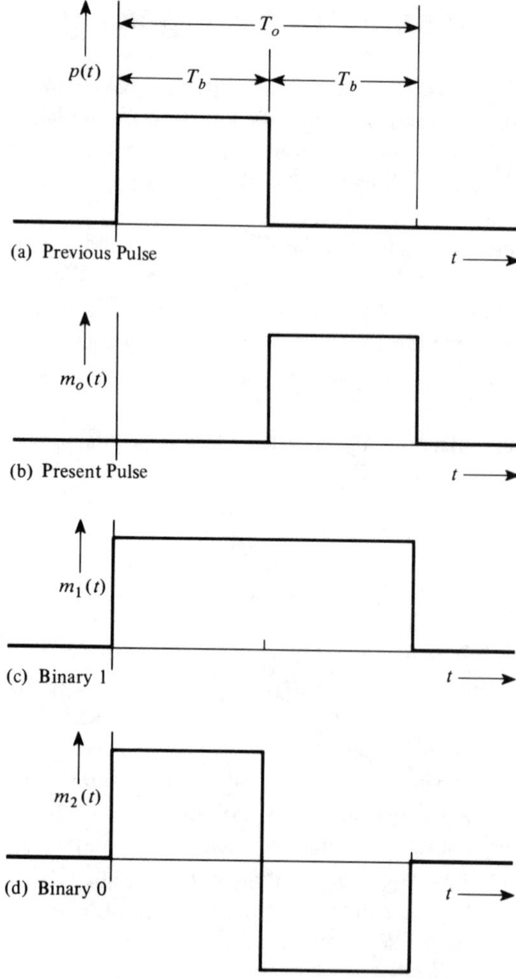

FIGURE 8-23 Typical DPSK waveforms.

These vectors are illustrated in Fig. 8-24. Obviously, $\mathbf{m}_1$ and $\mathbf{m}_2$ are orthogonal signals, each having energy $4E_b$. Thus

$$E_m = 4E_b, \qquad i = 1, 2 \tag{8-131}$$

where E_b is the energy per bit in the signal $s(t)$.

For noncoherent optimum (MAP) reception of these binary orthogonal signals over an AWGN channel, the probability of error is obtained by substituting (8-131) into (8-112). This gives the $P(E)$ for the case of equally likely signaling. Furthermore, since $M = 2$, $P(E) = P_e$, where P_e is the probability of bit error. Thus the BER for DPSK signaling over an AWGN channel is

$$P_e = \tfrac{1}{2}e^{-E_b/N_0} \tag{8-132}$$

for the optimum noncoherent receiver.

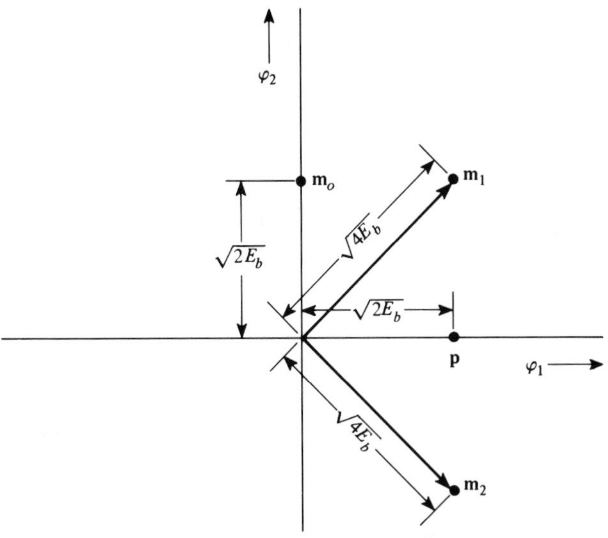

FIGURE 8-24 Vector signals for DPSK signaling.

The optimum receiver structure is obtained by implementing the MAP decision rule. From (8-104) the MAP receiver chooses $\tilde{m} = m_1$ when

$$(\mathbf{x} \cdot \mathbf{m}_1)^2 + (\mathbf{y} \cdot \mathbf{m}_1)^2 > (\mathbf{x} \cdot \mathbf{m}_2)^2 + (\mathbf{y} \cdot \mathbf{m}_2)^2 \tag{8-133}$$

x and **y** are the vectors for the quadrature components of the complex envelope of the input RF waveform $r(t)$. From (8-128),

$$\mathbf{m}_1 = \mathbf{p} + \mathbf{m}_0 \tag{8-134a}$$

and

$$\mathbf{m}_2 = \mathbf{p} - \mathbf{m}_0 \tag{8-134b}$$

Substituting (8-134) into (8-133), we get

$$[\mathbf{x} \cdot (\mathbf{p} + \mathbf{m}_0)]^2 + [\mathbf{y} \cdot (\mathbf{p} + \mathbf{m}_0)]^2 > [\mathbf{x} \cdot (\mathbf{p} - \mathbf{m}_0)]^2 + [\mathbf{y} \cdot (\mathbf{p} - \mathbf{m}_0)]^2$$

This reduces to choosing $\tilde{m} = m_1$ when

$$D \triangleq (\mathbf{x} \cdot \mathbf{p})(\mathbf{x} \cdot \mathbf{m}_0) + (\mathbf{y} \cdot \mathbf{p})(\mathbf{y} \cdot \mathbf{m}_0) > 0 \tag{8-135}$$

The MAP decision rule of (8-135) is implemented by using a receiver that contains two matched filters and a product (phase) detector, as shown in Fig. 8-25. The input to the receiver is

$$r(t) = \text{Re}\{g(t)e^{j(\omega_c t - \theta)}\}$$

where the complex envelope of the input is

$$g(t) = x(t) + jy(t)$$

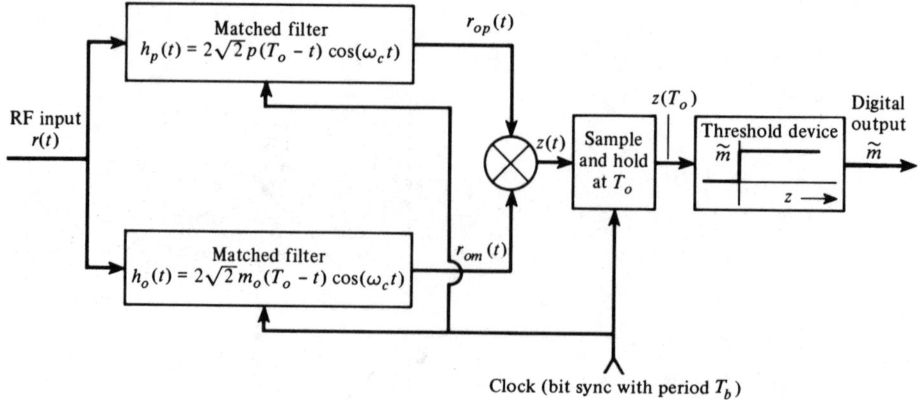

FIGURE 8-25 Optimum noncoherent receiver for DPSK signaling.

The complex envelope of the output of the upper matched filter is

$$g_{op}(t) = \tfrac{1}{2}[x(t) + jy(t)] * 2\sqrt{2}p(T_0 - t)$$

At the sampling time T_0, this complex envelope is

$$g_{op}(T_0) = \sqrt{2}\left[\int_0^{T_0} x(\lambda)p(\lambda)\,d\lambda + j\int_0^{T_0} y(\lambda)p(\lambda)\,d\lambda\right]$$

or

$$g_{op}(T_0) = \sqrt{2}[\mathbf{x}\cdot\mathbf{p} + j\mathbf{y}\cdot\mathbf{p}] \tag{8-136a}$$

Similarly, the complex envelope for the output of the lower matched filter at the sampling time is

$$g_{om}(T_0) = \sqrt{2}[\mathbf{x}\cdot\mathbf{m}_0 + j\mathbf{y}\cdot\mathbf{m}_0] \tag{8-136b}$$

The output of the multiplier (phase detector) is

$$\begin{aligned} z(t) &= r_{op}(t)r_{om}(t) \\ &= \text{Re}\{g_{op}(t)e^{j(\omega_c t-\theta)}\}\,\text{Re}\{g_{om}(t)e^{j(\omega_c t-\theta)}\} \\ &= \tfrac{1}{2}\,\text{Re}\{g_{op}(t)g_{om}^*(t)\} + \tfrac{1}{2}\,\text{Re}\{g_{op}(t)g_{om}(t)e^{j(2\omega_c t-2\theta)}\} \end{aligned}$$

In practice, the circuit passes only the baseband frequencies, so that the relevant multiplier output is

$$z(t) = \tfrac{1}{2}\,\text{Re}\{g_{op}(t)g_{om}^*(t)\}$$

This waveform is sampled at $t = T_0$, so the sample-and-hold output is

$$z(T_0) = \tfrac{1}{2}\,\text{Re}\{g_{op}(T_0)g_{om}^*(T_0)\} \tag{8-137a}$$

Substituting (8-136) into (8-137a), we get

$$z(T_0) = (\mathbf{x}\cdot\mathbf{p})(\mathbf{x}\cdot\mathbf{m}_0) + (\mathbf{y}\cdot\mathbf{p})(\mathbf{y}\cdot\mathbf{m}_0) \tag{8-137b}$$

Comparing this result with (8-135), we have

$$z(T_0) = D \tag{8-137c}$$

The threshold device chooses $\tilde{m} = m_1$ (high level out) when $z(T_0) = D > 0$ and $\tilde{m} = m_2$ (low level out) when $z(T_0) = D < 0$. Consequently, the MAP decision rule of (8-135) is realized by the receiver configuration of Fig. 8-25. This optimum receiver configuration may be compared with a suboptimum DPSK receiver that was given in Fig. 7-12a or the equivalent optimum receiver shown in Fig. 7-12b.

8-8 Summary

Any one of a number of criteria may be chosen to optimize the performance of a digital receiver. Some of these criteria are: the Bayes criterion, the minimax criterion, the Neyman–Pearson criterion, and the maximum a posteriori (MAP) criterion.

Vector signal space concepts were used to implement the MAP criterion. For the case of an AWGN channel, the MAP criterion is to choose $\tilde{m} = m_k$ when

$$D_i = \mathbf{r} \cdot \mathbf{s}_i + c_i, \qquad i = 1, 2, \ldots, M \tag{8-138}$$

is maximized for $i = k$. M is the number of available source messages. Here c_i is a constant that depends on the noise level, the probability of transmitting the ith signal, and the energy of the ith signal. $\mathbf{r} = (r_1, r_2, \ldots, r_N)$ is the N-dimensional vector corresponding to the received waveform $r(t)$.

$$r(t) = \sum_{j=1}^{N} r_j \varphi_j(t) \tag{8-139}$$

where $\{\varphi_j(t)\}$ is the set of orthonormal functions that is used to represent the signal waveforms $\{s_i(t)\}$. $\mathbf{s}_i = (s_{i1}, s_{i2}, \ldots, s_{iN})$ is the signal vector for the ith signal waveform that corresponds to the message m_i, where

$$s_i(t) = \sum_{j=1}^{N} s_{ij} \varphi_j(t) \tag{8-140}$$

The minimum number of dimensions needed to represent the M signal waveforms is $N \leq M$, where the minimum value for N can be realized by using the Gram–Schmidt orthogonalization procedure.

The optimum (MAP) receiver structure for the AWGN channel may be realized in any one of four canonical forms. They involve using either N correlators or matched filters, or either M correlators or matched filters. For bandpass signaling these optimum receivers require a synchronous (coherent) carrier phase reference. Bit synchronization is required for all cases.

The probability of error $P(E)$ was evaluated for MAP receivers that are designed for binary signals, QAM signals, and PCM signals. The union bound was used to obtain an upper bound on $P(E)$ for PPM (orthogonal) signaling.

Noncoherent MAP receivers were obtained by assuming that the carrier phase was unknown. Optimum noncoherent receiver structures were obtained that incorporated matched filters and envelope detectors. $P(E)$ was evaluated for optimum noncoherent detection of FSK.

Optimum receivers were also found for an AWGN channel with Rayleigh fading. The $P(E)$ for the fading case was found to be much poorer than that for the nonfading case. This suggests the use of diversity receivers on Rayleigh fading channels.

The BER (bit error rate) for DPSK signaling was evaluated using the MAP theory for unknown phase (noncoherent) signals. In addition, the optimum DPSK receiver structure was found.

PROBLEMS

8-1 The signal waveforms for $M = 3$ messages are

$$s_1(t) = \begin{cases} 5, & 0 < t \leq 10 \\ 0, & t \text{ elsewhere} \end{cases}$$

$$s_2(t) = \begin{cases} \cos(10\pi t), & 0 < t \leq 10 \\ 0, & t \text{ elsewhere} \end{cases}$$

and

$$s_3(t) = \begin{cases} [\cos(5\pi t)]^2, & 0 < t \leq 10 \\ 0, & t \text{ elsewhere} \end{cases}$$

Using the Gram–Schmidt procedure:

(a) Find an orthonormal function set $\{\varphi_j(t)\}$ that can be used to represent these signals.

(b) Find the corresponding signal vectors, $\mathbf{s}_1$, $\mathbf{s}_2$, and $\mathbf{s}_3$.

8-2 Using the results of Prob. 8-1,

(a) Find the energy in each of the signals by evaluating $E_{s_i} = |\mathbf{s}_i|^2$, $i = 1, 2, 3$.

(b) Compare these results by evaluating the signal energy directly from the waveforms, where

$$E_{s_i} = \int_0^{10} [s_i(t)]^2 \, dt, \qquad i = 1, 2, 3$$

8-3 Given the signaling waveforms as shown in Fig. P8-3:

(a) Find the minimum number of dimensions required to represent these signals in N-dimensional vector space.

(b) Find the values for the signal vectors $\mathbf{s}_1$, $\mathbf{s}_2$, $\mathbf{s}_3$, and $\mathbf{s}_4$.

(c) Find the maximum distance between the signal vectors.

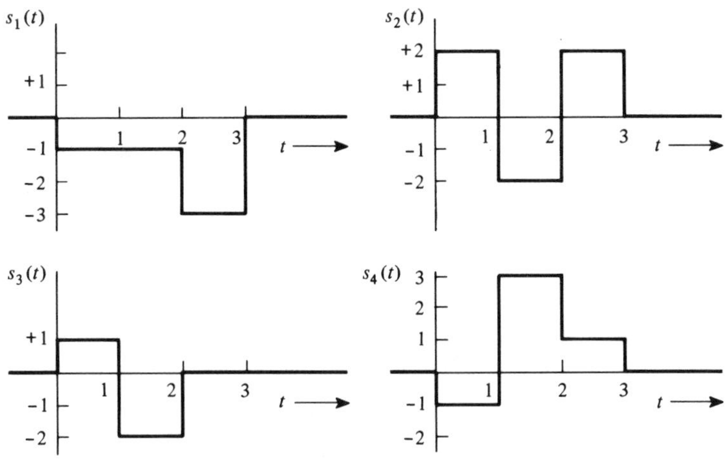

FIGURE P8-3

8-4 Show that (8-64) reduces to (7-33) for the case of OOK signaling.

8-5 Show that (8-64) reduces to (7-38) for the case of BPSK signaling.

8-6 Show that (8-64) reduces to (7-47) for the case of FSK signaling.

8-7 Show that (8-64) reduces to (7-69) for the case of QPSK signaling.

8-8 A BPSK signal is transmitted over an AWGN channel. The source probabilities are $P(m_1) = \frac{1}{3}$ and $P(m_2) = \frac{2}{3}$. The PSD of the noise is $\mathcal{P}_n(f) = N_0/2$.

(a) Find and sketch a block diagram for the optimum MAP receiver structure.

(b) Evaluate the BER for the MAP receiver in terms of E_b/N_0.

8-9 Work Prob. 8-8 for the case of FSK orthogonal signaling when a coherent MAP receiver is used.

8-10 A BPSK signal is transmitted over an AWGN channel. The source probabilities are $P(m_1) = \frac{1}{3}$ and $P(m_2) = \frac{2}{3}$. The PSD of the noise is $\mathcal{P}_n(f) = N_0/2$.

(a) Find the optimum *maximum-likelihood* decision rule.

(b) Find the maximum-likelihood receiver structure.

(c) Find the BER for this maximum-likelihood receiver in terms of E_b/N_0.

8-11 Evaluate $P(E)$ for an AWGN channel and the QAM signal set shown in Fig. P8-11. Assume that the signals are equally likely to occur.

(a) Find $P(E)$ as a function of d and N_0.

(b) Find $P(E)$ as a function of E_s and N_0, where E_s is the peak signal energy.

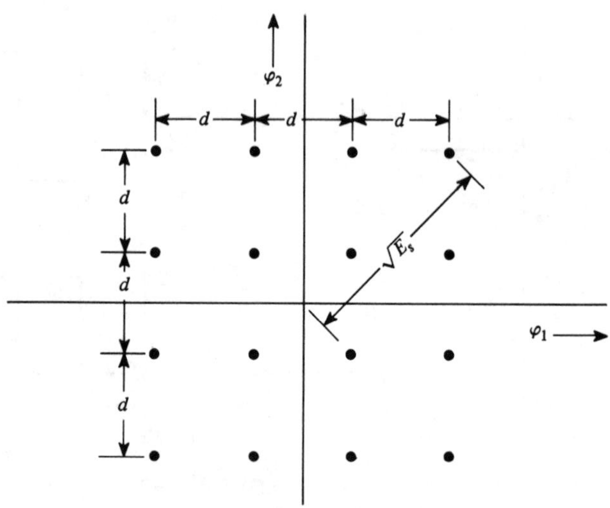

FIGURE P8-11

8-12 Show that the optimum MAP receiver that is designed for PCM signaling is also an optimum *minimax* receiver provided that the MAP receiver is designed for $P(m_i) = 1/M$, $i = 1, 2, \ldots, M$. Here it is assumed that an AWGN channel is used. (*Hint:* See Fig. 8-14 for the $M = 8$ vector signal set.)

8-13 Let $s(t)$ be an FSK orthogonal signal that is transmitted over an AWGN channel. Assume that a binary 1 and a binary 0 are equally likely to occur. Find the union bound for $P(E)$. Compare this with the exact result for $P(E)$.

8-14 A three-dimensional *biorthogonal* signaling scheme is shown in Fig. P8-14. Assuming that these signals are sent over an AWGN channel, find the union bound for the probability of error $P(E)$ as a function of E_s/N_0. Assume that any of the signals is equally likely to occur.

8-15 Plot the union bound for the $P(E)$ versus $(E_b/N_0)_{\text{dB}}$ for several cases of orthogonal signaling over an AWGN channel. Assume equally likely signaling. Specifically:

(a) Plot the union bound for $M = 2$, $M = 2^4$, $M = 2^8$, and $M = 2^{16}$.

(b) Discuss the implications of the threshold effect that occurs as $M \to$ large. [*Hint:* What is a minimum required $(E_b/N_0)_{\text{dB}}$ for communication?]

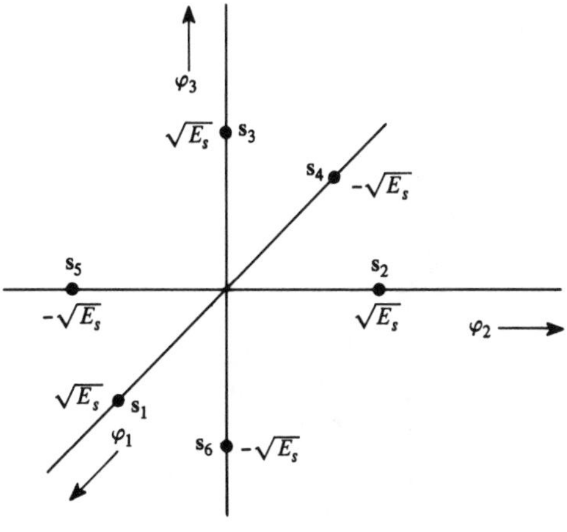

FIGURE P8-14

8-16 Show that (8-85) can be approximated by

$$P(E) < e^{-(K/2)[(E_b/N_0)-2\ln 2]}$$

As $K \rightarrow$ large, what is the minimum required value for E_b/N_0 to maintain communication? Compare this result with that of Shannon's channel capacity theorem.

8-17 A waveform signal set is described by

$$s_i(t) = \begin{Bmatrix} A_c \cos\left[\omega_c t + i\dfrac{2\pi}{M}\right], & 0 < t \le T_0 \\ 0, & t \text{ otherwise} \end{Bmatrix}, \qquad i = 1, 2, \ldots, M$$

This is called *M-ary PSK signaling*. Assume that $M = 6$, $P(m_i) = \frac{1}{6}$, and that the signals are transmitted over an AWGN channel.

(a) Find a set of orthonormal basis functions using the minimum N.
(b) Find the corresponding signal vectors and decision regions for MAP reception.
(c) Set up the equation for $P(E)$, but do not try to evaluate the integrals.
(d) Find a MAP receiver structure.

8-18 For the M-ary PSK signaling set described in Prob. 8-17, find the union bound for $P(E)$. Assume that $M = 6$.

8-19 An FM transmitter is modulated to a large FM index by a four-level, random data, digital waveform. A typical sample function of the modulation is shown in Fig. P8-19. The $\pm$1-V and $\pm$3-V levels occur with a probability of $\frac{1}{4}$ for each. The FM signal is transmitted over an AWGN channel. This is called *M-ary FSK,* where $M = 4$.

(a) Find the optimum coherent MAP receiver structure.

(b) Find $P(E)$ in terms of E_s/N_0, where E_s is the energy of the FM signal over T_0 sec. (*Note:* The union bound may be used if needed.)

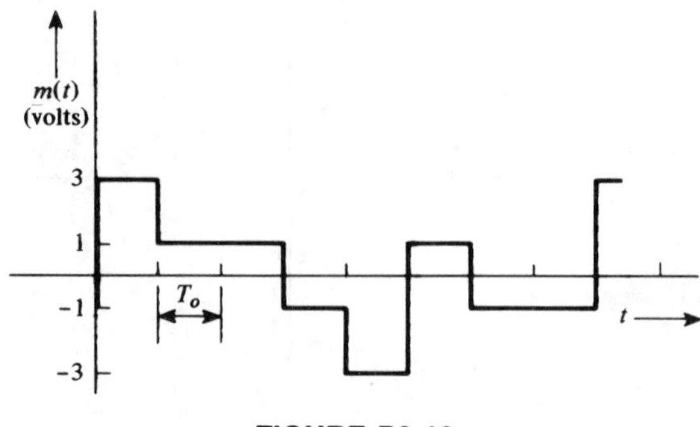

FIGURE P8-19

8-20 Referring to Prob. 8-19, let the carrier frequency be 450 MHz and the carrier level be A_c. The deviation constant of the FM modulator is D_f = 5000 rads/V-sec. Find the PSD for the transmitted digital signal if $T_0 = 1$ msec. (*Hint:* Use the PSD approximation that is valid for high-index FM signals.)

8-21 The information from a signal source with $M = 8$ equally likely messages is to be sent over a communication system. There is AWGN in the channel. The transmission rate is five messages per second and (E_s/N_0) is 10 dB. Compare the *error performance* for PCM and PPM signaling. That is:
(a) Find $P(E)$ when PCM signaling is used.
(b) Find the union bound on $P(E)$ when PPM signaling is used.

8-22 Refer to Prob. 8-21 and compare the *bandwidth performance* of PCM and PPM signaling techniques. [*Hint:* Use the dimensionality theorem (see Chapter 2).]
(a) Find the bandwidth required for PCM signaling.
(b) Find the bandwidth required for PPM signaling.

8-23 Rework Prob. 8-21 if $M = 256$.

8-24 Referring to Prob. 8-21, compare the *bandwidth performance* of PCM and PPM signaling techniques when $M = 256$. [*Hint:* Use the dimensionality theorem (see Chapter 2).]
(a) Find the bandwidth required for PCM signaling.
(b) Find the bandwidth required for PPM signaling.

8-25 An optimum noncoherent receiver for equal-energy signaling over an AWGN channel is shown in Fig. 8-19. Find a similar receiver structure that is an optimum noncoherent receiver for the case of *unequal*-energy signaling where the energies in the signals are E_{s_i}, i = 1, 2, . . . , M.

8-26 FSK signaling waveforms for $s_1(t)$ and $s_2(t)$ were described by (7-39). Find the corresponding waveforms for $m_1(t)$, $m_2(t)$, $\varphi_1(t)$, and $\varphi_2(t)$, as described by (8-103) for this case of FSK signaling.

8-27 If $f(z)$ is a Gaussian PDF with a mean m_z and a variance σ_z^2, show that $\int_{-\infty}^{\infty} e^{az^2} f(z)\, dz$ is given by the right-hand side of (8-110).

8-28 Starting with (8-118), show that (8-126) is correct.

8-29 The optimum receiver for DPSK signaling was given in Fig. 8-25. Examine the operation of this receiver as follows.

(a) Sketch representative waveforms for $r_{op}(t)$, $r_{om}(t)$, $z(t)$, $z(T_0)$, and $\tilde{m}$ for a typical $r(t)$ that contains a DPSK sequence plus noise.

(b) Explain exactly how the clock signal controls the matched filters and the sample-and-hold circuit.

8-30 Find an optimum MAP receiver for DPSK signaling that uses correlators instead of the matched filters shown in Fig. 8-25. Explain how the clock signal is used to control the circuits in the correlation DPSK receiver.

8-31 Assume that an $M = 4$ equally likely message source is connected to a digital communication system.

(a) Evaluate the WER, $P(E)$, for the case of BPSK signaling as a function of E_s/N_0 where the signal is corrupted by AWGN and is detected on a bit-by-bit basis.

(b) Evaluate the WER, $P(E)$, for the case of QPSK signaling as a function of E_s/N_0 where the signal is corrupted by AWGN and is detected on a message-by-message basis.

(c) Plot $P(E)$ versus E_s/N_0 for these two cases and discuss which signaling method is more desirable.

8-32 An $M = 4$ equally likely message source is connected to a QPSK digital communication system that has AWGN in the channel.

(a) Evaluate the WER, $P(E)$, as a function of E_s/N_0.

(b) Find the BER by using the results of part (a) and (8-2). Compare this BER result with the exact BER result of (7-69). Explain how these results agree or disagree.

8-33 Consider an M-ary amplitude shift keyed (MASK) signal set where $s_i(t) = A_i \cos(\omega_c t)$, $i = 1, 2, \ldots, M$, $0 < t < T$. Let $A_i = iA$, where the signals are equally likely.

(a) Find and sketch a block diagram for a MAP receiver for the case of an AWGN channel with a noise PSD of $N_0/2$.

(b) Find an expression for the probability of error, $P(E)$, as a function of A, T, and M.

(c) Find an expression for the average signal energy, E_s, over T seconds in terms of A, T, and M.

(d) On a log scale, plot $P(E)$ as a function of E_b/N_0 in decibel units over a range of 0 to 25 dB, where $E_b/N_0 = E_s/(N_0 \log_2 M)$ for the cases of $M = 4$, 8, 16, and 64.

8-34 Compare the performance of DPSK, DPSK with a Rayleigh fading channel, BPSK, and noncoherent detected FSK by plotting the BER for each as a function of E_b/N_0 in decibel units over a range of 0 to 20 dB. Use a log scale for the BER.

Mathematical Techniques, Identities, and Tables

A-1 Trigonometry

Definitions

$$\sin x = \frac{e^{jx} - e^{-jx}}{2j} \tag{A-1}$$

$$\cos x = \frac{e^{jx} + e^{-jx}}{2} \tag{A-2}$$

$$\tan x = \frac{\sin x}{\cos x} = \frac{e^{jx} - e^{-jx}}{j(e^{jx} + e^{-jx})} \tag{A-3}$$

Trigonometric Identities

$$e^{\pm jx} = \cos x \pm j \sin x \qquad \text{(Euler's theorem)} \tag{A-4}$$

$$\cos(x \pm y) = \cos x \cos y \mp \sin x \sin y \tag{A-5}$$

$$\sin(x \pm y) = \sin x \cos y \pm \cos x \sin y \tag{A-6}$$

$$\cos\left(x \pm \frac{\pi}{2}\right) = \mp \sin x \tag{A-7}$$

$$\sin\left(x \pm \frac{\pi}{2}\right) = \pm \cos x \tag{A-8}$$

$$\cos 2x = \cos^2 x - \sin^2 x \tag{A-9}$$

$$\sin 2x = 2 \sin x \cos x \tag{A-10}$$

$$2 \cos x \cos y = \cos(x - y) + \cos(x + y) \tag{A-11}$$

$$2 \sin x \sin y = \cos(x - y) - \cos(x + y) \tag{A-12}$$

$$2 \sin x \cos y = \sin(x - y) + \sin(x + y) \tag{A-13}$$

$$2 \cos^2 x = 1 + \cos 2x \tag{A-14}$$

$$2 \sin^2 x = 1 - \cos 2x \tag{A-15}$$

$$4\cos^3 x = 3\cos x + \cos 3x \tag{A-16}$$

$$4\sin^3 x = 3\sin x - \sin 3x \tag{A-17}$$

$$8\cos^4 x = 3 + 4\cos 2x + \cos 4x \tag{A-18}$$

$$8\sin^4 x = 3 - 4\cos 2x + \cos 4x \tag{A-19}$$

$$A\cos x - B\sin x = R\cos(x + \theta) \tag{A-20a}$$

$$\text{where } R = \sqrt{A^2 + B^2} \tag{A-20b}$$

$$\theta = \tan^{-1}(B/A) \tag{A-20c}$$

$$A = R\cos\theta \tag{A-20d}$$

$$B = R\sin\theta \tag{A-20e}$$

A-2 Differential Calculus

Definition

$$\frac{df(x)}{dx} = \lim_{\Delta x \to 0} \frac{f(x + (\Delta x/2)) - f(x - (\Delta x/2))}{\Delta x} \tag{A-21}$$

Differentiation Rules

$$\frac{du(x)v(x)}{dx} = u(x)\frac{dv(x)}{dx} + v(x)\frac{du(x)}{dx} \qquad \text{(products)} \tag{A-22}$$

$$\frac{d\left(\dfrac{u(x)}{v(x)}\right)}{dx} = \frac{v(x)\dfrac{du(x)}{dx} - u(x)\dfrac{dv(x)}{dx}}{v^2(x)} \qquad \text{(quotient)} \tag{A-23}$$

$$\frac{du[v(x)]}{dx} = \frac{du}{dv}\frac{dv}{dx} \qquad \text{(chain rule)} \tag{A-24}$$

Derivative Table

$$\frac{d[x^n]}{dx} = nx^{n-1} \tag{A-25}$$

$$\frac{d\sin ax}{dx} = a\cos ax \tag{A-26}$$

$$\frac{d\cos ax}{dx} = -a\sin ax \tag{A-27}$$

$$\frac{d \tan ax}{dx} = \frac{a}{\cos^2 ax} \tag{A-28}$$

$$\frac{d \sin^{-1} ax}{dx} = \frac{a}{\sqrt{1 - (ax)^2}} \tag{A-29}$$

$$\frac{d \cos^{-1} ax}{dx} = -\frac{a}{\sqrt{1 - (ax)^2}} \tag{A-30}$$

$$\frac{d \tan^{-1} ax}{dx} = \frac{a}{1 + (ax)^2} \tag{A-31}$$

$$\frac{d[e^{ax}]}{dx} = ae^{ax} \tag{A-32}$$

$$\frac{d[a^x]}{dx} = a^x \ln a \tag{A-33}$$

$$\frac{d(\ln x)}{dx} = \frac{1}{x} \tag{A-34}$$

$$\frac{d(\log_a x)}{dx} = \frac{1}{x} \log_a e \tag{A-35}$$

$$\frac{d\left[\int_{a(x)}^{b(x)} f(\lambda,x)\, d\lambda\right]}{dx} = f(b(x),x)\frac{db(x)}{dx} - f(a(x),x)\frac{da(x)}{dx} + \int_{a(x)}^{b(x)} \frac{\partial f(\lambda,x)}{\partial x}\, d\lambda \qquad \text{(Leibniz's rule)} \tag{A-36}$$

A-3 Indeterminate Forms

If $\lim_{x\to a} f(x)$ is of the form

$$\frac{0}{0}, \quad \frac{\infty}{\infty}, \quad 0 \cdot \infty, \quad \infty - \infty, \quad 0^0, \quad \infty^0, \quad 1^\infty$$

then

$$\lim_{x\to a} f(x) = \lim_{x\to a}\left[\frac{N(x)}{D(x)}\right] = \lim_{x\to a}\left[\frac{(dN(x)/dx)}{(dD(x)/dx)}\right] \qquad \text{(l'Hospital's rule)} \tag{A-37}$$

where $N(x)$ is the numerator of $f(x)$, $D(x)$ is the denominator of $f(x)$, $N(a) = 0$, and $D(a) = 0$.

A-4 Integral Calculus

Definition

$$\int f(x)\,dx = \lim_{\Delta x \to 0} \left\{ \sum_n [f(n\,\Delta x)]\,\Delta x \right\} \tag{A-38}$$

Integration Techniques

1. Change in variable. Let $v = u(x)$:

$$\int_a^b f(x)\,dx = \int_{u(a)}^{u(b)} \left(\frac{f(x)}{dv/dx} \bigg|_{x=u^{-1}(v)} \right) dv \tag{A-39}$$

2. Integration by parts

$$\int u\,dv = uv - \int v\,du \tag{A-40}$$

3. Integral tables
4. Complex variable techniques
5. Numerical methods

A-5 Integral Tables

Indefinite Integrals

$$\int (a + bx)^n\,dx = \frac{(a + bx)^{n+1}}{b(n + 1)}, \qquad 0 < n \tag{A-41}$$

$$\int \frac{dx}{a + bx} = \frac{1}{b} \ln|a + bx| \tag{A-42}$$

$$\int \frac{dx}{(a + bx)^n} = \frac{-1}{(n - 1)b(a + bx)^{n-1}}, \qquad 1 < n \tag{A-43}$$

$$\int \frac{dx}{c + bx + ax^2} = \begin{cases} \dfrac{2}{\sqrt{4ac - b^2}} \tan^{-1}\left(\dfrac{2ax + b}{\sqrt{4ac - b^2}} \right), & b^2 < 4ac \\ \dfrac{1}{\sqrt{b^2 - 4ac}} \ln \left| \dfrac{2ax + b - \sqrt{b^2 - 4ac}}{2ax + b + \sqrt{b^2 - 4ac}} \right|, & b^2 > 4ac \\ \dfrac{-2}{2ax + b}, & b^2 = 4ac \end{cases} \tag{A-44}$$

$$\int \frac{x\,dx}{c + bx + ax^2} = \frac{1}{2a} \ln|ax^2 + bx + c| - \frac{b}{2a} \int \frac{dx}{c + bx + ax^2} \tag{A-45}$$

$$\int \frac{dx}{a^2 + b^2x^2} = \frac{1}{ab} \tan^{-1}\left(\frac{bx}{a}\right) \tag{A-46}$$

$$\int \frac{x\, dx}{a^2 + x^2} = \frac{1}{2} \ln (a^2 + x^2) \tag{A-47}$$

$$\int \frac{x^2\, dx}{a^2 + x^2} = x - a \tan^{-1}\left(\frac{x}{a}\right) \tag{A-48}$$

$$\int \frac{dx}{(a^2 + x^2)^2} = \frac{x}{2a^2(a^2 + x^2)} + \frac{1}{2a^3} \tan^{-1}\left(\frac{x}{a}\right) \tag{A-49}$$

$$\int \frac{x\, dx}{(a^2 + x^2)^2} = \frac{-1}{2(a^2 + x^2)} \tag{A-50}$$

$$\int \frac{x^2\, dx}{(a^2 + x^2)^2} = \frac{-x}{2(a^2 + x^2)} + \frac{1}{2a} \tan^{-1}\left(\frac{x}{a}\right) \tag{A-51}$$

$$\int \frac{dx}{(a^2 + x^2)^3} = \frac{x}{4a^2(a^2 + x^2)^2} + \frac{3x}{8a^4(a^2 + x^2)} + \frac{3}{8a^5} \tan^{-1}\left(\frac{x}{a}\right) \tag{A-52}$$

$$\int \frac{x^2\, dx}{(a^2 + x^2)^3} = \frac{-x}{4(a^2 + x^2)^2} + \frac{x}{8a^2(a^2 + x^2)} + \frac{1}{8a^3} \tan^{-1}\left(\frac{x}{a}\right) \tag{A-53}$$

$$\int \frac{x^4\, dx}{(a^2 + x^2)^3} = \frac{a^2x}{4(a^2 + x^2)^2} - \frac{5x}{8(a^2 + x^2)} + \frac{3}{8a} \tan^{-1}\left(\frac{x}{a}\right) \tag{A-54}$$

$$\int \frac{dx}{(a^2 + x^2)^4} = \frac{x}{6a^2(a^2 + x^2)^3} + \frac{5x}{24a^4(a^2 + x^2)^2} + \frac{5x}{16a^6(a^2 + x^2)} + \frac{5}{16a^7} \tan^{-1}\left(\frac{x}{a}\right) \tag{A-55}$$

$$\int \frac{x^2\, dx}{(a^2 + x^2)^4} = \frac{-x}{6(a^2 + x^2)^3} + \frac{x}{24a^2(a^2 + x^2)^2} + \frac{x}{16a^4(a^2 + x^2)} + \frac{1}{16a^5} \tan^{-1}\left(\frac{x}{a}\right) \tag{A-56}$$

$$\int \frac{x^4\, dx}{(a^2 + x^2)^4} = \frac{a^2x}{6(a^2 + x^2)^3} - \frac{7x}{24(a^2 + x^2)^2} + \frac{x}{16a^2(a^2 + x^2)} + \frac{1}{16a^3} \tan^{-1}\left(\frac{x}{a}\right) \tag{A-57}$$

$$\int \frac{dx}{a^4 + x} = \frac{1}{4a^3\sqrt{2}} \ln\left(\frac{x^2 + ax\sqrt{2} + a^2}{x^2 - ax\sqrt{2} + a^2}\right) + \frac{1}{2a^3\sqrt{2}} \tan^{-1}\left(\frac{ax\sqrt{2}}{a^2 - x^2}\right) \tag{A-58}$$

$$\int \frac{x^2\,dx}{a^4 + x^4} = -\frac{1}{4a\sqrt{2}} \ln\left(\frac{x^2 + ax\sqrt{2} + a^2}{x^2 - ax\sqrt{2} + a^2}\right) + \frac{1}{2a\sqrt{2}} \tan^{-1}\left(\frac{ax\sqrt{2}}{a^2 - x^2}\right) \tag{A-59}$$

$$\int \cos x\,dx = \sin x \tag{A-60}$$

$$\int x \cos x\,dx = \cos x + x \sin x \tag{A-61}$$

$$\int x^2 \cos x\,dx = 2x \cos x + (x^2 - 2) \sin x \tag{A-62}$$

$$\int \sin x\,dx = -\cos x \tag{A-63}$$

$$\int x \sin x\,dx = \sin x - x \cos x \tag{A-64}$$

$$\int x^2 \sin x\,dx = 2x \sin x - (x^2 - 2) \cos x \tag{A-65}$$

$$\int e^{ax}\,dx = \frac{e^{ax}}{a}, \qquad a \text{ real or complex} \tag{A-66}$$

$$\int xe^{ax}\,dx = e^{ax}\left(\frac{x}{a} - \frac{1}{a^2}\right), \qquad a \text{ real or complex} \tag{A-67}$$

$$\int x^2e^{ax}\,dx = e^{ax}\left(\frac{x^2}{a} - \frac{2x}{a^2} + \frac{2}{a^3}\right), \qquad a \text{ real or complex} \tag{A-68}$$

$$\int x^3e^{ax}\,dx = e^{ax}\left(\frac{x^3}{a} - \frac{3x^2}{a^2} + \frac{6x}{a^3} - \frac{6}{a^4}\right), \qquad a \text{ real or complex} \tag{A-69}$$

$$\int e^{ax} \sin x\,dx = \frac{e^{ax}}{a^2 + 1}(a \sin x - \cos x) \tag{A-70}$$

$$\int e^{ax} \cos x\,dx = \frac{e^{ax}}{a^2 + 1}(a \cos x + \sin x) \tag{A-71}$$

Definite Integrals

$$\int_0^\infty \frac{x^{m-1}}{1 + x^n}\,dx = \frac{\pi/n}{\sin(m\pi/n)}, \qquad n > m > 0 \tag{A-72}$$

$$\int_0^\infty x^{\alpha-1}e^{-x}\,dx = \Gamma(\alpha), \qquad \alpha > 0 \tag{A-73a}$$

where $$\Gamma(\alpha + 1) = \alpha\Gamma(\alpha) \tag{A-73b}$$

$$\Gamma(1) = 1;\ \Gamma(\tfrac{1}{2}) = \sqrt{\pi} \tag{A-73c}$$

$$\Gamma(n) = (n-1)! \qquad \text{if } n \text{ is a positive integer} \tag{A-73d}$$

$$\int_0^\infty x^{2n} e^{-ax^2}\, dx = \frac{1 \cdot 3 \cdot 5 \cdot \cdots \cdot (2n-1)}{2^{n+1}a^n} \sqrt{\frac{\pi}{a}} \tag{A-74}$$

$$\int_{-\infty}^\infty e^{-a^2x^2+bx}\, dx = \frac{\sqrt{\pi}}{a} e^{b^2/(4a^2)}, \qquad a > 0 \tag{A-75}$$

$$\int_0^\infty e^{-ax} \cos bx\, dx = \frac{a}{a^2 + b^2}, \qquad a > 0 \tag{A-76}$$

$$\int_0^\infty e^{-ax} \sin bx\, dx = \frac{b}{a^2 + b^2}, \qquad a > 0 \tag{A-77}$$

$$\int_0^\infty e^{-a^2x^2} \cos bx\, dx = \frac{\sqrt{\pi} e^{-b^2/4a^2}}{2a}, \qquad a > 0 \tag{A-78}$$

$$\int_0^\infty x^{\alpha-1} \cos bx\, dx = \frac{\Gamma(\alpha)}{b^\alpha} \cos \tfrac{1}{2}\pi\alpha, \qquad 0 < \alpha < 1,\ b > 0 \tag{A-79}$$

$$\int_0^\infty x^{\alpha-1} \sin bx\, dx = \frac{\Gamma(\alpha)}{b^\alpha} \sin \tfrac{1}{2}\pi\alpha, \qquad 0 < |\alpha| < 1,\ b > 0 \tag{A-80}$$

$$\int_0^\infty x e^{-ax^2} I_k(bx)\, dx = \frac{1}{2a} e^{b^2/4a} \tag{A-81a}$$

where $$\mathrm{I}_k(bx) = \frac{1}{\pi} \int_0^\pi e^{bx\cos\theta} \cos k\theta\, d\theta \tag{A-81b}$$

$$\int_0^\infty \frac{\sin x}{x}\, dx = \int_0^\infty \mathrm{Sa}(x)\, dx = \frac{\pi}{2} \tag{A-82}$$

$$\int_0^\infty \left(\frac{\sin x}{x}\right)^2 dx = \int_0^\infty \mathrm{Sa}^2(x)\, dx = \frac{\pi}{2} \tag{A-83}$$

$$\int_0^\infty \frac{\cos ax}{b^2 + x^2}\, dx = \frac{\pi}{2b} e^{-ab}, \qquad a > 0,\ b > 0 \tag{A-84}$$

$$\int_0^\infty \frac{x \sin ax}{b^2 + x^2}\, dx = \frac{\pi}{2} e^{-ab}, \qquad a > 0,\ b > 0 \tag{A-85}$$

$$\int_{-\infty}^\infty e^{\pm j2\pi yx}\, dx = \delta(y) \tag{A-86}$$

A-6 Series Expansions

Finite Series

$$\sum_{n=1}^{N} n = \frac{N(N+1)}{2} \tag{A-87}$$

$$\sum_{n=1}^{N} n^2 = \frac{N(N+1)(2N+1)}{6} \tag{A-88}$$

$$\sum_{n=1}^{N} n^3 = \frac{N^2(N+1)^2}{4} \tag{A-89}$$

$$\sum_{n=0}^{N} a^n = \frac{a^{N+1}-1}{a-1} \tag{A-90}$$

$$\sum_{n=0}^{N} \frac{N!}{n!\,(N-n)!} x^n y^{N-n} = (x+y)^N \tag{A-91}$$

$$\sum_{n=0}^{N} e^{j(\theta+n\phi)} = \frac{\sin[(N+1)\phi/2]}{\sin(\phi/2)} e^{j[\theta+(N\phi/2)]} \tag{A-92}$$

$$\sum_{k=0}^{N} \binom{N}{k} a^{N-k} b^k = (a+b)^N \tag{A-93a}$$

where

$$\binom{N}{k} = \frac{N!}{(N-k)!\,k!} \tag{A-93b}$$

Infinite Series

$$f(x) = \sum_{n=0}^{\infty} \left(\frac{f^{(n)}(a)}{n!}\right)(x-a)^n \qquad \text{(Taylor's series)} \tag{A-94}$$

$$f(x) = \sum_{n=-\infty}^{\infty} c_n e^{jn\omega_0 x} \qquad \text{for } a \le x \le a+T \qquad \text{(Fourier series)} \tag{A-95a}$$

where

$$c_n = \frac{1}{T}\int_a^{a+T} f(x) e^{-jn\omega_0 x}\, dx \tag{A-95b}$$

and

$$\omega_0 = \frac{2\pi}{T} \tag{A-95c}$$

$$e^x = \sum_{n=0}^{\infty} \frac{x^n}{n!} \tag{A-96}$$

$$\sin x = \sum_{n=0}^{\infty} \frac{(-1)^n x^{2n+1}}{(2n+1)!} \tag{A-97}$$

$$\cos x = \sum_{n=0}^{\infty} \frac{(-1)^n x^{2n}}{(2n)!} \tag{A-98}$$

A-7 Hilbert Transform Pairs[†]

$$\hat{x}(t) \triangleq x(t) * \frac{1}{\pi t} = \frac{1}{\pi}\int_{-\infty}^{\infty} \frac{x(\lambda)}{t-\lambda}\,d\lambda \tag{A-99}$$

Function	Hilbert Transform
1. $x(at+b)$	$\hat{x}(at+b)$
2. $x(t)+y(t)$	$\hat{x}(t)+\hat{y}(t)$
3. $\dfrac{d^n x(t)}{dt^n}$	$\dfrac{d^n}{dt^n}\hat{x}(t)$
4. A constant	0
5. $\dfrac{1}{t}$	$-\pi\delta(t)$
6. $\sin(\omega_0 t+\theta)$	$-\cos(\omega_0 t+\theta)$
7. $\dfrac{\sin at}{at} = \mathrm{Sa}(at)$	$-\dfrac{1}{2\pi}\,at\,\mathrm{Sa}^2(at)$
8. $e^{\pm j\omega_0 t}$	$\mp je^{\pm j\omega_0 t}$
9. $\delta(t)$	$\dfrac{1}{\pi t}$
10. $\dfrac{a}{\pi(t^2+a^2)}$	$\dfrac{t}{\pi(t^2+a^2)}$
11. $\Pi\left(\dfrac{t}{T}\right) \triangleq \begin{cases}1, & \lvert t\rvert \le T/2\\ 0, & t \text{ elsewhere}\end{cases}$	$\dfrac{1}{\pi}\ln\left\lvert\dfrac{2t+T}{2t-T}\right\rvert$

A-8 The Dirac Delta Function

Definition: The *Dirac delta function* $\delta(x)$, also called the *unit impulse function,* satisfies *both* of the following conditions:

$$\int_{-\infty}^{\infty} \delta(x)\,dx = 1 \tag{A-100}$$

$$\delta(x) = \begin{cases}\infty, & x = 0\\ 0, & x \neq 0\end{cases} \tag{A-101}$$

Consequently, $\delta(x)$ is a "singular" function.[‡]

[†]Fourier transform theorems are given in Table 2-1, and Fourier transform pairs are given in Table 2-2.

[‡]The Dirac delta function is not an ordinary function since it is really undefined at $x = 0$. However, it is described by the mathematical theory of distributions [Bremermann, 1965].

Properties of Dirac Delta Functions

1. $\delta(x)$ can be expressed in terms of the limit of some ordinary functions such that (in the limit of some parameter) the ordinary function satisfies the definition for $\delta(x)$. For example,

$$\delta(x) = \lim_{\sigma \to 0} \left(\frac{1}{\sqrt{2\pi}\sigma} e^{-x^2/(2\sigma^2)} \right) \tag{A-102}$$

$$= \lim_{a \to \infty} \left[\frac{a}{\pi} \left(\frac{\sin ax}{ax} \right) \right] \tag{A-103}$$

For these two examples, $\delta(-x) = \delta(x)$, so for *these* cases $\delta(x)$ is said to be an *even-sided delta function*. The even-sided delta function is used throughout this text except when specifying the PDF of a discrete random variable.

$$\delta(x) = \begin{cases} \lim\limits_{a \to \infty} (ae^{ax}), & x \le 0 \\ 0, & x > 0 \end{cases} \tag{A-104}$$

This is an example of a single-sided delta function; in particular, this is a *left-sided delta function*. This type of delta function is used to specify the PDF of a discrete point of a random variable (see Appendix B).

2. Sifting property:

$$\int_{-\infty}^{\infty} w(x)\, \delta(x - x_0)\, dx = w(x_0) \tag{A-105}$$

3. For *even-sided* delta functions

$$\int_a^b w(x)\delta(x - x_0)\, dx = \begin{cases} 0, & x_0 < a \\ \frac{1}{2}w(a), & x_0 = a \\ w(x_0), & a < x_0 < b \\ \frac{1}{2}w(b), & x_0 = b \\ 0, & x_0 > b \end{cases} \tag{A-106}$$

where $b > a$.

4. For *left-sided* delta functions

$$\int_a^b w(x)\, \delta(x - x_0)\, dx = \begin{cases} 0, & x_0 \le a \\ w(x_0), & a < x_0 \le b \\ 0, & x_0 > b \end{cases} \tag{A-107}$$

where $b > a$.

5. $$\int_{-\infty}^{\infty} w(x)\, \delta^{(n)}(x - x_0)\, dx = (-1)^n\, w^{(n)}(x_0) \tag{A-108}$$

where the superscript (n) denotes the nth derivative with respect to x.

6. The Fourier transform of $\delta(x)$ is unity. That is,

$$\mathscr{F}[\delta(x)] = 1 \tag{A-109a}$$

Conversely,

$$\delta(x) = \mathscr{F}^{-1}[1] \tag{A-109b}$$

7. Scaling property:
$$\delta(ax) = \frac{1}{|a|}\,\delta(x) \tag{A-110}$$

8. For *even-sided* delta functions,

$$\delta(x) = \int_{-\infty}^{\infty} e^{\pm j2\pi xy}\,dy \tag{A-111}$$

A-9 Tabulation of Sa(x) = (sin x)/x

x	Sa(x)	Sa²(x)	x	Sa(x)	Sa²(x)
0.0	1.0000	1.0000	5.2	−0.1699	0.0289
0.2	0.9933	0.9867	5.4	−0.1431	0.0205
0.4	0.9735	0.9478	5.6	−0.1127	0.0127
0.6	0.9411	0.8856	5.8	−0.0801	0.0064
0.8	0.8967	0.8041	6.0	−0.0466	0.0022
1.0	0.8415	0.7081	6.2	−0.0134	0.0002
1.2	0.7767	0.6033	2π	0.0000	0.0000
1.4	0.7039	0.4955	6.4	0.0182	0.0003
1.6	0.6247	0.3903	6.6	0.0472	0.0022
1.8	0.5410	0.2927	6.8	0.0727	0.0053
2.0	0.4546	0.2067	7.0	0.0939	0.0088
2.2	0.3675	0.1351	7.2	0.1102	0.0122
2.4	0.2814	0.0792	7.4	0.1214	0.0147
2.6	0.1983	0.0393	7.6	0.1274	0.0162
2.8	0.1196	0.0143	7.8	0.1280	0.0164
3.0	0.0470	0.0022	8.0	0.1237	0.0153
π	0.0000	0.0000	8.2	0.1147	0.0132
3.2	−0.0182	0.0003	8.4	0.1017	0.0104
3.4	−0.0752	0.0056	8.6	0.0854	0.0073
3.6	−0.1229	0.0151	8.8	0.0665	0.0044
3.8	−0.1610	0.0259	9.0	0.0458	0.0021
4.0	−0.1892	0.0358	9.2	0.0242	0.0006
4.2	−0.2075	0.0431	9.4	0.0026	0.0000
4.4	−0.2163	0.0468	3π	0.0000	0.0000
4.6	−0.2160	0.0467	9.6	−0.0182	0.0003
4.8	−0.2075	0.0431	9.8	−0.0374	0.0014
5.0	−0.1918	0.0368	10.0	−0.0544	0.0030

A-10 Tabulation of $Q(z)$

$$Q(z) \triangleq \frac{1}{\sqrt{2\pi}} \int_z^\infty e^{-\lambda^2/2}\, d\lambda \tag{A-112}$$

For $z \geq 3$:

$$Q(z) \approx \frac{1}{\sqrt{2\pi}\, z} e^{-z^2/2} \qquad \text{(see Fig. B-7)} \tag{A-113}$$

Also,

$$Q(-z) = 1 - Q(z) \tag{A-114}$$

$$Q(z) = \tfrac{1}{2}\operatorname{erfc}\left(\frac{z}{\sqrt{2}}\right) = \tfrac{1}{2}\left[1 - \operatorname{erf}\left(\frac{z}{\sqrt{2}}\right)\right] \tag{A-115a}$$

where

$$\operatorname{erfc}(x) \triangleq \frac{2}{\sqrt{\pi}} \int_x^\infty e^{-\lambda^2}\, d\lambda \tag{A-115b}$$

$$\operatorname{erf}(x) = \frac{2}{\sqrt{\pi}} \int_0^x e^{-\lambda^2}\, d\lambda \tag{A-115c}$$

For $z \geq 0$, a rational function approximation is [Ziemer and Tranter, 1990]

$$Q(z) = \frac{e^{-z^2/2}}{\sqrt{2\pi}}(b_1 t + b_2 t^2 + b_3 t^3 + b_4 t^4 + b_5 t^5) \tag{A-116}$$

where $t = 1/(1 + pz)$, with $p = 0.2316419$,

and

$$\begin{aligned} b_1 &= 0.31981530 \\ b_2 &= -0.356563782 \\ b_3 &= 1.781477937 \\ b_4 &= -1.821255978 \\ b_5 &= 1.330274429 \end{aligned}$$

z	$Q(z)$	z	$Q(z)$
0.0	0.50000	2.0	0.02275
0.1	0.46017	2.1	0.01786
0.2	0.42074	2.2	0.01390
0.3	0.38209	2.3	0.01072
0.4	0.34458	2.4	0.00820
0.5	0.30854	2.5	0.00621
0.6	0.27425	2.6	0.00466
0.7	0.24196	2.7	0.00347
0.8	0.21186	2.8	0.00256
0.9	0.18406	2.9	0.00187
1.0	0.15866	3.0	0.00135
1.1	0.13567	3.1	0.00097
1.2	0.11507	3.2	0.00069
1.3	0.09680	3.3	0.00048
1.4	0.08076	3.4	0.00034
1.5	0.06681	3.5	0.00023
1.6	0.05480	3.6	0.00016
1.7	0.04457	3.7	0.00011
1.8	0.03593	3.8	0.00007
1.9	0.02872	3.9	0.00005
		4.0	0.00003

Also see Fig. B-7 for a plot of $Q(z)$.

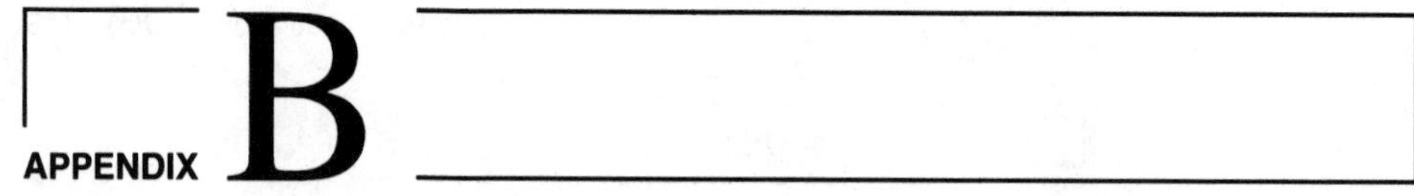

APPENDIX B

Probability and Random Variables

B-1 Introduction

The need for probability theory arises in every scientific discipline since it is impossible to be exactly sure of values that are obtained by measurement. For example, we might say that we are 90% sure that a voltage is within ± 0.1 V of a 5-V level. This is a statistical description of the voltage parameter as opposed to a deterministic description whereby we might define the voltage as being exactly 5 V.

This appendix is intended to be a short course in probability and random variables. Numerous excellent books cover this topic in more detail [Papoulis, 1984 and 1991; Peebles, 1987; Shanmugan and Breipohl, 1988]. However, this appendix will provide a good introduction to the topic for the new student or a quick review for the student who is already knowledgeable in this area.

If probability and random variables are new topics for you, you will soon realize that to understand them, you will need to master many new definitions that seem to be introduced all at once. It is important to memorize these definitions in order to have a vocabulary that can be used when conversing with others about statistical results. In addition, you must develop a feeling for the engineering application of these definitions and theorems. If you accomplish this at the beginning, you can grasp more complicated statistical ideas easily.

B-2 Sets

Definition: A *set* is a collection (or class) of objects.

The largest set or the all-embracing set of objects under consideration in an experiment is called the *universal set*. All other sets under consideration in the experiment are subsets or *events* of the universal set. This is illustrated in Fig. B-1a, where a *Venn diagram* is given. For example, M might denote the set containing all types of milk shakes and B might denote the subset of blueberry milk shakes. Thus B is contained in M, which is denoted by $B \subset M$. In parts b and c of this figure two sets, A and B, are shown. There are two basic ways of describing the combination of set A and set B. These are called *intersection* and *union*.

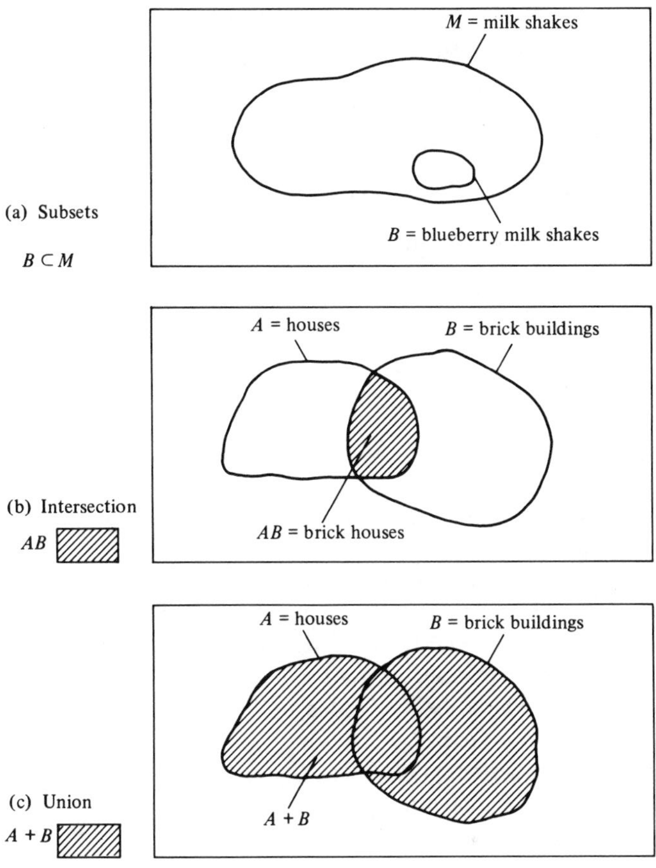

FIGURE B-1 Venn diagram.

Definition: The *intersection* of set A and set B, denoted by AB, is the set of elements that is common to A and B. (Mathematicians use the notation $A \cap B$.)

The intersection of A and B is analogous to the AND operation of digital logic. For example, if A denotes houses and B denotes brick buildings, then the event $C = AB$ would denote only brick houses. This is illustrated by Fig. B-1b.

Definition: The *union* of set A and set B, denoted by $A + B$, is the set that contains all the elements of A or all the elements of B or both. (Mathematicians use the notation $A \cup B$.)

The union of A and B is analogous to the OR operation of digital logic. Continuing with the example just given, $D = A + B$ would be the set that contains all brick buildings, all houses, as well as all brick houses. This is illustrated in Fig. B-1c.

The events A and B are called *simple events*, and the events $C = AB$ and $D = A + B$ are called *compound events* since they are logical functions of simple events.

B-3 Probability and Relative Frequency

Simple Probability

The probability of an event A, denoted by $P(A)$, may be defined in terms of the relative frequency of A occurring in n trials.

DEFINITION[†]

$$P(A) = \lim_{n\to\infty}\left(\frac{n_A}{n}\right) \tag{B-1}$$

where n_A is the number of times that A occurs in n trials.

In practice, n is taken to be some reasonable number such that a larger value for n would give approximately the same value for $P(A)$. For example, suppose that a coin is tossed 40 times and that the heads event, denoted by A, occurs 19 times. Then the probability of a head would be evaluated as approximately $P(A) = \frac{19}{40}$, where the true value of $P(A) = 0.5$ would have been obtained if $n = \infty$ had been used.

From the definition of probability, as given by (B-1), it is seen that all probability functions have the property

$$0 \le P(A) \le 1 \tag{B-2}$$

where $P(A) = 0$ if the event A is a *null event* (does not occur) and $P(A) = 1$ if the event A is a *sure event* (always occurs).

Joint Probability

DEFINITION: The probability of a *joint event*, AB, is

$$P(AB) = \lim_{n\to\infty}\left(\frac{n_{AB}}{n}\right) \tag{B-3}$$

where n_{AB} is the number of times that the event AB occurs in n trials.

In addition, two events, A and B, are said to be *mutually exclusive* if AB is a null event, which implies that $P(AB) \equiv 0$.

[†]Here an engineering approach is used to define probability. Strictly speaking, statisticians have developed probability theory based on three axioms: (1) $P(A) > 0$, (2) $P(S) = 1$ where S is the sure event, and (3) $P(A + B) = P(A) + P(B)$ provided that AB is a null event (i.e., $AB = \emptyset$). Statisticians define $P(A)$ as being any function of A that satisfies these axioms. The engineering definition is consistent with this approach since it satisfies these axioms.

EXAMPLE B-1 EVALUATION OF PROBABILITIES

Let event A denote an auto accident blocking a certain street intersection during a 1-min interval. Let event B denote that it is raining at the street intersection during a 1-min period. Then event $E = AB$ would be a blocked intersection while it is raining, as evaluated in 1-min increments.

Suppose that experimental measurements are tabulated continuously for 1 week, and it is found that $n_A = 25$, $n_B = 300$, $n_{AB} = 20$, and there are $n =$ 10,080 1-min intervals in the week of measurements. ($n_A = 25$ does not mean that there were 25 accidents in a 1-week period, but that the intersection was blocked for 25 1-min periods because of car accidents; similarly for n_B and n_{AB}.) These results indicate that the probability of having a blocked intersection is $P(A) = 0.0025$, the probability of rain is $P(B) = 0.03$, and the probability of having a blocked intersection while it is raining is $P(AB) = 0.002$.

The probability of the union of two events may be evaluated by measuring the compound event directly, or it may be evaluated from the probabilities of simple events as given by the following theorem.

THEOREM: *Let $E = A + B$; then*

$$P(E) = P(A + B) = P(A) + P(B) - P(AB) \tag{B-4}$$

Proof. Let the event A-only occur n_1 times out of n trials, the event B-only occur n_2 times out of n trials, and the event AB occur n_{AB} times. Thus

$$P(A + B) = \lim_{n\to\infty}\left(\frac{n_1 + n_2 + n_{AB}}{n}\right)$$

$$= \lim_{n\to\infty}\left(\frac{n_1 + n_{AB}}{n}\right) + \lim_{n\to\infty}\left(\frac{n_2 + n_{AB}}{n}\right) - \lim_{n\to\infty}\left(\frac{n_{AB}}{n}\right)$$

which is identical to (B-4) since

$$P(A) = \lim_{n\to\infty}\left(\frac{n_1 + n_{AB}}{n}\right),$$

$$P(B) = \lim_{n\to\infty}\left(\frac{n_2 + n_{AB}}{n}\right), \quad \text{and} \quad P(AB) = \lim_{n\to\infty}\left(\frac{n_{AB}}{n}\right)$$

EXAMPLE B-1 (Continued)

The probability of having a blocked intersection or rain occurring, or both, is then

$$P(A + B) = 0.0025 + 0.03 - 0.002 \approx 0.03 \tag{B-5}$$

Conditional Probabilities

DEFINITION: The probability that an event A occurs, given that an event B has also occurred, is denoted by $P(A|B)$ and is defined by

$$P(A|B) = \lim_{n_B \to \infty} \left(\frac{n_{AB}}{n_B} \right) \tag{B-6}$$

EXAMPLE B-1 (Continued)

The probability of a blocked intersection when it is raining is approximately

$$P(A|B) = \tfrac{20}{300} = 0.066 \tag{B-7}$$

THEOREM: *Let $E = AB$; then*

$$P(AB) = P(A)P(B|A) = P(B)P(A|B) \tag{B-8}$$

This is known as Bayes' theorem.

Proof

$$P(AB) = \lim_{n \to \infty} \left(\frac{n_{AB}}{n} \right) = \lim_{\substack{n \to \infty \\ n_A \to \text{large}}} \left(\frac{n_{AB}}{n_A} \frac{n_A}{n} \right) = P(B|A)P(A) \tag{B-9}$$

It is noted that the values obtained for $P(AB)$, $P(B)$, and $P(A|B)$ in Example B-1 can be verified by the use of (B-8).

DEFINITION: Two events, A and B, are said to be *independent* if either

$$P(A|B) = P(A) \tag{B-10}$$

or

$$P(B|A) = P(B) \tag{B-11}$$

Using this definition, we can easily demonstrate that events A and B of Example B-1 are not independent. On the other hand, if event A had been defined as getting a head on a coin toss, while B was the event that it was raining at the intersection, then A and B would be independent. Why?

Using (B-8) and (B-10), we can show that if a set of events $A_1, A_2, \ldots, A_n$, are independent; then a necessary condition is†

$$P(A_1 A_2 \cdots A_n) = P(A_1)P(A_2) \cdots P(A_n) \tag{B-12}$$

B-4 Random Variables

DEFINITION: A real *random variable* is a real-valued function defined on the events (elements) of the probability system.

†Equation (B-12) is not a sufficient condition for $A_1, A_2, \ldots, A_n$ to be independent [Papoulis, 1984, p. 34].

An understanding of why this definition is needed is fundamental to the topic of probability theory. So far we have defined probabilities in terms of events *A*, *B*, *C*, and so on. This method is awkward to use when the sets are objects (such as apples, oranges, etc.) instead of numbers. It is more convenient to describe sets by numerical values so that equations can be obtained as a function of numerical values instead of functions of alphanumeric parameters. This method is accomplished by using the random variable.

EXAMPLE B-2 RANDOM VARIABLE

Referring to Fig. B-2, we can show the mutually exclusive events *A, B, C, D, E, F, G,* and *H* by a Venn diagram. These are all the possible outcomes of an experiment, so the sure event is $S = A + B + C + D + E + F + G + H$. Each of these events is denoted by some convenient value of the random variable *x*, as

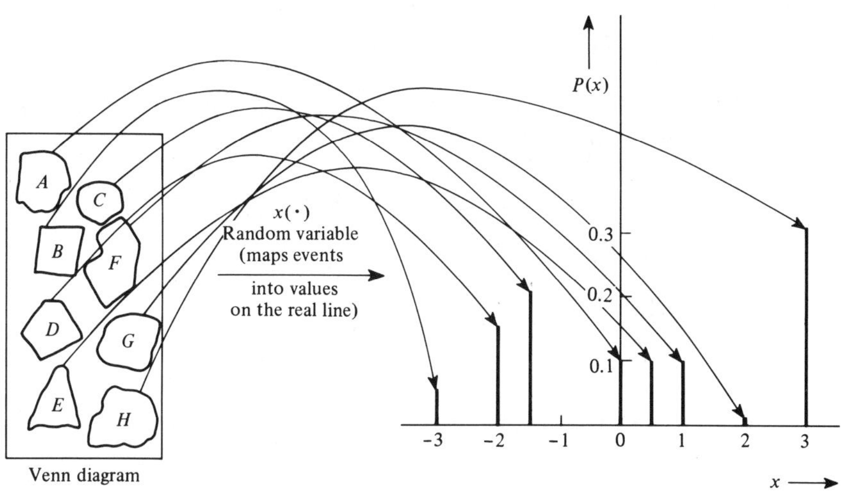

Event [·]	Value of Random Variable $x[\cdot]$	Probability of Event $P(x)$
A	0.0	0.10
B	−3.0	0.05
C	−1.5	0.20
D	−2.0	0.15
E	+0.5	0.10
F	+1.0	0.10
G	+2.0	0.00
H	+3.0	0.30
		Total = 1.00

FIGURE B-2 Random variable and probability functions for Example B-2.

shown in the table in this figure. The assigned values for x may be positive, negative, fractions, or integers as long as they are real numbers. Since all the events are mutually exclusive, using (B-4) yields

$$P(S) = 1 = P(A) + P(B) + P(C) + P(D) + P(E) + P(F) + P(G) + P(H) \tag{B-13}$$

That is, the probabilities have to sum to unity (the probability of a sure event), as shown in the table, and the individual probabilities have been given or measured as shown. For example, $P(C) = P(-1.5) = 0.2$. These values for the probabilities may be plotted as a function of the random variable x, as shown in the graph of $P(x)$. This is a *discrete* (or *point*) distribution since the random variable takes on only discrete (as opposed to continuous) values.

B-5 Cumulative Distribution Functions and Probability Density Functions

Definition: The *cumulative distribution function* (CDF) of the random variable x is given by $F(a)$, where

$$F(a) \triangleq P(x \le a) \equiv \lim_{n\to\infty}\left(\frac{n_{x\le a}}{n}\right) \tag{B-14}$$

where $F(a)$ is a unitless function.

Definition: The *probability density function* (PDF) of the random variable x is given by $f(x)$, where

$$f(x) = \left.\frac{dF(a)}{da}\right|_{a=x} = \left.\frac{dP(x\le a)}{da}\right|_{a=x} = \lim_{\substack{n\to\infty\\ \Delta x\to 0}}\left[\frac{1}{\Delta x}\left(\frac{n_{\Delta x}}{n}\right)\right] \tag{B-15}$$

where $f(x)$ has units of $1/x$.

EXAMPLE B-2 (Continued)

The CDF for this example that was illustrated in Fig. B-2 is easily obtained using (B-14). The resulting CDF is shown in Fig. B-3. Note that the CDF starts at a zero value on the left ($a = -\infty$), and the probability is accumulated until the CDF is unity on the right ($a = +\infty$).

Using (B-15), we obtain the PDF by taking the derivative of the CDF. The result is shown in Fig. B-4. The PDF consists of Dirac delta functions located at the assigned (discrete) values of the random variable and having weights equal to the probabilities of the associated event.[†]

[†]Left-sided delta functions are used here so that if $x = a$ happens to be a discrete point, $F(a) = P(x \le a)$ includes all the probability from $x = -\infty$ up to and *including* the point $x = a$. See Sec. A-8 (Appendix A) for properties of Dirac delta functions.

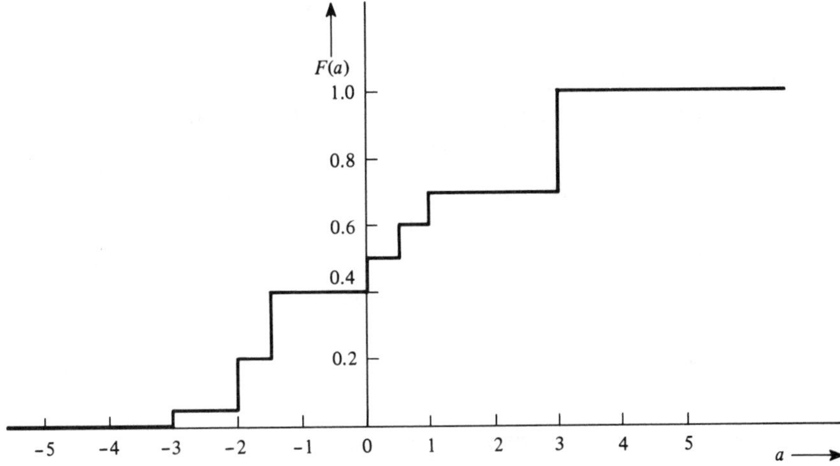

FIGURE B-3 CDF for Example B-2.

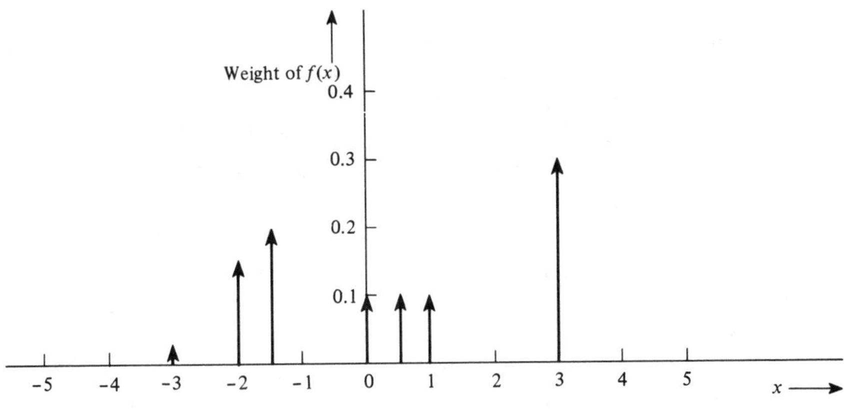

FIGURE B-4 PDF for Example B-2.

Properties of CDFs and PDFs

Some *properties of the CDF* are as follows:

1. $F(a)$ is a nondecreasing function.
2. $F(a)$ is *right-hand* continuous. That is,

$$\lim_{\substack{\epsilon \to 0 \\ \epsilon > 0}} F(a+\epsilon) = F(a)$$

3.

$$F(a) = \lim_{\substack{\epsilon \to 0 \\ \epsilon > 0}} \int_{-\infty}^{a+\epsilon} f(x)\, dx \tag{B-16}$$

4. $0 \leq F(a) \leq 1$.

5. $F(-\infty) = 0$.
6. $F(+\infty) = 1$.

Note that the ϵ is needed to account for a discrete point that might occur at $x = a$. If there is no discrete point at $x = a$, the limit is not necessary.

Some *properties of the PDF* are as follows.

1. $f(x) \geq 0$. That is, $f(x)$ is a nonnegative function.
2. $\int_{-\infty}^{\infty} f(x)\,dx = F(+\infty) = 1$. (B-17)

As we will see later, $f(x)$ may have values larger than unity; however, the area under $f(x)$ is equal to unity. These properties of the CDF and PDF are very useful in checking results of problems. That is, if a CDF or a PDF violates any of these properties, you *know* that an error has been made in the calculations.

Discrete and Continuous Distributions

Example B-2 was an example of a *discrete* (or *point*) *distribution*. That is, the random variable has M discrete values $x_1, x_2, x_3, \ldots, x_M = 7$ in this example. Consequently, the CDF increased only in jumps [i.e., $F(a)$ was discontinuous] as a increased, and the PDF consisted of delta functions located at the discrete values of the random variable. In contrast to this example of a discrete distribution, there are continuous distributions, one of which is illustrated in the following example. If a random variable is allowed to take on any value in some interval, it is a *continuously distributed* random variable in that interval.

EXAMPLE B-3 A CONTINUOUS DISTRIBUTION

Let the random variable denote the voltages that are associated with a collection of a large number of flashlight batteries (1.5-V cells). If the number of batteries in the collection were infinite, the number of different voltage values (events) that we could obtain would be infinite, so that the distributions (PDF and CDF) would be continuous functions. Suppose that, by measurement, the CDF is evaluated first by using $F(a) = P(x \leq a) = \lim_{n\to\infty} (n_{x\leq a}/n)$, where n is the number of batteries in the whole collection and $n_{x\leq a}$ is the number of batteries in the collection with voltages less than or equal to a volts, where a is a parameter. The CDF that might be obtained is illustrated in Fig. B-5a. The associated PDF is obtained by taking the derivative of the CDF, as shown in Fig. B-5b. Note that $f(x)$ exceeds unity for some values of x, but that the area under $f(x)$ is unity (a PDF property). (You might check to see that the other CDF and PDF properties are also satisfied.)

THEOREM

$$F(b) - F(a) = P(x \leq b) - P(x \leq a) = P(a < x \leq b)$$

$$= \lim_{\substack{\epsilon\to 0\\ \epsilon>0}}\left[\int_{a+\epsilon}^{b+\epsilon} f(x)\,dx\right] \tag{B-18}$$

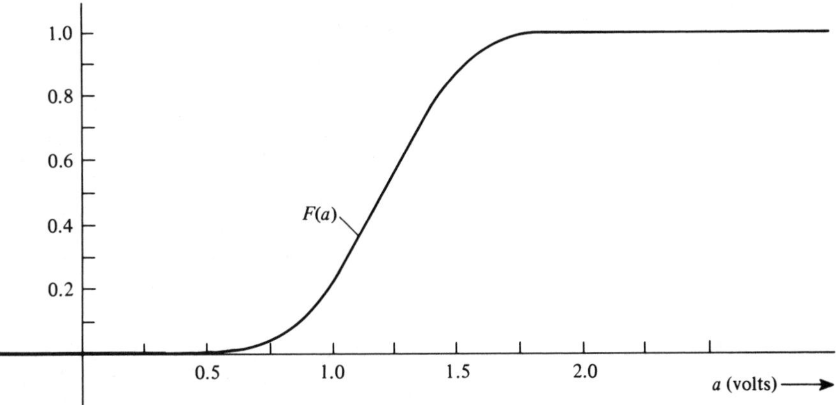

(a) Cumulative Distribution Function

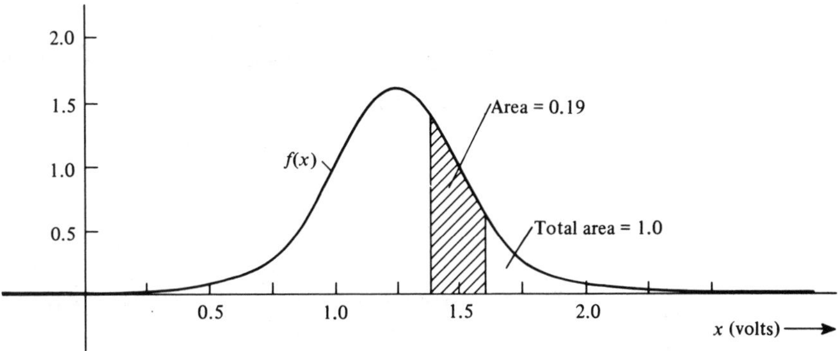

(b) Probability Density Function

FIGURE B-5 CDF and PDF for a continuous distribution (Example B-3).

Proof

$$F(b) - F(a) = \lim_{\substack{\epsilon \to 0 \\ \epsilon > 0}} \left[\int_{-\infty}^{b+\epsilon} f(x)\, dx - \int_{-\infty}^{a+\epsilon} f(x)\, dx \right]$$

$$= \lim_{\substack{\epsilon \to 0 \\ \epsilon > 0}} \left[\int_{-\infty}^{a+\epsilon} f(x)\, dx + \int_{a+\epsilon}^{b+\epsilon} f(x)\, dx - \int_{-\infty}^{a+\epsilon} f(x)\, dx \right]$$

$$= \lim_{\substack{\epsilon \to 0 \\ \epsilon > 0}} \left[\int_{a+\epsilon}^{b+\epsilon} f(x)\, dx \right]$$

EXAMPLE B-3 (Continued)

Suppose that we wanted to calculate the probability of obtaining a battery having a voltage between 1.4 and 1.6 V. Using this theorem, we compute

$$P(1.4 < x \leq 1.6) = \int_{1.4}^{1.6} f(x)\,dx = F(1.6) - F(1.4) = 0.19$$

(using Fig. B-5). We also realize that the probability of obtaining a 1.5-V battery is zero. Why? However, the probability of obtaining a 1.5 V ± 0.1 V battery is 0.19.

In communication systems we have digital signals that may have discrete distributions (one discrete value for each permitted level in a multilevel signal), and we have analog signals and noise that have continuous distributions. We may also have mixed distributions, which contain discrete as well as continuous values. These occur, for example, when a continuous signal and noise are clipped by an amplifier that is driven into saturation.

THEOREM: *If x is discretely distributed, then*

$$f(x) = \sum_{i=1}^{M} P(x_i)\,\delta(x - x_i) \tag{B-19}$$

where M is the number of discrete events and $P(x_i)$ is the probability of obtaining the discrete event x_i.

This theorem was illustrated by Example B-2, where the PDF of this discrete distribution is plotted in Fig. B-4.

THEOREM: *If x is discretely distributed, then*[†]

$$F(a) = \sum_{i=1}^{L} P(x_i) \tag{B-20}$$

where L is the largest integer such that $x_L \leq a$ and $L \leq M$, where M is the number of points in the discrete distribution. Here it is assumed that the discrete points x_i are indexed so that they occur in ascending order of the index. That is, $x_1 < x_2 < x_3 < \cdots < x_M$.

This theorem was illustrated in Fig. B-3, which is a plot of (B-20) for Example B-2.

In electrical engineering problems, the CDFs or PDFs of waveforms are relatively easy to obtain by using the relative frequency approach, as described by

[†]Equation (B-20) is identical to $F(a) = \Sigma_{i=1}^{M} P(x_i)u(a - x_i)$, where $u(y)$ is a unit step function [defined in (2-49)].

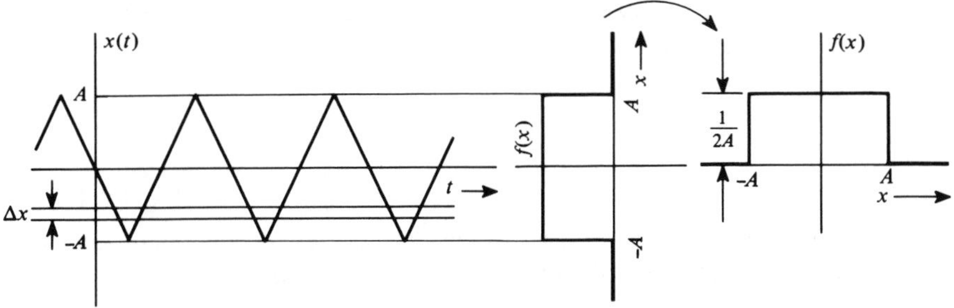

(a) Triangular Waveform and Its Associated PDF

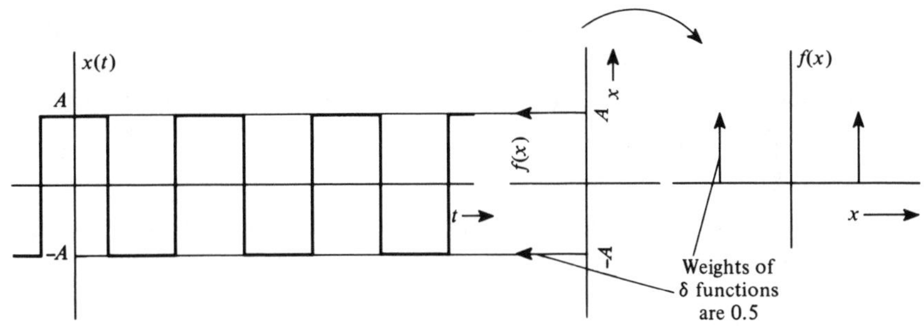

(b) Square Waveform and Its Associated PDF

FIGURE B-6 PDFs for triangular waves and square waves.

(B-14) and (B-15). For example, the PDFs for triangular and square waves are given in Fig. B-6. These are obtained by sweeping a narrow horizontal window, Δx volts wide, vertically across the waveforms and measuring the relative frequency of occurrence of voltages in the Δx window. The time axis is divided into n intervals, and the waveform appears $n_{\Delta x}$ times in these intervals in the Δx window. A rough idea of the PDF of a waveform can also be obtained by looking at the waveform on an oscilloscope. Using *no* horizontal sweep, the intensity of the presentation as a function of the voltage (y axis) gives the PDF. (This assumes that the intensity of the image is proportional to the time that the waveform dwells in the window Δy units wide.) The PDF is proportional to the intensity as a function of y (the random variable).

B-6 Ensemble Average and Moments

Ensemble Average

One of the primary uses of probability theory is to evaluate the average value of a random variable (which represents some physical phenomenon) or to evaluate the average value of some function of the random variable. In general, let the function of the random variable be denoted by $y = h(x)$.

DEFINITION: The *expected value,* which is also called the *ensemble average,* of $y = h(x)$ is given by

$$\overline{y} = \overline{[h(x)]} \triangleq \int_{-\infty}^{\infty} [h(x)]f(x)\,dx \tag{B-21}$$

This definition may be used for discrete as well as continuous random variables. Note that it is a linear operator where the operator is

$$\overline{[\cdot]} = \int_{-\infty}^{\infty} [\cdot]f(x)\,dx \tag{B-22}$$

Other authors denote the ensemble average of y by $E[y]$ or $\langle y \rangle$ as well as using this notation of $\overline{y}$. We will use the $\overline{y}$ notation since it is easier to write and more convenient to use when long equations are being evaluated with a large number of averaging operators.

THEOREM: *If x is a **discretely** distributed random variable, the expected value can be evaluated by using*

$$\overline{y} = \overline{[h(x)]} = \sum_{i=1}^{M} h(x_i)P(x_i) \tag{B-23}$$

where M is the number of discrete points in the distribution.

Proof. Using (B-19) in (B-21), we get

$$\begin{aligned}\overline{h(x)]} &= \int_{-\infty}^{\infty} h(x)\left[\sum_{i=1}^{M} P(x_i)\,\delta(x - x_i)\right] dx \\ &= \sum_{i=1}^{M} P(x_i) \int_{-\infty}^{\infty} h(x)\,\delta(x - x_i)\,dx \\ &= \sum_{i=1}^{M} P(x_i)h(x_i)\end{aligned}$$

EXAMPLE B-4 EVALUATION OF AN AVERAGE

We will now show that (B-23) and, consequently, the definition of expected value as given by (B-21) are consistent with the way in which we usually evaluate averages. Suppose that we have a class of $n = 40$ students who take a test. The resulting test scores are 1 paper with a grade of 100; 2 with scores of 95; 4,

90; 6, 85; 10, 80; 10, 75; 5, 70; 1, 65; and 1 paper with a score of 60. Then the class average will be

$$\bar{x} = \frac{100(1) + 95(2) + 90(4) + 85(6) + 80(10) + 75(10) + 70(5) + 65(1) + 60(1)}{40}$$

$$= 100\left(\frac{1}{40}\right) + 95\left(\frac{2}{40}\right) + 90\left(\frac{4}{40}\right) + 85\left(\frac{6}{40}\right) + 80\left(\frac{10}{40}\right) + 75\left(\frac{10}{40}\right)$$

$$+ 70\left(\frac{5}{40}\right) + 65\left(\frac{1}{40}\right) + 60\left(\frac{1}{40}\right)$$

$$= \sum_{i=1}^{9} x_i P(x_i) = 79.6 \tag{B-24}$$

Moments

Moments are defined as ensemble averages of some specific functions used for $h(x)$. For example, for the rth moment, defined subsequently, let $y = h(x) = (x - x_0)^r$.

Definition: The rth *moment* of the random variable x taken about the point $x = x_0$ is given by

$$\overline{(x - x_0)^r} = \int_{-\infty}^{\infty} (x - x_0)^r f(x)\, dx \tag{B-25}$$

Definition: The *mean* m is the first moment taken about the origin (i.e., $x_0 = 0$). Thus

$$m \triangleq \bar{x} = \int_{-\infty}^{\infty} x f(x)\, dx \tag{B-26}$$

Definition: The *variance* σ^2 is the second moment taken about the mean. Thus

$$\sigma^2 = \overline{(x - \bar{x})^2} = \int_{-\infty}^{\infty} (x - \bar{x})^2 f(x)\, dx \tag{B-27}$$

Definition: The *standard deviation* σ is the square root of the variance. Thus

$$\sigma = \sqrt{\sigma^2} = \sqrt{\int_{-\infty}^{\infty} (x - \bar{x})^2 f(x)\, dx} \tag{B-28}$$

As an engineer, you may recognize the integrals of (B-26) and (B-27) as related to applications in mechanical problems. The mean is equivalent to the center of gravity of a mass that is distributed along a single dimension, where $f(x)$ denotes the mass density as a function of the x axis. The variance is equivalent to the moment of inertia about the center of gravity. However, you might ask: What is the significance of the mean, variance, and other moments in electrical engineering problems? In Chapter 6 it is shown that if x represents a voltage or current waveform, the mean gives the dc value of the waveform. The second moment ($r = 2$) taken about the origin ($x_0 = 0$), which is $\overline{x^2}$, gives the normalized power. σ^2 gives the normalized power in the corresponding ac coupled waveform. Consequently, $\sqrt{\overline{x^2}}$ is the rms value of the waveform and σ is the rms value of the corresponding ac coupled waveform.

In statistical terms, m gives the *center of gravity* of the PDF and σ gives us the *spread* of the PDF about this center of gravity. For example, in Fig. B-5 the voltage distribution for a collection of flashlight batteries is given. The mean is $\overline{x} = 1.25$ V and the standard deviation is $\sigma = 0.25$ V. This figure illustrates a Gaussian distribution that will be studied in detail in Sec. B-7. For this Gaussian distribution the area under $f(x)$ from $x = 1.0$ to 1.5 V, which corresponds to the interval $\overline{x} \pm \sigma$, is 0.68. Thus we conclude that 68% of the batteries have voltages within one standard deviation of the mean value (Gaussian distribution).

There are several ways to specify a number that is used to describe the typical or most common value of x. The *mean* m is one such measure that gives the center of gravity. Another measure is the *median,* which corresponds to the value $x = a$, where $F(a) = \frac{1}{2}$. A third measure is called the *mode,* which corresponds to the value of x where $f(x)$ is a maximum, assuming that the PDF has only one maximum. For the Gaussian distribution, all these measures give the same number, namely, $x = m$. For other types of distributions, the values obtained for the mean, median, and mode will usually be nearly the same number. The variance is also related to the second moment about the origin and the mean, as described by the following theorem.

THEOREM

$$\sigma^2 = \overline{x^2} - (\overline{x})^2 \tag{B-29}$$

A proof of this theorem illustrates how the ensemble average operator notation is used.

Proof

$$\begin{aligned}\sigma^2 &= \overline{(x - \overline{x})^2} \\ &= \overline{[x^2]} - \overline{[2\, x\overline{x}]} + \overline{[(\overline{x})^2]}\end{aligned} \tag{B-30}$$

Since $\overline{[\cdot]}$ is a linear operator, $\overline{[2x\overline{x}]} = 2\overline{x}\,\overline{x} = 2(\overline{x})^2$. Moreover, $(\overline{x})^2$ is a constant and the average value of a constant is the constant itself. That is, for the constant c,

$$\overline{c} = \int_{-\infty}^{\infty} c\, f(x)\, dx = c \int_{-\infty}^{\infty} f(x)\, dx = c \tag{B-31}$$

So, using (B-31) in (B-30), we obtain

$$\sigma^2 = \overline{x^2} - 2(\bar{x})^2 + (\bar{x})^2$$

which is equivalent to (B-29).

Thus there are two ways to evaluate the variance: (1) by use of the definition as given by (B-27), or (2) by use of the theorem of (B-29).

B-7 Examples of Important Distributions

There are numerous types of distributions. Some of the more important ones used in communication and statistical problems are summarized in Table B-1. Here, the equations for their PDF and CDF, a sketch of the PDF, and the formula for the mean and variance are given. These distributions will be studied in more detail in the following paragraphs.

Binomial Distribution

The binomial distribution is useful for describing digital as well as other statistical problems. Its application is best illustrated by an example.

Assume that we have a binary word that is n bits long and that the probability of sending a binary 1 is p. Consequently, the probability of sending a binary 0 is $1 - p$. We want to evaluate the probability of obtaining n-bit words that contain k binary 1's. One such word is k binary 1's followed by $n - k$ binary 0's. The probability of obtaining this word is $p^k(1 - p)^{n-k}$. There are also other n-bit words that contain k binary 1's. In fact, the number of different n-bit words containing k binary 1's is

$$\binom{n}{k} = \frac{n!}{(n-k)!k!} \tag{B-32}$$

(This can be demonstrated by taking a numerical example such as $n = 8$ and $k = 3$.) The symbol $\binom{n}{k}$ is used in algebra to denote the operation described by (B-32) and is read "the combination of n things taken k at a time." Thus the probability of obtaining an n-bit word containing k binary 1's is

$$P(k) = \binom{n}{k} p^k(1 - p)^{n-k} \tag{B-33}$$

If we let the random variable x denote these discrete values, then $x = k$, where k can take on the values 0, 1, 2, . . ., n, and we obtain the *binomial PDF:*

$$f(x) = \sum_{k=0}^{n} P(k)\delta(x - k) \tag{B-34}$$

where $P(k)$ is given by (B-33).

TABLE B-1 Some Distributions and Their Properties

Name of Distribution	Type: $F(a)$ is:	Sketch of PDF	Equation for: Cumulative Distribution Function (CDF)	Equation for: Probability Density Function (PDF)	Mean	Variance
Binomial	Discrete	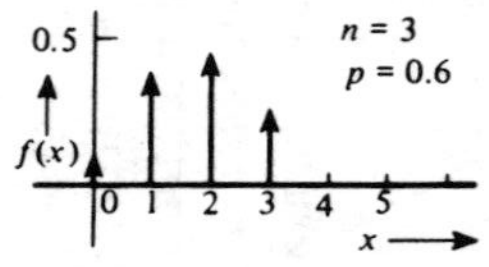 $n = 3$, $p = 0.6$	$F(a) = \sum_{\substack{k=0 \\ m \le a}}^{m} P(k)$ where $P(k) = \binom{n}{k} p^k (1-p)^{n-k}$	$f(x) = \sum_{k=0}^{n} P(k)\,\delta(x-k)$ where $P(k) = \binom{n}{k} p^k (1-p)^{n-k}$	np	$np(1-p)$
Poisson	Discrete	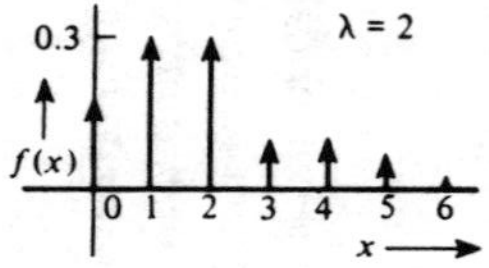 $\lambda = 2$	$F(a) = \sum_{\substack{k=0 \\ m \le a}}^{m} P(k)$ where $P(k) = \frac{\lambda^k}{k!} e^{-\lambda}$	$f(x) = \sum_{k=0}^{\infty} P(k)\,\delta(x-k)$ where $P(k) = \frac{\lambda^k}{k!} e^{-\lambda}$	λ	λ
Uniform	Continuous	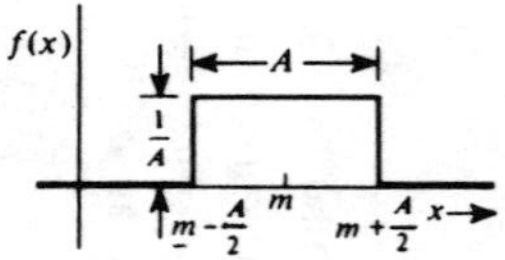	$F(a) = \begin{cases} 0, & a < \left(\frac{2m-A}{2}\right) \\ \frac{1}{A}\left[a - \left(\frac{2m-A}{2}\right)\right], & \lvert a-m \rvert \le \frac{A}{2} \\ 1, & a \ge \left(\frac{2m+A}{2}\right) \end{cases}$	$f(x) = \begin{cases} 0, & x < \left(\frac{2m-A}{2}\right) \\ \frac{1}{A}, & \lvert x-m \rvert \le \frac{A}{2} \\ 0, & x > \left(\frac{2m+A}{2}\right) \end{cases}$	m	$\frac{A^2}{12}$
Gaussian	Continuous	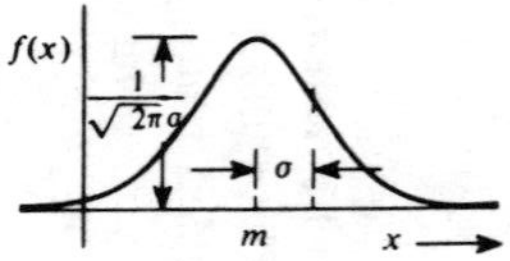	$F(a) = Q\left(\frac{m-a}{\sigma}\right)$ where $Q(\alpha) \triangleq \frac{1}{\sqrt{2\pi}} \int_{\alpha}^{\infty} e^{-x^2/2}\,dx$	$f(x) = \frac{1}{\sqrt{2\pi}\sigma} \exp[-(x-m)^2/2\sigma^2]$	m	σ^2
Sinusoidal	Continuous	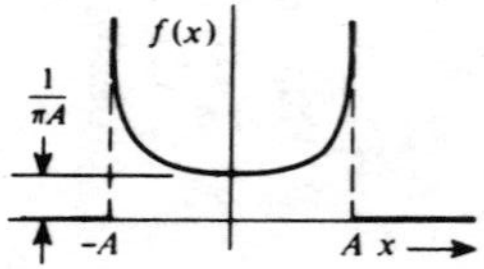	$F(a) = \begin{cases} 0, & a \le -A \\ \frac{1}{\pi}\left[\frac{\pi}{2} + \sin^{-1}\left(\frac{a}{A}\right)\right], & \lvert a \rvert \le A \\ 1, & a \ge A \end{cases}$	$f(x) = \begin{cases} 0, & x < -A \\ \frac{1}{\pi\sqrt{A^2 - x^2}}, & \lvert x \rvert \le A \\ 0, & x > A \end{cases}$	0	$\frac{A^2}{2}$

The name *binomial* comes from the fact that the $P(k)$ are the individual terms in a binomial expansion. That is, letting $q = 1 - p$,

$$(p + q)^n = \sum_{k=0}^{n} \binom{n}{k} p^k q^{n-k} = \sum_{k=0}^{n} P(k) \tag{B-35}$$

The combinations $\binom{n}{k}$, which are also the binomial coefficients, can be evaluated by using Pascal's triangle:

$n = 0$						1					
$n = 1$					1		1				
$n = 2$				1		2		1			
$n = 3$			1		3		3		1		
$n = 4$		1		4		6		4		1	
$n = 5$	1		5		10		10		5		1
⋮						⋮					

For a particular value of n, the combinations $\binom{n}{k}$, for $k = 0, 1, \ldots, n$, are the elements in the nth row. For example, for $n = 3$,

$$\binom{3}{0} = 1, \quad \binom{3}{1} = 3, \quad \binom{3}{2} = 3, \quad \text{and} \quad \binom{3}{3} = 1$$

The mean value of the binomial distribution is evaluated by the use of (B-23).

$$m = \bar{x} = \sum_{k=0}^{n} x_k P(x_k) = \sum_{k=0}^{n} kP(k)$$

or

$$m = \sum_{k=1}^{n} k \binom{n}{k} p^k q^{n-k} \tag{B-36}$$

Using the identity

$$k\binom{n}{k} = \frac{kn!}{(n-k)!k!} = \frac{n(n-1)!}{(n-k)!(k-1)!}$$
$$= n\left[\frac{(n-1)!}{((n-1)-(k-1))!(k-1)!}\right]$$

or

$$k\binom{n}{k} = n\binom{n-1}{k-1} \tag{B-37}$$

we have

$$m = \sum_{k=1}^{n} n \binom{n-1}{k-1} p^k q^{n-k} \tag{B-38}$$

Making a change in the index, let $j = k - 1$. Thus

$$m = \sum_{j=0}^{n-1} n \binom{n-1}{j} p^{j+1} q^{n-(j+1)}$$

$$= np \left[\sum_{j=0}^{n-1} \binom{n-1}{j} p^j q^{(n-1)-j} \right] = np\,[(p+q)^{n-1}] \tag{B-39}$$

Recalling that p and q are probabilities and $p + q = 1$, then we see that $(p+q)^{n-1} = 1$. Thus (B-39) reduces to

$$m = np \tag{B-40}$$

Similarly, it can be shown that the variance is $np(1-p)$ by using $\sigma^2 = \overline{x^2} - (\overline{x})^2$.

Poisson Distribution

The Poisson distribution (Table B-1) is obtained as a limiting approximation of a binomial distribution when n is very large and p is very small, but the product $np = \lambda$ is some reasonable size [Thomas, 1969].

Uniform Distribution

The uniform distribution is

$$f(x) = \begin{cases} 0, & x < \left(\dfrac{2m - A}{2}\right) \\ \dfrac{1}{A}, & |x - m| \le \dfrac{A}{2} \\ 0, & x > \left(\dfrac{2m + A}{2}\right) \end{cases} \tag{B-41}$$

where A is the peak-to-peak value of the random variable. This is illustrated by the sketch in Table B-1. The mean of this distribution is

$$\int_{-\infty}^{\infty} x\,f(x)\,dx = \int_{m-(A/2)}^{m+(A/2)} x \frac{1}{A}\,dx = m \tag{B-42}$$

and the variance is

$$\sigma^2 = \int_{m-(A/2)}^{m+(A/2)} (x - m)^2 \frac{1}{A}\,dx \tag{B-43}$$

Making a change in variable, let $y = x - m$.

$$\sigma^2 = \frac{1}{A} \int_{-A/2}^{A/2} y^2\,dy = \frac{A^2}{12} \tag{B-44}$$

The uniform distribution is useful in describing quantizing noise that is created when an analog signal is converted into a PCM signal, as discussed in Chapter 3.

In Chapter 6 it is shown that it also describes the noise out of a phase detector when the input is Gaussian noise (as described subsequently).

Gaussian Distribution

The Gaussian distribution, which is also called the *normal distribution,* is one of the most important distributions, if not *the* most important. As discussed in Chapter 5, thermal noise has a Gaussian distribution. Numerous other phenomena can also be described by Gaussian statistics, and many theorems have been developed by statisticians that are based on Gaussian assumptions. It cannot be overemphasized that the Gaussian distribution is very important in analyzing both communication problems and problems in statistics. It can also be shown that the Gaussian distribution can be obtained as the limiting form of the binomial distribution when n becomes large, while holding the mean $m = np$ finite, and letting the variance $\sigma^2 = np(1 - p)$ be much larger than unity [Feller, 1957; Papoulis, 1984].

DEFINITION: The Gaussian distribution is

$$f(x) = \frac{1}{\sqrt{2\pi}\,\sigma} e^{-(x-m)^2/(2\sigma^2)} \tag{B-45}$$

where m is the mean and σ^2 is the variance.

A sketch of (B-45) is given in Fig. B-5, together with the CDF for the Gaussian random variable where $m = 1.25$ and $\sigma = 0.25$. The Gaussian PDF is symmetrical about $x = m$, with the area under the PDF being $\frac{1}{2}$ for $(-\infty \le x \le m)$ and $\frac{1}{2}$ for $(m \le x \le \infty)$. The peak value of the PDF is $1/(\sqrt{2\pi}\,\sigma)$ so that as $\sigma \to 0$, the Gaussian PDF goes into a δ function located at $x = m$ (since the area under the PDF is always unity).

We will now show that (B-45) is properly normalized; that is, the area under $f(x)$ is unity. This can be accomplished by letting I represent the integral of the PDF:

$$I \triangleq \int_{-\infty}^{\infty} f(x)\, dx = \int_{-\infty}^{\infty} \frac{1}{\sqrt{2\pi}\,\sigma} e^{-(x-m)^2/(2\sigma^2)}\, dx \tag{B-46}$$

Making a change in variable, let $y = (x - m)/\sigma$, then

$$I = \frac{1}{\sqrt{2\pi}\,\sigma} \int_{-\infty}^{\infty} e^{-y^2/2}\, (\sigma\, dy) = \frac{1}{\sqrt{2\pi}} \int_{-\infty}^{\infty} e^{-y^2/2}\, dy \tag{B-47}$$

The integral I can be shown to be unity by showing that I^2 is unity.

$$\begin{aligned} I^2 &= \left[\frac{1}{\sqrt{2\pi}} \int_{-\infty}^{\infty} e^{-x^2/2}\, dx\right] \left[\frac{1}{\sqrt{2\pi}} \int_{-\infty}^{\infty} e^{-y^2/2}\, dy\right] \\ &= \frac{1}{2\pi} \int_{-\infty}^{\infty} \int_{-\infty}^{\infty} e^{-(x^2+y^2)/2}\, dx\, dy \end{aligned} \tag{B-48}$$

Make a change in the variables to polar coordinates from Cartesian coordinates. Let $r^2 = x^2 + y^2$ and $\theta = \tan^{-1}(y/x)$; then

$$I^2 = \frac{1}{2\pi}\int_0^{2\pi}\left[\int_0^{\infty} e^{-r^2/2}\, r\, dr\right] d\theta = \frac{1}{2\pi}\int_0^{2\pi} d\theta = 1 \tag{B-49}$$

Thus $I^2 = 1$ and, consequently, $I = 1$.

Up to now, we have *assumed* that the parameters m and σ^2 of (B-45) were the mean and the variance of the distribution. We need to *show* that indeed they are! This may be accomplished by writing the Gaussian form in terms of some arbitrary parameters α and β.

$$f(x) = \frac{1}{\sqrt{2\pi}\,\beta}\, e^{-(x-\alpha)^2/(2\beta^2)} \tag{B-50}$$

This function is still properly normalized, as demonstrated by (B-49). First, we need to show that the parameter α is the mean.

$$m = \int_{-\infty}^{\infty} x\, f(x)\, dx = \frac{1}{\sqrt{2\pi}\,\beta}\int_{-\infty}^{\infty} x e^{-(x-\alpha)^2/(2\beta^2)}\, dx \tag{B-51}$$

Making a change in variable, let $y = (x - \alpha)/\beta$; then

$$m = \frac{1}{\sqrt{2\pi}}\int_{-\infty}^{\infty} (\beta y + \alpha) e^{-y^2/2}\, dy$$

or

$$m = \frac{\beta}{\sqrt{2\pi}}\int_{-\infty}^{\infty} (y e^{-y^2/2})\, dy + \alpha\left(\int_{-\infty}^{\infty} \frac{1}{\sqrt{2\pi}}\, e^{-y^2/2}\, dy\right) \tag{B-52}$$

The first integral on the right of (B-52) is zero because the integrand is an odd function and the integral is evaluated over symmetrical limits. The second integral on the right is the integral of a properly normalized Gaussian PDF, so the integral has a value of unity. Thus (B-52) becomes

$$m = \alpha \tag{B-53}$$

and we have shown that the parameter α is the mean value.

The variance is

$$\sigma^2 = \int_{-\infty}^{\infty} (x - m)^2 f(x)\, dx$$

$$= \frac{1}{\sqrt{2\pi}\,\beta}\int_{-\infty}^{\infty} (x - m)^2 e^{-(x-m)^2/(2\beta^2)}\, dx \tag{B-54}$$

Similarly, we need to show that $\sigma^2 = \beta^2$. This will be left as a homework exercise.

The next question to answer is: What is the CDF for the Gaussian distribution?

THEOREM: *The cumulative distribution function (CDF) for the Gaussian distribution is*

$$F(a) = Q\left(\frac{m-a}{\sigma}\right) = \tfrac{1}{2}\,\text{erfc}\left(\frac{m-a}{\sqrt{2}\,\sigma}\right) \tag{B-55}$$

where the Q function is defined by

$$Q(z) \triangleq \frac{1}{\sqrt{2\pi}}\int_z^{\infty} e^{-\lambda^2/2}\,d\lambda \tag{B-56}$$

and the complementary error function (erfc) is defined as

$$\text{erfc}(z) \triangleq \frac{2}{\sqrt{\pi}}\int_z^{\infty} e^{-\lambda^2}\,d\lambda \tag{B-57}$$

It can also be shown that

$$\text{erfc}(z) = 1 - \text{erf}(z) \tag{B-58}$$

where the error function is defined as

$$\text{erf}(z) \triangleq \frac{2}{\sqrt{\pi}}\int_0^{z} e^{-\lambda^2}\,d\lambda \tag{B-59}$$

The Q function and the complementary error function, as used in (B-55), give the same curve for $F(a)$. Since neither of the corresponding integrals, given by (B-56) and (B-57), can be evaluated in closed form, math tables (see Sec. A-10, Appendix A), numerical integration techniques, or closed-form approximations must be used to evaluate them. The equivalence between the two functions is

$$Q(z) = \tfrac{1}{2}\,\text{erfc}\left(\frac{z}{\sqrt{2}}\right) \tag{B-60}$$

Communication engineers often prefer to use the Q function as compared to the erfc function since solutions to problems written in terms of the Q function do not require the writing of the $\frac{1}{2}$ and $1/\sqrt{2}$ factors. On the other hand, the advantage of using the erf(z) or erfc(z) functions is that they are one of the standard functions in the FORTRAN computer language and MathCAD, and these functions are also available on some hand calculators. However, textbooks in probability and statistics usually give a tabulation of the normalized CDF. This is a tabulation of $F(a)$ for the case of $m = 0$ and $\sigma = 1$, and it is equivalent to $Q(-a)$ and $\frac{1}{2}\,\text{erfc}(-a/\sqrt{2})$. Since $Q(z)$ and $\frac{1}{2}\,\text{erfc}(z/\sqrt{2})$ are equivalent, which one is used is a matter of personal preference. We use the Q-function notation in this book.

Proof of a Theorem for the Gaussian CDF

$$F(a) = \int_{-\infty}^{a} f(x)\,dx = \frac{1}{\sqrt{2\pi}\,\sigma}\int_{-\infty}^{a} e^{-(x-m)^2/(2\sigma^2)}\,dx \tag{B-61}$$

Making a change in variable, let $y = (m - x)/\sigma$.

$$F(a) = \frac{1}{\sqrt{2\pi}\,\sigma}\int_{\infty}^{(m-a)/\sigma} e^{-y^2/2}\,(-\sigma\,dy) \tag{B-62}$$

or

$$F(a) = \frac{1}{\sqrt{2\pi}}\int_{(m-a)/\sigma}^{\infty} e^{-y^2/2}\,dy = Q\left(\frac{m-a}{\sigma}\right) \tag{B-63}$$

Similarly, $F(a)$ may be expressed in terms of the complementary error function.

As mentioned earlier, it is unfortunate that the integrals for $Q(z)$ or $\text{erfc}(z)$ cannot be evaluated in closed form. However, for large values of z, very good closed-form approximations can be obtained, and for small values of z numerical integration techniques can be applied easily. A plot of $Q(z)$ is shown in Fig. B-7 for $z \geq 0$ and a tabulation of $Q(z)$ is given in Sec. A-10.

A relatively simple closed-form upper bound for $Q(z)$, $z > 0$, is

$$Q(z) < \frac{1}{\sqrt{2\pi}\,z}\,e^{-z^2/2}, \qquad z > 0 \tag{B-64}$$

This is also shown in Fig. B-7. It is obtained by evaluating the $Q(z)$ integral by parts.

$$Q(z) = \int_z^{\infty} \frac{1}{\sqrt{2\pi}}\,e^{-\lambda^2/2}\,d\lambda = \int_z^{\infty} u\,dv = uv\Big|_z^{\infty} - \int_z^{\infty} v\,du$$

where $u = 1/(\sqrt{2\pi}\lambda)$ and $dv = \lambda e^{-\lambda^2/2}\,d\lambda$. Thus

$$Q(z) = \left(\frac{1}{\sqrt{2\pi}\,\lambda}\right)\left(-e^{-\lambda^2/2}\right)\Big|_z^{\infty} - \int_z^{\infty}\left(-e^{-\lambda^2/2}\right)\left(-\frac{1}{\sqrt{2\pi}\,\lambda^2}\,d\lambda\right)$$

Dropping the integral, which is a positive quantity, we obtain the upper bound on $Q(z)$ as given by (B-64). If needed, a lower bound may also be obtained [Wozencraft and Jacobs, 1965]. A rational function approximation for $Q(z)$ is given by (A-116).

For values of $z \geq 3$ this upper bound has an error of less than 10% when compared to the actual value for $Q(z)$. That is, $Q(3) = 1.35 \times 10^{-3}$ and the upper bound has a value of 1.48×10^{-3} for $z = 3$. This is an error of 9.4%. For $z = 4$ the error is 5.6% and for $z = 5$ the error is 3.6%. In evaluating the probability of error for digital systems, as discussed in Chapters 6, 7, and 8, the result is often found to be a Q function. Since most useful digital systems have a probability of error of 10^{-3} or less, this upper bound becomes very useful for evaluating $Q(z)$. At any rate, if the upper bound approximation is used, we know that the value obtained indicates slightly poorer performance than is theoretically possible. In this sense, this approximation will give a worst-case result.

For the case of z negative, $Q(z)$ can be evaluated by using the identity

$$Q(-z) = 1 - Q(z) \tag{B-65}$$

where the Q value for positive z is used (as obtained from Fig. B-7) to compute the Q of the negative value of z.

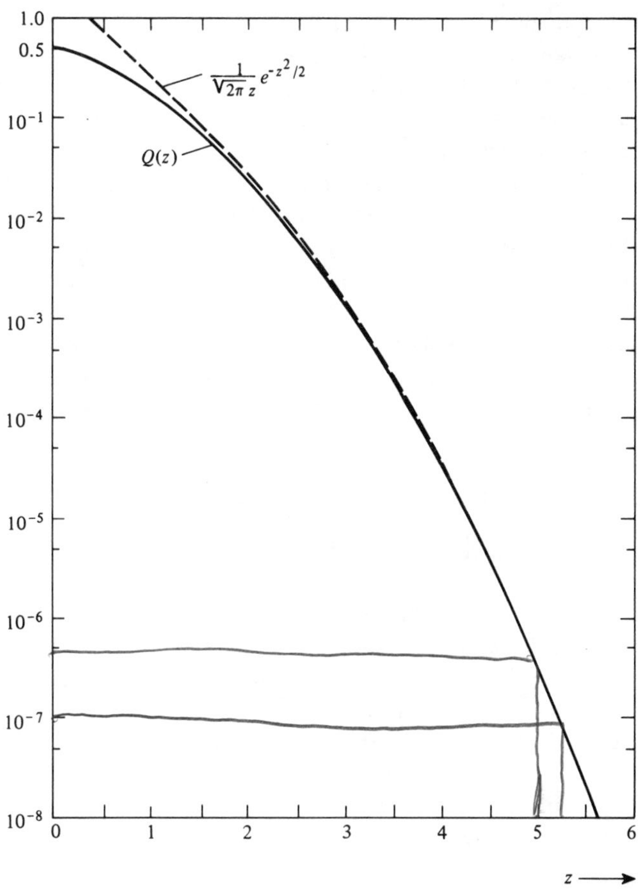

FIGURE B-7 The function $Q(z)$ and an overbound, $\dfrac{1}{\sqrt{2\pi}\,z}e^{-z^2/2}$.

Sinusoidal Distribution

Theorem: *If $x = A \sin \psi$, where ψ has the uniform distribution,*

$$f_\psi(\psi) = \begin{cases} \dfrac{1}{2\pi}, & |\psi| \le \pi \\ 0, & \text{elsewhere} \end{cases} \tag{B-66}$$

then the PDF for the sinusoid is given by

$$f_x(x) = \begin{cases} 0, & x < -A \\ \dfrac{1}{\pi\sqrt{A^2 - x^2}}, & |x| \le A \\ 0, & x > A \end{cases} \tag{B-67}$$

A proof of this theorem will be given in Sec. B-8.

A sketch of the PDF for a sinusoid is given in Table B-1, along with equations for the CDF and the variance. Note that the standard deviation, which is equivalent to the rms value as discussed in Chapter 6, is $\sigma = A/\sqrt{2}$. This should *not* be a surprising result.

The sinusoidal distribution can be used to model observed phenomena. For example, x might represent an oscillator voltage where $\psi = \omega_0 t + \theta_0$. Here the frequency of oscillation is f_0, and ω_0 and t are assumed to be deterministic values. θ_0 represents the random startup phase of the oscillator. (When the power of an unsynchronized oscillator is turned on, the oscillation builds up from a noise voltage that is present in the circuit.) In another oscillator model, time might be considered to be a uniformly distributed random variable where $\psi = \omega_0 t$. Here, once again, we would have a sinusoidal distribution for x.

B-8 Functional Transformations of Random Variables

As illustrated by the preceding sinusoidal distribution, we often need to evaluate the PDF for a random variable that is a function of another random variable for which the distribution is known. This is illustrated pictorially in Fig. B-8. Here the input random variable is denoted by x, and the output random variable is denoted by y. Since several PDFs are involved, *subscripts* (such as x in f_x) will be used to indicate with which random variable the PDF is associated. The arguments of the PDFs may change, depending on what substitutions are made, as equations are reduced.

THEOREM: *If $y = h(x)$, where $h(\cdot)$ is the output-to-input (transfer) characteristic of a device* ***without memory,***[†] *then the PDF of the output is*

$$f_y(y) = \sum_{i=1}^{M} \frac{f_x(x)}{|dy/dx|}\bigg|_{x=x_i=h_i^{-1}(y)} \tag{B-68}$$

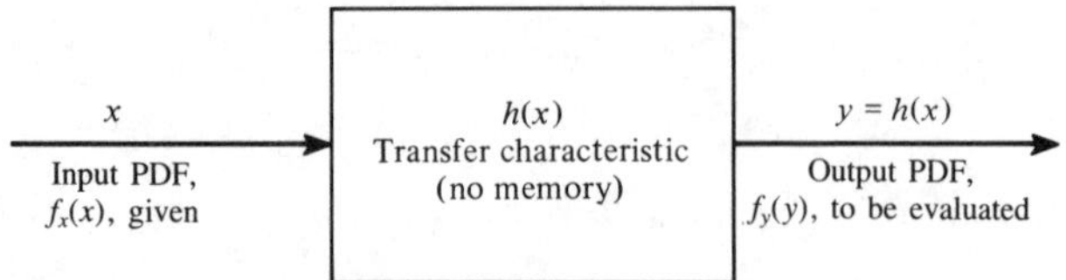

FIGURE B-8 Functional transformation of random variables.

[†]The output-to-input characteristic $h(x)$ should not be confused with the impulse response of a linear network, which was denoted by $h(t)$.

where $f_x(x)$ is the PDF of the input, x. M is the number of real roots of $y = h(x)$. That is, the inverse of $y = h(x)$ gives $x_1, x_2, \ldots, x_M$ for a single value of y. $|\cdot|$ denotes the absolute value and the single vertical line denotes the evaluation of the quantity at $x = x_i = h_i^{-1}(y)$.

Two examples will now be worked out to demonstrate the application of this theorem, and then the proof will be given.

EXAMPLE B-5 SINUSOIDAL DISTRIBUTION

Let

$$y = h(x) = A \sin x \tag{B-69}$$

where x is uniformly distributed over $-\pi$ to $+\pi$, as given by (B-66). This is illustrated in Fig. B-9. For a given value of y, say $-A < y_0 < A$, there are two possible inverse values for x, namely, x_1 and x_2, as shown in the figure. Thus $M = 2$, provided that $|y| < A$. Otherwise, $M = 0$. Evaluating the derivative of (B-69), we obtain

$$\frac{dy}{dx} = A \cos x$$

and for $0 \leq y \leq A$,

$$x_1 = \operatorname{Sin}^{-1}\left(\frac{y}{A}\right)$$

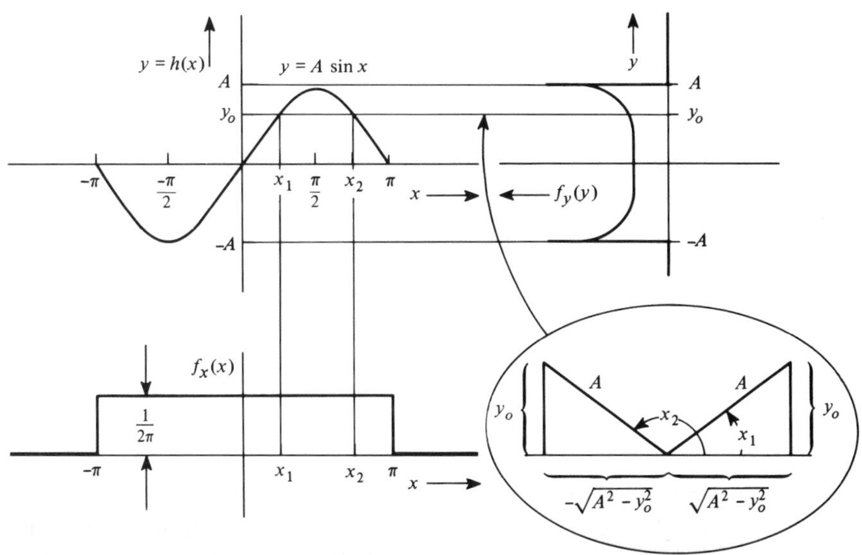

FIGURE B-9 Evaluation of the PDF of a sinusoid (Example B-5).

and

$$x_2 = \pi - x_1$$

where the uppercase S in $\text{Sin}^{-1}(\cdot)$ denotes the principal angle. A similar result is obtained for $-A \le y \le 0$. Using (B-68), we find that the PDF for y is

$$f_y(y) = \begin{cases} \dfrac{f_x(x_1)}{|A \cos x_1|} + \dfrac{f_x(x_2)}{|A \cos x_2|}, & |y| \le A \\ 0, & y \text{ elsewhere} \end{cases} \tag{B-70}$$

The denominators of these two terms are evaluated with the aid of the triangles shown in the insert of Fig. B-9. By using this result and substituting the uniform PDF for $f_x(x)$, (B-70) becomes

$$f_y(y) = \begin{cases} \dfrac{1/2\pi}{|\sqrt{A^2 - y^2}\,|} + \dfrac{1/2\pi}{|-\sqrt{A^2 - y^2}\,|}, & |y| \le A \\ 0, & y \text{ elsewhere} \end{cases}$$

or

$$f_y(y) = \begin{cases} 0, & y < -A \\ \dfrac{1}{\pi\sqrt{A^2 - y^2}}, & |y| \le A \\ 0, & y > A \end{cases} \tag{B-71}$$

which is the PDF for a sinusoid as given first by (B-67). This result is intuitively obvious since we realize that a sinusoidal waveform spends most of its time near its peak values and passes through zero relatively rapidly. Thus the PDF should peak up at $+A$ and $-A$ volts.

EXAMPLE B-6 PDF FOR THE OUTPUT OF A DIODE CHARACTERISTIC

Assume that a diode current–voltage characteristic is modeled by the ideal characteristic shown in Fig. B-10, where y is the current through the diode and x is the voltage across the diode. This type of characteristic is also called *half-wave linear rectification.*

$$y = \begin{cases} Bx, & x > 0 \\ 0, & x \le 0 \end{cases} \tag{B-72}$$

where $B > 0$. For $y > 0$, $M = 1$; and for $y < 0$, $M = 0$. However, if $y = 0$, there are an infinite number of roots for x (i.e., all $x \le 0$). Consequently, there will be a discrete point at $y = 0$ if the area under $f_x(x)$ is nonzero for $x \le 0$ (i.e., for the

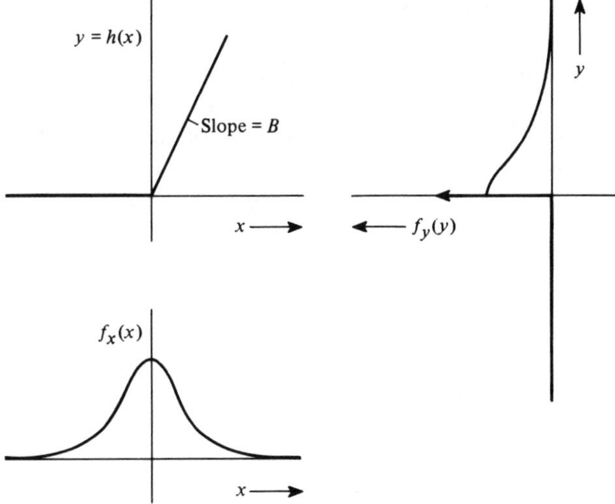

FIGURE B-10 Evaluation of the PDF out of a diode characteristic for Example B-6.

values of x that are mapped into $y = 0$). Using (B-68), we get

$$f_y(y) = \begin{Bmatrix} \dfrac{f_x(y/B)}{B}, & y > 0 \\ 0, & y < 0 \end{Bmatrix} + P(y = 0)\ \delta(y) \tag{B-73}$$

where

$$P(y = 0) = P(x \leq 0) = \int_{-\infty}^{0} f_x(x)\, dx = F_x(0) \tag{B-74}$$

Suppose that x has a Gaussian distribution with zero mean; then these equations reduce to

$$f_y(y) = \begin{Bmatrix} \dfrac{1}{\sqrt{2\pi}\, B\sigma} e^{-y^2/(2B^2\sigma^2)}, & y > 0 \\ 0, & y < 0 \end{Bmatrix} + \tfrac{1}{2}\ \delta(y) \tag{B-75}$$

A sketch of this result is shown in Fig. B-10. For $B = 1$, note that the output is the same as the input for x positive (i.e., $y = x > 0$), so that the PDF of the output is the same as the PDF of the input for $y > 0$. For $x < 0$, the values of x are mapped into the point $y = 0$, so that the PDF of y contains a δ function of weight $\frac{1}{2}$ at the point $y = 0$.

Proof of Theorem. We will demonstrate that (B-68) is valid by partitioning the x axis into intervals over which $h(x)$ is monotonically increasing, monotonically decreasing, or a constant. As we have seen in Example B-6, when $h(x)$ is a constant over some interval of x, a discrete point at y equal to that constant is possible. In addition, discrete points in the distribution of x will be mapped into discrete points in y even in regions where $h(x)$ is not a constant.

Now we will demonstrate that the theorem, as described by (B-68), is correct by taking the case, for example, where $y = h(x)$ is monotonically decreasing for $x < x_0$ and monotonically increasing for $x > x_0$. This is illustrated in Fig. B-11. The CDF for y is then

$$\begin{aligned} F_y(y_0) &= P(y \le y_0) = P(x_1 \le x \le x_2) \\ &= P[(x = x_1) + (x_1 < x \le x_2)] \end{aligned} \tag{B-76}$$

where the + sign denotes the union operation. Then using (B-4), we get

$$F_y(y_0) = P(x_1) + P(x_1 < x \le x_2)$$

or

$$F_y(y_0) = P(x_1) + F_x(x_2) - F_x(x_1) \tag{B-77}$$

The PDF of y is obtained by taking the derivative of both sides of this equation.

$$\frac{dF_y(y_0)}{dy_0} = \frac{dF_x(x_2)}{dx_2}\frac{dx_2}{dy_0} - \frac{dF_x(x_1)}{dx_1}\frac{dx_1}{dy_0} \tag{B-78}$$

where $dP(x_1)/dy_0 = 0$ was used since $P(x_1)$ is a constant. Because

$$dF_x(x_2)/dx_2 = f_x(x_2) \qquad \text{and} \qquad dF_x(x_1)/dx_1 = f_x(x_1)$$

(B-78) becomes

$$f_y(y_0) = \frac{f_x(x_2)}{dy_0/dx_2} + \frac{f_x(x_1)}{-dy_0/dx_1} \tag{B-79}$$

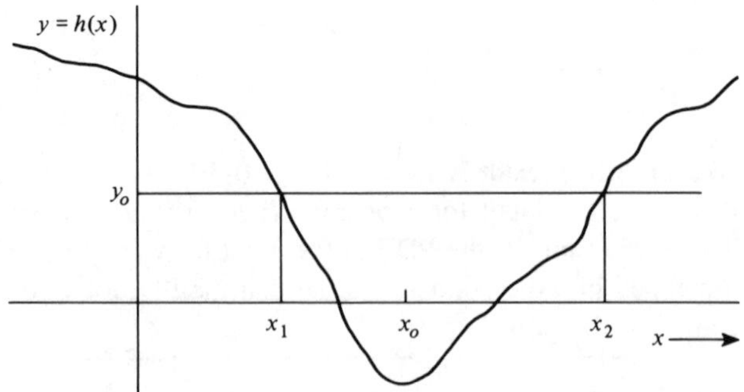

FIGURE B-11 Example of a function $h(x)$ that monotonically decreases for $x < x_0$ and monotonically increases for $x > x_0$.

At the point $x = x_1$, the slope of y is negative because the function is monotonically decreasing for $x < x_0$; thus $dy_0/dx_1 < 0$ and (B-79) becomes

$$f_y(y_0) = \sum_{i=1}^{M=2} \frac{f_x(x)}{|dy_0/dx|}\Bigg|_{x=x_i=h_i^{-1}(y_0)} \tag{B-80}$$

When there are more than two intervals during which $h(x)$ is monotonically increasing or decreasing, this procedure may be extended so that (B-68) is obtained.

In concluding this discussion on the functional transformation of a random variable, it should be emphasized that the description of the mapping function $y = h(x)$ assumes that the output y, at any instant, depends on the value of the input x only at that same instant and not on previous (or future) values of x. Thus this technique is applicable to devices that contain no memory (i.e., no inductance or capacitance) elements; however, the device may be nonlinear, as we have seen in the preceding examples.

B-9 Multivariate Statistics

In Sec. B-3 the probabilities of simple events, joint probabilities, and conditional probabilities were defined. In Secs. B-4 and B-5, using the probability of simple events, we developed the concepts of PDFs and CDFs. The PDFs and CDFs involved only one random variable, so these are one-dimensional problems. Similarly, the moments involved only one-dimensional integration.

In this section multiple-dimensional problems, also called *multivariate statistics,* will be developed. These involve PDFs and CDFs that are associated with probabilities of intersecting events and with conditional probabilities. In addition, multiple-dimensional moments will be obtained as an extension of one-dimensional moments that were studied in Sec. B-6. *If the reader clearly understands the one-dimensional case* (developed in the preceding sections), *there will be little difficulty in generalizing those results to the N-dimensional case.*

Multivariate CDFs and PDFs

Definition: The *N-dimensional CDF* is

$$F(a_1, a_2, \ldots, a_N) = P[(x_1 \le a_1)(x_2 \le a_2) \cdots (x_N \le a_N)]$$

$$= \lim_{n\to\infty}\left[\frac{n_{(x_1\le a_1)(x_2\le a_2)\cdots(x_N\le a_N)}}{n}\right] \tag{B-81}$$

where the notation $(x_1 \le a_1)(x_2 \le a_2)\cdots(x_N \le a_N)$ is the intersection event consisting of the intersection of the events associated with $x_1 \le a_1$, $x_2 \le a_2$, and so on.

DEFINITION: The *N-dimensional PDF* is

$$f(x_1, x_2, \ldots, x_N) = \left. \frac{\partial^N F(a_1, a_2, \ldots, a_N)}{\partial a_1 \partial a_2 \cdots \partial a_N} \right|_{\mathbf{a}=\mathbf{x}} \qquad \text{(B-82)}$$

where **a** and **x** are the row vectors; $\mathbf{a} = (a_1, a_2, \ldots, a_N)$ and $\mathbf{x} = (x_1, x_2, \ldots, x_N)$.

DEFINITION: The expected value of $y = h(\mathbf{x})$ is

$$\overline{[y]} = \overline{h(x_1, x_2, \ldots, x_N)}$$
$$= \int_{-\infty}^{\infty} \int_{-\infty}^{\infty} \cdots \int_{-\infty}^{\infty} h(x_1, x_2, \ldots, x_N)$$
$$\times f(x_1, x_2, \ldots, x_N)\, dx_1\, dx_2 \cdots dx_N \qquad \text{(B-83)}$$

Some *properties* of N-dimensional random variables are:

1. $f(x_1, x_2, \ldots, x_N) \geq 0$ (B-84a)
2. $\int_{-\infty}^{\infty} \int_{-\infty}^{\infty} \cdots \int_{-\infty}^{\infty} f(x_1, x_2, \ldots, x_N)\, dx_1\, dx_2 \cdots dx_N = 1$ (B-84b)
3. $F(a_1, a_2, \ldots, a_N)$

$$= \lim_{\substack{\epsilon \to 0 \\ \epsilon > 0}} \int_{-\infty}^{a_1+\epsilon} \int_{-\infty}^{a_2+\epsilon} \cdots \int_{-\infty}^{a_N+\epsilon} f(x_1, x_2, \ldots, x_N)\, dx_1\, dx_2 \cdots dx_N \qquad \text{(B-84c)}$$

4. $F(a_1, a_2, \ldots, a_N) \equiv 0$ if *any* $a_i = -\infty, \quad i = 1, 2, \ldots, N$ (B-84d)
5. $F(a_1, a_2, \ldots, a_N) = 1$ when *all* $a_i = +\infty, \quad i = 1, 2, \ldots, N$ (B-84e)
6. $P[(a_1 < x_1 \leq b_1)(a_2 < x_2 \leq b_2) \cdots (a_N < x_N \leq b_N)]$

$$= \lim_{\substack{\epsilon \to 0 \\ \epsilon > 0}} \int_{a_1+\epsilon}^{b_1+\epsilon} \int_{a_2+\epsilon}^{b_2+\epsilon} \cdots \int_{a_N+\epsilon}^{b_N+\epsilon} f(x_1, x_2, \ldots, x_N)\, dx_1\, dx_2 \cdots dx_N \qquad \text{(B-84f)}$$

The definitions and properties for multivariate PDFs and CDFs are based on the concepts of joint probabilities, as discussed in Sec. B-3. In a similar way, conditional PDFs and CDFs can be obtained [Papoulis, 1984]. Using the property $P(AB) = P(A)P(B|A)$ of (B-8), we find that the joint PDF of x_1 and x_2 is

$$f(x_1,x_2) = f(x_1)f(x_2|x_1) \qquad \text{(B-85)}$$

where $f(x_2|x_1)$ is the conditional PDF of x_2 given x_1. Generalizing further, we obtain

$$f(x_1,x_2|x_3) = f(x_1|x_3)f(x_2|x_1,x_3) \qquad \text{(B-86)}$$

Many other expressions for relationships between multiple-dimensional PDFs should also be apparent. When x_1 and x_2 are *independent*, $f(x_2|x_1) = f_{x_2}(x_2)$ and

$$f_{\mathbf{x}}(x_1,x_2) = f_{x_1}(x_1)f_{x_2}(x_2) \qquad \text{(B-87)}$$

where the subscript x_1 denotes the PDF associated with x_1, and the subscript x_2 denotes the PDF associated with x_2. For N independent random variables this becomes

$$f_{\mathbf{x}}(x_1, x_2, \ldots, x_N) = f_{x_1}(x_1)f_{x_2}(x_2)\cdots f_{x_N}(x_N) \tag{B-88}$$

THEOREM: *If the Nth-dimensional PDF of x is known, then the Lth-dimensional PDF of x can be obtained when L < N by*

$$f(x_1, x_2, \ldots, x_L) = \underbrace{\int_{-\infty}^{\infty}\int_{-\infty}^{\infty}\cdots\int_{-\infty}^{\infty}}_{N-L \text{ integrals}} f(x_1, x_2, \ldots, x_N)\, dx_{L+1}\, dx_{L+2}\cdots dx_N \tag{B-89}$$

This Lth-dimensional PDF, where L < N, is sometimes called the ***marginal PDF*** *since it is obtained from a higher-dimensional (Nth) PDF.*

Proof. First, show that this result is correct if $N = 2$ and $L = 1$.

$$\int_{-\infty}^{\infty} f(x_1,x_2)\, dx_2 = \int_{-\infty}^{\infty} f(x_1)f(x_2|x_1)\, dx_2$$

$$= f(x_1)\int_{-\infty}^{\infty} f(x_2|x_1)\, dx_2 = f(x_1) \tag{B-90}$$

since the area under $f(x_2|x_1)$ is unity. This procedure is readily extended to prove the Lth-dimensional case of (B-89).

Bivariate Statistics

Bivariate (or joint) distributions are the $N = 2$ dimensional case. In this section the definitions from the previous section will be used to evaluate two-dimensional moments. As shown in Chapter 6, bivariate statistics have some very important applications to electrical engineering problems, and some additional definitions need to be studied.

DEFINITION: The *correlation* (or *joint mean*) of x_1 and x_2 is

$$m_{12} = \overline{x_1 x_2} = \int_{-\infty}^{\infty}\int_{-\infty}^{\infty} x_1 x_2 f(x_1,x_2)\, dx_1\, dx_2 \tag{B-91}$$

DEFINITION: Two random variables x_1 and x_2 are said to be *uncorrelated* if

$$m_{12} = \overline{x_1 x_2} = \overline{x_1}\;\overline{x_2} = m_1 m_2 \tag{B-92}$$

If x_1 and x_2 are independent, it follows that they are also uncorrelated, but the converse is not generally true. However, as we will see, the converse is true for bivariate Gaussian random variables.

DEFINITION: Two random variables are said to be *orthogonal* if

$$m_{12} = \overline{x_1 x_2} \equiv 0 \tag{B-93}$$

Note the similarity of the definition of orthogonal random variables to that of orthogonal functions given by (2-73).

DEFINITION: The *covariance* is

$$\begin{aligned} u_{11} &= \overline{(x_1 - m_1)(x_2 - m_2)} \\ &= \int_{-\infty}^{\infty}\int_{-\infty}^{\infty} (x_1 - m_1)(x_2 - m_2) f(x_1,x_2)\, dx_1\, dx_2 \end{aligned} \tag{B-94}$$

It should be clear that if x_1 and x_2 are independent, the covariance is zero (and x_1 and x_2 are uncorrelated). The converse is not generally true, but it is true for the case of bivariate Gaussian random variables.

DEFINITION: The *correlation coefficient* is

$$\rho = \frac{u_{11}}{\sigma_1 \sigma_2} = \frac{\overline{(x_1 - m_1)(x_2 - m_2)}}{\sqrt{\overline{(x_1 - m_1)^2}}\,\sqrt{\overline{(x_2 - m_2)^2}}} \tag{B-95}$$

This is also called the *normalized covariance*. The correlation coefficient is always within the range

$$-1 \le \rho \le +1 \tag{B-96}$$

For example, suppose that $x_1 = x_2$; then $\rho = +1$. If $x_1 = -x_2$, then $\rho = -1$; and if x_1 and x_2 are independent, $\rho = 0$. Thus the correlation coefficient tells us, on the average, how likely a value of x_1 is to being proportional to the value for x_2. This subject is discussed in more detail in Chapter 6, where these results are extended to include random processes (time functions). There the dependence of the value of a waveform at one time is compared to the value of the waveform that occurs at another time. This will bring the concept of frequency response into the problem.

Gaussian Bivariate Distribution

A good example of a joint ($N = 2$) distribution that is of great importance is the bivariate Gaussian distribution. The bivariate Gaussian PDF is

$$f(x_1,x_2) = \frac{1}{2\pi\sigma_1\sigma_2\sqrt{1-\rho^2}}\, e^{-\frac{1}{2(1-\rho^2)}\left[\frac{(x_1-m_1)^2}{\sigma_1^2} - 2\rho\frac{(x_1-m_1)(x_2-m_2)}{\sigma_1\sigma_2} + \frac{(x_2-m_2)^2}{\sigma_2^2}\right]} \tag{B-97}$$

where σ_1^2 is the variance of x_1, σ_2^2 is the variance of x_2, m_1 is the mean of x_1, and m_2 is the mean of x_2. Examining (B-97), we see that if $\rho = 0$, $f(x_1, x_2) = f(x_1)f(x_2)$, where $f(x_1)$ and $f(x_2)$ are the one-dimensional PDFs of x_1 and x_2.

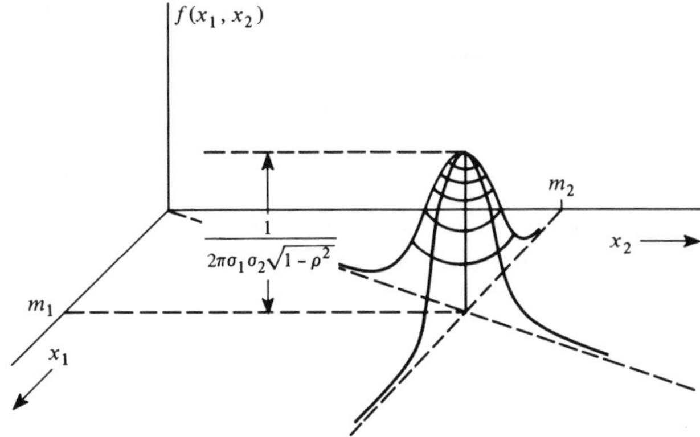

FIGURE B-12 Bivariate Gaussian PDF.

Thus if bivariate *Gaussian* random variables are uncorrelated (which implies that $\rho = 0$), they are independent.

A sketch of the bivariate (two-dimensional) Gaussian PDF is shown in Fig. B-12.

Multivariate Functional Transformation

Section B-8 will now be generalized for the multivariate case. Referring to Fig. B-13, we will obtain the PDF for **y**, denoted by $f_y(\mathbf{y})$ in terms of the PDF for **x**, denoted by $f_x(\mathbf{x})$.

Theorem: *Let* $\mathbf{y} = \mathbf{h}(\mathbf{x})$ *denote the transfer characteristic of a device (no memory) that has N inputs, denoted by* $\mathbf{x} = (x_1, x_2, \ldots, x_N)$*; N outputs, denoted by* $\mathbf{y} = (y_1, y_2, \ldots, y_N)$*; and* $y_i = h_i(\mathbf{x})$*. That is,*

$$\begin{aligned} y_1 &= h_1(x_1, x_2, \ldots, x_N) \\ y_2 &= h_2(x_1, x_2, \ldots, x_N) \\ &\vdots \\ y_N &= h_N(x_1, x_2, \ldots, x_N) \end{aligned} \tag{B-98}$$

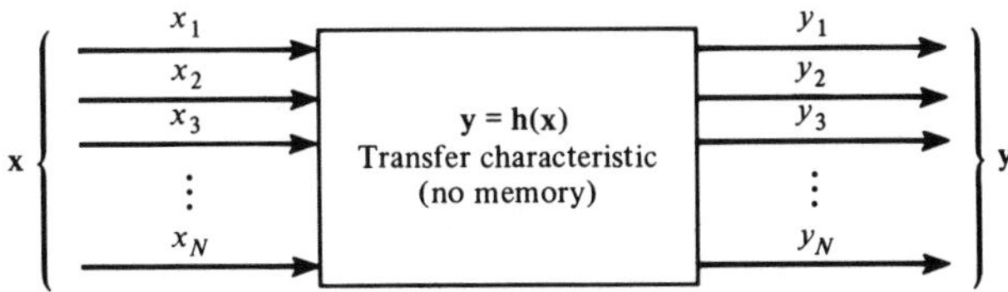

FIGURE B-13 Multivariate functional transformation of random variables.

Furthermore, let $\mathbf{x}_i$, $i = 1, 2, \ldots, M$, *denote the real roots (vectors) of the equation* $\mathbf{y} = \mathbf{h}(\mathbf{x})$. *The PDF of the output is then*

$$f_{\mathbf{y}}(\mathbf{y}) = \sum_{i=1}^{M} \frac{f_{\mathbf{x}}(\mathbf{x})}{|J(\mathbf{y}/\mathbf{x})|}\bigg|_{\mathbf{x}=\mathbf{x}_i=\mathbf{h}_i^{-1}(\mathbf{y})} \tag{B-99}$$

where $|\cdot|$ *denotes the absolute value operation and* $J(\mathbf{y}/\mathbf{x})$ *is the Jacobian of the coordinate transformation to* $\mathbf{y}$ *from* $\mathbf{x}$. *The Jacobian is*

$$J\left(\frac{\mathbf{y}}{\mathbf{x}}\right) = \mathrm{Det}\begin{bmatrix} \dfrac{\partial h_1(\mathbf{x})}{\partial x_1} & \dfrac{\partial h_1(\mathbf{x})}{\partial x_2} & \cdots & \dfrac{\partial h_1(\mathbf{x})}{\partial x_N} \\ \dfrac{\partial h_2(\mathbf{x})}{\partial x_1} & \dfrac{\partial h_2(\mathbf{x})}{\partial x_2} & \cdots & \dfrac{\partial h_2(\mathbf{x})}{\partial x_N} \\ \vdots & \vdots & & \vdots \\ \dfrac{\partial h_N(\mathbf{x})}{\partial x_1} & \dfrac{\partial h_N(\mathbf{x})}{\partial x_2} & \cdots & \dfrac{\partial h_N(\mathbf{x})}{\partial x_N} \end{bmatrix} \tag{B-100}$$

where $\mathrm{Det}[\cdot]$ *denotes the determinant of the matrix* $[\cdot]$.

A proof of this theorem will not be given, but it should be clear that it is a generalization of the theorem for the one-dimensional case that was studied in Sec. B-8. The coordinate transformation relates differentials in one coordinate system to those in another [Thomas, 1969]:

$$dy_1\, dy_2 \cdots dy_N = J\left(\frac{\mathbf{y}}{\mathbf{x}}\right) dx_1\, dx_2 \cdots dx_N \tag{B-101}$$

EXAMPLE B-7 PDF FOR THE SUM OF TWO RANDOM VARIABLES

Suppose that we have a circuit configuration (such as an operational amplifier) that sums two inputs, x_1 and x_2, to produce the output

$$y = A(x_1 + x_2) \tag{B-102}$$

where A is the gain of the circuit. Assume that $f(x_1,x_2)$ is known and that we wish to obtain a formula for the PDF of the output in terms of the joint PDF for the inputs.

We can use theorem (B-99) to solve this problem. However, for two inputs we need two outputs in order to satisfy the assumptions of the theorem. This is achieved by defining an auxiliary variable for the output (say, y_2). Thus

$$y_1 = h_1(\mathbf{x}) = A(x_1 + x_2) \tag{B-103}$$

$$y_2 = h_2(\mathbf{x}) = Ax_1 \tag{B-104}$$

The choice of the equation to use for the auxiliary variable, (B-104), is immaterial, provided that it is an independent equation so that the determinant, $J(\mathbf{y}/\mathbf{x})$, is not zero. However, the equation is usually selected so that the ensuing mathematics will be relatively simple. Using (B-103) and (B-104), we get

$$J = \text{Det}\begin{bmatrix} A & A \\ A & 0 \end{bmatrix} = -A^2 \tag{B-105}$$

Substituting this into (B-99) yields

$$f_{\mathbf{y}}(y_1,y_2) = \left.\frac{f_{\mathbf{x}}(x_1,x_2)}{|-A^2|}\right|_{\mathbf{x}=h^{-1}(\mathbf{y})}$$

or

$$f_{\mathbf{y}}(y_1,y_2) = \frac{1}{A^2} f_{\mathbf{x}}\left(\frac{y_2}{A}, \frac{1}{A}(y_1 - y_2)\right) \tag{B-106}$$

We want to find a formula for $f_{y_1}(y_1)$ since $y_1 = A(x_1 + x_2)$. This is obtained by evaluating the marginal PDF from (B-106).

$$f_{y_1}(y_1) = \int_{-\infty}^{\infty} f_{\mathbf{y}}(y_1,y_2)\, dy_2$$

or

$$f_{y_1}(y_1) = \frac{1}{A^2}\int_{-\infty}^{\infty} f_{\mathbf{x}}\left(\frac{y_2}{A}, \frac{1}{A}(y_1 - y_2)\right) dy_2 \tag{B-107}$$

This general result relates the PDF of $y = y_1$ to the joint PDF of $\mathbf{x}$ where $y = A(x_1 + x_2)$. *If* x_1 and x_2 are *independent* and $A = 1$, (B-107) becomes

$$f(y) = \int_{-\infty}^{\infty} f_{x_1}(\lambda) f_{x_2}(y - \lambda)\, d\lambda$$

or

$$f(y) = f_{x_1}(y) * f_{x_2}(y) \tag{B-108}$$

where $*$ denotes the convolution operation. Similarly, if we sum N independent random variables, the PDF for the sum is the $(N - 1)$-fold convolution of the one-dimensional PDFs for the N random variables.

Central Limit Theorem

If we have the sum of a number of independent random variables with arbitrary one-dimensional PDFs, the *central limit theorem* states that the PDF for the sum of these independent random variables approaches a Gaussian (normal) distribution under very general conditions. Strictly speaking, the central limit theorem does not hold for the PDF if the independent random variables are discretely distributed. In this case the PDF for the sum will consist of delta functions (not

Gaussian, which is continuous); however, if the delta functions are "smeared out" (e.g., replace the delta functions with rectangles that have corresponding areas), the resulting PDF will be approximately Gaussian. Regardless, the CDF for the sum will approach that of a Gaussian CDF.

EXAMPLE B-8 PDF FOR THE SUM OF THREE INDEPENDENT, UNIFORMLY DISTRIBUTED RANDOM VARIABLES

The central limit theorem will be illustrated by evaluating the exact PDF for the sum of three independent, uniformly distributed random variables. This exact result will be compared with the Gaussian PDF, as predicted by the central limit theorem.

Let each of the independent random variables x_i have a uniform distribution, as shown in Fig. B-14a. The PDF for $y_1 = x_1 + x_2$, denoted by $f(y_1)$, is obtained by the convolution operation described by (B-108) and Fig. 2-7. This result is shown in Fig. B-14b. It is seen that after only one convolution operation the PDF for the sum (which is a triangle) is going toward the Gaussian shape. The PDF

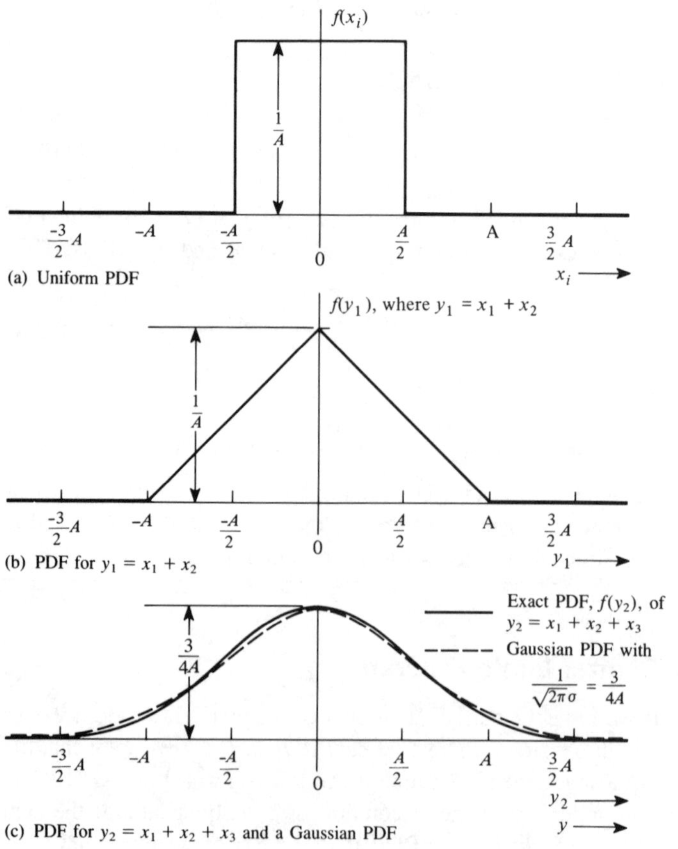

FIGURE B-14 Demonstration of the central limit theorem (Example B-8).

for $y_2 = x_1 + x_2 + x_3, f(y_2)$, is obtained by convolving the triangular PDF with another uniform PDF. The result is

$$f(y_2) = \begin{cases} 0, & y_2 \leq -\frac{3}{2}A \\ \dfrac{1}{2A^3}\left(\dfrac{3}{2}A + y_2\right)^2, & -\frac{3}{2}A \leq y_2 \leq -\frac{1}{2}A \\ \dfrac{1}{2A^3}\left(\dfrac{3}{2}A^2 - 2y_2^2\right), & |y_2| \leq \frac{1}{2}A \\ \dfrac{1}{2A^3}\left(\dfrac{3}{2}A - y_2\right)^2, & \frac{1}{2}A \leq y_2 \leq \frac{3}{2}A \\ 0, & y_2 \geq \frac{3}{2}A \end{cases} \tag{B-109}$$

This curve is plotted by the solid line in Fig. B-14c. For comparison purposes, a matching Gaussian curve, with $1/(\sqrt{2\pi}\,\sigma) = 3/(4A)$, is also shown by the dashed line. It is seen that $f(y_2)$ is very close to the Gaussian curve for $|y_2| < \frac{3}{2}A$, as predicted by the central limit theorem. Of course, $f(y_2)$ is certainly not Gaussian for $|y_2| > \frac{3}{2}A$ because $f(y_2) \equiv 0$ in this region, whereas the Gaussian curve is not zero except at $y = \pm\infty$. Thus we observe that the Gaussian approximation (as predicted by the central limit theorem) is not very good on the *tails* of the distribution. In Chapter 7 it is shown that the probability of bit error for digital systems is obtained by evaluating the area under the tails of a distribution. If the distribution is not known to be Gaussian, and the central limit theorem is used to approximate the distribution by a Gaussian PDF, the results are often not very accurate, as shown by this example. However, if the area under the distribution is needed near the mean value, the Gaussian approximation may be very good.

PROBLEMS

B-1 A long binary message contains 1428 binary 1's and 2668 binary 0's. What is the probability of obtaining a binary 1 in any received bit?

B-2 **(a)** Find the probability of getting an 8 in the toss of two dice.
(b) Find the probability of getting either a 5, 7, or 8 in the toss of two dice.

B-3 Show that

$$\begin{aligned} P(A + B + C) = {} & P(A) + P(B) + P(C) \\ & - P(AB) - P(AC) - P(BC) + P(ABC) \end{aligned}$$

B-4 A die is tossed. The probability of getting any face is $P(x) = \frac{1}{6}$, where $x = k = 1, 2, 3, 4, 5$, or 6.
(a) Find the probability of getting an odd-numbered face.
(b) Find the probability of getting a 4 when an even-numbered face is obtained on a toss.

B-5 Which of the following functions satisfy the properties for a PDF. Why?

(a) $f(x) = \frac{1}{\pi}\left(\frac{1}{1+x^2}\right)$

(b) $f(x) = \begin{cases} |x|, & |x| < 1 \\ 0, & x \text{ otherwise} \end{cases}$

(c) $f(x) = \begin{cases} \frac{1}{6}(8 - x), & 4 \leq x \leq 10 \\ 0, & x \text{ otherwise} \end{cases}$

(d) $f(x) = \sum_{k=0}^{\infty} \frac{3}{4}(\frac{1}{4})^k \, \delta(x - k)$

B-6 Show that all cumulative distribution functions must satisfy the properties given in Sec. B-5.

B-7 Let $f(x) = Ke^{-b|x|}$, where K and b are positive constants. Find the mathematical expression for the CDF and sketch the results.

B-8 Find the probability that $-\frac{1}{4}A \leq y_1 \leq \frac{1}{4}A$ for the triangular distribution shown in Fig. B-14b.

B-9 A triangular PDF is shown in Fig. B-14b.
(a) Find a mathematical expression that describes the CDF.
(b) Sketch the CDF.

B-10 Evaluate the PDFs for the two waveforms shown in Fig. PB-10.

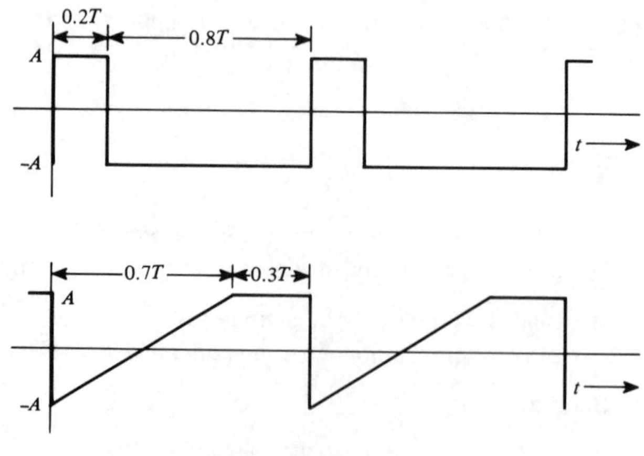

FIGURE PB-10

B-11 Let a PDF be given by $f(x) = Ke^{-bx}$ for $x \geq 0$ and zero for x negative where K and b are positive constants.
(a) Find the value required for K in terms of b.
(b) Find m in terms of b.
(c) Find σ^2 in terms of b.

B-12 A random variable x has a PDF

$$f(x) = \begin{cases} \frac{3}{32}(-x^2 + 8x - 12), & 2 < x < 6 \\ 0, & x \text{ elsewhere} \end{cases}$$

(a) Demonstrate that $f(x)$ is a valid PDF.
(b) Find the mean.
(c) Find the second moment.
(d) Find the variance.

B-13 Determine the standard deviation for the triangular distribution shown in Fig. B-14b.

B-14 (a) Find the terms in a binomial distribution for $n = 7$, where $p = 0.5$.
(b) Sketch the PDF for this binomial distribution.
(c) Find and sketch the CDF for this distribution.
(d) Find $\overline{x^3}$ for this distribution.

B-15 For a binomial distribution, show that $\sigma^2 = np(1 - p)$.

B-16 A binomial random variable x_k has values of k where

$$k = 0, 1, \ldots, n; \qquad P(k) = \binom{n}{k} p^k q^{n-k}; \qquad q = 1 - p$$

Assume that $n = 160$ and $p = 0.1$.
(a) Plot $P(k)$.
(b) Compare the plot of part (a) with a plot $P(k)$ using the Gaussian approximation

$$\binom{n}{k} p^k q^{n-k} \approx \frac{1}{\sqrt{2\pi}\,\sigma} e^{-(k-m)^2/2\sigma^2}$$

which is valid when $npq \gg 1$ and $|k - np|$ is in the neighborhood of $\sqrt{npq}$ where $\sigma = \sqrt{npq}$ and $m = np$.
(c) Also plot the Poisson approximation

$$\binom{n}{k} p^k q^{n-k} \approx \frac{\lambda^k}{k!} e^{-\lambda}$$

where $\lambda = np$, n is large, and p is small.

B-17 An order of $n = 3000$ transistors is received. The probability that each transistor is defective is $p = 0.001$. What is the probability that the number of defective transistors in the batch is 6 or less? (*Note:* The Poisson approximation is valid when n is large and p is small.)

B-18 In a fiber optic communication system, photons are emitted with a Poisson distribution as described in Table B-1. $m = \lambda$ is the average number of photons emitted in an arbitrary time interval, and $P(k)$ is the probability of k photons being emitted in the same interval.

(a) Plot the PDF for $\lambda = 0.5$.
(b) Plot the CDF for $\lambda = 0.5$.
(c) Show that $m = \lambda$.
(d) Show that $\sigma = \sqrt{\lambda}$.

B-19 Let x be a random variable that has a Laplacian distribution. The Laplacian PDF is $f(x) = (1/2b)e^{-|x-m|/b}$, where b and m are real constants and $b > 0$.
(a) Find the mean of x in terms of b and m.
(b) Find the variance of x in terms of b and m.

B-20 Given the Gaussian PDF

$$f(x) = \frac{1}{\sqrt{2\pi}\beta} e^{-(x-m)^2/(2\beta^2)}$$

show that the variance of this distribution is β^2.

B-21 In a manufacturing process for resistors, the values obtained for the resistors have a Gaussian distribution where the desired value is the mean value. If we want 95% of the manufactured 1-kΩ resistors to have a tolerance of $\pm 10\%$, what is the required value for σ?

B-22 Assume that x has a Gaussian distribution. Find the probability that:
(a) $|x - m| < \sigma$.
(b) $|x - m| < 2\sigma$.
(c) $|x - m| < 3\sigma$.
Obtain numerical results by using tables if necessary.

B-23 Show that:

(a) $Q(z) = \frac{1}{2} \operatorname{erfc}\left(\frac{z}{\sqrt{2}}\right)$.

(b) $Q(-z) = 1 - Q(z)$.

(c) $Q(z) = \frac{1}{2}\left[1 - \operatorname{erf}\left(\frac{z}{\sqrt{2}}\right)\right]$.

B-24 For a Gaussian distribution, show that:

(a) $F(a) = \frac{1}{2} \operatorname{erfc}\left(\frac{m-a}{\sqrt{2}\sigma}\right)$.

(b) $F(a) = \frac{1}{2}\left[1 + \operatorname{erf}\left(\frac{a-m}{\sqrt{2}\sigma}\right)\right]$.

B-25 A noise voltage has a Gaussian distribution. The rms value is 5 V and the dc value is 1.0 V. Find the probability of the voltage having values between -5 and $+5$ V.

B-26 Suppose that x is a Gaussian random variable with $m = 5$ and $\sigma = 0.6$.
(a) Find the probability that $x \leq 1$.
(b) Find the probability that $x \leq 6$.

B-27 The Gaussian random variable x has a zero mean and a variance of 2. Let A be the event such that $|x| < 3$.
(a) Find an expression for the conditional PDF $f(x|A)$.
(b) Plot $f(x|A)$ over the range $|x| < 5$.
(c) Plot $f(x)$ over the range $|x| < 5$ and compare these two plots.

B-28 Let x have a sinusoidal distribution with a PDF as given by (B-67). Show that the CDF is

$$F(a) = \begin{cases} 0, & a \le -A \\ \dfrac{1}{\pi}\left[\dfrac{\pi}{2} + \sin^{-1}\left(\dfrac{a}{A}\right)\right], & |a| \le A \\ 1, & a \ge A \end{cases}$$

B-29 **(a)** If x has a sinusoidal distribution with the peak value of x being A, show that the rms value is $\sigma = A/\sqrt{2}$. [*Hint:* Use (B-67).]
(b) If $x = A \cos \psi$, where ψ is uniformly distributed between $-\pi$ and $+\pi$, show that the rms value of x is $\sigma = A/\sqrt{2}$.

B-30 Given that $y = x^2$ and x is a Gaussian random variable with mean value m and variance σ^2, find a formula for the PDF of y in terms of m and σ^2.

B-31 x is a uniformly distributed random variable over the range $-1 \le x \le 1$ plus a discrete point at $x = \frac{1}{2}$ with $P(x = \frac{1}{2}) = \frac{1}{4}$.
(a) Find a mathematical expression for the PDF for x and plot your result.
(b) Find the PDF for y where

$$y = \begin{cases} x^2, & x \ge 0 \\ 0, & x < 0 \end{cases}$$

Sketch your result.

B-32 A saturating amplifier is modeled by

$$y = \begin{cases} Ax_0, & x > x_0 \\ Ax, & |x| \le x_0 \\ -Ax_0, & x < -x_0 \end{cases}$$

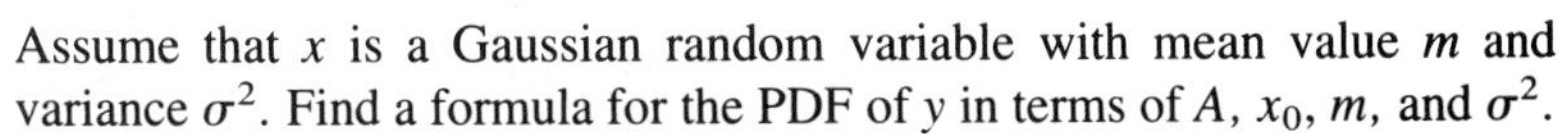

Assume that x is a Gaussian random variable with mean value m and variance σ^2. Find a formula for the PDF of y in terms of A, x_0, m, and σ^2.

B-33 A sinusoid with a peak value of 8 V is applied to the input of a quantizer. The quantizer characteristic is shown in Fig. 3-8a. Calculate and plot the PDF for the output.

B-34 A voltage waveform that has a Gaussian distribution is applied to the input of a full-wave rectifier circuit. The full-wave rectifier is described by $y(t) = |x(t)|$ where $x(t)$ is the input and $y(t)$ is the output. The input waveform has a dc value of 1 V and an rms value of 2 V.

(a) Plot the PDF for the input waveform.
(b) Plot the PDF for the output waveform.

B-35 Refer to Example B-6 and (B-75), which describe the PDF for the output of an ideal diode (half-wave rectifier) characteristic. Find the mean (dc) value of the output.

B-36 Given the joint density function,

$$f(x_1,x_2) = \begin{cases} e^{-(1/2)(4x_1+x_2)}, & x_1 \geq 0, x_2 \geq 0 \\ 0, & \text{otherwise} \end{cases}$$

(a) Verify that $f(x_1,x_2)$ is a density function.
(b) Show that x_1 and x_2 are either independent or dependent.
(c) Evaluate $P(1 \leq x_1 \leq 2,\ x_2 \leq 4)$.
(d) Find ρ.

B-37 A joint density function is

$$f(x_1,x_2) = \begin{cases} K(x_1 + x_1x_2), & 0 \leq x_1 \leq 1, \quad 0 \leq x_2 \leq 4 \\ 0, & \text{elsewhere} \end{cases}$$

(a) Find K.
(b) Determine if x_1 and x_2 are independent.
(c) Find $F_{x_1x_2}(0.5,2)$.
(d) Find $f_{x_2|x_1}(x_2|x_1)$.

B-38 Let $y = x_1 + x_2$ where x_1 and x_2 are uncorrelated random variables. Show that:
(a) $\overline{y} = m_1 + m_2$, where $m_1 = \overline{x_1}$ and $m_2 = \overline{x_2}$.
(b) $\sigma_y^2 = \sigma_1^2 + \sigma_2^2$, where $\sigma_1^2 = \overline{(x_1 - m_1)^2}$ and $\sigma_2^2 = \overline{(x_2 - m_2)^2}$.
[*Hint:* Use the ensemble operator notation similar to that used in the proof for (B-29).]

B-39 Let $x_1 = \cos\,\theta$ and $x_2 = \sin\,\theta$, where θ is uniformly distributed over $(0,2\pi)$. Show that:
(a) x_1 and x_2 are uncorrelated.
(b) x_1 and x_2 are not independent.

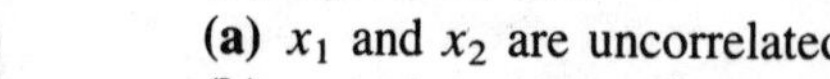

B-40 Two random variables x_1 and x_2 are jointly Gaussian. The joint PDF is described by (B-97) where $m_1 = m_2 = 1$, $\sigma_{x_1} = \sigma_{x_2} = 1$ and $\rho = 0.5$. Plot $f(x_1,x_2)$ for x_1 over the range $|x_1| < 5$ and $x_2 = 0$. Also give plots for $f(x_1,x_2)$ for $|x_1| < 5$ and $x_2 = 0.4$, 0.8, 1.2, and 1.6.

B-41 Show that the marginal PDF of a bivariate Gaussian PDF is a one-dimensional Gaussian PDF. That is, evaluate

$$f(x_1) = \int_{-\infty}^{\infty} f(x_1,x_2)\,dx_2$$

where $f(x_1,x_2)$ is given by (B-97). [*Hint:* Factor some terms outside the integral containing x_1 (but not x_2). Complete the square on the exponent of the remaining integrand so that a Gaussian PDF form is obtained. Use the property that the integral of a properly normalized Gaussian PDF is unity.]

B-42 **(a)** $y = A_1x_1 + A_2x_2$, where A_1 and A_2 are constants and the joint PDF of x_1 and x_2 is $f_x(x_1,x_2)$. Find a formula for the PDF of y in terms of the (joint) PDF of **x**.

(b) If x_1 and x_2 are independent, how can this formula be simplified?

B-43 Two independent random variables x and y have the PDFs $f(x) = 5e^{-5x}\, u(x)$ and $f(y) = 2e^{-2y}\, u(y)$. Plot the PDF for w where $w = x + y$.

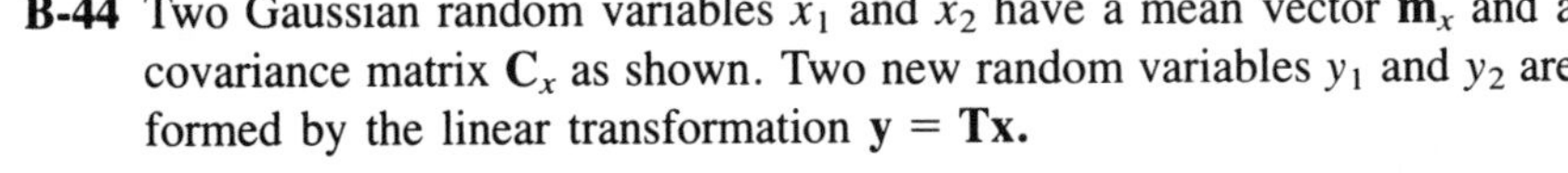

B-44 Two Gaussian random variables x_1 and x_2 have a mean vector $\mathbf{m}_x$ and a covariance matrix $\mathbf{C}_x$ as shown. Two new random variables y_1 and y_2 are formed by the linear transformation $\mathbf{y} = \mathbf{T}\mathbf{x}$.

$$\mathbf{m}_x = \begin{bmatrix} 2 \\ -1 \end{bmatrix} \qquad \mathbf{C}_x = \begin{bmatrix} 5 & -2/\sqrt{5} \\ -2/\sqrt{5} & 4 \end{bmatrix} \qquad \mathbf{T} = \begin{bmatrix} 1 & 1/2 \\ 1/2 & 1 \end{bmatrix}$$

(a) Find the mean vector for **y**, which is denoted by $\mathbf{m}_y$.

(b) Find the covariance matrix for **y**, which is denoted by $\mathbf{C}_y$.

(c) Find the correlation coefficient for y_1 and y_2. (*Hint:* See Sec. 6-6.)

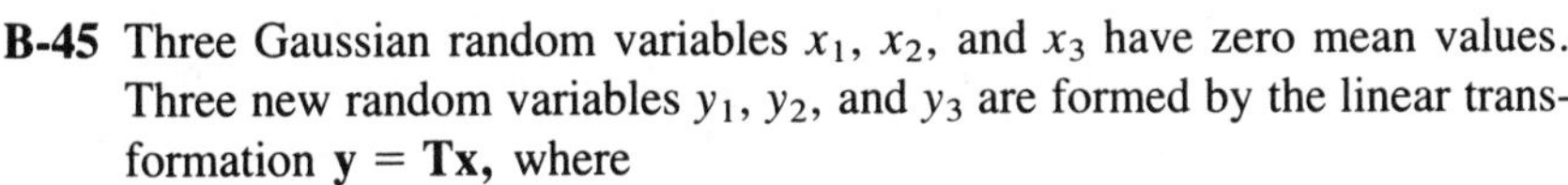

B-45 Three Gaussian random variables x_1, x_2, and x_3 have zero mean values. Three new random variables y_1, y_2, and y_3 are formed by the linear transformation $\mathbf{y} = \mathbf{T}\mathbf{x}$, where

$$\mathbf{C}_x = \begin{bmatrix} 6.0 & 2.3 & 1.5 \\ 2.3 & 6.0 & 2.3 \\ 1.5 & 2.3 & 6.0 \end{bmatrix} \qquad \mathbf{T} = \begin{bmatrix} 5 & 2 & -1 \\ -1 & 3 & 1 \\ 2 & -1 & 2 \end{bmatrix}$$

(a) Find the covariance matrix for **y**, which is denoted by $\mathbf{C}_y$.

(b) Write an expression for the PDF $f(y_1,y_2,y_3)$. (*Hint:* See Sec. 6.6.)

B-46 **(a)** Find a formula for the PDF of $y = Ax_1x_2$ where x_1 and x_2 are random variables having the joint PDF $f_{\mathbf{x}}(x_1,x_2)$.

(b) If x_1 and x_2 are independent, reduce the formula obtained in part (a) to a simpler result.

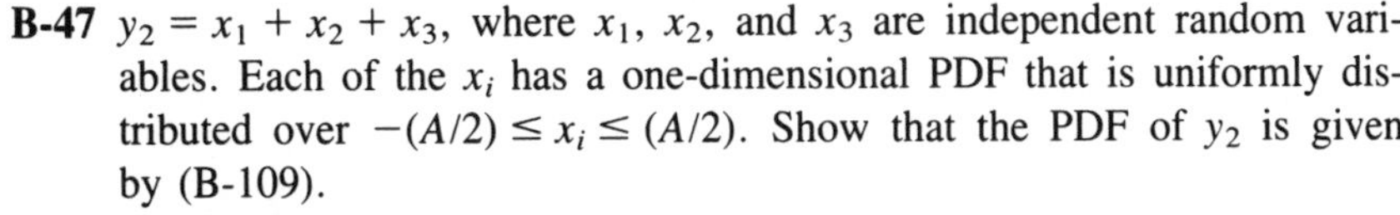

B-47 $y_2 = x_1 + x_2 + x_3$, where x_1, x_2, and x_3 are independent random variables. Each of the x_i has a one-dimensional PDF that is uniformly distributed over $-(A/2) \le x_i \le (A/2)$. Show that the PDF of y_2 is given by (B-109).

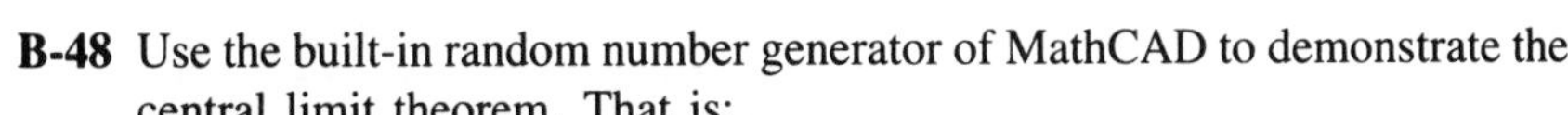

B-48 Use the built-in random number generator of MathCAD to demonstrate the central limit theorem. That is:

(a) Compute samples of the random variable y where $y = \Sigma\, x_i$ and the x_i values are obtained from the random number generator.

(b) Plot the PDF for y by using the histogram function of MathCAD.

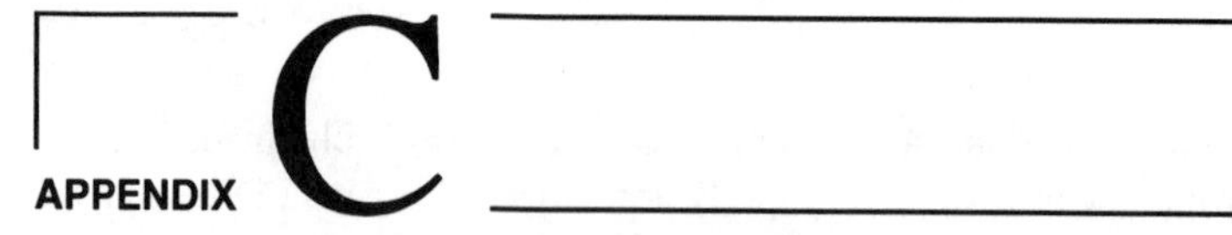

APPENDIX C

Standards and Terminology for Computer Communications

C-1 Codes

Morse Code

Although the Morse code is not used in computer communications, it is listed here because a form of this code was used in the first electrical digital communication. The first electrical communication system was the telegraph, which was patented in 1840 by Samuel F. B. Morse. Table C-1 lists the International Morse Code that is used today. The timing is such that the length of the dash is three dot units. Spaces between parts of a character are equal to one dot unit, spaces between characters are three dot units, and the space between words is five dot units.

TABLE C-1 International Morse Code

A	•—	N	—•	1	•— — — —
B	—•••	O	— — —	2	••— — —
C	—•—•	P	•— —•	3	•••— —
D	—••	Q	— —•—	4	••••—
E	•	R	•—•	5	•••••
F	••—•	S	•••	6	—••••
G	— —•	T	—	7	— —•••
H	••••	U	••—	8	— — —••
I	••	V	•••—	9	— — — —•
J	•— — —	W	•— —	0	— — — — —
K	—•—	X	—••—		
L	•—••	Y	—•— —		
M	— —	Z	— —••		

Period (.)	•—•—•—	Wait sign (AS)	•—•••
Comma (,)	— —••— —	Double dash (break)	—•••—
Interrogation (?)	••— —••	Error sign	••••••••
Quotation Mark (’’)	•—••—•	Fraction bar (/)	—••—•
Colon (:)	— — —•••	End of message (AR)	•—•—•
Semicolon (;)	—•—•—•	End of transmission (SK)	•••—•—
Parenthesis ( )	—•— —•—	International distress signal (SOS)	•••— — —•••

Punched Card

Historically, the punched card was used to enter data into mainframe computers, but today data are usually entered into mainframes from on-line video display terminals (VDT). The size of a standard IBM card is 3.75 in. × 7.375 in. × 0.0067 in. In the usual format it has 12 rows and 80 columns, for a total of 960

TABLE C-2 Punched Card Code

Character	Row Numbers Punched		Character	Row Numbers Punched			Character	Row Numbers Punched		
0	0		a	12	0	1	%	0	4	8
1	1		b	12	0	2	@	4	8	
2	2		c	12	0	3	#	3	8	
3	3		d	12	0	4	¢	12	2	8
4	4		e	12	0	5	<	12	4	8
5	5		f	12	0	6	(	12	5	8
6	6		g	12	0	7	†	12	6	8
7	7		h	12	0	8	|	12	7	8
8	8		i	12	0	9	!	11	2	8
9	9		j	12	11	1	)	11	5	8
A	12	1	k	12	11	2	;	11	6	8
B	12	2	l	12	11	3	¬	11	7	8
C	12	3	m	12	11	4	—	0	5	8
D	12	4	n	12	11	5	>	0	6	8
E	12	5	o	12	11	6	?	0	7	8
F	12	6	p	12	11	7	:	2	8	
G	12	7	q	12	11	8	'	5	8	
H	12	8	r	12	11	9	=	6	8	
I	12	9	s	11	0	2	"	7	8	
J	11	1	t	11	0	3	[	12	8	2
K	11	2	u	11	0	4	\	0	8	2
L	11	3	v	11	0	5	]	11	8	2
M	11	4	w	11	0	6	\	1	8	
N	11	5	x	11	0	7	{	12	0	
O	11	6	y	11	0	8	}	11	0	
P	11	7	z	11	0	9	~	11	0	1
Q	11	8	.	12	3	8				
R	11	9	□	12	4	8				
S	0	2	&	12						
T	0	3	space	BLANK						
U	0	4	$	11	3	8				
V	0	5	*	11	4	8				
W	0	6	-	11						
X	0	7	/	0	1					
Y	0	8	,	0	3					
Z	0	9								

hole positions. Each column designates a single character that is defined by one, two, or three punched holes in the 12 possible positions. The punched card code is given in Table C-2.

Baudot

In 1875, Emile Baudot developed a "multiplex telegraph" system. The Baudot code has evolved into what is now known as the *International Telegraph Alphabet (ITA) Number 2*. This Baudot code is given in Table C-3. The Baudot code is used for teleprinter service. It is used today for wire press and radio teletype press service, although it is being replaced by the ASCII code.

The Baudot code has two serious disadvantages: (1) it has no provisions for parity bits, and (2) it is a *sequential* code. That is, the Letters (down-shift) control character is sent to place the printer in the letter printing mode, and the Figure (up-shift) control character is used to shift the printer into the figures printing mode. If an error causes a control character to be received that shifts the printer into a false mode, the printer will print a wrong string of characters until a proper control-shift character is received.

EBCDIC

The Extended Binary-Coded Decimal Interchange Code (EBCDIC) was developed in 1962 and contains 8 bits with no parity. The EBCDIC code is given in Table C-4. It is used extensively in IBM mainframe computer systems.

TABLE C-3 Baudot Code[a]

Character Case		Bit Pattern					Character Case		Bit Pattern				
Lower	Upper	5	4	3	2	1	Lower	Upper	5	4	3	2	1
A	-	0	0	0	1	1	Q	1	1	0	1	1	1
B	?	1	1	0	0	1	R	4	0	1	0	1	0
C	:	0	1	1	1	0	S	'	0	0	1	0	1
D	$	0	1	0	0	1	T	5	1	0	0	0	0
E	3	0	0	0	0	1	U	7	0	0	1	1	1
F	!	0	1	1	0	1	V	;	1	1	1	1	0
G	&	1	1	0	1	0	W	2	1	0	0	1	1
H	#	1	0	1	0	0	X	/	1	1	1	0	1
I	8	0	0	1	1	0	Y	6	1	0	1	0	1
J	Bell	0	1	0	1	1	Z	"	1	0	0	0	1
K	(	0	1	1	1	1	Letters (shift) ↓		1	1	1	1	1
L	)	1	0	0	1	0	Figures (shift) ↑		1	1	0	1	1
M	.	1	1	1	0	0	Space (SP)		0	0	1	0	0
N	,	0	1	1	0	0	Carriage return		0	1	0	0	0
O	9	1	1	0	0	0	Line feed		0	0	0	1	0
P	0	1	0	1	1	0	Blank		0	0	0	0	0

[a] 1 = mark = punch hole and 0 = space = no punch hole for paper tape.

TABLE C-4 EBCDIC Code[a]

Bit Position				4	0	0	0	0	0	0	0	0	1	1	1	1	1	1	1	1
				3	0	0	0	0	1	1	1	1	0	0	0	0	1	1	1	1
				2	0	0	1	1	0	0	1	1	0	0	1	1	0	0	1	1
8	7	6	5	1	0	1	0	1	0	1	0	1	0	1	0	1	0	1	0	1
0	0	0	0		NUL	SOH	STX	ETX	PF	HT	LC	DEL			SMM	VT	FF	CR	SO	SI
0	0	0	1		DLE	DC1	DC2	DC3	RES	NL	BS	IL	CAN	EM	CC		IFS	IGS	IRS	IUS
0	0	1	0		DS	SOS	FS		BYP	LF	EOB	PRE			SM			ENQ	ACK	BEL
0	0	1	1				SYN		PN	RS	UC	EOT					DC4	NAK		SUB
0	1	0	0		SP										¢	.	<	(	+	|
0	1	0	1		&										!	$	*	)	;	¬
0	1	1	0		-	/									∧	,	%	-	>	?
0	1	1	1												:	#	@	'	=	"
1	0	0	0			a	b	c	d	e	f	g	h	i						
1	0	0	1			j	k	l	m	n	o	p	q	r						
1	0	1	0				s	t	u	v	w	x	y	z						
1	0	1	1																	
1	1	0	0			A	B	C	D	E	F	G	H	I						
1	1	0	1			J	K	L	M	N	O	P	Q	R						
1	1	1	0				S	T	U	V	W	X	Y	Z						
1	1	1	1		0	1	2	3	4	5	6	7	8	9						□

[a] ACK Acknowledge
BEL Bell, or alarm
BS Backspace
BYP Bypass
CAN Cancel
CC Unit backspace (word processing)
CR Carriage return
DC1 Device control 1
DC2 Device control 2
DC3 Device control 3
DC4 Device control 4
DEL Delete
DLE Data link escape
DS Digit select
EM End of medium
ENQ Enquiry
EOB End of block
EOT End of transmission
ETX End of text
EF Form feed
FS File separator
HT Horizontal tab
IFS Interchange file separator
IGS Interchange group separator
IL Idle
IRS Interchange record separator
IUS Interchange unit separator
LC Lowercase
LF Line feed (printer to next line)
NAK Negative acknowledge
NL New line (LF plus CR)
NUL Null, or all zeros
PF Punch off
PN Punch on
PRE Prefix
RES Restore
RS Record separator (or reader stop)
SI Shift in
SM Start message
SMM Repeat
SO Shift out
SOH Start of heading
SOS Start of significance
SP Space
STX Start of text
SUB Substitute
SYN Synchronous idle
UC Uppercase
VT Vertical tab

TABLE C-5 ASCII Code[a]

Bit Position				7	0	0	0	0	1	1	1	1
				6	0	0	1	1	0	0	1	1
4	3	2	1	5	0	1	0	1	0	1	0	1
0	0	0	0		NUL	DLE	SP	0	@	P	\	p
0	0	0	1		SOH	DC1	!	1	A	Q	a	q
0	0	1	0		STX	DC2	"	2	B	R	b	r
0	0	1	1		ETX	DC3	#	3	C	S	c	s
0	1	0	0		EOT	DC4	$	4	D	T	d	t
0	1	0	1		ENQ	NAK	%	5	E	U	e	u
0	1	1	0		ACK	SYN	&	6	F	V	f	v
0	1	1	1		BEL	ETB	'	7	G	W	g	w
1	0	0	0		BS	CAN	(	8	H	X	h	x
1	0	0	1		HT	EM	)	9	I	Y	i	y
1	0	1	0		LF	SUB	*	:	J	Z	j	z
1	0	1	1		VT	ESC	+	;	K	[	k	{
1	1	0	0		FF	FS	'	<	L	\	l	:
1	1	0	1		CR	GS	—	=	M	]	m	}
1	1	1	0		SO	RS	.	>	N	∧	n	~
1	1	1	1		SI	US	/	?	O	—	o	DEL

[a]ACK Acknowledge
BEL Bell, or alarm
BS Backspace
CAN Cancel
CR Carriage return
DC1 Device control 1
DC2 Device control 2
DC3 Device control 3
DC4 Device control 4
DEL Delete
DLE Data link escape
EM End of medium
ENQ Enquiry
EOT End of transmission
ESC Escape
ETB End of transmission block
ETX End of text
FF Form feed
FS File separator
GS Group separator
HT Horizontal tab
LF Line feed
NAK Negative acknowledge
NUL Null, or all zeros
RS Record separator
SI Shift in
SO Shift out
SOH Start of heading
SP Space
STX Start of text
SUB Substitute
SYN Synchronous idle
US Unit separator
VT Vertical tab

ASCII

The American National Standard Code for Information Interchange (ASCII) was first adopted in 1963 and updated in 1967. It was the first code specifically developed for computer communication systems. It is now the most popular code used by interactive computer terminals of both the printing and cathode ray tube (CRT) types. The code is shown in Table C-5. It has a total of 8 bits per character. Seven bits are completely specified as shown. The eighth bit is a parity bit that may be even or odd parity, depending on the choice selected for use in a particular installation.

C-2 DTE/DCE Interface Standards

A general block diagram for a computer communication system is shown in Fig. C-1. To facilitate the use of equipment supplied by different vendors, interface standards have been adopted for connection of data terminal equipment (DTE)

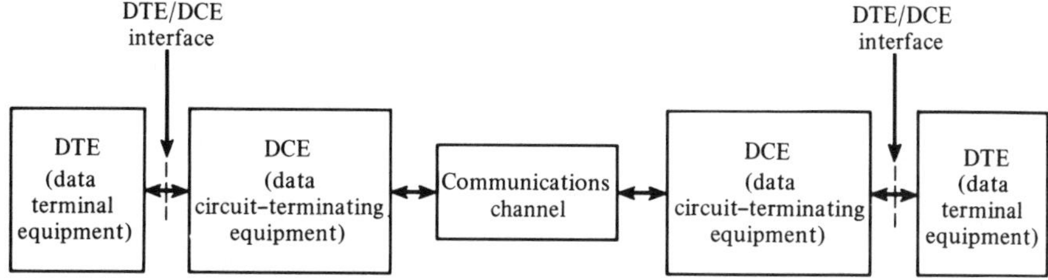

FIGURE C-1 Computer communication system.

and data communications equipment (DCE). The DTE may be a digital computer, a printer, a keyboard, and so on, and the DCE may be a modem. In addition, one type of DTE, say a printer, may be connected to another type of DTE, say a keyboard, by the same type of interface.

Numerous organizations are concerned with advancement of practical standards. Internationally, the International Telegraph and Telephone Consultative Committee (CCITT) and the International Organization for Standardization (ISO) are the best known.

In the United States, as well as internationally, four standards groups are the American National Standards Institute (ANSI), the Electronics Industries Association (EIA), the Institute of Electrical and Electronics Engineers (IEEE), and the National Communications System (NCS) of the federal government.

DTE/DCE standards are developed in terms of the following:

- The physical and electrical interface.
- The link control (software) interface that provides framing/synchronization and error detection/error correction functions.

Now several physical/electronic standards will be described.

Current Loop

Since early in the twentieth century, when mechanical teleprinter systems were developed, the current loop has been a popular interface for transmission of asynchronous serial data (see Fig. C-2). During the idle condition, the dc current

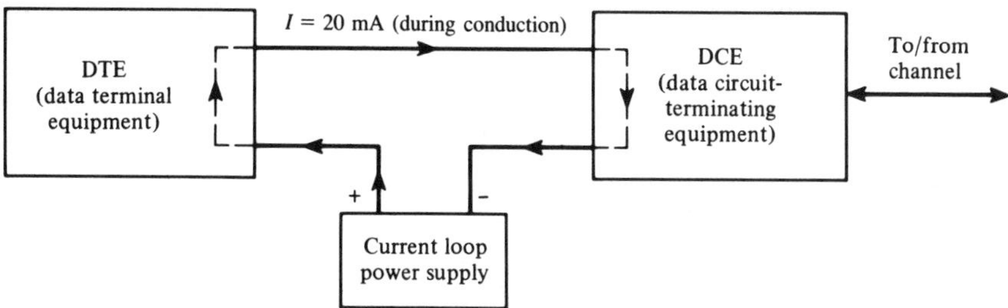

FIGURE C-2 20-mA current (local) loop interface.

TABLE C-6 EIA RS-232C, RS-422A, and RS-449 Interface Standards[a]

RS-449/422					RS-232 Interface			
9 Pin Aux.	37 Pin A	37 Pin B	RS-449 Circuit	RS-449 Description	25 Pin	EIA-RS-232C Circuit	CCITT-V.24 Circuit	Mnemonic Designation
1	1			Shield	1	AA	101	FG
5	19		SG	Signal ground	7	AB	102	SG
9	37		SC	Send common			102a	
6	20		RC	Receive common			102b	
	4	22	SD	Send data	2	BA	103	TD
	6	24	RD	Receive data	3	BB	104	RD
	7	25	RS	Request to send	4	CA	105	RTS
	9	27	CS	Clear to send	5	CB	106	CTS
	11	29	DM	Data mode	6	CC	107	DSR
	12	30	TR	Terminal ready	20	CD	108.2	DTR
	15		IC	Incoming call	22	CE	125	RI
	13	31	RR	Receiver ready	8	CF	109	DCD
	33		SQ	Signal quality	21	CG	110	SQ
	16		SR	Signaling rate selector	23	CH	111	
	2		SI	Signaling rate indicator	23	CI	112	
	17	35	TT	Terminal timing	24	DA	113	TC
	5	23	ST	Send timing	15	DB	114	TC
	8	26	RT	Receive timing	17	DD	115	RC
3			SSD	Secondary send data	14	SBA	118	STD
4			SRD	Secondary receive data	16	SBB	119	SRD
7			SRS	Secondary request to send	19	SCA	120	SRTS
8			SCS	Secondary clear to send	13	SCB	121	SCTS
2			SRR	Secondary receiver ready	12	SCF	122	SDCD
	10		LL	Local loopback			141	
	14		RL	Remote loopback			140	
	18		TM	Test mode			142	
	32		SS	Select standby			116	
	36		SB	Standby indicator			117	
	16		SF	Select frequency			126	
	28		IS	Terminal in service				
	34		NS	New signal				

Source: Black Box Corporation, Pittsburgh, PA. © Black Box Corporation, 1988. All rights reserved. Reprinted by permission.

[a]DB25 references the (25-pin) connector that is commonly used for RS-232 and V.24. Pins 9 and 10 are reserved for data set testing. Pins 11, 18, and 25 are undefined. Pins 3 and 21 of RS-449 interface connector are undefined. Lead 23 of RS-232 connector may be defined as CH or CI.

TABLE C-6 *(continued)*

RS-232 Interface	Signal Type and Direction						
		Data		Control		Timing	
RS-232 Description	GND	From DCE	To DCE	From DCE	To DCE	From DCE	To DCE
Protective frame ground	×						
Signal ground/common return	×						
DTE common	×						
DCE common	×						
Transmitted data			×				
Received data		×					
Request to send					×		
Clear to send				×			
Data set ready				×			
Data terminal ready					×		
Ring indicator				×			
Received line carrier signal detector				×			
Signal quality detector				×			
Data signal rate selector (DTE)					×		
Data signal rate selector (DCE)				×			
Transmitter clock signal element timing (DTE)							×
Transmitter clock signal element timing (DCE)						×	
Receiver clock signal element timing (DCE)						×	
Second transmitter data			×				
Secondary received data		×					
Secondary request to send					×		
Secondary clear to send				×			
Secondary received line signal detector				×			
Local loopback					×		
Remote loopback					×		
Test indicator				×			
Select standby					×		
Standby indicator				×			
Select transmit frequency				×			

in the local (interface) loop is specified to be some design value. In most teleprinters designed for computer terminal use, this value is 20 mA. To send an ASCII character, the current is interrupted at the transmitting device according to the binary word for that particular ASCII character. The presence of loop current denotes a mark, and the absence of loop current denotes a space. The interruption of the current is detected at the receiving point in the loop. If needed, a number of transmitting and receiving devices may be wired in series in this local loop (where only one transmits at a time).

RS-232C, RS-422A, RS-449, and RS-530 Interfaces

The RS-232C serial interface (also known as the *EIA standard interface*) was developed in 1969 by the EIA in cooperation with the Bell System and independent computer and modem manufacturers. The CCITT has adopted an interface standard, called V.24, that is very similar to the RS-232C standard. The military MIL-188c is also similar. The RS-232C pin connections are given in Table C-6, where the EIA and CCITT circuit equivalents are listed. The standard provides for serial data transmission by using a voltage level of $-V_0$ for a binary 1 (mark) and a level of $+V_0$ for a binary 0 (space), where $3 \leq V_0 \leq 25$ V. A typical value for V_0 is 6 V. The voltage appears on the appropriate lead measured with respect to the common (signal) ground lead. This is called *unbalanced signaling* and is also known as single-ended signaling.

The RS-232C standard defines the electrical characteristics, the functional description of interchange circuits, and a list of standard applications. The type of physical connector is not specified, but a DB25 connector (25 pins) is generally used in practice.

The RS-232C interface is intended for data rates up to 20 kbits/sec and cable lengths up to 50 ft. Longer lengths are possible if twisted-pair cable is used and the loading capacitance is kept below 2500 pF. This interface is one of the most popular serial interfaces used in computer communication equipment today.

For higher data rates and longer cable lengths, the RS-422A interface was developed. This interface uses *balanced signaling* (also known as differential signaling), and, consequently, two wires are required for each circuit. This method gives greater noise and crosstalk immunity because a common signal return path is not used. The balanced-signaling technique allows the transition region between the mark and space voltage levels to be reduced so that the signal levels are $\pm V_0$, where $0.2 \leq V_0 \leq 25$ V is used for balanced signaling. The RS-422A interface uses DB37 and DB9 connectors (37- and 9-pin) as specified in the ISO 4902 standard. The pin connections are given in Fig. C-6, where the two-wire connections for each balanced circuit are specified in columns A and B. The RS-422A standard allows for data rates up to 100 kbits/sec at 4000 ft or 40 Mbits/sec at 40 ft.

The RS-449 signal interface is intended to be the successor to the RS-232C interface. The RS-449 interface connections are also given in Table C-6. A 37-pin D connector is used for the primary channel and an auxiliary 9-pin D connector is used (when needed) for a secondary channel. The 37-pin connector

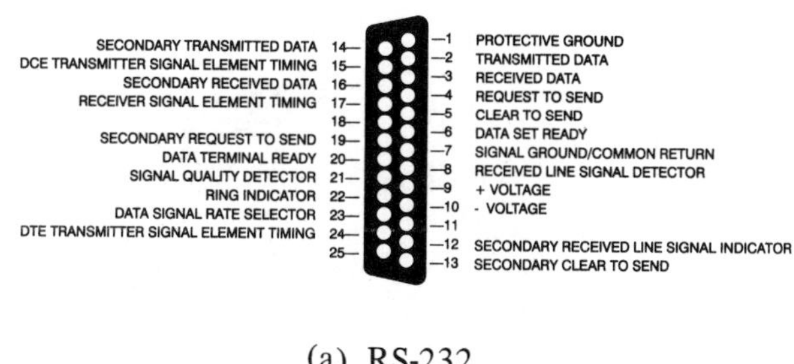

(a) RS-232

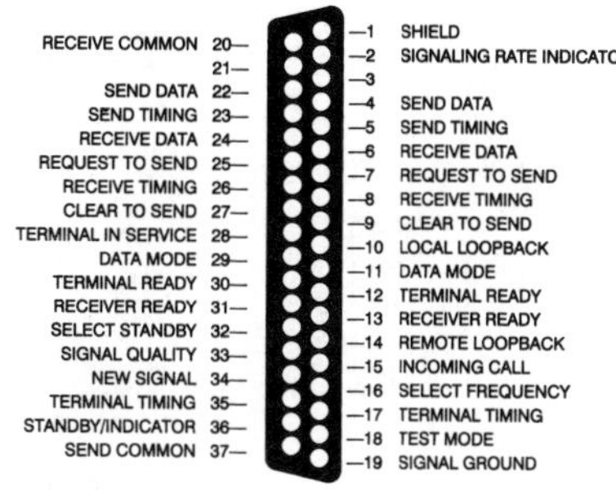

(b) RS-449/422

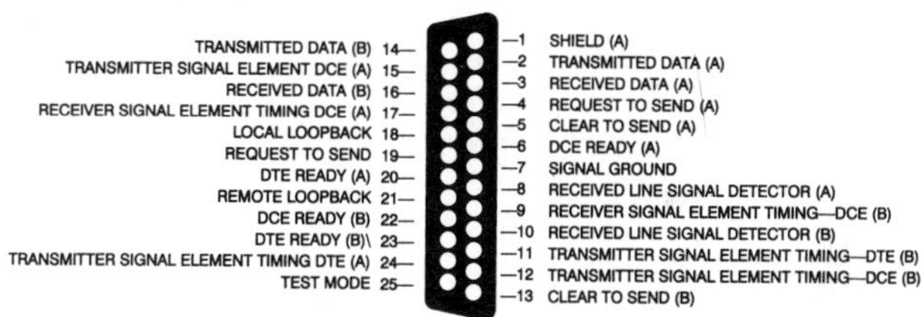

(c) RS-530

FIGURE C-3 RS-232C, RS-449/422, and RS-530 male connectors.

allows for unbalanced or (optionally) balanced signaling where the signaling levels are those for RS-232C or RS-222A, respectively. Balanced signaling allows the RS-449 interface to support data rates up to 2 Mbits/sec and cable lengths up to 667 ft. The RS-530 is a 25-pin (DB25 connector) version of the RS-449. Connectors for the RS-232C, RS-449/422, and RS-530 interfaces are shown in Fig. C-3.

Centronics Parallel Interface

Printers are often connected to computing equipment by either a RS-232C serial interface or a Centronics parallel interface. Figure C-4 and Table C-7 show the

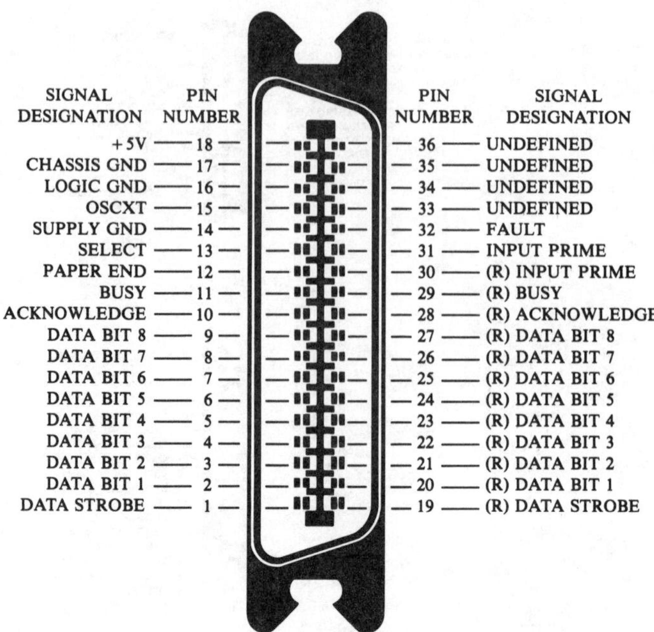

FIGURE C-4 Centronics connector. [Reprinted by permission of Black Box Corporation, Pittsburgh, PA. © Black Box Corporation, 1988. All rights reserved.]

36-pin connector for the Centronics parallel interface. This interface has eight data lines that provide eight bits of data in parallel. The data are controlled by the STROB pulse. This interface uses unbalanced signaling with TTL (transistor transistor logic) levels.

IEEE-488 Interface

The IEEE-488 interface is designed to connect computer equipment with programmable instruments such as programmable voltmeters, signal sources, and power supplies. It allows programmable instruments to be easily connected together to create an *automatic test equipment* (ATE) system. This interface was first developed in 1965 by Hewlett–Packard and was known as the HP Interface Bus (HP–IB) or the General Purpose Interface Bus (GPIB). In 1975 the IEEE adopted an evolved standard, which was named the IEEE-488 interface. Table C-8 presents some details on this parallel interface. The lines are held at +5 V for a binary 0 (false) and pulled to ground for a binary 1 (true). This standard allows the bus to extend over a distance of 67 ft. Every IEEE-488 device is a listener, a talker, or a controller. A *listener* is capable of receiving data from the bus, such as a printer or a programmable power supply. A *talker* is capable of

TABLE C-7 Centronics Parallel Interface[a]

Signal Pin No.	Return Pin No.	Signal	Direction (with Reference (to Printer)	Description
1	19	STROBE	In	STROBE pulse (negative going) enables reading data.
2	20	Data 1	In	1st to 8th bits of parallel data.
3	21	Data 2	In	Each signal is at "high" level
4	22	Data 3	In	when data are logical "1" and
5	23	Data 4	In	"low" when logical "0."
6	24	Data 5	In	
7	25	Data 6	In	
8	26	Data 7	In	
9	27	Data 8	In	
10	28	ACKNLG	Out	"Low" indicates that data have been received and that the printer is ready to accept other data.
11	29	BUSY	Out	"High" indicates that the printer cannot receive data.

Source: Black Box Corporation, Pittsburgh, PA.

[a] Pins 12, 13, 14, 15, 18, 31, 32, 34, 35, and 36 vary in function depending on application; they are commonly used for printer auxiliary controls, and error handling and indication. Pins 16 and 17 are commonly used for logic ground and chassis ground, respectively.

transmitting data over the bus, such as a frequency counter or a voltmeter. There can be only one active talker on the bus at a time. A *controller* is a computer-type device that controls the network (including itself) and specifies which devices are active talkers or listeners. A 24-pin IEEE-488 connector is illustrated in Fig. C-5.

CCITT X.25 Interface

The CCITT X.25 standard is entitled "Interface Between a DTE and a DCE for Terminals Operating in the Packet Mode on Public Data Networks." It involves the use of data packets as the basis for creating a virtual circuit. With a virtual circuit, two users appear to have a private data connection between themselves, although they are actually sharing the same physical channel with other users. This is accomplished by breaking the serial data stream of each user into "packets" of data of a certain length. Each packet has a header attached, with the destination address for the packet and other control information. The packets from each user are interspersed with those of other users on the X.25 line.

TABLE C-8 IEEE-488 Interface[a]

24-Pin A	24-Pin B	Circuit	Description	Data Line	Control Line	Handshake Line
	24	GND	Logic ground	×		
1	24	D101	Data bit 1	×		
2	24	D102	Data bit 2	×		
3	24	D103	Data bit 3	×		
4	24	D104	Data bit 4	×		
13	24	D105	Data bit 5	×		
14	24	D106	Data bit 6	×		
15	24	D107	Data bit 7	×		
16	24	D108	Data bit 8	×		
9	21	IFC	Interface Clear—used by controller to place devices into quiescent state		×	
10	22	SRQ	Service Request—used by device to request service		×	
17	24	REN	Remote Enable—used by controller to override device front panel controls		×	
5	24	EOI	End of Identity—(a) used by talker to end message, or (b) used by controller with ATN for parallel pole		×	
11	23	ATN	Attention—used by controller to end its data and listen for a command		×	
6	18	DAV	Data Valid—used by controller to indicate a control byte on data bus		×	
7	19	NRFD	Not Ready For Data—used by listener to indicate it is not ready			×
8	20	NDAC	No Data Accepted—used by listener to indicate it has not yet accepted last byte			×
	12		Shield			

[a]Column B pins are return leads to be grounded at the system ground point.

The virtual circuits may be set up temporarily or permanently. A temporary virtual circuit consists of three operations: (1) setting up the call, (2) transferring the data, and (3) taking down (disconnecting) the virtual connection. In setting up the call, a logical channel number is used. With a permanent virtual connection, the logical number is assigned permanently to the customer when the X.25 line is leased and the virtual connection is always available so that data can be

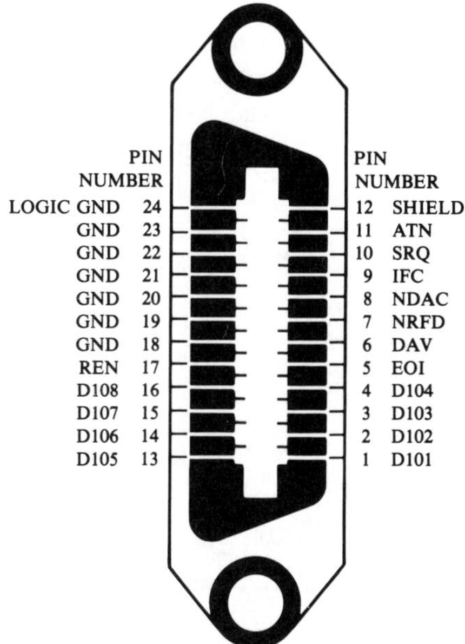

FIGURE C-5 IEEE-488 connector.

transferred at any instant. Thus the permanent virtual connection appears to be a private leased line to the end users.

The X.25 standard describes the DTE/DCE interface and does not specify how the virtual circuits are implemented. However, X.25 is actually concerned with three layers of protocols (see the ISO OSI protocol layers as described in Sec. C-3): (1) the DCE/CTE physical layer interface, which uses the X.21 standard for synchronous data transmission; (2) the data link level, which gives the method of attaching address information to the packets according to the HDLC protocol (described later); and (3) the network layer, which sets up the virtual circuits by the use of two different types of packets. The control packets contain control information such as CALL REQUEST (sent by the caller), CALL ACCEPT (sent by the called party), CALL CONNECT, CLEAR, and others. The data packets contain the information data that are exchanged between the users on a virtual circuit.

C-3 The ISO OSI Network Model

When users are connected together, many things must be considered for orderly, efficient data transmission. This complicated problem can be understood better and the network can be maintained more easily if the different tasks are divided into layered modules. These modules are implemented in both hardware and software. Most computer network designers and manufacturers follow the layered model recommended by the ISO in its Open System Interconnection (OSI) reference model. This model is shown in Fig. C-6. This figure presents an example of two end users (host A and host B) that are connected to a data communications network via the physical channel. The physical channel may be connections via telephone company (TELCO) lines, fiber optic link, microwave radio, or satellite links.

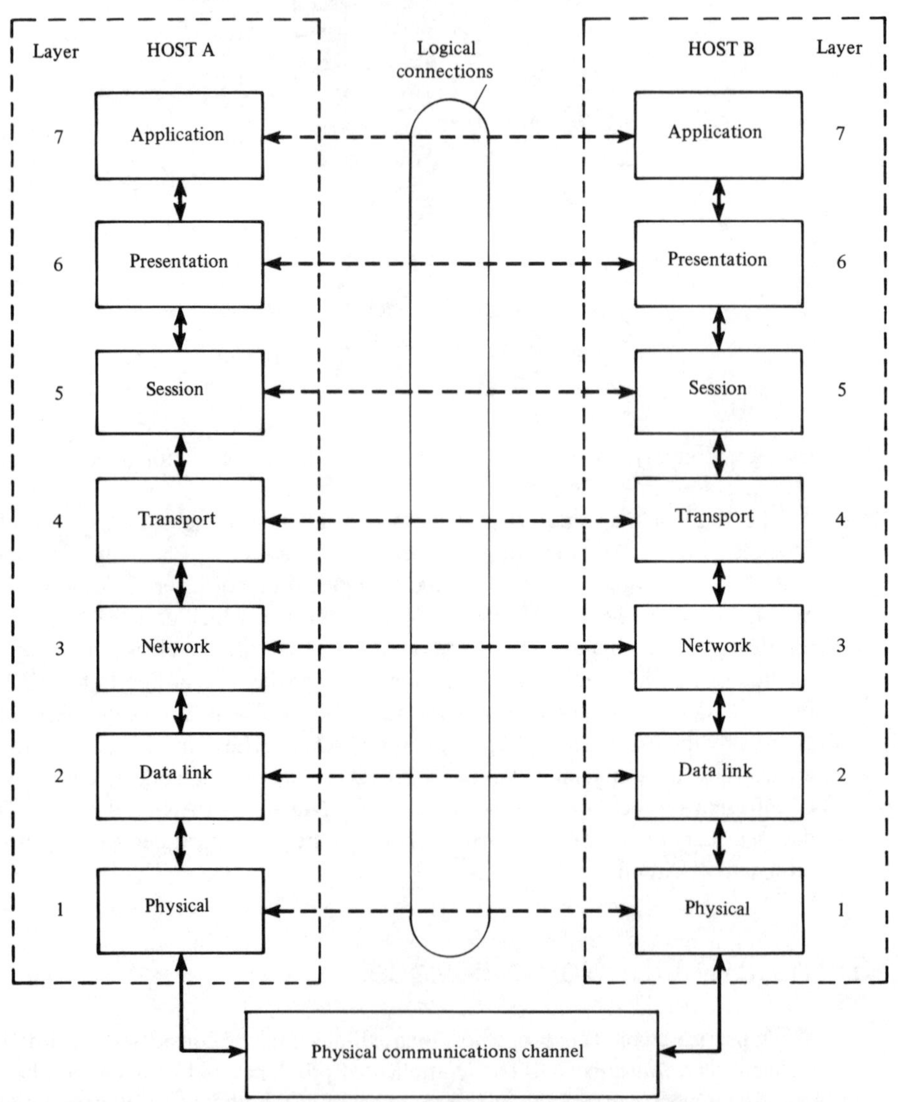

FIGURE C-6 ISO OSI reference model of physical and logical connections.

The layered approach has many attractive features [Stallings, 1985]. By adhering to the appropriate hardware and/or software specifications that define the layer boundaries, many different vendors can supply hardware/software that will work efficiently in the overall network. Furthermore, various parts of the system can be updated and maintained by replacing or modifying the hardware/software at each layer level instead of replacing the whole system.

The functions of the various layers may be described as follows:

Layer 1—Physical. Concerned with bit transmission. Standards specify signal levels, cable connectors, and cable. For multiple-access packet transmission systems, the methods of carrier sense and collision detection are also specified.

Layer 2—Data Link. Concerned with beginning message transmission, error detection and correction, and ending message transmission.

Layer 3—Network. Concerned with choosing the route taken by the message, flow control, and message priorities.

Layer 4—Transport. Concerned with end-to-end reliability (error detection and information recovery) and maps logical addresses to end-user devices.

Layer 5—Session. Starts and stops the communication session, transfers the user from one task to another, and provides recovery from communication problems (restart) without losing data.

Layer 6—Presentation. Converts data to the appropriate syntax for the display devices (character sets and graphics) and encodes/decodes compressed data.

Layer 7—Application. Provides a log-in procedure, checks passwords, allows file upload/download and remote job entry, and tabulates system resources used for billing purposes.

C-4 Data Link Control Protocols

Data link control (DLC) protocols, also called *line protocols,* are a set of rules for operating a computer communication system. They standardize the framing, addressing, and error control techniques used in the system.

BISYNC

The Binary Synchronous Communication (BISYNC) protocol was developed in 1968 by IBM for synchronous transmission between two computers or between a computer and batch terminals. (It has been superseded by SDLC, which we will discuss subsequently.) BISYNC is a character-oriented protocol. That is, it uses characters (usually EBCDIC or ASCII) to delineate the various fields of the message. It is also a half-duplex protocol, which means that the data flow in one direction at a time between two communication points. Periodically, an end-of-transmission-block (ETB) character is sent, which requires a response from the

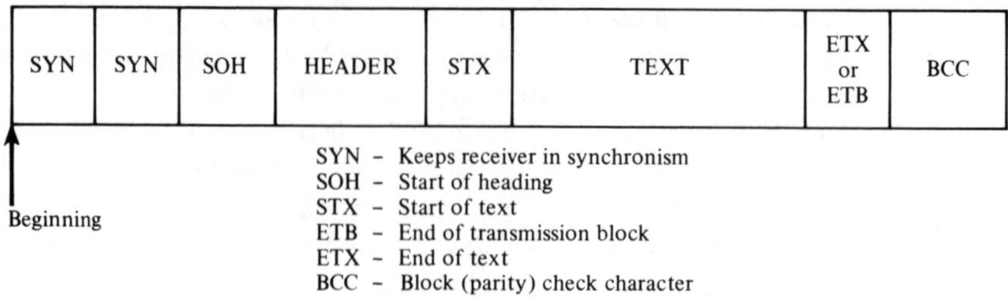

FIGURE C-7 BISYNC format.

receiving station indicating acknowledgment of reception of error-free data (ACK) or error in the previous received block (NAK). If a NAK is received by the transmitter end, the whole block is retransmitted. This is automatic request for repeat (ARQ) error control. A typical BISYNC message is shown in Fig. C-7. The text portion is variable in length.

SDLC

Synchronous Data Link Control (SDLC) is a bit-oriented protocol. That is, instead of control characters it uses a unique flag at the beginning and end of each frame. It was developed by IBM in 1974 to provide a more versatile protocol than BISYNC. It is a full-duplex protocol, which means that data can be transmitted in both directions simultaneously. Furthermore, SDLC permits the transmission to one remote location while receiving from a different remote location on a multidrop line. SDLC also uses ARQ error control, where a station has to acknowledge correct receipt of at least the eighth preceding frame if communication is to continue. The SDLC frame is shown in Fig. C-8.

HDLC

High-Level Data Link Control (HDLC) was approved by ISO in 1975 (ISO 3309) and has been updated since then by other ISO standards. It is a packet protocol that has been adopted worldwide by many vendors. IBM's SDLC is a subset of HDLC, and ANSI's Advanced Data Communications Control Procedure (ADCCP) is equivalent to HDLC. X.25 uses the HDLC protocol.

The general structure of the HDLC frame (packet) is shown in Fig. C-8. The address field may be extended recursively to allow more addresses. Two basic types of frames are allowed—control and information. In control frames there is

BEGINNING FLAG 01111110	ADDRESS 8 bits	CONTROL 8 bits	INFORMATION Any number of bits	FRAME CHECK 16 bits	ENDING FLAG 01111110

FIGURE C-8 SDLC and HDLC formats.

no information field, and the control frame is used to set up, take down, or check a virtual link between users. In the information frame, the information field may be of any size. The information frames are numbered in sequence (by bits in the control field), so that the sending station does not have to wait for an acknowledgment of the frame just transmitted before sending another frame. The acknowledgment can be received at a later time (for error detection and recovery) since the frames are numbered. In addition, the packets may arrive out of sequence at the receiver, and the numbering allows the receiver to place the information in the correct order.

HDLC was originally developed for unbalanced operation (ISO 4335), in which one primary station controls secondary stations by sending command frames. The secondary stations respond only when polled in the normal response mode. In the asynchronous mode (not to be confused with asynchronous bit transmission), the secondary stations are allowed to send unsolicited response frames. Newer modes (ISO 6256) allow balanced operation, in which any station may initiate the setting up of a virtual call.

C-5 Telephone Line Standards

Basically, a (twisted-pair) telephone line is capable of transmitting signals that have energy that is confined to the 300- to 3000-Hz frequency band.[†] (This does not mean that the signals will be undistorted when transmitted over the line.) There are two main categories of telephone service.

1. Dial-up lines, which use the normal telephone connections where calls are dialed up to almost anywhere in the world.
2. Private leased lines, which are either two-wire or four-wire lines that connect two points or multiple points on a dedicated 24-hr/day basis.

The quality of a dial-up line is difficult to predict, especially if it traverses several countries. One dial-up connection between two points may be excellent, and another, between the same two points, may be terrible.

The characteristics of private leased lines in the United States are guaranteed to meet certain criteria. There are two different sets of specifications.

1. *C-conditioning*. This specifies the frequency response and envelope delay (which are linear distortion) characteristics of the line. These specifications are given in Table C-9.
2. *D-conditioning*. This specifies the minimum signal-to-noise ratio and the second and third harmonic (nonlinear) distortion minimum signal-to-distortion ratios. These specifications are given in Table C-10.

When a customer orders a C-conditioned line, the telephone company checks to see if a line meets the particular C-conditioning class that has been ordered. If

[†] Short lines have a much wider frequency response. This is used to advantage by limited-distance modems that operate over this extended bandwidth to send data at high rates—up to 19,200 bits/sec.

TABLE C-9 C-Conditioning Specifications (Leased Lines)

Type	Frequency Range (Hz)	Frequency Response[a] (dB)	Frequency Range (Hz)	Variation in Envelope Delay (μsec)
Basic	500–2500	−2 to +8	800–2600	1750
(3002 line)	300–3000	−3 to +12		
C1	1000–2400	−1 to +3	1000–2400	1000
(Class E line)	300–2700	−2 to +6	800–2600	1750
	2700–3000	−3 to +12		
C2	500–2800	−1 to +3	1000–2600	500
(Class F line)	300–3000	−2 to +6	600–2600	1500
			500–2800	3000
C4	500–3000	−2 to +3	1000–2600	300
(Class H line)	300–3200	−2 to +6	800–2800	500
			600–3000	1500
			500–3000	3000
C5	500–2800	−0.5 to +1.5	1000–2600	100
	300–3000	−1 to +3	600–2600	300
			500–2800	600

[a]0-dB reference is 1004 Hz.

TABLE C-10 D-Conditioning Specifications (Leased Lines)

Type	S/N (dB)	S/D (dB)	
		Second	Third
Basic (3002)	24	25	28
D	28	35	40

the line does not meet the specifications, equalizers will be placed on the line so that they are met. By luck, the customer may get a much better line than he or she specifies (i.e., higher-order conditioning) when no equalization is required. The D-conditioning specifications are usually obtained by *line selection* so that the specifications are satisfied.

C-6 Modem Standards

Historically, the Bell System dominated the modem market. It calls its modems *data sets*. Since the Bell System divestiture, modems built to CCITT standards have also become popular. Table C-11 lists the modems that interface data lines

TABLE C-11 Modem Standards

Type	Data Rate (bits/sec) Normal	Data Rate (bits/sec) Fallback	Mode	Type of Line	Synch/ Asynch	Modulation[a]	Transmitting Frequencies (Hz)[b] Originate		Answer	
Bell 103/113	0-300		FDX	2W Dial-up	Async	FSK	1070S	1270M	2025S	2225M
CCITT V.21	0-300		FDX	2W Dial-up	Async	FSK	980M	1080S	1650M	1850S
Bell 202S	1,200		HDX	2W Dial-up	Async	FSK	1200M	2200S		
Bell 202T	1,200		FDX	4W Lease	Async	FSK	1200M	2200S		
CCITT V.23	1,200		HDX	2W Dial-up	Async	FSK	1300M	2100S		
		600	HDX	2W Dial-up	Async	FSK	1300M	1700S		
Bell 212A	1,200		FDX	2W Dial-up	Either	4DPSK	1200		2400	
		300	FDX	2W Dial-up	Either	FSK	Same as Bell 103			
CCITT V.22	1,200		FDX	2W Dial-up	Either	FSK	Same as Bell 212			
		600	FDX	2W Dial-up	Either	2DPSK	1200		2400	
Bell 201C	2,400		FDX	4W Lease	Sync	4DPSK	1800			
CCITT V.22bis	2,400		FDX	2W Dial-up	Either	16QAM	1200		2400	
		1200	FDX	2W Dial-up	Either	4DPSK	Same as CCITT V.22			
CCITT V.26	2,400		FDX	4W Lease	Sync	4DPSK	1800 (Same as Bell 201C)			
CCITT V.26bis	2,400		FDX	4W Lease	Sync	4DPSK	1800			
		1200	FDX	4W Lease	Sync	2DPSK	1800			
Bell 208A	4,800		FDX	4W Lease	Sync	8DPSK	1800			
Bell 208B	4,800		HDX	2W Dial-up	Sync	8DPSK	1800			
CCITT V.27bis	4,800		FDX	4W Lease	Sync	8DPSK	1800			
		2400	FDX	4W Lease	Sync	4DPSK	1800			
Bell 209	9,600		FDX	4W Lease D	Sync	16QAM	1700			
CCITT V.29	9,600	7200/4800	FDX	4W Lease D	Sync	16QAM	1700			
CCITT V.32	9,600	4800	FDX	2W Dial	Either	32QAM[c]	1800			
CCITT V.33	14,400		FDX	4W Lease C2	Sync	128QAM				
Bell 301B	40,800		FDX	Group Channel	Sync	BPSK				
Bell 303B	19,200		FDX	Half-group	Sync	VSB				
Bell 303C	50,000		FDX	Group Channel	Sync	VSB				
Bell 303D	230,400		FDX	Supergroup	Sync	VSB				

[a]FSK, frequency-shift keying; BPSK, binary-phase-shift keying; MDPSK, *M*-ary differential phase-shift keying; VSB, vestigial sideband; MQAM, *M*-ary quadratic amplitude modulation. [b]M, mark; S, space. [c]Trellis-coded modulation is used.

TABLE C-12 Popular Personal Modems (Tx, transmit; Rx receive)

	Bell System Type			CCITT
	103/113	**202**	**212A**	**V.22 bis**
Data	Serial binary Asynchronous Full duplex (two-wire line)	Dial-up or leased line Serial binary Asynchronous Half-duplex (two-wire line)	Serial binary Asynchronous or synchronous Full-duplex, dial-up lines (also contains a 103 type modem) 1200 bits/sec	Serial binary Asynchronous or synchronous Full-duplex, dial-up (also usually contains a 212A-type modem) 2400 bits/sec
Data rate	0 to 300 bits/sec	0 to 1200 bits/sec (dial-up) 0 to 1800 bits/sec (leased C2)	600 baud (symbols/sec)	600 baud (symbols/sec)
Modulation	FSK	FSK	QPSK ($M = 4$ phases)	QAM (16-point rectangular-type signal constellation)
Frequency/ phase assignment	*Originate end* / *Answer end* Tx: Space 1070 Hz / 2025 Hz Mark 1270 Hz / 2225 Hz Rx: Space 2025 Hz / 1070 Hz Mark 2225 Hz / 1270 Hz	Mark: 1200 Hz Space: 2200 Hz	Carrier frequencies Originate end / Answer end Tx: $f_c = 1200$ Hz / $f_c = 2400$ Hz Rx: $f_c = 2400$ / $f_c = 1200$ Hz *Message* (2 bits) / *Phase Angle* 00 / 90° 01 / 0° 10 / 180° 11 / 270°	Originate end / Answer end Tx: $f_c = 1200$ Hz / $f_c = 2400$ Hz Rx: $f_c = 2400$ Hz / $f_c = 1200$ Hz
Transmit level	0 to −12 dBm	0 to −12 dBm		
Receive level	0 to −50 dBm	0 to −50 dBm (dial-up line) 0 to −40 dBm (leased line)		

from computers to analog lines. Note that voice grade lines will support bit rates up to about 19.6 kbits/sec and that for higher rates, group and supergroup carrier lines are used. For large data rates, such as 56 kbits/sec or 1.544 Mbits/sec, it is usually more economical to use DS-0 or DS-1 digital lines (see Chapter 3) than to use modems in conjunction with FDM analog lines. Moreover, with ISDN (see Chapter 5), the telephone company provides 16- and 64-kbits/sec data lines directly to the user.

Table C-12 gives more details on some of the more popular modems. For personal computer data communication at 300, 1200, and 2400 bits/sec, the Bell 103, Bell 212A, and CCITT V.22bis modems, respectively, are used. Amateur radio operators use the Bell 103 standard for audio tones into and out of SSB transceivers for packet radio transmission on the HF amateur radio bands, and the Bell 202 standard is used for audio into and out of FM transceivers for packet transmission on the VHF and UHF bands.

C-7 Brief Computer Communications Glossary

Acoustic Coupler. A method used to couple modem signals to the telephone line via sound transducers (microphone and headphone) that are positioned on the telephone handset.

ACU (automatic calling unit). A unit that dials up the telephone line connection on receipt of the proper data signal from a computer.

Adaptive Equalizer. An electronic circuit that automatically compensates for the changing envelope delay and frequency response of a telephone line.

ARP (Address Resolution Protocol). A TCP/IP (Transmission Control Protocol/Internet Protocol) process that maps the IP address to the Ethernet address as required by TCP/IP for use with Ethernet.

ARQ (automatic retransmission request). An error correction technique whereby the message is retransmitted when the receiver detects an error.

ASCII (American Standard Code for Information Interchange). A 7-bit plus 1-parity bit code; see Table C-5.

ASR. Automatic sending–receiving.

Asynchronous Transmission. The time intervals between transmitted characters may be of unequal length. Start and stop bits are transmitted before and after each character to synchronize the receiver clock.

Baud. The data transmission rate, in units of symbols per second, whereby one symbol is counted for each possible modulation level. Baud is sometimes used incorrectly by modem manufacturers to express bits per second. A baud is the same as a bit per second if each symbol represents one bit (i.e., $M = 2$ possible levels).

Baudot Code. A 5-bit code set; see Table C-3. Also known as the *International Telegraph Alphabet (ITA) Number 2*.

BISYNC (Binary-Synchronous Communications Protocol). See Sec. C-4 and Fig. C-7.

Bridge. An electronic box that connects different local area networks at the data-link layer of the OSI model; see Fig. C-6 in Sec. C-3.

Broadband Channel. A channel that has a bandwidth much greater than that required to send one VF signal.

CCITT. The Consultative Committee for International Telephone and Telegraph, an international standards group.

Circuit Switching. A physically (or electronically) switched connection that connects the line of the calling party to that of the called party.

Concentrator. A device that connects a number of low-speed, usually asynchronous, devices to a high-speed, usually synchronous, device.

Control Character. A nonprinting functional character that provides some control function (e.g., see Table C-5 for the ASCII control characters).

CRC (cyclic redundancy check). Data are checked for errors by using a polynomial algorithm. The algorithm is used to compute a check field (usually 16 bits) that is appended to the frame of data at the transmitter and used by the receiver to verify the correctness of the data.

CSMA/CD (carrier sense multiple access/collision detection). Multiple users transmit data over a common channel. If two users transmit data at the same time, the resulting collision is detected and each tries to transmit again after a predetermined time interval.

CTS. Clear-to-send line on the RS-232C interface (see Table C-6).

CXR. Carrier.

Data Set. (1) A modem or (2) a collection of data records.

DCE (data communications equipment). A modem; see Fig. C-1.

Dial-Up Line. A switched-circuit telephone line connection (as opposed to a leased line).

DTE (data terminal equipment). The equipment that provides the data source or data sink; see Fig. C-1.

EBCDIC (Extended Binary-Coded Decimal Interchange Code). An 8-bit code used primarily in IBM equipment; see Table C-4.

Echo Check. A method of checking for errors in a data system by looping the data at the receive site back to the transmitter site for comparison.

Echo Suppressor. A device that suppresses the echoes on toll lines that are caused by transmission line reflections (usually at the hybrid interface).

Envelope Delay. In the context of telephone lines, the measure of the delay through the circuit for a sinusoidal signal at a certain frequency as compared to the minimum delay, which usually occurs at a frequency of about 1800 Hz.

Ethernet. A coaxial cable local-area network that transmits data at 10 Mbits/sec with a CSMA/CD technique. It was first developed by Xerox Corporation and is now standardized by the IEEE (IEEE 802.3).

Flow Control. Ability to stop data flow from a transmitter so that receiver buffers do not overflow.

Front-End Processor. A communications computer connected between the main computer and the remote terminals; provides line control, code conversion, data concentration, and error control.

Full Duplex. (1) A protocol that provides for simultaneous transmission of data in both directions; also, or alternatively, (2) in the context of telephone data lines, a four-wire (two-pair) circuit.

Gateway. An electronic box or logical network device that interconnects incompatible networks or devices by performing protocol conversion.

Half-Duplex. (1) A protocol that allows data to be transmitted in only one direction at a time; also, or alternatively, (2) in the context of telephone data lines, a two-wire (one-pair) circuit.

HDLC (High-Level Data Link Control Protocol). See Sec. C-4 and Fig. C-8.

Hybrid Network. A circuit that couples a two-wire telephone line to a four-wire circuit. (See Fig. 5-24.)

ISO. International Organization for Standardization.

ISO OSI Model. See Fig C-6.

LAN (local-area network). A connection of office or campus computer facilities to each other by twisted-pair wire, coaxial cable, or fiber optic cable.

Leased Line. A permanent line between two or more communication sites (no exchange switching) for leasing customers.

Message Switching. *See* Store-and-Forward.

Modem. An acronym for modulator–demodulator. A modem converts the baseband data signal to a bandpass signal for transmission over an analog telephone line, and vice versa. The Bell System calls a modem a *data set*.

Multipoint Line. A line that connects two or more communication sites and requires the use of some type of polling and addressing mechanism for each site.

Packet. A group of bits with appended address bits, sender identification bits, and/or other control bits.

Packet Switching. The transfer of data by means of addressed packets. The channel is occupied only during the transmission of the packet. In this sense, there is no permanent circuit between communication sites.

Parity Check. The addition of an extra noninformation bit to the data group so that the number of 1's in the group is either even or odd; permits the detection of single errors.

Polling. In a multipoint network, the process of inquiring which sites have data to transmit.

Protocol. The rules for control of data transmission over the communication network.

Repeater. An electronic device that receives an attenuated electrical or optical data signal with noise and regenerates a "noise-free" high-level signal (with, possibly, some bit errors) for retransmission.

Reverse Channel. A channel that allows the receiver in a modem to send slow-speed data to the transmitting modem for use in error control, circuit breaking, and network diagnostics.

RS-232C. A DCE/DTE interface. See Table C-6.

RTS. Request-to-send line on the RS-232C interface; see Table C-6.

SDLC (Synchronous Data Link Control Protocol). See Sec. C-4 and Fig. C-8.

Simplex Mode. The sending of data in one direction only.

Slicing Level. The threshold level of a comparator that determines whether the input level is converted into an output level corresponding to a binary 1 or a level corresponding to a binary 0.

Statistical Multiplexer. A time-division multiplexer that takes advantage of the fact that there is often ''dead time'' on the input data channels. Thus, by dynamically adapting the time-division algorithm, a larger number of input channels can be serviced than with a fixed (nonadaptive) time-division multiplexer.

Store-and-Forward. The process by which messages are accepted into memory and forwarded at a convenient time according to the address on the header of the message.

Switched Line. The process by which lines from different sites are connected (switched) when a communication session is desired (e.g., the dial-up telephone network).

Synchronous Transmission. The process by which data characters are transmitted at a fixed rate, with the transmitter and receiver being in synchronism. There are no start and stop bits, so this transmission technique is much more efficient than asynchronous transmission.

TCP/IP (Transmission Control Protocol/Internet Protocol). A layered set of protocols to allow high-speed (e.g., Ethernet) network connection between PCs and other computers. Different types of vendor equipment can communicate with each other over the TCP/IP network. TCP/IP corresponds to Layers 3 and 4 of the OSI model as shown in Fig. C-6.

Token Bus. A LAN that uses a unique word (a token) to be passed to a device before it can place data on the LAN for transmission.

Turnabout Time. Time required for a modem to reverse the direction of transmission when connected to a half-duplex line.

VF (Voice Frequency) Channel. A channel with a frequency response over the range of 300 to 3200 Hz.

X.25. A packet protocol. See Sec. C-2.

APPENDIX D

Introduction to MathCAD

As described in Chapter 1, MathCAD may be run on an IBM-compatible personal computer to solve communications problems. In this appendix we describe the use of the student edition of MathCAD (Version 2.0). A template floppy disk, which contains MathCAD programs for text equations and end-of-the chapter problems that are marked with a computer symbol is made available to your instructor from the publisher.

D-1 General Information

1. To start MathCAD type

   drive: \path \MCAD /M filename

   \MCAD calls up MathCAD. The option /M specifies the manual calculation mode, and filename is the name of the template MathCAD program file that you want to load into MathCAD for computation. The manual calculation option requires that you press F9 for the computation; without this option, the computation will start immediately when the file is loaded (before you have a chance to set values of parameters).
2. MathCAD will use a 8087, 80287, or a 80387 math coprocessor if you have one installed. The coprocessor is not required; however, it will certainly speed up the calculations and is strongly recommended.
3. Press the function key F10 to energize the pull-down command menu. This allows you to load files, edit files, print files, and save files. The escape key allows you to enter commands by typing them instead of using the pull-down menu.
4. Regions on the MathCAD screen are created (defined) to be any one of three types: (1) Equation—the default type; (2) Text—move cursor to a region of screen where the text is to begin, type ″, followed by the text; (3) Plots—move cursor to upper-left point where a plot region is to be defined, then type @ (shift 2).

5. Other useful function keys are F1 for help, F3 for deletion or cut of a region, and F4 to paste a region.
6. MathCAD begins calculating at the upper-left corner of the screen and proceeds to the right and down. All variables used by an equation, except for a Global variable, must be defined previously. A Global variable may be defined anywhere, even on the last line. If b is a Global variable that is equal to 5, it is displayed as b ≡ 5, where "≡" is entered by typing "~".
7. MathCAD is case sensitive. The variable A is not equal to the variable a. Built-in functions such as sin(z), log(z), and fft(v) are specified by lowercase letters, except for some functions such as Bessel functions.
8. MathCad defines an equation such as A:=B + C to mean $A = B + C$. The ":=" is obtained by typing ":". The "=" sign is used to give a computed answer. If a variable C is to have the values 0, 5, 10, 15, 20, 25, 30, this is specified by C:=0, 5 . . . 30, where the ". . ." is obtained by typing ";". A subscript is entered by typing "[" before the subscript. Elements of vectors are indexed to begin with the subscript 0 (not 1, which will give an "undefined" vector).
9. A plot region may be defined by moving the cursor to the upper-left point where the plot region is to begin and typing @. A plot grid with undefined values for the horizontal and vertical axes will appear. Use the tab key to position cursor at points on the plot so that variables to be plotted can be typed in as well as typing in minimum and maximum values for the vertical and horizontal axes. To change the size and scale of the plot, move the cursor inside the plot region and type "f". This will bring up the plot specification menu. The default scale is logs = 0,0, which specifies a linear vertical and horizontal scale. Logs = 0,2 specifies a linear vertical scale and a logarithmic horizontal scale of two cycles, and so on. Size = 6,15 denotes a plot region of 6 (horizontal) lines and 15 columns. Type = 1 specifies that the points on the plot will be connected by a line, type = d specifies that a dot will be plotted for each point, type = v specifies that the values will be shown by diamonds, and type = V specifies that diamonds connected by lines will be plotted. (For other possibilities, see the MathCAD manual.)
10. To display a table for a set of defined or computed variable values, just place the cursor at the upper-left point where a column of tabulated values are desired and type "A=", where A is the variable to be tabulated.
11. Predefined variables are $\pi = 3.14159$ and e = 2.71828. π is entered by typing Alt-P (hold down the Alternate key and press p).
12. The unit step function, $u(t)$, is a built-in function denoted in MathCAD by Φ(t), where Φ is obtained by typing Alt-H.
13. Some Greek letters are available. For example, for α, type Alt-A; for β, type Alt-B; for Φ = unit step, type Alt-H; for $\pi = 3.14159$, type Alt-P; and for ω, type Alt-W. (For other possibilities, see the MathCAD manual.)

See Table D-1 for a listing of popular MathCAD operators and Table D-2 for built-in functions. (For other possibilities, see the MathCAD manual.)

TABLE D-1 Some MathCAD Operators

Operator	Type This	Operator	Type This
$x - y$ (subtraction)	x−y	$\sum_i x$	i$x
$x + y$ (sum)	x+y	$\prod_i x$	i#x
$x \cdot y$ (multiplication)	x*y	$\int f(x)\, dx$	x&f(x)
x/y (division)	x/y	$\frac{df(x)}{dx}$	x?f(x)
$\mathbf{x} \cdot \mathbf{y}$ (dot product)	x*y	$x > y$	x > y
$\mathbf{x} \cdot \mathbf{y}$ (matrix multiply)	x*y	$x < y$	x < y
$\mathbf{x} \times \mathbf{y}$ (cross prod)	x Alt−* y	$x \geq y$	x Alt−) y
x^* (conjugate)	x″	$x \leq y$	x Alt−(y
$\mathbf{x}^T$ (transpose)	x Alt−!	$x \neq y$	x Alt−# y
x^y (power)	x^y	$x \approx y$	z Alt−> y
$\Sigma\, x$ (sum of values)	Alt−$ x		
$\sqrt{x}$ (square root)	\x		
$\lvert x \rvert$ (absolute value)	|x		

TABLE D-2 Built-in MathCAD Functions[a]

Trigonometric	Hyperbolic	Logs, exp, and complex	Besel	Fast Fourier Transform (See Sec. 2-7)	Statistics	Miscellaneous	
						Function	Definition
sin(z)	sinh(z)	exp(z)	Y0(x)	fft(v)	mean(v)	floor(x)	Greatest integer $\leq$ x
cos(z)	cosh(z)	ln(z)	Y1(x)	ifft(v)	stdev(v)	ceil(x)	Least integer $\geq$ x
tan(z)	tanh(z)	log(z)[b]	Yn(n,x)	cfft(v)	var(v)	Φ(x)	Unit step = u(x)
asin(z)	asinh(z)	Re(z)	J0(x)	icfft(v)	Γ(z)	mod(x1,x2)	Remainder of x1/x2
acos(z)	acosh(z)	Im(z)	J1(x)		erf(x)	until (x1,x2)	Returns x2 until x1 < 0
atan(z)	atanh(z)	arg(z)	Jn(n,x)		cnorm(x)[c]	if(c,x1,x2)	Returns x1 if c $\neq$ 0, x2 otherwise
					rnd(x)[d]		
					hist(v1,v2)[e]		

[a] Notation used: x and y are real, z may be complex, n is an integer, v is a vector, and M is a matrix.

[b] Base 10 logarithm.

[c] This is the CDF, $F(x)$, for the Gaussian distribution with a mean of 0 and a variance of 1.

[d] Returns a random number that is uniformly distributed between 0 and x.

[e] Histogram function. v1 is a vector of allowed discrete values and v2 is the data vector (elements consist of the data values).

D-2 MathCAD Example

A MathCAD program and the corresponding computed plot for (5-23) from Sec. 5-3 is shown in Table D-3. This is file 5-23.MCD from the template floppy disk. The plot shown corresponds to Fig. 5-9 for the case when the digital modulation index is selected to be $h = 0.7$.

TABLE D-3 MathCAD Example: File 5-23

```
The Student Edition of MathCAD 2.0                        For Educational Use Only

 Ac := 1          R := 1                     1              R
 ** SELECT Mod. Index, h, by ***     Tb := -     DF := h· -        Tb = 1    DF = 0.35
 ** Global on next screen. *****             R              2


  xs(f,n) := π·Tb·(f - DF·(2·n - 3))                 xf(f) := 2·π·f·Tb
             sin(xs(f,n))                            xF := 2·π·DF·Tb
  A(f,n) := ─────────────                            xc(n,m) := xF·(n + m - 3)
             xs(f,n)

               cos(xf(f) - xc(n,m)) - cos(xF)·cos(xc(n,m))
  B(f,n,m) := ──────────────────────────────────────────────
                                  2
                 1 + (cos(xF))  - 2·cos(xF)·cos(xf(f))


                   2                                   2
  P1(f) := A(f,1) ·(1 + B(f,1,1))      P2(f) := A(f,2) ·(1 + B(f,2,2))
  P12(f) := 2·B(f,1,2)·A(f,1)·A(f,2)

                   2
   P(f) := 0.5·Ac ·Tb·(P1(f) + P2(f) + P12(f))
              See NEXT SCREEN for plot.                      P(0) = 0.827

            R
 df := 1.5· ──          f := 0,df ..1.5·R
            50              ** Compare with Fig. 5-9 **
                        1.5

 h ≡ 0.7    <--SELECT

 DF = 0.35
 R = 1
 Tb = 1
                        P(f)


                          0
                          0                f                    1.5·R
```

References

ABRAMOWITZ, M., and I. A. STEGUN (Editors), *Handbook of Mathematical Functions,* National Bureau of Standards, Superintendent of Documents, U.S. Government Printing Office, Washington, DC, 1964. Also available in paperback from Dover Publications, New York, 1965.

AMOROSO, F., "The Bandwidth of Digital Data Signals," *IEEE Communications Magazine,* vol. 18, no. 6, November 1980, pp. 13–24.

AOYAMA, T., W. R. DAUMER, and G. MODENA (Editors), Special Issue on Voice Coding for Communications, *IEEE Journal on Selected Areas of Communications,* vol. 6, February 1988.

ARRL, *The ARRL Handbook for the Radio Amateur,* 68th ed., L. D. Wolfgang, C. L. Hutchinson, et al. (Editors), American Radio Relay League, Newington, CT, 1991.

BASCH, E. E., and T. G. BROWN, "Introduction to Coherent Optical Fiber Transmission," *IEEE Communications Magazine,* vol. 23, May 1985, pp. 23–30.

BEDROSIAN, E. B., and S. O. RICE, "Distortion and Crosstalk of Linearly Filtered Angle-Modulated Signals," *Proceedings of the IEEE,* vol. 56, January 1968, pp. 2–13.

BELL TELEPHONE LABORATORIES, *Transmission Systems for Communications,* 4th ed., Western Electric Company, Winston-Salem, NC, 1970.

BENDAT, J. S., and A. G. PIERSOL, *Random Data: Analysis and Measurement Procedures,* Wiley-Interscience, New York, 1971.

BENEDETTO, S., E. BIGLIERI, and V. CASTELLANI, *Digital Transmission Theory,* Prentice Hall, Englewood Cliffs, NJ, 1987.

BENNETT, W. R., and J. R. DAVEY, *Data Transmission,* McGraw-Hill Book Company, New York, 1965.

BENSON, K. B., and J. C. WHITAKER, *Television Engineering Handbook,* Rev. ed., McGraw-Hill Book Company, New York, 1992.

BERGLAND, G. D., "A Guided Tour of the Fast Fourier Transform," *IEEE Spectrum,* vol. 6, July 1969, pp. 41–52.

BHARGAVA, V. K., "Forward Error Correction Schemes for Digital Communications," *IEEE Communications Magazine,* vol. 21, January 1983, pp. 11–19.

BHARGAVA, V. K., D. HACCOUN, R. MATYAS, and P. P. NUSPL, *Digital Communications by Satellite,* Wiley-Interscience, New York, 1981.

BIGLIERI, E., D. DIVSALAR, P. J. MCLANE, and M. K. SIMON, *Introduction to Trellis-Coded Modulation with Applications,* Macmillan Publishing Company, New York, 1991.

BLACKMAN, R. B., and J. W. TUKEY, *The Measurement of Power Spectra,* Dover, New York, 1958.

BLAHUT, R. E., *Theory and Practice of Error Control Codes,* Addison-Wesley Publishing Company, Reading, MA, 1983.

BLANCHARD, A., *Phase-Locked Loops,* Wiley-Interscience, New York, 1976.

BOWRON, P., and F. W. STEPHENSON, *Active Filters for Communications and Instrumentation,* McGraw-Hill Book Company, New York, 1979.

BREMERMANN, H., *Distributions, Complex Variables and Fourier Transforms,* Addison-Wesley Publishing Company, Reading, MA, 1965.

BRILEY, B. E., *Introduction to Telephone Switching,* Addison-Wesley Publishing Company, Reading, MA, 1983.

BROADCASTING, *Broadcasting Yearbook,* Broadcasting Publications, Inc., Washington, DC, 1991.

BYLANSKI, P., and D. G. W. INGRAM, *Digital Transmission Systems,* Peter Peregrinus Ltd., Herts, England, 1976.

CAMPANELLA, M., U. LOFASO, and G. MAMOLA, "Optimum Pulse Shape for Minimum Spectral Occupancy in FSK Signals," *IEEE Transactions on Vehicular Technology,* vol. VT-33, May 1984, pp. 67–75.

CARLSON, A. B., *Communication Systems,* 3rd ed., McGraw-Hill Book Company, New York, 1986.

CATTERMOLE, K. W., *Principles of Pulse-code Modulation,* American Elsevier, New York, 1969.

CCITT STUDY GROUP XVII, "Recommendation V.32 for a Family of 2-wire, Duplex Modems Operating on the General Switched Telephone Network and on Leased Telephone-type Circuits," *Document AP VIII–43–E,* May 1984.

CHAKRABORTY, D., "VSAT Communication Networks—An Overview," *IEEE Communications Magazine,* vol. 26, no. 5, May 1988, pp. 10–24.

CHAPMAN, R. C. (Editor), "The SLC96 Subscriber Loop Carrier System," *AT&T Bell Laboratories Technical Journal,* vol. 63, no. 10, Part 2, December 1984, pp. 2273–2437.

CHILDERS, D., and A. DURLING, *Digital Filtering and Signal Processing,* West Publishing Company, New York, 1975.

CHORAFAS, D. N., *Telephony, Today and Tomorrow,* Prentice Hall, Englewood Cliffs, NJ, 1984.

CLARK, G. C., and J. B. CAIN, *Error–Correction Coding for Digital Communications,* Plenum Publishing Corporation, New York, 1981.

COOPER, G. R., and C. D. MCGILLEM, *Modern Communications and Spread Spectrum,* McGraw-Hill Book Company, New York, 1986.

COUCH, L. W., "A Study of a Driven Oscillator with FM Feedback by Use of a Phase-Lock-Loop Model," *IEEE Transactions on Microwave Theory and Techniques,* vol. MTT-19, no. 4, April 1971, pp. 357–366.

COURANT, R., and D. HILBERT, *Methods of Mathematical Physics,* Vol. 1, Wiley-Interscience, New York, 1953.

DAMMANN, C. L., L. D. MCDANIEL, and C. L. MADDOX, "D2 Channel Bank—Multiplexing and Coding," *Bell System Technical Journal,* vol. 51, October 1972, pp. 1675–1700.

DAVENPORT, W. B., JR., and W. L. ROOT, *An Introduction to the Theory of Random Signals and Noise,* McGraw-Hill Book Company, New York, 1958.

DAVIDOFF, M. R., *The Satellite Experimenter's Handbook,* American Radio Relay League, Newington, CT, 1984.

DAVIS, D. W., and D. L. A. BARBER, *Communication Networks for Computers,* John Wiley & Sons, New York, 1973.

DEANGELO, J., "New Transmitter Design for the 80's," *BM/E (Broadcast Management/Engineering),* vol. 18, March 1982, pp. 215–226.

DECINA, M., and G. MODENA, "CCITT Standards on Digital Speech Processing," *IEEE Journal on Selected Areas of Communications,* vol. 6, February 1988, pp. 227–234.

DEFFEBACH, H. L., and W. O. FROST, "A Survey of Digital Baseband Signaling Techniques," *NASA Technical Memorandum NASATM X-64615,* June 30, 1971.

DEJAGER, F., "Delta Modulation, A Method of PCM Transmission Using a 1-Unit Code," *Phillips Research Report,* no. 7, December 1952, pp. 442–466.

DEJAGER, F., and C. B. DEKKER, "Tamed Frequency Modulation: A Novel Method to Achieve Spectrum Economy in Digital Transmission," *IEEE Transactions on Communications,* vol. COM–26, May 1978, pp. 534–542.

DESHPANDE, G. S., and P. H. WITTKE, "Correlative Encoding of Digital FM," *IEEE Transactions on Communications,* vol. COM–29, February 1981, pp. 156–162.

DHAKE, A. M., *Television Engineering,* McGraw-Hill Book Company, New Delhi, India, 1980.

DIXON, R. C., *Spread Spectrum Systems,* 2nd ed., John Wiley & Sons, New York, 1984.

EDELSON, B. I., and A. M. WERTH, "SPADE System Progress and Application," *COMSAT Technical Review,* Spring 1972, pp. 221–242.

EILERS, C. G., "TV Multichannel Sound—The BTSC System," *IEEE Transactions on Consumer Electronics,* vol. CE–31, February 1985, pp. 1–7.

FASHANO, M., and A. L. STRODTBECK, "Communication System Simulation and Analysis with SYSTID," *IEEE Journal on Selected Areas of Communications,* vol. SAC–2, January 1984, pp. 8–29.

FEHER, K., *Digital Communications—Satellite/Earth Station Engineering,* Prentice Hall, Englewood Cliffs, NJ, 1981.

FELLER, W., *An Introduction to Probability Theory and Its Applications,* John Wiley & Sons, New York, 1957, p. 168.

FIKE, J. L., and G. E. FRIEND, *Understanding Telephone Electronics,* 2nd ed., Texas Instruments, Dallas, 1984.

FINK, D. G., and H. W. BEATY (Editors), *Standard Handbook for Electrical Engineers,* McGraw-Hill Book Company, New York, 1978.

FLANAGAN, J. L., M. R. SCHROEDER, B. S. ATAL, R. E. CROCHIERE, N. S. JAYANT, and J. M. TRIBOLET, "Speech Coding," *IEEE Transactions on Communications,* vol. COM–27, April 1979, pp. 710–737.

FORNEY, G. D., "The Viterbi Algorithm," *Proceedings of the IEEE,* vol. 61, March 1973, pp. 268–273.

GALLAGHER, R. G., *Information Theory and Reliable Communications,* John Wiley & Sons, New York, 1968.

GARDNER, F. M., *Phaselock Techniques,* 2nd ed., John Wiley & Sons, New York, 1979.

GARDNER, F. M., and W. C. LINDSEY (Guest Editors), Special Issue on Synchronization, *IEEE Transactions on Communications,* vol. COM–28, no. 8, August 1980.

GERSHO, A., "Charge-Coupled Devices: The Analog Shift Register Comes of Age," *IEEE Communications Society Magazine,* vol. 13, November 1975, pp. 27–32.

GLISSON, T. L., *Introduction to System Analysis,* McGraw-Hill Book Company, New York, 1986.

GOLDBERG, R. R., *Fourier Transforms,* Cambridge University Press, New York, 1961.

GOLOMB, S. W. (Editor), *Digital Communications with Space Applications,* Prentice Hall, Englewood Cliffs, NJ, 1964.

GRIFFITHS, J., *Radio Wave Propagation and Antennas,* Prentice Hall, Englewood Cliffs, NJ, 1987.

GROB, B., *Basic Television,* 4th ed., McGraw-Hill Book Company, New York, 1975.

GULSTAD, K., "Vibrating Cable Relay," *Electrical Review* (London), vol. 42, 1898; vol. 51, 1902.

GUPTA, S. C., "Phase Locked Loops," *Proceedings of the IEEE,* vol. 63, February 1975, pp. 291–306.

HA, T. T., *Digital Satellite Communications,* Macmillan Publishing Company, New York, 1986.

HA, T. T., *Digital Satellite Communications,* 2nd ed., John Wiley & Sons, New York, 1989.

HAMMING, R. W., "Error Detecting and Error Correcting Codes," *Bell System Technical Journal,* vol. 29, April 1950, pp. 147–160.

HARRIS, F. J., "On the Use of Windows for Harmonic Analysis with the Discrete Fourier Transform," *Proceedings of the IEEE,* vol. 66, January 1978, pp. 51–83.

HARRIS, F. J., "The Discrete Fourier Transform Applied to Time Domain Signal Processing," *IEEE Communications Magazine,* vol. 20, May 1982, pp. 13–22.

HARTLEY, R. V., "Transmission of Information," *Bell System Technical Journal,* vol. 27, July 1948, pp. 535–563.

HAUS, H. A., Chairman IRE Subcommittee 7.9 on Noise, "Description of Noise Performance of Amplifiers and Receiving Systems," *Proceedings of the IEEE,* vol. 51, no. 3, March 1963, pp. 436–442.

HAYKIN, S., *Communication Systems,* 2nd ed., John Wiley & Sons, New York, 1983.

HOLMES, J. K., *Coherent Spread Spectrum Systems,* Wiley-Interscience, New York, 1982.

HUANG, D. T., and C. F. VALENTI, "Digital Subscriber Lines: Network Considerations for ISDN Basic Access Standard," *Proceedings of the IEEE,* vol. 79, February 1991, pp. 125–144.

IRMER, T., "An Overview of Digital Hierarchies in the World Today," *IEEE International Conference on Communications,* San Francisco, June 1975, pp. 16–1 to 16–4.

IRWIN, J. D., *Basic Engineering Circuit Analysis,* Macmillan Publishing Company, New York, 1st ed., 1984, 2nd ed., 1987.

JACOBS, I., "Design Considerations for Long-Haul Lightwave Systems," *IEEE Journal on Selected Areas of Communications,* vol. 4, December 1986, pp. 1389–1395.

JAMES, R. T., and P. E. MUENCH, "A.T.&T. Facilities and Services," *Proceedings of the IEEE,* vol. 60, November 1972, pp. 1342–1349.

JAYANT, N. S., "Digital Encoding of Speech Waveforms," *Proceedings of the IEEE,* May 1974, pp. 611–632.

JAYANT, N. S., "Coding Speech at Low Bit Rates," *IEEE Spectrum,* vol. 23, August 1986, pp. 58–63.

JAYANT, N. S., and P. NOLL, *Digital Coding of Waveforms,* Prentice Hall, Englewood Cliffs, NJ, 1984.

JENKINS, M. G., and D. G. WATTS, *Spectral Analysis and Its Applications,* Holden-Day, San Francisco, CA, 1968.

JERRI, A. J., "The Shannon Sampling Theorem—Its Various Extensions and Applications: A Tutorial Review," *Proceedings of the IEEE,* vol. 65, no. 11, November 1977, pp. 1565–1596.

JORDAN, E. C. (Editor), *Reference Data for Engineers: Radio, Electronics, Computer, and Communications,* 7th ed., Howard W. Sams and Company, Indianapolis, 1985.

JORDAN, E. C., and K. G. BALMAIN, *Electromagnetic Waves and Radiating Systems,* 2nd ed., Prentice Hall, Englewood Cliffs, NJ, 1968.

JURGEN, R. K., "High-Definition Television Update," *IEEE Spectrum,* vol. 25, April 1988, pp. 56–62.

JURGEN, R. K., "The Challenges of Digital HDTV," *IEEE Spectrum,* vol. 28, April 1991, pp. 28–30 and 71–73.

KAUFMAN, M., and A. H. SEIDMAN, *Handbook of Electronics Calculations,* McGraw-Hill Book Company, New York, 1979.

KAY, S. M., *Modern Spectral Estimation—Theory and Applications,* Prentice Hall, Englewood Cliffs, NJ, 1986.

KAY, S. M., and S. L. MARPLE, "Spectrum-Analysis—A Modern Perspective," *Proceedings of the IEEE,* vol. 69, November 1981, pp. 1380–1419.

KAZAKOS, D., and P. PAPANTONI-KAZAKOS, *Detection and Estimation,* Computer Science Press, New York, 1990.

KLAPPER, J., and J. T. FRANKLE, *Phase-Locked and Frequency–Feedback Systems,* Academic Press, New York, 1972.

KRAUS, J. D., *Radio Astronomy,* 2nd ed., Cygnus-Quasar Books, Powell, OH, 1986.

KRAUSS, H. L., C. W. BOSTIAN, and F. H. RAAB, *Solid State Radio Engineering,* John Wiley & Sons, New York, 1980.

KRETZMER, E. R., "Generalization of a Technique for Binary Data Communications," *IEEE Transactions on Communication Technology,* vol. COM-14, February 1966, pp. 67–68.

LAM, S. S., "Satellite Packet Communications, Multiple Access Protocols and Performance," *IEEE Transactions on Communications,* vol. COM–27, October 1979, pp. 1456–1466.

LATHI, B. P., *Modern Digital and Analog Communication Systems,* Holt, Rinehart and Winston, New York, 1983 (2nd ed., 1989).

LEE, W. C. Y., "Elements of Cellular Mobile Radio Systems," *IEEE Transactions on Vehicular Technology,* vol. VT–35, May 1986, pp. 48–56.

LEE, W. C. Y., *Mobile Cellular Telecommunications Systems,* McGraw-Hill Book Company, New York, 1989.

LENDER, A., "The Duobinary Technique for High Speed Data Transmission," *IEEE Transactions on Communication Electronics,* vol. 82, May 1963, pp. 214–218.

LENDER, A., "Correlative Level Coding for Binary-Data Transmission," *IEEE Spectrum,* vol. 3, February 1966, pp. 104–115.

LIN, S., and D. J. COSTELLO, JR., *Error Control Coding,* Prentice Hall, Englewood Cliffs, NJ, 1983.

LINDSEY, W. C., and C. M. CHIE, "A Survey of Digital Phase-Locked Loops," *Proceedings of the IEEE,* vol. 69, April 1981, pp. 410–431.

LINDSEY, W. C., and M. K. SIMON, *Telecommunications System Engineering,* Prentice Hall, Englewood Cliffs, NJ, 1973.

LUCKY, R. W., and H. R. RUDIN, "Generalized Automatic Equalization for Communication Channels," *IEEE International Communication Conference,* vol. 22, 1966.

LUCKY, R. W., J. SALZ, and E. J. WELDON, *Principles of Data Communication,* McGraw-Hill Book Company, New York, 1968.

MANASSEWITSCH, V., *Frequency Synthesizers,* 3rd ed., Wiley-Interscience, New York, 1987.

MARKLE, R. E., "Single Sideband Triples Microwave Radio Route Capacity," *Bell Systems Laboratories Record,* vol. 56, no. 4, April 1978, pp. 105–110.

MARKLE, R. E., "Prologue, The AR6A Single-Sideband Microwave Radio System," *Bell System Technical Journal,* vol. 62, December 1983, pp. 3249–3253. (The entire December 1983 issue deals with the AR6A system.)

MARPLE, S. L., *Digital Spectral Analysis,* Prentice Hall, Englewood Cliffs, NJ, 1986.

McEliece, R. J., The Theory of Information and Coding (*Encyclopedia of Mathematics and Its Applications,* vol. 3), Addison-Wesley Publishing Company, Reading, MA, 1977.

Melsa, J. L., and D. L. Cohn, *Decision and Estimation Theory,* McGraw-Hill Book Company, New York, 1978.

Mennie, D., "AM Stereo: Five Competing Options," *IEEE Spectrum,* vol. 15, June 1978, pp. 24–31.

Miyaoka, S., "Digital Audio Is Compact and Rugged," *IEEE Spectrum,* vol. 21, March 1984, pp. 35–39.

Motorola, *Telecommunications Device Data,* Motorola Semiconductor Products, Austin, TX, 1985.

Mumford, W. W., and E. H. Scheibe, *Noise Performance Factors in Communication Systems,* Horizon House–Microwave, Dedham, MA, 1968.

Murphy, E., "Whatever Happened to AM Stereo?," *IEEE Spectrum,* vol. 25, March 1988, p. 17.

Nilsson, J. W., *Electric Circuits,* 3rd ed., Addison-Wesley Publishing Company, Reading, MA, 1990.

North, D. O., "An Analysis of the Factors Which Determine Signal/Noise Discrimination in Pulsed-Carrier Systems," *RCA Technical Report.* PTR–6–C, June 1943; reprinted in *Proceedings of the IEEE,* vol. 51, July 1963, pp. 1016–1027.

Nyquist, H., "Certain Topics in Telegraph Transmission Theory," *Transactions of the AIEE,* vol. 47, February 1928, pp. 617–644.

Okumura, Y., and M. Shinji (Guest Editors) Special Issue on Mobile Radio Communications, *IEEE Communications Magazine,* vol. 24, February 1986.

Olmsted, J.M.H., *Advanced Calculus,* Appleton-Century-Crofts, New York, 1961.

O'Neal, J. B., "Delta Modulation and Quantizing Noise, Analytical and Computer Simulation Results for Gaussian and Television Input Signals," *Bell System Technical Journal,* vol. 45, January 1966(a), pp. 117–141.

O'Neal, J. B., "Predictive Quantization (DPCM) for the Transmission of Television Signals," *Bell System Technical Journal,* vol. 45, May–June 1966(b), pp. 689–721.

Oppenheim, A. V., and R. W. Schafer, *Digital Signal Processing,* Prentice Hall, Englewood Cliffs, NJ, 1975.

Oppenheim, A. V., and R. W. Schafer, *Discrete-Time Signal Processing,* Prentice Hall, Englewood Cliffs, NJ, 1989.

Pandhi, S. N., "The Universal Data Connection," *IEEE Spectrum,* vol. 24, July 1987, pp. 31–37.

Panter, P. F, *Modulation, Noise and Spectral Analysis,* McGraw-Hill Book Company, New York, 1965.

Papoulis, A., *Probability, Random Variables and Stochastic Processes,* McGraw-Hill Book Company, New York, 2nd ed., 1984, 3rd ed., 1991.

Park, J. H., Jr., "On Binary DPSK Detection," *IEEE Transactions on Communications,* vol. COM–26, April 1978, pp. 484–486.

Pasupathy, S., "Correlative Coding," *IEEE Communications Society Magazine,* vol. 15, July 1977, pp. 4–11.

Peebles, P. Z., *Communication System Principles,* Addison-Wesley Publishing Company, Reading, MA, 1976.

Peebles, P. Z., *Digital Communication Systems,* Prentice Hall, Englewood Cliffs, NJ, 1987.

Peebles, P. Z., *Probability, Random Variables, and Random Signal Principles,* 2nd ed., McGraw-Hill Book Company, New York, 1987.

PEEK, J.B.H., "Communications Aspects of the Compact Disk Digital Audio System," *IEEE Communications Magazine,* vol. 23, February 1985, pp. 7–15.

PERSONICK, S. D., "Digital Transmission Building Blocks," *IEEE Communications Magazine,* vol. 18, January 1980, pp. 27–36.

PETERSON, W. W., and E. J. WELDON, *Error-Correcting Codes,* MIT Press, Cambridge, MA, 1972.

PETTIT, R. H., *ECM and ECCM Techniques for Digital Communication Systems,* Lifetime Learning Publications (a division of Wadsworth, Inc.), Belmont, CA, 1982.

PRATT, T., and C. W. BOSTIAN, *Satellite Communications,* John Wiley & Sons, New York, 1986.

PRITCHARD, W. L., and C. A. KASE, "Getting Set for Direct-Broadcast Satellites," *IEEE Spectrum,* vol. 18, August 1981, pp. 22–28.

PROAKIS, J. G., *Digital Communications,* McGraw-Hill Book Company, New York, 1983.

QURESHI, S., "Adaptive Equalization," *IEEE Communications Magazine,* vol. 20, March 1982, pp. 9–16.

RAMO, S., J. R. WHINNERY, and T. VANDUZER, *Fields and Waves in Communication Electronics,* 2nd ed., John Wiley & Sons, Inc., New York, 1984.

RAPPAPORT, T. S., "Characteristics of UHF Multipath Radio Channels in Factory Buildings," *IEEE Transactions on Antennas and Propagation,* vol. 37, August 1989, pp. 1058–1069.

RHEE, S. B., and W. C. Y. LEE (Editors), Special Issue on Digital Cellular Technologies, *IEEE Transactions on Vehicular Technology,* vol. 40, May 1991.

RICE, S. O., "Mathematical Analysis of Random Noise," *Bell System Technical Journal,* vol. 23, July 1944, pp. 282–333, and vol. 24, January 1945, pp. 46–156; reprinted in *Selected Papers on Noise and Stochastic Processes,* N. Wax (Editor), Dover Publications, New York, 1954.

RICE, S. O., "Statistical Properties of a Sine-Wave Plus Random Noise," *Bell System Technical Journal,* vol. 27, January 1948, pp. 109–157.

ROOT, W. L., "Remarks, Most Historical, on Signal Detection and Signal Parameter Estimation," *Proceedings of the IEEE,* vol. 75, November 1987, pp. 1446–1457.

ROSE, R. B., "MIMIMUF: A Simplified MUF–Prediction Program for Microcomputers," *QST,* vol. 66, December 1982, pp. 36–38.

ROSE, R. B., "Technical Correspondence—MIMIMUF Revisited," *QST,* vol. 68, March 1984, p. 46.

ROWE, H. E., *Signals and Noise in Communication Systems,* D. Van Nostrand Company, Princeton, NJ, 1965.

RYAN, J. S. (Editor), Special Issue on Telecommunications Standards, *IEEE Communications Magazine,* vol. 23, January 1985.

RYDER, J. D., and D. G. FINK, *Engineers and Electrons,* IEEE Press, New York, 1984.

SABIN, W. E., and E. O. SCHOENIKE, *Single-Sideband Systems and Circuits,* McGraw-Hill Book Company, New York, 1987.

SALTZBERG, B. R., T. R. HSING, J. M. CIOFFI, and D. W. LIN (Editors), Special Issue on High-Speed Digital Subscriber Lines, *IEEE Journal on Selected Areas in Communications,* vol. 9, August 1991.

SCHARF, L. L., *Statistical Signal Processing: Detection, Estimation, and Time Series Analysis,* Addison-Wesley Publishing Company, Reading, MA, 1991.

SCHILLING, D. L., R. L. PICKHOLTZ, and L. B. MILSTEIN, "Spread Spectrum Goes Commercial," *IEEE Spectrum,* vol. 27, August 1990, pp. 40–45.

SCHILLING, D. L., L. B. MILSTEIN, R. L. PICKHOLTZ, M. KULLBACK, and F. MILLER, "Spread Spectrum for Commercial Communications," *IEEE Communications Magazine,* vol. 29, April 1991, pp. 66–67.

SCHWARTZ, M., W. R. BENNETT, and S. STEIN, *Communication Systems and Techniques,* McGraw-Hill Book Company, New York, 1966.

SHANMUGAN, K. S., *Digital and Analog Communication Systems,* John Wiley & Sons, New York, 1979.

SHANMUGAN, K. S., and A. M. BREIPOHL, *Random Signals: Detection, Estimation and Data Analysis,* John Wiley & Sons, New York, 1988.

SHANNON, C. E., "A Mathematical Theory and Communication," *Bell System Technical Journal,* vol. 27, July 1948, pp. 379–423, and October 1948, pp. 623–656.

SHANNON, C. E., "Communication in the Presence of Noise," *Proceedings of the IRE,* vol. 37, January 1949, pp. 10–21.

SIMON, M. K., "Comments on 'On Binary DPSK Detection'," *IEEE Transactions on Communications,* vol. COM–26, October 1978, pp. 1477–1478.

SINNEMA, W., and R. MCPHERSON, *Electronic Communications,* Prentice Hall Canada, Inc., Scarborough, Ontario, 1991.

SIPERKO, C. M., "LaserNet—A Fiber Optic Intrastate Network (Planning and Engineering Considerations)," *IEEE Communications Magazine,* vol. 23, May 1985, pp. 31–45.

SKLAR, B., *Digital Communications,* Prentice Hall, Englewood Cliffs, NJ, 1988.

SLEPIAN, D., "On Bandwidth," *Proceedings of the IEEE,* vol. 64, No. 3, March 1976, pp. 292–300.

SMITH, B., "Instantaneous Companding of Quantized Signals," *Bell System Technical Journal,* vol. 36, May 1957, pp. 653–709.

SMITH, J. R., *Modern Communication Circuits,* McGraw-Hill Book Company, New York, 1986.

SOLOMON, L., "The Upcoming New World of TV Reception," *Popular Electronics,* vol. 15, no. 5, May 1979, pp. 49–62.

SPILKER, J. J., *Digital Communications by Satellite,* Prentice Hall, Englewood Cliffs, NJ, 1977.

STALLINGS, W., *Data and Computer Communications,* Macmillan Publishing Company, New York, 1985.

STARK, H., F. B. TUTEUR, and J. B. ANDERSON, *Modern Electrical Communications,* Prentice Hall, Englewood Cliffs, NJ, 1988.

STIFFLER, J. J., *Theory of Synchronous Communication,* Prentice Hall, Englewood Cliffs, NJ, 1971.

STUMPERS, F. L., "Theory of Frequency–Modulation Noise," *Proceedings of the IRE,* vol. 36, September 1948, pp. 1081–1092.

SUNDE, E. D., *Communications Engineering Technology,* John Wiley & Sons, New York, 1969.

SWEENEY, P., *Error Control Coding,* Prentice Hall, Englewood Cliffs, NJ, 1991.

TAUB, H., and D. L. SCHILLING, *Principles of Communication Systems,* 2nd ed., McGraw-Hill Book Company, New York, 1986.

THOMAS, J. B., *An Introduction to Statistical Communication Theory,* John Wiley & Sons, New York, 1969.

TURIN, G., "An Introduction to Digital Matched Filters," *Proceedings of the IEEE,* vol. 64, no. 7, July 1976, pp. 1092–1112.

UNGERBOECK, G., "Channel Coding with Multilevel/Phase Signals," *IEEE Transactions on Information Theory,* vol. IT–28, January 1982, pp. 55–67.

UNGERBOECK, G., "Trellis-coded Modulation with Redundant Signal Sets," Parts 1 and 2, *IEEE Communications Magazine,* vol. 25, no. 2, February 1987, pp. 5–21.

VAN DER ZIEL, A., *Noise in Solid State Devices and Circuits,* Wiley-Interscience, New York, 1986.

VITERBI, A. J., "When not to Spread Spectrum—A Sequel," *IEEE Communications Magazine,* vol. 23, April 1985, pp. 12–17.

VITERBI, A. J., and J. K. OMURA, *Principles of Digital Communication and Coding,* McGraw-Hill Book Company, New York, 1979.

WEI, L., "Rotationally Invariant Convolutional Channel Coding with Expanded Signal Space—Part II: Nonlinear Codes," *IEEE Journal on Selected Areas in Communications,* vol. SAC-2, no. 2, September 1984, pp. 672–686.

WHALEN, A. D., *Detection of Signals in Noise,* Academic Press, New York, 1971.

WHITTAKER, E. T., "On Functions Which Are Represented by the Expansions of the Interpolation Theory," *Proceedings of the Royal Society* (Edinburgh), vol. 35, 1915, pp. 181–194.

WIENER, N., *Extrapolation, Interpolation, and Smoothing of Stationary Time Series with Engineering Applications,* MIT Press, Cambridge, MA, 1949.

WOZENCRAFT, J. M., and I. M. JACOBS, *Principles of Communication Engineering,* John Wiley & Sons, New York, 1965.

WYLIE, C. R., JR., *Advanced Engineering Analysis,* John Wiley & Sons, New York, 1960.

YOUNG, P. H., *Electronic Communication Techniques,* Merrill Publishing Co., Columbus, OH, 1990.

ZIEMER, R. E., and R. L. PETERSON, *Digital Communications and Spread Spectrum Systems,* Macmillan Publishing Company, New York, 1985.

ZIEMER, R. E., and W. H. TRANTER, *Principles of Communications,* 3rd ed., Houghton Mifflin Company, Boston, 1990.

ZIEMER, R. E., W. H. TRANTER, and D. R. FANNIN, *Signals and Systems,* Macmillan Publishing Company, New York, 1st ed., 1983, 2nd ed., 1989.

ZOU, W. Y., "Comparison of Proposed HDTV Terrestrial Broadcasting Systems," *IEEE Transactions on Broadcasting,* vol. 37, December 1991, pp. 145–147.

Answers to Selected Problems

Chapter 1

1-4 1.97 bits

1-10 $H = 3.32$ bits, $T = 1.66$ sec

1-14 **(a)** $R = 3.6 \times 10^7$ bits/sec **(b)** $C = 5.23 \times 10^7$ bits/sec

(c) Allow 10 intensity levels for each pixel and allow each pixel to be one of three colors.

Chapter 2

2-9 36 dB

2-10 **(a)** 4.08×10^{-14} W **(b)** -103.9 dBm **(c)** $1.75\ \mu$V

2-16 $S(f) = -\dfrac{A}{\omega^2} + Ae^{-j\omega T_0}\left(\dfrac{1}{\omega^2} + j\dfrac{T_0}{\omega}\right)$, where $\omega = 2\pi f$.

Note that $S(0) = AT_0^2/2$.

2-40 **(a)** $(\sin 4t)/(4t)$ **(b)** 7

2-45 **(a)** $-0.4545A$ **(b)** $0.9129A$

2-46 **(a)** The expression for $v(t)$ over one period is

$$v(t) = \begin{cases} \dfrac{3A}{T}t, & 0 < t < \frac{1}{3}T \\ \dfrac{-3A}{T}\left(t - \dfrac{2}{3}T\right), & \frac{1}{3}T < t < \frac{5}{6}T \\ \dfrac{3A}{T}(t - T), & \frac{5}{6}T < t < T \end{cases}$$

(b) $$c_n = \begin{cases} \frac{1}{4}A, & n = 0 \\ \dfrac{3Ae^{-j2\pi n/3}}{(\pi n)^2}, & n = \text{odd} \\ 0, & \text{other } n \end{cases}$$

(c) $V_{\text{rms}} = A/2$ **(d)** BW = $(1/T)$ Hz

2-48 **(a)** $(A_1 + A_2)/\sqrt{2}$ **(b)** $\left(\sqrt{A_1^2 + A_2^2}\,\right)/\sqrt{2}$ **(c)** $|A_2 - A_1|/\sqrt{2}$

(d) $\left(\sqrt{A_1^2 + A_2^2}\,\right)/\sqrt{2}$ **(e)** $\left(\sqrt{A_1^2 + A_2^2}\,\right)/\sqrt{2}$

2-58
$$c_n = \begin{cases} \dfrac{AT^2}{2T_0}, & n = 0 \\[2ex] \dfrac{A\{e^{-j2\pi nT/T_0}[1 + j2\pi n(T/T_0)] - 1\}}{(2\pi n)^2/T_0}, & n \neq 0 \end{cases}$$

2-65 **(a)**
$$c_n = \begin{cases} 0, & n = \text{even} \\[1ex] \dfrac{4}{n^2\pi^2}, & n = \text{odd} \end{cases}$$
(b) $\frac{1}{3}$ W

2-84 **(a)** 6.28 msec **(b)** 160

2-86 **(a)** 4 bits/word **(b)** 5 kbits/sec

Chapter 3

3-3 **(a)**

$$W_s(f) = d \sum_{n=-\infty}^{\infty} \frac{\sin \pi nd}{\pi nd} \begin{cases} 1, & |f - nf_s| \leq 2500 \\[1ex] \dfrac{-1}{1500}(|f - nf_s| - 4000), & 2500 \leq |f - nf_s| \leq 4000 \\[1ex] 0, & f \text{ elsewhere} \end{cases}$$

where $d = 0.5$ and $f_s = 10{,}000$

(b)

$$W_s(f) = 0.5\left(\frac{\sin \pi\tau f)}{\pi\tau f}\right) \sum_{k=-\infty}^{\infty} \begin{cases} 1, & |f - kf_s| \leq 2500 \\[1ex] \dfrac{-1}{1500}(|f - kf_s| - 4000), & 2500 \leq |f - kf_s| \leq 4000 \\[1ex] 0, & f \text{ elsewhere} \end{cases}$$

where $\tau = 50 \times 10^{-6}$ and $f_s = 10{,}000$

3-7
$$W_s(f) = -j\left(\frac{1 - \cos(\pi f/f_s)}{\pi f/f_s}\right)$$
$$\times \sum_{k=-\infty}^{\infty} \begin{cases} 1, & |f - kf_s| \leq 2500 \\[1ex] \dfrac{-1}{1500}(|f - kf_s| - 4000), & 2500 \leq |f - kf_s| \leq 4000 \\[1ex] 0, & f \text{ elsewhere} \end{cases}$$

3-9 (**a**) 200 samples/sec (**b**) 9 bits/word (**c**) 1.8 kbits/sec (**d**) 900 Hz

3-15 (**a**) 5 bits (**b**) 27 kHz

3-24 (**a**) $\mathcal{P}(f) = \frac{A^2T_b}{4}\left[\frac{\sin(\frac{1}{2}\pi f T_b)}{\frac{1}{2}\pi f T_b}\right]^2$

(**b**) $\mathcal{P}(f) = \frac{A^2T_b}{4}\left[\frac{\sin(\frac{1}{4}\pi f T_b)}{\frac{1}{4}\pi f T_b}\right]^2 [\sin(\frac{1}{4}\pi f T_b)]^2$

3-34 (**a**) 32.4 kbits/sec (**b**) 10.8 ksymbols/sec (**c**) 5.4 kHz

3-35 (**a**) 3.2 ksymbols/sec (**b**) 0.5

3-46 (**a**) 5.33 kbits/sec (**b**) 667 Hz

3-47 (**a**) 10.7 kbits/sec (**b**) 1.33 kHz

3-64 (**a**) 0.00534 (**b**) 0.427

3-72 (**a**) 11 bits/character (**b**) 20 terminals

3-75 (**a**) 13,896 (**b**) 5095 (**c**) 1273 (**d**) 159 (**e**) 23

3-82 (**a**) 40 Hz (**b**) 1280 Hz

Chapter 4

4-9 (**a**) $g(t) = 500 + 200 \sin \omega_a t$, AM, $m(t) = 0.4 \sin \omega_a t$

(**b**) $x(t) = 500 + 200 \sin \omega_a t$, $y(t) = 0$

(**c**) $R(t) = 500 + 200 \sin \omega_a t$, $\theta(t) = 0$

(**d**) 2.7 kW

4-20 48.3%

4-23 $K\sqrt{m^2(t) + [\hat{m}(t)]^2}$, yes

4-28 (**a**) $(j2\pi f)/[j2\pi f + K_d K_v F_1(f)]$

(**b**) $(j2\pi f)(f_1 + jf)/[j2\pi f(f_1 + jf) + K_d K_v f_1]$

4-35 (**a**) 107.6 MHz (**b**) RF: At least 96.81–96.99 MHz; IF: 10.61–10.79 MHz

(**c**) 118.3 MHz

4-42 (**a**) 6.99 dBk (**b**) 707 V (**d**) 7025 W (**e**) 18,050 W

4-47 $$S(f) = \frac{1}{2}\left[\sum_{n=-\infty}^{\infty} c_n \delta(f - f_c - nf_m) + \sum_{n=-\infty}^{\infty} c_n^* \delta(f + f_c + nf_m)\right]$$

where $c_n = \frac{A_c}{2\pi}\left[2\theta_1\left(\frac{\sin n\theta_1}{n\theta_1}\right) + 2A_m \frac{\cos n\theta_1 \sin \theta_1 - n \sin n\theta_1 \cos \theta_1}{1 - n^2}\right]$,

$A_m = 1.2$, and $\theta_1 = 146.4°$

4-48 (**a**) $s(t) = (\cos \omega_1 t + 2 \cos 2\omega_1 t) \cos \omega_c t$, where $\omega_1 = 1000\pi$

(**b**) $S(f) = \frac{1}{4}[\delta(f - f_c + f_1) + \delta(f + f_c - f_1) + \delta(f - f_c - f_1) + \delta(f + f_c + f_1)] + \frac{1}{2}[\delta(f - f_c + 2f_1) + \delta(f + f_c - 2f_1) + \delta(f - f_c - 2f_1) + \delta(f + f_c + 2f_1)]$

(**c**) 1.25 W (**d**) 4.5 W

4-55 (**b**) $s(t) = \Pi(t) \cos \omega_c t - \dfrac{1}{\pi} \ln\left(\dfrac{|t + \frac{1}{2}|}{|t - \frac{1}{2}|}\right) \sin \omega_c t$

(**c**) $\max[s(t)] = \infty$

4-65 (**a**) $f_{\text{BPF}} = 12.96$ MHz, $B_{\text{BPF}} = 48.75$ kHz (**b**) 7.96 or 17.96 MHz

(**c**) 9.38 kHz

4-67 (**a**) $m(t) = 0.2 \cos(2000\pi t)$, $M_p = 0.2$ V, $f_m = 1$ kHz

(**b**) $m(t) = -0.13 \sin(2000\pi t)$, $M_p = 0.13$ V, $f_m = 1$ kHz

(**c**) $P_{\text{AV}} = 2500$ W, $P_{\text{PEP}} = 2500$ W

4-72 $S(f) = \frac{1}{2}[G(f - f_c) + G^*(-f - f_c)]$, where $G(f) = \sum_{n=-\infty}^{\infty} J_n(0.7854)\, \delta(f - nf_m)$, $f_c = 146.52$ MHz, and $f_m = 1$ kHz. $B = 3.57$ kHz

4-76 $S(f) = \frac{1}{2}[\sum_{n=-\infty}^{\infty} c_n\, \delta(f - f_c - nf_m) + \sum_{n=-\infty}^{\infty} c_n^*\, \delta(f + f_c + nf_m)]$, where $c_n = 5(e^{j\beta} - 1)\left(\dfrac{\sin(n\pi/2)}{n\pi/2}\right)$, $f_m = 1$ kHz, $\beta = 0.7854$, and $f_c = 60$ MHz

4-80 (**a**) 5.5 Hz

(**b**) $$S(f) = \frac{A_c}{2}\left\{\delta(f - f_c) + \delta(f + f_c) + \frac{D_f}{2\pi} \sum_{\substack{n=-\infty \\ n\neq 0}}^{\infty} \left(\frac{c_n}{nf_m}\right) \times [\delta(f - f_c - nf_m) - \delta(f + f_c - nf_m)]\right\},$$

where $c_n = 5e^{-jn\pi/2}\left(\dfrac{\sin(n\pi/2)}{n\pi/2}\right)$ $(n \neq 0)$,

$D_f = 6.98$ rads/V-sec, $f_m = 100$ Hz, and $f_c = 30$ MHz

4-83 $\mathcal{P}(f) \approx 0.893[2\delta(f - f_c + 5f_0) + \delta(f - f_c + 3f_0) + \delta(f - f_c - f_0) + 2\delta(f - f_c - 3f_0) + \delta(f - f_c - 5f_0) + 2\delta(f + f_c - 5f_0) + \delta(f + f_c - 3f_0) + \delta(f + f_c + f_0) + 2\delta(f + f_c + 3f_0) + \delta(f + f_c + 5f_0)]$

where $f_0 \triangleq \dfrac{D_f}{2\pi} = 15.9$ kHz and $f_c = 2$ GHz

4-86 (**a**) 20 kHz

(**b**) $S(f) = \frac{1}{2}[M_1(f - f_c) + M_1(f + f_c)] + \dfrac{j}{2}[M_2(f + f_c) - M_2(f - f_c)]$

Chapter 5

5-1 **(a)** $S(f) = \frac{1}{2}[X(f - f_c) + X^*(-f - f_c)]$,

$$\text{where } X(f) = \frac{A_c}{2} \sum_{n=-\infty}^{\infty} \left(\frac{\sin(n\pi/2)}{n\pi/2}\right) \delta(f - \tfrac{1}{2}nR), \quad R = 2400$$

(b) 4.8 kHz

5-8 **(a)** 3.6 kHz **(b)** 8.6 kHz

5-31 **(a)** 0.000126 μV **(b)** 8 **(c)** 4%

5-35 **(a)** 25.6 dB **(b)** 36.3 W **(c)** −54.9 dBm

5-39

$$G_a = \frac{h_{fe}^2 R_s}{h_{oe}(R_s + \mathrm{h}_{ie})^2}$$

5-43 **(a)** 129 K **(b)** −76.6 dBm

5-45 $G_a = 46$ dB, $F = 3.23$ dB

5-47 **(a)** 19.8 dB **(b)** 7.52 dB **(c)** 4.67 dB

5-59 12.84-m diameter

5-70 20.6 kW

Chapter 6

6-2 **(a)** $\bar{x} = (2\sqrt{2}\,A/\pi)\cos(\omega_0 t + \pi/4)$ **(b)** $x(t)$ is not stationary

6-4 $2/\sqrt{\pi} = 1.128$

6-8 **(a)** 15 W **(b)** 11 W **(c)** 19 W

6-10 **(a)** No **(b)** Yes **(c)** Yes **(d)** No

6-17 **(a)** $\sqrt{B}$ **(b)** $R_x(\tau) = B[(\sin \pi B\tau)/(\pi B\tau)]^2$

6-22 **(a)** $N_0 K^2/(8\pi^2 f^2)$ **(b)** ∞

6-25 $1/(2\pi f_0)$

6-30 **(a)** $H(f) = 10\Big/\left[\left(1 + j\frac{f}{f_0}\right)^2\right]$

(b) $0.690\, f_0$, where $f_0 = 1/(2\pi RC)$

6-35 **(a)** Not WSS **(b)** $\mathcal{P}_x(f) = \dfrac{A_0^2}{4}[\delta(f - f_0) + \delta(f + f_0)]$ **(c)** Yes

6-42 $x(t) = m_1(t) + m_2(t)$, $y(t) = \hat{m}_1(t) - \hat{m}_2(t)$

$$\mathcal{P}_{m_1}(f) = A\Pi\left(\frac{f}{2(f_3 - f_c)}\right), \quad \mathcal{P}_{m_2}(f) = A\Lambda\left(\frac{f}{f_c - f_1}\right)$$

6-44 $\mathcal{P}_v(f) = \frac{1}{4}[\mathcal{P}_x(f - f_c) + \mathcal{P}_x(-f - f_c)]$,
where $\mathcal{P}_x(f) = T_b\left(\frac{1 - \cos \pi f T_b}{\pi f T_b}\right)^2$

6-50 **(a)** $y(t) = \begin{cases} \frac{1}{2}, & t = 0 \\ 1, & 0 < t < T \\ \frac{1}{2}, & t = T \\ 0, & t \text{ elsewhere} \end{cases}$ **(b)** Same pulse shape answer for part (a)

Chapter 7

7-1 $P_e = \frac{1}{2}e^{-\sqrt{2}A/\sigma_0}$

7-11 $P_e = Q\left(\sqrt{0.7992(E_b/N_0)}\right)$

7-14 **(a)** $H(f) = Te^{-j\pi f T}\left(\frac{\sin \pi f T}{\pi f T}\right)$ **(b)** $B_{eq} = 1/(2T)$

7-16 $P_e = P(1)Q\left(\sqrt{\frac{2(-V_T + A)^2 T}{N_0}}\right) + P(0)Q\left(\sqrt{\frac{2(V_T + A)^2 T}{N_0}}\right)$

7-21 **(a)** 5×10^{-7} **(b)** 4.42×10^{-2}

7-22 **(a)** 2400 bits/sec **(b)** $(P_e)_{\text{overall}} = 1.11 \times 10^{-14}$

7-27 3.2×10^{-7}

7-31 **(a)** 4800 bits/sec **(b)** 19,943 bits/sec

7-32 **(a)** 9.09×10^{-4} **(b)** 29.1 dB

7-35 11.3 dB inferior

7-42 −100.9 dBm

7-47 **(a)** 651% **(b)** Low-frequency components predominate

7-49 $(S/N)_{\text{out}} = (5 \times 10^{-5})(P_s/N_0)$ for all five channels

Chapter 8

8-1 **(a)** $\varphi_1(t) = 0.3162, \quad 0 < t \le 10$; $\varphi_2(t) = 0.4472 \cos 10\pi t, \quad 0 < t \le 10$

(b) $\mathbf{s}_1 = (15.81, 0)$; $\mathbf{s}_2 = (0, 2.236)$; $\mathbf{s}_3 = (1.581, 1.118)$

8-3 **(a)** $N = 2$

(b) $\mathbf{s}_1 = (3.317, 0)$; $\mathbf{s}_2 = (-1.811, 2.954)$; $\mathbf{s}_3 = (0.3015, 2.216)$;
$\mathbf{s}_4 = (-1.508, -2.954)$

(c) 5.916

8-11 **(a)** $P(C) = \frac{1}{4}(2 - 3p)^2$, where $p = Q\left(d/\sqrt{2N_0}\right)$

(b) Same as (a) except $p = Q\left(\frac{1}{3}\sqrt{E_s/N_0}\right)$

8-14 $P(E) = 4Q\left(\sqrt{E_s/N_0}\right) + Q\left(\sqrt{2E_s/N_0}\right)$

8-16 For PPM, using union bound: $(E_b/N_0) > 1.42$ dB for $P(E) \rightarrow 0$
From Shannon's theorem: $(E_b/N_0) > -1.59$ dB for $P(E) \rightarrow 0$

8-21 **(a)** 0.0147 **(b)** 0.0056

8-22 **(a)** 7.5 Hz **(b)** 20 Hz

8-26 $\phi_1(t) = \dfrac{1}{\sqrt{T}} e^{j\Delta\omega t}, \quad 0 < t \leq T; \; \varphi_2(t) = \dfrac{1}{\sqrt{T}} e^{-j\Delta\omega t}, \quad 0 < t \leq T$

$E_m = A^2T$, where $\Delta\omega = 2\pi\Delta F = \pi(f_1 - f_2)$

Appendix B

B-1 0.3486

B-2 **(a)** 5/36 **(b)** 15/36

B-4 **(a)** 1/2 **(b)** 1/3

B-8 0.4375

B-11 **(a)** b **(b)** $1/b$ **(c)** $1/b^2$

B-21 51.0

B-26 **(a)** 1.337×10^{-11} **(b)** 0.9520

B-30 $$f(y) = \begin{cases} \dfrac{1}{2\sigma\sqrt{2\pi y}} \left[e^{-(\sqrt{y}+m)^2/(2\sigma^2)} + e^{-(\sqrt{y}-m)^2/(2\sigma^2)}\right], & y \geq 0, \\ 0, & y < 0 \end{cases}$$

B-35 $B\sigma/\sqrt{2\pi}$

B-37 **(a)** 1/6 **(b)** Yes **(c)** 1/12

(d) $$f(x_2|x_1) = \begin{cases} \frac{1}{12}(1 + x_2), & 0 \leq x_1 \leq 1, \quad 0 \leq x_2 \leq 4 \\ 0, & \text{otherwise} \end{cases}$$

B-46 **(a)** $f_y(y) = \displaystyle\int_{-\infty}^{\infty} \frac{1}{|Ax|} f_x\left(\frac{y}{Ax}, x\right) dx$ **(b)** $f_y(y) = \displaystyle\int_{-\infty}^{\infty} \frac{1}{|Ax|} f_{x_1}\left(\frac{y}{Ax}\right) f_{x_2}(x)\, dx$

Index

A

D

G

H

I

J

K

L

N

O

P

Q

R

S

T

U

V

W

X

Z

TABLE 2-1 Some Fourier Transform Theorems[a]

Operation	Function	Fourier Transform
Linearity	$a_1 w_1(t) + a_2 w_2(t)$	$a_1 W_1(f) + a_2 W_2(f)$
Time delay	$w(t - T_d)$	$W(f) e^{-j\omega T_d}$
Scale change	$w(at)$	$\frac{1}{\lvert a \rvert} W\left(\frac{f}{a}\right)$
Conjugation	$w^*(t)$	$W^*(-f)$
Duality	$W(t)$	$w(-f)$
Real signal frequency translation [$w(t)$ is real]	$w(t)\cos(\omega_c t + \theta)$	$\frac{1}{2}[e^{j\theta} W(f - f_c) + e^{-j\theta} W(f + f_c)]$
Complex signal frequency translation	$w(t) e^{j\omega_c t}$	$W(f - f_c)$
Bandpass signal	$\text{Re}\{g(t) e^{j\omega_c t}\}$	$\frac{1}{2}[G(f - f_c) + G^*(-f - f_c)]$
Differentiation	$\frac{d^n w(t)}{dt^n}$	$(j2\pi f)^n W(f)$
Integration	$\int_{-\infty}^{t} w(\lambda)\, d\lambda$	$(j2\pi f)^{-1} W(f) + \frac{1}{2} W(0)\, \delta(f)$
Convolution	$w_1(t) * w_2(t) = \int_{-\infty}^{\infty} w_1(\lambda) \cdot w_2(t - \lambda)\, d\lambda$	$W_1(f) W_2(f)$
Multiplication[b]	$w_1(t) w_2(t)$	$W_1(f) * W_2(f) = \int_{-\infty}^{\infty} W_1(\lambda)\, W_2(f - \lambda)\, d\lambda$
Multiplication by t^n	$t^n w(t)$	$(-j2\pi)^{-1} \frac{d^n W(f)}{df^n}$

[a] $\omega_c = 2\pi f_c$.

[b] $*$ denotes convolution as described in detail by (2-62).